AN INTRODUCTION TO
Physical Science

Thirteenth Edition

SHIPMAN · WILSON · HIGGINS

AN INTRODUCTION TO

Physical Science

Thirteenth Edition

JAMES T. SHIPMAN
Ohio University

JERRY D. WILSON
Lander University

CHARLES A. HIGGINS, Jr.
Middle Tennessee State University

BROOKS/COLE
CENGAGE Learning

Australia · Brazil · Japan · Korea · Mexico · Singapore · Spain · United Kingdom · United States

BROOKS/COLE
CENGAGE Learning

An Introduction to Physical Science,
Thirteenth Edition
James T. Shipman
Jerry D. Wilson
Charles A. Higgins, Jr.

Publisher, Physical Sciences: Mary Finch

Publisher, Physics and Astronomy:
Charles Hartford

Development Editors: Brandi Kirksey,
Margaret Pinette

Editorial Assistant: Brendan Killion

Senior Media Editor: Rebecca Berardy Schwartz

Marketing Manager: Jack Cooney

Marketing Coordinator: Julie Stefani

Marketing Communications Manager:
Darlene Macanan

Senior Content Project Manager: Cathy Brooks

Senior Art Director: Cate Rickard Barr

Rights Acquisition Specialist: Shalice
Shah-Caldwell

Production Service: Lachina Publishing Services

Text Designer: Real Time Design

Cover Designer: Real Time Design

Cover Image: "Moon and Venus Over
Switzerland." ©: David Kaplan

Compositor: Lachina Publishing Services

For product information and technology assistance, contact us at
Cengage Learning Customer & Sales Support, 1-800-354-9706
For permission to use material from this text or product,
submit all requests online **www.cengage.com/permissions.**
Further permissions questions can be emailed to
permissionrequest@cengage.com.

Library of Congress Control Number: 2011937665

Student Edition:
ISBN-13: 978-1-133-10409-4
ISBN-10: 1-133-10409-6

Paper Edition:
ISBN-13: 978-1-133-10909-9
ISBN-10: 1-133-10909-8

Instructor's Edition:
ISBN-13: 978-1-133-10933-4
ISBN-10: 1-133-10933-0

Brooks/Cole
20 Channel Center Street
Boston, MA 02210
USA

Cengage Learning is a leading provider of customized learning solutions with office locations around the globe, including Singapore, the United Kingdom, Australia, Mexico, Brazil and Japan. Locate your local office at **international.cengage.com/region**

Cengage Learning products are represented in Canada by Nelson Education, Ltd.

For your course and learning solutions, visit **www.cengage.com**

Purchase any of our products at your local college store or at our preferred online store **www.cengagebrain.com**

Instructors: Please visit **login.cengage.com** and log in to access instructor-specific resources.

Printed in the United States of America
1 2 3 4 5 6 7 15 14 13 12 11

Brief Contents

Contents

Preface

Science and technology are the driving forces of change in our world today. They have created a revolution in all aspects of our lives, including communication, transportation, medical care, the environment, and education. Because the world is rapidly being transformed, it is important that today's students advance their knowledge of science. While increasing their understanding of the principles of science, students must also know how science is conducted, and when, where, and to what it is applied. Equipped with this knowledge, they can better adapt to their environment and make informed decisions that ultimately affect their lives and the lives of others.

The primary goal of the thirteenth edition of *An Introduction to Physical Science* is in keeping with that of previous editions: to stimulate students' interest in science and to build a solid foundation of general knowledge in the physical sciences. Additionally, we continue to present the content in such a way that students develop the critical reasoning and problem-solving skills that are needed in our ever-changing technological world.

An *Introduction to Physical Science,* Thirteenth Edition, as for previous editions, is intended for an introductory course for college nonscience majors. The five divisions of physical science are covered: physics, chemistry, astronomy, meteorology, and geology. Each division of physical science is discussed in the context of real-world examples. The textbook is readily adaptable to either a one- or two-semester course, or a two- to three-quarter course, allowing the instructor to select topics of his or her choice.

Approach

One of the outstanding features of this textbook continues to be its emphasis on fundamental concepts. These concepts are built on as we progress through the chapters. For example, Chapter 1, which introduces the concepts of measurement, is followed by chapters on the basic topics of physics: motion, force, energy, heat, wave motion, electricity and magnetism, atomic physics, and nuclear physics. This foundation in physics is useful in developing the principles of chemistry, astronomy, meteorology, and geology in the chapters that follow. We hope that this will lead to more students choosing careers in the sciences, engineering, and mathematics.

Organizational Updates in the Thirteenth Edition

The thirteenth edition of *An Introduction to Physical Science* retains its 24-chapter format. There have been several organizational changes in chapter order and title. Each chapter has several new Conceptual Questions and Answers features. These highlight not only important material but important nonmathematical concepts as well. A short section, Chapter 10.8, Elementary Particles, has been added so students may be familiar with items often heard in the news, such as quarks, the Large Hadron Collider (LHC), the Higgs boson (the God particle), and antimatter.

Chapter 13.5 has been expanded slightly, incorporating an example and exercises to emphasize the importance of Avogadro's number. Chapter 14 is reorganized slightly by adding section 14.6, Biochemistry, which groups all the information about biological materials together as they relate to chemistry.

Chapters 16 and 17 have updated sections because of the 2006 reorganization of solar system bodies by the International Astronomical Union (IAU), as well as new discoveries and classifications of objects beyond Neptune. These are section 16.6, The Dwarf Planets, and section 17.5, Moons of the Dwarf Planets. Also, the title of section 17.6 has been changed to Small Solar-System Bodies to conform to the new definition. Section 18.7, Cosmology, has been reorganized and updated slightly to incorporate the new findings regarding dark matter and dark energy. Chapters 16, 17, and 18 each have several new photos from the Hubble Space Telescope.

Chapters 19 and 20 have been updated to include recent tornado outbreaks, hurricanes, pollution, and global warming.

Chapters 21 and 22 retain their organization to provide a top-down discussion of geologic processes. The largest scale processes are presented first and work towards smaller ones. For example, the Earth's interior, continental drift, and plate tectonics are introduced in Chapter 21, and smaller-scale rock and mineral formations are discussed in Chapter 22. Finally, Chapter 24 retains its emphasis on the use of the scientific method to define the absolute geologic time scale and the age of the Earth.

Math Coverage and Support

Each discipline is treated both descriptively and quantitatively. To make the thirteenth edition user-friendly for students who are not mathematically inclined, we continue to introduce concepts to be treated mathematically as follows. First, the concept is defined, as briefly as possible, using words. The definition is then presented, where applicable, as an equation in word form. And, finally, the concept is expressed in symbolic notation.

The level of mathematics in the textbook continues to be no greater than that of general high school math. Appendixes I though VII provide a review of the math skills needed to deal with the mathematical exercises in this textbook. It may be helpful for students to begin their study by reading through these seven appendixes. This will help identify and remediate any mathematical weaknesses and thereby build confidence and ability for working the mathematical exercises in the textbook. Practice Exercises for mathematical concepts and skills appear in Cengage Learning's CourseMate.

Assistance is also offered to students by means of in-text worked *Examples* and follow-up *Confidence Exercises* (with answers at the end of the book). However, the relative emphasis, whether descriptive or quantitative, is left to the discretion of the instructor. To those who wish to emphasize a descriptive approach, the *Exercises* may be omitted, and the other end-of-chapter sections may be used for assignments.

Outstanding Pedagogical Features in the Thirteenth Edition

▶ New to the thirteenth edition are chapter Conceptual Questions and Answers. Conceptual in nature (no mathematics), the questions are designed to pique student interest in associated chapter material—and answers are given. A few example questions (see text for answers):
  • At night, a glass windowpane acts as a mirror when viewed from inside a lighted room. Why isn't it a mirror during the day?
  • Why do wet clothes or water spots on clothes appear to be a darker color? Does the color change?
  • Microwave glass oven doors have a metal mesh with holes. What is the purpose of this?
  • We have a periodic table of elements. Why not a periodic table of compounds?
  • Why do onions make you cry?
  • Why is NO_2 called "laughing gas"?
  • Does it ever get too cold to snow?
  • Will the Sun turn into a black hole?
  • What is one global Earth process we have studied that drives the rock cycle?

▶ Each chapter begins with a list of *Facts*—a brief description of interesting, pertinent, and user-friendly items regarding concepts and topics to be covered in the chapter.

▶ Each section begins with *Preview Questions* that ask about principles and concepts that should be learned in studying the section. The questions are also designed to introduce important topics to pique the curiosity of the student.

▶ Each section ends with *Did You Learn?* statements that remind and emphasize the answers to the *Preview Questions* and important section topics that should have been noted.

▶ The acclaimed *Highlight* feature has been retained, in detailed features of pertinent chapter topics.

▶ All worked-out *Examples* within a chapter give step-by-step solutions and are followed by related *Confidence Exercises* that give students immediate practice in solving that specific type of problem. *Answers to Confidence Exercises* may be found at the back of the book, so students can judge immediately their degree of comprehension.

▶ Nine *Appendixes* and a back-of-the-book *Glossary* of all chapter Key Terms and other associated terms are included to further aid student learning.

End-of-Chapter Features

For the thirteenth edition, the end-of-chapter material continues to include important features in the following order:

1. *Key Terms* that summarize the important boldface chapter terms, all of which are defined in the back-of-the-book *Glossary*.
2. Matching Questions following the *Key Terms* are designed to test students' ability to match an appropriate statement with each key term. For immediate feedback, answers to these questions are provided at the back of the book.
3. *Multiple-Choice Questions* follow the *Matching Questions*. The questions are keyed to the appropriate chapter section should the student need help, and the answers are given at the back of the book.
4. Next come *Fill-in-the-Blank Questions*. These questions are keyed to the appropriate chapter section should the student need help, and the answers are given at the back of the book.
5. *Short-Answer Questions* test students' knowledge of important concepts by section.
6. Following the *Short-Answer Questions* section is a colorful, associative *Visual Connection* of chapter terms. The answers to the *Visual Connection* can be found at the back of the book.
7. Next is the *Applying Your Knowledge* section. These questions involve conceptual and practical applications of material covered in the chapter and everyday topics relevant to the subject matter and challenge the student to apply the concepts learned.
8. When the chapter contains mathematical equations, a list of *Important Equations* is given as a helpful review tool for students and for quick reference when needed in working Exercises.
9. The *Exercises* section follows, but only for those chapters with mathematical content. As in previous editions, these exercises are *paired,* with the answer being provided to the first exercise (odd number) of each pair.
10. *On the Web* is the last end-of-chapter feature that challenges students to answer questions related to concepts discussed in each chapter. This feature is available and integrated with the textbook at: **www.cengagebrain.com/shop/ISBN/9781133104094** and also within CourseMate. It includes suggested answers and recommended hot-linked websites.

Design, Photo, and Illustration Program

Recognizing that many students are visual learners, we have increased the visual appeal and accessibility of this edition with new and more color photos and an updated art program.

Complete Text Support Package

CourseMate with eBook Interested in a simple way to complement your text and course content with study and practice materials? Cengage Learning's CourseMate brings course concepts to life with interactive learning, study, and exam preparation tools that support the printed textbook. Watch student comprehension soar as your class works with the printed textbook and the textbook-specific website.

CourseMate goes beyond the book to deliver what you need!

Enhanced WebAssign for Shipman, Wilson, Higgins's *An Introduction to Physical Science* 13th Edition Exclusively from Cengage Learning, Enhanced WebAssign® combines the exceptional mathematics, physics, and astronomy content that you know and love with the most powerful online homework solution, WebAssign. Enhanced WebAssign engages students with immediate feedback and an interactive eBook, helping students to develop a deeper conceptual understanding of their subject matter. Online assignments can be built by selecting from hundreds of text-specific problems or supplemented with problems from any Cengage Learning textbook.

Enhanced WebAssign includes the Cengage YouBook: an engaging and customizable new eBook that lets you tailor the textbook to fit your course and connect with your students. You can remove and rearrange chapters in the table of contents and tailor assigned readings that match your syllabus exactly. Powerful editing tools let you change as much as you'd like–or leave it just like it is. You can highlight key passages or add sticky notes to pages to comment on a concept in the reading, and then share any of these individual notes and highlights with your students, or keep them personal. You can also edit narrative content in the textbook by adding a text box or striking out text. With a handy link tool, you can drop in an icon at any point in the eBook that lets you link to your own lecture notes, audio summaries, video lectures, or other files on a personal website or anywhere on the web. A simple YouTube widget lets you easily find and embed videos from YouTube directly into eBook pages.

The Cengage YouBook helps students go beyond just reading the textbook. Students can also highlight the text, add their own notes, and bookmark the text. Animations and videos play right on the page at the point of learning so that they're not speed bumps to reading but true enhancements.

WebTutor Rich with content for your physical sciences course, this Web-based teaching and learning tool integrates with your school's learning management system and includes course management, study/mastery, and communication tools. A wealth of student learning activities includes animations, quizzing, and flashcards. Chapter-based practice quizzes offer immediate feedback and link to the interactive eBook so students can focus their efforts where they need to. Reduce your preparation time with a fully customizable test bank and PowerPoint lectures. Use WebTutor™ to provide virtual office hours, post your syllabus, and track student progress—all directly from your learning management system.

PowerLecture for Shipman, Wilson, Higgins's *An Introduction To Physical Science*, 13th Edition This DVD provides the instructor with dynamic media tools for teaching. This dual-platform digital library and presentation tool provides text art, photos, and tables in a variety of easily exportable electronic formats. This enhanced DVD also contains animations to supplement your lectures, as well as lecture outlines. In addition, you can customize your presentations by importing your personal lecture slides or any other material you choose. Microsoft® PowerPoint® lecture slides and figures from the book are also included on this DVD. Turn your lectures into an interactive learning environment that promotes conceptual understanding with "clicker" content. With these slides you can pose text-specific questions to a large group, gather results, and display students' answers seamlessly. Using Diploma, instructors have all of the tools they need to create, author/edit, customize, and deliver multiple types of tests. Instructors can import questions directly from the test bank, create their own questions, or edit existing algorithmic questions, all within Diploma's powerful electronic platform.

Acknowledgments

We wish to thank our colleagues and students for the many contributions they continue to make to this textbook through correspondence, questionnaires, and classroom testing of the material. We would also like to thank all those who have helped us greatly in shaping this text over the years and especially the following reviewers of this thirteenth edition:

Jennifer Cash, South Carolina State University

Richard Holland, Southeastern Illinois College

Mark Holycross, Spartanburg Methodist College

Trecia Markes, University of Nebraska—Kearney

Eric C. Martell, Millikin University

Robert Mason, Illinois Eastern Community College

Edgar Newman, Coastal Carolina University

Michael J. O'Shea, Kansas State University

Kendra Sibbernsen, Metropolitan Community College

Todd Vaccaro, Francis Marion University

While they are not official reviewers for the book, we would like to acknowledge the contributions to the chemistry chapters by Lynn Deanhardt, Lander University, and Allison Wind, Middle Tennessee State University.

We are grateful to those individuals and organizations who contributed photographs, illustrations, and other information used in the text. We are also indebted to the Cengage Learning staff and several others for their dedicated and conscientious efforts in the production of *An Introduction to Physical Science*. We especially would like to thank Mary Finch, Publisher, Physical Sciences; Charles Hartford, Publisher, Astronomy and Physics; Brandi Kirksey, Development Editor; Margaret Pinette, Development Editor; Brendan Killion, Editorial Assistant; Cathy Brooks, Senior Content Project Manager; Jack Cooney, Marketing Manager; and Cate Barr, Senior Art Director.

As in previous editions, we continue to welcome comments from students and instructors of physical science and invite you to send us your impressions and suggestions.

About the Authors

James T. Shipman passed away July 10, 2009. (See descriptive note on Jim below.)

Jerry D. Wilson received his physics degrees from: B.S., Ohio University; M.S., Union College (Schenectady, NY); and Ph.D., Ohio University. He is one of the original authors of the first edition of *An Introduction to Physical Science,* published nationally in 1971, and has several physics textbooks to his credit. Wilson is currently Emeritus Professor of Physics at Lander University, Greenwood, SC. Email: *jwilson@greenwood.net*

Charles A. (Chuck) Higgins received his B.S. degree in physics from the University of Alabama in Huntsville, and his M.S. and Ph.D. degrees in astronomy from the University of Florida in 1996. Areas of interest and research include planetary radio astronomy, astronomy education, and public outreach. He is currently an Associate Professor in the Department of Physics and Astronomy at Middle Tennessee State University in Murfreesboro, Tennessee. Email: *chiggins@mtsu.edu*

A note about our coauthor, James T. Shipman, who passed away July 10, 2009, at the age of 90. Jim initiated the first of three sections of *An Introduction to Physical Science* in the late 1960s. He wrote the first physics section and collaborated with Jerry L. Adams on the second section, modern physics and chemistry. Being elected chair of a large physics department at Ohio University and his time limited, he asked Jerry D. Wilson and Jack Baker to write the third meteorology and geology section. The textbook was published locally until the national publication by D.C. Heath and Co. in 1971. Jim contributed directly to nine editions until his retirement but continued to revise the accompanying laboratory manual with Clyde Baker. He will be missed and fondly remembered.

A note about a previous coauthor, *Aaron W. Todd.* Aaron passed away April 5th, 2007, after a valiant battle against brain cancer. His contributions to five editions of *An Introduction to Physical Science* were many and appreciated. His input and friendship are greatly missed.

Measurement

Getty Images

It is a capital mistake to theorize before one has data. Insensibly one begins to twist the facts to suit the theories, instead of the theories to suit the facts.

•

Sherlock Holmes
(Arthur Conan Doyle, 1859–1930)

< Bring in the chain for a measurement. First and 10!

Science is concerned with the description and understanding of our environment. A first step in understanding our environment is to measure and describe the physical world. Over the centuries, humans have developed increasingly sophisticated methods of measurement, and scientists make use of the most advanced of these.

We are continually making measurements in our daily lives. Watches and clocks are used to measure the time it takes for events to take place. A census is taken every 10 years to determine (measure) the population. Money, calories, and the days and years of our lives are counted.

It was once thought that all things could be measured with exact certainty, but as smaller and smaller objects were measured, it became evident that the act of

PHYSICS FACTS

▶ Tradition holds that in the twelfth century, King Henry I of England decreed that 1 yard should be the distance from his royal nose to the thumb of his outstretched arm. (Had King Henry's arm been 3.37 inches longer, the yard and the meter would have been equal in length.)

▶ The abbreviation for the pound, lb, comes from the Latin word *libra,* which was a Roman unit of weight approximately equal to 1 pound. The word *pound* comes from the Latin *pondero,* meaning "to weigh." Libra is also a sign of the zodiac, symbolized by a set of scales (used for weight measurement).

Chapter Outline

measuring distorted the measurement. This uncertainty in making measurements of the very small is discussed in more detail in Chapter 9.5. (Note that "Chapter 9.5" means "Chapter 9, Section 5." This format will be used throughout this book to call your attention to further information in another part of the book.)

Measurement is crucial to understanding our physical environment, but first let's discuss the physical sciences and the methods of scientific investigation.

1.1   The Physical Sciences

Preview Questions*

● What are the two major divisions of natural science?
● What are the five major divisions of physical science?

Think about the following:

▶ *Hung up.* A basketball player leaping up to make a shot seems to "hang" in the air before he slam-dunks a basketball.

▶ *Spot you one.* Driving in the summer, you may see what looks like water or a "wet spot" on the road ahead, but you never get to it.

▶ *All stuck up.* The professor rubs a balloon on his sweater and touches it to the ceiling, and the balloon stays there.

▶ *Mighty small.* There are pictures of individual atoms.

▶ *It doesn't add up.* Exactly 100 cc of ethanol alcohol is mixed with exactly 100 cc of water, and the resulting mixture is less than 200 cc.

▶ *Get in line.* There won't be a total solar eclipse visible from the United States until 2017, but there will be six or more visible elsewhere before then.

▶ *Dark Moon.* The dark side of the Moon isn't dark all the time.

▶ *A bolt from the blue.* You don't have to be in a thunderstorm for lightning to strike.

▶ *No blow.* One continent has no hurricanes, and a particular latitude has none either.

▶ *All shook up.* An earthquake with a magnitude of 8.0 on the Richter scale is not twice as energetic as one with a magnitude of 4.0 (but about a million times more).

▶ *Keep an eye on the sky.* There is evidence that a meteorite caused dinosaurs to become extinct.

Would you like to know how or why such things occur, or how they are known? All these statements are explained in this book. Most people are curious about such topics, and explanations of these and many other phenomena are obtained through scientific observations. The above statements pertain to physical science, but there are several other branches of science as well.

Science (from the Latin *scientia,* meaning "knowledge") may be defined as an organized body of knowledge about the natural universe and the processes by which that knowledge is acquired and tested. In general, there are *social sciences,* which deal with human society and individual relationships, and *natural sciences,* which investigate the natural universe. In turn, the natural sciences are divided into the *biological sciences* (sometimes called *life sciences*), which are concerned with the study of living matter, and the *physical sciences,* which involve the study of nonliving matter.

*Preview Questions are listed at the beginning of each section. The answers to these questions are found in the related Did You Learn at the end of the section.

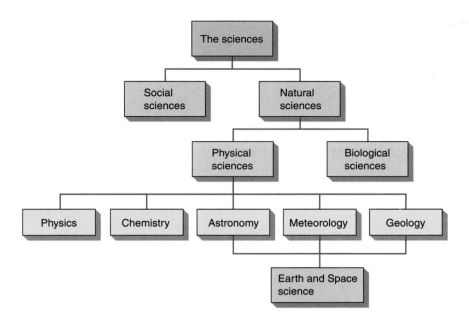

Figure 1.1 The Major Physical Sciences
A diagram showing the five major physical sciences and how they fit into the various divisions of the sciences. (See text for discussion.)

This book introduces the various disciplines of physical science, the theories and laws fundamental to each, some of the history of their development, and the effect each has on our lives. Physical science is classified into five major divisions (● Fig. 1.1):

Physics, the most fundamental of the divisions, is concerned with the basic principles and concepts of matter and energy.

Chemistry deals with the composition, structure, and reactions of matter.

Astronomy is the study of the universe, which is the totality of all matter, energy, space, and time.

Meteorology is the study of the atmosphere, from the surface of the Earth to where it ends in outer space.

Geology is the science of the planet Earth: its composition, structure, processes, and history.

(The last three physical sciences are sometimes combined as *Earth and Space Science.*)

Physics is considered the most fundamental of these divisions because each of the other disciplines applies the principles and concepts of matter and energy to its own particular focus. Therefore, our study of physical science starts with physics (Chapters 1–10); then moves on to chemistry (Chapters 11–14), astronomy (Chapters 15–18), and meteorology (Chapters 19 and 20); and ends with geology (Chapters 21–24).

This exploration will enrich your knowledge of the physical sciences and give you perspective on how science has grown throughout the course of human history; how science influences the world we live in today; and how it is employed through *technology* (the application of scientific knowledge for practical purposes).

Although the earliest humans had no sophisticated means to make measurements, they did have curiosity about the world around them, along with a compelling need to survive in a harsh environment that resulted in the making of tools and the harnessing of fire. The desire to understand the movement of the stars, the passing of the seasons, and the hope of predicting the weather by using the clues of the wind and the clouds grew out of such curiosity and all were addressed by observations of the Earth and sky.

Indeed, observation forms the basis of all scientific knowledge, even in the modern world. Scientific knowledge is cumulative, and if our predecessors had not asked questions and made observations, our own knowledge of the physical sciences would be far less extensive. Each new discovery yields the possibility for more.

Did You Learn?*

- Biology (life) and physical sciences make up the natural sciences.
- The major divisions of physical science are physics, chemistry, astronomy, meteorology, and geology.

1.2   Scientific Investigation

Preview Questions

- What does the scientific method say about the description of nature?
- Do scientific laws and legal laws have anything in common?

Theory guides. Experiment decides. Johannes Kepler (1571–1630)

Today's scientists do not jump to conclusions as some of our ancestors did, which often led to superstitious results. Today, measurements are the basis of scientific investigation. Phenomena are observed, and questions arise about how or why these phenomena occur. These questions are investigated by the **scientific method.**

The scientific method can be broken down into the following elements:

1. *Observations* and *measurements* (quantitative data).

2. *Hypothesis.* A possible explanation for the observations; in other words, a tentative answer or an educated guess.

3. *Experiments.* The testing of a hypothesis under controlled conditions to see whether the test results confirm the hypothetical assumptions, can be duplicated, and are consistent. If not, more observations and measurements may be needed.

4. *Theory.* If a hypothesis passes enough experimental tests and generates new predictions that also prove correct, then it takes on the status of a theory, a well-tested explanation of observed natural phenomena. (Even theories may be debated by scientists until experimental evidence decides the debate. If a theory does not withstand continued experimentation, then it must be modified, rejected, or replaced by a new theory.)

5. *Law.* If a theory withstands the test of many well-designed, valid experiments and there is great regularity in the results, then that theory may be accepted by scientists as a *law.* A law is a concise statement in words or mathematical equations that describes a fundamental relationship of nature. Scientific laws are somewhat analogous to legal laws, which may be repealed or modified if inconsistencies are later discovered. Unlike legal laws, scientific laws are meant to describe, not regulate.

The bottom line on the scientific method is that no hypothesis, theory, or law of nature is valid unless its predictions are in agreement with experimental (quantitative measurement) results. See ● Fig. 1.2 for a flowchart representing the scientific method.

The **Highlight: The "Face" on Mars,** which follows, illustrates the need for the scientific method.

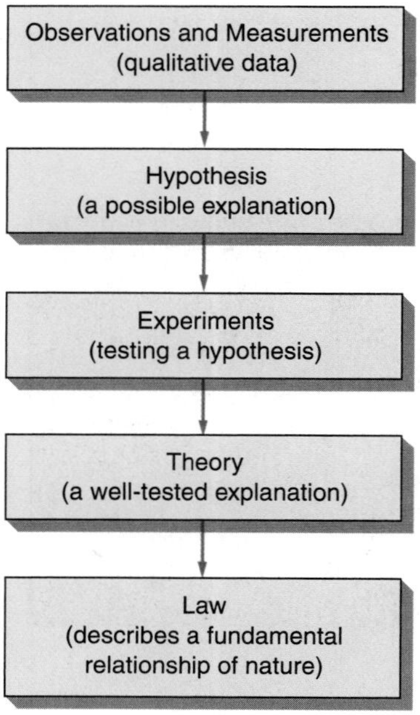

Figure 1.2 The Scientific Method
A flowchart showing the elements of the scientific method. If experiments show that a hypothesis is not consistent with the facts, more observations and measurements may be needed.

The flowchart contains the following boxes connected by downward arrows:

Observations and Measurements (qualitative data)

Hypothesis (a possible explanation)

Experiments (testing a hypothesis)

Theory (a well-tested explanation)

Law (describes a fundamental relationship of nature)

Did You Learn?

- No hypothesis, theory, or law of nature is valid unless its predictions are in agreement with experimental results.
- Scientific laws describe nature, and legal laws regulate society.

*Did You Learn? notes are listed at the end of each section and relate to the Preview Questions at the beginning of each section.

1.3 The Senses

Preview Questions

- Which two senses give us the most information about our environment?
- How may our senses be enhanced?

Our environment stimulates our senses, either directly or indirectly. The five senses (sight, hearing, touch, taste, and smell) make it possible for us to know about our environment. Therefore, the senses are vitally important in studying and understanding the physical world.

Most information about our environment comes through sight. Hearing ranks second in supplying the brain with information about the external world. Touch, taste, and smell, although important, rank well below sight and hearing in providing environmental information.

All the senses have limitations. For example, the unaided eye cannot see the vast majority of stars and galaxies. We cannot immediately distinguish the visible stars of our galaxy from the planets of our solar system, which all appear as points of light. The limitations

Highlight   The "Face" on Mars

In 1976, NASA's *Viking 1* spacecraft was orbiting Mars. When snapping photos, the spacecraft captured the shadowy likeness of an enormous head, 2 miles from end to end and located in a region of Mars called Cydonia (Fig. 1a).

The surprise among the mission controllers at NASA was quickly tempered as planetary scientists decided that the "face" was just another Martian mesa, a geologic landform common in the Cydonia region. When NASA released the photo to the public a few days later, the caption noted a "huge rock formation . . . which resembles a human head . . . formed by shadows giving the illusion of eyes, nose, and mouth." NASA scientists thought that the photo would attract the public's attention to its Mars mission, and indeed it did!

The "face" on Mars became a sensation, appearing in newspapers (particularly tabloids), in books, and on TV talk shows. Some people thought that it was evidence of life on Mars, either at present or in the past, or perhaps that it was the result of a visit to the planet by aliens. As for NASA's contention that the "face" could be entirely explained as a combination of a natural landform and unusual lighting conditions, howls arose from some of the public about "cover-up" and "conspiracy." Other people, with a more developed scientific attitude, gave provisional acceptance to NASA's conclusion, realizing that extraordinary claims (witty aliens) need extraordinary proof.

Twenty-two years later, in 1998, the *Mars Global Surveyor* (MGS) mission reached Mars, and its camera snapped a picture of the "face" 10 times sharper than the 1976 *Viking* photo. Thousands waited for the image to appear on NASA's website. The photo revealed a natural landform, not an alien monument. However, the image was taken through wispy clouds, and some people were still not convinced that the object was just a plain old mesa.

Not until 2001 did the MGS camera again pass over the object. This time there were no clouds, and the high-resolution picture was clearly that of a mesa similar to those common in the Cydonia region and the American West (Fig. 1b).

Why would so many articles and books be written extolling the alien origin of the "face"? Perhaps many authors were trading on the gullibility and ignorance of part of our population to line their own pockets or to gain attention. If so, the best ways to deal with similar situations in the future would be to improve the standard of education among the general public and to emphasize the importance of a well-developed scientific method.

Most of the information for this Highlight came from Tony Phillips, "Unmasking the Face on Mars," NASA, May 24, 2001.

NASA/JPL

(a)

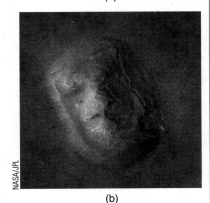

NASA/JPL

(b)

Figure 1   The Face on Mars
(a) In 1976, at the low resolution of the *Viking I* camera, the appearance of a sculpted face can be seen. (b) In 2001, at the high resolution of the *Mars Global Surveyor* camera, the object is seen to be a common mesa.

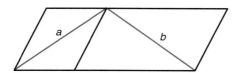

(a) Is the diagonal line *b* longer than the diagonal line *a*?

(b) Are the horizontal lines parallel or do they slope?

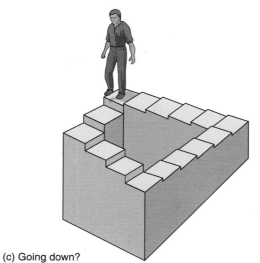

(c) Going down?

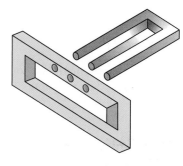

(d) Is something dimensionally wrong here?

Figure 1.3 Some Optical Illusions
We can be deceived by what we see. Answer the questions under the drawings.

of the senses can be reduced by using measuring instruments such as microscopes and telescopes. Other examples of limitations are our temperature sense of touch being limited to a range of hotness and coldness before injury and our hearing being limited to a certain frequency range (Chapter 6.4).

Not only do the senses have limitations, but they also can be deceived, thus providing false information about our environment. For example, perceived sight information may not always be a true representation of the facts because the brain can be fooled. There are many well-known optical illusions, such as those shown in ● Fig. 1.3. Some people may be quite convinced that what they see in such drawings actually exists as they perceive it. However, we can generally eliminate deception by using instruments. For example, the use of rulers to answer the questions in Fig. 1.3a and b.

> ### Did You Learn?
>
> - Sight and hearing give us the greatest amount of information about our environment.
> - The limitations of the senses can be reduced by using instruments such as microscopes and telescopes.

1.4   Standard Units and Systems of Units

Preview Questions

- What is a standard unit?
- What are the standards units of length, mass, and time in the SI?

To describe nature, we make measurements and express these measurements in terms of the magnitudes of units. Units enable us to describe things in a concrete way, that is, numerically. Suppose that you are given the following directions to find the way to cam-

Figure 1.4 A Mostly Metric World
This Canadian sign warns drivers that the metric system is in use. Notice the differences in the magnitudes of the speed limit. You'd better not go 90 mi/h.

pus when you first arrive in town: "Keep going on this street for a few blocks, turn left at a traffic light, go a little ways, and you're there." Certainly some units or numbers would be helpful.

Many objects and phenomena can be described in terms of the *fundamental* physical quantities of length, mass, and time (*fundamental* because they are the most basic quantities or properties we can imagine). In fact, the topics of *mechanics*—the study of motion and force—covered in the first few chapters of this book require *only* these physical quantities. Another fundamental quantity, electric charge, will be discussed in Chapter 8. For now, let's focus on the units of length, mass, and time.

To measure these fundamental quantities, we compare them with a reference, or standard, that is taken to be a standard unit. That is, a **standard unit** is a fixed and reproducible value for the purpose of taking accurate measurements. Traditionally, a government or international body establishes a standard unit.

A group of standard units and their combinations is called a **system of units**. Two major systems of units in use today are the **metric system** and the **British system**. The latter is used primarily in the United States but is gradually being replaced in favor of in favor of the metric system, which is used throughout most of the world (● Fig. 1.4). The United States is the only major country that has not gone completely metric.

Length

The description of space might refer to a location or to the size of an object (amount of space occupied). To measure these properties, we use the fundamental quantity of **length**, the measurement of space in any direction.

Space has three dimensions, each of which can be measured in terms of length. The three dimensions are easily seen by considering a rectangular object such as a bathtub (● Fig. 1.5). It has length, width, and height, but each dimension is actually a length. The dimensions of space are commonly represented by a three-dimensional Cartesian coordinate system (named in honor of French mathematician René Descartes, 1596–1650, who developed the system).

The standard unit of length in the metric system is the **meter** (m), from the Greek *metron*, "to measure." It was defined originally as one ten-millionth of the distance from the Earth's equator to the geographic North Pole (● Fig. 1.6a). A portion of the meridian between Dunkirk, France, and Barcelona, Spain, was measured to determine the meter length, and

Figure 1.5 Space Has Three Dimensions
(a) The bathtub has dimensions of length (*l*), width (*w*), and height (*h*), but all are actually measurements of length. (b) The dimensions of space are commonly represented by a three-dimensional Cartesian coordinate system (*x, y, z*) with the origin as the reference point.

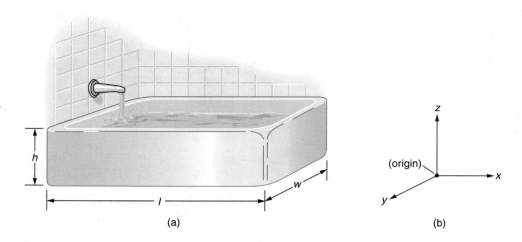

(a)

(b)

that unit was first adopted in France in the 1790s. One meter is slightly longer than 1 yard, as illustrated in Fig. 1.6b.

From 1889 to 1960, the standard meter was defined as the length of a platinum–iridium bar kept at the International Bureau of Weights and Measures in Paris, France. However, the stability of the bar was questioned (for example, length variations occur with temperature changes), so new standards were adopted in 1960 and again in 1983. The current definition links the meter to the speed of light in a vacuum, as illustrated in Fig. 1.6c. Light travels at a speed of 299,792,458 meters/second (usually listed as 3.00×10^8 m/s). So, by definition, 1 meter is the distance light travels in 1/299,792,458 of a second.

The standard unit of length in the British system is the *foot*, which historically was referenced to the human foot. As noted in the Physics Facts at the beginning of this chapter, King Henry I used his arm to define the yard. Other early units commonly were referenced to parts of the body. For example, the *hand* is a unit that even today is used in measuring the heights of horses (1 hand is 4 in.).

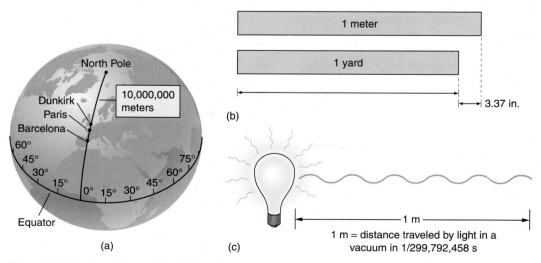

(a)

(b)

(c)

Figure 1.6 The Metric Length Standard: The Meter
(a) The meter was originally defined such that the distance from the North Pole to the equator would be 10 million meters. (b) One meter is a little longer than one yard, about 3.4 in. longer (not to scale). (c) The meter is now defined by the distance light travels in a vacuum in a small fraction of a second.

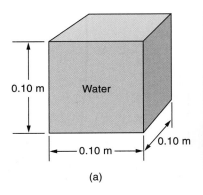

(a)

(b)

Figure 1.7 The Metric Mass Standard: The Kilogram
(a) The kilogram was originally defined in terms of a specific volume of water, that of a cube 0.10 m on a side (at 4°C, the temperature at which water has its maximum density). As such, the mass standard was associated with the length standard. (b) Prototype kilogram number 20 is the U.S. standard unit of mass. The prototype is a platinum–iridium cylinder 39 mm in diameter and 39 mm high.

Mass

Mass is the amount of matter an object contains. The more massive an object, the more matter it contains. (More precise definitions of mass in terms of force and acceleration, and in terms of gravity, will be discussed in Chapter 3.)

The standard metric unit of mass is the **kilogram** (kg). Originally, this amount of matter was related to length and was defined as the amount of water in a cubic container 0.10 m, or 10 cm, on a side (● Fig. 1.7a). However, for convenience, the mass standard was referenced to a material standard (an artifact or a human-made object). Currently, the kilogram is defined to be the mass of a cylinder of platinum–iridium kept at the International Bureau of Weights and Measures in Paris. The U.S. prototype (copy) is kept at the National Institute of Standards and Technology (NIST) in Washington, D.C. (Fig. 1.7b).

This standard is based on an artifact rather than on a natural phenomenon. Even though the cylinder is kept under controlled conditions, its mass is subject to slight changes because of contamination and loss from surface cleaning. A property of nature, by definition, is always the same and in theory can be measured anywhere. Scientists have yet to agree on an appropriate fundamental constant, such as the speed of light for the meter, on which to base the kilogram.

The unit of mass in the British system is the *slug,* a rarely used unit. We will not use this unit in our study because a quantity of matter in the British system is expressed in terms of weight on the surface of the Earth and in units of pounds. (The British system is sometimes said to be a gravitational system.) Unfortunately, weight is not a *fundamental* quantity, and its use often gives rise to confusion. Of course, a fundamental quantity should be the same and not change. However, weight is the gravitational attraction on an object by a celestial body, and this attraction is different for different celestial bodies. (The gravitational attraction of a body depends on its mass).

For example, on the less massive Moon, the gravitational attraction is $\frac{1}{6}$ that on the Earth, so an object on the Moon weighs $\frac{1}{6}$ less than on the Earth. For example, a suited astronaut who weighs 300 pounds on the Earth will weigh $\frac{1}{6}$ that amount, or 50 pounds, on the Moon, but the astronaut's mass will be the same (● Fig. 1.8).

A fundamental quantity does not change at different locations. The astronaut has the same mass, or quantity of matter, wherever he or she is. As will be learned in Chapter 3.3, mass and weight are related, but they are not the same. For now, keep in mind that *mass, not weight, is the fundamental quantity*.

Time

Each of us has an idea of what time is, but when asked to define it, you may have to ponder a bit.

Some terms often used when referring to time are *duration, period,* and *interval.* A common descriptive definition is that **time** is the continuous, forward flow of events. Without

Figure 1.8 Mass Is the Fundamental Quantity
The weight of an astronaut on the Moon is $\frac{1}{6}$ the astronaut's weight on the Earth, but the astronaut's mass is the same in any location.

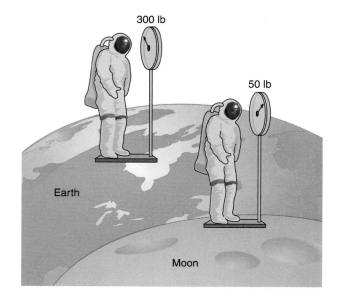

Figure 1.9 Time and Events
Events mark intervals of time. Here, at the New York City Marathon, after starting out (beginning event), a runner crosses the finish line (end event) in a time of 2 hours, 12 minutes, and 38 seconds.

events or happenings of some sort, there would be no perceived time (● Fig. 1.9). The mind has no innate awareness of time, only an awareness of events taking place in time. In other words, we do not perceive time as such, only events that mark locations in time, similar to marks on a meterstick indicating length intervals.

Note that time has only one direction—forward. Time has never been observed to run backward. That would be like watching a film run backward in a projector.

Conceptual Question and Answer

Time and Time Again

Q. What is time?

A. Time is a difficult concept to define. The common definition that *time is the continuous, forward flow of events* is more of an observation that a definition.

Others have thought about time. Marcus Aurelius, the Roman emperor and philosopher, wrote:

Time is a strong river of passing events, and strong is its current.

St. Augustine pondered this question, too:

What is time? If no one asks me, I know; if I want to explain it to a questioner, I do not know.

The Mad Hatter in Lewis Carroll's *Alice in Wonderland* thought he knew time:

If you know time as well as I do, you wouldn't talk about wasting it. It's him. . . . Now, if you only kept on good terms with him, he'd do almost anything you liked with the clock. For instance, suppose it were nine o'clock in the morning, just time to begin lessons; you'd only have to whisper a hint to Time, and around goes the clock in a twinkling: Half past one, time for dinner.

A safe answer is: *Time is a fundamental property or concept.* This definition masks our ignorance, and physics goes on from there, using the concept to describe and explain what we observe.

Time and space seem to be linked. In fact, time is sometimes considered a fourth dimension, along with the other three dimensions of space. If something exists in space, it also must exist in time. But for our discussion, space and time will be regarded as separate quantities.

Fortunately, the standard unit of time is the same in both the metric and British systems: the **second**. The second was originally defined in terms of observations of the Sun, as a certain fraction of a solar day (● Fig. 1.10a).

In 1967, an atomic standard was adopted. The second was defined in terms of the radiation frequency of the cesium-133 atom. This "atomic clock" used a beam of cesium atoms to maintain our time standard with a variation of about 1 second in 300 years (Fig. 1.10b). In 1999, another cesium-133 clock was adopted. This atomic "fountain clock," as its name implies, uses a fountain of cesium atoms (Fig. 1.10c). The variation of this timepiece is within 1 second per 100 million years.

NIST is currently working on a "quantum logic" clock that makes use of the oscillations of a single ion of aluminum. It is expected to not gain or lose more than 1 second in about 3.7 billion years!

The standard units for length, mass, and time in the metric system give rise to an acronym, the **mks** system. The letters *mks* stand for *meter, kilogram,* and *second.* They are also standard units for a modernized version of the metric system, called the International System of Units (abbreviated **SI**, from the French *Système International d'Unités*).

When more applicable, smaller units than those standard in the mks system may be used. Although the mks system is the *standard* system, the smaller *cgs system* is sometimes used, where *cgs* stands for *centimeter, gram,* and *second.* For comparison, the units for length, mass, and time for the various systems are listed in ● Table 1.1.

Table 1.1	Units of Length, Mass, and Time for the Metric and British Systems of Measurement		
	Metric		
Fundamental Quantity	*SI or mks*	*cgs*	**British**
Length	meter (m)	centimeter (cm)	foot (ft)
Mass	kilogram (kg)	gram (g)	slug
Time	second (s)	second (s)	second (s)

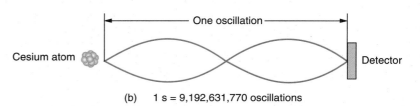

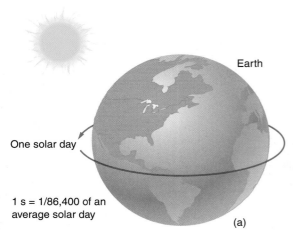

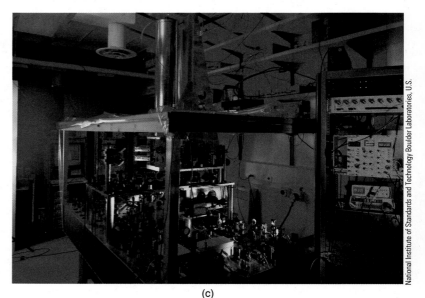

Figure 1.10 A Second of Time
(a) The second was defined originally in terms of a fraction of the average solar day. (b) One second is currently defined in terms of the frequency of radiation emitted from the cesium atom. (c) A clock? Yes, the National Institute of Standards and Technology (NIST) "fountain" cesium atomic clock. Such a clock provides the time standard for the United States.

Did You Learn?

● A standard unit is a fixed and reproducible value for accurate measurements.

● The SI standard units for length, mass, and time are the meter, kilogram, and second, respectively.

1.5   More on the Metric System

Preview Questions

● What are the four most common metric prefixes?

● What is the difference between a cubic centimeter (cm³) and a milliliter (mL)?

The SI was established in 1960 to make comprehension and the exchange of ideas among the people of different nations as simple as possible. It now contains seven base units: the meter (m), the kilogram (kg), the second (s), the ampere (A) to measure the flow of electric charge, the kelvin (K) to measure temperature, the mole (mol) to measure the amount of a substance, and the candela (cd) to measure luminous intensity. A definition of each of these units is given in Appendix I. However, we will be concerned with only the first three of these units for several chapters.

One major advantage of the metric system is that it is a decimal (base-10) system. The British system is a duodecimal (base-12) system, as in 12 inches equals a foot. The base-10 system allows easy expression and conversion to larger and smaller units. A series of *metric prefixes* is used to express the multiples of 10, but you will only need to know a few common ones:

mega- (M)   1,000,000 (million, 10^6)
kilo- (k)   1000 (thousand, 10^3)
centi- (c) $\frac{1}{100}$ = 0.01 (hundredth, 10^{-2})
milli- (m) $\frac{1}{1000}$ = 0.001 (thousandth, 10^{-3})

Examples of the relationships of these prefixes follow.

1 meter is equal to 100 centimeters (cm) or 1000 millimeters (mm).
1 kilogram is equal to 1000 grams (g).
1 millisecond (ms) is equal to 0.001 second (s).
1 megabyte (Mb) is equal to a million bytes.
(See ● Table 1.2 for more metric prefixes.)

You are familiar with another base-10 system: our currency. A cent is $\frac{1}{100}$ of a dollar, or a centidollar. A dime is $\frac{1}{10}$ of a dollar, or a decidollar. Tax assessments and school bond levies are sometimes given in mills. Although not as common as a cent, a mill is $\frac{1}{1000}$ of a dollar, or a millidollar.

As using factors of 10 demonstrates, the decimal metric system makes it much simpler to convert from one unit to another than in the British system. For example, it is easy to see that 1 kilometer is 1000 meters and 1 meter is 100 centimeters. In the British system, though, 1 mile is 5280 feet and 1 foot is 12 inches, making this system unwieldy compared with the metric system.

Some nonstandard metric units are in use. One of the most common is a unit of fluid volume or capacity. Recall that the kilogram was originally defined as the mass of a cube of water 0.10 m, or 10 cm, on a side (Fig. 1.7a). This volume was defined to be a **liter** (L). Hence, 1 L has a volume of 10 cm × 10 cm × 10 cm = 1000 cm^3 (cubic centimeters, sometimes abbreviated as cc, particularly in chemistry and biology). Because 1 L is 1000 cm^3 and 1 kg of water is 1000 g, it follows that 1 cm^3 of water has a mass of 1 g. Also, because 1 L contains 1000 milliliters (mL), 1 cm^3 is the same volume as 1 mL (● Fig. 1.11).*

Table 1.2 Some Metric Prefixes

Prefix	Symbol	Example: meter (m)	Pronunciation
giga- (billion)	G	Gm (gigameter, 1,000,000,000 m or 10^9 m)*	JIG-a (*jig* as in *jiggle*, *a* as in *about*)
mega- (million)	M	Mm (megameter, 1,000,000 m or 10^6 m)	MEG-a (as in *megaphone*)
kilo- (thousand)	k	km (kilometer, 1000 m or 10^3 m)	KIL-o (as in *kilowatt*)
hecto- hundred)	h	hm (hectometer, 100 m or 10^2 m)	HEK-to (*heck-toe*)
deka- (ten)	da	dam (dekameter, 10 m or 10^1 m)	DEK-a (*deck* plus *a* as in *about*)
		meter (m)	
deci- (one-tenth)	d	dm (decimeter, 0.10 m or 10^{-1} m)	DES-I (as in *decimal*)
centi- (one-hundredth)	c	cm (centimeter, 0.01 m or 10^{-2} m)	SENT-i (as in *sentimental*)
milli- (one-thousandth)	m	mm (millimeter, 0.001 m or 10^{-3} m)	MIL-li (as in *military*)
micro- (one-millionth)	μ	μm (micrometer, 0.000001 m or 10^{-6} m)	MI-kro (as in *microphone*)
nano- (one-billionth)	n	nm (nanometer, 0.000000001 m or 10^{-9} m)†	NAN-oh (*an* as in *annual*)

*Powers-of-10, or scientific, notation (10^x) is often used instead of decimals. If you are not familiar with this notation, or if you need to review it, see Appendix VI.
†You will be hearing more about nano- in terms of *nanotechnology*, which is any technology on the nanometer scale. To get an idea of this size, a human hair is about 50,000 nm across, and it takes 10 hydrogen atoms in a line to make 1 nanometer.

---

*The liter is sometimes abbreviated with a lowercase "ell" (l) as in ml, but a capital "ell" (L) is preferred so that the abbreviation is less likely to be confused with the numeral one (1). In type, 1 L is much clearer than 1 l.

Figure 1.11 Mass and Volume (the Liter)
(a) The kilogram was originally related to length. The mass of the quantity of water in a cubic container 10 cm on a side was taken to be 1 kg. As a result, 1 cm³ of water has a mass of 1 g. The volume of the container was defined to be 1 liter (L), and 1 cm³ = 1 mL. (b) One liter is slightly larger than 1 quart.

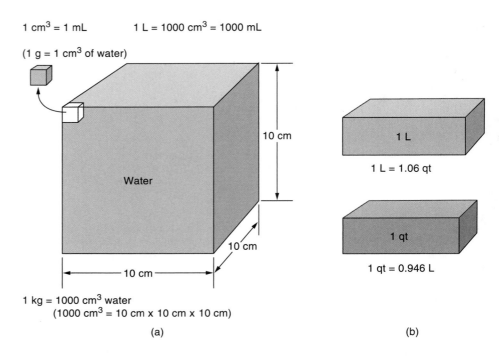

$1 \text{ cm}^3 = 1 \text{ mL}$ $1 \text{ L} = 1000 \text{ cm}^3 = 1000 \text{ mL}$

$(1 \text{ g} = 1 \text{ cm}^3 \text{ of water})$

10 cm

Water

10 cm

10 cm

$1 \text{ kg} = 1000 \text{ cm}^3 \text{ water}$
$(1000 \text{ cm}^3 = 10 \text{ cm} \times 10 \text{ cm} \times 10 \text{ cm})$

(a)

1 L
1 L = 1.06 qt

1 qt
1 qt = 0.946 L

(b)

You may wonder why a nonstandard volume such as the liter is used when the standard metric volume would use the standard meter length, a cube 1 m on a side. This volume is rather large, but it too is used to define a unit of mass. The mass of a quantity of water in a cubic container 1 m on a side (1 m³) is taken to be a *metric ton* (or *tonne*) and is a relatively large mass. One cubic meter contains 1000 L (can you show this?), so 1 m³ of water = 1000 kg = 1 metric ton.

The liter is now used commonly for soft drinks and other liquids, having taken the place of the quart. One liter is a little larger than 1 quart: 1 L = 1.06 qt (● Fig. 1.12).

Figure 1.12 Liter and Quart
(a) The liter of drink on the right contains a little more liquid than 1 quart of milk. (b) One quart is equivalent to 946 mL, or slightly smaller than 1 liter (1 L = 1.06 qt).

(a)

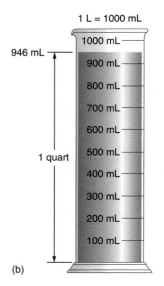

$1 \text{ L} = 1000 \text{ mL}$
1000 mL

946 mL

900 mL
800 mL
700 mL
600 mL
500 mL
400 mL
300 mL
200 mL
100 mL

1 quart

Jerry D. Wilson

(b)

1.6 Derived Units and Conversion Factors

Preview Questions

- What are derived units?
- How can you tell which is longer, 1 kilometer or 1 mile?

Derived Units

How are most physical quantities generally described using *only* the three basic units of length, mass, and time? We use **derived units**, which are multiples or combinations of units. The various derived units will become evident to you during the course of your study. Some examples of derived units follow.

Derived Quantity	Unit
Area (length)^2	m^2, cm^2, ft^2, etc.
Volume (length)^3	m^3, cm^3, ft^3, etc.
Speed (length/time)	m/s, cm/s, ft/s, etc.

Let's focus on a particular quantity with derived units, density, which involves mass and volume.

The **density** of a substance reflects the compactness of the matter or mass of a substance. In more formal language, *density,* commonly represented by the lowercase Greek letter rho (ρ), is the amount of mass located in a definite volume, or simply the mass per volume.

$$\text{density} = \frac{\text{mass}}{\text{volume}} = \frac{\text{mass}}{(\text{length})^3}$$

or

$$\rho = \frac{m}{V} \qquad\qquad 1.1$$

Figure 1.13 **Mass, Volume, and Density**
Both the weight and the pillow have the same mass, but they have very different volumes and hence have different densities. The metal weight is much denser than the pillow. If the mass of the metal were distributed uniformly throughout its volume (homogeneous), then the density would be constant. This distribution would not be the case for the pillow, and an average density would be expressed.

Thus, if sample A has a mass of 20 kg that occupies a volume of 5.0 m^3, then it has a density of 20 kg/5.0 m^3 = 4.0 kg/m^3. If sample B has a mass of 20 kg that occupies a volume of 4.0 m^3, then it has a density of 20 kg/4.0 m^3 = 5.0 kg/m^3. So sample B is denser (has greater density) and its mass is more compact than sample A.

Also, if mass is distributed uniformly throughout a volume, then the density will be constant. Such would not be the case for the pillow in ● Fig. 1.13.

Although the standard units of density are kg/m^3, it is often convenient to use g/cm^3 (grams per cubic centimeter) for more manageable numbers. For example, by our original definition, 1 L (1000 cm^3) of water has a mass of 1 kg (1000 g), so water has a density of $\rho = m/V = $ 1000 g/1000 cm^3 = 1.0 g/cm^3. If density is expressed in units of grams per cubic centimeter, then the density of a substance can be easily compared with that of water. For example, a rock with a density of 3.3 g/cm^3 is 3.3 times as dense as water. Iron has a density of 7.9 g/cm^3, and the Earth as a whole has an average density of 5.5 g/cm^3. The planet Saturn, with an average density of 0.69 g/cm^3, is less dense than water.

| EXAMPLE 1.1 | Determining Density |

Density can be useful in identifying substances. For example, suppose a chemist had a sample of solid material that is determined to have a mass of 452 g and a volume of 20.0 cm³. What is the density of the substance?

Solution

Density is easily computed using Eq. 1.1:

$$\rho = \frac{m}{V} = \frac{452 \text{ g}}{20.0 \text{ cm}^3} = 22.6 \text{ g/cm}^3$$

This density is quite large, and by looking up the known densities of substances, the chemist would suspect that the material is the chemical element osmium, the densest of all elements. (Gold has a density of 19.3 g/cm³, and silver has a density of 10.5 g/cm³.)

Confidence Exercise 1.1

A sample of gold has the same mass as that of the osmium sample in Example 1.1. Which would have the greater volume? Show by comparing the volume of the gold with that of the osmium. (The density of gold is given in Example 1.1.)

The answers to Confidence Exercises may be found at the back of the book.

Densities of liquids such as blood and alcohol can be measured by means of a *hydrometer,* which is a weighted glass bulb that floats in the liquid (● Fig. 1.14). The higher the bulb floats, the greater the density of the liquid.

When a medical technologist checks a sample of urine, one test he or she performs is for density. For a healthy person, urine has a density of 1.015 to 1.030 g/cm³; it consists mostly of water and dissolved salts. When the density is greater or less than this normal range, the urine may have an excess or deficiency of dissolved salts, usually caused by an illness.

A hydrometer is also used to test the antifreeze in a car radiator. The closer the density of the radiator liquid is to 1.00 g/cm³, the closer the antifreeze and water solution is to being pure water, and more antifreeze may be needed. The hydrometer is usually calibrated in degrees rather than density and indicates the temperature to which the amount of anti-freeze will protect the radiator.

Finally, when a combination of units becomes complicated, it is often given a name of its own. For example, as discussed in Chapter 3.3, the SI unit of force is the newton, which in terms of standard units is*

$$\text{newton (N)} = \text{kg} \cdot \text{m/s}^2$$

It is easier to talk about a newton (N) than about a kg·m/s². As you might guess, the newton unit is named in honor of Sir Isaac Newton. The abbreviation of a unit named after an individual is capitalized, but the unit name itself is not: newton (N). We will encounter other such units during the course of our study.

Conversion Factors

Often we want to convert units within one system or from one system to another. For example, how many feet are there in 3 yards? The immediate answer would be 9 feet, because it is commonly known that there are 3 feet per yard. Sometimes, though, we may want to make comparisons of units between the metric and the British systems. In general, to convert units, a **conversion factor** is used. Such a factor relates one unit to another. Some convenient conversion factors are listed on the inside back cover of this book. For instance,

$$1 \text{ in.} = 2.54 \text{ cm}$$

Figure 1.14 Measuring Liquid Density
A hydrometer is used to measure the density of a liquid. The denser the liquid, the higher the hydrometer floats. The density can be read from the calibrated stem.

Edward Kinsman/Photo Researchers, Inc.

_____
*The centered dot means that the quantities for these units are multiplied.

Although it is commonly written in equation form, this expression is really an *equivalence statement;* that is, 1 in. has an equivalent length of 2.54 cm. (To be a true equation, the expression must have the same magnitudes and units or dimensions on both sides.) However, in the process of expressing a quantity in different units, a conversion relationship in ratio or factor form is used:

$$\frac{1 \text{ in.}}{2.54 \text{ cm}} \quad \text{or} \quad \frac{2.54 \text{ cm}}{1 \text{ in.}}$$

For example, suppose you are 5 ft 5 in., or 65.0 in., tall, and you want to express your height in centimeters. Then

$$65.0 \text{ in.} \times \frac{2.54 \text{ cm}}{1 \text{ in.}} = 165 \text{ cm}$$

Note that the in. units cancel and the cm unit is left on both sides of the equation, which is now a true equation, 65.0 × 2.54 cm = 165 cm (equal on both sides). In the initial example of converting units in the same system, 3 yd to feet, a conversion factor was actually used in the form:

$$3 \text{ yd} \times \frac{3 \text{ ft}}{1 \text{ yd}} = 9 \text{ ft}$$

where 3 ft/1 yd, or 3 ft per yard. In this case, because the conversion is so common, the mathematics can be done mentally.

The steps may be summarized as follows:

Steps for Converting from One Unit to Another

Step 1

Use a conversion factor, a ratio that may be obtained from an equivalence statement. (Often it is necessary to look up these factors or statements in a table; see the inside back cover of this book.)

Step 2

Choose the appropriate form of conversion factor (or factors) so that the unwanted units cancel.

Step 3

Check to see that the units cancel and that you are left with the desired unit. Then perform the multiplication or division of the numerical quantities. Here is an example done in stepwise form.

EXAMPLE 1.2   Conversion Factors: One-Step Conversion

The length of a football field is 100 yards. In constructing a football field in Europe, the specifications have to be given in metric units. How long is a football field in meters?

Solution

Question: 100 yd is equivalent to how many meters? That is, 100 yd = ? m.

Step 1

There is a convenient, direct equivalence statement between yards and meters given inside the back cover of the textbook under Length:

$$1 \text{ yd} = 0.914 \text{ m}$$

The two possible conversion factor ratios are

$$\frac{1 \text{ yd}}{0.914 \text{ m}} \quad \text{or} \quad \frac{0.914 \text{ m}}{\text{yd}} \quad \text{(conversion factors)}$$

For convenience, the number 1 is commonly left out of the denominator of the second conversion factor; that is, we write 0.914 m/yd instead of 0.914 m/1 yd.

Step 2

The second form of this conversion factor, 0.914 m/yd, would allow the yd unit to be canceled. (Here yd is the unwanted unit in the denominator of the ratio.)

Step 3

Checking this unit cancellation and performing the operation yields

$$100 \text{ yd} \times \frac{0.914 \text{ m}}{\text{yd}} = 91.4 \text{ m}$$

Confidence Exercise 1.2

In a football game, you often hear the expression "first and 10" (yards). How would you express this measurement in meters to a friend from Europe?

The answers to Confidence Exercises may be found at the back of the book.

As the use of the metric system in the United States expands, unit conversions and the ability to do such conversions will become increasingly important. Automobile speedometers showing speeds in both miles per hour (mi/h) and kilometers per hour (km/h) are common. Also, road signs comparing speeds can be seen. Some are designed to help drivers with metric conversion (● Fig. 1.15).

In some instances, more than one conversion factor may be used, as in Example 1.3.

Figure 1.15 Unit Conversions
(a) The speedometers of automobiles may be calibrated in both British and metric units. The term mph is a common, nonstandard abbreviation for miles per hour; the standard abbreviation is mi/h. Note that 60 mi/h is about 100 km/h. (b) Road signs in Canada, which went metric, are designed to help drivers with the conversion, particularly U.S. drivers going into Canada. Notice the highlighted letters in Think*m*etric.

EXAMPLE 1.3   Conversion Factors: Multistep Conver

A computer printer has a width of 18 in. What is its width in meters?

Solution

Question: How many meters are there in 18 inches? 18 in. = ? m. Suppose you didn't know and couldn't look up the conversion factor for inches to meters but remembered that 1 in. = 2.54 cm (which is a good length equivalence statement to remember between the British and metric systems). Then, using this information and another well-known equivalence statement, 1 m = 100 cm, you could do the multiple conversion as follows:

$$\text{inches} \times \text{centimeters} \times \text{meters}$$

$$18 \text{ in.} \times \frac{2.54 \text{ cm}}{\text{in.}} \times \frac{1 \text{ m}}{100 \text{ cm}} = 0.46 \text{ m}$$

(a)

(b)

Highlight   Is Unit Conversion Important? It Sure Is.

In 1999, the *Mars Climate Orbiter* spacecraft reached its destination after having flown 670 million km (415 million mi) over a 9.5-month period (Fig. 1). As the spacecraft was to go into orbit around the red planet, contact between it and personnel on the Earth was lost, and the *Orbiter* was never heard from again.

What happened? Investigations showed that the loss of the *Orbiter* was primarily a problem of unit conversion, or a lack thereof. At Lockheed Martin Astronautics, which built the spacecraft, engineers calculated the navigational information in British units. When flight control at NASA's Jet Propulsion Laboratory received the data, it was assumed that the information was in metric units, as called for in the mission specifications.

Unit conversions were not made, and as a result, the *Orbiter* approached Mars at a far lower altitude than planned. It either burned up in the Martian atmosphere or crashed to the planet's surface. Because of a lack of unit conversion, a $125 million spacecraft was lost on the red planet, causing more than a few red faces.

This incident underscores the importance of using appropriate units, making correct conversions, and working consistently in the same system of units.

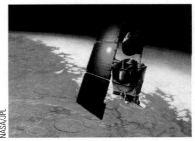

NASA/JPL

Figure 1   An Artist's Conception of the *Mars Climate Orbiter*

Notice that the units cancel correctly.

Let's check this result directly with the equivalence statement 1 m = 39.4 in.

$$18 \text{ in.} \times \frac{1 \text{ m}}{39.4 \text{ in.}} = 0.46 \text{ m}$$

Confidence Exercise 1.3

How many seconds are there in 1 day? (Use multiple conversion factors, starting with 24 h/day.)

Equivalence statements that are not dimensionally or physically correct are sometimes used; an example is 1 lb = 2.2 kg. This equivalence statement may be used to determine the weight of an object in pounds, given its mass in kilograms. It means that 1 kg is *equivalent* to 2.2 lb; that is, a 1-kg *mass* has a *weight* of 2.2 lb. For example, a person with a mass of 60 kg ("kilos") would have a weight in pounds of 60 kg × 2.2 lb/kg = 132 lb.

For an example of the kind of problem that can result from the concurrent use of both the British and metric systems, see the **Highlight: Is Unit Conversion Important? It Sure Is.**

Did You Learn?

- Derived units are multiples or combination of units. For example, density $\rho = kg/m^3$.
- Looking at the equivalence statements (inside back cover) 1 mi = 1.61 km and 1 km = 0.62 mi, it can be seen that 1 mi is longer than 1 km.

1.7   Significant Figures

Preview Questions

- What is the purpose of significant figures?
- Why are mathematical results rounded?

Jerry D. Wilson

Figure 1.16 Significant Figures and Insignificant Figures
Performing the division operation of 6.8/1.67 on a calculator with a floating decimal point gives many figures. However, most of these figures are insignificant, and the result should be rounded to the proper number of significant figures, which is two. (See text for further explanation.)

When working with quantities, hand calculators are often used to do mathematical operations. Suppose in an exercise you divided 6.8 cm by 1.67 cm and got the result shown in ● Fig. 1.16. Would you report 4.0718563? Hopefully not—your instructor might get upset.

The reporting problem is solved by using what are called **significant figures** (or *significant digits*), a method of expressing measured numbers properly. This method involves the accuracy of measurement and mathematical operations.

Note that in the multiplication example, 6.8 cm has two figures or digits and 1.67 has three. These figures are significant because they indicate a magnitude read from some measurement instrument. In general, more digits in a measurement implies more accuracy or the greater fineness of the scale of the measurement instrument. That is, the smaller the scale (or the more divisions), the more numbers you can read, resulting in a better measurement. The 1.67-cm reading is more accurate because it has one more digit than 6.8 cm.

The number of significant figures in 6.8 cm and 1.67 cm is rather evident, but some confusion may arise when a quantity contains one or more zeros. For example, how many significant figures does the quantity 0.0254 have? The answer is three. Zeros at the beginning of a number are not significant, but merely locate the decimal point. Internal or end zeros are significant; for example, 104.6 and 3705.0 have four and five significant figures, respectively. (An end or "trailing" zero must have a decimal point associated with it. The zero in 3260 would not be considered significant.)

However, a mathematical operation cannot give you a better "reading" or more significant figures than your original quantities. Thus, as general rules,

1. When multiplying and dividing quantities, leave as many significant figures in the answer as there are in the quantity with the least number of significant figures.

2. When adding or subtracting quantities, leave the same number of decimal places in the answer as there are in the quantity with the least number of significant places.*

Applying the first rule to the example in Fig. 1.16 indicates that the result of the division should have two significant figures (abbreviated s.f.). Hence, rounding the result:

$$6.8 \text{ cm}/1.67 \text{ cm} = 4.1$$

↑   ↑

limiting term   4.0718563 is
has 2 s.f.   rounded to 4.1 (2 s.f.)

If the numbers were to be added, then, by the addition rule,

$$6.8 \text{ cm (least number of decimal places)}$$
$$+ \ 1.67 \text{ cm}$$
$$\overline{8.47 \text{ cm}} \ \rightarrow \ 8.5 \text{ cm (final answer rounded to one decimal place)}$$

Clearly, it is necessary to round numbers to obtain the proper number of significant figures. The following rules will be used to do this.

Rules for Rounding

1. If the first digit on the right to be dropped is less than 5, then leave the preceding digit as is.

2. If the first digit to be dropped is 5 or greater, then increase the preceding digit by one.

EXAMPLE 1.4   Rounding Numbers

Round each of the following:
(a) 26.142 to three significant figures.
The 4 is the first digit to be removed and is less than 5. Then, by rule 1,

$$26.142 \rightarrow 26.1$$

---

*See Appendix VII for more on significant figures.

(b) 10.063 to three significant figures.

The 6 is the first digit to be removed. (Here, the zeros on each side of the decimal point are significant because they are internal and have digits on both sides. Then, by rule 2,

$$10.063 \rightarrow 10.1$$

(c) 0.09970 to two significant figures.

In this case, the first nondigit to be removed is the 7. (The zeros to the immediate left and right of the decimal point are not significant but merely serve to locate the decimal point.) Because 7 is greater than 5, by rule 2,

$$0.0997 \rightarrow 0.10$$

(d) The result of the product of the measured numbers 5.0×356.

Performing the multiplication,

$$5.0 \times 356 = 1780$$

Because the result should have only two significant figures, as limited by the quantity 5.0, we round

$$1780 \rightarrow 1800$$

A problem may exist here. Standing alone, it might not be known whether the "trailing" zeros in the 1800 result are significant or not. This problem may be remedied by using *powers-of-ten (scientific) notation*. The 1800 may be equivalently written as

$$1800 = 1.8 \times 10^3$$

and there is no doubt that there are two significant figures. See Appendix VI if you are not familiar with powers-of-ten notation. More information on this notation usage is given in the next chapter.

Confidence Exercise 1.4

Multiply 2.55 by 3.14 on a calculator and report the result in the proper number of significant figures.

Did You Learn?

- Significant figures are a method of expressing measured numbers properly.
- A mathematical result is rounded so as to express the proper number of significant figures.

KEY TERMS

1. science (1.1)
2. scientific method (1.2)
3. standard unit (1.4)
4. system of units
5. metric system
6. British system
7. length
8. meter
9. mass
10. kilogram
11. time
12. second
13. mks
14. SI
15. mega- (1.5)
16. kilo-
17. centi-
18. milli-
19. liter
20. derived units (1.6)
21. density
22. conversion factor
23. significant figures (1.7)

MATCHING

For each of the following items, fill in the number of the appropriate Key Term from the preceding list. Compare your answers with those at the back of the book.

a. _____ The most widely used system of units

b. _____ One-hundredth

c. _____ Method of properly expressing measured numbers

d. _____ The measurement of space in any direction

e. _____ The relationship of one unit to another

f. _____ Thousand

g. _____ Standard unit of mass

h. _____ A valid theory of nature must be in agreement with experimental results.

i. _____ Defined in terms of the radiation frequency of a cesium atom

j. _____ An organized body of knowledge about the natural universe

k. _____ The amount of matter an object contains

l. _____ A group of standard units and their combinations

m. _____ One-thousandth

n. _____ A system of units that is slowly being phased out

o. _____ The continuous forward flowing of events

p. _____ A fixed and reproducible value for making measurements

q. _____ Multiples and combinations of standard units

r. _____ Defined in terms of the speed of light

s. _____ Million

t. _____ Modernized version of the metric system

u. _____ Compactness of matter

v. _____ Acronym for metric standard units

w. _____ $V = 10 \text{ cm} \times 10 \text{ cm} \times 10 \text{ cm}$

MULTIPLE CHOICE

Compare your answers with those at the back of the book.

1. Which is the most fundamental of the physical sciences? (1.1)
   (a) astronomy   (b) chemistry   (c) physics   (d) meteorology

2. Which one of the following is a concise statement about a fundamental relationship in nature? (1.2)
   (a) hypothesis   (b) law   (c) theory   (d) experiment

3. Which human sense is first in supplying the most information about the external world? (1.3)
   (a) touch   (b) taste   (c) sight   (d) hearing

4. Which is the standard unit of mass in the metric system? (1.4)
   (a) gram   (b) kilogram   (c) slug   (d) pound

5. Which one of the following is *not* a fundamental quantity? (1.4)
   (a) length   (b) weight   (c) mass   (d) time

6. Which metric prefix means "one-thousandth"? (1.5)
   (a) centi-   (b) milli-   (c) mega-   (d) kilo-

7. Which metric prefix means "thousand"? (1.5)
   (a) centi-   (b) milli-   (c) mega-   (d) kilo-

8. Which of the following metric prefixes is the smallest? (1.5)
   (a) micro-   (b) centi-   (c) nano-   (d) milli-

9. How many base units are there in the SI? (1.5)
   (a) four   (b) five   (c) six   (d) seven

10. Which combination of units expresses density? (1.6)
    (a) mass/(time)^3 (b) mass/(kg)^3
    (c) mass/(length)^3 (d) mass/m^2

11. What is the expression 1 in. = 2.54 cm properly called? (1.6)
    (a) equation   (b) conversion factor
    (c) SI factor   (d) equivalence statement

12. A student measures the length and width of a rectangle to be 49.4 cm and 0.590 cm, respectively. Wanting to find the area (in cm^2) of this rectangle, the student multiplies on a calculator and obtains a result of 2.9146. The area should be reported as ___. (1.7)
    (a) 2914.6 cm^2 (b) 2915 cm^2
    (c) 2.9×10^3 cm^2 (d) 2.91×10^2 cm^2

13. Which of the following numbers has the greatest number of significant figures? (1.6)
    (a) 103.07   (b) 124.5   (c) 0.09914   (d) 5.048×10^5

FILL IN THE BLANK

Compare your answers with those at the back of the book.

1. The natural sciences, in which scientists study the natural universe, are divided into physical and ___ sciences. (1.1)

2. A(n) ___ is used to test a hypothesis. (1.2)

3. According to the ___, no hypothesis or theory of nature is valid unless its predictions are in agreement with experimental results. (1.2)

4. Most information about our environment reaches us through the sense of ___. (1.3)

5. All the human senses have ___. (1.3)

6. One liter is slightly ___ than 1 quart. (1.4)

7. One yard is slightly ___ than 1 meter. (1.4)

8. Unlike mass, weight is not a(n) ___ property. (1.4)

9. The standard unit of ___ is the same in all measurement systems. (1.4)

10. The metric prefix *mega-* means ___. (1.5)

11. A common nonstandard metric unit of fluid volume or capacity is the ___. (1.5)

12. If A is denser than B, then A contains more ___ per unit volume. (1.6)

What is the first step in understanding our environment?

1.1 The Physical Sciences

1. What is the definition of *science?*
2. What are the five major divisions of physical science?

1.2 Scientific Investigation

3. What is the first element of the scientific method?
4. Which generally comes first in solving problems, hypothesis or experiment?
5. What is the difference between a law and a theory?
6. What does the controversy over the "face" on Mars illustrate?

1.3 The Senses

7. How do the five senses rank in importance in yielding information about our environment?
8. The senses cannot be completely relied on. Why?
9. Answer the questions that accompany ● Fig. 1.17.

1.4 Standard Units and Systems of Units

10. Why are some quantities called "fundamental"?
11. A standard unit must have what characteristics?
12. What makes up a system of units?
13. For a given speed limit, would the numerical value be greater in mi/h or in km/h?
14. Is the United States officially on the metric system? Explain.
15. Initially, how many metric standard units were referenced to artifacts? How many are still referenced to artifacts?
16. Which is a fundamental property, weight or mass? Why?
17. Explain the acronyms mks, SI, and cgs.

1.5 More on the Metric System

18. What are the metric prefixes for million and millionth?
19. What are the four most common metric prefixes?
20. What is a metric ton, and how is it defined?
21. What is the standard unit of volume in the SI?

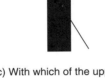

(a) Is the upper horizontal line longer?

(b) Are the diagonal lines parallel?

(c) With which of the upper lines does the line on the right connect?

Figure 1.17   Seeing Is Believing
See Short-Answer Question 10.

1.6 Derived Units and Conversion Factors

22. How many standard units are generally used to describe the mechanics of nature?
23. What does density describe?
24. In general, when a derived unit becomes complicated (involves too many standard units), what is done?
25. Is an equivalence statement a true equation? Explain.
26. In Fig. 1.15a, the abbreviation mph. is used. Is this a correct abbreviation? Why might it be confusing to some people? (*Hint: miles, meters.*)

1.7 Significant Figures

27. Why are significant figures used?
28. How are significant figures obtained?
29. If you multiplied 9874 m by 36 m, how many significant figures should you report in your answer?

Visualize the connections and give answers for the blanks. Compare your answers with those at the back of the book.

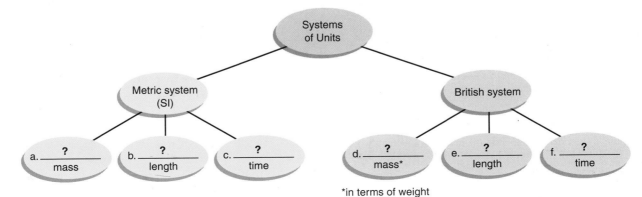

*in terms of weight

APPLYING YOUR KNOWLEDGE

1. Give two ways in which a scientific law and a legal law differ.

2. In general, common metric units are larger than their British counterparts; for example, 1 m is a little longer than 1 yd. Give two other examples, as well as one notable exception.

3. In the original definition of the kilogram as a volume of water, it was specified that the water be at its maximum density, which occurs at 4°C (39°F). Why? What would be the effect if another temperature were used? (*Hint:* See the Highlight: Freezing from the Top Down in Chapter 5.2.)

4. Suppose you could buy a quart of a soft drink and a liter of the soft drink at the same price. Which would you choose, and why?

5. In the opening scenes of the movie *Raiders of the Lost Ark,* Indiana Jones tries to remove a gold idol from a booby-trapped pedestal. He replaces the idol with a bag of sand of approximately the same volume. (Density of gold = 19.32 g/cm³ and density of sand ≈ 2 g/cm³.)
   (a) Did he have a reasonable chance of not activating the booby trap? Explain.
   (b) In a later scene, he and an unscrupulous guide play catch with the idol. Assume that the volume of the idol is about 1.0 L. If it were solid gold, then what mass would the idol have? Do you think playing catch with it is plausible?

6. Currently, the tallest building in the world is the Burj Khalifa (with 160 stories) in Dubai, United Arab Emirates, which is 828 m tall. Previously, the tallest was the Taipei 101 skyscraper (with 101 floors) in Taiwan, which is 508 m tall. A friend wants you to describe the process by which you could find how much taller, in feet, the Burj is compared with the Taipei 101. What would you tell him?

IMPORTANT EQUATION

density: $\rho = \dfrac{m}{V}$ (1.1)

EXERCISES

1.5 More on the Metric System

1. What is your height in meters?

   Answer: varies with person (5.0 ft = 1.5 m)

2. What is your height in centimeters?

3. What is the volume of a liter in cubic millimeters?

   Answer: 1,000,000 or 10⁶ mm³

4. Show that 1 cubic meter contains 1000 L.

5. Water is sold in half-liter bottles. What is the mass, in kilograms and in grams, of the water in such a full bottle?

   Answer: 0.50 kg, 500 g

6. A rectangular container measuring 10 cm × 20 cm × 25 cm is filled with water. What is the mass of this volume of water in kilograms and in grams?

7. Write the following quantities in standard units.
   (a) 0.55 Ms   (c) 12 mg
   (b) 2.8 km   (d) 100 cm

   Answer: (a) 550,000 s (b) 2800 m (c) 0.000012 kg (d) 1 m

8. Fill in the blanks with the correct numbers for the metric prefixes.
   (a) 40,000,000 bytes = ___ Mb
   (b) 0.5722 L = ___ mL
   (c) 540.0 m = ___ cm
   (d) 5,500 bucks = ___ kilobucks

1.6 Derived Units and Conversion Factors

9. Compute, in centimeters and in meters, the height of a basketball player who is 6 ft 5 in. tall.

   Answer: 196 cm, 1.96 m

10. Compute, in both feet and inches, the height of a woman who is 165 cm tall.

11. In Fig 1.15b, is the conversion on the sign exact? Justify your answer.

    Answer: Yes

12. If we changed our speed limit signs to metric, what would probably replace (a) 55 mi/h and (b) 65 mi/h?

13. Is the following statement reasonable? (Justify your answer.) It took 300 L of gasoline to fill up the car's tank.

    Answer: No, 300 L is about 300 qt, or 75 gal.

14. Is the following statement reasonable? (Justify your answer.) The area of a dorm room is 49 m².

15. Referring to number 6 in Applying Your Knowledge, how much shorter in feet is the Taipei 101 than the Burj Khalifa?

    Answer: 1050 ft

16. The new Hoover Dam Bridge connecting Arizona and Nevada opened in October 2010 (● Fig. 1.18). It is the highest and longest arched concrete bridge in the Western Hemisphere, rising 890 ft above the Colorado River and extending 1900 ft in length. What are these dimensions in meters?

17. A popular saying is "Give him an inch, and he'll take a mile.". What would be the equivalent saying using comparable metric units?

    Answer: "Give him a centimeter and he'll take a kilometer."

18. A metric ton is 1000 kg, and a British ton is 2000 lb. Which has the greater weight and by how much?

19. Compute the density in g/cm³ of a piece of metal that has a mass of 0.500 kg and a volume of 63 cm³.

    Answer: 7.9 g/cm³ (the density of iron)

20. What is the volume of a piece of iron (ρ = 7.9 g/cm³) that has a mass of 0.50 kg?

Figure 1.18 High and Wide
An aerial view of the new four-lane Hoover Dam Bridge between Arizona and Nevada with the Colorado River beneath (as seen from behind the dam). See Exercise 16.

Figure 1.19 How Many Significant Figures?
See Exercise 26.

23. Round the following numbers to three significant figures.
    (a) 0.9986   (c) 0.01789
    (b) 7384.38   (d) 47.645
    Answer: (a) 0.999 (b) 7380 or 7.38 × 10³ (c) 0.0179 (d) 47.6

24. Round the following numbers to four significant figures.
    (a) 3.1415926   (c) 483.5960
    (b) 0.00690745   (d) 0 .0234973

25. What is the result of (3.15 m × 1.53 m)/0.78 m with the proper number of significant figures?
    Answer: 6.2 m

26. The calculator result of multiplying 2.15 × π is shown in
    ● Fig. 1.19. Round the result to the proper number of significant figures.

1.7 Significant Figures

21. Round the following numbers to two significant figures.
    (a) 7.66   (c) 9438
    (b) 0.00208   (d) 0.000344
    Answer: (a) 7.7 (b) 0.0021 (c) 9400 or 94 × 10² (d) 0.00034

22. Round the following numbers to three significant figures.
    (a) 0.009995   (c) 0.010599
    (b) 644.73   (d) 8429.55

ON THE WEB

1. Method to the Madness

Outline the three steps of problem solving as presented in this textbook. Visit the student website for this textbook at **www.cengagebrain.com/shop/ISBN/1133104096** to verify your answers.

2. The Measurement of Time

How fast is the new optical clock? What is the basis of atomic clocks? How does the NIST compact atomic clock differ from the atomic clock? What effect might atomic clocks have on your life? Explore answers to these questions on the student website at **www.cengagebrain.com/shop/ISBN/1133104096.**

Motion

Give me matter and motion, and I will construct the universe.

•

René Descartes
(1596–1650)

Strobe lighting captures a > bullet tearing through a card at about 900 m/s (about 2000 mi/h)

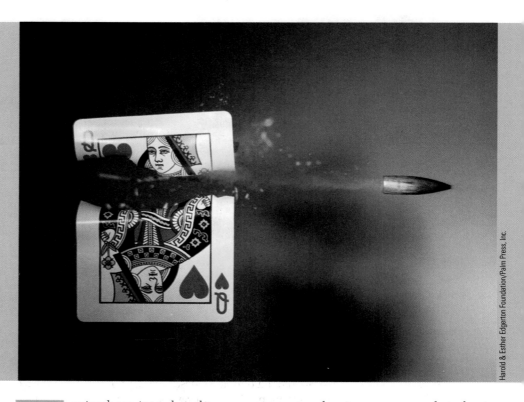

Harold & Esther Edgerton Foundation/Palm Press, Inc.

PHYSICS FACTS

- It takes 8.33 minutes for light to travel from the Sun to the Earth. (See Example 2.2.)

- Electrical signals between your brain and muscles travel at about 435 km/h (270 mi/h).

- A bullet from a high-powered rifle travels at a speed of about 2900 km/h (1800 mi/h).

- NASA's X43A uncrewed jet flew at a speed of 7700 km/h (4800 mi/h), faster than a speeding bullet.

- Nothing can exceed the speed of light (in vacuum), 3.0×10^8 m/s (186,000 mi/s).

- *Velocity* comes from the Latin *velocitas*, meaning "swiftness."

- *Acceleration* comes from the Latin *accelerare*, meaning "hasten."

H aving been introduced to measurement and units, you are ready to begin your study of physics. **Physics**, the most fundamental physical science, is concerned with the basic principles and concepts that describe the workings of the universe. It deals with matter, motion, force, and energy.

There are various areas of physics:

- *Classical mechanics* is the study of the motion of objects moving at relatively low speeds.

- *Waves and sound* is the study of wave motion and its application to sound.

- *Thermodynamics* is the study of temperature, heat, and thermal processes.

- *Electromagnetism* is the study of electricity and magnetism.

Chapter Outline

▶ *Optics* is the study of the properties of light, mirrors, and lenses.

▶ *Quantum mechanics* is the study of the behavior of particles on the microscopic and submicroscopic levels.

▶ *Atomic and nuclear physics* is the study of the properties of atoms and nuclei.

▶ *Relativity* is the study of objects traveling at speeds approaching the speed of light.

We will delve into all of these areas, except relativity, and begin with classical mechanics and the study of motion.

Motion is everywhere. Students walk to class. You drive to the store. Birds fly. The wind blows the trees. Rivers flow. Even the continents slowly drift. On a larger scale, the Earth rotates on its axis; the Moon revolves around the Earth, which revolves around the Sun; the Sun moves in its galaxy; and the galaxies move with respect to one another.

This chapter focuses on describing motion and on defining and discussing terms such as *speed*, *velocity*, and *acceleration*. These concepts will be considered without the forces that produce motion. The discussion of forces is reserved for Chapter 3.

[*Note:* If you are not familiar with powers-of-10 notation (scientific notation) or if you need a math review, then see the appropriate sections in the Appendix.]

2.1   Defining Motion

Preview Questions

● What is needed to designate a position?

● What is motion?

The term **position** refers to the location of an object. To designate the position of an object, a *reference point* and a *measurement scale* are needed. For example, the entrance to campus is 1.6 km (1 mi) from the *intersection* with a particular traffic light. The book is 15 cm from the *corner* of the table. The Cartesian coordinates of the point on a graph are $(x, y) = (2.0 \text{ cm}, 3.0 \text{ cm})$. Here the reference point is the *origin* of the coordinate system (Chapter 1.4).

If an object changes position, we say that motion has occurred. That is, an object is in **motion** when it is undergoing a continuous change in position.

Consider an automobile traveling on a straight highway. The motion of the automobile may or may not be occurring at a constant rate. In either case, the motion is described by using the fundamental units of *length* and *time*.

The description of motion by length and time is evident in running. For example, as shown in ● Fig. 2.1, the cheetah runs a certain distance at full speed in the shortest possible time. Combining length and time to give the *time rate of change of position* is the basis of describing motion in terms of speed and velocity, as discussed in the following section.

Did You Learn?

● To designate a position or location, both a reference point and a measurement scale are needed.

● Motion involves a continuous change of position.

2.2   Speed and Velocity

Preview Questions

● Between two points, which may be greater in magnitude, distance or displacement?

● What is the difference between speed and velocity?

Figure 2.1 Motion
Motion is described in terms of time and distance. Here, a running cheetah appears to be trying to run a distance in the shortest time possible. The cheetah is the fastest of all land animals, capable of reaching speeds of up to 113 km/h, or 70 mi/h. (The slowest land creature is the snail, which can move at a speed of 0.05 km/h, or 0.03 mi/h.)

Steve Umland/The Image Works

The terms *speed* and *velocity* are often used interchangeably; however, in physical science, they have different distinct meanings. Speed is a *scalar* quantity, and velocity is a *vector* quantity. Let's distinguish between scalars and vectors now, because other terms will fall into these categories during the course of our study. The distinction is simple.

A **scalar** quantity (or a *scalar*, for short) has only magnitude (numerical value and unit of measurement). For example, you may be traveling in a car at 90 km/h (about 55 mi/h). Your speed is a scalar quantity; the magnitude has the numerical value of 90 and unit of measurement km/h.

A **vector** quantity has magnitude *and* direction. For example, suppose you are traveling at 90 km/h *north*. This describes your velocity, which is a vector quantity because it consists of magnitude *plus* direction. By including direction, a vector quantity (or a *vector*, for short) gives more information than a scalar quantity. No direction is associated with a scalar quantity.

Vectors may be graphically represented by arrows (● Fig. 2.2). The length of a vector arrow is proportional to the magnitude and may be drawn to scale. The arrowhead indicates the direction of the vector. Notice in Fig. 2.2 that the red car has a negative velocity vector, $-v_c$, that is equal in magnitude (length of arrow) but opposite in direction, to the positive velocity vector, $+v_c$, for the blue car. The velocity vector for the man, v_m, is shorter than the vectors for the cars because he is moving more slowly in the positive (+) direction. (The + sign is often omitted as being understood.)

Speed

Now let's look more closely at speed and velocity, which are basic quantities used in the description of motion. The **average speed** of an object is the total distance traveled divided by the time spent in traveling the total distance. **Distance** (*d*) is the actual length of the path that is traveled. In equation form,

Figure 2.2 Vectors
Vectors may be graphically represented by arrows. The length of a vector arrow is proportional to the magnitude of the quantity it represents, and the arrowhead indicates the direction. (See text for description.)

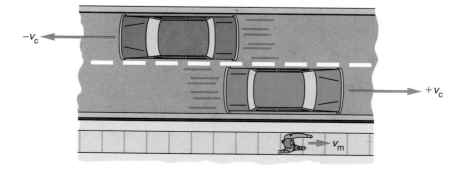

$$\text{average speed} = \frac{\text{distance traveled}}{\text{time to travel the distance}}$$

$$\bar{v} = \frac{d}{t} \qquad\qquad 2.1$$

where the bar over the symbol ($\bar{v}$, "vee-bar") indicates that it is an average value.

Note that length d and time t are *intervals*. They are sometimes written Δd and Δt to indicate explicitly that they are intervals ($\Delta d / \Delta t$). The Δ (Greek delta) means "change in" or "difference in." For example, $\Delta t = t - t_o$, where t_o and t are the original (or initial) and final times (on a clock), respectively. If $t_o = 0$, then $\Delta t = t$. Speed has the standard units of meters per second (m/s) or feet per second (ft/s)—length/time. Other common nonstandard units are mi/h and km/h.

Taken over a time interval, speed is an average. This concept is somewhat analogous to an average class grade. Average speed gives only a general description of motion. During a long time interval like that of a car trip, you may speed up, slow down, and even stop. The average speed, however, is a single value that represents the average rate of motion for the entire trip.

The description of motion can be made more specific by taking smaller time intervals such as a few seconds or even an instant. The speed of an object at any instant of time may be quite different from its average speed, and it gives a more accurate description of the motion. In this case, we have an instantaneous speed.

The **instantaneous speed** of an object is its speed at that instant of time (Δt becoming extremely small). A common example of nearly instantaneous speed is the speed registered on an automobile speedometer (● Fig. 2.3). This value is the speed at which the automobile is traveling at that moment, or instantaneously.

Velocity

Now let's look at describing motion with velocity. Velocity is similar to speed, *but* a direction is involved. **Average velocity** is the displacement divided by the total travel time. **Displacement** is the straight-line distance between the initial and final positions, with direction toward the final position, and is a vector quantity (● Fig. 2.4).

For straight-line motion in one direction, speed and velocity have something in common. Their magnitudes are the same because the lengths of the distance and the displacement are the same. The distinction between them in this case is that a direction must be specified for the velocity.

As you might guess, there is also **instantaneous velocity**, which is the velocity at any instant of time. For example, a car's instantaneous speedometer reading and the direction

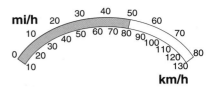

Figure 2.3 Instantaneous Speed
The speed on an automobile speedometer is a practical example of instantaneous speed, or the speed of the car at a particular instant. Here it is 80 km/h, or 50 mi/h.

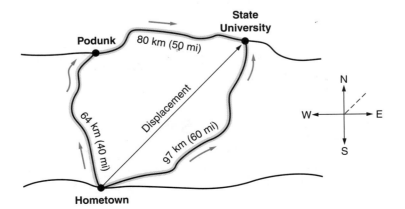

Figure 2.4 Displacement and Distance
Displacement is a vector quantity and is the straight-line distance between two points (initial and final), plus direction. In traveling from Hometown to State University, the displacement would be so many kilometers NE (northeast). Distance is a scalar quantity and is the actual path length traveled. Different routes have different distances. Two routes are shown in the figure, with distances of 97 km and 144 km (64 km + 80 km = 144 km).

in which it is traveling at that instant give the car's instantaneous velocity. Of course, the speed and direction of the car may and usually do change. This motion is then *accelerated motion*, which is discussed in the following section.

If the velocity is *constant*, or *uniform*, then there are no such changes. Suppose an airplane is flying at a constant speed of 320 km/h (200 mi/h) directly eastward. Then the airplane has a constant velocity and flies in a straight line. (Why?)

For this special case, you should be able to convince yourself that the constant velocity and the average velocity are the same ($\bar{v} = v$). By analogy, think about everyone in your class getting the same (constant) test score. How do the class average and the individual scores compare under these circumstances?

A car traveling with a constant velocity is shown in ● Fig. 2.5. Examples 2.1 through 2.3 illustrate the use of speed and velocity.

EXAMPLE 2.1   Finding Speed and Velocity

Describe the motion (speed and velocity) of the car in Fig. 2.5.

Solution

Step 1

The data are taken from the figure.
Given: $d = 80$ m and $t = 4.0$ s
(The car travels 80 m in 4.0 s.)

Step 2

Wanted: Speed and velocity. The units of the data are standard.

Step 3

The car has a constant (uniform) speed and travels 20 m each second. When a speed is constant, it is equal to the average speed; that is, $v = \bar{v}$. (Why?) Calculating the average speed using Eq. 2.1 yields

$$\bar{v} = \frac{d}{t} = \frac{80 \text{ m}}{40 \text{ s}} = 20 \text{ m/s}$$

If the motion is in one direction (straight-line motion), then the car's velocity is also constant and is 20 m/s *in the direction of the motion.*

Confidence Exercise 2.1

How far would the car in Example 2.1 travel in 10 s?
    The answers to Confidence Exercises may be found at the back of the book.

Example 2.1 was worked in stepwise fashion, as suggested in the approach to problem solving in Appendix II. This stepwise approach will be used in the first example in early chapters as a reminder. Thereafter, examples generally will be worked directly, unless a stepwise solution is considered helpful.

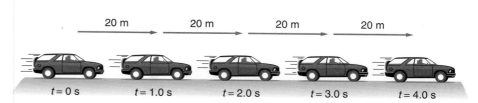

Figure 2.5 Constant Velocity
The car travels equal distances in equal periods of time in straight-line motion. With constant speed and constant direction, the velocity of the car is constant, or uniform.

EXAMPLE 2.2   Finding the Time It Takes for Sunlight to Reach the Earth

The speed of light in space (vacuum) is $c = 3.00 \times 10^8$ m/s. (The speed of light is given the special symbol c.) How many seconds does it take light from the Sun to reach the Earth if the distance from the Sun to the Earth is 1.50×10^8 km?

Solution

Given: $v = c = 3.00 \times 10^8$ m/s, $d = 1.5 \times 10^8$ km. Converting the distance to meters (the standard unit) yields

$$d = 1.50 \times 10^8 \text{ km} \left(\frac{10^3 \text{ m}}{\text{km}} \right) = 1.50 \times 10^{11} \text{ m}$$

Wanted: t (time for light to travel from the Sun)
Rearranging Eq. 2.1 for the unknown t, we have $t = d/v$, where the bar over the v is omitted because the speed is constant, and $\bar{v} = v$. Then

$$t = \frac{d}{v} = \frac{1.50 \times 10^{11} \text{ m}}{3.00 \times 10^8 \text{ m/s}}$$

$$= 5.00 \times 10^2 \text{ s} = 500 \text{ s}$$

From this example, it can be seen that although light travels very rapidly, it still takes 500 seconds, or 8.33 minutes, to reach the Earth after leaving the Sun (● Fig. 2.6). Again we are working with a constant speed and velocity (straight-line motion).

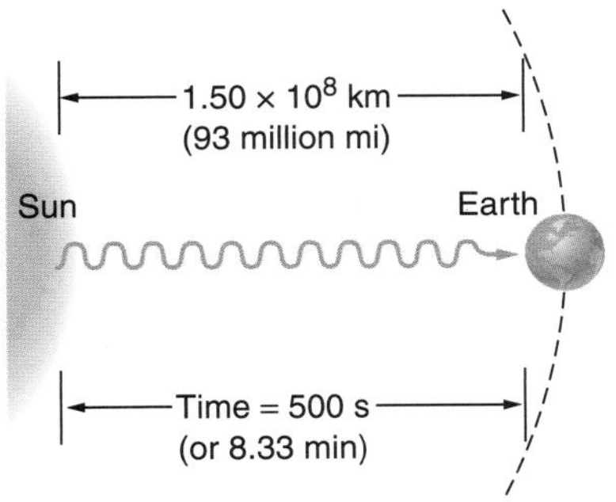

Confidence Exercise 2.2

A communications satellite is in a circular orbit about the Earth at an an altitude of 3.56×10^4 km. How many seconds does it take a signal from the satellite to reach a television receiving station? (Radio signals travel at the speed of light, 3.00×10^8 m/s.)*
The answers to Confidence Exercises may be found at the back of the book.

Figure 2.6 Traveling at the Speed of Light
Although light travels at a speed of 3.00×10^8 m/s, it still takes more than 8 minutes for light from the Sun to reach the Earth. (See Example 2.2.)

EXAMPLE 2.3   Finding the Orbital Speed of the Earth

What is the average speed of the Earth in miles per hour as it makes one revolution about the Sun? (Consider the Earth's orbit to be circular.)

Solution

Our planet revolves once around the Sun in an approximately circular orbit in a time of 1 year. The distance it travels in one revolution is the circumference of this circular orbit. (The orbit is actually slightly elliptical, as discussed in Chapter 16.1.)
    The circumference of a circle is given by $2\pi r$, where r is the radius of the orbit. In this case, r is taken to be 93.0 million miles (9.30×10^7 mi), which is the mean distance of the Earth from the Sun (see Fig. 2.6). The time it takes to travel this distance is 1 year, or about 365 days.
    Putting these data into Eq. 2.1 and multiplying the time by 24 h/day to convert to hours gives

$$\bar{v} = \frac{d}{t} = \frac{2\pi r}{t} = \frac{2(3.14)(9.30 \times 10^7 \text{ mi})}{365 \text{ days } (24 \text{ h/day})}$$

$$= 6.67 \times 10^4 \text{ mi/h (or about 18.5 mi/s)}$$

---

*The greatest speed that has been achieved *on* the Earth is Mach 9.6 (9.6 times the speed of sound; see Chapter 6.4), which is about 7000 mi/h (1.9 ft/s) or 11,000 km/h (3.1 km/s). This speed record was set by NASA's X43A, an experimental, pilotless, scram-jet-powered aircraft.

You are traveling through space with the Earth at this speed. In 1 second (the time it takes to say, "one-thousand-one"), you travel 18.5 mi (about 30 km). That's pretty fast!

Confidence Exercise 2.3

What is the average speed in mi/h of a person at the equator as a result of the Earth's rotation? (Take the radius of the Earth to be $R_E = 4000$ mi.)

The solution to Example 2.3 shows that the Earth *and all of us on it* are traveling through space at a speed of 66,700 mi/h (or 18.5 mi/s). Even though this speed is exceedingly large and relatively constant, we do not sense this great speed visually because of the small relative motions (apparent motions) of the stars and because our atmosphere moves along with the Earth (which is a good thing). Think about how you know you are in motion when traveling in a perfectly smooth-riding and quiet car. You see trees and other fixed objects "moving" in relative motion.

Also, there is no sensing of any change in velocity if the change is small. Changes in motion can be sensed if they are appreciable. Think about being in a smooth-riding car and being blindfolded. Minor changes in velocity would go unnoticed, but you would be able to tell if the car suddenly sped up, slowed down, or went around a sharp curve, all of which are changes in velocity. A change in velocity is called an *acceleration*, the topic of the following sections.

Did You Learn?

- Distance (actual path length) is always greater than or equal to the magnitude of displacement (straight-line distance).
- Speed is a scalar (magnitude), and velocity is a vector (magnitude and direction).

2.3 Acceleration

Preview Questions

- What motional changes produce an acceleration?
- Is a negative acceleration ($-a$) necessarily a deceleration?

When you are riding in a car on a straight interstate highway and the speed is suddenly increased—say, from 20 m/s to 30 m/s (45 mi/h to 56 mi/h)—you feel as though you are being forced back against the seat. If the car then whips around a circular cloverleaf, you feel forced to the outside of the circle. These experiences result from changes in velocity.

Because velocity is a vector quantity, with both magnitude and direction, a change in velocity involves either or both of these factors. Therefore, an acceleration may result from (1) a change in *speed* (magnitude), (2) a change in *direction*, or (3) a change in *both speed and direction*. When any of these changes occur, an object is accelerating. Examples are (1) a car speeding up (or slowing down) while traveling in a straight line, (2) a car rounding a curve at a constant speed, and (3) a car speeding up (or slowing down) while rounding a curve.

Acceleration is defined as *the time rate of change of velocity*. Taking the symbol Δ (delta) to mean "change in," the equation for **average acceleration** ($\bar{a}$) can be written as

$$\text{average acceleration} = \frac{\text{change in velocity}}{\text{time for change to occur}}$$

$$\bar{a} = \frac{\Delta v}{\Delta t} = \frac{v_f - v_o}{t}$$

2.2

The change in velocity (Δv) is the final velocity v_f minus the original velocity v_o. Also, the interval Δt is commonly written as t ($\Delta t = t - t_o = t$), with t_o taken to be zero (starting the clock at zero, and t is understood to be an interval).

The units of acceleration in the SI are meters per second per second, (m/s)/s, or meters per second squared, m/s². These units may be confusing at first. Keep in mind that an acceleration is a measure of a *change* in velocity during a given time period.

Consider a constant acceleration of 9.8 m/s². This value means that the velocity changes by 9.8 m/s *each* second. Thus, for straight-line motion, as the number of seconds increases, the velocity goes from 0 to 9.8 m/s during the first second, to 19.6 m/s (that is, 9.8 m/s + 9.8 m/s) during the second second, to 29.4 m/s (that is, 19.6 m/s + 9.8 m/s) during the third second, and so forth, adding 9.8 m/s each second. This sequence is illustrated in ● Fig. 2.7 for an object that falls with a constant acceleration of 9.8 m/s². Because the velocity increases, the distance traveled by the falling object each second also increases, but *not* uniformly.

Our discussion will be limited to such constant accelerations. For a constant acceleration, $\bar{a} = a$ (we omit the overbar because the average acceleration is equal to the constant value). Equation 2.2 may be rearranged to give an expression for the final velocity of an object in terms of its original velocity, time, and constant acceleration (a):

$$v_f - v_o = at$$

or

$$v_f = v_o + at \qquad\qquad \textbf{2.3}$$

This Equation 2.3 is useful for working problems in which the quantities a, v_o, and t are all known and v_f is wanted. If the original velocity $v_o = 0$, then

$$v_f = at \qquad\qquad \textbf{2.3a}$$

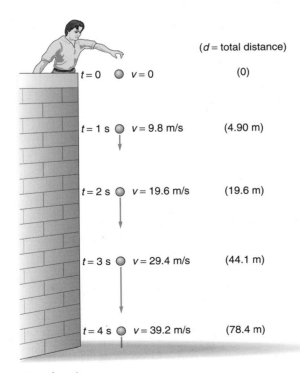

(d = total distance)

$t = 0$ ○ $v = 0$		(0)
$t = 1$ s ○ $v = 9.8$ m/s		(4.90 m)
$t = 2$ s ○ $v = 19.6$ m/s		(19.6 m)
$t = 3$ s ○ $v = 29.4$ m/s		(44.1 m)
$t = 4$ s ○ $v = 39.2$ m/s		(78.4 m)

Figure 2.7 Constant Acceleration
For an object with a constant downward acceleration of 9.8 m/s², its velocity increases 9.8 m/s each second. The increasing lengths of the velocity arrows indicate the increasing velocity of the ball. The distances do not increase uniformly. (The distances of fall are obviously not to scale.)

It should be noted that the use of Eq. 2.3 requires a *constant* acceleration, one for which the velocity changes uniformly.

EXAMPLE 2.4   Finding Acceleration

A race car starting from rest accelerates uniformly along a straight track, reaching a speed of 90 km/h in 7.0 s. What is the magnitude of the acceleration of the car in m/s²?

Solution

Let's work this example stepwise for clarity.

Step 1

Given: $v_o = 0$, $v_f = 90$ km/h, $t = 7.0$ s
(Because the car is initially at rest, $v_o = 0$.)

Step 2

Wanted: a (acceleration in m/s²)
Because the acceleration is wanted in m/s², 90 km/h is converted to m/s using the conversion factor from the inside back cover of this book:

$$v_f = 90 \ \cancel{\text{km/h}} \left(\frac{0.278 \ \text{m/s}}{\cancel{\text{km/h}}} \right) = 25 \ \text{m/s}$$

Step 3

The acceleration may be calculated using Eq. 2.2:

$$a = \frac{v_f - v_o}{t} = \frac{25 \ \text{m/s} - 0}{7.0 \ \text{s}} = 3.6 \ \text{m/s}^2$$

Confidence Exercise 2.4

If the car in the preceding example continues to accelerate at the same rate for an additional 3.0 s, what will be the magnitude of its velocity in m/s at the end of this time?

Because velocity is a vector quantity, acceleration is also a vector quantity. For an object in straight-line motion, the acceleration (vector) may be in the same direction as the velocity (vector), as in Example 2.4, *or* the acceleration may be in the direction opposite the velocity (● Fig. 2.8). In the first instance, the acceleration causes the object to speed up and the velocity increases. If the velocity and acceleration are in opposite directions, however, then the acceleration slows the object and the velocity decreases, which is sometimes called a *deceleration*.*

Figure 2.8 Acceleration and Deceleration
When the acceleration is in the same direction as the velocity of an object in straight-line motion, the speed increases. When the acceleration is in the opposite direction of the velocity (beginning at the vertical dashed line), there is a deceleration and the speed decreases.

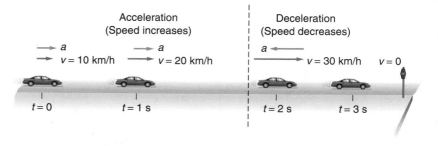

Acceleration
(Speed increases)

a
$v = 10$ km/h

a
$v = 20$ km/h

Deceleration
(Speed decreases)

a
$v = 30$ km/h $v = 0$

$t = 0$ $t = 1$ s $t = 2$ s $t = 3$ s

---

*Note that a deceleration is *not* necessarily a negative acceleration ($-a$). If the motion is in the negative direction ($-v$) and the velocity and the acceleration are in the *same* direction, then the object speeds up; that is, its velocity increases in the negative direction.

Conceptual Question and Answer

Putting the Pedal to the Metal

Q. Why is the gas pedal of an automobile commonly called the "accelerator"? What might the brake pedal be called? How about the steering wheel?

A. When you push down on the gas pedal or accelerator, the car speeds up (increasing magnitude of velocity), but when you let up on the accelerator, the car slows down or decelerates.

  Putting on the brakes would produce an even greater deceleration, so perhaps the brake pedal should be called a "decelerator." An acceleration results from a change in the velocity's magnitude and/or direction. Technically, then, the steering wheel might also be called an "accelerator," since it changes a speeding car's direction.

In general, we will consider only constant, or uniform, accelerations. A special constant acceleration is associated with the acceleration of falling objects. The **acceleration due to gravity** at the Earth's surface is directed downward and is denoted by the letter g. Its magnitude in SI units is

$$g = 9.80 \text{ m/s}^2$$

This value corresponds to 980 cm/s^2, or about 32 ft/s^2.

  The acceleration due to gravity varies slightly depending on such factors as distance from the equator (latitude) and altitude. However, the variations are very small, and for our purposes, g will be taken to be the same everywhere on or near the Earth's surface.

  Italian physicist Galileo Galilei (1564–1642), commonly known just as Galileo, was one of the first scientists to assert that all objects fall with the same acceleration. Of course, this assertion assumes that frictional effects are negligible. To exclude frictional and any other effects, the term *free fall* is used. Objects in motion solely under the influence of gravity are said to be in **free fall**. One can illustrate the validity of this statement experimentally by dropping a small mass, such as a coin, and a larger mass, such as a ball, at the same time from the same height. They will both hit the floor, as best as can be judged, at the same time (negligible air resistance). Legend has it that Galileo himself performed such experiments. (See the **Highlight: Galileo and the Leaning Tower of Pisa.**) More modern demonstrations are shown in ● Fig. 2.9.

  On the Moon, there is no atmosphere, so there is no air resistance. In 1971, while on the lunar surface, astronaut David Scott dropped a feather and a hammer simultaneously. They

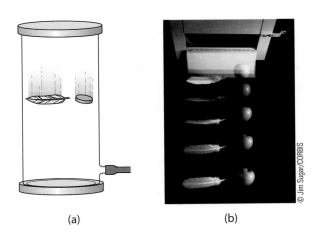

(a)   (b)

© Jim Sugar/CORBIS

Figure 2.9 Free Fall and Air Resistance
(a) When dropped simultaneously from the same height, a feather falls more slowly than a coin because of air resistance. But when both objects are dropped in an evacuated container with a good partial vacuum, where air resistance is negligible, the feather and the coin fall together with a constant acceleration. (b) An actual demonstration with multiflash photography: An apple and a feather are released simultaneously through a trap door into a large vacuum chamber, and they fall together, almost. Because the chamber has a partial vacuum, there is still some air resistance. (How can you tell?)

Highlight   Galileo and the Leaning Tower of Pisa

One of Galileo's greatest contributions to science was his emphasis on experimentation, a basic part of the scientific method (Chapter 1.2). See Fig. 1. However, it is not certain whether he actually carried out a now-famous experiment. There is a popular story that Galileo dropped stones or cannonballs of different masses from the top of the Tower of Pisa to determine experimentally whether objects fall with the same acceleration (Fig. 2).

Galileo did indeed question Aristotle's view that objects fell because of their "earthiness" and that the heavier, or more earthy, an object, the faster it would fall in seeking its "natural" place toward the center of the Earth. His ideas are evident in the following excerpts from his writings.*

How ridiculous is this opinion of Aristotle is clearer than light. Who ever would believe, for example, that . . . if two stones were flung at the same moment from a high tower, one stone twice the size of the other, . . . that when the smaller was half-way down the larger had already reached the ground?

And Aristotle says that "an iron ball of one hundred pounds falling a height of one hundred cubits reaches the ground before a one-pound ball has fallen a single cubit." I say that they arrive at the same time.

Although Galileo refers to a *high tower*, the Tower of Pisa is not mentioned in his writings, and there is no independent record of such an experiment. Fact or fiction? No one really knows. What we do know is that all freely falling objects near the Earth's surface fall with the same acceleration.

The Granger Collection, NYC

Figure 1 Galileo Galilei (1564–1642)
The motion of objects was one of Galileo's many scientific inquiries.

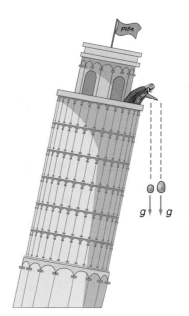

Figure 2 Free Fall
All freely falling objects near the Earth's surface have the same constant acceleration. Galileo is alleged to have proved this by dropping cannonballs or stones of different masses from the Tower of Pisa. Over short distances, the air resistance can be neglected, so objects dropped at the same time will strike the ground together.

---

*From L. Cooper, *Aristotle, Galileo, and the Tower of Pisa* (Ithaca, NY: Cornell University Press, 1935).

---

both hit the surface of the Moon at the same time because neither the feather nor the hammer was slowed by air resistance. This experiment shows that Galileo's assertion applies on the Moon as well as on the Earth. Of course, on the Moon all objects fall at a slower rate than do objects on the Earth's surface because the acceleration due to gravity on the Moon is only $\frac{1}{6}$ of the acceleration due to gravity on the Earth.

The velocity of a freely falling object on the Earth increases 9.80 m/s each second, so its magnitude increases uniformly with time, but how about the distance covered each second? The distance covered is not uniform because the object speeds up. The distance a dropped object ($v_\text{o} = 0$) travels downward with time can be computed from the equation $d = \frac{1}{2}at^2$, with g substituted for a.

$$d = \tfrac{1}{2}gt^2$$

2.4

EXAMPLE 2.5   **Finding How Far a Dropped Object Falls**

A ball is dropped from the top of a tall building. Assuming free fall, how far does the ball fall in 1.50 s?

Solution

With $t = 1.50$ s, the distance is given by Eq. 2.4, where it is known that $g = 9.80$ m/s^2.

$$d = \tfrac{1}{2}gt^2 = \tfrac{1}{2}(9.80 \text{ m/s}^2)(1.50 \text{ s})^2 = 11.0 \text{ m}$$

Confidence Exercise 2.5

What is the speed of the ball 1.50 s after it is dropped?

Review Fig. 2.7 for the distances and velocities of a falling object with time. Note that the increase in distance is directly proportional to the time squared ($d \propto t^2$, Eq. 2.4), and the increase in velocity is directly proportional to time ($v_f \propto t$, Eq. 2.3a).

Using these equations, it can be shown that at the end of the third second ($t = 3.00$ s), the object has fallen 44.1 m, which is about the height of a 10-story building. It is falling at a speed of 29.4 m/s, which is about 65 mi/h. Pretty fast. Look out below!

When an object is thrown straight upward, it slows down while traveling. (The velocity decreases 9.80 m/s each second.) In this case, the velocity and acceleration are in opposite directions, and there is a deceleration (● Fig. 2.10). The object slows down (its velocity decreases) until it stops *instantaneously* at its maximum height. Then it starts to fall downward as though it had been dropped from that height. The travel time downward to the original starting position is the same as the travel time upward.

The object returns to its starting point with the same speed it had initially upward. For example, if an object is thrown upward with an initial speed of 29.4 m/s, it will return to the starting point with a speed of 29.4 m/s. You should be able to conclude that it will travel 3.00 s upward, to a maximum height of 44.1 m, and return to its starting point in another 3.00 s.

The preceding numbers reflect free-fall motion, but air resistance generally retards or slows the motion. For a good example of air resistance, consider skydiving, jumping out of an airplane and falling toward the Earth (with a parachute). Before the parachute opens,

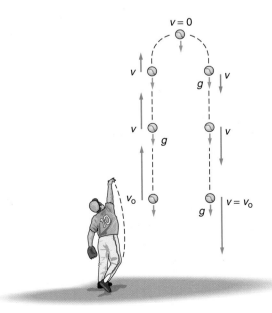

Figure 2.10 Up and Down
An object projected straight upward slows down because the acceleration due to gravity is in the opposite direction of the velocity, and the object stops ($v = 0$) for an instant at its maximum height. Note the blue acceleration arrow at $v = 0$. That is $a \neq 0$. Gravity still acts and $a = g$. The ball then accelerates downward, returning to the starting point with a velocity equal in magnitude and opposite in direction to the initial velocity, $-v_o$. (The downward path is displaced to the right in the figure for clarity.)

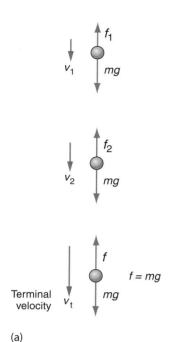

Getty Images

(a)

(b)

Figure 2.11 Air Resistance and Terminal Velocity

(a) As the velocity of a falling object increases, so does the upward force f of air resistance. At a certain velocity, called the *terminal velocity*, the force f is equal to the weight of the object (mg). The object then falls with a constant velocity. (b) Skydivers assume a "spread-eagle" position to increase air resistance and reach terminal velocity more quickly, allowing for a greater time of fall.

the falling skydiver is said to be in "free fall," but it is not free fall by our definition. Here, there is air resistance, which skydivers use to their advantage.

Air resistance is the result of a moving object colliding with air molecules. Therefore, air resistance (considered a type of friction) depends on an object's size and shape, as well as on its speed. The larger the object (or the more downward area exposed) and the faster it moves, the more collisions and the more air resistance there will be. As a skydiver accelerates downward, his or her speed increases, as does the air resistance. At some point, the upward air resistance balances the downward weight of the diver and the acceleration goes to zero. The skydiver then falls with a constant velocity, which is called the **terminal velocity**, v_t (● Fig. 2.11a).

Wanting to maximize the time of fall, skydivers assume a "spread-eagle" position to provide greater surface area and maximize the air resistance (Fig. 2.11b). The air resistance then builds up faster and terminal velocity is reached sooner, giving the skydiver more fall time. This position is putting air resistance to use. The magnitude of a skydiver's terminal velocity during a fall is reached at about 200 km/h (125 mi/h).

Conceptual Question and Answer

And the Winner Is . . .

Q. Suppose two metal balls of different materials, but the same size and shape, were dropped simultaneously from a high-altitude balloon. Let one ball be more massive (heavier) than the other. Which ball will strike the ground first, or would they hit at the same time?

A. Here air resistance is a factor. As the falling balls gain speed, the upward air resistance would increase the same on each (same size and shape). However, when the air resistance is sufficient to balance the weight of the lighter ball, it would reach terminal velocity and cease to accelerate.

The magnitude of this air resistance does not balance the weight of the heavier ball, which would continue to accelerate downward until it reached its greater terminal velocity. So, the heavier ball would be ahead of the lighter ball and falling faster, reaching the ground first.

Did You Learn?

- Acceleration results from a change in velocity, which may result from a change in magnitude, a change in direction, or both.
- A negative acceleration ($-a$) will speed up an object with a negative velocity ($-v$).

2.4   Acceleration in Uniform Circular Motion

Preview Questions

- What is needed for uniform circular motion?
- For an object in uniform circular motion, are both the speed and the velocity constant?

An object in *uniform* circular motion has a constant speed. A car goes around a circular track at a uniform speed of 90 km/h (about 55 mi/h). However, the velocity of the car is *not* constant because the direction is continuously changing, giving rise to a change in velocity and an acceleration.

This acceleration cannot be in the direction of the instantaneous motion or velocity, because otherwise the object would speed up and the motion would not be uniform. In what direction, then, is the acceleration? Because the object must continually change *direction* to maintain a circular path, the acceleration is actually perpendicular, or at a right angle, to the velocity vector.

Consider a car traveling in uniform circular motion, as illustrated in ● Fig. 2.12. At any point, the instantaneous velocity is *tangential* to the curve (at an angle of 90° to a radial line at that point). After a short time, the velocity vector has changed (in direction). The change in velocity (Δv) is given by a vector triangle, as illustrated in the figure.

This change is an average over a time interval Δt, but notice how the Δv vector generally points inward toward the center of the circle. For instantaneous measurement, this generalization is true, so for an object in uniform circular motion, the acceleration is toward the center of the circle. This acceleration is called **centripetal acceleration**.

Even when traveling at a constant (tangential) speed, an object in uniform circular motion must have an inward acceleration. For a car, this acceleration is supplied by friction on the tires. When a car hits an icy spot on a curved road, it may slide outward if the centripetal acceleration is not great enough to keep it in a circular path.

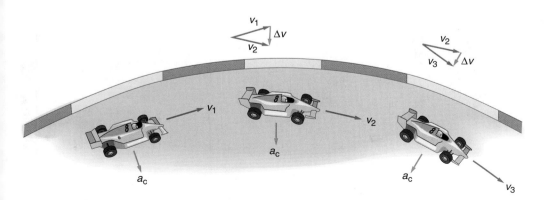

Figure 2.12 Centripetal Acceleration
A car traveling with a constant speed on a circular track is accelerating because its velocity is changing (direction change). This acceleration is toward the center of the circular path and is called *centripetal* ("center-seeking") *acceleration*.

In general, whenever an object moves in a circle of radius r with a constant speed v, the magnitude of the centripetal acceleration a_c in terms of tangential speed is given by the equation

$$\text{centripetal acceleration} = \frac{(\text{speed})^2}{\text{radius}}$$

$$a_c = \frac{v^2}{r} \qquad\qquad 2.5$$

The centripetal acceleration increases as the square of the speed. That is, when the speed is doubled, the acceleration increases by a factor of 4. Also, the smaller the radius, the greater the centripetal acceleration needed to keep an object in circular motion for a given speed.

EXAMPLE 2.6   Finding Centripetal Acceleration

Determine the magnitude of the centripetal acceleration of a car going 12 m/s (about 27 mi/h) on a circular cloverleaf with a radius of 50 m (● Fig. 2.13).

Solution

Given: $v = 12$ m/s, $r = 50$ m
The centripetal acceleration is given by Eq. 2.5:

$$a_c = \frac{v^2}{r} = \frac{(12\ \text{m/s})^2}{50\ \text{m}}$$

$$= 2.9\ \text{m/s}^2$$

The value of 2.9 m/s² is about 30% of the acceleration due to gravity, $g = 9.8$ m/s², and is a fairly large acceleration.

Confidence Exercise 2.6

Using the result of Example 2.3, compute the centripetal acceleration in m/s² of the Earth in its nearly circular orbit about the Sun ($r = 1.5 \times 10^{11}$ m).

Did You Learn?

● A centripetal (center-seeking) acceleration is needed for uniform circular motion.

● In uniform circular motion, there is a change in velocity (direction), but the tangential speed is constant.

Figure 2.13 Frictional Centripetal Acceleration
The inward acceleration necessary for a vehicle to go around a level curve is supplied by friction on the tires. On a banked curve, some of the acceleration is supplied by the inward component of the acceleration due to gravity.

2.5   Projectile Motion

Preview Questions

- Neglecting air resistance, why would a ball projected horizontally and another ball dropped at the same time from the same initial height hit the ground together?
- On what does the range of a projectile depend? (Neglect air resistance.)

Another common motion in two dimensions is that of an object thrown or projected by some means. For example, the motion of a thrown ball or a golf ball driven off a tee is called **projectile motion**. A special case of projectile motion is that of an object projected horizontally, or parallel, to a level surface. Suppose a ball is projected horizontally, while at the same time another ball is dropped from the same height (● Fig. 2.14a). You might be surprised that both balls hit the ground at the same time (neglecting air resistance). An object thrown horizontally will fall at the same rate as an object that is dropped (both fall with acceleration g). The velocity in the horizontal direction does not affect the velocity and acceleration in the vertical direction.

The multiflash photo in Fig. 2.14b shows a dropped ball and one projected horizontally at the same time. Notice that the balls fall vertically together as the projected ball moves to the right. Neglecting air resistance, a horizontally projected object essentially travels in a horizontal direction with a constant velocity (no acceleration in that direction) while falling vertically under the influence of gravity. The resulting path is a curved arc, as shown in Fig. 2.14. Occasionally, a sports announcer claims that a hard-throwing quarterback can throw a football so many yards "on a line," meaning a straight line. This statement, of course, must be false. All objects thrown horizontally begin to fall as soon as they leave the thrower's hand.

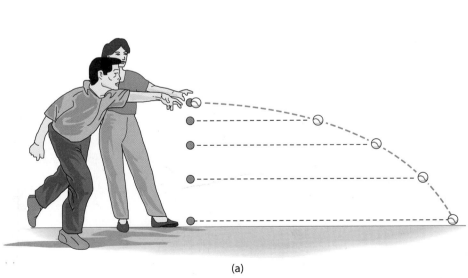
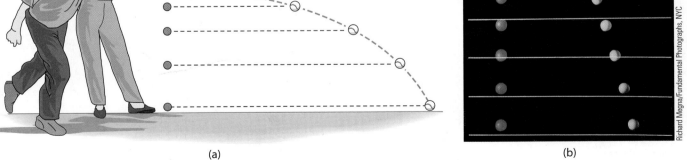

(a)

(b)

Figure 2.14 Same Vertical Motions
(a) When one ball is thrown horizontally and another ball is dropped simultaneously from the same height, both will hit the ground at the same time (with air resistance neglected), because the vertical motions are the same. (Diagram not to scale.) (b) A multiflash photograph of two balls, one of which was projected horizontally at the same time the other ball was dropped. Notice from the horizontal lines that both balls fall vertically at the same rate.

Figure 2.15 Projectile Motion
(a) The curved path of a projectile is a result of the combined motions in the vertical and horizontal directions. As the ball goes up and down, it travels to the right. The combined effect is a curved path. (b) Neglecting air resistance, the projected football has the same horizontal velocity (v_x) throughout its flight, but its vertical velocity (v_y) changes in the same way as that of an object thrown upward.

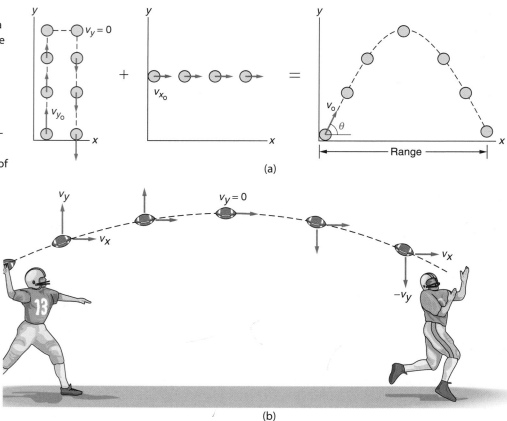

(a)

(b)

If an object is projected at an angle θ (lowercase Greek theta) to the horizontal, then it will follow a symmetric curved path, as illustrated in ● Fig. 2.15, where air resistance is again neglected. The curved path is essentially the result of the combined motions in the vertical and horizontal directions. The projectile goes up and down vertically while at the same time traveling horizontally with a constant velocity.

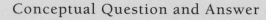

Conceptual Question and Answer

Hanging in There

Q. When a basketball player drives in and jumps to shoot for a goal (or "slam dunk"), he seems to momentarily "hang in the air" (● Fig. 2.16). Why?

A. When running and jumping for the shot, the player is in near projectile motion. Notice in Fig. 2.10 that the velocity is small for the vertical motion near the maximum height, decreasing to zero and then increasing slowly downward. During this time, the combination of the slow vertical motions and the constant horizontal motion gives the illusion of the player "hanging" in midair.

When a ball or other object is thrown, the path that it takes depends on the projection angle. Neglecting air resistance, each path will resemble one of those in ● Fig. 2.17. As shown, the *range*, or horizontal distance the object travels, is maximum when the object is projected at an angle of 45° relative to level ground. Notice in the figure that for a given initial speed, projections at complementary angles (angles that add up to 90°)—for example, 30° and 60°—have the same range as long as there is no air resistance.

With little or no air resistance, projectiles have symmetric paths, but when a ball or object is thrown or hit hard, air resistance comes into effect. In such a case, the projectile

Figure 2.16 Hanging in There
Carmelo Anthony seems to "hang" in the air at the top of his "projectile" jump path. (See text for description.)

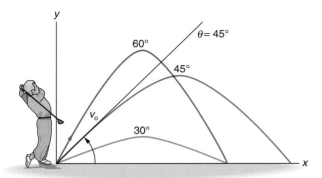

Figure 2.17 Maximum Range
A projectile's maximum range on a horizontal plane is achieved with a projection angle of 45° (in the absence of air resistance). Projectiles with the same initial speed and projection angles the same amount above and below 45° have the same range, as shown here for 30° and 60°.

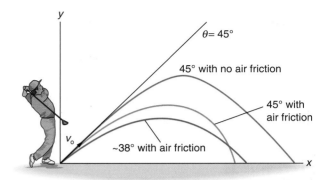

Figure 2.18 Effects of Air Resistance on Projectiles
Long football passes, hard-hit baseballs, and driven golf balls follow trajectories similar to those shown here. Frictional air resistance reduces the range.

John Cumming/Jupiter Images

Figure 2.19 Going for the Distance
Because of air resistance, the athlete hurls the javelin at an angle of less than 45° for maximum range.

path is no longer symmetric and resembles one of those shown in ● Fig. 2.18. Air resistance reduces the velocity of the projectile, particularly in the horizontal direction. As a result, the maximum range occurs at an angle less than 45°.

Athletes such as football quarterbacks and baseball players are aware of the best angle at which to throw to get the maximum distance. A good golfing drive also depends on the angle at which the ball is hit. Of course, in most of these instances there are other considerations, such as spin. The angle of throw is also a consideration in track and field events such as discus and javelin throwing. ● Figure 2.19 shows an athlete hurling a javelin at an angle of less than 45° in order to achieve the maximum distance.

Did You Learn?

● A horizontally projected ball and another ball dropped from the same height have the same downward motion because the vertical acceleration (g) is the same (neglecting air resistance).

● The range of a projectile depends on the initial velocity and angle of projection (neglecting air resistance).

KEY TERMS

1. physics (intro)
2. position (2.1)
3. motion
4. scalar (2.2)
5. vector
6. average speed
7. distance
8. instantaneous speed
9. average velocity
10. displacement
11. instantaneous velocity
12. acceleration (2.3)
13. average acceleration
14. acceleration due to gravity
15. free fall
16. terminal velocity
17. centripetal acceleration (2.4)
18. projectile motion (2.5)

MATCHING

For each of the following items, fill in the number of the appropriate Key Term from the preceding list. Compare your answers with those at the back of the book.

a. _____ Displacement/travel time
b. _____ Has magnitude only
c. _____ Velocity at an instant of time
d. _____ Directed toward the center of circular motion
e. _____ Actual path length
f. _____ Straight-line directed distance
g. _____ A continuous change of position
h. _____ Time rate of change of velocity
i. _____ Has magnitude and direction

j. _____ Motion solely under the influence of gravity
k. _____ Motion of a thrown object
l. _____ Speed at an instant of time
m. _____ The location of an object
n. _____ 9.8 m/s^2
o. _____ Distance traveled/travel time
p. _____ The most fundamental of the physical sciences
q. _____ Difference between final and initial velocities divided by time
r. _____ Zero acceleration in free fall

MULTIPLE CHOICE

Compare your answers with those at the back of the book.

1. What is necessary to designate a position? (2.1)
   (a) fundamental units   (b) motion
   (c) a direction   (d) a reference point

2. Which one of the following describes an object in motion? (2.1)
   (a) A period of time has passed.
   (b) Its position is known.
   (c) It is continuously changing position.
   (d) It has reached its final position.

3. Which one of the following is always true about the magnitude of a displacement? (2.2)
   (a) It is greater than the distance traveled.
   (b) It is equal to the distance traveled.
   (c) It is less than the distance traveled.
   (d) It is less than or equal to the distance traveled.

4. Distance is to displacement as ___. (2.2)
   (a) centimeters is to meters   (b) a vector is to a scalar
   (c) speed is to velocity   (d) distance is to time

5. Acceleration may result from what? (2.3)
   (a) an increase in speed
   (b) a decrease in speed
   (c) a change in direction
   (d) all of the preceding

6. For a constant linear acceleration, what changes uniformly? (2.3)
   (a) acceleration
   (b) velocity
   (c) distance
   (d) displacement

7. Which one of the following is true for a deceleration? (2.3)
   (a) The velocity remains constant.
   (b) The acceleration is negative.
   (c) The acceleration is in the direction opposite to the velocity.
   (d) The acceleration is zero.

8. Which is true for an object in free fall? (2.3)
   (a) It has frictional effects.
   (b) It has a constant velocity.
   (c) It has a constant displacement.
   (d) It increases in distance proportionally to t^2.

9. If the speed of an object in uniform circular motion is tripled and the radial distance remains constant, then the magnitude of the centripetal acceleration increases by what factor? (2.4)
   (a) 2   (b) 3   (c) 4   (d) 9

10. Neglecting air resistance, which of the following is true for a ball thrown at an angle θ to the horizontal? (2.5)
   (a) It has a constant velocity in the $+x$ direction.
   (b) It has a constant acceleration in the $-y$ direction.
   (c) It has a changing velocity in the $+y$ direction.
   (d) All of the preceding are true.

11. In the absence of air resistance, a projectile launched at an angle of 28° above the horizontal will have the same range as a projectile launched at which of the following angles? (2.5)
   (a) 45°   (b) 57°   (c) 62°   (d) 180° − 33° = 147°

12. A football is thrown on a long pass. Compared to the ball's initial horizontal velocity, the velocity at the highest point is ___. (2.5)
   (a) greater   (b) less   (c) the same

FILL IN THE BLANK

Compare your answers with those at the back of the book.

1. An object is in motion when it undergoes a continuous change of ___. (2.1)
2. Speed is a(n) ___ quantity. (2.2)
3. Velocity is a(n) ___ quantity. (2.2)
4. ___ is the actual path length. (2.2)
5. A car's speedometer reads instantaneous ___. (2.2)
6. Speed and direction do not change for a(n) ___ velocity. (2.2)
7. The distance traveled by a dropped object increases with ___. (2.3)
8. Objects in motion solely under the influence of gravity are said to be in ___. (2.3)
9. The metric units associated with acceleration are ___. (2.3)
10. An object in uniform circular motion has a constant ___. (2.4)
11. If the speed of one object in uniform circular motion is two times that of another such object (same radius), then its centripetal acceleration is ___ times greater. (2.4)
12. Neglecting air resistance, a horizontally thrown object and an object dropped from the same height fall with the same constant ___. (2.5)

SHORT ANSWER

1. What area of physics involves the study of objects moving at relatively low speeds?

2.1 Defining Motion

2. What is necessary to designate the position of an object?
3. How are position and motion related?

2.2 Speed and Velocity

4. Distinguish between scalar and vector quantities.
5. Distinguish between distance and displacement. How are these quantities associated with speed and velocity?
6. How is average speed analogous to an average class grade?
7. A jogger jogs two blocks directly north.
   (a) How do the jogger's average speed and the magnitude of the average velocity compare?
   (b) If the jogger's return trip follows the same path, then how do average speed and magnitude of the average velocity compare for the total trip?

2.3 Acceleration

8. What changes when there is acceleration? Give an example.
9. The gas pedal of a car is commonly referred to as the accelerator. Would this term be appropriate for (a) the clutch in a stick-shift car? (b) the gears in a stick-shift car? Explain.
10. Does a negative acceleration always mean that an object is slowing down? Explain.
11. A ball is dropped. Assuming free fall, what is its initial speed? What is its initial acceleration? What is the final acceleration?

12. A vertically projected object has zero velocity at its maximum height, but the acceleration is not zero. What would be implied if the acceleration were zero?

2.4 Acceleration in Uniform Circular Motion

13. Can a car be moving at a constant speed of 60 km/h and still be accelerating? Explain.
14. What does the term *centripetal* mean?
15. Are we accelerating as a consequence of the Earth spinning on its axis? Explain.
16. What is the direction of the acceleration vector of a person on the spinning Earth if the person is (a) at the equator? (b) at some other latitude? (c) at the poles?

2.5 Projectile Motion

17. For projectile motion, what quantities are constant? (Neglect air resistance.)
18. How do the motions of horizontal projections with the same initial speed compare on the Earth and on the Moon?
19. On what does the range of a projectile depend?
20. Can a baseball pitcher throw a fastball in a straight, horizontal line? Why or why not?
21. Figure 2.14(b) shows a multiflash photograph of one ball dropped from rest and, at the same time, another ball projected horizontally from the same height. The two hit the floor at the same time. Explain.
22. Taking into account air resistance, how do you throw a ball to get the maximum range? Why?

VISUAL CONNECTION

Visualize the connections and give answers for the blanks. Compare your answers with those at the back of the book.

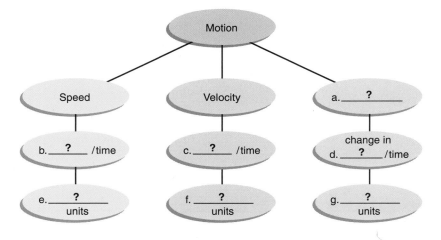

APPLYING YOUR KNOWLEDGE

1. Do highway speed limit signs refer to average speeds or to instantaneous speeds? Explain.

2. (a) If we are moving at a high speed as the Earth revolves about the Sun (18.5 m/s, Example 2.3), then why don't we generally feel this motion? (b) Similarly, we are traveling through space because of the Earth's rotation (1000 mi/h at the equator, Confidence Exercise 2.3). We don't generally feel this motion either, but we do easily sense it otherwise. How?

3. What is the direction of the acceleration vector of a person on the spinning Earth if the person is (a) at the equator? (b) at some other latitude? (c) at the poles?

4. Is an object projected vertically upward in free fall? Explain.

5. A student sees her physical science professor approaching on the sidewalk that runs by her dorm. She gets a water balloon and waits. When the professor is 2.0 s from being directly under her window 11 m above the sidewalk, she drops the balloon. You finish the story.

6. How would (a) an updraft affect a skydiver in reaching terminal velocity? (b) a downdraft?

7. A skydiver uses a parachute to slow the landing speed. Parachutes generally have a hole in the top. Why? Wouldn't air going through the hole deter the slowing?

8. Tractor-trailer rigs often have an airfoil on top of the cab, as shown in ● Fig. 2.20. What is the purpose of this airfoil?

Figure 2.20 Air Foil
See Question 8 in Applying Your Knowledge.

IMPORTANT EQUATIONS

Average Speed: $\bar{v} = \dfrac{d}{t}$ (2.1)

Acceleration: $\bar{a} = \dfrac{\Delta v}{t} = \dfrac{v_f - v_o}{t}$ (2.2)

$v_f = v_o + at$ (constant acceleration)   (2.3)

or with $v_o = 0$,

$v_f = at$ (2.3a)

Distance Traveled by a Dropped Object: $d = \frac{1}{2}gt^2$ (2.4)

Centripetal Acceleration: $a_c = \dfrac{v^2}{r}$ (2.5)

EXERCISES

2.2 Speed and Velocity

1. A gardener walks in a flower garden as illustrated in ● Fig. 2.21. What distance does the gardener travel?

   Answer: 7 m

2. What is the gardener's displacement (Fig. 2.21)? Give a general direction, such as south or west. (*Hint:* Think of a 3-4-5 triangle, Pythagorean theorem.)

3. At a track meet, a runner runs the 100-m dash in 12 s. What was the runner's average speed?

   Answer: 8.3 m/s

4. A jogger jogs around a circular track with a diameter of 300 m in 10 minutes. What was the jogger's average speed in m/s?

5. A space probe on the surface of Mars sends a radio signal back to the Earth, a distance of 7.86×10^7 km. Radio waves travel at the speed of light (3.00×10^8 m/s). How many seconds does it take the signal to reach the Earth?

   Answer: 2.62×10^2 s

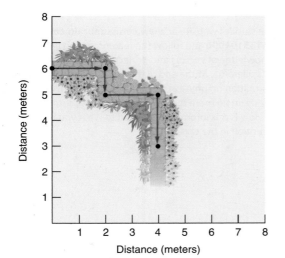

Figure 2.21 A Walk through the Garden
See Exercises 1 and 2.

6. A group of college students eager to get to Florida on a spring break drove the 750-mi trip with only minimum stops. They computed their average speed for the trip to be 55.0 mi/h. How many hours did the trip take?

7. A student drives the 100-mi trip back to campus after spring break and travels with an average speed of 52 mi/h for 1 hour and 30 minutes for the first part of the trip.
   (a) What distance was traveled during this time?
   (b) Traffic gets heavier, and the last part of the trip takes another half-hour. What was the average speed during this leg of the trip?
   (c) Find the average speed for the total trip.

   Answer: (a) 78 mi (b) 44 mi/h (c) 50 mi/h

8. Joe Cool drives to see his girlfriend who attends another college.
   (a) The 130-km trip takes 2.0 h. What was the average speed for the trip?
   (b) The return trip over the same route takes 3.0 h because of heavy traffic. What is the average speed for the return trip?
   (c) What is the average speed for the entire trip?

9. An airplane flying directly eastward at a constant speed travels 300 km in 2.0 h.
   (a) What is the average velocity of the plane?
   (b) What is its instantaneous velocity?

   Answer: (a) 150 km/h, east (b) same

10. A race car traveling northward on a straight, level track at a constant speed travels 0.750 km in 20.0 s. The return trip over the same track is made in 25.0 s.
    (a) What is the average velocity of the car in m/s for the first leg of the run?
    (b) What is the average velocity for the total trip?

2.3 Acceleration

11. A sprinter starting from rest on a straight, level track is able to achieve a speed of 12 m/s in 6.0 s. What is the sprinter's average acceleration?

    Answer: 2.0 m/s²

12. Modern oil tankers weigh more than a half-million tons and have lengths of up to one-fourth mile. Such massive ships require a distance of 5.0 km (about 3.0 mi) and a time of 20 min to come to a stop from a top speed of 30 km/h.

(a) What is the magnitude of such a ship's average acceleration in m/s^2 in coming to a stop?

(b) What is the magnitude of the ship's average velocity in m/s? Comment on the potential of a tanker running aground.

13. A motorboat starting from rest travels in a straight line on a lake.
    (a) If the boat achieves a speed of 8.0 m/s in 10 s, what is the boat's average acceleration?
    (b) Then, in 5.0 more seconds, the boat's speed is 12 m/s. What is the boat's average acceleration for the total time?

    Answer: (a) 0.80 m/s^2 in the direction of motion (b) same

14. A car travels on a straight, level road.
    (a) Starting from rest, the car is going 44 ft/s (30 mi/h) at the end of 5.0 s. What is the car's average acceleration in ft/s^2?
    (b) In 4.0 more seconds, the car is going 88 ft/s (60 mi/h). What is the car's average acceleration for this time period?
    (c) The car then slows to 66 ft/s (45 mi/h) in 3.0 s. What is the average acceleration for this time period?
    (d) What is the overall average acceleration for the total time?
    (Note these convenient British unit conversions: 60 mi/h = 88 ft/s, 45 mi/h = 66 ft/s, and 30 mi/h = 44 ft/s.)

15. A ball is dropped from the top of an 80-m-high building. Does the ball reach the ground in 4.0 s? (See Figure 2.7.)

    Answer: No, it falls only 78 m in 4.0 s.

16. What speed does the ball in Exercise 15 have in falling 3.5 s?

17. Figure 1.18 (Chapter 1) shows the Hoover Dam Bridge over the Colorado River at a height of 271 m. If a heavy object is dropped from the bridge, how much time passes before the object makes a splash?

    Answer: 7.4 s

18. A student drops an object out the window of the top floor of a high-rise dormitory.
    (a) Neglecting air resistance, how fast is the object traveling when it strikes the ground at the end of 3.0 s? Express the speed in mi/h for a familiar comparison. Can things dropped from high places be dangerous to people below?
    (b) How far, in meters, does the object fall during the 3.0 s? Comment on how many floors the dormitory probably has.

2.4 Acceleration in Uniform Circular Motion

19. A person drives a car around a circular, level cloverleaf with a radius of 70 m at a uniform speed of 10 m/s.
    (a) What is the acceleration of the car?
    (b) Compare this answer with the acceleration due to gravity as a percentage. Would you be able to sense the car's acceleration if you were riding in it?

    Answer: (a) 1.4 m/s^2, toward the center (b) 14%, yes

20. A race car goes around a circular, level track with a diameter of 1.00 km at a constant speed of 90.0 km/h. What is the car's centripetal acceleration in m/s^2?

2.5 Projectile Motion

21. If you drop an object from a height of 1.5 m, it will hit the ground in 0.55 s. If you throw a baseball horizontally with an initial speed of 30 m/s from the same height, how long will it take the ball to hit the ground? (Neglect air resistance.)

    Answer: 0.55 s

22. A golfer on a level fairway hits a ball at an angle of 42° to the horizontal that travels 100 yd before striking the ground. He then hits another ball from the same spot with the same speed, but at a different angle. This ball also travels 100 yd. At what angle was the second ball hit? (Neglect air resistance.)

ON THE WEB

1. Galileo and the Leaning Tower of Pisa

The Highlight: Galileo and the Leaning Tower of Pisa talks about Galileo and the "experiments" that he may or may not have conducted. Follow the recommended links on the student website at **www.cengagebrain.com/shop/ISBN/1133104096** to perform simulations of three experiments. After you are finished, note what you found out about gravity.

2. Vectors and Projectiles: Can You Help the Zookeeper with His Monkey?

Visit the student website at **www.cengagebrain.com/shop/ISBN/1133104096** and follow the recommended links to expand your knowledge of vectors and projectiles by considering the zookeeper's dilemma. At the end of the online exercise are various links for information on physical descriptions of motion. Click on the links and briefly (in no more than a sentence or two) summarize what was stated. How is this information related to what you learned about projectile motion in the text?

Force and Motion

Syracuse Newspapers/Li-Hua Lan/The Image Works

> The whole burden of philosophy seems to consist of this—from the phenomena of motions to investigate the forces of nature, and from these forces to explain the other phenomena. [Physics was once called natural philosophy.]
>
> Sir Isaac Newton
> (1642–1727)

< A force is needed to produce a change in motion, as the pusher well knows.

I n Chapter 2 the description of motion was discussed but not the cause of motion. What, then, causes motion? You might say a push causes something to start moving or to speed up or slow down, but in more scientific terms, the push is the application of a *force*. In this chapter we will go one step further and study force *and* motion. The discussion will consider Newton's three laws of motion, his law of universal gravitation, and later the laws of conservation of linear and angular momentum, making it a very legal sounding chapter with all these laws. In Chapter 3.6 buoyant force is considered. Will an object float or sink?

During the sixteenth and seventeenth centuries, a "scientific revolution" occurred. The theories of motion, which were handed down from the Greeks for almost 2000 years, were reexamined and changed. Galileo (1564–1642) was one

PHYSICS FACTS

▶ "If I have seen further [than certain other men] it is by standing upon the shoulders of giants." Isaac Newton, in a letter to Robert Hooke, February 5, 1675, in reference to work in physics and astronomy done by Galileo and Kepler.

Chapter Outline

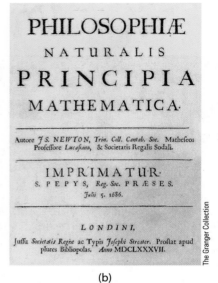

(a)   (b)

Figure 3.1 The Man and the Book
(a) Sir Isaac Newton (1642–1727). (b) The title page from the *Principia.* Can you read the Roman numerals at the bottom of the page that give the year of publication?

of the first scientists to experiment on moving objects. It remained for Sir Isaac Newton, who was born the year Galileo died, to devise the laws of motion and explain the phenomena of moving objects on the Earth and the motions of planets and other celestial bodies.

Newton was only 25 years old when he formulated most of his discoveries in mathematics and mechanics. His book *Mathematical Principles of Natural Philosophy* (commonly referred to as the *Principia*) was published in Latin in 1687 when he was 45. It is considered by many to be the most important publication in the history of physics and certainly established Newton as one of the greatest scientists of all time. (See ● Fig. 3.1.)

3.1   Force and Net Force

Preview Questions

● Does a force always produce motion?

● What is the condition for motion when more than one force acts?

First let's look at the meaning of force. It is easy to give examples of forces, but how would you define a force? A force is defined in terms of what it does, and as you know from experience, a force can produce changes in motion. A force can set a stationary object into motion. It can also speed up or slow down a moving object, or it can change the direction of the motion. In other words, a force can produce a *change in velocity* (speed and/or direction) or an *acceleration*. Therefore, an observed change in motion, including motion starting from rest, is evidence of a force, which leads to the following definition: A **force** is a vector quantity capable of producing motion or a change in motion, that is, a change in velocity or an acceleration (Chapter 2.3).

The word *capable* here is significant. A force may act on an object, but its capability to produce a change in motion may be balanced or canceled by one or more other forces. The net effect is then zero. Thus, a force does not *necessarily* produce a change in motion. It follows, though, that if a force acts *alone*, then the object on which it acts will exhibit a change in velocity, or an acceleration.

(a)   (b)

Figure 3.2 Balanced and Unbalanced Forces
(a) When two applied forces are equal in magnitude and acting in opposite directions, they are said be *balanced*, and there is no net force and no motion if the system is initially at rest. (b) When F_2 is greater than F_1, there is an *unbalanced*, or net force to the right, and motion occurs.

To take into account the application of more than one force on an object, we speak of an **unbalanced** or **net force**. To understand the difference between balanced and unbalanced forces, consider the tug of war shown in ● Fig. 3.2a. Forces are applied, but there is no motion. The forces in this case are balanced; they are equal in magnitude and opposite in direction. In effect, the forces cancel each other, and the net force is zero because the forces are "balanced." Motion occurs when the forces are *unbalanced*, when they are not equal and there is a *net* force F_{net} to the right, and motion occurs (Fig. 3.2b).

Because forces have directions as well as magnitudes, they are *vector* quantities. Several forces may act on an object, but for there to be a change in velocity or for an acceleration to occur, there must be an unbalanced, or net, force. With this understanding, let's next take a look at both force and motion.

Did You Learn?

● A single or net applied force can produce an acceleration.

● When more than one force acts, a net or unbalanced force is needed to produce a change in velocity or an acceleration.

3.2 Newton's First Law of Motion

Preview Questions

● If you were moving with a constant velocity in deep space, how far would you travel?

● How can the inertias of objects be compared?

Long before the time of Galileo and Newton, scientists had asked themselves, "What is the natural state of motion?" The early Greek scientist Aristotle had presented a theory that prevailed for almost 20 centuries after his death. According to this theory, an object required a force in order to remain in motion. The natural state of an object was one of rest, with the exception of celestial bodies, which were naturally in motion. It is easily observed that moving objects such as a rolling ball tend to slow down and come to rest, so a natural state of being at rest must have seemed logical to Aristotle.

Galileo studied motion using a ball rolling down an inclined plane onto a level surface. The smoother he made the surface, the farther the ball would roll (● Fig. 3.3). He reasoned that if a very long surface could be made perfectly smooth, there would be nothing to stop the ball, so it would continue to slide in the absence of friction indefinitely or until

Figure 3.3 Motion without Resistance
If the level surface could be made perfectly smooth, then how far would the ball travel?

something stopped it. Thus, contrary to Aristotle, Galileo concluded that objects could *naturally* remain in motion rather than coming to rest.

Newton also recognized this phenomenon, and Galileo's result is incorporated in **Newton's first law of motion**:

> An object will remain at rest or in uniform motion in a straight line unless acted on by an external, unbalanced force.

Uniform motion in a straight line means that the velocity is constant. An object at rest has a constant velocity of zero. An *external* force is an applied force, one applied on or to the object or system. There are also internal forces. For example, suppose the object is an automobile and you are a passenger traveling inside. You can push (apply a force) on the floor or the dashboard, but this has no effect on the car's velocity because your push is an internal force.

Because of the ever-present forces of friction and gravity on the Earth, it is difficult to observe an object in a state of uniform motion, but in free space, where there is no friction and negligible gravitational attraction, an object initially in motion maintains a constant velocity. For example, after being projected on its way, an interplanetary spacecraft approximates this condition quite well. Upon going out of the solar system where there is negligible gravitational influence, as two *Voyager* spacecraft have done (Chapter 16.6), a spacecraft will travel with a constant velocity until an external, unbalanced force alters this velocity.

Motion and Inertia

Galileo also introduced another important concept. It appeared that objects had a property of maintaining a state of motion; there was a resistance to changes in motion. Similarly, if an object was at rest, it seemed to "want" to remain at rest. Galileo called this property *inertia*. **Inertia** is the natural tendency of an object to remain in a state of rest or in uniform motion in a straight line.

Newton went one step further and related the concept of inertia to something that could be measured: mass. **Mass** is a measure of inertia. The greater the mass of an object, the greater its inertia, and vice versa.

To help understand the relationship between mass and inertia, suppose you horizontally pushed two different people on swings initially at rest, one of them a very large man and the other a small child (● Fig. 3.4a). You would quickly find that it was more difficult to get the adult moving; there would be a noticeable difference in the resistance to a change in motion between the man and the child. Also, once you got them swinging and then tried to stop their motions, you would notice a difference in the resistance to a change in motion again. Being more massive, the man has greater inertia and a greater resistance to a change in motion.*

Another example of inertia is shown in Fig. 3.4b. The stack of coins has inertia and resists being moved when at rest. If the paper is jerked quickly, then the inertia of the coins

---

*A suggested inertia example: When you enter a supermarket with an empty shopping cart, the cart is easy to maneuver, but when you leave with a full cart, making changes in motion (speed and direction) requires much greater effort even though the cart rolls freely with little friction. (Courtesy of Dr. Philip Blanco, Grossmont College.)

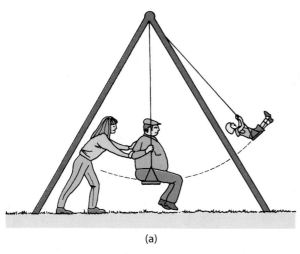

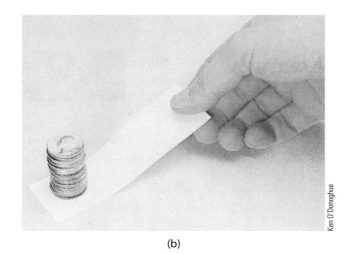

Ken O'Donoghue

(a)   (b)

Figure 3.4 Mass and Inertia
(a) An external, applied force is necessary to set an object in motion. The man has more mass and more inertia than the child and hence has more resistance to motion. (b) Because of inertia, it is possible to remove the paper strip from beneath the stack of quarters without toppling it, by giving the paper a quick jerk. Try it.

will prevent them from toppling. You may have pulled a magazine from the bottom of a stack with a similar action.

Because of the relationship between motion and inertia, Newton's first law is sometimes called the *law of inertia*. This law can be used to describe some of the observed effects in everyday life. For example, suppose you were in the front seat of a car traveling at a high speed down a straight road and the driver suddenly put on the brakes for an emergency stop. What would happen, according to Newton's first law, if you were not wearing a seat belt? The friction on the seat of your pants would not be enough to change your motion appreciably, so you'd keep on moving while the car was coming to a stop. The next external, unbalanced force you'd experience would not be pleasant. Newton's first law makes a good case for buckling up.

Conceptual Question and Answer

You Go Your Way, I'll Go Mine

Q. An air-bubble level on a surface, as illustrated in ● Fig. 3.5, is given an applied horizontal force toward the left. Which way does the bubble move?

A. Some might say the bubble stays behind and "moves" to the right, but it actually moves to the left in the direction of the force. The incorrect answer arises because we are used to observing the bubble instead of the liquid. The bubble is chiefly air, which has little mass or inertia, and readily moves. The correct answer is explained by Newton's first law. Because of inertia, the liquid resists motion and "piles up" toward the rear of the level, forcing the bubble forward. Think about giving a stationary pan of water on a table a push. What happens to the water?

Did You Learn?

● An object in uniform motion would travel in a straight line until acted upon by some external unbalanced force, so in free space with negligible gravity an object would travel indefinitely.

● Objects' relative inertias can be compared by their masses.

Figure 3.5 Pushing a Level on a Table
Which way does the bubble go? See the Conceptual Question and Answer.

3.3 Newton's Second Law of Motion

Preview Questions

- How are force and motion related?
- Which is generally greater, static friction or kinetic friction?

In our initial study of motion (Chapter 2), acceleration was defined as the time rate of the change of velocity ($\Delta v / \Delta t$). However, nothing was said about what *causes* acceleration, only that a change in velocity was required to have an acceleration. So, what causes an acceleration? The answer follows from Newton's first law: *If an external, unbalanced force is required to produce a change in velocity, then an external, unbalanced force causes an acceleration.*

Newton was aware of this result, but he went further and also related acceleration to inertia or mass. Because inertia is the tendency not to undergo a *change* in motion, a reasonable assumption is that the greater the inertia or mass of an object, the smaller the change in motion or velocity (acceleration) when a force is applied. Such insight was typical of Newton in his many contributions to science.

Summarizing

1. The acceleration produced by an unbalanced force acting on an object (or mass) is directly proportional to the magnitude of the force ($a \propto F$) and in the direction of the force (the $\propto$ symbol is a proportionality sign). In other words, the greater the unbalanced force, the greater the acceleration.

2. The acceleration of an object being acted on by an unbalanced force is inversely proportional to the mass of the object ($a \propto 1/m$). That is, for a given unbalanced force, the greater the mass of an object, the smaller the acceleration.

Combining these effects of force and mass on acceleration gives

$$\text{acceleration} = \frac{\text{unbalanced force}}{\text{mass}}$$

When appropriate units are used, the effects of force and mass on acceleration can be written in equation form as $a = F/m$. Or, as commonly written in terms of force in magnitude form, we have **Newton's second law of motion**:

$$\text{force} = \text{mass} \times \text{acceleration}$$

$$F = ma \qquad\qquad 3.1$$

These relationships are illustrated in ● Fig. 3.6. Note that if the force acting on a mass is doubled, then the acceleration doubles (direct proportion, $a \propto F$). However, if the same

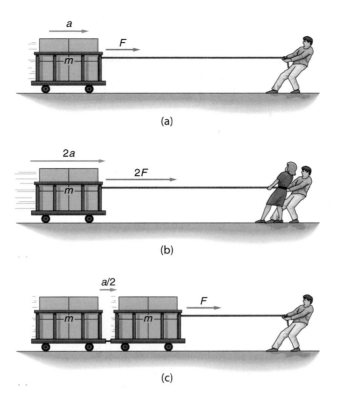

(a)

(b)

(c)

Figure 3.6 Force, Mass, and Acceleration
(a) An unbalanced force *F* acting on a mass *m* produces an acceleration *a*. (b) If the mass remains the same and the force is doubled, then the acceleration is doubled. (c) If the mass is doubled and the force remains the same, then the acceleration is reduced by one-half. The friction of the cars is neglected in all cases.

force is applied to twice as much mass, the acceleration is one-half as great (inverse proportion, $a \propto 1/m$).

Notice from Fig. 3.6 that the mass *m* is the *total* mass of the system or all the mass that is accelerated. A system may be two or more separate masses, as will be seen in Example 3.1. Also, *F* is the net, or unbalanced, force, which may be the vector sum of two or more forces. Unless otherwise stated, a general reference to force means an *unbalanced* force.

In the SI metric system, the unit of force is appropriately called the **newton** (abbreviated N), This is a derived unit. The standard unit equivalent may be seen from Eq. 3.1 by putting it in standard units: force = mass × acceleration = kg × m/s² = kg·m/s² = N.

In the British system, the unit of force is the *pound* (lb). This unit also has derived units of mass multiplied by acceleration (ft/s²). The unit of mass is the *slug*, which is rarely used and will not be employed in this textbook. Recall that the British system is a gravitational or force system and that in this system, objects are weighed in pounds (force). As will be seen shortly, *weight* is a force and is expressed in newtons (N) in the SI and in pounds (lb) in the British system. If you are familiar with the story that Newton gained insight by observing (or being struck by) a falling apple while meditating on the concept of gravity, it is easy to remember that an average-size apple weighs about 1 newton (● Fig. 3.7).

Example 3.1 uses $F = ma$ to illustrate that *F* is the unbalanced or net force and *m* is the total mass.

EXAMPLE 3.1   **Finding Acceleration with Two Applied Forces**

Forces are applied to blocks connected by a string and resting on a frictionless surface, as illustrated in ● Fig. 3.8. If the mass of each block is 1.0 kg and the mass of the string is negligible, then what is the acceleration of the system?

Solution

Step 1
Given: $m_1 = 1.0$ kg, $F_1 = -5.0$ N (left, negative direction)
               $m_2 = 1.0$ kg, $F_2 = +8.0$ N (right, positive direction)
Keep in mind that force (*F*) is a vector with direction, here + or −.

Figure 3.7 About a Newton
An average-size apple weighs about 1 newton (or 0.225 lb).

Figure 3.8 F = ma
Net force and total mass. (See
Example 3.1.)

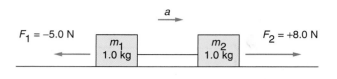

Step 2

Wanted: *a* (acceleration)
(The units are standard in the metric system.)

Step 3

The acceleration may be calculated using Eq. 3.1, $F = ma$, or $a = F/m$. Note, however, that F is the unbalanced (net) force, which in this case is the vector sum of the forces. Effectively, F_1 cancels part of F_2, and there is a net force $F_{net} = F_2 + F_1 = 8.0 \text{ N} - 5.0 \text{ N}$ in the direction of F_2. The total mass of the system being accelerated is $m = m_1 + m_2$. Hence, we have an acceleration in the direction of the net force (to the right).

$$a = \frac{F}{m} = \frac{F_{net}}{m_1 + m_2} = \frac{8.0 \text{ N} - 5.0 \text{ N}}{1.0 \text{ kg} + 1.0 \text{ kg}} = \frac{3.0 \text{ N}}{2.0 \text{ kg}} = 1.5 \text{ m/s}^2$$

Question: What would be the case if the surface were not frictionless? *Answer:* There would be another (frictional) force in the direction of F_1 opposing the motion.

Confidence Exercise 3.1

Given the same conditions as in Example 3.1, suppose $F_1 = -9.0 \text{ N}$ and $F_2 = 6.0 \text{ N}$ in magnitude. What would be the acceleration of the system in this case?

The answers to Confidence Exercises may be found at the back of the book.

Because Newton's second law is so general, it can be used to analyze many situations. A dynamic example is *centripetal force*. Recall from Chapter 2.4 that the centripetal acceleration for uniform circular motion is given by $a_c = v^2/r$ (Eq. 2.5). The magnitude of the centripetal force that supplies such an acceleration is given by Newton's second law, $F = ma_c = mv^2/r$.

Mass and Weight

This is a good place to make a clear distinction between mass and weight. As learned previously, *mass* is the amount of matter an object contains (it is also a measure of inertia). **Weight** is related to the force of gravity (related to the gravitational force acting on a mass or object). The quantities are related, and Newton's second law clearly shows this relationship.

On the surface of the Earth, where the acceleration due to gravity is relatively constant ($g = 9.80 \text{ m/s}^2$), the weight w on an object with a mass m is given by

$$\text{weight} = \text{mass} \times \text{acceleration due to gravity}$$

$$w = mg \qquad\qquad 3.2$$

Note that this equation is a special case of $F = ma$ where different symbols, w and g, have been used for force and acceleration.

Conceptual Question and Answer

Fundamental is Fundamental

Q. Is weight a fundamental quantity?

A. No, and here's why. A fundamental quantity is constant or has the same value no matter where it is measured. In general, a physical object always has the same amount of matter, so it has a constant mass. The weight of an object, on the other hand, may differ, depending on the value of the acceleration due to gravity, g.

For example, on the surface of the Moon, the acceleration due to gravity (g_M) is one-sixth the acceleration due to gravity on the surface of the Earth [$g_M = g/6 = (9.8$ m/s$^2)/6 = 1.6$ m/s^2], because of the Moon's mass and size. (See Chapter 3.5.) Thus, an object will have the *same* mass on the Moon as on the Earth, but its weight will be *different*. Mass, not weight, is the fundamental quantity. (A review of an important item from Chapter 1.4. See Fig. 1.8.)

EXAMPLE 3.2   **Computing Weight**

What is the weight (in newtons) of a 1.0-kg mass on (a) the Earth and (b) the Moon?

Solution

(a) Using Eq. 3.2 and the Earth's $g = 9.8$ m/s^2,

$$w = mg = (1.0 \text{ kg})(9.8 \text{ m/s}^2) = 9.8 \text{ N}$$

(b) On the Moon, where $g_M = g/6 = 1.6$ m/s^2,

$$w = mg_M = (1.0 \text{ kg})(1.6 \text{ m/s}^2) = 1.6 \text{ N}$$

Although the mass is the same in both cases, the weights are different because of different *g*'s.

Confidence Exercise 3.2

On the surface of Mars, the acceleration due to gravity is 0.39 times that on the Earth. What would a kilogram weigh in newtons on Mars?

The answers to Confidence Exercises may be found at the back of the book.

An unknown mass or object may be "weighed" on a scale. The scale can be calibrated in mass units (kilograms or grams) or in weight units (newtons or pounds). See ● Fig. 3.9. In Europe, weight is expressed in terms of mass. People "weigh" themselves in kilograms or "kilos."

Finally, an object in free fall has an unbalanced force of $F = w = mg$ acting on it (downward).* Why, then, do objects in free fall all descend at the same rate, as stated in the last chapter? Even Aristotle thought a heavy object would fall faster than a lighter one. Newton's laws explain.

Suppose two objects were in free fall, one having twice the mass of the other, as illustrated in ● Fig. 3.10. According to Newton's second law, the more massive object would have two times the weight, or gravitational force of attraction. By Newton's first law, however, the more massive object has twice the inertia, so it needs *twice the force to fall at the same rate*.

Figure 3.9 Mass and Weight
A mass of 1.0 kg suspended on a scale calibrated in newtons shows the weight to be 9.8 N, which is equivalent to 2.2 lb.

Tom Pantages

---

*Recall from Chapter 2.3 that an object falling solely under the influence of gravity (no air resistance) is said to be in *free fall*.

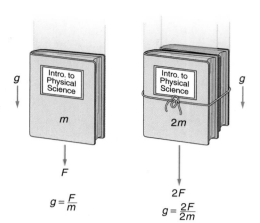

$$g = \frac{F}{m}$$

$$g = \frac{2F}{2m}$$

Figure 3.10 Acceleration Due to Gravity
The acceleration due to gravity is independent of the mass of a freely falling object. Thus the acceleration is the same for all such falling objects. An object with two times the mass of another has twice the gravitational force acting on it, but it also has twice the inertia and so falls at the same rate.

Friction

The force of friction is commonplace in our everyday lives. Without it, we would not be able to walk (our feet would slip), pick things up, and so on. **Friction** is the ever-present resistance to relative motion that occurs whenever two materials are in contact with each other, whether they are solids, liquids, or gases. In some instances, we want to increase friction by, for example, putting sand on an icy road or sidewalk to improve traction. In other instances, it is desired to reduce friction. Moving machine parts are lubricated to allow them to move freely, thereby reducing wear and the expenditure of energy. Automobiles would not run without friction-reducing oils and greases. With friction, energy is generally lost in the form of heat.

This section is concerned with friction between solid surfaces. (Air friction or resistance was discussed in Chapter 2.3.) All surfaces, no matter how smooth they appear or feel, are microscopically rough. It was originally thought that friction between solids arose from an interlocking of surface irregularities, or high and low spots, but research has shown that friction between contacting surfaces (particularly metals) is due to local adhesion. When surfaces are pressed together, local bonding or welding occurs in a few small patches. To overcome this local adhesion, a force great enough to pull apart the bonded regions must be applied.

Friction is characterized by a force of friction that opposes an applied force. Between solid surfaces, friction is generally classified into two types: static friction and sliding (or kinetic) friction.* *Static friction* occurs when the frictional force is sufficient to prevent relative motion between surfaces. Suppose you want to slide a filing cabinet across the floor. You push on it, but it doesn't move. The force of static friction between the bottom of the cabinet and the floor is opposite and at least equal to your applied force, so there is no motion; it is a static condition.

To get the job done, you push harder, but still the cabinet does not move. This indicates that the static frictional force increased when you increased the applied force. Increasing the applied force even more, you finally get the cabinet moving, but there is still a great deal of resistance—called kinetic friction—between the cabinet bottom and the floor. *Kinetic (or sliding) friction* occurs when there is a relative sliding motion between the surfaces in contact. You may notice that it is easier to keep the cabinet sliding than it is to get it moving. This is because sliding friction is generally less than the maximum static friction.

As you might imagine, the frictional force depends on how hard the surfaces are pressed together, which usually depends on the weight of an object on a surface. Because contacting surfaces can be of various materials, *coefficients of friction*—coefficients of static friction and coefficients of kinetic friction—are used to characterize particular situations. ● Table 3.1 lists some approximate values of these for various materials. The coefficient of static fric-

Table 3.1 Approximate Values for Coefficients of Static and Kinetic Friction between Certain Surfaces

Friction between Materials	μ_s	μ_k
Aluminum on aluminum	1.90	1.40
Aluminum on steel	0.61	0.47
Teflon on steel	0.04	0.04
Rubber on concrete		
dry	1.20	0.85
wet	0.80	0.60
Wood on wood	0.58	0.40
Lubricated ball bearings	< 0.01	< 0.01

*There is also rolling friction, such as occurs between a train wheel and a rail. This type of friction is difficult to analyze and will not be discussed here.

tion, μ_s, is proportional to the maximum applied force needed to just overcome static friction. The coefficient of kinetic friction, μ_k, is proportional to the applied force necessary to keep an object moving (at a constant velocity). Notice that $\mu_s > \mu_k$, which reflects that it takes more force to get an object moving than it does to keep it moving.

Did You Learn?

- According to Newton's second law, an external net force will produce a change in velocity or an acceleration.
- The maximum force of static friction is generally greater than the force of kinetic friction. It takes less force to keep an object moving than to get it going.

3.4 Newton's Third Law of Motion

Preview Questions

- What's the difference between an action and a reaction?
- Is the net force of Newton's third-law force pair equal to zero?

Newton's third law of motion is of great physical significance. We commonly think of single forces, but Newton recognized that it is *impossible* to have a single force. He observed that in the application of a force there is always a mutual reaction, so forces always occur in pairs. As Newton illustrated, "If you press a stone with your finger, the finger is also pressed upon by the stone." Another example is when you ride in a car that is braked to a quick stop. If you push against your seat belt, your seat belt pushes against you. This is Newton's third law in action.

Newton's third law of motion is sometimes called the *law of action and reaction*. **Newton's third law of motion** states the following:

For every action there is an equal and opposite reaction.

In this statement, the words *action* and *reaction* refer to forces. The law is also stated another way:

For every force there is an equal and opposite force.

A more descriptive statement would be the following:

Whenever one object exerts a force on a second object, the second object exerts an equal and opposite force on the first object.

Expressed in equation form, Newton's third law may be written as

$$\text{action} = \text{opposite reaction}$$
$$F_1 = -F_2 \qquad\qquad 3.3$$

where

$$F_1 = \text{force exerted on object 1 by object 2}$$
$$-F_2 = \text{force exerted on object 2 by object 1}$$

The negative sign in Eq. 3.3 indicates that F_2 is in the opposite direction from F_1.

Jet propulsion is an example of Newton's third law. In the case of a rocket, the exhaust gas is accelerated from the rocket, and the rocket accelerates in the opposite direction (● Fig. 3.11a). A common misconception is that on launch the exhaust gas pushes against the launch pad to accelerate the rocket. If this were true, then there would be no space travel because there is nothing to push against in space. The correct explanation is one of action (gas being forced backward by the rocket) and reaction (the rocket being propelled forward by the escaping gas). The gas (or gas particles) exerts a force on the rocket, and

(a)

NASA/Johnson Space Center

(b)

Arnoz Eckerson/Visuals Unlimited

Figure 3.11 Newton's Third Law in Action
(a) The rocket and exhaust gas exert equal and opposite forces on each other and so are accelerated in opposite directions. (b) The equal and opposite forces are obvious here. Notice the distortion of both the tennis ball and the racquet.

the rocket exerts a force on the gas. The equal and opposite actions of Newton's third law should be evident in Fig. 3.11b.

Let's take a look at the third law in terms of the second law ($F = ma$). Writing Eq. 3.3 in the form

$$F_1 = -F_2$$

or

$$m_1a_1 = -m_2a_2$$

shows that if m_2 is much greater than m_1, then a_1 is much greater than a_2.

Consider dropping a book on the floor. As the book falls, it has a force acting on it (the Earth's gravity) that causes it to accelerate. What is the equal and opposite force? It is the force of the book's gravitational pull on the Earth. Technically, the Earth accelerates upward to meet the book. However, because our planet's mass is so huge compared with the book's mass, the Earth's acceleration is so minuscule it cannot be detected.

An important distinction to keep in mind is that Newton's third law relates two equal and opposite forces that act on two *separate* objects. Newton's second law concerns how forces acting on a *single* object cause an acceleration. If two forces acting on a single object are equal and opposite, there will be no net force and no acceleration, but these forces are *not* the third-law force pair. (Why?)

Let's look at one more example that illustrates the application of Newton's laws. Imagine that you are a passenger in a car traveling down a straight road and entering a circular curve at a constant speed. As you know, there must be a centripetal force to provide the centripetal acceleration necessary to negotiate the curve (Chapter 2.4). This force is supplied by friction on the tires, and the magnitude of this frictional force (f) is given by $f = ma_c = mv^2/r$. Should the frictional force not be great enough—say, if you hit an icy spot (reduced friction)—the car would slide outward because the centripetal force would not be great enough to keep the car in a circular path (● Fig. 3.12).

You have no doubt experienced a lack of centripetal force when going around a curve in a car and have had a feeling of being "thrown" outward. Riding in the car before entering the curve, you tend to go in a straight line in accordance with Newton's first law. As the car makes the turn, you continue to maintain your straight-line motion until the car turns "into you."

It may feel that you are being thrown outward toward the door, but actually the door is coming toward you because the car is turning, and when the door gets to you, it exerts a force on you that supplies the centripetal force needed to cause you to go around the curve with the car. The friction on the seat of your pants is not enough to do this, but if you are buckled up, then the requisite force may be supplied by a seat belt instead of the door.

Another practical application involving Newton's laws is discussed in the **Highlight: The Automobile Air Bag**.

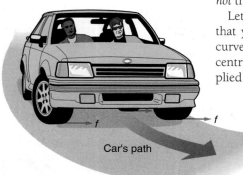

f   *f*

Car's path

Figure 3.12 Newton's First Law in Action
Friction supplies the centripetal force necessary for the car to negotiate the curve. When a car goes around a sharp curve or corner, the people in it tend to have the sensation of being thrown outward as the car rounds the curve, but that is not so. See the text for a description.

Did You Learn?

- A reaction is an equal, but opposite, force to an action force.
- The equal and opposite force pair of Newton's third law do not cancel each other as they act on different objects.

3.5   Newton's Law of Gravitation

Preview Questions

- What keeps the Moon in orbit around the Earth?
- Are astronauts seen floating in the International Space Station really weightless?

Highlight   The Automobile Air Bag

A major automobile safety feature is the air bag. As noted earlier, seat belts restrain you so that you don't keep going per Newton's first law when the car comes to a sudden stop. Where, though, does the air bag come in, and what is its principle?

When a car has a head-on collision with another vehicle or hits an immovable object such as a tree, it stops almost instantaneously. Even with seat belts, the impact of a head-on collision could be such that seat belts might not restrain a person completely and injuries could occur.

Enter the automobile air bag. This balloon-like bag inflates automatically on hard impact and cushions the driver. Front air bags are mounted in the center of the steering wheel on the driver's side and in the dashboard on the passenger side.

The air bag tends to increase the contact time in stopping a person, thereby reducing the impact force, as compared with the force experienced when hitting the dashboard or steering column (Fig. 1). Also, the impact force is spread over a large general area and not applied to certain parts of the body, as in the case of seat belts.

You might wonder what causes an air bag to inflate and what inflates it. Keep in mind that to do any good, the inflation must occur in a fraction of a second, a shorter time than the time between the driver's reaction and collision contact. The air bag's inflation is initiated by an electronic sensing unit. This unit contains sensors that detect rapid decelerations such as those that occur in high-impact collisions. The sensors have threshold settings so that normal hard braking does not activate them.

Sensing an impact, a control unit sends an electric current to an igniter in the air-bag system that sets off a chemical explosion. The gases (mostly nitrogen) rapidly inflate the thin nylon bag. The total process of sensing to complete inflation takes about 30 milliseconds, or 0.030 second. Pretty fast, and that's a good thing, too! A car's occupant hits the air bag about 50 milliseconds after a collision. (For more on the chemical explosion, see the Highlight: The Chemistry of Air Bags in Chapter 13.2.)

In a front-end collision, a car's battery and alternator are among the first things to go, so an air bag's sensing unit is equipped with its own electrical power source. Such front automobile air bags offer protection only for front-end collisions, in which the car's occupants are thrown forward (excuse me, *continue to travel forward*, per Newton's first law).

Automobile side air bags are also available to protect occupants from side-impact collisions, which account for 30% of all accidents. Side air bags must be engineered to deploy much more quickly than those that protect from front-end collisions because only a few inches of door separate the occupant from the colliding vehicle. Side air bags are mounted in the doors, in the roof (from which they deploy downward), and in seat backs (from which they deploy toward the door).

Unfortunately, injuries and deaths have resulted from the deployment of front air bags. An air bag is not a soft, fluffy pillow. When activated, it deploys at speeds of up to 320 km/h (200 mi/h) and could hit a person close by with enough force to cause severe injury and even death. Adults are advised to sit at least 25 cm (10 in.) from the air-bag cover. Seats should be adjusted to allow for the proper safety distance. Probably the most serious concern is associated with children. Children may get close to the dashboard if they are not buckled in or are not buckled in securely. Another dangerous practice is using a rear-facing child seat in the front passenger seat. On deployment, the air bag could force the child into the back of the front seat, causing injury.

The next generation of "smart" air bags will be designed to sense the severity of an accident and inflate accordingly, as well as to base the inflation on the weight of the occupant. For example, if the sensors detect the small weight of a child in the passenger seat, the air bag won't deploy.

Specific problems may exist, but air bags save many lives. Even if your car is equipped with air bags, *always* remember to buckle up. (Maybe we should make that Newton's "fourth law of motion.")

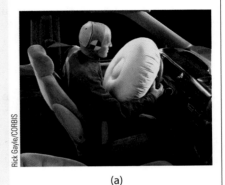

(a)

(b)

Figure 1 Softening the Impact
(a) Automobile air bags tend to increase the contact time, thereby reducing the impact force, (b) compared with hitting the dashboard, windshield, or steering column.

Having gained a basic understanding of forces, let's take a look at a common fundamental force of nature—gravity. Gravity is a fundamental force because we do not know what causes it and can only observe and describe its effects. Gravity is associated with mass and causes the mutual attraction of mass particles.

The law describing the gravitational force of attraction between two particles was formulated by Newton from his studies on planetary motion. Known as **Newton's law of universal gravitation** (universal because it is believed that gravity acts everywhere in the universe), this law may be stated as follows:

> Every particle in the universe attracts every other particle with a force that is directly proportional to the product of their masses and inversely proportional to the square of the distance between them.

Suppose the masses of two particles are designated as m_1 and m_2, and the distance between them is r (● Fig. 3.13a). Then the statement of the law is written in symbol form as

$$F \propto \frac{m_1 m_2}{r^2}$$

where $\propto$ is a proportionality sign. Notice in the figure that F_1 and F_2 are equal and opposite; they are a mutual interaction *and* a third-law force pair.

When an appropriate constant (of proportionality) is inserted, the equation form of Newton's law of universal gravitation is

$$F = \frac{Gm_1 m_2}{r^2} \qquad \text{3.4}$$

where **G** is the universal gravitational constant and has a value of $G = 6.67 \times 10^{-11}$ N·m²/kg².

The gravitational force between two masses is said to have an infinite range. That is, the only way for the force to approach zero is for the masses to be separated by a distance approaching infinity $r \to \infty$. (You can't escape gravity.)

An object is made up of a lot of point particles, so the gravitational force on an object is the vector sum of all the particle forces. This computation can be quite complicated, but one simple and convenient case is a homogeneous sphere.* In this case, the net force acts as though all the mass were concentrated at the center of the sphere, and the separation distance for the mutual interaction with another object is measured from there. If it is assumed that the Earth, other planets, and the Sun are spheres with uniform mass distributions, then to a reasonable approximation the law of gravitation can be applied to such bodies (Fig. 3.13b).

In using such an approximation, your weight (the gravitational attraction of the Earth on your mass, m) is computed as though all the mass of the Earth, M_E, were concentrated at its center, and the distance between the masses is the radius of the Earth, R_E. That is, $w = mg = GmM_E/R_E^2$.

Figure 3.13 Newton's Law of Gravitation
(a) Two particles attract each other gravitationally, and the magnitude of the forces is given by Newton's law of gravitation. The forces are equal and opposite: Newton's third-law force pair. (b) For a homogeneous or uniform sphere, the force acts as though all the mass of the sphere were concentrated as a particle at its center.

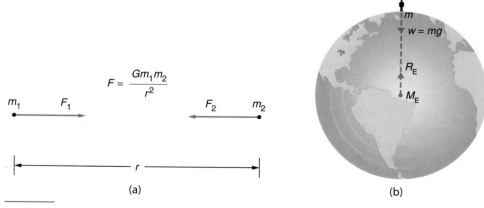

$$F = \frac{Gm_1 m_2}{r^2}$$

$m_1 \quad F_1 \qquad\qquad F_2 \quad m_2$

$\longleftarrow\quad r\quad \longrightarrow$

(a)

(b)

*_____
Homogeneous means that the mass particles are distributed uniformly throughout the object.

In Eq. 3.4, as r increases, the force of gravity becomes less. Hence, gravity and the acceleration due to gravity become less with altitude above the Earth.

Conceptual Question and Answer

A Lot of Mass

Q. If you look inside the back cover of this book, you will find the mass of the Earth is listed as 6.0×10^{24} kg. How could the Earth be "weighed" or "massed"?

A. It would take some big scales, and of course isn't possible. Like many quantities in physics, values are gained indirectly. If you look at the previous equation for your weight on the surface of the Earth using Newton's law of gravitation, then you will note that the m's cancel and you are left with $g = GM_E/R_E^2$. With a little mathematical manipulation, the mass of the Earth is given by $M_E = gR_E^2/G$. The values of g, G, and R_E are known. (The circumference of the Earth was measured in the first century BCE, from which the radius R_E was calculated.) So, plug in the numbers and you have the mass of the Earth.

Newton used his law of gravitation and second law of motion to show that gravity supplies the centripetal acceleration and force required for the Moon to move in its nearly circular orbit about the Earth, but he did not know or experimentally measure the value of G. This very small value was measured some 70 years after Newton's death by the English scientist Henry Cavendish, who used a very delicate balance to measure the force between two masses.

When an object is dropped, the force of gravity is made evident by the acceleration of the falling object, but if there is a gravitational force of attraction between every two objects, why don't you feel the attraction between yourself and this textbook? (No pun intended.) Indeed there *is* a force of attraction between you and this textbook, but it is so small that you don't notice it. Gravitational forces can be very small, as illustrated in Example 3.3.

EXAMPLE 3.3   Applying Newton's Law of Gravitation

Two objects with masses of 1.0 kg and 2.0 kg are 1.0 m apart (● Fig. 3.14). What is the magnitude of the gravitational force between these masses?

Solution

The magnitude of the force is given by Eq. 3.4:

$$F = \frac{Gm_1 m_2}{r^2}$$

$$= \frac{(6.67 \times 10^{-11}\,\text{N} \cdot \text{m}^2/\text{kg}^2)(1.0\,\text{kg})(2.0\,\text{kg})}{(1.0\,\text{m})^2}$$

$$= 1.3 \times 10^{-10}\,\text{N}$$

This number is very, very small. A grain of sand would weigh more. For an appreciable gravitational force to exist between two masses, at least one of the masses must be relatively large (see Fig. 3.14).

Confidence Exercise 3.3

If the distance between the two masses in Fig. 3.14 were tripled, by what factor would the mutual gravitational force change? (*Hint*: Use a ratio.)

With regard to astronauts in space, we hear the (incorrect) terms *zero g* and *weightlessness* (● Fig. 3.15). These terms are not true descriptions. *Microgravity* and *apparent*

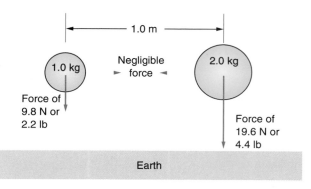

Figure 3.14 The Amount of Mass Makes a Difference
A 1.0-kg mass and a 2.0-kg mass separated by a distance of 1.0 m have a negligible mutual gravitational attraction (about 10^{-10} N). However, because the Earth's mass is quite large, the masses are attracted to the Earth with forces of 9.8 N and 19.6 N, respectively. These forces are the weights of the masses.

Figure 3.15 Floating Around
Astronauts in a space shuttle orbiting the Earth are said to "float" around because of "zero g" or "weightlessness." Actually, the gravitational attraction of objects keeps them in orbit with the spacecraft. Because gravity is acting, by definition an astronaut has weight. Here, astronaut pilot Susan L. Still enters data into an onboard computer.

weightlessness are more applicable terms. Gravity certainly acts on an astronaut in an orbiting spacecraft to provide the necessary centripetal force. Without gravity, the spacecraft (and astronaut) would not remain in orbit but instead would fly off tangentially in a straight line (analogous to swinging a ball on a string about your head and the string breaks). Because gravity is acting, the astronaut by definition has weight.

The reason an astronaut floats in the spacecraft and feels "weightless" is that the spacecraft and the astronaut are both "falling" toward the Earth. Imagine yourself in a freely falling elevator standing on a scale. The scale would read zero because it is falling just as fast as you are. You are not weightless, however, and g is not zero, as you would discover upon reaching the bottom of the elevator shaft.

Did You Learn?

- The gravitational attraction between the Earth and the Moon supplies the necessary centripetal force to keep the Moon in its orbit around the Earth.

- By definition, astronauts in an Earth-orbiting spacecraft are not "weightless." Gravity acts on them, so they have weight.

3.6 Archimedes' Principle and Buoyancy

Preview Questions

- What is the magnitude of the buoyant force on an object in a fluid?

- What determines if an object will float or sink in water?

Let's take a look at another common force associated with fluids. (Unlike solids, fluids can "flow," so liquids and gases are fluids.) Objects float in fluids because they are buoyant or are buoyed up. For example, if you immerse a cork in water and release it, the cork will be buoyed up to the surface and remain there. From your knowledge of forces, you know that such motion requires an upward net force. For an object to come to the surface, there must be an upward force acting on it that is greater than the downward force of its weight. When the object is floating, these forces must balance each other, and we say the object is in *equilibrium* (zero net force).

The upward force resulting from an object being wholly or partially immersed in a fluid is called the **buoyant force**. The nature of this force is summed up by **Archimedes' principle**:*

> An object immersed wholly or partially in a fluid experiences a buoyant force equal in magnitude to the weight of the *volume of fluid* that is displaced.

We can see from Archimedes' principle that the buoyant force depends on the weight of the volume of fluid displaced. Whether an object will sink or float depends on the density of the object (ρ_o) relative to that of the fluid (ρ_f). There are three conditions to consider:

▶ An object will float in a fluid if its average density is less than the density of the fluid ($\rho_o < \rho_f$).

▶ An object will sink if its average density is greater than the density of the fluid ($\rho_o > \rho_f$).

▶ An object will be in equilibrium at any submerged depth in a fluid if the average density of the object and the density of the fluid are equal ($\rho_o = \rho_f$).

(Average density implies that the object does not have a uniform mass, that is, that it has more mass in one area than another.) An example of the application of the first condition is shown in ● Fig. 3.16.

These three conditions also apply to a fluid within a fluid, provided that the two are immiscible (do not mix). For example, you might think cream is "heavier" than skim milk, but that's not so. Cream floats on milk, so it is less dense.

Figure 3.16 Fluid Buoyancy
The air is a fluid in which objects, such as this dirigible, float. Because of the helium gas inside, the average density of the blimp is less than that of the surrounding air. The weight of the volume of air displaced is greater than the weight of the blimp, so the blimp floats, supported by a buoyant force. (The ship is powered so that it can maneuver and change altitude.) It is sometimes said that helium is "lighter" than air, but it is more accurate to say that helium is *less dense* than air.

Conceptual Question and Answer

Float the Boat

Q. A 1.0-lb piece of iron or steel readily sinks in water, yet ocean liners made of iron and steel weigh thousands of tons and float. Why?

A. Because an ocean liner floats, its average density must be less than that of seawater. An ocean liner is made of iron and steel, but overall most of its volume is occupied by air. Thus, its average density is less than that of seawater. Displacing a huge volume of water, it floats. Similarly, the human body has air-filled spaces, so most of us float in water.

In some instances the overall density is purposefully varied to change the depth. For example, a submarine submerges by flooding its tanks with seawater (called "taking on ballast"). This flooding increases the sub's average density, and it sinks and dives. (Propulsion power enables it to maneuver.) When the sub is to surface, water is pumped out of the tanks. The average density becomes less than that of the surrounding seawater, and up it goes.

When an object floats, some of it is submerged, displacing enough volume so that the buoyant force equals the weight force. The volume submerged depends on the weight of the object. In some instances the submerged volume can be appreciable. For example, the

---

*Archimedes (287–212 BCE), a Greek scholar, was given the task of determining whether a gold crown made for the king was pure gold or contained a quantity of some other metal. Legend has it that the solution to the problem came to Archimedes while he was in the bathtub. It is said that he was so excited he jumped out of the tub and ran through the streets of the city (unclothed) shouting, "Eureka! Eureka!" (Greek for "I have found it!")

submerged volume of an iceberg is on the order of 90% (● Fig. 3.17). Thus, the 10% that remains above water is "only the tip of the iceberg."

Another important property of liquids is discussed in the **Highlight: Surface Tension, Water Striders, and Soap Bubbles**.

Did You Learn?

● The buoyant force is equal in magnitude to the weight of the volume of fluid an object displaces.

● An object with an average density greater than 1.0 g/cm³ (the density of water) will sink in water; if it is less than 1.0 g/cm³, then the object will float.

3.7 Momentum

Preview Questions

● When is the linear momentum of a system conserved?

● What gives rise to a change in angular momentum?

Another important quantity in the description of motion is *momentum*. This term is commonly used; for example, it is said that a sports team has a lot of momentum or has lost its momentum. Let's see what momentum means scientifically. There are two types of momentum: linear and angular.

Linear Momentum

Stopping a speeding bullet is difficult because it has a high velocity. Stopping a slowly moving oil tanker is difficult because it has a large mass. In general, the product of mass and velocity is called **linear momentum**, the magnitude of which is

$$\text{linear momentum} = \text{mass} \times \text{velocity}$$
$$p = mv \qquad\qquad 3.5$$

where v is the instantaneous velocity.

Because velocity is a vector, momentum is also a vector with the same direction as the velocity. Both mass and velocity are involved in momentum. A small car and a large truck both traveling at 50 km/h in the same direction have the same velocity, but the truck has more momentum because it has a much larger mass. For a system of masses, the linear momentum of the system is found by adding the linear momentum vectors of all the individual masses.

The linear momentum of a system is important because if there are no external unbalanced forces, then the linear momentum of the system is *conserved;* it does not change with time. In other words, with no unbalanced forces, no acceleration occurs, so there is no change in velocity and no change in momentum. This property makes linear momentum extremely important in analyzing the motion of various systems. The **law of conservation of linear momentum** may be stated as follows:

The total linear momentum of an isolated system remains the same if there is no external, unbalanced force acting on the system.

Even though the internal conditions of a system may change, the vector sum of the momenta remains constant.

$$\text{total final momentum} = \text{total initial momentum}$$
or
$$P_f = P_i$$
where
$$P = p_1 + p_2 + p_3 + \ldots \qquad\qquad 3.6$$
(sum of individual momentum vectors)

Figure 3.17 The Tip of the Iceberg The majority of a floating iceberg is beneath the water, as illustrated here. Approximately 90% of its bulk is submerged. The displacement of water by the submerged portion gives rise to a buoyant force that equals the iceberg's weight.

Surface Tension, Water Striders, and Soap Bubbles

Have you ever noticed how drops of water bead up on a clean kitchen surface or on a newly waxed car? What causes this beading? The answer is tension. The molecules of a liquid exert a small attractive force on one another (due to an asymmetry of electrical charge; see Chapter 12). Within a liquid, any molecule is completely surrounded by other molecules, and the net force is zero (Fig. 1). For molecules at the surface, however, there is no attractive force acting on them from above the surface. As a result, net forces act on molecules of the surface layer due to the attraction of neighboring molecules just below the surface. The inward pull on the surface molecules causes the surface of a liquid to contract, giving rise to what is called *surface tension*.

If a sewing needle is carefully placed on the surface of a bowl of water, the surface acts like an elastic membrane because of surface tension. There is a slight depression in the surface, and molecular forces along the depression act at an angle to the surface (Fig. 2). The vertical component of these forces balances the weight (*mg*) of the needle, and it "floats" on the surface. Similarly, surface tension supports the weight of a water strider on water.

The net effect of surface tension is to make the surface area of a liquid as small as possible. A volume of liquid tends to assume the shape that has the least surface area. As a result, drops of water and soap bubbles have spherical shapes because a sphere has the smallest surface area for a given volume (Fig. 3). In forming a drop or a bubble, surface tension pulls the molecules together to minimize the surface area.

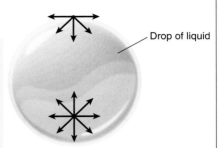

Figure 1 Surface Tension
The net force on a molecule in the interior of a liquid is zero because the molecule is surrounded by other molecules. At the surface, though, there are no molecules above, and an inward force thus acts on a molecule due to the attractive forces of the neighboring molecules just below the surface. This gives rise to surface tension.

Figure 3 Surface Tension at Work
Because of surface tension, soap bubbles assume the shape of a sphere to minimize the surface area. Beads of water on a newly waxed car also assume spherical shapes.

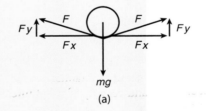
(a)

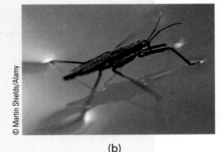
(b)

Figure 2 Walk on Water
(a) A light object, such as a needle, forms a depression in the liquid, and the surface area acts like a stretched membrane, with the weight (*mg*) of the object supported by the upward components of the surface tension. (b) Some insects, such as water striders, can walk on water because of surface tension, as you might walk on a large trampoline. Notice the depressions in the liquid's surface where the legs touch it.

Suppose you are standing in a boat near the shore, and you and the boat are the system (● Fig. 3.18). Let the boat be stationary, so the total linear momentum of the system is zero (no motion for you or the boat, so zero linear momentum). On jumping toward the shore, you will notice immediately that the boat moves in the opposite direction. The boat moves because the force you exerted in jumping is an *internal* force. Thus, the total linear momentum of the system is conserved and must remain zero (water resistance neglected). You have momentum in one direction, and to cancel this vectorially so that the total momentum remains zero, the boat must have an equal and opposite momentum. Remember that momentum is a vector quantity, and momentum vectors can add to zero.

Figure 3.18 Conservation of Linear Momentum
When the system is at rest, the total momentum of the system (man and boat) is zero. When the man jumps toward the shore (an internal force), the boat moves in the opposite direction, conserving linear momentum.

Man jumps forward

Boat moves backward

EXAMPLE 3.4   Applying the Conservation of Linear Momentum

Two masses at rest on a frictionless surface have a compressed spring between them but are held together by a light string (● Fig. 3.19). The string is burned, and the masses fly apart. If m_1 has a velocity $v_1 = 1.8$ m/s (to the right), then what is the velocity of m_2?

Solution

Let's solve this example stepwise for clarity.

Step 1
Given: $m_1 = 1.0$ kg, $v_1 = 1.8$ m/s (positive, to the right)
$m_2 = 2.0$ kg (masses from figure)

Step 2
Wanted: v_2 (velocity of m_2)

Step 3

Reasoning: The total momentum of the system is initially zero ($P_i = 0$) before the string is burned. (The reason for releasing the masses in this manner is to ensure that no external forces are applied as there might be if the masses were held together with hands.)

The spring is part of the system and so applies an *internal* force to each of the masses. Hence, the linear momentum is conserved. After leaving the spring, the moving masses have nonzero momenta, and the total momentum of the system is $P_f = p_1 + p_2$. (The spring is assumed to be motionless.)

With the total linear momentum being conserved (no unbalanced *external* forces acting), using Eq. 3.6,

$$P_f = P_i$$

or

$$p_1 + p_2 = 0$$

Rearranging gives

$$p_1 = -p_2$$

which tells us that the momenta are equal and opposite. Then, in terms of mv,

$$m_1 v_1 = -m_2 v_2$$

Solving for v_2 yields

$$v_2 = -\frac{m_1 v_1}{m_2} = -\frac{(1.0 \text{ kg})(1.8 \text{ m/s})}{2.0 \text{ kg}} = -0.90 \text{ m/s}$$

to the left in Fig. 3.19, because v_1 was taken to be positive to the right.

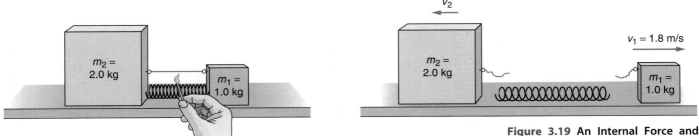

Figure 3.19 An Internal Force and Conservation of Linear Momentum When the string is burned, the compressed spring applies an internal force to the system. See Example 3.4.

Confidence Exercise 3.4

Suppose you were not given the values of the masses but only that $m_1 = m$ and $m_2 = 3m$. What could you say about the velocities in this case?

Earlier we looked at the jet propulsion of rockets in terms of Newton's third law. This phenomenon can also be explained in terms of linear momentum. The burning of the rocket fuel gives energy by which *internal* work is done and hence internal forces act. As a result, the exhaust gas goes out the back of the rocket with momentum in that direction, and the rocket goes in the opposite direction to conserve linear momentum. Here the many, many exhaust gas molecules have small masses and large velocities, whereas the rocket has a large mass and a relatively small velocity.

You can demonstrate this rocket effect by blowing up a balloon and letting it go. The air comes out the back and the balloon is "jet" propelled, but without a guidance system, the balloon zigzags wildly.

Angular Momentum

Another important quantity Newton found to be conserved is angular momentum. Angular momentum arises when objects go in paths around a center of motion or axis of rotation. Consider a comet going around the Sun in an elliptical orbit, as illustrated in ● Fig. 3.20. The magnitude of **angular momentum** (*L*) is given by

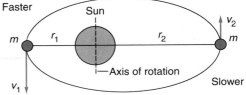

Figure 3.20 Angular Momentum The angular momentum of a comet going around the Sun in an elliptical orbit is given at the two opposite points in the orbit by mv_1r_1 and mv_2r_2. Angular momentum is conserved in this case, and $mv_1r_1 = mv_2r_2$. As the comet comes closer to the Sun, the radial distance *r* decreases, so the speed *v* must increase. Similarly, the speed decreases when *r* increases. Thus, a comet moves fastest when it is closest to the Sun and slowest when it is farthest from the Sun, which is also true for the Earth. (The orbit here is exaggerated to show radial differences.)

> angular momentum = mass × velocity × object distance from axis of rotation
>
> or $\qquad\qquad L = mvr$ $\qquad\qquad$ 3.7

An external, unbalanced force can change the linear momentum. Similarly, angular momentum can be changed by an external, unbalanced (net) *torque*. Such a torque gives rise to a twisting or rotational effect. Basically, a force produces linear motion, and a torque produces rotational motion. For example, in ● Fig. 3.21, a net torque on the steering wheel is produced by two equal and opposite forces acting on different parts of the wheel. These forces give rise to two torques, resulting in a net torque that causes the steering wheel to turn or rotate. (Would there be a net torque [and rotation] if both forces were upward?)

Note that these forces are at a distance *r* (called the *lever arm*) from the center of motion or axis of rotation. When *r* and *F* vectors are perpendicular, the magnitude of the **torque** (τ, Greek tau) is given by

> torque = force × lever arm
>
> $\qquad\qquad \tau = Fr$ $\qquad\qquad$ 3.8

with the units N · m.

Figure 3.21 Torque
A torque is a twisting action that produces rotational motion or a change in rotational motion. Torque is analogous to a force producing linear motion or a change in linear motion. The forces F_1 and F_2 supply the torque.

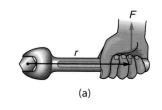

(a)

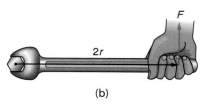

(b)

Figure 3.22 Torque and Lever Arm
(a) Torque varies with the length of the lever arm r. (b) When the length of the lever arm is doubled for a given force, the torque is doubled. Thus, by using a longer wrench, more torque can be applied to a bolt or nut.

Torque varies with r, so for a given force, the greater r is, the greater the torque. You have probably used this fact in trying to produce rotational motion to loosen something, such as a bolt or a nut (● Fig. 3.22). Increasing the lever arm r increases the torque, making it easier to loosen the nut. For the same reason, doorknobs are placed far from the hinges. Have you ever tried to push open a door near the hinges? It's very difficult; there is not enough torque because the lever arm is too short.

There is also a conservation law for angular momentum. The **law of conservation of angular momentum** states that the angular momentum of an object remains constant if there is no external, unbalanced torque acting on it.

That is, the magnitudes of the angular momenta are equal at times 1 and 2:

$$L_1 = L_2$$

or
$$m_1 v_1 r_1 = m_2 v_2 r_2 \qquad \textbf{3.9}$$

where the subscripts 1 and 2 denote the angular momentum of the object at different times.

In our example of a comet, the angular momentum mvr remains the same because the gravitational attraction is internal to the system. As the comet gets closer to the Sun, r decreases, so the speed v increases. For this reason, a comet moves more rapidly when it is closer to the Sun than when it is farther away (see Fig. 3.20). Comet orbits are highly elliptical, so they move at very different speeds in different parts of their orbits. Planets have relatively less elliptical orbits but do move with different speeds.

EXAMPLE 3.5   Applying the Conservation of Angular Momentum

In its orbit at the farthest point from the Sun, a certain comet is 900 million miles away and traveling at 6000 mi/h. What is its speed at its closest point to the Sun at a distance of 30 million miles?

Solution

We are given v_2, r_2, and r_1, so v_1 can be calculated (Eq. 3.9):

$$m v_1 r_1 = m v_2 r_2$$
$$v_1 r_1 = v_2 r_2$$

or
$$v_1 = \frac{v_2 r_2}{r_1}$$

$$= \frac{(6.0 \times 10^3 \text{ mi/h})(900 \times 10^6 \text{ mi})}{30 \times 10^6 \text{ mi}}$$

$$= 1.8 \times 10^5 \text{ mi/h, or } 180{,}000 \text{ mi/h}$$

Thus, the comet moves much more rapidly when it is close to the Sun than when it is far away.

Confidence Exercise 3.5

The Earth's orbit about the Sun is not quite circular. At its closest approach, our planet is 1.47×10^8 km from the Sun, and at its farthest point, it is 1.52×10^8 km from the Sun. At which of these points does the Earth have the greater orbital speed and by what factor? (*Hint:* Use a ratio.)

Another example of the conservation of angular momentum is demonstrated in ● Fig. 3.23. Ice-skaters use the principle to spin faster. The skater extends both arms and perhaps one leg and obtains a slow rotation. Then, drawing the arms in and above the head (making r smaller), the skater achieves greater angular velocity and a more rapid spin because of the decrease in the radial distance of the mass.

(a)

Figure 3.23 Conservation of Angular Momentum: Ice-Skater Spin
(a) An ice-skater starts with a slow rotation, keeping the arms and a leg extended. (b) When the skater stands and draws the arms inward and above the head, the average radial distance of mass decreases and the angular velocity increases to conserve angular momentum, producing a rapid spin.

(b)

Angular momentum also affects the operation of helicopters. What would happen when a helicopter with a single rotor tried to get airborne? To conserve angular momentum, the body of the helicopter would have to rotate in the direction opposite that of the rotor. To prevent such rotation, large helicopters have two oppositely rotating rotors (● Fig. 3.24a). Smaller helicopters instead have small "antitorque" rotors on the tail (Fig. 3.24b). These rotors are like small airplane propellers that provide a torque to counteract the rotation of the helicopter body.

Did You Learn?

- Linear momentum is conserved when there is no external, unbalanced force acting on the system.

- An external, unbalanced torque causes a change in angular momentum.

(a)

(b)

Figure 3.24 Conservation of Angular Momentum in Action
(a) Large helicopters have two overhead rotors that rotate in opposite directions to balance the angular momentum. (b) Small helicopters with one overhead rotor have an "antitorque" tail rotor to balance the angular momentum and prevent rotation of the helicopter body.

KEY TERMS

1. force (3.1)
2. unbalanced, or net, force
3. Newton's first law of motion (3.2)
4. inertia
5. mass
6. Newton's second law of motion (3.3)
7. newton

8. weight
9. friction
10. Newton's third law of motion (3.4)
11. Newton's law of universal gravitation (3.5)
12. *G*
13. buoyant force (3.6)

14. Archimedes' principle
15. linear momentum (3.7)
16. law of conservation of linear momentum
17. angular momentum
18. torque
19. law of conservation of angular momentum

MATCHING

For each of the following items, fill in the number of the appropriate Key Term from the preceding list. Compare your answers with those at the back of the book.

a. _____ Law of inertia
b. _____ Changes angular momentum
c. _____ *mg*
d. _____ Required for an object to float
e. _____ *mvr*
f. _____ Tendency of an object to remain at rest or in uniform, straight-line motion
g. _____ Mass × velocity
h. _____ *F* = *ma*
i. _____ Conservation law requiring the absence of an unbalanced torque

j. _____ Action and reaction
k. _____ Capable of producing motion or a change in motion
l. _____ Universal constant
m. _____ SI unit of force
n. _____ Resistance to relative motion
o. _____ A nonzero vector sum of forces
p. _____ Occurs in the absence of an unbalanced force
q. _____ Describes the force of gravity
r. _____ A measure of inertia
s. _____ Gives the magnitude of the buoyant force

MULTIPLE CHOICE

Compare your answers with those at the back of the book.

1. Mass is related to an object's ___. (3.1)
   (a) weight   (b) inertia
   (c) density   (d) all of the preceding

2. What is a possible state of an object in the absence of a net force? (3.2)
   (a) at rest   (b) constant speed
   (c) zero acceleration (d) all of the preceding

3. What term refers to the tendency of an object to remain at rest or in uniform, straight-line motion? (3.2)
   (a) mass   (b) force
   (c) inertia   (d) external force

4. What is necessary for a change in velocity? (3.3)
   (a) inertia   (b) an unbalanced force
   (c) a zero net force (d) a change in direction

5. According to Newton's second law of motion, when an object is acted upon by an unbalanced force, what can be said about the acceleration? (3.3)
   (a) It is inversely proportional to the object's mass.
   (b) It is zero.
   (c) It is inversely proportional to the net force.
   (d) It is independent of mass.

6. A net force ___.
   (a) can produce motion
   (b) is a scalar quantity
   (c) is capable of producing a change in velocity
   (d) both (a) and (c)

7. For every action force, there is which of the following? (3.4)
   (a) a net force
   (b) a friction force
   (c) an unbalanced force
   (d) an equal and opposite force

8. Which is true of the force pair of Newton's third law? (3.4)
   (a) The two forces never produce an acceleration.
   (b) The two forces act on different objects.
   (c) The two forces always cancel each other.
   (d) The two forces are in the same direction.

9. Which is true about the acceleration due to gravity? (3.5)
   (a) It is a universal constant.
   (b) It is a fundamental property.
   (c) It decreases with increasing altitude.
   (d) It is different for different objects in free fall.

10. What is true about the constant *G*? (3.5)
    (a) It is a very small quantity.   (b) It is a force.
    (c) It is the same as *g*.   (d) It decreases with altitude.
11. A child's toy floats in a swimming pool. The buoyant force exerted on the toy depends on the volume of ___. (3.6)
    (a) water in the pool   (b) the pool
    (c) the water displaced   (d) the toy under water
12. If a submerged object displaces an amount of liquid with a weight less than its own, when the object is released, it will ___. (3.6)
    (a) sink
    (b) remain submerged in equilibrium
    (c) float
    (d) pop up out of the surface

13. If a submerged object displaces a volume of liquid of greater weight than its own and is then released, what will the object do?
    (a) sink
    (b) rise to the surface and float
    (c) remain at its submerged position
14. A change in linear momentum requires which of the following? (3.7)
    (a) a change in velocity   (b) an unbalanced force
    (c) an acceleration   (d) all of these
15. Angular momentum is conserved in the absence of which of the following? (3.7)
    (a) inertia   (b) gravity
    (c) a net torque   (d) linear momentum

FILL IN THE BLANK

Compare your answers with those at the back of the book.

1. A force is a quantity that is ___ of producing motion or a change in motion. (3.1)
2. Forces are ___ quantities. (3.1)
3. Galileo concluded that objects ___ (could/could not) remain in motion without a net force. (3.2)
4. An object will *not* remain at rest or in uniform, straight-line motion if acted upon by a(n) ___ force. (3.2)
5. The inertia of an object is related to its ___. (3.2)
6. According to Newton's second law, an object's acceleration is ___ proportional to its mass. (3.3)
7. The newton unit of force is equal to ___ in standard units. (3.3)

8. The coefficient of ___ friction is generally greater than the coefficient of ___ friction. (3.3)
9. Newton's third-law force pair acts on ___ objects. (3.4)
10. The universal gravitational constant is believed to be constant ___. (3.5)
11. An object will sink in a fluid if its average density is ___ than that of the fluid. (3.7)
12. Milk is ___ dense than the cream that floats on top. (3.7)
13. The total linear momentum is *not* conserved if there is a(n) ___ force acting on the system. (3.7)
14. The angular momentum of an object is *not* conserved if the object is acted upon by unbalanced ___. (3.7)

SHORT ANSWER

3.1 Force and Net Force

1. Does a force always produce motion? Explain.
2. Distinguish between a net force and an unbalanced force.

3.2 Newton's First Law of Motion

3. Consider a child holding a helium balloon in a closed car at rest. What would the child observe the balloon to do when the car (a) accelerates from rest and (b) brakes to a stop? (The balloon does not touch the roof of the car.)
4. An old party trick is to pull a tablecloth out from under dishes and glasses on a table. Explain how this trick is done without pulling the dishes and glasses with the cloth.
5. To tighten the loose head of a hammer, the base of the handle is sometimes struck on a hard surface (● Fig. 3.25). Explain the physics behind this maneuver.
6. When a paper towel is torn from a roll on a rack, a jerking motion tears the towel better than a slow pull. Why? Does this method work better when the roll is large or when it is small and near the end? Explain.

3.3 Newton's Second Law of Motion

7. Describe the relationship between (a) force and acceleration, and (b) mass and acceleration.

Figure 3.25 Make It Tight
See Short-Answer Question 5.

8. Can an object be at rest if forces are being applied to it? Explain.

9. If no forces are acting on an object, can the object be in motion? Explain.

10. What is the unbalanced force acting on a moving car with a constant velocity of 25 m/s (56 mi/h)?

11. The coefficient of kinetic friction is generally less than than the coefficient of static friction, Why?

12. A 10-lb rock and a 1-lb rock are dropped simultaneously from the same height.
    (a) Some say that because the 10-lb rock has 10 times as much force acting on it as the 1-lb rock, it should reach the ground first. Do you agree?
    (b) Describe the situation if the rocks were dropped by an astronaut on the Moon.

3.4 Newton's Third Law of Motion

13. When a rocket blasts off, is it the fiery exhaust gases "pushing against" the launch pad that cause the rocket to lift off? Explain.

14. There is an equal and opposite reaction for every force. Explain how an object can be accelerated when the vector sum of these forces is zero.

15. Suppose your physical science textbook is lying on a table. How many forces are acting on it? (Neglect air pressure.)

16. Explain the kick of a rifle in terms of Newton's third law. Do the masses of the gun and the bullet make a difference?

3.5 Newton's Law of Gravitation

17. Show that the universal gravitational constant G has units of $N \cdot m^2/kg^2$.

18. The gravitational force is said to have an infinite range. What does that mean?

19. Explain why the acceleration due to gravity on the surface of the Moon is one-sixth that of the acceleration due to gravity on the Earth's surface.

20. An astronaut has a mass of 70 kg when measured on the Earth. What is her weight in deep space far from any celestial object? What is her mass there?

21. Is "zero g" possible? Explain.

3.6 Archimedes' Principle and Buoyancy

22. In Chapter 1.6 in the discussion of the hydrometer, it is stated: "The higher the bulb floats, the greater the density of the liquid." Why is this? (See Fig. 1.14.)

23. What is a major consideration in constructing a life jacket that will keep a person afloat?

24. As learned in Chapter 1.5, 1 L of water has a mass of 1 kg. A thin, closed plastic bag (negligible weight) with 1 L of water in it is lowered into a lake by means of a string and submerged. When fully submerged, how much force would you have to exert on the string to prevent the bag from sinking more?

25. A large piece of iron with a volume of 0.25 m³ is lowered into a lake by means of a rope until it is just completely submerged. It is found that the support on the rope is less than when the iron is in air. How does the support vary as the iron is lowered more?

26. Is it easier for a large person to float in a lake than a small person? Explain.

27. Why must a helium balloon be held with a string?

3.7 Momentum

28. What are the units of linear momentum?

29. Explain how the conservation of linear momentum follows directly from Newton's first law of motion.

30. In Example 3.4 there are external forces acting on the masses. (a) Identify these forces. (b) Why is the linear momentum still conserved?

31. When a high diver in a swimming event springs from the diving board and "tucks in," a rapid spin results. Why?

VISUAL CONNECTION

Visualize the connections and give answers for the blanks. Compare your answers with those at the back of the book.

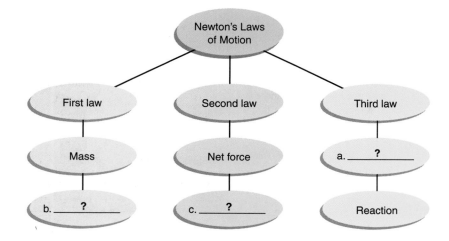

APPLYING YOUR KNOWLEDGE

1. Astronauts walking on the Moon are seen "bounding" rather than walking normally. Why?

2. A person places a bathroom scale in the center of the floor and stands on the scale with his arms at his sides (● Fig. 3.26a). If he keeps his arms *rigid* and quickly raises them over his head (Fig. 3.26b), he notices the scale reading increases as he brings his arms upward. Why? Then, with his arms over his head (Fig. 3.26c), he quickly lowers his arms to his side. How does the scale reading change and why? (Try it yourself.)

3. Using Eq. 3.4 show that the acceleration due to gravity is given by:

$$g = GM_E/R_E^2 \quad (g \text{ on the Earth's surface})$$

where M_E is the mass of the Earth and R_E its radius (Fig. 3.13b) Note that g is independent of m, so it is the same for all objects on the surface of the Earth. (If you are adventurous, plug in the values of G, M_E, and R_E from the back cover of the book and see what you get for g.) Why is the value of g less on the surface of the Moon?

4. People can easily float in the Great Salt Lake in Utah and in the Dead Sea in Israel and Jordan. Why?

5. In a washing machine, water is extracted from clothes by a rapidly spinning drum (● Fig. 3.27). Explain the physics behind this process.

6. When you push on a heavy swinging door to go into a store, why is it harder to push the door open if you mistakenly push on the side closer to the hinges?

7. When unable to loosen the lug nut on an automobile tire, a mechanic may put a piece of pipe on the handle of the tire wrench so as to extend its length. How does that help?

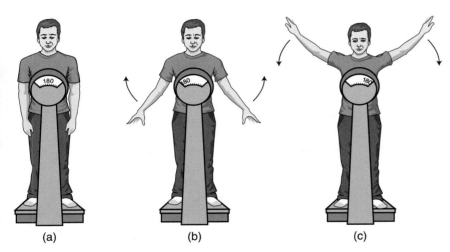

(a)   (b)   (c)

Figure 3.26 Up and Down
A quick way to gain or lose weight. See Applying Your Knowledge Question 2.

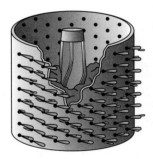

Figure 3.27 Get the Water Out
See Applying Your Knowledge Question 5.

IMPORTANT EQUATIONS

Newton's Second Law: $F = ma$ (3.1)

for weight: $w = mg$ (3.2)

Newton's Third Law: $F_1 = -F_2$ (3.3)

Newton's Law of Gravitation: $F = \dfrac{Gm_1m_2}{r^2}$ (3.4)

(Universal gravitational constant: $G = 6.67 \times 10^{-11} \text{ N} \cdot \text{m}^2/\text{kg}^2$)

Linear Momentum: $p = mv$ (3.5)

Conservation of Linear Momentum:

$$P_f = P_i$$

where $P = p_1 + p_2 + p_3 + \dots$ (3.6)

Angular Momentum: $L = mvr$ (3.7)

Torque: $\tau = Fr$ (3.8)

Conservation of Angular Momentum:

$$L_1 = L_2$$
$$m_1v_1r_1 = m_2v_2r_2 \quad (3.9)$$

EXERCISES

3.1 Force and Net Force

1. What is the net force of a 5.0-N force and an 8.0-N force acting on an object for each of the following conditions?
   (a) The forces act in opposite directions.
   (b) The forces act in the same direction.

   Answer: (a) 3.0 N, (b) 13 N

2. A horizontal force of 250 N is applied to a stationary wooden box in one direction, and a 600-N horizontal force is applied in the opposite direction. What additional force is needed for the box to remain stationary?

3.3 Newton's Second Law of Motion

3. Determine the net force necessary to give an object with a mass of 3.0 kg an acceleration of 5.0 m/s².

   Answer: 15 N

4. A force of 2.1 N is exerted on a 7.0-g rifle bullet. What is the bullet's acceleration?

5. A 1000-kg automobile is pulled by a horizontal tow line with a net force of 950 N. What is the acceleration of the auto? (Neglect friction.)

   Answer: 0.95 m/s²

6. A constant net force of 1500 N gives a toy rocket an acceleration of 2.5 m/s². What is the mass of the rocket?

7. What is the weight in newtons of a 6.0-kg package of nails?

   Answer: 59 N

8. What is the force in newtons acting on a 4.0-kg package of nails that falls off a roof and is on its way to the ground?

9. (a) What is the weight in newtons of a 120-lb person? (b) What is your weight in newtons?

   Answer: (a) 534 N (b) Answers will vary

10. What is the weight in newtons of a 250-g package of breakfast cereal?

3.5 Newton's Law of Gravitation

11. Two 3.0-kg physical science textbooks on a bookshelf are 0.15 m apart. What is the magnitude of the gravitational attraction between the books?

    Answer: 2.7×10^{-8} N

12. (a) What is the force of gravity between two 1000-kg cars separated by a distance of 25 m on an interstate highway? (b) How does this force compare with the weight of a car?

13. How would the force of gravity between two masses be affected if the separation distance between them were (a) doubled? (b) decreased by one-half?

    Answer: (a) $F_2 = F_1/4$ (b) $F_2 = 4F_1$

14. The separation distance between two 1.0-kg masses is (a) decreased by two-thirds and (b) increased by a factor of 3. How is the mutual gravitational force affected in each case?

15. (a) Determine the weight on the Moon of a person whose weight on the Earth is 180 lb. (b) What would be your weight on the Moon?

    Answer: (a) 30 lb (b) Answers will vary.

16. Suppose an astronaut has landed on Mars. Fully equipped, the astronaut has a mass of 125 kg, and when the astronaut gets on a scale, the reading is 49 N. What is the acceleration due to gravity on Mars?

3.6 Archimedes' Principle and Buoyancy

17. A child's cubic play block has a mass of 120 g and sides of 5.00 cm. When placed in a bathtub full of water, will the cube sink or float? (*Hint*: See Chapter 1.6.)

    Answer: float, $p = 0.96$ g/cm³

18. A ball with a radius of 8.00 cm and a mass of 600 g is thrown into a lake. Will the ball sink or float?

3.7 Momentum

19. Calculate the linear momentum of a pickup truck that has a mass of 1000 kg and is traveling eastward at 20 m/s.

    Answer: 2.0×104 kg · m/s, east

20. A small car with a mass of 900 kg travels northward at 30 m/s. Does the car have more or less momentum than the truck in Exercise 19, and how much more or less? (Is direction a factor in this exercise?)

21. Two ice-skaters stand together as illustrated in ● Fig. 3.28. They "push off" and travel directly away from each other, the boy with a velocity of 0.50 m/s to the left. If the boy weighs 735 N and the

Figure 3.28 Pushing Off
See Exercises 21 and 22.

(a) Before

0.50 m/s ←   → ?

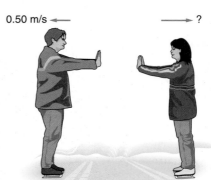

(b) After

girl weighs 490 N, what is the girl's velocity after they push off? (Consider the ice to be frictionless.)

Answer: −0.75 m/s

22. For the couple in Fig. 3.28, suppose you were told that the girl's mass was three-fourths that of the boy's mass. What would be the girl's velocity in this case?

23. A comet goes around the Sun in an elliptical orbit. At its farthest point, 600 million miles from the Sun, it is traveling with a speed of 15,000 mi/h. How fast is it traveling at its closest approach to the Sun, at a distance of 100 million miles?

Answer: 90,000 mi/h

24. An asteroid in an elliptical orbit about the Sun travels at 1.2×10^6 m/s at perihelion (the point of closest approach) at a distance of 2.0×10^8 km from the Sun. How fast is it traveling at aphelion (the farthest point), which is 8.0×10^8 km from the Sun?

ON THE WEB

1. Newton's First Law of Motion

How many "laws of motion" did Sir Isaac Newton propose? What is Newton's first law of motion (and what is it sometimes called)? What do the two parts of the law predict? How does the law apply to your own life? Name at least two ways in which understanding this law can help keep you safe. To see this law in motion, go to the student website for this text at **www.cengagebrain.com/shop/ISBN/1133104096**.

2. Newton's Law of Gravitation

Follow the recommended links on the student website at **www.cengagebrain.com/shop/ISBN/1133104096** to explore antigravity further by creating an antigravity environment. How is this subject related to what you have learned in the text about Newton's law of gravitation?

CHAPTER

4

Work and Energy

Courtesy Cedar Point Amusement Park/Resort, Sandusky Ohio

Courtesy Cedar Point Amusement Park/Resort, Sandusky Ohio

I like work; it fascinates me. I can sit and look at it for hours.

•

Jerome K. Jerome
(1859–1927)

A lot of work going up, and a > lot of energy at the top. The roller coaster at Cedar Point Amusement Park in Sandusky, Ohio, is shown here. The roller coaster reaches speeds of 120 miles per hour within 4 seconds.

PHYSICS FACTS

▶ *Kinetic* comes from the Greek *kinein*, meaning "to move."

▶ *Energy* comes from the Greek *energeia*, meaning "activity."

▶ The United States has about 5% of the world's population and consumes about 26% of the world's energy supply.

▶ Muscles are used to propel the human body by turning stored (potential) energy into motion (kinetic energy).

▶ The human body operates within the limits of the conservation of total energy. The sum of dietary input energy minus the energy expended in the work of daily activities, internal activities, and system heat losses equals zero.

The commonly used terms *work* and *energy* have general meanings for most people. For example, work is done to accomplish some task or job. To get the work done, energy is expended. Hence, work and energy are related. After a day's work, one is usually tired and requires rest and food to regain one's energy.

The scientific meaning of work is quite different from the common meaning. A student standing and holding an overloaded book bag is technically doing no work, yet he or she will feel tired after a time. Why does the student do no work? As will be learned in this chapter, mechanical work involves force *and* motion.

Energy, one of the cornerstones of science, is more difficult to define. Matter and energy make up the universe. Matter is easily understood; in general, we can touch

Chapter Outline

it and feel it. Energy is not actually tangible; it is a concept. We are aware of energy only when it is being used or transformed, such as when it is used to do work. For this reason, energy is sometimes described as stored work.

Our main source of energy is the Sun, which constantly radiates enormous amounts of energy into space. Only a small portion of this energy is received by the Earth, where much of it goes into sustaining plant and animal life. On the Earth, energy exists in various forms, including chemical, electrical, nuclear, and gravitational energies. However, energy may be classified more generally as either *kinetic energy* or *potential energy*. Read on.

4.1 Work

Preview Questions

● Is work a vector quantity? In other words, does it need a direction associated with it?

● What are the units of work?

Mechanically, work involves force and motion. One can apply a force all day long, but if there is no motion, then there is technically no work. The work done by a *constant force F* is defined as follows:

The **work** done by a constant force *F* acting on an object is the product of the magnitude of the force (or parallel component of the force) and the parallel distance *d* through which the object moves while the force is applied.

In equation form,

$$\text{work} = \text{force} \times \text{parallel distance}$$
$$W = Fd \qquad\qquad 4.1$$

In this form, it is easy to see that work involves motion. If $d = 0$, then the object has not moved and no work is done.

Figures 4.1 and 4.2 illustrate the difference between the application of force without work resulting and the application of force that results in work. In ● Fig. 4.1, a force is being applied to the wall, but no work is done because the wall doesn't move. After a while the man may become quite tired, but no mechanical work is done. ● Figure 4.2 shows an object being moved through a distance *d* by an applied force *F*. Note that the force and the directed distance are parallel to each other and that the force *F* acts through the parallel distance *d*. The work is then the product of the force and distance, $W = Fd$.

Another important consideration is shown in ● Fig. 4.3. When the force and the distance are not parallel to each other, only a component or part of the force acts through the parallel distance. When a lawn mower is pushed at an angle to the horizontal, only the component of the force that is parallel to the level lawn (horizontal component F_h) moves through a parallel distance and does work ($W = F_h d$). The vertical component of the force (F_v) does no work because this part of the force does not act through a distance. It only tends to push the lawn mower against the ground.

Work is a *scalar* quantity. Both force and parallel distance (actually, displacement) have directions associated with them, but their product, work, does not. Work is expressed only as a magnitude (a number with proper units). There is no direction associated with it.

Because work is the product of force and distance, the units of work are those of force times length ($W = Fd$). The SI unit of work is thus the newton-meter (N·m, force × length). This unit combination is given the special name **joule** (abbreviated J and pronounced "jool") in honor of nineteenth-century English scientist James Prescott Joule, whose research concerned work and heat.

Figure 4.1 No Work Done
A force is applied to the wall, but no work is done because there is no motion ($d = 0$).

$W = Fd = 0$ $\qquad F$

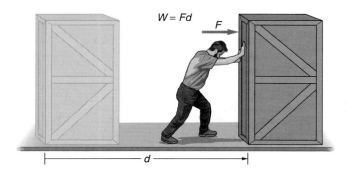

$W = Fd$

F

d

Figure 4.2 Work Being Done
An applied constant force *F* acts through a parallel distance *d*. When the force and the displacement are in the same direction, the amount of work equals the force times the distance the object moves, $W = Fd$.

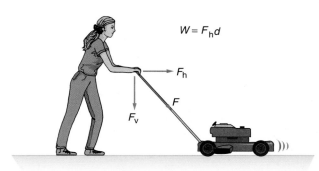

$W = F_h d$

F_h

F

F_v

Figure 4.3 Work and No Work
Only the horizontal component F_h does work because only it is in the direction of the motion. The vertical component F_v does no work because $d = 0$ in that direction.

One joule is the amount of work done by a force of 1 N acting through a distance of 1 m. Similarly, the unit of force multiplied by length in the British system is the pound-foot. However, English units are commonly listed in the reverse order, and work is expressed in **foot-pound** (ft · lb) units. A force of 1 lb acting through a distance of 1 ft does 1 ft · lb of work. The units of work are summarized in ● Table 4.1.

When doing work, you apply a force and feel the other part of Newton's third-law force pair acting against you. Therefore, we say that work is done *against* something, such as work against gravity or friction. When something is lifted, a force must be applied to overcome the force of gravity (as expressedw by an object's weight $w = mg$), so work is done *against* gravity. This work is given by $W = Fd = wh = mgh$, where *h* is the height to which the object was lifted.

Similarly, work is done *against* friction. Friction opposes motion. Hence, to move something on a surface in a real situation, you must apply a force. In doing so, you do work against friction. As illustrated in ● Fig. 4.4a, if an object is moved with a constant velocity, then the applied force *F* is equal and opposite to the frictional force *f* (zero net force on the block and no acceleration). The work done by the applied force against friction is $W = Fd$.

Notice in Fig. 4.4(a) that the frictional force acts through the distance and therefore does work. The frictional force and the displacement are in opposite directions. Taking right to be positive and expressing directions by plus and minus signs, we have $+d$ and $-f$. The work done by the frictional force is then $-fd = -W$; that is, it is negative work. The negative frictional work is equal to the positive work done by the applied force, so the total work is zero, $W_t = Fd - fd = 0$. Otherwise, there would be net work and an energy transfer such that the block would not move with a constant velocity. If the applied force were removed, then the block would slow down (decrease in energy) and eventually stop because of frictional work.

When walking, there must be friction between our feet and the floor; otherwise, we would slip. In this case, though, no work is done against friction because the frictional force prevents the foot from slipping (Fig. 4.4b). No motion of the foot, no work. Of course, other (muscle) forces do work because in walking there is motion. It is interesting

Table 4.1 Work Units (Energy Units)		
System	Force × Distance Units $W = F \times d$	Special or Common Name
SI	newton × meter (N · m)	joule (J)
British	pound × foot	foot-pound (ft · lb)

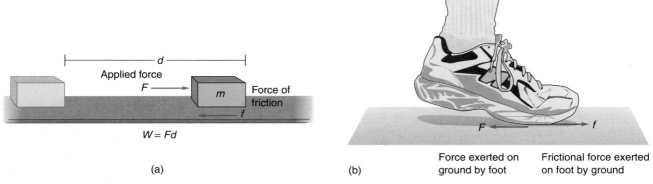

Figure 4.4 Work and No Work Done Against Friction
(a) The mass is moved to the right with a constant velocity by a force F, which is equal and opposite to the frictional force f. (b) When you are walking, there is friction between your feet and the floor. This example is a static case, and the frictional force prevents the foot from moving or slipping. No motion, no work.

to note in Fig. 4.4b that to walk forward, one exerts a backward force on the floor. The frictional force is in the direction of the walking motion and opposes the motion (slipping) of the foot on the ground.

Did You Learn?

- Work is a scalar quantity. Both force and parallel distance (actually, displacement) have directions associated with them, but their product, work, does not.

- Work = $F \times d$ and so has the units newton-meter (N · m), which is called a joule (J).

4.2 Kinetic Energy and Potential Energy

Preview Questions

- By what process is energy transferred from one object to another?
- To find the difference in gravitational potential energies, the difference in heights is taken. What is taken to find the difference in kinetic energies?

When work is done on an object, what happens? When work is done against gravity, an object's height is changed, and when work is done against friction, heat is produced. Note that in these examples some physical quantity changes when work is done.

The concept of energy helps unify all the possible changes. Basically, when work is done there is a change in energy, and the amount of work done is equal to the change in energy. But what is energy? Energy is somewhat difficult to define because it is abstract. Like force, it is a concept: easier to describe in terms of what it can do rather than in terms of what it is.

Energy, one of the most fundamental concepts in science, may be described as a property possessed by an object or system (a group of objects). **Energy** is the ability to do work; that is, an object or system that possesses energy has the ability or capability to do work. That is how the notion of energy as stored work arises. When work is done by a system, the amount of energy of the system decreases. Conversely, when work is done on a system, the system gains energy. (Remember, however, that not all the energy possessed by a system may be available to do work.)

Hence, work is the process by which energy is transferred from one object or system to another. An object with energy can do work on another object and give it energy. That being the case, it should not surprise you to learn that work and energy have the same units. In the SI, energy is measured in joules, as is work. Also, both work and energy are scalar quantities, so no direction is associated with them.

Energy occurs in many forms—chemical energy, heat energy, and so on (Chapter 4.5). Here the focus will be on *mechanical energy*, which has two fundamental forms: *kinetic energy* and *potential energy*.

Kinetic Energy

As noted previously, when a constant net force acts and work is done on an object, the object's velocity changes. This can be seen by using equations from Chapter 2, where $d = \frac{1}{2} at^2$ and $v = at$. Then, with work given by $W = Fd$ and with $F = ma$, we have

$$W = Fd = mad$$
$$= ma(\tfrac{1}{2} at^2) = \tfrac{1}{2}m(at)^2$$
$$= \tfrac{1}{2} mv^2$$

This amount of work is now *energy of motion*, and it is defined as *kinetic energy*.

Kinetic energy is the energy an object possesses because of its motion, or simply stated, it is the energy of motion. The amount of kinetic energy an object has when traveling with a velocity v is given by*

$$\text{kinetic energy} = \tfrac{1}{2} \times \text{mass} \times (\text{velocity})^2$$

$$E_k = \tfrac{1}{2} mv^2 \qquad\qquad \textbf{4.2}$$

As an example of the relationship between work and kinetic energy, consider a pitcher throwing a baseball (● Fig. 4.5). The amount of work required to accelerate a baseball from rest to a speed v is equal to the baseball's kinetic energy, $\frac{1}{2} mv^2$.

Suppose work is done on a moving object. Because the object is moving, it already has some kinetic energy, and *the work done goes into changing the kinetic energy*. Hence,

$$\text{work} = \text{change in kinetic energy}$$
$$W = \Delta E_k = E_{k_2} - E_{k_1} = \tfrac{1}{2} mv_2^2 - \tfrac{1}{2} mv_1^2 \qquad\qquad \textbf{4.3}$$

If an object is initially at rest ($v_1 = 0$), then the change in kinetic energy is equal to the kinetic energy of the object. Also keep in mind that to find the change in kinetic energy, you must first find the kinetic energy for each velocity and then subtract, *not* find the change or difference in velocities and then compute to find the change in kinetic energy.

EXAMPLE 4.1   Finding the Change in Kinetic Energy

A 1.0-kg ball is fired from a cannon. What is the change in the ball's kinetic energy when it accelerates from 4.0 m/s to 8.0 m/s?

Solution

Given: $m = 1.0$ kg and $v_1 = 4.0$ m/s
　　　　　　　　　　　$v_2 = 8.0$ m/s

Equation 4.3 can be used directly to compute the change in kinetic energy. Notice that the kinetic energy is calculated for each velocity.

$$\Delta E_k = E_{k_2} - E_{k_1} = \tfrac{1}{2} mv_2^2 - \tfrac{1}{2} mv_1^2$$
$$= \tfrac{1}{2}(1.0 \text{ kg})(8.0 \text{ m/s})^2 - \tfrac{1}{2}(1.0 \text{ kg})(4.0 \text{ m/s})^2$$
$$= 32 \text{ J} - 8.0 \text{ J} = 24 \text{ J}$$

---

*Although velocity is a vector, the product of $v \times v$, or v^2, gives a scalar, so kinetic energy is a scalar quantity. Either instantaneous velocity (magnitude) or speed may be used to determine kinetic energy.

Figure 4.5 Work and Energy
The work necessary to increase the velocity of a mass is equal to the increase in kinetic energy; here it is that of the thrown ball. (It is assumed that no energy is lost to friction, that is, converted to heat.)

Confidence Exercise 4.1

In working the preceding example, suppose a student first subtracts the velocities and says $\Delta E_k = \frac{1}{2} m(v_2 - v_1)^2$. What would the answer be, and would it be correct? Explain and show your work.

The answers to Confidence Exercises may be found at the back of the book.

Work is done in getting a stationary object moving, and the object then has kinetic energy. Suppose you wanted to stop a moving object such as an automobile. Work must be done here, too, and the amount of work needed to stop the automobile is equal to its change in kinetic energy. The work is generally supplied by brake friction.

In bringing an automobile to a stop, we are sometimes concerned about the braking distance, which is the distance the car travels after the brakes are applied. On a level road the work done to stop a moving car is equal to the braking force (f) times the braking distance ($W = fd$). As has been noted, the required work is equal to the kinetic energy of the car ($fd = \frac{1}{2} mv^2$). Assuming that the braking force is constant, the braking distance is directly proportional to the square of the velocity ($d \propto v^2$).

Squaring the velocity makes a big difference in the braking distances for different speeds. For example, if the speed is doubled, then the braking distance is increased by a factor of 4. (What happens to the braking distance if the speed is tripled?)

This concept of braking distance explains why school zones have relatively low speed limits, commonly 32 km/h (20 mi/h). The braking distance of a car traveling at this speed is about 8.0 m (26 ft). For a car traveling twice the speed, 64 km/h (40 mi/h), the braking distance is four times that distance, or $4 \times 8.0 \text{ m} = 32 \text{ m}$ (105 ft). See ● Fig. 4.6. The driver's reaction time is also a consideration. This simple calculation shows that if a driver exceeds the speed limit in a school zone, then he or she may not be able to stop in time to avoid hitting a child who darts into the street. Remember v^2 the next time you are driving through a school zone.

Potential Energy

An object doesn't have to be in motion to have energy. It also may have energy by virtue of where it is. **Potential energy** is the energy an object has because of its position or location, or more simply, it is the energy of position. Work is done in changing the position of an object; hence, there is a change in energy.

For example, when an object is lifted at a slow constant velocity, there is no net force on it because it is not accelerating. The weight mg of the object acts downward, and there is an equal and opposite upward force. The distance parallel to the applied upward force is the height h to which the

Figure 4.6 Energy and Braking Distance
Given a constant braking force, if the braking distance of a car traveling 32 km/h (20 mi/h) is 8.0 m (26 ft), then for a car traveling twice as fast, or 64 km/h (40 mi/h), the braking distance is four times greater, or 32 m (105 ft), that is, $d \propto v^2$.

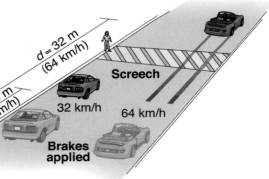

Figure 4.7 Work Against Gravity
In lifting weights (*mg*) to a height *h*, a weight lifter applies an upward lifting force *F*. The work done in lifting the weights is *mgh*. While standing there with the weights overhead, is the weight lifter doing any work?

object is lifted (● Fig. 4.7). Thus, the *work done against gravity* is, in equation form,

$$\text{work} = \text{weight} \times \text{height}$$
$$W = mgh \tag{4.4}$$
$$(W = Fd)$$

Suppose you lift a 1.0-kg book from the floor to a tabletop 1.0 m high. The amount of work done in lifting the book is

$$W = mgh$$
$$= (1.0 \text{ kg})(9.8 \text{ m/s}^2)(1.0 \text{ m}) = 9.8 \text{ J}$$

With work being done, the energy of the book changes (increases), and the book on the table has energy and the ability to do work because of its height or position. This energy is called **gravitational potential energy.** If the book were allowed to fall back to the floor, it could do work; for example, it could crush something.

As another example, the water stored behind a dam has potential energy because of its position. This gravitational potential energy is used to generate electrical energy. Also, when walking up stairs to a classroom, you are doing work. On the upper floor you have more gravitational potential energy than does a person on the lower floor. (Call down and tell the person so.)

The gravitational potential energy E_p is equal to the work done, and this is equal to the weight of the object multiplied by the height(Eq. 4.4). That is,

$$\text{gravitational potential energy} = \text{weight} \times \text{height}$$
$$E_p = mgh \tag{4.5}$$

When work is done by or against gravity, the potential energy changes, and

$$\text{work} = \text{change in potential energy}$$
$$= E_{p_2} - E_{p_1}$$
$$= mgh_2 - mgh_1$$
$$= mg(h_2 - h_1)$$
$$= mg\,\Delta h$$

Similar to an object having kinetic energy for a particular velocity, an object has a potential energy for each particular height or position. When work is done there is a *change* in position, so the (Δh) is really a height *difference*. The h in Eqs. 4.4 and. 4.5 is also a height difference, with $h_1 = 0$.

The value of the gravitational potential energy at a particular position depends on the reference point or the reference or zero point from which the height is measured. Near the surface of the Earth, where the acceleration due to gravity (g) is relatively constant, the designation of the zero position or height is arbitrary. Any point will do. Using an arbitrary zero point is like using a point other than the zero mark on a meterstick to measure length (● Fig. 4.8). This practice may give rise to negative positions, such as the minus (−) positions on a Cartesian graph.

Heights (actually, displacements or directed lengths) may be positive or negative relative to the zero reference point. However, note that the height *difference*, or change in the potential energy between two positions, is the same in any case. For example, in Fig. 4.8 the top ball is at a height of $h = y_2 - y_1 = 50$ cm − 0 cm = 50 cm according to the scale on the left, and $h = 100$ cm − 50 cm = 50 cm according to the meterstick on the right. Basically, you can't change a length or height by just changing scales.

A negative (−) h gives a *negative* potential energy. A negative potential energy is analogous to a position in a hole or a well shaft because we usually designate $h = 0$ at the Earth's surface. Negative energy "wells" will be important in the discussion of atomic theory in Chapter 9.3.

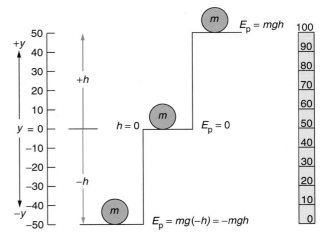

Figure 4.8 Reference Point
The reference point for measuring height is arbitrary. For example, the zero reference point may be that on a Cartesian axis (*left*) or that at one end of the meterstick (*right*). For positions below the chosen zero reference point on the Cartesian *y*-axis, the potential energy is negative because of negative displacement. However, the potential-energy values measured from the zero end of the meterstick would be positive. The important point is that the energy *differences* are the same for any reference.

Conceptual Question and Answer

Double Zero

Q. A fellow student tells you that she has both zero kinetic energy and zero gravitational potential energy. Is this possible?

A. Yes. If she is sitting motionless ($v = 0$), then her kinetic energy is zero. The value of the gravitational potential energy depends on the reference or zero point. If this point is taken at the student's position ($h = 0$), then she will have both zero kinetic and potential energies. (Some work would change the situation.)

There are other types of potential energy besides gravitational. For example, when a spring is compressed or stretched, work is done (against the spring force) and the spring has potential energy as a result of the change in length (position). Also, work is done when a bowstring is drawn back. The bow and the bowstring bend and acquire potential energy. This potential energy is capable of doing work on an arrow, thus producing motion and kinetic energy. Note again that work is a process of transferring energy.

Did You Learn?

● Energy is the ability or capability to do work, and work is the process by which energy is transferred from one object to another.

● To find the difference in kinetic energies, the difference in the squares of the velocity is taken, $\Delta E = \frac{1}{2}m(v_2^2 - v_1^2)$, *not* the difference in velocities squared $(v_2 - v_1)^2$.

4.3 Conservation of Energy

Preview Questions

● Overall, can energy be created or destroyed?

● What is the difference between total energy and mechanical energy?

Energy may change from one form to another and does so without a net loss or net gain. That is to say, energy is *conserved*, and the total amount remains constant. The study of energy transformations has led to one of the most basic scientific principles, the law of conservation of energy. Although the meaning is the same, the *law of conservation of energy* (or simply the conservation of energy) can be stated in different ways. For example, "energy can be neither created nor destroyed" and "in changing from one form to another, energy is always conserved." are both ways of stating this law.

The law of **conservation of total energy** may also be conveniently stated as follows:

The total energy of an isolated system remains constant.

Thus, although energy may be changed from one form to another, energy is not lost from the system, and so it is conserved. A *system* is something enclosed within boundaries, which may be real or imaginary, and *isolated* means that nothing from the outside affects the system (and vice versa).

For example, the students in a classroom might be considered a system. They may move around in the room, but if no one leaves or enters, then the number of students is conserved (the "law of conservation of students"). It is sometimes said that the total energy of the universe is conserved, which is true. The universe is the largest system of which we can think, and all the energy that has ever been in the universe is still there somewhere in some form or other.

To simplify the understanding of the conservation of energy, we often consider *ideal* systems in which the energy is in only two forms, kinetic and potential. In this case, there is **conservation of mechanical energy,** which may be written in equation form as

$$\text{initial energy} = \text{final energy}$$
$$(E_k + E_p)_1 = (E_k + E_p)_2$$
$$(\tfrac{1}{2} mv^2 + mgh)_1 = (\tfrac{1}{2} mv^2 + mgh)_2 \qquad \textbf{4.6}$$

where the subscripts indicate the energy at different times. There are initial (subscript 1) and final (subscript 2) times. Here it is assumed that no energy is lost in the form of heat because of frictional effects (or any other cause). This unrealistic situation is nevertheless instructive in helping us understand the conservation of energy.

EXAMPLE 4.2   Finding Kinetic and Potential Energies

A 0.10-kg stone is dropped from a height of 10.0 m. What will be the kinetic and potential energies of the stone at the heights indicated in ● Fig. 4.9? (Neglect air resistance.)

Solution

With no frictional losses, by the conservation of mechanical energy the total energy ($E_T = E_k + E_p$) will be the same at all heights above the ground. When the stone is released, the total energy is all potential energy ($E_T = E_p$) because $v = 0$ and $E_k = 0$.

At any height h, the potential energy will be $E_p = mgh$. Thus, the potential energies at the heights of 10 m, 7.0 m, 3.0 m, and 0 m are

$$h = 10 \text{ m}: E_p = mgh = (0.10 \text{ kg})(9.8 \text{ m/s}^2)(10.0 \text{ m}) = 9.8 \text{ J}$$
$$h = 7.0 \text{ m}: E_p = mgh = (0.10 \text{ kg})(9.8 \text{ m/s}^2)(7.0 \text{ m}) = 6.9 \text{ J}$$
$$h = 3.0 \text{ m}: E_p = mgh = (0.10 \text{ kg})(9.8 \text{ m/s}^2)(3.0 \text{ m}) = 2.9 \text{ J}$$
$$h = 0 \text{ m}: E_p = mgh = (0.10 \text{ kg})(9.8 \text{ m/s}^2)(0 \text{ m}) = 0 \text{ J}$$

Because the total mechanical energy is conserved, or constant, the kinetic energy (E_k) at any point can be found from the equation $E_T = E_k + E_p$. That is, $E_k = E_T - E_p = 9.8 \text{ J} - E_p$ because the total energy is all potential energy at the 10-m height. ($E_T = E_p = 9.8$ J. Check to see whether the equation gives $E_k = 0$ at the release height.) A summary of the results is given in ● Table 4.2.

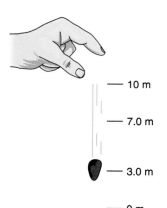

— 10 m

— 7.0 m

— 3.0 m

— 0 m

Figure 4.9 Changing Kinetic and Potential Energies
See Example 4.2.

Table 4.2 Energy Summary for Example 4.2

Height (m)	E_T (J)	E_p (J)	E_k (J)	v (m/s)
10.0	9.8	9.8	0	0
7.0	9.8	6.9	2.9	7.7
3.0	9.8	2.9	6.9	12
0	9.8	0	9.8	14
(decreases)	(constant)	(decreases)	(increases)	(increases)

Confidence Exercise 4.2

Find the kinetic energy of the stone in the preceding example when it has fallen 5.0 m. The answers to Confidence Exercises may be found at the back of the book.

Note from Table 4.2 that as the stone falls, the potential energy becomes less (decreasing h) and the kinetic energy becomes greater (increasing v). Potential energy is converted into kinetic energy. Just before the stone hits the ground ($h = 0$), all the energy is kinetic, and the velocity is a maximum.

This relationship can be used to compute the magnitude of the velocity, or the speed, of a falling object released from rest. The potential energy lost is $mg\Delta h$, where Δh is the change or decrease in height measured *from the release point down*. This is converted into kinetic energy, $\frac{1}{2}mv^2$. By the conservation of mechanical energy, these quantities are equal:

$$\tfrac{1}{2}mv^2 = mg\,\Delta h$$

Then, canceling the m's and solving for v, *the speed of a dropped object after it has fallen a distance h from the point of release is*

$$v = \sqrt{2g\,\Delta h} \qquad\qquad 4.7$$

This equation was used to compute the v's in Table 4.2. For example, at a height of 3.0 m, $\Delta h = 10.0 \text{ m} - 3.0 \text{ m} = 7.0 \text{ m}$ (a decrease of 7.0 m), and

$$v = \sqrt{2g\Delta h} = \sqrt{2(9.8 \text{ m/s}^2)(7.0 \text{ m})} = 12 \text{ m/s}$$

Conceptual Question and Answer

The Race Is On

Q. They're off! Two identical balls are released simultaneously from the same height on individual tracks as shown in ● Fig. 4.10. Which ball will reach the end of its track first (or do they arrive together)?

A. The Track B ball finishes first and wins the race. Obviously, the B ball has a longer distance to travel, so how can this be? Although the B ball must travel farther, it gains greater speed by giving up potential energy for kinetic energy on the downslope, which allows it to cover a greater distance in a shorter time. The B ball has a greater average speed on both the lower downslope and upslope of the track, and so it pulls ahead of the Track A ball and gets to the finish line first. (On the upslope, ball B is decelerating, but it is still traveling faster than ball A.)

Question: Are their speeds the same at the finish line? *Hint:* Look at their overall changes in potential energies. Did you get it? Note that the balls overall have the same decrease in height. Hence, they have the same decrease in potential energy and the same increase in kinetic energy, and they arrive at the finish line with the same speed.

Figure 4.10 The Race Is On
Two identical balls are released simultaneously from the same height on individual tracks. Which ball will reach the end of its track first? (See Conceptual Question and Answer.)

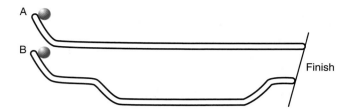

Did You Learn?

● In changing energy from one form to another, the total energy is always conserved.

● Total energy is all the energy of an isolated system, which may be of any form. Mechanical energy is that of an ideal system having only kinetic and potential energies.

4.4 Power

Preview Questions

● What is the difference in the operations of a 2-hp motor and a 1-hp motor?

● Electric bills from power companies charge for so many kilowatt-hours (kWh). What are we paying for?

When a family moves into a second-floor apartment, a lot of work is done in carrying their belongings up the stairs. In fact, each time the steps are climbed, the movers must not only carry up the furniture, boxes, and so on, but they also raise their own weights.

If the movers do all the work in 3 hours, then they will not have worked as rapidly as if the job had been done in 2 hours. The same amount of work would have been done in each case, but there's something different: the *time rate* at which the work is done.

To express how fast work is done, the concept of power is used. **Power** is the time rate of doing work. Power is calculated by dividing the work done by the time required to do the work. In equation form,

$$\text{power} = \frac{\text{work}}{\text{time}}$$

or

$$P = \frac{W}{t} \tag{4.8}$$

Because work is the product of force and distance ($W = Fd$), power also may be written in terms of these quantities:

$$\text{power} = \frac{\text{force} \times \text{distance}}{\text{time}}$$

$$P = \frac{Fd}{t} \tag{4.9}$$

In the SI, work is measured in joules, so power (W/t) has the units of joules per second (J/s). This unit is given the special name **watt** (W) after James Watt, a Scottish engineer who developed an improved steam engine, and 1 W = 1 J/s (● Fig. 4.11).

Light bulbs are rated in watts such as a 100-W bulb, which means that the bulb uses 100 J of electrical energy each second (100 W = 100 J/s). You have been introduced to several SI units in a short time. For convenience, they are summarized in ● Table 4.3.

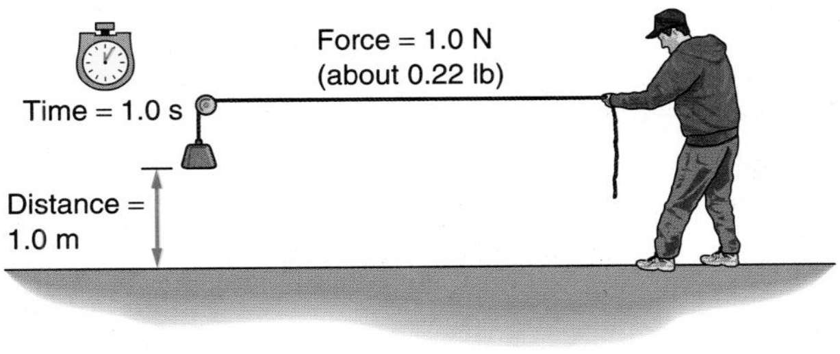

Figure 4.11 The Watt
In applying a force of 1.0 N to raise a mass a distance of 1.0 m, the amount of work done (*Fd*) is 1.0 J. If this work is done in a time of 1.0 s, then the power, or time rate of doing work, is 1.0 W. ($P = W/t = 1.0$ J/1.0 s $= 1.0$ W)

Table 4.3 SI Units of Force, Work, Energy, and Power

Quantity	Unit	Symbol	Equivalent Units
Force	newton	N	$kg \cdot m/s^2$
Work	joule	J .	$N \cdot m$
Energy	joule	J	$N \cdot m$
Power	watt	W	J/s

One should be careful not to be confused by the two meanings of the capital letter W. In the equation $P = W/t$, the (italic) *W* stands for work. In the statement $P = 25$ W, the (roman) W stands for watts. In equations, letters that stand for variable quantities are in italic type, whereas letters used as abbreviations for units are in regular (roman) type.

In the British system, the unit of work is the foot-pound and the unit of power is the foot-pound per second (ft · lb/s). Technically, by equation form the unit of work ($W = Fd$) would be pound-foot. However, this is the same unit as torque (Chapter 3.7, $\tau = Fr$, pound-foot), so to distinguish, work is written foot-pound.

A larger unit, the **horsepower** (hp), is commonly used to rate the power of motors and engines, and

$$1 \text{ horsepower (hp)} = 550 \text{ ft} \cdot \text{lb/s} = 746 \text{ W}$$

The horsepower unit was originated by James Watt, after whom the SI unit of power is named. In the 1700s horses were used in coal mines to bring coal to the surface and to power water pumps. In trying to sell his improved steam engine to replace horses, Watt cleverly rated the engines in horsepower to compare the rates at which work could be done by an engine and by an average horse.

The greater the power of an engine or motor, the faster it can do work; that is, it can do more work in a given time. For instance, a 2-hp motor can do twice as much work as a 1-hp motor in the same amount of time, or a 2-hp motor can do the same amount of work as a 1-hp motor in half the time.

Example 4.3 shows how power is calculated.

EXAMPLE 4.3   Calculating Power

A constant force of 150 N is used to push a student's stalled motorcycle 10 m along a flat road in 20 s. Calculate the power expended in watts.

Solution

First we list the given data and what is to be found in symbol form:

Given: $F = 150$ N   *Wanted*: P (power)
                $d = 10$ m
                $t = 20$ s

Equation 4.9 can be used to find the power with the work expressed explicitly as *Fd*:

$$P = \frac{W}{t} = \frac{Fd}{t} = \frac{(150 \text{ N})(10 \text{ m})}{20 \text{ s}} = 75 \text{ W}$$

Notice that the units are consistent, N · m/s = J/s = W. The given units are all SI, so the answer will have the SI unit of power, the watt.

Bob Daemmrich/The Image Works

Figure 4.12 Energy Consumption
Electrical energy is consumed as the motor of the grinder does work and turns the grinding wheel. Notice the flying sparks and that the operator wisely is wearing a face shield rather than just goggles, as the sign in the background suggests. An electric *power* company is actually charging for *energy* in units of kilowatt-hours (kWh).

Confidence Exercise 4.3

A student expends 7.5 W of power in lifting a textbook 0.50 m in 1.0 s with a constant velocity. (a) How much work is done, and (b) how much does the book weigh (in newtons)?

As we have seen, work produces a change in energy. Thus <u>power may be thought of as energy produced or consumed divided by the time taken to do so.</u> That is,

$$\text{power} = \frac{\text{energy produced or consumed}}{\text{time taken}}$$

or

$$P = \frac{E}{t} \qquad\qquad 4.10$$

Rearranging the equation yields

$$E = Pt \qquad\qquad 4.10a$$

Equation 4.10a is useful in computing the amount of electrical energy consumed in the home. Because energy is power times time ($P \times t$), it has units of watt-second (W·s). Using the larger units of kilowatt (kW) and hour (h) gives the larger unit of **kilowatt-hour** (kWh).

When paying the power company for electricity, in what units are you charged? That is, what do you pay for? If you check an electric bill, you will find that the bill is for so many kilowatt-hours (kWh). Hence, people actually pay the power company for the amount of energy consumed, which is used to do work (● Fig. 4.12). Example 4.4 illustrates how the energy consumed can be calculated when the power rating is known.

EXAMPLE 4.4   Computing Energy Consumed

A 1.0-hp electric motor runs for 10 hours. How much energy is consumed (in kilowatt-hours)?

Solution

Given: $P = 1.0$ hp   *Wanted:* E (energy in kWh)
            $t = 10$ h

Note that the time is given in hours, which is what is wanted, but the power needs to be converted to kilowatts. With 1 hp = 746 W, we have

$$1.0 \text{ hp} = 746 \text{ W} (1 \text{ kW}/1000 \text{ W}) = 0.746 \text{ kW}$$
$$= 0.75 \text{ kW (rounding)}$$

Then, using Eq. 4.10a,

$$E = Pt = (0.75 \text{ kW})(10 \text{ h}) = 7.5 \text{ kWh}$$

This is the electrical energy consumed when the motor is running (doing work).

We often complain about our electric bills. In the United States, the cost of electricity ranges from about 7¢ to 17¢ per kWh, depending on location. Thus, running the motor for 10 hours at a rate of 12¢ per kWh costs 90¢. That's pretty cheap for 10 hours of work output. (Electrical energy is discussed further in Chapter 8.2.)

Confidence Exercise 4.4

A household uses 2.00 kW of power each day for 1 month (30 days). If the charge for electricity is 8¢ per kWh, how much is the electric bill for the month?

Conceptual Question and Answer

Payment for Power

Q. Some factory workers are paid by the hour. Others are paid on a piecework basis (paid according to the number of pieces or items they process or produce.) Is there a power consideration in either of these methods of payment?

A. For hourly payment, there is little consideration for worker incentive or power consumed. A worker gets paid no matter how much work or power is expended. For piecework, on the other hand, the more work done in a given time or the more power expended, the more items produced and the greater the pay.

Did You Learn?

- A 2-hp engine can do twice as much work as a 1-hp engine in the same time, or the same amount of work in half the time.
- The kilowatt-hour (kWh) is a unit of energy ($E = Pt$).

4.5 Forms of Energy and Consumption

Preview Questions

- How many common forms of energy are there, and what are they?
- What are the two leading fuels consumed in the United States, and which is used more in the generation of electricity?

Forms of Energy

We commonly talk about various *forms* of energy such as chemical energy and electrical energy. Many forms of energy exist, but the main unifying concept is the conservation of energy. Energy cannot be created or destroyed, but it can change from one form to another.

In considering the conservation of energy to its fullest, there has to be an accounting for *all* the energy. Consider a swinging pendulum. The kinetic and potential energies of the pendulum bob change at each point in the swing. Ideally, the *sum* of the kinetic and potential energies—the total mechanical energy—would remain constant at each point in the swing and the pendulum would swing indefinitely.

In the real world, however, the pendulum will eventually come to a stop. Where did the energy go? Of course, friction is involved. In most practical situations, the kinetic and potential energies of objects eventually end up as heat. *Heat*, or *thermal energy*, will be examined at some length in Chapter 5.2, but for now let's just say that heat is transferred energy that becomes associated with kinetic and potential energies on a molecular level.

We have already studied *gravitational potential energy*. The gravitational potential energy of water is used to generate electricity in hydroelectric plants. Electricity may be described in terms of electrical force and *electrical energy* (Chapter 8.2). This energy is associated with the motions of electric charges (electric currents). It is electrical energy that runs numerous appliances and machines that do work for us.

Electrical forces hold or bond atoms and molecules together, and there is potential energy in these bonds. When fuel is burned (a chemical reaction), a rearrangement of the electrons and atoms in the fuel occurs, and energy—*chemical energy*—is released. Our

Table 4.4	Common Forms of Energy
Chemical energy	
Electrical energy	
Gravitational (potential) energy	
Nuclear energy	
Radiant (electromagnetic) energy	
Thermal energy	

main fossil fuels (wood, coal, petroleum, and natural gas) are indirectly the result of the Sun's energy. This *radiant energy*, or light from the Sun, is electromagnetic radiation. When electrically charged particles are accelerated, electromagnetic waves are "radiated" (Chapter 6.3). Visible light, radio waves, TV waves, and microwaves are examples of electromagnetic waves.

A more recent entry into the energy sweepstakes is *nuclear energy*. Nuclear energy is the source of the Sun's energy. Fundamental nuclear forces are involved, and the rearrangement of nuclear particles to form different nuclei results in the release of energy as some of the mass of the nuclei is converted into energy. In this case mass is considered to be a form of energy (Chapter 10). See ● Table 4.4 for a summary of the common forms of energy.

As we go about our daily lives, each of us is constantly using and giving off energy from body heat. The source of this energy is food (● Fig. 4.13). An average adult radiates heat energy at about the same rate as a 100-W light bulb. This explains why it can become uncomfortably warm in a crowded room. In winter extra clothing helps keep our body heat from escaping. In summer the evaporation of perspiration helps remove heat and cool our bodies.

The commercial sources of energy on a national scale are mainly coal, oil (petroleum), and natural gas. Nuclear and hydroelectric energies are about the only other significant commercial sources. ● Figure 4.14 shows the current percentage of energy supplied by each of these resources. Over one-half of the oil consumption in the United States comes from imported oil. The United States does have large reserves of coal, but there are some pollution problems with this resource (see Chapter 20.4). Even so, it is the major energy source for the generation of electricity (● Fig. 4.15).

Perhaps you're wondering where all this energy goes and who consumes it. ● Figure 4.16 gives a general breakdown of energy use by economic sector.

Energy Consumption

All these forms of energy go into satisfying a growing demand. Although the United States has less than 5% of the world's population, it accounts for approximately 26% of the world's annual energy consumption of fossil fuels: coal, oil, and natural gas. With increasing world population (now about 7 billion), there is an ever-increasing demand for energy. Where will it come from?

Of course, fossil fuels and nuclear processes will continue to be used, but increasing use gives rise to pollution and environmental concerns. Fossil fuels contribute to greenhouse gases and possible global warming (Chapter 20.5). Nuclear reactors do not have gaseous

Fgure 4.13 Refueling
The source of human energy is food. Food is the fuel our bodies convert into energy that is used in performing tasks and carrying out body functions. Also, energy is given off as heat and may be stored for later use.

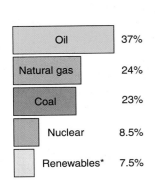

Figure 4.14 Comparative Fuel Consumption
The bar graph shows the approximate relative percentages of current fuel consumption in the United States.

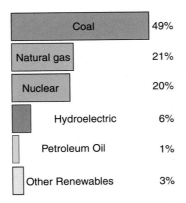

Figure 4.15 Fuels for Electrical Generation
The bar graph shows the relative percentages of fuels used for generating electricity in the United States.

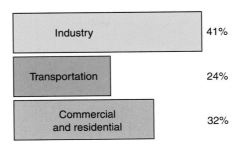

Figure 4.16 Energy Consumption by Sector
The bar graph shows the relative consumption of energy by three main sectors of the United States economy.

*Renewables (hydro, biofuels, solar, and wind).

emissions, but nuclear wastes are a problem. (See the Chapter 10.7 Highlight: Nuclear Power and Waste Disposal.)

Research is being done on so-called alternative or renewable fuels and energy sources, which would be nonpolluting supplements to our energy supply. These sources will be addressed in the next section. Before then, however, let's consider a relatively new and potential alternative *form* of "fossil" fuel in the **Highlight: Ice That Burns**.

Highlight Ice That Burns

A frozen substance, *methane hydrate*, is described as "ice that burns" (Fig. 1). Found under the ocean floors and below polar regions, methane hydrate is a crystalline form of natural gas and water. (Methane is the major constituent of natural gas.) Methane hydrate resembles ice, but it burns if ignited. Until recently, it was looked upon as a nuisance because it sometimes plugged natural gas lines in polar regions.

Now this frozen gas–water combination is the focus of research and exploration. Methane hydrate occupies as much as 50% of the space between sediment particles in samples obtained by exploratory drilling. It has been estimated that the energy locked in methane hydrate amounts to twice the global reserves of conventional sources (coal, oil, and natural gas).

Methane hydrate may be an energy source in the future, but for now, there are many problems to solve, such as finding and drilling into deposits of methane hydrate and separating the methane from the water. Care must be taken to ensure that methane does not escape into the atmosphere. Methane, a "greenhouse" gas, is 10 times more effective than carbon dioxide in causing climate warming (Chapter 20.5).

Leibniz-Institute of Marine Sciences at Kiel University, Germany, Dr. Jens Greinert

Figure 1 Burning Ice
Found under the ocean floors and below polar regions, methane hydrate is a crystalline form of natural gas and water. It resembles ice and will burn if ignited.

4.6   Alternative and Renewable Energy Sources

Preview Questions

● What is the difference between alternative and renewable energy sources?

● Why are solar power and wind power somewhat unreliable?

Fossil fuels will be our main source of energy for some time. However, fossil-fuel combustion contributes to pollution and possible climate change: greenhouse gases and global warming (Chapter 20.5). The amount of fossil fuels is limited. They will be depleted someday. For example, it is estimated that at our present rate of consumption, known world oil reserves will last perhaps only 50 years.

Let's distinguish between alternative energy and renewable energy. **Alternative energy sources** are energy sources that are not based on the burning of fossil fuels and nuclear processes. **Renewable energy sources** are energy sources from natural processes that are constantly replenished. In large part these energy sources overlap. It is estimated that they account for about 7–8% of the energy consumption in the United States. See whether you would classify each of the following as alternative or renewable energy sources or as both.

▶ *Hydropower* Hydropower is used widely to produce electricity (● Fig. 4.17). We would like to increase this production because falling water generates electricity cleanly and efficiently. However, most of the best sites for dams have already been developed. There are over 2000 hydroelectric dams in the United States. The damming of rivers usually results in the loss of agricultural land and alters ecosystems.

▶ *Wind power* Wind applications have been used for centuries. If you drive north from Los Angeles into the desert, you will suddenly come upon acres and acres of windmills

Figure 4.17 Grand Coulee Dam
The potential energy of dammed water can be used to generate electricity. Shown here, the Grand Coulee Dam across the Columbia River in Washington state is the largest facility in the United States producing hydroelectric power and the fifth largest in the world.

Harald Sund/Getty Images

Figure 4.18 Wind Energy
Wind turbines in Tehachapi Pass, California, generate electricity using the desert wind.

(Fig. 4.18). Windmills for pumping water were once common on American farms. There have been significant advances in wind technology, and modern wind turbines, as shown in Fig. 4.18, generate electricity directly. The wind is free and nonpolluting. However, the limited availability of sites with sufficient wind (at least 20 km/h or 12 mi/h) prevents widespread development of wind power. And the wind does not blow continuously.

One projected solution is a wind farm of floating wind turbines offshore in the ocean. This technology requires the development of an undersea power cable network to bring the electricity ashore.

▶ *Solar power* The Sun is our major source of energy and one of the most promising sources of energy for the future. Solar power is currently put to use, but more can be done. Solar heating and cooling systems are used in some homes and businesses, and other technologies focus on concentrating solar radiation for energy production.

One of the most environmentally promising solar applications is the photovoltaic cell (or photocell, for short). These cells convert sunlight directly into electricity (Fig. 4.19).

Figure 4.19 Solar Energy
Photocells convert solar energy directly into electrical energy. (a) Solar panels at the Nellis Air Force Base near Las Vegas, Nevada. Occupying 140 acres, 70,000 solar panels use a solar-tracking system for greater capacity. (b) The world's largest solar farm is in southern Spain. The farm consists of 120,000 solar panels occupying 247 acres.

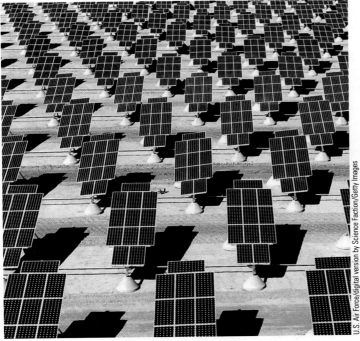

(a)

(b)

The light meter used in photography is a photocell, and photocell arrays are used on Earth satellites. Efficiency has been a problem with photocells, but advanced technology has boosted it to about 30–35%. Even so, electricity from photocells costs approximately 30¢ per kWh, which is not economically competitive with electricity produced from fossil fuels (on average, 10¢ to 12¢ per kWh). Photocell arrays could be put on the roofs of buildings to reduce the need for additional land, but electrical backup systems would be needed because the photocells could be used only during the daylight hours. Like windmills, photocell energy production depends on the weather. And clouds would reduce its efficiency.

Perhaps we will one day learn to mimic the Sun and produce energy by nuclear fusion. You'll learn more about nuclear fusion in Chapter 10.

▶ *Geothermal* Geothermal hotspots and volcanic features can be found around the world (Chapter 22.4). The Earth's interior is hot, and heat is energy. At the surface, geysers and hot springs are evidence of this extreme heat. Steam from geysers in California and Italy are used to generate electricity (● Fig. 4.20). In Iceland, water from hot springs is used to heat the capital of Reykjavik. Work is being done on geothermal systems that pump hot water into underground hotspots and then use the resulting steam to generate electricity. Such systems are relatively inefficient and costly.

▶ *Tides* Unlike the weather problems with wind and solar energy production, tidal energy production is steady, reliable, and predictable. There is constant water motion and thus a constant energy source. In France, electrical power is generated by strong tides going in and out of the Rance River. Underwater generators are planned to take advantage of the tide going in and out. Surface generators are also being developed to take advantage of surface wave action (● Fig. 4.21).

▶ *Biofuels* Because of the agricultural capacity of the United States, large amounts of corn can be produced, from which ethanol (an alcohol) is made. A mixture of gasoline and ethanol, called "gasohol," is used to run cars. Ethanol is advertised as reducing air pollution (less carbon dioxide) when mixed and burned with gasoline. Some pollution is reduced, but there are still emissions. Also, there is the disposal of waste by-products from the ethanol production to consider, and more fossil-fuel energy is actually used in ethanol production than the use of ethanol saves.

In some places a variety of biofuels are available. ● Fig. 4.22 shows a filling station in San Diego, CA advertising fuel choices, including ethanol and biodiesel. (Biodiesel is typically made by chemically reacting vegetable oil with an alcohol).

Figure 4.20 The Geysers
Located 72 miles north of San Francisco, the "Geysers" is a naturally occurring steam field reservoir below the Earth's surface used to generate electricity. As the largest complex of geothermal power plants in the world, the net generation capacity is enough to provide electricity for 750,000 homes.

Figure 4.21 Wave-Action Electrical Generation
The Aguçadoura wave in Peru converts the energy of ocean surface waves into electrical power. The snake-like structures float in the water, where they arc and bend, forcing oil to be pumped through high-pressure motors that in turn drive electrical generators. The power is then transferred to shore. (Underwater turbines generators also take advantage of the in-and-out motions of ocean water due to the daily rise and fall of tides.)

Figure 4.22 Take Your Choice
A sign advertising gasoline (3 grades), ethanol, diesel, biodiesel, natural gas, and propane at a San Diego (CA) filling station.

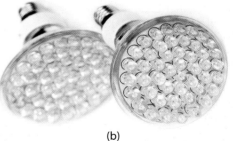

Work is being done on algae-based biofuels. Algae are organisms that grow in aquatic environments. A green layer of algae is commonly seen on ponds in the summer. Algae use photosynthesis (Chapter 19.1) to produce energy for rapid growth and can double in weight several times on a sunny day. As part of the photosynthesis process, algae produce oils which can be harvested as biofuels.

▶ *Biomass* Biomass is any organic matter available on a renewable basis. It includes agricultural crops, wood and wood wastes, animal wastes, municipal wastes, and aquatic plants. Processed and capable of being burned, biomass constitutes a source of energy, some of which can be used in transportation fuels or chemicals.

For other energy topics, see the **Highlight: Hybrids and Hydrogen**.

In addition to alternative energy resources, emphasis is placed on using our available energy more efficiently. Appliances come with Energy Guide labels that compare energy costs or usage. Also, there are more efficient light sources. Compact fluorescent bulbs are coming into increasing use (● Fig. 4.23a). The common incandescent bulbs are relatively inefficient, producing most of the radiation in the infrared region, whereas fluorescent light is more in the visible region. There is some environmental concern because fluorescent lamps and bulbs contain mercury, which is potentially environmentally hazardous (see Chapter 23.3). Compact bulbs use less energy (watts or energy per time) and last up to 5 years, saving not only energy but replacement costs.

More recently, light-emitting diode, or LED, bulbs have been introduced (Fig. 4.23b). The common little Christmas tree lights are LEDs. These bulbs use just 12 W to produce the same amount of light as a 60-W incandescent bulb, an 80% savings of energy. LED bulbs are reported to last 25 times as long as an incandescent bulb and have no mercury. However, they are much more expensive. LEDs are also becoming common in flashlights and automobile headlights.

Figure 4.23 More Efficient Lightning
(a) A compact fluorescent light (CFL) bulb. Note that the 15-W CFL is as efficient as a 60-W incandescent bulb. (b) For even more efficiency and longer life, LED (light emitting diode) bulbs are coming into use.

Highlight   Hybrids and Hydrogen

Automobiles have been powered primarily by gasoline combustion for 100 years or so. With concerns about petroleum (gasoline) supply and environmental pollution (Chapter 20.4), alternative power sources are being sought. You have probably heard about electric cars that are powered by batteries. However, these cars have a limited range, an average of 120 km (75 mi), before the batteries need to be recharged. Recharging requires the generation of electricity, which can be polluting.

Two top candidates that offer some improvement are *hybrid cars* and an *alternative fuel engine* that uses hydrogen as a fuel. First let's look at the gasoline–electric hybrid car, which is a combination of a gasoline-powered car and an electric car. These hybrids are becoming increasingly common.

A hybrid car has a small gasoline engine and an electric motor that assists the engine when accelerating. The electric motor is powered by batteries that are recharged when the car is running on gasoline. For more efficiency (and less pollution), the engine temporarily shuts off when stopped in traffic and restarts automatically when put back in gear. The electric motor can also assist the car in slowing. The combo hybrid saves fuel in getting more miles per gallon (mpg) of gasoline. Now, all-electric cars are on the market.

Hydrogen (derived from Greek words meaning "maker of water") is the most abundant element in the universe. It is the third most abundant on the Earth's surface, where it is found primarily in water and organic compounds. There are vast quantities of hydrogen in the water of the oceans, rivers, lakes, polar ice caps, and atmosphere (humidity). Hydrogen is produced from sources such as natural gas, methanol, and water. A common chemistry lab experiment is the electrolysis of water. When an electric current is supplied to platinum plates in water (H_2O), hydrogen gas (H_2) is formed at one plate and oxygen (O_2) at the other.

Hydrogen is used in fuel cells, which work somewhat like a battery. A fuel cell, though, does not run down or need recharging. As long as hydrogen is supplied, a fuel cell produces electricity (which can power an automobile) and heat. We will not go into the details of the operation of a fuel cell. Suffice it to say that just as electricity can be used in the electrolysis process described above to separate water into hydrogen and oxygen, electricity is *produced* when hydrogen and oxygen are combined in a fuel cell (hydrogen fuel and oxygen from the air). Fuel cells combine hydrogen and oxygen without combustion to yield water, heat, and electricity with no pollution.

Automobiles with fuel-cell engines have been built, but it will probably be some time before they become commonplace. One problem is the large-scale distribution of hydrogen fuel (Fig. 1). Hydrogen is normally a gas, but when compressed it liquefies, and the very cold liquid hydrogen can be pumped. As you can imagine, adding liquid-hydrogen pumps to filling stations would be very costly. Another concern is that hydrogen is very flammable (as is gasoline). Even so, hydrogen is as safe for transport, storage, and use as many other fuels. There is, however, a historical reluctance to use hydrogen known as the *Hindenburg syndrome*. Check out Chapter 11.6 to see why.

Figure 1 Fill It Up . . . with Hydrogen
Automobiles with fuel-cell engines have been built, but a major problem for commercial use is the distribution of hydrogen fuel. Imagine the enormous task of adding liquid-hydrogen pumps to filling stations.

Henning Bock/Bransch

Did You Learn?

- Alternative energy sources are those not based on fossil fuels and nuclear processes; biofuels are an example. Renewable energy sources are those that cannot be exhausted; solar and wind are examples.

- Solar power (sunlight) varies because of weather conditions and seasonal changes. Wind power varies because the wind does not blow continuously.

KEY TERMS

1. work (4.1)
2. joule
3. foot-pound
4. energy (4.2)
5. kinetic energy

6. potential energy
7. gravitational potential energy
8. conservation of total energy (4.3)
9. conservation of mechanical energy
10. power (4.4)

11. watt
12. horsepower
13. kilowatt-hour
14. alternative energy sources (4.6)
15. renewable energy sources

MATCHING

For each of the following items, fill in the number of the appropriate Key Word from the preceding list. Compare your answers with those at the back of the book.

a. _____ The ability to do work

b. _____ Time rate of doing work

c. _____ SI unit of energy

d. _____ British unit of power

e. _____ Equal to work done against gravity

f. _____ Energy of position

g. _____ A process of transferring energy

h. _____ British unit of work

i. _____ Energy sources that cannot be exhausted

j. _____ SI unit of power

k. _____ $E_k + E_p$ = a constant

l. _____ Energy of motion

m. _____ Unit of electrical energy

n. _____ Requires an isolated system

o. _____ Energy sources other than fossil fuels and nuclear reactions

MULTIPLE CHOICE

Compare your answers with those at the back of the book.

1. Work is done on an object when it is (4.1)
   (a) moved
   (b) stationary
   (c) acted on by a balanced force
   (d) none of the preceding

2. Which of the following is a unit of work? (4.1)
   (a) W   (b) J·s   (c) N/s   (d) N·m

3. What is the SI unit of energy? (4.2)
   (a) ft·lb   (b) newton   (c) watt   (d) joule

4. Which of the following objects has the greatest kinetic energy? (4.2)
   (a) an object with a mass of 4m and a velocity of v
   (b) an object with a mass of 3m and a velocity of 2v
   (c) an object with a mass of 2m and a velocity of 3v
   (d) an object with a mass of m and a velocity of 4v

5. When negative work is done on a moving object, its kinetic energy (4.2)
   (a) increases   (b) decreases   (c) remains constant

6. The reference point for gravitational potential energy may be which of the following? (4.2)
   (a) zero
   (b) negative
   (c) positive
   (d) all of the preceding

7. When the height of an object is changed, the gravitational potential energy (4.2)
   (a) increases
   (b) decreases
   (c) depends on the reference point
   (d) remains constant

8. Energy *cannot* be ____ .
   (a) created
   (b) conserved
   (c) transferred
   (d) in more than one form

9. On which of the following does the speed of a falling object depend? (4.3)
   (a) mass
   (b) $\sqrt{\Delta h}$
   (c) $\frac{1}{2}mv^2$
   (d) parallel distance

10. Power is expressed by which of the following units? (4.4)
    (a) J/s   (b) N·m   (c) W·s   (d) W/m

11. If one motor has three times as much power as another, then the less powerful motor ____ . (4.4)
    (a) can do the same work in three times the time
    (b) can do the same work in the same time
    (c) can do the same work in one-third the time
    (d) can never do the same work as the larger motor

12. In the United States, which one of the following sectors consumes the most energy? (4.5)
    (a) residential
    (b) commercial
    (c) industry
    (d) transportation

13. Which one of the following would not be classified as a total alternative fuel source? (4.6)
    (a) photocells   (b) gasohol   (c) windmills   (d) wood

14. Which of the following renewable energy sources currently produces the most energy?
    (a) wind power
    (b) solar power
    (c) hydropower
    (d) tidal power

FILL IN THE BLANK

Compare your answers with those at the back of the book.

1. Work is equal to the force times the ___ distance through which the force acts. (4.1)
2. Work is a ___ quantity. (4.1)
3. The unit N · m is given the special name of ___ . (4.1)
4. When energy is transferred from one object to another, ___ is done. (4.2)
5. Mechanical energy consists of ___ and potential energy. (4.2)
6. The stopping distance of an automobile on a level road depends on the ___ of the speed. (4.2)
7. Work is a process of ___ energy. (4.2)
8. The total energy of a(n) ___ system remains constant. (4.3)
9. Power is the time rate of doing ___ . (4.4)
10. A horsepower is equal to about ___ kW. (4.4)
11. The kilowatt-hour (kWh) is a unit of ___ . (4.4)
12. In the United States, ___ is the most consumed fuel in generating electricity. (4.5)
13. Renewable energy sources cannot be ___ . (4.6)
14. Gasohol is gasoline mixed with ___ . (4.6)

SHORT ANSWER

4.1 Work

1. What is required to do work?
2. Do all forces do work? Explain.
3. How much work is required to lift a bucket of water from a well?
4. A weight lifter holds 900 N (about 200 lb) over his head. Is he doing work on the weights? Did he do any work on the weights? Explain.
5. For the situation in Fig. 4.4a, if the applied force is removed, then the frictional force will continue to do work, so there is an energy transfer. Explain this transfer.

4.2 Kinetic Energy and Potential Energy

6. Car B is traveling twice as fast as car A, but car A has three times the mass of car B. Which car has the greater kinetic energy?
7. Why are water towers very tall structures and often placed on high elevations?
8. If the speed of a moving object is doubled, how many times more work is required to bring it to rest?
9. A book sits on a library shelf 1.5 m above the floor. One friend tells you the book's total mechanical energy is zero, and another says it is not. Who is correct? Explain.
10. (a) A car traveling at a constant speed on a level road rolls up an incline until it stops. Assuming no frictional losses, comment on how far up the hill the car will roll.
    (b) Suppose the car rolls back down the hill. Again, assuming no frictional losses, comment on the speed of the car at the bottom of the hill.
11. An object is said to have a negative potential energy. Because it is preferable not to work with negative numbers, how can you change the value of the object's potential energy without moving the object?

4.3 Conservation of Energy

12. Distinguish between total energy and mechanical energy.
13. A ball is dropped from a height at which it has 50 J of potential energy. How much kinetic energy does the ball have just before hitting the ground?
14. When is total energy conserved? When is mechanical energy conserved?

15. A simple pendulum as shown in ● Fig. 4.24 oscillates back and forth. Use the letter designations in the figure to identify the pendulum's position(s) for the following conditions. (There may be more than one answer. Consider the pendulum to be ideal with no energy losses.)
    (a) Position(s) of instantaneous rest ___
    (b) Position(s) of maximum velocity ___
    (c) Position(s) of maximum E_k ___
    (d) Position(s) of maximum E_p ___
    (e) Position(s) of minimum E_k ___
    (f) Position(s) of minimum E_p ___
    (g) Position(s) after which E_k increases ___
    (h) Position(s) after which E_p increases ___
    (i) Position(s) after which E_k decreases ___
    (j) Position(s) after which E_p decreases ___
16. Two students throw identical snowballs from the same height, both snowballs having the same initial speed v_o (● Fig. 4.25). Which snowball has the greater speed on striking the level ground at the bottom of the slope? Justify your answer using energy considerations.
17. A mass suspended on a spring is pulled down and released. It oscillates up and down as illustrated in ● Fig. 4.26. Assuming the total energy to be conserved, use the letter designations to

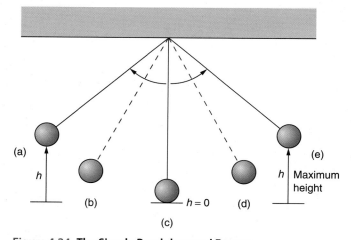

Figure 4.24 The Simple Pendulum and Energy
See Short-Answer Question 15.

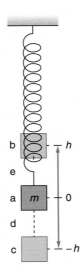

Figure 4.26 Energy Transformation
See Short-Answer Question 17.

Figure 4.25 Away They Go!
See Short-Answer Question 16.

identify the spring's position(s) as listed in Question 15. (There may be more than one answer.)

18. When you throw an object into the air, is its return speed just before hitting your hand the same as its initial speed? (Neglect air resistance.) Explain by applying the conservation of mechanical energy.

4.4 Power

19. (a) What is the SI unit of power?
    (b) Show that in terms of fundamental units, the units of power are $kg \cdot m^2/s^3$.

20. Persons A and B do the same job, but person B takes longer. Who does the greater amount of work? Who is more "powerful"?

21. What does a greater power rating mean in terms of (a) the amount of work that can be done in a given time and (b) how fast a given amount of work can be done?

22. What do we pay the electric company for, power or energy? In what units?

23. A 100-W light bulb uses how much more energy than a 60-W light bulb?

4.5 Forms of Energy and Consumption

24. Which fuel is consumed the most in the United States?

25. Which fuel is used the most in the generation of electricity in the United States?

26. On average, how much energy do you radiate each second?

27. List some different general forms of energy (other than kinetic energy and potential energy).

4.6 Alternative and Renewable Energy Sources

28. What are two examples of alternative energy sources?

29. What are two examples of renewable energy sources?

30. Which two renewable energy sources are affected by the weather?

VISUAL CONNECTION

Visualize the connections and give answers for the blanks. Compare your answers with those at the back of the book.

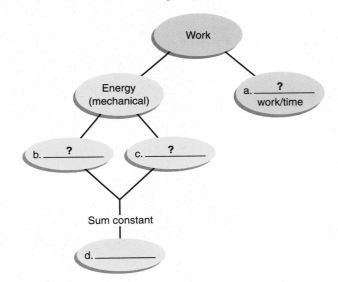

APPLYING YOUR KNOWLEDGE

1. A fellow student tells you that she has both zero kinetic energy and zero potential energy. Is this possible? Explain.

2. Two identical stones are thrown from the top of a tall building. Stone 1 is thrown vertically downward with an initial speed v, and stone 2 is thrown vertically upward with the same initial speed. Neglecting air resistance, how do their speeds compare on hitting the ground?

3. A person on a trampoline can go higher with each bounce. How is this possible? Is there a maximum height to which the person can go?

4. With which of our five senses can we detect energy?

5. What are three common ways to "save" electricity to reduce electric bills?

IMPORTANT EQUATIONS

Work: $W = Fd$ (4.1)

Kinetic Energy: $E_k = \frac{1}{2} mv^2$ (4.2)

and work: $W = \Delta E_k = E_{k_2} - E_{k_1}$

$\qquad = \frac{1}{2} mv_2^2 - \frac{1}{2} mv_1^2$ (4.3)

Potential Energy (Gravitational): $E_p = mgh$ (4.4)

Conservation of Mechanical Energy:

$$(E_k + E_p)_1 = (E_k + E_p)_2$$

or $(\frac{1}{2} mv^2 + mgh)_1 = (\frac{1}{2} mv^2 + mgh)_2$ (4.6)

Speed and Height (from rest): $v = \sqrt{2g\,\Delta h}$ (4.7)

Power: $P = \dfrac{W}{t} = \dfrac{Fd}{t}$ (4.8–4.9)

(or, in terms of energy, $P = \dfrac{E}{t}$ and $E = Pt$) (4.10)

EXERCISES

4.1 Work

1. A worker pushes horizontally on a large crate with a force of 250 N, and the crate is moved 3.0 m. How much work was done?

   Answer: 750 J

2. While rearranging a dorm room, a student does 400 J of work in moving a desk 2.0 m. What was the magnitude of the applied horizontal force?

3. A 5.0-"kilo" bag of sugar is on a counter. How much work is required to put the bag on a shelf a distance of 0.45 m above the counter?

   Answer: 22 J

4. How much work is required to lift a 4.0-kg concrete block to a height of 2.0 m?

5. A man pushes a lawn mower on a level lawn with a force of 200 N. If 40% of this force is directed downward, then how much work is done by the man in pushing the mower 6.0 m?

   Answer: 7.2×10^2 J

6. If the man in Exercise 5 pushes the mower with 40% of the force directed horizontally, then how much work is done?

7. How much work does gravity do on a 0.150-kg ball falling from a height of 10.0 m? (Neglect air resistance.)

   Answer: 14.7 J

8. A student throws the same ball straight upward to a height of 7.50 m. How much work did the student do?

4.2 Kinetic Energy and Potential Energy

9. (a) What is the kinetic energy in joules of a 1000-kg automobile traveling at 90 km/h?

   (b) How much work would have to be done to bring a 1000-kg automobile traveling at 90 km/h to a stop?

   Answer: (a) 3.1×10^5 J (b) same

10. A 60-kg student traveling in a car with a constant velocity has a kinetic energy of 1.2×10^4 J. What is the speedometer reading of the car in km/h?

11. What is the kinetic energy of a 20-kg dog that is running at a speed of 9.0 m/s (about 20 mi/h)?

    Answer: 8.1×10^2 J

12. Which has more kinetic energy, a 0.0020-kg bullet traveling at 400 m/s or a (6.4×10^7)-kg ocean liner traveling at 10 m/s (20 knots)? Justify your answer.

13. A 0.50-kg block initially at rest on a frictionless, horizontal surface is acted upon by a force of 8.0 N for a distance of 4.0 m. How much kinetic energy does the block gain?

    Answer: 32 J

14. If the force in Exercise 13 had acted for a distance of 8.0 m, then what would be the block's velocity?

15. What is the potential energy of a 3.00-kg object at the bottom of a well 10.0 m deep as measured from ground level? Explain the sign of the answer.

    Answer: −294 J

16. How much work is required to lift a 3.00-kg object from the bottom of a well 10.0 m deep?

4.3 Conservation of Energy

17. An object is dropped from a height of 12 m. At what height will its kinetic energy and its potential energy be equal?

    Answer: 6.0 m

18. A 1.0-kg rock is dropped from a height of 6.0 m. At what height is the rock's kinetic energy twice its potential energy?

19. A sled and rider with a combined weight of 60 kg are at rest on the top of a hill 12 m high.
    (a) What is their total energy at the top of the hill?
    (b) Assuming there is no friction, what would the total energy be on sliding halfway down the hill?

    Answer: (a) 7.1×10^3 J (b) same

20. A 30.0-kg child starting from rest slides down a water slide with a vertical height of 10.0 m. What is the child's speed (a) halfway down the slide's vertical distance and (b) three-fourths of the way down? (Neglect friction.)

4.4 Power

21. If the man in Exercise 5 pushes the lawn mower 6.0 m in 10 s, how much power does he expend?

    Answer: 72 W

22. If the man in Exercise 5 expended 60 W of power in pushing the mower 6.0 m, how much time is spent in pushing the mower this distance?

23. A student who weighs 556 N climbs a stairway (vertical height of 4.0 m) in 25 s.
    (a) How much work is done?
    (b) What is the power output of the student?

    Answer: (a) 2.2×10^3 J (b) 89 W

24. A 125-lb student races up stairs with a vertical height of 4.0 m in 5.0 s to get to a class on the second floor. How much power in watts does the student expend in doing work against gravity?

25. On a particular day, the following appliances are used for the times indicated: a 1600-W coffee maker, 10 min, and a 1100-W microwave oven, 4.0 min. With these power requirements, find how much it costs to use these appliances at an electrical cost of 8¢ per kWh.

    Answer: 3¢

26. A microwave oven has a power requirement of 1250 W. A frozen dinner requires 4.0 min to heat on full power.
    (a) How much electrical energy (in kWh) is used?
    (b) If the cost of electricity is 12¢ per kWh, then how much does it cost to heat the dinner?

ON THE WEB

1. Forms and Sources of Energy: To Conserve or Not to Conserve

If energy can be neither created nor destroyed, why is there discussion about energy conservation? Where do the fossil fuels—coal, oil, and natural gas—come from? Why is the loss of these energy sources problematic? How can we conserve these sources of energy? Follow the recommended links on our student website at **www.cengage brain.com/shop/ISBN/1133104096** to discover more about "The Energy Story" and to answer the questions above.

2. Clean Energy

What is renewable energy? Why is it important? Why is energy efficiency important? What does "clean energy" have to do with you? From our student website at **www.cengagebrain.com/shop/ISBN/1133104096**, visit the Clean Energy site to answer these questions.

CHAPTER

5

Temperature and Heat

If you can't stand the heat, stay out of the kitchen.

•

Harry S. Truman
(1884–1972)

Molten steel being poured at a steel mill. >

Mark Joseph/Getty Images

PHYSICS FACTS

- Daniel Gabriel Fahrenheit (1686–1736), a German instrument maker, constructed the first alcohol thermometer (1709) and the first mercury thermometer (1714). Fahrenheit used temperatures of 0° and 96° for reference points. The freezing and boiling points of water were then measured to be 32°F and 212°F.
- Anders Celsius (1701–1744), a Swedish astronomer, invented the Celsius temperature scale with a 100° interval between the freezing and boiling points of water (0°C and 100°C). Celsius' original scale was reversed, 100°C (freezing) and 0°C (boiling), but it was later changed.
- The Golden Gate Bridge across San Francisco Bay varies in length about 1 m (3.3 ft)

Both *temperature* and *heat* are commonly used terms when referring to hotness or coldness, but they are not the same thing. They have different and distinct meanings, as we will discover in this chapter. The concepts of temperature and heat play an important part in our daily lives. We like hot coffee and cold ice cream. We heat (or cool—remove heat from) our living and working spaces to adjust the temperature to body comfort. The daily temperature forecast is an important part of a weather report. How cold or how hot it will be affects the clothes we wear and the plans we make.

How the Sun provides heat to the Earth will be discussed in detail in Chapter 19.2. The heat balance between various parts of the Earth and its atmosphere gives rise to wind, rain, and other weather phenomena. On a cosmic scale, the temperature

Chapter Outline

of various stars gives clues to their ages and to the origin of the universe. More locally, the study of temperature and heat will enable the many phenomena that occur in our environment to be explained.

5.1 Temperature

Preview Questions

● We talk about temperature, but what does it physically represent?

● Are there any limits on the lowest and highest temperatures?

Temperature tells us whether something is hot or cold. That is, temperature is a *relative* measure, or indication, of hotness or coldness. For example, if water in one cup has a higher temperature than water in another cup, then we know the water in the first cup is hotter, but it would be colder than a cup of water with an even higher temperature. Thus, hot and cold are *relative* terms; that is, they are comparisons.

On the molecular level, temperature depends on the kinetic energy of the molecules of a substance. The molecules of all substances are in constant motion. This observation is true even for solids, in which the molecules are held together by intermolecular forces that are sometimes likened to springs. The molecules move back and forth about their equilibrium positions.

In general, the greater the temperature of a substance, the greater the motion of its molecules. On this basis, we say that **temperature** is a measure of the average kinetic energy of the molecules of a substance.

Humans have temperature perception in the sense of touch, but this perception is unreliable and may vary a great deal among different people. Our sense of touch doesn't enable us to measure temperature accurately or quantitatively. The quantitative measurement of temperature may be accomplished through the use of a thermometer. A thermometer is an instrument that uses the physical properties of materials to determine temperature accurately. The temperature-dependent property most commonly used to measure temperature is thermal expansion. Nearly all substances expand with increasing temperature and contract with decreasing temperature.

The change in length or volume of a substance is quite small, but the effects of thermal expansion can be made evident by using special arrangements. For example, a bimetallic strip is made of pieces of different metals bonded together (● Fig. 5.1). When it is heated, one metal expands more than the other and the strip bends toward the metal with the smaller thermal expansion. As illustrated in Fig. 5.1, the strip can be calibrated with a scale to measure temperature. Bimetallic strips in the form of a coil or helix are used in dial-type thermometers and thermostats.

The most common type of thermometer is the *liquid-in-glass thermometer*, with which you are no doubt familiar. It consists of a glass bulb on a glass stem with a capillary bore and a sealed upper end. A liquid in the bulb (usually mercury, or alcohol colored with a red dye to make it visible) expands on heating, forcing a column of liquid up the capillary tube. The glass also expands, but the liquid expands much more.*

Thermometers are calibrated so that numerical values can be assigned to different temperatures. The calibration requires two reference, or fixed, points and a choice of unit. By analogy, think of constructing a stick to measure length. You need two marks, or reference points, and then you divide the interval between the marks into sections or units. For example, you might use 100 units between the reference marks to calibrate the length of a meter in centimeters.

Two common reference points for a temperature scale are the ice and steam points of water. The *ice point* is the temperature of a mixture of ice and water at normal atmospheric

PHYSICS FACTS *cont.*

between summer and winter due to thermal expansion. The bridge is not golden in color, but rather, orange vermillon (Fig. 5.3b).

▶ The temperatures on the Fahrenheit and Celsius scales are equal at −40°. That is, −40°F = −40°C.

▶ During a bicycle race on a hot day, a professional cyclist can evaporate as much as 7 L (7.4 qt) of perspiration in 3 hours in getting rid of the heat generated by this vigorous activity.

---

*There are a variety of thermometers. One digital type monitors the infrared radiation in the ear, which is proportional to the body temperature.

Figure 5.1 Bimetallic Strip and Thermal Expansion
(a) and (b) Because of different degrees of thermal expansion, a bimetallic strip of two different metals bends when heated. The degree of deflection of the strip is proportional to the temperature, and a calibrated scale could be added for temperature readings. Such a scale is shown but not calibrated. (c) Bimetallic coils are used as shown here. An indicator arrow is attached directly to the coil. See also Figure 5.21.

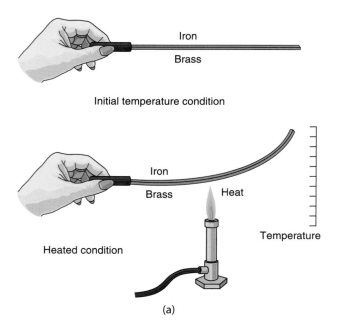

(a)

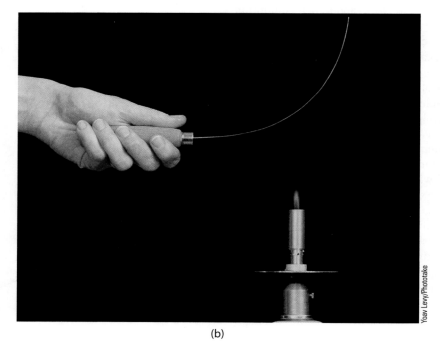

(b)

(c)

pressure. The *steam point* is the temperature at which pure water boils at normal atmospheric pressure. The ice point and the steam point are commonly called the *freezing point* and the *boiling point*, respectively.

Two familiar temperature scales are the Fahrenheit and Celsius scales (● Fig. 5.2). The **Fahrenheit scale** has an ice point of 32° (read "32 degrees") and a steam point of 212°. The interval between the ice point and the steam point is evenly divided into 180 units, and each unit is called a *degree*. Thus, a *degree Fahrenheit*, abbreviated °F, is $\frac{1}{180}$ of the temperature change between the ice point and the steam point.

The **Celsius scale** is based on an ice point of 0° and a steam point of 100°, and there are 100 equal units or divisions between these points. So, a *degree Celsius*, abbreviated °C, is $\frac{1}{100}$ of the temperature change between the ice point and the steam point. Thus, a degree

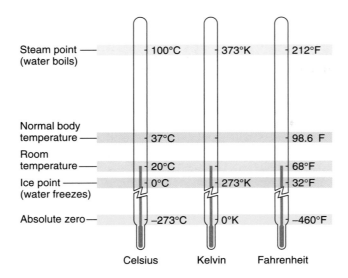

Figure 5.2 Temperature Scales
The common temperature scales are the Fahrenheit and Celsius scales. They have 180-degree and 100-degree intervals, respectively, between the ice and steam points. A third scale, the Kelvin (absolute) scale, is used primarily in scientific work. The unit or interval on the absolute Kelvin scale is the kelvin (K). This scale takes zero as the lower limit of temperature, or absolute zero (0 K).

Celsius is 1.8 times (almost twice) as large as a degree Fahrenheit (100°C and 180°F for the same temperature interval). The Celsius temperature scale is used predominantly in "metric countries" and hence throughout most of the world (Chapter 1.4).

There is no known upper limit of temperature, but there is a lower limit. It is about −273°C (actually −273.15°C), or −460°F, and is called *absolute zero*. Another temperature scale, the **Kelvin scale**, has its zero temperature at this absolute limit (see Fig. 5.2).* It is sometimes called the *absolute temperature scale*. The unit on the Kelvin scale is the **kelvin**, abbreviated K (*not* °K), and is the same size as a degree Celsius. Because the Kelvin scale has absolute zero as its lowest reading, it can have no negative temperatures.

Since the kelvin and the degree Celsius are equal intervals, it is easy to convert from the Celsius scale to the Kelvin scale: simply add 273 to the Celsius temperature. In equation form,

$$T_K = T_C + 273 \quad \text{(Celsius } T_C \text{ to Kelvin } T_K) \qquad 5.1$$

As examples, a temperature of 0°C equals 273 K, and a Celsius temperature of 27°C is equal to 300 K: $T_K = T_C + 273 = 27 + 273 = 300$ K.

Converting from Fahrenheit to Celsius and vice versa is also quite easy. The equations for these conversions are

$$T_F = \tfrac{9}{5} T_C + 32 \qquad 5.2a$$

or

$$T_F = 1.8 T_C + 32$$

(Celsius T_C to Fahrenheit T_F)

and

$$T_C = \tfrac{5}{9}(T_F - 32) \qquad 5.2b$$

or

$$T_C = \frac{T_F - 32}{1.8}$$

(Fahrenheit T_F to Celsius T_C)

The two equations in 5.2a and 5.2b are the same; they are just different arrangements for scale conversion.

As examples, try converting 100°C and 32°F to their equivalent temperatures on the other scales. (You already know the answers.) Remember that on these scales there are negative temperatures, whereas there are no negative values on the Kelvin scale.

---

*The Kelvin scale is named for Lord Kelvin (William Thomson, 1824–1907), the British physicist who developed it.

EXAMPLE 5.1   Converting Temperatures between Scales

The normal human body temperature is usually taken as 98.6°F. What is the equivalent temperature on the Celsius scale?

Solution

With $T_F = 98.6°F$ and using Eq. 5.2b to find T_C,

$$T_C = \tfrac{5}{9}(T_F - 32) = \tfrac{5}{9}(98.6 - 32) = \tfrac{5}{9}(66.6) = 37.0°C$$

So, on the Celsius scale, normal body temperature is 37.0°C (a nice, round number).

Confidence Exercise 5.1

Show that a temperature of −40° is the same on both the Fahrenheit and the Celsius scales.
   Answers to Confidence Exercises may be found at the back of the book.

In converting temperatures between Celsius and Fahrenheit, it is sometimes difficult to remember whether one multiplies by 9/5 or by 5/9 and whether one adds or subtracts 32. Keep the following in mind. When going from T_C to T_F, you get a larger number (for example, 100°C to 212°F), so the larger fraction, 9/5, is used. (This reflects that a degree Celsius is larger than a degree Fahrenheit.) Similarly, in going from T_F to T_C, a smaller number is obtained (for example, 212°F to 100°C), so the smaller fraction, 5/9, is used.

There is another convenient conversion procedure that eliminates the 32 question you might care to use knowing the 9/5 and 5/9 distinction. Try this method:

Celsius to Fahrenheit: $T_C + 40$, multiply by 9/5, subtract 40 ($= T_F$)

Fahrenheit to Celsius: $T_F + 40$, multiply by 5/9, subtract 40 ($= T_C$)

(Remember the sequence: add 40, multiply, subtract 40.)

For example, converting $T_C = 100°C$ to T_F (where 9/5 is used to convert to a larger number):

$$100° + 40 = 140 \times (9/5) = 252 - 40 = 212°F.$$

Conceptual Question and Answer

The Easy Approximation

Q. In most countries temperatures are given in Celsius. Is there a simple way to convert these temperatures to Fahrenheit without using the regular conversion equation and a calculator?

A. A quick way to do Celsius to Fahrenheit conversions for ambient temperatures is to use an approximation of Eq. 5.2a ($T_F = 1.8T_C + 32$) that you can do in your head. This is, $T_F \approx 2T_C + 30$ (increasing $1.8T_C$ to $2T_C$ and decreasing 32 to 30). That is, double the Celsius temperature and add 30. For example, for 20°C we have $T_F \approx 2(20) + 30 = 70°F$. This answer is not exactly 68°F, but it is close enough to give you an idea of the Fahrenheit temperature. (Remember this simple conversion on your next trip abroad.)

Did You Learn?

● Temperature is a measure of the average kinetic energy of the molecules of a substance.

● There is no known upper limit to temperature, and the lower limit is absolute zero (0 K, −273°C, or −460°F).

5.2 Heat

Preview Questions

- Why is heat called "energy in transit"?
- Most substances contract with decreasing temperature. Is this true for water?

We commonly describe heat as a form of energy. However, this definition can be made more descriptive. The molecules of a substance may vibrate back and forth, rotate, or move from place to place. They have kinetic energy. As stated previously, the average kinetic energy of the molecules of a substance is related to its temperature. For example, if an object has a high temperature, the average kinetic energy of the molecules is relatively high. (The molecules move relatively faster.) In kinetic theory of gases (Chapter 5.6), the kinetic energy is actually the average *translational* kinetic energy of the molecules. (Translation means that the molecule moves linearly as a whole.)

For a diatomic gas, however, besides having translational "temperature" kinetic energy, the molecules may also have kinetic energy due to vibrations and rotations. This kinetic energy is associated with the vibrations or rotational modes of the atoms within molecules and the molecular bonds, which simplistically may be thought of as "springs." There is also potential energy due to intramolecular interactions of the bond "springs" in stretching and compressing. The total energy (sum of kinetic plus potential) is called the *internal energy*.

When heat is transferred from one body to another, we often say that heat "flows" from a region of higher temperature to a region of lower temperature.* Actually, **heat** is the net energy transferred from one object to another because of a temperature difference. In other words, heat is energy in transit because of a temperature difference. When heat is added to a body, its internal energy increases. Some of the transferred energy may go into the translational kinetic energy of the molecules, which is manifested as a temperature increase, and some may go into the kinetic-potential energy of the internal energy.

Because heat is energy, it has the SI unit of joule (J), but a common and traditional unit for measuring heat energy is the calorie. A **calorie** (cal) is defined as the amount of heat necessary to raise one *gram* of pure water by one Celsius degree at normal atmospheric pressure. In terms of the SI energy unit,

$$1 \text{ cal} = 4.186 \text{ J} (\approx 4.2 \text{ J})$$

The calorie just defined is not the same as the one used when discussing diets and nutrition. This is the kilocalorie (kcal), and 1 kcal is equal to 1000 cal. A **kilocalorie** is the amount of heat necessary to raise the temperature of one *kilogram* of water one Celsius degree. A food calorie (Cal) is equal to 1 kcal and is commonly written with a capital C to avoid confusion. We sometimes refer to a "big" (kilogram) Calorie and a "little" (gram) calorie to make this distinction.

1 food Calorie = 1000 calories (1 kcal)

1 food Calorie = 4186 joules (≈ 4.2 kJ)

Food Calories indicate the amount of energy produced when a given amount of the particular food is completely burned.

The unit of heat in the British system is the British thermal unit (Btu). One **Btu** is the amount of heat necessary to raise one *pound* of water one Fahrenheit degree at normal atmospheric pressure. Air conditioners and heating systems are commonly rated in Btu's. These ratings are actually the Btu's removed or supplied per hour. If food energies were rated in Btu's instead of Calories (kcal), then the values would be greater. Because 1 kcal is about 4 Btu, a 100-Cal soft drink would have 400 Btu.

As we saw in the measurement of temperature, one effect of heating a material is expansion. As a general rule, nearly all matter—solids, liquids, and gases—expands when heated

---

*Heat was once thought to be a fluid called "caloric," and fluids flow.

Highlight   Freezing from the Top Down

As a general rule, a substance expands when heated and contracts when cooled. An important exception to this rule is water. A volume of water does contract when cooled, to a point (about 4°C). When cooled from 4°C to 0°C, however, a volume of water *expands*. Another way of looking at it is in terms of density ($\rho = m/V$, Chapter 1.6). This behavior is illustrated in Fig. 1. As a volume of water is cooled to 4°C, its density increases (its volume decreases), but from 4°C to 0°C, the density decreases (the volume increases). Hence, water has its maximum density at 4°C (actually 3.98°C).

The reason for this unique behavior is molecular structure. When water freezes, the water molecules bond together in an open hexagonal (six-sided) structure, as evident in (six-sided) snowflakes (Fig. 2). The open space in the molecular structure explains why ice is less dense than water and therefore floats (see Chapter 3.6). Again, this property is nearly unique to water. The solids of almost all substances are denser than their liquids.

Water has its maximum density at 4°C, which explains why lakes freeze from the top down. Most of the cooling takes place at the top open surface. As the temperature of the top layer of water is cooled toward 4°C, the cooler water at the surface is denser than the water below and the cooler water sinks to the bottom. This process takes place until 4°C is reached. Below 4°C, the surface layer is less dense than the water below and so remains on top, where it freezes at 0°C.

Thus, because of water's very unusual properties in terms of density versus temperature, lakes freeze from the top down. If the water thermally contracted and the density increased all the way to 0°C, then the coldest layer would be on the bottom and freezing would begin there. Think of what that would mean to aquatic life, let alone ice-skating.

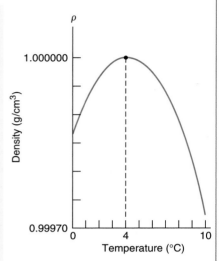

Figure 1 Strange Behavior
As is true of most substances, the volume of a quantity of water decreases (and its density increases) with decreasing temperature, but in the case of water, that is true only down to 4°C. Below this temperature, the volume increases slightly. With a minimum volume at 4°C, the density of water is a maximum at this temperature and decreases at lower temperatures.

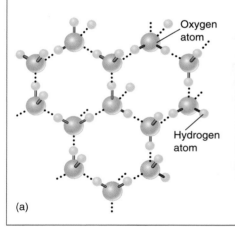

(a)

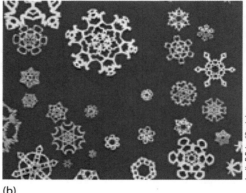

(b)

Figure 2 Structure of Ice
(a) An illustration of the open hexagonal (six-sided) molecular structure of ice. (b) Six-sided snowflakes.

and contracts (negative expansion) when cooled. *The most important exception to this rule is water in the temperature range near its freezing point.* When water is frozen, it expands. That is, ice at 0°C occupies a larger volume than the same mass of water at 0°C. The underlying reason is hydrogen bonding (Chapter 12.6), which leads to some interesting environmental effects as discussed in the **Highlight: Freezing from the Top Down**.

The change in length or volume of a substance as a result of heat and temperature changes is a major factor in the design and construction of items ranging from steel bridges and automobiles to watches and dental cements. The cracks in some highways are designed so that the concrete will not buckle as it expands in the summer heat. Expansion joints are designed into bridges for the same reason (● Fig. 5.3). The Golden Gate Bridge across San Francisco Bay varies in length by about 1 m (3.3 ft) between summer and winter.

Heat expansion characteristics of metals are used to control such things as the flow of water in car radiators and the flow of heat in homes through the operation of metallic thermostats (see Fig. 5.21). Electronic thermostats are now coming into common use.

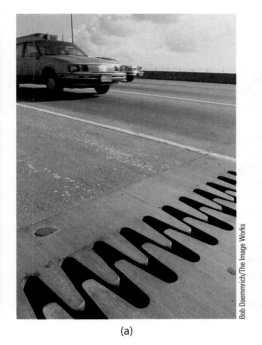

(a)   (b)

Figure 5.3 Thermal-Expansion Joints
(a) Expansion joints are built into bridges and connecting roadways to allow for the expansion and contraction of the steel girders caused by the addition and removal of heat. Were the girders allowed to come into contact when expanding, serious damage could result. (b) Thermal expansion causes the Golden Gate Bridge in San Francisco to vary over 1 m (3.3 ft) in length between summer and winter. (The "Golden" Gate is actually orange vermilion in color.)

Did You Learn?

● Heat is the net energy transfer from one object to another because of a temperature difference.

● In general, water contracts on cooling, but from 4°C to 0°C, it expands.

5.3 Specific Heat and Latent Heat

Preview Questions

● What is specific about specific heat?
● Why is latent heat referred to as "hidden" heat?

Specific Heat

Heat and temperature, although different, are intimately related. When heat is added to a substance, the temperature generally increases. For example, suppose you added equal amounts of heat to equal masses of iron and aluminum. How do you think their temperatures would change? You might be surprised to find that if the temperature of the iron increased by 100°C, then the corresponding temperature change in the aluminum would be only 48°C. You would have to add more than twice the amount of heat to the aluminum to get the same temperature change as for an equal mass of iron.

This result reflects that the internal forces of the materials are different (different inter-molecular "springs," so to speak). In iron, more of the energy goes into translational kinetic energy than into intramolecular potential energy, so the iron has a higher temperature. (Both have the same change in internal energy.)

This difference is expressed in terms of specific heat. The **specific heat** of a substance is the amount of heat necessary to raise the temperature of one kilogram of the substance one Celsius degree.

By definition, 1 kcal is the amount of heat that raises the temperature of 1 kg of water 1°C, so it follows that water has a specific heat of 1.00 kcal/kg·°C (that is, 1.00 kcal per kilogram per degree Celsius). For ice and steam, the specific heats are nearly equal, about 0.50 kcal/kg·°C. Other substances require different amounts of heat to raise the temperature of 1 kg of the substance by 1°C. A specific heat is *specific* to a particular substance.

The SI units of specific heat are J/kg·°C, but kcal is sometimes used for energy, kcal/kg·°C. We will work in both kcal and J. (The latter is generally preferred, but the larger kcal unit makes the specific heat values smaller and more manageable mathematically, particularly for water.) The specific heats of a few common substances are given in ● Table 5.1.

The greater the specific heat of a substance, the greater the amount of heat necessary to raise the temperature per unit mass. Put another way, the greater the specific heat of a substance, the greater its capacity for heat (given equal masses and temperature change). In fact, the full technical name for specific heat is *specific heat capacity*. With regard to the previous example of aluminum and iron, note in Table 5.1 that the specific heat of aluminum is slightly greater than twice that of iron. Hence, for equal masses, it takes a little more than twice as much heat to raise the temperature of aluminum as to raise the temperature of iron.

We often say that certain materials "hold their heat" because such materials have relatively high specific heats. Since it takes more heat per unit mass to raise their temperatures, or "heat them up," they have more stored energy, as is sometimes painfully evident when one eats a baked potato or cheese on a pizza. Other food and pizza toppings and crust may have cooled, but you still might burn your mouth. This is because water, which has one of the highest specific heats, makes up large portions of both potato and cheese.

Water has one of the highest specific heats and so can store more heat energy for a given temperature change. Because of this, water is used in solar energy applications. Solar energy is collected during the day and is used to heat water, which can store more energy than most liquids without getting overly hot. At night, the warm water may be pumped through a home to heat it. The high specific heat of water also has a moderating effect on temperature around bodies of water, such as large lakes.

Table 5.1 Specific Heats of Some Common Substances

Substance	Specific Heat (20°C)	
	kcal/kg·°C	J/kg·°C
Air (0°C, 1 atm)	0.24	1000
Alcohol (ethyl)	0.60	2510
Aluminum	0.22	920
Copper	0.092	385
Ice	0.50	2100
Iron	0.105	440
Mercury	0.033	138
Steam (at 1 atm)	0.50	2100
Water (liquid)	1.00	4186
Wood (average)	0.40	1700

The specific heat, or the amount of heat necessary to change the temperature of a given amount of a substance, depends on three factors: the mass (m), the specific heat (designated by c), and the temperature change (ΔT). In equation form,

amount of heat = mass $\times$ specific $\times$ temperature
to change   heat   change
temperature

or $\qquad\qquad\qquad\qquad H = mc\,\Delta T \qquad\qquad\qquad\qquad\qquad$ 5.3

Equation 5.3 applies when heat is added to (or removed from, $-\Delta T$, negative ΔT) a substance and it does *not* undergo a phase change (such as changing from a solid to a liquid). When heat is added to, or removed from, a substance that is changing phase, the temperature does not change and a different equation, which will be presented shortly, must be used.

EXAMPLE 5.2   Using Specific Heat

How much heat in kcal does it take to heat 80 kg of bathwater from 12°C (about 54°F) to 42°C (about 108°F)?

Solution

Step 1

Given: $m = 80$ kg

$\qquad \Delta T = 42°C - 12°C = 30°C$

$\qquad\quad c = 1.00$ kcal/kg·°C (for water, known)

Step 2

Wanted: H (heat)

(The units are consistent, and the answer will come out in kilocalories. If the specific heat c were expressed in units of J/kg·°C, then the answer would be in joules.)

Step 3

The amount of heat required may be computed directly from Eq. 5.3:

$H = mc\,\Delta T = (80\text{ kg})(1.00\text{ kcal/kg·°C})(30°C) = 2.4 \times 10^3$ kcal

Let's get an idea how much it costs to heat the bathwater electrically. Each kilocalorie corresponds to 0.00116 kWh (kilowatt-hour; recall from Chapter 4.4 that we pay for electrical energy in these units). Hence, this amount of heat in kWh is

$$H = 2.4 \times 10^3\text{ kcal}\left(\frac{0.00116\text{ kWh}}{\text{kcal}}\right) = 2.8\text{ kWh}$$

At 10¢ per kWh, the cost of the electricity to heat the bathwater is 2.8 kWh $\times$ 10¢/kWh = 28¢. For four people each taking a similar bath each day for 1 month (30 days), it would cost 4 $\times$ 30 $\times$ \$0.28 = \$33.60 to heat the water.

Confidence Exercise 5.2

A liter of water at room temperature (20°C) is placed in a refrigerator with a temperature of 5°C. How much heat in kcal must be removed from the water for it to reach the refrigerator temperature?

    Answers to Confidence Exercises may be found at the back of the book.

Latent Heat

Substances in our environment are normally classified as solids, liquids, or gases. These forms are called *phases of matter*. When heat is added to (or removed from) a substance, it may undergo a change of phase. For example, when water is heated sufficiently, it boils and changes to steam, and when enough heat is removed, water changes to ice.

Water changes to steam at a temperature of 100°C (or 212°F) under normal atmospheric pressure. If heat is added to a quantity of water at 100°C, it continues to boil as the liquid is converted into gas, but the temperature remains constant. Here is a case of adding heat to a substance without a resulting temperature change. Where does the energy go? The heat associated with a phase change is called **latent heat** (*latent* means "hidden"). Latent heat is sometimes called hidden heat because it is not reflected by a temperature change.

On a molecular level, when a substance goes from a liquid to a gas, work must be done to break the intermolecular bonds and separate the molecules. The molecules of a gas are farther apart than the molecules in a liquid, relatively speaking. Hence, during such a phase change, the heat energy must go into the work of separating the molecules and not into increasing the molecular kinetic energy, which would increase the temperature. (Phase changes are discussed in more detail in Chapter 5.5.)

Referring to ● Fig. 5.4, let's go through the process of heating a substance (water in the figure) and changing phases from solid to liquid to gas. In the lower left-hand corner, the substance is represented in the solid phase. As heat is added, the temperature rises. When point A is reached, adding more heat does *not* change the temperature (horizontal line, constant temperature). Instead, the heat energy goes into changing the solid into a liquid. The amount of heat necessary to change 1 kg of a solid into a liquid at the same temperature is called the *latent heat of fusion* of the substance. In Fig. 5.4 this heat is simply the amount of heat necessary to go from point A to point B (assuming 1 kg of the substance).

When point B has been reached, the substance is all liquid. The temperature of the substance at which this change from solid to liquid takes place is known as the *melting point* (or *freezing point* when going from liquid to solid). After point B, further heating again causes a rise in temperature. As heat is added, the temperature continues to rise until point C is reached.

Figure 5.4 Graph of Temperature versus Heat for Water
The solid, liquid, and gas phases are ice, water, and steam, as shown over the sloping graph lines. During a phase change (A to B and C to D) heat is added, but the temperature does not change. Also, the two phases exist together.

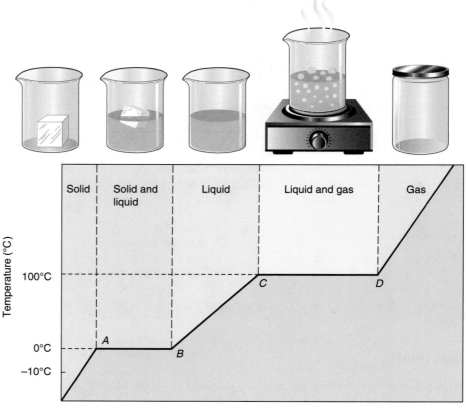

From point *B* to point *C*, the substance is in the liquid phase. When point *C* is reached, adding more heat does *not* change the temperature. The added heat now goes into changing the liquid into a gas. The amount of heat necessary to change one kilogram of a liquid into a gas is called the *latent heat of vaporization* of the substance. In Fig. 5.4 it is the amount of heat necessary to go from point *C* to point *D*.

When point *D* has been reached, the substance is all in the gaseous phase. The temperature of the substance at which this change from liquid to gas phase occurs is known as the *boiling point* (or *condensation point* when going from a gas to a liquid). After point *D* is reached, further heating again causes a rise in the temperature (superheated steam for water).

In some instances a substance can change directly from the solid to the gaseous phase. This change is called *sublimation*. Examples include dry ice (solid carbon dioxide, CO_2), mothballs, and solid air fresheners. The reverse process of changing directly from the gaseous to the solid phase is called *deposition*. The formation of frost is an example (Chapter 20.1). The processes of phase changes are illustrated in ● Fig. 5.5.

The temperature of a body rises as heat is added only when the substance is not undergoing a change in phase. When heat is added to ice at 0°C, the ice melts without a change in its temperature. The more ice there is, the more heat is needed to melt it. In general, the heat required to change a solid into a liquid at the melting point can be found by multiplying the mass of the substance by its latent heat of fusion. Thus, we may write

$$\begin{array}{c} \text{heat needed to} \\ \text{melt a substance} \end{array} = \text{mass} \times \begin{array}{c} \text{latent heat} \\ \text{of fusion} \end{array}$$

or $$H = mL_f \qquad\qquad 5.4$$

Similarly, at the boiling point, the amount of heat necessary to change a mass of liquid into a gas can be written as

$$\begin{array}{c} \text{heat needed to} \\ \text{boil a substance} \end{array} = \text{mass} \times \begin{array}{c} \text{latent heat of} \\ \text{vaporization} \end{array}$$

or $$H = mL_v \qquad\qquad 5.5$$

For water, the latent heat of fusion (L_f) and the latent heat of vaporization (L_v) are

$$L_f = 80 \text{ kcal/kg} = 3.35 \times 10^5 \text{ J/kg}$$
$$L_v = 540 \text{ kcal/kg} = 2.26 \times 10^6 \text{ J/kg}$$

These values are rather large numbers compared with the specific heats. Note that it takes 80 times as much heat energy (latent heat) to melt 1 kg of ice at 0°C as to raise the temperature of 1 kg of water by 1°C (specific heat). Similarly, changing 1 kg of water to steam at 100°C takes 540 times as much energy as raising its temperature by 1°C (definition of kcal). Also note that it takes almost seven times as much energy to change 1 kg of water at 100°C to steam as it takes to change 1 kg of ice at 0°C to water.

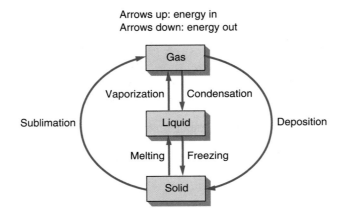

Figure 5.5 Phase Changes
This diagram illustrates the various phase changes for solids, liquids, and gases. See text for description.

Table 5.2 Temperatures of Phase Changes and Latent Heats for Some Substances (normal atmospheric pressure)

Substance	Latent Heat of Fusion, L_f (kcal/kg)	Melting Point	Latent Heat of Vaporization, L_v (kcal/kg)	Boiling Point
Alcohol, ethyl	25	−114°C	204	78°C
Helium*	—	—	377	−269°C
Lead	5.9	328°C	207	1744°C
Mercury	2.8	−39°C	65	357°C
Nitrogen	6.1	−210°C	48	−196°C
Water	80	0°C	540	100°C

*Not a solid at 1 atm pressure; melting point −272°C at 26 atm.

See ● Table 5.2 for the latent heats and the boiling and melting points of some other substances.

EXAMPLE 5.3   Using Latent Heat

Calculate the amount of heat necessary to change 0.20 kg of ice at 0°C into water at 10°C.

Solution

The total heat necessary is found in two steps: ice melting at 0°C and water warming from 0°C to 10°C.

$$H = H_{\text{melt ice}} + H_{\text{change } T}$$
$$= mL_f + mc\,\Delta T$$
$$= (0.20 \text{ kg})(80 \text{ kcal/kg}) + (0.20 \text{ kg})(1.00 \text{ kcal/kg} \cdot °C)(10°C - 0°C)$$
$$= 18 \text{ kcal}$$

Confidence Exercise 5.3

How much heat must be removed from 0.20 kg of water at 10°C to form ice at 0°C? (Show your calculations.)

Pressure has an effect on the boiling point of water. The boiling point increases with increasing pressure as would be expected. *Boiling* is the process by which energetic molecules escape from a liquid. This energy is gained from heating. If the pressure is greater above the liquid, then the molecules must have more energy to escape, and the liquid has to be heated to a higher temperature for boiling to take place.

Normally, when heated water approaches the boiling point in an open container, pockets of energetic molecules form gas bubbles. When the pressure due to the molecular activity in the bubbles is great enough (when it is greater than the pressure on the surface of the liquid), the bubbles rise and break the surface. The water is then boiling. In this sense boiling is a cooling mechanism for the water. Energy is removed, and the water's temperature does not exceed 100°C.

The increase in the boiling point of water with increasing pressure is the principle of the pressure cooker (● Fig. 5.6). In a sealed pressure cooker, the pressure above the liquid is increased, causing the boiling point to increase. The extra pressure is regulated by a pressure valve, which allows vapor to escape. (There is also a safety valve in the lid in case the pressure valve gets stuck.) Hence, the water content of the cooker boils at a temperature greater than 100°C, and the cooking time is reduced.

At mountain altitudes, the boiling point of water may be several degrees lower than at sea level. For example, at the top of Pikes Peak (elevation 4300 m, or 14,000 ft), the atmospheric pressure is reduced to the point where water boils at about 86°C rather than at 100°C. Pressure cookers come in handy at high altitudes, especially if you want to eat on time. It is interesting that cake mixes designed for use at high altitudes contain less baking powder than those used at or near sea level. The baking powder supplies gas to "raise" a cake. If normal cake mixes were used at high altitudes, then the cake would rise too much and could explode.

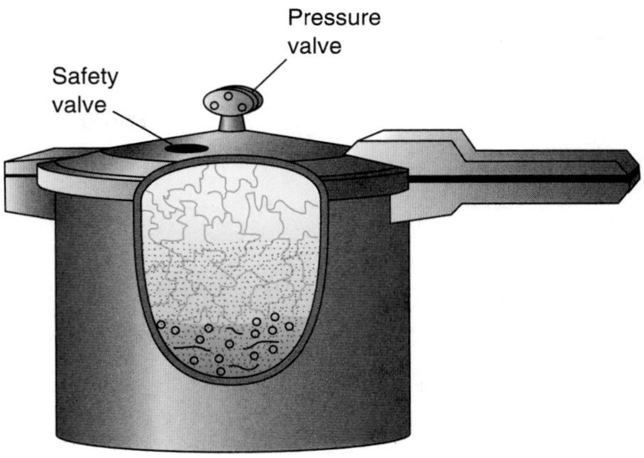

Figure 5.6 The Pressure Cooker Because of the increased pressure in the cooker, the boiling point of water is raised and food cooks faster at the higher temperature.

Conceptual Question and Answer

Under Pressure

Q. Automobile engine cooling systems operate under pressure. What is the purpose of this pressurizing?

A. Under pressure, the boiling point of the water coolant is raised and the engine can operate at a higher temperature, which makes it more efficient in removing heat. (Never remove a radiator cap immediately after turning off a hot engine. Why?)

Finally, let's consider a type of phase change that is important to our personal lives. *Evaporation* is a relatively slow phase change of a liquid to a gas, which is a major cooling (heat removal) mechanism of our bodies. When hot, we perspire, and the evaporation of perspiration has a cooling effect because energy is lost. This cooling effect of evaporation is quite noticeable on the bare skin when one gets out of a bath or shower or has a rubdown with alcohol.

The comforting evaporation of perspiration is promoted by moving air. When you are perspiring and standing in front of a blowing electric fan, you might be tempted to say that the air is cool, but it is the same temperature as the other air in the room. The motion of the air promotes evaporation by carrying away molecules (and energy) and thus has a cooling effect.

On the other hand, air can hold only so much moisture at a given temperature. The amount of moisture in the air is commonly expressed in terms of *relative humidity* (Chapter 19.3). When it is quite humid, there is little evaporation of perspiration and we feel hot.

Did You Learn?

- Different amounts of heat are required to raise the temperature of 1 kg of a substance 1°C. Specific heat is substance specific (varies with substance).

- Latent heat is associated with a phase change, and there is no change in temperature when heat is added (seemingly "hidden").

5.4   Heat Transfer

Preview Questions

- What are the three methods of heat transfer?
- Which type of heat transfer involves mass transfer?

Table 5.3 Thermal Conductivities of Some Common Substances

Substance	W/°C · m*
Copper	390
Iron	80
Brick	3.5
Floor tile	0.7
Water	0.6
Glass	0.4
Wood	0.2
Cotton	0.08
Styrofoam	0.033
Air	0.026
Vacuum	0

*W/°C = (J/s)/°C is the rate of heat flow per temperature difference, where W represents the watt. The length unit (m) comes from considering the dimensions (area and thickness) of the conductor.

Because heat is energy in transit as a result of a temperature difference, how the transfer is done is an important consideration. Heat transfer is accomplished by three methods: conduction, convection, and radiation.

Conduction is the transfer of heat by molecular collisions. The kinetic energy of molecules is transferred from one molecule to another through collisions. How well a substance conducts heat depends on its molecular bonding. Solids, especially metals, are generally the best thermal conductors.

In addition to undergoing molecular collisions, metals contain a large number of "free" (not permanently bound) electrons that can move around. These electrons contribute significantly to heat transfer, or thermal conductivity. (These electrons also contribute to electrical current; see Chapter 8.1.) The *thermal conductivity* of a substance is a measure of its ability to conduct heat. As shown in ● Table 5.3, metals have relatively high thermal conductivities.

In general, liquids and gases are relatively poor thermal conductors. Liquids are better conductors than gases because their molecules are closer together and consequently collide more often. Gases are relatively poor conductors because their molecules are farther apart, so conductive collisions do not occur as often. Substances that are poor thermal conductors are sometimes referred to as *thermal insulators*.

Cooking pots and pans are made of metals so that heat will be readily conducted to the foods inside. Looking at Table 5.3, you can see why some cooking pots have copper bottoms; they provide better thermal conductivity (heat transfer) and faster cooking. Most pot holders, on the other hand, are made of cloth, which is a poor thermal conductor for obvious reasons. Many solids, such as cloth, wood, and plastic foam (Styrofoam), are porous and have large numbers of air (gas) spaces that add to their poor conductivity. For example, foam coolers depend on this property, as does the fiberglass insulation used in the walls and attics of our homes.

Conceptual Question and Answer

Hug the Rug

Q. In a bedroom with a tile or vinyl floor (common in dorms), when you rise and shine and your bare feet hit the floor, you might remark, "Oh, that floor is cold!" A throw rug is often used to prevent this discomfort. How does this help, when the rug and floor are in contact all night and are at the same temperature?

A. The rug and floor are at the same temperature, but their thermal conductivities are different. A tile floor *feels* colder because it conducts heat from the feet faster. See Table 5.3 for the conductivities of cotton (rug) and tile. To be more accurate, you should say, "Oh, that floor has a high thermal conductivity!"

Convection is the transfer of heat by the movement of a substance, or mass, from one place to another. The movement of heated air or water is an example. Many homes are heated by convection (movement of hot air). In a forced-air system, air is heated at the furnace and circulated throughout the house by way of metal ducts. When the air temperature has dropped, it passes through a cold-air return on its way back to the furnace to be reheated and recirculated (● Fig. 5.7).

The warm-air vents in a room are usually in the floor (under windows). Cold-air return ducts are in the floor too, but on opposite sides of a room. The warm air entering the room rises (being "lighter," or less dense, and therefore buoyant). As a result, cooler air is forced toward the floor and convection cycles that promote even heating are set up in the room. Heat is distributed in the Earth's atmosphere (Chapter 19.4) in a manner similar to the transfer of heat by convection currents set up in a room.

The transfer of heat by convection and conduction requires a material medium for the process to take place. The heat from the Sun is transmitted through the vacuum of space and this occurs by radiation. **Radiation** is the process of transferring energy by means of electromagnetic waves (Chapter 6.3). Electromagnetic waves carry energy and can travel through a vacuum. These waves include visible light, infrared, and ultraviolet radiations.

Another example of heat transfer by radiation occurs in an open fire or a fireplace. The warmth of the fire can be felt on exposed skin, yet air is a poor conductor. Moreover, the air warmed by the fire is rising (by convection and and most goes up the chimney in a fireplace). Therefore, the only mechanism for appreciable heat transfer via a fireplace is radiation. The three methods of heat transfer are illustrated in ● Fig. 5.8.

In general, dark objects are good absorbers of radiation, whereas light-colored objects are poor absorbers and good reflectors. For this reason, we commonly wear light-colored clothing in summer to be cooler. In winter, we generally wear dark-colored clothes to take advantage of the absorption of solar radiation.

A device that involves all three methods of heat transfer is the thermos bottle, which is used to keep liquids either hot or cold (● Fig. 5.9). Actually, knowledge of the methods of heat transfer is used to *minimize* the transfer in this case. The space between the double glass walls of the bottle is partially evacuated and then sealed. Glass is a relatively poor conductor, and any heat conducted through a wall (from the outside in or the inside out) will encounter a partial vacuum, an even greater thermal insulator. This vacuum also reduces heat transfer from one glass wall to the other by convection. Finally, the inner surface of the glass bottle is silvered for reflection to prevent heat transfer through the glass by radiation. Thus, a hot (or cold) drink in the bottle remains hot (or cold) for some time.

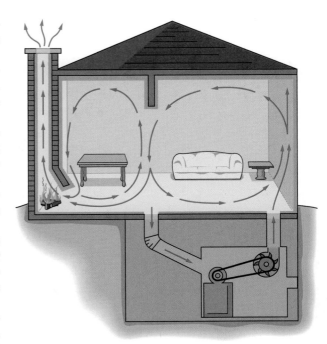

Figure 5.7 Convection Cycles
In a forced-air heating system, warm air is blown into a room. The warm air rises, the cold air descends, and a convection cycle is set up that promotes heat distribution. Some of the cold air returns to the furnace for heating and recycling. Note that a great deal of heat is lost up the chimney of the fireplace.

Did You Learn?

- Heat may be transferred by conduction, convection, or radiation.

- Convection is the transfer of heat by the movement of a substance, or mass, from one place to another.

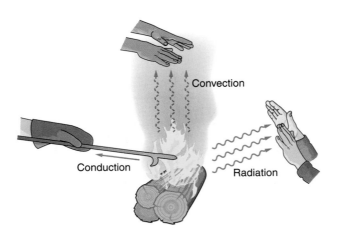

Figure 5.8 Conduction, Convection, and Radiation
The gloved hand is warmed by conduction. Above, the hands are warmed by the convection of the rising hot air (and some radiation). To the right, the hands are warmed by radiation.

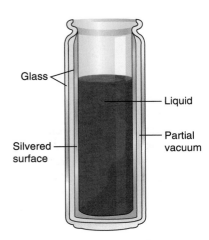

Figure 5.9 A Vacuum Bottle
A vacuum thermos bottle is designed to keep hot drinks hot and cold drinks cold. The bottle usually has a stopper and protective case, which are not shown. (See text for description.)

5.5   Phases of Matter

Preview Questions

● What are the three common phases of matter?

● How are the three phases of matter defined in terms of shape and volume?

As learned in Chapter 5.3, the addition (or removal) of heat can cause a substance to change phase. The three common **phases of matter** are solid, liquid, and gas. All substances exist in each phase at some temperature and pressure. At normal room temperature and atmospheric pressure, a substance will be in one of the three phases. For instance, at room temperature, oxygen is a gas, water is a liquid, and copper is a solid.*

The principal distinguishing features of solids, liquids, and gases are easier to understand if we look at the phases from a *molecular* point of view. Most substances are made up of very small particles called *molecules*, which are chemical combinations of atoms (see Chapter 11.3). For example, two hydrogen atoms attached to an oxygen atom form a water molecule, H_2O.

A solid has relatively fixed molecules and a definite shape and volume. In a *crystalline* solid, the molecules are arranged in a particular repeating pattern. This orderly arrangement of molecules is called a *lattice*. ● Figure 5.10(a) illustrates a lattice structure in three dimensions. The molecules (represented by small circles in the figure) are bound to one another by electrical forces.

Upon heating, the molecules gain kinetic energy and vibrate about their equilibrium positions in the lattice. The more heat that is added, the greater the vibrations become. Thus, the molecules move farther apart, and as shown diagrammatically in Fig. 5.10b, the solid expands.

When the melting point of a solid is reached, additional energy (the latent heat of fusion, 80 kcal/kg for water) breaks apart the bonds that hold the molecules in place. As bonds break, holes are produced in the lattice and nearby molecules can move toward the holes. As more and more holes are produced, the lattice becomes significantly distorted and the solid becomes a liquid.

Solids that lack an ordered molecular structure are said to be *amorphous*. Examples are glass and asphalt. They do not melt at definite temperatures but do gradually soften when heated.

● Figure 5.11 illustrates an arrangement of the molecules in a liquid. A liquid has only a slight, if any, lattice structure. Molecules are relatively free to move. A liquid is an arrange-

Figure 5.10 Crystalline Lattice Expansion
(a) In a crystalline solid the molecules are arranged in a particular repeating pattern. This orderly array is called a *lattice*. (b) A schematic diagram of a solid crystal lattice in two dimensions (left). Heating causes the molecules to vibrate with greater amplitudes in the lattice, thereby increasing the volume of the solid (right). The arrows represent the molecular bonds, and the drawing is obviously not to scale.

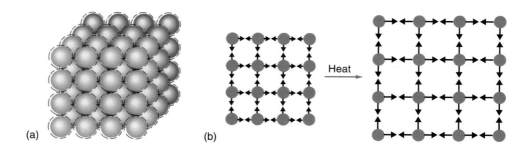

(a)   (b)   Heat

*Solids, liquids, and gases are sometimes referred to as states of matter rather than phases of matter. Thermodynamically, however, the *state* of a quantity of matter is a particular condition defined by certain variables such as, for a gas, pressure (*p*), volume (*V*), and temperature (*T*). See Chapter 5.6. Thus, a quantity of gas can have many states. The term *phases of matter* is preferred and is used here to avoid confusion.

ment of molecules that may move and assume the shape of the container. A liquid has a definite volume but no definite shape.

When a liquid is heated, the individual molecules gain kinetic energy. The result is that the liquid expands. When the boiling point is reached, the heat energy is sufficient to break the molecules completely apart from one another. The latent heat of vaporization is the heat per kilogram necessary to free the molecules completely from one another. Because the electrical forces holding the different molecules together are quite strong, the latent heat of vaporization is fairly large (540 kcal/kg for water). When the molecules are completely free from one another, the gaseous phase is reached.

A gas is made up of rapidly moving molecules and assumes the size and shape of its container; a gas has no definite shape or volume. The molecules exert little or no force on one another except when they collide. The distance between molecules in a gas is quite large compared with the size of the molecules (● Fig. 5.12). The pressure, volume, and temperature of a gas are closely related, as will be seen in the next section.

Continued heating of a gas causes the molecules to move more and more rapidly. Eventually, at ultrahigh temperatures the molecules and atoms are ripped apart by collisions with one another. Inside hot stars such as our Sun, atoms and molecules do not exist and another phase of matter, called a plasma, occurs (no relationship to blood plasma). A plasma is a hot gas of electrically charged particles.

Although plasma is not generally listed as a common phase of matter, there are plasmas all around. One of them—a dense plasma that surrounds the Earth in a layer of the atmosphere called the *ionosphere* (Chapter 19.1)—is *literally* all around. Plasmas exist in the paths of lightning strikes, where the air is heated up to 30,000°C (54,000°F), and in neon and fluorescent lamps. Another phase of matter has been reported—a Bose-Einstein condensate. Jagadis Chandra Bose was an Indian physicist who, with Albert Einstein, predicted the theoretical possibility of the existence of this phase in the 1920s.

By extreme cooling of atoms, a condensate is formed that has completely different properties from all other phases of matter. This extreme cooling was first accomplished in 1995, but this new phase of matter isn't very common. Most recently, rubidium atoms had to be cooled to less than 170 billionths of a degree above absolute zero before condensation occurred.

Did You Learn?

- The three common phases of matter are solid, liquid, and gas.

- A solid has a definite shape and volume. A liquid has a volume but no definite shape. A gas has no definite shape or volume.

5.6   The Kinetic Theory of Gases

Preview Questions

- In the ideal gas law, pressure is directly proportional to what temperature?
- In kinetic theory, how are the temperature and molecular speed of a gas related?

Unlike solids and liquids, gases take up the entire volume and shape of any enclosing container and are easily compressed. Gas pressure rises as the temperature increases, and gases *diffuse* (travel) slowly into the air when their containers are opened. These observations and others led to a model called the *kinetic theory of gases*.

The **kinetic theory** describes a gas as consisting of molecules moving independently in all directions at high speeds; the higher the temperature, the higher the average speed. The molecules collide with one another and with the walls of the container. The distance between molecules is, on average, large compared to the size of the molecules themselves. Theoretically, an *ideal gas* (or *perfect gas*) is one in which the molecules are point particles

Figure 5.11 Liquids and Molecules This illustration depicts the arrangement of molecules in a liquid. The molecules are packed closely together and form only a slight lattice structure. Some surface molecules may acquire enough energy through collisions to break free of the liquid. This process is called *evaporation*. When a liquid is heated and surface molecules break free from the boiling liquid, it is called *vaporization*.

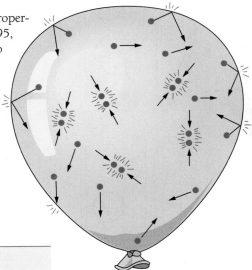

Figure 5.12 Gases and Molecules Gas molecules, on average, are relatively far apart. They move randomly at high speeds, colliding with one another and the walls of the container. The force of their collisions with the container walls causes pressure on the walls.

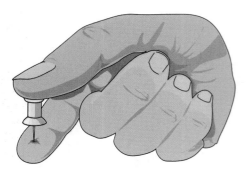

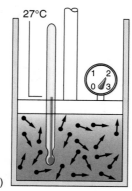

Figure 5.13 Force and Pressure
When holding a tack as shown, the thumb and the finger experience equal forces (Newton's third law). However, at the sharp end of the tack, the area is smaller and the pressure is greater. It can hurt you if you apply enough force. (Ouch!)

(have no size at all) and interact only by collision. A *real gas* behaves somewhat like an ideal gas unless it is under so much pressure that the space between its molecules becomes small relative to the size of the molecules or unless the temperature drops to the point at which attractions among the molecules can be significant.

Each molecular collision with a wall exerts only a tiny force. However, the frequent collisions of billions of gas molecules with the wall exert a steady average force per unit area, or pressure, on the wall. **Pressure** is defined as the force per unit area, $p = F/A$. ● Figure 5.13 illustrates how these three quantities are related.

The SI unit of pressure is newtons per square meter, or N/m^2 (force per area), which is called *a pascal* (Pa) in honor of Blaise Pascal, a seventeenth-century French scientist who was one of the first to develop the concept of pressure.

In the British system, the unit of pressure is pound per inches squared (lb/in^2). A nonstandard unit of pressure used when dealing with gases is the *atmosphere* (atm), where 1 atm is the atmospheric pressure at sea level and 0°C (1.01×10^5 Pa, or 14.7 lb/in^2).

Pressure and Number of Molecules

To see how pressure (p), volume (V), *Kelvin temperature* (T), and number of molecules (N) are related, let's examine the effect of each of V, T, and N on pressure (p) when the other two are held constant.

If the temperature and volume (T and V) are held constant for a gas, then pressure is directly proportional to the number of gas molecules present: $p \propto N$ (● Fig. 5.14). It is logical that the greater the number of molecules, the greater the number of collisions with the sides of the container.

Pressure and Kelvin Temperature

If the volume and number of molecules (V and N) are held constant for a gas, pressure is directly proportional to the Kelvin temperature: $p \propto T$ (● Fig. 5.15). As T increases, the molecules move faster and strike the container walls harder and more frequently. No wonder the pressure increases.

Be aware that many accidental deaths have resulted from lack of knowledge of how pressure builds up in a closed container when it is heated. An explosion can result. A discarded spray can is a good example of such a dangerous container.

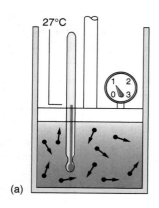

27°C

(a)

27°C

(b)

Figure 5.14 Pressure and Number of Molecules
In both containers the temperature and volume are constant. However, in the container in (b), there are twice as many molecules as in the container in (a). This causes the pressure to be twice as great, as indicated on the gauge. (The more molecules, the more collisions and the greater the pressure.)

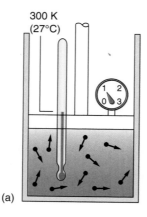

300 K
(27°C)

(a)

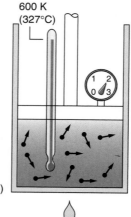

600 K
(327°C)

(b)

Burner

Figure 5.15 Pressure and Kelvin Temperatures
In both containers the number of molecules and the volume are constant. However, the gas in (b) has been heated to twice the Kelvin temperature of that in (a), that is, to 600 K (327°C) versus 300 K (27°C). This temperature causes the pressure to be twice as great as shown on the gauge (the higher the temperature, the greater the kinetic energy, the more collisions, and the greater the pressure).

Pressure and Volume

If the number of molecules and the Kelvin temperature (N and T) are held constant for a gas, then pressure and volume are found to be inversely proportional: $p \propto 1/V$ (● Fig. 5.16). As the volume decreases, the molecules do not have so far to travel and have a smaller surface area to hit. It is logical that they exert more pressure (more force per unit area) than they did before. This relationship was recognized in 1662 by Robert Boyle (Chapter 11.2) and is called *Boyle's law*.

The Ideal Gas Law

Summarizing the factors affecting the pressure of a confined gas, we find that pressure (p) is directly proportional to the number of molecules (N) and to the Kelvin temperature (T) and inversely proportional to the volume (V); that is,

$$p \propto \frac{NT}{V} \qquad\qquad 5.6$$

This proportion can be used to make a useful equation for a given amount of gas (N is constant). In this case the relationship $p \propto T/V$ applies at any time, and we may write the **ideal gas law** in ratio form as

$$\frac{p_2}{p_1} = \left(\frac{V_1}{V_2}\right)\left(\frac{T_2}{T_1}\right)$$

(*N*, number of molecules, constant)   5.7

where the subscripts indicate conditions at different times. Examples of some gas law relationships are shown in ● Fig. 5.17.

EXAMPLE 5.4   Changing Conditions

A closed, rigid container holds a particular amount of gas that behaves like an ideal gas. If the gas is initially at a pressure of 1.80×10^6 Pa at room temperature (20°C), then what will be the pressure when the gas is heated to 40°C? (See Fig. 5.15.)

Solution

Let's work this example in steps for clarity.

Step 1

Given: $V_1 = V_2$ (rigid container)
$p_1 = 1.80 \times 10^6$ Pa (or N/m^2)
$T_1 = 20°C + 273 = 293$ K
$T_2 = 40°C + 273 = 313$ K

The temperatures here were converted to kelvins. This step is a *must* when the ideal gas law is used.

Step 2

Wanted: p_2 (new pressure)

Step 3

Equation 5.7 may be used directly. Because $V_1 = V_2$, the volumes cancel, and

$$p_2 = \left(\frac{T_2}{T_1}\right)p_1 = \left(\frac{313\ \text{K}}{293\ \text{K}}\right)(1.80 \times 10^6\ \text{Pa})$$

$$= 1.92 \times 10^6\ \text{Pa}$$

The pressure increases, as would be expected, because $p \propto T$.

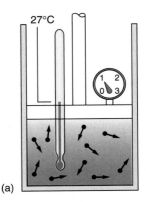

(a)

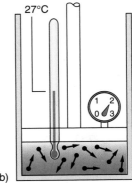

(b)

Figure 5.16 Pressure and Volume
In both containers the temperature and the number of molecules are constant, but the container in (b) has only half the volume of the container in (a). This difference causes the pressure to be twice as great as shown on the gauge (same average kinetic energy but less distance to travel, on average, in a smaller volume, so more collisions).

Figure 5.17 Up They Go!
(top) In a cold climate with low temperatures, the pressure in the weather balloon is low and is increased by adding more gas, $p \propto N$ (number of molecules), as the penguins watch. (bottom) For a hot-air balloon, the balloon is inflated and a flame is used to heat the air in the balloon, which increases the temperature and results in increased pressure and volume. The increased volume makes the air less dense and buoyant in the surrounding cooler air.

BIOS/Peter Arnold, Inc./PhotoLibrary

John Elk III/Bruce Coleman, Inc.

Confidence Exercise 5.4

Suppose the initial pressure of the gas in this example doubled. What would be the final temperature in kelvins in this case?

Some applications of the ideal gas law are given in the **Highlight: Hot Gases: Aerosol Cans and Popcorn.**

Did You Learn?

- The temperature used in the ideal gas law is the Kelvin or absolute temperature.
- In the kinetic theory, the higher the temperature of a gas, the greater the average speed of the molecules.

Highlight Hot Gases: Aerosol Cans and Popcorn

Suppose a trapped gas (constant volume) is heated. What happens? As the temperature of the gas increases, its molecules become more active and collide with the walls of the container more frequently, and the pressure increases. This process may be seen from the ideal gas law in the form $p \propto NT/V$ (Eq. 5.6), which tells us that with the number of molecules (N) and the volume (V) constant, the pressure (p) will change in direct proportion to the change in Kelvin temperature (T). That is,

$$p \propto T$$

Continued heating and temperature increase may cause the gas pressure to build up to the point at which the container is ruptured or explodes. This situation could be dangerous, and warnings to this effect are printed on aerosol can labels (Fig. 1a).

A more beneficial hot-gas explosion occurs when we make popcorn. When heated, moisture inside the popcorn kernel is vaporized and trapped therein. Continued heating raises the gas (steam) pressure until it becomes great enough to rupture the kernel (Fig. 1b). This "explosion" causes the cornstarch inside to expand up to 40 times its original size. (Butter and salt, anyone?)

Another example of using hot gases appeared in the Highlight: The Automobile Air Bag, in Chapter 3.5. Here, a chemical explosion occurred, and the rapidly expanding hot gas inflated the automobile air bag. Could you analyze this situation in terms of the ideal gas law?

(a)

(b)

Figure 1 Hot Gases
Hot gases can be dangerous in some situations (a), but beneficial in others (b). (See text for description.)

5.7 Thermodynamics

Preview Questions

- What is the basis of the first law of thermodynamics?
- What is the difference between a heat engine and a heat pump?

The term **thermodynamics** means "the dynamics of heat," and its study includes the production of heat, the flow of heat, and the conversion of heat to work. Heat energy is used either directly or indirectly to do most of the work that is performed in everyday life. The operation of heat engines, such as internal-combustion engines, is based on the laws of thermodynamics. These laws are important because of the relationships among heat energy, work, and the directions in which thermodynamic processes may occur.

First Law of Thermodynamics

Because one aspect of thermodynamics is concerned with heat energy transfer, accounting for the energy involved in a thermodynamic process is important. This accounting is done by the principle of the conservation of energy, which states that energy can be neither created nor destroyed. The first law of thermodynamics is simply the principle of conservation of energy applied to thermodynamic processes.

Suppose some heat (*H*) is added to a system. Where does it go? One possibility is that it goes into increasing the system's internal energy (ΔE_i). Another possibility is that it results in work (*W*) being done by the system. Or both possibilities could occur. Thus, heat added to a closed system goes into the internal energy of the system, into doing work, or both.

For example, consider heating an inflated balloon. As heat is added to the system (the balloon and the air inside), the temperature increases and the system expands. The temperature of the air inside the balloon increases because some of the heat goes into the internal energy of the air. The gas does work in expanding the balloon (work done by the system).

The **first law of thermodynamics**, which expresses this and other such energy balances, may be written in general as follows:

heat added to (or removed from) a system	=	change in internal energy of the system	+	work done by (or on) the system

or in equation form

$$H = \Delta E_i + W \qquad\qquad 5.8$$

For this equation, a positive value of heat (+*H*) means that heat is *added to* the system, and a positive value of work (+*W*) means that work is *done by* the system (as in the case of the heated balloon). Negative values indicate the opposite conditions.

Heat Engines

Another good example of the first law is the heat engine. A **heat engine** is a device that converts heat into work. Many types of heat engines exist: gasoline engines on lawn mowers and in cars, diesel engines in trucks, and steam engines in old-time locomotives. They all operate on the same principle. Heat input (for example, from the combustion of fuel) goes into doing useful work, but some of the input energy is lost or wasted.

Thermodynamics is concerned not with the components of an engine, but rather with its general operation. A heat engine may be represented schematically as illustrated in ● Fig. 5.18.

A heat engine operates between a high-temperature reservoir and a low-temperature reservoir. These reservoirs are systems from which heat can be readily absorbed and to which heat can be readily expelled. In the process, the engine uses some of the input energy to do work. Notice that the widths of the heat and work paths in the figure are in keeping with the conservation of energy. Actual heat engines usually operate in cycles.

Second Law of Thermodynamics

The first law of thermodynamics is concerned with the conservation of energy. As long as the energy check sheet is balanced, the first law is satisfied. Suppose a hot body at 100°C and a colder body at room temperature (20°C) are placed in contact and heat flows from the colder body to the hotter body. The energy is easily accounted for and the first law is satisfied. but something is physically wrong. Heat does not spontaneously flow from a colder body to a hotter body (● Fig. 5.19). That would be like

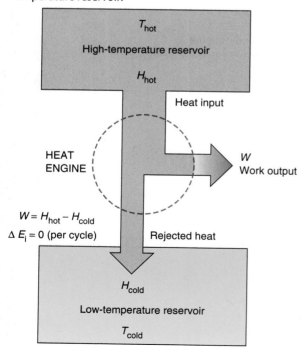

Figure 5.18 Schematic Diagram of a Heat Engine
A heat engine takes heat, H_{hot} from a high-temperature reservoir, converts some into useful work, *W*, and rejects the remainder, H_{cold}, to a low-temperature reservoir.

T_{hot}

High-temperature reservoir

H_{hot}

Heat input

HEAT ENGINE

W Work output

$W = H_{hot} - H_{cold}$

$\Delta E_i = 0$ (per cycle)

Rejected heat

H_{cold}

Low-temperature reservoir

T_{cold}

heat flowing up a "temperature hill," a situation analogous to a ball spontaneously rolling *up* a hill.

Something more than the first law is needed to describe a thermodynamic process. As you might have guessed, it is the direction of the process or whether or not something actually occurs. The **second law of thermodynamics** specifies what can and what cannot happen thermodynamically. This law can be stated in several ways, depending on the situation. A common statement of the second law as applied to our preceding example of heat flow is as follows:

It is impossible for heat to flow spontaneously from a colder body to a hotter body.

Another statement of the second law applies to heat engines. Suppose a heat engine operated in such a way that *all* the heat input was converted into work. Such an engine doesn't violate the first law, but it would have a thermal efficiency of 100%, which has never been observed. Then, the second law as applied to heat engines may be stated as follows:

No heat engine operating in a cycle can convert thermal energy completely into work.

Third Law of Thermodynamics

Another law is associated with absolute zero (Chapter 5.1). Absolute zero (0 K) is the lower limit of temperature. The temperature of absolute zero cannot be attained physically because to do so would require virtually all the heat to be taken from an object. Therefore,

It is impossible to attain a temperature of absolute zero.

This law is sometimes called the **third law of thermodynamics**.

Scientists try to get close to absolute zero. In one attempt, the temperature of a sodium gas was cooled to below 1 nK (nanokelvin, or one-billionth, 10^{-9}, of a kelvin), setting a record.

The third law has still never been violated experimentally. It becomes more difficult to lower the temperature of a material (to pump heat from it) the closer the temperature gets to absolute zero. The difficulty increases with each step to the point at which an infinite amount of work would be required to reach the very bottom of the temperature scale.

Conceptual Question and Answer

Common Descriptions

Q. Do the following statements have any general association with the laws of thermodynamics? (a) You can't get something for nothing. (b) You can't even break even. (c) You can't sink that low.

A. (a) First law. Energy cannot be created (or destroyed). (b) Second law. Thermal energy cannot convert completely into work. (c) Third law. Absolute zero cannot be reached.

Heat Pumps

The second law of thermodynamics states that heat will not flow *spontaneously* from a colder body to a hotter body, or up a "temperature hill," so to speak. The analogy of a ball spontaneously rolling up a hill was used. Of course, we can get a ball to roll up a hill by applying a force and doing work on it. Similarly, heat will flow up the "temperature hill" when work is done. This is the principle of a heat pump.

A **heat pump** is a device that uses work input to transfer heat from a low-temperature reservoir to a high-temperature reservoir (● Fig. 5.20). Work input is required to "pump" heat from a low-temperature reservoir to a high-temperature reservoir. Theoretically, a heat pump is the reverse of a heat engine.

A refrigerator is an example of a heat pump. Heat is transferred from the inside volume of a refrigerator to the outside by the compressor doing work on a gas (by the expenditure of electrical energy). The heat is transferred to the room (high-temperature reservoir). Similarly, an automobile air conditioner transfers heat from the inside of a car to the outside.

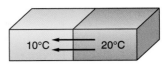

Occurs spontaneously

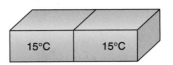

Thermal equilibrium at a later time

(a)

Does **not** occur spontaneously

(b)

Figure 5.19 Heat Flow
(a) When objects are in thermal contact, heat flows spontaneously from a hotter object to a colder object until both objects are at the same temperature, or come to thermal equilibrium. (b) Heat never flows spontaneously from a colder object to a hotter one. That is, a cold object never spontaneously gets colder when placed in thermal contact with a warm object.

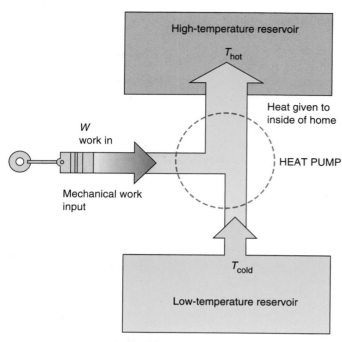

Figure 5.20 Schematic Diagram of a Heat Pump
Work input is necessary to pump heat from a low-temperature reservoir to a high-temperature reservoir—for example, a refrigerator. The inside volume of the refrigerator (low-temperature reservoir) is kept cool by pumping heat into the room (high-temperature reservoir). The pumping requires work input. Notice that the widths of the heat and work paths are in keeping with the conservation of energy.

The "heat pumps" used for home heating and cooling are aptly named. In summer they operate as air conditioners by pumping air from the inside (low-temperature reservoir) to the outside (high-temperature reservoir). In winter heat is extracted from the outside air or from water in a reservoir such as a well or a system of underground coils (low-temperature reservoir) and pumped inside the home (high-temperature reservoir) for heating. In the United States, heat pumps are used extensively in the South where the climate is mild. On very cold days, an auxiliary heating unit in the heat pump (generally an electric heater) is used to supply extra heat when needed.

A heat pump is generally more expensive than a regular furnace, but it has long-term advantages. For example, it has no associated fuel costs (other than the cost of the electricity to supply the work input) because it takes heat from the air.

Entropy

You may have heard the term *entropy* and wondered what it means. **Entropy** is a mathematical quantity; thermodynamically speaking, its change tells whether or not a process can take place naturally. Hence, entropy is associated with the second law.

In terms of entropy, the second law of thermodynamics can be stated as follows:

The entropy of an isolated system never decreases.

To facilitate understanding this idea, entropy is sometimes expressed as a measure of the disorder of a system. When heat is added to an object, its entropy increases because the added energy increases the disordered motion of the molecules. As a natural process takes place, the disorder increases. For example, when a solid melts, its molecules are freer to move in a random motion in the liquid phase than in the solid phase. Similarly, when evaporation takes place, the result is greater disorder and increased entropy.

Systems that are left to themselves (isolated systems) tend to become more and more disordered, but never the reverse. By analogy, a student's dormitory room or room at home naturally becomes disordered, never the reverse. Of course, the room can be cleaned and items put in order, and in this case the entropy *of the room system* decreases. To put things back in order, someone must expend energy resulting in a greater overall entropy increase than the room's entropy decrease. Another statement of the second law is:

The total entropy of the universe increases in every natural process.

This statement has long-term implications. Heat naturally flows from a region of higher temperature to one of lower temperature. In terms of order, heat energy is more "orderly" when it is concentrated. When transferred naturally to a region of lower temperature, it is "spread out" or more "disorderly," and the entropy increases. Hence, the universe—the stars and the galaxies—eventually should cool down to a final common temperature when the entropy of the universe has reached a maximum. This possible fate, billions of years from now, is sometimes referred to as the "heat death" of the universe.

Did You Learn?

● The first law of thermodynamics is the principle of conservation of energy applied to thermodynamic processes.

● A heat engine converts heat to work. A heat pump uses work to transfer heat from a low-temperature reservoir to a high-temperature reservoir. (A refrigerator is an example.)

KEY TERMS

1. temperature (5.1)
2. Fahrenheit scale
3. Celsius scale
4. Kelvin scale
5. kelvin
6. heat (5.2)
7. calorie
8. kilocalorie
9. Btu

10. specific heat (5.3)
11. latent heat
12. conduction (5.4)
13. convection
14. radiation
15. phases of matter (5.5)
16. kinetic theory (5.6)
17. pressure
18. ideal gas law

19. thermodynamics (5.7)
20. first law of thermodynamics
21. heat engine
22. second law of thermodynamics
23. third law of thermodynamics
24. heat pump
25. entropy

MATCHING

For each of the following items, fill in the number of the appropriate Key Word from the preceding list. Compare your answers with those at the back of the book.

a. _____ Water has one of the highest
b. _____ Food Calorie
c. _____ Transfer of heat by electromagnetic waves
d. _____ Can never attain absolute zero
e. _____ Transfer of heat by molecular collisions
f. _____ A measure of average molecular kinetic energy
g. _____ Solid, liquid, and gas
h. _____ Never decreases in an isolated system
i. _____ Scale based on absolute zero
j. _____ Heat associated with a phase change
k. _____ Common temperature scale in the United States
l. _____ Describes gases in terms of moving molecules
m. _____ Raises the temperature of one gram of water one Celsius degree

n. _____ Same size as a degree Celsius
o. _____ Transfer of heat by mass movement
p. _____ Thermodynamic conservation of energy
q. _____ $p \propto NT/V$
r. _____ Common temperature scale worldwide
s. _____ Uses work to transfer heat to a high-temperature reservoir
t. _____ Raises the temperature of one pound of water one Fahrenheit degree
u. _____ Force per unit area
v. _____ Tells what can and what cannot happen thermodynamically
w. _____ Energy transferred because of temperature difference
x. _____ The dynamics of heat
y. _____ Converts heat into work

MULTIPLE CHOICE

Compare your answers with those at the back of the book.

1. Temperature is _____ . (5.1)
   (a) a measure of heat
   (b) a relative measure of hotness and coldness
   (c) internal energy in transit
   (d) both (b) and (c)

2. Which unit of the following is smaller? (5.2)
   (a) a degree Fahrenheit
   (b) a kelvin
   (c) a degree Celsius

3. Which of the following is the largest unit of heat energy? (5.2)
   (a) kilocalorie   (b) calorie   (c) joule   (d) Btu

4. The specific heat of substance A is 10 times that of substance B. If equal amounts of heat are added to equal masses of the substances, then the temperature increase of substance A would be _____ . (5.3)
   (a) the same as that of B   (b) 10 times that of B
   (c) one-tenth that of B   (d) none of the preceding

5. Which of the following methods of heat transfer generally involves mass movement? (5.4)
   (a) conduction   (b) convection   (c) radiation

6. The heat we get from the Sun is transferred through space by which process? (5.4)
   (a) conduction   (b) convection
   (c) radiation   (d) all of the preceding

7. In which of the following is intermolecular bonding greatest? (5.5)
   (a) solids   (b) liquids   (c) gases

8. Which of the following has a definite volume but no definite shape? (5.5)
   (a) solid   (b) liquid   (c) gas   (d) plasma

9. Pressure is defined as (5.6)
   (a) force   (b) force times area
   (c) area divided by force   (d) force divided by area

10. When we use the ideal gas law, the temperature must be in which of the following units? (5.6)
    (a) °C   (b) °F   (c) K

11. When heat is added to a system, it goes into which of the following? (5.7)
    (a) doing work only
    (b) adding to the internal energy only
    (c) doing work, increasing the internal energy, or both

12. The direction of a natural process is indicated by which of the following? (5.7)
    (a) conservation of energy
    (b) change in entropy
    (c) thermal efficiency
    (d) specific heat

FILL IN THE BLANK

Compare your answers with those at the back of the book.

1. When a bimetallic strip is heated, it bends toward the metal with the ___ thermal expansion. (5.1)

2. A ___ difference is necessary for net heat transfer. (5.2)

3. The food Calorie is equal to ___ calories. (5.2)

4. From the equation $H = mc\,\Delta T$, it can be seen that the SI units of specific heat are ___. (5.3)

5. The latent heat of vaporization for water is almost ___ times as great as its latent heat of fusion. (5.3)

6. The boiling point of water may be increased by increasing ___. (5.3)

7. Electrons contribute significantly to the ___ heat transfer process in metals. (5.4)

8. The ___ phase of matter has no definite shape, and no definite volume. (5.4)

9. Pressure is defined as force per ___. (5.5)

10. In the ideal gas law, pressure is ___ proportional to volume. (5.5)

11. The second law of thermodynamics essentially gives the ___ of a process. (5.7)

12. The household refrigerator is a heat ___. (5.7)

SHORT ANSWER

5.1 Temperature

1. When the temperature changes during the day, which scale, Fahrenheit or Celsius, will have the greater degree change?

2. On which temperature scale would a temperature change of 10 degrees be the largest?

3. The two common liquids used in liquid-in-glass thermometers are alcohol (ethanol) and mercury, which have melting points and boiling points of −114°C, 79°C and −39°C, 357°C, respectively. Would either one of these thermometers be better for low-temperature or high-temperature measurements?

4. An older type of thermostat used in furnace and heat pump control is shown in ● Fig. 5.21. The glass vial tilts back and forth so that electrical contacts are made via the mercury (an electrically conducting liquid metal), and the furnace or heat pump is turned off and on. Explain why the vial tilts back and forth. (Newer thermostats are electronic.)

5.2 Heat

5. Heat may be thought of as the "middleman" of energy. Why?

6. When one drinking glass is stuck inside another, an old trick to unstick them is to put water in one of them and run water of a different temperature on the outside of the other. Which water should be hot, and which water should be cold?

7. Heat always flows from a high-temperature body in contact with one with a low temperature. Does heat always flow from a body with more internal energy to one with less internal energy? (Hint: Think about dropping a hot BB into a tub of water.)

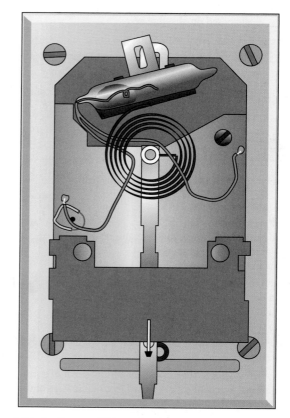

Figure 5.21 An Exposed View of a Thermostat
See Short-Answer Question 4.

Figure 5.22 A Cold Warning
See Short-Answer Question 15.

Figure 5.23 Keeping Warm
See Short-Answer Question 16.

5.3 Specific Heat and Latent Heat

8. What does the specific heat of a substance tell you when you compare it with the specific heat of another substance?

9. When eating a piece of hot apple pie, you may find that the crust is only warm but the apple filling burns your mouth. Why?

10. If equal amounts of heat are added to two containers of water and the resultant temperature change of the water in one container is twice that of the water in the other container, then what can you say about the quantities of water in the containers?

11. When you exhale outdoors on a cold day, you can "see your breath." Why?

12. Compare the SI units of specific heat and latent heat and explain any differences.

5.4 Heat Transfer

13. Give two examples each of good thermal conductors and good thermal insulators. In general, what makes a substance a conductor or an insulator?

14. Which would feel colder on your bare feet, a tile floor or a wooden floor (at the same temperature)? (*Hint*: See Table 5.3.)

15. Highway road signs give warnings, as shown in ● Fig. 5.22. Why would a bridge freeze or ice before the road?

16. Thermal underwear is made to fit loosely. (● Fig. 5.23). What is the purpose of this?

5.5 Phases of Matter

17. What determines the phase of a substance?

18. Give descriptions of a solid, a liquid, and a gas in terms of shape and volume.

19. Name the processes of going (a) from a solid to a gas and (b) from a gas to a solid.

5.6 The Kinetic Theory of Gases

20. How does the kinetic theory describe a gas?

21. What is meant by an *ideal* gas?

22. When does the behavior of a real gas approximate that of an ideal gas?

23. On the molecular level, what causes gas pressure?

24. In terms of kinetic theory, explain why a basketball stays inflated.

5.7 Thermodynamics

25. An inflated balloon is put in a refrigerator, and it shrinks. How does the first law of thermodynamics apply to this case?

26. What does the first law of thermodynamics tell you about a thermodynamic process? What does the second law tell you?

27. Explain the following statements in terms of the laws of thermodynamics.
    (a) Energy can be neither created nor destroyed.
    (b) Entropy can be created but not destroyed.

28. Why can't entropy be destroyed?

29. What can be said about the total entropy of the universe? Why is it true?

30. Can absolute zero be attained? Explain.

31. According to the ideal gas law, what would be the pressure of a gas at absolute zero?

VISUAL CONNECTION

Visualize the connections and give answers for the blanks. Compare your answers with those at the back of the book.

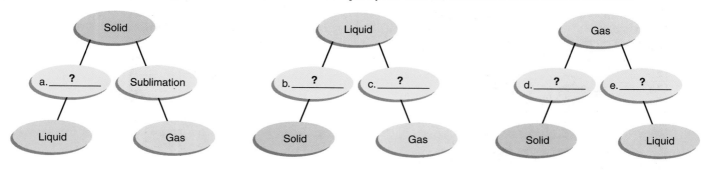

APPLYING YOUR KNOWLEDGE

1. When someone leaves an outside door open on a cold, still day, it is often said that the person is letting in cold air. If so, how? (*Hint:* Think of convection.)

2. Why are steam burns more severe than hot water burns? (Consider the same mass of steam and hot water.)

3. Your automobile has a "radiator" as part of its cooling system. Is radiation the only method of heat transfer in this system? Explain.

4. When a thermometer is inserted into hot water, the mercury or alcohol is sometimes observed to fall slightly before it rises. Why?

5. Could the Earth be considered a heat engine? Explain.

6. When you freeze ice cubes in a tray, there is a decrease in entropy because there is more order in the crystalline lattice of the ice than in the water you started with. Are you violating the second law? Explain.

7. In the Conceptual Question and Answer in Chapter 5.1, an approximation for Celsius to Fahrenheit conversion was given: $T_F \sim 2T_C + 30$. What might be an approximation equation for Fahrenheit to Celsius conversion?

IMPORTANT EQUATIONS

Kelvin–Celsius Conversion:

$$T_K = T_C + 273 \tag{5.1}$$

Fahrenheit–Celsius Conversion:

$$T_F = \tfrac{9}{5}T_C + 32 \qquad (\text{or } T_F = 1.8T_C + 32) \tag{5.2a}$$

$$T_C = \tfrac{9}{5}(T_F - 32) \qquad [\text{or } T_C = (T_F - 32)/1.8] \tag{5.2b}$$

Celsius to Fahrenheit: $T_C + 40$, multiply by 9/5, subtract 40.

Fahrenheit to Celsius: $T_F + 40$, multiply by 5/9, subtract 40.

(Remember: add, multiply, subtract.)

Specific Heat: $H = mc\,\Delta T$ $\qquad$ (5.3)

$$c_{water} = 1.00 \text{ kcal/kg} \cdot {}^\circ C$$

$$c_{ice} = c_{steam} = 0.50 \text{ kcal/kg} \cdot {}^\circ C$$

Latent Heat: $\qquad H = mL$ $\qquad$ (5.4–5.5)

$$(\text{water}) \; L_f = 80 \text{ kcal/kg}$$

$$L_v = 540 \text{ kcal/kg}$$

Changing Conditions for a Gas (Ideal Gas Law):

$$p \propto \frac{NT}{V} \tag{5.6}$$

Ideal Gas Law:

$$\frac{p_2}{p_1} = \left(\frac{V_1}{V_2}\right)\left(\frac{T_2}{T_1}\right) \tag{5.7}$$

First Law of Thermodynamics: $H = \Delta E_i + W$ $\qquad$ (5.8)

EXERCISES

(Assume temperatures in the Exercises to be exact numbers.)

5.1 Temperature

1. While in Europe, a tourist hears on the radio that the temperature that day will reach a high of 17°C. What is this temperature on the Fahrenheit scale?

   Answer: 63°F

2. Which is the lower temperature: (a) 245°C or 245°F? (b) 200°C or 375°F?

3. Normal room temperature is about 68°F. What is the equivalent temperature on the Celsius scale?

   Answer: 20°C

4. A person running a fever has a temperature of 39.4°C. What is this temperature on the Fahrenheit scale?

5. Researchers in the Antarctic measure the temperature to be −40°F. What is this temperature (a) on the Celsius scale? (b) on the Kelvin scale?

Answer: (a) −40°C (b) 233 K

6. The temperature of outer space is about 3 K. What is this temperature on (a) the Celsius scale? (b) on the Fahrenheit scale?

5.2 Heat

7. A college student produces about 100 kcal of heat per hour on the average. What is the rate of energy production in joules?

Answer: 420 kJ

8. How many kilocalories of heat does an expenditure of 250 kJ produce?

9. A pound of body fat stores an amount of chemical energy equivalent to 3500 Cal. When sleeping, the average adult burns or expends about 0.45 Cal/h for every pound of body weight. How many Calories would a 150-lb person burn during 8 hours of sleep?

Answer: 540 Cal (kcal)

10. How long would the person have to sleep continuously to lose 1 lb of body fat?

11. On a brisk walk, a person burns about 325 Cal/h. At this rate, how many hours of brisk walking would it take to lose 1 lb of body fat? (See Exercise 9.)

Answer: 10.8 h

12. If the brisk walk were done at 4.0 mi/h, how far would a person have to walk to burn off 1 lb of body fat?

5.3 Specific Heat and Latent Heat

13. How much heat in kcal must be added to 0.50 kg of water at room temperature (20°C) to raise its temperature to 30°C?

Answer: 5.0 kcal

14. How much heat in joules is needed to raise the temperature of 1.0 L of water from 0°C to 100°C? (*Hint*: Recall the original definition of the liter.)

15. (a) How much energy is necessary to heat 1.0 kg of water from room temperature (20°C) to its boiling point? (Assume no energy loss.)

(b) If electrical energy were used, how much would it cost at 12¢ per kWh?

Answer: (a) 80 kcal (b) 1.1¢

16. Equal amounts of heat are added to equal masses of aluminum and copper at the same initial temperature. Which metal will have the higher final temperature, and how much greater will that temperature change be than the temperature change of the other metal?

17. How much heat is necessary to change 500 g of ice at −10°C to water at 20°C?

Answer: 52.5 kcal

18. A quantity of steam (300 g) at 110°C is condensed, and the resulting water is frozen into ice at 0°C. How much heat was removed?

5.6 The Kinetic Theory of Gases

19. A sample of neon gas has its volume quadrupled and its temperature held constant. What will be the new pressure relative to the initial pressure?

Answer: $p_2 = p_1/4$

20. A fire breaks out and increases the Kelvin temperature of a cylinder of compressed gas by a factor of 1.2. What is the final pressure of the gas relative to its initial pressure?

21. A cylinder of gas is at room temperature (20°C). The air conditioner breaks down, and the temperature rises to 40°C. What is the new pressure of the gas relative to its initial pressure?

Answer: $p_2 = 1.07p_1$

22. A cylinder of gas at room temperature has a pressure p_1. To what temperature in degrees Celsius would the temperature have to be increased for the pressure to be $1.5p_1$?

23. A quantity of gas in a piston cylinder has a volume of 0.500 m³ and a pressure of 200 Pa. The piston compresses the gas to 0.150 m³ in an isothermal (constant-temperature) process. What is the final pressure of the gas?

Answer: 667 Pa

24. If the gas in Exercise 23 is initially at room temperature (20°C) and is heated in an isobaric (constant-pressure) process, then what will be the temperature of the gas in degrees Celsius when it has expanded to a volume of 0.700 m³?

ON THE WEB

1. To the Boiling Point

What happens to the temperature of water when it is boiling? Find out! Follow the recommended links on the student website at **www.cengagebrain.com/shop/ISBN/1133104096** to carry out an experiment in latent heat. You must be careful to take necessary safety precautions when performing this experiment.

2. Thermodynamic Equilibrium

Perform a series of thermodynamic experiments to give you some hands-on experience. How can you use this information in your daily life? How could you use this information to design a product to sell to the consuming public? What would that product be, and how would it work? Follow the recommended links on the student website at **www.cengagebrain.com/shop/ISBN/1133104096** to get started.

Waves and Sound

You can't just cut out the perfect wave and take it home with you. It's constantly moving all the time.

•

Jimi Hendrix
(1942–1970)

Ride the wave! Surfers ride the >
wave of a breaking surf.

Rick Doyle/CORBIS

PHYSICS FACTS

▶ Waves of different types can travel through solids, liquids, gases, and vacuum.

▶ Brain waves are tiny oscillating electrical voltages in the brain.

▶ Tidal waves are not related to tides. More appropriate is the Japanese name *tsunami*, which means "harbor wave." Tsunamis are generated by subterranean earthquakes and can cross the ocean at speeds up to 960 km/h (595 mi/h) with little surface evidence. (See Chapter 21.5.)

▶ Sound is (a) the physical propagation of a disturbance (energy) in a medium and (b) the physiological and psychological response, generally to pressure waves in air.

T he word *wave* brings to mind different things for different people. Probably most would associate it with ocean waves (a big one is shown in the opening photo) or their smaller relatives, those on the surface of a lake or pond. Others might think of sound waves or light waves. Music is made up of sound waves, and we see beautiful rainbows because of light waves.

Indeed, waves are all around us, and understanding their properties is essential to describing our physical environment. Our eyes and ears are the two main wave-detecting devices that link us to our world (Chapter 1.3). This and the next chapter explore the important roles that waves play in our lives.

Chapter Outline

6.1 Waves and Energy Propagation

Preview Questions

- What causes waves, and how and what do they propagate?
- Is matter propagated by waves?

Since beginning the study of energy, much has been said about its forms, its relationship to work, and its conservation. It was learned that heat is energy in transit because of a temperature difference, but the transfer and *propagation* of energy in matter are not limited to temperature differences. In many common cases such as a disturbance in water, energy is propagated in media as **waves**.

For example, when a stone is thrown into a still pond or when you dive into a swimming pool, there is a disturbance. In such instances, water waves propagate outward, propagating energy from the disturbance source (● Fig. 6.1). When you are swimming in a lake and a high-speed boat goes by a short distance away, waves propagate the energy of the boat's disturbance of the water toward you, and when they hit you, it is clear that the waves have energy. Wind can also generate wave disturbances on a lake or pond. This transfer of energy is called *wave motion*. As a wave propagates outward from a disturbance, energy is transmitted from one particle to another in the medium.

In general, only energy, not matter, is transferred by waves. When you are fishing and water waves come toward and reach your floating bobber, the bobber bobs up and down but generally stays in the same place (unless there is a current). Only energy is propagated, not matter (the water or the bobber).

A similar situation can occur in a solid. For example, during an earthquake a disturbance takes place because of a slippage along a fault, and the energy is transmitted through the Earth by waves (Chapter 21.5). Again, the disturbance is a transfer of energy, not of matter. On a more local scale, suppose someone is playing loud music in an adjacent closed room. You can hear sound, so wave energy must propagate through the wall.

We know that waves propagate in gases, with air being the most common example. Disturbances give rise to sound waves such as an explosion, which produces a shock wave, or a vibrating guitar string, which produces continuous waves (● Fig. 6.2). Another common source of sound waves—a whistle—is also shown in Fig. 6.2.

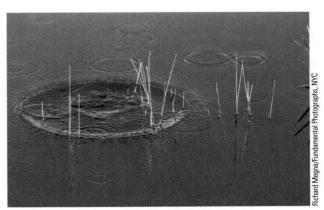

Richard Megna/Fundamental Photographs, NYC

Figure 6.1 Waves and Energy Waves propagate energy outward from disturbances.

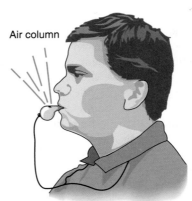

Air column

Steel or nylon string

Figure 6.2 Vibrations Examples of vibrational disturbances that produce sound waves. The vibrations of guitar strings produce waves that are propagated in the air. Blowing a whistle disturbs the small ball inside the whistle, causing it to vibrate and produce sound waves.

However, some waves can propagate without a medium (in vacuum), such as light from the Sun and radio waves from space probes. If light needed a medium to propagate, then the Earth would receive no sunlight.

After discussing the general properties of waves in the next section, light waves and sound waves will be considered in more detail.

Did You Learn?

● From an originating disturbance, energy is transmitted from one particle to another in a medium as the wave propagates outward.

● In general, energy is propagated by waves, not matter.

6.2   Wave Properties

Preview Questions

● What is the distinguishing difference between longitudinal and transverse waves?

● How are wave frequency and period related, and what is the unit of frequency?

A disturbance that generates a wave may be a simple pulse or shock, such as the clapping of hands or a book hitting the floor. A disturbance may also be periodic, repeated again and again at regular intervals. A vibrating guitar string or a whistle sets up periodic waves, and the waves are continuous as long as the disturbances are too (Fig. 6.2).

In general, waves may be classified as *longitudinal* or *transverse* based on particle motion and wave direction. In a **longitudinal wave,** the particle motion and the wave velocity are parallel to each other. For example, consider a stretched spring, as illustrated in ● Fig. 6.3. When several coils at one end are compressed and released, the disturbance is propagated along the length of the spring with a certain wave velocity. Notice that the displacements of the spring "particles" and the wave velocity vector are parallel to each other. The directions of the "particle" oscillations can be seen by tying a small piece of ribbon to the spring. The ribbon will oscillate similarly to the way the particles of the spring oscillate.

The single disturbance in the spring is an example of a longitudinal wave pulse. Periodic longitudinal waves are common. An example of a longitudinal wave is sound, as will be seen shortly. In a **transverse wave,** the particle motion is perpendicular to the direction of the wave velocity. A transverse wave may be generated by shaking one end of a stretched cord up and down or side to side (● Fig. 6.4). Notice how the cord "particles" oscillate perpendicularly (at an angle of 90°) to the direction of the wave velocity vector. The particle motion may again be demonstrated by tying a small piece of ribbon to the cord. Another example of a transverse wave is light. Recall that light and other such waves can propagate through the vacuum of space without a medium.

Certain wave characteristics are used in describing periodic wave motion (● Fig. 6.5). The wave velocity describes the speed and direction of the wave motion. The **wavelength** (λ, Greek lambda) is the distance from any point on the wave to the adjacent point with similar oscillation: the distance of one complete "wave," or where it starts to repeat itself. For example, the wavelength distance may be measured from one wave crest to an adjacent crest (or from one wave trough to the next wave trough).

The wave **amplitude** (A) is the maximum displacement of any part of the wave (or wave particle) from its equilibrium position. The energy transmitted by a wave is directly proportional to the square of its amplitude (A^2). However, the amplitude does not affect the wave speed. (See amplitude in Fig. 6.5.)

The oscillations of a wave may be characterized in terms of frequency. The wave **frequency** (f) is the number of oscillations or cycles that occur

Figure 6.3 Longitudinal Wave
In a longitudinal wave, as illustrated here for a stretched spring, the wave velocity, or direction of the wave propagation, is parallel to the (spring) particle displacements (back and forth). There are regions of compression and stretching (rarefaction) in the spring.

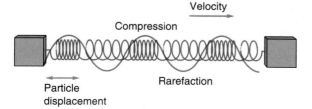

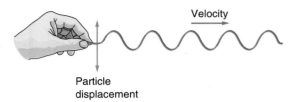

Figure 6.4 Transverse Wave
In a transverse wave as illustrated here for a stretched cord, the wave velocity (vector), or the direction of the wave propagation, is perpendicular to the (cord) particle displacements (up and down).

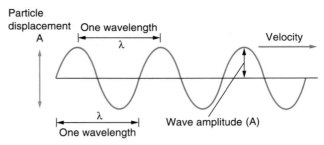

Figure 6.5 Wave Description
Some terms used to describe wave characteristics. (See text for description.)

during a given period of time, usually one second. Frequency is the number of cycles per second, but this unit is given the name **hertz** (Hz).* One hertz is one cycle per second. (In standard units, Hz = 1/s. The unitless "cycle" is carried along for descriptive convenience.)

For example, as illustrated in ● Fig. 6.6a, if four complete wavelengths pass a given point in 1 second, then the frequency of the wave is four cycles per second, or 4 Hz. The more "wiggles" or wavelengths per time period, the greater the frequency. A wave with a frequency of 8 Hz is illustrated in Fig. 6.6b.

There is another quantity used to characterize a wave. The wave **period** (T) is the time it takes the wave to travel a distance of one wavelength (Fig. 6.6a). Looking at the dashed-line portion, it can be seen that a particle in the medium makes one complete oscillation in a time of one period.

The frequency and period are inversely proportional. In equation form

$$\text{frequency} = \frac{1}{\text{period}}$$

or

$$f = \frac{1}{T}$$

6.1

Examine the units. The frequency is given in cycles per second, and the period in seconds per cycle.

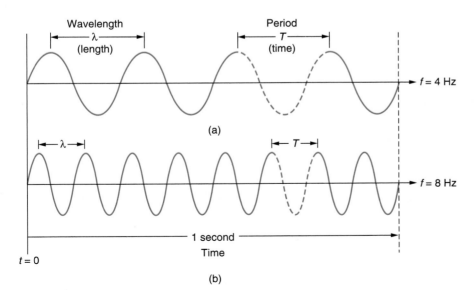

(a)

(b)

Figure 6.6 Wave Comparison
Wave (a) has a frequency of 4 Hz, which means that 4 wavelengths pass by a point in 1 second. With a wave of 8 Hz, there would be 8 wavelengths passing by in 1 second. Hence, the wavelength of wave (a) is twice as long as the wavelength of wave (b). The period (T) is the time it takes one wavelength to pass by, so the period of wave (a) is also twice as long as that of wave (b). The relationship between the frequency and the period is $f = 1/T$.

---

*In honor of Heinrich Hertz (1857–1894), a German scientist and an early investigator of electromagnetic waves.

Suppose a wave has a frequency of $f = 4$ Hz (Fig. 6.6a). Then, four wavelengths pass by a point in 1 second, and one wavelength passes in $\frac{1}{4}$ second ($T = 1/f = \frac{1}{4}$ s). That is, the period, or time for one cycle, is $\frac{1}{4}$ second. What, then, would be the period for the wave in Fig. 6.6b? (Right, $\frac{1}{8}$ s.)

Another simple relationship for wave characteristics relates the **wave speed** (v), the wavelength, and the period (or frequency). Because speed is the distance divided by time and because a wave moves one wavelength in a time of one period, this may be expressed in equation form as

$$\text{wave speed} = \frac{\text{wavelength}}{\text{period}}$$

$$v = \frac{\lambda}{T} \qquad \qquad 6.2$$

or, by Eq. 6.1 ($f = 1/T$),

$$\text{wave speed} = \text{wavelength} \times \text{frequency}$$
$$v = \lambda f \qquad \qquad 6.3$$

EXAMPLE 6.1   Calculating Wavelengths

Consider sound waves with a speed of 344 m/s and frequencies of (a) 20 Hz and (b) 20 kHz. Find the wavelength of each of these sound waves.

Solution

Step 1

Given: $v = 344$ m/s

(a) $f = 20$ Hz
(b) $f = 20$ kHz $= 20 \times 10^3$ Hz

(Note that kHz was converted to Hz, the standard unit, and recall that Hz = 1/s.)

Step 2

Wanted: λ (wavelength)

Step 3

(a) Rearrange Eq. 6.3 and solve for λ:

$$v = \lambda f$$

$$\lambda = \frac{v}{f}$$

$$= \frac{344 \text{ m/s}}{20 \text{ Hz}} = \frac{344 \text{ m/s}}{20 \ (1/\text{s})} = 17 \text{ m}$$

(b)
$$\lambda = \frac{v}{f} = \frac{344 \text{ m/s}}{20 \times 10^3 \text{ Hz}}$$

$$= 17 \times 10^{-3} \text{ m} = 0.017 \text{ m}$$

Confidence Exercise 6.1

A sound wave has a speed of 344 m/s and a wavelength of 0.500 m. What is the frequency of the wave?

Answers to Confidence Exercises may be found at the back of the book.

As will be learned in Chapter 6.4, the frequencies of 20 Hz and 20,000 Hz given in Example 6.1 define the general range of audible sound wave frequencies. Thus, the wavelengths of audible sound cover the range from about 1.7 cm for the highest-frequency sound to about 17 m for the lowest-frequency sound. In British units sound waves range from wavelengths of approximately $\frac{1}{2}$ in. up to about 50 ft.

Did You Learn?

- Longitudinal and transverse waves are distinguished in terms of particle motion and wave velocity. They are parallel in longitudinal waves and perpendicular (90°) in transverse waves.

- The frequency and period are inversely related ($f = 1/T$), and the unit of frequency is 1/s or hertz (Hz).

6.3 Light Waves

Preview Questions

- How is the electromagnetic spectrum arranged?
- What is the speed of light in vacuum?

Light belongs to a family of waves that include radio waves, microwaves, and X-rays. Technically, these are *electromagnetic waves*. When charged particles such as electrons are accelerated, energy is radiated away in the form of waves. Electromagnetic waves consist of oscillating electric and magnetic fields (Chapter 8). For now, we will consider electromagnetic waves only in the context of waves.

Accelerated charged particles produce waves of various frequencies or wavelengths that form a continuous **electromagnetic (EM) spectrum** (● Fig. 6.7). Waves at one end (right) of the spectrum that have relatively low frequencies (10^4 Hz to 10^8 Hz) and long wavelengths are known as *radio* waves. Waves with frequencies greater than radio waves but less than visible light, from 10^{11} Hz to about 4.3×10^{14} Hz, include the *microwave* and *infrared* regions.

With increasing frequency come the visible and ultraviolet regions. Note the small portion of the spectrum that is visible to the human eye, 4.3×10^{14} Hz to 7.5×10^{14} Hz (or $\lambda = 700$ nm to 400 nm). It lies between the infrared (IR) and ultraviolet (UV) regions. It is the ultraviolet in sunlight that tans and burns our skin.

At still higher frequencies is the more energetic *X-ray* region. X-rays are widely used in medical and dental applications to take "pictures" of bones and teeth. Too much X-ray radiation can cause cell damage. Therefore, lead aprons are used when taking dental X-rays to protect other parts of the body. X-rays cannot penetrate lead. *Gamma rays* occupy a

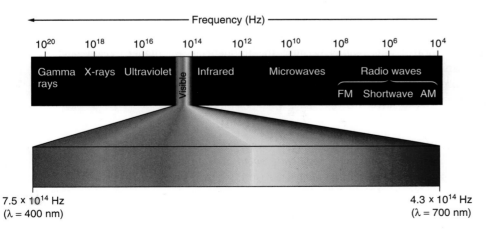

Figure 6.7 Electromagnetic (EM) Spectrum
Different frequency (or wavelength) regions are given names. Notice that the visible region forms only a very small part of the EM spectrum (much smaller than shown in the figure).

still higher frequency range. This energetic radiation is emitted in nuclear decay processes (discussed in Chapter 10.3).

The term *light* is commonly used for electromagnetic regions in or near the visible region; for example, we say *ultraviolet light*. Only the frequency (or wavelength) distinguishes visible light from the other portions of the spectrum. Our human eyes are sensitive only to certain frequencies or wavelengths, but other instruments can detect other portions of the spectrum. For example, a radio receiver can detect radio waves.

Keep in mind that radio waves are *not* sound waves. If they were, then you would "hear" many stations at once. Radio waves are detected and distinguished by frequency. Then, they are processed electronically and sent to the speaker system that produces sound waves.

Electromagnetic radiation consists of transverse waves. These waves can travel through a vacuum. For instance, light from the Sun travels through the vacuum of space before arriving at the Earth. This speed, which is called the **speed of light**, is designated by the letter *c* and has a value of

$$c = 3.00 \times 10^8 \text{ m/s}$$

(To a good approximation, this value is also the speed of light in air.*)

Equation 6.3 with $v = c$, that is, $c = \lambda f$, can be used to find the wavelength of light or any electromagnetic wave in a vacuum and to a good approximation in air. For example, let's calculate the wavelength of a typical radio wave.

EXAMPLE 6.2   Computing the Wavelength of a Radio Wave

What is the wavelength of the radio waves produced by an AM station with an assigned frequency of 600 kHz?

Solution

First, convert the frequency from kilohertz (kHz) into hertz. This conversion can be done directly, because *kilo-* is 10^3.

$$f = 600 \text{ kHz} = 600 \times 10^3 \text{ Hz} = 6.00 \times 10^5 \text{ Hz}$$

Then, using Eq. 6.3 in the form $c = \lambda f$ and rearranging for λ yields

$$\lambda = \frac{c}{f} = \frac{3.00 \times 10^8 \text{ m/s}}{6.00 \times 10^5 \text{ Hz}}$$

$$= 0.500 \times 10^3 \text{ m} = 500 \text{ m}$$

Confidence Exercise 6.2

The station in this example is an AM station, which generally uses kHz frequencies. FM stations have MHz frequencies. What is the wavelength of an FM station with an assigned frequency of 90.0 MHz?

The answers to Confidence Exercises may be found at the back of the book.

The wavelengths of AM radio waves are quite long compared to FM radio waves, which have shorter wavelengths because the frequencies are higher. Visible light, with frequencies on the order of 10^{14} Hz, has relatively short wavelengths, as can be seen in the approximation

$$\lambda = \frac{c}{f} \approx \frac{10^8 \text{ m/s}}{10^{14} \text{ Hz}} = 10^{-6} \text{ m}$$

Thus, the wavelength of visible light is on the order of one millionth of a meter (micrometer, μm). To avoid using negative powers of 10, wavelengths are commonly expressed in a smaller unit called the *nanometer* (nm, where 1 nm = 10^{-9} m) so as to have whole

---

*In this sense, *light* means all electromagnetic waves, all of which travels at the speed of light in vacuum. Why the letter *c* symbol for speed? It is adopted from *celeritas*, Latin for speed.

numbers with powers of 10. Using the values given in Fig. 6.7, you should be able to show that the approximate wavelength range for the visible region is between 4×10^{-7} m and 7×10^{-7} m. This range corresponds to a span of 400 nm to 700 nm.

For visible light, different frequencies or wavelengths are perceived by the eye as different colors, and the brightness depends on the energy of the wave (Chapter 7.4).

Did You Learn?

- The continuous electromagnetic spectrum is arranged in sequences of wave frequencies or wavelengths.

- The speed of light in vacuum is 3.00×10^{8} m/s (about 186,000 mi/s).

6.4 Sound Waves

Preview Questions

- What is the frequency range of human hearing?
- When the decibel (dB) level is doubled, does the sound intensity double?

Technically, **sound** is the propagation of *longitudinal* waves through matter. Sound waves involve longitudinal particle displacements in all kinds of matter: solid, liquid, or gas.

Probably most familiar are sound waves in air, which affect our sense of hearing, but sound also travels in liquids and solids. When while swimming underwater someone clicks two rocks together also underwater, you can hear this disturbance. Also, sound can be heard through thin (but solid) walls.

The wave motion of sound depends on the elasticity of the medium. A longitudinal disturbance produces varying pressures and stresses in the medium. For example, consider a vibrating tuning fork, as illustrated in ● Fig. 6.8.

As an end of the fork moves outward, it compresses the air in front of it, and a *compression* is propagated outward. When the fork end moves back, it produces a region of decreased air pressure and density called a *rarefaction*. With continual vibration, a series of high- and low-pressure regions travels outward, forming a longitudinal sound wave. The waveform may be displayed electronically on an oscilloscope, as shown in ● Fig. 6.9.

Sound waves may have different frequencies and so form a spectrum analogous to the electromagnetic spectrum (● Fig. 6.10). However, the **sound spectrum** has much lower frequencies and is much simpler, with only three frequency regions. These regions are defined in terms of the audible range of human hearing, which is about 20 Hz to 20 kHz (20,000 Hz) and constitutes the *audible region* of the spectrum.

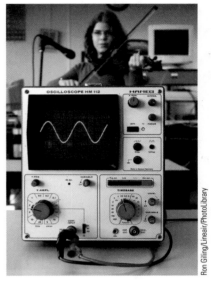

Ron Giling/Lineair/PhotoLibrary

Figure 6.9 Waveform
The waveform of a tone from a violin is displayed on an oscilloscope by using a microphone to convert the sound wave into an electric signal.

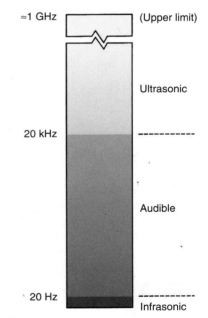

Frequency

Figure 6.10 Sound Spectrum
The sound spectrum consists of three regions: the infrasonic region ($f < 20$ Hz), the audible region (20 Hz $< f < 20$ kHz), and the ultrasonic region ($f > 20$ kHz).

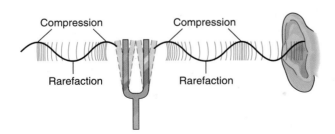

Figure 6.8 Sound Waves
Sound waves consist of a series of compressions (high-pressure regions) and rarefactions (low-pressure regions), as illustrated here for waves in air from a vibrating tuning fork. These regions can be described by a waveform.

Below the audible region is the *infrasonic region* and above is the *ultrasonic region*. (Note the analogy to infrared and ultraviolet light.) The sound spectrum has an upper limit of about a billion hertz (1 GHz, gigahertz) because of the elastic limitations of materials.

Waves in the infrasonic region, which humans cannot hear, are found in nature. Earthquakes produce infrasonic waves, and earthquakes and their locations are studied by using these waves (Chapter 21.5). Infrasound is also associated with tornados and weather patterns. Elephants and cattle have hearing response in the infrasonic region and may get advance warnings of earthquakes and weather disturbances. Aircraft, automobiles, and other rapidly moving objects produce infrasound.

The audible region is of prime importance in terms of hearing. Sound is sometimes defined as those disturbances perceived by the human ear, but as Fig. 6.10 shows, this definition would omit a majority of the sound spectrum. Indeed, ultrasound, which we cannot hear, has many practical applications, some of which are discussed shortly

We hear sound because propagating disturbances cause the eardrum to vibrate, and sensations are transmitted to the auditory nerve through the fluid and bones of the inner ear. The characteristics associated with human hearing are physiological and can differ from their physical counterparts.

For example, *loudness* is a relative term. One sound may be louder than another, and as you might guess, this property is associated with the energy of the wave. The measurable physical quantity is **intensity** (I), which is the rate of sound energy transfer through a given area. Intensity may be given as so many joules per second (J/s) through a square meter (m^2). Recall, though, that a joule per second is a watt (W), so sound intensity has units of W/m^2.

The loudness or intensity of sound decreases the farther one is from the source. As the sound is propagated outward, it is "spread" over a greater area and so has less energy per unit area. This characteristic is illustrated for a point source in ● Fig. 6.11. In this case the intensity is inversely proportional to the square of the distance from the source ($I \propto 1/r^2$); that is, it is an inverse-square relationship. This relationship is analogous to painting a larger room with the same amount of paint. The paint (energy) must be spread thinner and so is less "intense."

The minimum sound intensity that can be detected by the human ear (called the *threshold of hearing*) is about 10^{-12} W/m^2. At a greater intensity of about 1 W/m^2, sound becomes painful to the ear. Because of the wide range, the sound intensity level is commonly measured on a logarithmic scale, which conveniently compresses the linear scale. This logarithmic scale is called the decibel (dB) scale and is illustrated in ● Fig. 6.12. Notice that *sound intensity* is measured in W/m^2 and that *sound intensity level* is expressed in decibels (dB).

Figure 6.11 Sound Intensity
An illustration of the inverse-square law for a point source ($I \propto 1/r^2$). Note that when the distance from the source doubles—for example, from 1 m to 2 m—the intensity decreases to one-fourth its former value because the sound must pass through four times the area.

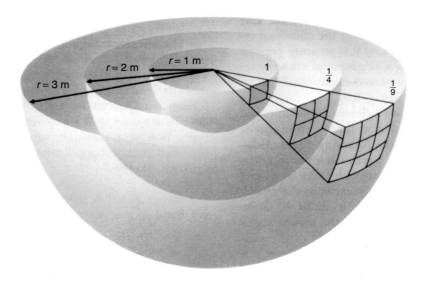

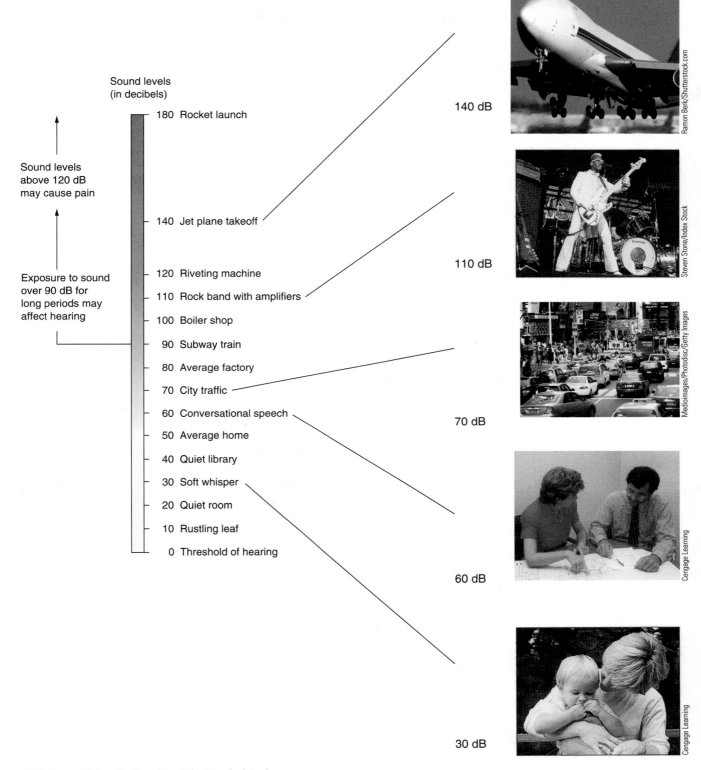

Sound levels
(in decibels)

180 Rocket launch

140 Jet plane takeoff

120 Riveting machine

110 Rock band with amplifiers

100 Boiler shop

90 Subway train

80 Average factory

70 City traffic

60 Conversational speech

50 Average home

40 Quiet library

30 Soft whisper

20 Quiet room

10 Rustling leaf

0 Threshold of hearing

Sound levels
above 120 dB
may cause pain

Exposure to sound
over 90 dB for
long periods may
affect hearing

140 dB

110 dB

70 dB

60 dB

30 dB

Figure 6.12 Sound Intensity Level and the Decibel Scale
The decibel scale of sound intensity levels, with examples of typical sources.

Table 6.1	Sound Intensity Levels and Decibel Differences		
Source of Sound	Sound Intensity Levels (dB)	Times Louder Than Threshold	Decibel Difference (ΔdB)
Riveting machine	120	1,000,000,000,000	—
Rock band with amplifiers	110	100,000,000,000	—
Boiler shop	100	10,000,000,000	—
Subway train	90	1,000,000,000	—
Average factory	80	100,000,000	—
City traffic	70	10,000,000	— (and so on)
Conversational speech	60	1,000,000	Δ60 dB 1,000,000 increase*
Average home	50	100,000	Δ50 dB 100,000 increase
Quiet library	40	10,000	Δ40 dB 10,000 increase
Soft whisper	30	1,000	Δ30 dB 1,000 increase
Quiet room	20	100	Δ20 dB 100 increase
Rustling leaf	10	10	Δ10 dB 10 increase
Threshold of hearing	0	0	Δ3 dB 2 increase

*Similar decreases in intensity occur for $-\Delta$dB.

A **decibel (dB)** is one-tenth of a bel (B). The bel unit was named in honor of Alexander Graham Bell, who received the first patent for the telephone. Because the decibel scale is not linear with intensity, the dB level does not double when the sound intensity is doubled. Instead, the intensity level increases by only 3 dB. In other words, a sound with an intensity level of 63 dB has twice the intensity of a sound with an intensity level of 60 dB.

Comparisons are conveniently made on the dB scale in terms of decibel differences and factors of 10. Sound intensity levels and decibel differences (ΔdB) are shown in ● Table 6.1.

Exposure to loud sounds or noise can be damage one's hearing, as discussed in the **Highlight: Noise Exposure Limits**.

Loudness is related to intensity, but it is subjective, and estimates differ from person to person. Also, the ear does not respond equally to all frequencies. For example, two sounds with different frequencies but the same intensity levels are judged by the ear to differ in loudness.

The frequency of a sound wave may be physically measured, whereas *pitch* is the *perceived* highness or lowness of a sound. For example, a soprano has a high-pitched voice compared with a baritone. Pitch is related to frequency. If a sound with a single frequency is heard at two intensity levels, then nearly all listeners will agree that the more intense sound has a lower pitch.

Conceptual Question and Answer

A Tree Fell

Q. Here's an old one. If a tree falls in the forest where there is no one to hear it, then is there sound?

A. Physically, sound is simply wave disturbances (energy) that propagate in solids, liquids, and gases. When perceived by our ears, sound is interpreted as speech, music, noise, and so forth. The answer to the question depends on the distinction between physical and sensory sound. The answer is no if thinking in terms of sensory hearing, but yes if considering physical waves.

Highlight   Noise Exposure Limits

Sounds with intensity levels of 120 dB and higher can be painfully loud to the ear. Brief exposures to even higher sound intensity levels can rupture eardrums and cause permanent hearing loss. Long exposure to relatively lower sound (noise) levels can also cause hearing problems. (*Noise* is defined as unwanted sound.)

Examples of such loud sounds are motorcycles and rock bands with amplifiers (Fig. 1). Another more recent concern is a problem with loudness involving listening devices with earphones. Whether hearing is affected depends primarily on the loudness and the length of time one is exposed to the sound. (Pitch or frequency may also have a contributing effect.) Long exposure to loud sounds can damage the sensitive hair cells of the inner ear as well as the hearing nerve. Typical symptoms include feeling pressure in the ear, hearing speech as muffled, and hearing a ringing sound (tinnitus) in the ear. When the damage is not severe, these symptoms may go away in minutes or hours after the sound exposure ends.

Occupational noise hazards can also present problems. Workers in a noisy factory and members of an airplane ground crew, for example, may be affected. When danger to hearing exists on such jobs, ear protectors should be required (Fig. 2). Ear protectors are available at hardware stores, and people are wise to wear them when mowing grass or using a chain saw.

Federal standards now set permissible noise exposure limits for occupational loudness. These limits are listed in Table 1. Note that a person can safely work on a subway train (90 dB, Fig. 6.13) for 8 hours but should not play in (or listen to) an amplified rock band (110 dB) continuously for more than half an hour.

Table 1	Permissible Noise Exposure Limits
Maximum Duration per Day (h)	Intensity Level (dB)
8	90
6	92
4	95
3	97
2	100
$1\frac{1}{2}$	102
1	105
$\frac{1}{2}$	110
$\frac{1}{4}$ or less	115

Figure 1 A Lot of Decibels
Loud rock music may cause a ringing in the ears (tinnitus); long exposure could lead to permanent damage.

Bill Gallery/Stock Boston

Figure 2 Sound Intensity Safety
An airport cargo worker wears ear protectors to prevent hearing damage from the high-intensity levels of jet plane engines.

Ultrasound is the term used for sound waves with frequencies greater than 20,000 Hz or 20 kHz. These waves cannot be detected by the human ear, but the audible frequency range for other animals includes ultrasound frequencies. For example, dogs can hear ultrasound, and ultrasonic whistles used to call dogs don't disturb humans. Bats use ultrasonic sonar to navigate at night and to locate and catch insects. This is called *echolocation* (● Fig. 6.13).

An important use of ultrasound is in examining parts of the body. Ultrasound is an alternative to potentially harmful X-rays. The ultrasonic waves allow different tissues, such as organs and bone, to be "seen" or distinguished by bouncing waves off the object examined. The reflected waves are detected, analyzed, and stored in a computer. An *echogram* is then reconstructed, such as the one of a fetus shown in ● Fig. 6.14. Energetic X-rays might

Figure 6.13 Ultrasonic Sonar
Bats use the reflections of ultrasound for navigation and to locate food (echolocation). The emitted sound waves (blue) are reflected, and the echoes (red) enable the bat to locate the wall and an insect.

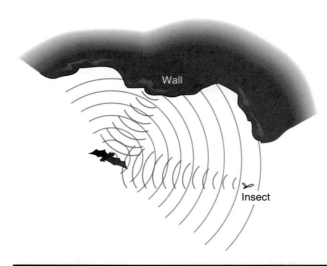

Figure 6.14 Echogram
An echogram in which the outline of a fetus at 21 weeks can be clearly seen.

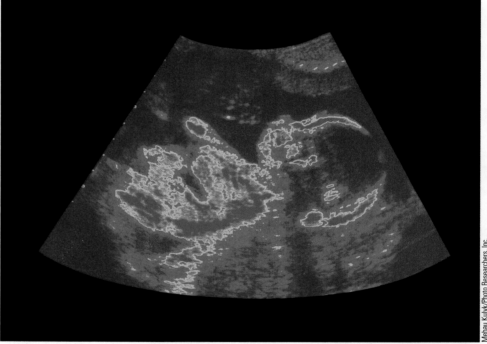

Mehau Kulyk/Photo Researchers, Inc.

harm the fetus and cause birth defects, but ultrasonic waves have less energetic vibrations and less chance of harming a fetus.

Ultrasound is also used as a cleaning technique. Minute foreign particles can be removed from objects placed in a liquid bath through which ultrasound is passed. The wavelength of ultrasound is on the same order of magnitude as the particle size, and the wave vibrations can get into small crevices and "scrub" particles free. Ultrasound is especially useful in cleaning objects with hard-to-reach recesses, such as rings and other jewelry. Ultrasonic cleaning baths for false teeth are also commercially available, as are ultrasonic toothbrushes. The latter transmit 1.6-MHz wave action to the teeth and gums to help remove bacterial plaque. There are also "sonic" toothbrushes that vibrate with a frequency of 18 kHz.

The **speed of sound** in a particular medium depends on the makeup of the material. A common medium is air. The speed of sound (v_s in air at 20°C) is

$$v_s = 344 \text{ m/s} \ (770 \text{ mi/h})$$

or approximately $\frac{1}{3}$ km/s, or $\frac{1}{5}$ mi/s.

The speed of sound increases with increasing temperature. For example, it is 331 m/s at 0°C and 344 m/s at 20°C. The speed of sound in air is much, much less than the speed of light.

The relatively slow speed of sound in air may be observed at a baseball game. A spectator sitting far from home plate may see a batter hit the ball but not hear the "crack" of the bat until slightly later. Similarly, a lightning flash is seen almost instantaneously, but the resulting thunder comes rumbling along afterward at a speed of about $\frac{1}{5}$ mi/s.

By counting the seconds between seeing a lightning flash and hearing the thunder, you can estimate your distance from the lightning or the storm's center (where lightning usually occurs). For example, if 5 seconds elapsed, then the storm's center is at a distance of approximately $\frac{1}{3}$ km/s × 5 s (= 1.6 km), or $\frac{1}{5}$ mi/s × 5 s (= 1.0 mi).

In general, as the density of the medium increases, the speed of sound increases. The speed of sound in water is about 4 times greater than in air, and in general, the speed of sound in solids is about 15 times greater than in air.

By using the speed of sound and frequency, the wavelength of a sound wave can easily be computed.

EXAMPLE 6.3   Computing the Wavelength of Ultrasound

What is the wavelength of a sound wave in air at 20°C with a frequency of 22 MHz?

Solution

The speed of sound at 20°C is $v_s = 344$ m/s (given previously), and the frequency is $f = 22$ MHz $= 22 \times 10^6$ Hz, which is in the ultrasonic region.

To find the wavelength λ, Eq. 6.3 can be used with the speed of sound:

$$\lambda = \frac{v_s}{f}$$

$$= \frac{344 \text{ m/s}}{22 \times 10^6 \text{ Hz (1/s)}} = 1.6 \times 10^{-5} \text{ m}$$

This wavelength of ultrasound is on the order of particle size and can be used in ultrasonic cleaning baths, as described earlier.

Confidence Exercise 6.3

What is the wavelength of an infrasonic sound wave in air at 20°C with a frequency of 10.0 Hz? (How does this wavelength compare with ultrasound wavelengths?)

Did You Learn?

- The human hearing frequency range is 20 Hz to 20 kHz.
- The dB scale is nonlinear, and an increase of 3 dB doubles the sound intensity. For example, a sound with 63 dB has twice the intensity of a sound with 60 dB.

6.5   The Doppler Effect

Preview Questions

- What is the Doppler effect?
- What is necessary for a jet aircraft to generate a sonic boom?

When watching a race and a racing car with a loud engine approaches, one hears an increasing high-frequency "whee." When the car passes by, the frequency suddenly shifts

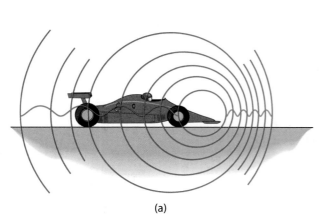

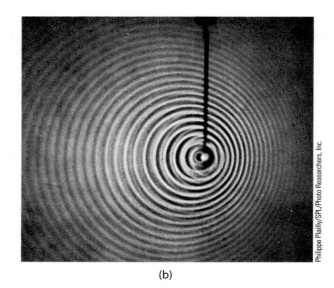

Philippe Plailly/SPL/Photo Researchers, Inc.

(a)   (b)

Figure 6.15 Doppler Effect
(a) Because of the motion of a sound source, illustrated here as a racing car, sound waves are "bunched up" in front and "spread out" in back. The result is shorter wavelengths (higher frequencies) in front of the sources and longer wavelengths (lower frequencies) behind the source. (b) The Doppler effect in water waves in a ripple tank. The source of the disturbance is moving to the right.

lower and a low-pitched "whoom" sound is heard. Similar frequency changes may be heard when a large truck passes by. The apparent change in the frequency of the moving source is called the **Doppler effect**.* The reason for the observed change in frequency (and wavelength) of a sound is illustrated in ● Fig. 6.15.

As a moving sound source approaches an observer, the waves are "bunched up" in front of the source. With the waves closer together (shorter wavelength), an observer perceives a higher frequency. Behind the source the waves are spread out, and with an increase in wavelength, a lower frequency is heard ($f = v/\lambda$). If the source is stationary and the observer moves toward and passes the source, then the shifts in frequency are also observed. Hence, the Doppler effect depends on the *relative* motion of the source and the observer.

Waves propagate outward in front of a source as long as the speed of sound, v_s, is greater than the speed of the source, v (● Fig. 6.16a). However, as the speed of the source approaches the speed of sound in a medium, the waves begin to bunch up closer and closer. When the speed of the source exceeds the speed of sound in the medium, a V-shaped bow wave is formed. Such a wave is readily observed for a motorboat traveling faster than the wave speed in water.

In air, when a jet aircraft travels at a supersonic speed (a speed greater than the speed of sound in air), a bow wave is in the form of a conical shock wave that trails out and downward from the aircraft. When this high-pressure, compressed wave front passes over an observer, a *sonic boom* is heard. The bow wave travels with the supersonic aircraft and does not occur only at the instant the aircraft "breaks the sound barrier" (first exceeds the speed of sound). There are actually two booms, because shock waves are formed at both the front and the tail of the plane (Fig. 6.16c).

Conceptual Question and Answer

Faster Than Sound

Q. If a jet pilot is flying faster than the speed of sound, is he or she able to hear sound?

A. Yes. The air in the cockpit is traveling with the pilot and is relatively stationary, so sound would be heard in the cockpit if not obstructed by communication headsets.

*Named after Christian Doppler (1803–1853), an Austrian physicist who first described the effect.

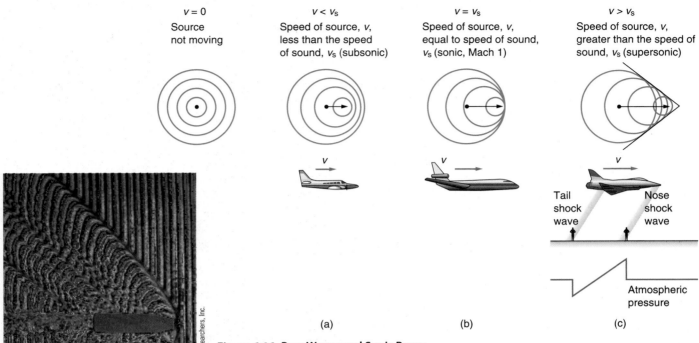

Figure 6.16 **Bow Waves and Sonic Boom**
Just as a moving boat forms a bow wave in water, a moving aircraft forms a bow wave in the air.
The sound waves bunch up in front of the airplane for increasing sonic speeds [(a) and (b)]. A
plane traveling at supersonic speeds forms a high-pressure shock wave in the air (c) that is heard
as a sonic boom when the plane passes over the observer. Actually, shock waves are formed
at both the nose and the tail of the aircraft. (d) This bullet is traveling at the speed of 500 m/s.
Notice the shock waves produced (and the turbulence behind the bullet).

You may hear jet aircraft speeds expressed as Mach 1 or Mach 2. The number Mach 1 is
equal to the speed of sound, Mach 2 is equal to twice the speed of sound, and so on. This
Mach number is named after Ernst Mach (1838–1916), an Austrian physicist who studied
supersonics.

On a smaller scale, you have probably heard a "mini" sonic boom, the crack of a whip.
When a whip is given a flick of the wrist, a wave pulse travels down the length of the whip.
Whips generally taper down from the handle to the tip, and the pulse speed increases the
thinner the whip. Traveling the length of the whip, the speed increases until it is greater
than the speed of sound. The "crack" is made by the air rushing back into the region of
reduced pressure created by the final supersonic flip of the whip's tip, much as the sonic
boom emanates from a shock trail behind a supersonic aircraft.

The Doppler effect is general and occurs for all kinds of waves, including water waves,
sound waves, and light waves. Because the Doppler effect can be used to detect and provide information on moving objects, it is used to examine blood flow in arteries and veins.
In this application ultrasound reflects from moving red blood cells with a change in frequency according to the speed of the cells. Information about the speed of blood flow helps
physicians to diagnose such things as blood clots and arterial closing.

For example, the Doppler effect can be used to assess the risk of stroke. Accumulations
of plaque deposits on the inner walls of blood vessels can restrict blood flow. A major cause
of stroke is obstruction of the carotid artery in the neck, which supplies blood to the brain.
The presence and severity of such obstructions may be detected by using ultrasound. An
ultrasonic generator (transducer), which generates high-frequency ultrasonic pulses, is
placed on the neck. Reflections from the red blood cells moving through the artery are
monitored to determine the rate of blood flow, providing an indication of and the severity
of any blockage.

In the Doppler effect for visible light, the frequency is shifted toward the blue end of the spectrum when the light source (such as a star) is approaching. (Blue light has a shorter wavelength, or higher frequency.) In this case we say that a Doppler *blueshift* has occurred.

When a stellar light source is moving away, the frequency is shifted toward the red (longer wavelength) end of the spectrum and a Doppler **redshift** occurs. The magnitude of the frequency shift is related to the speed of the source. The rotations of the planets and stars can be established by looking at the Doppler shifts from opposite sides; one is receding (redshift), and the other approaching (blueshift). Also, Doppler shifts of light from stars in our Milky Way galaxy indicate that the galaxy is rotating.

Light from other galaxies shows redshifts, which indicate that they are moving away from us according to the Doppler effect. By modern interpretations, however, it is not a Doppler shift but rather a shift in the wavelength of light influenced by the expansion of the universe. The wavelength of light expands along with the universe, giving a *cosmological redshift* (see Chapter 18.7).

The Doppler shift of waves also is applied here on the Earth. You may have experienced one such application. Radar (radio waves), which police officers use in determining the speed of moving vehicles, uses the Doppler effect.

Did You Learn?

- The Doppler effect is the apparent change in sound frequency of a moving source and a stationary observer (or a moving observer and stationary source).

- Jet aircraft with a speed greater than the speed of sound in air ($M > 1$) can generate a sonic boom.

6.6 Standing Waves and Resonance

Preview Questions

- Are the particles in a standing wave really "standing" or stationary?

- What does resonance mean in terms of a system's energy?

You may have shaken one end of a stretched cord or rope and observed wave patterns that seem to "stand" along the rope when it is shaken just right. These "stationary" waveforms are referred to as **standing waves** (● Fig. 6.17).

Standing waves are caused by the interference of waves traveling down and reflected back along the rope. When two waves meet, they interfere, and the combined waveform of the superimposed waves is the sum of the waveforms or particle displacements of the medium. Some points on the rope, called *nodes*, remain stationary at all times. At these points, the displacements of the interfering waves are equal and opposite and cancel each other completely such that the rope has zero displacement there. At other points, called *antinodes*, the individual amplitudes add to give the maximum amplitudes of the combined wave forms.

Notice that the string in Fig. 6.17 vibrates in standing wave modes only at *particular* frequencies (as evidenced by the number of loops or half wavelengths). These frequencies are referred to as the *characteristic*, or *natural*, *frequencies* of the stretched string.

When a stretched string or an object is acted on by a periodic driving force with a frequency equal to one of the natural frequencies, the oscillations have large amplitudes. This phenomenon is called **resonance**, and in this case there is maximum energy transfer to the system.

A common example of driving a system in resonance is pushing a swing. A swing is essentially a pendulum with only one natural frequency, which depends on the rope

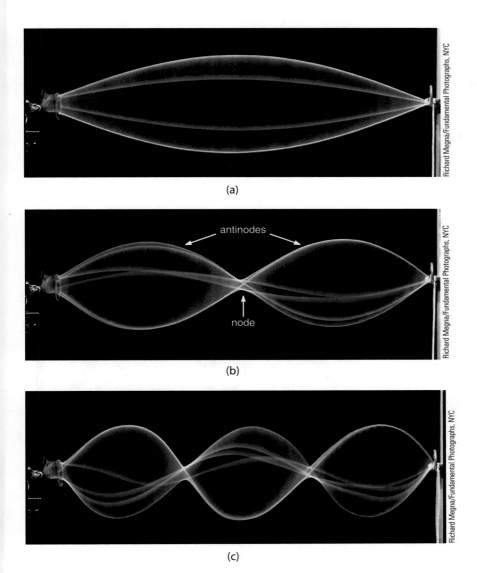

(a)

antinodes

node

(b)

(c)

Richard Megna/Fundamental Photographs, NYC

Figure 6.17 Standing Waves
Actual standing waves in a stretched rubber string. Standing waves are formed only when the string is vibrated at particular frequencies. Note the differences in wavelength; in (a), one-half wavelength ($\lambda/2$); in (b), one wavelength (λ); and in (c), a wavelength and a half ($3\lambda/2$).

length. When a swing is pushed periodically with a period of $T = 1/f$, energy is transferred to the swing and its amplitude gets larger (higher swings). If the swing is not pushed at its natural frequency, then the pushing force may be applied as the swing approaches or after it has reached its maximum amplitude. In either case the swing is not driven in resonance (at its natural frequency).

A stretched string has many natural frequencies, not just one like a pendulum. When the string is shaken, waves are generated, but when the driving frequency corresponds to one of the natural frequencies, the amplitude at the antinodes will be larger because of resonance.

There are many examples of resonance. The structures of the throat and nasal cavities give the human voice a particular tone as a result of resonances. Tuning forks of the same frequency can be made to resonate, as illustrated in ● Fig. 6.18a.

A steel bridge or most any structure is capable of vibrating at natural frequencies, sometimes with dire consequences (Fig. 6.18b). Soldiers marching in columns across bridges are told to break step and not march at a periodic cadence that might correspond to a natural frequency of the bridge and result in resonance and large oscillations that could cause structural damage. In 1850 in France, about 500 soldiers marching across a suspension bridge over a river caused a resonant vibration that rose to such a level that the bridge collapsed. More than 200 of the soldiers drowned.

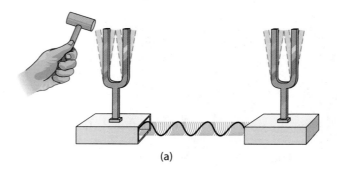

(a)

(b)

ETPM0096816   Tacoma, WA: November 7, 1940. Large section of the concrete roadway in the center span of the new Tacoma Narrows Bridge, aka "Galloping Gertie" hurtled into Puget Sound. High winds caused the bridge to sway, undulate and finally collapse under the strain. ©Topham / The Image Works   NOTE: The copyright notice must include "The Image Works" DO NOT SHORTEN THE NAME OF THE COMPANY

(c)

Figure 6.18 Resonance
(a) When one tuning fork is activated, the other tuning fork of the same frequency is driven in resonance and starts to vibrate. (b) and (c) Unwanted resonance. The famous Tacoma Narrows Bridge in the state of Washington collapsed in 1940 after wind drove the bridge into resonance vibrations.

Conceptual Question and Answer

It Can Be Shattering

Q. Opera singers are said to be able to shatter crystal glasses with their voices. How is this possible?

A. A few opera singers with powerful voices have been able to shatter glasses (● Fig. 6.19). If you wet your finger and move it around the rim of a crystal glass with a little pressure, you can get the glass to "sing" or vibrate at its natural frequency. The wavelength of the sound is the same as the distance around the rim of the glass. (If you ping the glass with your finger, then you'll hear the resonance frequency.)

   If a singer is able to sustain an intense tone with the proper frequency, then the resonance energy transfer may increase the amplitude of the vibrations of the glass to the point it shatters. (Try this resonance with your finger, not your singing.)

Musical instruments use standing waves and resonance to produce different tones. On stringed instruments such as the guitar, violin, and piano, standing waves are formed on strings fixed at both ends. When a stringed instrument is tuned, a string is tightened or loosened, which adjusts the tension and the wave speed in the string. This adjustment changes the frequency with the length of the string being fixed. Different musical notes (frequencies) on a guitar or violin are obtained by placing a finger on a string, which effectively shortens its length (● Fig. 6.20).

A vibrating string does not produce a great disturbance in air, but the body of a stringed instrument such as a violin acts as a sounding board and amplifies the sound. Thus, the body of such an instrument acts as a resonance cavity, and sound comes out through holes in the top surface.

In wind instruments standing waves are set up in air columns. Organ pipes have fixed lengths just as fixed strings do, so only a certain number of wavelengths can be fitted in the tube. However, the length of an air column and the frequency or tone can be varied in some instruments, such as a trombone, by varying the length of the column.

The *quality* (or timbre) of a sound depends on the waveform or the number of waveforms present. For example, you can sing the same note as a famous singer, but a different combination of waveforms gives the singer's voice a different quality, perhaps a pleasing "richness." It is the quality of our voices that gives them different sounds.

Figure 6.19 Shattering Glass
Some singers with the right frequency and intensity can cause a glass to vibrate in resonance and shatter.

Did You Learn?

- In a standing wave, opposite moving particles continually interfere, with the nodes being the only stationary points.
- At resonance frequencies there is maximum energy transfer to a system.

Figure 6.20 Different Notes
Placing the fingers on the strings at different locations effectively shortens the lengths and changes the musical notes (frequencies).

KEY TERMS

1. waves (6.1)
2. longitudinal wave (6.2)
3. transverse wave
4. wavelength
5. amplitude
6. frequency
7. hertz
8. period
9. wave speed
10. electromagnetic spectrum (6.3)
11. speed of light
12. sound (6.4)
13. sound spectrum
14. intensity
15. decibel
16. ultrasound
17. speed of sound
18. Doppler effect (6.5)
19. redshift
20. standing waves (6.6)
21. resonance

MATCHING

For each of the following items, fill in the number of the appropriate Key Word from the preceding list. Compare your answers with those at the back of the book.

a. _____ Particle motion and wave velocity parallel

b. _____ In air, its value is 344 m/s at 20°C

c. _____ A spectrum of waves, including visible light

d. _____ Maximum displacement of wave particle

e. _____ Indicates movement of a receding source

f. _____ Time to travel one wavelength

g. _____ Waveforms caused by wave interference

h. _____ Longitudinal waves that propagate through matter

i. _____ Propagation of energy after a disturbance

j. _____ Maximum energy transfer to a system

k. _____ Rate of transfer of sound energy through a given area

l. _____ Number of oscillations per time

m. _____ $f > 20,000$ Hz

n. _____ 3.00×10^8 m/s

o. _____ Particle motion perpendicular to wave velocity

p. _____ Equal to λ/T

q. _____ Regions of sound

r. _____ Unit of sound intensity level

s. _____ Apparent change of frequency because of relative motion

t. _____ Unit equivalent to 1/s

u. _____ Distance between two wave maxima

MULTIPLE CHOICE

Compare your answers with those at the back of the book.

1. A wave with particle oscillation parallel to the direction of propagation is a(n) _____ . (6.2)
   (a) transverse wave
   (b) longitudinal wave
   (c) light wave
   (d) none of the preceding

2. If a piece of ribbon were tied to a stretched string carrying a transverse wave, then how is the ribbon observed to oscillate? (6.2)
   (a) perpendicular to wave direction
   (b) parallel to wave direction
   (c) neither (a) nor (b)
   (d) both (a) and (b)

3. The energy of a wave is related to the square of which of the following? (6.2)
   (a) amplitude
   (b) frequency
   (c) wavelength
   (d) period

4. How fast do electromagnetic waves travel in vacuum? (6.3)
   (a) 3.00×10^8 m/s
   (b) 9.8 m/s²
   (c) 344 m/s
   (d) 3.44×10^6 m/s

5. Which of the following is true for electromagnetic waves? (6.3)
   (a) They have different speeds in vacuum for different frequencies.
   (b) They are longitudinal waves.
   (c) They require a medium for propagation.
   (d) None of the preceding is true.

6. Which one of the following regions has frequencies just slightly less than the visible region in the electromagnetic frequency spectrum? (6.3)
   (a) radio wave
   (b) ultraviolet
   (c) microwave
   (d) infrared

7. The speed of sound is generally greatest in _____ . (6.4)
   (a) gases   (b) liquids   (c) solids   (d) vacuum

8. Which of the following sound frequencies would not be heard by the human ear? (6.4)
   (a) 25 Hz
   (b) 900 Hz
   (c) 20 kHz
   (d) 25 kHz

9. A sound with an intensity level of 30 dB is how many times louder than the threshold of hearing? (6.4)
   (a) 10   (b) 3000   (c) 100   (d) 1000

10. A moving observer approaches a stationary sound source. What does the observer hear? (6.5)
    (a) an increase in frequency
    (b) a decrease in frequency
    (c) the same frequency as the source

11. Which of the following properties does *not* change in the Doppler effect?
    (a) wavelength
    (b) speed
    (c) frequency
    (d) period

12. If an astronomical light source were moving toward us, then what would be observed? (6.5)
    (a) a blueshift
    (b) a shift toward longer wavelengths
    (c) a shift toward lower frequencies
    (d) a sonic boom

13. Which of the following occur(s) when a stretched string is shaken at one of its natural frequencies? (6.6)
    (a) standing waves
    (b) resonance
    (c) maximum energy transfer
    (d) all of the preceding

FILL IN THE BLANK

Compare your answers with those at the back of the book.

1. Waves involve the propagation of ___. (6.1)
2. Wave velocity and particle motion are ___ in transverse waves. (6.2)
3. The distance from one wave crest to an adjacent wave crest is called a(n) ___. (6.2)
4. Wave speed is equal to the wavelength times the ___. (6.2)
5. In vacuum electromagnetic waves travel at the speed of ___. (6.3)
6. Light waves are ___ waves. (6.3)
7. Sound waves are ___ waves. (6.4)
8. The audible region is above a frequency of ___ Hz. (6.4)
9. Decibels are used to measure the sound property of ___. (6.4)
10. To double the loudness, or sound intensity, a dB difference of ___ is needed. (6.4)
11. In the Doppler effect, when a moving sound source approaches a stationary observer, the apparent shift in frequency is ___. (6.5)
12. A Doppler blueshift in light from a star indicates that the star is ___. (6.5)
13. Resonance occurs at ___ frequencies. (6.6)

SHORT ANSWER

6.1 Waves and Energy Propagation

1. What is meant by a "wave"?
2. Do all waves require a medium to propagate? Explain.

6.2 Wave Properties

3. A wave travels upward in a medium (vertical wave velocity). What is the direction of particle oscillation for (a) a longitudinal wave and (b) a transverse wave?
4. What are the SI units for (a) wavelength, (b) frequency, (c) period, and (d) amplitude?
5. How many values of amplitude are there in one wavelength of a wave, and how is the amplitude related to the energy of a wave?
6. A displacement-versus-time graph for a wave form is shown in ● Fig. 6.21. What are the (a) amplitude and (b) frequency of the wave?

6.3 Light Waves

7. With what speed do electromagnetic waves propagate in vacuum?

8. Which end (blue or red) of the visible spectrum has the longer wavelength? Which has the higher frequency?
9. Are radio waves sound waves? Explain.
10. What is the range of wavelengths of visible light? How do these wavelengths compare with those of audible sound?

6.4 Sound Waves

11. What is a rarefaction?
12. What happens to the energy when a sound "dies out"?
13. Referring to Fig. 6.11, indicate over how many squares the sound waves would spread for $r = 5$ m. The sound intensity would decrease to what fraction in value?
14. What is the chief physical property that describes (a) pitch, (b) loudness, and (c) quality?
15. Why does the music coming from a band marching in a spread-out formation on a football field sometimes sound discordant?
16. What is the difference between sound wave energy and intensity?
17. Does doubling the decibels of a sound intensity level double the intensity? Explain.
18. Why lightning seen before thunder is heard?

6.5 The Doppler Effect

19. How is the wavelength of sound affected when (a) a sound source moves toward a stationary observer and (b) the observer moves away from a stationary sound source?
20. Under what circumstances would sound have (a) a Doppler "blueshift"? (b) a Doppler "redshift"?

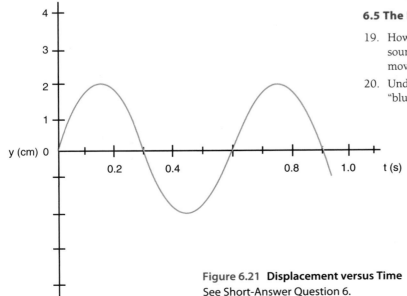

Figure 6.21 Displacement versus Time
See Short-Answer Question 6.

21. On a particular day the speed of sound in air is 340 m/s. If a plane flies at a speed of 680 m/s, is its Mach number (a) 1.5, (b) 2.0, (c) 2.5, or (d) 2.7?

22. Radar and sonar are based on similar principles. Sonar (which stands for *sound navigation and ranging*) uses ultrasound, and radar (which stands for *radio detecting and ranging*) uses radio waves. Explain the principle of detecting and ranging in these applications.

6.6 Standing Waves and Resonance

23. What is the effect when a system is driven in resonance? Is a particular frequency required? Explain.

24. Would you expect to find a node or an antinode at the end of a plucked guitar string? Explain.

25. What determines the pitch or frequency of a string on a violin or a guitar? How does a musician get a variety of notes from one string?

VISUAL CONNECTION

Visualize the connections and give answers for the blanks. Compare your answers with those at the back of the book.

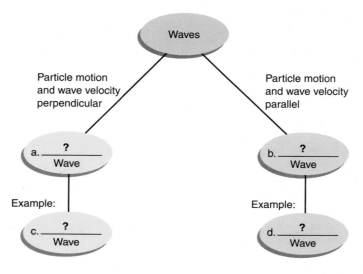

APPLYING YOUR KNOWLEDGE

1. The speed of sound in air is given for still air. How would wind affect the speed? What would happen if you shouted into the wind?

2. Were an astronaut on the Moon to drop a hammer, would there be sound? Explain. (*Follow-up*: How do astronauts communicate with one another and with mission control?)

3. We cannot see ultraviolet (UV) radiation, which can be dangerous to the eyes, but we can detect some UV radiations (without instruments). How is this done?

4. How fast would a "jet" fish have to swim to create an "aquatic" boom?

5. When one sings in the shower, the tones sound full and rich. Why is this?

6. Standing waves are set up in organ pipes. Consider an organ pipe with a fixed length, which may be open or closed. An open pipe is open at both ends, and a closed pipe is closed at one end and open at the other. A standing wave will have an antinode at an open end and a node at a closed end (● Fig. 6.22). In the figure determine (draw) the wavelengths of the first three characteristic frequencies for the pipes (that is, the first three wavelength segments that will "fit" into the pipes). Express your answers in terms of the length L of the pipe, such as $L = \lambda/3$ or $\lambda = 3L$ (not correct values).

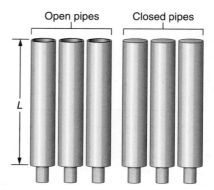

Figure 6.22 Organ Pipes
See Question 6 in Applying Your Knowledge.

IMPORTANT EQUATIONS

Frequency–Period Relationship: $f = \dfrac{1}{T}$ (6.1)

Wave Speed: $v = \dfrac{\lambda}{T} = \lambda f$ (6.2–6.3)

Speed of Light: $c = 3.00 \times 10^8$ m/s

EXERCISES

6.2 Wave Properties

1. A periodic wave has a frequency of 5.0 Hz. What is the wave period?

   Answer: 0.20 s

2. What is the period of the wave motion for a wave with a frequency of 0.25 kHz?

3. Waves moving on a lake have a speed of 2.0 m/s and a distance of 1.5 m between adjacent crests. (a) What is the frequency of the waves? (b) Find the period of the wave motion.

   Answer: (a) 1.3 Hz (b) 0.77 s

4. A sound wave has a frequency of 2000 Hz. What is the distance between crests or compressions of the wave? (Take the speed of sound to be 344 m/s.)

6.3 Light Waves

5. Compute the wavelength of the radio waves from (a) an AM station operating at a frequency of 650 kHz and (b) an FM station with a frequency of 95.1 MHz.

   Answer: (a) 4.62×10^2 m (b) 3.2 m

6. Compute the wavelength of an X-ray with a frequency of 10^{18} Hz.

7. What is the frequency of blue light that has a wavelength of 420 nm?

   Answer: 7.14×10^{14} Hz

8. An electromagnetic wave has a wavelength of 6.00×10^{-6} m. In what region of the electromagnetic spectrum is this radiation?

9. How far does light travel in 1 year? [This distance, known as a light-year (ly), is used in measuring astronomical distances (Ch. 18.1).]

   Answer: $9.4.8 \times 10^{12}$ km (about 6 trillion mi)

10. Approximately how long would it take a telephone signal to travel 3000 mi from coast to coast across the United States? (Telephone signals travel at about the speed of light.)

6.4 Sound Waves

11. Compute the wavelength in air of ultrasound with a frequency of 50 kHz if the speed of sound is 344 m/s.

    Answer: 6.9×10^{-3} m

12. What are the wavelength limits of the audible range of the sound spectrum? (Use the speed of sound in air.)

13. The speed of sound in a solid medium is 15 times greater than that in air. If the frequency of a wave in the solid is 20 kHz, then what is the wavelength?

    Answer: 0.26 m

14. A sound wave in a solid has a frequency of 15.0 kHz and a wavelength of 0.333 m. What would be the wave speed, and how much faster is this speed than the speed of sound in air?

15. During a thunderstorm, 4.5 s elapses between observing a lightning flash and hearing the resulting thunder. Approximately how far away, in kilometers and miles, was the lightning flash?

    Answer: 1.5 km or 0.90 mi

16. Picnickers see a lightning flash and hear the resulting thunder 9.0 s later. If the storm is traveling at a rate of 15 km/h, then how long, in minutes, do the picnickers have before the storm arrives at their location?

17. A subway train has a sound intensity level of 90 dB, and a rock band has a sound intensity level of about 110 dB. How many times greater is the sound intensity of the band than that of the train?

    Answer: 100 times

18. A loudspeaker has an output of 70 dB. If the volume of the sound is turned up so that the output intensity is 10,000 times greater, then what is the new sound intensity level?

ON THE WEB

1. Introduction to Waves

Do you now have a grasp of the fundamentals of waves? What is wave motion? Can you distinguish between a longitudinal wave and a transverse wave? What is the speed of light? Visit the student website at **www.cengagebrain.com/shop/ISBN/1133104096** and follow the recommended links to answer these questions. Also, use the worksheets on Vibration and Waveform Graph and Wave Vocabulary to reinforce what you know.

2. The Doppler Effect

Have you ever wondered about that shift in sound frequency as another vehicle is approaching and then passing yours? How does the speed of sound differ when the sound wave travels through the three different media of gas, liquid, and solid? How does the speed of an airplane affect the sound? How do stationary and moving sound sources differ? What creates the sonic boom as a plane breaks through the sound barrier? To explore the Doppler effect and sonic booms and to answer the questions above, follow the recommended links on the student website at **www.cengagebrain.com/shop/ISBN/1133104096**.

Optics and Wave Effects

*Music is the arithmetic
of sound as optics is
the geometry of light.*

•

Claude Debussy
(1862–1918)

Because of refraction, the pencil >
appears to be severed.

Bill Beatty/Visuals Unlimited

PHYSICS FACTS

▶ Some emergency vehicles have AMBULANCE printed on the front. This is so that the word AMBULANCE can be read in the rearview mirrors of vehicles ahead ("right–left" reversal).

▶ The "right–left" reversal of a plane mirror is really a "front–back" reversal (Chapter 7.1).

You look into a mirror and see your image. Many of us wear eyeglasses (lenses) to be able to see more clearly. These common applications involve light and vision, which are the basis for the branch of physical science known as *optics*. To be able to describe light phenomena is of great importance because most of the information we receive about our physical environment involves light and sight (see Chapter 1.3). Optics is generally divided into two areas: *geometrical (ray) optics* and *physical (wave) optics*.

Geometrical optics uses lines, or "light rays," to explain phenomena such as reflection and refraction, which are the principles of mirrors and lenses, respectively. The majority of our mirrors are plane mirrors, like the flat mirrors used in bathrooms and dressing rooms. There are other types of mirrors as well, such as

Chapter Outline

cosmetic mirrors that produce magnified images. This type of mirror is curved, so geometry comes into play in describing mirrors.

The same goes for lenses. There are many different shapes of lenses that produce different effects. An optometrist must know these effects when prescribing lenses for eyeglasses (or contact lenses). And lenses are central components of a most important optical instrument: the human eye. Geometrical optics also explains such things as fiber optics and the brilliance of diamonds.

Physical optics, on the other hand, takes into account the wave effects that geometrical optics ignores. Wave theory leads to satisfactory explanations of such phenomena as polarization, interference, and diffraction. You may wear polarizing sunglasses to reduce glare. When you see a colorful display in a soap bubble, interference is occurring. When you speak loudly, a person out of sight around the corner of a doorway can hear you, indicating that sound waves "bend," or are diffracted around corners. You can be heard in the next room, but you are not seen. Light is diffracted also but not enough for us to observe in most cases.

Also discussed in this chapter are liquid crystal displays (LCDs), which you probably have on a calculator or watch and may have on a computer or TV monitor. So let's get started with the fascinating study of optics.

7.1   Reflection

Preview Questions

- What is reflection?
- Does the law of reflection apply to both specular (regular) and diffuse (irregular) reflections?

Light waves travel through space in a straight line and will continue to do so unless diverted from the original direction. A change in direction takes place when light strikes and rebounds from a surface. A change in direction by this method is called **reflection**.

Reflection may be thought of as light "bouncing off" a surface. However, it is much more complicated and involves the absorption and emission of complex atomic vibrations of the reflecting medium. To describe reflection simply, the reflection of rays is considered, which ignores the wave nature of light. A **ray** is a straight line that represents the path of light with a directional arrowhead.

An incident light ray is reflected from a surface in a particular way. As illustrated in ● Fig. 7.1, the angles of the incident and reflected rays (θ_i and θ_r) are measured relative to the *normal*, a line perpendicular to the reflecting surface. These angles are related by the **law of reflection**:

The angle of reflection θ_r is equal to the angle of incidence θ_i.

Also, the reflected and incident rays are in the same plane.

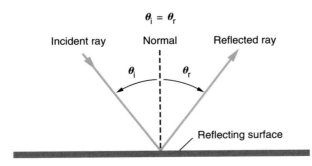

Figure 7.1 Law of Reflection
The angle of reflection θ_r is equal to the angle of incidence θ_i relative to the normal (a line perpendicular to the reflecting surface). The rays and the normal line lie in the same plane.

Regular reflection

(a)

Diffuse reflection

(b)

Figure 7.2 Reflection

(a) A smooth (mirror) surface produces specular (or regular) reflection. (b) A rough surface produces diffuse (or irregular) reflection. Both reflections obey the law of reflection.

The reflection from very smooth (mirror) surfaces is called **specular reflection** (or regular reflection, ● Fig. 7.2a). In specular reflection, incident parallel rays are parallel on reflection. In contrast, rays reflected from relatively rough surfaces are not parallel but scattered. This reflection is called **diffuse reflection** (or irregular reflection, Fig. 7.2b). The reflection from the pages of this book is diffuse. The law of reflection applies to both types of reflection, but the rough surface causes the incoming light rays to be reflected in different directions.

Conceptual Question and Answer

No Can See

Q. Some tractor-trailers have a sign on the back, "If you can't see my mirror, I can't see you." What does this mean?

A. For the trucker to see you in his rearview mirror, you must be in a position to see the mirror. The law of reflection works both ways; the rays are just reversed.

Rays may be used to determine the image formed by a mirror. A *ray diagram* for determining the apparent location of an image formed by a plane mirror is shown in ● Fig. 7.3. The image is located by drawing two rays emitted by the object and applying the law of reflection. Where the rays intersect, or appear to intersect, locates the image. For a plane mirror, the image is located behind, or "inside," the mirror at the same distance as the object is in front of the mirror.

Figure 7.3 Ray Diagram

By tracing the reflected rays, it is possible to locate a mirror image where the rays intersect, or appear to intersect, behind the mirror.

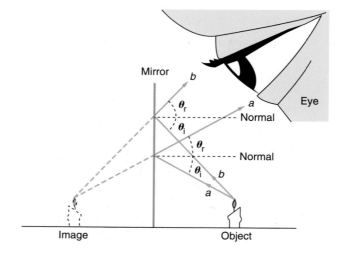

Image

Object

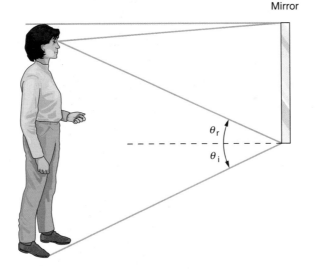

Mirror

θ_r

θ_i

Figure 7.4 Complete Figure
For a person to see his or her complete figure in a plane mirror, the height of the mirror must be at least one-half the height of the person, as can be demonstrated by ray tracing.

● Figure 7.4 shows a ray diagram for the light rays involved when a person sees a complete head-to-toe image. Applying the law of reflection, it can be shown that the total image is seen in a plane mirror that is only half the person's height. Also, the distance from the mirror is not a factor (see Exercise 3).

It is the reflection of light that enables us to see things. Look around you. What you see is light reflected from the walls, ceiling, floor, and other objects, which in general is diffuse reflection. Some beautiful specular reflections are seen in nature, as shown in ● Fig. 7.5.

Conceptual Question and Answer

Nighttime Mirror

Q. At night, a glass windowpane acts as a mirror viewed from the inside a lighted room. Why isn't it a mirror during the day?

A. When light strikes a transparent medium, most of it is transmitted but some is reflected. During the day, the light reflected from the inside of the window is overwhelmed by the light coming through the window from the outside. At night, though, when the light transmitted from the outside is greatly reduced, the inside reflections can be discerned and the windowpane acts as a mirror. (Can you now explain the principle of one-way mirrors and reflective sunglasses?)

Coco McCoy/Rainbow

Figure 7.5 Natural Reflection
Beautiful reflections, such as the one shown here on a still lake, are often seen in nature. Is the picture really printed right-side up? Turn it upside down and take a look. Without the tree limbs on the left, would you be able to tell?

Before leaving plane mirrors, let's consider another interesting aspect, commonly called a *right–left reversal*. When you raise your right hand when looking at a plane (bathroom-type) mirror, your image raises its left hand. This is really a *front–back reversal*. To understand, suppose you are facing north. Then your mirror image "faces" south. That is, your image has its front to the south and its back to the north: a front–back reversal. This reversal can be demonstrated by asking a friend to stand facing you (without a mirror). When your friend raises a right hand, you can see that the hand is actually on your left side.

Did You Learn?

● Reflection is a change in direction that takes place when light strikes a surface and rebounds.

● The law of reflection applies to both specular (parallel) reflection and diffuse (scattered) reflection.

7.2 Refraction and Dispersion

Preview Questions

● What causes light refraction, and what does the index of refraction (*n*) express?

● Why is white light separated into colors when passing through a prism?

Refraction

When light strikes a transparent medium, some light is reflected and some is transmitted. This transmission is illustrated in ● Fig. 7.6 for a beam of light incident on the surface of a body of water. Upon investigation, you will find that the transmitted light has changed direction in going from one medium to another. The deviation of light from its original path arises because of a change in speed in the second medium. This effect is called **refraction**. You have probably observed refraction effects for an object in a glass of water. For example, a spoon or pencil in a glass of water will appear to be displaced and perhaps severed (see the chapter-opening photo).

The directions of the incident and refracted rays are expressed in terms of the angle of incidence θ_1 and the angle of refraction θ_2, which are measured relative to the normal, a line perpendicular to the surface boundary of the medium (● Fig. 7.7).* The different

Figure 7.6 Refraction in Action
A beam of light is refracted—that is, its direction is changed—on entering the water.

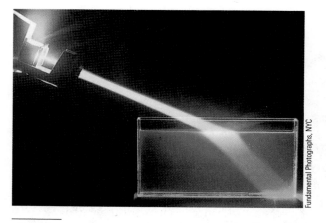

*For refraction, subscripts 1 and 2 are used for angles to distinguish from the i and r subscripts used for reflection.

speeds in the different media are expressed in terms of a ratio relative to the speed of light. This is known as the **index of refraction** n:

$$\text{index of refraction} = \frac{\text{speed of light in vacuum}}{\text{speed of light in medium}}$$

or

$$n = \frac{c}{c_m} \qquad\qquad 7.1$$

The index of refraction is a pure number (it has no units) because c and c_m are measured in the same units, which cancel. The indexes of refraction of some common substances are given in ● Table 7.1. Note that the index of refraction for air is close to that for vacuum.

When light passes obliquely ($0° < \theta_1 < 90°$) into a denser medium—for example, from air into water or glass—the light rays are refracted, or bent toward the normal ($\theta_2 < \theta_1$). It is the slowing of the light that causes this deviation. Complex processes are involved, but intuitively, we might expect the passage of light by atomic absorption and emission through a denser medium to take longer. For example, the speed of light in water is about 75% of its speed in air or vacuum, as shown in the following example.

EXAMPLE 7.1 Finding the Speed of Light in a Medium

What is the speed of light in water?

Solution

Step 1

Given: Nothing directly; therefore, quantities are known or available from tables.

Step 2

Wanted: Speed of light in a medium (water). This value may be found from Eq. 7.1. The speed of light c is known, $c = 3.00 \times 10^8$ m/s, and from Table 7.1, $n = 1.33$ for water.

Step 3

Rearranging Eq. 7.1 and doing the calculation,

$$c_m = \frac{c}{n} = \frac{3.00 \times 10^8 \text{ m/s}}{1.33}$$

$$= 2.26 \times 10^8 \text{ m/s}$$

This is about 75% of the speed of light in a vacuum (c), as can be seen by the ratio

$$\frac{c_m}{c} = \frac{1}{n} = \frac{1}{1.33} = 0.752 \ (= 75.2\%)$$

Confidence Exercise 7.1

According to Eq. 7.1, what is the speed of light in (a) a vacuum and (b) air, and how do they compare?

Answers to the Confidence Exercises may be found at the back of the book.

To help understand how light is bent, or refracted, when it passes into another medium, consider a band marching across a field and entering a wet, muddy region obliquely (at an angle), as illustrated in ● Fig. 7.8a. Those marchers who first enter the muddy region keep marching at the same cadence (frequency). Because they are slipping in the muddy earth, though, they don't cover as much ground and are slowed down (smaller wave speed).

The marchers in the same row on solid ground continue on with the same stride, and as a result, the direction of the marching column is changed as it enters the muddy region. This change in direction with change in marching speed is also seen when a marching band turns a corner and the inner members mark time.

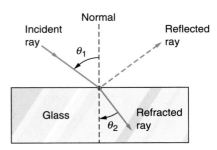

Figure 7.7 Refraction
When light enters a transparent medium at an angle, it is deviated, or refracted, from its original path. As illustrated here, when light passes from air into a denser medium such as glass, the rays are refracted, or "bent," toward the normal ($\theta_2 < \theta_1$; that is, the angle of refraction is less than the angle of incidence). Some of the light is also reflected from the surface, as indicated by the dashed ray.

Table 7.1	Indexes of Refraction of Some Common Substances
Substance	**n**
Water	1.33
Crown glass	1.52
Diamond	2.42
Air (0°C, 1 atm)	1.00029
Vacuum	1.00000

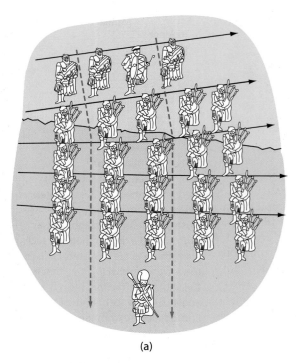

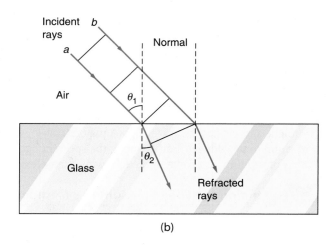

(a)   (b)

Figure 7.8 Refraction Analogy
(a) Marching obliquely into a muddy field causes a band column to change direction. The cadence (or frequency) remains the same, but the marchers in the mud slip and travel shorter distances (shorter wavelengths). This is analogous to the refraction of a wave front (b).

Wave fronts may be thought of as analogous to marching rows (Fig. 7.8b). In the case of light, the wave frequency (cadence) remains the same, but the wave speed is reduced, as is the wavelength ($c_m = \lambda_m f$, Eq. 6.3). The wavelength may be thought of as the distance covered in each step (shorter when slipping).

Conceptual Question and Answer

Twinkle, Twinkle

Q. On a clear night, stars are observed to "twinkle." What causes this?

A. The index of refraction of a gas varies with its density, which varies with temperature. At night, starlight passes through the atmosphere, which has temperature density variations and turbulence. As a result, the refraction causes the star's image to appear to move and vary in brightness, making the stars "twinkle."

A couple of other refraction effects are shown in ● Fig. 7.9. You probably have seen a "wet spot" mirage on a road on a hot day (Fig. 7.9a). No matter how long or how rapidly you travel toward the apparent water, the "spot" is never reached. As illustrated in the figure, the mirage is caused by refraction of light in the hot air near the road surface. This "water," which is the same illusion of thirst-quenching water that appears to desert travelers, is really a view of the sky via refracted sky light. Also, the variation in the density of the rising hot air causes refractive variations that allow hot air to be "seen" rising from the road surface. (Stop and think. You can't see air.)

Have you ever tried to catch a fish underwater and missed? Figure 7.9b shows why. We tend to think of our line of sight as a straight line, but light bends as a result of refraction at the air–water interface. Unless refraction is taken into account, the fish is not where it appears to be.

Finally, you may have noticed that the setting Sun appears flattened, as shown in ● Fig. 7.10a. This is a refractive effect. Light coming from the top and bottom portions is refracted differently as it passes through different atmospheric densities near the horizon, giving rise to a flattening effect. Light from the sides of the Sun is refracted the same, so there is no apparent side difference.

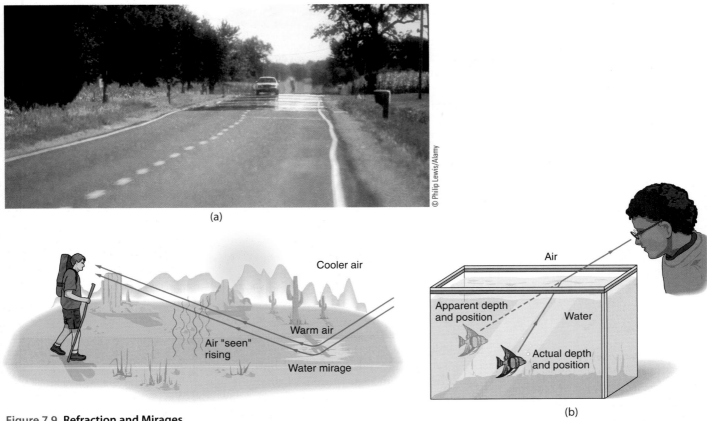

Figure 7.9 **Refraction and Mirages**

(a) A "wet spot" mirage is produced when light from the sky is refracted by warm air near the road surface (the index of refraction varies with temperature). Such refraction also enables us to perceive heated air "rising." Although air cannot be seen, the refraction in the turbulent updrafts causes variations in the light passing through. (b) Try to catch a fish. We tend to think of light as traveling in straight lines, but because light is refracted, the fish is not where it appears to be.

Figure 7.10 **Refraction Effects**

(a) The Sun appears flattened because light from the top and bottom portions is refracted as it passes through different atmospheric densities near the horizon. (b) The Sun is seen before it actually rises above the eastern horizon and after it sets below the western horizon because denser air near the Earth refracts the light over the horizon. (Drawing exaggerated for clarity.)

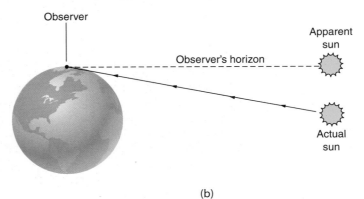

Atmospheric effects also lengthen the day in a sense. Because of refraction, the Sun is seen before it actually rises above the eastern horizon and after it sets below the western horizon. It may take as much as 20 minutes for rising and setting. The denser air near the Earth refracts the light over the horizon (Fig. 7.10b). The same effect occurs for the rising and setting of the Moon.

Reflection by Refraction

When light goes from a denser medium into a less dense medium (for example, from water or glass into air), the ray is refracted and bent away from the normal. This refraction can be seen by reverse ray tracing of the light shown in Fig. 7.7, this time as it goes from glass into air. ● Figure 7.11 shows this type of refraction for a light source in water.

Note that an interesting thing happens as the angle of incidence becomes larger. The refracted ray is bent farther from the normal, and at a particular *critical angle* θ_c, the refracted ray is along the boundary of the two media. For angles greater than θ_c, the light is reflected and not refracted. This phenomenon is called **total internal reflection**.

Refraction and total internal reflection are illustrated in ● Fig. 7.12a. With total internal reflection, a prism can be used as a mirror.

Internal reflection is used to enhance the *brilliance* of diamonds. In the so-called brilliant cut, light entering a diamond is totally reflected (Fig. 7.12b). The light emerging from the upper portion gives the diamond a beautiful sparkle.

Another example of total internal reflection occurs when a fountain of water is illuminated from below. The light is totally reflected within the streams of water, providing a spectacular effect. Similarly, light can travel along transparent plastic tubes called "light pipes." When the incident angle for light in the tube is greater than the critical angle, the light undergoes a series of internal reflections down the tube (● Fig. 7.13a).

Light can also travel along thin fibers, and bundles of such fibers are used in the field of *fiber optics* (Fig. 7.13b). You may have seen fiber optics used in decorative lamps. An important use of the flexible fiber bundle is to pipe light to hard-to-reach places. Light may also be transmitted down one set of fibers and reflected back through another so that an image of the illuminated area is seen. This illuminated area might be a person's stomach or heart in medical applications (Fig. 7.13c).

Fiber optics is also used in telephone communications. In this application electronic signals in wires are replaced by light (optical) signals in fibers. The fibers can be drawn out thinner than copper wire, and more fibers can be bundled in a cable for a greater number of calls.

Dispersion

The index of refraction for a material actually varies slightly with wavelength. When light is refracted, the different wavelengths of light are bent at slightly different angles. This

Figure 7.11 Internal Reflection
When light goes from a denser medium into a less dense medium, such as from water into air, as illustrated here, it is refracted away from the normal. At a certain critical angle θ_c, the angle of refraction is 90°. For incidence greater than the critical angle, the light is reflected internally.

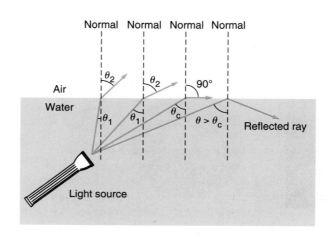

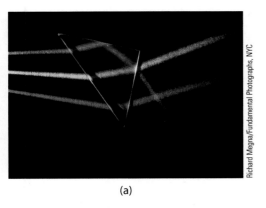

(a)

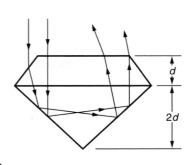

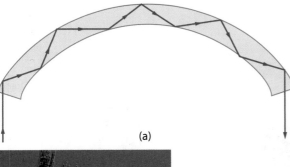

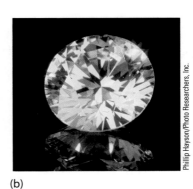

(b)

Figure 7.12 Refraction and Internal Reflection
(a) Beams of colored light are incident on a piece of glass from the left. As light passes from air into the glass, the beams are refracted. The incident angle of the blue beam at the glass–air interface exceeds the critical angle θ_c, and the beam is internally reflected. (b) Refraction and internal reflection give rise to the brilliance of a diamond. In the so-called brilliant cut, a diamond is cut with a certain number of faces, or facets, along with the correct depth to give the proper refraction and internal reflection.

(a)

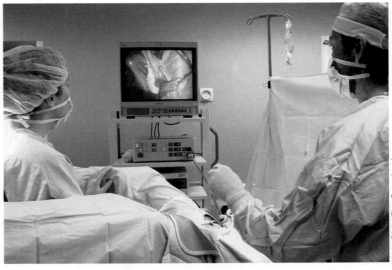

(b)

(c)

Figure 7.13 Fiber Optics
(a) When light that is incident on the end of a transparent tube exceeds the critical angle, it is internally reflected along the tube, which acts as a "light pipe." (b) A fiber-optic bundle held between a person's fingers. The ends of the fibers are lit up because of transmission of light by multiple internal reflections. (c) A fiber-optic application. Endoscopes are used to view various internal body parts. Light traveling down some fibers is reflected back through others, making it possible to view otherwise inaccessible places on a monitor.

phenomenon is called **dispersion**. When white light (light containing all wavelengths of the visible spectrum) passes through a glass prism, the light rays are refracted upon entering the glass (● Fig. 7.14). With different wavelengths refracted at slightly different angles, the light is *dispersed* into a spectrum of colors (wavelengths).

That the amount of refraction is a function of wavelength can be seen by combining the two equations $n = c/c_m$ and $c_m = \lambda f$. Substituting the second into the first for c_m, we have $n = c/\lambda f$, and the index of refraction n varies inversely with wavelength λ. So, shorter wavelengths have greater indexes of refraction and are diverted from their path by a greater amount. Blue light has a shorter wavelength than red light; hence, blue light is refracted more than red light, as shown in Fig. 7.14.

As an example of dispersion, a diamond is said to have "fire" because of colorful dispersion. This is in addition to having brilliance because of internal reflection. Rainbows, a natural phenomenon involving dispersion and internal reflection, are discussed in the **Highlight: The Rainbow: Dispersion and Internal Reflection**.

The fact that light can be separated into its component wavelengths provides an important investigative tool. This process is illustrated in ● Fig. 7.15a. Light from a source goes through a narrow slit, and when the light is passed through a prism, the respective wavelengths are separated into "line" images of the slit. The line images, representing definite wavelengths, appear as bright lines. A scale is added for measurement of the wavelengths of the lines in the spectrum, providing an instrument called a *spectrometer*. The line spectra of four elements are shown in Fig. 7.15b.*

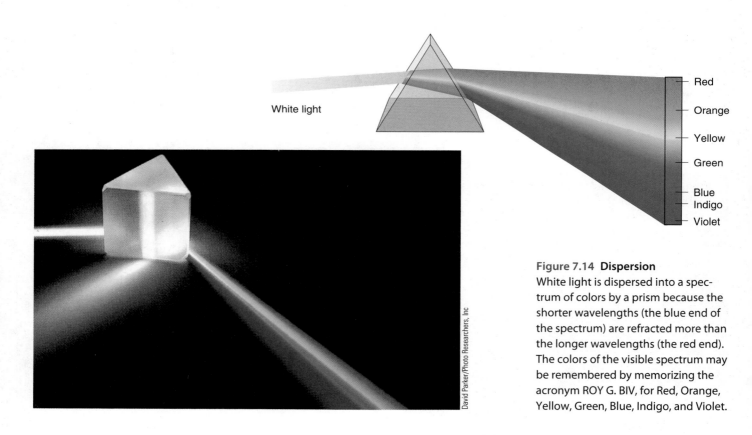

White light

Red
Orange
Yellow
Green
Blue
Indigo
Violet

David Parker/Photo Researchers, Inc

Figure 7.14 Dispersion
White light is dispersed into a spectrum of colors by a prism because the shorter wavelengths (the blue end of the spectrum) are refracted more than the longer wavelengths (the red end). The colors of the visible spectrum may be remembered by memorizing the acronym ROY G. BIV, for Red, Orange, Yellow, Green, Blue, Indigo, and Violet.

*In a spectrometer a diffraction grating (Chapter 7.6), which produces sharper lines, is commonly used instead of a prism.

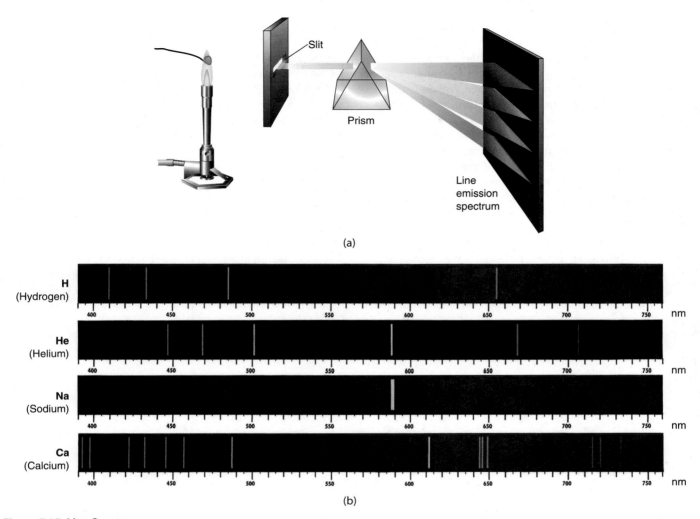

(a)

(b)

Figure 7.15 Line Spectra
(a) A line spectrum is generated when light from a heated source passes through a slit to produce a sharp beam that passes through a prism, which disperses the beam into line images of the slit. (b) Line spectra of various elements in the visible region. Note that each spectrum is unique to, or characteristic of, that element.

Notice in Fig. 7.15b that the spectra are different, or characteristic. Every substance, when sufficiently heated, gives off light of characteristic frequencies. Spectra can be studied and substances identified by means of spectrometers. Astronomers, chemists, physicists, and other scientists have acquired much basic information from the study of light. In fact, the element helium was first identified in the spectrum of sunlight (hence the name *helium* from *helios*, the Greek word for "Sun").

Did You Learn?

- Refraction occurs for transmitted light when the speeds are different in the different media. The index of refraction expresses the speed of light in a medium, relative to the speed of light in vacuum.

- The index of refraction varies with the wavelengths of light, and different wavelengths (colors) are refracted at different angles (dispersion).

Highlight   The Rainbow: Dispersion and Internal Reflection

A beautiful atmospheric phenomenon often seen after rain is the rainbow. The colorful arc across the sky is the result of several optical effects: refraction, internal reflection, and dispersion. The conditions must be just right. As is well known, a rainbow is seen after rain but not after *every* rain.

Following a rainstorm, the air contains many tiny water droplets. Sunlight incident on the droplets produces a rainbow, but whether a rainbow is visible depends on the relative positions of the Sun and the observer. The Sun is generally behind you when you see a rainbow.

To understand the formation and observation of a rainbow, consider what happens when sunlight is incident on a water droplet. On entering the droplet, the light is refracted and dispersed into component colors as it travels in the droplet (Fig. 1a). If the dispersed light strikes the water–air interface of the droplet at greater than the critical angle, then it is internally reflected and the component colors emerge from the droplet at slightly different angles. Because of the conditions for refraction and internal reflection, the component colors lie in a narrow range of 40° to 42° for an observer on the ground.

Thus, a display of colors is seen only when the Sun is positioned such that the dispersed light is reflected to you through these angles. With this condition satisfied and an abundance of water droplets in the air, you see the colorful arc of a *primary rainbow,* with colors running vertically upward from violet to red (Fig. 1a and c).

Occasionally, conditions are such that sunlight undergoes *two* internal reflections in water droplets. The result is a vertical inversion of colors in a higher, fainter, and less frequently seen *secondary rainbow* (Fig. 1b and c). Note the bright region below the primary rainbow. Light from the rainbows combines to form this illuminated region.

The arc length of a rainbow you see depends on the altitude (angle above the horizon) of the Sun. As the altitude of the Sun increases, less of the rainbow is seen. On the ground, you cannot see a (primary) rainbow if the altitude of the Sun is greater than 42°. The rainbow is below the horizon in this case. However,

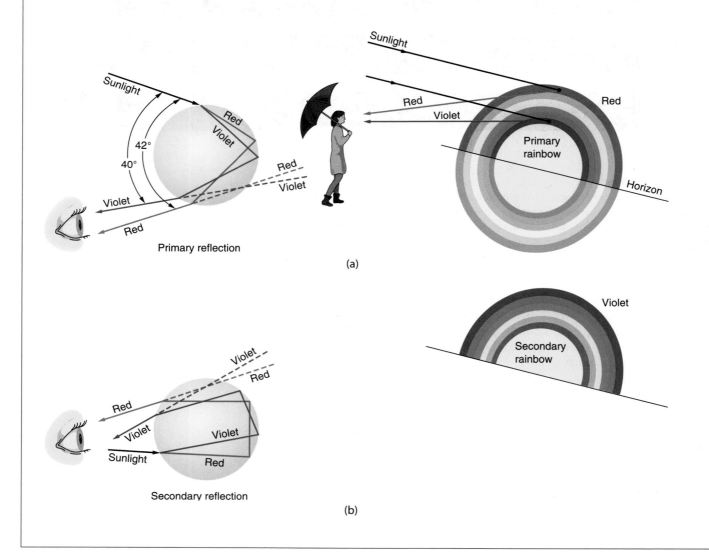

Primary reflection

(a)

Secondary reflection

(b)

Highlight

if your elevation is increased, then more of the rainbow arc is seen. For instance, airplane passengers sometimes view a completely circular rainbow, similar to the miniature one that can be seen in the mist produced by a lawn sprayer. (No need for a pot of gold.)

(c)

Figure 1 Rainbow Formation
Sometimes two rainbows can be seen, a primary rainbow and a secondary rainbow. (a) For the more common primary rainbow, there are single internal reflections in the water droplets (top, previous page). Note that this separates the colors such that the red component appears above the violet component (dashed lines), and a primary rainbow's colors run sequentially upward from violet to red. (b) For the secondary rainbow, the sunlight enters the water droplets such that there are two internal reflections (bottom, previous page). This causes the sequence of colors in the secondary rainbow to be the reverse of that in the primary rainbow. Secondary rainbows are higher, fainter, and less frequently seen than primary rainbows. (c) Both the Sun and an observer must be properly positioned for the observer to see a rainbow. For the photo, the observer was positioned in such a way that both the primary and secondary rainbows were seen.

7.3 Spherical Mirrors

Preview Questions

- What are the shapes of converging and diverging spherical mirrors?
- What is the difference between real and virtual images?

Spherical surfaces can be used to make practical mirrors. The geometry of a spherical mirror is shown in ● Fig. 7.16. A spherical mirror is a section of a sphere of radius R. A line drawn through the center of curvature C perpendicular to the mirror surface is called the *principal axis*. The point where the principal axis meets the mirror surface is called the *vertex* (V in Fig. 7.16).

Another important point in spherical mirror geometry is *the focal point F*. The distance from the vertex to the focal point is called the **focal length** f. (What is "focal" about the focal point and the focal length will become evident shortly.) For a spherical mirror, the focal length is one-half the value of the radius of curvature of the spherical surface. Expressed in symbols, the *focal length of a spherical mirror* is

$$f = \frac{R}{2}$$

7.2

Figure 7.16 Spherical Mirror Geometry
A spherical mirror is a section of a sphere with a center of curvature *C*. The focal point *F* is halfway between *C* and the vertex *V*. The distance from *V* to *F* is called the *focal length f*. The distance from *V* to *C* is the radius of curvature *R* (the radius of the sphere). And $R = 2f$, or $f = R/2$.

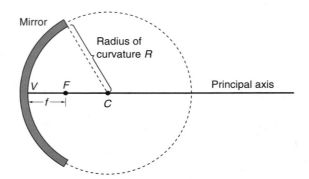

where f is the focal length and R is the radius of curvature for the spherical mirror. Equation 7.2 can be used to locate the focal point C or the center of the curvature C when f or R is known.

The inside surface of a spherical section is said to be *concave* (as though looking into a recess or cave), and when it has a mirrored surface, it is a **concave (converging) mirror**. The reason for "converging" is illustrated in ● Fig. 7.17a. Reflecting light rays parallel to the principal axis converge and pass through the focal point. The rays are "focused" at the focal point. (Off-axis parallel rays converge in the focal plane.)

Similarly, the outside surface of a spherical section is said to be *convex*, and when it has a mirrored surface, it is a **convex (diverging) mirror**. Parallel rays along the principal axis are reflected in such a way that they *appear* to diverge from the focal point (Fig. 7.17b).

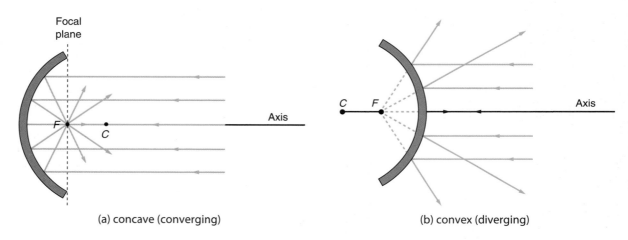

(a) concave (converging)   (b) convex (diverging)

Figure 7.17 Spherical Mirrors
(a) Rays parallel to the principal axis of a concave or converging spherical mirror converge at the focal point. Rays not parallel to the principal axis converge in the focal plane so as to form extended images. (b) Rays parallel to the axis of a convex or diverging spherical mirror are reflected so as to appear to diverge from the focal point inside the mirror. (c) The divergent property of a diverging mirror is used to give an expanded field of view, as shown here for a department store.

(c)

Aldo Mastrocola/Lightwave

In regard to reverse ray tracing, light rays coming to the mirror from the surroundings are made parallel, and an expanded field of view is seen in the diverging mirror. Diverging mirrors are used on side mirrors of cars and trucks to give drivers a wider rear view of traffic and in stores to monitor aisles (Fig. 7.17c).

Ray Diagrams

The images formed by spherical mirrors can be found graphically using *ray diagrams*. An arrow is commonly used as the object, and the location and size of the image are determined by drawing two rays:

1. Draw a ray parallel to the principal axis that is reflected through the focal point.

2. Draw a ray through the center of curvature C that is perpendicular to the mirror surface and reflected back along the incident path.

The intersection of these rays (tip of arrow) locates the position of the image.

These rays are shown in ● Fig. 7.18 for an object at various positions in front of a concave mirror. In the ray diagrams, D_o is the *object distance* (distance of the object from the vertex) and D_i is the *image distance* (distance of the image from the vertex). The object distance in a ray diagram can be determined relative to the focal point (F) or the center of curvature (C), which is usually known from f or R.

The characteristics of an image are described as being (1) *real or virtual*, (2) *upright (erect) or inverted*, and (3) *magnified or reduced* (smaller than the object). If the diagram is drawn to scale, then the magnification (greater or less than 1) may be found by comparing the heights of the object and image arrows in Fig. 7.18. A **real image** is one for which the light rays converge so that an image can be formed on a screen. A **virtual image** is one for which

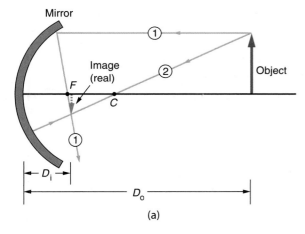

(a)

Figure 7.18 Ray Diagrams
Illustrations of rays 1 and 2 (see text) in ray diagrams. D_o and D_i are the object distance and image distance, respectively. (a) A ray diagram for an object beyond the center of curvature C for a concave spherical mirror shows where the image is formed. (b) A ray diagram for a concave mirror for an object located between F and C. In (a) and (b) the images are real and thus could be seen on a screen placed at the image distances. Note that the image moves out and grows larger as the object moves toward the mirror. (c) A ray diagram for a concave mirror with the object inside the focal point F. In this case, the image is virtual and is formed behind, or "inside," the mirror.

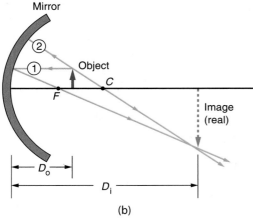

(b)

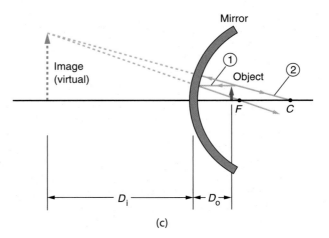

(c)

Figure 7.19 Magnification
Concave cosmetic mirrors give magnification so that facial features can be seen better.

the light rays diverge and cannot be formed on a screen. Both real and virtual images are formed by a concave (converging) spherical mirror, depending on the object distance. An example of a virtual image is shown in ● Fig. 7.19, where magnification is put to use.

Note that the real images are formed in front of the mirror where a screen could be positioned. Virtual images are formed behind or "inside" the mirror where the light rays *appear* to converge. For a converging concave mirror, a virtual image always results when the object is inside the focal point. For a diverging mirror, a virtual image always results wherever the object is located. (What type of image is formed by a plane mirror?)

A convex mirror may also be treated by ray diagrams. The rays of a ray diagram are drawn by using the law of reflection, but they are extended through the focal point and the center of curvature inside or behind the mirror, as shown in ● Fig. 7.20. A virtual image is formed where these extended rays intersect. As Fig 7.20 suggests, even though the object distance may vary, the image of a diverging convex mirror is always virtual, upright, and smaller than the object.

EXAMPLE 7.2   Finding the Images of Spherical Mirrors Using Ray Diagrams

A spherical concave mirror has a radius of curvature of 20 cm.

(a) An object is placed 25 cm in front of the concave mirror. Draw a ray diagram for this situation. Estimate the image distance and give the image characteristics.

(b) An object is placed 15 cm in front of the concave mirror. Draw a ray diagram, estimate the image distance, and give the image characteristics for this case.

Solution

(a) With $R = 20$ cm, then $f = R/2 = 20$ cm/2 $= 10$ cm, which gives the locations of C and F, respectively. Locating the object and drawing the two rays to locate the image gives the ray diagram shown in ● Fig. 7.21a. (When not asked to draw to scale, you may use a sketch to give the approximate image distance, as well as the image characteristics.) As can be seen from the sketch, the image distance is about midway between F and C, or 15 cm (mathematical calculations reveal that it is actually 16.7 cm), and the image characteristics are real, inverted, and reduced.

(b) Using the same procedure, from Fig. 7.21b the image distance is about 30 cm (mathematically it is actually 30 cm), and the image is real, inverted, and magnified.

The image is reduced when the object is beyond the center of curvature (C) but magnified when the object is inside the center of curvature. This result is true in general.

Confidence Exercise 7.2

Suppose the object in part (a) of Example 7.2 were placed 5 cm from the mirror. What would be the image characteristics in this case?

Answers to Confidence Exercises may be found at the back of the book.

Figure 7.20 Ray Diagram for a Diverging Mirror
A ray diagram for a convex spherical mirror with an object in front of the mirror. A convex mirror always forms a reduced, virtual image.

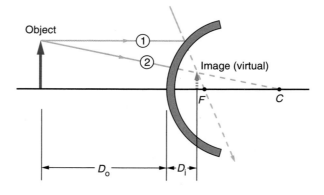

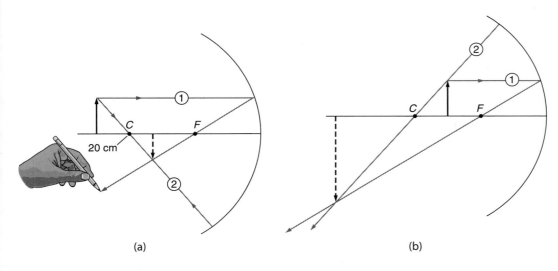

Figure 7.21 Ray Diagrams for Example 7.2 See text for description.

(a)   (b)

Conceptual Question and Answer

Up and Down

Q. When you look into the front side of a shiny spoon, you will see an inverted image of yourself. When you look at the back side of the spoon, your image is upright. Why?

A. You are alternately looking at concave and convex mirrors. Looking into the front of the spoon, or a concave mirror, as an object you are outside the focal length and the image is inverted. As you move the spoon away from you, the inverted image becomes smaller. [Compare with the ray diagrams in Fig. 7.18a and b.]

Looking at the back of the spoon, or a convex mirror, your image is upright as all images are for convex mirrors (see Fig. 7.20). Get a spoon and check it out.

Did You Learn?

- A converging mirror is concave, and a diverging mirror is convex.
- A real image is one for which light rays converge so that an image can be formed on a screen. For a virtual image, the light rays diverge and an image cannot be formed.

7.4 Lenses

Preview Questions

- What is the general difference in shapes between a converging lens and a diverging lens?
- What are the functions of the rods and cones in the retina of the human eye?

A lens consists of material such as a transparent piece of glass or plastic that refracts light waves to give an image of an object. Lenses are extremely useful and are found in eyeglasses, telescopes, magnifying glasses, cameras, and many other optical devices.

In general, there are two main types of lenses. A **converging lens** is thicker at the center than at the edges. A **diverging lens** is thinner at the center than at the edges. These two types and some of the possible shapes for each are illustrated in ● Fig. 7.22. In general, we

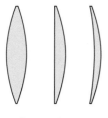

Converging, or
convex, lenses;
greatest thickness
at center

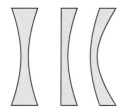

Diverging, or
concave, lenses;
greatest thickness
at edge

Figure 7.22 Lenses
Different types of converging and diverging lenses. Note that converging lenses are thicker at the center than at the edges, whereas diverging lenses are thinner at the center and thicker at the edges.

will investigate the spherical biconvex and biconcave lenses at the left of each group in the figure (*bi-* because they have similar spherical surfaces on both sides).

Light passing through a lens is refracted twice, once at each surface. The lenses most commonly used are known as *thin lenses*. Thus, when constructing ray diagrams, the thickness of the lens can be neglected.

The principal axis for a lens goes through the center of the lens (● Fig. 7.23). Rays coming in parallel to the principal axis are refracted toward the principal axis by a converging lens. For a converging lens, the rays are focused at point *F*, the focal point. For a diverging lens, the rays are refracted away from the principal axis and appear to emanate from the focal point on the incident side of the lens.

Ray Diagrams

How lenses refract light to form images can be shown by drawing graphic ray diagrams similar to those applied to mirrors.

1. The first ray is drawn parallel to the principal axis and then refracted by the lens along a line drawn through a focal point of the lens.

2. The second ray is drawn through the center of the lens without a change in direction.

The intersection of these rays (tips of arrows) locates the position of the image (tip of the image arrow).

Examples of this procedure are shown in ● Fig. 7.24. Only the focal points for the respective surfaces are shown—just focal points are needed. The lenses do have radii of curvature, but for spherical lenses, $f \neq R/2$ in contrast to $f = R/2$ for spherical mirrors. The characteristics of the images formed by a converging or convex lens change, similar to the way those of a converging mirror change as an object is brought toward the mirror from a distance. Beyond the focal point, an inverted, real image is formed, which becomes larger as the object approaches the focal point.

The magnification becomes greater than 1 when the object distance is less than *2f*. Once inside the focal point of a converging mirror, an object always forms a virtual image. For lenses, a *real image* is formed on the opposite side of a lens from the object and can be seen on a screen (Fig. 7.23a). A *virtual image* is formed on the object side of the lens (Fig. 7.24b).

For a diverging or concave lens, the image is always upright and reduced, or smaller than the object. When looking through a concave lens, one sees images as shown in ● Fig. 7.25. Also, a concave lens forms only virtual objects.

EXAMPLE 7.3   **Finding the Images of Converging Lenses Using Ray Diagrams**

A convex lens has a focal length of 12 cm. Draw ray diagrams for objects at (a) 18 cm and (b) 8 cm from the lens. Estimate the image distances and give the image characteristics for each case.

Figure 7.23 Lens Focal Points
For a converging spherical lens, rays parallel to the principal axis and incident on the lens converge at the focal point on the opposite side of the lens. Rays parallel to the axis of a diverging lens appear to diverge from a focal point on the incident side of the lens.

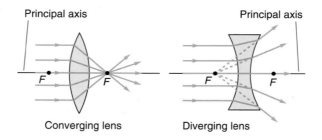

Principal axis   Principal axis

F   *F*   *F*   *F*

Converging lens   Diverging lens

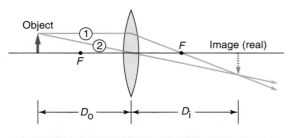

(a)

Figure 7.24 Ray Diagrams
(a) A ray diagram for a converging lens with the object outside the focal point. The image is real and inverted and can be seen on a screen placed at the image distance, as shown in the photo. (b) A ray diagram for a converging lens with the object inside the focal point. In this case, an upright, virtual image is formed on the object side of the lens.

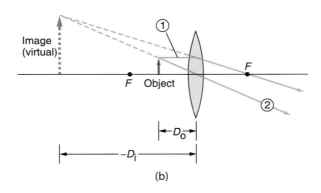

(b)

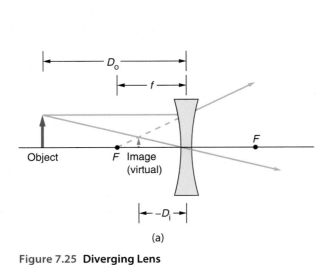

(a)

(b)

Figure 7.25 Diverging Lens
(a) A ray diagram for a diverging, concave lens. A virtual image is formed on the object side of the lens. Diverging lenses form only virtual images. (b) Like a diverging mirror, a diverging (concave) lens gives an expanded field of view.

Solution

(a) The focal length locates the focal point F. Locating the object and drawing the two rays to locate the image gives the ray diagram shown in ● Fig. 7.26a. It shows the image distance to be about twice (36 cm) that of the object distance (18 cm), and the image is real, inverted, and magnified.

(b) Using the same procedure, from Fig. 7.26b we see that the image is behind the object (with an image distance of about 24 cm) and that the image is virtual (on the object side of the lens), upright, and magnified.

Confidence Exercise 7.3

Suppose a diverging (concave) lens were used in part (b) of Example 7.3, with the same focal length and object distance. What would be the image characteristics in this case?

The Human Eye

The human eye contains a convex lens along with other refractive media in which most of the light refraction occurs. Even so, a great deal can be learned about the optics of the eye by considering only the focusing action of the lens. As illustrated in ● Fig. 7.27, the lens focuses the light entering the eye on the *retina*. The photoreceptors of the retina, called *rods* and *cones*, are connected to the optic nerve, which sends signals to the brain. The rods are more sensitive than the cones and are responsible for light and dark "twilight" vision; the cones are responsible for color vision. The retina contains about 120 million rods and 6 million cones. (Remember "c"ones for "c"olor.)

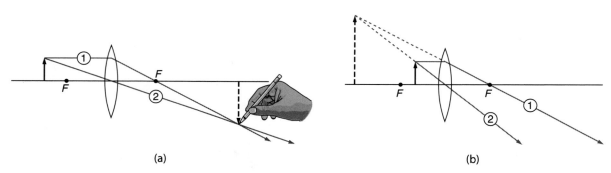

(a)   (b)

Figure 7.26 Ray Diagrams for Example 7.3
See text for description.

Figure 7.27 The Human Eye
The lens of the human eye forms an image on the retina, which contains rod and cone cells. The rods are more sensitive than the cones and are responsible for light and dark "twilight" vision; the cones are responsible for color vision. (The image is drawn vertically for clarity.)

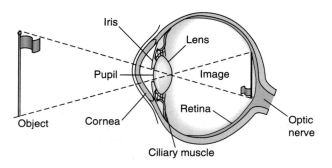

Conceptual Question and Answer

Right Side Up from Upside Down

Q. Note in Fig. 7.27 that an image focused on the retina is upside down. Why don't we see the world that way?

A. The brain learns early in life to interpret the inverted images of the world right side up. Experiments have been done with persons wearing special glasses that give them an inverted view of the world. After some initial run-ins, they become accustomed to and function quite well in their upside-down world.

Because the distance between the lens and retina does not vary, in this situation D_i is constant. Because D_o varies for different objects, the focal length of the lens of the eye must vary for the image to be on the retina. The lens is called the *crystalline lens* and consists of glassy fibers. By action of the attached ciliary muscles, the shape and focal length of the lens vary as the lens is made thinner and thicker. The optical adjustment of the eye is truly amazing. Objects can be seen quickly at distances that range from a few centimeters (the near point) to infinity (the far point).

Although the far point is infinity, objects can be resolved only for certain distances. When one is viewing distant objects, they appear smaller and eventually become indistinguishable (cannot be resolved). It is sometimes said that the Great Wall of China is the only human construction that can be seen by the unaided eye of an astronaut orbiting the Earth, but this statement is false. The Great Wall is on the order of 8.0 m (26 ft) wide at the base and 4.0 m (13 ft) wide at the top. The Los Angeles Freeway is much wider. Astronauts orbit the Earth at altitudes on the order of 400 km (250 mi). It can be shown that to see (visually resolve) the Great Wall, an astronaut would have to be at an altitude of 35 km (22 mi) or lower. This height is within the Earth's atmosphere. (See Chapter 19.1.)

Speaking of a "normal" eye implies that visual defects exist in some eyes, which is readily apparent from the number of people who wear glasses or contact lenses. Many people have trouble seeing objects at certain distances. These individuals have one of the two most common visual defects: nearsightedness and farsightedness.

Nearsightedness (myopia) is the condition of being able to see nearby objects clearly but not distant objects. This occurs when, for some reason, the distant image is focused in front of the retina (● Fig. 7.28a). Glasses with diverging lenses that move the image back to the retina can be used to correct this defect.

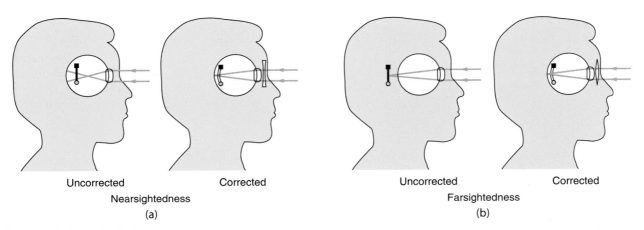

Uncorrected   Corrected
Nearsightedness
(a)

Uncorrected   Corrected
Farsightedness
(b)

Figure 7.28 Vision Defects
Two common vision defects arise because the image is not focused on the retina. (a) Nearsightedness occurs when the image is formed in front of the retina. This condition is corrected by wearing glasses with diverging lenses. (b) Farsightedness occurs when the image is formed behind the retina. This condition is corrected by wearing glasses with converging lenses.

Farsightedness (hyperopia) is the condition of being able to see distant objects clearly but not nearby objects. The images of such objects are focused behind the retina (Fig. 7.28b). The near point is the position closest to the eye at which objects can be seen clearly. (Bring your finger toward your nose. The position where the tip of the finger goes out of focus is your near point.) For farsighted people, the near point is not at the normal position but at some point farther from the eye.

Children can see sharp images of objects as close as 10 cm (4 in.) to their eyes. The crystalline lens of the normal young-adult eye can be deformed to produce sharp images of objects as close as 12 to 15 cm (5 to 6 in.). However, at about the age of 40, the near point normally moves beyond 25 cm (10 in.).

You may have noticed older people holding reading material at some distance from their eyes so as to see it clearly. When the print is too small or the arm too short, reading glasses with converging lenses are the solution (Fig. 7.28b). The recession of the near point with age is not considered an abnormal defect of vision. It proceeds at about the same rate in all normal eyes. (You too may need reading glasses someday.)

Did You Learn?

● A converging (convex) lens is thicker at the center than at the edges. A diverging (concave) lens is thinner at the center than at the edges.

● The rods in the retina of your eye are responsible for light and dark "twilight" vision, and the cones are responsible for color vision.

7.5   Polarization

Preview Questions

● What does the polarization of light experimentally prove about light?

● What is the principle of polarizing sunglasses?

The wave nature of light gives rise to an interesting and practical optical phenomenon. Light waves are transverse waves with oscillations perpendicular to the direction of propagation (electromagnetic waves, see Figure 8.24). The atoms of a light source generally emit light waves that are randomly oriented, and a beam of light has transverse oscillations in all directions. Viewing a beam of light from the front, the transverse oscillations may be indicated by vector arrows as shown in ● Fig. 7.29a.

In the figure the oscillations are in planes perpendicular to the direction of propagation. Such light is said to be *unpolarized*, with the oscillations randomly oriented. **Polarization** refers to the preferential orientation of the oscillations. If the oscillations have some partial preferential orientation, then the light *is partially polarized* (Fig. 7.29b). If the oscillations are in a single plane, then the light is **linearly polarized**, or *plane polarized* (Fig. 7.29c).

Figure 7.29 Polarization
(a) When the electric field vectors are randomly oriented, as viewed in the direction of propagation, the light is unpolarized. (b) With preferential orientation, the light is partially polarized. (c) If the field vectors lie in a plane, then the light is linearly polarized.

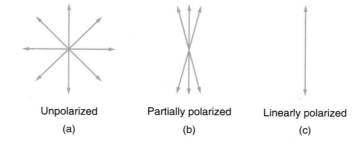

Unpolarized
(a)

Partially polarized
(b)

Linearly polarized
(c)

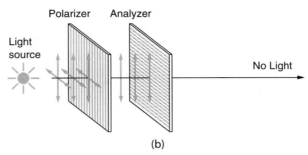

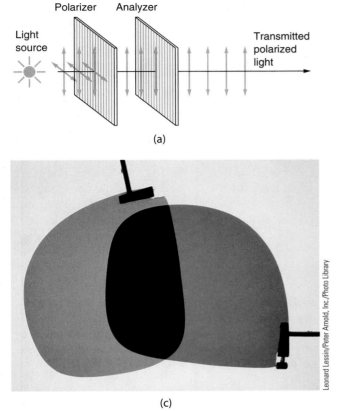

(c)

Figure 7.30 Polarized Light
(a) Light is linearly polarized when it passes through a polarizer. (The lines on the polarizer indicate the direction of polarization.) The polarized light passes through the analyzer if it is similarly oriented. (b) When the polarization direction of the analyzer is perpendicular (90°) to that of the polarizer ("crossed Polaroids"), little or no light is transmitted. (c) A photo showing the condition of (b), "crossed Polaroids," using polarizing sunglass lenses.

A light wave can be polarized by several means. A common method uses a polymer sheet or "Polaroid."* Polaroid sheets have a polarization direction associated with long, oriented molecular chains of the polymer film. The *polarizer* allows only the components in a specific plane to pass through, as illustrated in ● Fig. 7.30a. The other field vectors are absorbed and do not pass through the polarizer.

The human eye cannot detect polarized light, so an *analyzer*, another polarizing sheet, is needed. When a second polarizer is placed in front of the first polarizer, as illustrated in Fig. 7.30b, little light (theoretically, no light) is transmitted and the sheets appear dark. In this case, the polarization directions of the sheets are at 90° and the polarizing sheets are said to be "crossed" (Fig. 7.30c).

The polarization of light is experimental proof that light is a transverse wave. Longitudinal waves, such as sound, cannot be polarized.

Sky light is partially polarized as a result of atmospheric scattering by air molecules. When unpolarized sunlight is incident on air molecules, the light waves set the electrons of the molecules into vibration. The accelerating charges emit radiation, similar to the vibrating charges in the antenna of a broadcast station. The radiated, or "scattered," sky light has polarized components, as may be observed with an analyzer. (The best direction to look to observe the polarization depends on the location of the Sun. At sunset and sunrise, the best direction is directly overhead.) It is believed that some insects, such as bees, use polarized sky light to determine navigational directions relative to the Sun.

A common application of polarization is in polarizing sunglasses. The lenses of these glasses are polarizing sheets oriented such that the polarization direction is vertical. When sunlight is reflected from a surface, such as water or a road, the light is partially polarized in the horizontal direction. Because the reflected light is scattered in a preferred direction, the intensity increases, which an observer sees as glare. Polarizing sunglasses allow only

*Named after the first commercial polarizing sheet, called Polaroid, developed by Edwin Land around 1930.

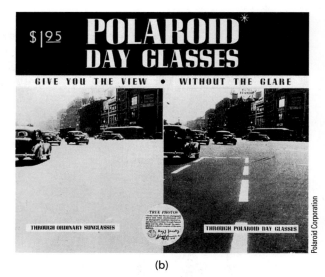

Polaroid Corporation

(a)   (b)

Figure 7.31 Polarizing Sunglasses
(a) Light reflected from a surface is partially polarized in the horizontal direction of the plane of the surface. Because the polarizing direction of the sunglasses is oriented vertically, the horizontally polarized component of the reflected light is blocked, thereby reducing the intensity or glare. (b) Old-time glare-reduction advertisement from the Polaroid Corporation Archives. Note the vintage of the cars and the price of the glasses.

the vertical component of light to pass through. The horizontal component is blocked out, which reduces the glare (● Fig. 7.31).

Another common but not as well-known application of polarized light is discussed in the **Highlight: Liquid Crystal Displays (LCDs)**.

Did You Learn?

● The polarization of light proves that light is a transverse wave.

● The lenses of polarizing sunglasses have the polarizing direction vertical so as to reduce the horizontally polarized reflective glare.

7.6   Diffraction and Interference

Preview Questions

● On what does diffraction depend?

● What is the difference between constructive interference and destructive interference?

Diffraction

Water waves passing through slits are shown in ● Fig. 7.32. Note how the waves are bent, or deviated, around the corners of the slit as they pass through. All waves (sound, light, and so on) show this type of bending as they go through relatively small slits or pass by the corners of objects. The deviation of waves in such cases is referred to as **diffraction**.

In Fig. 7.32 there are different degrees of bending, or diffraction. The degree of diffraction depends on the wavelength of the wave and the size of the opening or object. In general, the longer the wavelength compared to the width of the opening or object, the greater the diffraction.

Highlight Liquid Crystal Displays (LCDs)

When a crystalline solid melts, the resulting liquid generally has no orderly arrangement of atoms or molecules. However, some organic compounds have an intermediate state in which the liquid retains some orderly molecular arrangement, hence the name liquid crystal (LC).

Some liquid crystals are transparent and have an interesting property. When an electrical voltage is applied, the liquid crystal becomes opaque. The applied voltage upsets the orderly arrangement of the molecules, and light is scattered, making the LC opaque.

Another property of some liquid crystals is how they affect linearly polarized light by "twisting" or rotating the polarization direction 90°. This twisting, however, does not occur if a voltage is applied, causing molecular disorder.

A common application of these properties is in liquid crystal displays (LCDs), which are found on wristwatches, calculators, and TV and computer screens. How LCDs work is illustrated in Fig. 1.

Trace the incident light in the top diagram of Fig. 1. Unpolarized light is linearly polarized by the first polarizer. The LC then rotates the polarization direction, and the polarized light passes through the second polarizer (which is "crossed" with the first) and then is reflected by the mirror. On the reverse path, the rotation of the polarization direction in the LC allows the light to emerge from the LCD, which consequently appears bright or white.

However, if a voltage is applied to the LC such that it loses its rotational property, then light is not passed by the second polarizer. With no reflected light, the display appears dark. Thus, by applying voltages to segments of numeral and letter displays, it is possible to form dark regions on a white background (Fig. 2). The white background is the reflected, polarized light, which can be demonstrated by using an analyzer. The LCD display on your calculator may appear dark if you are wearing polarizing sunglasses.

(a)

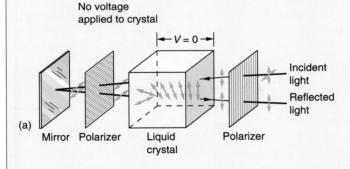

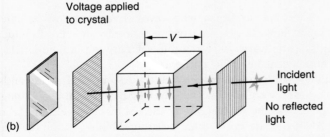

Figure 1 Liquid Crystal Display (LCD)
(a) An illustration of how a liquid crystal "twists" the light polarization through 90°. The light passes through the polarizer and is reflected back and out of the crystal with another twist.
(b) When a voltage is applied to the crystal, there is no twisting and light does not pass through the second polarizer. In this case, the light is absorbed and the crystal appears dark.

Figure 2 LCDs and Polarization
(a) The light from the bright regions of an LCD is polarized, as can be shown by using polarizing sunglasses as an analyzer. Here the numbers can still be seen, (b) but not when the glasses have been rotated 90°.

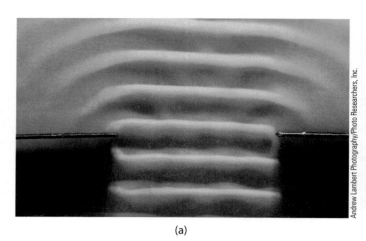

(a)

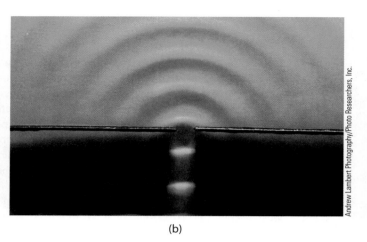

(b)

Figure 7.32 Diffraction
(a) The diffraction, or bending, of the water waves passing through the slit can be seen in a ripple tank, with the bending at the edges of the waves. (b) When the slit is made smaller, the diffraction becomes greater. Note the greater bending of the waves.

As was shown in Example 6.1, audible sound waves have wavelengths on the order of centimeters to meters. Visible light waves, on the other hand, have wavelengths of about 10^{-6} m (a millionth of a meter). Ordinary objects (and slits) have dimensions of centimeters to meters. Thus, the wavelengths of sound are larger than or about the same size as objects, and diffraction readily occurs for sound. For example, when you are standing in one room, you can talk through a doorway into another room in which people are standing around the corner unseen on each side, and they can hear you.

However, the dimensions of ordinary objects or slits are much greater than the wavelengths of visible light, so the diffraction of light is not commonly observed. For instance, when you shine a beam of light at an object, there will be a shadow zone behind the object with very sharp boundaries.

Some light diffraction does occur at corners, but it goes largely unnoticed because it is difficult to see. Very close inspection reveals that the shadow boundary is blurred or fuzzy, and there is actually a pattern of bright and dark regions (● Fig. 7.33). This is an indication that some diffraction has occurred.

Think about this: When you sit in a lecture room or movie theater, sound is easily diffracted around a person directly in front of you, but light is not; you cannot see anything directly in front (other than the back of a head). What does that say about wavelengths and the size of people's heads?

As another example, radio waves are electromagnetic waves of very long wavelengths, in some cases hundreds of meters long. In this case, ordinary objects and slits are much smaller than the wavelength, so radio waves are easily diffracted around buildings, trees, and so on, making radio reception generally quite efficient.

You may have noticed a difference in reception between the AM and FM radio bands, which have different frequencies and wavelengths. The wavelengths of the AM band range from about 180 m to 570 m, whereas the wavelengths of the FM band range from 2.8 m to 3.4 m. Hence, the longer AM waves are easily diffracted around buildings and the like, whereas FM waves may not be. As a result, AM reception may be better than FM reception in some areas.

Interference

Figure 7.33 Diffraction Pattern
Using special lighting, it is possible to see diffraction patterns clearly in the opening of a razor blade.

When two or more waves meet, they are said to *interfere*. For example, water waves from two disturbances on a lake or pond commonly interfere with each other. The resultant waveform of the interfering waves is a combination of the individual waves. Specifically, the waveform is given by the **principle of superposition**:

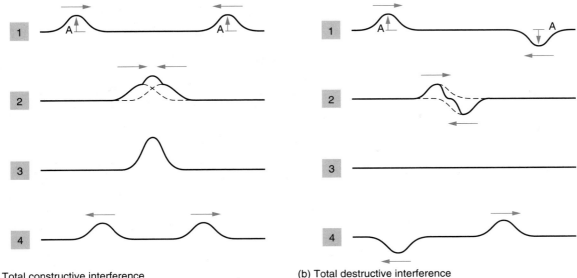

(a) Total constructive interference   (b) Total destructive interference

Figure 7.34 Interference
(a) When wave pulses of equal amplitude that are in phase meet and overlap, there is total constructive interference. At the instant of the overlap (a, 3), the amplitude of the combined pulse is
twice that of the individual ones ($A + A = 2A$). (b) When two wave pulses of equal amplitude are
out of phase, the waveform disappears for an instant when waves exactly overlap (b, 3); that is,
the combined amplitude of the waveform is zero ($A - A = 0$).

**At any time, the combined waveform of two or more interfering waves is given by
the sum of the displacements of the individual waves at each point in the medium.**

The displacement of the combined waveform of two waves at any point is given by
$y = y_1 + y_2$, where the directions are indicated by plus and minus signs. The waveform of
the interfering waves changes with time, and after interfering, the waves pass on with their
original forms.

It is possible for the waves to reinforce one another when they overlap, or interfere,
causing the amplitude of the combined waveform to be greater than either pulse. This is
called **constructive interference**. On the other hand, if two waves tend to cancel each
other when they overlap or interfere (that is, if one wave has a negative displacement), then
the amplitude of the combined waveform is smaller than that of either pulse. This is called
destructive interference.

Special cases of constructive and destructive interference are shown in ● Fig. 7.34 for
pulses with the same amplitude A. When the interfering pulses are exactly in phase (crest
coincides with crest when the pulses are overlapped), the amplitude of the combined
waveform is twice that of either individual pulse ($y = A + A = 2A$), and this is referred
to as *total constructive interference* (Fig. 7.34a). However, when two pulses are completely
out of phase (crest coincides with trough when the pulses are overlapped), the waveforms
disappear; that is, the amplitude of the combined waveform is zero ($y = A - A = 0$). This
is referred to as *total destructive interference* (Fig. 7.34b).

The word *destructive* is misleading. Do not get the idea that the energy of the pulses
is destroyed. The waveform is destroyed, but the propagating energy is still there in the
medium, in the form of potential energy (conservation of energy). After interfering, the
individual pulses continue on with their original waveforms.

The colorful displays seen in oil films and soap bubbles can be explained by interference.
Consider light waves incident on a thin film of oil on the surface of water or on a wet road.
Part of the light is reflected at the air–oil interface, and part is transmitted. The part of the
light in the oil film is then reflected at the oil–water interface (● Fig. 7.35).

The two reflected waves may be in phase, totally out of phase, or somewhere in between.
In Fig. 7.35a the waves are shown in phase, but this result will occur only for certain

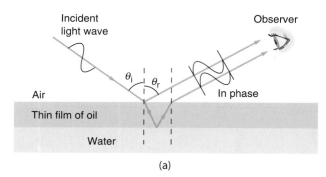

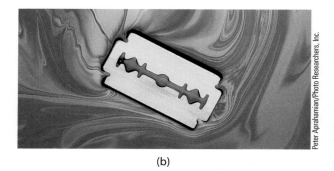

(a)

(b)

Figure 7.35 Thin-Film Interference
(a) When reflected rays from the top and bottom surfaces of an oil film are in phase, constructive interference occurs and an observer sees only the color of light for a certain angle and film thickness. When the reflected rays are out of phase, destructive interference occurs, which means that light is transmitted at the oil–water interface rather than reflected, and this area appears dark.
(b) Because the thickness of the oil film varies, a colorful display is seen for different wavelengths of light.

angles of observation, wavelengths of light (colors), and thicknesses of oil film. At certain angles and oil thicknesses, only one wavelength of light shows constructive interference. The other visible wavelengths interfere destructively, and these wavelengths are transmitted and not reflected.

Hence, different wavelengths interfere constructively for different thicknesses of oil film, and an array of colors is seen (Fig. 7.35b). In soap bubbles the thickness of the soap film moves and changes with time, and so does the array of colors.

Diffraction can also give rise to interference. This interference can arise from the bending of light around the corners of a single slit. An instructive technique employs two narrow double slits that can be considered point sources, as illustrated in ● Fig. 7.36. When the slits are illuminated with monochromatic light (light of only one wavelength), the diffracted light through the slits spreads out and interferes constructively and destructively at different points where crest meets crest and where crest meets trough, respectively. By placing a screen a distance from the slits, an observer can see an interference pattern of alternating bright and dark fringes. A double-slit experiment done in 1801 by English scientist Thomas Young demonstrated the wave nature of light. Such an experiment makes it possible to compute the wavelength of the light from the geometry of the experiment.

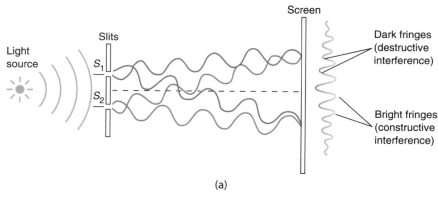

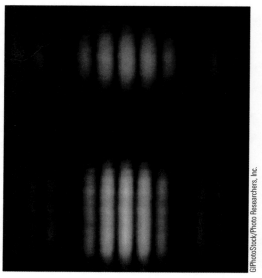

(a)

(b)

Figure 7.36 Double-Slit Interference
(a) Light waves interfere as they pass through two narrow slits that act as point sources, giving rise to regions of constructive interference, or bright fringes, and regions of destructive interference, or dark fringes. (b) Actual diffraction patterns of different colors of laser light through a double slit. Note that the fringe spacing is smaller for the shorter (green) wavelength.

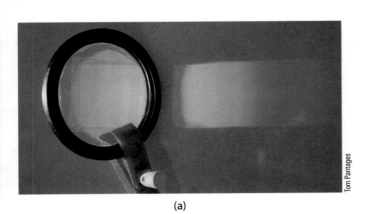

(a)

(b)

Figure 7.37 Diffraction Grating Interference
(a) The many slits of a diffraction grating produce a very sharp interference pattern compared with those of only two slits. The photo shows the colorful separation of colors (wavelengths) of white light passing through a transmission grating. (b) Diffraction is now readily observed. The grooves of a compact disc (CD) form a reflection diffraction grating, and the incident light is separated into a spectrum of colors.

   This double-slit experiment may be extended. The intensity of the lines becomes less when the light has to pass through a number of narrow slits, but this produces sharp lines that are useful in the analysis of light sources and other applications. A *diffraction grating* consists of many narrow, parallel lines spaced very close together. When light is transmitted through a grating, it is called a *transmission grating*. Such gratings are made by using a laser to etch fine lines on a photosensitive material. The interference of waves passing through such a diffraction grating produces an interference pattern, as shown in ● Fig. 7.37a.
   Reflection gratings (reflecting lines) are made by etching lines on a thin film of aluminum deposited on a flat surface. The narrow grooves of a compact disc (CD) act as a reflection diffraction grating, producing colorful displays (Fig. 7.37b).
   Diffraction gratings are more effective than prisms for separating the component wavelengths of light emitted by stars (including our Sun) and other light sources (see Fig. 7.15).

Did You Learn?

● In general, the longer the wavelength compared to the size of the opening or object, the greater the diffraction (bending).

● In constructive interference, when two waves interfere, the combined wave form is greater in amplitude than either wave. In destructive interference the opposite occurs.

KEY TERMS

1. reflection (7.1)
2. ray
3. law of reflection
4. specular reflection
5. diffuse reflection
6. refraction (7.2)
7. index of refraction
8. total internal reflection
9. dispersion
10. focal length (7.3)
11. concave (converging) mirror
12. convex (diverging) mirror
13. real image
14. virtual image
15. converging lens (7.4)
16. diverging lens
17. polarization (7.5)
18. linearly polarized light
19. diffraction (7.6)
20. principle of superposition
21. constructive interference
22. destructive interference

MATCHING

For each of the following items, fill in the number of the appropriate Key Term from the preceding list. Compare your answers with those at the back of the book.

a. _____ Reflection from a very smooth surface

b. _____ Image for which light rays diverge and cannot form an image

c. _____ Parallel light rays appear to diverge from mirror focal point

d. _____ Amplitude of combined wave form is greater

e. _____ An image that can be formed on a screen

f. _____ A change in direction when going from one medium into another

g. _____ Combined waveform is given by sum of individual displacements

h. _____ A change in the direction of light at a surface

i. _____ Parallel light rays converge and pass through mirror focal point

j. _____ Amplitude of combined wave form is smaller

k. _____ Reflection from a relatively rough surface

l. _____ Lens thicker at the edge than at the center

m. _____ Distance from vertex to focal point

n. _____ Reflection back into same medium

o. _____ Preferential orientation of field vectors

p. _____ A straight line that represents the path of light

q. _____ Bending of light waves passing through slits and around corners

r. _____ Ratio of light speeds in vacuum and medium

s. _____ Refraction of wavelengths at slightly different angles

t. _____ Light that is plane polarized

u. _____ $\theta_i = \theta_r$

v. _____ Lens thicker at the center than at the edge

MULTIPLE CHOICE

Compare your answers with those at the back of the book.

1. For ray reflections from a surface, which statement is true? (7.1)
   (a) The angle of reflection is equal to the angle of incidence.
   (b) The reflection angle is measured from a line perpendicular to the reflecting surface.
   (c) The rays lie in the same plane.
   (d) All the preceding are true.

2. To what does the law of reflection apply? (7.1)
   (a) regular reflection
   (b) specular reflection
   (c) diffuse reflection
   (d) all the preceding

3. What is the case when the angle of refraction is smaller than the angle of incidence? (7.2)
   (a) The critical angle is exceeded.
   (b) The first medium is less dense.
   (c) The second medium has a smaller index of refraction.
   (d) The speed of light is greater in the second medium.

4. In refraction, which of the following wave properties is unchanged? (7.2)
   (a) frequency   (b) wavelength
   (c) speed   (d) both (a) and (b)

5. What is the unit of the index of refraction? (7.2)
   (a) m   (b) none; it is unitless
   (c) m/s   (d) 1/s

6. Which is true of a convex mirror? (7.3)
   (a) It has a radius of curvature equal to f.
   (b) It is a converging mirror.
   (c) It forms only virtual images.
   (d) It forms magnified and reduced images.

7. Which is true for a real image? (7.3)
   (a) It is always magnified.
   (b) It is formed by converging light rays.

   (c) It is formed behind a mirror.
   (d) It occurs only for $D_i = D_o$.

8. Which of the following is true of a concave lens? (7.4)
   (a) It is a converging lens.
   (b) It is thicker at the center than at the edge.
   (c) It is a lens that forms virtual images for $D_o > f$.
   (d) It is a lens that forms real images for $D_o < f$.

9. Which is true of a virtual image? (7.4)
   (a) It is always formed by a convex lens.
   (b) It can be formed on a screen.
   (c) It is formed on the object side of a lens.
   (d) It cannot be formed by a concave lens.

10. What happens when the polarization directions of two polarizing sheets are at an angle of 90° to each other? (7.5)
    (a) No light gets through.
    (b) There is maximum transmission.
    (c) Maximum transmission is reduced by 50%.
    (d) None of the preceding.

11. Which is true of diffraction? (7.6)
    (a) It occurs best when the slit width is less than the wavelength of a wave.
    (b) It depends on refraction.
    (c) It is caused by interference.
    (d) It does not occur for light.

12. When does total constructive interference occur? (7.6)
    (a) when waves are in phase
    (b) at the same time as total destructive interference
    (c) when the waves are equal in amplitude and are completely out of phase
    (d) when total internal reflection occurs

FILL IN THE BLANK

Compare your answers with those at the back of the book.

1. Light rays are used in ___ optics. (Intro)
2. Reflection from a rough surface is referred to as ___ reflection. (7.1)
3. The index of refraction is the ratio of the speed of light in a medium to the speed of light in a(n) ___. (7.2)
4. When light passes obliquely into a denser medium, the light rays are bent ___ the normal. (7.2)
5. When light is reflected and none is refracted at an interface, it is called ___ reflection. (7.2)
6. A concave mirror is commonly called a ___ mirror. (7.3)
7. A virtual image ___ be formed on a screen. (7.3)
8. A diverging lens is ___ at the center than at the edge. (7.4)
9. A virtual image is always formed by a(n) ___ lens. (7.4)
10. Polarization is proof that light is a(n) ___ wave. (7.5)
11. The larger the wavelength compared to the size of an opening or object, the ___ the diffraction. (7.6)
12. The resultant waveform of combining waves is described by the ___. (7.6)

SHORT ANSWER

7.1 Reflection

1. For specular reflection, what is the situation with an angle of incidence of (a) 0° and (b) 90°?
2. Dutch painter Vincent van Gogh was emotionally troubled and once cut off part of his own ear. His *Self Portrait with Bandaged Ear* (1889) is shown in ● Fig. 7.38. Which ear did he cut? (*Hint:* How do you paint a self-portrait?)
3. When you walk toward a full-length plane mirror, what does your image do? How fast does the image move? Is the image in step with you?
4. How long does the image of a 12-in. ruler appear in a plane mirror? Does it depend on the distance the ruler is from the mirror?
5. Where would an observer see the image of the arrow shown in ● Fig. 7.39? (Draw in the image the observer would see.)

7.2 Refraction and Dispersion

6. Is there refraction for incident angles of (a) 0° and (b) 90°?
7. From the Earth, stars are seen to "twinkle." What does an astronaut see from the International Space Station (ISS) orbiting the Earth?
8. Explain why the pencil appears severed in the chapter-opening photo.
9. For any substance, the index of refraction is always greater than 1. Why?
10. Explain why cut diamonds have brilliance and why they have "fire."
11. Does atmospheric refraction affect the length of the day? Would the daylight hours be longer or shorter if the Earth had no atmosphere?

7.3 Spherical Mirrors

12. What relationships exist between the center of curvature, the focal point, the focal length, and the vertex of a spherical mirror?
13. Distinguish between real images and virtual images for spherical mirrors.
14. Explain when real and virtual images are formed by (a) a convex mirror and (b) a concave mirror.

Erich Lessing/Art Resource, NY

Figure 7.38 *Self Portrait with Bandaged Ear* **(1889) by Vincent Van Gogh**
See Short-Answer Question 2.

Mirror

Figure 7.39 Reflection and Image
See Short-Answer Question 5.

15. What happens to a light ray that passes through the focal point at an angle to the optic axis of a concave mirror?

16. When a light ray parallel to the optic axis is reflected from a concave mirror, where does it go?

17. Why are the back surfaces of automobile headlights curved?

18. What type of mirror would be used for the solar heating of water?

7.4 Lenses

19. Where is a diverging lens thickest?

20. Explain when real and virtual images are formed by (a) a convex lens and (b) a concave lens.

21. Why are slides put into a slide projector upside down, and where is the slide relative to the projector lens?

22. A magnifying glass is a convex lens. Sunlight can be focused to a small spot using such a lens. What is the small spot an image of? Why are holes burnt in pieces of paper or leaves when the small spot is focused on them?

23. Why do some people wear bifocal or trifocal eyeglasses?

7.5 Polarization

24. Give two examples of practical applications of polarization.

25. Is it possible for the human eye to detect polarized light? If not, why not?

26. How could you use polarization to distinguish between longitudinal and transverse waves?

27. While you are looking through two polarizing sheets, one of the sheets is rotated 180°. Will there be any change in what you observe? Explain.

7.6 Diffraction and Interference

28. How can it be shown or proved that light is diffracted?

29. Why do sound waves bend around everyday objects, whereas the bending of light is not generally observed?

30. Which are more easily diffracted by ordinary objects, AM radio waves or FM radio waves? Explain why.

31. Describe the interference of two wave pulses with different amplitudes if they are (a) in phase and (b) completely out of phase.

32. What optical phenomenon causes soap bubbles and oil slicks to show colorful displays?

VISUAL CONNECTION

Visualize the connections and give answers for the blanks. Compare your answers with those at the back of the book.

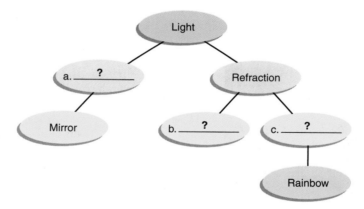

APPLYING YOUR KNOWLEDGE

1. If the Moon's spherical surface gave specular reflection, what would it resemble at full moon?

2. When you look at a window from the inside at night, two similar images, one behind the other, are often seen. Why?

3. On most automobile passenger-side rearview mirrors, a warning is printed: "Objects in mirror are closer than they appear." Why, and what makes the difference? (*Hint*: The mirrors are convex mirrors.)

4. How would a fish see the above-water world when looking up at various angles? (*Hint*: Think in terms of the critical angle and the "cone" of light coming in from above the water in reverse-ray tracing.)

5. You wish to buy a second pair of polarizing sunglasses. How can you check to make certain that the new glasses are indeed polarizing?

6. While you are looking through two polarizing sheets, one of the sheets is slowly rotated 180°. Describe what you will observe.

IMPORTANT EQUATIONS

Index of Refraction: $n = \dfrac{c}{c_m}$ (7.1)

Spherical Mirror Radius and Focal Length Equation: (7.2)

$$f = \frac{R}{2}$$

EXERCISES

7.1 Reflection

1. Light is incident on a plane mirror at an angle of 30° relative to the normal. What is the angle of reflection?

    Answer: 30°

2. Light is incident on a plane mirror at an angle of 30° relative to its surface. What is the angle of reflection?

3. Show that for a person to see his or her complete (head-to-toe) image in a plane mirror, the mirror must have a length (height) of at least one-half a person's height (see Fig. 7.4). Does the person's distance from the mirror make a difference? Explain.

    Answer: Bisecting triangles in the figure give one-half height. This is true for any distance.

4. How much longer must the minimum length of a plane mirror be for a 6-ft 4-in. man to see his complete head-to-toe image than for a 5-ft 2-in. woman to do so?

7.2 Refraction and Dispersion

5. What is the speed of light in a diamond?

    Answer: 1.24×10^8 m/s

6. The speed of light in a particular type of glass is 1.60×10^8 m/s. What is the index of refraction of the glass?

7. What percentage of the speed of light in vacuum is the speed of light in crown glass?

    Answer: 65.8%

8. The speed of light in a certain transparent material is 41.3% of the speed of light in vacuum. What is the index of refraction of the material? (Can you identify the material?)

7.3 Spherical Mirrors

(Assume significant figures to 0.1 cm.)

9. Sketch a ray diagram for a concave mirror with an object at $D_o = R$ and describe the image characteristics.

    Answer: Real, inverted, and same size

10. Sketch ray diagrams for a concave mirror showing objects at (a) $D_o > R$, (b) $D_o > f$, and (c) $D_o < f$. Describe how the image changes as the object is moved toward the mirror.

11. An object is placed 15 cm from a convex spherical mirror with a focal length of 10 cm. Estimate where the image is located and what its characteristics are.

    Answer: $D_i = 6.0$ cm, virtual, upright, and reduced

12. A reflecting, spherical Christmas tree ornament has a diameter of 8.0 cm. A child looks at the ornament from a distance of 15 cm. Describe the image she sees.

7.4 Lenses

13. Sketch a ray diagram for a spherical convex lens with an object at $D_o = 2f$ and describe the image characteristics.

    Answer: Real, inverted, and same size

14. Sketch ray diagrams for a spherical convex lens with objects at (a) $D_o > 2f$, (b) $2f > D_o > f$, and (c) $D_o < f$. Describe how the image changes as the object is moved closer to the lens.

15. An object is placed 45 cm in front of a converging lens with a focal length of 20 cm. Draw a ray diagram. Estimate the image distance and give the image characteristics.

    Answer: $D_i = 36$ cm, real, inverted, reduced

16. An object is placed in front of a converging lens at an object distance of twice the focal length of the lens. Sketch a ray diagram and compare the image and object distances. Repeat with two more ray diagrams, using different focal lengths and still making the object distance twice the focal length. Can you draw any conclusions by comparing the object distance and the image distance?

17. A particular convex lens has a focal length of 15 cm, and an object is placed at the focal point. Draw a ray diagram and comment on where the image is formed.

    Answer: Parallel rays never meet, or meet at infinity. (Thus, there really is no image.)

18. A spherical concave lens has a focal length of 20 cm, and an object is placed 15 cm from the lens. Draw a ray diagram. Estimate the image distance and give the image characteristics.

ON THE WEB

1. The Cause of Refraction

What is light? What happens when light hits an object? Why does light refract? What causes this behavior? Why is there one angle of incidence at which no refraction occurs? What is this angle? Does refractive behavior always occur? Explore answers to these questions by following the recommended links on the student website at **www.cengagebrain.com/shop/ISBN/1133104096.**

2. Let's Wish on a Rainbow

Have you ever wondered about the "physics" of a rainbow? Have you ever wondered exactly what a rainbow is? What will affect whether you see a rainbow (or two)? What do dispersion, refraction, and reflection have to do with rainbows? To answer these questions, follow the recommended links on the student website at **www.cengagebrain.com/shop/ISBN/1133104096.**

Electricity and Magnetism

Like charges repel, and unlike charges attract each other, with a force that varies inversely with the square of the distance between them. . . . Frictional forces, wind forces, chemical bonds, viscosity, magnetism, the forces that make the wheels of industry go round—all these are nothing but Coulomb's law.

•

J. R. Zacharias
(1905–1986)

Electrical transmission lines > transport electrical energy over long distances. Here the towers run through a field of sunflowers.

© Lester Lefkowitz/CORBIS

PHYSICS FACTS

▶ Electric eels can kill or stun prey by producing voltages up to 650 V, more than 50 times the voltage of a car battery.

▶ A tooth hurts when aluminum foil touches an amalgam (metal) filling because a voltage is generated when two different metals are separated by a conducting liquid, in this case saliva.

O urs is indeed an electrical society. Think of how your life might be without electricity. Some idea of this is obtained during extended power outages. Yet when asked to define electricity, many people have difficulty. The terms *electric charge* and *electric current* come to mind, but what are they?

You may recall from Chapter 1 that *electric charge* was mentioned as a fundamental quantity. That is, we really don't know what it is, so our chief concern is what it does, which is the description of electrical phenomena.

As you will learn in this chapter, electric charge is associated with certain particles that have interacting forces. With a force, there is motion of electric charges (current) as well as electrical energy and power. Understanding these principles

Chapter Outline

makes the benefits of electricity available to us. Electricity runs motors, heats food, provides lighting, powers our televisions and stereos, and so on.

But the *electric force* is even more basic than "electricity." It keeps atoms and molecules—even the ones that make up our bodies—together. It may be said that the electric force holds matter together, whereas the gravitational force (Chapter 3.5) holds our solar system and galaxies together.

Closely associated with electricity is *magnetism*. In fact, we refer to *electromagnetism* because these phenomena are basically inseparable. For example, without magnetism, there would be no generation of electrical power. As children (and perhaps as adults), most of us have been fascinated with the properties of small magnets. Have you ever wondered what causes magnets to attract and repel each other?

This chapter introduces the basic properties of electricity and magnetism. Examples of these exciting phenomena are everywhere around us.

8.1 Electric Charge, Electric Force, and Electric Field

Preview Questions

- What is the difference between the law of charges and Coulomb's law?
- What is static electricity?

Electric charge is a fundamental quantity. The property of electric charge is associated with certain subatomic particles, and experimental evidence leads to the conclusion that there are two types of charges, *positive* (+) and *negative* (−). In general, all matter is made up of small particles called *atoms*, which are composed in part of negatively charged particles called **electrons**, positively charged particles called **protons**, and particles called *neutrons* that have no electric charge (they are electrically neutral) and are slightly more massive than protons. ● Table 8.1 summarizes the fundamental properties of these atomic particles, which are discussed in more detail in Chapters 9 and 10.

As the table indicates, all three particles have certain masses, but only electrons and protons possess electric charges. The magnitudes of the electric charges on the electron and the proton are equal, but their natures are different, as expressed by the plus and minus signs. When there is the same number of electrons and protons, the *total* charge is zero (same number of positive and negative charges of equal magnitude), and the atom as a whole is electrically *neutral*.

The unit of electric charge is the *coulomb* (C), after Charles Coulomb (1736–1806), a French scientist who studied electrical effects. Electric charge is usually designated by the letter q. The symbol $+q$ indicates that an object has an excess number of positive charges, or fewer electrons than protons; and $-q$ indicates an excess of negative charges, or more electrons than protons.

Electric Force

An electric force exists between any two charged particles. On investigation, it is found that the mutual forces on the particles may be either attractive or repulsive, depending on the types of charges (+ or −). In fact, it is because of these different force interactions that

PHYSICS FACTS *cont.*

- The voltage and current in an average lightning strike are 5,000,000 V and 25,000 A, respectively. The air in the vicinity is heated to a temperature of 20,000°C, about three times the temperature of the Sun's surface.

- For a current in a metal wire, the electric field travels near the speed of light (in the wire), which is much faster than the speed of the charge carriers themselves. The speed of the latter is only about a millimeter per second.

- During the twentieth century, the north magnetic pole drifted northwest at an average speed of 10 km (6 mi) per year. Now it appears to be moving faster.

Table 8.1 Some Properties of Atomic Particles

Particle	Symbol	Mass	Charge
Electron	e^-	9.11×10^{-31} kg	-1.60×10^{-19} C
Proton	p^+	1.673×10^{-27} kg	$+1.60 \times 10^{-19}$ C
Neutron	n	1.675×10^{-27} kg	0

we know there are two different types of charges. Recall from Chapter 3.5 that for gravitation there is only one type of mass and that the force interaction between masses is always attractive.

The attraction and repulsion between different types of charges are described by the **law of charges**:

Like charges repel; unlike charges attract.

In other words, two negative charges (charged particles) or two positive charges experience repulsive electric forces: forces equal and opposite (Newton's third law). A positive charge and a negative charge experience attractive forces: forces toward each other. (Newton's third law still applies. Why?)

The law of charges gives the direction of an electric force, but what about its magnitude? In other words, how strong is the electric force between charged particles or bodies? Charles Coulomb derived a relationship for the magnitude of the electric force between two charged bodies that is appropriately known as **Coulomb's law**:

The force of attraction or repulsion between two charged bodies is directly proportional to the product of the two charges and inversely proportional to the square of the distance between them.

Written in equation form,

$$\text{force} = \frac{\text{constant} \times \text{charge 1} \times \text{charge 2}}{(\text{distance between charges})^2}$$

$$F = \frac{kq_1q_2}{r^2} \qquad\qquad 8.1$$

where F is the magnitude of the force in newtons, q_1 the magnitude of the first charge in coulombs, q_2 the magnitude of the second charge in coulombs, and r the distance between the charges in meters.

Here k is a proportionality constant with the value of

$$k = 9.0 \times 10^9 \text{ N} \cdot \text{m}^2/\text{C}^2$$

Coulomb's law is similar in form to Newton's law of universal gravitation (Chapter 3.5, $F = Gm_1m_2/r^2$). Both forces depend on the inverse square of the separation distance. One obvious difference between them is that Coulomb's law depends on charge, whereas Newton's law depends on mass.

Two other important differences exist. One is that Coulomb's law can have either an attractive or a repulsive force, depending on whether the two charges are different or the same (law of charges). The force of gravitation, on the other hand, is *always* attractive.

The other important difference is that the electric force is comparatively much stronger than the gravitational force. For example, an electron and a proton are attracted to each other both electrically and gravitationally. However, the gravitational force is so relatively weak that it can be ignored, and only the electric forces of attraction and repulsion are considered. (See Exercise 4.)

An object with an excess of electrons is said to be *negatively charged*, and an object with a deficiency of electrons is said to be *positively charged*. A negative charge can be placed on a rubber rod by stroking the rod with fur. (Electrons are transferred from the fur to the rod by friction. This process is called *charging by friction*.) In ● Fig. 8.1a, a rubber rod that has been stroked with fur and given a net charge is shown suspended by a thin thread that allows the rod to swing freely. The charge on the rod is negative. When another rubber rod that has been negatively charged is brought close to the suspended rod, the one that is free to move will swing away; the charged rods repel each other (like charges repel).

Doing the same thing with two glass rods that have been stroked with silk has similar results (Fig. 8.1b). Here electrons are transferred from the rods to the silk, leaving a positive charge on each rod.

The experiments show repulsion in both cases, but as shown in Fig. 8.1c, the charges on the stroked rubber rod and those on the glass rod attract one another (unlike charges

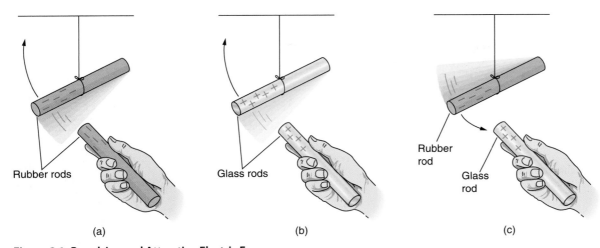

(a)   (b)   (c)

Figure 8.1 Repulsive and Attractive Electric Forces
(a) Two negatively charged objects repel each other. (b) Two positively charged objects repel each other. (c) A negatively charged object and a positively charged object attract each other.

attract). The charge on the glass rod is positive, and the charge on the rubber rod is negative. A charged object is said to have a static, or electrostatic, charge. *Electrostatics* is the study of charge at rest.

Static charge can be a problem. After walking across a carpet, you have probably been annoyingly zapped by a spark when reaching for a metal doorknob. You were charged by friction in crossing the carpet, and the electric force was strong enough to cause the air to ionize and conduct charge to the metal doorknob. This zapping is most likely to occur on a dry day. With high humidity (a lot of moisture in the air), there is a thin film of moisture on objects, and charge is conducted away before it can build up. Even so, such sparks are undesirable in the vicinity of flammable materials such as in an operating room with explosive gases or around gas pumps at a filling station (● Fig. 8.2).

From Coulomb's law (Eq. 8.1), it can be seen that as two charges get closer together, the force of attraction or repulsion increases. This effect can give rise to regions of charge, as illustrated in ● Fig. 8.3a. When a negatively charged rubber comb is brought near small pieces of paper, the charges in the paper molecules are acted on by electric forces—positive charges are attracted, negative charges repelled—and the result is an effective separation of charge. The molecules are then said to be *polarized*: they possess definite regions of charge.

Because the positive-charge regions are closer to the comb than the negative-charge regions, the attractive forces are stronger than the repulsive forces. Thus, a net attraction exists between the comb and the pieces of paper. Small bits of paper can be picked up by the comb, which indicates that the attractive electric force is greater than the paper's weight (the gravitational force on it).

Keep in mind, however, that overall the paper is uncharged; it is electrically neutral. Only molecular regions within the paper are charged. This procedure is termed *charging by induction*.

Conceptual Question and Answer

Defying Gravity

Q. Why does a balloon stick to a ceiling or wall after being rubbed on a person's hair or clothing (Fig. 8.3b)?

A. The balloon is charged by the frictional rubbing, which causes a transfer of charge. When the balloon is placed on a wall, the charge on the surface of the balloon induces regions of charge in the molecules of the wall material, attracting and holding the balloon the ceiling.

Figure 8.2 Static Danger
Warnings on gas pumps note the danger of static electric spark and how to prevent it.

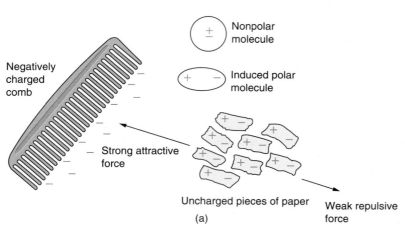

Figure 8.3 Polarization of Charge
(a) When a negatively charged comb is brought near small pieces of paper, the paper molecules are polarized with definite regions of charge, giving rise to a net attractive force. As a result, the bits of paper are attracted to and cling to the comb. (b) The force associated with electrical charges causes hair to stand on end and balloons to stick to the ceiling.

Another demonstration of electric force is shown in ● Fig. 8.4. When a charged rubber rod is brought close to a thin stream of water, the water is attracted toward the rod and the stream is bent. Water molecules have a permanent separation of charge, or regions of different charges. Such molecules are called *polar molecules* (Chapter 12.5).

Electric Field

The electric force, like the gravitational force, is an "action-at-a-distance" force. Since the electric force has an infinite range ($F \propto 1/r^2$) and approaches zero only as r approaches infinity, a charge can have an effect on any additional charge placed anywhere.

The idea of a force acting at a distance through space was difficult for early investigators to accept, and the concept of a *field* was introduced. In this approach, only the effect of the electrical interaction is considered, not the cause. We think of a charge interacting with an electric field rather than another charge responsible for it. An **electric field** surrounds a charge and represents the physical effect of a particular charge in nearby space. When another charge is placed in the field, the field will exert an electric force on that charge. This approach could have been used in Chapter 3.5 with the interaction of a mass with the gravitation field of another mass.

You can imagine determining or mapping out an electric field by using a small positive charge (a *test charge*) at a location near the charge of field interest. The force (magnitude and direction) on the test charge is recorded. When the electric force is determined at many locations, we have a vector "map" of the electric force field. The force is divided by a unit positive charge, and the electric field ($E = F/q_+$) is the force per unit charge. (Like force, the electric field E is a vector.) In this manner, if an arbitrary charge is placed in the field, then the magnitude of the force on it can be found by $F = q_+E$, with the direction depending on the sign of the charge or how it would react to a positive test charge.

The electric fields for some charges are illustrated in ● Fig. 8.5. For a single positive charge in Fig. 8.5a, the vectors point away from the charge. (Why?) Note that the vectors get shorter the farther from the charge as the force diminishes with distance.

Two configurations of charge are shown in Fig. 8.5b. When the field vectors are connected, we have *lines of force*, where the arrowheads indicate the force direction on a positive charge. If an arbitrary charge were put in a field and released, it would follow one of these lines, the field acting on the charge. The lines between the positive and negative charges begin and end on the charges, respectively. This indicates that the charges are

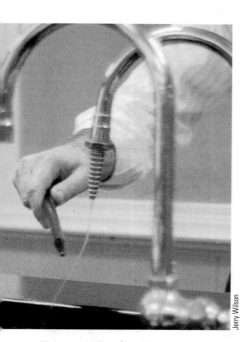

Figure 8.4 Bending Water
A charged rod brought close to a small stream of water will bend the stream because of the polarization of the water molecules.

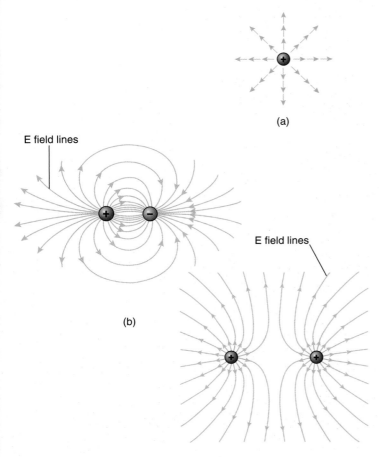

Figure 8.5 Electric Field Lines
(a) Electric field lines near a positive charge. (b) Electric field lines between unlike and like charges.

attractive, and the closer together the lines of force, the stronger the field. For the two posi-tive charges (or two negative charges), there are no lines of force between, which indicates that the charges are repulsive.

Did You Learn?

- Coulomb's law gives the magnitude of the electrical forces between two charges, but not the direction. The law of charges indicates attraction or repulsion between differ-ent types of charges, giving the direction of the electric force.

- Static electricity, or electrostatics, is the study of charge at rest.

8.2 Current, Voltage, and Electrical Power

Preview Questions

- What is electric current, and how is it expressed?
- What is joule heat?

Current

When a net charge flows or moves in one direction, there is a direct electric current. **Current** is defined as the time rate of flow of electric charge:

$$\text{current} = \frac{\text{charge}}{\text{time}}$$

$$\text{or} \qquad\qquad I = \frac{q}{t} \qquad\qquad\qquad 8.2$$

where I is the electric current in amperes, q the electric charge flowing past a given point in coulombs, and t the time for the charge to move past the point in seconds.

As Eq. 8.2 indicates, the unit of current is *coulomb per second* (q/t). This combination of units is called an *ampere* (A) in honor of French physicist André Ampère (1775–1836), an early investigator of electric and magnetic phenomena. The word *ampere* is commonly shortened to *amp*; for example, a current of 5 A is read as "five amps."

Early theories considered electrical phenomena to be due to some type of fluid in materials, which is probably why we say a current *flows*. Actually, it is electric charge that "flows." Electrical *conductors* are materials in which an electric charge flows readily. Metals are good conductors. Metal wires are widely used to conduct electric currents. This conduction is due primarily to the outer, loosely bound electrons of the atoms. (Recall from Chapter 5.4 that electrons also contribute significantly to thermal conduction in metals.)

Materials in which electrons are more tightly bound do not conduct electricity very well, and they are referred to as electrical *insulators*. Examples are wood, glass, and plastics. Electric cords are coated with rubber or plastic so that they can be handled safely. Materials that are neither good conductors nor good insulators are called *semiconductors;* graphite (carbon) and silicon are examples.

In the definition of current in Eq. 8.2, we speak of an amount of charge flowing past a given point. This is *not* a flow of charge in a manner similar to fluid flow. In a metal wire, for example, the electrons move randomly and chaotically, colliding with the metal atoms. Some go in one direction past a point, and others go in the opposite direction. An electric current exists when more electrons move in one direction than in the other. Thus, q is really a *net* charge in Eq. 8.2 (analogous to a *net* force in Chapter 3.1).

The net flow of charge is characterized by an average velocity called the *drift velocity*. The drift velocity is much smaller than the random velocities of the electrons themselves. Typically, the magnitude of the drift velocity is on the order of 1 mm/s. At this speed, it would take an electron about 17 min to travel 1 m in the wire.

You may be wondering how electrical signals in wires, such as telephone signals, can be transferred almost instantaneously across the country. The answer is that it is the electrical field that is transmitted at near the speed of light (in the wire), not the charge.

Voltage

The effects produced by moving charges give rise to what is generally called *electricity*. For charges to move, they must be acted on by other positive or negative charges. Consider the situation shown in ● Fig. 8.6. Start out with some unseparated charges and then begin to separate them. It takes very little work to pull the first negative charge to the left and the first positive charge to the right. When the next negative charge is moved to the left, it is repelled by the negative charge already there, so more work is needed. Similarly, it takes more work to move the second positive charge to the right. And as more and more charges are separated, it takes more and more work.

Because work is done in separating the charges, there is **electric potential energy**. If a separated charge were free to move, then it would move toward the charges of opposite sign. For example, a negative charge, as shown in Fig. 8.6, would move toward the positive charges. Electric potential energy would then be converted into kinetic energy as required by the conservation of energy.

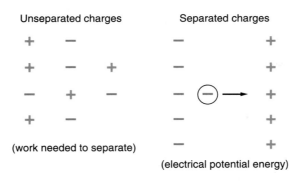

Unseparated charges   Separated charges

(work needed to separate)

(electrical potential energy)

Figure 8.6 Electric Potential Energy Work must be done to separate positive and negative charges. The work is done against the attractive electric force. When separated, the charges have electric potential energy and would move if free to do so.

Instead of electric potential energy, a related but different quantity called *potential difference*, or *voltage*, is most often considered. Voltage is defined as the amount of work it would take to move a charge between two points divided by the value of the charge. In other words, **voltage** (V) is the work (W) per unit charge (q), or the electric potential energy per unit charge.

$$\text{voltage} = \frac{\text{work}}{\text{charge}}$$

or $\qquad\qquad V = \frac{W}{q}$ $\qquad\qquad$ 8.3

The *volt* (V) is the unit of voltage and is equal to one joule per coulomb (J/q). Voltage is caused by a separation of charge. When work is done in separating the charges, there is electric potential energy, which may be used to set up a current. The symbol for voltage is an italic "vee" (V), whereas the symbol for the volt unit is a roman "vee" (V).*

When there is a current, it meets with some opposition because of collisions within the conducting material. This opposition to the flow of charge is called **resistance** (R). The unit of resistance is the *ohm* (Ω, the Greek letter omega). A simple relationship involving voltage, current, and resistance was formulated by Georg Ohm (1787–1854), a German physicist, and applies to many materials. It is called **Ohm's law** and in equation form may be written

$$\text{voltage} = \text{current} \times \text{resistance}$$

or $\qquad\qquad V = IR$ $\qquad\qquad$ 8.4

From this equation, it can be seen that one ohm resistance is one volt per ampere ($R = V/I$).

An example of a simple electric circuit is shown in ● Fig. 8.7a, together with a circuit diagram. The water circuit analogy given in Fig. 8.7b may help you better understand the components of the electric circuit. The battery provides the voltage to drive the circuit through chemical activity (chemical energy). This is analogous to the pump driving the water circuit. When the switch is *closed* (when the valve is opened in the water circuit), there is a current in the circuit. Electrons move away from the negative terminal of the battery toward the positive terminal.

The light bulb in the circuit offers resistance, and work is done in lighting it, with electrical energy being converted into heat and radiant energy. The waterwheel in the water circuit provides analogous resistance to the water flow and uses gravitational potential energy to do work. Note that there is a voltage, or potential, difference (drop) across the bulb, similar to the gravitational potential difference across the waterwheel. The components of an electric circuit are represented by symbols in a circuit diagram, as shown in the figure.

The switch in the circuit allows the path of the electrons to be open or closed. When the switch is open, there is not a complete path or circuit through which charge can flow and there is no current. (This is called an *open* circuit.) When the switch is closed, the circuit is completed, and there is a current. (The circuit is then said to be *closed*.) A sustained electric current requires a closed path or circuit.

*The volt unit is named in honor of Alessando Volta (1745–1827), an Italian scientist who constructed the first battery.

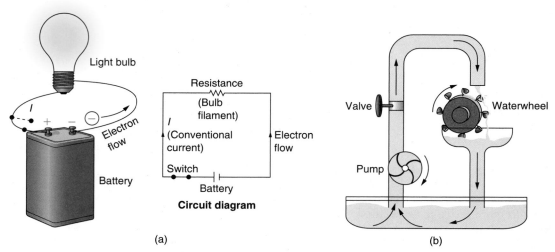

Figure 8.7 Simple Electric Circuit and a Water Analogy
(a) A simple electric circuit in which a battery supplies the voltage and a lightbulb supplies the resistance is shown. When the switch is closed, electrons flow from the negative terminal of the battery toward the positive terminal. Electrical energy is expended in heating the bulb filament. A circuit diagram with the component symbols is at the right. (b) In the water "circuit," the pump is analogous to the battery, the valve is analogous to the switch, and the waterwheel is analogous to the light bulb in furnishing resistance. Energy is expended, or work is done, in turning the waterwheel. (See text for more details.)

Note in the circuit diagram in Fig. 8.7a that the current (I) is in the opposite direction around the circuit to that of the electron flow. Even though electron charges are flowing in the circuit, it is customary to designate the *conventional current I* in the direction in which positive charges would flow. This designation is a historical remnant from Ben Franklin, who once advanced a fluid theory of electricity. All bodies supposedly contained a certain normal amount of this mysterious fluid, a surplus or deficit of which gave rise to electrical properties. With an excess resulting from a fluid flow, a body was positively "excited."

This theory later gave rise to the idea that it was the positive charges that flowed or moved. (Electrons were unknown at the time.) In any case, the current direction is still designated in the conventional sense or in the direction in which the positive charges would flow in the circuit, that is, away from the positive terminal of the battery and toward the negative terminal.

Electric Power

When current exists in a circuit, work is done to overcome resistance and power is expended. Recall from Eq. 4.8 that one definition of power (P) is

$$P = \frac{W}{t}$$

From Eq. 8.3, $W = qV$, and substituting for W yields

$$P = \frac{q}{t} V$$

Substituting $q/t = I$ gives an equation in terms of current and voltage for **electric power**:

$$\text{power} = \text{current} \times \text{voltage}$$
$$P = IV \qquad\qquad 8.5$$

Using Ohm's law ($V = IR$) for V, we have $P = I(IR)$, and

$$\text{power} = (\text{current})^2 \times \text{resistance}$$
$$P = I^2R \qquad\qquad 8.6$$

The power that is dissipated in an electric circuit is frequently in the form of heat. This heat is called *joule heat*, or I^2R *losses* (read *I* squared *R* losses), as given by Eq. 8.6. This heating effect is used in electric stoves, heaters, cooking ranges, hair dryers, and so on. Hair dryers have heating coils of low resistance so as to get a large current for large I^2R losses. When a light bulb is turned on, much of the power goes to produce heat as well as light. The unit of power is the watt, and light bulbs are rated in watts (● Fig. 8.8).

● Table 8.2 gives some typical power requirements for a few common household appliances.

(a)

(b)

Figure 8.8 Wattage (Power) Ratings
(a) A 60-W light bulb dissipates 60 J of electrical energy each second. (b) The curling iron uses 13 W at 120 V. Given the wattage and voltage ratings, you can find the current drawn by an appliance by using $I = P/V$.

EXAMPLE 8.1   Finding Current and Resistance

Find the current and resistance of a 60-W, 120-V light bulb in operation.

Solution

Step 1

Given: P = 60 W (power)
            V = 120 V (voltage)

Step 2

Wanted: *I* (current)
             R (resistance)

Step 3

The current is obtained using Eq. 8.5, $P = IV$. Rearranging yields

$$I = \frac{P}{V} = \frac{60 \text{ W}}{120 \text{ V}} = 0.50 \text{ A}$$

Equation 8.4 (Ohm's law) can be rearranged to solve for resistance:

$$R = \frac{V}{I} = \frac{120 \text{ V}}{0.50 \text{ A}} = 240 \ \Omega$$

Note that Eq. 8.6 could also be used to solve for *R*:

$$R = \frac{P}{I^2} = \frac{60 \text{ W}}{(0.50 \text{ A})^2} = 240 \ \Omega$$

Confidence Exercise 8.1

A coffeemaker draws 10 A of current operating at 120 V. How much romelectrical energy does the coffeemaker use each second?

Answers to Confidence Exercises may be found at the back of the book.

There are two principal forms of electric current and voltage. In a battery circuit, such as that shown in Fig. 8.7, the electron flow is always in one direction, away from the negative terminal and toward the positive terminal. This type of current is called **direct current**, or **dc**. Direct current is used in battery-powered devices such as flashlights, portable radios, and automobiles. Dc voltage usually has a steady, constant value. Batteries are rated in this voltage.

The other common type of current is **alternating current**, or **ac**, which is produced by a constantly changing (alternating) voltage from positive (+) to negative (−) to positive (+), and so on. (Although the usage is redundant, we commonly say "ac current" and "ac voltage.") Alternating current is produced by electric companies and is used in the home. (Alternating current and voltage generation are discussed in Chapter 8.5.)

The frequency of changing from positive to negative voltages is usually at the rate of 60 cycles per second (cps) or 60 Hz (see Fig. 8.8b). The average voltage varies from 110 V to 120 V, and household ac voltage is commonly listed as 110 V, 115 V, or 120 V. The equations for Ohm's law (Eq. 8.4) and power (Eqs. 8.5 and 8.6) apply to both dc and ac circuits containing only resistances.

Table 8.2 Typical Power Requirements of Some Household Appliances

Appliance	Power (W)
Air conditioner	
Room	1500
Central	5000
Coffeemaker	1650
Dishwasher	1200
Water heater	4500
Microwave oven	1250
Refrigerator	500
Stove	
Range top	6000
Oven	4500
Television (color)	100

For the most part, we use 110-V ac voltage in household circuits. However, in Europe, the common household voltage is 220-V ac. To learn why the difference exists, see the **Highlight: United States and Europe: Different Voltages**.

Did You Learn?

● When a net charge flows in one direction, there is a direct electric current (I). Current is expressed as the time rate of flow of electric charge ($I = q/t$).

● Power is commonly dissipated in an electrical circuit in the form of heat, which is referred to as joule heat (or I^2R losses, where $P = IV = I^2R$).

8.3 Simple Electric Circuits and Electrical Safety

Preview Questions

● How do the currents in a series circuit and a parallel circuit differ?

● What happens when more resistance is added to (a) a series circuit and (b) a parallel circuit?

Once electricity enters the home or business, it is used in circuits to power various appliances and other items. Plugging appliances, lamps, and other electrical applications into an outlet places them in a circuit. There are two basic ways to connect elements in a circuit: in *series* and *in parallel*.

An example of a series circuit is shown in ● Fig. 8.9. The lamps are conveniently represented as resistances in the circuit diagram. In a **series circuit**, the same current passes through all the resistances. This is analogous to a liquid circuit with a single line connecting several components. The total resistance is simply the sum of the individual resistances. As with different height potentials, the total voltage is the sum of the individual voltage drops, and

$$V = V_1 + V_2 + V_3 + \cdots$$
$$V = IR_1 + IR_2 + IR_3 + \cdots$$
or
$$V = I(R_1 + R_2 + R_3 + \cdots)$$

where the equation is written for three or more resistances. Writing $V = IR_s$, where R_s is the *total equivalent series resistance*, then by comparison,

total equivalent series resistance = summation of individual resistors
$$R_s = R_1 + R_2 + R_3 + \cdots \tag{8.7}$$
resistances in series

Equation 8.7 means that all the resistances in series could be replaced with a single resistance R_s. The same current would flow, and the same power would be dissipated. For

Figure 8.9 Series Circuit
The light bulbs are connected in series, and the current is the same through each bulb.

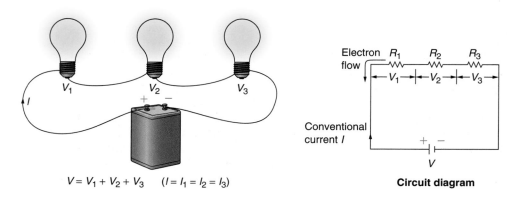

$V = V_1 + V_2 + V_3$ $(I = I_1 = I_2 = I_3)$

Circuit diagram

Highlight United States and Europe: Different Voltages

Many travelers to Europe have difficulty with their electrical appliances because the United States and a handful of other countries in the Americas use 110 V/60 Hz electricity. Europe and most other countries of the world use a 220 V/50 Hz system.* As a result, using appliances in Europe designed to operate on U.S. voltage can cause some real problems. For example, a 110-V hairdryer used in a 220-V European outlet would quickly burn up.

Why the different voltages? There is an historical explanation. In 1879 Thomas Edison invented an improved incandescent light bulb. He realized the need for an electrical distribution system to provide power for lighting and built a 110-V dc system that initially provided power in Manhattan. The dc system, with large wire conductors and big voltage drops, was somewhat cumbersome.

George Westinghouse introduced a distribution system based on 110-V alternating current (ac). Alternating current allows the voltage to be changed through the use of transformers. This allowed transmission at higher voltages and less current (see Chapter 8.5), thereby reducing line losses due to conductor resistance (joule heat) and making for greater transmission distances.

Alternating voltage generation took over from dc voltage generation, and power plants built in the early 1900s used 110 V/60 Hz voltage. There was some voltage variation, from 110 V to 115 V to 120 V, as there is today.

By the time most European countries got around to developing electrical distribution systems, engineers had figured out how to make 220-V bulbs. These bulbs did not burn out as quickly as 110-V bulbs, making them more economical. So, in Germany around the beginning of the twentieth century, 220-V/50 Hz (or 220 V–240 V) generation was adopted and spread throughout Europe. The United States stayed with 110 V because of the big investment in 110-V equipment. However, 220 V–240 V voltage is available on the U.S. three-wire system. (See Fig. 8.11 and text discussion there.) There are different plugs for 110-V and 220-V outlets, so the voltages cannot be mistaken.

Because of the voltage difference, travelers to Europe usually need to take a voltage converter with them. Nowadays, some devices, such as hair dryers, are designed to operate on either 110 V/60 Hz or 220 V/50 Hz by switching from one to the other. Equipment rated at 50 Hz or 60 Hz will usually operate on either cycle. However some devices, such as electric plug-in clocks that use the ac frequency for timing, may not function well.

One final note: Many countries use different plugs for normal outlets (Fig. 1). So, if you are going abroad, think about taking a plug adapter kit for connecting to foreign plugs.

———
*The 60 Hz and 50 Hz are the frequencies of the alternating voltages.

Stephen Kirschenmann/iStockphoto.com

Figure 1 Different Plugs
Different electrical plugs are used in different parts of the world. Going counterclockwise from the lower left; vertical prongs (North/South America), round prongs (Europe), three prongs (Great Britain), and slanted prongs (Australia).

---

example, $P = I^2 R_s$ is the power used in the whole circuit. (The resistances of the connecting wires are considered negligible.)

The example of lamps or resistances in series in Fig. 8.9 could just as easily have been an early string of Christmas tree lights, which used to be connected in a simple series circuit. When a bulb burned out, the whole string of lights went out because there was no longer a complete path for the current and the circuit was "open." Having a bulb burn out was like opening a switch in the circuit to turn off the lights. However, in most strings of lights purchased today, one light can burn out but the others remain lit. Each bulb has a secondary "shunt" resistor that takes over if the main filament resistor burns out.

The other type of simple circuit is called *a parallel circuit*, as illustrated in ● Fig. 8.10. In a **parallel circuit**, the voltage across each resistance is the same, but the current through each resistance may vary (different resistances, different currents). Note that the current from the voltage source (battery) divides at the junction where all the resistances are connected. This arrangement is analogous to liquid flow in a large pipe coming to a junction where it divides into several smaller pipes.

Figure 8.10 Parallel Circuit
The light bulbs are connected in parallel, and the current from the battery divides at the junction (where the three bulbs are connected). The amount of current in each parallel branch is determined by the relative values of the resistance in the branches; the greatest current is in the path of least resistance.

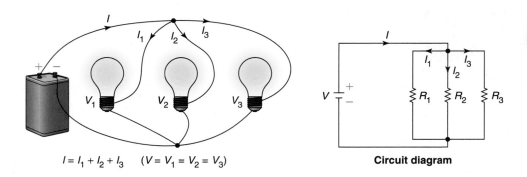

$I = I_1 + I_2 + I_3$ $(V = V_1 = V_2 = V_3)$

Circuit diagram

Because there is no buildup of charge at a junction, the charge leaving the junction must equal the charge entering the junction (law of conservation of charge), and in terms of current,

$$I = I_1 + I_2 + I_3 + \cdots$$

Using Ohm's law (in the form $I = V/R$), we have for the different resistances,

$$I = \frac{V}{R_1} + \frac{V}{R_2} + \frac{V}{R_3} + \cdots$$

or

$$I = V\left(\frac{1}{R_1} + \frac{1}{R_2} + \frac{1}{R_3} + \cdots\right)$$

The voltage V is the same across each resistance R because the voltage source is effectively connected "across" each resistance, and each gets the same voltage effect or drop.

If Ohm's law is written as $I = V/R_p$, where R_p is the *total equivalent parallel resistance*, then, by comparison,

> reciprocal of total equivalent = summation of the reciprocals
> parallel resistance   of the individual resistances
>
> $$\frac{1}{R_p} = \frac{1}{R_1} + \frac{1}{R_2} + \frac{1}{R_3} + \cdots$$ 8.8
>
> resistances in parallel

For a circuit with only two resistances in parallel, this equation can be conveniently written as

> $$R_p = \frac{R_1 R_2}{R_1 + R_2}$$ 8.9
>
> two resistances in parallel

As in the case of the series circuit, all the resistances could be replaced by a single resistance R_p without affecting the current from the battery or the power dissipation.

EXAMPLE 8.2   Resistances in Parallel

Three resistors have values of $R_1 = 6.0\ \Omega$, $R_2 = 6.0\ \Omega$, and $R_3 = 3.0\ \Omega$. What is their total resistance when connected in parallel, and how much current will be drawn from a 12-V battery if it is connected to the circuit?

Solution

Let's first combine R_1 and R_2 into a single equivalent resistance, using Eq. 8.9:

$$R_{p_1} = \frac{R_1 R_2}{R_1 + R_2} = \frac{(6.0\ \Omega)(6.0\ \Omega)}{6.0\ \Omega + 6.0\ \Omega} = 3.0\ \Omega$$

Hence, an equivalent circuit is a resistance R_{p_1} connected in parallel with R_3. Apply Eq. 8.9 to these parallel resistances to find the total resistance:

$$R_p = \frac{R_{p_1} R_3}{R_{p_1}} + R_3 = \frac{(3.0\ \Omega)(3.0\ \Omega)}{3.0\ \Omega + 3.0\ \Omega} = 1.5\ \Omega$$

The same current will be drawn from the source for a 1.5-Ω resistor as for the three resistances in parallel.

The problem also can be solved by using Eq. 8.8. In this case, R is found by using the lowest common denominator for the fractions (zeros and units initially omitted for clarity):

$$\frac{1}{R_p} = \frac{1}{R_1} + \frac{1}{R_2} + \frac{1}{R_3} = \frac{1}{6} + \frac{1}{6} + \frac{1}{3}$$

$$= \frac{1}{6} + \frac{1}{6} + \frac{2}{6} = \frac{4}{6\,\Omega}$$

or
$$R_p = \frac{6\,\Omega}{4} = 1.5\,\Omega$$

The current drawn from the source is then given by Ohm's law using R_p:

$$I = \frac{V}{R_p} = \frac{12\,V}{1.5\,\Omega} = 8.0\,A$$

Confidence Exercise 8.2

Suppose the resistances in Example 8.2 were wired in series and connected to the 12-V battery. Would the battery then supply more or less current than it would for the parallel arrangement? What would be the current in the circuit in this case?

The answers to Confidence Exercises may be found at the back of the book.

Conceptual Question and Answer

Series or Parallel

Q. Are automobile headlights wired in series or in parallel? How do you know?

A. Headlights are wired in parallel. When one headlight goes out, the other remains lit, which indicates a circuit path of a parallel circuit. You have probably seen a car approaching with only one headlight. If wired in series, then both headlights would go out.

An interesting fact:

> For resistances connected in parallel, the total resistance is always less than the smallest parallel resistance.

Such is the case in Example 8.2. Try to find a parallel circuit that proves otherwise. (Or better yet, forget it. You'd be wasting your time!)

Home appliances are wired in parallel (● Fig. 8.11). There are two major advantages of parallel circuits:

1. The same voltage (110–120 V) is available throughout the house, which makes it much easier to design appliances. (The 110–120 V voltage is obtained by connecting across the "hot," or high-voltage, side of the line to *ground*, or zero potential. This gives a voltage *difference* of 120 V, even if one of the "high" sides is at a potential of −120 V. The voltage for large appliances, such as central air conditioners and heaters, is 220–240 V, which is available by connecting across the two incoming potentials, as shown in Fig. 8.11. This potential is analogous to a height difference between two positions, one positive and one negative, for gravitational potential energy. See Fig. 4.8.)

2. If one appliance fails to operate, then the others in the circuit are not affected because their circuits are still complete. In a series circuit, when one component fails, none of the others will operate because the circuit is incomplete, or "open."

Figure 8.11 Household Circuits
As illustrated here, household circuits are wired in parallel. For small appliances, the circuit voltage is 120 V. Because there are independent branches, any particular circuit element can operate when others in the same circuit do not. For large appliances, such as a central air conditioner or electric stove, the connection is between the +120V and the −120V potential wires to give a voltage difference of 240 V.

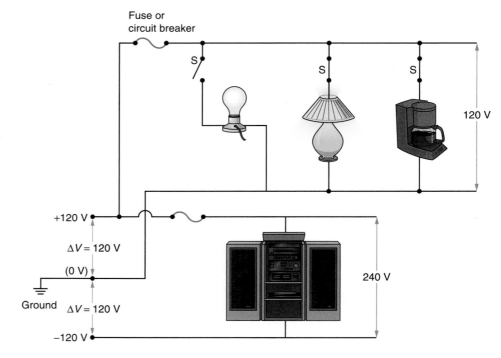

Resistances can be connected in *series-parallel circuits*, which give intermediate equivalent total resistances, but we will not examine these.

Conceptual Question and Answer

More Resistance, More Current

Q. When you turn on more lights and appliances in your home, the current demand in the circuit is greater. Why is this, given that you are adding more resistance with each component?

A. Household circuits are wired in parallel, and for resistances connected in parallel, the total resistance is always less than the smallest parallel resistance. Resistances may be added, but the total resistance is restricted to less than the smallest resistance, so there is more current.

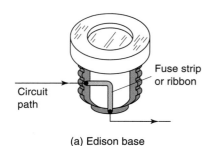

Figure 8.12 Fuses
(a) An Edison-base fuse. If the current exceeds the fuse rating, then joule heat causes the fuse strip or ribbon to burn out and the circuit is opened. (b) Type S fuses. Edison-base fuses have the same screw thread for different ratings, so a 30-A fuse could be put into a 15-A circuit, which would be dangerous. (Why?) Type S fuses have different threads for different fuse ratings and cannot be interchanged.

Electrical Safety

Electrical safety for both people and property is an important consideration in using electricity. For example, in household circuits as more and more appliances are turned on, there is more and more current and the wires get hotter and hotter (joule heat). The fuse or circuit breaker shown in the circuit diagram in Fig. 8.11 is a safety device that prevents the wires from carrying too much current, getting too hot, and possibly starting a fire. When the preset amount of current is reached, the fuse filament gets so hot that it melts and opens the circuit. Fuses are primarily found in older homes. Circuit breakers (described below) are installed in newer homes.

Two types of fuses are generally used in household circuits. The *Edison-base fuse* has a base with threads similar to those on a light bulb (● Fig. 8.12a). Thus. they will fit into any socket, which can create a problem. For example, a 30-A fuse might be screwed into a socket that should have a 15-A fuse. Such a mix-up could be dangerous, so *type S fuses* are often used instead (Fig. 8.12b). With type S fuses, a threaded adapter specific to a particular fuse is put into the socket. Fuses that are rated differently have different threads, and a 30-A fuse cannot be screwed into a 15-A socket. Household fuses are being phased out, but small fuses are still used in automobile and other circuits.

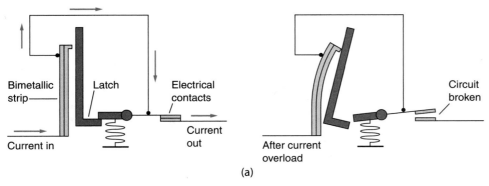

Figure 8.13 Circuit Breaker
(a) Thermal type. As the current through the bimetallic strip increases, it becomes warmer (joule heat) and bends. When the current-rated value is reached, the strip bends sufficiently to open the circuit. (b) Circuit breakers in a home electrical service entrance.

The more popular *circuit breaker* has generally replaced fuses and is required in newer homes. It serves the same function as fuses. A thermal type of circuit breaker, illustrated in ● Fig. 8.13a, uses a bimetallic strip (see Chapter 5.1). As the current through the strip increases, it becomes warmer (joule heat) and bends. When the current-rated value is reached, the strip bends sufficiently to open the circuit. The strip quickly cools, and the breaker can be reset. However, a blown fuse or a tripped circuit breaker indicates that the circuit is drawing or attempting to draw too much current. *You should find and correct the problem before replacing the fuse or resetting the circuit breaker.* Some circuit breakers operate on magnetic principles: the more current, the greater the magnetic attraction, and the circuit is opened at some preset current value. (See electromagnets in Chapter 8.4.)

Switches, fuses, and circuit breakers are always placed in the "hot," or high-voltage, side of the line. If they placed in the ground side, there would be no current when a circuit was opened. But there would still be a 120-V potential to the appliance, which could be dangerous if someone came in contact with it.

However, even when wired properly, fuses and circuit breakers may not always give protection from electrical shock. A hot wire inside an appliance or power tool may break loose and come into contact with its housing or casing, putting it at a high voltage. The fuse does not blow unless there is a large current. If a person touches a casing that is conductive, as illustrated in ● Fig. 8.14a, a path is provided to ground and the person receives an electric shock.

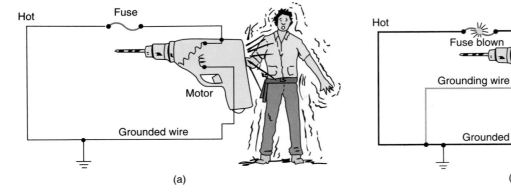

Figure 8.14 Electrical Safety by Use of Dedicated Grounding
(a) Suppose an internal "hot" wire broke and came in contact with the metal casing of an appliance. Without a dedicated ground wire, the casing would be at a high potential without a fuse being blown or a circuit breaker being tripped. If someone were to touch the casing, then a dangerous shock could result. (b) If the case were grounded with a dedicated ground wire through a third prong on the plug, then the circuit would be opened and the casing would be at zero potential.

(a)

(b)

Ken O'Donoghue

Figure 8.15 Electric Plugs
(a) A three-prong plug and socket. The third, rounded prong is connected to a dedicated grounding wire used for electrical safety. (b) A two-prong polarized plug and socket. Note that one blade, or prong, is larger than the other. The ground, or neutral (zero-potential) side of the line, is wired to the large-prong side of the socket. This distinction, or polarization, permits paths to ground for safety purposes.

This condition is prevented by *grounding* the casing, as shown in Fig. 8.14b. If a hot wire touches the casing, there would be a large current that would blow the fuse (or open a circuit breaker). Grounding is the purpose of the three-prong plugs found on many electrical tools and appliances (● Fig. 8.15a).

A *polarized* plug is shown in Fig. 8.15b. You have probably noticed that some plugs have one blade or prong larger than the other and will fit into a wall outlet only one way. Polarized plugs are an older type of safety feature. Being polarized, or directional, one side of the plug is always connected to the ground side of the line. The casing of an appliance can be connected to ground in this way, with an effect similar to that of the three-wire system. Because the polarized system depends on the circuit and the appliance being wired properly, there is a chance of error. So a dedicated grounding wire is better. Also, even though a polarized plug is wired to the ground side of the line, it is still a current-carrying wire, whereas the dedicated ground wire carries no current.

An electric shock can be very dangerous, and *touching exposed electric wires should always be avoided.* Many injuries and deaths occur from electric shock. The effects of electric shock are discussed in the **Highlight: Electrical Effects on Humans**.

Did You Learn?

● In a series circuit, the current is the same through all resistances. In a parallel circuit, the current through each resistance may vary (for different resistances).

● Resistance added to a series circuit increases the total resistance and less current flows for a given voltage. Resistance added to a parallel circuit reduces the total resistance and more current flows for a given voltage.

8.4 Magnetism

Preview Questions

● How are the law of poles and the law of charges similar?

● Where is the Earth's north magnetic pole?

Like most people, you have probably been fascinated by the attractive and repulsive forces between magnets: a hands-on example of force at a distance. Magnets are readily available today because we know how to make them, but they were once quite scarce and were found only as rocks in nature, or natural magnets.

Natural magnets, called lodestones, were discovered as early as the sixth century BCE in ancient Greece in the province of Magnesia, from which magnetism derives its name. Lodestones could attract pieces of iron and other lodestones. For centuries the attractive properties of natural magnets were attributed to supernatural forces. Early Greek philosophers believed that a magnet had a "soul" that caused it to attract pieces of iron. Now, we know otherwise.

Some time around the first century BCE, the Chinese learned to make artificial magnets by stroking pieces of iron with natural magnets. This led to one of the first practical applications of magnets, the compass, which implied that the Earth has magnetic properties. (The Chinese are said to have developed the compass, but several other peoples claim this invention, too.)

Probably the most familiar magnets are the common bar magnet and the horseshoe magnet (a bar magnet bent in the form of a horseshoe). One of the first things one notices when examining a bar magnet is that it has two regions of magnetic strength or concentration—one at each end of the magnet—which are called *poles*. One is designated as the north pole, N, and the other as the south pole, S. This is because the N pole of a magnet, when used as a compass, is the "north-seeking pole" (it points north) and the S pole is the "south-seeking pole."

Highlight Electrical Effects on Humans

When working with electricity, common sense and knowledge of fundamental electrical principles are important. Electric shocks can be very dangerous, and they kill and injure many people every year. A major problem can be poor maintenance (Fig. 1).

The danger is proportional to the amount of electric current that goes through the body. The amount of current going through the body is given by Ohm's law as

$$I = \frac{V}{R_{body}} \qquad (1)$$

where R_{body} is the resistance of the body.

A current of 5 mA (milliamp) can be felt as a shock (Table 1). A current of 100 mA is nearly always fatal.

The amount of current, as indicated in Eq. 1, depends critically on a person's body resistance. Human body resistance varies considerably, mainly as a result of whether the skin is wet or dry. Because our bodies are mostly water, skin resistance makes up most of the body's resistance.

A dry body can have a resistance as high as 500,000 Ω, and the current from a 110-V source would be only 0.00022 A, or 0.22 mA. Danger arises when the skin is moist or wet. Then, the resistance of the body can go as low as 100 Ω, and the current will rise to 1.1 A, or 1100 mA. Injuries and death from shocks usually occur when the skin is wet. Therefore, appliances such as radios should not be used near a bathtub. If a plugged-in radio happened to fall into the bathtub, then the whole tub, including the person in it, might be plugged into 110 V.

Despite the availability of various electrical safety devices, it does not take much current to cause human injuries and fatalities. Only milliamps of current are needed. Were you to come into contact with a hot wire and become part of a circuit, it is not only your body resistance that is important (as discussed) but also how you are connected in the circuit. If the circuit is completed through your hand (finger to thumb), then a shock and a burn can result. However, if the circuit is completed through the body from hand to hand or hand to foot, then the resulting effects can be much more serious, depending on the amount of current.

Note in Table 1 that a current of 15 mA to 25 mA can cause muscular freeze, so the person affected may not be able to let go of a hot wire. Muscles are controlled by nerves, which in turn are controlled by electrical impulses. Slightly larger currents can cause breathing difficulties, and just slightly larger currents can cause ventricular fibrillation, or uncontrolled contractions of the heart. A current greater than 100 mA, or 0.10 A, generally results in death. Keep in mind, for your personal electrical safety, that a little current goes a long way.

Figure 1 Electrical Hazards
Electrical hazards, such as this frayed wire, can be dangerous and cause fire and injury.

Table 1	Effects of Electric Currents on Humans	
Current (mA)	Effect*
1	Barely perceived
5–10	Mild shock
10–15	Difficulty in releasing
15–25	Muscular freeze, cannot release or let go
50–100	May stop breathing, ventricular fibrillation
100	Death

*Effects vary with individuals.

When examining two magnets, you will notice that there are attractive and repulsive forces between them that are specific to the poles. These forces are described by the **law of poles**:

Like poles repel; unlike poles attract.

In other words, N and S poles (N-S) attract, and N-N and S-S poles repel each other (● Fig. 8.16a). The strength of the attraction or repulsion depends on the strength of the magnetic poles. Also, in a manner similar to Coulomb's law, the strength of the magnetic

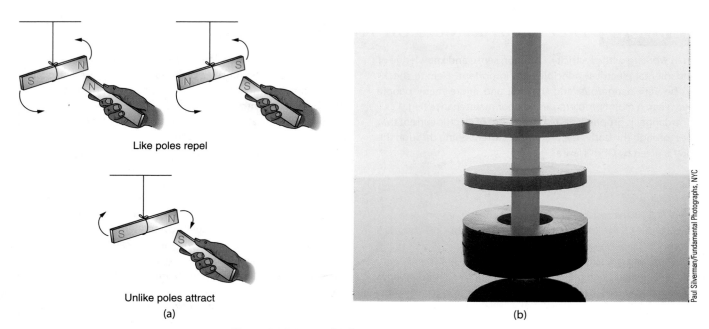

Like poles repel

Unlike poles attract

(a)

(b)

Figure 8.16 Laws of Poles

(a) Like poles repel, and unlike poles attract. (b) The adjacent poles of the circular magnets must be like poles. (Why?)

force is inversely proportional to the square of the distance between the poles. Figure 8.16b shows some toy magnets that seem to defy gravity as a result of their magnetic repulsion.

All magnets have two poles, so magnets are dipoles (*di*- means "two"). Unlike electric charge, which occurs in single charges, magnets are always dipoles. A *magnetic monopole* would consist of a single N or S pole without the other. There is no known physical reason for magnetic monopoles not to exist, but so far, their existence has not been confirmed experimentally. The discovery of a magnetic monopole would be an important fundamental development.

Every magnet produces a force on every other magnet. To discuss these effects, let's discuss the concept of a magnetic field, similar to an electric field in Chapter 8.1. A **magnetic field** (*B*) is a set of imaginary lines that indicates the direction in which a small compass needle would point if it were placed near a magnet. Hence, the field lines are indications of the magnetic force, or a force field. ● Figure 8.17a shows the magnetic field lines around a simple bar magnet. The arrows in the field lines indicate the direction in which the north pole of a compass needle would point. The closer together the field lines, the stronger the magnetic force.

Magnetic field patterns can be "seen" by using iron filings. The iron filings become magnetized and act as small compass needles. The outline of the magnetic field produced in this manner is shown for the bar magnet in Fig. 8.17b. The field concept is analogous to an *electric field* around charges, but with iron filings the magnetic field is more easily visualized. The electric and magnetic fields are vector quantities, and electromagnetic waves, as discussed in Chapter 6.3, are made up of electric and magnetic fields that vary with time.

Electricity and magnetism are generally discussed together because they are linked. (Electromagnetism is the topic of Chapter 8.5.) In fact, *the source of magnetism is moving and "spinning" electrons.* Hans Oersted, a Danish physicist, first discovered in 1820 that a compass needle is deflected by a wire carrying electric current. When a compass is placed near a wire in a simple battery circuit and the circuit is closed, there is current in the wire and the compass needle is deflected from its north-seeking direction. When the circuit is opened, the compass needle goes back to pointing north again.

The strength of the magnetic field is directly proportional to the magnitude of the current: the greater the current, the greater the strength of the magnetic field. Hence, a current produces a magnetic field that can be turned off and on at will.

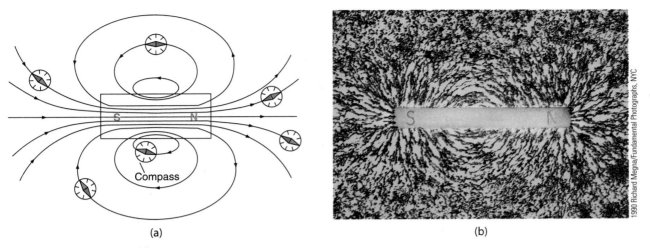

Figure 8.17 Magnetic Field
(a) Magnetic field lines may be plotted by using a small compass. The N pole of the compass needle (blue) points in the direction of the field at any point, (b) Iron filings become induced magnets and conveniently outline the pattern of the magnetic field.

Different configurations of current-carrying wires give different magnetic field configurations. For example, a straight, dc current–carrying wire produces a field in a circular pattern around the wire (● Fig. 8.18a). A single loop of wire gives a field comparable to that of a small bar magnet (Fig. 8.18b), and the field of a coil of wire with several loops is very similar to that of a bar magnet.

But what produces the magnetic field of a permanent magnet such as a bar magnet? In a basic model of the atom, electrons are pictured as going around the nucleus (Chapter 9.3). This is electric charge in motion, or a current loop, and it might be expected that it would be a source of a magnetic field. However, the magnetic field produced by orbiting atomic electrons is very small. Also, the atoms of a material are distributed such that the magnetic fields would be in various directions and would generally cancel each other, giving a zero net effect.

Modern theory predicts the magnetic field to be associated with electron "spin." This effect is pictured classically as an electron spinning on its axis in the same way that the

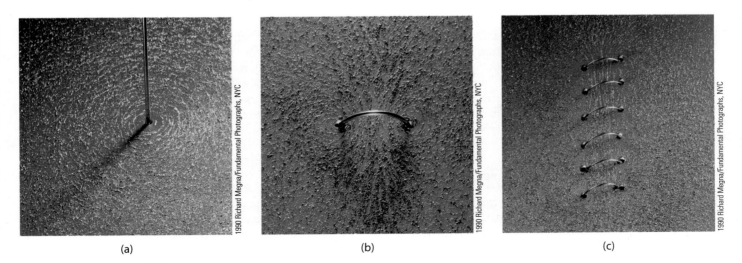

Figure 8.18 Magnetic Field Patterns
Iron filing patterns near current-carrying wires outline the magnetic fields for (a) a long, straight wire; (b) a single loop of wire; and (c) a coil of wire (a solenoid).

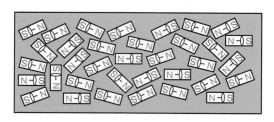

(a) Unmagnetized material

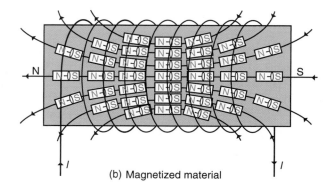

(b) Magnetized material

Figure 8.19 Magnetization

(a) In a ferromagnetic material, the magnetic domains are generally unaligned, so there is no magnetic field. For simplicity, the domains are represented by small bar magnets. (b) In a magnetic field produced by a current-carrying loop of wire, the domains become aligned with the field (the aligned domains may grow at the expense of others) and the material becomes magnetized.

Earth rotates on its axis.* A material has many atoms and electrons, and the magnetic spin effects of all these electrons usually cancel each other out. Therefore, most materials are not magnetic or are only slightly magnetic. In some instances, however, the magnetic effect can be quite strong.

Materials that are highly magnetic are called **ferromagnetic**. Ferromagnetic materials include the elements iron, nickel, and cobalt, as well as certain alloys of these and a few other elements. In ferromagnetic materials, the magnetic fields of many atoms combine to give rise to **magnetic domains**, or local regions of alignment. A single magnetic domain acts like a tiny bar magnet.

In iron the domains can be aligned or nonaligned. A piece of iron with the domains randomly oriented is not magnetic. This effect is illustrated in ● Fig. 8.19. When the iron is placed in a magnetic field, such as that produced by a current-carrying loop of wire, the domains line up or those parallel to the field grow at the expense of other domains, and the iron is magnetized. When the magnetic field is removed, the domains tend to return to a mostly random arrangement because of heat effects that cause disordering. The amount of domain alignment remaining after the field is removed depends on the strength of the applied magnetic field.

An application of this effect is an *electromagnet*, which consists of a coil of insulated wire wrapped around a piece of iron (● Fig. 8.20). Because a magnetic field can be turned on and off by turning an electric current on and off, it can be controlled whether or not the iron will be magnetized. When the current is on, the magnetic field of the coil magnetizes the iron. The aligned domains add to the field, making it about 2000 times stronger.

Electromagnets have many applications. Large ones are used routinely to pick up and transfer scrap iron, and small ones are used in magnetic relays and solenoids, which act as magnetic switches. Solenoids are used in automobiles to engage the starting motor. One type of circuit breaker uses an electromagnetic switch. The strength of an electromagnet is directly proportional to the current in its coils. When there is a certain amount of current in the breaker circuit, an electromagnet becomes strong enough to attract and "trip" a metallic conductor, thus opening the circuit.

The iron used in electromagnets is called "soft" iron. This does not mean that it is physically soft; rather, it means that this type of iron can be magnetized but quickly becomes demagnetized. Certain types of iron, along with nickel, cobalt, and a few other elements, are known as "hard" magnetic materials. Once magnetized, they retain their magnetic properties for a long time.

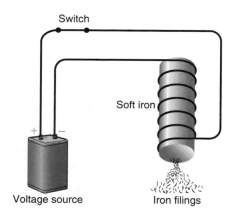

Figure 8.20 Electromagnet
A simple electromagnet consists of an insulated coil of wire wrapped around a piece of iron. When the switch is closed, there is a current in the wire, which gives rise to a magnetic field that magnetizes the iron and thus creates a magnet. When the switch is open, there is no current in the coil and the iron is not magnetized.

---

* This is not actually the case. Electron spin is a quantum mechanical effect (Chapter 9) with no classical analog. However, the classical model is useful in understanding the effect.

Hard iron is used for permanent magnets such as bar magnets. When permanent magnets are heated or struck, the domains are shaken from their alignment and the magnet becomes weaker. Also, there is a temperature effect. In fact, above a certain temperature called the *Curie temperature*, a material ceases to be ferromagnetic. The Curie temperature of iron is 770°C.*

A permanent magnet is made by "permanently" aligning the domains inside the material. One way to do so is to heat a piece of hard ferromagnetic material above its Curie temperature and then apply a strong magnetic field. The domains line up with the field, and as the material cools, the domain alignment is frozen in, so to speak, producing a permanent magnet.

Conceptual Question and Answer

Coin Magnet

Q. Are coins (money) magnetic?

A. Some are, and some are not. It depends on the metal content. In general, U.S. "silver" coins are not magnetic, but Canadian coins are because Canadian coins contain nickel (a ferromagnetic material). Get a magnet and a few coins and see for yourself.

Have you ever put a Canadian coin in a coin-operated vending machine and had it refuse to drop down? This happens because the machines are equipped with magnets in the coin path to prevent odd-sized coins, metal slugs, or washers (usually iron) from entering the coin mechanism and causing damage.

The Earth's Magnetic Field

At the beginning of the seventeenth century, William Gilbert, an English scientist, suggested that the Earth acts as a huge magnet. Today we know that such a magnetic effect does exist for our planet. It is the Earth's magnetic field that causes compasses to point north. Experiments have shown that a magnetic field exists within the Earth and extends many hundreds of miles out into space. The aurora borealis and aurora australis (the northern lights and southern lights, respectively), common sights in higher latitudes near the poles, are associated with the Earth's magnetic field. This effect is discussed in Chapters 9.3 and 19.1.

The origin of the Earth's magnetic field is not known, but the most widely accepted theory is that it is caused by the Earth's rotation, which produces internal electrical currents in the liquid part of the core deep within the planet. It is not due to some huge mass of magnetized iron compound within the Earth. The Earth's interior is well above the Curie temperature, so materials are not ferromagnetic. Also, the magnetic poles slowly change their positions, a phenomenon that suggests changing currents.

The Earth's magnetic field does approximate that of a current loop or a huge imaginary bar magnet, as illustrated in ● Fig. 8.21. Note that the magnetic south pole is near the geographic North (N) Pole. For this reason, the north pole of a compass needle, which is a little magnet, points north (law of poles). Because the "north-seeking" pole of the compass needle points north along magnetic field lines, this direction is referred to as *magnetic north*. That is, magnetic north is in the direction of a magnetic "north" pole (actually a magnetic south pole), which is near the geographic North

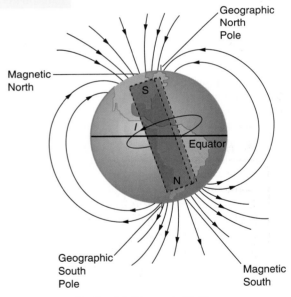

Figure 8.21 The Earth's Magnetic Field
The Earth's magnetic field is thought to be caused by internal currents in the liquid outer core, in association with the planet's rotation. The magnetic field is similar to that of a giant current loop or giant bar magnet within the Earth (such a bar magnet does not and could not really exist). Note that magnetic north (toward which the compass points) and the geographic North Pole (the Earth's axis of rotation) do not coincide.

---

* The Curie temperature is named after Pierre Curie (1859–1906), the French scientist who discovered the effect. Pierre Curie was the husband of Marie (Madame) Curie. They both did pioneering work in radioactivity (Chapter 10.3).

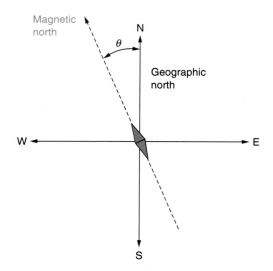

Figure 8.22 Magnetic Declination
The angle of magnetic declination is the angle between geographic (true) north and magnetic north (as indicated by a compass). The declination is measured in degrees east or west of geographic north.

Pole.* The direction of the geographic pole is called *true north*, or in the direction of the Earth's north spin axis. The magnetic and geographic poles do not coincide. The magnetic north pole is about 1000 km (620 mi) from the geographic North Pole. In the Southern Hemisphere the magnetic "south" pole (actually a magnetic north pole) is displaced even more from its corresponding geographic pole.

Hence, the compass does not point toward true north but toward magnetic north. The variation between the two directions is expressed in terms of **magnetic declination**, which is the angle between geographic (true) north and magnetic north (● Fig. 8.22). The declination may vary east or west of a geographic meridian (an imaginary line running from pole to pole).

It is important in navigation to know the magnetic declination at a particular location so that the magnetic compass direction can be corrected for true north. Magnetic declination is provided on navigational maps that show lines of declination expressed in degrees east and west, as illustrated in ● Fig. 8.23. Because the magnetic north pole moves, the magnetic declination varies with time. (See the **Highlight: Magnetic North Pole**.)

Notice that the magnetic field lines around the Earth are curved (Fig. 8.21). For north–south direction, a compass needle lines up with the horizontal components of the lines. As a result of being curved, however, there is a downward component giving rise to a *magnetic dip* or *inclination*, which is the angle made by a compass needle with the horizontal at any point on the Earth's surface. If a compass were held vertically, then the needle would dip downward in the Northern Hemisphere. What would be the inclination in the Southern Hemisphere?

The magnetic field of the Earth is relatively weak compared with that of magnets used in the laboratory, but the field is strong enough to be used by certain animals (including ourselves) for orientation and direction. For instance, it is believed that migratory birds and

Figure 8.23 United States Magnetic Declination
The map shows isogonic (same magnetic declination) lines for the conterminous United States. For locations on the 0° line, magnetic north is in the same direction as geographic (true) north. On either side of this line, a compass has an easterly or westerly variation. For example, on the 20°W line, a compass has a westerly declination of 20°; that is, magnetic north is 20° west of true north.

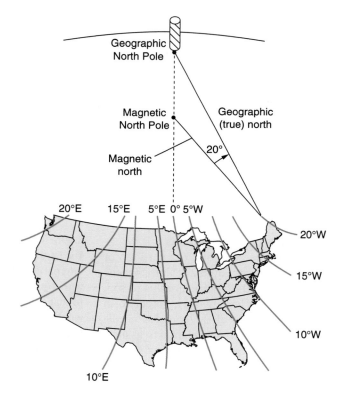

*These designations sometimes cause confusion. From the law of poles, the north pole of a compass points toward a south pole. However, it has become common to say that the magnetic north direction of a compass is toward "magnetic north" in the north polar regions, keeping everything north, particularly in navigation.

Highlight Magnetic North Pole

The magnetic north pole is on the move. Its location was first determined in 1831 to be on the west coast of the Boothia Peninsula in northern Canada by British explorer James Clark Ross. The next determination was in 1904 by Norwegian explorer Roald Amundsen, who found that the magnetic pole had moved some 50 km (30 mi) from its first determined location (Fig. 1).

The magnetic pole's position was next determined by the Canadian government in 1948. This time it was located some 250 km (155 mi) northwest of where Amundsen determined. Other observations in 1962, 1972, 1984, and 1994 showed that the magnetic pole was continuing to move northwest with an average speed of 10 km (6.2 mi) per year. A 2001 measurement found the pole to be located in the Arctic Ocean. The movement of the magnetic north pole had sped up to about 40 km (25 mi) per year. The estimated position of the pole in 2005 is shown in Fig. 1 and has probably moved since then.

Where will the magnetic north pole be in the future? If the northwest path continues, then it will skirt south of the geo-graphic North Pole and be near or in Siberia later in this century. Of course, this is conjecture. The movement of the magnetic pole may change direction or become erratic because of changes within the Earth.

If the magnetic north pole *did* get to Siberia, then there would be notable changes. One would be that the aurora borealis ("northern lights") would not be seen predominantly in Alaska and Canada, as they are now. The auroras are caused by charged particles from the Sun, which are trapped in the Earth's magnetic field and deflected toward the poles (see Chapter 19.1). With the magnetic north pole in Siberia, the aurora would be observed predominantly in Siberia and northern Europe.

Another change would be in the magnetic declination, and this change would make compass navigation in the United States quite difficult. However, we now have the GPS (global position-ing system) that references directions to the fixed geographic North Pole. [See the Highlight: Global Positioning System (GPS) in Chapter 15.2.]

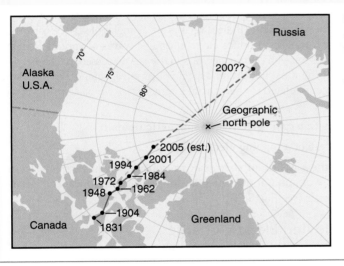

Figure 1 Magnetic North Pole
Measurements show that the magnetic north pole has been mov-ing to the northwest. If this northwest path continues, then the pole will skirt south of the geographic North Pole and will be near or in Siberia later in the century.

homing pigeons use the Earth's magnetic field to directionally orient themselves in their homeward flights. Iron compounds have been found in their brains.

Did You Learn?

- Like poles (charges) repel, and unlike poles (charges) attract.
- The Earth's magnetic north pole is near the geographic South Pole; the magnetic south pole is near the geographic North Pole, which is why compass needles point north. The direction the compass points is referred to as magnetic north.

8.5 Electromagnetism

Preview Questions

- What are the two basic principles of electromagnetism?
- What's the difference between a motor and a generator?

Figure 8.24 Electromagnetic Wave
An illustration of the vector components of an electromagnetic wave. The wave consists of two force fields, electric (*E*) and magnetic (*M*), oscillating perpendicularly to each other and to the direction of wave propagation (velocity vector).

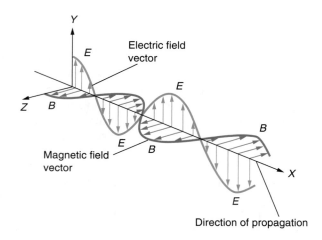

The interaction of electrical and magnetic effects is known as **electromagnetism**. One example of this intimate relationship is electromagnetic waves. When charged particles such as electrons are accelerated, energy is radiated away in the form of waves. *Electromagnetic waves* consist of oscillating electric and magnetic fields, and the field energy radiates outward at the speed of light, 3.00×10^8 m/s in vacuum.

An illustration of an electromagnetic wave is shown in ● Fig. 8.24. The wave is traveling in the *x* direction. The electric (*E*) and magnetic (*B*) field vectors are at angles of 90° to each other. Accelerated, oscillating charged particles produce electromagnetic waves of various frequencies or wavelengths. These waves form an *electromagnetic spectrum* as discussed in Chapter 6.3.

Electromagnetism is one of the most important aspects of physical science, and most of our current technology is directly related to this crucial interaction. Two basic principles of this interaction are as follows:

1. **Moving electric charges (current) give rise to magnetic fields.**

2. **A magnetic field may deflect a moving electric charge.**

The first principle forms the basis of an *electromagnet* considered previously. Electromagnets are found in a variety of applications, such as doorbells, telephones, and devices used to move magnetic materials (see Fig. 8.20).

Magnetic Force on Moving Electric Charge

The second of the electromagnetic principles may be described in a qualitative way: A magnetic field can be used to deflect moving electric charges. A stationary electric charge in a magnetic field experiences no force, but when a moving charge enters a magnetic field as shown in ● Fig. 8.25, it experiences a force. This magnetic force (F_{mag}) is perpendicular to the plane formed by the velocity vector (*v*) and the magnetic field (*B*).

In the figure, the force initially would be out of the page, and with an extended field, the negatively charged particle would follow a circular arc path. If the moving charge were positive, then it would be deflected in the opposite direction, or into the page. Also, if a charge, positive or negative, is moving parallel to a magnetic field, there is no force on the charge.

This effect can be demonstrated experimentally as shown in ● Fig. 8.26. A beam of electrons is traveling in the tube from left to right and is made visible by a piece of fluorescent paper in the tube. In Fig. 8.26a the beam is undeflected in the absence of a magnetic field. In Fig. 8.26b the magnetic field of a bar magnet causes the beam to be deflected downward. In Fig. 8.26c the opposite magnetic pole deflects the beam in the opposite direction.

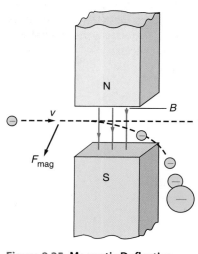

Figure 8.25 Magnetic Deflection
Electrons entering a vertical magnetic field as shown experience a force F_{mag} that deflects them out of the page. (See text for description.)

Motors and Generators

The electrons in a conducting wire also experience force effects caused by magnetic fields. Hence, a current-carrying wire in a magnetic field can experience a force. With such a force available, it might quickly come to mind that the force could be used to do work, and this

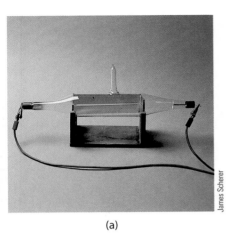

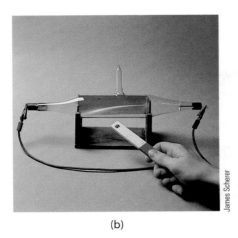

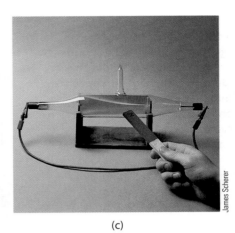

(a)   (b)   (c)

Figure 8.26 Magnetic Force on a Moving Charge
(a) The presence of a beam of electrons is made evident by a fluorescent strip in the tube, which allows the beam to be seen. (b) The magnetic field of a bar magnet gives rise to a force on the electrons, and the beam is deflected (downward). (c) The opposite pole of the bar magnet deflects the beam the other way.

is what is happens in electric motors. Basically, a **motor** is a device that converts electrical energy into mechanical energy. The mechanical rotation of a motor's shaft is used to do work. To help understand the electromagnetic-mechanical interaction of motors (of which there are many types), consider the diagram of a simple *dc motor* shown in ● Fig. 8.27a. Real motors have many loops or windings, but for simplicity only one is shown here. The battery supplies current to the loop, which is free to rotate in the magnetic field between the pole faces. The force on the current-carrying loop produces a torque, causing it to rotate.

Figure 8.27 A dc Motor
(a) An illustration of a loop of a coil in a dc motor. When carrying a current in a magnetic field, the coil experiences a torque and rotates the attached shaft. The split-ring commutator effectively reverses the loop current each half cycle so that the coil will rotate continuously. (b) The forces on the coil show why the reversal of current is necessary.

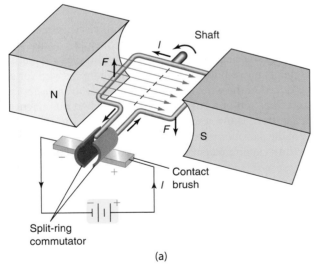

(a)

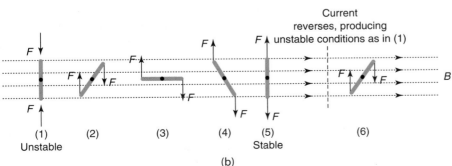

(b)

Figure 8.28 Electromagnetic Induction
An illustration of Faraday's experiment showing electromagnetic induction. (a) When the magnet is moved down through the coil of wire, the reading on the meter indicates a current in the circuit. (b) When the magnet is moved upward, the current in the circuit is reversed.

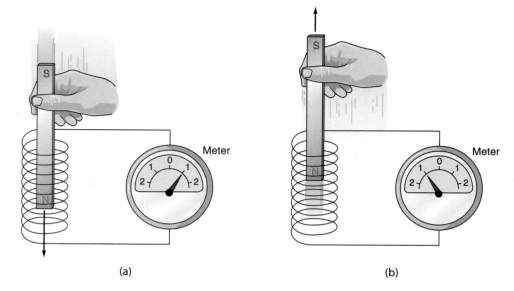

(a)   (b)

Continuous rotation requires a *split-ring commutator* that reverses the polarity and the current in the loop each half cycle so that the loop has the appropriate force to rotate continuously (Fig. 8.27b). The inertia of the loop carries it through the positions where unstable conditions exist. When a rotating armature has many windings (loops), the effect is enhanced. The rotating loops cause a connected shaft to rotate, and this motion is used to do mechanical work. The conversion of electrical energy into mechanical energy is enhanced by many loops of wire and stronger magnetic fields.

One might ask whether the reverse is possible. That is, is the conversion of mechanical energy into electrical energy possible? Indeed it is, and this principle is the basis of electrical generation. Have you ever wondered how electricity is generated? This crucial process is based on electromagnetic principles.

A **generator** is a device that converts mechanical work or energy into electrical energy. A generator operates on what is called *electromagnetic induction*. This principle was discovered in 1831 by Michael Faraday, an English scientist. An illustration of his experiment is shown in ● Fig. 8.28a. When a magnet is moved downward through a loop of wire (or a coil for enhancement), a current is induced in the wire, as indicated on the *galvanometer* (a meter that detects the magnitude and direction of a current). Similarly, when the magnet is moved upward through the coil, a current is induced, but in the opposite direction, as indicated by the opposite deflection of the galvanometer needle in Fig. 8.28b. Investigation shows that this effect is caused by a time-varying magnetic field through the loop because of the motion of the magnet.

The same effect is obtained by using a stationary magnetic field and rotating the loop in the field. The magnetic field through the loop varies with time, and a current is induced. A simple *ac generator* is illustrated in ● Fig. 8.29. When the loop is mechanically rotated, a voltage and current are induced in the loop that vary in magnitude and alternate back and forth, changing direction each half cycle. Hence, there is an alternating current (ac). There are also dc generators, which are essentially dc motors operated in reverse. However, most electricity is generated as ac and then converted, or *rectified*, to dc.

Generators are used in power plants to convert other forms of energy into electrical energy. For the most part, fossil fuels and nuclear energy are used to heat water to generate steam, which is used to turn turbines that supply mechanical energy in the generation process. The electricity is carried to homes and businesses, where it is either converted back into mechanical energy to do work or converted into heat energy.

Transformers

But how is electrical energy or power transmitted? You have probably seen towers with high-voltage (or high-tension) transmission lines running across the land, as shown in the chapter-

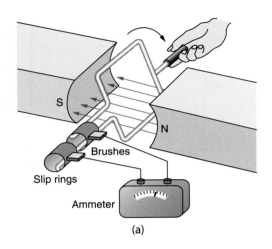
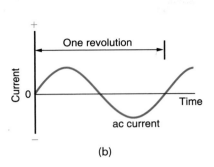

(a)

(b)

Figure 8.29 **An ac Generator**
(a) An illustration of a coil loop in an ac generator. When the loop is mechanically turned, a current is induced, as indicated by the ammeter. (b) The current varies in direction each half cycle and hence is called ac, or alternating current.

opening photo. The voltage for transmission is stepped up by means of **transformers**, simple devices based on electromagnetic induction (● Fig. 8.30). A transformer consists of two insulated coils of wire wrapped around an iron core, which concentrates the magnetic field when there is current in the input coil. With an ac current in the input or primary coil, there is a time-changing magnetic field as a result of the current going back and forth. The magnetic field goes through the secondary coil and induces a voltage and current.

Conceptual Question and Answer

No Transformation

Q. Will transformers operate on dc current? If not, why not?

A. Transformers operate on the induction of a time-changing magnetic field, which is obtained using ac current. Direct current produces a static magnetic field and so cannot be used for transformer operations (see Fig. 8.18).

Because the secondary coil has more windings than the primary coil, the induced ac voltage is greater than the input voltage, and this type of transformer is called a *step-up transformer*. However, when the voltage is stepped up, the secondary current is stepped down (by the conservation of energy, because $P = IV$). The factor of voltage step-up depends on the ratio of the numbers of windings on the two coils, which can be easily controlled.

So why step up the voltage? Actually, it is the step-down in current that is really of interest. Transmission lines have resistance and therefore I^2R losses. Stepping down the current reduces these losses, and energy is saved that would otherwise be lost as joule heat. If the voltage is stepped up by a factor of 2, then the current is reduced by a factor of 2, or by $\frac{1}{2}$. With half the current, there is only one-fourth the I^2R losses. (Why?) Thus, for power transmission, the voltage is stepped up to a very high voltage to get the corresponding current step-down and thereby avoid joule heat losses.

Of course, such high voltages cannot be used in our homes which, in general, have 220- to 240-V service entries. Therefore, to get the voltage back down (and the current up), a *step-down transformer* is used, which steps down the voltage and steps up the current. This is done by simply reversing the input and output coils. If the primary coil has more windings than the secondary coil, then the voltage is stepped down.

The voltage change for a transformer is given by

$$\text{secondary} = \left(\frac{\text{number of secondary coils}}{\text{number of rimary coils}}\right) \times \text{primary voltage}$$

$$V_2 = \left(\frac{N_2}{N_1}\right)V_1 \qquad \qquad 8.10$$

transformer voltage change

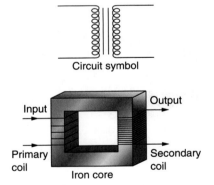

Circuit symbol

Input   Output

Primary   Secondary
coil   coil
Iron core

Figure 8.30 **The Transformer**
The basic features of a transformer and the circuit symbol for it are shown. A transformer consists of two coils of insulated wire wrapped around a piece of iron. Alternating current in the primary coil creates a time-varying magnetic field, which is concentrated by the iron core, and the field passes through the secondary coil. The varying magnetic field in the secondary coil produces an alternating-current output.

| EXAMPLE 8.3 | Finding Voltage Output for a Transformer |

A transformer has 500 windings in its primary coil and 25 in its secondary coil. If the primary voltage is 4400 V, what is the secondary voltage?

Solution

Using Eq. 8.10 with $N_1 = 500$, $N_2 = 25$, and $V_1 = 4400$ V,

$$V_2 = \left(\frac{N_2}{N_1}\right)V_1 = \left(\frac{25}{500}\right)(4400 \text{ V}) = 220\text{V}$$

This result is typical of a *step-down* transformer on a utility pole near residences. The voltage is stepped down by a factor of 25/500 = 1/20, so the current is stepped up by an inverse factor of 20 (that is, 500/25 = 20).

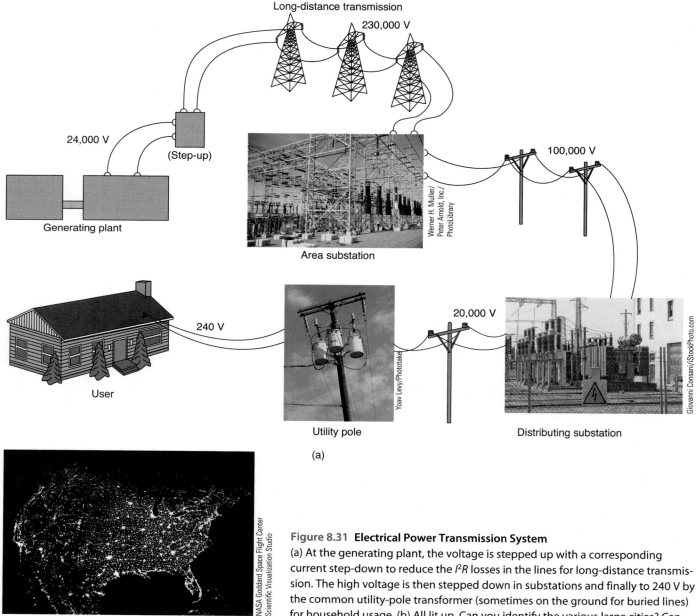

Figure 8.31 Electrical Power Transmission System
(a) At the generating plant, the voltage is stepped up with a corresponding current step-down to reduce the I^2R losses in the lines for long-distance transmission. The high voltage is then stepped down in substations and finally to 240 V by the common utility-pole transformer (sometimes on the ground for buried lines) for household usage. (b) All lit up. Can you identify the various large cities? Can you find your hometown?

Confidence Exercise 8.3

Suppose the transformer in Example 8.3 were reversed and used as a step-up transformer with a voltage input of 100 V. What would the voltage output be?

An illustration of stepping up and stepping down the voltage in electrical transmission is shown in ● Fig. 8.31a. The voltage step-up and the corresponding current step-down, which reduces joule heat losses, are major reasons we use ac electricity in power transmission. Note how efficient electrical transmission is in Fig. 8.31b.

Indeed, we are now in an electrical age of high technology and electronics. *Electronics* is the branch of physics and engineering that deals with the emission and control of electrons. Our electronic instruments have become smaller and smaller, primarily because of the development of solid-state *diodes* and *transistors*. These devices control the direction of electron flow, and the transistor allows for the amplification of an input signal (as in a transistor radio). Solid-state diodes and transistors offer great advantages over the older vacuum tubes and have all but replaced them. Major advantages are smaller size, lower power consumption, and material economy (● Fig. 8.32).

Even further miniaturization has come about through *integrated circuits* (ICs). An integrated circuit may contain millions of transistors on a silicon *chip*. Current technology can produce transistors just 45 nm (10^{-9} m) wide and pack more than four hundred million transistors on a chip. Chips have the dimensions of only a few millimeters (called a *microchip or microprocessor*, ● Fig. 8.33).

Such chips can have many logic circuits and are the "brains" of computers. Microchips have many applications in processors that perform tasks almost instantaneously. For example, they are in our automobiles, computing the miles (or kilometers) per gallon of gasoline, applying antilock brakes, signaling the inflation of air bags when needed, and indicating location and direction to your destination with global positioning satellites. [See the Highlight: Global Positioning System (GPS) in Chapter 15.2.] What will the future bring? Think about it.

Did You Learn?

- Moving electric charges (current) give rise to a magnetic field, and a magnetic field may deflect moving electric charge.
- A motor converts electrical energy into mechanical energy; a generator converts mechanical energy or work into electrical energy.

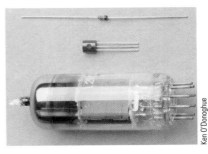

Figure 8.32 Solid-State Diode, Transistor, and Vacuum Tube A diode (two leads, top) and a transistor (three leads) are shown above a vacuum tube. Solid-state diodes and transistors have all but replaced vacuum tubes in most applications.

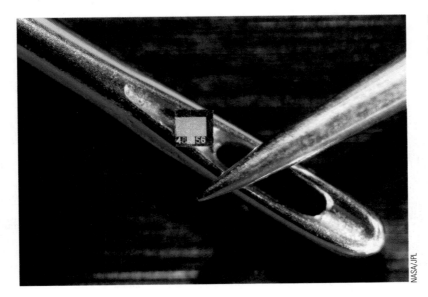

Figure 8.33 Integrated Circuits A microprocessor integrated circuit, or chip, lies near the eye of a needle.

KEY TERMS

1. electric charge (8.1)
2. electrons
3. protons
4. law of charges
5. Coulomb's law
6. electric field
7. current (8.2)
8. electric potential energy
9. voltage
10. resistance
11. Ohm's law
12. electric power
13. direct current (dc) (8.3)
14. alternating current (ac)
15. series circuit
16. parallel circuit
17. law of poles (8.4)
18. magnetic field
19. ferromagnetic
20. magnetic domains
21. magnetic declination
22. electromagnetism (8.5)
23. motor
24. generator
25. transformer

MATCHING

For each of the following items, fill in the number of the appropriate Key Term from the preceding list. Compare your answers with those at the back of the book.

a. _____ Positively charged particles
b. _____ Angle between true north and magnetic north
c. _____ *IV*
d. _____ Unlike charges attract; like charges repel
e. _____ Interaction of electrical and magnetic effects
f. _____ A fundamental quantity
g. _____ Steps voltages up and down
h. _____ Relates voltage and current to resistance
i. _____ Converts electrical energy into mechanical energy
j. _____ Negatively charged particles
k. _____ Imaginary lines indicating magnetic direction
l. _____ Describes the force between charged particles
m. _____ Highly magnetic material

n. _____ Produced by a constantly changing voltage
o. _____ Time rate of flow of electric charge
p. _____ Unlike poles attract; like poles repel
q. _____ Equals W/q
r. _____ Same current through all resistances
s. _____ Opposition to charge flow
t. _____ Local regions of alignment
u. _____ Net electron flow in one direction
v. _____ Converts mechanical energy into electrical energy
w. _____ Results from separating electrical charges
x. _____ Voltage across each resistance is the same
y. _____ $E = F/q_+$ mapping

MULTIPLE CHOICE

Compare your answers with those at the back of the book.

1. What can be said about the electric force between two charged particles?
   (a) it is repulsive for unlike charges.
   (b) it varies as $1/r$.
   (c) it depends only on the magnitudes of the charges.
   (d) it is much, much greater than the attractive gravitational force.

2. Two equal positive charges are placed equidistant on either side of another positive charge. What would the middle positive charge experience? (8.1)
   (a) a net force to the right
   (b) a net force to the left
   (c) a zero net force

3. In a dc circuit, how do electrons move? (8.2)
   (a) with a slow drift velocity.
   (b) in alternate directions.
   (c) near the speed of light.
   (d) none of the preceding.

4. What is a unit of voltage? (8.2)
   (a) joule
   (b) joule/coulomb
   (c) amp-coulomb
   (d) amp/coulomb

5. In electrical terms, power has what units? (8.2)
   (a) joule/coulomb
   (b) amp/ohm
   (c) amp-coulomb
   (d) amp-volt

6. Appliances with heating elements require which of the following? (8.3)
   (a) a large current
   (b) a large resistance
   (c) a low joule heat

7. The greatest equivalent resistance occurs when resistances are connected in which type of arrangement? (8.3)
   (a) series   (b) parallel   (c) series–parallel

8. Given three resistances, the greatest current occurs in a battery circuit when the resistances are connected in what type of arrangement? (8.3)
   (a) series   (b) parallel   (c) series–parallel

9. When two bar magnets are near each other, the north pole of one of the magnets experiences what type of force from the other magnet? (8.4)
    (a) an attractive force
    (b) a repulsive force
    (c) a Coulomb force
    (d) both (a) and (b)

10. What is the variation in the location of the Earth's magnetic north pole from true north given by? (8.4)
    (a) the law of poles   (b) the magnetic field
    (c) magnetic domains   (d) the magnetic declination

11. What type of energy conversion does a motor perform? (8.5)
    (a) chemical energy into mechanical energy
    (b) mechanical energy into electrical energy
    (c) electrical energy into mechanical energy
    (d) mechanical energy into chemical energy

12. What type of energy conversion does a generator perform? (8.5)
    (a) chemical energy into mechanical energy
    (b) mechanical energy into electrical energy
    (c) electrical energy into mechanical energy
    (d) mechanical energy into chemical energy

13. Which of the following is true of a step-up transformer? (8.5)
    (a) It has an equal number of windings on the primary and secondary coils.
    (b) It has fewer windings on the secondary coil.
    (c) It has fewer windings on the primary coil.
    (d) None of the preceding statements is true.

14. A transformer with more windings on the primary coil than on the secondary coil does which of the following? (8.4)
    (a) Steps up the voltage.
    (b) Steps up the current.
    (c) Steps up both current and voltage.
    (d) Will operate off dc current.

FILL IN THE BLANK

Compare your answers with those at the back of the book.

1. An object with a deficiency of electrons is ___ charged. (8.1)
2. The unit of electric current is the ___. (8.1)
3. ___ are neither good conductors nor good insulators. (8.1)
4. Voltage is defined as work per ___. (8.2)
5. An electric circuit that is not a complete path is called a(n) ___ circuit. (8.2)
6. The unit of resistance is the ___. (8.2)
7. Another name for joule heat is ___ losses. (8.2)
8. A battery-powered device uses ___ current. (8.3)
9. In a circuit with resistances connected in parallel, the total resistance is always less than the ___ resistance. (8.3)
10. A material ceases to be ferromagnetic above the ___ temperature. (8.4)
11. Magnetic north is generally in the direction of the ___ north pole. (8.4)
12. A step-up transformer has more windings on the ___ coil. (8.5)

SHORT ANSWER

8.1 Electric Charge, Electric Force, and Electric Field

1. Which two particles that make up atoms have about the same mass? Which two have the same magnitude of electric charge?
2. A large charge $+Q$ and a small charge $-q$ are a short distance apart. How do the electric forces on each charge compare? Is this comparison described by another physical law? (*Hint:* See Chapter 3.4.)
3. Explain how a charged rubber comb attracts bits of paper and how a charged balloon sticks to a wall or ceiling.
4. Why do clothes sometimes stick together when removed from a dryer?

8.2 Current, Voltage, and Electrical Power

5. Distinguish between electric potential energy and voltage.
6. What are two things that are required for there to be a current in a circuit?
7. How does joule heat vary with increasing and decreasing resistances?

8. If the drift velocity in a conductor is so small, then why does an auto battery influence the starter as soon as you turn the ignition switch?

8.3 Simple Electric Circuits and Electrical Safety

9. Distinguish between alternating current and direct current.
10. Why are home appliances connected in parallel rather than in series?
11. Compare the safety features of (a) fuses, (b) circuit breakers, (c) three-prong plugs, and (d) polarized plugs.
12. Is it safe to stay inside a car during a lightning storm (● Fig. 8.34)? Explain.
13. Sometimes resistances in a circuit are described as being connected "head to tail" and "all heads and all tails" connected together. What is being described?

8.4 Magnetism

14. Why do iron filings show magnetic field patterns?
15. Compare the law of charges and the law of poles.

Figure 8.34 Safe Inside a Car?
See Short-Answer Question 12.

16. (a) What is a ferromagnetic material? (b) Why does a permanent magnet attract pieces of ferromagnetic materials? What would happen if the pieces were above their Curie temperature?

17. What is the principle of an electromagnet?

18. (a) What does the Earth's magnetic field resemble, and where are its poles? (b) Why do airline pilots and ship navigators guided by compasses have to make corrections to stay on a course charted on a map?

8.5 Electromagnetism

19. Describe the basic principle of a dc electric motor.

20. What happens (a) when a proton moves parallel to a magnetic field and (b) when a proton moves perpendicular to a magnetic field?

21. What is the principle of a transformer, and how are transformers used?

Figure 8.35 High Voltage or High Current?
See Short-Answer Question 24.

22. What would happen if electric power were transmitted from the generating plant to your home at 120 V?

23. Why can birds perch on high-voltage power lines and not get hurt?

24. Body injury from electricity depends on the magnitude of the current and its path (see the Highlight: Electrical Effects on Humans, Chapter 8.4). However, signs warning "Danger. High Voltage" are commonly seen (● Fig. 8.35). Shouldn't the signs refer to high *current*? Explain.

VISUAL CONNECTION

Visualize the connections and give answers for the blanks. Compare your answers with those at the back of the book.

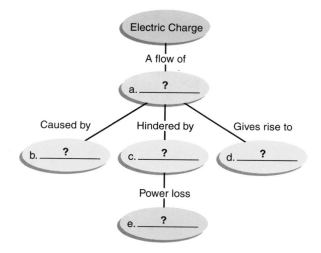

APPLYING YOUR KNOWLEDGE

1. What is the cause of "static cling" in clothes, and what is an economical way to get rid of it?

2. Two negative charges of -1 C each are placed at opposite ends of a meterstick. (a) Could a free electron be placed somewhere on the meterstick so that it would be in static equilibrium (zero net force)? How about a proton? (b) Could an electron or a proton be placed anywhere beyond the ends of the meterstick so that it would be in static equilibrium? Explain.

3. Answer both parts of Question 2 for a charge of $+1$ C placed at one end of the meterstick and a charge of -1 C placed at the other.

4. An old saying about electrical safety states that you should keep one hand in your pocket when working with high voltage electricity. What does this mean?

5. The following are incandescent light bulb questions. (Several billion bulbs are used in the United States each year.)
   (a) The gas in a light bulb is not air, but is a mixture of argon and nitrogen gases at low pressure. Why isn't air used? Wouldn't it be cheaper?

(b) After long periods of use, a gray spot develops on the inside of a bulb. What is this gray spot? (*Hint:* It is metallic.)

6. Show that the resistance that is equivalent to two resistances in parallel can be written

$$R_p = \frac{R_1 R_2}{R_1 + R_2}$$

7. What happens when you cut a bar magnet in half? If it were continually cut in half, then would it finally yield two magnetic monopoles? Explain.

8. Suppose you are on an expedition to locate the magnetic north pole. (Assume it is still on land to make it easier to plant a flag.) How would you go about locating it? (*Hint:* Think of inclination.)

IMPORTANT EQUATIONS

Coulomb's Law: $F = \dfrac{kq_1q_2}{r_2}$ (8.1)

$(k = 9.0 \times 10^9 \text{ N} \cdot \text{m}^2/\text{C}^2)$

Current: $I = \dfrac{q}{t}$ (8.2)

Voltage: $V = \dfrac{W}{q}$ (8.3)

Ohm's Law: $V = IR$ (8.4)

Electric Power: $P = IV = I^2R$ (8.5–8.6)

Resistance in Series: $R_S = R_1 + R_2 + R_3 + \cdots$ (8.7)

Resistances in Parallel: $\dfrac{1}{R_p} = \dfrac{1}{R_1} + \dfrac{1}{R_2} + \dfrac{1}{R_3} \cdots$ (8.8)

Two Resistances in Parallel: $R_p = \dfrac{R_1 R_2}{R_1 + R_2}$ (8.9)

Transformer (Voltages and Turns): $V_2 = \left(\dfrac{N_2}{N_1}\right)V_1$ (8.10)

EXERCISES

8.1 Electric Charge, Electric Force, and Electric Field

1. How many electrons make up one coulomb of charge? (*Hint:* $q = ne$.)

   Answer: 6.25×10^{18}

2. An object has one million more electrons than protons. What is the net charge of the object?

3. What are the forces on two charges of $+0.60$ C and $+2.0$ C, respectively, if they are separated by a distance of 3.0 m?

   Answer: 1.0×10^9 N, mutually repulsive

4. Find the force of electrical attraction between a proton and an electron that are 5.3×10^{-11} m apart (the arrangement in the hydrogen atom). Compare this force to the gravitational force between these particles (see Chapter 3.5).

5. There is a net passage of 4.8×10^{18} electrons by a point in a wire conductor in 0.25 s. What is the current in the wire?

   Answer: 3.1 A

6. A current of 1.50 A flows in a conductor for 6.5 s. How much charge passes a given point in the conductor during this time?

8.2 Current, Voltage, and Electrical Power

7. To separate a 0.25-C charge from another charge, 30 J of work is done. What is the electric potential energy of the charge?

   Answer: 30 J

8. What is the voltage of the 0.25-C charge in Exercise 7?

9. If an electrical component with a resistance of 50 Ω is connected to a 120-V source, then how much current flows through the component?

   Answer: 2.4 A

10. What battery voltage is necessary to supply 0.50 A of current to a circuit with a resistance of 20 Ω?

11. A car radio draws 0.25 A of current in the auto's 12-V electrical system.
    (a) How much electric power does the radio use?
    (b) What is the effective resistance of the radio?

    Answer: (a) 3.0 W (b) 48 Ω

12. A flashlight uses batteries that add up to 3.0 V and has a power output of 0.50 W.
    (a) How much current is drawn from the batteries?
    (b) What is the effective resistance of the flashlight?

13. How much does it cost to run a 1500-W hair dryer 30 minutes each day for a month (30 days) at a cost of 12¢ per kWh?

    Answer: $2.70

14. A refrigerator using 1000 W runs one-eighth of the time. How much does the electricity cost to run the refrigerator each month at 10¢ per kWh?

15. A 24-Ω component is connected to a 12-V battery. How much energy is expended per second?

    Answer: 6.0 W

16. In Exercise 15, using the equation given in the chapter, you probably found the power in two steps. Show that $P = V^2/R$, which requires only one step.

17. The heating element of an iron operates at 110 V with a current of 10 A.
    (a) What is the resistance of the iron?
    (b) What is the power dissipated by the iron?

    Answer: (a) 11 Ω (b) 1100 W

18. A 100-W light bulb is turned on. It has an operating voltage of 120 V.
    (a) How much current flows through the bulb?
    (b) What is the resistance of the bulb?
    (c) How much energy is used each second?

19. Two resistors with values of 25 Ω and 35 Ω, respectively, are connected in series and hooked to a 12-V battery.
    (a) How much current is in the circuit?
    (b) How much power is expended in the circuit?

    Answer: (a) 0.20 A (b) 2.4 W

20. Suppose the two resistors in Exercise 19 were connected in parallel. What would be (a) the current and (b) the power in this case?

21. A student in the laboratory connects a 10-Ω resistor, a 15-Ω resistor, and a 20-Ω resistor in series and then connects the arrangement to a 50-V dc source.
    (a) How much current is in the circuit?
    (b) How much power is expended in the circuit?

    Answer: (a) 1.1 A (b) 55 W

22. The student in Exercise 21 repeats the experiment but with the resistors connected in parallel.
    (a) What is the current?
    (b) What is the power?

23. A 30.0-Ω resistor and a 60.0-Ω resistor in series are connected to a 120-V dc source.
    (a) What is (are) the current(s) through the resistors?
    (b) What is the voltage drop across each resistor?

    Answer: (a) 1.33 A (b) 40 V and 80 V

24. A 30.0-Ω resistor and a 60.0-Ω resistor in parallel are connected to a 120-V dc source.
    (a) What is (are) the current(s) thorough the resistors?
    (b) What is the voltage drop across each resistor?

8.5 Electromagnetism

25. A transformer has 300 turns on its secondary and 100 turns on its primary. The primary is connected to a 12-V source.
    (a) What is the voltage output of the secondary?
    (b) If 2.0 A flows in the primary coil, then how much current is there in the secondary coil?

    Answer: (a) 36 V (b) 0.67 A

26. A transformer has 500 turns on its primary and 200 turns on its secondary.
    (a) Is it a step-up or a step-down transformer?
    (b) If a voltage of 100 V is applied to the primary and a current of 0.25 A flows in these windings, then what are the voltage output of the secondary and the current in the secondary?

27. A transformer with 1000 turns in its primary coil has to decrease the voltage from 4400 V to 220 V for home use. How many turns should there be on the secondary coil?

    Answer: 50

28. A power company transmits current through a 240,000-V transmission line. This voltage is stepped down at an area substation to 40,000 V by a transformer that has 900 turns on the primary coil. How many turns are on the secondary of the transformer?

ON THE WEB

1. Static Electricity

In winter, why do you get a shock when you touch a light switch or even another person or pet? Why do balloons cling to the wall at birthday parties? Describe the process called "induction" and explain how it is related to static electricity. Explain briefly how knowing about static electricity can help you understand other aspects of electricity. Visit the student website at **www.cengagebrain.com/shop/ISBN/1133104096** to answer these questions.

2. Electromagnetic Waves

What are force fields? What role do vibrating charges play in electromagnetic processes? Can you visualize classical and electromagnetic waves? What do you know about lines of force? Follow the recommended links on the student website at **www.cengagebrain.com/shop/ISBN/1133104096** to better understand the various terms and components of electromagnetic waves and how electromagnetism works.

Atomic Physics

Courtesy IBM Corporation. Research Division, Almaden Research Center. Comp.

All things are made of atoms—little particles that move around in perpetual motion, attracting each other when they are a little distance apart, but repelling upon being squeezed into one another.

•

Richard Feynman,
physics Nobel Laureate
(1918–1988)

< Individual atoms in the "stadium corral"—iron (Fe) and copper (Cu).

T he development of physics prior to about 1900 is termed *classical physics* or *Newtonian physics*. It was generally concerned with the *macrocosm*, that is, with the description and explanation of large-scale observable phenomena such as the movements of projectiles and planets.

As the year 1900 approached, scientists thought the field of physics was in fairly good order. The principles of mechanics, wave motion, sound, and optics were reasonably well understood. Electricity and magnetism had been combined into electromagnetism, and light had been shown to be electromagnetic waves. Certainly some rough edges remained, but it seemed that only a few refinements were needed.

PHYSICS FACTS

▶ Theoretically, a sample of hot hydrogen gas can give off many different wavelengths, but only four are in the visible range.

▶ The basis of the microwave oven was discovered by a scientist working with radar

Chapter Outline

As scientists probed deeper into the submicroscopic world of the atom (the *microcosm*), however, they observed strange things, strange in the sense that they could not be explained by the classical principles of physics. These discoveries were unsettling because they made clear that physicists would need radical new approaches to describe and explain submicroscopic phenomena. One of these new approaches was *quantum mechanics*, which describes the behavior of matter and interactions on the atomic and subatomic levels.

The development of physics since about 1900 is called *modern physics*. This chapter is concerned with *atomic physics*, the part of modern physics that deals mainly with phenomena involving the electrons in atoms. Chapter 10 will then examine *nuclear physics*, which deals with the central core, or nucleus, of the atom.

An appropriate way to start the examination of atomic physics is with a brief history of the concept of the atom. How that concept has changed provides an excellent example of how scientific theories adapt as new experimental evidence is gathered.

9.1   Early Concepts of the Atom

Preview Questions

● What was John Dalton's hypothesis about the make-up of matter?

● How did Rutherford's nuclear model picture the atom?

Around 400 BCE, Greek philosophers were debating whether matter was continuous or discrete (particulate). For instance, if one could repeatedly cut a sample of gold in half, would an ultimate particle of gold theoretically be reached that could not be divided further? If an ultimate particle could be reached, matter would be discrete; if not, matter would be continuous.

Most of these philosophers, including the renowned Aristotle, decided that matter was continuous and could be divided again and again, indefinitely. A few philosophers thought that an ultimate, *indivisible* (Greek: *atomos*) particle would indeed be reached. The question was purely philosophical, since neither side could present scientific evidence to support its viewpoint. Modern science differs from the approach of the ancient Greeks by relying not only on logic but also on the scientific method (Chapter 1.2): the systematic gathering of facts by observation and experimentation and the rigorous testing of hypotheses.

The "continuous" model of matter postulated by Aristotle and his followers prevailed for about 2200 years until an English scientist, John Dalton, in 1807 presented evidence that matter is discrete and exists as particles. Dalton's evidence for the atomic theory will be examined in Chapter 12.3. For now, it is sufficient to point out that Dalton's major hypothesis was that each chemical element is composed of tiny, indivisible particles called **atoms**, which are identical for that element but different (particularly in masses and chemical properties) from atoms of other elements. Dalton's concept of the atom has been called the "billiard ball model," because he thought of atoms as essentially featureless, indivisible spheres of uniform density (● Fig. 9.1a).

Dalton's model had to be refined about 90 years later when the electron was discovered by J. J. Thomson at Cambridge University in England in 1897. Thomson studied electrical discharges in tubes of low-pressure gas called *gas-discharge tubes* or *cathode-ray tubes*. He discovered that when high voltage was applied to the tube, a "ray" was produced at the negative electrode (the cathode) and sped toward the positive electrode (the anode). Unlike electromagnetic radiation, the ray was deflected by electric and magnetic fields. Thomson concluded that the ray consisted of a stream of negatively charged particles, which are now called **electrons**. (See Chapter 8.5, Fig. 8.26.)

Further experiments by Thomson and others showed that an electron has a mass of 9.11×10^{-31} kg and a charge of -1.60×10^{-19} C and that the electrons were being produced by the voltage "tearing" them away from atoms of gas in the tube. Because identical electrons were being produced regardless of what gas was in the tube, it became apparent that atoms of all types contain electrons.

Dalton's model

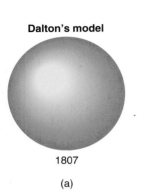

1807

(a)

Thomson's model

Spherical cloud
of positive
charge

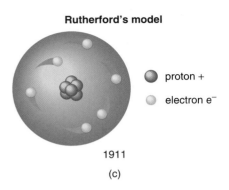

Electrons   1903

(b)

Rutherford's model

1911

(c)

proton +

electron e⁻

Figure 9.1 Dalton's, Thomson's, and Rutherford's Models of the Atom
(a) John Dalton's 1807 "billiard ball model" pictured the atom as a tiny, indivisible, uniformly dense, solid sphere. (b) J. J. Thomson's 1903 "plum pudding model" of the atom was a sphere of positive charge in which negatively charged electrons were embedded. (c) Ernest Rutherford's 1911 "nuclear model" depicted the atom as having a dense center of positive charge, called the *nucleus*, around which electrons orbited. (The sizes of the internal particles relative to the atom are *greatly* exaggerated in the figure.)

As a whole, an atom is electrically neutral. Therefore, some other part of the atom must be positively charged. Indeed, further experiments detected positively charged particles (now called *positive ions*) as they sped to and passed through holes in the negative electrode of a cathode-ray tube. In 1903, Thomson concluded that an atom was much like a sphere of plum pudding; electrons were the raisins stuck randomly in an otherwise homogeneous mass of positively charged pudding (Fig. 9.1b).

Thomson's "plum pudding model" of the atom was modified only 8 years later in 1911. Ernest Rutherford discovered that 99.97% of the mass of an atom is concentrated in a very tiny core which he called the *nucleus*. (The classic experiment that Rutherford performed to discover the nucleus is discussed in Chapter 10.2.)

Rutherford's nuclear model of the atom pictured the electrons as circulating in some way in the otherwise empty space around this very tiny, positively charged core (Fig. 9.1c). However, as will be discussed in the following sections, even Rutherford's model has undergone modification.

Did You Learn?

● John Dalton proposed that each chemical element was composed of tiny, indivisible particles called "atoms."

● Rutherford's model of the atom pictured electrons circulating around a tiny, positively charged core which he called the nucleus.

9.2 The Dual Nature of Light

Preview Questions

● What is a photon, and on what does its energy depend?

● What is meant by the "dual nature of light"?

Before discussing the next improvement in the model of the atom, a radical development about the nature of light should be considered. Even before the turn of the twentieth century, scientists knew that visible light of all frequencies was emitted by the atoms of an incandescent (glowing hot) solid, such as the filament of a light bulb. For example, ● Fig. 9.2

Figure 9.2 Red-Hot Steel
The radiation component of maximum intensity determines a hot solid's color, as shown here by orange-red hot steel coming out of a furnace.

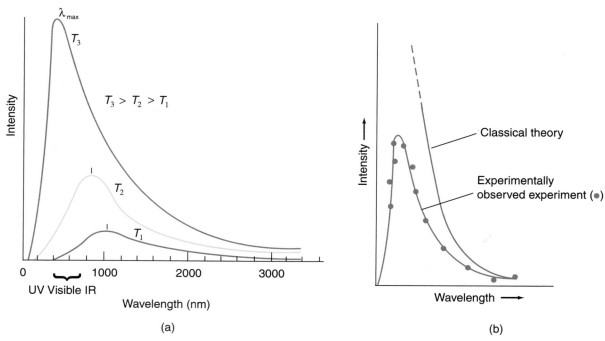

(a)

(b)

Figure 9.3 Thermal Radiation
(a) Intensity versus wavelength curves for thermal radiation. The wavelength associated with the maximum intensity (λ_{max}) becomes shorter with increasing temperature. (b) Classical theory predicts that the intensity of thermal radiation should be inversely related to the wavelength of the emitted radiation. If that were true, then the intensity would be much greater for shorter wavelengths than is actually observed.

Figure 9.4 Max Planck (1858–1947)
While a professor of physics at the University of Berlin in 1900, Planck proposed that the energy of thermal oscillators exists in only discrete amounts, or quanta. The important small constant *h* is called *Planck's constant.* Planck was awarded the Nobel Prize in physics in 1918 for his contributions to quantum physics. Planck's second son was executed by the Nazis for conspiring to assassinate Adolf Hitler.

shows that the radiation emitted by a hot ingot of steel as it comes from the furnace has its maximum intensity in the orange-red region of the visible spectrum.

Although there is a dominant orange-red color perceived by our eyes, in actuality there is a continuous spectrum. As illustrated in ● Fig. 9.3a, the intensity of the emitted radiation depends on the wavelength (λ). Practically all wavelengths are present, but there is a dominant color (wavelength), which depends on temperature. Consequently, a hot solid heated to a higher and higher temperature appears to go from a dull red to a bluish white. This outcome is expected because the hotter the solid, the greater the vibrations in the atoms and the higher frequency (shorter wavelength) of the emitted radiation. The wavelength of the maximum intensity component (λ_{max}) becomes shorter.

According to classical wave theory, the intensity (*I*) of the radiation spectrum is inversely related to the wavelength (actually, $I \propto 1/\lambda^4$). This relationship predicts that the intensity should increase without limit as the wavelengths get shorter, as illustrated in Fig. 9.3b. This prediction is sometimes called the *ultraviolet catastrophe*: *ultraviolet* because the difficulty occurs with wavelengths shorter than the ultraviolet end of the spectrum and *catastrophe* because it predicts emitted energy growing without bounds at these wavelengths.

The dilemma was resolved in 1900 by Max Planck (pronounced "plonck"), a German physicist (● Fig. 9.4). He introduced a radical idea that explained the observed distribution of thermal radiation intensity. In doing so, Planck took the first step toward a new theory called *quantum physics.* Classically, an electron oscillator may vibrate at any frequency or have any energy up to some maximum value, but *Planck's hypothesis* stated that the energy is *quantized.* That is, an oscillator can have only discrete, or specific, amounts of energy. Moreover, Planck concluded that the energy (*E*) of an oscillator depends on its frequency (*f*) in accordance with the following equation:

energy = Planck's constant × frequency

$$E = hf \tag{9.1}$$

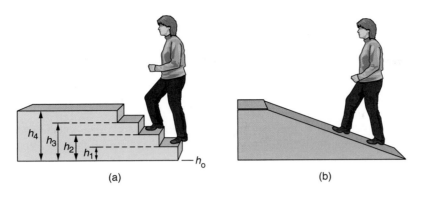

Figure 9.5 The Concept of Quantized Energy
(a) The woman on the staircase can stop only on one of the steps, so she can have only four potential-energy values. (b) When on the ramp, she can be at any height and thus can have a continuous range of potential-energy values. (See Conceptual Question and Answer.)

where h is a constant, called *Planck's constant*, that has the very small value of 6.63×10^{-34} J·s.

Planck's hypothesis correctly accounted for the observed radiation curve shown in Fig. 9.3b. Thus, Planck introduced the idea of a **quantum**, a discrete amount of energy (or a "packet" of energy).

Conceptual Question and Answer

Step Right Up

Q. Can mechanical energy be quantized?

A. Quantized energy in quantum physics is analogous to the potential energy of a person on a staircase who can have only discrete potential-energy values determined by the height of each particular step (● Fig. 9.5a). Continuous energy, on the other hand, is like the potential energy of a person on a ramp who can stand at any height and thus can have any value of potential energy (Fig. 9.5b).

In the latter part of the nineteenth century, scientists observed that electrons are emitted when certain metals are exposed to light. The phenomenon was called the **photoelectric effect**. This direct conversion of light (radiant energy) into electrical energy now forms the basis of photocells used in calculators, in the automatic door openers at your supermarket, and in the "electric eye" beam used in many automatic garage door openers to protect children and pets from a descending door, as in ● Fig. 9.6.

As in the case of the ultraviolet catastrophe, certain aspects of the photoelectric effect could not be explained by classical theory. For example, the amount of energy necessary to

Nonvisible light beam

Electric eye with photocell

Figure 9.6 Photoelectric Effect Application: The Electric Eye
When light strikes a photocell, electrons are freed from the atoms and a current is set up in an "electric eye" circuit. When the garage door starts to move downward, any interruption of the electric eye beam (usually infrared, IR), which affects the circuit current and signals the door to stop, protects anything that might be under it.

free an electron from a photomaterial could be calculated. According to classical theory, in which light is considered a wave with a continuous flow of energy, it would take an appreciable time for electromagnetic waves to supply the energy needed for an electron to be emitted. However, electrons flow from photocells almost immediately upon being exposed to light. Also, it was observed that only light above a certain frequency would cause electrons to be emitted. According to classical theory, light of any frequency should be able to provide the needed energy.

In 1905, Albert Einstein solved the problems of the photoelectric effect. Applying Planck's hypothesis, Einstein postulated that light (and, in fact, all electromagnetic radiation) was quantized and consisted of "particles," or "packets," of energy, rather than waves. Einstein coined the term **photon** to refer to such a quantum of electromagnetic radiation. He used Planck's relationship ($E = hf$) and stated that a quantum, or photon, of light contains a discrete amount of energy (E) equal to Planck's constant (h) times the frequency (f) of the light. The higher the frequency of the light, the greater the energy of its photons. For example, because blue light has a higher frequency (shorter wavelength) than red light, photons of blue light have more energy than photons of red light. Example 9.1 illustrates how photon energy can be determined.

EXAMPLE 9.1   Determining Photon Energy from the Frequency

Find the energy in joules of the photons of red light of frequency 5.00×10^{14} Hz. (Recall from Chapter 6.2 that the unit hertz is equivalent to reciprocal second, 1/s.)

Solution

Given the frequency, the energy can be found directly by using Eq. 9.1.

$$E = hf = (6.63 \times 10^{-34} \, \text{J} \cdot \text{s})(5.00 \times 10^{14} \, 1/\text{s})$$
$$= 33.2 \times 10^{-20} \, \text{J} = 3.32 \times 10^{-19} \, \text{J}$$

Confidence Exercise 9.1

Find the energy in joules of the photons of blue light of frequency 7.50×10^{14} Hz.
Answers to Confidence Exercises may be found at the back of the book.

By considering light to be composed of photons, Einstein was able to explain the photoelectric effect. The classical time delay necessary to get enough energy to free an electron is not a problem when using the concept of photons of energy. A photon with the proper amount of energy could deliver the release energy instantaneously in a "packet." (An analogy to illustrate the difference between the delivery of wave energy and that of quantum energy is shown in ● Fig. 9.7.) The concept of the photon also explains why light with greater than a certain minimum frequency is required for emission of an electron. Because $E = hf$, a photon of light with a frequency smaller than this minimum value would not have enough energy to free an electron. (See the **Highlight: Albert Einstein.**)

How can light be composed of photons (discrete packets of energy) when it shows wave phenomena such as polarization, diffraction, and interference (Chapters 7.5 and

Figure 9.7 Wave and Quantum Analogy
A wave supplies a continuous flow of energy, somewhat analogous to the stream of water from the garden hose. A quantum supplies its energy all at once in a "packet" or "bundle," somewhat like each bucketful of water thrown on the fire.

Wave nature   Quantum nature

Highlight   Albert Einstein

Isaac Newton's only rival for the accolade of "greatest scientist of all time" is Albert Einstein, who was born in Ulm, Germany, in 1879 (Fig. 1). In high school, Einstein did poorly in Latin and Greek and was interested only in mathematics. His teacher told him, "You will never amount to anything, Einstein."

Einstein attended college in Switzerland, graduated in 1901, and accepted a job as a junior official at the patent office in Berne, Switzerland. He spent his spare time working in theoretical physics. In 1905, five of his papers were published in the *German Yearbook of Physics*, and in that same year he earned his Ph.D. at the age of 26. One paper explained the photoelectric effect (Chapter 9.2), and he was awarded the 1921 Nobel Prize in physics for that contribution.

Another of Einstein's 1905 papers put forth his ideas on what came to be called the *special theory of relativity*. This paper dealt with what would happen to an object as it approached the speed of light. Einstein asserted that to an outside observer, the object would get shorter in the direction of motion, it would become more massive, and a clock would run more slowly in the object's system. These predictions are against "common sense," but common sense is based on limited experience with objects of ordi-

nary size moving at ordinary speeds. All three of Einstein's predictions have now been verified experimentally.

One other important result came from the special theory of relativity: Energy and matter are related by what has become one of the most famous equations in scientific history: $E = mc^2$. This relationship and its use will be discussed in Chapter 10.6.

In 1915, Einstein published his *general theory of relativity*, which deals mainly with the effect of a gravitational field on the behavior of light and has profound implications with regard to the structure of the universe. Once again, Einstein's predictions were verified experimentally, and the general theory of relativity is a cornerstone of modern physics and astronomy.

Of course, Einstein was not infallible. He thought quantum mechanics bordered on the absurd, and throughout the 1930s he fought a friendly battle with Niels Bohr on the subject. (See Chapter 9.3 for information on Bohr.) Bohr won. Predictions of quantum mechanics are demonstrated by experiments.

One famous exchange between Einstein and Bohr went like this:

Einstein: "God doesn't play dice with the cosmos."

Bohr: "Einstein, don't tell God what to do."

Figure 1 Albert Einstein (1879–1955)
Einstein is shown here during a visit to Caltech in the 1930s.

7.6)? Such behavior is explained by assuming the wave nature of light, but it cannot be explained by the photon (particle) concept. On the other hand, the photoelectric effect cannot be explained by invoking the wave nature of light; for this effect, the photon concept is necessary. Hence, there is a confusing situation. Is light a wave, or is it a particle? The situation is expressed by the term **dual nature of light**, which means that to explain various phenomena, light must be described sometimes as a wave and sometimes as a particle (● Fig. 9.8).

Our idea that something is *either* a wave *or* a particle breaks down here. Light is not really a wave, nor is it really a particle. It has characteristics of both, and we simply have no good, single, macroscopic analogy that fits the combination. Therefore, scientists (renowned for being pragmatic) use whichever model of light works in a specific type of experiment. In

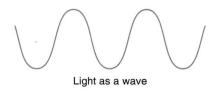

Light as a wave

Light as a stream of photons
(packets of energy)

Figure 9.8 The Wave-Particle Duality of Light
Electromagnetic radiation (a beam of light) can be pictured in two ways: as a wave (top) or as a stream of individual packets of energy called *photons* (bottom).

some experiments the wave model does the job; in other experiments the particle model is necessary.

The concept of the dual nature of light has been around for almost a century now. It actually does an excellent job of explaining both known and newly discovered phenomena and of making valid predictions that have served as the basis of new technologies. But it is still puzzling.

Did You Learn?

● A photon is a quantum or "packet" of electromagnetic energy (light), and its energy depends on the frequency, $E = hf$.

● In explaining phenomena, light sometimes behaves as a wave and sometimes as a particle, giving it a dual nature.

9.3   Bohr Theory of the Hydrogen Atom

Preview Questions

● What does the principal quantum number n in the Bohr theory designate?

● When does a hydrogen atom emit or absorb radiant energy?

Now that you have an idea of what is meant by quantum theory and photons, it is time to discuss the next advance in the understanding of the atom. Recall from Chapter 7.2 that when light from incandescent sources, such as light bulb filaments, is analyzed with a spectrometer, a *continuous spectrum* (continuous colors) is observed (● Fig. 9.9a).

In addition to continuous spectra, there are two types of *line spectra*. In the late 1800s much experimental work was being done with gas-discharge tubes, which contain a little neon (or another element) and emit light when subjected to a high voltage. When the light from a gas-discharge tube is analyzed with a spectrometer, a **line emission spectrum**, rather than a continuous spectrum, is observed (Fig. 9.9b). In other words, only spectral lines of certain frequencies or wavelengths are found, and each element gives a different set of lines (see Fig. 7.15). Spectroscopists at that time did not understand why only discrete, characteristic wavelengths of light were emitted by atoms in various excited gases.

The second type of line spectrum is found when visible light of all wavelengths is passed through a sample of a cool, gaseous element before entering the spectrometer. The **line absorption spectrum** that results has dark lines of missing colors (Fig. 9.9c). The dark lines are at exactly the same wavelengths as the bright lines of the *line emission spectrum* for that particular element. Compare, for instance, the wavelengths of the lines in the emission and absorption spectra of hydrogen in Fig. 9.9b and c.

An explanation of the spectral lines observed for hydrogen was advanced in 1913 by Danish physicist Niels Bohr. The hydrogen atom is the simplest atom, which is no doubt why Bohr chose it for study. Its nucleus is a single proton, and Bohr's theory assumes that its one electron revolves around the nuclear proton in a circular orbit in much the same way a satellite orbits the Earth or a planet orbits the Sun. (In fact, Bohr's model is often called *the planetary model* of the atom.)

However, in the atom, the electric force instead of the gravitational force supplies the necessary centripetal force. The revolutionary part of the theory is that Bohr assumed that the angular momentum (Chapter 3.7) of the electron is quantized. As a result, an electron can have only specific energy values in an atom. He correctly reasoned that a discrete line spectrum must be the result of a quantum effect. This assumption led to the prediction that the hydrogen electron could exist only in discrete (specific) orbits with particular radii. (This is not the case for Earth satellites. With proper maneuvering, a satellite can have any orbital radius.)

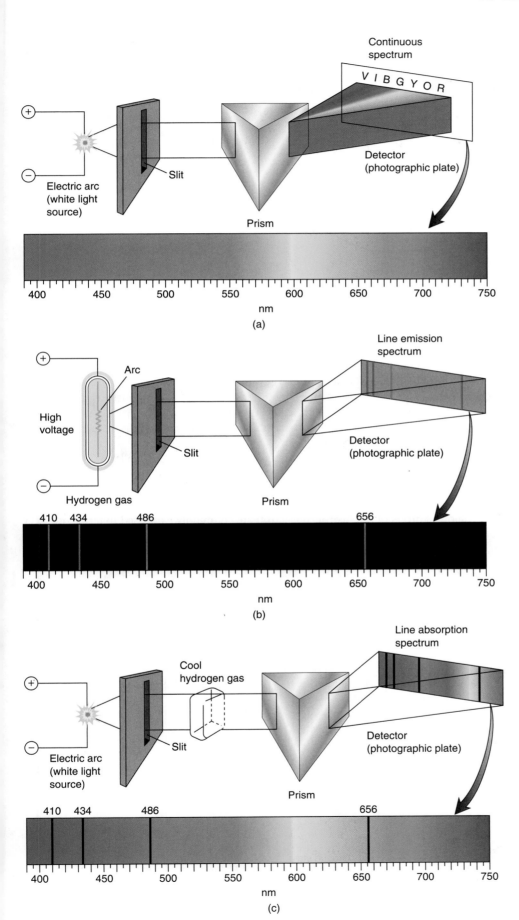

Figure 9.9 Three Types of Spectra (a) A continuous spectrum containing all wavelengths of visible light (indicated by the initial letters of the colors of the rainbow). (b) The line emission spectrum for hydrogen consists of four discrete wavelengths of visible radiation. (c) The line absorption spectrum for hydrogen consists of four discrete "missing" wavelengths that appear as dark lines against a rainbow-colored background.

Figure 9.10 Bohr Electron Orbits
The Bohr hypothesis predicts only certain discrete orbits for the hydrogen electron. Each orbit is indicated by a principal quantum number *n*. The orbit shown in blue is the ground state (*n* = 1). The orbits move outward and get farther apart with increasing values of *n*.

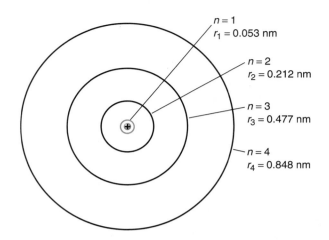

Bohr's possible electron orbits are characterized by whole-number values, $n = 1, 2, 3, \ldots$, where *n* is called the **principal quantum number** (● Fig. 9.10). The lowest-value orbit (*n* = 1) has the smallest radius, and the radii increase as the principal quantum number increases. Note in Fig. 9.10 that the orbits are not evenly spaced; as *n* increases, the distances from the nucleus increase, along with the distances *between* orbits.

A problem remained with Bohr's hypothesis. According to classical theory, an accelerating electron radiates electromagnetic energy. An electron in circular orbit has centripetal acceleration (Chapter 2.4) and hence should radiate energy continuously as it circles the nucleus. Such a loss of energy would cause the electron to spiral into the nucleus, in a manner similar to the death spiral of an Earth satellite in a decaying orbit because of energy losses due to atmospheric friction.

However, atoms do not continuously radiate energy, nor do they collapse. Accordingly, Bohr made another nonclassical assumption: the hydrogen electron does not radiate energy when in an allowed, discrete orbit, but does so only when it makes a *quantum jump*, or *transition*, from one allowed orbit to another.

The allowed orbits of the hydrogen electron are commonly expressed in terms of *energy states*, or *energy levels*, with each state corresponding to a specific orbit (● Fig. 9.11). We characterize the energy levels as states in *a potential well* (Chapter 4.2). Just as energy must be expended to lift a bucket in a water well, energy is necessary to lift the electron to a

Figure 9.11 Orbits and Energy Levels of the Hydrogen Atom
The Bohr theory predicts that the hydrogen electron can occupy only certain orbits having discrete radii. Each orbit has a particular energy value, or energy level. The lowest level, which has the quantum number *n* = 1, is called the *ground state*. The higher energy levels, which have *n* values greater than 1, are called *excited states*.

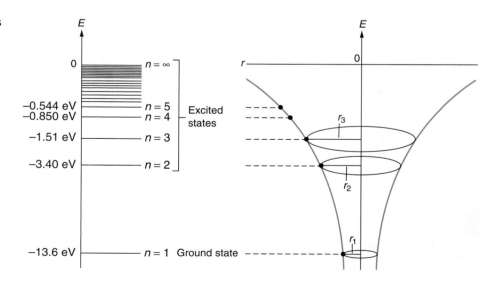

higher level. If the top of the potential well is taken as the zero reference level, then the energy levels in the well all have negative values.

In a hydrogen atom, the electron is normally at the bottom of the "well," or in the **ground state** ($n = 1$) and must be given energy, or "excited," to raise it up in the well to a higher energy level, or orbit. The states above the ground state ($n = 2, 3, 4, \ldots$) are called **excited states**. The level for $n = 2$ is the *first excited state*, the level for $n = 3$ is the *second excited state*, and so on.

The levels in the energy well resemble the rungs of a ladder except that the energy level "rungs" are not evenly spaced. Just as a person going up and down a ladder must do so in discrete steps on the ladder rungs, so a hydrogen electron must be excited (or de-excited) from one energy level to another by discrete amounts. If enough energy is applied to excite the electron to the top of the well, then the electron is no longer bound to the nucleus and the atom is *ionized*, becoming a positive ion (H^+).

A mathematical development of Bohr's theory is beyond the scope of this textbook. However, the important results are its prediction of the radii and the energies of the allowed orbits. The radius of a particular orbit is given by

Table 9.1	Allowed Values of the Hydrogen Electron's Radius and Energy for Low Values of n	
n	r_n	E_n
1	0.053 nm	−13.60 eV
2	0.212 nm	−3.40 eV
3	0.477 nm	−1.51 eV
4	0.848 nm	−0.85 eV

Bohr radius = 0.053 × the square of the principal quantum number n

$$r_n = 0.053 \, n^2 \text{ nm} \qquad \text{9.2}$$

where r is the orbit radius, measured in nanometers (1 nm $= 10^{-9}$ m), and n is the principal quantum number of an orbit ($n = 1, 2, 3, \ldots$).

Allowed values of r are listed in ● Table 9.1 for some low values of n. Note that the radii indeed get farther apart with increasing n, as indicated in Fig. 9.10.

EXAMPLE 9.2   Determining the Radius of an Orbit in a Hydrogen Atom

Determine the radius in nanometers (nm) of the first orbit ($n = 1$, the ground state) in a hydrogen atom.

Solution

The principal quantum number is given, so the orbital radius can be found directly using Eq. 9.2.
$$r_1 = 0.053 \, n^2 \text{ nm} = 0.053 \, (1)^2 \text{ nm} = 0.053 \text{ nm}$$

Confidence Exercise 9.2

Determine the radii in nm of the second and third orbits ($n = 2, 3$, the first and second excited states) in a hydrogen atom. Compare the values you compute with those in Table 9.1.

Answers to Confidence Exercises may be found at the back of the book.

The total energy (in eV) of an electron in an allowed orbit is given by

$$E_n = \frac{-13.60}{n^2} \text{ eV} \qquad \text{9.3}$$

where E is energy measured in electron volts (eV) and n is the orbit's principal quantum number.*

---

*An *electron volt* is the amount of energy an electron acquires when it is accelerated through an electric potential of 1 volt. The eV is a small, common, nonstandard unit of energy in atomic and nuclear physics, and 1 eV $= 1.60 \times 10^{-19}$ J. Its size makes it an appropriate unit on these levels.

EXAMPLE 9.3   Determining the Energy of an Orbit
in the Hydrogen Atom

Determine the energy of an electron in the first orbit ($n = 1$, the ground state) in a hydrogen atom.

Solution

The energy of a particular orbit, or energy level, in a hydrogen atom is calculated by using Eq. 9.3 and the n value for that orbit. For $n = 1$, then,

$$E_1 = \frac{-13.60}{(1)^2} \text{ eV} = -13.60 \text{ eV}$$

Confidence Exercise 9.3

Determine the energies of an electron in the second and third orbits ($n = 2, 3$, the first and second excited states) in a hydrogen atom. Compare your answers with the values given in Table 9.1.

Table 9.1 shows the energies for the hydrogen electron for low values of n (orbits nearest the nucleus). These values correspond to the energy levels shown in Fig. 9.11. Note that, unlike the distances between orbits, the energy levels get *closer* together as n increases. Recall that the minus signs, indicating negative energy values, show that the electron is in a potential-energy well. Because the energy value is -13.60 eV for the ground state, it would require that much energy input to ionize a hydrogen atom. Thus, we say that the hydrogen electron's *binding energy* is 13.60 eV.

How did the hypothesis stand up to experimental verification? Recall that Bohr was trying to explain discrete line spectra. According to his hypothesis, an electron can make transitions only between two allowed orbits, or energy levels. In these transitions, the total energy must be conserved. If the electron is initially in an excited state, then it will lose energy when it "jumps down" to a less excited (lower n) state. In this case, the electron's energy loss will be carried away by a photon, or a quantum of light.

By the conservation of energy, the total initial energy (E_{n_i}) must equal the total final energy ($E_{n_f} + E_{\text{photon}}$):

$$E_{n_i} = E_{n_f} + E_{\text{photon}}$$

(energy before = energy after)

Or, rearranging, the energy of the emitted photon (E_{photon}) is the difference between the energies of the initial and final states:

photon energy = energy of initial orbit minus energy of final orbit

$$E_{\text{photon}} = E_{n_i} - E_{n_f} \qquad \textbf{9.4}$$

A schematic diagram of the process of *photon emission* is shown in ● Fig. 9.12a. Hydrogen's line emission spectrum results from the relatively few allowed energy transitions as the electron de-excites. Figure 9.12b illustrates the reverse process of *photon absorption* to excite the electron. Hydrogen's line absorption spectrum results from exposing hydrogen atoms in the ground state to visible light of all wavelengths (or frequencies). The hydrogen electrons absorb only those wavelengths that can cause electron transitions "up." These wavelengths are taken out of the incoming light, whereas the other, inappropriate wavelengths pass through to produce a spectrum of color containing dark lines.

Of course, in a hydrogen atom, a photon of the same energy emitted in a "down" transition will have been absorbed in an "up" transition between the same two levels. Therefore, the dark lines in the hydrogen absorption spectrum exactly match up with the bright lines in the hydrogen emission spectrum (see Figs. 9.9b and 9.9c).

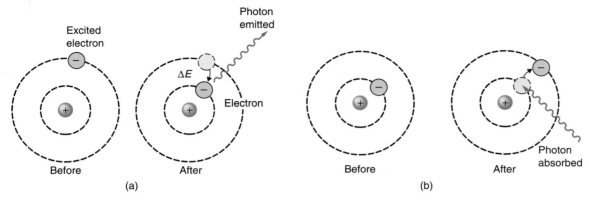

Figure 9.12 Photon Emission and Absorption
(a) When an electron in an excited hydrogen atom makes a transition to a lower energy level, or orbit, the atom loses energy by emitting a photon. (b) When a hydrogen atom absorbs a photon, the electron is excited into a higher energy level, or orbit.

The transitions for photon emissions in the hydrogen atom are shown on an energy level diagram in ● Fig. 9.13. The electron may "jump down" one or more energy levels in becoming de-excited. That is, the electron may go down the energy "ladder" using adjacent "rungs," or it may skip "rungs." (To show photon absorption and jumps to higher energy levels, the arrows in Fig. 9.13 would point up.)

EXAMPLE 9.4   Determining the Energy of a Transition in a Hydrogen Atom

Use Table 9.1 and Eq. 9.4 to determine the energy of the photon emitted as an electron in the hydrogen atom jumps down from the $n = 2$ level to the $n = 1$ level.

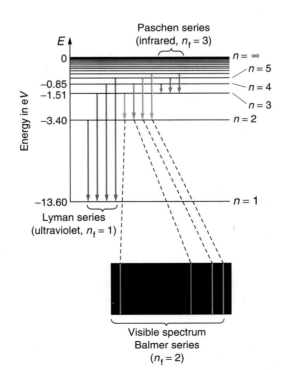

Figure 9.13 Spectral Lines for Hydrogen

The transitions among discrete energy levels by the electron in the hydrogen atom give rise to discrete spectral lines. For example, transitions down to $n = 2$ from $n = 3$, 4, 5, and 6 give the four spectral lines in the visible region that form the Balmer series. Bohr correctly predicted the existence of both the Lyman series and the Paschen series.

Solution

Table 9.1 shows values of -3.40 eV for the $n = 2$ level and -13.60 eV for $n = 1$. Then, using Eq. 9.4,

$$E_{photon} = E_2 - E_1$$
$$= -3.40 \text{ eV} - (-13.60 \text{ eV}) = 10.20 \text{ eV}$$

The positive value indicates that the 10.20-eV photon is *emitted*.

Confidence Exercise 9.4

Use Table 9.1 and Eq. 9.4 to determine the energy of the photon absorbed as an electron in the hydrogen atom jumps up from the $n = 1$ level to the $n = 3$ level.

Thus, the Bohr hypothesis predicts that an excited hydrogen atom will emit light with discrete frequencies (and hence wavelengths) corresponding to discrete "down" transitions. Conversely, a hydrogen atom in the ground state will absorb light with discrete frequencies corresponding to discrete "up" transitions. The theoretical frequencies for the four allowed transitions that give lines in the visible region were computed using the energy values for the various initial and final energy levels. These were compared with the frequencies of the lines of the hydrogen emission and absorption spectra. The theoretical and the experimental values were identical, a triumph for the Bohr theory.

Transitions to a particular lower level form a *transition series*. These series were named in honor of early spectroscopists who discovered or experimented with the hydrogen atom's spectral lines that belonged to particular regions of the electromagnetic spectrum. For example, the series of lines in the *visible* spectrum of hydrogen—which corresponds to transitions from $n = 3, 4, 5,$ and 6 down to $n = 2$—is called the *Balmer series* (Fig. 9.13).

Bohr's calculations predicted the existence of a series of lines of specific energies in the *ultraviolet* (UV) region of the hydrogen spectrum and another series of lines in the *infrared* (IR) region. Lyman and Paschen, who like Balmer were spectroscopists, discovered these series, the lines of which were exactly at the wavelengths Bohr had predicted. These series bear their names (Fig. 9.13).

So quantum theory and the quantum nature of light scored another success. As you might imagine, the energy level arrangements for atoms other than hydrogen—those with more than one electron—are more complex. Even so, the line spectra for atoms of various elements are indicative of their energy level spacings and provide characteristic line emission and line absorption "fingerprints" by which atoms may be identified using spectroscopy.

The compositions of distant stars can be determined via analysis of the dark absorption lines in their spectra. In fact, the element helium was discovered in the Sun before it was found on the Earth. In 1868 a dark line was detected in the solar spectrum that did not match the absorption line of any known element. It was concluded that the line must be that of a new element. The element was named *helium* (after the Greek word for Sun, *helios*). Twenty-seven years later, helium was found on the Earth, trapped in a uranium mineral. The major source of helium today is from natural gas.

Another interesting quantum phenomenon is *auroras*, or what are often called the *northern lights* and *southern lights*. Auroras are caused by charged particles from the Sun entering the Earth's atmosphere close to the magnetic north and south poles (see Fig. 19.6). These particles interact with molecules in the air and excite some of their electrons to higher energy levels. When the electrons fall back down, some of the absorbed energy is emitted as visible radiation.

A similar transition phenomenon occurs in the production of light by fluorescent lamps—those long, white tubes that probably light your classroom—or by the relatively new compact fluorescent bulbs (Chapter 4.6). The primary radiation emitted by electrically excited mercury atoms in a fluorescent tube is in the ultraviolet (UV) region. The UV radiation is absorbed by the white fluorescent material that coats the inside of the tube. This material reradiates at frequencies in the visible region from various "down" transitions of lesser energy, providing white light for reading and other purposes.

- The principal quantum number (*n*) designates the possible electron orbits (energy levels or states) in the Bohr theory, *n* = 1, 2, 3, ...

- Photons are emitted and absorbed by a hydrogen atom when its electron changes energy levels, "down" and "up" transitions, respectively.

9.4 Microwave Ovens, X-Rays, and Lasers

Preview Questions

- What is the main reason foods heat up in microwave ovens?
- What does the acronym "laser" stand for?

In this section three important technological applications that involve quantum theory will be considered: microwave ovens, X-rays, and lasers.

Microwave Ovens

Large parts of modern physics and chemistry are based on the study of energy levels of various atomic and molecular systems. When light is emitted or absorbed, scientists study the emission or absorption spectrum to learn about the energy levels of the system, as shown in Fig. 9.13 for the hydrogen atom.

Some scientists do research in *molecular spectroscopy*, the study of the spectra and energy levels of molecules (combinations of atoms). Molecules of one substance produce a spectrum different from that produced by molecules of another substance. Molecules can have quantized energy levels because of molecular vibrations or rotations or because they contain excited atoms.

The water molecule has some rotational energy levels spaced very closely together. The energy differences are such that microwaves, which have relatively low frequencies and energies (see Figure 6.7), are absorbed by the water molecules. This principle forms the basis of the *microwave oven*.* Because all foods contain moisture, their water molecules absorb microwave radiation, thereby gaining energy and rotating more rapidly; thus, the food is heated and cooked. Molecules of fats and oils in a food also are excited by microwave radiation, so they, too, contribute to the cooking. The interior metal sides of the oven reflect the radiation and remain cool.

Because it is the water content of foods that is crucial in microwave heating, objects such as paper plates and ceramic or glass dishes do not get hot immediately in a microwave oven. However, they often become warm or hot after being in contact with hot food (heat transfer by conduction, Chapter 5.4).

Some people assume that the microwaves penetrate the food and heat it throughout, but that is not the case. Microwaves penetrate only a few centimeters before being completely absorbed, so the interior of a large mass of food must be heated by conduction as in a regular oven. For this reason, microwave oven users are advised to let foods sit for a short time after microwaving. Otherwise, the center may be disagreeably cool even though the outside of the food is quite hot.

---

*"Percy Spencer didn't know better than to bring candy with him into his microwave lab in 1946. When the American engineer, who was developing radar components for the Raytheon Corp., let his chocolate bar get too close to a piece of equipment, it turned into chocolate goo. Cooking would never be the same. Within a year, Raytheon had introduced the first commercial microwave oven."—*Time,* March 29, 1999

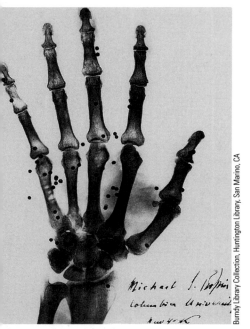

Burndy Library Collection, Huntington Library, San Marino, CA

Figure 9.14 X-Rays Quickly Found Practical Use
Roentgen discovered X-rays in December 1895. By February 1896 they were being put to practical use. X-rays can penetrate flesh relatively easily and leave a skeletal image on film. The black spots in this "X-ray," or radiograph, are bird shot embedded in the subject's hand from a hunting accident.

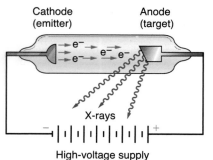

Figure 9.15 X-Ray Production
X-rays are produced in a tube in which electrons from the cathode are accelerated toward the anode. Upon interacting with the atoms of the anode material, the electrons are slowed down, and the atoms emit energy in the form of X-rays.

Conceptual Question and Answer

Can't Get Through

Q. Microwave oven glass doors have a metal mesh with holes. What is the purpose of this mesh?

A. The oven door glass window is for easy viewing, but the window has a perforated metal mesh over it for safety. Because of the size of the perforations (holes) in the mesh are much less than microwave wavelengths [on the order of 12 cm (4.7 in.)], most of the microwave radiation cannot pass through the door, whereas light with wavelengths on the order of 10^{-11} cm can. For microwaves, the mesh behaves as a continuous metal sheet that reflects the radiation back into the oven.

X-Rays

X-rays are another example of the technological use of quantum phenomena. **X-rays** are high-frequency, high-energy electromagnetic radiation (Chapter 6.3). These rays were discovered accidentally in 1895 by German physicist Wilhelm Roentgen (pronounced "RUNT-gin"). While working with a gas-discharge tube, Roentgen noticed that a piece of fluorescent paper across the lab was glowing, apparently from being exposed to some unknown radiation emitted from the tube. He called it *X-radiation*, with the X standing for "unknown." X-rays are now widely used in industrial and medical fields (● Fig. 9.14).

In a modern X-ray tube, electrons are accelerated through a large electrical voltage toward a metal target (● Fig. 9.15). When the electrons strike the target, they interact with the electrons in the target material and the electrical repulsion decelerates the incident electrons. The result is an emission of high-frequency X-ray photons (quanta). In keeping with their mode of production, X-rays are called *Bremsstrahlung* ("braking rays") in German.

The wise use of X-rays wasn't always the rule. Large doses of X-rays can cause skin "burns," cancer, and other conditions. Chest and dental X-rays once exposed patients to large doses of X-rays. These X-rays are now less intense and are monitored for appropriate safety levels.

X-ray imagery, along with magnetic resonance imagery, is discussed in the Highlight: X-Ray CAT Scan and MRI.

Lasers

Another device based on energy levels is the *laser*, the development of which was a great success for modern science. Scientific discoveries, such as X-rays and microwave heating, have often been made accidentally. X-rays were put to practical use before anyone understood the how or why of the X-ray phenomenon. Similarly, early investigators often applied a trial-and-error approach until they found something that worked. Edison's improvement of the incandescent light is a good example of this approach. He tried various materials before settling on a carbonized filament. In contrast to X-rays and the light bulb, the idea of the laser was first developed "on paper" from theory around 1965. The laser was then built with the full expectation that it would work as predicted.

The word **laser** is an acronym for *light amplification by stimulated emission of radiation*. The amplification of light provides an intense beam. Ordinarily, when an electron in an atom is excited by a photon, it emits a photon and then returns to its ground state immediately. In this process, which is called *spontaneous emission*, one photon goes in and one photon comes out (● Fig. 9.16a and b).

However, some substances, such as ruby crystals and carbon dioxide gas, and some combinations of substances, such as a mixture of the gases helium and neon, have *metastable* excited states. That is, some of their electrons can jump up into these excited energy levels and remain there briefly.

When many of the atoms or molecules of a substance have been excited into a metastable state by the input of the appropriate energy, we say that a *population inversion* has occurred

Highlight X-Ray CAT Scan and MRI

CAT scan refers to an X-ray medical imaging method, not a feline glance. In conventional medical and dental X-ray photography, the rays emerging from an X-ray tube are detected on film. The X-rays themselves may expose the film or excite some fluorescent material that produces light for the film exposure. The latter method reduces the amount of X-rays needed. The difference in the absorption by different structures in the body gives rise to the image production: the less the absorption, the greater the transmission and darker the film. In a sense, the image is a "shadow" of what the rays have passed through.

In the 1970s a new technique called computer tomography (CT) was developed. In conventional X-ray images, the entire thickness of the body is projected on the film. As a result, one structure may obscure the view of another. A tomographic image, on the other hand, is an image of a "slice" through the body. (*Tomography* comes from the Greek words *tomo*, meaning "slice," and *graphon*, meaning "written.")

Together, an X-ray source and detector rotate around the body, scanning it at a great number of points to produce an image slice. Obtaining data for a complete picture would take some time, but fan beams and multiple digital detectors for a computerized image speed up the process (Fig. 1). Dental X-rays are now digitized and can be viewed almost instantaneously.

*Resonance is discussed in Chapter 6.6.

What does *CAT* stand for? Because the image of slice is perpendicular to the body axis, CAT stands for *computerized axial tomography* or *computer-assisted tomography,* but it is usually shortened to just CT for a CT scan. The computer assists in reconstructing the angular slices with resolution that cannot be achieved by conventional X-ray photographs.

Like CT scans, MRI (magnetic resonance imaging) is a medical imagery technique used to view detailed internal body structures. The good contrast it provides between soft tissues make it particularly useful in brain, heart, spinal cord, and cancer imaging.

Unlike CT scans or regular X-rays, which use potentially dangerous radiation, MRI uses magnets and radio waves as the patient passes through a chamber, much like a CT chamber. The human body is largely composed of water molecules (H_2O), with each molecule having two protons (hydrogen nuclei). A magnet creates a powerful magnetic field that aligns the protons of the hydrogen atoms, which are then exposed to radio waves. When the radio waves have just the right frequency (resonance), the protons "flip" in alignment and energy (a photon) is emitted.* This energy is detected by the scanner and sent to a computer. With multiple scans, a slice image of an organ or part of the body is formed, and these slices can be combined into a three-dimensional image (Fig. 2).

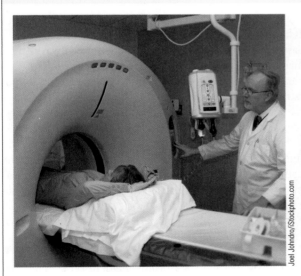

Figure 1 CAT (CT) Scan
A patient entering the scanning chamber.

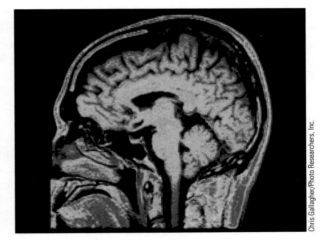

Figure 2 MRI Image
A color-enhanced MRI scan of a 38-year-old male showing the cranial anatomy. (Can you identify any of the organs?)

(● Fig. 9.17a). In such a condition, an excited atom can be stimulated to emit a photon (Fig. 9.16c and b). In a **stimulated emission**, the key process in a laser, an excited atom is struck by a photon of the same energy as the allowed transition and two photons are emitted (one in, two out: amplification). Of course, in this process, the laser is not giving something for nothing; energy is needed to excite the atom initially.

The light intensity is amplified because the emitted photon is in phase with the stimulating photon and thus interferes constructively to produce maximum intensity (Chapter 7.6).

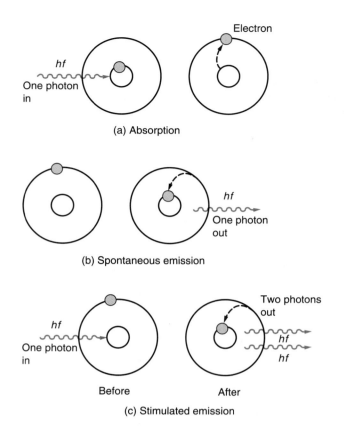

(a) Absorption

(b) Spontaneous emission

(c) Stimulated emission

Each of the two photons can then stimulate the emission of yet another identical photon. The result of many stimulated emissions and reflections in a laser tube is a narrow, intense beam of laser light. The beam consists of photons having the same energy and wavelength (*monochromatic*) and traveling in phase in the same direction (*coherent*). See Fig. 9.17c.

Light from sources such as an incandescent bulb consists of many wavelengths and is *incoherent* because the excitation occurs randomly. The atoms emit randomly at different wavelengths, and the waves have no particular directional or phase relationships to one another.

Because a laser beam is so directional, it spreads very little as it travels. This feature has enabled scientists to reflect a laser beam back to the Earth from a mirror placed on the Moon by astronauts. This technique makes accurate measurements of the distance to the Moon possible so that small fluctuations in the Moon's orbit can be studied. Similar measurements of light reflected from mirrors on the Moon are used to help determine the rate of plate tectonic movement on the Earth (see Chapter 21.3).

Lasers are used in an increasing number of applications. For instance, long-distance communications use laser beams in space and in optical fibers for telephone conversations. Lasers are also used in medicine as diagnostic and surgical tools (● Fig. 9.18).

In industry the intense heat produced by focused laser beams incident on a small area can drill tiny holes in metals and can weld machine parts. Laser "scissors" cut cloth in the garment industry. Laser printers produce computer printouts. Other applications occur in surveying, weapons systems, chemical processing, photography, and holography (the process of making three-dimensional images).

Another common laser application is found at supermarket checkout counters. You have probably noticed the reddish glow that is produced by a helium–neon laser in an optical scanner used for reading the product codes on items in supermarkets and other stores.

You may own a laser in a compact disc (CD) player. A laser "needle" is used to read the information (sound) stored on the disc in small dot patterns. The dots produce reflection patterns that are read by photocells and converted into electronic signals, which are changed to sound waves by a speaker system.

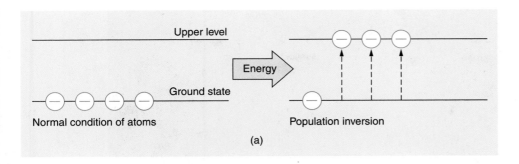

Upper level

Energy

Ground state

Normal condition of atoms   Population inversion

(a)

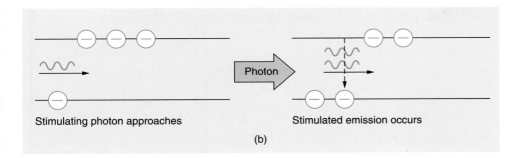

Photon

Stimulating photon approaches   Stimulated emission occurs

(b)

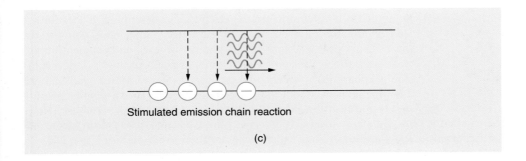

Stimulated emission chain reaction

(c)

Figure 9.17 Steps in the Action of a Laser
(a) Atoms absorb energy and move to a higher energy level (a population inversion). (b) A photon approaches, and stimulated emission occurs. (c) The photons that are emitted cause other stimulated emissions in a chain reaction. The photons are of the same wavelength, are in phase, and are all moving in the same direction.

Did You Learn?

- The absorption of microwaves mainly by water molecules in food causes it to heat up.

- Laser is an acronym for *l*ight *a*mplification by *s*timulated *e*mission of *r*adiation.

9.5 Heisenberg's Uncertainty Principle

Preview Questions

- How does Heisenberg's uncertainty principle affect the determination of a particle's location and velocity?

- Where does Heisenberg's uncertainty principle have practical importance in measurement?

There is another important aspect of quantum mechanics. According to classical mechanics, there is no limit to the accuracy of a measurement. Theoretically, accuracy can be continually improved by refinement of the measuring instrument or procedure to the point where the measurement contains no uncertainty. This notion resulted in a deterministic view of nature. For example, it implied that if you either know or measure the exact

Figure 9.18 Eye Surgery by Laser
A laser beam can be used to "weld" a detached retina into its proper place. In other surgical operations, a laser beam can serve as a scalpel, and the immediate cauterization prevents excessive bleeding.

position and velocity of a particle at a particular time, then you can determine where it will be in the future and where it was in the past (assuming no future or past unknown forces).

However, quantum theory predicts otherwise and sets limits on the accuracy of measurement. This idea, developed in 1927 by German physicist Werner Heisenberg, is called **Heisenberg's uncertainty principle**:

It is impossible to know a particle's exact position and velocity simultaneously.

This concept is often illustrated with a simple example. Suppose you want to measure the position and velocity of an electron, as illustrated in ● Fig. 9.19. If you are to see the electron and determine its location, at least one photon must bounce off the electron and come to your eye. In the collision process, some of the photon's energy and momentum are transferred to the electron. (This situation is analogous to a classical collision of billiard balls, which involves a transfer of momentum and energy.)

At the moment of collision, the electron recoils. The very act of measuring the electron's position has altered its velocity. Hence, the very effort to locate the position accurately causes an uncertainty in knowing the electron's velocity.

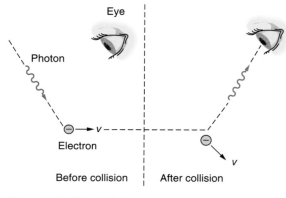

Figure 9.19 Uncertainty
Imagine trying to determine accurately the location of an electron with a single photon, which must strike the electron and come to the detector. The electron recoils at the moment of collision, which introduces a great deal of uncertainty in knowing the electron's velocity or momentum.

Further investigation led to the conclusion that when the mass (m) of the particle, the minimum uncertainty in velocity (Δv), and the minimum uncertainty in position (Δx) are multiplied, a value on the order of Planck's constant ($h = 6.63 \times 10^{-34}$ J·s) is obtained;

$$m(\Delta v)(\Delta x) \approx h$$

The bottom line of Heisenberg's uncertainty principle: There *is* a limit on measurement accuracy that is philosophically significant, but it is of practical importance only when dealing with particles of atomic or subatomic size. As long as the mass is relatively large, Δv and Δx will be very small.

Did You Learn?

● If you know the position of a moving particle, then there is uncertainty in knowing its velocity and vice versa.

● The act of microscopic measurement causes uncertainty.

9.6 Matter Waves

Preview Questions

● What are matter waves or de Broglie waves?

● How do mass and speed affect the wavelength of matter waves?

As the concept of the dual nature of light developed, what was thought to be a wave was sometimes found to act as a particle. Can the reverse be true? In other words, can particles have a wave nature? This question was considered by French physicist Louis de Broglie, who in 1925 hypothesized that matter, as well as light, has properties of both waves and particles.

According to de Broglie's hypothesis, any moving particle has a wave associated with it whose wavelength is given by

$$\text{wavelength} = \frac{\text{Planck's constant}}{\text{mass} \times \text{speed}}$$

$$\lambda = \frac{h}{mv} \qquad\qquad 9.5$$

where λ is the wavelength of the moving particle, m the mass of the particle, v its speed, and h is Planck's constant (6.63×10^{-34} J·s). The waves associated with moving particles are called **matter waves** or **de Broglie waves**.

In Eq. 9.5, the wavelength (λ) of a matter wave is inversely proportional to the mass of the particle or object; that is, the smaller the mass, the larger (longer) the wavelength. Thus, the longest wavelengths are generally for particles with little mass. (Speed is also a factor, but particle masses have more effect because they can vary over a much wider range than speeds can.) However, Planck's constant is such a small number (6.63×10^{-34} J·s) that *any* wavelengths of matter waves are quite small. Let's use an example to see how small.

EXAMPLE 9.5   **Finding the de Broglie Wavelength**

Find the de Broglie wavelength for an electron ($m = 9.11 \times 10^{-31}$ kg) moving at 7.30×10^{5} m/s.

Solution

The mass and speed are given, so the wavelength can be found by using Eq. 9.5.

$$\lambda = \frac{h}{mv}$$

$$= \frac{6.63 \times 10^{-34}\,\text{J·s}}{(9.11 \times 10^{-31}\,\text{kg})(7.30 \times 10^{5}\,\text{m/s})}$$

$$= 1.0 \times 10^{-9}\,\text{m} = 1.0\,\text{nm (nanometer)}$$

This wavelength is several times larger than the diameter of the average atom, so although small, it is certainly significant relative to the size of an electron. (See whether the units in the equation are correct and actually cancel to give meters.)

Confidence Exercise 9.5

Find the de Broglie wavelength for a 1000-kg car traveling at 25 m/s (about 56 mi/h).

Conceptual Question and Answer

A Bit Too Small

Q. If moving masses have wave properties, why aren't the waves normally observed?

A. For normal objects and speeds, the wavelengths are so small that the wave properties go unnoticed. Your answer to Confidence Exercise 9.5 should

Highlight   Electron Microscopes

Can atoms and molecules be seen? You bet. They can even be moved around. Welcome to the world of electron microscopy. Electron microscopes use the wave-like properties of electrons to image objects that are difficult or impossible to see with the unaided eye or with ordinary microscopes, which rely on beams of light (Fig. 1).

Unlike light photons, electrons are charged particles and so can be focused by the use of electric and magnetic fields. Such focusing was done routinely to form images on TV screens using older cathode ray tubes. By accelerating electrons to very high speeds, wavelengths as small as 0.004 nm (nanometer, 10^{-9} m) can be obtained. According to the laws of physics, it would be theoretically possible to image objects as small as 0.002 nm. Compare this value

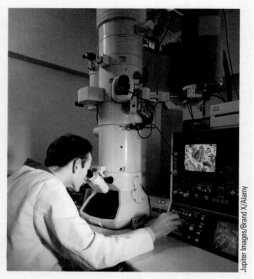

Figure 1 Observing through an electron microscope.

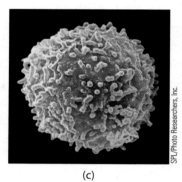

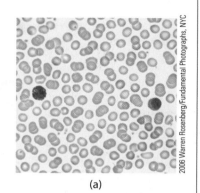

(a)

Figure 2 (a) Two white blood cells in a field of red blood cells seen under an optical (light) microscope. (The white blood cells are stained for distinction.) (b) A lymphocyte (type of white blood cell) seen under a tunneling electron microscope (TEM). (c) A lymphocyte seen under a scanning electron microscope (SEM). Note the distinct surface features.

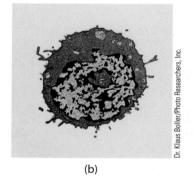

(b)

(c)

confirm that a moving, relatively massive object has a short wavelength. The 1000-kg car traveling at 56 mi/h has a wavelength on the order of 10^{-38} m, which is certainly not significant relative to the size of the car.

Recall that wave properties, such as diffraction (Chapter 7.6), are observed when the wavelength is on the same order as the size of an object or opening. A wavelength of 10^{-38} m is just too small for wave effects to be observed.

De Broglie's hypothesis was met with skepticism at first, but it was verified experimentally in 1927 by G. Davisson and L. H. Germer in the United States. They showed that a beam of electrons exhibits a diffraction pattern. Because diffraction is a wave phenomenon, a beam of electrons must have wave-like properties.

For appreciable diffraction to occur, a wave must pass through a slit with a width smaller than the wavelength (Chapter 7.6). Visible light has wavelengths from about 400 nm to 700 nm, and slits with widths of these sizes can be made quite easily. As Example 9.5 showed, however, a fast-moving electron has a wavelength of about 1 nm. Slits of this width cannot be manufactured.

Fortunately, nature has provided suitably small slits in the form of crystal lattices. The atoms in these crystals are arranged in rows (or some other orderly arrangement), and spaces

Highlight

with the imaging power of visible light, where the limit on the smallest object that can be seen is a relatively whopping 200 nm.

The first two types of electron microscopes developed were the transmission electron microscope (TEM) and the scanning electron microscope (SEM). In TEMs the electron beam passes through a very thin slice of material and probes its interior structure. In SEMs the beam reflects off the surface of the material and reveals its exterior details (Fig. 2).

A more recent development, the scanning tunneling microscope (STM), probes surfaces with a tungsten needle that is only a few atoms wide at its tip. At very short distances, electrons "tunnel" (a quantum phenomenon) from the needle across the gap and through the surface being examined, which produces a tiny current that can be converted into images of individual atoms. Incredibly, IBM scientists then found that STMs could apply a voltage to the needle tip that would allow the atoms (or small molecules) on the surface to be moved around (see the chapter-opening photo).

Figure 3 shows a stunning image of the 5-nm-tall "molecular man," which is formed from 28 carbon monoxide (CO) molecules on a platinum surface. Chemists have even been able to take two molecules, break the appropriate chemical bonds, and rearrange the parts into new molecules. The various types of electron microscopes are among today's most powerful tools in scientific research and have led to the development of the field called *nanotechnology* (any technology done on a nanometer scale). There is no question that exciting discoveries involving "nanotech" lie in our future.

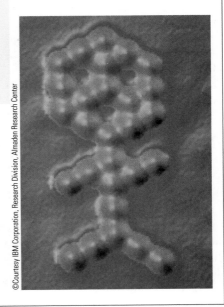

©Courtesy IBM Corporation, Research Division, Almaden Research Center

Figure 3 Molecular Man
This "molecular man" was crafted by moving 28 individual molecules, one at a time. Each of the gold-colored peaks is the image of a carbon monoxide (CO) molecule. The molecules rest on a single-crystal platinum surface (represented in blue).

between the rows provide natural "slits." Davisson and Germer bombarded nickel (Ni) crystals with electrons and obtained a diffraction pattern on a photographic plate. A diffraction pattern made by X-rays (electromagnetic radiation) and one made by an electron beam incident on a thin aluminum (Al) foil are shown in ● Fig. 9.20. The similarity in the diffraction patterns from the electromagnetic *waves* and from the electron *particles* is evident.

Electron diffraction demonstrates that moving matter has not only particle characteristics but also wave characteristics. Remember that the wave nature of ordinary-size moving particles is too small to be measurable. The wave nature of matter becomes of practical importance only with small particles such as electrons and atoms.

The *electron microscope* is based on the theory of matter waves. This instrument uses a beam of electrons, rather than a beam of light, to view an object. See the **Highlight: Electron Microscopes**.

Did You Learn?

- De Broglie or matter waves are waves associated with moving particles (moving particles have a wave nature).

- The wavelength of a matter or de Broglie wave of a moving particle is inversely proportional to its mass and speed.

Figure 9.20 Diffraction Patterns
(a) The diffraction pattern produced by X-rays can be explained using a wave model of the X-rays. (b) The appearance of the diffraction pattern made by a beam of electrons shows that electrons have a wave nature, which can be explained using de Broglie's concept of matter waves.

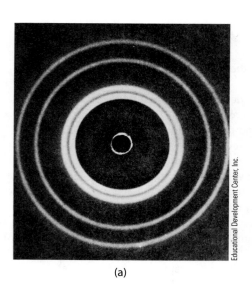

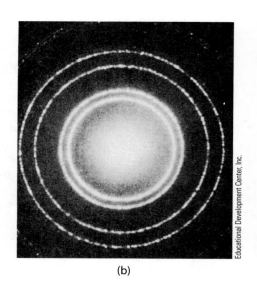

(a)   (b)

9.7   The Electron Cloud Model of the Atom

Preview Questions

● In quantum mechanics, what replaces the classical view of mechanics?
● What is the principle of the electron cloud model?

Niels Bohr chose to analyze the hydrogen atom because it's the simplest atom (Chapter 9.3). It is increasingly difficult to analyze atoms with two or more electrons (*multielectron atoms*) and determine their electron energy levels. The difficulty arises because in multi-electron atoms, more electrical interactions exist than in the hydrogen atom. Forces exist among the various electrons, and in large atoms, electrons in outer orbits are partially shielded from the attractive force of the nucleus by electrons in inner orbits. Bohr's theory, so successful for the hydrogen atom, did not give the correct results when applied to multielectron atoms.

Recall from Chapter 9.3 that Bohr deduced from the small number of lines in the hydrogen emission spectra that electron energy levels are quantized. But he was unable to say *why* they are quantized. Bohr also stated that although classical physics says that the electron should radiate energy as it travels in its orbit, that does not occur. Here, too, he could not offer an explanation. A better model of the atom was needed.

As a result of the discovery of the dual nature of waves and particles, a new kind of physics called **quantum mechanics** or *wave mechanics*, based on the synthesis of wave and quantum ideas, was born in the 1920s and 1930s. In accordance with Heisenberg's uncertainty principle, the concept of *probability* replaced the view of classical mechanics that everything moves according to *exact* laws of nature.

De Broglie's hypothesis showed that waves are associated with moving particles and somehow govern or describe the particle behavior. In 1926 Erwin Schrödinger, an Austrian physicist, presented a widely applicable mathematical equation that gave new meaning to de Broglie's matter waves. Schrödinger's equation is a formulation of the conservation of energy. The detailed form of the equation is quite complex, but it is written in simple form as

$$(E_k + E_p)\Psi = E\Psi$$

where E_k, E_p, and E are the kinetic energy, potential energy, and total energy, respectively, and Ψ (the Greek letter psi) is a wave function.

Schrödinger's *electron cloud model* (or *quantum model*) of the atom focuses on the wave nature of the electron and treats it as a spread-out wave, its energy levels being a consequence of the wave, requiring a *whole number* of wavelengths to form standing waves (Chapter 6.6) in orbits around the nucleus (● Fig. 9.21a and b). Any orbit between two adjacent permissible orbits would require a fractional number of wavelengths and would not produce a standing wave (Fig. 9.21c).

This requirement of a whole number of wavelengths explains the quantization that Bohr had to assume. Furthermore, standing waves do not move from one place to another, so an electron in a standing wave is not accelerating and would not have to radiate light, which explains why Bohr's second assumption was correct.

In Schrödinger's equation the symbol Ψ is called the *wave function* and mathematically represents the wave associated with a particle. At first, scientists were not sure how Ψ should be interpreted. For the hydrogen atom, they concluded that Ψ^2 (the wave function squared, psi squared), when multiplied by the square of the radius r, represents the *probability* that the hydrogen electron will be at a certain distance r from the nucleus. (In Bohr's theory, the electron can be only in circular orbits with discrete radii given by $r = 0.053 \, n^2$ nm.)

A plot *of* $r^2\Psi^2$ versus r for the hydrogen electron (● Fig. 9.22a) shows that the most probable radius for the hydrogen electron is $r = 0.053$ nm, which is the same value Bohr calculated in 1913 for the ground state orbit of the hydrogen atom. (In fact, all the energy levels for the hydrogen atom were found to be exactly the same as those Bohr had calculated.) The electron might be found at other radii, but with less likelihood, that is, with lower probability. This idea gave rise to the model of an *electron cloud* around the nucleus, where the cloud's density reflects the probability that the electron is in that region (Fig. 9.22b).

Thus, Bohr's simple planetary model was replaced by a more sophisticated, highly mathematical model that treats the electron as a wave and can explain more data and predict more accurately. The electron cloud model, or quantum model, of the atom is more difficult to visualize than the Bohr model. The location of a specific electron becomes more vague than in the Bohr model and can be expressed only in terms of probability. The important point is that the quantum mechanical model enables us to determine the energy of the electrons in multielectron atoms. For scientists, knowing the electron's energy is much more important than knowing its exact location.

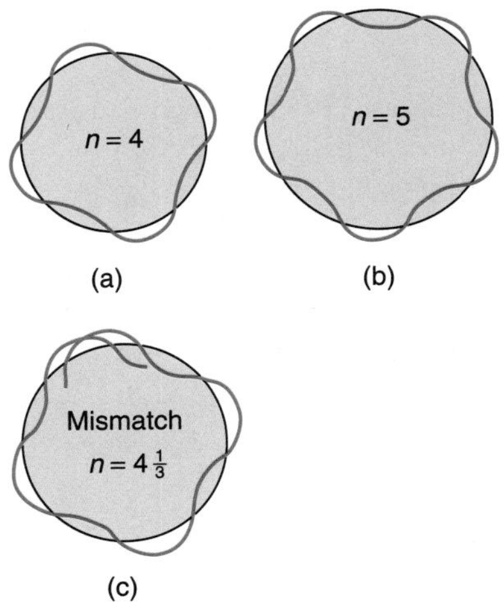

Figure 9.21 The Electron as a Standing Wave
The hydrogen electron can be treated as a standing wave in a circular orbit around the nucleus. For the wave to be stable, however, the circumference must accommodate a *whole number* of wavelengths, as shown in (a) and (b). In (c), the wave would destructively interfere with itself, so this orbit is forbidden. This restriction on the orbits, or energies, explains why the atom is quantized.

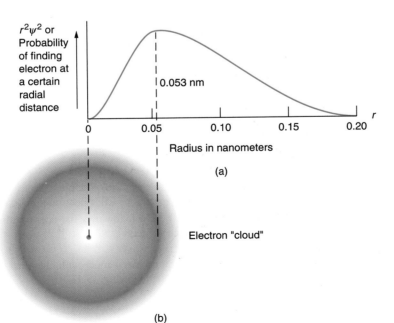

(b)

Figure 9.22 $r^2\Psi^2$ Probability
The square of the wave function (Ψ^2) multiplied by the square of the radius (r^2) gives the probability of finding the electron at that particular radius. As shown here, the radius of a hydrogen atom with the greatest probability of containing the electron is 0.053 nm, which corresponds to the first Bohr radius. (b) The probability of finding the electron at other radii gives rise to the concept of an "electron cloud," or probability distribution.

Did You Learn?

● The concept of probability replaced the classical view that everything moves according to *exact* laws of nature.

● The density of an electron cloud around the nucleus reflects the probability that an electron will be in that region.

KEY TERMS

1. atoms (9.1)
2. electrons
3. quantum (9.2)
4. photoelectric effect
5. photon
6. dual nature of light

7. line emission spectrum (9.3)
8. line absorption spectrum
9. principal quantum number
10. ground state
11. excited states
12. X-rays (9.4)

13. laser
14. stimulated emission
15. Heisenberg's uncertainty principle (9.5)
16. matter (de Broglie) waves (9.6)
17. quantum mechanics (9.7)

MATCHING

For each of the following items, fill in the number of the appropriate Key Term from the preceding list. Compare your answers with those at the back of the book.

a. _____ The electron level of lowest energy in an atom

b. _____ Type of spectrum given by light from a gas-discharge tube

c. _____ A discrete amount, or packet, of energy

d. _____ A device based on light amplification by stimulated emission of radiation

e. _____ Type of spectrum obtained when white light is passed through a cool gas

f. _____ Branch of physics based on synthesis of wave and quantum ideas

g. _____ The basic particles of elements

h. _____ It is impossible to know a particle's exact position and velocity simultaneously

i. _____ Electron emission by some metals when exposed to light

j. _____ Negatively charged particles that are components of all atoms

k. _____ High-frequency electromagnetic radiation produced when high-speed electrons strike a metal target

l. _____ Light has both wave-like and particle-like characteristics

m. _____ Integer that identifies a Bohr orbit

n. _____ Waves associated with moving particles

o. _____ A quantum of electromagnetic radiation

p. _____ Process wherein an excited atom is struck by a photon and emits additional photons

q. _____ Electron levels of higher-than-normal energy

MULTIPLE CHOICE

Compare your answers with those at the back of the book.

1. Who championed the idea of the atom about 400 BCE? (9.1)
   (a) Aristotle   (b) Plato
   (c) Democritus   (d) Archimedes

2. Which scientist is associated with the "plum pudding model" of the atom? (9.1)
   (a) Thomson   (b) Rutherford   (c) Bohr   (d) Dalton

3. Planck developed his quantum hypothesis to explain which of these phenomena? (9.2)
   (a) the ultraviolet catastrophe   (b) line spectra
   (c) the photoelectric effect   (d) uncertainty

4. Light of which of the following colors has the greatest photon energy? (9.2)
   (a) red   (b) orange   (c) yellow   (d) violet

5. The Bohr theory was developed to explain which of these phenomena? (9.3)
   (a) energy levels   (b) the photoelectric effect
   (c) line spectra   (d) quantum numbers

6. In which of the following states does a hydrogen electron have the greatest energy? (9.3)
   (a) $n = 1$ (b) $n = 3$ (c) $n = 5$ (d) $n = 7.5$

7. Bombarding a metal anode with high-energy electrons produces which of the following? (9.4)
   (a) laser light   (b) X-rays   (c) microwaves   (d) neutrons

8. The "s" in the acronym laser stands for (9.4)
   (a) simple   (b) specific
   (c) spontaneous   (d) stimulated

9. Which of the following does a laser do?
   (a) amplifies light
   (b) produces monochromatic light
   (c) produces coherent light
   (d) all the preceding

10. Limitations on measurements are described by which of the following? (9.5)
    (a) Heisenberg's uncertainty principle
    (b) de Broglie's hypothesis
    (c) Schrödinger's equation
    (d) Einstein's special theory of relativity

11. Which of the following pairs of particle properties is it impossible to determine exactly and simultaneously? (9.5)
    (a) charge and mass
    (b) position and velocity
    (c) charge and position
    (d) velocity and momentum

12. What scientist first hypothesized matter waves? (9.6)
    (a) Schrödinger
    (b) de Broglie
    (c) Heisenberg
    (d) Einstein

13. According to the de Broglie hypothesis, how is the wavelength associated with a moving particle?
    (a) It is independent of mass.
    (b) It is longer the greater the speed of the particle.
    (c) It easily shows diffraction effects.
    (d) None of the preceding.

14. Why did the Bohr model need improvement? (9.7)
    (a) It worked only for the hydrogen atom.
    (b) It did not explain why the atom is quantized.
    (c) It did not explain why an electron does not emit radiation as it orbits.
    (d) All these answers are correct.

FILL IN THE BLANK

Compare your answers with those at the back of the book.

1. The subatomic particle called the ___ was discovered by J. J. Thomson. (9.1)

2. The scientist associated with the "nuclear model" of the atom is ___. (9.1)

3. In the equation $E = hf$, the h is called ___. (9.2)

4. A quantum of electromagnetic radiation is commonly called a(n) ___. (9.2)

5. In the Bohr model, as n increases, the distance of the electron from the nucleus ___. (9.3)

6. When analyzed with a spectrometer, light from an incandescent source produces a(n) ___ spectrum. (9.3)

7. A photon is absorbed when an electron makes a transition from one energy level to a(n) ___ one. (9.3)

8. Microwave ovens heat substances mainly by exciting molecules of ___. (9.4)

9. The X in X-ray stands for ___. (9.4)

10. Heisenberg's ___ principle is of practical importance only with particles of atomic or subatomic size. (9.5)

11. According to de Broglie's hypothesis, a moving particle has a(n) ___ associated with it. (9.6)

12. In the electron cloud model of the atom, the electron's location is stated in terms of ___. (9.7)

SHORT ANSWER

9.1 Early Concepts of the Atom

1. What is the basic difference between classical mechanics and quantum mechanics?

2. How did Thomson know that electrons are negatively charged?

3. What major change was made in Thomson's model of the atom after Rutherford's discovery?

9.2 The Dual Nature of Light

4. How does the radiation from a hot object change with temperature?

5. Name a phenomenon that can be explained only by light having a wave nature and one that can be explained only by light having a particle nature.

6. If electromagnetic radiation is made up of quanta, then why don't we hear a radio intermittently as discrete packets of energy arrive?

7. Distinguish between a proton and a photon.

8. How are the frequency and wavelength of an electromagnetic wave related?

9. Explain the difference between a photon of red light and one of violet light in terms of energy, frequency, and wavelength.

10. Light shining on the surface of a photomaterial causes the ejection of electrons if the frequency of the light is above a certain minimum value. Why is there a certain minimum value?

11. What scientist won the Nobel Prize for explaining the photoelectric effect? Name another theory for which that scientist is famous.

9.3 Bohr Theory of the Hydrogen Atom

12. How does the number of lines in the emission spectrum for an element compare with the number of lines in the absorption spectrum?

13. Does light from a neon sign have a continuous spectrum? Explain.

14. In the Bohr theory, principal quantum numbers are denoted by what letter?

15. Why was it necessary for Bohr to assume that a bound electron in orbit did not emit radiation?

16. Distinguish between a *ground state* for an electron and an *excited state*.

17. How many visible lines make up the emission spectrum of hydrogen? What are their colors?

18. How does the Bohr theory explain the discrete lines in the *emission* spectrum of hydrogen?

19. How does the Bohr theory explain the discrete lines in the *absorption* spectrum of hydrogen?

20. In which transition is the photon of greater energy emitted, $n = 3$ to $n = 1$ or $n = 2$ to $n = 1$?

21. A hydrogen electron is in the excited state $n = 3$. How many photons of different frequencies could possibly be emitted in the electron's return to the ground state?

9.4 Microwave Ovens, X-Rays, and Lasers

22. Why does a microwave oven heat a potato but not a ceramic plate?

23. What does the acronym *laser* stand for?

24. What is unique about light from a laser source?

25. Why should you never look directly into a laser beam or into its reflection?

26. Why are X-rays called "braking rays" in German?

9.5 Heisenberg's Uncertainty Principle

27. State Heisenberg's uncertainty principle.

28. Why isn't Heisenberg's uncertainty principle relevant to everyday observations?

9.6 Matter Waves

29. What is a matter wave, and when is the associated wavelength significant?

30. Niels Bohr was never able to actually explain why a hydrogen electron is limited to certain orbits. How did de Broglie explain it?

31. How was a beam of electrons shown to have wave-like properties?

32. What useful instrument takes advantage of the wave properties of electrons?

9.7 The Electron Cloud Model of the Atom

33. What scientist is primarily associated with the electron cloud model of the atom?

34. What other name is often used to refer to the electron cloud model of the atom?

VISUAL CONNECTION

Visualize the connections and give answers for the blanks. Compare your answers with those at the back of the book.

Models of the Atom

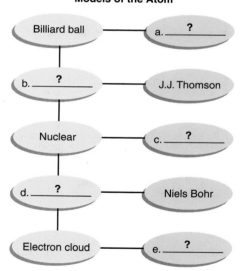

APPLYING YOUR KNOWLEDGE

1. Why are microwave ovens constructed so that they will not operate when the door is open?

2. While you are shopping for a new car, a salesperson tells you that the newest technology is a *quantized* cruise control. What might this statement mean?

3. Your friend says that atoms do not exist because no one has ever seen an atom. What would be your reply?

4. Why at night, under the mercury or sodium vapor lights in a mall parking lot, do cars seem to be peculiar colors?

5. Explain why Heisenberg's uncertainty principle wouldn't pose a problem for police officers using radar to determine a car's speed.

IMPORTANT EQUATIONS

Photon Energy: $$E = hf$$
$$(h = 6.63 \times 10^{-34}\,\text{J} \cdot \text{s}) \tag{9.1}$$

Hydrogen Electron Orbit Radii:
$$r_n = 0.053\,n^2\,\text{nm}\ (n = 1, 2, 3, \dots) \tag{9.2}$$

Hydrogen Electron Energy:
$$E_n = \frac{-13.60}{n^2}\,\text{eV}\ (n = 1, 2, 3, \dots) \tag{9.3}$$

Photon Energy for Transition: $E_{\text{photon}} = E_{n_i} - E_{n_f}$ (9.4)

de Broglie Wavelength: $\lambda = \dfrac{h}{mv}$ (9.5)

EXERCISES

9.2 The Dual Nature of Light

1. The human eye is most sensitive to yellow-green light having a frequency of about 5.45×10^{14} Hz (a wavelength of about 550 nm). What is the energy in joules of the photons associated with this light?

   Answer: 3.61×10^{-19} J

2. Light that has a frequency of about 5.00×10^{14} Hz (a wavelength of about 600 nm) appears orange to our eyes. What is the energy in joules of the photons associated with this light?

3. Photons of a certain ultraviolet light have an energy of 6.63×10^{-19} J. (a) What is the frequency of this UV light? (b) Use $\lambda = c/f$ to calculate its wavelength in nanometers (nm).

   Answer: (a) 1.00×10^{15} Hz (b) 300 nm

4. Photons of a certain infrared light have an energy of 1.50×10^{-19} J. (a) What is the frequency of this IR light? (b) Use $\lambda = c/f$ to calculate its wavelength in nanometers (nm).

9.3 Bohr Theory of the Hydrogen Atom

5. What is the radius in nm of the electron orbit of a hydrogen atom for $n = 3$?

   Answer: 0.48 nm

6. What is the radius in nm of the electron orbit of a hydrogen atom for $n = 4$?

7. What is the energy in eV of the electron of a hydrogen atom for the orbit designated $n = 3$?

   Answer: -1.51 eV

8. What is the energy in eV of the electron of a hydrogen atom for the orbit designated $n = 4$?

9. Use Table 9.1 to determine the energy in eV of the photon emitted when an electron jumps down from the $n = 4$ orbit to the $n = 2$ orbit of a hydrogen atom.

   Answer: 2.55 eV

10. Use Table 9.1 to determine the energy in eV of the photon absorbed when an electron jumps up from the $n = 1$ orbit to the $n = 4$ orbit of a hydrogen atom.

9.6 Matter Waves

11. Calculate the de Broglie wavelength of a 0.50-kg ball moving with a constant velocity of 26 m/s (about 60 mi/h).

    Answer: 5.1×10^{-35} m

12. Estimate your de Broglie wavelength when you are running. (Recall that $h \sim 10^{-34}$ in SI units and 1 lb is equivalent to 0.45 kg.) For the computation, estimate how fast you can run in meters per second.

ON THE WEB

1. Atoms with Attitude

What does the earliest model of the atom look like? What quandary does it present? What do you know about the quantum atom and spectral lines? What can you say about Niels Bohr and Bohr's atom? Where does the term *quantum leap* come from, and how has its real meaning changed? Explore answers to these questions by following the links at **www.cengagebrain.com/shop/ISBN/1133104096**.

2. You're Cookin' with Gas? Nope, the Microwave.

How do microwaves work? Are they dangerous? Can you explain the relationship between water and microwave cooking? What happens when you cook frozen foods in the microwave, and why don't such foods always cook evenly? What might you do to deal with that? Why do microwave manufacturers warn the user about the dangers of heating *only* water in the microwave oven? Visit the student website at **www.cengagebrain.com/shop/ISBN/1133104096** and follow the links to learn more about cooking with microwaves and answer the above questions.

Nuclear Physics

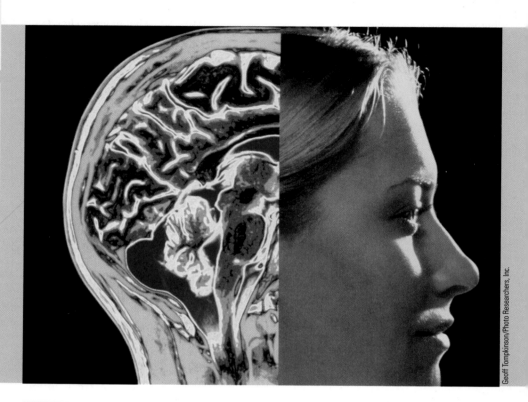

Geoff Tompkinson/Photo Researchers, Inc.

*It is a source of grati-
fication to us all that
we have been able
to contribute a little
to an understanding
of the nucleus of the
atom.*

•

Ernest Lawrence,
physics Nobel Laureate
(1901–1958)

Technetium-99, a laboratory- >
produced radioactive isotope, is
often used in brain scans. Here,
a color-enhanced scan has been
superimposed on the back of a
woman's head.

PHYSICS FACTS

▶ The elements technetium (Tc)
and promethium (Pm) are
not found in nature, but they
can be artificially made. They
don't exist naturally because
their half-lives are on the order
of hours and minutes, and if
originally present, they would
have decayed away.

T he atomic nucleus and its properties have had an important impact on
our society. For example, the nucleus is involved with archeological
dating, diagnosis and treatment of cancer and other diseases, chemical
analysis, radiation damage and nuclear bombs, and the generation of electricity.
This chapter discusses these topics and includes Highlights on The Discovery of
Radioactivity and Nuclear Power and Waste Disposal.

An appropriate way to begin the study of nuclear physics is with a brief history
of how the concept of the element arose and how elements and their nuclei are
expressed by symbols.

Chapter Outline

10.1   Symbols of the Elements

Preview Questions

● What original "elements" did Aristotle think composed all matter on the Earth?

● Why are some element symbols very different from their names? For example, carbon is C, but silver is Ag.

The Greek philosophers who lived during the period from about 600 to 200 BCE were apparently the first people to speculate about what basic substance or substances make up matter. In the fourth century BCE, the Greek philosopher Aristotle developed the idea that all matter on the Earth is composed of four "elements": earth, air, fire, and water. He was wrong on all four counts, and in Chapter 11 the discovery and properties of true elements will be discussed. We will consider some of these elements in terms of their atomic nuclei.

The symbol notation used to designate the different elements was first introduced in the early 1800s by Swedish chemist Jöns Jakob Berzelius ("bur-ZEE-lee-us"). He used one or two letters of the Latin name to represent each element. For example, sodium was designated Na for *natrium* and silver Ag for *argentum* (● Table 10.1).

Since Berzelius' time, most elements have been symbolized by the first one or two letters of the English name. Examples include C for carbon, O for oxygen, and Ca for calcium. The first letter of a chemical symbol is always capitalized, and the second is lowercase. Inside the front cover of this book, you will find a periodic table of the elements showing the positions, names, and symbols of the elements presently known. There are 112 officially named elements. Elements 113 through 118 have designations but are unnamed. (An international committee assigns new element names. For more on the elements and periodic table, see Chapter 11.4.)

Although you are not expected to learn the names and symbols of all the elements, you should become familiar with most of the names and symbols of the elements listed in ● Table 10.2.

Did You Learn?

● Aristotle's first "periodic table" consisted of earth, air, fire, and water.

● Some chemical symbols are derived from their Latin names, for example, silver, Ag (argentum) or sodium, Na (natrium).

PHYSICS FACTS *cont.*

❱ You are radioactive because your body contains carbon-14.

❱ A lengthy plane flight at high altitude can expose passengers to an amount of radiation energy (from cosmic rays) comparable to that of a chest X-ray.

❱ "Moonshine! Pure moonshine!!" That's what Ernest Rutherford, the discoverer of the proton and the atomic nucleus, said about the possibility of atomic (nuclear) energy.

❱ Spent nuclear fuel rods from nuclear reactors contain many radioactive isotopes. The disposal of this waste is a problem. (See the Highlight: Nuclear Power and Waste Disposal.)

❱ More radioactive isotopes are released into the atmosphere from power plants burning coal and oil than from nuclear power plants.

Table 10.1	Some Chemical Symbols from Latin Names	
Modern Name	Symbol	Latin Name
Copper	Cu	*Cuprum*
Gold	Au	*Aurum*
Iron	Fe	*Ferrum*
Lead	Pb	*Plumbum*
Mercury	Hg	*Hydrargyrum*
Potassium	K	*Kalium*

Table 10.2 Names and Symbols of Some Common Elements

Name	Symbol	Name	Symbol	Name	Symbol
Aluminum	Al	Gold	Au	Phosphorus	P
Argon	Ar	Helium	He	Platinum	Pt
Barium	Ba	Hydrogen	H	Potassium	K
Boron	B	Iodine	I	Radium	Ra
Bromine	Br	Iron	Fe	Silicon	Si
Calcium	Ca	Lead	Pb	Silver	Ag
Carbon	C	Magnesium	Mg	Sodium	Na
Chlorine	Cl	Mercury	Hg	Sulfur	S
Chromium	Cr	Neon	Ne	Tin	Sn
Cobalt	Co	Nickel	Ni	Uranium	U
Copper	Cu	Nitrogen	N	Zinc	Zn
Fluorine	F	Oxygen	O		

10.2   The Atomic Nucleus

Preview Questions

● How was it determined that atoms have nuclei?

● What is the atomic number, and how is it related to an element?

Matter is made up of atoms. An atom is composed of negatively charged particles, called **electrons**, which surround a positively charged nucleus. The **nucleus** is the central core of an atom. It consists of positively charged **protons** and electrically neutral **neutrons**. An electron and a proton have the same magnitude of electric charge, but the charges are different. The charge on the electron is designated negative ($-$) and that on the proton positive ($+$). (See Chapter 8 1.)

Protons and neutrons have almost the same mass and are about 2000 times more massive than an electron. Nuclear protons and neutrons are collectively called **nucleons**. ● Table 10.3 summarizes the basic properties of electrons, protons, and neutrons.

In 1911, British scientist Ernest Rutherford discovered that the atom consists of a nucleus surrounded by orbiting electrons. He was curious about what would happen when energetic alpha particles (helium nuclei) bombarded a very thin sheet of gold.* J. J. Thomson's "plum pudding model" (Fig. 9.1b) predicted that the alpha particles would pass through the evenly distributed positive charges in the gold atoms with little or no deflection from their original paths.

Rutherford's experiment was conducted using a setup such as that illustrated in ● Fig. 10.1. The behavior of the alpha particles was determined by using a movable screen coated with zinc sulfide. When an alpha particle hit the screen, a small flash of light was emitted that could be observed with a low-power microscope. (A similar phenomenon causes cathode-ray TV screens to glow when hit by moving electrons.) Rutherford found that the vast

Table 10.3 Major Constituents of an Atom

Particle (symbol)	Charge (C)	Electronic Charge	Mass (kg)	Location
Electron (e)	-1.60×10^{-19}	-1	9.109×10^{-31}	Outside nucleus
Proton (p)	$+1.60 \times 10^{-19}$	$+1$	1.673×10^{-27}	Nucleus
Neutron (n)	0	0	1.675×10^{-27}	Nucleus

*Alpha particles (doubly positively charged helium nuclei) come from the radioactive decay of certain elements, as will be discussed in Chapter 10.3.

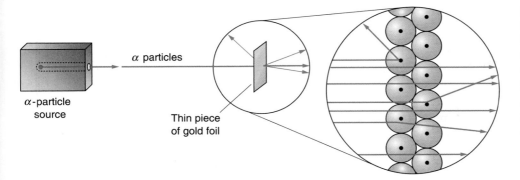

Figure 10.1 Rutherford's Alpha-Scattering Experiment
Nearly all the alpha particles striking the gold foil went through. A few were deflected, and some bounced back. These results led to the discovery of the nucleus. (The zinc sulfide screen used to monitor the alpha particles is not shown.)

majority of the alpha particles went through the gold foil as though it were not even there. A few of the positively charged alpha particles were deflected, however, and about 1 out of 20,000 actually bounced back. Rutherford could only explain this behavior by assuming that each gold atom had its positive charge concentrated in a small core rather than distributed throughout the atom. He called the core the atomic *nucleus* and assumed that electrons move around the nucleus "like bees around a hive."*

The alpha-scattering experiment showed that a nucleus has a diameter of about 10^{-14} m (● Fig. 10.2). In contrast, the atom's outer electrons have orbits with diameters of about 10^{-10} m (Chapter 9.3). Thus, the diameter of an atom is approximately 10,000 times the diameter of its nucleus, and most of an atom's volume consists of empty space. Imagine that the nucleus of an atom were the size of a peanut. If you place the nut in the middle of a baseball stadium, then the stadium itself would be about the relative size of the atom.

The electrical repulsion between an atom's electrons and those of adjacent atoms keeps matter from collapsing. Electron orbits determine the size (volume) of atoms, but the nucleus contributes more than 99.97% of the mass.

The particles in an atom are designated by certain numbers. The **atomic number**, symbolized by the letter Z, is the number of protons in the nucleus of each atom of that element.

An **element** is a substance in which all the atoms have the same number of protons (the same atomic number, Z). For an atom to be electrically neutral (have a total net charge of zero), the numbers of electrons and protons must be the same. Therefore, the atomic number also indicates the number of electrons in a neutral atom.

Electrons may be gained or lost by an atom, and the resulting particle, called an *ion*, will be electrically charged. However, because the number of protons has not changed, the particle is a positively charged ion of that same element. For instance, if a *sodium atom* (Na) loses an electron, then it becomes a *sodium ion* (Na^+), not an atom or ion of some other element. The sodium ion still has the same number of protons (11) as the sodium atom.

▶ The **neutron number** (N) is the number of neutrons in a nucleus.

▶ The **mass number** (A) is the number of protons plus neutrons in the nucleus: the total number of nucleons.

The general designation for a specific nucleus places the mass number (A) to the upper left of the chemical symbol (shown here as X for generality), and the atomic number (Z) goes at the lower left.†

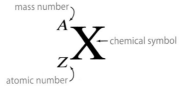

mass number
atomic number
← chemical symbol

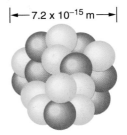

|←— 7.2×10^{-15} m —→|

Figure 10.2 A Representation of the Nucleus
The nucleus of an aluminum-27 atom consists of 13 protons (blue) and 14 neutrons (yellow), for a total of 27 nucleons. The diameter of this nucleus is 7.2×10^{-15} m, close to the 10^{-14}-m diameter of an average nucleus.

*Rutherford described the backscattering as "almost as incredible as if you had fired a 15-inch shell at a piece of tissue paper, and it came back and hit you."

†Why are the letters Z and A used? "Atomic number" in German is *"Atomzahl,"* so the Z probably comes from *zahl* (number). *M* is sometimes used for mass number (*Massenzahl* in German), but the symbol A is recommended by the American Chemical Society (ACS) Style Guide.

Sometimes the neutron number N is placed at the lower right ($^A_Z X_N$). However, the number of neutrons (N) in a nucleus is easily determined by subtracting the atomic number (Z) from the mass number (A).

$$\text{neutron number} = \text{mass number} - \text{atomic number}$$

$$N = A - Z$$

10.1

It is common to write the symbol for a uranium nucleus as $^{238}_{92}U$. But because it is a simple matter to obtain an element's atomic number from the periodic table, a nucleus of an element is sometimes represented by just the mass number and the chemical symbol (for example, ^{238}U) or by the name of the element followed by a hyphen and the mass number (uranium-238). The chemical symbols and names for all the elements are given in the periodic table inside the front cover of this book.

EXAMPLE 10.1   Determining the Composition of an Atom

Determine the number of protons, electrons, and neutrons in the fluorine atom $^{19}_{9}F$.

Solution

The atomic number Z is 9, so the number of protons is 9 (as is the number of electrons for a neutral atom). The mass number A is 19, so the number of neutrons is $N = A - Z$ $= 19 - 9 = 10$. The answer is 9 protons, 9 electrons, and 10 neutrons.

Confidence Exercise 10.1

Determine the number of protons, electrons, and neutrons in the carbon atom ^{131}I (iodine-131), a radioactive nucleus used in the treatment of thyroid cancer.

Answers to Confidence Exercises may be found at the back of the book.

Atoms of the same element can be different because of different numbers of neutrons in their nuclei. Atoms that have the same number of protons (same Z, same element) but differ in their numbers of neutrons (different N, and therefore different A) are known as the **isotopes** of that element. *Isotope* literally means "same place" (*iso-*, Greek for same, and *tropos*, meaning place), and it designates atoms that occupy the same place in the periodic table of elements. Isotopes are like members of a family. They all have the same atomic number (Z) and the same element name (surname), but they are distinguishable by the number of neutrons (N) in their nuclei (the equivalent of their given name). For example, three isotopes of carbon are $^{12}_{6}C_6$, $^{13}_{6}C_7$, and $^{14}_{6}C_8$: carbon-12, carbon-13, and carbon-14, respectively.

The isotopes of an element have the same chemical properties because they have the same number of electrons, which determines chemical activity and reactions. But they differ somewhat in physical properties because they have different masses. ● Figure 10.3 illustrates the atomic composition of the three isotopes of hydrogen. They even have their own names: ^1_1H is *protium* (or just *hydrogen*); ^2_1H is *deuterium* (D); and ^3_1H is *tritium* (T). The atomic nuclei in these cases are referred to as *protons*, *deuterons*, and *tritons*, respectively. That is, the proton is the nucleus of a protium atom, and so on.

In a given sample of naturally occurring hydrogen, about 1 atom in 6000 is deuterium and about 1 atom in 10,000,000 is tritium. Protium and deuterium are stable atoms, whereas tritium is unstable (that is, radioactive; Chapter 10.3). Deuterium is sometimes called heavy hydrogen. It combines with oxygen to form *heavy water* (D_2O).

The Atomic Mass

Generally, each element occurs naturally as a combination of its isotopes. The weighted average mass of an atom of the element in a naturally occurring sample is called the

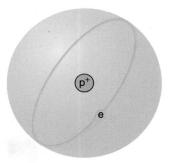

Protium
^1_1H

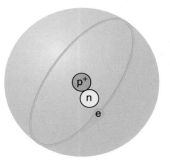

Deuterium
^2_1H

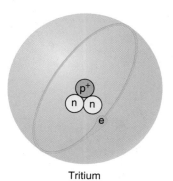

Tritium
^3_1H

Figure 10.3 The Three Isotopes of Hydrogen

Each atom has one proton and one electron, but they differ in the number of neutrons in the nucleus. (*Note*: This figure is not drawn to scale; the nucleus is shown much too large relative to the size of the whole atom.)

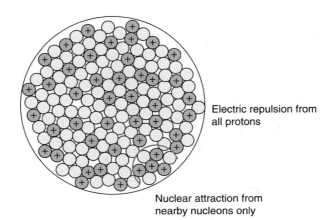

Figure 10.4 A Multinucleon Nucleus
The protons on the surface of the nucleus, such as those shown in the red semicircle, are attracted by the strong nuclear force of only the six or seven closest nucleons, but they are electrically repelled by all the other protons. When the number of protons exceeds 83, the electrical repulsion overcomes the nucleon attraction and the nucleus is unstable.

Electric repulsion from all protons

Nuclear attraction from nearby nucleons only

atomic mass and is given under its symbol in the periodic table (in *atomic mass units*, symbolized u).*

All atomic masses are based on the ^{12}C atom, which is assigned a relative atomic mass of *exactly* 12 u. Naturally occurring carbon has an atomic mass slightly greater than 12.0000 u because it contains not only ^{12}C but also a little ^{13}C and a trace of ^{14}C. An isotope's mass number closely approximates its atomic mass (its actual mass in u).

The Strong Nuclear Force

In previous chapters, two fundamental forces were considered: the electric force and gravitational force. The electric force between a proton and an electron in an atom is about 10^{39} times the corresponding gravitational force. The electric force is the only important force on the electrons in an atom and is responsible for the structure of atoms, molecules, and matter in general. In a nucleus the positively charged protons are packed closely together. According to Coulomb's law (Chapter 8.1), like charges repel each other, so the repulsive electric forces in a nucleus are huge. Then, why doesn't the nucleus fly apart?

Obviously, because nuclei generally remain intact, there must be something else: a third fundamental force. This **strong nuclear force** (or just *strong force* or *nuclear force*) acts between nucleons: between two protons, between two neutrons, and between a proton and a neutron. It holds the nucleus together. The exact equation describing the nucleon–nucleon interaction is unknown. However, for very short nuclear distances of less than about 10^{-14} m, the interaction is strongly attractive; in fact, it is the strongest fundamental force known. At distances greater than about 10^{-14} m, however, the nuclear force is zero.

A multinucleon nucleus is illustrated in ● Fig. 10.4. A proton on the surface of the nucleus is attracted only by the six or seven nearest nucleons. Because the strong nuclear force is a short-range force, only the nearby nucleons contribute to the attractive force.

On the other hand, the repulsive electric force is a long-range force and acts between any two protons, no matter how far apart they are in the nucleus. As nuclei of different elements contain more and more protons, the electric repulsive forces increase, yet the attractive nuclear forces remain constant because they are determined by nearest neighbors only.

When the nucleus has more than 83 protons, the electric forces of repulsion overcome the nuclear attractive forces and the nucleus is subject to spontaneous disintegration, or *decay*. That is, particles are emitted to adjust the neutron–proton imbalance.

There is also a *weak nuclear force*. It is a short-range force that reveals itself principally in beta decay (Chapter 10.3). The weak nuclear force is stronger than the gravitational

*Because the masses are so small in relation to the SI standard kilogram, another unit of appropriate size, the *atomic mass unit* (u), is used, where 1 u = 1.66054 × 10^{-27} kg.

force but very much weaker than the electromagnetic force and the strong nuclear force. Physicists seek to combine three of the known forces—electromagnetic, strong nuclear, and weak nuclear—into a single underlying theory called the *grand unified model (GUT)*, but incorporating the gravitational force into the model has yet to be accomplished (Chapter 10.8).

Did You Learn?

● Using alpha particle scattering, Rutherford determined that atoms have a small electrically positive core, which he called the nucleus.

● The atomic number is the number of protons in each atom of an element. An element is a substance in which all the atoms have the same atomic number (Z).

10.3   Radioactivity and Half-Life

Preview Questions

● What are the three common processes of radioactive decay, and what is emitted in each?

● What is half-life?

A particular species or isotope of any element is called a *nuclide*. A nuclide is a nucleus characterized by a definite atomic number and mass number, such as ^1_1H, $^{14}_{12}C$, and $^{238}_{92}U$. Nuclides whose nuclei undergo spontaneous decay (disintegration) are called **radioactive isotopes** (or radioisotopes for short, or radionuclides). The spontaneous process of nuclei undergoing a change by emitting particles or rays is called *radioactive decay*, or **radioactivity**. Substances that give off such radiation are said to be *radioactive*. (The **Highlight: The Discovery of Radioactivity** discusses the discovery of radioactivity by Becquerel and the discovery of two new radioactive elements by the Curies.)

Radioactive nuclei can disintegrate in three common ways: *alpha decay*, *beta decay*, and *gamma decay*; see ● Fig. 10.5. (Fission, another important decay process, will be discussed in Chapter 10.5.) In all decay processes, energy is given off, usually in the form of energetic particles that produce heat. Equations for radioactive decay are generally written in the form

$$A \rightarrow B + b$$

The original nucleus (A) is sometimes called the *parent* nucleus, and the resulting nucleus (B) is referred to as the *daughter* nucleus. The b in the equation represents the emitted particle or ray.

Figure 10.5 The Three Components of Radiation from Radioactive Isotopes
An electric field separates the rays from a sample of a heavy radioisotope, such as uranium, into alpha (α) particles (positively charged helium nuclei), beta (β) particles (negatively charged electrons), and neutral gamma (γ) rays (high-energy electromagnetic radiation). The electrically charged particles are deflected toward oppositely charged plates.

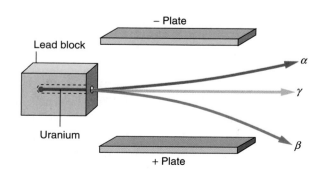

Highlight | The Discovery of Radioactivity

In Paris in 1896, Henri Becquerel ("beh-KREL") heard of Wilhelm Roentgen's recent discovery of X-rays (Chapter 9.4). While experimenting with photographic plates to determine whether any of the fluorescent materials he was investigating might emit X-rays, Becquerel discovered that a mineral containing the element uranium was emitting radiation that had nothing to do with fluorescence. What he discovered was a new type of radiation resulting from radioactivity.

In 1897, also in Paris, Marie Curie began a search for naturally radioactive elements. In 1898, she and her husband, physicist Pierre Curie (discoverer of the magnetic Curie temperature, Chapter 8.4), isolated a minute amount of a new element from tons of uranium ore. Named *polonium* (Po) after Marie's native country, Poland, it was hundreds of times more radioactive than uranium. Later, they found an even more radioactive element, which they named *radium* (Ra) because of the intense radiation it emitted (Fig. 1).

In 1903, the Curies shared the Nobel Prize in physics with Becquerel for their work on radioactivity. Marie Curie was also awarded the Nobel Prize in chemistry in 1911 for further work on radium and the study of its properties. (Pierre was killed in a horse-drawn carriage accident in 1906.) Madame Curie (as she is commonly known) was the first person to win two Nobel Prizes in two different fields, being one of only two persons to do so. The other was American chemist Linus Pauling, who won Nobel prizes in chemistry (1954) and peace (1962).

The rest of her career was dedicated to establishing and supervising laboratories for research on radioactivity and the use of radium in the treatment of cancer. In 1921, Madame Curie toured the United States, and President Warren Harding, on behalf of the women of the United States, presented her with a gram of radium (a relatively large and expensive amount) in recognition of her services to science.

Madame Curie died in 1934 of leukemia, probably caused by overexposure to radioactive substances. She carried test tubes containing radioactive isotopes in her pocket and stored them in a desk drawer. Her death came shortly before an event that would no doubt have made her very proud. In 1935, the Curies' daughter Irene Joliot-Curie and her husband, Frederic Joliot, were awarded the Nobel Prize in chemistry.

Time & Life Pictures/Getty Images

Figure 1 The Curies
Marie (1867–1934) and Pierre (1859–1906) Curie discovered polonium and radium in 1898. They were awarded the Nobel Prize in physics in 1903 for their work on radioactivity. (The prize was shared with Henri Becquerel.)

Alpha decay is the disintegration of a nucleus into a nucleus of another element with the emission of an *alpha particle*, which is a helium nucleus ($^4_2$He). An alpha particle with two protons has a positive charge of $+2e$. Alpha decay is common for elements with atomic numbers greater than 83. An example of alpha decay is that of thorium (Th) into radium (Ra):

$$^{232}_{90}\text{Th} \rightarrow {}^{228}_{88}\text{Ra} + {}^4_2\text{He}$$

In this decay equation, the sum of the mass numbers is the same on each side of the arrow; that is, $232 = 228 + 4$. Also, the sum of the atomic numbers is the same on each side; that is, $90 = 88 + 2$. This principle holds for all nuclear decays and involves the conservation of nucleons and the conservation of charge, respectively.

In a nuclear decay equation, the sum of the mass numbers will be the same on both sides of the arrow, as will the sum of the atomic numbers.

EXAMPLE 10.2   Finding the Products of Alpha Decay

$^{238}_{92}\text{U}$ undergoes alpha decay. Write the equation for the process.

Solution

Step 1

Write the symbol for the parent nucleus followed by an arrow.

$$^{238}_{92}\text{U} \rightarrow$$

Step 2

Because alpha decay involves the emission of $^{4}_{2}\text{He}$, this symbol can be written to the right of the arrow and preceded by a plus sign, leaving space for the symbol for the daughter nucleus.

$$^{238}_{92}\text{U} \rightarrow \underline{} + {}^{4}_{2}\text{He}$$

Step 3

Determine the mass number, atomic number, and chemical symbol for the daughter nucleus. The sum of the mass numbers on the left is 238. The sum on the right must also be 238, and so far, only the 4 for the alpha particle shows. Thus, the daughter must have a mass number of $238 - 4 = 234$. By similar reasoning, the atomic number of the daughter must be $92 - 2 = 90$. From the periodic table (see inside front cover), it can be seen that the element with atomic number 90 is Th (thorium). The complete equation for the decay is

$$^{238}_{92}\text{U} \rightarrow {}^{234}_{90}\text{Th} + {}^{4}_{2}\text{He}$$

Confidence Exercise 10.2

Write the equation for the alpha decay of the radium isotope $^{226}_{88}\text{Ra}$.

Answers to the Confidence Exercises may be found at the back of the book.

Beta decay is the disintegration of a nucleus into a nucleus of another element with the emission of a *beta particle*, which is an electron $(_{-1}^{0}e)$. An example of beta decay is

$$^{14}_{6}\text{C} \rightarrow {}^{14}_{7}\text{N} + {}^{0}_{-1}e$$

A beta particle, or electron, is assigned a mass number of 0 (because it contains no nucleons) and an atomic number of -1 (because its electric charge is opposite that of a proton's $+1$ charge). The sums of the mass numbers and atomic numbers on both sides of the arrow are equal: $14 = 14 + 0$ and $6 = 7 - 1$. In beta decay, with a decrease in the neutron number, a neutron $(_{0}^{1}n)$ is transformed into a proton and an electron $(_{0}^{1}n \rightarrow {}_{1}^{1}p + {}_{-1}^{0}e)$. The proton remains in the nucleus, and the electron is emitted.

Gamma decay occurs when a nucleus emits a *gamma ray* (γ) and becomes a less energetic form of the same nucleus. A gamma ray is a photon of high-energy electromagnetic radiation and has no mass number and no atomic number. Gamma rays are similar to X-rays but are more energetic. An example of gamma decay is

$$^{204}_{82}\text{Pb*} \rightarrow {}^{204}_{82}\text{Pb} + \gamma$$

The asterisk (*) following the lead (Pb) symbol means that the nucleus is in an excited state, analogous to an atom being in an excited state with an electron in a higher energy level (Chapter 9.3). When the nucleus de-excites, one or more gamma rays are emitted. The nucleus is left in a state of lower excitation and ultimately in the "ground (stable) state" of the same nuclide. Gamma decay generally occurs when a nucleus is formed in an excited state, as a product of alpha or beta decay.

Table 10.4	Nuclear Radiations		
Name	Symbol	Charge	Mass Number
Alpha	^4_2He	2+	4
Beta	$^0_{-1}\text{e}$	1−	0
Gamma	γ	0	0
Positron	$^0_{+1}\text{e}$	1+	0
Neutron	^1_0n	0	1

In addition to alpha, beta, and gamma radiation, certain nuclear processes (generally involving artificial radioisotopes) emit *positrons* $\left(^0_{+1}\text{e}\right)$. For example,

$$^{17}_{9}\text{F} \rightarrow {}^{17}_{8}\text{O} + {}^0_{+1}\text{e}$$

Positrons are sometimes referred to as *beta-plus* particles because they are the *antiparticle* of the electron, having the same mass but an electric charge of +1. ● Table 10.4 lists five common forms of nuclear radiation. (Neutrons are also the product of some nuclear reactions.)

A nucleus with atomic number greater than 83 is always radioactive and commonly undergoes a series of alpha, beta, and gamma decays until a stable nucleus is produced. For example, the series of decays beginning with uranium-238 and ending with stable $^{206}_{82}\text{Pb}$ is illustrated in ● Fig. 10.6. Note how the alpha (α) and beta (β) transitions are indicated in Fig. 10.6. The gamma decays that accompany the alpha and beta decays in the series are not apparent on the diagram because the neutron and proton numbers do not change in gamma decay.

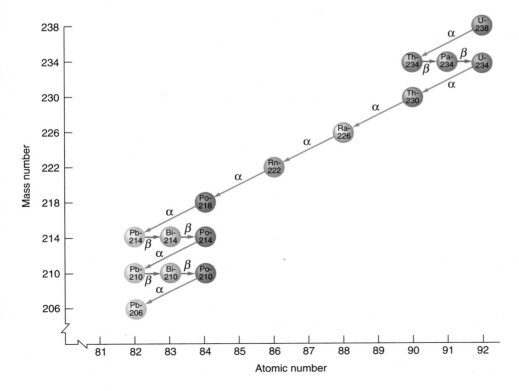

Figure 10.6 The Decay of Uranium-238 to Lead-206
Each radioactive nucleus in the series undergoes either alpha (α) decay or beta (β) decay. Finally, stable lead-206 is formed as the end product. The gamma decays, which change only the energy of nuclei, are not shown.

Conceptual Question and Answer

A Misprint?

Q. Suppose you picked up a newspaper and saw a story reporting the discovery of a new radioactive element. The story states that the element was formed from radioactive decay: $^{274}_{120}X \rightarrow ^{276}_{121}Y$. Did the newspaper make a misprint?

A. Yes. The increase in the proton number in the decay (120 to 121) would imply beta decay with an additional proton. However, in beta decay, the mass number does not change, so if the proton number is correct, then the daughter element should be $^{274}_{121}Y$.

Identifying Radioactive Nuclei

Which nuclei are unstable (radioactive) and which are stable? When the number of protons (Z) versus the number of neutrons (N) for each stable nucleus is plotted, the points (red dots) form a narrow band called the *band of stability* (● Fig. 10.7). For comparison, the straight red line in the figure represents equal numbers of protons and neutrons. The increasing divergence of the band from the N = Z line shows that there are more neutrons than protons.

The blue dots in Fig. 10.7 represent known radioactive nuclei. Note that the nuclei cluster on each side of the band of stability and sometimes are found within it. No stable nuclides (red dots) are found past Z = 83, but numerous radioactive nuclei with more than 83 protons are known.

An inventory of the number of protons and the number of neutrons in stable nuclei reveals an interesting pattern (● Table 10.5). Most of the stable nuclides have both an even

Figure 10.7 A Plot of Number of Neutrons (N) versus Number of Protons (Z) for Nuclei
The red dots representing stable nuclei trace out a *band of stability*. It begins on a line where the number of *neutrons* (N) and the number of *protons* (Z) are equal and gradually diverges from the line as the *number of protons* gets greater. Because all nuclei with more than 83 protons are radioactive, the band ends at this number of protons. The blue dots represent known radioisotopes.

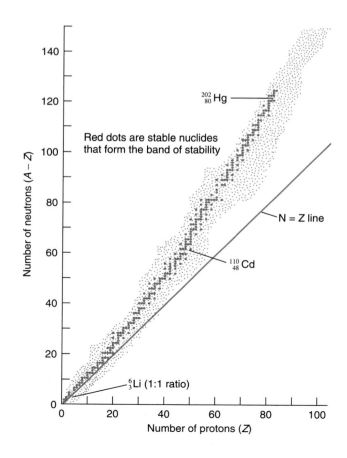

Table 10.5 The Pairing Effect in Stabilizing Nuclei

Proton Number, Z	Neutron Number, N	Number of Stable Nuclides	Example
Even	Even	160	$^{24}_{12}Mg$
Even	Odd	52	$^{13}_{6}C$
Odd	Even	52	$^{6}_{5}Be$
Odd	Odd	4	$^{14}_{7}N$

number of protons (Z) and an even number of neutrons (N) in their nuclei and are referred to as *even–even* nuclei. Practically all the other stable nuclei are either *even–odd* or *odd–even*. Nature dislikes *odd–odd* nuclei (only four stable ones exist) apparently because of the existence of energy levels in the nucleus that favor the pairing of two protons or two neutrons.

Such descriptions as *odd–odd* refer to the number of protons and neutrons, respectively, not to the atomic number and mass number. For example, because $N = A - Z$, an *odd* atomic number (say, 9) coupled with an *even* mass number (say, 20) means an *odd* number of protons (9) but also an *odd* number of neutrons (11).

A nucleus will be radioactive if it meets any of the following criteria.

1. Its atomic number is greater than 83.

2. It has fewer neutrons, n, than protons, p (except for $^{1}_{1}H$ and $^{3}_{2}He$).

3. It is an odd–odd nucleus (except for $^{2}_{1}H$, $^{6}_{3}Li$, $^{10}_{5}B$, and $^{14}_{7}N$).*

EXAMPLE 10.3   Identifying Radioactive Isotopes

Identify the radioactive nucleus in each pair and state your reasoning.

(a) $^{208}_{82}Pb$ and $^{222}_{86}Rn$

(b) $^{19}_{10}Ne$ and $^{20}_{10}Ne$

(c) $^{63}_{29}Cu$ and $^{64}_{29}Cu$

Solution

(a) $^{222}_{86}Rn$ (Z above 83)

(b) $^{19}_{10}Ne$ (fewer n than p)

(c) $^{64}_{29}Cu$ (odd–odd)

Confidence Exercise 10.3

Predict which two of the following nuclei are radioactive.

$$^{232}_{90}Th \quad ^{24}_{12}M \quad ^{40}_{19}K \quad ^{31}_{15}P$$

Half-Life

Some samples of radioisotopes take a long time to decay; others decay very rapidly. In a sample of a given isotope, the decay of an individual nucleus is a random event. It is impossible to predict which nucleus will be the next to undergo a nuclear change. However, given a large number of nuclei, it is possible to predict how many will decay in a certain length of time. The rate of decay of a given radioisotope is described by the term **half-life**,

---

*A fourth criterion, which will not be used because it is difficult to apply, is that unless the mass number of a nucleus is relatively close to the element's atomic mass, the nucleus will be radioactive.

Figure 10.8 Radioactive Decay and Half-Life

Starting with the number of nuclei N_o of a radioactive sample, after one half-life has elapsed, only one-half $\left(\frac{1}{2}N_o\right)$ of the original nuclei will remain undecayed (as indicated by the shading in the box above the curve). The other half of the sample consists of nuclei of the decay product (white portion of the box). After two half-lives, only one-quarter of the original nuclei will remain, and so on.

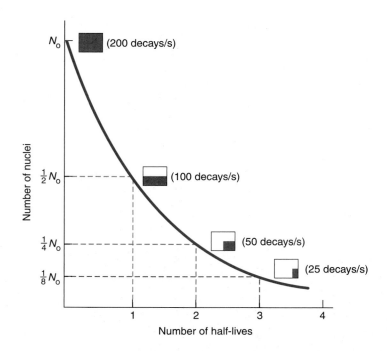

which is the time it takes for half of the nuclei of a given radioactive sample to decay. In other words, after one half-life has gone by, one-half of the original amount of isotope remains undecayed; after two half-lives $\left(\frac{1}{2}\times\frac{1}{2}\right)=$ one-fourth $\left(\frac{1}{4}\right)$ of the original amount is undecayed; and so on (● Fig. 10.8).*

To determine the half-life of a radioisotope, the *activity* (the rate of emission of decay particles) is monitored. The activity is commonly measured in counts per minute (cpm). This measurement may be done with an instrument such as a *Geiger counter* (● Fig. 10.9). Hans Geiger, who developed the counter in 1913, was one of Rutherford's assistants. If half

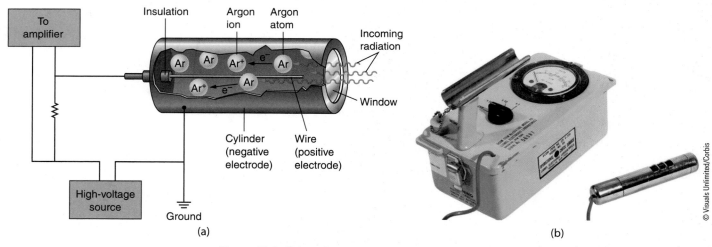

Figure 10.9 Geiger Counter
(a) A Geiger counter detects ions formed as a result of a high-energy particle from a radioactive source entering the window and ionizing argon atoms along its path. The ions and electrons formed produce a pulse of current, which is amplified and counted. (b) A portable (battery-operated) Geiger counter.

*A radioactive isotope's half-life can be measured without waiting throughout the duration of one half-life. The half-life can be calculated by measuring the rate of decay of a known quantity of the isotope.

of the original nuclei of sample decay in one half-life, then the activity decreases to one-half of its original counts during that time.

If a radio isotope's half-life is, say, 12 y, then keep the units straight by putting that information into your calculations as 12 y/half-life (12 years per half-life).

For simplicity, we will work with only a number of half-lives that are a small whole number. In a given exercise, the quantity solved for will be one of the following: the number of half-lives, the final sample amount, or the elapsed time.

EXAMPLE 10.4   Finding the Number of Half-Lives and the Final Amount

What fraction and mass of a 40-mg sample of iodine-131 (half-life = 8 d) will remain after 24 days?

Solution

Step 1

Find the number of half-lives that have passed in 24 d.

$$\frac{24 \text{ d}}{8 \text{ d/half-life}} = 3 \text{ half-lives}$$

Step 2

Starting with the defined original amount N_o, halve it three times (because three half-lives have passed).

Thus, three half-lives have passed, and with $N_o/8$ remaining, the final amount of iodine-131 is $\frac{1}{8}$ of 40 mg, or 5 mg.

Confidence Exercise 10.4

Strontium-90 (half-life = 29 y) is one of the worst components of fallout from atmospheric testing of nuclear bombs because it concentrates in the bones. The last such bomb was tested in 1963. In the year 2021, how many half-lives will have gone by for the strontium-90 produced in the blast? What fraction of the strontium-90 will remain in that year?

EXAMPLE 10.5   Finding the Elapsed Time

How long would it take a sample of ^{14}C to decay to one-fourth of its original activity? The half-life of ^{14}C is 5730 y.

Solution

The ^{14}C will decay for a time period equal to two half-lives, as shown by the number of arrows in the sequence

$$N_o \rightarrow \frac{N_o}{2} \rightarrow \frac{N_o}{4}$$

To find the elapsed time, multiply the number of half-lives by the half-life.

$$(2 \text{ half-lives})(5730 \text{ y/half-life}) = 11{,}460 \text{ y}$$

Confidence Exercise 10.5

Technetium-99 is often used as a radioactive tracer to assess heart damage. Its half-life is 6.0 h. How long would it take a sample of technetium-99 to decay to one-sixteenth of its original amount?

Carbon-14 Dating

Because their decay rates are constant, radioisotopes can be used as nuclear "clocks." Half-life can be used to determine how much of a radioactive sample will exist in the future (see Fig. 10.8). Similarly, by using the half-life to calculate backward in time, scientists can determine the ages of objects that contain known radioisotopes. Of course, some idea of the initial amount of the isotope must be known.

An important dating procedure commonly used in archeology involves the radioisotope ^{14}C. **Carbon-14 dating** is used on materials that were once part of living things, such as wood, bone, and parchment. The process depends on the fact that living things (including you) contain a known amount of radioactive ^{14}C, which has a half-life of 5730 years.

The ^{14}C nuclei exist in living things because the isotope is continually being produced in the atmosphere by cosmic rays. *Cosmic rays* are high-speed charged particles that reach the Earth from various sources, like the Sun. The "rays" are primarily protons, and on entering the upper atmosphere, they can cause reactions that produce neutrons (● Fig. 10.10). These neutrons react with the nuclei of nitrogen atoms in the atmosphere to produce ^{14}C and a proton ($^{1}_{1}H$):

$$^{14}_{7}N + ^{1}_{0}n \rightarrow ^{14}_{6}C + ^{1}_{1}H$$

It is assumed that the intensity of incident cosmic rays has been relatively constant over thousands of years because of atmospheric mixing. Changes in solar activity and the

Figure 10.10 Carbon-14 Dating
An illustration of how carbon-14 forms in the atmosphere and enters the biosphere. See text for description.

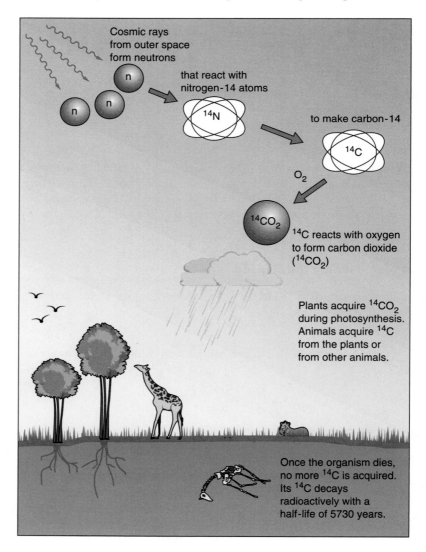

Cosmic rays from outer space form neutrons

that react with nitrogen-14 atoms

^{14}N

to make carbon-14

^{14}C

O_2

$^{14}CO_2$

^{14}C reacts with oxygen to form carbon dioxide ($^{14}CO_2$)

Plants acquire $^{14}CO_2$ during photosynthesis. Animals acquire ^{14}C from the plants or from other animals.

Once the organism dies, no more ^{14}C is acquired. Its ^{14}C decays radioactively with a half-life of 5730 years.

Earth's magnetic field may have caused it to vary somewhat, but the dating is a good approximation.

The newly formed ^{14}C reacts with oxygen in the air to form radioactive carbon dioxide, $^{14}CO_2$ (Fig. 10.10). This, along with ordinary carbon dioxide, $^{12}CO_2$, is used by plants in photosynthesis (Chapter 19.1). About one out of every trillion (10^{12}) carbon atoms in plants is ^{14}C. Animals eat the plants containing ^{14}C, and animals eat the animals that ate the plants.

Thus, all living matter has about the same level of radioactivity due to ^{14}C, an activity of about 16 counts (beta emissions) per minute per gram of total carbon ($^{14}_{6}C \rightarrow {}^{14}_{7}N + {}^{0}_{-1}e$). Once an organism dies, it ceases to take in ^{14}C, but the original amount of ^{14}C continues to undergo radioactive decay. The longer an organism has been dead, the lower the radioactivity is of each gram of carbon in its remains.

The limit of radioactive carbon dating depends on the ability to measure the very low activity in old samples. Current techniques give an age-dating limit of about 40,000–50,000 years, depending on the size of the sample. After about ten half-lives, the radioactivity is barely measurable.

EXAMPLE 10.6   Carbon-14 Dating

An old scroll (like the Dead Sea Scrolls) is found in a cave that has a carbon-14 activity of 4 counts/min per gram of total carbon. Approximately how old is the scroll?

Solution

First the number of half-lives of ^{14}C that have passed is determined. The plant, from which the parchment was made, originally had an activity of 16 counts/min per gram of carbon. Then, to get the current 4 counts:

$$16 \text{ counts} \rightarrow \left(t_{\frac{1}{2}}\right) 8 \text{ counts} \rightarrow \left(t_{\frac{1}{2}}\right) 4 \text{ counts}$$

so two half-lives have elapsed.

Knowing that the half-life of ^{14}C is 5730 years,

$$2 \times 5730 \text{ years} = 11,460 \text{ years}$$

so the parchment is between approximately 11,000 and 12,000 years old.

Confidence Exercise 10.6

An archeology dig unearths a skeleton. Analysis shows that there is a ^{14}C activity of 1 count/min per gram of total carbon. Approximately how old is the skeleton?

Did You Learn?

● Radioactivity is the spontaneous decay of nuclei undergoing a change by emitting particles or rays: alpha decay (alpha particle or helium nucleus), beta decay (beta particle or electron), or gamma decay (gamma ray or photon).

● Half-life is the time it takes for one-half of a given radioactive sample to decay.

10.4 Nuclear Reactions

Preview Questions

● What quantities are conserved in nuclear reactions?

● What are transuranium elements?

Through the emission of alpha and beta particles, radioactive nuclei spontaneously change (undergo *transmutation*) into nuclei of other elements. Scientists wondered whether the reverse process was possible. Could a particle be added to a nucleus to change it into that of another element? The answer is yes.

Rutherford produced the first *nuclear reaction* in 1919 by bombarding nitrogen (^{14}N) gas with alpha particles from a radioactive source. Other particles were observed coming from the gas and were identified as protons. Rutherford reasoned that an alpha particle colliding with a nitrogen nucleus can occasionally knock out a proton. The result is an *artificial transmutation* of a nitrogen nucleus into an oxygen nucleus. The equation for the reaction is

$$^4_2\text{He} + {}^{14}_7\text{N} \rightarrow {}^{17}_8\text{O} + {}^1_1\text{H}$$

The conservation of mass number and the conservation of atomic number hold in nuclear reactions, just as in nuclear decay.

The general form of a nuclear reaction is

$$a + A \rightarrow B + b$$

where a is the particle that bombards nucleus A to form nucleus B and an emitted particle b. In addition to the particles listed in Table 10.4, the particles commonly encountered in nuclear reactions are protons (^1_1H), deuterons (^2_1H), and tritons (^3_1H).

EXAMPLE 10.7   Completing an Equation for a Nuclear Reaction

Complete the equation for the proton bombardment of lithium-7.

$$^1_1\text{H} + {}^7_3\text{Li} \rightarrow \underline{\quad} + {}^1_0\text{n}$$

Solution

The sum of the mass numbers on the left is 8. So far, only a mass number of 1 shows on the right, so the missing particle must have a mass number of $8 - 1 = 7$.

The sum of the atomic numbers on the left is 4. The total showing on the right is 0. Thus, the missing particle must have an atomic number of $4 - 0 = 4$. The atom with mass number 7 and atomic number 4 is an isotope of Be (beryllium, $Z = 4$; see the periodic table inside the front cover). The completed equation is

$$^1_1\text{H} + {}^7_3\text{Li} \rightarrow {}^7_4\text{Be} + {}^1_0\text{n}$$

Confidence Exercise 10.7

Complete the equation for the deuteron bombardment of aluminum-27.

$$^2_1\text{H} + {}^{27}_{13}\text{Al} \rightarrow \underline{\quad} + {}^4_2\text{He}$$

The reaction in Rutherford's experiment was discovered almost by accident because it took place so infrequently. One proton is produced for about every one million alpha particles that shoot through the nitrogen gas. Consider the implications of its discovery: One element had been changed into another. It was the age-old dream of the alchemists, the original researchers into transmutation, although their main concern was to change common metals, such as lead, into gold.

Such artificial transmutations are now common. Large machines called *particle accelerators* use electric fields to accelerate charged particles to very high energies. The energetic particles are used to bombard nuclei and initiate nuclear reactions. Different reactions require different particles and different bombarding energies. One nuclear reaction that occurs when a proton strikes a nucleus of mercury-200 is

$$^1_1\text{H} + {}^{200}_{80}\text{Hg} \rightarrow {}^{197}_{79}\text{Au} + {}^4_2\text{He}$$

Highlight Number of Naturally Occurring Elements: A Quandary

We currently have 118 known elements. (See the periodic table inside the front cover.) Each element has nuclei with the same atomic number Z (same number of protons). As noted, some of the elements do not occur naturally and are made artificially by nuclear reactions. The question then arises, how many *naturally* occurring elements are there? That is, how many of the 118 known elements are found in nature?

The transuranium elements ($Z > 92$) are all artificially made, so that eliminates 16 elements. It would seem logical that the first 92 elements in the periodic table, from hydrogen ($_1$H) to uranium ($_{92}$U), would be naturally occurring, but a couple of elements, technetium ($_{43}$Tc) and promethium ($_{61}$Pm), have only been created artificially. Technetium (Tc) was the first synthetic element produced (in 1937) by nuclear bombardment of the element molybdenum (Mo) with deuterons, $_1^2$H:

$$_{42}^{96}\text{Mo} + _1^2\text{H} \rightarrow _{43}^{98}\text{Tc}$$

The Tc and Pm natural absence can be understood from their half-lives, which are about 8 hours and about 20 minutes, respectively. If formed at the beginning of the universe, these elements would have long since decayed away and not occur in nature today.

So, we are down to 90 elements. It is commonly said that there are 88 naturally occurring elements. The difference involves element 85, astatine ($_{85}$At), and element 87, francium ($_{87}$Fr). These elements do appear in nature, but only briefly and in trace amounts. They are found in radioactive decay series (see Exercise 10), and their half-lives are on the order of 2 seconds and 20 minutes, respectively.

Some people believe that even though only small amounts of astatine and francium are present at any given time, they are "naturally occurring" because they occur spontaneously in nature, making the total 90 elements. Others argue that such elements should not really be considered "naturally occurring" since they are not in nature in the sense of other elements (thereby making the total 88 elements).

What do you think? Are there 88 or 90 naturally occurring elements? (Most go with 88.)

Gold (Au) can indeed be made from another element. Unfortunately, making gold by this process would cost millions of dollars an ounce, much more than the gold is worth.

Neutrons produced in nuclear reactions can be used to induce other nuclear reactions. Because they have no electric charge, neutrons do not experience repulsive electrical interactions with nuclear protons as would alpha particle and proton projectiles, both of which have positive charges. As a result, neutrons are especially effective at penetrating the nucleus and inducing a reaction. For example,

$$_0^1\text{n} + _{21}^{45}\text{Sc} \rightarrow _{19}^{42}\text{K} + _2^4\text{He}$$

The *transuranium elements*, which have an atomic number greater than uranium-92, are all artificially made as a result of induced reactions, as are Tc (43) and Pm (61). The elements At (85) and Fr (87) can also be made artificially, but there is some question about their occurring naturally. (See the **Highlight: Number of Naturally Occurring Elements**.) Elements 93 (neptunium, Np) to 101 (mendelevium, Md) can be made by bombarding a lighter nucleus with alpha particles or neutrons. For example,

$$_0^1\text{n} + _{92}^{238}\text{U} \rightarrow _{93}^{239}\text{Np} + _{-1}^0\text{e}$$

Beyond mendelevium, heavier bombarding particles are required. For example, element 109, meitnerium (Mt), is made by bombarding bismuth-209 with iron-58 nuclei.

$$_{26}^{58}\text{Fe} + _{83}^{209}\text{Bi} \rightarrow _{109}^{266}\text{Mt} + _0^1\text{n}$$

Figure 10.11 A Smoke Detector
In most smoke detectors, a weak radioactive source ionizes the air and sets up a small current. If smoke particles enter the detector, then the current is reduced, causing an alarm to sound.

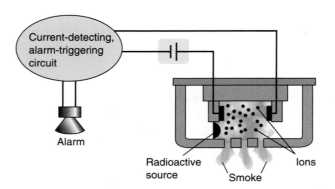

Atoms of hydrogen ($_1$H), helium ($_2$He), and lithium ($_3$Li) are thought to have been formed in the Big Bang theory of the universe (Chapter 18.7), whereas atoms of beryllium ($_4$Be) up through iron ($_{26}$Fe) are made in the cores of stars by nuclear reactions (Chapter 18.4). Atoms of elements heavier than iron are believed to be formed during supernova explosions of stars, when neutrons are in abundance and can enter into nuclear reactions with medium-size atoms to form larger ones (Chapter 18.5).

Some Uses of Radioactive Isotopes

Conceptual Question and Answer

Around the House

Q. Do you have any radioactive sources in your residence?

A. Probably so, if you are prudent enough to have a smoke detector. Americium-241, an artificial transuranium radioactive isotope (half-life = 432 y), is used in the most common type of residential smoke detector. As the americium-241 decays, the alpha particles that are emitted ionize the air inside part of the detector (● Fig. 10.11).

The ions form a small current that allows a battery (or house voltage) to power a closed circuit. If smoke enters the detector, then the ions become attached to the smoke particles and slow down, causing the current in the circuit to decrease and an alarm to sound. (An older model of smoke detector uses a light path and photocell. When smoke dims the light path and less current is supplied by the photocell, an alarm is sounded.)

Radioactive isotopes have many uses in medicine, chemistry, biology, agriculture, and industry. For example, a radioactive isotope of iodine, ^{123}I, is used in a diagnostic measurement connected with the thyroid gland. The patient is administered a prescribed amount of ^{123}I, which, like regular iodine in the diet, is absorbed by the thyroid gland. Doctors can monitor the iodine uptake of the thyroid by measuring the absorbed radioactive iodine.

Radiation is used in many other types of medical diagnoses, like the brain scans pictured in the chapter-opening photo. Plutonium-238 powers a tiny battery used in heart pacemakers. Nuclear radiation also can be used to treat diseased cells, which generally can be destroyed by radiation more easily than healthy cells. For example, focusing an intense beam of radiation from cobalt-60 on a cancerous tumor destroys its cells and thus impairs or halts its growth.

In chemistry and biology, radioactive "tracers," such as ^{14}C (radiocarbon) and ^{3}H (tritium), are used to tag an atom in a certain part of a molecule so that it can be followed through a series of reactions. In this way, the reaction pathways of hormones, drugs, and other substances can be determined.

In industry, tracer radioisotopes help manufacturers test the durability of mechanical components and identify structural weaknesses in equipment. In environmental studies,

Figure 10.12 Is It Safe?
Airport baggage being screened using neutron activation analysis. Almost all explosives contain nitrogen, which when bombarded with neutrons gives off gamma rays that can be detected. Note the radiation caution sign over the opening.

© Justin Sullivan/ Getty Images

small amounts of radioisotopes help detect groundwater movement through soil and trace the paths of industrial air and water pollutants.

In agriculture, food is irradiated. Meat, poultry, and egg products are irradiated with gamma rays to reduce the number of harmful bacteria and parasites present. The Food and Drug Administration (FDA) found irradiation to be safe and approved the process in the 1960s. Irradiation is an important food safety tool.

Neutron activation analysis is one of the most sensitive analytical methods in science. A beam of neutrons irradiates the sample, and each constituent element forms a specific radioisotope that can be identified by the characteristic energies of the gamma rays it emits. Neutron activation analysis has the advantage over chemical and spectral identification of elements because it needs only minute samples. It can be used to identify and measure 50 different elements in amounts as small as 1 picogram (10^{-12} g).

Neutron activation analysis is used as an antiterrorist tool in airports. Virtually all explosives contain nitrogen. By using neutron activation analysis and analyzing the energy of any emission of gamma rays from airport baggage, it is possible to detect the presence of nitrogen and then check the baggage manually to investigate any suspicious finding (● Fig. 10.12).

The next two sections address the controlled and uncontrolled release of nuclear energy.

Did You Learn?

● The mass number (*A*) and the atomic number (*Z*) are conserved in nuclear reactions.

● Transuranium elements are those with atomic numbers greater than 92, and all are artificially made.

10.5 Nuclear Fission

Preview Questions

● Is mass conserved in fission reactions?

● What is a self-sustaining chain reaction, and how can it grow in energy release?

Figure 10.13 Fission and Chain Reaction
(a) In a fission reaction such as that shown for uranium-235, a neutron is absorbed and the unstable ^{236}U nucleus splits into two lighter nuclei with the emission of energy and two or more neutrons. (b) If the emitted neutrons cause increasing numbers of fission reactions, then an expanding *chain reaction* occurs.

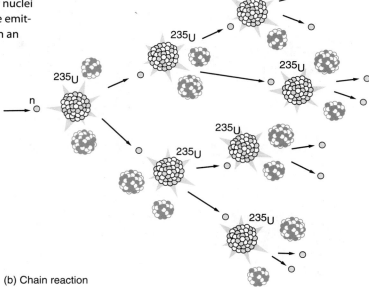

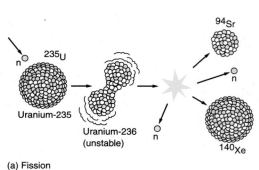

(a) Fission

(b) Chain reaction

Fission is the process in which a large nucleus "splits" (fissions) into two intermediate-size nuclei, with the emission of neutrons and the conversion of mass into energy. For example, consider the fission decay of ^{236}U. If ^{235}U is bombarded with low-energy neutrons, then ^{236}U is formed:

$$^{1}_{0}n + ^{235}_{92}U \rightarrow ^{236}_{92}U$$

The ^{236}U immediately fissions into two smaller nuclei, emits several neutrons, and releases energy. ● Figure 10.13a illustrates the following typical fission of ^{236}U:

$$^{236}_{92}U \rightarrow ^{140}_{54}Xe + ^{94}_{38}Sr + 2^{1}_{0}n$$

This is just one of many possible fission decays of ^{236}U. Another is

$$^{236}_{92}U \rightarrow ^{132}_{50}Sn + ^{101}_{42}Mo + 3^{1}_{0}n$$

EXAMPLE 10.8   Completing an Equation for Fission

Complete the following equation for fission.

$$^{236}_{92}U \rightarrow ^{88}_{36}Kr + ^{144}_{56}Ba + \underline{}$$

Solution

The atomic numbers are balanced (92 = 36 + 56), so the other particle must have an atomic number of 0. The mass number on the left is 236, and the sum of the mass numbers on the right is 88 + 144 = 232. Hence, if the mass numbers are to balance, then there must be four additional units of mass on the right-hand side of the equation. Because no particle with an atomic number of 0 and a mass number of 4 exists, the missing "particle" is actually four neutrons. The reaction is then

$$^{236}_{92}U \rightarrow ^{88}_{36}Kr + ^{144}_{56}Ba + 4^{1}_{0}n$$

Both the atomic numbers and the mass numbers are now balanced or conserved.

Confidence Exercise 10.8

Complete the following equation for fission.

$$^{236}_{92}U \rightarrow ^{90}_{38}Sr + \underline{} + 2^{1}_{0}n$$

The fast-fissioning ^{236}U is an intermediate nucleus and is often left out of the equation for the neutron-induced fission of ^{235}U; the equation for the reaction in Example 10.8 is usually written

$$^1_0n + {}^{235}_{92}U \rightarrow {}^{88}_{36}Kr + {}^{144}_{56}Ba + 4{}^1_0n$$

Nuclear fission reactions have three important features:

1. The fission products are always radioactive. Some have half-lives of thousands of years, which lead to major problems in the disposal of nuclear waste.

2. Relatively large amounts of energy are produced.

3. Neutrons are released.

In an *expanding* **chain reaction**, one initial reaction triggers a growing number of subsequent reactions. In the case of fission, one neutron hits a nucleus of ^{235}U and forms ^{236}U, which can fission and emit two (or more) neutrons. These two neutrons can then hit another two ^{235}U nuclei, causing them to fission and release energy and four neutrons. These four neutrons can cause four more fissions, releasing energy and eight neutrons, and so on (Fig. 10.13b). Each time a nucleus fissions, energy is released; as the chain expands, the energy output increases.

For a *self-sustaining* chain reaction, each fission event needs to cause only one more fission event, which leads to a steady release of energy, not a growing release. The process of energy production by fission is not as simple as just described. For a self-sustaining chain reaction to proceed, a sufficient amount and concentration of fissionable material (^{235}U) must be present. Otherwise, too many neutrons would escape from the sample before reaction with a ^{235}U nucleus. The chain would be broken. The minimum amount of fissionable material necessary to sustain a chain reaction is called the **critical mass**. The critical mass for pure ^{235}U is about 4 kg, which is approximately the size of a baseball. With a *subcritical mass*, no chain reaction occurs. With a *supercritical mass*, the chain reaction grows, and under certain conditions an explosion occurs.

Natural uranium is 99.3% ^{238}U, which does not undergo fission. Only the remaining 0.7% is the fissionable ^{235}U isotope. So that more fissionable ^{235}U nuclei will be present in a sample, the ^{235}U is concentrated, or "enriched." The enriched uranium used in U.S. nuclear reactors for the production of electricity is about 3% ^{235}U. Weapons-grade uranium is enriched to 90% or more; this percentage provides many fissionable nuclei for a large and sudden release of energy.

In a fission bomb, or "atomic bomb," a supercritical mass of highly enriched fissionable material must be formed and held together for a short time to get an explosive release of energy. Subcritical segments of the fissionable material in a fission bomb are kept separated before detonation so that a critical mass does not exist for the chain reaction. A chemical explosive is used to bring the segments together in an interlocking, supercritical configuration that holds them long enough for a large fraction of the material to undergo fission. The result is an explosive release of energy.

Nuclear Reactors

A nuclear (atomic) bomb is an example of *uncontrolled* fission. A nuclear reactor is an example *of controlled* fission, in which the growth of the chain reaction and the release of energy are controlled. The first commercial fission reactor for generating electricity went into operation in 1957 at Shippingport, Pennsylvania. That reactor was shut down in 1982 after 25 years of operation.

The basic design of a nuclear reactor vessel is shown in ● Fig. 10.14. Enriched uranium oxide fuel pellets are placed in metal tubes to form long *fuel rods*, which are placed in the reactor core where fission takes place. Also in the core are *control rods* made of neutron-absorbing materials such as boron (B) and cadmium (Cd). The control rods are adjusted (inserted or withdrawn) so that only a certain number of neutrons are absorbed, ensuring that the chain reaction releases energy at the rate desired. For a steady rate of energy release, one neutron from each fission event should initiate only one additional fission event. If more energy is needed, then the rods are withdrawn farther. When fully inserted

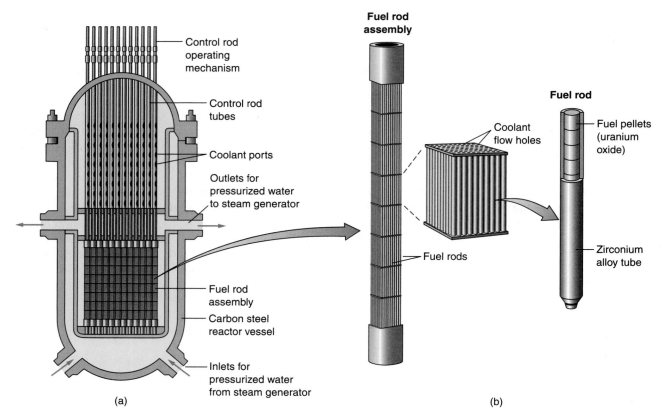

Figure 10.14 Nuclear Reactor Diagram
(a) A schematic diagram of a reactor vessel. (b) A fuel rod and its assembly. A typical reactor contains fuel rod assemblies of approximately 200 rods each, and the fuel core can have up to 3000 fuel assemblies. One of the uranium nuclear fuel pellets the size of the tip of your little finger inside a rod can provide energy equivalent to 1780 lb of coal or 149 gallons of oil. That's quite a bit of energy.

into the core, the control rods absorb enough neutrons to stop the chain reaction, and the reactor shuts down.

A reactor's core is a heat source, and the heat energy is removed by a coolant flowing through the core. Reactors in the United States are light-water reactors in which the coolant is commonly H_2O (light water, as opposed to heavy water, D_2O, which is used in some Canadian reactors and requires lower uranium fuel enrichment). The coolant flowing through the hot fuel assemblies removes heat that is used to produce steam to drive a turbogenerator, which produces electricity. ● Fig. 10.15 shows a *pressurized water reactor* (PWR). Pressure prevents the reactor water from boiling (giving it a higher temperature, Chapter 5.3), and the hot water is sent to a steam generator that produces steam to drive a turbine. The generation of steam is accomplished without radioactive water coolant reaching the turbine. In the United States, 78% of the reactors are PWRs. An older type, the *boiling water reactor* (BWR), uses boiling water in the reactor to produce steam that directly drives the turbine. The water is then cooled, condensed, and returned to the reactor.

In addition, the coolant acts as a *moderator*. The ^{235}U nuclei react best with "slow" neutrons. The neutrons emitted from the fission reactions are relatively "fast," with energies that are not best suited for ^{235}U fission. The fast neutrons are slowed down, or moderated, by transferring energy to the water molecules in collision processes. After only a few collisions, the neutrons are slowed down to the point at which they efficiently induce fission in the ^{235}U nuclei.

With a continuous-fission chain reaction, the possibility of a nuclear accident is always present. The word *meltdown* is commonly used when discussing such accidents. The

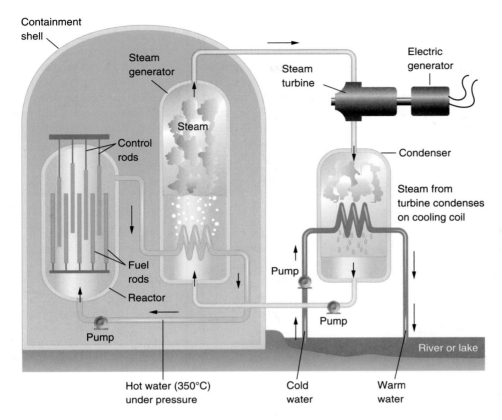

Figure 10.15 Diagram of a Nuclear Reactor
The nuclear reactor consists of fuel rods with interspersed control rods. By raising or lowering the control rods, an operator can increase or decrease the rate of energy release from the fuel rods. Heat from the fuel rods raises the temperature of the liquid water in the reactor. A pump circulates the hot water to a steam generator, and the resulting steam passes through a turbine that operates an electric generator. The steam leaves the turbine and goes into the condenser, where it liquefies on the cooling coil. A nearby river or lake provides the cold water for the condenser.

coolant must be maintained. If heat energy is not removed continuously and the core becomes partially exposed, then some of the fuel rods may "melt," or fuse. This is called a *partial meltdown*. With the loss of more coolant and the exposure of more of the fuel rods, there is a fissioning mass of fuel pellets becoming extremely hot, fusing, dropping down, melting, and breaching the containing vessel. At this stage, there is a *total meltdown* and radioactive material enters the environment. (This situation is sometimes called the *China Syndrome* because the "melt" is heading downward through the center of the Earth "toward China." Of course, this description is inaccurate because China is not on the other side of the Earth from the United States; both countries are in the Northern Hemisphere).

Well-known accidents occurred in 1979 at the Three Mile Island nuclear plant in Pennsylvania, in 1986 with the reactor at Chernobyl in the Ukraine in Russia, and in 2011 at the Fukushima Daiichi nuclear plant in Japan. At Three Mile Island, a partial meltdown occurred as the result of an accidental shutdown of cooling water. There was a slight fusing of the fuel pellets and the release of large amounts of radioactive material inside the containment building. However, only a small amount of radioactive gases escaped into the environment. At Chernobyl, poor human judgment, including the disconnection of several emergency safety systems, led first to a meltdown, then to an explosion in the reactor core, and finally to a fire. This particular type of reactor used graphite (carbon) blocks for a moderator. Gas explosions caused the carbon to catch fire, radioactive material escaped with the smoke, and weather conditions caused radioactive fallout to spread over many European countries. Several hundred deaths occurred in the immediate region, and it is estimated that as many as 50,000 additional cancer deaths will occur from the long-term effects of the radioactive fallout.

In March 2011, a severe earthquake and tsunami occurred off the northeast coast of Japan (see Chapter 21.5). The boiling water nuclear reactors at the Fukushima Daiichi generating facility were affected when external electrical power was lost in the earthquake and the tsunami waves disabled the backup generators that were to keep the water coolant flowing in the reactors. Without sufficient water, three reactor cores were partially exposed.

This partial exposure generated steam and caused the metal rod cladding of the fuel to oxidize, resulting in the production of hydrogen. When the gases were vented from the containment vessel to prevent pressure buildup and the hydrogen united with oxygen in the air, explosions occurred, releasing some radioactive material. A fourth generator had been shut down for maintenance, but its stored spent fuel rods, which remain hot, were exposed and caught on fire. Sea water was pumped in to replace the coolant flowing in the three reactors and to control the fire. Even so, partial meltdowns occurred, and water leaking from containment vessels contaminated the environment. With continued efforts, conditions were stabilized, and fortunately, there were no complete meltdowns.

Conceptual Question and Answer

Out of Control

Q. Were the control-rod mechanism in a nuclear reactor to malfunction and the chain reaction proceeded uncontrolled, could the reactor explode like a nuclear bomb?

A. No. A nuclear reactor cannot explode like an atomic or nuclear bomb. Recall that reactor-grade uranium contains only about 3% fissionable U-235, whereas weapons-grade uranium contains more than 90%. Also, the high-grade material must be held together briefly in critical mass to achieve an explosive release of energy. A meltdown and gaseous explosions could be possible, but not a tremendous explosion like a nuclear bomb.

In addition to ^{235}U, the other fissionable nuclide of importance is plutonium, ^{239}Pu (half-life of 2.4×10^4 y). This plutonium isotope is produced by bombardment of ^{238}U with fast neutrons, meaning that ^{239}Pu is produced as nuclear reactors operate because not all the neutrons are moderated or slowed down. Being fissionable, ^{239}Pu extends the time before the refueling of a reactor is necessary.

In a *breeder reactor*, the process of producing fissionable ^{239}Pu from ^{238}U is promoted. ^{238}U is otherwise useless for energy production. Breeder reactors are currently being operated in France and Germany, but not in the United States. The ^{239}Pu can be chemically separated from the fission by-products and used as the fuel in an ordinary nuclear reactor.

In addition to concerns about the safety of operating nuclear fission power plants, the other major problem is what to do with the radioactive waste generated. This subject is discussed later after some other considerations.

Did You Learn?

● In nuclear fission, some mass is converted to energy.

● In a self-sustaining chain reaction, each fission event causes one or more fission events, leading to a steady release of energy, and the reaction may grow with sufficient amounts and concentration of fissionable material (supercritical mass).

10.6 Nuclear Fusion

Preview Questions

● Where does nuclear fusion occur "naturally"?

● What are the major advantages of fusion energy production over fission energy production?

Fusion is the process in which smaller nuclei combine to form larger ones with the release of energy. Fusion is the source of energy of the Sun and other stars. In the Sun, the fusion process produces a helium nucleus from four protons (hydrogen nuclei). Also produced are two positrons. The thermonuclear process takes place in several steps (see Chapter 18.2), the net result being

$$4{}^1_1\text{H} \rightarrow {}^4_2\text{He} + 2({}^0_{+1}e) + \text{energy}$$

In the Sun, about 600 million tons of hydrogen is converted into 596 million tons of helium *every second*. The other 4 million tons of matter are converted into energy. Fortunately, the Sun has enough hydrogen to produce energy at its present rate for several billion more years.

Two other examples of fusion reactions are

$$ {}^2_1\text{H} + {}^2_1\text{H} \rightarrow {}^3_1\text{H} + {}^1_1\text{H}$$

and

$$ {}^2_1\text{H} + {}^3_1\text{H} \rightarrow {}^4_2\text{He} + {}^1_0\text{n}$$

In the first reaction, two deuterons fuse to form a triton and a proton. This is termed a D-D (deuteron–deuteron) reaction. In the second example (a D-T reaction), a deuteron and a triton form an alpha particle and a neutron (● Fig. 10.16).

Fusion involves no critical mass or size because there is no chain reaction to maintain. However, the repulsive force between two positively charged nuclei opposes fusing. This force is smallest for hydrogen fusion because the nuclei contain only one proton. To overcome the repulsive forces and initiate fusion, the kinetic energies of the particles must be increased by raising the temperature to about 100 million kelvins. At such high temperatures, the hydrogen atoms are stripped of their electrons and a **plasma** (a gas of free electrons and positively charged ions) results. To achieve fusion, a high temperature is necessary, and the plasma must be confined at a high enough density for the protons (or other nuclei) to collide frequently.

Large amounts of fusion energy have been released in an uncontrolled manner in a hydrogen bomb (H-bomb), where a fission bomb is used to supply the energy needed to initiate the fusion reaction (● Fig. 10.17). Unfortunately, controlled fusion for commercial use remains elusive. Controlled fusion might be accomplished by steadily adding fuel

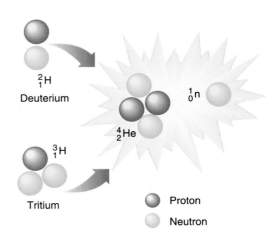

Figure 10.16 A D-T Fusion Reaction The combination of a deuteron (D) and a triton (T) to produce an alpha particle and a neutron is one example of fusion. Nuclei of other elements of low atomic mass also can undergo fusion to produce heavier, more stable nuclei and release energy in the process.

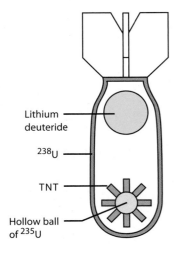

Figure 10.17 H-Bomb
The diagram shows the basic elements of a hydrogen bomb. To detonate an H-bomb, the TNT is exploded, forcing the ^{235}U together to get a supercritical mass and a fission explosion (a small atomic bomb, so to speak). The fusionable material is deuterium in the lithium deuteride (LiD). When it is heated to a plasma, D-D fusion reactions occur. Neutrons from the fission explosion react with the lithium to give tritium (^{3}H), and then D-T fusion reactions also take place. The bomb is surrounded with ^{238}U, which tops off the explosion with a fission reaction. The result is shown in the photo.

in small amounts to a fusion reactor. Since the D-T reaction has the lowest temperature requirement (about 100 million K) of any fusion reaction, it is likely to be the first fusion reaction developed as an energy source.

Major problems arise in reaching such temperatures and in confining the high-temperature plasma. If the plasma were to touch the reactor walls, then it would cool rapidly, but the walls would not melt. Even though the plasma is nearly 100 million K above the melting point of any material, the total quantity of heat that could be transferred from the plasma is very small because the plasma's concentration is extremely low.

One approach to controlled fusion is *inertial confinement*, a technique in which simultaneous high-energy laser pulses from all sides cause a fuel pellet containing deuterium and tritium to implode, resulting in compression and high temperatures. If the pellet stays intact for a sufficient time, then fusion is initiated.

Another approach to controlled fusion is *magnetic confinement*. Because a plasma is a gas of charged particles, it can be controlled and manipulated with electric and magnetic fields. A nuclear fusion reactor called a *tokomak* uses a doughnut-shaped magnetic field to hold the plasma away from any material. Electric fields produce currents that raise the temperature of the plasma.

Plasma temperatures, densities, and confinement times have been problems with magnetic and inertial confinements, and no one knows when commercial energy production via nuclear fusion can be expected. Even so, fusion is a promising energy source because of three advantages over fission:

1. The low cost and abundance of deuterium, which can be extracted inexpensively from water. Scientists estimate that the deuterium in the top 2 inches of water in Lake Erie could provide fusion energy equal to the combustion energy in all the world's oil reserves. On the other hand, uranium for fission is scarce, expensive, and hazardous to mine.

2. Dramatically reduced nuclear waste disposal problems. Some fusion by-products are radioactive because of nuclear reactions involving the neutrons that are formed in the D-T reactions, but they have relatively short half-lives compared with those of fission wastes.

3. Fusion reactors could not get out of control. In the event of a system failure in a fusion plant, the reaction chamber would immediately cool down and energy production would halt.

The two disadvantages of fusion compared with fission are that (1) fission reactors are presently operational (commercial fusion reactors are at least decades away) and (2) fusion plants will probably be more costly to build and operate than fission plants.

Nuclear Reactions and Energy

In 1905, Albert Einstein published his *special theory of relativity*, which deals with the changes that occur in mass, length, and time as an object's speed approaches the speed of light (c). The theory also predicted that mass (m) and energy (E) are not separate quantities but are related by the equation

$$\text{energy} = \text{mass} \times \text{the speed of light squared}$$
$$E = mc^2 \hspace{4cm} \text{10.2}$$

The prediction proved correct. Scientists have indeed changed mass into energy and, on a very small scale, have converted energy into mass.

For example, a mass of 1.0 g (0.0010 kg) has an equivalent energy of

$$E = mc^2 = (0.0010 \text{ kg})(3.00 \times 10^8 \text{ m/s})^2$$
$$= 90 \times 10^{12} \text{ J}$$

This 90 *trillion* joules is the same amount of energy that is released by the explosion of about 20,000 *tons* of TNT. Such calculations convinced scientists that nuclear reactions, in which just a very small amount of mass was "lost," were a potential source of vast amounts of energy.

The units of mass and energy commonly used in nuclear physics differ from those discussed in preceding chapters. Mass is usually given in atomic mass units (u), and energy is usually given in mega electron volts, MeV (1 MeV = 1.60×10^{-13} J). With these units, Einstein's equation reveals that 1 u of mass has the energy equivalent of 931 MeV, so there are 931 MeV/u (931 MeV per atomic mass unit).

To determine the change in mass and hence the energy released or absorbed in any nuclear process, just add up the masses of all reactant particles and from that sum subtract the total mass of all product particles. If an increase in mass has taken place, then the reaction is *endoergic* (absorbs energy) by that number of atomic mass units times 931 MeV/u. As is more common, if a decrease in mass has resulted, then the reaction is *exoergic* (releases energy) by that number of atomic mass units times 931 MeV/u. The decrease in mass in a nuclear reaction is called the **mass defect**.

EXAMPLE 10.9   Calculating Mass and Energy Changes in Nuclear Reactions

Calculate the mass defect and the corresponding energy released during this typical fission reaction:*

$$\underset{(236.04556\ \text{u})}{^{236}_{92}\text{U}} \rightarrow \underset{(87.91445\ \text{u})}{^{88}_{36}\text{Kr}} + \underset{(143.92284\ \text{u})}{^{144}_{56}\text{U}} + \underset{(4 \times 1.00867\ \text{u})}{4^1_0\text{n}}$$

Solution

The total mass on the left of the arrow is 236.04556 u. Adding the masses of the particles on the right gives 235.87197 u. The difference (0.17359 u) is the mass defect, which has been converted to

$$(0.17359\ \text{u})(931\ \text{MeV/u}) = 162\ \text{MeV of energy}$$

Thus, during the reaction 0.17359 u of mass is converted to 162 MeV of energy.

Confidence Exercise 10.9

Calculate the mass defect and the corresponding energy released during a D-T fusion reaction:

$$\underset{(2.0140\ \text{u})}{^2_1\text{H}} + \underset{(3.0161\ \text{u})}{^3_1\text{H}} \rightarrow \underset{(4.0026\ \text{u})}{^4_2\text{He}} + \underset{(1.0087\ \text{u})}{^1_0\text{n}}$$

When nuclear fuels are compared, it is apparent that *kilogram for kilogram*, more energy is available from fusion than from fission. Comparing fission and fusion with energy production by ordinary chemical reactions, the fission of 1 kg of uranium-235 provides energy equal to burning 2 million kg of coal, whereas the fusion of 1 kg of deuterium releases the same amount of energy as burning of 40 million kg of coal.

● Figure 10.18 plots the relative stability of various nuclei as a function of the mass number (the number of nucleons) and shows that energy can be released in both nuclear fission *and* nuclear fusion. Note that fission of heavy nuclei at the far right of the curve to intermediate-size nuclei in the middle leads upward on the curve. Fusion of small nuclei on the left to larger nuclei farther to the right also leads upward on the curve. Any reaction that leads *upward* on the curve in Fig. 10.18 releases energy because such a reaction is accompanied by a mass defect. Basically, each nucleon in the reactant nucleus (or nuclei) loses a little mass in the process.

Any nuclear reaction in which the products are lower on the curve than the reactants can proceed only with a net increase in mass and a corresponding net absorption of energy.

---

*Because we are dealing with *differences* in mass, either the masses of the atoms or the masses of just their nuclei can be used. The number of electrons is the same on each side of the equation and thus does not affect the mass difference. The masses of the atoms are used because they are more easily found in handbooks.

Figure 10.18 The Relative Stability of Nuclei
To obtain more stability (with the release of energy), the fission of more massive nuclei (right) to less massive nuclei is upward on the curve. The fusion of less massive nuclei into more massive nuclei (left) is upward on the curve. Any reaction that leads upward on the curve releases energy.

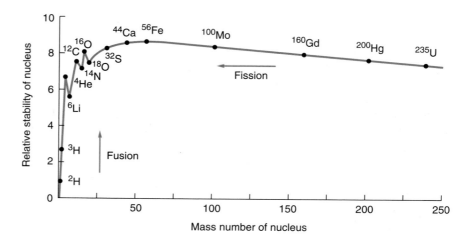

One type of nucleus cannot give a net release of nuclear energy either by fission or by fusion. Of course, it is the one at the top of the curve, ^{56}Fe (iron-56). You cannot go higher than the top, so no net energy will be released either by splitting ^{56}Fe into smaller nuclei or by fusing several ^{56}Fe into a larger nucleus.

Did You Learn?

● The Sun's energy results from the fusion process of converting hydrogen nuclei into helium nuclei.

● Fusion advantages over fission: (1) low cost and abundance of fuel (deuterium), (2) reduced nuclear waste, and (3) little chance of reaction going out of control.

10.7   Effects of Radiation

Preview Questions

● What is "ionizing" radiation, and why is it harmful?

● How do particles and rays rate in protective shielding?

It is a common misconception that radioactivity is something new in the environment. Radioactivity has been around far longer than humans as a natural part of the environment. Still, we must be aware of the dangers associated with radiation.

Radiation that is energetic enough to knock electrons out of atoms or molecules and form ions is classified as *ionizing radiation*. Alpha particles, beta particles, neutrons, gamma rays, and X-rays all fall into this category. Such radiation can damage or even kill living cells, and it is particularly harmful when it affects protein and DNA molecules involved in cell reproduction. Ionizing radiation is especially dangerous because you cannot see, smell, taste, or feel it.

Radiation doses are measured in *rads* (short for radiation *a*bsorbed *d*ose), where 1 rad corresponds to 0.01 joule of energy deposited per kilogram of tissue. Because alpha, beta, and gamma radiations differ in penetrating ability and ionizing capabilities, both the energy dose of the radiation *and* its effectiveness in causing human tissue damage must be considered. The *rem* (short for *r*oentgen *e*quivalent for *m*an), which takes into account both the dosage and its relative biological effectiveness, is the unit we generally use when discussing biological effects of radiation.*

---

*The SI unit of radiation biological effectiveness is the sievert (Sv), where 1 Sv = 100 rem (named after Rolf Sievert, a Swedish medical physicist).

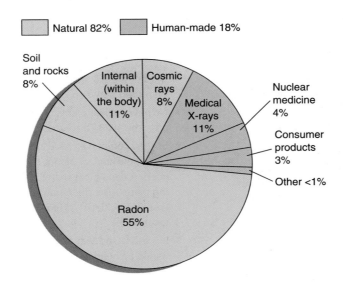

Natural 82%   Human-made 18%

Soil and rocks 8%
Internal (within the body) 11%
Cosmic rays 8%
Medical X-rays 11%
Nuclear medicine 4%
Consumer products 3%
Other <1%
Radon 55%

Figure 10.19 Sources of Exposure to Radiation
On average, each person in the United States receives a yearly radiation exposure of 0.2 rem, of which 82% is from the natural sources. The other 18% is from human-made sources.

The average U.S. citizen receives about 0.2 rem of natural and human-made background radiation each year. ● Figure 10.19 shows that about 82% of this background radiation comes from natural sources and about 18% from human activities. However, individual exposures vary widely, depending on location, occupation, and personal habits.

Sources of natural radiation include cosmic rays from outer space. Travel in high-flying jetliners and living in high-altitude cities, such as Denver, Colorado, entail more exposure to this part of the background radiation. Other sources of natural background radiation include radionuclides in the rocks and minerals in our environment. One of the decay products of ^{238}U is radon-222, an inert noble gas that quickly alpha decays (half-life = 3.8 days) into an isotope of polonium (^{218}Po). Uranium-238 occurs naturally in the environment, as does its by-product radon-222. As a gas, radon can seep from the ground into the air and into a building, primarily thorough the foundation with cracks in the basement floor and drains.

Radon-laden air breathed into the lungs can decay into polonium-218. This isotope can settle in the lungs, where it begins (half-life = 3 min) a relatively short decay series with a sequence of alpha, beta, and gamma emissions, leading to mutations in the lung tissue and causing cancer.

The Environmental Protection Agency estimates that nearly 1 out of 15 homes in the United States has radon at or above the recommended levels. There are relatively inexpensive, do-it-yourself kits to check on radon levels in a residence.

Human-made sources of radiation include X-rays and radioisotopes used in medical procedures, fallout from nuclear testing, tobacco smoke, nuclear wastes, and emissions from power plants. Ironically, because fossil fuels contain traces of uranium, thorium, and their daughter nuclei, more radioactive isotopes are released into the atmosphere from power plants burning coal and oil than from nuclear power plants.

Shielding is used to decrease radiation exposure. Clothing and our skin are sufficient protection against alpha particles, unless the particles are ingested or inhaled. Beta particles can burn the skin and penetrate into body tissue, but not far enough to affect internal organs. Heavy clothing can shield against beta particles. Gamma rays, X-rays, and neutrons are more difficult to stop (● Fig. 10.20). Protective shielding from these rays requires thick lead, concrete, or earth.

So, with the knowledge gained in this chapter, let's consider a final, but very important, nuclear topic. See the **Highlight: Nuclear Power and Waste Disposal**.

Did You Learn?

● Ionizing radiation is radiation energetic enough to knock electrons out of atoms or molecules and form ions. Such radiation can damage cells and produce harmful health hazards.

Highlight   Nuclear Power and Waste Disposal

In making reasoned judgments about nuclear power, there are many aspects to consider. This includes nuclear waste disposal. Rather than present a narrative on these, salient facts and predictions will be presented for your consideration.

Nuclear Power

▸ Unlike coal, oil, and natural gas, nuclear power doesn't emit carbon dioxide and other greenhouse gases (Chapter 19.2).

▸ About 20% of the electricity in the United States is generated by nuclear power. (In France, it is 80%.) By 2035, the U.S. demand for electricity is expected to increase 28%.

▸ In the United States, there are 104 nuclear generating plants at 65 locations in 31 states (Fig. 1).

▸ Existing nuclear plants are aging and are normally retired after 40 to 60 years of service, but some have been granted

an extension because of a good safety record. More than one-third of the U.S. generating capacity of nuclear plants should be retired between 2029 and 2035.

▸ No new nuclear plants have been built in the United States after the 1978 Three Mile Island partial meltdown.

▸ Several companies have applied for regulatory approval to build new nuclear plants, but the process and building takes years. It costs more than $2 billion to build a nuclear generating plant. Also, the Japanese Fukushima Daiichi earthquake-tsunami incident has led to a re-evaluation of safe building locations.

▸ It is estimated that as many as 29 new generating plants would be needed to meet the 2035 electricity demand.

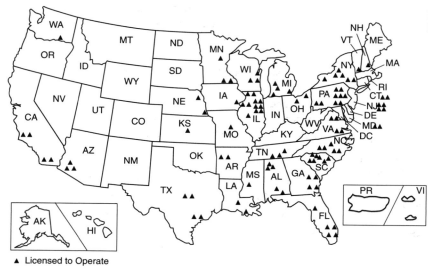

Figure 1 Location of Nuclear Power Reactors in the U.S.

▲ Licensed to Operate

● Alpha particles and beta particles are easily shielded, but gamma rays, X-rays, and neutrons are not. Alpha particles can be shielded by clothing and skin, and beta particles can be shielded by heavy clothing.

10.8   Elementary Particles

Preview Questions

● Why are some subatomic particles called "elementary" particles?

● What are "exchange" particles?

The search for the *fundamental* "building blocks" of nature is as old as science itself. The simple picture of indivisible atoms gave way to a model of the atom with subatomic particles. By the 1930s, scientists had identified four of these particles: the electron, proton,

Highlight

Nuclear Waste

▶ Nuclear generating plants generate nuclear wastes classified as "low level" and "high level."

▶ Low-level nuclear waste usually comes from material used to handle highly radioactive parts of a nuclear reactor, such as cooling water pipes. Waste from medical procedures involving radioactive treatment is also low level.

▶ High-level radioactive waste is generally material from nuclear reactor cores and nuclear weapons. The waste includes uranium, plutonium, and other highly radioactive isotopes made during fission. These isotopes have extremely long half-lives, sometimes longer than 100,000 years.

▶ Generating plants in the United States produce on the order of 32,000 tons of nuclear waste annually.

How do we get rid of or dispose of nuclear waste?

▶ The most promising disposal method at this time is a geologic repository. In this process, waste is packaged and is stored deep below the Earth's surface in underground tunnels or chambers.

▶ The location of such a repository must be geologically "safe," without the possibility of groundwater seepage and with low earthquake probability.

▶ A proposed site for a deep geological repository was Yucca Mountain in southwest Nevada, 100 miles from Las Vegas. It was scheduled to accept waste in 1998. However, the opening was delayed for more than a decade, and in 2010, after more than $10 million was spent, the project was cancelled.

▶ There is one deep geological repository in the United States, the Waste Isolation Pilot Plant (or WIPP), 26 miles east of Carlsbad, New Mexico. It uses tunnels in salt formations. WIPP began operation in 1999 and is expected to continue disposal until 2070. Most of the waste comes from government storage sites.

Some Alternatives?

▶ Because of delays with adequate underground repositories, a number of power plants in the United States have resorted to onsite "dry cast storage" where waste is stored in steel and concrete aboveground casts.

▶ Weapons-grade plutonium destined for disposal can be mixed with uranium, producing a fuel that can be used in nuclear reactors.

▶ Nuclear waste could be stored in large salt domes. Once the waste is stored, the salt oozes around it, becoming geologically stable for 50 million to 100 million years, but there is no further access to the waste.

▶ It has even been suggested to load the waste on a rocket and shoot it into deep space. However, what would happen if the rocket malfunctioned and exploded in the Earth's atmosphere?

▶ More realistic is a subductive waste disposal method. Subduction refers to a process in which one tectonic plate slides underneath another and is absorbed in the Earth's mantle (see Chapter 21.1). Were a repository located on a subducting plate, both the waste and the plate would be absorbed into the mantle. The best site for a plate repository would be on the ocean floor where plate subduction occurs. However, the waste would have to be packaged for a long stay because subduction progresses at only a few centimeters per year.

Have you got any ideas about possibilities for nuclear waste disposal?

---

neutron, and photon. Since then, more than 200 various particles have been discovered coming from the nucleus. They are called **elementary particles** because it is not certain that they are fundamental. You may have heard of some of them, such as neutrinos, mesons, bosons, lambda particles, and others. (Only a few particles and theories will be considered here.)

Scientists believe that most elementary particles do not exist outside the nucleus but rather are created when the nucleus is disrupted. Some elementary particles are thought to be the *exchange particles* responsible for the four fundamental interactions: electromagnetic, strong, weak, and gravitational. For example, in one theory the "exchange" particle of the strong nuclear interaction is the pi meson, or pion. Nuclear particles with a strong interaction are viewed as interacting by exchanging pions back and forth (analogous to the way you interact with someone by tossing a beach ball back and forth). All strongly interacting particles, which include the nuclear protons and neutrons, are called *hadrons*.

In 1964, Murray Gell-Mann at the California Institute of Technology and George Zweig in Switzerland independently suggested that all hadrons, the elementary particles with the strong interaction, were made up of no more than three subatomic particles with fractional

Figure 10.20 Penetration of Radiation
Alpha particles cannot penetrate clothing or skin. Beta particles can slightly penetrate body tissue. However, gamma rays, X-rays, and neutrons can penetrate an arm easily.

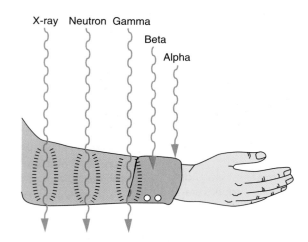

electronic charges, called *quarks.** For instance, a nuclear proton is thought to consist of three quarks (two with a $+\frac{2}{3}$ and one with $-\frac{1}{3}$ electronic charge).

Quarks apparently do not exist as free particles but are permanently bound inside hadrons, so we never see these fractional electronic charges in nature. However, there is indirect experimental evidence for their existence.

Other exchange particles, W and Z bosons, were discovered for the nuclear weak force that is important in governing the stability of basic matter particles. This discovery generated an *electroweak theory* that unified the electromagnetic and weak interactions. The electroweak theory and the strong force hadron theory then combined into a *grand unified theory* (GUT) or *standard model*. Thus, three forces (electromagnetic, strong, and weak) have been combined into a single theory. The gravitational force is not included in this theory. Scientists would like to believe that all forces could be combined and manifest a single *superforce*.

Elementary particles are currently studied in the Large Hadron Collider (LHC), a huge particle accelerator built on the French–Swiss border (● Fig. 10.21). It became operational in 2010, after almost 20 years of building its 10-km-diameter, 27-km-circumference circular tunnels. Scientists build particle accelerators to collide particles at high energies in hope of releasing elementary particles from the nucleus, and the LHC is the biggest particle accelerator built so far. Two counterorbiting proton beams make 11,000 round trips per second at speeds close to the speed of light, and more than 1200 superconducting magnets provide the centripetal magnetic force.

An inconsistency exists in the standard model due to a large mass difference between the exchange particles. This inconsistency led to the proposal of the *Higgs boson* particle by British physicist Peter Higgs in the 1960s. If the Higgs boson exists, then it would explain why particles have mass and would resolve the wide differences in the mass of the standard model's exchange particles. If the energy of the LHC is enough to produce a free Higgs boson particle, then it could provide the fundamental knowledge about the basic makeup of matter. The Higgs boson is often popularized as the "God particle."

In Chapter 10.3, it was mentioned that the positron ($_{+1}^{0}e$) was the *antiparticle* of the electron. *Antimatter* is believed to be composed of antiparticles in the same way that normal matter is composed of particles. For example, a positron and an antiproton can form an antihydrogen atom in the same way that an electron and a proton form a normal hydrogen atom. The mixing of particles and their antiparticles or matter–antimatter causes an annihilation of the particles with a release of energy.

Antiparticles have been created and quickly annihilated in particle accelerators, but they tend to disappear so fast that scientists do not have time to study them. Antihydrogen atoms

---

*The name is from a line in James Joyce's novel *Finnegans Wake,* "Three quarks for Muster Mark!" The "three quarks" denote the children of Mr. Finn, who sometimes appear as Mister (Muster) Mark. It is a bit strange, but there are also strange quarks.

Figure 10.21 Large Hadron Collider (LHC)
(a) A safety inspector riding a bicycle along the LHC tunnel at CERN (the European particle physics laboratory) near Geneva, Switzerland. The LHC is a 27-kilometer (17 mile) -long-underground ring of superconducting magnets housed in a pipe-like structure. (b) Aerial view of countryside outside Geneva, Switzerland. Beneath the fields lies a 27-kilometer-circumference tunnel that houses the large particle accelerator used to study elementary particles.

have been created in the LHC that may be able to stay around long enough for possible study. Such study would be an exciting new field for energy with antimatter–matter propulsion.

Conceptual Question and Answer

Star Trek Adventure

Q. When antimatter comes in contact with regular matter, annihilation occurs with the release of energy. In the science fiction series *Star Trek*, antimatter–matter energy was used as propulsion for the starship *Enterprise*, with the antimatter being stored on board in antimatter pods. Since antimatter reacts with matter, what could be used to contain the antimatter?

A. Obviously the container couldn't be a regular matter container or tank. What is a container without ordinary walls? Plasma is contained by magnetic fields (magnetic confinement) (Chapter 10.6). Charged antiparticles might be contained in a magnetic field "container."

The study of elementary particles may yield unknown information about forces and matter. In addition to exploring the Higgs boson theory, others theories predict that the LHC has the potential to provide information on other forms of matter, such as dark matter (Chapter 18.7), black holes (Chapter 18.8), and magnetic monopoles (Chapter 8.4). Stay tuned for some interesting results.

Did You Learn?

- Particles are called elementary when it is not certain if they are fundamental.
- Exchange particles are believed to be responsible for the interaction of fundamental forces. Some theoretical exchange particles are the pi meson (pion) for the strong nuclear force, and W and Z bosons for the nuclear weak force.

KEY TERMS

1. electrons (10.2)
2. nucleus
3. protons
4. neutrons
5. nucleons
6. atomic number
7. element
8. neutron number
9. mass number
10. isotopes
11. atomic mass
12. strong nuclear force
13. radioactive isotope (10.3)
14. radioactivity
15. alpha decay
16. beta decay
17. gamma decay
18. half-life
19. carbon-14 dating
20. fission (10.5)
21. chain reaction
22. critical mass
23. fusion (10.6)
24. plasma
25. mass defect
26. elementary particle

MATCHING

For each of the following items, fill in the number of the appropriate Key Term from the preceding list. Compare your answers with those at the back of the book.

a. _____ Fundamental force that holds the nucleus together

b. _____ Process in which smaller nuclei combine to form larger ones

c. _____ Neutral particles in atoms

d. _____ Spontaneous process of nuclei changing by emitting particles or rays

e. _____ Negatively charged particles in atoms

f. _____ Radioactive decay in which electrons are emitted

g. _____ Specific types of nuclei that are unstable

h. _____ A, the number of protons plus neutrons in a nucleus

i. _____ The central core of an atom

j. _____ Process in which a large nucleus splits and emits neutrons

k. _____ Used to date organic objects

l. _____ Positively charged particles in atoms

m. _____ A hot gas of electrons and ions

n. _____ Z, the number of protons in an atom

o. _____ The minimum amount of fissionable material needed to sustain a chain reaction

p. _____ N, the number of neutrons in a nucleus

q. _____ The weighted average mass of atoms of an element in a naturally occurring sample

r. _____ Disintegration of a nucleus, with the emission of a helium nucleus

s. _____ A substance in which all the atoms have the same number of protons

t. _____ The time it takes for the decay of half the atoms in a sample

u. _____ Process in which one initial reaction triggers a growing number of subsequent reactions

v. _____ Collective term for nuclear protons and neutrons

w. _____ Forms of atoms having the same number of protons but differing in number of neutrons

x. _____ The decrease in mass during a nuclear reaction

y. _____ Disintegration of a nucleus, with the emission of a high-energy photon

z. _____ Do not exist outside of the nucleus

MULTIPLE CHOICE

Compare your answers with those at the back of the book.

1. Which scientist devised the symbol notation we now use for elements? (10.1)
   (a) Newton
   (b) Berzelius
   (c) Dalton
   (d) Einstein

2. What is the symbol notation for the element potassium? (10.1)
   (a) P   (b) Po   (c) Pt   (d) K

3. How many neutrons are in the nucleus of the atom $^{35}_{17}$Cl? (10.2)
   (a) 35   (b) 17   (c) 18   (d) 52

4. Is a nucleon (a) a proton, (b) a neutron, (c) an electron, or (d) both a proton and a neutron? (10.2)

5. Which radioactive decay mode does not result in a different nuclide? (10.3)
   (a) alpha   (b) beta
   (c) gamma   (d) all the preceding

6. What is the missing particle in the nuclear decay
   $^{179}_{79}$Au → $^{175}_{77}$Ir + __ ? (10.3)
   (a) deuteron
   (b) neutron
   (c) beta particle
   (d) alpha particle

7. The majority of stable isotopes belong to which category? (10.3)
   (a) odd–odd
   (b) even–even
   (c) even–odd
   (d) odd–even

8. Which of the following scientists discovered radioactivity? (10.3)
   (a) Rutherford
   (b) Heisenberg
   (c) Becquerel
   (d) Pierre Curie

9. How many half-lives would it take for a sample of a radio-active isotope to decrease its activity to $\frac{1}{32}$ of the original amount? (10.3)
   (a) 5   (b) 16   (c) 6   (d) 32

10. Which of the following is not conserved in all nuclear reactions?
    (a) nucleons
    (b) mass number,
    (c) atomic number
    (d) neutron number

11. Which of the following completes the reaction
    $_{1}^{2}\text{H} + _{42}^{98}\text{Mo} \rightarrow$ ___ $+ _{0}^{1}\text{n}$? (10.4)
    (a) $_{42}^{97}\text{Mo}$ (b) $_{44}^{100}\text{Ru}$
    (c) $_{43}^{99}\text{Tc}$ (d) $_{41}^{93}\text{Nb}$

12. What is the appropriate procedure to decrease the heat output of a fission reactor core during a crisis? (10.5)
    (a) Insert the control rods farther.
    (b) Remove fuel rods.
    (c) Increase the level of coolant.
    (d) Decrease the amount of moderator.

13. What is a very hot gas of nuclei and electrons called? (10.6)
    (a) tokomak
    (b) laser
    (c) plasma
    (d) ideal gas

14. Which unit is most closely associated with the biological effects of radiation? (10.7)
    (a) the curie
    (b) the rem
    (c) the becquerel
    (d) the cpm

15. What is the theoretical exchange particle for the nuclear weak force? (10.8)
    (a) Z particle
    (b) pion
    (c) graviton
    (d) gluon

FILL IN THE BLANK

Compare your answers with those at the back of the book.

1. K is the symbol for the element ___. (10.1)

2. All atomic masses are based on an atom of ___. (10.2)

3. The collective name for neutrons and protons in a nucleus is ___. (10.2)

4. Carbon-12, carbon-13, and carbon-14 are ___. (10.2)

5. No stable nuclides exist that have Z greater than ___. (10.3)

6. The nuclear notation $_{-1}^{0}\text{e}$ refers to a(n) ___. (10.3)

7. The amount of a radioactive isotope will have dropped to 12.5% of what it was originally after ___ half-lives have elapsed. (10.3)

8. The proton number of the daughter nucleus in beta decay is ___ than that of the parent nucleus. (10.3)

9. In nuclear reactions, both atomic number and ___ are conserved. (10.4)

10. For an atomic bomb to explode, a(n) ___ mass is necessary. (10.5)

11. In discussions of nuclear fusion reactions, the letter D stands for ___. (10.6)

12. Of the average amount of radiation received by a person in the United States, ___% comes from natural sources. (10.7)

13. All hadrons are believed to be made up of ___.

SHORT ANSWER

10.1 Symbols of the Elements

1. What are the chemical symbols for (a) carbon, (b) chlorine, and (c) lead?

2. What are the names of the elements with the symbols (a) N, (b) He, and (c) Fe?

3. Why is the symbol for the transuranic element Rutherfordium Rf instead of Ru?

10.2 The Atomic Nucleus

4. Why is the neutron number N equal to the mass number A minus the atomic number Z?

5. What is the collective name of the two particles that make up a nucleus?

6. What evidence is there to support the idea that the strong nuclear force is stronger than the electric force?

7. The diameter of the uranium atom is about the same as that of a hydrogen atom. What does that imply about the structure of the atom?

8. About what percentage of the mass of an atom is contained in the nucleus?

9. What do the letters Z, A, and N in nuclear notation stand for?

10. State the special names by which the isotopes ^{1}H, ^{2}H, and ^{3}H are known.

11. What are the nuclear notations for the isotopes of hydrogen, using H, D, and T? What names are given to these nuclei?

12. On which atom are all atomic masses based? What mass is assigned to that atom?

13. Name the force that holds a nucleus together. At what distance does this force drop to zero?

10.3 Radioactivity and Half-Life

14. What convincing evidence is there that radioactivity is a nuclear effect and not an atomic effect?

15. How does the charge of a nucleus change with beta decay?

16. How does the mass of a nucleus change with alpha decay?

17. Indicate which of the three types of radiation—alpha, beta, or gamma—each of the following phrases describes.
    (a) is not deflected by a magnet
    (b) has a negative charge
    (c) consists of ions
    (d) is similar to X-rays
    (e) has a positive charge

18. After three half-lives have gone by, what fraction of a sample of a radioactive isotope remains?

19. Why can't carbon-14 dating be used for ages 50,000 years and older?

10.4 Nuclear Reactions

20. Use the letters a, A, B, and b to write the general form of a nuclear reaction.

21. How does the common household smoke alarm detect smoke?

22. What is the principle of neutron activation analysis, and how is it used?

10.5 Nuclear Fission

23. What subatomic particle is emitted when a nucleus fissions?

24. In terms of a chain reaction, explain what is meant by critical mass, subcritical mass, and supercritical mass.

25. What percentage of natural uranium is fissionable ^{235}U? To what percentage must it be enriched for use in a U.S. nuclear reactor? For use in nuclear weapons?

26. What is a "meltdown?" How is this related to the "China syndrome"?

27. Both control rods and moderators are involved with neutrons in a nuclear reactor. What is the role of each?

28. Why can't nuclear reactors accidentally cause a nuclear explosion like an atomic bomb?

29. Fission was once referred to as "splitting the atom." Is this terminology correct?

30. During operation, breeder reactors make fissionable fuel from nonfissionable material. What is the fuel made, and what is the nonfissionable material used?

10.6 Nuclear Fusion

31. The fusion of what element allows the Sun to emit such enormous amounts of energy?

32. In discussions of nuclear fusion, what do the letters D and T stand for?

33. What is a plasma? Distinguish between magnetic and inertial confinements. Why is confinement such a problem for fusion?

34. Briefly list three advantages and two disadvantages of energy production by fusion compared with by fission.

35. The equation $E = mc^2$ was developed by which scientist? Identify what each of the letters in the equation stands for.

36. Why is an atomic bomb needed to start a hydrogen bomb?

37. What term is applied to a nuclear reaction if the products have less mass than the reactants? What term is applied to the difference in mass in such a case?

38. Show by means of sketches the nuclear processes by which energy can be released by both fission and fusion.

10.7 Effects of Radiation

39. What is ionizing radiation, and what problems can it cause?

40. How do nuclear radiations vary in terms of shielding?

10.8 Elementary Particles

41. What are exchange particles, and how do they relate to the fundamental forces?

42. How many quarks are believed to make up a proton?

VISUAL CONNECTION

Visualize the connections and give answers for the blanks. Compare your answers with those given at the back of the book.

Constituents of the Atom

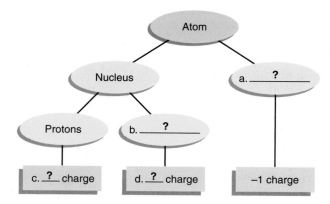

APPLYING YOUR KNOWLEDGE

1. Suppose a magnetic field instead of an electric field were used in Fig. 10.5 to distinguish nuclear radiations. What would be observed?

2. The technique of carbon-14 dating relies on the assumption that the cosmic-ray intensity has been constant for at least the last 50,000 years. Suppose it was discovered that the cosmic-ray intensity was much greater 10,000 years ago. How would this discovery affect the ages of samples that have already been dated?

3. Would you rather live downwind of a nuclear power plant or downwind of a coal-burning power plant?

4. What steps might you take to lower your exposure to ionizing radiation?

5. If the Higgs boson were confirmed, what effect would this have on the theory of matter?

IMPORTANT EQUATIONS

Neutron Number = Mass Number − Atomic Number:

$$N = A - Z \qquad (10.1)$$

Mass–Energy: $E = mc^2$ $\qquad$ (10.2)

EXERCISES

10.2 The Atomic Nucleus

1. Fill in the nine gaps in this table.

Element symbol	B		
Protons		9	
Neutrons		10	
Electrons			18
Mass number	11		40

Answer: First col.: 5, 6, 5; second col.: F, 9, 19; third col.: Ar, 18, 22

2. Fill in the nine gaps in this table.

Element symbol	Ga		
Protons		15	
Neutrons	39		20
Electrons			17
Mass number		31	

3. Oxygen ($_8$O) has three stable isotopes with consecutive mass numbers. Write the complete nuclear symbol (with neutron number) for each.

Answer: $^{16}_{8}O_8$, $^{17}_{8}O_9$, $^{18}_{8}O_{10}$

4. Show that the neutron number for the following nuclides generally exceeds the atomic number for nuclei with atomic numbers greater than 20: lithium-6, bromine-80, silicon-28, titanium-48, fluorine-19, and platinum-179.

10.3 Radioactivity and Half-Life

5. Complete the following equations for nuclear decay, and state whether each process is alpha decay, beta decay, or gamma decay.

(a) $^{46}_{21}Sc^* \rightarrow {}^{46}_{21}Sc +$ ____

(b) $^{232}_{90}Th \rightarrow$ ____ $+ {}^{4}_{2}He$

(c) $^{47}_{21}Sc \rightarrow {}^{47}_{22}Ti +$ ____

Answer: (a) γ, gamma (b) $^{228}_{88}Ra$, alpha (c) $^{0}_{-1}e$, b, beta

6. Complete the following equations for nuclear decay, and state whether each process is alpha decay, beta decay, or gamma decay.

(a) $^{8}_{5}B \rightarrow {}^{8}_{4}B + {}^{0}_{-1}e$

(b) $^{210}_{84}Po \rightarrow {}^{206}_{82}Pb +$ ____

(c) $^{207}_{84}Po^* \rightarrow {}^{207}_{84}Po +$ ____

7. Write the equation for each of the following.

(a) alpha decay of $^{226}_{88}Ra$

(b) beta decay of $^{60}_{27}Co$

Answer: (a) $^{226}_{88}Ra \rightarrow {}^{222}_{86}Rn + {}^{4}_{2}He$ (b) $^{60}_{27}Co \rightarrow {}^{60}_{28}Ni + {}^{0}_{-1}e$

8. Actinium-225 ($^{225}_{89}Ac$) undergoes alpha decay.

(a) Write the equation.

(b) The daughter formed in part (a) undergoes beta decay. Write the equation.

9. Radon-222 ($^{222}_{86}Rn$), a radioactive gas, undergoes alpha decay. (a) What is the daughter nucleus in this process? (b) The daughter nucleus undergoes both beta and alpha decay. What is the "granddaughter" nucleus in each case?

Answer: (a) $^{218}_{84}Po$, (b) $^{218}_{85}At$ and $^{214}_{82}Pb$

10. A radioactive decay series called the neptunium-237 series is shown in ● Fig. 10.22. (a) What is the decay mode for each of the sequential decays? Notice the double decay mode toward the end of the series. (b) Determine the daughter nucleus at the end of each decay. (You can find information on astatine and francium given in the Highlight: Number of Naturally Occurring Elements.)

11. Pick the radioactive isotope in each set. Explain your choice.

(a) $^{249}_{98}Cf$ $^{12}_{6}C$

(b) $^{79}_{35}Br$ $^{76}_{33}As$

(c) $^{15}_{8}O$ $^{17}_{8}O$

Answer: (a) $^{249}_{98}Cf$ (Z > 83), (b) $^{76}_{33}As$ (odd-odd), (c) $^{15}_{8}O$ (fewer n than p)

12. Pick the radioactive isotope in each set. Explain your choice.

(a) $^{17}_{9}F$ $^{32}_{16}S$

(b) $^{209}_{83}Bi$ $^{226}_{88}Ra$

(c) $^{23}_{11}Na$ $^{20}_{9}F$

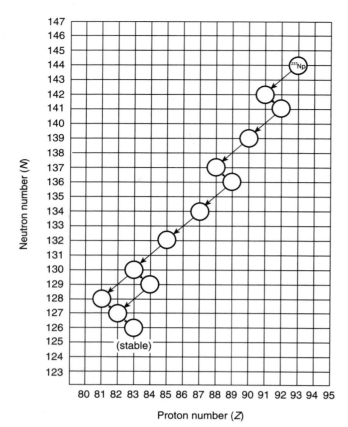

Figure 10.22 The neptunium-237 series. See Exercise 10.

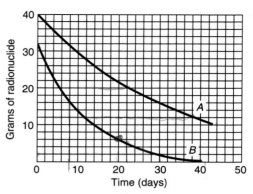

Figure 10.23 See Exercise 15.

13. Technetium-99 (half-life = 6.0 h) is used in medical imaging. How many half-lives would go by in 36 h?

Answer: six half-lives

14. How many half-lives would have to elapse for a sample of a radioactive isotope to decrease from an activity of 160 cpm to an activity of 5 cpm?

15. A thyroid cancer patient is given a dosage of ^{131}I (half-life = 8.1 d). What fraction of the dosage of ^{131}I will still be in the patient's thyroid after 24.3 days?

Answer: $\frac{1}{8}$

16. A clinical technician finds that the activity of a sodium-24 sample is 480 cpm. What will be the activity of the sample 75 h later if the half-life of sodium-24 is 15 h?

17. Tritium (half-life = 12.3 y) is used to verify the age of expensive brandies. If an old brandy contains only $\frac{1}{16}$ of the tritium present in new brandy, then how long ago was it produced?

Answer: 49 y

18. What is the half-life of thallium-206 if the activity of a sample drops from 2000 cpm to 250 cpm in 21.0 min?

19. Use the graph in ● Fig. 10.23 to find the half-life of isotope A. All you need is a sound understanding of the definition of half-life.

Answer: For A, half-life = 22 d, because half of 40 g is 20 g, and 20 g is reached after 22 d.

20. Use the graph in Exercise 15 to find the half-life of isotope B.

21. Use the graph in Exercise 15 to find how long it would take the original sample of isotope A to decrease to 12 g.

Answer: 38 days

22. Use the graph in Exercise 15 to find how long it would take the original sample of isotope B to decrease to 2 g.

10.4 Nuclear Reactions

23. Complete the following nuclear reaction equations.
    (a) $^{4}_{2}He + ^{14}_{7}N \rightarrow ^{17}_{8}O +$ ____
    (b) $^{4}_{2}He + ^{27}_{13}Al \rightarrow ^{30}_{15}P +$ ____
    (c) ____ $+ ^{66}_{29}Cu \rightarrow ^{67}_{30}Zn + ^{1}_{0}n$
    (d) $^{1}_{0}n + ^{235}_{92}U \rightarrow ^{138}_{54}Xe +$ ____ $+ 5^{1}_{0}n$

Answer: (a) $^{1}_{1}H$ (b) $^{1}_{0}n$ (c) $^{2}_{1}H$ (d) $^{93}_{38}H$

24. Complete the following nuclear reaction equations.
    (a) $^{16}_{8}O + ^{20}_{10}Ne \rightarrow$ ____ $+ ^{12}_{6}C$
    (b) $^{1}_{0}n + ^{28}_{14}Si \rightarrow$ ____ $+ ^{1}_{1}H$
    (c) ____ $+ ^{230}_{90}Th \rightarrow ^{223}_{87}Fr + 2^{4}_{2}He$
    (d) $^{246}_{96}Cm + ^{12}_{6}C \rightarrow$ ____ $+ 4^{1}_{0}n$

10.5 Nuclear Fission

25. Complete the following equation for fission.

$$^{240}_{94}Pu \rightarrow ^{97}_{38}Sr + ^{140}_{56}Ba +$$ ____

Answer: $3^{1}_{0}n$

26. Complete the following equation for fission.

$$^{252}_{98}Cf \rightarrow$$ ____ $+ ^{142}_{55}Cs + 4^{1}_{0}n$

10.6 Nuclear Fusion

27. One of the fusion reactions that takes place as a star ages is called the triple alpha process.

$$3^{4}_{2}He \rightarrow ^{12}_{6}C$$

Calculate the mass defect (in u) and the energy produced (in MeV) each time the reaction takes place. (Atomic mass of ^{4}He = 4.00260 u; atomic mass of ^{12}C = 12.00000 u.)

Answer: 0.00780 u, 7.26 MeV

28. Calculate the mass defect (in u) and the energy produced (in MeV) in the D-D reaction shown.

$$\begin{array}{cccc} ^{2}_{1}H & ^{2}_{1}H & ^{3}_{1}H & ^{1}_{1}H \\ (2.0140\ u) & + (2.0140\ u) & \rightarrow (3.0161\ u) & + (1.0078\ u) \end{array}$$

ON THE WEB

1. It's All Quite Elemental, Dear Hydrogen Atom

Follow the recommended links on the student website at **www.cengagebrain.com/shop/ISBN/1133104096** to "interact" with the periodic table. What basic information about hydrogen are you given there? Notice that hydrogen is a nonmetal. Describe the characteristics of nonmetals and compare them with metals. What other elements are considered to be nonmetallic?

2. Beam Me Up, EBIT

Commercial use of controlled fusion is presently elusive. Have you heard of the Electron Beam Ion Trap (EBIT)? How close are we really to being able to control production of those highly charged ions necessary for commercial use? What exactly is the EBIT, and what are its practical applications? If you were to become a physicist interested in nuclear fusion, then how might you be able to use this device? What future applications can you see in the field of nuclear fusion, and how could you develop them? Explore answers to these questions by following the recommended links on the student website at **www.cengagebrain.com/shop/ISBN/1133104096**.

I have always wanted to know as much as possible about the world.

•

Linus Pauling, Nobel Laureate in chemistry (1901–1994)

The beautiful element gold coats > *these cathedral domes in Russia.*

© Morton Beebe, S.F./CORBIS

CHEMISTRY FACTS

▶ In 1855, pure aluminum sold for about $100,000 a pound, and samples were exhibited with the crown jewels of France. One year after graduating from Oberlin College in Ohio in 1885, Charles Martin Hall succeeded in finding an inexpensive way to manufacture aluminum. By 1890 aluminum sold for $2 a pound.

▶ In 1894, British scientists Lord Rayleigh and Sir William Ramsey discovered a new element that they named argon ("the lazy or idle one"). Ramsey went on to find other gaseous elements—neon ("new"), krypton ("hidden"), and xenon ("stranger")—giving rise to a new column of elements in the periodic table.

▶ There is no elemental lead (Pb) in pencils; pencil "lead" is

The division of physical science called **chemistry** is the study of the composition and structure of matter (anything that has mass) and the chemical reactions by which substances are changed into other substances. Chemistry had its beginnings early in history, when humans mastered fire more than 100,000 years ago. Egyptian hieroglyphs from 3400 BCE show wine making, which requires a chemical fermentation process. By about 2000 BCE, the Egyptians and Mesopotamians produced and worked metals, and the ancient Egyptians and Chinese prepared dyes, glass, pottery, and embalming fluids, all using chemical processes.

From about 500 to 1600, *alchemy* flourished. Its main objectives (never attained) were to change common metals into gold and to find an "elixir of life" to prevent

Chapter Outline

aging. Modern chemistry began in 1774, with the work of Antoine Lavoisier (see the Highlight: Lavoisier, "The Father of Chemistry" in Chapter 12.1) and differs from alchemy by using the scientific method (Chapter 1.2), which requires reasonable objectives and avoids mysticism, superstition, and secrecy.

Chemistry has five major divisions. *Physical chemistry*, the most fundamental, applies the theories of physics (especially thermodynamics) to the study of chemical systems in general. *Analytical chemistry* identifies what substances are present in a material and determines how much of each substance is present. *Organic chemistry* is the study of compounds that contain carbon and hydrogen. The study of all other chemical compounds is called *inorganic chemistry*. *Biochemistry*, where chemistry and biology meet, deals with the chemical reactions that occur in living organisms. These major divisions overlap, and smaller divisions exist, such as polymer chemistry and nuclear chemistry.

The 88 naturally occurring elements in our environment, either singly or in chemical combination, are the components of virtually all matter. We are affected by the physical and chemical properties of the various elements; our bodies and everything we normally encounter are made of chemical elements. In this chapter we will examine how matter is classified by chemists, discuss elements, develop an understanding of the periodic table, and learn how compounds (chemical combinations of elements) are named.

11.1   Classification of Matter

Preview Questions

- What is the difference between a compound and a mixture?
- How does the solubility of a substance vary with the temperature of solutions?

In Chapter 5 matter was classified by its physical phase: solid, liquid, gas, or plasma. However, the classification scheme summarized in ● Fig. 11.1 is also useful in chemistry. It first divides matter into pure substances and mixtures.

A *pure substance* is a type of matter in which all samples have fixed composition and identical properties. Pure substances are divided into elements and compounds. An **element** is a pure substance in which all the atoms have the same number of protons; that is, they have the same atomic number (Chapter 10.2).

A **compound** is a pure substance composed of two or more elements chemically bonded in a definite, fixed ratio by mass. All samples of a given compound have identical properties that are usually different from the properties of the elements of which the compound is

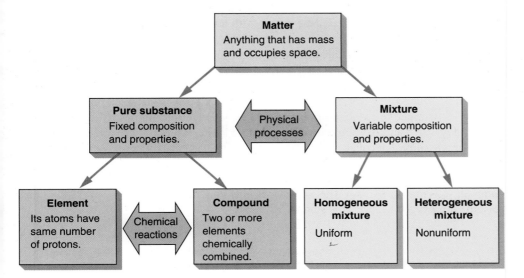

Figure 11.1 A Chemical Classification of Matter (See text for description.)

Table 11.1 The Properties of a Compound Compared with Those of Its Component Elements

Property	Zinc Sulfide	Zinc	Sulfur
Appearance	White powder	Silvery metal	Yellow powder or crystals
Density (g/cm³)	3.98	7.14	2.07
Melting point (°C)	1700	281	113
Conducts electricity as a solid	No	Yes	No
Conducts electricity as a liquid	Yes	Yes	No
Soluble in carbon disulfide	No	No	Yes

composed. ● Table 11.1 compares the properties of the compound zinc sulfide (ZnS) with those of zinc (Zn) and sulfur (S), the two elements into which it decomposes in a fixed ratio by mass of 2.04 parts zinc to 1 part sulfur.

A compound can be decomposed into its component elements only by chemical processes, such as the passage of electricity through melted zinc sulfide. The formation of a compound from elements is also a chemical process.

Conceptual Question and Answer

A Compound Question

Q. Is a compound uniquely different from the elements from which it is made?

A. Yes. For example, water (H_2O) is a liquid, whereas the elements that compose it, hydrogen (H_2) and oxygen (O_2), are gases. Similarly, common table salt (NaCl) is composed of two dangerous elements, metallic sodium (Na) and chlorine gas (Cl_2).

A **mixture** is a type of matter composed of varying proportions of two or more substances that are just physically mixed, *not* chemically bonded. Different samples of a particular mixture can have variable composition and properties. For example, a mixture of zinc and sulfur could consist of any mass ratio of zinc to sulfur; it would not be restricted to the 2.04-to-1 mass ratio found in the compound zinc sulfide. Mixtures are formed and separated by physical processes such as dissolving and evaporation. For example, the mixture of iron and sulfur could be separated simply by using a magnet.

A *heterogeneous mixture* is one in which at least two components can be observed. This means it is nonuniform. A pizza, zinc mixed with sulfur, a salad, and salad dressing are examples (● Fig. 11.2).

A mixture that is uniform throughout is called a *homogeneous mixture* or a **solution**, which looks like it might be just one substance.* Most solutions we encounter consist of one or more substances (such as coffee crystals or salt) dissolved in water. A solution need not be a liquid, however; it may also be a solid or a gas. A metal *alloy* such as brass, a mixture of copper and zinc, is an example of a solid solution. Air, a mixture composed mainly of nitrogen and oxygen, is an example of a gaseous solution. Each appears uniform, but different samples of coffee, saltwater, brass, and air can have different compositions. In a solution containing two or more substances, the substance present in the larger amount is called the *solvent* and the other substance is called the *solute*. For example, water is a common solvent in which salt (the solute) is readily dissolved. In air, nitrogen (78%) is the solvent, and oxygen (21%) is the solute.

Figure 11.2 Heterogeneous Mixtures Both the salad and the Italian dressing are heterogeneous mixtures.

_____

*Technically, in a true solution, the components must be mixed on the atomic or molecular level, so even a magnifying glass or microscope would not reveal that the sample was several substances mixed.

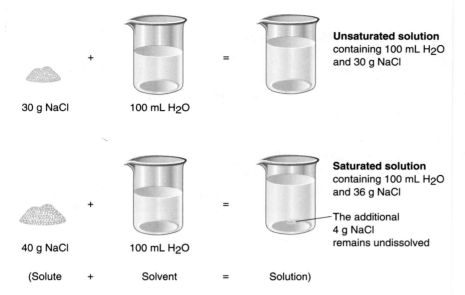

Figure 11.3 Comparison of
Unsaturated and Saturated Solutions
(Top) The 30 g of NaCl will dissolve
completely in 100 mL of water at 20°C,
giving an unsaturated solution. (Bottom) When 40 g of NaCl is stirred into
100 mL of H_2O, only 36.0 g dissolves
at 20°C, leaving 4 g of the crystalline
solid on the bottom of the beaker.
This solution is saturated.

Aqueous Solutions

A solution in which water is the solvent is called an *aqueous solution* (abbreviated *aq*).
When a solute dissolves in water and is mixed thoroughly, the distribution of its particles
(molecules or ions) is the same throughout the solution. If more solute can be dissolved
in the solution at the same temperature, then the solution is called an *unsaturated solution*
(● Fig. 11.3). As more solute is added, the solution becomes more and more concentrated.
Finally, when the maximum amount of solute is dissolved in the solvent, it is a *saturated
solution* (Fig 11.3). Usually, some undissolved solute remains on the bottom of the container, and a dynamic (active) equilibrium is set up between solute dissolving and solute
crystallizing so the solution remains saturated (● Fig. 11.4).

The **solubility** of a particular solute is the amount of solute that will dissolve in a specified volume or mass of solvent (at a given temperature) to produce a saturated solution.
Solubility depends on the temperature of the solution. The solubility of practically all solids is directly proportional to the temperature (● Fig. 11.5). Hot water dissolves more
solute than the same amount of cold water.

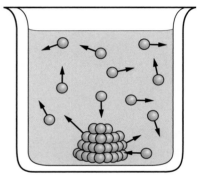

Figure 11.4 A Saturated Solution
In a saturated solution containing
excess solute, a dynamic equilibrium
exists between solute dissolving and
solute crystallizing.

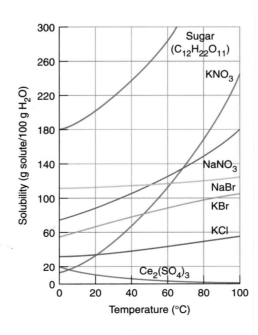

**Figure 11.5 The Effect of
Temperature on Solubilities
of Solids in Water**
As the graph indicates, the solubilities of solids generally increase as the
temperature of the solution rises.

(a)

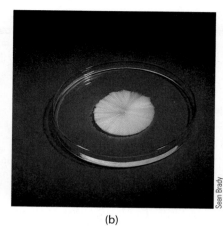

(b)

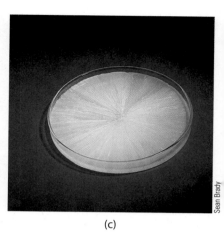

(c)

Figure 11.6 Supersaturated Solution
When a seed crystal is added to a supersaturated solution of sodium acetate, $NaC_2H_3O_2$ (a), the excess solute quickly crystallizes (b), forming a saturated solution (c).

When unsaturated solutions are prepared at high temperatures and then cooled, solubility decreases and may reach the saturation point, where excess solute normally begins crystallizing from the solution. However, if no crystals of the solid are already present in the solution (and no scratches are on the inside surface of the container), then crystallization may not take place if the solution is cooled carefully. This kind of solution, one that contains more than the normal maximum amount of dissolved solute at its temperature, is a *supersaturated solution*. Such solutions are unstable, and the introduction of a "seed" crystal will cause the excess solute to crystallize immediately (● Fig. 11.6).

This is the principle behind seeding clouds to cause rain. If the air is supersaturated with water vapor, then the introduction of certain types of crystals into the clouds greatly increases the probability that the water vapor will form raindrops (see Chapter 20.1).

Consider the solubility of gases in water: The solubility of gases is directly proportional to pressure. This principle is used in the preparation of soft drinks when carbon dioxide (CO_2) is forced into the beverage under high pressure. The beverage is then bottled and capped tightly to maintain pressure on it. Once the bottle is opened, the pressure inside the bottle is reduced to normal atmospheric pressure and the CO_2 starts escaping from the liquid, as evidenced by rising bubbles. After the bottle has been open for some time, most of the CO_2 escapes and the drink tastes flat.

The solubility of gases in water is inversely proportional to temperature. If an unopened soft drink is allowed to warm, the solubility of CO_2 decreases. When the bottle is opened, CO_2 may escape so fast that the beverage shoots out of the bottle (particularly if shaken before).

Did You Learn?

- The components of a compound are chemically bonded. Those of a mixture are not; they are just physically mixed.

- The solubility of practically all solids increases with temperature. (Hot water dissolves more solute than cold water.)

11.2   Discovery of the Elements

Preview Questions

- How many naturally occurring elements are there?
- How many known elements are there?

It was noted in Chapter 10.1 that Aristotle's erroneous idea of four elements prevailed for almost 2000 years. Then, in 1661, Robert Boyle (● Fig. 11.7), an Irish-born chemist, developed a modern definition of *element* by making it subject to laboratory testing. (In this same era, Isaac Newton was developing the laws of physics.) In his book *The Skeptical Chemist*, Boyle proposed that the designation *element* be applied only to those substances that could not be separated into components by any method.

Boyle championed the need for experimentation in science. It was he who initiated the practice of carefully and completely describing experiments so that anyone might repeat and confirm them. This procedure became universal in science and caused great improvements in scientific progress. In addition, Boyle first performed truly quantitative physical experiments, finding the inverse relationship between the pressure and volume of a gas (Chapter 5.6).

In the earliest civilizations, twelve substances that later proved to be elements were isolated: gold, silver, lead, copper, tin, iron, carbon, sulfur, antimony, arsenic, bismuth, and mercury. Phosphorus is the first element whose date of discovery is known. It was isolated in 1669 from urine by Hennig Brandt, a German chemist, eight years after Boyle's definition of the term *element*. By 1746, platinum, cobalt, and zinc had been discovered. The rest of the 1700s saw the discovery of ten more metals, the nonmetal tellurium, and the gaseous elements hydrogen, oxygen, nitrogen, and chlorine. ● Figure 11.8 shows how many elements were known at various times in history.

About 1808, Humphry Davy, an English chemist, used electricity from a huge battery (a recent invention at the time) to break down compounds, thereby discovering six elements—sodium, potassium, magnesium, calcium, barium, and strontium—a record for one person. Davy's work is a good example of how advances in technology can result in advances in basic science.

By 1895, a total of 73 elements were known. From 1895 to 1898, the noble gases helium, neon, krypton, and xenon ("ZEE-non") were discovered. Marie and Pierre Curie discovered polonium and radium in 1898. (See the Highlight: The Discovery of Radioactivity in Chapter 10.3.)

In total, there are 88 naturally occurring elements. Synthetic elements are unstable and are created by nuclear bombardment using particle accelerators. The first synthetic element, technetium (Tc), was produced in 1937 by nuclear bombardment of the element molybdenum (Mo) with deuterons, ^2_1H (see Chapter 10.4):

$$^{96}_{42}Mo + {}^2_1H \rightarrow {}^{98}_{43}Tc$$

Figure 11.7 Robert Boyle (1627–1691)
Boyle wrote *The Skeptical Chemist* in 1661. In 1662, he discovered what is known as Boyle's law of gases: The volume and pressure of a gas are inversely proportional (Chapter 5.6).

© Bettmann/CORBIS

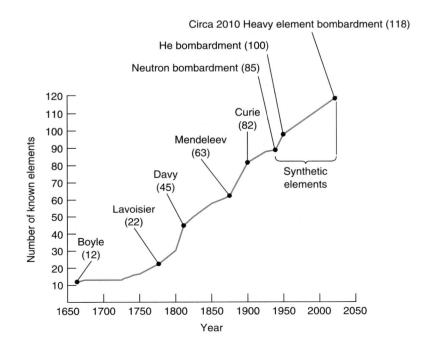

Figure 11.8 The Discovery of the Elements
The graph shows how many elements were known at various points in history. Note the number of elements known at the times when several well-known scientists made their greatest contributions.

Approximately 30 more synthetic elements have followed, and the present total is 118 known elements.

In Chapter 10.1 it was stated that Jöns Jakob Berzelius was responsible for our present method of designating elements by using one or two letters of their names. (See the **Highlight: Berzelius and How New Elements Are Named**.) Table 10.2 lists the names and symbols of common elements.

Did You Learn?

● There are 88 naturally occurring elements.

● There are a total of 118 known elements. (Synthetic elements are created by nuclear reactions in labs.)

11.3 Occurrence of the Elements

Preview Questions

● What is the most common element in the Earth's crust?

● What are the two allotropes of oxygen?

● Figure 11.9 shows that about 74% of the mass of the Earth's crust is composed of only two elements, oxygen (47%) and silicon (27%). The Earth's core is thought to be about 85% iron and 15% nickel, and its atmosphere close to the surface consists of 78% nitrogen, 21% oxygen, and almost 1% argon, together with traces of other elements and compounds (see Fig. 19.1 and Chapters 21.1 and 22.1). The human body consists primarily of oxygen (65%) and carbon (18%).

Analysis of electromagnetic radiation from stars, galaxies, and nebulae (interstellar clouds of gas and dust) indicates that hydrogen, the simplest element of all, accounts for about 75% of the mass of elements in the universe, with about 24% being helium, the next simplest element. All the other elements account for only 1%. Chapters 10.4 and 18.4 briefly discuss how elements are formed in stars, a process called *nucleosynthesis*.

Figure 11.9 Relative Abundance (by Mass) of Elements in the Earth's Crust
Two elements, oxygen and silicon, account for about 74% of the mass of elements in the Earth's crust.

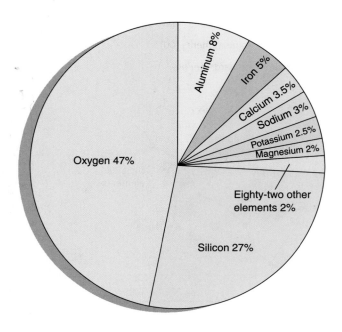

Highlight   Berzelius and How New Elements Are Named

Swedish chemist Jöns Jakob Berzelius (1779–1848) was probably the best experimental chemist of his generation, and given the crudeness of his laboratory equipment, he may have been the best experimental chemist of all time (Fig. 1). Unlike his contemporary French nobleman Antoine Lavoisier (Chapter 12.1), who could afford to buy the best laboratory equipment, Berzelius worked with minimal equipment in very plain surroundings: two rooms without furnaces, hoods, or running water. He had a closet with chemicals and a furnace next door in the kitchen.

In these simple facilities Berzelius performed more than 2000 experiments over a 10-year period to determine accurate atomic masses for the 50 elements then known. His success can be seen from the data in the table below. These remarkably accurate values attest to his experimental skills and patience.

In addition to his experimental measurements, Berzelius discovered the elements cerium, thorium, selenium, and silicon. Of these elements, selenium and silicon are particularly important in today's world because they are connected to human health and computers, respectively. Berzelius discovered selenium in 1817 in connection with his studies of sulfuric acid. Berzelius prepared silicon in its pure form in 1824 by heating silicon tetrafluoride (SiF_4) with potassium metal. He is also responsible for the chemical notation that is still used today (O for oxygen, Si for silicon, and so on).

Once an element is discovered, how do scientists name the new element? Before and during the time of Berzelius, new elements were named by their discoverer, usually from a Greek root relating the element to something commonly known. Berzelius named the new element selenium (from the Greek goddess Selene for "Moon") because it was associated with the element tellurium (from the Latin word *tellus* for "Earth"). In 1777, Lavoisier named the newly discovered element oxygen from the Greek word "oxys," which meant "acid," because Lavoisier thought that oxygen was common to all acids.

Modern science is not as simple. During the time lapse between a new element's confirmed discovery and its official naming, the element is symbolized by the initial letters for the Greek prefixes that designate its atomic number. For example, element 117, discovered in Russia in 2010, is temporarily named ununseptium, symbolized Uus (from the initial letters of *una* for one, *una* for one, and *septa* for seven). (See the periodic table inside the front cover.)

Using temporary names has been standard since 1990 because the naming process can be contentious and take a very long time. Chemists that make up the International Union of Pure and Applied Chemists (IUPAC) come from 80 different countries, and each country argues for honoring its scientists with a new elemental name. According to the IUPAC guidelines, "Elements can be named after a mythological concept, a mineral, a place or country, a property, or a scientist." And, for consistency, all new names must end in -*ium*. For example, element 112 was discovered in 1996, and it took 13 years for scientists to agree on the final name: copernicium, after Polish astronomer Nicolaus Copernicus, who first proposed the Sun-centered model of the solar system (see Chapter 16.1).

Although scientists don't always agree on the names, at least they agree on the pursuit of finding new elements. Naming elements was certainly much easier in Berzelius' day.

Source (in part): Steven S. Zumdahl, *Chemistry*, seventh ed. Copyright © 2007 by Houghton Mifflin Company. Adapted with permission.

Comparison of Several of Berzelius' Atomic Masses with the Modern Values

Element	Atomic Mass (atomic mass units)	
	Berzelius' Value	Current Value
Chlorine (Cl)	35.41	35.45
Copper (Cu)	63.00	63.55
Hydrogen (H)	1.00	1.01
Lead (Pb)	207.12	207.2
Nitrogen (N)	14.05	14.01
Oxygen (O)	16.00	16.00
Potassium (K)	39.19	39.10
Silver (Ag)	108.12	107.87
Sulfur (S)	32.18	32.07

© Bettmann/CORBIS

Figure 1   Jöns Jakob Berzelius
This portrait was painted by O. J. Soedermark in 1843, after Berzelius had been secretary of the Royal Swedish Academy of Sciences for 25 years. Courtesy of the Royal Swedish Academy of Sciences.

Figure 11.10 Atoms in Iron, Neon, and Hydrogen
(a) Iron and other metallic elements are represented by just the symbol such as Fe, because the individual atoms pack together in a repeating pattern. (b) Neon and some other gases exist as single atoms, so they also are represented by just the symbol such as Ne. (c) Hydrogen gas is composed of diatomic molecules, so the element is written as H_2.

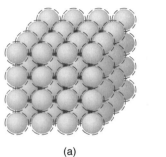
(a)

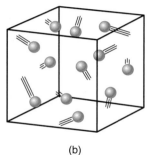
(b)

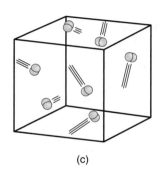
(c)

Atoms and Molecules

The fundamental units of elements and all matter are atoms (see Chapter 9.1). Each type of atom uniquely identifies the element by the number of protons and associated neutrons and electrons to make a stable, neutral element. An atom with one proton and one electron is hydrogen (^1_1H). Adding one proton (and associated neutrons and electrons) makes a new atom with a new, unique elemental name: helium (^4_2He) in this case. In another example, the atoms in a large sample of iron are packed together as shown in ● Fig. 11.10a, with each atom bonding equally to all its nearest neighbors. Each atom of iron is the same. Thus, for a metallic element such as iron, just writing its symbol (Fe) represents its composition adequately.

The composition of a single (monatomic) gas is individual atoms because the atoms are stable by themselves. Therefore a monatomic gas, such as neon (Ne), is identified by the element's symbol. (Fig. 11.10b).

With certain elements, however, individual atoms are too reactive to exist independently; these elements, such as hydrogen and oxygen, are only stable when two atoms are bonded chemically and are called *diatomic molecules* (Fig. 11.10c). A **molecule** is an electrically neutral particle composed of two or more atoms chemically bonded (● Fig. 11.11). If the atoms are of the same element, then it is a molecule of the element (for example, H_2 or N_2). If the atoms are of different elements, then it is a molecule of a compound (for example, H_2O or NH_3).

● Figure 11.12 illustrates the seven common elements that exist as diatomic molecules. Note that six of the seven are close together and form a "7" shape at the right of the periodic table, making them easy to recall. (See the periodic table inside the front cover.) When chemical equations are written, as will be done in Chapter 13, the formulas of these seven elements are written in diatomic form (such as $H_2 + Cl_2 \rightarrow 2 HCl$).

Allotropes

Two or more forms of the same element that have different bonding structures in the same physical phase are called **allotropes**. The three allotropes of carbon are shown in

Figure 11.11 Representations of Molecules
Ball-and-stick models (*top*) and space-filling models (*bottom*) of the element hydrogen and the three common compounds water, ammonia, and methane.

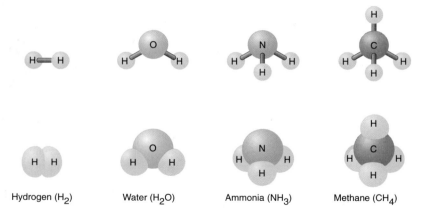

Hydrogen (H_2)   Water (H_2O)   Ammonia (NH_3)   Methane (CH_4)

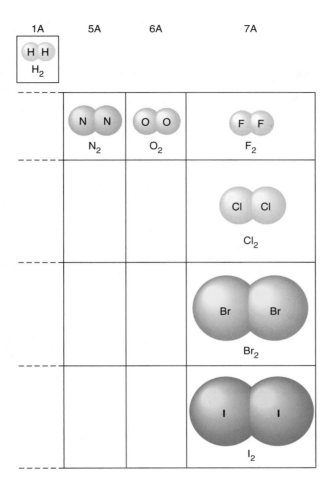

Figure 11.12 Elements That Exist as Diatomic Molecules
Hydrogen, nitrogen, oxygen, fluorine, chlorine, bromine, and iodine exist in nature as diatomic (two-atom) molecules.

● Fig. 11.13. Diamonds are composed of pure carbon, and ● Figure 11.14a shows that each carbon atom in diamond bonds to four of its neighbors. The carbon atom is in the middle of a geometric structure called a *regular tetrahedron*, a figure having four faces that are identical equilateral triangles. The four bonds of the carbon atom point toward the four corners of the tetrahedron. (See also Fig. 14.4.) This geometry leads to a three-dimensional network that helps make diamond the hardest substance known.

Another form of carbon is graphite, which is a black, slippery solid that is a major component of pencil "lead." (Remember that pencil "lead" contains no lead, Pb.) Figure 11.14b shows that the carbon atoms in graphite are bonded in a network of flat hexagons, giving an entirely different structure from that of diamond. Each carbon atom is bonded to three

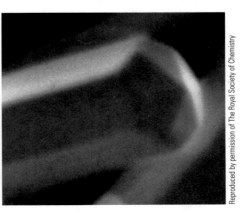

Figure 11.13 Allotropes of Carbon
The diamonds (near the pencil tip in the left image) were produced by heating graphite (black powder) under high pressure. C-60 fullerene, seen as hexagonal rods under high magnification (right image) can be made by vaporizing graphite with a laser.

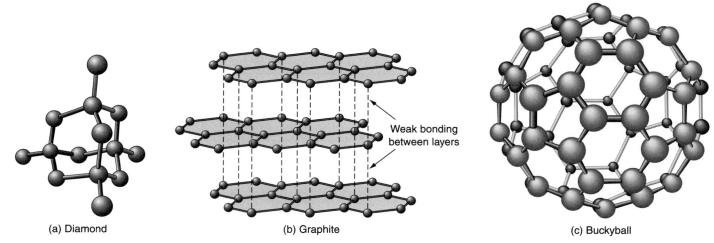

(a) Diamond   (b) Graphite   (c) Buckyball

Figure 11.14 Models of Three Allotropes of Carbon
(a) Diamond consists of a network of carbon atoms in which each atom bonds to four others in a tetrahedral fashion. (b) Graphite consists of a network of carbon atoms in which each atom bonds to three others and forms sheets of flat, interlocking hexagons. The sheets are held to each other only by weak electrical forces. (c) Buckminsterfullerene ("buckyball"), C-60, is a soccer-ball-like arrangement of interlocking hexagons and pentagons formed by carbon atoms.

other carbon atoms that lie in the same plane, and thus each forms part of three hexagons. Each carbon atom is left with one loosely held outer electron, which enables graphite to conduct an electric current, whereas diamond does not. The weak forces between the planes of hexagons allow them to slide easily over one another, thus giving graphite its characteristic slipperiness. As a result, graphite is used as a lubricant.

By applying intense heat and pressure, it is possible to form diamonds from graphite in the laboratory. Conversely, when heated to 1000°C in the absence of air, diamond changes to graphite.

The *fullerenes*, a whole class of ball-like substances (such as C_{32}, C_{60}, C_{70}, and C_{240}), make up the third allotropic form of carbon. The most stable fullerene was first prepared in the laboratory in 1985 by vaporizing graphite with a laser. It is a 60-carbon-atom, hollow, soccer-ball-shaped molecule named buckminsterfullerene, or "buckyball" (Fig. 11.14c). It is named for American engineer and philosopher Buckminster Fuller, who invented the geodesic dome, the architectural principle of which underlies the structure of a buckyball. Scientists succeeded in isolating C_{60} in bulk in 1990 and found that it forms naturally in sooty flames such as those of candles. Chemists have prepared molecules of C_{60} having attached groups of atoms, and they also have made C_{60} molecules with metal atoms trapped inside their hollow structure. Fullerenes containing argon, neon, and helium inside their "cages" have even been recovered from 4.6-billion-year-old meteorites that date back to the solar system's origin.

Procedures similar to those used to prepare fullerenes led to the discovery of nanotubes, which are long structures of carbon atoms in a mesh pattern (reminiscent of chicken wire) wound into a cylinder capped on each end by half a buckyball (● Fig. 11.15). These nanotubes can have single walls or have walls several layers thick. Nanotubes with varying

Figure 11.15 Sketch of a Short Nanotube
The term *nanotube* arose because the pores of these tubes of carbon are about 1 nanometer (10^{-9} m) wide.

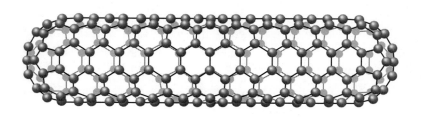

properties can be made. For example, some conduct electricity, yet others are insulators. It is possible that these nanotubes can function as tiny wires in electronic devices built on a molecular level.

Oxygen has two allotropes. Oxygen usually occurs as a gas of diatomic O_2 molecules, but it can also exist as gaseous, very reactive, triatomic O_3 molecules named *ozone*, ● Fig. 11.16. (Ozone is an important constituent of the atmosphere; see Chapter 19.1.) Phosphorus, sulfur, tin, and several other elements also exist in allotropic forms.

Did You Learn?

● Oxygen (47%) and silicon (27%) are the two most abundant elements found in the Earth's crust.

● The allotropes of oxygen are O_2 and O_3 (ozone).

11.4 The Periodic Table

Preview Questions

● What are groups and periods on the periodic table?

● What is the relationship between valence electrons and chemical groups?

By 1869, a total of 63 elements had been discovered, but a system for classifying the elements had not been established. The first detailed and useful periodic table aligned the elements in rows and columns in order of increasing atomic mass. It was published in 1869 by Dmitri Mendeleev ("men-duh-LAY-eff"), a Russian chemist (● Fig. 11.17).

Mendeleev's table is useful because it demonstrates how elements with similar physical and chemical properties recur at regular intervals (that is, *periodically*). Using this method, Mendeleev predicted the properties of three unknown elements, and he even left vacancies in his table for them. He lived to witness the discovery of all three of these elements: gallium (Ga), scandium (Sc), and germanium (Ge). They had the properties that Mendeleev had predicted and filled the vacancies that he had left for them.

Mendeleev was unable to explain why elements with similar properties occurred at regular intervals with increasing atomic mass. He realized that in several positions in his table, the increasing atomic mass and the properties of the elements did not exactly coincide. For example, on the basis of atomic mass, iodine (I) should come *before* tellurium (Te), but iodine's properties indicate that it should *follow* tellurium. Mendeleev thought that perhaps in these cases the atomic masses were incorrect. However, it is now known that the periodic law is a function of the atomic number (number of protons) and the corresponding electron configuration of an element, not its atomic mass. Fortunately for Mendeleev, the atomic numbers and atomic masses generally increase together.

Thus, the *periodic table* organizes the elements, on the basis of atomic numbers, into seven horizontal rows called **periods**. As a result, the properties of the elements show regular trends, and similar properties occur *periodically*, at definite intervals. Later in this section we will discuss some examples of these regular trends. The modern statement of the **periodic law** is

The properties of elements are periodic functions of their atomic numbers.

The vertical columns in the periodic table are called **groups**. At present, some debate exists about the designation of the groups. In 1986, the IUPAC decreed that the groups be labeled 1 through 18 from left to right, as shown in the periodic table inside the front cover. For many years in the United States, however, the groups have been divided into A and B subgroups, as is also shown in the periodic table inside the front cover and in ● Fig. 11.18. Which designation will ultimately prevail is still in question. The A and B designation will be used in our discussion.

The elements in the periodic table are classified in several ways, one of which is into representative, transition, and inner transition elements.

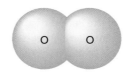

(a) Oxygen molecule

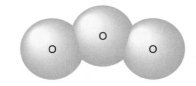

(b) Ozone molecule

Figure 11.16 Oxygen and Ozone
The element oxygen normally exists as gaseous diatomic molecules (O_2). Oxygen also can exist as gaseous, highly reactive, triatomic molecules (O_3) called *ozone*.

Figure 11.17 Dmitri Mendeleev (1834–1907)
Mendeleev, a Russian chemist, developed the first useful periodic table of the chemical elements.

Figure 11.18 Names of Specific Portions of the Periodic Table
The representative elements are shown in green, the transition elements in blue, and the inner transition elements in purple. Note the location of the alkali metals (Group 1A, except H), the alkaline earth metals (Group 2A), the halogens (Group 7A), the noble gases (Group 8A), and the lanthanides and actinides.

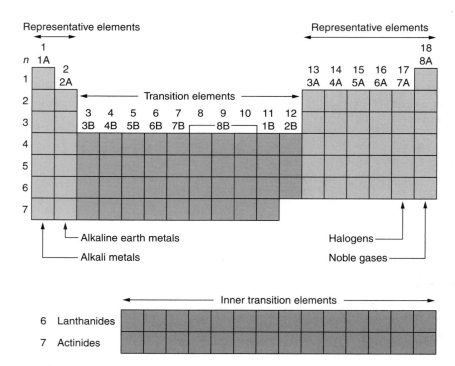

1. **Representative (or Main group) elements** are those in Groups 1A through 8A, shown in green in Fig. 11.18 and in your periodic table. Four of the groups of representative elements have commonly used special names: *alkali metals* (Group 1A), *alkaline earth metals* (Group 2A), *halogens* (Group 7A), and *noble gases* (Group 8A).

2. **Transition elements**, shown in blue, are the groups designated by a numeral and the letter B. Such familiar elements as iron, copper, and gold are transition elements.

3. **Inner transition elements**, shown in purple, are placed in two rows at the bottom of the periodic table. Each row has its own name. The elements cerium (Ce) through lutetium (Lu) are called the *lanthanides*, and thorium (Th) through lawrencium (Lr) make up the *actinides*. Except for uranium, few of the inner transition elements are well known. However, the phosphors in color TV and computer screens are primarily compounds of lanthanides.

Metals and Nonmetals

Another useful method for classifying elements (namely, *metals* and *nonmetals*) was originally done on the basis of certain distinctive properties (● Table 11.2). Our modern definitions are as follows:

> **A metal is an element whose atoms tend to lose electrons during chemical reactions.**
> **A nonmetal is an element whose atoms tend to gain (or share) electrons.**

As shown in ● Fig. 11.19, the metallic character of the elements increases as you go down a group and decreases across a period (from left to right). As you might expect, the nonmetallic character of the elements shows the opposite trend, decreasing down a group and increasing across a period. Cesium (Cs) is the most metallic naturally occurring element, and fluorine (F) is the most nonmetallic. (Francium, Fr, which technically might be the most metallic element, is a synthetic element and is unavailable in amounts greater than a few atoms.)

Most elements are metals. The actual dividing line between metals and nonmetals cuts through the periodic table like a staircase (Fig. 11.19). The elements boron, silicon, germanium, arsenic, antimony, and tellurium are called *semimetals* or *metalloids*. They are located next to the staircase line and display properties of both metals and nonmetals. Several

Table 11.2 Some General Properties of Metals and Nonmetals

Metals	Nonmetals
Good conductors of heat and electricity	Poor conductors of heat and electricity
Malleable; can be beaten into thin sheets	Brittle if a solid
Ductile; can be stretched into wire	Nonductile
Possess metallic luster	Do not possess metallic luster
Solids at room temperature (exception: Hg)	Solids, liquids, or gases at room temperature
Usually have 1 to 3 valence electrons	Usually have 4 to 8 valence electrons
Lose electrons, forming positive ions	Gain electrons to form negative ions or share electrons

semimetals are of crucial importance as semiconductors in the electronics industry (● Fig. 11.20).

Elements can also be classified according to whether they are solids, liquids, or gases at room temperature and atmospheric pressure. Only two, bromine and mercury, occur as liquids. Eleven occur as gases: hydrogen, nitrogen, oxygen, fluorine, chlorine, and the six noble gases (Chapter 11.6). All the rest are solids, the vast majority being metallic solids.

Electron Configuration and Valence Electrons

In Chapter 9.3 it was stated that the main electron energy levels in atoms are designated by the principal quantum number n, which can have the values 1, 2, 3, and so on. We often refer to these energy levels as *shells*. For example, the first shell is the energy level where $n = 1$.

The chemical reactivity of the elements depends on the **electron configuration**, or the order of the electrons in the energy levels in their atoms. The transition elements are a block comprising 10 columns, and the inner transition elements have 14 columns. Our interest will be learning how the electrons are arranged in the various shells, learning the *shell electron configurations*. To keep things simple yet still show the basic ideas, only the

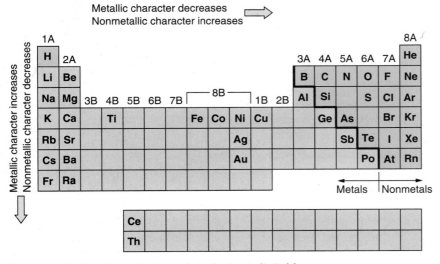

Figure 11.19 Metals and Nonmetals in the Periodic Table
The elements shown in blue, at the left of the "staircase" line, are metals. The nonmetals are shown in pink. Fluorine (F) is the most reactive nonmetal, and cesium (Cs) and francium (Fr) are the most reactive metals. The semimetals are found next to the staircase line and have some metallic and some nonmetallic properties. Note that hydrogen is generally considered a nonmetal.

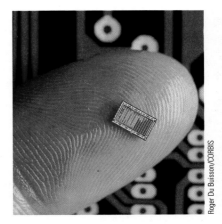

Figure 11.20 A Silicon Microchip
The metalloid element silicon (Si) was discovered by Berzelius in 1824. Silicon forms the basis for the modern microelectronics industry. (See also Fig. 8.32.)

Figure 11.21 Shell Distribution of Electrons for Periods 1, 2, and 3
Each atom's nucleus is shown in light blue. The number of shells containing electrons is the same as the period number. Each shell of electrons is shown as a partial circle, with a numeral showing the number of electrons in the shell. The number of valence electrons, those in the outer shell, is the same for all the atoms in the group and is equal to the group number (exception: helium, He).

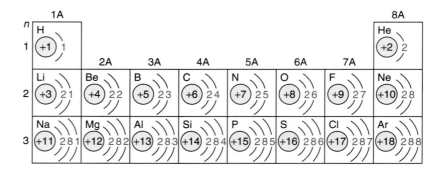

first 18 elements will be considered. However, it is important to be able to determine how many electrons are in the outermost shell of an atom of *any* representative element.

The outer shell of an atom is known as the **valence shell**, and the electrons in it are called the **valence electrons**. The valence electrons are extremely important because they are the ones involved in forming chemical bonds. Elements in a given group have the same number of valence electrons and therefore have similar chemical properties.

Here are the rules for writing shell electron configurations of the representative elements:

1. The number of electrons in an atom of a given element is the same as the element's atomic number. For example, the periodic table shows that lithium (Li) has atomic number 3, so there are three electrons to "place" in the proper shells.

2. For atoms of a given element, the number of shells that contain electrons will be the same as the period number. For example, lithium is in Period 2, so the three electrons are in two shells.

3. For the representative (Group A) elements, the number of valence electrons (the electrons in the outer shell) is the same as the group number. (The only exception is helium, which has just two electrons in the outer shell, although it is in Group 8A.)

For example, lithium is in Group 1A, so it has one valence electron. Clearly, if lithium's three electrons are to be distributed among two shells and the outer shell has one electron, the first shell must contain the other two. Thus, the shell electron configuration of lithium is written 2,1 (that is, two electrons in the first shell and one in the second shell). In fact, the first shell of all atoms holds a maximum of two electrons, whereas the second shell holds a maximum of eight. For the representative elements, the number of valence electrons increases by one as you proceed across a given period. ● Figure 11.21 shows the shell electron configurations for the first 18 elements.

EXAMPLE 11.1	**Identifying Some Properties of an Element**

● Figure 11.22 shows the very reactive element sodium (Na). Find Na in the periodic table and answer the following questions.

(a) What are its atomic number (Z) and atomic mass?
(b) Is it a representative, transition, or inner transition element? Is it a metal or a nonmetal?
(c) What period is it in, and what is its group number?
(d) How many total electrons are in an atom of sodium? How many protons?
(e) How many valence electrons are in it?
(f) How many shells of a Na atom contain electrons?
(g) What is sodium's shell electron configuration?

Solution

(a) $Z = 11$, and the atomic mass is 23.0 u.
(b) Representative (its group number has "A" in it); metal (at *left* of staircase line).
(c) Period 3 (it is in the third horizontal row down) and Group 1A.
(d) Eleven electrons and 11 protons (same as Z).

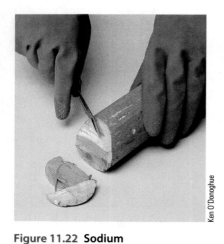

Figure 11.22 Sodium
This metallic element is so soft that it can be sliced with a knife. It is so reactive that an explosion or chemical burn can occur, so it must be handled with caution.

Ken O'Donoghue

(e) One valence electron (same as the group number, for representative elements).

(f) Three shells contain electrons (same as the period number).

(g) Two electrons are always in the first shell, and the third shell of sodium must contain one valence electron, so that leaves 8 of the 11 for the second shell (which we already know can hold a maximum of 8). Thus, sodium's shell electron configuration is 2,8,1.

Confidence Exercise 11.1

Find the element phosphorus (P) in the periodic table, and answer the following questions.

(a) What are its atomic number and atomic mass?

(b) Is it a representative, transition, or inner transition element? Is it a metal or a nonmetal?

(c) What period is it in, and what is its group number?

(d) How many total electrons are in an atom of phosphorus? How many protons?

(e) How many valence electrons are in a P atom?

(f) How many shells of a P atom contain electrons?

(g) What is the shell electron configuration of phosphorus?

Answers to Confidence Exercises may be found in the back of the book.

Atomic Size: Another Periodic Characteristic

Several periodic characteristics—such as metallic and nonmetallic character, electron configuration, and number of valence electrons—have been discussed. Another periodic nature of elements is atomic size, which ranges from a diameter of about 0.064 nm for helium (He) to about 0.47 nm for cesium (Cs).

● Figure 11.23 illustrates that <u>atomic size increases down a group</u>. This statement is logical because each successive element of the group has an additional shell containing electrons. Additional shells shield the outer electron from the increasing charge on the nucleus.

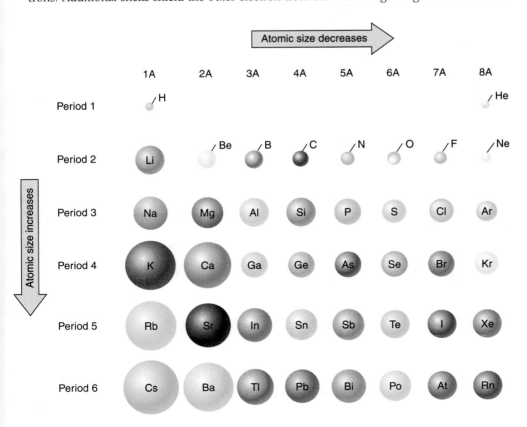

Figure 11.23 The Relative Sizes of Atoms of the Representative Elements
In general, atomic size decreases across a period and increases down a group. The diameter of the smallest atom, He, is about 0.064 nm, whereas that of the largest atom shown, Cs, is about 0.47 nm.

Figure 11.24 Ionization Energy: A Periodic Trend
The ionization energy (in kilojoules per mole) increases in a generally regular fashion across a period and decreases down a group.

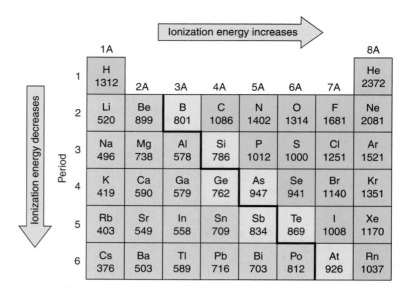

Figure 11.23 also shows that atomic size decreases across a period. Notice that each Group 1A atom is large with respect to atoms of the other elements of that period. That occurs because its one outer electron is loosely bound to the nucleus. As the charge on the nucleus increases (more protons) without adding an additional shell of electrons, the outer electrons are more tightly bound, thus decreasing the atomic size from left to right across the period.

Ionization Energy: Yet Another Periodic Characteristic

When an atom gains or loses electrons, it acquires a net electric charge and becomes an *ion*. The amount of energy that it takes to remove an electron from an atom is called its **ionization energy**. In general, the ionization energy increases across a period (● Fig. 11.24). The Group 1A elements have the lowest ionization energies. Their one valence electron is in an outer shell, so it does not take much energy to remove the electron. Elements located to the right of the Group 1A element in a particular period have additional protons and electrons, but the electrons are added to the same shell. The added protons bind the electrons more and more strongly until the shell is completely filled.

Ionization energy decreases down a group because the electron to be removed is shielded from the attractive force of the nucleus by each additional shell of electrons. Another periodic characteristic—*electronegativity*—will be encountered in Chapter 12.5.

Did You Learn?

● In the periodic table, groups are vertical columns of elements that have similar properties. Periods are horizontal rows that show similar properties periodically.

● Elements of a given group have the same number of valence electrons (electrons in the outer atomic shell) and therefore have similar properties.

11.5   Naming Compounds

Preview Questions

● How are chemical formulas written?

● What is an ion?

Our discussion from this point on requires a basic understanding of compound *nomenclature* (from the Latin words for "name" and "to call"). Elements combine chemically to form compounds. Each compound is represented by a *chemical formula*, which is written by putting the elements' symbols adjacent to each other, usually with the more metallic element first.

A *subscript* (subscripted numeral) following each symbol designates the number of atoms of the element in the formula. For an element with only one atom in the formula, the subscript 1 is not written; it is understood. For instance, the formula for water is written H_2O, which means that a single molecule of water consists of two atoms of hydrogen and one atom of oxygen. Chemical formulas, such as H_2O, are often used to identify specific chemical compounds, but *names* that unambiguously identify them are also needed.

Conceptual Question and Answer

A Table of Compounds?

Q. We have a periodic table of elements. Why not a table of compounds?

A. With millions of known compounds, it would be quite a table. Also, new compounds are being created or developed. Elements have periodicity, meaning that you can line them up by functional groups, such as alkali metals and halogens. There is some grouping of compounds by reactions (see Chapter 14) but generally only by elements that they contain, such as H, C, and N. The *Chemical Abstract Service* maintains a database of more than 55 million different inorganic and organic chemical compounds.

Compounds with Special Names

Some compounds have such well-established special names that no systematic nomenclature can compete. Their names are just learned individually. Common examples to know are given in ● Table 11.3.

Naming a Compound of a Metal and a Nonmetal

For simplicity, we will learn to name compounds of metals that form only one ion, which are mainly those of Groups 1A (ionic charge 1+) and 2A (ionic charge 2+), along with Al, Zn, and Ag (ionic charges of 3+, 2+, and 1+ , respectively). Chapter 12.4 discusses how to name compounds of metals that form more than one ion.

To name a binary (two-element) compound of a metal combined with a nonmetal, first give the name of the metal and then give the name of the nonmetal with its ending changed to *-ide* (● Table 11.4). For example:

NaCl	sodium chloride
Al_2O_3	aluminum oxide
Ca_3N_2	calcium nitride

Naming Compounds of Two Nonmetals

In a compound of two nonmetals, the element with more metallic character (the one farther left or farther down in the periodic table) is usually written first in the formula and named first. The second element is named using its *-ide* ending. Generally, two nonmetallic elements form several binary compounds, which are distinguished by using Greek prefixes to designate the number of atoms of the element that occur in the molecule (● Table 11.5). The prefix *mono-* is always omitted from the name of the first element in the compound and is usually omitted from the second (with the common exception of carbon monoxide, CO). For example:

HCl	hydrogen chloride
CS_2	carbon disulfide
PBr_3	phosphorus tribromide
IF_7	iodine heptafluoride

Table 11.3 Eleven Compounds with Special Names

Name	Formula
Water	H_2O
Ammonia	NH_3
Methane	CH_4
Nitrous oxide	N_2O
Nitric oxide	NO
Hydrochloric acid	HCl(*aq*)
Nitric acid	HNO_3(*aq*)
Acetic acid	$HC_2H_3O_2$(*aq*)
Sulfuric acid	H_2SO_4(*aq*)
Carbonic acid	H_2CO_3(*aq*)
Phosphoric acid	H_3PO_4(*aq*)

Table 11.4 The *-ide* Nomenclature for Common Nonmetals

Element Name	*-ide* Name
Bromine	Bromide
Chlorine	Chloride
Fluorine	Fluoride
Hydrogen	Hydride
Iodine	Iodide
Nitrogen	Nitride
Oxygen	Oxide
Sulfur	Sulfide

Table 11.5 Greek Prefixes

Prefix	Number	Prefix	Number
mono-	1	penta-	5
di-	2	hexa-	6
tri-	3	hepta-	7
tetra-	4	octa-	8

Naming Compounds That Contain Polyatomic Ions

An **ion** is an atom, or chemical combination of atoms, having a net electric charge because of a gain or loss of electrons. An ion formed from a single atom is a *monatomic ion*, whereas an electrically charged combination of atoms is a *polyatomic ion*. The names and formulas of eight common polyatomic ions are given in ● Table 11.6.

For a compound of a metal combined with a polyatomic ion, simply name the metal and then name the polyatomic ion. For example:

$ZnSO_4$	zinc sulfate
$NaC_2H_3O_2$	sodium acetate
$Mg(NO_3)_2$	magnesium nitrate
K_3PO_4	potassium phosphate

When hydrogen is combined with a polyatomic ion, the compound is generally named as an acid as is shown in Table 11.3.

The only common positive polyatomic ion present in compounds is the ammonium ion, NH_4^+. If the ammonium ion is combined with a nonmetal, then change the ending of the nonmetal to -*ide*. If it is combined with a negative polyatomic ion, then simply name each ion. For example:

$(NH_4)_3P$	ammonium phosphide
$(NH_4)_3PO_4$	ammonium phosphate

The basic nomenclature rules are summarized in ● Fig. 11.25. Note the first question to ask: Is it one of the 11 compounds with a special name? If the answer is yes, then just name it. If the answer is no, then ask the second question: Is it a binary compound?

The naming of the complex compounds found in organic chemistry follows its own set of rules, as will be seen in Chapter 14.3. And by the way, don't be upset if a few compounds seem to be named in violation of the rules given above. Space limitations prohibit complete coverage of the rules.

Table 11.6 Some Common Polyatomic Ions

Name	Formula
Acetate	$C_2H_3O_2^-$
Hydrogen carbonate	HCO_3^-
Hydroxide	OH^-
Nitrate	NO_3^-
Carbonate	CO_3^{2-}
Sulfate	SO_4^{2-}
Phosphate	PO_4^{3-}
Ammonium	NH_4^+

Note: The names of positive polyatomic ions end in -*ium*. Many negative polyatomic ions have names ending in -*ate*. Another name for the hydrogen carbonate ion is *bicarbonate*.

EXAMPLE 11.2   Naming Compounds

Name each of these six compounds: (a) $H_2SO_4(aq)$, (b) $ZnCO_3$, (c) Na_2S, (d) SiO_2, (e) NH_3, (f) NH_4NO_3.

Solution

(a) $H_2SO_4(aq)$: Sulfuric acid (a compound with a special name).
(b) $ZnCO_3$: Zinc carbonate (a compound of a metal with the carbonate polyatomic ion, so name the metal and then name the polyatomic ion).
(c) Na_2S: Sodium sulfide (a binary compound of a metal with a nonmetal, so the metal is named first, and then the -*ide* name of the nonmetal is given).
(d) SiO_2: Silicon dioxide (a binary compound of two nonmetals, so the Greek prefix system is preferred).
(e) NH_3: Ammonia (a common compound that is always called by its special name).
(f) NH_4NO_3: Ammonium nitrate (a compound that contains an ammonium ion with another polyatomic ion, so just name each one).

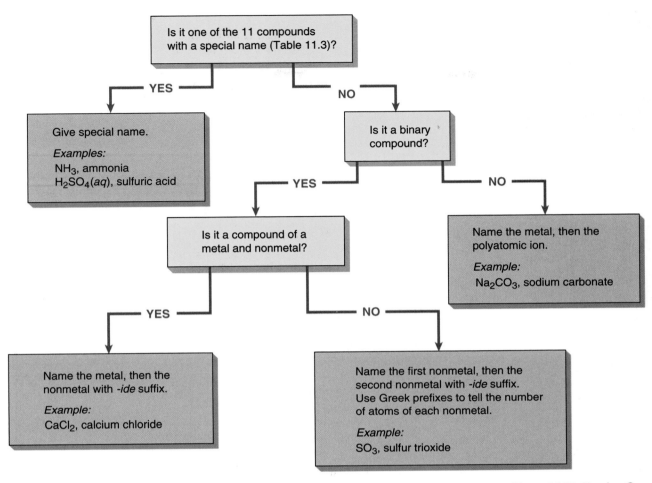

Figure 11.25 Naming Compounds
This flowchart summarizes the basic nomenclature rules for naming compounds.

Confidence Exercise 11.2

Name the compounds AsF_5 and $CaCl_2$.
  Answers to Confidence Exercises may be found at the back of the book.

Did You Learn?

- Chemical formulas are written by putting the symbols of the elements adjacent to each other with a subscript following each symbol to designate the number of that atom.

- A ion is an atom, or chemical combination of atoms, having a net electric charge because of a gain or loss of electron(s).

11.6 Groups of Elements

Preview Questions

- Why do the elements of a group react similarly?

- Which group is composed of monatomic gases, and which group is composed of diatomic gases?

Figure 11.26 "Neon" Lights
When the atoms of a small amount of gas are subjected to an electric current, the gas glows, or fluoresces. The color of a particular light depends on the identity of the gas whose atoms are being excited. Neon gas glows a beautiful orange-red, whereas argon emits a blue light.

Recall that a row of the periodic table is called *a period* and that a column of the table is termed a *group*. As illustrated in Fig. 11.21, all the elements in a group have the same number of valence electrons. If one element in a group reacts with a given substance, then the other elements in the group usually react in a similar manner with that substance. The formulas of the compounds produced also will be similar in form. Four of these groups of elements are discussed in this section: the noble gases (8A), the alkali metals (1A), the halogens (7A), and the alkaline earth metals (2A).

The Noble Gases

The elements of Group 8A are known as the **noble gases**. They are monatomic; that is, they exist as single atoms. The noble gases have the unusual chemical property of almost never forming compounds with other elements or even bonding to themselves. That is why the noble gas argon is used inside light bulbs, where the intense heat would cause the tungsten (W) filament to react with most gases and quickly deteriorate.

We can conclude that noble gas electron configurations with eight electrons in the outer shell (or two if the element, like helium, uses only the first shell) are quite stable. This conclusion is of crucial importance when considering the bonding characteristics of atoms of other elements, as will be learned in Chapter 12.

Helium is found in natural gas, and radon (Rn) is a radioactive by-product of the decay of radium. Other noble gases are found only in the air. Recall that argon makes up almost 1% of air. A common use of some of the noble gases is in "neon" signs, which contain minute amounts of various noble gases or other gases in a sealed glass vacuum tube (● Fig. 11.26). When an electric current is passed through the tube, the gases glow, giving a color that is characteristic of the specific gas present (see Chapter 9.3).

The most striking physical property of the noble gases is their low melting and boiling points. For example, gaseous helium liquefies at −269°C and solidifies at −272°C, just four degrees and one degree, respectively, above absolute zero (−273°C). For this reason, liquid helium is often used in low-temperature research.

The Alkali Metals

The elements in Group 1A, except for hydrogen, are called the **alkali metals**. (Hydrogen, a gaseous *nonmetal*, is discussed at the end of this section.) Each alkali metal atom has only one valence electron. The atom tends to lose this outer electron quite easily, so the alkali metals react readily with other elements and are said to be *active* metals.

Sodium and potassium are abundant in the Earth's crust, but lithium, rubidium, and cesium are rare. The alkali metals are all soft (Fig. 11.22) and are so reactive with oxygen and moisture that they must be stored under oil. The most common compound containing an alkali metal is table salt (sodium chloride, NaCl). Other alkali metal compounds known even to ancient civilizations are *potash* (potassium carbonate, K_2CO_3) and *washing soda* (sodium carbonate, Na_2CO_3). Two other common compounds of sodium are *lye* (sodium hydroxide, NaOH) and *baking soda* (sodium hydrogen carbonate, or sodium bicarbonate, $NaHCO_3$).

By seeing these formulas of sodium compounds and by knowing that all the elements in a group produce compounds with similar formulas, the formulas of compounds of the other alkali metals can be predicted.

EXAMPLE 11.3   Chemical Formulas for Groups

Give the chemical formula for the compounds (a) potassium chloride and (b) lithium carbonate.

Solution

(a) KCl: Potassium is a Group 1A nonmetal, so it is written similar to sodium chloride.

(b) Li_2CO_3: Lithium is a Group 1A nonmetal, so it is written similar to sodium carbonate.

Confidence Exercise 11.3

Predict the chemical formula for lithium chloride and potassium hydroxide.

The Halogens

The Group 7A elements are called the **halogens**. Their atoms have seven electrons in their valence shell and have a strong tendency to gain one more electron. They are active non-metals and are present in nature only in the form of their compounds. As shown in Fig. 11.12, the halogens consist of diatomic molecules (F_2, Cl_2, Br_2, and I_2).

Fluorine, a pale yellow, poisonous gas, is the most reactive of all the elements. It corrodes even platinum, a metal that withstands most other chemicals. Because it is so reactive, a stream of fluorine gas causes wood, rubber, and even water to burst into flame. Fluorine was responsible for the unfortunate deaths of several very able chemists before it was finally isolated.

Chlorine, a pale green, poisonous gas, was used as a chemical weapon in World War I (● Fig. 11.27). In small amounts, it is used as a purifying agent in swimming pools and in public water supplies.

Bromine is a reddish brown, foul-smelling, poisonous liquid used as a disinfectant, whereas iodine is a violet-black, brittle solid (Fig. 11.27). The "iodine" found in a medicine cabinet is not the pure solid element; it is a *tincture* of iodine dissolved in alcohol. Most table salt is now iodized; that is, it contains 0.02% sodium iodide (NaI), which is added to supplement the human diet because an iodine deficiency causes thyroid problems (● Fig. 11.28). The last halogen in the group, astatine (At), is only found in minute amounts, is radioactive, and is currently being studied for nuclear medicine.

Some formulas and names for other halogen compounds are:

$AlCl_3$	aluminum chloride
NH_4F	ammonium fluoride
$CaBr_2$	calcium bromide

By analogy with these formulas, we can predict the correct formulas of many other halogen compounds, such as AlF_3, NH_4Cl, and CaI_2.

The Alkaline Earth Metals

The elements in Group 2A are called the **alkaline earth metals**. Their atoms contain two valence electrons. They are active metals but are not as chemically reactive as the alkali metals. They have higher melting points and are generally harder and stronger than their Group 1A neighbors.

Beryllium occurs in the mineral beryl, some varieties of which make beautiful gemstones such as aquamarines and emeralds. Magnesium is used in safety flares because of its ability to burn with an intensely bright flame (see Fig. 13.1). Both magnesium and beryllium are used to make lightweight, high-strength, metal alloys like MgAlZn or BeCu. The common over-the-counter medicine called *milk of magnesia* is magnesium hydroxide, $Mg(OH)_2$.

Calcium is used by vertebrates in the formation of bones and teeth, which are mainly calcium phosphate, $Ca_3(PO_4)_2$. Calcium carbonate ($CaCO_3$) is a mineral present in nature in many forms. The shells of marine creatures, such as clams, are made of calcium carbonate, as are coral reefs. Limestone, a sedimentary rock often formed from seashells and thus from calcium carbonate, is used to build roads and to neutralize soils that are acidic. Tums tablets are mainly calcium carbonate. Also composed of calcium carbonate are the impressive formations of stalactites and stalagmites in underground caverns (see Fig. 22.28). Compounds of calcium are also used to outline baseball and softball fields.

Strontium compounds produce the red colors in fireworks, and barium compounds produce the green (see the Chapter 12 opening photo). Radioactive strontium-90 is

Figure 11.27 The Halogens
Chlorine is the pale-green gas in the left-hand flask. Bromine is the reddish brown liquid in the center. Iodine is the violet-black solid in the right-hand flask. Note how readily bromine vaporizes and iodine sublimes.

Figure 11.28 Iodine in Food
In general, table salt is iodized for health.

Figure 11.29 The *Hindenburg* Disaster, Lakehurst, New Jersey, May 6, 1937
The 800-ft-long airship burst into flame just before landing after a trip from Germany. The disaster killed 35 of the 96 passengers and is responsible for the "*Hindenburg* syndrome," a reluctance to use hydrogen as a fuel. For size comparison, the Goodyear blimps are about 200 ft long. Airships nowadays use nonflammable helium for buoyancy.

synthesized in nuclear labs and nuclear reactors; if ingested, it goes into the bone marrow because of its similarity to calcium and may cause cancer.

Radium is an intensely radioactive element that glows in the dark. Its radioactivity is a deterrent to practical uses, although a few decades ago watch dials were painted with radium chloride ($RaCl_2$) so that they could be read in darkness.

Hydrogen

Although hydrogen is commonly listed in Group 1A, it is not considered an alkali metal. Hydrogen is a nonmetal that usually reacts like an alkali metal, forming HCl, H_2S, and so forth (similar to NaCl, Na_2S, and so forth). Sometimes, though, hydrogen reacts like a halogen, forming such compounds as NaH (sodium hydride) and CaH_2 (compare with NaCl and $CaCl_2$). As a light gas, it was used in early dirigibles (● Fig. 11.29). At room temperature, hydrogen is a colorless, odorless gas that consists of diatomic molecules (H_2).

Although hydrogen is combustible in air, it can be used as a fuel source without combustion. Hydrogen and oxygen gas are combined in a fuel cell to produce electricity and water. Therefore, it is environmentally cleaner than gasoline. It may be the main fuel used in transportation in the future, which would greatly help our air pollution problems. (See the Chapter 4.6 Highlight: Hybrids and Hydrogen.)

Did You Learn?

● All the elements of a group react similarly because they have the same number of valence electrons.

● The noble gases are monatomic, and the halogens are diatomic.

KEY TERMS

1. chemistry
2. element (11.1)
3. compound
4. mixture
5. solution
6. solubility
7. molecule (11.3)
8. allotropes
9. periods (11.4)
10. periodic law
11. groups
12. representative elements
13. transition elements
14. inner transition elements
15. metal
16. nonmetal
17. electron configuration
18. valence shell
19. valence electrons
20. ionization energy
21. ion (11.5)
22. noble gases (11.6)
23. alkali metals
24. halogens
25. alkaline earth metals

MATCHING

For each of the following items, fill in the number of the appropriate Key Term from the preceding list. Compare your answers with those at the back of the book.

a. _____ The elements in Groups 1A through 8A

b. _____ The lanthanides and actinides

c. _____ The elements in Groups 1B through 8B

d. _____ An atom or chemical combination of atoms with a net electric charge

e. _____ The outer shell of an atom

f. _____ Physical science dealing with the composition and structure of matter

g. _____ An electrically neutral particle composed of two or more chemically bonded atoms

h. _____ The properties of elements are periodic functions of their atomic numbers

i. _____ A pure substance in which all atoms have the same number of protons

j. _____ Can be broken into its component elements only by chemical processes

k. _____ The elements of Group 1A, except hydrogen

l. _____ The elements magnesium, calcium, and barium are part of this group

m. _____ Fluorine, chlorine, and iodine are part of this group

n. _____ The elements of Group 8A

o. _____ Matter composed of varying proportions of two or more physically mixed substances

p. _____ Forms of the same element that have different bonding structures in the same phase

q. _____ An element whose atoms tend to gain or share electrons

r. _____ An element whose atoms tend to lose electrons

s. _____ The amount of solute that will dissolve in a specified amount of solvent at a given temperature

t. _____ A homogeneous mixture

u. _____ The electrons involved in forming chemical bonds

v. _____ The vertical columns of the periodic table

w. _____ The horizontal rows of the periodic table

x. _____ The order of electrons in the energy levels of an atom

y. _____ The amount of energy it takes to remove an electron from an atom

MULTIPLE CHOICE

Compare your answers with those at the back of the book.

1. Which scientist is often referred to as the "Father of Chemistry"? (Intro)
   (a) Boyle (b) Lavoisier
   (c) Berzelius (d) Mendeleev

2. A solute crystal dissolves when added to a solution. What type was the original solution? (11.1)
   (a) saturated (b) supersaturated
   (c) unsaturated (d) presaturated

3. Which of these scientists in 1661 defined *element* in a manner that made it subject to laboratory testing? (11.2)
   (a) Boyle (b) Davy
   (c) Berzelius (d) Mendeleev

4. Which of these is a synthetic element? (11.2)
   (a) oxygen (8) (b) roentgenium (111)
   (c) uranium (92) (d) iodine (53)

5. Which one of these elements normally exists as a gas of diatomic molecules? (11.3)
   (a) iodine (b) argon
   (c) sulfur (d) chlorine

6. Which element is the most common in the Earth's crust? (11.3)
   (a) silicon (b) magnesium
   (c) oxygen (d) aluminum

7. What is the Group IIIA element in Period 2? (11.4)
   (a) Mg (b) Sc
   (c) B (d) Ga

8. Which one of these elements has the greatest atomic size? (11.4)
   (a) lithium (b) fluorine
   (c) potassium (d) bromine

9. Which one of these elements has the greatest metallic character? (11.4)
   (a) nickel (b) barium
   (c) fluorine (d) bromine

10. Which of these is the preferred name for Na_2SO_4? (11.5)
    (a) sodium sulfide (b) disodium sulfide
    (c) disodium sulfate (d) sodium sulfate

11. Which of these is the preferred name for Na_2S? (11.5)
    (a) sodium sulfide
    (b) disodium sulfide
    (c) disodium sulfate
    (d) sodium sulfate

12. Which one of these is a halogen compound? (11.6)
    (a) $Mg(OH)_2$ (b) Na_2S
    (c) $AlCl_3$ (d) $(NH_4)_3P$

FILL IN THE BLANK

Compare your answers with those at the back of the book.

1. _____ is the study of compounds that contain carbon and hydrogen. (Intro)

2. In a solution, the substance present in the largest amount is called the ___. (11.1)

3. Hot water dissolves _____ solute than the same amount of cold water. (11.1)

4. Boyle championed the need for ___ in science. (11.2)

5. The individual units making up substances such as water and methane are called ___. (11.3)

6. The most abundant *metal* in the Earth's crust is ___. (11.3)

7. The allotrope of oxygen called ozone has the chemical formula ___. (11.3)

8. In 1869, a scientist named ___ developed the periodic table. (11.4)

9. Atoms of elements of Group 4A all contain ___ valence electrons. (11.4)

10. The metallic character of the elements ___ as you go down a group and decreases across a period.

11. The name of the NO_3^- polyatomic ion is ___. (11.5)

12. The common name for the compound NaOH is ___. (11.6)

SHORT ANSWER

1. What do those in the field of chemistry study?

2. Name the five major divisions of chemistry.

11.1 Classification of Matter

3. Which illustrations in ● Fig. 11.30 represent mixtures? A compound? Only one element? Only diatomic molecules?

4. When salad dressing is poured on a salad, what kind of mixture results?

5. What type of process is involved in going from mixtures to pure substances and vice versa?

6. What characteristic distinguishes an element from a compound?

7. What type of process is involved in going from elements to compounds and vice versa?

8. Give one example each of a solid solution, a liquid solution, and a gaseous solution.

9. When sugar is dissolved in iced tea, what is the solvent and what is the solute? When would you become aware that the solution was saturated?

11.2 Discovery of the Elements

10. Name 6 of the first 12 elements that were isolated. Are most of these metals, or are they nonmetals? List some of the first gaseous elements discovered.

11. Name and describe how the first synthetic element was produced.

12. About how many elements are known at present? About how many occur naturally in our environment?

11.3 Occurrence of the Elements

13. Name the two elements that predominate in (a) the Earth's crust, (b) the Earth's core, (c) the human body, and (d) the air. In each case, list the more predominant element first.

14. What are the shapes of the three allotropes of carbon?

11.4 The Periodic Table

15. Name the scientist who receives the major credit for the development of the periodic table. In what year was his work published?

16. At first, the periodic table placed the elements in order of increasing atomic mass. What property is now used instead of atomic mass?

17. What characteristics of the elements are periodic? There are at least five.

18. What formal term is applied to (a) the horizontal rows of the periodic table and (b) the vertical columns?

19. Why are chemists so interested in the number of valence electrons in atoms?

20. How do metals differ from nonmetals with regard to (a) number of valence electrons, (b) conductivity of heat and electricity, and (c) phase?

21. How does metallic character change (a) across a period and (b) down a group?

22. What are semimetals? What are their properties, and what industry are they primarily used for?

23. List the two elements that are liquids and the five (other than the six noble gases) that are gases at room temperature and atmospheric pressure.

24. How does the size of atoms change (a) across a period and (b) down a group?

11.5 Naming Compounds

25. What are the most common names for these compounds and ions: NH_3, HCl, OH^-, SO_4^{2-}?

26. Distinguish between (a) an atom and a molecule, (b) an atom and an ion, and (c) a molecule and a polyatomic ion.

27. Why is it necessary to use Greek prefixes when naming compounds containing two nonmetals?

11.6 Groups of Elements

28. Briefly, why do elements in a given group have similar chemical properties?

29. What are the interesting properties of noble gases? How are noble gases used in lighting?

30. Name the four common halogens, and tell the normal phase of each. Which one is the most reactive of all elements?

31. Explain why sodium iodide is added to table salt.

32. What is the formula for the common household compound known as *baking soda*?

33. List some common uses for alkaline earth metals and their compounds.

34. What properties of hydrogen make it very useful in chemistry?

Figure 11.30
See Short Answer
Question 3.

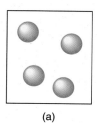 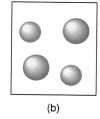 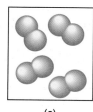 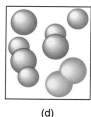 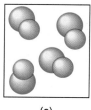

(a)   (b)   (c)   (d)   (e)

VISUAL CONNECTION

Visualize the connections and give answers for the blanks. Compare your answers with those at the back of the book.

Match the correct elements with the group name below:
K, Zn, Mg, Br, Ar, V, Li, Ca, I, Ne

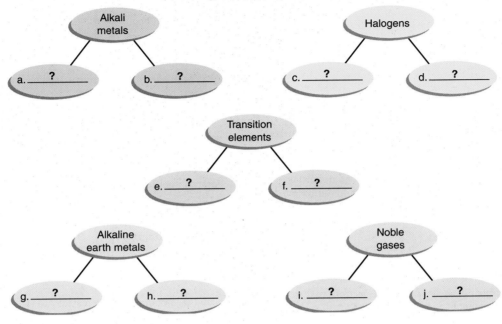

APPLYING YOUR KNOWLEDGE

1. Element 118 was reportedly synthesized in 1999, but the claim was later retracted. As a consequence of that experiment, element 116 was discovered. To what period and group would each of these two elements belong, and what known elements should each resemble?

2. Do you know your periodic table ABC's? What is the only letter that does not appear in the periodic table?

3. Homogenized milk is composed of microscopic globules of fat suspended in a watery medium. Is homogenized milk a true solution (homogeneous mixture)? Explain.

4. Why might a locksmith put graphite in a car's door lock?

5. An article in *Science News* stated, "Noble gases are snobs." What did the author mean?

6. Figure 11.27 on page 317 shows a photograph of chlorine, bromine, and iodine. Why should neither you nor the photographer attempt to take a picture of fluorine?

7. Consider the hypothetical case in which the charge on the electron is twice that of a proton. If the neutral atom must maintain a charge of 0, then how many protons and electrons would calcium-20 contain? Is it possible to make hydrogen this way?

8. In late 2001, news reports said anthrax-contaminated buildings were being treated with chlorine dioxide. What is the formula for this compound?

EXERCISES

11.1 Classification of Matter

1. Classify each of the following materials as an element, compound, heterogeneous mixture, or homogeneous mixture: (a) air, (b) water, (c) diamond, (d) soil.

   Answer: (a) homogeneous mixture (b) compound (c) element (d) heterogeneous mixture

2. Classify each of the following materials as an element, compound, heterogeneous mixture, or homogeneous mixture: (a) a fried egg, (b) ozone, (c) brass, (d) carbon dioxide.

3. Refer to Fig. 11.5 on page 299. Find the approximate solubility (g/100 g water) of KBr at 20°C.

   Answer: about 65 g/100 g H_2O

4. Refer to Fig. 11.5. How many times more soluble (g/100 g water) is KNO_3 at 80°C compared to 40°C?

5. Refer to Fig. 11.5. Would the resulting solution be saturated or unsaturated if 100 g of sugar were stirred into 100 g of water at 40°C?

   Answer: unsaturated, about 240 g of sugar is soluble per 100 g H_2O at 40°C

6. Refer to Fig. 11.5. Would the resulting solution be saturated or unsaturated if 150 g of $NaNO_3$ were stirred into 100 g of water at 60°C?

11.2 Discovery of the Elements

7. Give the symbol for each element: (a) sulfur, (b) sodium, (c) aluminum. (Refer to Table 10.2 if you have difficulty answering Questions 7 through 10.)

   Answer: (a) S (b) Na (c) Al

8. Give the symbol for each element: (a) lithium, (b) lead, (c) xenon.

9. Give the name of each element: (a) N, (b) K, (c) Zn.

   Answer: (a) nitrogen (b) potassium (c) zinc

10. Give the name of each element: (a) Ba, (b) W, (c) Sn.

11.4 The Periodic Table

11. Refer to the periodic table, and give the period and group number for each of these elements: (a) magnesium, (b) zinc, (c) tin.

    Answer: (a) 3, 2A (b) 4, 2B (c) 5, 4A

12. Refer to the periodic table and give the period and group number for each of these elements: (a) neon, (b) barium, (c) iodine.

13. Classify each of these elements as representative, transition, or inner transition; state whether each is a metal or a nonmetal; and give its normal phase: (a) krypton, (b) iron, (c) uranium.

    Answer: (a) representative, nonmetal, gas (b) transition, metal, solid (c) inner transition, metal, solid

14. Classify each of these elements as representative, transition, or inner transition; state whether each is a metal or a nonmetal; and give its normal phase: (a) nickel, (b) magnesium, (c) bromine.

15. Give the total number of electrons, the number of valence electrons, and the number of shells containing electrons in (a) a silicon atom and (b) an argon atom.

    Answer: (a) 14, 4, 3 (b) 38, 8, 3

16. Give the total number of electrons, the number of valence electrons, and the number of shells containing electrons in (a) a beryllium atom and (b) a sulfur atom.

17. Use the periodic table to find the atomic mass, atomic number, number of protons, and number of electrons for an atom of (a) lithium and (b) gold.

    Answer: (a) 6.94 u, 3, 3, 3 (b) 197.0 u, 79, 79, 79

18. Use the periodic table to find the atomic mass, atomic number, number of protons, and number of electrons for an atom of (a) argon and (b) strontium.

19. Give the shell electron configuration for (a) carbon and (b) aluminum.

    Answer: (a) 2, 4 (b) 2, 8, 3

20. Give the shell electron configuration for (a) lithium and (b) phosphorus.

21. Arrange in order of increasing *nonmetallic* character (a) the Period 4 elements Se, Ca, and Mn; and (b) the Group 6A elements Se, Po, and O.

    Answer: (a) Ca, Mn, Se (b) Po, Se, O

22. Arrange in order of increasing *nonmetallic* character (a) the Period 3 elements P, Cl, and Na; and (b) the Group 7A elements F, Br, and Cl.

23. Arrange in order of increasing ionization energy (a) the Period 5 elements Sn, Sr, and Xe; and (b) the Group 8A elements Ar, He, and Ne.

    Answer: (a) Sr, Sn, Xe (b) Ar, Ne, He

24. Arrange in order of increasing ionization energy (a) the Group 1A elements Na, Cs, and K; and (b) the Period 4 elements As, Ca, and Br.

25. Arrange in order of increasing atomic size (a) the Period 4 elements Ca, Kr, and Br, and (b) the Group 1A elements Cs, Li, and Rb.

    Answer: (a) Kr, Br, Ca (b) Li, Rb, Cs

26. Arrange in order of increasing atomic size (a) the Period 2 elements C, Ne, and Be, and (b) the Group 4A elements Si, Pb, and Ge.

11.5 Naming Compounds

27. Name each of these common acids: (a) $H_2SO_4(aq)$, (b) $HNO_3(aq)$, (c) $HCl(aq)$.

    Answer: (a) sulfuric acid (b) nitric acid (c) hydrochloric acid

28. Name each of these common acids: (a) $H_3PO_4(aq)$, (b) $HC_2H_3O_2(aq)$, (c) $H_2CO_3(aq)$.

29. Give the preferred names for (a) $CaBr_2$, (b) N_2S_5, (c) $ZnSO_4$, (d) KOH, (e) $AgNO_3$, (f) IF_7, (g) $(NH_4)_3PO_4$, and (h) Na_3P.

    Answer: (a) calcium bromide (b) dinitrogen pentasulfide (c) zinc sulfate (d) potassium hydroxide (e) silver nitrate (f) iodine heptafluoride (g) ammonium phosphate (h) sodium phosphide

30. Give the preferred names for (a) $Al_2(CO_3)_3$, (b) $(NH_4)_2SO_4$, (c) Li_2S, (d) SO_3, (e) Ba_3N_2, (f) $Ba(NO_3)_2$, (g) SiF_4, and (h) S_2Cl_2.

11.6 Groups of Elements

31. Sodium forms the compounds Na_2S, Na_3N, and $NaHCO_3$. What are the formulas for lithium sulfide, lithium nitride, and lithium hydrogen carbonate?

    Answer: Li_2S, Li_3N, $LiHCO_3$

32. Magnesium forms the compounds $Mg(NO_3)_2$, $MgCl_2$, and $Mg_3(PO_4)_2$. What are the formulas for barium nitrate, barium chloride, and barium phosphate?

ON THE WEB

1. Light Up the Sky!

What important elements make fireworks work? Go to the student website at **www.cengagebrain.com/shop/ISBN/1133104096**, follow the recommended links, and compile your list.

2. Everything You've Ever Wanted to Know about the Periodic Table of Elements

What is a "neutral" atom, and what happens when the number of electrons changes? How is the periodic table built up? What do the letters *s*, *p*, and *d* mean? How would you explain the rules for electron configurations? What are the origins of the periodic table? Have you ever thought about how you could possibly "weigh" an atom? How did scientists arrive at the name "periodic table"? To answer these questions, follow the recommended links on the student website at **www.cengagebrain.com/shop/ISBN/1133104096**.

Chemical Bonding

*See plastic nature
working to this end
The single atoms each
to other tend
Attract, attracted to,
the next in place
Form'd and impell'd its
neighbor to embrace.*

•

Alexander Pope
(1688–1744)

< The breaking and formation of
chemical bonds are responsible
for the vivid colors in this fireworks display over Capitol Hill.

I n this chapter the focus will be on chemical bonding and its role in compound formation. As the chapter-opening quotation implies, virtually
everything in nature depends on chemical bonds. The proteins, carbohydrates, fats, and nucleic acids that make up living matter are complex molecules
held together by chemical bonds. Simpler molecules such as the ozone, carbon
dioxide, and water in the air have bonds that allow them to absorb radiant energy,
thus keeping the Earth at a livable temperature.

Various molecules and ions bond together to form the compounds that make
up the minerals and rocks of the Earth. Potassium ions in our heart cells help
maintain the proper contractions, and in the extracellular fluids they help control
nerve transmissions to muscles. Were it not for the hydrogen bonding that causes

CHEMISTRY FACTS

▶ Snowflakes are generally sym-
metrical, hexagonal (six-sided)
structures with shapes like
ornate plates, needles, and
columns (see the Chapter 5.2
Highlight). As water crystallizes the symmetry results from
the maximization of attractive
electrical forces between water
molecules (hydrogen bonds).

▶ The yellow-brown color of the
noxious photochemical smog
that gathers over large cities
is due to nitrogen oxides, NO_X
(Chapter 20.4). These pollutants are typically formed when
nitrogen and oxygen combine
in high-temperature combustion reactions like those in
automobile engines.

▶ Each year approximately
21,000 toddlers visit emergency rooms in the United
States after ingesting coins

Chapter Outline

one water molecule to attract four others, water would not be a liquid at room temperature and life as we know it would not exist on the Earth.

Chemical bonding results from the electromagnetic forces among the various electrons and nuclei of the atoms involved, and chemists often must use quantum mechanics (Chapter 9.7) to solve the complicated submicroscopic problems that arise. As will be seen, however, relatively simple concepts can correlate and explain much of the information encountered in chemical bonding.

Our study of chemical bonding begins with a discussion of two laws that describe mass relationships in compounds and helped lead John Dalton to the atomic theory.

12.1   Law of Conservation of Mass

Preview Questions

● Does a change in chemical mass occur during a chemical reaction?

● Why is Antoine Lavoisier called "The Father of Chemistry"?

If the total mass involved in a chemical reaction is precisely measured before and after the reaction takes place, the most sensitive balances cannot detect any change (● Fig. 12.1). This generalization is known as the **law of conservation of mass**:

> No detectable change in the total mass occurs during a chemical reaction.

As discussed in the **Highlight: Lavoisier, "The Father of Chemistry,"** this law was discovered in 1774 by Antoine Lavoisier ("lah-vwah-ZHAY"). Now, two centuries later, Lavoisier's law of conservation of mass is still valid and useful.

EXAMPLE 12.1   Using the Law of Conservation of Mass

The complete burning of 4.09 g of carbon in oxygen produces 15.00 g of carbon dioxide as the only product. How many grams of oxygen gas must have reacted?

Solution

The total mass before and that after reaction must be equal. Because carbon dioxide was the only product and weighed 15.00 g, the total mass of carbon and oxygen that combined to produce it must have been 15.00 g. Because carbon's mass was 4.09 g, the oxygen that reacted with it must have contributed 15.00 g minus 4.09 g, or 10.91 g.

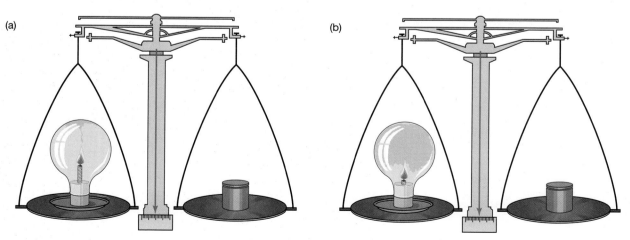

FIGURE 12.1 The Law of Conservation of Mass
When a candle is burned in an airtight container with oxygen, there is no detectable change in the candle's mass, as illustrated by the balance's pointer being in the same place (a) before the reaction and (b) after the reaction.

Highlight   Lavoisier, "The Father of Chemistry"

In the seventeenth and eighteenth centuries, the phenomenon of combustion was studied intensely. Earlier hypotheses were discarded thanks to the quantitative experiments of the wealthy French nobleman Antoine Lavoisier (Fig. 1). In 1774, he made chemistry a modern science, just as Galileo and Newton had done for physics more than a century earlier.

Lavoisier performed the first quantitative chemical experiments to explain combustion and to settle the question of whether mass was gained, lost, or unchanged during a chemical reaction. Unlike others who had tried to answer the same question, Lavoisier understood the nature of gases and took care to do his experiments in closed containers so that no substances could enter or leave. In addition, he used the most precise balance that had ever been built.

After many experiments, he was able to formulate the following law: *No detectable change in the total mass occurs during a chemical reaction* (the law of conservation of mass). During the course of these investigations, Lavoisier established conclusively that when things burn, they are not losing something to the air but gaining *oxygen* from it. Sometimes the burned materials gained mass by forming solid oxides, and sometimes they lost mass by forming gaseous oxides. The successes of Lavoisier caused the importance of measurement to be widely recognized by chemists.

In addition to introducing quantitative methods into chemistry, discovering the role of oxygen in combustion, and finding the law of conservation of mass, Lavoisier established the principles for naming chemicals. In 1789, he wrote the first modern chemistry textbook. Lavoisier is justly referred to as "The Father of Chemistry."

Because Lavoisier was wealthy and because he had made an enemy of a leader of the French Revolution, Lavoisier was guillotined in 1794 during the last months of the Revolution. When he objected upon arrest that he was a scientist, not a "tax-farmer," the arresting officer replied that "the republic has no need of scientists." Within two years of Lavoisier's death, the regretful French were unveiling busts of him.

?PRISMA/V&W/The Image Works

FIGURE 1 The Lavoisiers
Antoine Lavoisier (1743–1794), known as "The Father of Chemistry," and his wife, Marie-Anne, who helped him with experiments and produced the illustrations in his textbook.

Confidence Exercise 12.1

If 111.1 g of calcium chloride is formed when 40.1 g of calcium is reacted with chlorine, then what mass of chlorine combined with the calcium?

Answers to Confidence Exercises may be found at the back of the book.

Did You Learn?

- No detectable change in the total mass occurs during a chemical reaction; this is the law of conservation of mass.

- Lavoisier performed the first quantitative experiments that help prove the conservation of mass and wrote the first modern chemistry textbook.

12.2   Law of Definite Proportions

Preview Questions

- What is the sum of the atomic masses given in the chemical formula of a substance called?

- What is the proportion of elements in different samples of a compound?

Formula Mass

Recall from Chapter 10.2 that the *atomic mass* (abbreviated here AM) of an element is the *average* mass assigned to each atom in its naturally occurring mixture of isotopes. The masses of elements are based on a scale that assigns the ^{12}C atom the value of exactly 12 u. The atomic masses for most elements have been determined to several decimal places, as shown in the periodic table inside the front cover of this textbook, but for convenience these values are rounded to the nearest 0.1 u here. For example, the atomic masses of hydrogen, oxygen, and calcium are considered to be 1.0 u, 16.0 u, and 40.1 u, respectively.

The **formula mass** (abbreviated FM) of a compound or element is the sum of the atomic masses given in the formula of the substance. For example, the formula mass of O_2 is 16.0 u + 16.0 u = 32.0 u, and the formula mass of methane (swamp gas, CH_4) is 12.0 u + (4 × 1.0 u) = 16.0 u. If the formula of an element is only given by the element's *symbol* (for example Fe or Xe), then the formula mass is simply the atomic mass.*

EXAMPLE 12.2   Calculating Formula Masses

Find the formula mass of lead chromate, $PbCrO_4$, the bright yellow compound used in paint for the yellow lines on streets.

Solution

Find the atomic masses of Pb, Cr, and O in the periodic table. The formula shows one atom of Pb (207.2 u), one atom of Cr (52.0 u), and four atoms of O (16.0 u), so FM = 207.2 u + 52.0 u + (4 × 16.0 u) = 323.2 u.

Confidence Exercise 12.2

Find the formula mass of hydrogen sulfide, H_2S, the gas that gives rotten eggs their offensive odor.

Answers to Confidence Exercises may be found at the back of the book.

Law of Definite Proportions

In 1799, the French chemist Joseph Proust ("Proost") discovered the **law of definite proportions**:

Different samples of a pure compound always contain the same elements in the same proportion by mass.

For example,

9 g H_2O is composed of 8 g of oxygen and 1 g of hydrogen.

18 g H_2O is composed of 16 g of oxygen and 2 g of hydrogen.

36 g H_2O is composed of 32 g of oxygen and 4 g of hydrogen.

In each case, the ratio by mass of oxygen to hydrogen is 8 to 1.

Because the elements in a compound are present in a specific proportion (ratio) by mass, they are also present in a specific percentage by mass. The general equation for calculating the percentage of any component X in a total is

$$\% \ X \ \text{component} = \frac{\text{amount of component } X}{\text{total amount}} \times 100\% \qquad 12.1$$

*Chemists often use the term *atomic weight* in place of *atomic mass, molecular weight* or *formula weight* instead of *formula mass*, and *amu* rather than *u*.

Equation 12.1 can be used to find the percentage by mass of an element if the total mass of the compound and the mass contribution of the element are known. The percentage by mass of an element X in a compound can be calculated from the compound's formula by using Equation 12.2.

$$\% \ X \text{ by mass} = \frac{\text{mass of component } X}{\text{formula mass of compound}} \times 100\% \qquad \textbf{12.2}$$

EXAMPLE 12.3 Finding the Percent Component or Percent Mass in a Compound

(a) A compound having 44 total atoms consists of 11 atoms of element X and 33 atoms of element Y. Find the percent element X in the compound.

(b) Dry ice is solid carbon dioxide (● Fig. 12.2). Find the percent mass of carbon and oxygen in CO_2.

Solution

(a) Using Eq. 12.1, it is found that

$$\% \ X \text{ component} = \frac{11}{44} \times 100\% = 25\%$$

(b) The periodic table shows that the atomic masses of C and O are 12.0 u and 16.0 u, respectively. The total mass of C is 12.0 u and for O_2 is $(2 \times 16.0 \text{ u}) = 32.0 \text{ u}$. Therefore, the formula mass of CO_2 is $12.0 \text{ u} + (2 \times 16.0 \text{ u}) = 44.0 \text{ u}$. Using Eq. 12.2, it is found that

$$\% \ C \text{ by mass} = \frac{\text{total mass of C}}{\text{formula mass of } CO_2} \times 100\% = \frac{12.0 \text{ u}}{44.0 \text{ u}} \times 100\% = 27.3\%$$

Because oxygen is the only other component, the percent O and the percent C must add up to 100%. Thus, the % O by mass must be 100.0% minus 27.3%, or 72.7%.

Confidence Exercise 12.3

Find the percent mass of aluminum and oxygen in aluminum oxide, Al_2O_3, the major compound in rubies.

When a compound is broken down, its elements are found to be in a definite proportion by mass. Conversely, when the same compound is made from its elements, the elements will combine in that same proportion by mass. If the elements that are combined to form a compound are not mixed in the correct proportion, then one of the elements, called the **limiting reactant**, will be used up completely. The other, the **excess reactant**, will be only partially used up; some of it will remain unreacted.

In ● Fig. 12.3a, 10.00 g of Cu wire reacts completely with 5.06 g S to form 15.06 g of CuS, copper(II) sulfide. None of either reactant is left over. In Fig. 12.3b, 10.00 g of Cu reacts with 7.06 g of S, but this is not the proper ratio.

In accordance with the law of definite proportions, only 5.06 g of S can combine with 10.00 g of Cu. Therefore, 15.06 g of CuS is again formed, and the other 2.00 g of sulfur is in excess. Notice that the law of conservation of mass is satisfied; the total mass before and after reaction is 17.06 g. In Fig. 12.3c, it is the copper that is the excess reactant, and both the law of conservation of mass and the law of definite proportions again hold.

Did You Learn?

- The formula mass of a compound or element is the sum of the atomic masses given in the formula of a substance.

- The law of definite proportions states that different samples of a given compound always contain the same elements in the same proportion by mass.

FIGURE 12.2 Dry Ice
Solid carbon dioxide has a temperature of about $-78°C$, so cold that it should never touch the bare skin. Like mothballs, it sublimes (passes directly from the solid phase into the gaseous phase). The white fog is water vapor condensing from the air in contact with the cold CO_2.

Sean Brady

FIGURE 12.3 The Law of Definite Proportions

(a) The law of definite proportions indicates that when Cu and S react to form a specific compound, they will always react in the same ratio by mass. (b) and (c) If the ratio in which they are mixed is different, then part of one reactant will be left over.

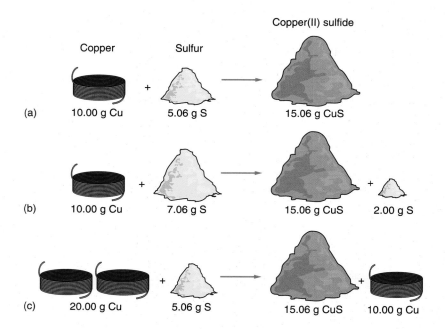

12.3 Dalton's Atomic Theory

Preview Questions

● According to Dalton's theory, are atoms lost or gained during chemical reactions?

● Which two laws did Dalton's hypotheses explain?

In 1803, atomic scientist John Dalton (Chapter 9.1) proposed the following hypotheses to explain the laws of conservation of mass and definite proportions.

1. Each element is composed of tiny, indivisible particles called *atoms,* which are identical for that element but are different (particularly in their masses and chemical properties) from atoms of other elements.

2. Chemical combination is simply the bonding of a definite, small whole number of atoms of each of the combining elements to make one molecule of the formed compound. A given compound always has the same relative numbers and types of atoms.

3. No atoms are gained, lost, or changed in identity during a chemical reaction; they are just rearranged to produce new substances.

If a chemical reaction is just a rearrangement of atoms, then it is easy to see that the law of conservation of mass is explained.

These hypotheses also explain the law of definite proportions. If different samples of a particular compound are made up only of various numbers of the same basic molecule, then it is clear that the mass ratio of the elements will have to be the same in every sample of that compound.

For example, each molecule of H_2O is composed of one atom of oxygen (AM 16.0 u) and two atoms of hydrogen (AM 1.0 u, total mass of hydrogen = 2 × 1.0 u = 2.0 u). Thus, each molecule of H_2O is composed of 16.0 parts oxygen by mass and 2.0 parts hydrogen by mass, or a ratio of 8.0 to 1.0. Because every pure sample of a molecular compound is simply a very large collection of identical molecules, the proportion by mass of any element in every sample of the compound will be the proportion by mass that it has in an individual molecule of that compound.

Because Dalton's hypotheses explained these two laws, it is no wonder that he believed that atoms exist and behave as he stated. In addition, Dalton also determined that the *mass ratio* of two molecules is also equal to the small whole-number *atom ratio* of each element in two molecules (carbon in CO and oxygen in CO_2 exist in a 1-to-2 mass ratio and in a 1-to-2 atom ratio). Experiments by other scientists verified Dalton's prediction (now called the *law of multiple proportions*).

More and more supporting evidence for Dalton's concept of the atom accumulated. Some modification of his original ideas occurred as new evidence arose, but modification is expected with all scientific ideas. Dalton's *atomic theory* is the cornerstone of chemistry. Whenever an explanation of a chemical occurrence is sought, our thoughts immediately turn to the concept of atoms.

Did You Learn?

- No atoms are gained or lost during a chemical reaction; they are just rearranged to produce a new substance.

- Dalton's hypotheses explained the law of conservation of mass and the law of definite proportions.

12.4 Ionic Bonding

Preview Questions

- In forming compounds, how many electrons do atoms tend to get in the outer shell?
- What are ionic bonds?

You learned in Chapter 11 that elements in the same group have the same number of valence (outer) electrons and form compounds with similar formulas; for example, the chlorides of Group 1A are LiCl, NaCl, KCl, RbCl, and CsCl, and all have a 1-to-1 ratio of atoms. Because of this behavior, we conclude that the valence electrons are the ones involved in compound formation.

Recall from Chapter 11.6 that the noble gases (Group 8A) are unique in that, except for several compounds of Xe and of Kr, they do not bond chemically with atoms of other elements. Also, all noble gases are monatomic; that is, their atoms do not bond to one another to form molecules. Thus, a noble gas electron configuration (eight electrons in the outer shell, but just two for He) is uniquely stable. It seems to provide a balanced structural arrangement that minimizes the tendency of an atom to react with other atoms.

The formation of the vast majority of compounds is explained by combining these two conclusions into the **octet rule**:

In forming compounds, atoms tend to gain, lose, or share valence electrons to achieve electron configurations of the noble gases; that is, they tend to get eight electrons (an octet) in the outer shell. Hydrogen is the main exception; it tends to get two electrons in the outer shell, like the configuration of the noble gas helium.

Individual atoms can achieve a noble gas electron configuration in two ways: by transferring electrons or by sharing electrons. Bonding by transfer of electrons is discussed in this section, and bonding by sharing of electrons is treated in Chapter 12.5.

In the transfer of electrons, one or more atoms lose their valence electrons, and another one or more atoms gain these same electrons to achieve noble gas electron configurations. Compounds formed by this electron transfer process are called **ionic compounds**. This name is used because the loss or gain of electrons destroys the electrical neutrality of the atom and produces the net positive or negative electric charge that characterizes an ion.

Atoms of metals generally have low ionization energies and thus tend to lose electrons and form positive ions (Chapter 11.4). On the other hand, atoms of nonmetals tend to gain electrons and form negative ions. A nonmetal atom that needs only one or two electrons to fill its outer shell can easily acquire the electrons from atoms that have low ionization

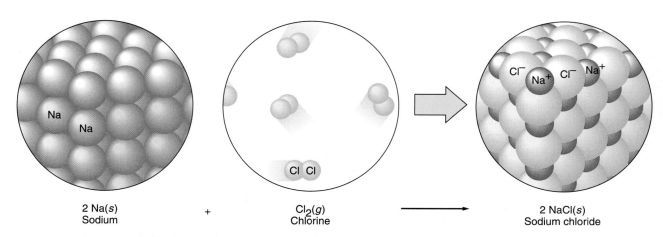

FIGURE 12.4 The Formation of Sodium Chloride
Metallic sodium (*left*) and chlorine gas (*middle*) react to form the ionic compound *sodium chloride* (*right*).

energy, that is, from atoms with only one or two electrons in the outer shell, such as those of Groups 1A and 2A. Atoms of some elements gain or lose as many as three electrons.

To illustrate, consider how common table salt, NaCl, is formed. When the nonmetal chlorine, a pale green gas composed of Cl_2 molecules, is united with the metal sodium, an energetic chemical reaction forms the ionic compound sodium chloride (● Fig. 12.4).

The loss of the valence electron from a neutral sodium atom to form a sodium ion with a 1+ charge is illustrated in ● Fig. 12.5a. The neutral atom has the same number of protons (11) as electrons (11). Thus, the net electric charge on the atom is zero. After the loss of an electron, the number of protons (11) is one more than the number of electrons (10), leaving a net charge on the sodium ion of 11 minus 10, or 1+. Similarly, Fig. 12.5b illustrates the gain of an electron by a neutral chlorine atom (17 p, 17 e) to form a chloride ion (17 p, 18 e) with a charge of 17 minus 18, or 1−. The net electric charge on an ion is the number of protons minus the number of electrons.

As Na^+ and Cl^- ions are formed, they are bonded by the attractive electric forces among the positive and negative ions. ● Figure 12.6 shows a model of sodium chloride that illustrates how the sodium ions and chloride ions arrange themselves in an orderly, three-dimensional lattice to form a cubic crystal. (A *crystal* is a solid whose external symmetry reflects an orderly, geometric, internal arrangement of atoms, molecules, or ions.)

No neutral units of fixed size are shown in Fig. 12.6; it is impossible to associate any one Na^+ with one specific Cl^-. Thus, it is somewhat inappropriate to refer to a "molecule" of sodium chloride or of any other ionic compound. Instead, we generally refer to one sodium ion and one chloride ion as being a *formula unit* of sodium chloride, the smallest combination of ions that gives the formula of the compound.

FIGURE 12.5 The Formation of (a) a Sodium Ion and (b) a Chloride Ion
After the transfer of one electron from the Na atom to the Cl atom, each ion has the electron configuration of a noble gas (eight electrons in the outer shell).

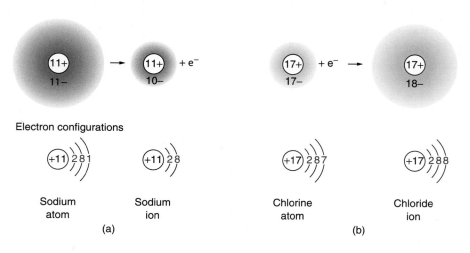

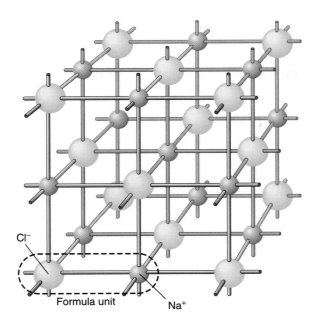

FIGURE 12.6 Sodium Chloride (NaCl) Schematic diagram of the sodium chloride crystal. The oval of dashes indicates what is meant by a *formula unit* of NaCl.

Similarly, the formula CaF_2 means that the compound calcium fluoride has one calcium ion (Ca^{2+}) for every two fluoride ions ($2F^-$). The three ions constitute a formula unit of calcium fluoride.

These ideas of compound formation emerged largely from the work of American chemist Gilbert Newton Lewis (1875–1946), who in 1916 developed *electron dot symbols* to help explain chemical bonding. In a **Lewis symbol**, the nucleus and the inner electrons of an atom or ion are represented by the element's symbol, and the valence electrons are shown as dots arranged in four groups of one or two dots around the symbol (● Table 12.1). It makes no difference on which side of the symbol various electron dots are placed, but they are left unpaired to the greatest extent possible.

Lewis structures use Lewis symbols to show valence electrons in molecules and ions of compounds. In a Lewis structure, a shared electron pair is indicated by two dots halfway between the atoms or, more often, by a dash connecting the atoms. Unshared pairs of valence electrons called *lone pairs*, those not used in bonding, are shown as belonging to the individual atom or ion. Lewis structures are two-dimensional representations, so they do not show the actual three-dimensional nature of molecules.

The sodium and chlorine atoms of Fig. 12.5 are represented in Lewis symbols as

$$Na \cdot \qquad \cdot \overset{\cdot\cdot}{\underset{\cdot\cdot}{Cl}} :$$

The sodium and chloride ions of Fig. 12.5 are represented in Lewis symbols as

$$Na^+ \qquad : \overset{\cdot\cdot}{\underset{\cdot\cdot}{Cl}} :^-$$

Table 12.1	Lewis Symbols for the First Three Periods of Representative Elements						
1A							8A
H•	2A	3A	4A	5A	6A	7A	He:
Li•	•Be•	•Ḃ•	•Ċ•	•N̈•	:Ö•	:F̈•	:N̈e:
Na•	•Mg•	•Ȧl•	•Ṡi•	•P̈•	:S̈•	:C̈l•	:Är:

FIGURE 12.7 Pattern of Ionic Charges
Most of the representative elements exhibit a regular pattern of ionic charges when they form ions.

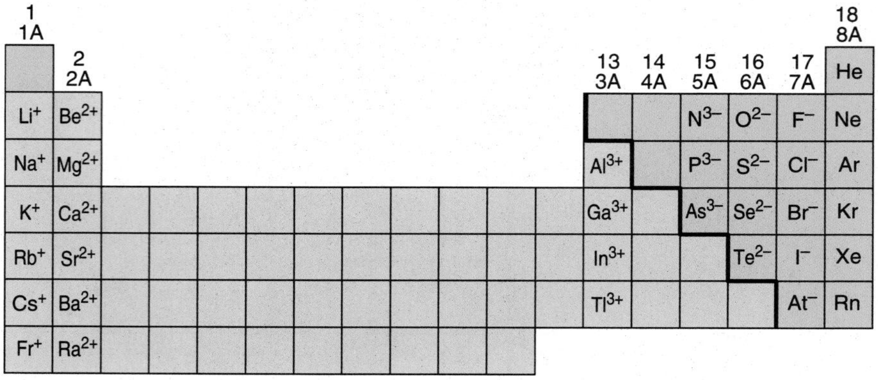

where the + and − indicate the same charge as that on a proton and an electron, respectively. If the charge of the ion is greater than 1+ or 1−, then it is represented by a numeral followed by the appropriate sign, as shown in the Lewis symbols for the magnesium ion and the oxide ion:

$$Mg^{2+} \qquad :\ddot{O}:^{2-}$$

The charge on most ions of the representative elements can be determined easily. First, metals form positive ions, or **cations** ("CAT-eye-ons"); nonmetals form negative ions, or **anions** ("AN-eye-ons"). Second, the positive charge on the metal's ion will be equal to the atom's number of valence electrons (its group number), whereas the negative charge on the nonmetal's ion will be the atom's number of valence electrons (its group number) minus 8 (● Fig. 12.7).

Group 8A elements already have eight valence electrons and thus form no ions in chemical reactions. For Group 4A, carbon and silicon do not form ions during chemical reactions because it is so hard to either lose or gain four electrons. (Only the ions shown in Fig. 12.7 and a few transition metals in Table 12.1 will be considered in this text.)

The pattern of charges in Fig. 12.7 arises because valence electrons are being lost by 12 to 17 metals and gained by nonmetals, generally to the extent necessary to get eight electrons in the outer shell; in other words, to acquire the same electron configuration, termed *isoelectronic,* as a noble gas.

For example, an atom of aluminum, a metal in Group 3A, has three valence electrons. When the atom loses three electrons, an aluminum ion with a charge of 3+ is formed. The Al^{3+} is isoelectronic with an atom of the noble gas neon (Ne) because both have a 2,8 electron configuration (see Fig. 11.21). Sulfur, a nonmetal in Group 6A, has six valence electrons. When forming an ion, a sulfur atom gains two additional electrons. The sulfide ion (S^{2-}) thus has a negative charge of 6 − 8, or 2−, and it is isoelectronic with the noble gas argon (Ar) because both have a 2,8,8 electron configuration.

In the formation of simple ionic compounds, one element loses its electrons and the other element gains them, resulting in ions. The ions are then held together in the crystal lattice by the electrical attractions among them, the **ionic bonds**. Using Lewis symbols and structures, we can represent the formation of NaCl as

$$Na\cdot + \cdot\ddot{\underset{..}{C}l}: \longrightarrow Na^{+} \quad :\ddot{\underset{..}{C}l}:^{-}$$

In this example, the sodium atom lost one electron and the chlorine atom gained one. In every ionic compound, the total charge in the formula adds to zero and the compound exhibits electrical neutrality. Thus, in the case of NaCl, the ratio of Na^{+} to Cl^{-} must be 1 to 1 so that the compound will exhibit net electrical neutrality.

Another example of an ionic compound is calcium oxide, CaO. Its formation is represented as

$$\cdot Ca\cdot + \cdot\ddot{O}: \longrightarrow Ca^{2+} \quad :\ddot{O}:^{2-}$$

Table 12.2 Formulas of Ionic Compounds

General Cation Symbol	General Anion Symbol	Cation-to-Anion Ratio for Neutrality	General Compound Formula	Specific Example of Compound
M^+	X^-	1 to 1	MX	NaF
M^{2+}	X^{2-}	1 to 1	MX	MgO
M^{3+}	X^{3-}	1 to 1	MX	AlN
M^+	X^{2-}	2 to 1	M_2X	Na_2O
M^{2+}	X^-	1 to 2	MX_2	MgF_2
M^+	X^{3-}	3 to 1	M_3X	Na_3N
M^{3+}	X^-	1 to 3	MX_3	AlF_3
M^{2+}	X^{3-}	3 to 2	M_3X_2	Ca_3N_2
M^{3+}	X^{2-}	2 to 3	M_2X_3	Al_2O_3

The calcium atom loses two electrons and the oxygen atom gains two. Thus, to get a net electrically neutral compound, we again need a 1-to-1 ratio of ions of each element.

Now consider what happens when calcium and chlorine react. A calcium atom has two valence electrons to lose, but each atom of chlorine can gain only one. There must be two atoms of chlorine to accept the two electrons of the calcium atom and give a net electrically neutral compound:

$$\cdot Ca \cdot + \cdot \ddot{\underset{..}{Cl}} : + \cdot \ddot{\underset{..}{Cl}} : \longrightarrow Ca^{2+} \quad : \ddot{\underset{..}{Cl}} :^- \quad : \ddot{\underset{..}{Cl}} :^-$$

Thus, the formula for calcium chloride is $CaCl_2$. All the ions in the compound have noble gas configurations, and the total charge is zero.

You can now understand how formulas of ionic compounds arise. The numbers of atoms of the various elements involved in the compound are determined by the requirements (1) that the total electrical charge be zero and (2) that all the atoms have noble gas electron configurations.

For ionic compounds to be electrically neutral, each formula unit must have an equal number of positive and negative charges. Thus, for the formulas for ionic compounds to be written correctly, the anions and cations must be shown in the smallest whole-number ratio that will equal zero charge. A simple way to determine the formula of an ionic compound is as follows: Use the anion charge (no sign) as the subscript for the cation and the cation charge (no sign) as the subscript for the anion. Convert the subscripts to the smallest whole-number ratio if necessary. This principle, which works for both monatomic and polyatomic ions, can be mastered by studying ● Table 12.2, Example 12.4, and Confidence Exercise 12.4.

The only other information you need to become adept at writing the formulas of ionic compounds is knowledge of the charges on common polyatomic ions (given in Table 11.6) and on ions of the representative elements (follow the pattern in Fig. 12.7).

EXAMPLE 12.4   Writing Formulas for Ionic Compounds

Write the formula for calcium phosphate, the major component of bones.

Solution

Calcium is in Group 2A, so this cation has an ionic charge of 2+. The phosphate ion is the anion PO_4^{3-} (Table 11.6). Use the anion charge number (3) as the subscript for Ca and the cation charge number (2) as the subscript for PO_4^{3-}. Therefore, neutrality can be achieved with three Ca^{2+} and two PO_4^{3-}; the correct formula is $Ca_3(PO_4)_2$. (The parentheses around the phosphate ion show that the subscript 2 applies to the entire polyatomic ion.)

Confidence Exercise 12.4

In the nine blank spaces in the matrix below, write the formulas for the ionic compounds formed by combining each metal ion (*M*) with each nonmetal ion (*X*). This matrix covers the general formulas for all possible compounds formed from ions having charges of magnitudes one through three.

	X^-	X^{2-}	X^{3-}
M^+			
M^{2+}			
M^{3+}			

Very strong forces of attraction exist among oppositely charged ions, so ionic compounds are usually crystalline solids with high melting and boiling points. Recently, chemists have prepared ionic substances that have unusual properties due to their large, nonspherical cations that lead to especially weak ionic bonding. Often these ionic substances are liquid at room temperature. Some of these *ionic liquids* promise to be "supersolvents," and because of their environmental benefits, they may replace some organic solvents. For an interesting application of ionic bonding, see the **Highlight: Photochromic Sunglasses**.

Another important property of ionic compounds is their behavior when an electric current is passed through them. If a lightbulb is connected to a battery by two wires, then the bulb glows. Electrons flow from the negative terminal of the battery, through the lightbulb, and back to the positive terminal. If one wire is cut, then the electrons cannot flow and the lightbulb will not glow. If the ends of the cut wire are inserted into a solid ionic compound, then the bulb does not light, because the ions are held in place and cannot move (● Fig. 12.8a). If the cut wires are inserted into a *melted* ionic compound, such as molten (liquid) NaCl, however, then the bulb lights (Fig. 12.8b).

Ionic compounds in the liquid phase conduct an electric current because the ions are now free to move and carry charge from one wire to the other. This is the crucial test of whether a compound is ionic: when melted, does it conduct an electric current? If so, then it is ionic. ● Figure 12.9 illustrates how the conduction takes place.

Many ionic compounds dissolve in water, thus forming aqueous solutions in which the ions are free to move. Like molten salts, such solutions conduct an electric current.

FIGURE 12.8 Ionic Compounds Conduct Electricity When Melted (but Not When Solid)
(a) When dry salt, an ionic compound, is used to connect the electrodes, the bulb in the conductivity tester does not light. (b) When the salt is melted, the ions are free to move, the circuit is completed, and the bulb lights. (A solution of salt dissolved in water also conducts electricity.)

Highlight Photochromic Sunglasses

Sunglasses can be troublesome. It seems that they are often getting lost or sat on. One solution to this problem for people who wear glasses is photochromic glass, glass that darkens automatically in response to intense light. Recall that glass is a complex, noncrystalline material that is composed of polymeric silicates. Of course, glass transmits visible light; its transparency is its most useful property.

Glass can be made photochromic by adding tiny silver chloride crystals that get trapped in the glass matrix as the glass solidifies. Silver chloride has the unusual property of darkening when struck by light; it is the property that makes silver halide salts so useful for photographic films. This darkening occurs because light causes an electron transfer from Cl^- to Ag^+ in the silver chloride crystal, forming a silver atom and a chlorine atom. The silver atoms formed in this way tend to migrate to the surface of the silver chloride crystal, where they aggregate to form a tiny crystal of silver metal, which is opaque to light.

In photography the image defined by the grains of silver is fixed by chemical treatment so that it remains permanent. In photochromic glass, however, this process must be reversible; the glass must become fully transparent again when the person goes back indoors. The secret to the reversibility of photochromic glass is the presence of Cu^+ ions. The added Cu^+ ions serve two important functions. First, they reduce the Cl atoms formed in the light-induced reaction, which prevents them from escaping from the crystal:

$$Ag^+ + Cl^- \xrightarrow{\text{light}} Ag + Cl$$
$$Cl + Cu^+ \longrightarrow Cu^{2+} + Cl^-$$

Second, when the exposure to intense light ends (the person goes indoors), the Cu^{2+} ions migrate to the surface of the silver chloride crystal, where they accept electrons from silver atoms as the tiny crystal of silver atoms disintegrates:

$$Cu^+ + Ag \longrightarrow Cu^+ + Ag^+$$

The Ag^+ ions are re-formed in this way and then return to their places in the silver chloride crystal, making the glass transparent once again.

Typical photochromic glass decreases to about 20% transmittance (transmits 20% of the light that strikes it) in strong sunlight and then, over a period of a few minutes, returns to about 80% transmittance indoors (normal glass has 92% transmittance).

FIGURE 1 Glasses with Photochromic Lenses
The right lens has been exposed to light; the left one has not.

AP Photo/PRNewsFoto/Transitions Optical, Inc.

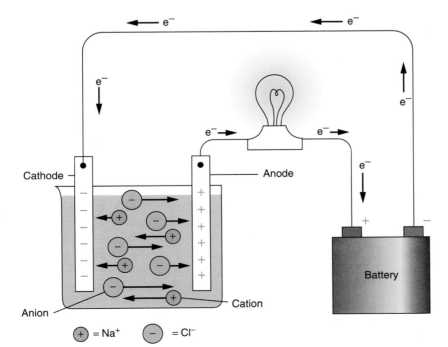

+ = Na$^+$ − = Cl$^-$

FIGURE 12.9 How Melted Salt (NaCl) Conducts Electricity
The negative ions (anions) move toward the anode, and the positive ions (cations) move toward the cathode. The movement of the ions closes the electric circuit and allows electrons to flow in the wire, as indicated by the lighted bulb.

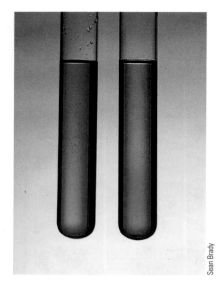

FIGURE 12.10 The Ions of Chromium Solutions containing Cr^{2+} are usually blue; those of Cr^{3+} are generally green.

Sean Brady

(Table 12.5 summarizes the properties of ionic compounds and compares them with those of covalent compounds.)

The Stock System

The rules for naming compounds are discussed in Chapter 11.5, where it is stated that the rules must be expanded for metals that form two (or more) types of ions. Such metals form more than one compound with a given nonmetal or polyatomic ion. To distinguish the compounds, the **Stock system** is used. In this system, a Roman numeral giving the value of the metal's ionic charge is placed in parentheses directly after the metal's name. For example,

$CrCl_2$ chromium(II) chloride

$CrCl_3$ chromium(III) chloride

Because the two chloride ions have a 1− charge and the compound must be electrically neutral, the chromium ion in $CrCl_2$ must have a charge of 2+. Similar reasoning indicates that the chromium ion in $CrCl_3$ must have a charge of 3+. The two compounds, $CrCl_2$ and $CrCl_3$, have entirely different properties (● Fig. 12.10).

For simplicity, only the four metals listed in ● Table 12.3 (copper, gold, iron, and chromium) will be considered here. The Stock system names of the metal ions are given in the table.

EXAMPLE 12.5   Naming Compounds of Metals That Form Several Ions

A certain compound of gold and sulfur has the formula Au_2S. As shown in Table 12.3, gold is a metal that forms several ions, so it would be preferable to name Au_2S using the Stock system. What is its Stock system name?

Solution

Sulfur has a 2− ionic charge (Fig. 12.7). Thus, the two Au atoms must contribute a total of 2+, or 1+ each. Therefore, gold's ionic charge in this compound is 1+, and the compound's Stock system name is gold(I) sulfide.

Confidence Exercise 12.5

Some copper and fluoride compounds have formulas CuF and CuF_2. What would be their Stock system names?

Did You Learn?

● In forming compounds, atoms tend to get eight electrons (an octet) in the outer shell (the configuration of the noble gases).

● When one element loses electrons and another element gains them, ions result and ionic compounds are held together by electrical ionic bonds.

12.5   Covalent Bonding

Preview Questions

● What are covalent compounds?

● How does electronegativity vary in the periodic table?

When electrons are *transferred*, ions are produced and an ionic compound is formed. When electrons are *shared*, molecules are produced and a **covalent compound** is formed. When a

Table 12.3	Some Common Metals That Form Two Ions
Ion	**Stock System Name**
Cu^+	copper(I)
Cu^{2+}	copper(II)
Au^+	gold(I)
Au^{3+}	gold(III)
Fe^{2+}	iron(II)
Fe^{3+}	iron(III)
Cr^{2+}	chromium(II)
Cr^{3+}	chromium(III)

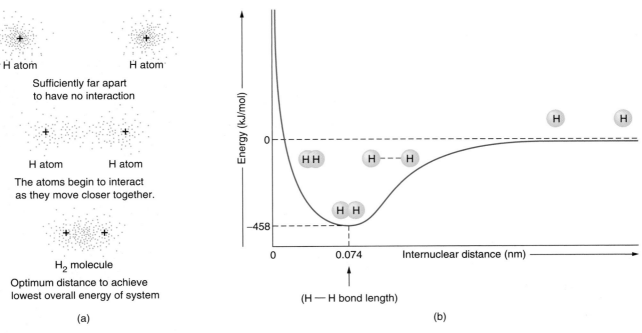

(a)

(b)

FIGURE 12.11 Bonding in the H$_2$ Molecule

(a) The interaction of two hydrogen atoms shown by electron cloud model. (b) Energy of the system is plotted as a function of distance between the nuclei of the two hydrogen atoms. As the atoms approach one another, the energy decreases until the distance reaches 0.074 nm. This is the distance at which the system is most stable. At lesser distances, the energy begins to increase as a consequence of increased repulsion between the two nuclei.

pair of electrons is shared by two atoms, a **covalent bond** exists between the atoms. If the covalent bond is between atoms of the same element, then the molecule formed is that of an element. As examples, let's examine the H$_2$ and Cl$_2$ molecules.

Hydrogen gas, H$_2$, is the simplest example of a molecule with a single covalent bond. When two hydrogen atoms are brought together, the result is attraction between opposite electrons and protons and repulsion between the two electrons and between the two protons. Once the nuclei are 0.074 nm apart, the net attraction is at a maximum (● Fig. 12.11). The two electrons no longer orbit individual nuclei. They are shared equally by both nuclei, which holds the atoms together.

In Lewis symbols, the separated hydrogen atoms are written as

$$\text{H}\cdot \qquad \cdot\text{H}$$

The Lewis structure of the H$_2$ molecule is written as

$$\text{H}:\text{H}$$

The two dots between the hydrogen atoms indicate that these electrons are being shared, giving each atom a share in the electron configuration of the noble gas helium. (Shared electrons are counted as belonging to each atom.) The single covalent bond (one shared pair of electrons) can also be represented by a dash.

$$\text{H—H}$$

Not all atoms will share electrons. For instance, a helium atom, with two electrons already filling its valence shell, does not form a stable molecule with another helium atom or with any other atom. The molecule H$_2$ is stable, but He$_2$ is not. Stable covalent molecules are formed when the atoms share electrons in such a way as to give all atoms a share in a noble gas configuration (in other words, whenever the octet rule is followed).

Now consider the bonding in Cl$_2$. In Lewis symbols, two chlorine atoms are shown as

$$:\ddot{\text{C}}\text{l}\cdot \qquad \cdot\ddot{\text{C}}\text{l}:$$

Each chlorine atom needs one electron to complete its octet. If each shares its unpaired electron with the other, then a Cl$_2$ molecule is formed.

$$:\ddot{\text{C}}\text{l}:\ddot{\text{C}}\text{l}: \qquad :\ddot{\text{C}}\text{l}—\ddot{\text{C}}\text{l}:$$

In the Cl_2 molecule, each chlorine atom has six electrons (three lone pairs) plus two shared electrons, giving each an octet of electrons. (Remember that the shared electrons are counted as belonging to each atom.)

Covalent bonds between atoms of different elements form molecules of compounds. Consider the gas HCl, hydrogen chloride. The H and Cl atoms have the Lewis symbols

$$H\cdot \qquad \cdot\ddot{\underset{\cdot\cdot}{C}}l:$$

If each shares its one unpaired electron with the other, then they both acquire noble gas configurations, and the Lewis structure for hydrogen chloride is

$$H:\ddot{\underset{\cdot\cdot}{C}}l: \quad \text{or} \quad H\!\!-\!\!\ddot{\underset{\cdot\cdot}{C}}l:$$

Up to this point, we have seen that the noble gases tend to form zero covalent bonds, that H forms one bond, and that Cl and the other elements of Group 7A form one bond. Let's take a look at the covalent bonding tendencies of oxygen, nitrogen, and carbon, representatives of Groups 6A, 5A, and 4A, respectively.

One of the most common molecules containing an oxygen atom is water, H_2O. The Lewis symbols of the individual atoms are

$$H\cdot \qquad \cdot\ddot{\underset{\cdot}{O}}:$$
$$\dot{H}$$

When they combine, two single bonds are formed to the oxygen atom as each hydrogen atom shares its electron with the oxygen atom, which shares its two unpaired valence electrons. The Lewis structure of water is

$$H:\ddot{\underset{\cdot\cdot}{O}}: \quad \text{or} \quad H\!\!-\!\!\overset{\cdot\cdot}{\underset{|}{O}}:$$
$$\dot{H} \qquad\qquad\qquad\quad H$$

Oxygen and the other Group 6A elements tend to form two covalent bonds.

One of the most common compounds containing a nitrogen atom is ammonia, NH_3, an important industrial compound. The nitrogen atom has five valence electrons $(\cdot\dot{N}\cdot)$ and shares a different one of its three unpaired electrons with three different H atoms $H\cdot$. The Lewis structure of ammonia is

$$H:\overset{\cdot\cdot}{\underset{\cdot}{N}}:H \quad \text{or} \quad H\!\!-\!\!\overset{\cdot\cdot}{\underset{|}{N}}\!\!-\!\!H$$
$$\dot{H} \qquad\qquad\qquad\quad H$$

Nitrogen and the other Group 5A elements tend to form three covalent bonds.

A common and simple compound of carbon is methane, CH_4. By now, you can predict that a carbon atom with its four valence electrons $(\cdot\dot{C}\cdot)$ will share a different valence electron with four different hydrogen atoms $(H\cdot)$. The Lewis structure of methane is

$$\qquad\qquad\qquad\qquad H$$
$$\qquad\qquad\qquad\qquad |$$
$$\overset{H}{\underset{\dot{H}}{H:\overset{\cdot}{C}:H}} \quad \text{or} \quad H\!\!-\!\!\overset{}{\underset{|}{C}}\!\!-\!\!H$$
$$\qquad\qquad\qquad\qquad |$$
$$\qquad\qquad\qquad\qquad H$$

Carbon and the other Group 4A elements tend to form four covalent bonds.

Covalent bonding is encountered mainly, but not exclusively, in compounds of nonmetals with other nonmetals. Writing Lewis structures for covalent compounds requires an understanding of the number of covalent bonds normally formed by common nonmetals. ● Table 12.4 summarizes our discussion of the number of covalent bonds to be expected from the elements of Groups 4A through 8A. (Exceptions are uncommon in Periods 1 and 2 but occur with more frequency starting with Period 3.)

When an element has two, three, or four *unpaired* valence electrons, its atoms will sometimes share more than one of them with another atom. Thus, double bonds and triple bonds between two atoms are possible (but, for geometric reasons, quadruple bonds are

Table 12.4	Number of Covalent Bonds Formed by Common Nonmetal Groups			
4A	5A	6A	7A	8A
$\cdot \overset{\cdot}{\underset{\cdot}{X}} \cdot$	$\cdot \overset{\cdot \cdot}{\underset{\cdot}{X}} \cdot$	$: \overset{\cdot \cdot}{\underset{\cdot}{X}} \cdot$	$: \overset{\cdot \cdot}{\underset{\cdot \cdot}{X}} \cdot$	$: \overset{\cdot \cdot}{\underset{\cdot \cdot}{X}} :$
4 bonds	3 bonds	2 bonds	1 bond	0 bonds
$-\overset{|}{\underset{|}{C}}-$	$-\overset{\cdot \cdot}{N}-$	$: \overset{\cdot \cdot}{\underset{|}{O}}-$	$: \overset{\cdot \cdot}{\underset{\cdot \cdot}{F}}-$	$: \overset{\cdot \cdot}{\underset{\cdot \cdot}{Ne}} :$

not). Consider the bonding in carbon dioxide, CO_2. The Lewis symbols for two oxygen atoms and one carbon atom are

$$: \overset{\cdot \cdot}{O} \cdot \qquad \cdot \overset{\cdot}{C} \cdot \qquad \cdot \overset{\cdot \cdot}{O} :$$

To get a stable molecule, eight electrons are needed around each atom. Therefore, we must have four electrons shared between the carbon atom and each oxygen atom. The Lewis structure of the CO_2 molecule is

$$\overset{\cdot \cdot}{O} :: C :: \overset{\cdot \cdot}{O} \quad \text{or} \quad \overset{\cdot \cdot}{O} = C = \overset{\cdot \cdot}{O}$$

Each atom in the CO_2 molecule has an octet of electrons around it. Perhaps you can see it more clearly if each atom and its octet of electrons are enclosed in a circle.

A sharing of two pairs of electrons between two atoms produces a *double bond,* represented by a double dash. A sharing of three pairs of electrons between two atoms produces a *triple bond*, represented by a triple dash. Nitrogen gas, N_2, is an example of a molecule with a triple bond. Two nitrogen atoms may be represented as

$$: \overset{\cdot}{N} \cdot \qquad \cdot \overset{\cdot}{N} :$$

To satisfy the octet rule, each nitrogen atom must share its three unpaired valence electrons with the other, forming a triple bond.

The Lewis structure for the N_2 molecule is

$$: N ::: N : \quad \text{or} \quad : N \equiv N :$$

In general, when drawing Lewis structures for covalent compounds, write the Lewis symbol for each atom in the formula. Realize that atoms forming only one bond can never connect two other atoms (just as a one-arm person cannot grab *two* people and serve as a "connection" between them). Try connecting the atoms using single bonds, remembering how many covalent bonds each atom normally forms. If each atom gets a noble gas configuration, then the job is done. If not, then see whether a double or triple bond will work.

EXAMPLE 12.6 Drawing Lewis Structures for Simple Covalent Compounds

Draw the Lewis structure for chloroform, $CHCl_3$, a covalent compound that was once used as an anesthetic but now has been replaced by less toxic compounds.

Solution

A carbon atom (Group 4A) forms four bonds, H forms one bond, and Cl (Group 7A) forms one bond. Only C can be the central atom; thus,

$$\text{H}\cdot \quad \cdot\overset{\cdot}{\text{C}}\cdot \quad \cdot\ddot{\text{C}}\overset{\cdot\cdot}{\text{l}}\text{:} \quad \cdot\ddot{\text{C}}\overset{\cdot\cdot}{\text{l}}\text{:} \quad \cdot\ddot{\text{C}}\overset{\cdot\cdot}{\text{l}}\text{:}$$

gives

$$\overset{\text{H}}{\underset{\text{:Cl:}}{\text{:Cl:C:Cl:}}} \quad \text{or} \quad \overset{\text{H}}{\underset{\text{:Cl:}}{\text{:Cl—C—Cl:}}}$$

Confidence Exercise 12.6

Dilute solutions of hydrogen peroxide, H_2O_2, are used as cleaners, and concentrated solutions have been used as rocket propellants. Draw the Lewis structure for H_2O_2.

Because of the nature of the bonding involved, covalent compounds have quite different properties from those of ionic compounds. Unlike ionic compounds, covalent compounds are composed of individual molecules with a specific molecular formula. For example, carbon tetrachloride consists of individual CCl_4 molecules, each composed of one carbon atom and four chlorine atoms. Although each covalent bond is strong *within* a molecule, the various molecules in a sample of the compound only weakly attract *one another*. Therefore, the melting points and boiling points of covalent compounds are generally much lower than those of ionic compounds. For instance, CCl_4 melts at $-23°C$ and is a liquid at room temperature, whereas the ionic compound NaCl is a solid that melts at $801°C$. Many covalent compounds occur as liquids or gases at room temperature.

Covalent compounds do not conduct electricity well, no matter their phase. The general properties of ionic and covalent compounds are summarized in ● Table 12.5.

Some compounds contain both ionic and covalent bonds. Sodium hydroxide (lye, NaOH) is an example. In Chapter 11.5 we saw that it is common for several atoms to form a polyatomic ion, such as the hydroxide ion, OH^-. The atoms *within* the polyatomic ion are covalently bonded, but the whole OH^- aggregation behaves like an ion in forming compounds. Because strong covalent bonds are present between atoms within polyatomic ions, it is difficult to break them up. Therefore, in chemical reactions they frequently act as a single unit, as will be seen in Chapter 13.3. The Lewis structure of sodium hydroxide is

$$\text{Na}^+[\text{:}\ddot{\text{O}}\text{:H}]^-$$

A covalent bond exists between the O and the H in each hydroxide ion, but the hydroxide ions and the sodium ions are bound together in a crystal lattice by ionic bonds, so NaOH is an ionic compound.

Follow these rules to predict whether a particular compound is ionic or covalent.

Table 12.5	Comparison of Properties of Ionic and Covalent Compounds
Ionic Compounds	**Covalent Compounds**
Crystalline solids (made of ions)	Gases, liquids, or solids (made of molecules)
High melting and boiling points	Low melting and boiling points
Conduct electricity when melted	Poor electrical conductors in all phases
Many soluble in water but not in nonpolar liquids	Many soluble in nonpolar liquids but not in water

1. Compounds formed of only nonmetals are covalent (except ammonium compounds).
2. Compounds of metals and nonmetals are generally ionic (especially for Group 1A or 2A metals).
3. Compounds of metals with polyatomic ions are ionic.
4. Compounds that are gases, liquids, or low-melting-point solids are covalent.
5. Compounds that conduct an electric current when melted are ionic.

EXAMPLE 12.7 **Predicting Bonding Type**

Predict which compounds are ionic and which are covalent: (a) KF; (b) SiH_4; (c) $Ca(NO_3)_2$; (d) compound X, a gas at room temperature; (e) compound Y, which melts at 900 ° C and then conducts an electric current.

Solution

(a) KF is ionic (Group 1A metal and a nonmetal).
(b) SiH_4 is covalent (only nonmetals).
(c) $Ca(NO_3)_2$ is ionic (metal and polyatomic ion).
(d) X is covalent (as are all substances that are gases or liquids at room temperature).
(e) Y is ionic (the compound has a high melting point, and the liquid conducts electricity).

Confidence Exercise 12.7

Is PCl_3 ionic or covalent in bonding? What about MgF_2?

Polar Covalent Bonding

In covalent bonding, the electrons involved in the bond between two atoms are shared. However, unless the atoms are of the same element, the bonding electrons will spend more time around the more nonmetallic element; that is, the sharing is unequal. Such a bond is called a **polar covalent bond**, indicating that it has a slightly positive end and a slightly negative end. We can think of an electron pair in a polar covalent bond as the object of a tug of war, where one atom "tugs" harder on the shared electron(s) than the other.

Electronegativity (abbreviated EN) is a measure of the ability of an atom in a molecule to draw bonding electrons to itself. ● Figure 12.12 gives the numerical values calculated for the electronegativities of the representative elements. Electronegativity increases across (from left to right) a period and decreases down a group, just as does the nonmetallic character (Chapter 11.4).

Consider the covalent bond in HCl. The chlorine atom is more electronegative (EN = 3.0) than the hydrogen atom (EN = 2.1). Although the two bonding electrons are shared between the two atoms, they tend to spend more time at the chlorine end than at the hydrogen end of the bond. A polar bond is obtained in which the polarity can be represented by an arrow:

$$H \xrightarrow{\;+\!\!\!\!\to\;} \ddot{\underset{\cdot\cdot}{Cl}}:$$

The head of the arrow points to the more electronegative atom and denotes the negative end of the bond. The "feathers" of the arrow make a plus sign (at left above) that indicates the positive end of the bond. ● Figure 12.13 summarizes our discussion of ionic, covalent, and polar covalent bonding.

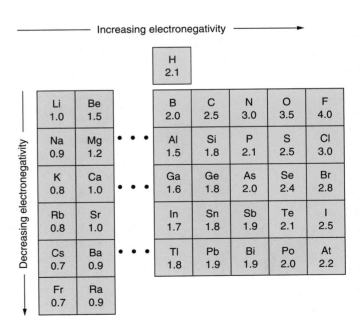

FIGURE 12.12 Electronegativity Values

In general, electronegativity increases across a period and decreases down a group. Fluorine (F) is the most electronegative element, and Francium (Fr) is the least. Note that hydrogen is generally considered a nonmetal.

FIGURE 12.13 A Summary of Ionic and Covalent Bonding
Bonding electrons may be either transferred (ionic bonding) or shared (covalent bonding). If shared, then they may be shared equally (nonpolar covalent bonding) or unequally (polar covalent bonding).

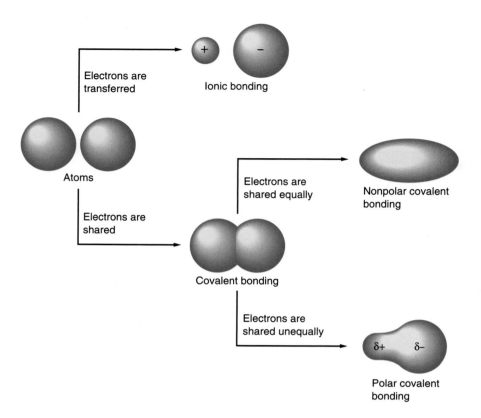

EXAMPLE 12.8   Showing the Polarity of Bonds

Use arrows to show the polarity of the covalent bonds in (a) water, H_2O; and (b) carbon tetrachloride, CCl_4.

Solution

(a) Oxygen (EN = 3.5) is more electronegative than hydrogen (EN = 2.1), so the arrows denoting the polarity of the bonds would point as shown here.

$$H \overset{\longleftrightarrow}{} \ddot{\text{O}}:$$
$$\big| \updownarrow$$
$$H$$

(b) Chlorine (EN = 3.0) is more electronegative than carbon (EN = 2.5), so the arrows denoting the polarity of the bonds would point as shown.

$$:\ddot{\text{Cl}}:$$
$$\updownarrow$$
$$:\ddot{\text{Cl}} \overset{\longleftarrow}{} \text{C} \overset{\longleftrightarrow}{} \ddot{\text{Cl}}:$$
$$\updownarrow$$
$$:\ddot{\text{Cl}}:$$

Confidence Exercise 12.8

Use arrows to show the polarity of the covalent bonds in ammonia, NH_3.

Conceptual Question and Answer

A Matter of Purity

Q. Are bonds purely ionic or purely covalent?

A. Pure ionic bonds do not exist because in all ionic compounds the electrons are not completely transferred; there is some partial sharing due to quantum mechanical effects. For example, NaCl has a strong ionic bond, but it has a few percent of covalent bonding characteristics. In the case of a covalent bond, if the sharing of electrons is equal, then we can say we have a purely covalent bond (like diatomic molecules, H_2, Cl_2, and so on). When a mostly covalent bond shows some ionic character, we call these polar covalent molecules, like H_2O.

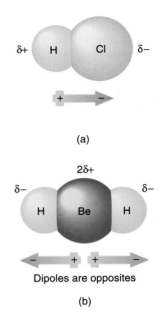

(a)

Dipoles are opposites

(b)

FIGURE 12.14 Polar Bonds and Polar Molecules
To predict the polarity of a molecule, both the bond polarities and the molecular shape must be known. (a) The arrow shown here for the one bond in an HCl molecule is a common symbol for the dipole of a polar bond. By convention, the arrow points to the more electronegative end of the bond. With only one polar bond present, the HCl molecule is polar overall. (b) Beryllium hydride, BeH_2, has two polar bonds, but the equal-magnitude dipoles are oriented in opposite directions, thereby canceling each other and making the molecule nonpolar overall.

Polar Bonds and Polar Molecules

Molecules as well as bonds can have polarity. A molecule, as a whole, is polar if electrons are more attracted to one end of the molecule than to the other end. Such a molecule has a slightly negative end and a slightly positive end, and we say it has a *dipole* or is a **polar molecule**. The slightly negative end of the polar molecule is denoted by a $\delta-$ (delta minus) and the slightly positive end by a $\delta+$ (delta plus). Consider the HCl molecule with its one polar bond (● Fig. 12.14a). With only one polar bond present, it should be obvious that the chlorine end of this molecule must be slightly negative and the hydrogen end slightly positive, resulting in a polar molecule.

Now consider the linear molecule beryllium hydride, BeH_2 (Fig. 12.14b). It has two equal-magnitude polar bonds, but the dipoles are oriented in opposite directions and thus cancel one another, making the molecule nonpolar overall.

As yet another example, consider the water molecule. If it were linear (it's not), then it would be nonpolar. The dipoles of the two bonds would cancel because the center point of the positive charges of the bonds would be in exactly the same place as the center of the negative charges, right in the middle of the molecule. No center-of-charge separation means no molecular dipole.

$$H \xrightarrow{\quad} \overset{..}{O} \xleftarrow{\quad} H$$

The water molecule is actually angular (105° between the two bonds), however, so the bond polarities reinforce one another instead of canceling. The center of positive charge is midway between the two hydrogen atoms. The center of negative charge is at the oxygen atom. Thus, the centers of charge are separated, and water molecules are polar.

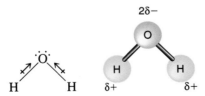

Then, as a prerequisite for determining whether or not a molecule containing polar bonds has a molecular dipole (is polar), you must know the molecule's shape, or geometry. Space does not allow for full discussion of the factors affecting molecular geometry, but a useful theory is that of valence-shell-electron-pair repulsion, or VSEPR (pronounced "vesper"). This theory states that the shape of many molecules is determined largely by the efforts of the valence shell electrons to stay out of each other's way to the greatest extent possible.

The polarity of the molecules of a liquid compound can be tested by a simple experiment. Bring an electrically charged rod close to the compound and see whether attraction occurs. If it does, then the molecules are polar. Because water molecules are polar, a thin

FIGURE 12.15 Polar and Nonpolar Liquids
(a) A stream of polar water molecules is deflected toward an electrically charged rod.
(b) A stream of nonpolar carbon tetrachloride (CCl_4) molecules is not deflected.

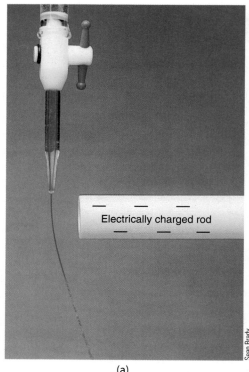

(a)
Water (H_2O)

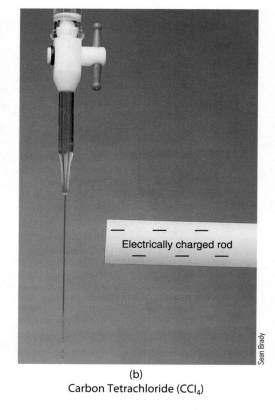

(b)
Carbon Tetrachloride (CCl_4)

stream of water is attracted to a charged rod, but a similar stream of nonpolar carbon tetrachloride is not attracted (● Fig. 12.15; see also Fig. 8.4 for water).

In summary,

1. If the bonds in a molecule are nonpolar, then the molecule can only be *nonpolar*.

2. A molecule with only one polar bond has to be *polar*.

3. A molecule with more than one polar bond will be nonpolar if the shape of the molecule causes the polarities of the bonds to cancel (● Table 12.6). If the bond polarities do not cancel, then it will be a polar molecule.

Table 12.6 Types of Molecules with Polar Bonds but No Resulting Dipole

Type		Cancellation of Polar Bonds	Example	Ball-and-Stick Model
Linear molecules with two identical bonds	B — A — B	←—+ +—→	CO_2	
Planar molecules with three identical bonds 120° apart	B, A, B B 120°		SO_3	
Tetrahedral molecules with four identical bonds 109.5° apart	B, A, B, B, B		CCl_4	

EXAMPLE 12.9 **Predicting the Polarity of Molecules**

The carbon tetrachloride molecule is tetrahedral, as shown in the accompanying sketch. All the bond angles are 109.5°, and the bonds are polar. Is the molecule polar?

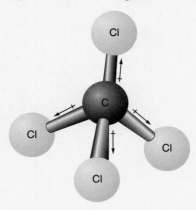

Solution

Each bond has the same degree of polarity, the center of positive charge is at the C atom, and the center of negative charge is also there. The polar bonds cancel; thus, the molecule is nonpolar.

Confidence Exercise 12.9

Boron trifluoride, BF_3, is an exception to the octet rule. A single bond extends from each fluorine atom to the boron atom, which has a share in only six electrons. The three bonds are polar, yet the molecule itself is nonpolar. What must be the geometry of the molecule and the angle between the bonds?

> **Did You Learn?**

- In covalent compounds, pairs of electrons are shared by atoms, producing molecules with covalent bonding.

- Electronegativity increases across (from left to right) a period and decreases down a group.

12.6 Hydrogen Bonding

Preview Questions

- What does the phase "like dissolves like" mean?
- How does hydrogen bonding occur in water (H_2O), and what effect does it have?

Why does water dissolve table salt but does not dissolve oil? The polar nature of water molecules causes them to interact with an ionic substance such as salt. The positive ends of the water molecules attract the negative ions, and the negative ends attract the positive ions (● Fig. 12.16). If the attraction of the water molecules overcomes the attractions among the ions in the crystal, the salt dissolves. As Fig. 12.16 shows, the negative ions move into solution surrounded by several water molecules with their positive ends pointed toward the ion. Just the opposite is true for the positive ions. Such attractions are called *ion–dipole interactions*.

As you would expect, the molecules of two polar substances have a *dipole–dipole interaction* and tend to dissolve in one another. Similarly, it would not be surprising to find that

FIGURE 12.16 Sodium Chloride Dissolving in Water
The negative ends of the polar water molecules attract and surround the positive sodium ions (purple spheres). The positive ends of the water molecules attract and surround the negative chloride ions (green spheres).

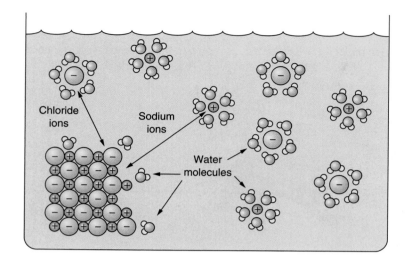

two nonpolar substances, such as oil and gasoline, mix well. The nonpolar molecules in oil have no more affinity for one another than they do for the nonpolar molecules in gasoline. These examples illustrate the well-known solubility principle of *like dissolves like*. In general, polar substances and nonpolar substances (*unlike* substances) do not dissolve in one another because the polar molecules tend to gather together and exclude the nonpolar molecules.

A **hydrogen bond** is a special kind of dipole–dipole interaction that can occur whenever a compound contains hydrogen atoms covalently bonded to small, highly electronegative atoms (only O, F, and N meet these criteria). Because the bond is so polar and a hydrogen atom is so small, the partial positive charge on hydrogen is highly concentrated. Thus, a hydrogen atom has an electrical attraction for nearby O, F, or N atoms that may be in the same or neighboring molecules, because these atoms have highly concentrated partial negative charges.

For small molecules, such as H_2O, HF, and NH_3, hydrogen bonding is a weak force of attraction between a hydrogen atom in *one* molecule and an O, F, or N atom in *another* molecule (● Fig. 12.17). Such hydrogen bonds are *inter*molecular forces, not *intra*molecular forces.*

Conceptual Question and Answer

Hydrogen Bond Highways

Q. How are hydrogen bonds similar to *inter*state and *intra*state highways?

A. An *intra*state highway only exists within one state, whereas an *inter*state highway passes through multiple states and "connects" them together. Hydrogen bonds are defined the same way: When the forces are within the same molecule, they are intramolecular bonds, but when the forces "connect" two molecules together, the hydrogen bonds are intermolecular.

Hydrogen bonds are strong enough (about 5–10% of the strength of a covalent bond) to have a pronounced effect on the properties of the substance. In a given group, the boiling points of similar compounds generally increase with increasing formula mass. But note in ● Fig. 12.18 that the hydrogen bonding in H_2O, HF, and NH_3 causes those compounds

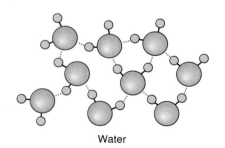

Water

FIGURE 12.17 Hydrogen Bonding in Water
Weak forces of attraction (symbolized here by red dots) between the hydrogen atom of one water molecule and the oxygen atom of another molecule cause water to be a liquid. At room temperature, about 80% of the molecules are hydrogen bonded at any one time.

---

*For large molecules such as DNA and proteins, hydrogen bonds exist between some of the H atoms and O or N atoms in other parts of the same molecule that have twisted around into close proximity. The famous "unzipping" of DNA molecules is actually a breaking of intramolecular hydrogen bonds.

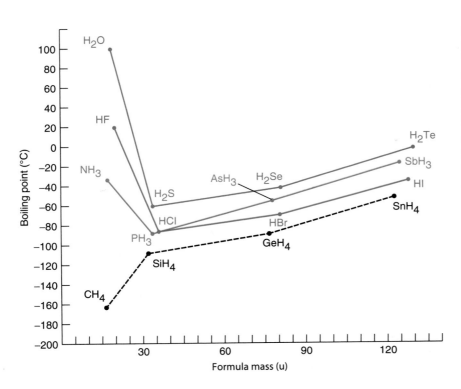

FIGURE 12.18 **Hydrogen Bonding at Work**
The boiling points of the hydrogen compounds of elements in Groups 4A, 5A, 6A, and 7A show that extensive hydrogen bonding exists and causes inordinately high boiling points in samples of H_2O, HF, and NH_3, but not in CH_4. Without hydrogen bonding, the boiling points of compounds in each group should all increase with increasing formula mass.

to have much higher boiling points relative to the other hydrogen compounds in their respective groups. However, because hydrogen bonding does not occur in CH_4, its boiling point shows a normal pattern relative to those of the other hydrogen compounds of Group 4A elements. Most substances with a formula mass as low as water's 18 u are gases. Without hydrogen bonding, water would have a boiling point of about $-80°$ C and thus at room temperature would be a gas, not a liquid, because there would be so little attraction among the water molecules.

Nearly all solids sink to the bottom in a sample of their liquid. Why, then, does ice float in water (● Fig. 12.19)? As the temperature of a sample of water drops, the hydrogen bonds align more and more in a hexagonal manner. This alignment gives ice an open structure, increases its volume, and makes it less dense than liquid water. (The Chapter 5.2 Highlight: Freezing from the Top Down gives more details.)

FIGURE 12.19 **Hydrogen Bonding and Density**
Hydrogen bonding and the shape of the water molecule cause ice to have an open structure, making it less dense than liquid water. Thus, ice floats, as shown on the left. For virtually every other substance, such as the benzene (C_6H_6) shown on the right, the solid is more dense than the liquid and thus does not float.

Did You Learn?

- Two polar substances tend to dissolve each other, whereas polar and nonpolar substances do not. Like dissolves like.
- Weak hydrogen bonding occurs between the hydrogen atom of one water molecule and an oxygen atom of another molecule. This bonding causes the water to be liquid at room temperature.

KEY TERMS

1. law of conservation of mass (12.1)
2. formula mass (12.2)
3. law of definite proportions
4. limiting reactant
5. excess reactant
6. octet rule (12.4)
7. ionic compounds
8. Lewis symbol
9. Lewis structures

10. cations
11. anions
12. ionic bonds
13. Stock system

14. covalent compound (12.5)
15. covalent bond
16. polar covalent bond

17. electronegativity
18. polar molecule
19. hydrogen bond (12.6)

MATCHING

For each of the following items, fill in the number of the appropriate Key Term from the preceding list. Compare your answers with those at the back of the book.

a. _____ General term for positive ions

b. _____ A molecule having an end that is slightly positive and an end that is slightly negative

c. _____ Compounds tend to achieve electron configurations of the noble gases

d. _____ Compounds in which pairs of electrons are shared

e. _____ A starting material that is completely used up in a reaction

f. _____ A measure of an atom's ability to attract bonding electrons to itself

g. _____ A pair of electrons shared between two atoms

h. _____ Shows, as dots, only the valence electrons of an atom or ion

i. _____ General term for negative ions

j. _____ The attraction that occurs between an atom of H and an atom of O, F, or N in another molecule

k. _____ No detectable change in mass occurs during a chemical reaction

l. _____ A starting material that is *not* completely used up in a reactant

m. _____ Electrical attractions that hold ions together

n. _____ Different samples of a pure compound contain the same elements in the same proportion by mass

o. _____ Nomenclature used to give the value of the ionic charge of a metal, such as $CrCl_2$ chromium(II) chloride

p. _____ Shows the structure of a molecule or formula unit using dots and dashes for electrons

q. _____ The sum of the atomic masses given in the formula of a substance

r. _____ A bond formed by unequal sharing of an electron pair

s. _____ Compounds formed by an electron transfer process

MULTIPLE CHOICE

Compare your answers with those at the back of the book.

1. Which one of these three forces is responsible for chemical bonding? (Intro)
    (a) gravitational   (b) electromagnetic   (c) strong nuclear

2. What quantity remains unchanged in a chemical reaction? (12.1)
    (a) mass   (b) crystal shape   (c) color   (d) phase

3. How many total atoms would be in one formula unit of $(NH_4)_3PO_4$? (12.2)
    (a) 4   (b) 16   (c) 18   (d) 20

4. A sample of compound AB decomposes to 48 g of A and 12 g of B. Another sample of the same compound AB decomposes to 24 g of A. Predict the number of grams of B obtained. (12.2)
    (a) 12   (b) 8   (c) 6   (d) 3

5. Which scientist is responsible for our first atomic theory? (12.3)
    (a) Lavoisier   (b) Proust   (c) Dalton   (d) Lewis

6. Which of the following statements comparing the compounds CO and CO_2 is true? (12.3)
    (a) The atom of oxygen weighs less in CO_2 than in CO.
    (b) Both compounds have the same formula mass.
    (c) The atom of carbon weighs more in CO_2 than in CO.
    (d) The mass ratio of oxygen to carbon is equal to the atom ratio in both compounds.

7. What is the normal charge on an ion of sulfur? (12.4)
    (a) 6−   (b) 6+   (c) 2−   (d) 2+

8. An ionic compound formed between a Group 2A element M and a Group 7A element X would have what general formula? (12.4)
    (a) M_2X (b) M_7X_2 (c) MX_2 (d) M_2X_7

9. Which of the following is the formula for iron(III) bromide? (12.4)
    (a) FeBr   (b) Fe_3Br (c) Fe_2Br_3 (d) $FeBr_3$

10. Carbon is a Group 4A element. How many covalent bonds are there in methane, CH_4? (12.5)
    (a) 4   (b) 5   (c) 6   (d) 8

11. How many shared pairs of electrons are in an ammonia molecule NH_3? (12.5)
    (a) 1   (b) 2   (c) 3   (d) 4

12. Which one is definitely a covalent compound? (12.5)
    (a) $TiCl_2$ (b) NO_2 (c) Na_2O (d) $CaSO_4$

13. Which element in the periodic table has the highest electronegativity value? (12.5)
    (a) fluorine   (b) oxygen
    (c) lithium   (d) bismuth

14. In which one of these compounds does hydrogen bonding occur? (12.6)
    (a) H_2S (b) HF   (c) PH_3 (d) CH_4

FILL IN THE BLANK

Compare your answers with those at the back of the book.

1. The total ___ before and after a reaction must be equal. (12.1)
2. "Dry ice" has the chemical formula ___. (12.2)
3. Dalton's atomic theory explains the laws of conservation of mass and ___. (12.3)
4. Atomic theory is the cornerstone of ___. (12.3)
5. The formula for copper(II) oxide is ___. (12.4)
6. An atom that has gained an electron is called a(n) ___. (12.4)
7. When bonding, a hydrogen atom tends to get ___ electrons, not eight, in its valence shell. (12.4)

8. The formula of an ionic compound of a Group 1A element M and a Group 6A element X is ___. (12.4)
9. Compounds that conduct an electric current when melted are ___. (12.5)
10. A covalent bond in which two atoms share two pairs of electrons is called a(n) ___ bond. (12.5)
11. Electronegativity increases across a period and ___ down a group. (12.5)
12. The highly electronegative atoms that are involved in hydrogen bonding are oxygen, fluorine, and ___. (12.6)

SHORT ANSWER

12.1 Law of Conservation of Mass (and the Highlight)

1. State the law of conservation of mass, and give an example.
2. Give four reasons Lavoisier is generally designated "The Father of Chemistry."
3. What techniques did Lavoisier use to determine that mass was unchanged in a combustion reaction?

12.2 Law of Definite Proportions

4. The atomic masses of the elements in the periodic table are measured relative to an atomic mass standard. Name the isotope that is used as the atomic mass standard, and give its assigned mass in atomic mass units (u).
5. What are the two most striking physical properties of dry ice?
6. When a sample of A reacts with one of B, a new substance AB is formed, and no B but a little of A is left. Which is the limiting reactant?

12.3 Dalton's Atomic Theory

7. State the three parts of Dalton's atomic theory.
8. Dalton thought that all atoms of a given element were alike. We now know that most elements consist of a mixture of isotopes (Table 10.4 on page 265). Explain why this has no effect on the explanation of the law of definite proportions.

12.4 Ionic Bonding

9. State the octet rule. What common element is an exception to this rule, and why?
10. What are valence electrons, and how are they used in the formation of compounds?
11. How is the number of valence electrons in an atom related to its tendency to gain or lose electrons during compound formation?
12. Explain the process in which one atom of sodium bonds with one atom of chlorine.
13. A formula unit of potassium sulfide (K_2S) would consist of what particles?
14. In a Lewis symbol, what determines the number of dots used?
15. What is the difference between a Lewis symbol and a Lewis structure?

16. What is the basic difference between ions formed by metals and ions formed by nonmetals?
17. What do we mean when we say that F^-, Ne, and Na^+ are *isoelectronic*?
18. State the two principles used in writing the formulas of ionic compounds.
19. What are some characteristics of ionic compounds?
20. Why does an aqueous solution of table salt conduct electricity, whereas an aqueous solution of table sugar does not?
21. When is the use of the Stock system of nomenclature preferred?

12.5 Covalent Bonding (and the Chemistry Fact)

22. Briefly, how are covalent compounds formed?
23. What is the basic difference distinguishing single, double, and triple bonds?
24. How does electronegativity vary, in general, left to right across a period and down a group?
25. What element has the highest electronegativity? The second highest?
26. A covalent bond in which the electron pair is unequally shared is called by what name?
27. Could a molecule composed of two atoms joined by a polar covalent bond ever be nonpolar? Explain.
28. To predict whether a molecule consisting of two or more polar bonds is polar, you must know what else about the molecule?
29. Would you expect a liquid solution of carbon tetrachloride (CCl_4) to conduct electricity?
30. A certain compound has a boiling point of $-10°C$. Is it probably ionic or probably covalent?
31. Since 1982, pennies are composed of 98% of what element? Does that pose a problem?

12.6 Hydrogen Bonding

32. State the short general principle of solubility, and explain what it means.
33. The density of a solid substance is usually greater than its density in a liquid form, yet the density of liquid water is greater than the density of solid water (ice). Explain.

VISUAL CONNECTION

Visualize the connections and give answers for the blanks. Compare your answers with those at the back of the book.

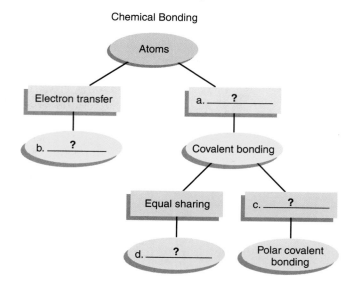

APPLYING YOUR KNOWLEDGE

1. You decide to have hot dogs for dinner. In the grocery store, you find that buns come only in packages of 12, whereas the hot dogs come in packages of 8. How can your purchases lead to having no buns or hot dogs left over? (How does this problem fit in with chemical bonding and other parts of this chapter?)

2. Why can't we destroy bothersome pollutants by just dissolving them in the ocean?

3. You are the supervisor in a car assembly plant where each car produced requires one body and four wheels. You discover that the plant has 200 bodies and 900 wheels. How many cars can be made from this inventory? (How does this problem fit in with the law of definite proportions?)

4. When you use a bottle of vinegar-and-oil salad dressing, you have to shake it up for mixing. Why?

5. Your friend has glasses with photochromic lenses; they darken and lighten as he goes outdoors and then back indoors. Explain to him that there is a lot of ionic chemistry going on inside the glass to cause the lenses to change.

6. Look up the electronegativity values of elements in the periodic table. There are several exceptions to the trend that electronegativity decreases down a group. Find several examples, and determine how these exceptions relate to the overall atomic size of each atom compared with their group. How do atomic size and electronegativity compare?

IMPORTANT EQUATIONS

Percentage of Component in a Sample:

$$\% \ X \text{ component} = \frac{\text{amount of component } X}{\text{total amount}} \times 100\% \quad (12.1)$$

Percentage by Mass of Element X from Compound's Formula:

$$\% \ X \text{ by mass} = \frac{\text{mass of component } X}{\text{formula mass of compound}} \times 100\% \quad (12.2)$$

EXERCISES

12.1 Law of Conservation of Mass

1. Volcanoes emit much hydrogen sulfide gas, H_2S, which reacts with the oxygen in the air to form water and sulfur dioxide, SO_2. Every 68 tons of H_2S reacts with 96 tons of oxygen and forms 36 tons of water. How many tons of SO_2 are formed?

   Answer: 128 tons

2. An antacid tablet weighing 0.942 g contained calcium carbonate as the active ingredient, in addition to an inert binder. When an acid solution weighing 56.413 g was added to the tablet, carbon dioxide gas was released, producing a fizz. The resulting solution weighed 57.136 g. How many grams of carbon dioxide were produced?

3. Calculate (to the nearest 0.1 u) the formula mass of these compounds.
   (a) carbon dioxide, CO_2
   (b) methane, CH_4
   (c) sodium phosphate, Na_3PO_4

   Answer: (a) 44.0 u (b) 16.0 u (c) 164.0 u

4. Calculate (to the nearest 0.1 u) the formula mass of these compounds.
   (a) lithium chloride, LiCl
   (b) silver carbonate, Ag_2CO_3
   (c) zinc nitrate, $Zn(NO_3)_2$

12.2 Law of Definite Proportions

5. Find the percentage by mass of Cl in $MgCl_2$ if it is 25.5% Mg by mass.

   Answer: 74.5%

6. Find the percentage by mass of sodium (Na) in Na_2S if it is 41.1% sulfur (S) by mass.

7. Determine the percentage by mass of each element in (a) salt, NaCl; and (b) sugar, $C_{12}H_{22}O_{11}$.

   Answer: (a) 39.3% Na, 60.7% Cl (b) 42.1% C, 6.4% H, 51.5% O

8. Determine the percentage by mass of each element in (a) lime, $CaCO_3$; and (b) milk of magnesia, $Mg(OH)_2$.

9. In a lab experiment, 6.1 g of Mg reacts with sulfur to form 14.1 g of magnesium sulfide. (a) Use Equation 12.1 to calculate the percentage by mass of Mg in magnesium sulfide. (b) How many grams of sulfur reacted, and how do you know?

   Answer: (a) 43% (b) 8.0 g, law of conservation of mass

10. In a lab experiment, 7.75 g of phosphorus reacts with bromine to form 67.68 g of phosphorus tribromide. (a) Calculate the percentage by mass of P in phosphorus tribromide. (b) How many grams of bromine reacted, and how do you know?

11. Refer to Exercise 9. How much magnesium sulfide would be formed if 6.1 g of Mg were reacted with 10.0 g of S, and how do you know?

    Answer: Still 14.1 g. The law of definite proportions indicates that the proper ratio is 6.1 g Mg to 8.0 g S, so $10.0 - 8.0 = 2.0$ g of S is in excess.

12. Refer to Exercise 10. How much phosphorus tribromide would be formed if 10.00 g of phosphorus reacted with 59.93 g of bromine, and how do you know?

12.4 Ionic Bonding

13. Referring only to a periodic table, give the ionic charge expected for each of these representative elements: (a) S, (b) K, (c) Br, (d) N, (e) Mg, (f) Ne, (g) C, and (h) Al.

    Answer: (a) 2− (b) 1+ (c) 1− (d) 3− (e) 1+ (f) 0 (g) 0 (h) 3+

14. Referring only to a periodic table, give the ionic charge expected for each of these representative elements: (a) I, (b) Na, (c) Se, (d) Ca, (e) In, (f) As, (g) Ar, and (h) Si.

15. Write the Lewis symbols and structures that show how Na_2O forms from sodium and oxygen atoms.

    Answer: $Na^+ \, Na^+ \, :\ddot{O}:^{2-}$

16. Write the Lewis symbols and structures that show how $BaBr_2$ forms from barium and bromine atoms.

17. Predict the formula for each ionic compound. (Recall the polyatomic ions in Table 11.6.)
    (a) cesium iodide
    (b) barium fluoride
    (c) aluminum nitrate
    (d) lithium sulfide
    (e) beryllium oxide
    (f) ammonium sulfate

    Answer: (a) CsI (b) BaF_2 (c) $Al(NO_3)_3$ (d) Li_2S (e) BeO (f) $(NH_4)_2SO_4$

18. Predict the formula for each ionic compound. (Recall the polyatomic ions in Table 11.6.)
    (a) lithium iodide
    (b) sodium nitride
    (c) sodium sulfate
    (d) aluminum bromide
    (e) ammonium acetate
    (f) calcium phosphate

19. Give the Stock system name for (a) $Fe(OH)_2$, (b) $CuCl_2$, and (c) AuI_3.

    Answer: (a) iron(II) hydroxide (b) copper(II) chloride (c) gold(III) iodide

20. Each of the following names is incorrect. Give the correct names for (a) $Zn(NO_3)_2$, zinc(II) nitrate, (b) CrS, chromium(I) sulfide, and (c) Cu_2O, copper(II) oxide.

12.5 Covalent Bonding

21. Referring only to a periodic table, give the number of covalent bonds expected for each of these representative elements: (a) S, (b) Ne, (c) Br, (d) N, and (e) C.

    Answer: (a) 2 (b) 0 (c) 1 (d) 3 (e) 4

22. Referring only to a periodic table, give the number of covalent bonds expected for each of these representative elements: (a) H, (b) Ar, (c) F, (d) Si, and (e) P.

23. Draw the Lewis structure for the rocket fuel hydrazine, N_2H_4. Show a structure with dots only, and then show one with both dots and dashes.

    Answer: $H:\ddot{N}:\ddot{N}:H$ and $H-\overset{|}{\underset{|}{N}}-\overset{|}{\underset{|}{N}}-H$
       $\quad\; H\,H \qquad\qquad\quad H\,H$

24. Draw the Lewis structure for formaldehyde, H_2CO, a compound whose odor is known to most biology students because of its use as a preservative. Show a structure with dots only, and then show one with both dots and dashes.

25. Use your knowledge of the correct number of covalent bonds to predict the formula for a simple compound formed between chlorine and (a) hydrogen, (b) nitrogen, (c) sulfur, and (d) carbon.

    Answer: (a) HCl (b) NCl_3 (c) SCl_2 (d) CCl_4

26. Use your knowledge of the correct number of covalent bonds to predict the formula for a simple compound formed between carbon and (a) hydrogen, (b) bromine, (c) sulfur, and (d) helium.

27. Predict which of these compounds are ionic and which are covalent. State your reasoning.

    (a) N_2H_4 (d) CBr_4
    (b) NaF   (e) $C_{12}H_{22}O_{11}$
    (c) $Ca(NO_3)_2$ (f) $(NH_4)_3PO_4$

    Answer: (a) covalent, two nonmetals (b) ionic, Group 1A metal and nonmetal
    (c) ionic, metal and polyatomic ion (d) covalent, two nonmetals
    (e) covalent, all nonmetals (f) ionic, two polyatomic ions

28. Predict which of these compounds are ionic and which are covalent. State your reasoning.

    (a) SF_6 (d) BaS
    (b) KBr   (e) $CaCO_3$
    (c) N_2O_3 (f) C_6H_5Cl

29. Use arrows to show the polarity of each bond in (a) SCl_2 and (b) CO_2. (*Hint*: Use the general periodic trends in electronegativity.)

    Answer: (a) S and Cl are in the same period, but Cl is farther right than S, so it is the more electronegative. Thus, the arrows would point to the two Cl atoms. (b) C and O are in the same period, but O is farther right, so it is the more electronegative. Thus, the arrows would point to the two oxygen atoms.

30. Use arrows to show the polarity of each bond in (a) BrCl and (b) NI_3. (*Hint*: Use the general periodic trends in electronegativity.)

ON THE WEB

1. A Tale of Killing the Golden Goose

Why was Antoine Lavoisier considered "The Father of Chemistry," and what was his fate? One early theory of combustion was that an unknown substance was released into the air when it burned. How did Lavoisier rectify this early theory with his law of conservation of mass? What was the effect of Lavoisier's solution? With regard to combustion, what did Lavoisier and his friends find? How were his experiments conducted? How might you be able to use what you've learned about Antoine Lavoisier in your own life? Discover more about Lavoisier and answer these questions by following the recommended links on the student website at **www.cengagebrain.com/shop/ISBN/1133104096**.

2. The Ties That Bind, or Bond

What is chemical bonding? How do you identify types of bonds? Who was Linus Pauling, and why do you think he was the only person to have won two Nobel Prizes (at least without having to share with someone else)? What does chemical bonding have to do with your life? Visit the student website at **www.cengagebrain.com/shop/ISBN/1133104096** to explore answers to these questions.

Chemical Reactions

© Gary Hansen/Phototake

< Green plants use chemical reactions to store the Sun's energy.

Having discussed elements, compounds, and chemical bonding, we are ready to examine chemical reactions. Our environment is composed of atoms and molecules that undergo chemical changes to produce the many substances we need and use. In producing new products, energy is released or absorbed.

For example, green plants absorb carbon dioxide from the air and, with energy from the Sun and chlorophyll as a catalyst, react with water from the soil to form carbohydrates and oxygen. (See the chapter-opening photo.) This complex chemical reaction is called *photosynthesis*, and the Sun's energy is stored in chemical bonds as *chemical energy*. Animal life inhales the oxygen and ingests the plants' carbohydrates to obtain energy. Chemical changes in the animals' cells then returns water and carbon dioxide to the environment (Chapter 19.1).

CHEMISTRY FACTS

▶ Glass has interesting physical properties; it is an amorphous solid material. This means that glass has no melting point; it is a rigid structure, but the molecules are disordered like a liquid. Glass is made from heating silica sand (SiO_2) to high temperatures to make a thick, syrupy mixture that hardens on cooling.

▶ Glove-warmer heat packs contain powdered iron (and other components) moistened with water. When a pack's plastic cover is removed, air can penetrate the paper packet, producing a common exothermic reaction through the rusting of iron, which generates a comfortable amount of heat.

▶ Pearls are layers of calcium carbonate, $CaCO_3$, deposited around a grain of sand that has

Chapter Outline

entered the shell of an oyster. ($CaCO_3$ is the major component of blackboard chalk.)

▶ "Skywriting" results from the spraying of titanium(IV) chloride ($TiCl_4$) from an airplane. It reacts with moisture in the air to form the compound titanium dioxide (TiO_2), which constitutes the white "writing" in the air.

▶ Catalytic converters in automobiles use a catalyst to clean the unwanted gases produced in engine combustion. Harmful nitric oxide is converted to nitrogen, and poisonous carbon monoxide is converted to carbon dioxide. Unfortunately, sulfur dioxide is converted to the more reactive sulfur trioxide, which reacts with water to form sulfuric acid, the main component in acid rain (see Ch 20.4).

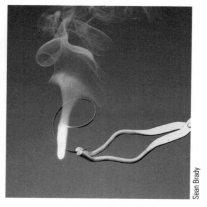

Figure 13.1 The Reaction of a Wire of Magnesium with Oxygen
A wire of magnesium metal (Mg) and oxygen (O_2) in the air combine to form magnesium oxide (MgO, the "smoke"). Magnesium is used in fireworks because of the bright light produced by the reaction.

Sean Brady

In this chapter, general types of chemical reactions and the major principles that underlie chemical change will be considered. There are two Highlights: The Chemistry of Air Bags and The Chemistry of Tooth Decay show the importance of chemical reactions in our lives.

13.1 Balancing Chemical Equations

Preview Questions

● Which properties change in a chemical reaction?

● How is a chemical equation balanced?

The characteristics of a substance are its properties. *Physical properties* are those that do not describe the chemical reactivity of the substance, and they can be measured without causing new substances to form. Among them are density, hardness, phase, color, melting point (m.p.), electrical conductivity, and specific heat.

Chemical properties reflect the ways in which a substance can be transformed into another; that is, they describe the substance's chemical reactivity. A chemical property of wood is that it burns when heated to ignition temperature in the presence of air, a chemical property of iron is that it rusts, and a chemical property of water is that it can be decomposed electrically into hydrogen and oxygen.

Changes that do not alter the substance's chemical composition are classified as *physical changes*. Examples include the freezing of water (it is still H_2O), the dissolving of table salt in water (it is still sodium ions and chloride ions), the heating of an iron bar (still Fe), and the evaporation of isopropyl alcohol (still C_3H_8O).

A change that alters the chemical composition of a substance and hence forms one or more new substances is called a *chemical change* or, more often, a **chemical reaction**. The decomposition of water, the rusting of iron, and the burning of magnesium in oxygen to form magnesium oxide are examples of chemical changes (● Fig. 13.1). ● Table 13.1 summarizes the distinctions between physical and chemical properties and changes.

A chemical reaction is a rearrangement of atoms in which some of the original chemical bonds are broken and new bonds are formed to give different chemical structures (● Fig. 13.2). Generally, only an atom's valence electrons are involved directly in a chemical reaction. The nucleus, and hence the atom's identity as a particular element (number of protons), is unchanged.

Consider the reaction shown by the generalized chemical equation

$$A + B \longrightarrow C + D$$
$$\text{Reactants} \qquad \text{Products}$$

The arrow indicates the direction of the reaction and has the meaning of "reacts to form" or "yields." (See ● Table 13.2 for the meanings of other symbols commonly seen in chemical equations.) Thus, the equation is read, "Substances *A* and *B* react to form substances *C* and *D*." The original substances *A* and *B* are called the **reactants**, and the new substances *C* and *D* are called the **products**. In any chemical reaction, three things take place:

Table 13.1 Physical and Chemical Properties and Changes		
	Physical	**Chemical**
Property	*Description* such as size, color, odor, density, and melting point	*Description* that tells how a substance reacts, or fails to react, chemically
Change	*Change* in which no new substance is formed, only a different form of the original substance	*Change* in which the original substances disappear and new substances are formed

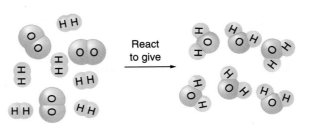

Figure 13.2 A Chemical Reaction Is a Rearrangement of Atoms When hydrogen and oxygen form water, bonds are broken in the reactants and new bonds are formed to give the products. No atoms can be lost, gained, or changed in identity.

Reactants   Products

1. The reactants disappear or are diminished.

2. New substances appear as products that have different chemical and physical properties from the original reactants. See ● Table 13.3 and note that in some reactions only one product is formed.

3. Energy (heat, light, electricity, sound) is either released or absorbed, although sometimes the energy change is too small to be detected.

A *chemical equation* can be written for each chemical reaction. The correct chemical formulas for the reactants and products must be used and *cannot* be changed. For example, the decomposition of hydrogen iodide is initially written $HI \longrightarrow H_2 + I_2$. However, until the equation is *balanced,* it does not express the actual *ratio* in which the substances react and form. Most chemical reactions can be balanced by trial and error using three simple principles:

1. The same number of atoms of each element must be represented on each side of the reaction arrow, because no atoms can be gained, lost, or changed in identity during a chemical reaction.

The equation $HI \longrightarrow H_2 + I_2$ is unbalanced, because two atoms of both H and I are represented on the right side but only one of each is shown on the left side.

2. Only the *coefficients* may be manipulated. The coefficients are the numbers in front of the formulas that designate the relative amounts of the substances; the *subscripts*, which denote the correct formulas of the substances, cannot be manipulated.

$$\boxed{\text{Coefficients}}$$
$$2\,H_2 + O_2 \longrightarrow 2\,H_2O$$
$$\boxed{\text{Subscripts}}$$

Table 13.2 Common Symbols in Chemical Equations

Symbol	Meaning
+	Plus, or and
→	Reacts to form, or yields
(g)	Gas
(l)	Liquid
(s)	Solid
(aq)	Aqueous (water) solution
$\xrightarrow{MnO_2}$	Catalyst (MnO_2, in this case)
⇌	Equilibrium (equal reaction rates)

Table 13.3 Clues That a Chemical Reaction Has Occurred

1. The color changes.
2. The odor changes.
3. Gas bubbles form.
4. Solid particles form in solution (*precipitate*).
5. Heat is produced or absorbed.

Thus, the equation in item 1 cannot be balanced by changing the formula of HI to H_2I_2. However, a coefficient of 2 can be placed *before* the formula of HI. The notation 2 HI represents two molecules of hydrogen iodide, each made up of one hydrogen atom and one iodine atom. This representation gives 2 HI $\longrightarrow$ H_2 + I_2 and balances the equation with the same number of each kind of atom on opposite sides of the equation. (As with subscripts, a coefficient of 1 is not written, it is understood.)

3. The final set of coefficients should be whole numbers (not fractions) and should be the smallest whole numbers possible.

For example, 2 HI $\longrightarrow$ H_2 + I_2 is appropriate, but *not* HI $\longrightarrow$ $\frac{1}{2}$ H_2 + $\frac{1}{2}$ I_2 *or* 4 HI $\longrightarrow$ 2 H_2 + 2 I_2.

The 5 following tips will help.

1. Count the atoms. Consider 4 $Al_2(SO_4)_3$. The subscript 2 multiplies the Al, the subscript 4 multiplies the O, the subscript 3 multiplies everything in parentheses, and the coefficient 4 multiplies the whole formula. Therefore, there is a total of 8 Al atoms, 12 S atoms, and 48 O atoms.

2. Start with an element that is present in only one formula on each side of the arrow. For example, when balancing C + SO_2 $\longrightarrow$ CS_2 + CO, start with S or O, not C.

3. Find the lowest common denominator (the smallest whole number that is divisible by each denominator) for each element that is present in only one place on each side. Insert coefficients in such a way as to get the same number of atoms of that element on each side.

   For example, in C + SO_2 $\longrightarrow$ CS_2 + CO, two atoms of sulfur show on the product side and only one shows on the reactant side. The lowest common denominator is 2, so put a coefficient 2 in front of the SO_2, giving C + 2 SO_2 $\longrightarrow$ CS_2 + CO. Next, take care of the oxygen. Four oxygen atoms show on the reactant side (in 2 SO_2) and only one on the product side (in CO). Putting a 4 before the CO gives C + 2 SO_2 $\longrightarrow$ CS_2 + 4 CO. Finally, balance the carbons. Five atoms of carbon now show on the product side and only one appears on the reactant side, so place a 5 before the C on the reactant side to get

$$5 \text{ C} + 2 \text{ SO}_2 \longrightarrow \text{CS}_2 + 4 \text{ CO}$$

   A quick recheck shows that all is in balance (two S, four O, five C).

4. Balance *polyatomic ions* as a unit when they remain intact during the reaction. For example, in Al + H_2SO_4 $\longrightarrow$ $Al_2(SO_4)_3$ + H_2, you would balance Al atoms, H atoms, and SO_4^{2-} (sulfate ions). (What coefficient would you put before the H_2SO_4? A 3 would make a good choice.)

5. If you come to a point where everything would be balanced if it weren't for a *fractional* coefficient that has to be used in one place, then multiply all the coefficients by whatever number is in the denominator of the fraction.

   For example, in C_2H_2 + O_2 $\longrightarrow$ CO_2 + H_2O, putting a 2 in front of the CO_2 and leaving an understood 1 in front of both H_2O and C_2H_2 would require a $\frac{5}{2}$ in front of the O_2 (the oxygen is left for last because it is present in two places on the product side). This would give C_2H_2 + $\frac{5}{2}$ O_2 $\longrightarrow$ 2 CO_2 + H_2O, but we generally don't want fractional coefficients, so multiply the whole equation by 2 (the number in the denominator). Thus, the final balanced equation is

$$2 \text{ C}_2\text{H}_2 + 5 \text{ O}_2 \longrightarrow 4 \text{ CO}_2 + 2 \text{ H}_2\text{O}$$

Basic Reactions

For practice in this fundamental chemical skill, let's go through the process of balancing two more equations in Example 13.1. Each example reaction also illustrates a type of reaction that we want to recognize.

1. A **combination reaction** (● Fig. 13.3a) occurs when at least two reactants combine to form just one product: $A + B \longrightarrow AB$.

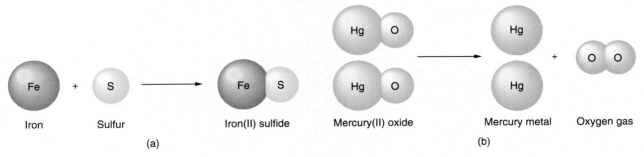

Iron   Sulfur   Iron(II) sulfide   Mercury(II) oxide   Mercury metal   Oxygen gas

(a)   (b)

Figure 13.3 A Combination Reaction and a Decomposition Reaction (a) Combination reactions have the format $A + B \longrightarrow AB$, as shown in this schematic of iron and sulfur combining to give iron(II) sulfide. (The FeS figure represents an ion pair, or formula unit, not a molecule.) (b) Decomposition reactions have the format $AB \longrightarrow A + B$, as shown in this schematic of mercury(II) oxide decomposing to give mercury metal and oxygen gas. (The HgO figure represents an ion pair, or formula unit, not a molecule.)

2. A **decomposition reaction** (Fig. 13.3b) occurs when only one reactant is present and decomposes into two (or more) products: $AB \longrightarrow A + B$.

EXAMPLE 13.1   Balancing Equations

(a) An example of a *combination reaction* is the reaction of magnesium and oxygen to form magnesium oxide (see Fig. 13.1). Balance the equation. (Recall that elemental oxygen exists as diatomic molecules.)

$$Mg + O_2 \longrightarrow MgO$$

(b) Air bags in cars and trucks are inflated by the nitrogen gas produced by electrical ignition of sodium azide, NaN_3. Balance the equation for this *decomposition reaction*.

$$NaN_3(s) \longrightarrow Na(s) + N_2(g)$$

Solution

(a) The magnesium atoms are balanced (one on the left and one on the right), but two oxygens show on the left and only one on the right. Because the lowest common denominator of 2 and 1 is 2, to balance the oxygen, place a 2 in front of the MgO.

$$Mg + O_2 \longrightarrow 2\,MgO$$

This step balances the oxygen, but now the magnesium is unbalanced (two on the right and one on the left). To rebalance the magnesium, place a 2 in front of the Mg. The balanced equation is

$$2\,Mg + O_2 \longrightarrow 2\,MgO$$

(b) Each element is present in only one place on each side, so it does not matter which is addressed first. The sodium atoms are balanced already, but the nitrogen is not balanced. With three atoms of N on the left and two on the right, the lowest common denominator is 6. Therefore, put a 2 in front of the NaN_3 and a 3 in front of the N_2, which gives

$$2\,NaN_3(s) \longrightarrow Na(s) + 3\,N_2(g)$$

The nitrogen is balanced with six atoms on each side, but the sodium is now unbalanced. However, it can be balanced by putting a 2 in front of the Na. The balanced equation is

$$2\,NaN_3(s) \longrightarrow 2\,Na(s) + 3\,N_2(g)$$

Confidence Exercise 13.1

The passage of an electric current through water (the *electrolysis* of water) forms hydrogen gas and oxygen gas at the electrodes, as shown in ● Fig. 13.4. Is it a combination reaction or a decomposition reaction? Balance the equation.

$$H_2O \longrightarrow H_2 + O_2$$

Answers to Confidence Exercises may be found at the back of the book.

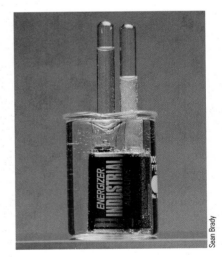

Sean Brady

Figure 13.4 The Electrolysis of Water Bubbles of oxygen gas (the tube on the left) and hydrogen gas (the tube on the right) form as the battery provides an electric current that decomposes water into its component elements. This is an example of *electrolysis,* the use of an electric current to cause a chemical reaction.

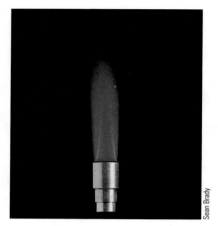

Sean Brady

Figure 13.5 A Common Exothermic Reaction
Natural gas consists mostly of methane (CH_4), with smaller amounts of other hydrocarbons. When hydrocarbons burn in abundant oxygen, the products are water and carbon dioxide. Energy is given off in the form of heat and light.

Did You Learn?

● A chemical reaction occurs when reactants are used, new substances appear, and energy is consumed or released.

● A balanced chemical equation has an equal number of the same atoms on each side.

13.2   Energy and Rate of Reaction

Preview Questions

● What is a combustion reaction?

● Which factors affect the rate of a reaction?

All chemical reactions involve a change in energy, as can be seen by the bright light in Fig. 13.1. The energy change is related to the bonding energies between the atoms that form the molecules. During a chemical reaction, some chemical bonds are broken and others are formed. Energy must be absorbed to break bonds, and energy is released when bonds are formed. The energy is released or absorbed in the form of heat, light, electrical energy, or sound. When a net release of energy to the surroundings occurs in a chemical reaction, it is called an **exothermic reaction**. An example of a common exothermic reaction is the burning of natural gas, which is composed primarily of methane, CH_4 (● Fig. 13.5). The reaction is

$$CH_4 + 2\,O_2 \longrightarrow CO_2 + 2\,H_2O\ (+\ energy)$$

When methane burns in air, the chemical energy of the bonds in the products is less than the chemical energy of the bonds in the reactants. More energy is given off when the new bonds form than is absorbed in breaking the old bonds (● Fig. 13.6).

When a net absorption of energy from the surroundings occurs during a chemical reaction, it is called an **endothermic reaction**. An example of an endothermic reaction is the production of nitric oxide (NO) from nitrogen and oxygen.

$$N_2 + O_2\ (+\ energy) \longrightarrow 2\,NO$$

Energy is released when the bonds are formed in the nitric oxide molecules, but the amount is less than that absorbed in breaking the oxygen and nitrogen molecule bonds

Figure 13.6 Exothermic Reaction
The reaction between methane and oxygen results in a net release of energy (E_R) because the bonds in the products (CO_2 and H_2O) have less total energy than the bonds in the reactants (CH_4 and O_2).

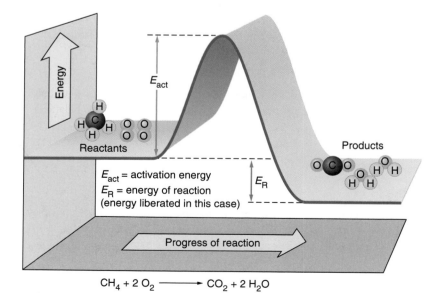

$$CH_4 + 2\,O_2 \longrightarrow CO_2 + 2\,H_2O$$

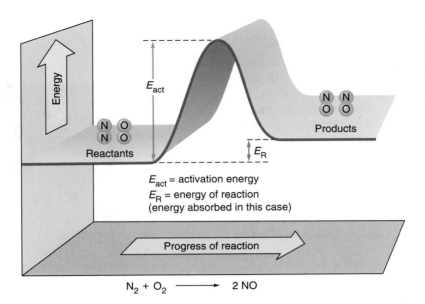

$$N_2 + O_2 \longrightarrow 2\,NO$$

Figure 13.7 Endothermic Reaction
In the oxygen–nitrogen reaction, a net absorption of energy (E_R) occurs because the bonds in the oxygen and nitrogen molecules (reactants) have a lower total energy than the bonds in the two nitric oxide molecules (products).

($\bullet$ Fig. 13.7). Thus, there is a net absorption of energy from the surroundings, and the reaction is endothermic.

The necessity of striking a match to ignite it is well known ($\bullet$ Fig. 13.8). One must contribute some energy—through friction—to initiate the chemical reaction. When one burns methane gas in a gas stove, a flame or spark is necessary to ignite the methane because the C—H and O—O bonds must be broken initially. Once the gas is ignited, the net energy released breaks the bonds of still more CH_4 and O_2 molecules, and the reaction proceeds continuously, giving off energy in the form of heat and light. The energy necessary to start a chemical reaction is called the **activation energy**.

The activation energy is a measure of the minimum kinetic energy that colliding molecules must possess in order to react chemically. Once the activation energy is supplied, however, an exothermic reaction can release more energy than was supplied. Think of a boulder resting next to a low wall at the top edge of a cliff; your goal is to drop the boulder off the cliff. An initial input of energy is required to raise the boulder to the top of the wall, but then a larger amount of energy is released as the boulder crashes below.

An analogous situation exists for an exothermic chemical reaction. Once the activation energy is supplied, the formation of new chemical bonds can release more energy than was absorbed in breaking the original bonds. This energy is usually perceived on a large scale as the surroundings becoming warmer ($\bullet$ Fig. 13.9a).

In endothermic reactions, more energy is needed (absorbed) to break chemical bonds than is generated when new bonds are formed. In this case, there is a net intake of energy of reaction (E_R) from the surroundings, which is usually perceived on a large scale as a cooling of the surroundings (Fig. 13.9b). The activation energy (E_{act}) is not fully recovered.

Figure 13.8 Activation Energy
Rubbing the head of a match against a rough surface (friction) provides the activation energy necessary for the match to ignite.

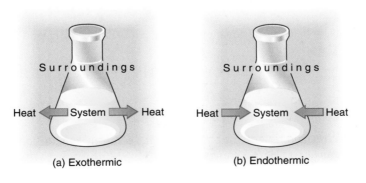

(a) Exothermic   (b) Endothermic

Figure 13.9 Heat Flow in Exothermic and Endothermic Reactions
(a) During exothermic reactions, the reaction vessel heats up and the heat flows to the surroundings, which also get warmer. (b) During endothermic reactions, the reaction vessel gets cold and heat flows in from the surroundings, which also cool down.

Highlight   The Chemistry of Air Bags

Most experts agree that air bags represent a very important advance in automobile safety. These bags, which are stored in the auto's steering wheel, dash, or doors, are designed to inflate rapidly (within about 40 milliseconds) in the event of a crash, cushioning the occupants against impact (Fig. 1). The bags then deflate immediately to allow vision and movement after the crash. Air bags are activated when a severe deceleration (an impact) causes a sensing control unit to send a signal that electrically ignites a detonator, which, in turn, causes sodium azide (NaN_3) to decompose explosively, forming sodium and nitrogen gas:

$$2 \, NaN_3(s) \longrightarrow 2 \, Na(s) + 3 \, N_2(g)$$

This system works very well and requires a relatively small amount of sodium azide [100 g yields 56 L $N_2(g)$ at 25°C and 1.0 atm]. Because the sodium metal is dangerously reactive (see Fig. 11.22), KNO_3 and SiO_2 are mixed in with the reactants using the produced Na metal to form a harmless sodium silicate glass.

When a vehicle containing air bags reaches the end of its useful life, the sodium azide present in the activators must be given proper disposal. Sodium azide, besides being explosive, has a toxicity roughly equal to that of sodium cyanide. It also forms hydrazoic acid (HN_3), a toxic and explosive liquid, when treated with strong acids.

The air bag represents an application of chemistry that has saved thousands of lives.

Figure 1 Inflated air bags.
(See also the Highlight: The Automobile Air Bag in Chapter 3.5.)

Thus, the "humps" (called *energy barriers*) in Figs. 13.6 and 13.7 indicate the activation energy that must be supplied for the reaction to proceed.

An explosion occurs when an exothermic chemical reaction liberates its energy almost instantaneously, simultaneously producing large volumes of gaseous products (see the **Highlight: The Chemistry of Air Bags**). A **combustion reaction** is a reaction in which a substance reacts with oxygen to burst into flame and form an oxide. It proceeds more slowly than an explosion and yet is still quite rapid. A common example of combustion is the burning of natural gas, coal, paper, and wood. <u>All carbon-hydrogen compounds (hydrocarbons) and carbon–hydrogen–oxygen compounds produce energy and yield carbon dioxide and water when burned completely.</u> We will focus on hydrocarbon combustion reactions.

EXAMPLE 13.2   Complete Hydrocarbon Combustion

One of the components of gasoline is the hydrocarbon named *heptane,* C_7H_{16}. Write the balanced equation for its complete combustion.

Solution

Write the formula for heptane plus that of oxygen gas, O_2, followed by a reaction arrow.

$$C_7H_{16} + O_2 \longrightarrow$$

The products of complete hydrocarbon combustion are always CO_2 and H_2O, so write their formulas on the product side.

$$C_7H_{16} + O_2 \longrightarrow CO_2 + H_2O$$

Now balance the equation, starting with either C or H, and leaving O until last (why?). The answer is

$$C_7H_{16} + 11 \, O_2 \longrightarrow 7 \, CO_2 + 8 \, H_2O$$

Confidence Exercise 13.2

Write and balance the equation for the complete combustion of the hydrocarbon named *propane,* C_3H_8, a common fuel gas.

Answers to Confidence Exercises may be found at the back of the book.

From Example 13.2, when the heptane (C_7H_{16}) in gasoline is burned completely, the combustion reaction is

$$C_7H_{16} + 11\ O_2 \longrightarrow 7\ CO_2 + 8\ H_2O$$

If there is insufficient oxygen or time for complete combustion, sooty black carbon (C) and the poisonous gas carbon monoxide (CO) will be products as well. The automobile engine has insufficient oxygen to burn the hydrocarbon completely, so the reaction is

$$C_7H_{16} + 9\ O_2 \longrightarrow 4\ CO_2 + 2\ CO + C + 8\ H_2O$$

The black color of some exhaust gases indicates the presence of large amounts of carbon (C) and shows that oxidation is incomplete. Because of the significant amounts of CO formed, running an automobile engine in a closed garage can be fatal. When tobacco burns, CO is among the harmful gases that are released.

Rate of Reaction

The rate of a reaction depends on (1) temperature, (2) concentration, (3) surface area, and (4) the possible presence of a catalyst (a substance that can speed up a reaction). Let's examine the first factor, *temperature*.

To react, molecules (or atoms or ions) must collide in the proper orientation and with enough kinetic energy to break bonds; this is the activation energy. The kinetic energy involved in any given collision may or may not be great enough for reaction. Molecules in reactions that have low activation energies react readily because of the greater number of *effective collisions*. These are collisions in which the molecules collide with at least the minimum energy and proper orientation to break bonds and form new substances. Recall from Chapter 5.1 that if heat is added to a substance and raises its temperature, the average speed and kinetic energy of the molecules will be increased. For chemical reactions, this increased temperature causes more collisions *and* more violent collisions, and the reaction rate increases dramatically. A biochemical example of this occurs when snakes, lizards, and other cold-blooded creatures warm themselves in sunlight to speed up their metabolism and become more active.

What role does the second factor, *concentration,* play in the rates of chemical reactions? Generally, if the concentration of the reactants is greater, then the rate of the reaction is greater. Because the molecules are packed more closely, more collisions occur each second and the reaction rate should increase (● Fig. 13.10). For example, a glowing coal will burst into flame in pure oxygen.

Astronauts Gus Grissom, Roger Chaffee, and Ed White lost their lives in 1967 when fire broke out in the Project Apollo spacecraft in which they were training on the ground. The environment inside the capsule was 100% oxygen, and everything burned furiously. Since then, an environment with a much smaller oxygen concentration has been used in spacecraft.

Surface area, the third factor, can play a surprisingly important role in the rate of a reaction. You would be startled to see a lump of coal or a pile of grain explode when a match was held to it. Get finely divided coal dust or grain dust in the air, however, and its enormous surface area can cause a combustion reaction that takes place with explosive speed (● Fig. 13.11). Knowledge of basic scientific principles can sometimes mean the difference between life and death. The second "Chemistry Fact" at the beginning of the chapter offers another example of surface area.

(a)

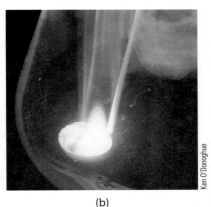

(b)

Figure 13.10 Concentration and Reaction Rate
The effect of concentration on reaction rate is apparent from the lower intensity of light coming from phosphorus burning in air's 21% oxygen (a) compared with burning in 100% oxygen (b).

Figure 13.11 A Grain Dust Explosion
A grain elevator exploded in 1982 in Council Bluffs, Iowa, killing five people. When finely divided grain dust is suspended in air in a confined space, the enormous surface area of the dust particles can cause such a fast combustion reaction that an explosion occurs.

Conceptual Question and Answer

Burning Iron!

Q. How would you get iron to burn in a Bunsen burner flame?

A. Iron pellets will not burn in a Bunsen burner flame. Why? The activation energy is too high. How, then, would you lower the activation energy? You could grind the iron pellets into a fine iron powder, because small particles have more surface area per unit mass (show this is true by comparing the surface area of a whole apple with the two halves of the apple). This increased surface area lowers the activation energy enough to cause iron powder to burn.

The fourth and final factor in reaction rate is the possible presence of a **catalyst**, a substance that increases the rate of reaction but is not itself consumed in the reaction.* Some catalysts act by providing a surface on which the reactants are concentrated. The majority of catalysts work by providing a new reaction pathway with lower activation energy; in effect, they lower the energy barrier (● Fig. 13.12).

By definition, catalysts are not consumed in the reaction, but they *are* involved in it. They unite with a reactant to form an intermediate substance that takes part in the chemical process and then decomposes to release the catalyst in its original form. For example, in the manufacture of sulfuric acid, which is a chemical produced and used in the largest quantity in industry, sulfur dioxide must react with oxygen to form sulfur trioxide:

$$2\,SO_2 + O_2 \longrightarrow 2\,SO_3 \quad \text{(slow)}$$

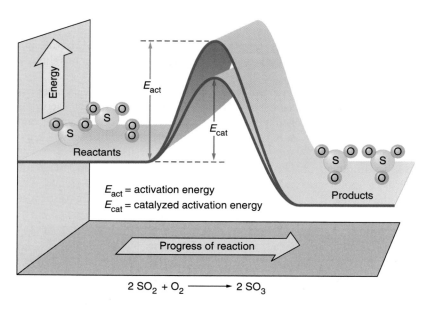

Figure 13.12 How a Catalyst Works
A catalyst generally operates by providing a new reaction pathway with a lower activation energy requirement ($E_{cat} < E_{act}$). Thus, more collisions possess enough energy to break bonds, and the reaction rate increases. Is this reaction exothermic or endothermic?

---

*A catalyst never slows a reaction. A substance that reduces the rate of a chemical reaction is called an *inhibitor*. Such a substance generally acts by tying up a catalyst for the reaction or by interacting with a reactant to reduce its concentration. Unlike a catalyst, an inhibitor is used up in a reaction.

This reaction is very slow unless nitric oxide (NO) is added as a catalyst. Once the NO is added, the reaction then proceeds in two fast steps. The NO combines with oxygen to form nitrogen dioxide (NO_2). The NO_2 then reacts with SO_2 to form SO_3, releasing the NO catalyst, which can be used again. Adding together the two fast reactions shows that the net result is the same as that of the one slow reaction:

$$2 \cancel{NO} + O_2 \longrightarrow 2 \cancel{NO_2} \quad \text{(fast)}$$
$$\underline{2 SO_2 + 2 \cancel{NO_2} \longrightarrow 2 SO_3 + 2 \cancel{NO} \quad \text{(fast)}}$$
$$2 SO_2 + O_2 \longrightarrow 2 SO_3 \quad \text{net reaction (fast)}$$

Another example of the use of a catalyst occurs in the decomposition of hydrogen peroxide, H_2O_2. At room temperature, a solution of H_2O_2 decomposes, slowly producing water and releasing oxygen.

$$2 H_2O_2 \longrightarrow 2 H_2O + O_2 \quad \text{(slow)}$$

However, if a small amount of manganese(IV) oxide, MnO_2 (often called *manganese dioxide*), is mixed with the H_2O_2, the reaction takes place rapidly at room temperature (● Fig. 13.13). The manganese(IV) oxide is not consumed in the reaction but acts only as a catalyst. The presence of a catalyst is indicated by placing its formula over the reaction arrow.

$$2 H_2O_2 \xrightarrow{MnO_2} 2 H_2O + O_2 \quad \text{(fast)}$$

A common example of catalysts is the use of catalytic converters in cars. Beads of a platinum (Pt), rhodium (Rh), or palladium (Pd) catalyst are packed into a chamber through which the exhaust gases must pass before they leave a car's tailpipe. During the passage through the converter, noxious CO and NO are changed to CO_2 and N_2, which are normal components of the atmosphere (● Fig. 13.14). The result is a great decrease in air pollution. The amount of CO_2 being discharged into the atmosphere is now a concern, however (see Chapter 20.5).

Catalysts are used extensively in manufacturing, and they also play a crucial role in biochemical processes. The human body has thousands of biological catalysts called *enzymes* that act to control various physiologic reactions. The names of enzyme catalysts usually end in *-ase*. During digestion, lactose (milk sugar) is broken down in a reaction catalyzed by the enzyme lactase. Many infants and adults, particularly those of African and Asian descent, have a deficiency of lactase and thus are unable to digest the lactose in milk. ● Figure 13.15 summarizes the factors that influence reaction rates.

(a)

(b)

Figure 13.13 A Catalyst at Work
Aqueous hydrogen peroxide (H_2O_2) decomposes to O_2 and H_2O at an imperceptible rate at room temperature. When a lump of manganese(IV) oxide is lowered into the solution, the reaction takes place rapidly because of the catalytic effect of the MnO_2.

Conceptual Question and Answer

No Pineapple

Q. Why should you not put pineapple in Jell-O?

A. Pineapple contains an enzyme called bromelin that breaks down bonds in proteins. Gelatin is a protein mixture with trapped pockets of flavored and colored liquid. If pineapple is added, then the bromelin enzyme cuts the protein chain and keeps the gelatin from jelling properly. Warnings are usually found on the gelatin box about pineapple, along with kiwi, papaya, figs, and guava.

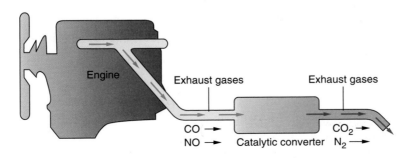

Figure 13.14 Catalytic Converters
A catalytic converter changes the poisonous carbon monoxide (CO) and nitric oxide (NO) gases coming from the car's engine into less harmful carbon dioxide (CO_2) and harmless nitrogen (N_2). Catalytic converters help reduce air pollution.

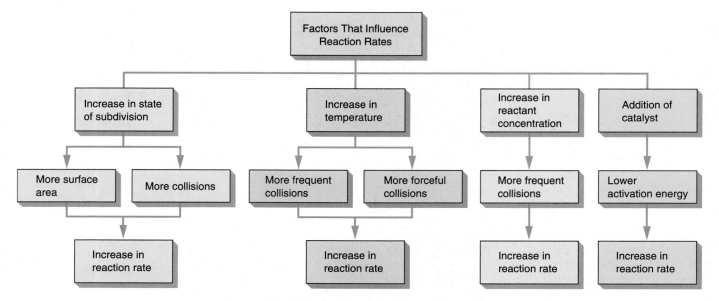

Figure 13.15 Reaction Rates
This diagram summarizes the four factors that influence the rate of reactions and how they do so.

Did You Learn?

● Combustion reactions are reactions in which a substance reacts with oxygen to burst into flame and form an oxide.

● Four factors affect reaction rates: temperature, concentration, surface area, and catalysts.

13.3   Acids and Bases

Preview Questions

● What is the pH scale?

● How is a salt defined in chemistry?

The classification of substances as acids or bases originated early in the history of chemistry. An acid, when dissolved in water, has the following properties:

1. It conducts electricity.

2. It changes the color of litmus dye from blue to red.

3. It tastes sour. (But *never* taste an acid or anything else in a lab!)

4. It reacts with a base to neutralize its properties.

5. It reacts with active metals to liberate hydrogen gas.

A base, when dissolved in water, has the following properties:

1. It conducts electricity.

2. It changes the color of litmus dye from red to blue.

3. It reacts with an acid to neutralize its properties.

One of the first theories formulated to explain acids and bases was presented in 1887 by Svante Arrhenius ("ar-RAY-nee-us"), a Swedish chemist. He proposed that the characteristic properties of aqueous solutions of acids and bases are due to the hydrogen ion (H^+) and the hydroxide ion (OH^-), respectively.

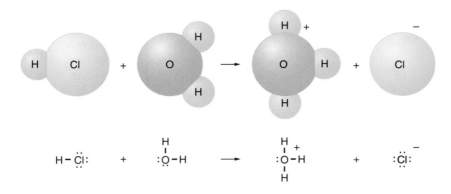

Figure 13.16 Hydrogen Chloride Reacts with Water
When gaseous hydrogen chloride (HCl) molecules are added to water (H_2O), hydronium ions (H_3O^+) and chloride ions (Cl^-) are formed.

A more advanced definition of acids and bases comes from chemists Johannes Brønsted and Thomas Lowry (1923), who define an acid to be a proton donor, and a base to be a proton acceptor. American chemist Gilbert Lewis (1923) developed a more general definition removing the necessity of the hydrogen ion. A Lewis acid is defined to be any species that *accepts* lone pair electrons, and a Lewis base is any species that *donates* lone pair electrons. In both definitions, water need not be the solvent. However, as this discussion is an introduction to the subject, we will limit it to the simplest definition of acids and bases.

According to the *Arrhenius acid–base concept,* when a substance such as colorless, gaseous hydrogen chloride (HCl) is added to water, virtually all the HCl molecules ionize into H^+ ions and Cl^- ions.

$$HCl \rightarrow H^+ + Cl^-$$

The acidic properties of HCl are due to the H^+ ions. Actually, when hydrogen chloride is placed in water, hydrogen ions are transferred from the HCl to the water molecules, as shown by the following equation:

$$HCl + H_2O \longrightarrow H_3O^+ + Cl^-$$

The H_3O^+, called the *hydronium ion,* and the Cl^- are formed in this reaction (● Fig. 13.16). An Arrhenius **acid** is a substance that gives hydrogen ions, H^+ (or hydronium ions, H_3O^+), in water.

Acids are classified as strong or weak. A *strong acid* is one that ionizes almost completely in solution (● Fig. 13.17a). For example, hydrochloric acid (HCl), nitric acid (HNO_3), and sulfuric acid (H_2SO_4) are common strong acids. In water, they ionize virtually completely.

Every chemical reaction is to some extent reversible; while the reactants are forming the products, some of the products are reacting to form the reactants. When in a given reaction the reverse reaction is significant, a *double arrow* ($\rightleftharpoons$) is placed between the reactants and products. When two competing reactions or processes are occurring at the same rate, we say that the system is in dynamic **equilibrium** (● Fig. 13.18). In a chemical reaction, the number of molecules on each side of the reaction does not have to be equal; the molecules just have to be changing from one side to the other at the same rate.

A *weak acid* is one that does not ionize to any great extent; at equilibrium, only a small fraction of its molecules react with H_2O to form H_3O^+ ions (Fig. 13.17b). Acetic acid, $HC_2H_3O_2$, is a common weak acid. In aqueous solution, we have

$$HC_2H_3O_2 + H_2O \rightleftharpoons H_3O^+ + C_2H_3O_2^-$$

Notice that the shorter, right-pointing equilibrium arrow indicates that ionization is not substantial for acetic acid in water.

Figure 13.17 Strong and Weak Acids
(a) A strong acid such as HCl ionizes almost completely in water, so the bulb glows brightly. Note the abundance of ions in the "circle."
(b) A weak acid such as acetic acid ionizes only slightly, so the bulb glows dimly. Notice the scarcity of ions in the "circle."

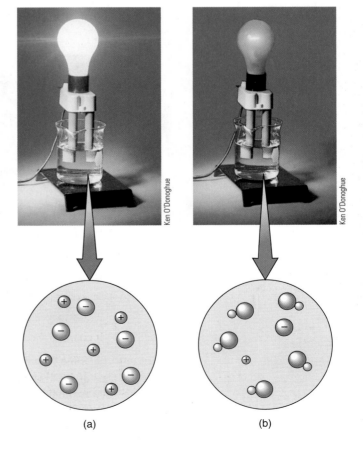

Ken O'Donoghue

(a)   (b)

Figure 13.18 Dynamic Equilibrium
Each juggler throws clubs to the other at the same rate at which he receives clubs. Because clubs are thrown continuously in both directions (hence the term *dynamic*), the number of clubs moving in each direction is constant, and the number of clubs each juggler has at a given time remains constant. *Note:* In a chemical reaction, the numbers of "clubs"(molecules) on the two sides of the reaction do not have to be equal; they just have to be changing from one side to another at the same rate.

Acids are useful compounds. For example, sulfuric acid is used in refining petroleum, processing steel, and manufacturing fertilizers and numerous other products. A dilute solution of hydrochloric acid is present in the human stomach to help digest food. Many weak acids are present in our foods, such as citric acid ($H_3C_6H_5O_7$) and ascorbic acid (vitamin C, $C_6H_8O_6$) in citrus fruits, carbonic acid (H_2CO_3) and phosphoric acid (H_3PO_4) in soft drinks, and acetic acid ($HC_2H_3O_2$) in vinegar (● Fig. 13.19).

When pure sodium hydroxide (NaOH), a white solid commonly known as *lye*, is added to water, it dissolves and releases Na^+ and OH^- into the solution. The reason NaOH solutions are basic is due to the hydroxide ions, OH^-. An Arrhenius **base** is a substance that produces hydroxide ions, OH^-, in water. However, a substance need not *initially* contain hydroxide ions to have the properties of a base. For example, ammonia (NH_3) contains no OH^- in its formula, but in water solutions it is a weak base. Its molecules react to a slight extent with water to form OH^-.

$$NH_3 + H_2O \rightleftharpoons NH_4^+ + OH^-$$

Common household bases are Drano, which contains NaOH, Windex, which has NH_3 in it, and baking soda, $NaHCO_3$ (● Fig. 13.20).

Water ionizes, but only slightly, as shown in the equation

$$H_2O + H_2O \rightleftharpoons H_3O^+ + OH^-$$

Thus, all aqueous solutions contain both H_3O^+ and OH^-. For pure water, the concentrations of H_3O^+ and OH^- are equal, and the liquid is *neutral*. An *acidic solution* contains a higher concentration of H_3O^+ than OH^-. If the concentration of OH^- is higher than that of H_3O^+, then it is a *basic solution* (● Fig. 13.21).

Figure 13.19 Some Common Acids
Vinegar, citrus fruits and juices, some cleaners, aspirin, and soft drinks are examples of common household substances that contain acids.

Figure 13.20 Some Common Bases
Ammonia, bleach, soap, cleaning supplies, and Maalox are examples of common household substances that contain bases.

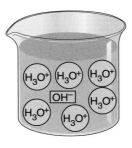 

Figure 13.21 Acidic and Basic Solutions
An acidic solution (*left*) has a higher concentration of H_3O^+ than of OH^-. In a basic solution (*right*), the situation is the reverse.

Conceptual Question and Answer

Crying Time

Q. Why do onions make you cry?

A. Cutting into an onion breaks the cells of the onion, producing sulfenic acids from the contents that are released. Enzymes from the cells mix with the sulfenic acids to produce propanethiol S-oxide, a gaseous sulfur compound. Upon getting into your eyes, the gas reacts with the water in your eyes to form sulfuric acid (H_2SO_4). The acid burns, causing your eyes to release tears to wash the irritant away. Chilling the onion first helps reduce these reactions. See why in Chapter 13.2.

It is common practice to designate the relative acidity or basicity of a solution by citing its **pH** (power of hydrogen), which is a logarithmic scale measure of the concentration of hydrogen ions (or hydronium ions) in the solution. Developed in 1909 by the Danish chemist Søren Sørensen, the pH scale has a value of 7 to indicate a neutral solution. Values of pH from 6 down to 0 indicate increasing acidity, with each drop of 1 in value meaning a *tenfold* increase in acidity. Similarly, pH values from 8 up to 14 indicate increasing basicity, with each increase of 1 in value meaning a tenfold increase in basicity. ● Figure 13.22 illustrates this concept and shows the pH values of some common solutions.

Most body fluids have a normal pH range, and a continued deviation from normal usually indicates some disorder in a body function. Thus, the pH value can be used as a means

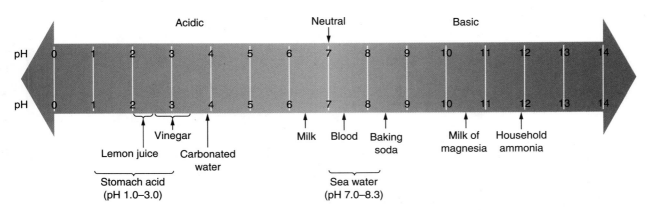

Figure 13.22 The pH Scale
A solution with a pH of 7 is neutral, a solution with pH less than 7 is acidic, and a solution with pH greater than 7 is basic.

Figure 13.23 Two pH Meters
Each pH meter's digital display gives the pH of the solution in the beaker. Which solution is acidic?

of diagnosis. For example, a blood pH that is not between 7.35 and 7.45 indicates an illness.

The pH of a solution is usually measured by a pH meter that can find the value to several decimal places (● Fig. 13.23). Approximate values can be found by the use of an acid–base indicator, which is a chemical that changes color over a narrow pH range. For example, *litmus* is red below pH 5 and blue above pH 8.

An important property of an acid is the disappearance of its characteristic properties when brought into contact with a base, and vice versa. This property is known as an **acid–base reaction**: The H^+ of an acid unites with the OH^- of a base to form water, whereas the cation of the base combines with the anion of the acid to form a salt. Thus, we can generalize: An acid and a hydroxide base react to give water and a salt. (The **Highlight: The Chemistry of Tooth Decay** discusses an example of the importance of equilibrium and acid–base reactions.)

Of course, NaCl is commonly called "salt," but actually there are many salts. A **salt** is an ionic compound composed of any cation except H^+ and any anion except OH^-. Examples are potassium chloride (KCl) and calcium phosphate, $Ca_3(PO_4)_2$. The salt may remain dissolved in water (KCl does), or it may form solid particles (a *precipitate*) if it is insoluble like $Ca_3(PO_4)_2$.

Some salts occur in our environment as *hydrates,* salts that contain molecules of water bonded in their crystal lattices. A common example is the blue, crystalline hydrate named copper(II) sulfate pentahydrate, $CuSO_4 \cdot 5H_2O$ (● Fig. 13.24). When this hydrate is heated, *anhydrous* $CuSO_4$, a white powder, is formed as water is expelled from the crystals. The dunes of White Sands National Monument in New Mexico are composed of a hydrate named gypsum, $CaSO_4 \cdot 2H_2O$.

EXAMPLE 13.3 Acid–Base Reactions

If you have an "acid stomach" (excess HCl), then you might take a milk of magnesia tablet, $Mg(OH)_2$, to neutralize it. Write the balanced chemical equation for the reaction.

Solution

You know that HCl is an acid and that $Mg(OH)_2$ is a base (note the hydroxide ion, OH^-). Therefore, the products of this acid–base reaction must be water and a salt. Write the formulas for both reactants, put in the reaction arrow, and then write the formula for water.

$$HCl + Mg(OH)_2 \longrightarrow H_2O + \text{(a salt)}$$

Now determine the correct formula for the salt. (*Note*: If you ever have trouble balancing the equation for an acid–base reaction, then you probably have an incorrect formula for the salt and may need to refer to Chapter 12.4.) Write the cation from the base first in the salt formula and then follow it with the anion of the acid: $Mg^{2+}Cl^-$, in this case. You can see that a 1-to-1 ratio of ions does not give electrical neutrality; two Cl^- are needed for each Mg^{2+}. Thus, the correct formula for the salt is $MgCl_2$. Add $MgCl_2$ to the product side and then balance the equation. The answer is

$$2 HCl + Mg(OH)_2 \longrightarrow 2 H_2O + MgCl_2$$

Confidence Exercise 13.3

Aluminum hydroxide, found in Di-Gel and Mylanta, is another popular antacid ingredient. Complete and balance the equation for the reaction of this base with stomach acid.

$$HCl + Al(OH)_3 \longrightarrow$$

A common chemical in many households is sodium hydrogen carbonate ($NaHCO_3$), known as *baking soda,* an excellent odor absorber.

Highlight   The Chemistry of Tooth Decay

Tooth enamel is composed of fibrous protein and a calcium mineral [hydroxyapatite, $Ca_5(PO_4)_3OH$]. This compound constantly dissolves (demineralizes) and reforms (mineralizes) in the saliva at the surface of a tooth. If these processes occur at the same rate, then there is no net loss of enamel. However, if demineralization gains the upper hand, then tooth decay can occur (Fig. 1).

This problem is caused by deposits called *plaque* that collect on the surfaces of teeth. Plaque occurs because of food debris containing sugars and the bacteria normally present in the mouth, which convert the sugars into acids. If plaque is allowed to collect, then it hardens over time into what is called *tartar*.

It is the acids in plaque and tartar that can dissolve the protective, hard enamel coating of a tooth (demineralization) into a porous and spongy surface. Such tooth decay can give rise to cavities that delve into the sensitive tooth interior, and then you're off to the dentist with a toothache. The acids generated by bacterial food breakdown can begin attacking the tooth enamel within 20 minutes.

The best prevention is brushing, flossing, and a visit to a dentist for a checkup every 6 months to a year. Most toothpaste contains a mild abrasive, or polishing agent, and stannous (tin) fluoride. The latter is a source of fluorine that helps combat tooth decay. If the saliva contains small amounts of fluoride ions, F^-, then the $Ca_5(PO_4)OH$ mineral in tooth enamel is changed to $Ca_5(PO_4)3F$, which is more resistant to future decay.

Fluorides are also gained from many municipal water supplies. About two-thirds of the public water in the United States has undergone fluoridation. The history of fluoridation is interesting. In the early 1900s, children in the area of Pikes Peak, Colorado, were noted to have tooth stain. Investigations found the staining to be due to high levels of fluoride ions (from a local mineral) in the drinking water. It was also observed that these children had fewer dental cavities than normal. Lower fluoride levels were found not to cause staining but did help prevent tooth decay.

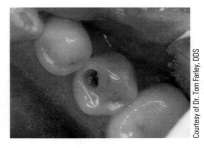

Courtesy of Dr. Tom Farley, DDS

Figure 1 Tooth Decay
Photo showing decay (dark area) on the top of the tooth.

© Richard Megna/Fundamental Photographs, NYC

Figure 13.24 A Hydrate Salt
Copper(II) sulfate pentahydrate ($CuSO_4 \cdot 5H_2O$) is the deep blue, crystalline hydrate salt on the left. When the hydrate is heated, the water is driven off and white, powdery anhydrous copper(II) sulfate is formed (*right*). As shown at the right, when water is added to the white anhydrous salt, the blue hydrated salt forms again.

Conceptual Question and Answer

Odors, Be Gone!

Q. How does baking soda eliminate odors, and where does an odor go?

A. Odors are usually generated from acidic compounds (like those found in spoiled milk) or from basic compounds (like those found in spoiled meats). The equations below show that the bicarbonate ion in baking soda can react with acids, forming carbon dioxide gas, or it can react with bases, forming

Figure 13.25 An Acid–Carbonate Reaction on a Large Scale
Nitric acid (20,000 gal) was spilled from a railroad tank car in Denver in 1983. Firefighters used an airport snowblower to throw sodium carbonate (Na_2CO_3) onto the HNO_3 acid and neutralize it.

the carbonate ion. So, both acidic and basic odor compounds are removed or neutralized in an acid–base reaction with baking soda. The odors become salt, water, and sometimes a gas, as shown in the following reactions.

$$NaHCO_3 + H^+ \text{ (from an acid)} \longrightarrow Na^+ + H_2O + CO_2(g)$$

$$NaHCO_3 + OH^- \text{ (from a base)} \longrightarrow Na^+ + H_2O + CO_3^-$$

Acids acting on baking soda are also involved in the leavening process in baking. Baking powders contain baking soda plus an acidic substance such as $KHC_4H_4O_6$ (cream of tartar). When this combination is dry, no reaction occurs, but when water is added, CO_2 is given off. This property is called an **acid–carbonate reaction**: An acid and a carbonate (or hydrogen carbonate) react to give carbon dioxide, water, and a salt.

● Figure 13.25 shows how this type of reaction can be used to neutralize acid spills.

EXAMPLE 13.4   Acid–Carbonate Reactions

Another way of relieving an overacid stomach is to take an antacid tablet (such as Tums) that contains calcium carbonate, $CaCO_3$ (● Fig. 13.26). Write the balanced equation for the reaction between stomach acid (HCl) and such an antacid ($CaCO_3$).

Solution

This is an acid–carbonate reaction, so the products are carbon dioxide, water, and a salt. Start by writing the formulas for both reactants, the reaction arrow, and the formulas for carbon dioxide and water. Leave a space for the formula of the salt.

$$HCl + CaCO_3 \longrightarrow CO_2 + H_2O + \text{ (a salt)}$$

As in the preceding example, determine the correct formula for the salt. The Ca^{2+} and Cl^- will form $CaCl_2$. Add the formula of the salt and then balance the equation.

$$2\ HCl + CaCO_3 \longrightarrow CO_2 + H_2O + CaCl_2$$

Confidence Exercise 13.4

Complete and balance the equation for the reaction of nitric acid with sodium carbonate, the reaction shown in Fig. 13.25.

$$HNO_3 + Na_2CO_3 \longrightarrow$$

Double-Replacement Reactions

Acid–base and acid–carbonate reactions are types of **double-replacement reactions**, those in which the positive and negative components of the two compounds "change partners" (● Fig. 13.27). The general format is

$$AB + CD \longrightarrow AD + CB$$

Figure 13.26 An Acid–Carbonate Reaction on a Small Scale
Calcium carbonate in a Tums antacid tablet reacts with an acidic solution of HCl to give CO_2 gas, H_2O, and dissolved $CaCl_2$.

| Potassium chloride | Silver nitrate | | Potassium nitrate | Silver chloride |

Figure 13.27 The Format of a Double-Replacement Reaction
Double-replacement reactions have the format $AB + CD \longrightarrow AD + CB$, as shown in this schematic of aqueous solutions of potassium chloride and silver nitrate "changing partners" to form potassium nitrate and silver chloride. (The silver chloride is insoluble in water and thus precipitates.) Note that each compound sketch represents an ion pair, or formula unit, not a molecule.

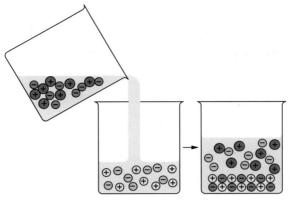

Figure 13.28 A Double-Replacement Precipitation Reaction
The positive ions of one soluble substance combine with the negative ions of another soluble substance to form a precipitate when the solutions are mixed. The other positive and negative ions generally remain dissolved.

For example, in the acid–base reaction of HCl and NaOH, the positive part of the acid (H^+) attaches to the negative part of the base (OH^-) to form HOH (another way of writing H_2O). The positive part of the base (Na^+) attaches to the negative part of the acid (Cl^-).

$$HCl + NaOH \rightarrow HOH + NaCl$$

As another example, in the acid–carbonate reaction of H_2SO_4 and K_2CO_3, the positive part of the acid (H^+) attaches to the negative part of the carbonate (CO_3^{2-}) to form H_2CO_3 (which immediately decomposes to CO_2 and H_2O). The positive part of the carbonate (K^+) attaches to the negative part of the acid (SO_4^{2-}).

$$H_2SO_4 + K_2CO_3 \longrightarrow CO_2 + H_2O + K_2SO_4$$

When aqueous solutions of two salts are mixed, the "changing of partners" often results in the formation of a **precipitate** (abbreviated *s* or *ppt*), an insoluble solid that appears when two liquids (usually aqueous solutions) are mixed. ● Figure 13.28 illustrates how a double-replacement precipitation reaction might look if we could see the ions. The formation of a precipitate of insoluble lead(II) iodide and soluble potassium nitrate from aqueous solutions of potassium iodide and lead(II) nitrate is shown in ● Fig. 13.29 and discussed in Example 13.5. ● Table 13.4 contains a useful, but not complete, list of water-soluble salts.

EXAMPLE 13.5   Double-Replacement Precipitation Reactions

Write the equation for the double-replacement reaction shown in Fig. 13.29.

Solution

First, write the correct formulas for the reactants, using your knowledge of nomenclature and ionic charges.

$$KI(aq) + Pb(NO_3)_2(aq) \longrightarrow$$

Now switch ion partners and put them together so that the positive ion is first and in such a ratio that each compound will be net electrically neutral. To show that the lead(II) iodide is insoluble, place (*s*), for solid, after its formula. To show that potassium nitrate is soluble (it contains nitrate ions), put (*aq*) after its formula. Finally, balance the equation. The answer is

$$2\,KI(aq) + Pb(NO_3)_2(aq) \longrightarrow 2\,KNO_3(aq) + PbI_2(s)$$

Confidence Exercise 13.5

Complete and balance the equation for the double-replacement reaction between aqueous solutions of sodium sulfate and barium chloride.

$$Na_2SO_4(aq) + BaCl_2(aq) \longrightarrow$$

Figure 13.29 The Yellow Precipitate Is Lead(II) Iodide
When a clear, colorless solution of potassium iodide is poured into a clear, colorless solution of lead(II) nitrate, yellow lead(II) iodide precipitates. The solute potassium nitrate remains in the solution.

Table 13.4 Some Water-Soluble Salts

Salts Containing	Examples
Alkali metal ions (Li^+, Na^+, K^+, Rb^+, Cs^+)	$NaCl$, K_2SO_4
Ammonium ions (NH_4^+)	NH_4Br, $(NH_4)_2CO_3$
Nitrate ions (NO_3^-)	$AgNO_3$, $Mg(NO_3)_2$
Acetate ions ($C_2H_3O_2^-$)	$Al(C_2H_3O_2)_3$

Did You Learn?

● The pH scale is a measure of the concentration of the hydrogen ion in a solution.

● A salt is an ionic compound with any cation except H^+ and any anion except OH^-.

13.4 Single-Replacement Reactions

Preview Questions

● What is oxidation?

● What determines the activity of a metal?

It was shown in Chapter 13.2 that combustion reactions involve the addition of oxygen to a substance. When oxygen combines with a substance or when an atom or ion loses electrons, we call the process **oxidation**. Conversely, when oxygen is removed from a compound or when an atom or ion gains electrons, the process is called **reduction**. For example, an important step in the steel-making industry is the reduction of hematite ore (Fe_2O_3) with coke (C) in a blast furnace:

$$2\ Fe_2O_3 + 3\ C \longrightarrow 4\ Fe + 3\ CO_2$$

Oxygen is removed from the iron(III) oxide, so we say that the iron oxide has been reduced. At the same time, oxygen has reacted with the carbon to form carbon dioxide; the carbon has been oxidized. We call such a chemical change an *oxidation-reduction reaction,* or, for short, a *redox reaction.*

Redox reactions do not always have to involve the gain and loss of oxygen. So that the term will apply to a larger number of reactions, redox reactions are also identified in terms of electrons lost (oxidation) or gained (reduction). Because all the electrons lost by atoms or ions must be gained by other atoms or ions, it follows that oxidation and reduction occur at the same time and at the same rate.

The relative *activity* of any metal is its tendency to lose electrons to ions of another metal or to hydrogen ions. It is determined by placing the metal in a solution that contains the ions of another metal and observing whether the test metal replaces the one in solution. If it does, then it is more active, because it has given electrons to the ions of the metal in solution. For example, ● Fig. 13.30 shows that when zinc metal (Zn) is placed in a solution that contains Cu^{2+} ions, the Zn loses electrons to the Cu^{2+} ions, producing copper metal (Cu) and Zn^{2+} ions.

$$Zn + Cu^{2+} \longrightarrow Cu + Zn^{2+}$$

Thus, the Zn has lost electrons and has been oxidized; the Cu^{2+} has gained electrons and has been reduced.

(a)

(b)

Figure 13.30 Zinc Replaces Copper Ions
(a) A strip of Zn is about to be placed into blue $CuSO_4$ solution. (b) A few minutes later, the single-replacement redox reaction has formed metallic copper and colorless $ZnSO_4$ solution.

Table 13.5 Activity Series

Metals	Ion Found
Lithium	Li^+
Potassium	K^+
Calcium	Ca^{2+}
Sodium	Na^+
Magnesium	Mg^{2+}
Aluminum	Al^{3+}
Zinc	Zn^{2+}
Chromium	Cr^{3+}
Iron	Fe^{2+}
Nickel	Ni^{2+}
Tin	Sn^{2+}
Lead	Pb^{2+}
HYDROGEN*	H^+
Copper	Cu^{2+}
Silver	Ag^+
Platinum	Pt^{2+}
Gold	Au^{3+}

Increasing activity

*Hydrogen is in capital letters because the activities of the metals are often determined in relation to the activity of hydrogen.

On the other hand, copper metal placed in a solution of Zn^{2+} leads to no reaction. Thus, zinc is more active than copper. If similar experiments are carried out for all metals (and hydrogen), then an **activity series** can be obtained (● Table 13.5).

If element A is listed *above* element B in the activity series, then A is more active than B and will replace B in a compound BC. This property is called a **single-replacement reaction** and has the general format

$$A + BC \longrightarrow B + AC$$

Only the most common type of single-replacement reaction is examined, the type in which element A is a metal (● Fig. 13.31). In such cases, A will lose its valence electrons and thus be oxidized, and B will gain these electrons and be reduced. All single-replacement reactions are also redox reactions.

Zinc metal   Copper(II) sulfate   Copper metal   Zinc sulfate

Figure 13.31 The Format of a Single-Replacement Reaction
Single-replacement reactions have the format $A + BC \longrightarrow B + AC$, as shown in this schematic of zinc reacting with an aqueous solution of copper(II) sulfate. The copper precipitates, but the other product, zinc sulfate, is soluble. The sketches of $CuSO_4$ and $ZnSO_4$ represent an ion pair, or formula unit, not a molecule. See Fig. 13.30 for photos of the actual reaction.

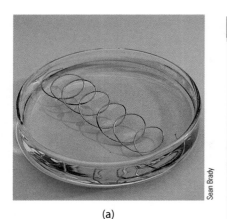

(a)

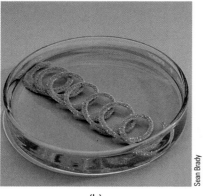

(b)

Figure 13.32 A Single-Replacement Reaction between Copper Wire and Silver Nitrate
A copper wire is shown immediately after placement in a solution of silver nitrate. (b) Later, the reaction is complete. One product, metallic silver, is evident on the surface of the wire. Copper(II) nitrate, the other product, remains dissolved in solution (notice the characteristic blue color of Cu^{2+}).

EXAMPLE 13.6　Single-Replacement Reactions

Refer to the activity series (Table 13.5) and predict whether placing copper metal in a solution of silver nitrate will lead to a reaction. If so, then complete and balance the equation.

Solution

Copper is above silver in the activity series; thus, a reaction will occur, as shown in ● Fig. 13.32. The copper atoms will be oxidized to Cu^{2+} ions, and the Ag^+ ions will be reduced to Ag atoms. The nitrate ions, NO_3^-, stay intact and dissolved in solution, but it is written as though they were joined to the Cu^{2+} ions to give $Cu(NO_3)_2(aq)$. The *unbalanced* equation for the reaction is

$$Cu + AgNO_3(aq) \longrightarrow Ag + Cu(NO_3)_2(aq)$$

Because the nitrate ions stay intact during the reaction, they can be balanced as a unit. Two nitrates are showing on the product side, so place a 2 before $AgNO_3$ on the left. Complete the balancing by adding a 2 before the Ag on the right.

$$Cu + 2\,AgNO_3(aq) \rightarrow 2\,Ag + Cu(NO_3)_2(aq)$$

Confidence Exercise 13.6

A strip of aluminum metal is placed in a solution of copper(II) sulfate, $CuSO_4(aq)$, and a strip of copper metal is placed in a solution of aluminum sulfate, $Al_2(SO_4)_3(aq)$. Refer to Table 13.5 and predict in which case a single-replacement reaction will take place; then complete and balance the equation for the reaction.

The metals above hydrogen in Table 13.5 will undergo a single-replacement reaction with acids to give hydrogen gas and a salt of the metal (● Fig. 13.33). For example,

$$Fe + H_2SO_4(aq) \longrightarrow H_2 + FeSO_4(aq)$$

In fact, metals above magnesium in Table 13.5 react vigorously with water and produce hydrogen gas and the metal hydroxide (● Fig. 13.34). The metal hydroxide is a base, or *alkali*, and that is why Groups 1A and 2A have *alkali* and *alkaline* in their names.

At this point, let's summarize the reaction types we have covered in this chapter. Please examine ● Table 13.6 and ● Fig. 13.35 carefully.

Did You Learn?

● Oxidation occurs when oxygen combines with a substance, or when atoms lose electrons.

● The activity of a metal depends on its ability to lose electrons to ions of another metal.

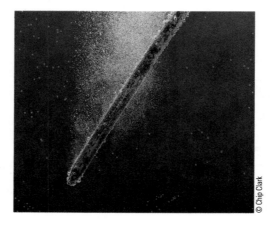

Figure 13.33 Metals and Acids
A single-replacement reaction occurs when an iron nail reacts with an aqueous solution of sulfuric acid to form H_2 gas and aqueous $FeSO_4$.

Table 13.6 A Summary of Reaction Types

Reaction Type	Example
Combination	$2\ Mg + O_2 \longrightarrow 2\ MgO$
Decomposition	$2\ HgO \rightarrow 2\ Hg + O_2$
Hydrocarbon combustion (complete)	$C_2H_4 + 3\ O_2 \longrightarrow 2\ CO_2 + 2\ H_2O$
Single-replacement	
(a) two metals	$Zn + CuSO_4 \longrightarrow Cu + ZnSO_4$
(b) metal and acid	$Fe + 2\ HCl \rightarrow H_2 + FeCl_2$
Double-replacement	
(a) precipitation	$BaCl_2 + Na_2SO_4 \longrightarrow BaSO_4(s) + 2\ NaCl$
(b) acid–base	$2\ HCl + Ca(OH)_2 \longrightarrow 2\ H_2O + CaCl_2$
(c) acid–carbonate	$H_2SO_4 + Na_2CO_3 \longrightarrow H_2O + CO_2 + Na_2SO_4$ (The H_2CO_3 that is initially formed decomposes.)

Ken O'Donoghue

Figure 13.34 Active Metals and Water
Metals, such as calcium, that are above magnesium in the activity series react with water to form hydrogen gas and the metal hydroxide. Notice that no reaction is occurring between magnesium and water (*left beaker*) but that hydrogen gas is being formed at a rapid rate as calcium reacts (*right beaker*).

Figure 13.35 A Summary of Reaction Types
The five basic types of reactions are displayed. A picture of each reaction is shown along with its generic equation and an example equation.

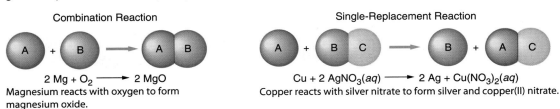

Combination Reaction

$2\ Mg + O_2 \longrightarrow 2\ MgO$
Magnesium reacts with oxygen to form magnesium oxide.

Single-Replacement Reaction

$Cu + 2\ AgNO_3(aq) \longrightarrow 2\ Ag + Cu(NO_3)_2(aq)$
Copper reacts with silver nitrate to form silver and copper(II) nitrate.

$C_2H_4 + 3\ O_2 \longrightarrow 2\ CO_2 + 2\ H_2O$
Hydrocarbon Combustion (complete)
A hydrocarbon reacts with oxygen to form carbon dioxide and water.

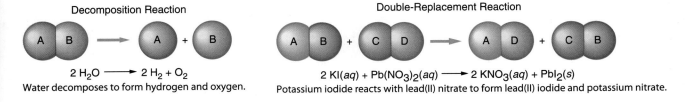

Decomposition Reaction

$2\ H_2O \longrightarrow 2\ H_2 + O_2$
Water decomposes to form hydrogen and oxygen.

Double-Replacement Reaction

$2\ KI(aq) + Pb(NO_3)_2(aq) \longrightarrow 2\ KNO_3(aq) + PbI_2(s)$
Potassium iodide reacts with lead(II) nitrate to form lead(II) iodide and potassium nitrate.

13.5 Avogadro's Number

Preview Questions

● What is Avogadro's number?

● How is molarity determined?

Just as bakers often speak of their products in terms of dozens, chemists often speak of moles of chemicals. The SI defines the **mole** (abbreviated mol) as the quantity of a substance that contains as many elementary units as there are atoms in exactly 12 g of carbon-12. That turns out to be 6.02×10^{23} carbon-12 atoms, and this huge number is referred to as **Avogadro's number**.

Ken O'Donoghue

Figure 13.36 One Mole of Six Substances
Each sample consists of 6.02×10^{23} formula units. *Clockwise from the top:* 63.5 g of copper, 27.0 g of aluminum, 55.8 g of iron, 32.1 g of sulfur, 253.8 g of iodine, and (in the center) 200.6 g of mercury.

The mole is one of the most important concepts in chemistry. The concept of mole is like that of dozen. Just as when you hear *dozen* you think 12 units, when you hear *mole* you should think 6.02×10^{23} units. Also, a mole of any substance has a mass equal to the same number of grams as the formula mass (see Chapter 12.2) of the substance, called the *molar mass* of the substance. For example, a mole of copper atoms (FM 63.5 u) is 6.02×10^{23} atoms and has a mass of 63.5 g, and a mole of water (FM 18.0 u) is 6.02×10^{23} water molecules and has a mass of 18.0 g (● Fig. 13.36). Alternatively, we can say that the molar mass of copper is 63.5 g/mol and the molar mass of water is 18.0 g/mol.

As discussed in Chapter 13.1, the coefficients in a balanced chemical equation indicate the ratio in which moles of the substances react or form. Thus, if you know the moles (or grams) of one substance in a balanced equation, then it is easy to calculate the moles (or grams) of every other substance in the equation.

EXAMPLE 13.7   Converting Grams to Moles and Moles to Grams

(a) Determine the mass in grams of 3.50 moles of lead chromate, $PbCrO_4$.

(b) How many moles are in 2.35 g of the salt magnesium chloride, $MgCl_2$?

Solution

(a) Step 1

Compute the molar mass of 1 mole of $PbCrO_4$ (see Example 12.2).

$$1\ Pb\ (207.2\ g) + 1\ Cr\ (52.0\ g) + 4\ O\ (16.0\ g) = 323.2\ g/mol$$

Step 2

Multiply the number of moles given by the molar mass.

$$3.50\ mol\ PbCrO_4 \times \frac{323.2\ g}{mol} = 1131.2\ g\ \text{or}\ 1.13\ kg$$

(b) Step 1

Compute the molar mass of $MgCl_2$.

$$1\ Mg\ (24.3\ g) + 2\ Cl\ (35.45\ g) = 95.2\ g/mol$$

Step 2

Multiply the number of grams given by the conversion factor for molar mass.

$$2.35\ g\ MgCl_2 \times \frac{1\ mol}{95.2\ g} = 0.025\ mol$$

Confidence Exercise 13.7

How many grams are in 1.70 moles of potassium permanganate, $KMnO_4$? How many moles are in 25.0 g of NaCl?

The number 6.02×10^{23} was named *Avogadro's number* not because Italian physicist Amedeo Avogadro discovered it, but in his honor 50 years after his death. Avogadro was the first person to use the term *molecule*. In 1811, he used the concept of molecules of elements to explain the newly discovered law of combining gas volumes and proved later that the correct formula for hydrogen gas was not H, but H_2.

How can such a large number as 6.02×10^{23} be comprehended? Pour out 6.02×10^{23} BBs on the United States, and they would cover the entire country to a depth of about 4 miles!

How is it known how many particles are in 1 mole? From Chapter 8.1, the unit for measuring electric charge is the coulomb (C), and it was determined that 96,485 C reduces 1 mole of singly charged ions to atoms. For example, by using an electric current to cause a chemical change (*electrolysis*), it takes 96,485 C to produce 1 mole (23.0 g) of sodium metal from molten sodium chloride (see Fig. 12.4). It takes one electron to reduce each sodium ion to a sodium atom:

$$Na^+ + e^- \longrightarrow Na$$

Thus, 96,485 C must be the total charge on 1 mole of electrons. A single electron has a negative charge of 1.6022×10^{-19} C. Therefore, Avogadro's number must be

$$\frac{96,485 \text{ C/mol}}{1.6022 \times 10^{-19} \text{ C/electron}} = 6.0220 \times 10^{23} \text{ electrons/mol}$$

This is the number of electrons in a mole of electrons and the number of units in 1 mole of anything.

Molarity

When the solutes of solutions take part in reactions, chemists want to know the number of reacting particles present. For this purpose, chemists use **molarity** (*M*), which expresses solution concentration by stating the number of moles of solute per every one liter of solution.

$$\text{molarity } (M) = \frac{\text{number of moles of solute}}{\text{volume of solution (in L)}} \qquad \text{13.1}$$

$$M = \frac{n_{\text{solute}}}{V_{\text{solution}}}$$

Figure 13.37 Making 1.0 L of 1.0 *M* Sucrose Solution
Dissolving 1.0 mol (342 g) of sucrose in enough water to make 1.0 L of solution gives a concentration of 1.0 *M*. The sucrose is added to the volumetric flask and dissolved in some water, and then more water is added to fill the flask to the circle partway up the neck (the 1.0-L mark).

For example, a 1.0 *M* (one molar) solution of sucrose (table sugar, $C_{12}H_{22}O_{11}$) could be made by dissolving 1.0 mol (342 g) of sucrose in enough water to make 1.0 L of solution (● Fig. 13.37). Notice that the same *concentration* of solution could also be obtained by dissolving half that amount of sucrose in enough water to make 0.50 L of solution.

EXAMPLE 13.8   **Finding the Molarity of a Solution**

What is the molarity of a sucrose solution in which 0.400 mol of the sugar is dissolved in water to give 1.80 L of solution?

Solution

Step 1

Given: $n_{\text{solute}} = 0.400$ mol, $V_{\text{solution}} = 1.80$ L

Step 2

Wanted: Molarity, *M*

Step 3

Use Equation 13.1.

$$M = \frac{n_{\text{solute}}}{V_{\text{solution}}} = \frac{0.400 \text{ mol}}{1.80 \text{ L}} = 0.222 \frac{\text{mol}}{\text{L}} = 0.222 \text{ M}$$

Confidence Exercise 13.8

What is the molarity of a salt (NaCl) solution in which 1.60 mol of salt is dissolved in enough water to give 2.00 L of solution?

Did You Learn?

- A mole of a substance contains Avogadro's number of elementary units.
- Molarity is the concentration of a solution determined by the ratio of the number of moles of solute and the solution volume.

KEY TERMS

1. chemical reaction (13.1)
2. reactants
3. products
4. combination reaction
5. decomposition reaction
6. exothermic reaction (13.2)
7. endothermic reaction
8. activation energy
9. combustion reaction

10. catalyst
11. acid (13.3)
12. equilibrium
13. base
14. pH
15. acid–base reaction
16. salt
17. acid–carbonate reaction
18. double-replacement reactions

19. precipitate
20. oxidation (13.4)
21. reduction
22. activity series
23. single-replacement reaction
24. mole (13.5)
25. Avogadro's number
26. molarity

MATCHING

For each of the following items, fill in the number of the appropriate Key Term from the preceding list. Compare your answers with those at the back of the book.

a. _____ Any substance that produces hydroxide ions in water

b. _____ The burning of wood is a common reaction of this type

c. _____ Term applied to any change that alters the composition of a substance

d. _____ A measure of the concentration of hydrogen ion in a solution

e. _____ A listing of metals in order of their tendency to lose electrons to ions of another metal

f. _____ An insoluble solid that appears when two liquids are mixed

g. _____ Reaction type when two or more substances combine to form only one product

h. _____ Term applied to a gain of oxygen or a loss of electrons

i. _____ Any reaction that has a net release of energy to the surroundings

j. _____ The general format of this type of reaction: $A + BC \rightarrow B + AC$

k. _____ The new substances formed in a chemical reaction

l. _____ Term applied to a loss of oxygen or a gain of electrons

m. _____ The energy necessary to start a chemical reaction

n. _____ An ionic compound that includes neither hydrogen ions nor hydroxide ions

o. _____ A substance that increases the rate of a reaction but is not itself consumed

p. _____ The number of particles in 1 mole

q. _____ Reaction that produces water, carbon dioxide, and a salt

r. _____ Any substance that gives hydrogen ions in water

s. _____ The general format of this type of reaction: $AB \rightarrow A + B$

t. _____ The original substances in a chemical reaction

u. _____ Reaction that has a net absorption of energy from the surroundings

v. _____ Term applied to two opposing reactions or processes occurring at the same rate

w. _____ Involves the combination of a hydrogen ion with a hydroxide ion

x. _____ The general format of this type of reaction: $AB + CD \rightarrow AD + CB$

y. _____ The quantity of a substance that contains as many elementary units as are in 12 g of carbon-12

z. _____ Moles of solute per liter of solution

MULTIPLE CHOICE

Compare your answers with those at the back of the book.

1. When iron rusts in the presence of oxygen and water, which of the following is occurring? (13.1)
   (a) physical property
   (b) chemical property
   (c) physical change
   (d) chemical change

2. How many aluminum atoms are indicated by "3 Al_2O_3"? (13.1)
   (a) 2   (b) 3   (c) 6   (d) 18

3. What is the coefficient in front of Ag when the following equation is properly balanced? (13.1) $Ag + H_2S + O_2 \longrightarrow Ag_2S + H_2O$
   (a) 2   (b) 4   (c) 6   (d) 8

4. What does a catalyst do to the activation energy of a chemical reaction? (13.2)
   (a) forms a precipitate in the reaction
   (b) lowers the activation energy to start the reaction
   (c) alters the chemical activity of one of the reactants
   (d) changes the phase of the chemical products

5. In a combustion reaction, which products are typically produced? (13.2)
   (a) CO_2 and CO
   (b) H_2O and C_7H_{16}
   (c) CH_4 and CO_2
   (d) CO_2 and H_2O

6. Which of the following statements does *not* correctly describe a basic solution? (13.3)
   (a) The solution has a pH less than 7.
   (b) The solution conducts electricity.
   (c) The solution changes the color of litmus dye from red to blue.
   (d) The solution contains a higher concentration of OH^- ions than H^+ ions.

7. What is removed from a hydrated salt to make an anhydrous product? (13.3)
   (a) carbon
   (b) oxygen
   (c) water
   (d) hydrogen

8. What is the pH of a solution ten times as acidic as one of pH 4? (13.3)
   (a) 3   (b) 14   (c) 5   (d) −6

9. *Oxidation* can be defined as which of the following? (13.4)
   (a) a gain of electrons
   (b) a loss of oxygen
   (c) a loss of electrons
   (d) both (a) and (b)

10. The reaction $3 Zn + 2 Au(NO_3)_3(aq) \longrightarrow$ $2 Au + 2 Zn(NO_3)_2(aq)$ will occur if Zn is where relative to Au in the activity series? (13.4)
    (a) above
    (b) to the right of it
    (c) below
    (d) to the left of it

11. One mole of hydrogen peroxide, H_2O_2, would consist of how many molecules? (13.5)
    (a) 6.02×10^{23} (b) 1
    (c) $34.0 \times 6.02 \times 10^{23}$ (d) 34.0

12. One mole of hydrogen peroxide, H_2O_2, would consist of how many grams? (13.5)
    (a) 6.02×10^{23} (b) 17.0
    (c) $34.0 \times 6.02 \times 10^{23}$ (d) 34.0

FILL IN THE BLANK

Compare your answers with those at the back of the book.

1. Density and boiling point are ___ properties. (13.1)

2. A balanced chemical reaction has ___ number of atoms of each element on each side of the equation. (13.1)

3. A combination reaction has the format $A + B \longrightarrow$ ___. (13.1)

4. In living systems, biological catalysts are called ___. (13.2)

5. When the total bond energy of the products is ___ than that of the reactants, the reaction will be exothermic. (13.2)

6. A(n) ___ can increase the rate of reaction by providing a surface for the reaction. (13.2)

7. An Arrhenius base is a substance that produces ___ ions in water. (13.3)

8. A pH of 7 indicates a(n) ___ solution. (13.3)

9. A(n) ___ reaction is used to neutralize acid spills. (13.3)

10. A reaction in which both oxidation and reduction are occurring is usually called a(n) ___ reaction. (13.4)

11. If element *A* is below element *B* in the activity series, then the reaction $A + BC \longrightarrow B + AC$ ___ occur. (13.4)

12. Thirty-six grams of water contain ___ moles of water. (13.5)

SHORT ANSWER

1. Name the reactants, products, and catalyst for photosynthesis. What is the source of the necessary energy?

13.1 Balancing Chemical Equations

2. Iodine is a (a) blue-black, (b) crystalline, and (c) solid. It (d) sublimes to a violet-colored gas and (e) reacts with aluminum and many other metals. Which of these five properties are physical, and which are chemical?

3. Classify each of the following as either a physical change or a chemical change.
   (a) dissolving sugar in water
   (b) crushing rock salt
   (c) burning sulfur
   (d) digesting a chili dog

4. What are some clues that a chemical reaction has occurred?

5. In a chemical reaction, explain what happens to the reactants, products, and energy.

6. What is the difference between coefficients and subscripts in a balanced chemical equation?

7. The following reaction occurs when a butane cigarette lighter is operated: $C_4H_{10} + O_2 \longrightarrow CO_2 + H_2O$. When balancing the equation, you should *not* start with which element?

8. What is inappropriate about each of the "balanced" equations shown?
   (a) $C_4H_{10} + \frac{13}{2} O_2 \longrightarrow 4 CO_2 + 5 H_2O$
   (b) $4 H_2O \longrightarrow 2 O_2 + 4 H_2$
   (c) $Na + H_2O \longrightarrow NaOH + H$
   (d) $He + Br_2 \longrightarrow HeBr_2$

9. Tell what is indicated by each of these six symbols sometimes seen in chemical reactions: (aq), (s), (l), (g), $\longrightarrow$, and $\rightleftharpoons$, and by a chemical formula written above a reaction arrow.

10. Describe the type of reaction for each reaction shown.
    (a) $2 H_2O \longrightarrow 2 H_2 + O_2$
    (b) $2 Ca + O_2 \longrightarrow 2 CaO$
    (c) $2 NaN_3(s) \longrightarrow 2 Na(s) + 3 N_2(g)$

13.2 Energy and Rate of Reaction

11. Why does a reaction vessel feel warm during an exothermic reaction and feel cold during an endothermic reaction?

12. What is absorbed during bond breaking but liberated during bond formation?

13.

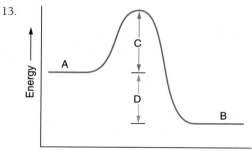

    (a) What term applies to the substances at point *A*?
    (b) What term applies to the substances at point *B*?
    (c) What term applies to the arrow designated *C*?
    (d) What term applies to the arrow designated *D*?
    (e) Is the reaction exothermic or endothermic?

14. Name the two products of the complete combustion of hydrocarbons. What is the other reactant, and how do the products change when this reactant is in short supply?

15. A collision between two molecules that have the potential to react may or may not result in a reaction. Explain.

16. List the four major factors that influence the rate of a chemical reaction.

17. Explain why chemical reactions proceed faster (a) as the temperature is increased and (b) as the concentrations of the reactants are increased.

18. Explain why heating a mixture of lumps of sulfur and zinc does not lead to reaction nearly as rapidly as heating a mixture of powdered sulfur and zinc.

19. What is the role of a catalyst in a chemical reaction? Briefly describe how it accomplishes its role.

20. What do sucrase and cholinesterase have in common?

13.3 Acids and Bases

21. What is the pH of a neutral aqueous solution? How many times as acidic is a solution of pH 2 than one of pH 6?

22. What color will litmus be in a solution of pH 9? A solution of pH 3?

23. What is the name of the electronic instrument used to measure pH? Which solution shown in Fig. 13.23 is basic?

24. List five general properties of acids and three of bases.

25. In the Arrhenius theory, how are acids and bases defined? Distinguish between hydrogen ions, hydronium ions, and hydroxide ions.

26. What are the chemical formulas of (a) stomach acid, (b) milk of magnesia, (c) lye, (d) baking soda, and (e) vinegar?

27. The reaction of an acid with a hydroxide base gives what two products? What is the most common mistake made in writing an equation for such a reaction?

28. What do we call salts that contain molecules of water bonded in their crystal lattices?

29. The reaction of an acid with a carbonate or hydrogen carbonate gives what three products?

30. Use the letters *A, B, C,* and *D* to illustrate the general format of a double-replacement reaction.

31. Describe what is seen in a precipitation reaction. What is happening on the atomic level to cause what is observed?

13.4 Single-Replacement Reactions

32. Describe oxidation and reduction from the standpoint of gain and loss of (a) oxygen and (b) electrons.

33. Explain the relative activity of a metal.

34. Metals above hydrogen in the activity series react with acids to give what two products? What general type of reaction does this exemplify?

35. Why are the metals gold and silver found as elements in nature, whereas the metals sodium and magnesium are found in nature only in compounds?

13.5 Avogadro's Number

36. What is the relationship between the terms *mole* and *Avogadro's number*?

37. What is meant by the "molarity" of a solution?

38. Suppose you are given the volume (in liters) of a salt (NaCl) solution and its molarity. Explain how you would determine the moles of salt in this solution.

VISUAL CONNECTION

Visualize the connections and give answers for the blanks. Compare your answers with those at the back of the book.

Types of Chemical Reactions

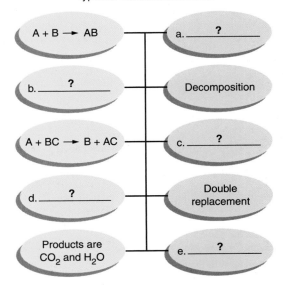

APPLYING YOUR KNOWLEDGE

1. Match the following chemical reactions with their appropriate name.
   a. $2\ Mg + O_2 \longrightarrow 2\ MgO$ 1. decomposition
   b. $C_2H_4 + 3O_2 \longrightarrow 2\ CO_2 + H_2O$ 2. double replacement
   c. $HCl + NaOH \longrightarrow HOH + NaCl$ 3. single replacement
   d. $2\ HgO \longrightarrow 2\ Hg + O_2$ 4. combination
   e. $Cu + 2\ AgNO_3(aq) \longrightarrow 2\ Ag + Cu(NO_3)_2$ 5. combustion

2. The best pH for a swimming pool is between 7.2 and 7.6. If your swimming pool has a pH of 8.0, then do you need to add an acidic or a basic substance to bring the pH into the proper range?

3. Explain why a bag of charcoal briquettes contains the following warning: *Do not use for indoor heating or cooking unless ventilation is provided for exhausting fumes to the outside.*

4. An Alka-Seltzer tablet contains solid citric acid and sodium hydrogen carbonate. What happens, and why, when the tablet is dropped into water?

5. The human body converts sugar into carbon dioxide and water at body temperature (98.6°F, or 37.0°C). Why are much higher temperatures required for the same conversion in the laboratory?

6. Why are sticks, rather than logs, used to start a fire?

7. Why is an open box of baking soda often placed in the refrigerator?

8. The equation $2\ NaHCO_3 \longrightarrow Na_2CO_3 + H_2O + CO_2$ is important in the preparation of what basic food item?

9. Sometimes water is seen dripping from the tailpipe of a car as it is running. Does this mean that someone bought watered-down gasoline? Explain.

10. Using atomic physics (Chapter 9.3), explain the difference between strong and weak acids and bases.

IMPORTANT EQUATION

Molarity: $M = \dfrac{n_{solute}}{V_{solution}}$ (13.1)

EXERCISES

13.1 Balancing Chemical Equations

1. Balance these chemical equations. (Each answer shows the correct coefficients in order.)
   (a) $CuCl_2 + H_2 \longrightarrow Cu + HCl$
   (b) $Fe + O_2 \longrightarrow Fe_2O_3$
   (c) $Al + H_2SO_4 \longrightarrow Al_2(SO_4)_3 + H_2$
   (d) $CaC_2 + H_2O \longrightarrow Ca(OH)_2 + C_2H_2$
   (e) $KNO_3 \longrightarrow KNO_2 + O_2$
   (f) $C_6H_6 + O_2 \longrightarrow CO_2 + H_2O$
   Answer: (Each answer shows the correct coefficients in order.) (a) 1, 1, 1, 2 (b) 4, 3, 2 (c) 2, 3, 1, 3 (d) 1, 2, 1, 1 (e) 2, 2, 1 (f) 2, 15, 12, 6

2. Balance these chemical equations.
   (a) $SO_2 + O_2 \longrightarrow SO_3$
   (b) $NH_3 + O_2 \longrightarrow N_2 + H_2O$
   (c) $C_4H_{10} + O_2 \longrightarrow CO_2 + H_2O$
   (d) $Pb(NO_3)_2 \longrightarrow PbO + NO_2 + O_2$
   (e) $Al + Fe_3O_4 \longrightarrow Al_2O_3 + Fe$
   (f) $Sr + H_2O \longrightarrow Sr(OH)_2 + H_2$

3. Identify any combination or decomposition reactions in Exercise 1.
   Answer: (b) combination, (e) decomposition

4. Identify any combination or decomposition reactions in Exercise 2.

5. (a) Nitrogen and hydrogen react to give ammonia in a combination reaction. Write and balance the equation. (b) Electrolysis can decompose KCl into its elements. Write and balance the equation.
   Answer: (a) $N_2 + 3 H_2 \rightarrow 2 NH_3$, (b) $2 KCl \rightarrow 2 K + Cl_2$

6. (a) Calcium and chlorine react to give calcium chloride in a combination reaction. Write and balance the equation.
   (b) Heating decomposes Na_2CO_3 to sodium oxide and carbon dioxide. Write and balance the equation.

13.2 Energy and Rate of Reaction

7. Write and balance the reaction for the complete combustion of pentane, C_5H_{12}.
   Answer: $C_5H_{12} + 8 O_2 \rightarrow 6 H_2O + 5 CO_2$

8. Write and balance the reaction for the complete combustion of butene, C_4H_8.

13.3 Acids and Bases

9. Complete and balance the following acid–base and acid–carbonate reactions:
   (a) $HNO_3 + KOH \longrightarrow$
   (b) $HC_2H_3O_2 + K_2CO_3 \longrightarrow$
   (c) $H_3PO_4 + NaOH \longrightarrow$
   (d) $H_2SO_4 + CaCO_3 \longrightarrow$
   Answer: (a) $HNO_3 + KOH \rightarrow H_2O + KNO_3$ (b) $2 HC_2H_3O_2 + K_2CO_3 \rightarrow H_2O + CO_2 + 2 KC_2H_3O_2$ (c) $H_3PO_4 + 3 NaOH \rightarrow 3 H_2O + Na_3PO_4$ (d) $H_2SO_4 + CaCO_3 \rightarrow H_2O + CO_2 + CaSO_4$

10. Complete and balance the following acid–base and acid–carbonate reactions:
    (a) $HCl + Ba(OH)_2 \longrightarrow$
    (b) $HCl + Al_2(CO_3)_3 \longrightarrow$
    (c) $H_3PO_4 + LiHCO_3 \longrightarrow$
    (d) $Al(OH)_3 + H_2SO_4 \longrightarrow$

11. Complete and balance the following double-replacement reactions. (*Hint:* Information in Table 13.4 will help you identify the precipitate.)
    (a) $AgNO_3(aq) + HCl(aq) \longrightarrow$
    (b) $BaCl_2(aq) + K_2CO_3(aq) \longrightarrow$
    Answer: (a) $AgNO_3(aq) + HCl(aq) \rightarrow AgCl(s) + HNO_3(aq)$ (b) $BaCl_2(aq) + K_2CO_3(aq) \rightarrow BaCO_3(s) + 2 KCl(aq)$

12. Complete and balance the following double-replacement reactions. (*Hint:* Information in Table 13.4 will help you identify the precipitate.)
    (a) $K_2SO_4(aq) + Pb(NO_3)_2(aq) \longrightarrow$
    (b) $K_3PO_4(aq) + CaBr_2(aq) \longrightarrow$

13.4 Single-Replacement Reactions

13. Refer to the activity series (Table 13.5) and predict in each case whether the single-replacement reaction shown will, or will not, actually occur.
    (a) $Na + KCl(aq) \longrightarrow K + NaCl(aq)$
    (b) $Ni + CuBr_2(aq) \longrightarrow Cu + NiBr_2(aq)$
    (c) $2 Al + 6 HCl(aq) \longrightarrow 3 H_2 + 2 AlCl_3(aq)$
    Answer: (a) will not, (b) will, (c) will

14. Refer to the activity series (Table 13.5) and predict in each case whether the single-replacement reaction shown will, or will not, actually occur.
    (a) $Zn + Fe(NO_3)_2(aq) \longrightarrow Fe + Zn(NO_3)_2(aq)$
    (b) $Pb + FeCl_2(aq) \longrightarrow Fe + PbCl_2(aq)$
    (c) $2 Ag + 2 HNO_3(aq) \longrightarrow H_2 + 2 AgNO_3(aq)$

15. Refer to the activity series (Table 13.5). Complete and balance the equation for the following single-replacement reactions.
    (a) $Ni + Pt(NO_3)_2(aq) \longrightarrow$
    (b) $Zn + H_2SO_4(aq) \longrightarrow$
    Answer: (a) $Ni + Pt(NO_3)_2(aq) \rightarrow Pt + Ni(NO_3)_2(aq)$ (b) $Zn + H_2SO_4(aq) \rightarrow H_2 + ZnSO_4(aq)$

16. Refer to the activity series (Table 13.5). Complete and balance the equation for the following single-replacement reactions.
    (a) $Mg + HCl(aq) \longrightarrow$
    (b) $Al + AgNO_3(aq) \longrightarrow$

13.5 Avogadro's Number

17. Two moles of hydrogen sulfide, H_2S, would consist of how many molecules?
    Answer: 12.04×10^{23} molecules

18. Four moles of sulfur dioxide, SO_2, would consist of how many molecules?

19. Fill in the blanks in the following table for the element sodium, Na.

Moles	Atoms	Mass
1.00		
	18.06×10^{23}	
		46.0 g

Answers: 6.02×10^{23} atoms, 23.0 g; 3.00 mol, 69.0 g; 2.00 mol, 12.04×10^{23} atoms

20. Fill in the blanks in the following table for the compound carbon dioxide, CO_2.

Moles	Molecules	Mass
3.00		
		44.0 g
	3.01×10^{23}	

21. Determine the mass in grams of 2.5 moles of $Ca(OH)_2$.

Answer: 185.25 g

22. How many moles are in 15.0 g of copper sulfate ($CuSO_4$)?

23. What would be the molarity of a solution in which 0.50 mole of NaCl is dissolved in enough water to make 2.0 L of solution?

Answer: 0.25 M

24. What would be the molarity of a solution in which 1.0 mole of KOH is dissolved in enough water to make 0.25 L of solution?

ON THE WEB

1. Chemical Equations

All life is constantly affected by countless types of chemical reactions. What was Proust's "law of definite proportions" (Chapter 12.2)? What do you know about chemical reactions? Follow the links on the student website at **www.cengagebrain.com/shop/ISBN/1133104096** to try an experiment to reinforce what you know.

2. Splitting Water

Has it ever occurred to you that you could "split water"? Visit the links on the student website at **www.cengagebrain.com/shop/ISBN/1133104096** and give it a try. What reaction did you expect? What reaction do you get? How can you apply practically what you learned from this experiment?

Organic Chemistry

Organic chemistry nowadays almost drives me mad. To me it appears like a primeval tropical forest full of the most remarkable things, a dreadful and endless jungle into which one dares not enter for there seems to be no way out.

•

Friedrich Wöhler, German chemist (1800–1882)

Skydivers trust their lives to parachutes made of nylon, a polymer, described in this chapter. >

George H.H. Huey/Corbis

CHEMISTRY FACTS

▶ The reaction of glycerol (glycerin) with nitric acid forms a thick, pale yellow liquid that is extremely explosive (nitroglycerin). With careful experiments and a bit of luck, Alfred Nobel, a Swedish inventor, formulated dynamite in 1866 by mixing nitroglycerin with porous material to form a paste. This invention led Nobel to great wealth and enabled him to establish the Nobel Prizes.

▶ A biochemical family of proteins called keratin is the fibrous material in human skin, hair, and nails. These materials are also found in claws of reptiles, beaks and feathers of birds, the shells of turtles, and the webs of spiders.

▶ Ethanol is added to gasoline to help oxygenate the gasoline,

Organic chemistry is the study of carbon compounds. The importance of carbon compounds to life cannot be overemphasized. What if all carbon compounds were removed from the Earth? Certainly there would be no animals, plants, or other forms of life, because nucleic acids, proteins, fats, carbohydrates, enzymes, vitamins, and hormones are organic (carbon) compounds. The food, fuel, and clothing that are so essential to our lives contain organic chemicals. The paper, ink, and glue that constitute this textbook are all organic compounds.

Scientists had originally defined organic chemistry as the study of compounds of animal or plant origin, and these compounds were thought to contain a "vital force" based on their natural origin. However, the concept of a "vital force" was disproved by German chemist Friedrich Wöhler in 1828 (the early days of organic

Chapter Outline

chemistry), who showed that urea, a carbon compound found in human urine, could be made in the laboratory from a mineral. This finding led to the present definition of organic chemistry. The division of chemistry called *biochemistry* was developed later to study the chemical compounds and reactions that occur in living cells.

Organic chemistry is a gigantic field. In this chapter, enough of the fundamental concepts, compounds, and reactions are introduced to impart a basic appreciation and comprehension of this fascinating area.

14.1 Bonding in Organic Compounds

Preview Questions

- What common elements other than carbon are in organic compounds?
- How are organic compounds bonded?

In addition to carbon, the other most common elements in organic compounds are hydrogen, oxygen, nitrogen, sulfur, and halogens. Because they are all nonmetals, organic compounds are covalent in bonding (Chapter 12.5). Examination of the Lewis symbols and application of the octet rule show that these elements should bond as summarized in ● Table 14.1.

Any structural formula that follows the bonding rules probably represents a known or possible compound. Any structure drawn that breaks one of these rules is unlikely to represent a real compound.

EXAMPLE 14.1 Identifying Valid and Incorrect Structural Formulas

Two structural formulas are shown. Which one does not represent a real compound? Why?

(a)   (b)

Solution

In structure (a), each hydrogen and halogen has one bond, each carbon has four, and the oxygen has two. It is a valid structure. Structure (b) cannot be correct. Although the nitrogen has three bonds, the hydrogens one each, and the carbons four each, the oxygen has three bonds when it should have only two.

Confidence Exercise 14.1

Which of the following two structural formulas is not correct? Why?

(a)   (b)

Answers to Confidence Exercises may be found at the back of the book.

Table 14.1 Numbers and Types of Bonds for Common Elements in Organic Compounds

Element	Total Number of Bonds	Distribution of Total Number of Bonds and Examples		
C	4	4 singles $-\overset{\|}{\underset{\|}{C}}-$	2 singles, 1 double $-\overset{\|}{C}=$	1 single, 1 triple $-C\equiv$
N	3	3 singles $-\overset{\|}{N}-$	1 single, 1 double $-N=$	1 triple $N\equiv$
O (or S)	2	2 singles $\overset{\|}{O}-$	1 double $O=$	
H or halogens	1	1 single H—, Cl—, etc.		

Did You Learn?

● Hydrogen, oxygen, nitrogen, and other nonmetals are common elements in organic compounds.

● Organic bonds are covalent bonds and follow the octet rule.

14.2 Aromatic Hydrocarbons

Preview Questions

● What is the most important aromatic compound?

● What are some uses of aromatic hydrocarbons?

The simplest organic compounds are **hydrocarbons**, which contain only carbon and hydrogen. For purposes of classification, all other organic compounds are considered to be *derivatives* of hydrocarbons. Thus, the first class of compounds discussed in organic chemistry is the hydrocarbons, which are first divided into aromatic hydrocarbons or aliphatic ones (● Fig. 14.1).

Aromatic hydrocarbons are hydrocarbons that possess one or more benzene rings (many have pungent aromas). The most important one is benzene (C_6H_6), a clear, colorless liquid with a distinct odor. It is a *carcinogen* (a cancer-causing agent) and has the following structural formulas and shorthand symbols:

Benzene
(Lewis structure)

Benzene
(Kekulé symbol)

Benzene
(modern symbol)

German chemist August Kekulé ("KECK-you-lay") discovered the ring-like structure of benzene in 1865. The Lewis structure and Kekulé symbol imply that the ring has alternating single and double bonds between the carbon atoms. However, the properties of the benzene molecule and advanced bonding theory show that six electrons are shared by *all* the carbon atoms in the ring, forming an electron cloud that extends above and below the

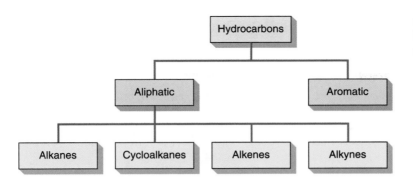

Figure 14.1 Classification of Hydrocarbons
(See text for description.)

plane of the ring (● Fig. 14.2). This sharing of six "delocalized" electrons by all the ring atoms lends a special stability to benzene and its relatives. The modern shorthand symbol for benzene is a hexagon with a circle inside.

Benzene is obtained mainly from petroleum. Other aromatic hydrocarbons are commonly obtained from coal tar, a by-product of soft coal. When other atoms or groups of atoms are substituted for one or more of the hydrogen atoms in the benzene ring, a vast number of different compounds can be produced, including such things as perfumes, explosives, drugs, solvents, insecticides, and lacquers. Examples of other aromatic hydrocarbons, whose structures are shown below, are toluene (or methylbenzene, used in model airplane glue), naphthalene (used in mothballs), and the explosive called TNT.

CH₃

Toluene
(methylbenzene)

Naphthalene

CH₃
O₂N   NO₂

NO₂

TNT (2,4,6-trinitrotoluene)

To draw the structures for simple benzene derivatives, draw a benzene ring and attach the substituent to the ring in the position indicated by the number before the substituent's name. (If only one substituent is on the ring, then the number 1 is omitted.)

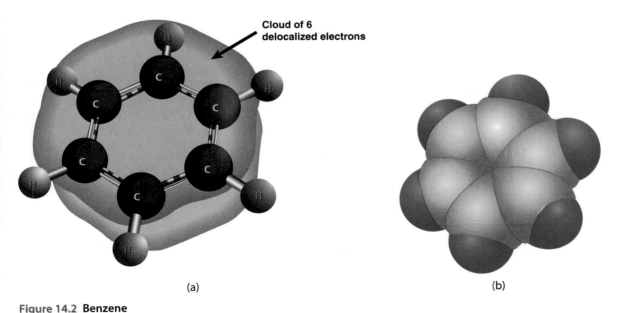

Cloud of 6 delocalized electrons

(a)

(b)

Figure 14.2 Benzene
(a) A representation of benzene shows that it is a flat molecule with six delocalized electrons forming an electron cloud above and below the plane of the ring. (b) A space-filling model of benzene.

EXAMPLE 14.2 **Drawing Structures for Benzene Derivatives**

Draw the structural formula for 1,3-dibromobenzene.

Solution

First, draw a benzene ring. Second, attach a bromine atom (*bromo* means "bromine atom") to the carbon atom at the ring position you choose to be number 1. Third, attach a second bromine atom (the name says *di*, meaning "two") to ring position 3 (you may number either clockwise or counterclockwise from carbon 1), and you have the answer.

Step 1   Step 2   Step 3

Confidence Exercise 14.2

Draw the structural formula for 1-chloro-2-fluorobenzene.
Answers to Confidence Exercises may be found at the back of the book.

Did You Learn?

● Benzene is the most important aromatic hydrocarbon; it has a ring-like structure, is clear and colorless, and has a distinct odor.

● Aromatic hydrocarbons are used in perfumes, explosives, drugs, solvents, and insecticides.

14.3 Aliphatic Hydrocarbons

Preview Questions

● What are saturated hydrocarbons?

● Which type of organic material is most used as an energy source?

Hydrocarbons having no benzene rings are **aliphatic hydrocarbons** and are divided into four major classes: alkanes, cycloalkanes, alkenes, and alkynes (Fig. 14.1). Each major class will be examined in turn.

Alkanes

Hydrocarbons that contain only single bonds are called **alkanes**. They are *saturated hydrocarbons* because their hydrogen content is at a maximum. Alkanes have a composition that satisfies the general formula

$$C_nH_{2n+2}$$

where n = the number of carbon atoms.

$$2n + 2 = \text{the number of hydrogen atoms}$$

The number of hydrogen atoms present in a particular alkane is twice the number of carbon atoms, plus two. The general formula applies to the alkanes listed in ● Table 14.2.

Methane (CH_4) is the first member ($n = 1$) of the alkane series. Ethane (C_2H_6) is the second member, propane (C_3H_8) the third, and butane (C_4H_{10}) the fourth. After butane, the number of carbon atoms is indicated by Greek prefixes such as *penta-, hexa-,* and so

Table 14.2 The First Eight Members of the Alkane Series

Name	Molecular Formula	Condensed Structural Formula
Methane	CH_4	CH_4
Ethane	C_2H_6	CH_3CH_3
Propane	C_3H_8	$CH_3CH_2CH_3$
Butane	C_4H_{10}	$CH_3(CH_2)_2CH_3$
Pentane	C_5H_{12}	$CH_3(CH_2)_3CH_3$
Hexane	C_6H_{14}	$CH_3(CH_2)_4CH_3$
Heptane	C_7H_{16}	$CH_3(CH_2)_5CH_3$
Octane	C_8H_{18}	$CH_3(CH_2)_6CH_3$

on. The names of alkanes always end in -*ane*. Methane through butane are gases, pentane through about $C_{17}H_{36}$ are liquids, and the rest are solids. The alkanes generally are color-less, and, being nonpolar, they do not dissolve in water.

A hydrocarbon's structure is easy to visualize when we write its *structural formula* (a graphical representation of the way the atoms are connected to one another) instead of its *molecular formula* (which tells only the type and number of atoms). For example, the full structural formulas for methane, ethane, and pentane are shown in color in the first row in ● Fig. 14.3. Each dash represents a covalent bond (two shared electrons). To save time and space, *condensed* structural formulas are often used, as shown in Fig. 14.3 (and the right-hand column of Table 14.2).

Figure 14.3 shows photos of ball-and-stick models of methane, ethane, and pentane. The four single bonds of each carbon point to the corners of a regular tetrahedron (a geo-metric figure with four identical equilateral triangles as faces). Carbon's four single bonds form angles of 109.5°, not 90°, as it may appear from two-dimensional structural formulas. ● Figure 14.4, which shows a ball-and-stick model and a space-filling model of methane, emphasizes the tetrahedral geometry of four single bonds to a carbon atom.

Alkanes are highly combustible. Like all hydrocarbons, when ignited they react with the oxygen in air, forming carbon dioxide and water and releasing heat (Chapter 13.2).

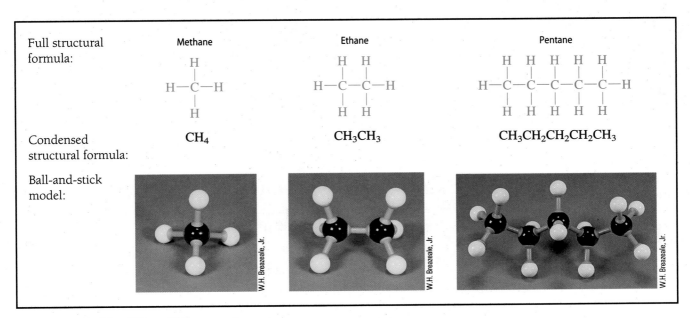

Figure 14.3 Models of Three Alkanes
Structural formulas, condensed structural formulas, and ball-and-stick models for methane, ethane, and pentane.

Figure 14.4 Methane
The tetrahedral geometry of four single bonds to a carbon atom is emphasized in the ball-and-stick and space-filling models for methane.

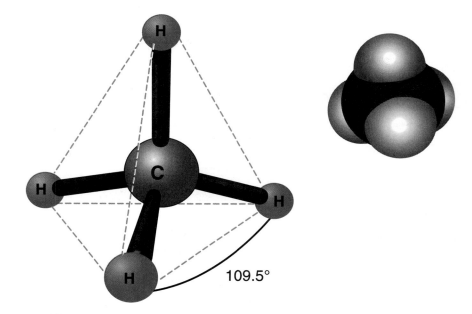

109.5°

Figure 14.5 Catalytic Cracking Unit at a Petroleum Refinery
Catalytic "cracking" breaks larger hydrocarbons into smaller molecules to increase the yield of gasoline. The hot, hydrocarbon vapor is passed over a catalyst of tin(IV) oxide and aluminum oxide.

Courtesy American Petroleum Institute

Otherwise, alkanes are not very reactive, because any reaction would involve breaking the strong C—H or C—C single bonds.

The alkanes make up many well-known products. Methane is the principal component of natural gas, which is used in many homes for heating and cooking (see Fig. 13.5). Propane and butane are also used for that purpose. Petroleum is made up chiefly of alkanes but also contains other classes of hydrocarbons. The crude oil must be refined; it must be separated into fractions by distillation. Each fraction is still a very complex mixture of hydrocarbons. Gasoline consists of the alkanes from pentane to decane ($n = 5$ to $n = 10$). At oil refineries, additional gasoline is made by catalytic "cracking" of larger alkanes into smaller ones (● Fig. 14.5).

Kerosene contains the alkanes with $n = 10$ to 16. The alkanes with higher values of n make up other products, such as diesel fuel, fuel oil, petroleum jelly, paraffin wax, and lubricating oil. The largest alkanes make up asphalt. Alkanes are also used as starting materials for products such as paints, plastics, drugs, detergents, insecticides, and cosmetics. (See the **Highlight: Tanning in the Shade**.) However, only about 6% of the petroleum consumed daily goes into making such useful materials. The other 94% is burned for energy.

A substituent that contains one less hydrogen atom than the corresponding alkane is called an **alkyl group** and is given the general symbol R. The name of the alkyl group is obtained by dropping the *-ane* suffix and adding *-yl*. For example, *methane* becomes *methyl,* and *ethane* becomes *ethyl.* The open bonds in the methyl and ethyl groups indicate that these groups are bonded to another atom; they do not have an independent existence (see structural formulas).

Methane   Methyl group   or   CH₃—

Ethane   Ethyl group   or   CH₃CH₂—

Highlight   Tanning in the Shade

Among today's best selling cosmetics are self-tanning lotions (see Fig. 1 for some examples). Many light-skinned people want to look like they have just spent a vacation at the beach, but they recognize the dangers of too much sun. It causes premature aging and may lead to skin cancer. Chemistry has come to the rescue in the form of lotions that produce an authentic-looking tan. All these lotions have the same active ingredient: dihydroxyacetone (DHA), which has the structure

$$H-O-\overset{\displaystyle H}{\underset{\displaystyle H}{C}}-\overset{\displaystyle O}{\overset{\displaystyle \|}{C}}-\overset{\displaystyle H}{\underset{\displaystyle H}{C}}-OH$$

DHA is a nontoxic, simple sugar that occurs as an intermediate in carbohydrate metabolism in higher-order plants and animals.

The tanning effects of DHA were discovered by accident in the 1950s; when DHA was accidentally spilled on the skin, it produced brown spots.

The mechanism of the browning process involves the Maillard reaction, which was discovered by French chemist Louis-Camille Maillard in 1912. In this process, amino acids react with sugars to create brown or golden brown products. The same reaction is responsible for much of the browning that occurs during the manufacture and storage of foods.

The browning of skin occurs in the stratum corneum—the outermost layer—where the DHA reacts with free amino ($-NH_2$) groups of the proteins found there.

DHA is present in most tanning lotions at concentrations between 2% and 5%, although some products designed to give deeper tan are more concentrated.

Thanks to these new products, tanning is now both safe and easy.

 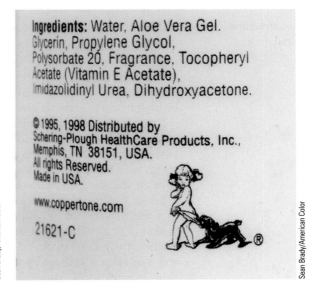

Figure 1 Tanning without Fear
(a) Self-tanning products. (b) A label showing the contents.

The way chains are built up in three dimensions is illustrated by butane, C_4H_{10} (● Fig. 14.6). In the structure of the compound named *n*-butane ("normal" butane), each end carbon atom has three hydrogen atoms attached, and two hydrogens are bonded to each of the middle two carbon atoms (Fig. 14.6a). Another arrangement of the atoms of the butane molecule is also possible. Isobutane (2-methyl-propane) also has the molecular formula C_4H_{10} (Fig. 14.6b).

Examination of the two-dimensional representations of the *n*-butane and isobutane molecules and their three-dimensional ball-and-stick models in Fig. 14.6 shows that the structures are indeed different. The structure of *n*-butane shows a *continuous chain* (or *straight chain*) in which no carbon is bonded directly to more than two others, whereas isobutane has a *branched-chain* structure in which one carbon atom is bonded directly to three others. Although they have the same molecular formula, C_4H_{10}, these compounds have different physical and chemical properties (for example, *n*-butane boils at $-0.5°C$, whereas isobutane boils at $-11.6°C$).

(a) n-Butane

(b) Isobutane

Figure 14.6 Different Structures of Butane
Butane, C_4H_{10}, has two structural arrangements: (a) *n*-butane and (b) isobutane.

Table 14.3	Number of Possible Isomers of Alkanes
Molecular Formula	**Total Isomers**
CH_4	1
C_2H_6	1
C_3H_8	1
C_4H_{10}	2
C_5H_{12}	3
C_6H_{14}	5
C_7H_{16}	9
C_8H_{18}	18
C_9H_{20}	35
$C_{10}H_{22}$	75
$C_{15}H_{32}$	4,347
$C_{20}H_{42}$	366,319
$C_{30}H_{62}$	4.11×10^9

Isobutane and *n*-butane are **constitutional isomers** (or *structural isomers*), compounds that have the same *molecular* formula but different *structural* formulas. In other words, they have the same number and type of each atom but differ in how these atoms are connected to one another. Constitutional isomers exist whenever two or more structural formulas can be built from the same number and type of each atom without violating the octet rule.

The phenomenon of constitutional isomerism is somewhat akin to using the same amounts of wood and brick to build houses that are entirely different in structure. Because of the ability of carbon atoms to bond to many other carbon atoms and atoms of other elements in so many different ways, the number of possible organic compounds is incredibly large, as ● Table 14.3 indicates.

Because of the number and complexity of organic compounds, a consistent method of nomenclature was developed so that communication would be effective. The IUPAC system of nomenclature for organic compounds begins with the rules for alkanes.* Let's examine the general rules and see how they enable us to write a structure from the name, and vice versa. Don't worry if these rules seem confusing the first time you read them. Some examples will make them clear.

1. The longest continuous chain of carbon atoms (the "backbone") is found, and the compound is named as a derivative of the alkane with this number of carbon atoms.

2. The positions and names of the substituents (single atoms or groups; see ● Table 14.4) that have replaced hydrogen atoms on the backbone chain are added. When more

Table 14.4	Substituents in Organic Compounds
Formula of Substituent	**Name of Substituent**
Br—	Bromo
Cl—	Chloro
F—	Fluoro
I—	Iodo
CH_3—	Methyl
CH_3CH_2—	Ethyl

---

*IUPAC ("EYE-you-pack") stands for International Union of Pure and Applied Chemistry.

than one type of substituent is present, either on the same carbon atom or on different carbon atoms, the substituents are listed in alphabetical order. When more than one of the same type of substituents are present, the prefixes *di-, tri-, tetra-, penta-,* and so forth are used to indicate how many.

3. The carbon atoms on the backbone chain are numbered by counting from the end of the chain nearest the substituents. The position of attachment of each substituent is identified by giving the number of the carbon atom in the chain. Each substituent must have a number. Commas are used to separate numbers from other numbers, and hyphens are used to separate numbers from names.

For example, consider the structure

$$\overset{5}{CH_3}-\overset{4}{CH_2}-\overset{3}{CH_2}-\overset{2}{CH}-\overset{1}{CH_3}$$
$$|$$
$$CH_3$$

2-Methylpentane

The longest continuous chain of carbon atoms is five, so the compound is named as a pentane derivative. Attached to the pentane backbone is a methyl group (see Table 14.4). Numbering the backbone from right to left gives a smaller number for the methyl substituent than numbering from left to right and so would be the correct way. The compound's name is 2-methylpentane.

Rather than use these rules to assign names to the structural formulas for compounds, they will be used in reverse to draw structural formulas from the names. This method is somewhat simpler and still gets the major points across about the relationship between structure and IUPAC name.

EXAMPLE 14.3 **Drawing a Structure from a Name**

Draw the structural formula for 2,3-dimethylhexane.

Solution

The end of the name is *hexane,* so draw a continuous chain of six carbon atoms joined by five single bonds and add enough bonds to each C atom so that all have four.

$$-\overset{|}{\underset{|}{C}}-\overset{|}{\underset{|}{C}}-\overset{|}{\underset{|}{C}}-\overset{|}{\underset{|}{C}}-\overset{|}{\underset{|}{C}}-\overset{|}{\underset{|}{C}}-$$

In your mind, number the carbons 1 through 6, starting from either end you wish, and attach a methyl group (CH_3-) to carbon 2 and another to carbon 3. Hydrogen atoms are added to the remaining bonds, giving

$$\begin{array}{ccccccc} & H & H & H & CH_3 & H & & H \\ & | & | & | & | & | & & | \\ H- & C- & C- & C- & C- & C- & - & C-H \\ & | & | & | & | & | & & | \\ & H & H & H & H & CH_3 & & H \end{array}$$

Confidence Exercise 14.3

The octane rating for gasoline assigns a value of 100 to the combustion of the "octane" whose IUPAC name is 2,2,4-trimethylpentane. Draw the structure of this important hydrocarbon.

Cycloalkanes

The **cycloalkanes,** members of a second series of saturated hydrocarbons, have the general molecular formula C_nH_{2n} and possess *rings* of carbon atoms, with each carbon atom

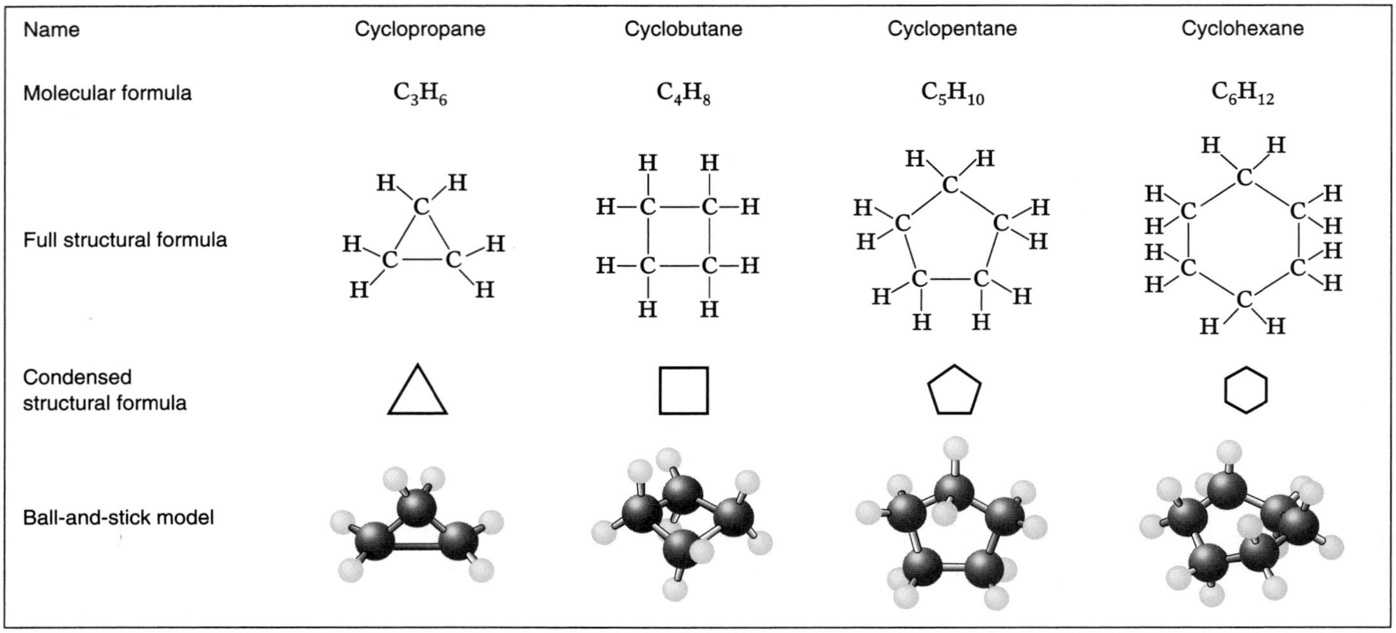

Name	Cyclopropane	Cyclobutane	Cyclopentane	Cyclohexane
Molecular formula	C_3H_6	C_4H_8	C_5H_{10}	C_6H_{12}
Full structural formula				
Condensed structural formula	△	□	⬠	⬡
Ball-and-stick model				

Figure 14.7 The First Four Cycloalkanes
Shown are saturated aliphatic hydrocarbons characterized by rings of carbon atoms.

bonded to a total of four carbon or hydrogen atoms. The smallest possible ring occurs with cyclopropane, C_3H_6; then come cyclobutane, cyclopentane, and so forth. Note that the prefix *cyclo-* is included when naming cycloalkanes. ● Figure 14.7 shows the names, molecular formulas, structural formulas, condensed structural formulas, and ball-and-stick models for the first four cycloalkanes.

To draw the structure of a cycloalkane derivative, draw the geometric figure that has the number of sides indicated by the compound's name. Then place each substituent on the ring in the numbered position indicated in the name. (If there is only one substituent, then the number 1 is omitted.) For example, the structure of 1-chloro-2-ethylcyclopentane is

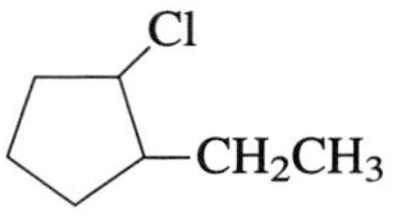

Alkenes

Hydrocarbons that have a double bond between two carbon atoms are called **alkenes**. Imagine that a hydrogen atom has been removed from each of two adjacent carbon atoms in an alkane, thereby allowing these C atoms to form an additional bond between them. Thus, the general formula for the alkene series is C_nH_{2n} (the same as for cycloalkanes). The series begins with ethene, C_2H_4 (more commonly known as *ethylene*), shown by the structural formula

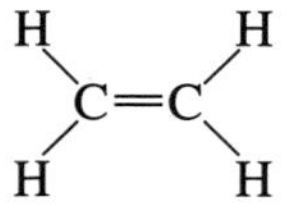

Ethene (ethylene)

Some of the simpler alkenes are listed in ● Table 14.5. The *-ane* suffix for alkane names is changed to *-ene* for alkenes. A number preceding the name indicates the carbon atom on which the double bond starts. The carbons in the chain are numbered starting at the end

Table 14.5 Some Members of the Alkene Series

Name	Molecular Formula	Condensed Structural Formula
Ethene (ethylene)	C_2H_4	$CH_2{=}CH_2$
Propene	C_3H_6	$CH_3CH{=}CH_2$
1-Butene	C_4H_8	$CH_3CH_2CH{=}CH_2$
2-Butene	C_4H_8	$CH_3CH{=}CHCH_3$
1-Pentene	C_5H_{10}	$CH_3(CH_2)_2CH{=}CH_2$

that gives the double bond the lower number. For example, 1-butene and 2-butene have the structural formulas

1-Butene   2-Butene

Alkenes are more reactive than alkanes because of the double bond. They are termed *unsaturated hydrocarbons* because a characteristic reaction is the *addition of hydrogen* to form a corresponding alkane. This reaction is important in the manufacturing of saturated fats like solid shortenings.

Alkynes

Hydrocarbons that have a triple bond between two carbon atoms are called **alkynes**. Imagine that two hydrogen atoms have been removed from each of two adjacent carbon atoms in an alkane, thereby allowing these C atoms to form two additional bonds between them. The general formula for the alkyne series is C_nH_{2n-2}. The simplest alkyne is ethyne, C_2H_2, more commonly called *acetylene* (● Fig. 14.8):

$$H{-}C{\equiv}C{-}H$$

Ethyne (acetylene)

Figure 14.8 An Ironworker Cutting Steel with an Oxyacetylene Torch The combustion of acetylene (ethyne) produces the intense heat needed to fuse metals. Oxyacetylene welding can even be used under water, because the necessary oxygen is supplied along with the acetylene.

Some members of the alkyne series are listed in ● Table 14.6. Note that the nomenclature for alkynes follows rules similar to those for the alkenes. (Cycloalkenes and cycloalkynes do exist, but they will not be discussed.)

Alkynes are also reactive, unsaturated hydrocarbons and, like alkenes, add hydrogen. Alkynes can add *two* molecules of hydrogen across the triple bond to form a corresponding alkane. Alkynes are not common in nature; they are used in refineries and in some pharmaceuticals.

● Figure 14.9 summarizes the names and characteristic structural features of the hydrocarbons as well as the general formulas for derivatives of hydrocarbons, which we will study next.

Table 14.6 Some Members of the Alkyne Series

Name	Molecular Formula	Condensed Structural Formula
Ethyne (acetylene)	C_2H_2	$HC{\equiv}CH$
Propyne	C_3H_4	$CH_3C{\equiv}CH$
1-Butyne	C_4H_6	$CH_3CH_2C{\equiv}CH$
2-Butyne	C_4H_6	$CH_3C{\equiv}CCH_3$
1-Pentyne	C_5H_8	$CH_3(CH_2)_2C{\equiv}CH$

Figure 14.9 A Summary of Hydrocarbons and Their Derivatives
Shown is a summary chart of the names and characteristic structural features of the hydrocarbons as well as the general formulas for derivatives of hydrocarbons.

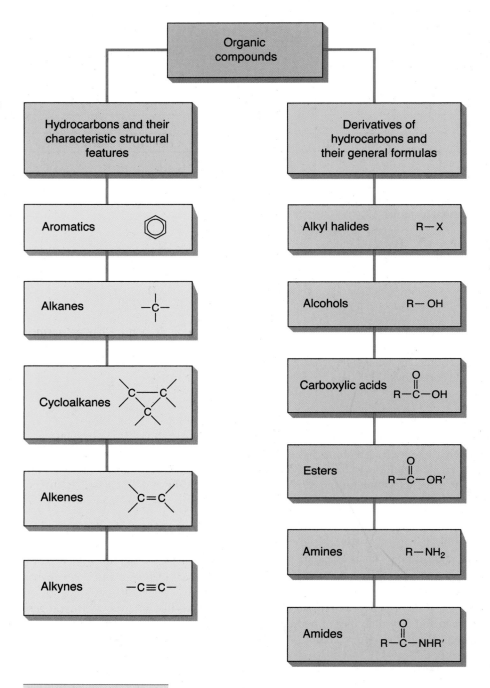

Did You Learn?

- Alkanes are saturated hydrocarbons because their hydrogen content is at a maximum.

- Most alkanes are used as energy sources.

14.4   Derivatives of Hydrocarbons

Preview Questions

- What are some common hydrocarbon derivatives?

- Which types of hydrocarbon derivatives have foul odors? Which have pleasing odors?

The characteristics of organic molecules depend on the number, type, and arrangement of their atoms. Any atom, group of atoms, or organization of bonds that determines specific properties of a molecule is called a **functional group**. <u>Generally, the functional group is the reactive part of a molecule, and its presence signifies certain predictable chemical properties.</u>

The double bond in an alkene and the triple bond in an alkyne are functional groups. Other functional groups (for example, a chlorine atom or an —OH group) may be attached to a carbon atom in place of a hydrogen atom. (See Fig. 14.9.) Organic compounds that contain elements other than carbon and hydrogen are called *derivatives* of hydrocarbons.

In the general structural formula for a derivative, the letter R usually represents an alkyl group to which the functional group is attached. Let's look at some interesting and useful derivatives of hydrocarbons.

Alkyl Halides

The general formula for an **alkyl halide** is R—X, or just RX, where X is a halogen atom and R is an alkyl group. Recall that in the naming of organic compounds, a fluorine atom is designated by *fluoro*, a chlorine atom by *chloro*, and so forth (see Table 14.4).

The alkyl halides called **CFCs** are chloroflurocarbons, such as dichlorodifluoromethane (Freon-12), and have been used commonly in air conditioners, refrigerators, and heat pumps. They are unreactive gases at the Earth's surface.

$$
\begin{array}{c}
F \\
| \\
F—C—Cl \\
| \\
Cl
\end{array}
$$

Dichlorodifluoromethane (a CFC)

The major problem with the use of CFCs is that they travel upward into the stratosphere, where the Earth's protective ozone (O_3) layer is formed by the action of sunlight on ordinary oxygen (O_2). The ozone layer absorbs much of the solar ultraviolet rays (UV) that can harm plant and animal life on the Earth's surface. The manner in which CFCs react with and destroy ozone is discussed in Chapter 20.5, which has the **Highlight: The Ozone Hole and Global Warming**.

Simple alkyl halides are generally colorless, odorless liquids at room temperature and are insoluble in water. Many are toxic, but they are still used as solvents, in pharmaceuticals, in chemical production, and in pesticides. Other examples of alkyl halides are chloroform ($CHCl_3$), carbon tetrachloride (CCl_4), and iodomethane (CH_3I). Oxygen is extremely soluble in liquid fluorocarbons, and those called perfluorocarbons (PFCs) are used in medicine for eye surgery, lung therapy for premature babies, diagnostic imaging, and artificial blood (● Fig. 14.10).

Figure 14.10 Liquid Fluorocarbons
Artificial blood, or blood substitute, uses fluorocarbons to transport oxygen. Still under clinical trials, these products could potentially be very useful for blood transfusions and for rapid treatment in trauma situations.

EXAMPLE 14.4   **Drawing Constitutional Isomers**

To understand constitutional isomerism better, draw the structural formulas for the two alkyl halide isomers that have the molecular formula $C_2H_4Cl_2$.

Solution

Carbon atoms form four bonds, but H and Cl can form only one each. Thus, H and Cl can never connect two other atoms; they can only stick to the "backbone." Therefore, first draw the two-carbon backbone using a single bond and then add enough bonds so that each carbon has four.

$$
\begin{array}{c}
| \quad | \\
—C—C— \\
| \quad |
\end{array}
$$

Count the open bonds. They total six, which is just right for the attachment of four H atoms and two Cl atoms. (If there were too many open bonds, then an alkene, alkyne, or ring structure would be tried.)

Figure 14.11 Constitutional Isomers
The two constitutional isomers of dichloroethane ($C_2H_4C_2$) are shown in these models.

Ignore the H atoms for now and see how many different ways you can put on the two Cl atoms. You will find two different ways. (*Note:* You must remember the tetrahedral geometry of the four bonds to C. <u>There is no difference in the bonds you are drawing "out," "up," or "down." Each bond is really at 109.5° from another, not 90° or 180°.</u>)

$$
\begin{array}{cc}
\text{Cl} & \text{Cl} \\
| & | \\
-\text{C}-\text{C}- \\
| & | \\
\end{array}
\qquad
\begin{array}{cc}
& \text{Cl} \\
& | \\
-\text{C}-\text{C}-\text{Cl} \\
| & | \\
\end{array}
$$

Add the four hydrogens to each structural formula and the question is answered (● Fig. 14.11). The names of the two isomers denote their structures.

$$
\begin{array}{cc}
\text{Cl} & \text{Cl} \\
| & | \\
\text{H}-\text{C}-\text{C}-\text{H} \\
| & | \\
\text{H} & \text{H} \\
\end{array}
\qquad
\begin{array}{cc}
\text{H} & \text{Cl} \\
| & | \\
\text{H}-\text{C}-\text{C}-\text{Cl} \\
| & | \\
\text{H} & \text{H} \\
\end{array}
$$

1,2-Dichloroethane   1,1-Dichloroethane

Confidence Exercise 14.4

Two constitutional isomers of C_3H_7F exist. Draw the structure for each.

Alcohols

Alcohols are organic compounds containing a *hydroxyl group*, —OH, attached to an alkyl group. The general formula for an alcohol is R—OH, or just ROH, and the IUPAC names of alcohols end in *-ol*. Thousands of alcohols exist, some with only one hydroxyl group and others with two or more. The simplest alcohol is methanol, which is also called *methyl alcohol* or *wood alcohol*. It is poisonous, but useful (● Fig. 14.12).

$$
\begin{array}{c}
\text{H} \\
| \\
\text{H}-\text{C}-\text{O}-\text{H} \\
| \\
\text{H} \\
\end{array}
\qquad \text{or} \qquad \text{CH}_3\text{OH}
$$

Methanol (methyl alcohol)

Ethanol (CH_3CH_2OH) is also called *ethyl alcohol* or *grain alcohol*. It is a colorless liquid that mixes with water in all proportions and is the least toxic and most economically important of all alcohols. Ethanol is found in alcoholic beverages and used in the production of many substances, including perfumes, dyes, and varnishes.

$$
\begin{array}{cc}
\text{H} & \text{H} \\
| & | \\
\text{H}-\text{C}-\text{C}-\text{O}-\text{H} \\
| & | \\
\text{H} & \text{H} \\
\end{array}
\qquad \text{or} \qquad \text{CH}_3\text{CH}_2\text{OH}
$$

Ethanol (ethyl alcohol)

Figure 14.12 Methanol (CH_3OH), the Simplest Alcohol
Methanol is used for fuel in some types of racing cars.

Rubbing alcohol (isopropyl alcohol, 2-hydroxypropane) is often used to swab skin before a needle injection, because it acts like an antiseptic, dissolves grime, and evaporates quickly.

$$CH_3CHCH_3$$
$$|$$
$$OH$$

2-Hydroxypropane (isopropyl alcohol)

Ethylene glycol (1,2-ethanediol, $OHCH_2CH_2OH$), a compound widely used as an antifreeze and coolant, is an example of an alcohol with two hydroxyl groups. It causes kidney failure if ingested and its sweet taste can act as a deadly lure to children and animals.

Amines

Amines ("ah-MEANS") are organic compounds that contain nitrogen and are basic (alkaline). The general formula for an amine is $R-NH_2$, or just RNH_2, but one or two additional alkyl groups could be attached to the nitrogen atom of the *amino group*, $-NH_2$, in place of one or more hydrogen atoms. (Recall that a nitrogen atom forms three bonds.) Examples are methylamine, dimethylamine, and trimethylamine. A condensed formula is shown for dimethylamine, and an even more condensed formula for trimethylamine.

Methylamine   Dimethylamine $(CH_3)_3N$
                                                     Trimethylamine

The simple amines have strong odors. The odor of raw fish comes from the amines it contains. Decaying flesh forms putresine (1,4-diaminobutane) and cadaverine (1,5-diaminopentane), two amines with especially foul odors, as their names vividly suggest.

Amines have many applications as medicinals and as starting materials for synthetic fibers. Aniline (aminobenzene) is the starting material for a whole class of synthetic dyes, as well as for many other useful compounds. Amines occur widely in nature as drugs, such as nicotine, the addictive substance in tobacco, and coniine, the poison that killed Socrates (● Fig. 14.13). A *drug* is a chemical that can produce a physiologic change in a human.

Amphetamines, such as Benzedrine, are synthetic amines that are powerful stimulants of the central nervous system. They raise the level of glucose in the blood, thereby fighting fatigue and reducing appetite. Although these drugs have legitimate medical usages, they are addictive and can lead to insomnia, excessive weight loss, and paranoia.

Figure 14.13 *The Death of Socrates*
The hemlock that Socrates drank contained the deadly alkaloid coniine. In 1787, French artist Jacques David painted *The Death of Socrates*.

Carboxylic Acids

The **carboxylic acids** ("CAR-box-ILL-ic") contain a *carboxyl group* and have the general formula RCOOH, in which the bonding is

$$\underset{\text{Carboxyl group}}{\overset{\displaystyle O}{-\overset{\|}{C}-O-H}} \qquad \underset{\substack{\text{Carboxylic acid} \\ \text{(general formula)}}}{\overset{\displaystyle O}{R-\overset{\|}{C}-O-H}}$$

The simplest carboxylic acid, formic acid (methanoic acid), is the cause of the painful discomfort from insect bites or bee stings. Vinegar is a 5% solution of acetic acid (ethanoic acid) in water. The structural formulas for formic and acetic acids are

$$\underset{\substack{\text{Formic acid} \\ \text{(methanoic acid)}}}{\overset{\displaystyle O}{H-\overset{\|}{C}-O-H}} \qquad \underset{\substack{\text{Acetic acid} \\ \text{(ethanoic acid)}}}{\overset{\displaystyle O}{CH_3-\overset{\|}{C}-O-H}}$$

Esters

An **ester** is a compound that has the general formula

$$\overset{\displaystyle O}{R-\overset{\|}{C}-O-R'}$$

where R and R′ (read "R prime") are any alkyl groups. R and R′ may be identical, but they are usually different.

Unlike amines, most esters possess pleasant odors. The fragrances of many flowers and the pleasing tastes of ripe fruits are due to one or more esters (● Fig. 14.14). Wintergreen mints and Pepto-Bismol get their characteristic fragrance from the ester named *methyl salicylate*, commonly called *oil of wintergreen*.

Figure 14.14 Esters Smell Good!
The distinctive aroma and flavor of oranges are due in part to the ester octyl acetate.

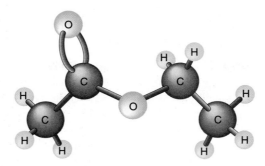

Figure 14.15 The Ethyl Acetate Molecule
Ethyl acetate, an ester, is used as a solvent in lacquers and other protective coatings.

Methyl salicylate (oil of wintergreen)

A carboxylic acid and an alcohol react to give an ester and water. This process is referred to as *ester formation*. The reaction mixture must be heated, and sulfuric acid is used as a catalyst. For example, acetic acid and ethanol react to give water and the ester called *ethyl acetate*, $CH_3COOCH_2CH_3$ (● Fig. 14.15). You would recognize its odor as that of fingernail polish remover.

Amides

Amides ("AM-eyeds"), another class of nitrogen-containing organic compounds, have the general formula

$$R-\overset{\overset{\displaystyle O}{\|}}{C}-\overset{\overset{\displaystyle H}{|}}{N}-R'$$

The reaction for *amide formation* is very similar to that of ester formation. A carboxylic acid and an amine have a net reaction that yields water and an amide. The simplest amide, ethanamide (CH_3CONH_2, also called acetamide), is formed from ammonia and acetic acid. Amides are commonly used as solvents and plasticizers.

Review Figure 14.9, which summarizes the names and general formulas of the derivatives of the hydrocarbons we have studied.

Did You Learn?

● Common hydrocarbon derivatives are chlorofluorocarbons (CFCs), alcohols, and many types of drugs.

● Amines have foul odors, whereas esters have pleasant odors.

14.5 Synthetic Polymers

Preview Questions

● How are polymers formed?

● What are some useful polymers?

Highlight   Veggie Gasoline?

Gasoline and other fuel consumption is high and will no doubt increase with world population increase, and petroleum supplies will eventually dwindle. One possible alternative to petroleum as a source of fuels and lubricants is vegetable oil such as soybean oil, one of the vegetable oils now used to cook french fries. Researchers believe that the oils from soybeans, corn, canola, and sunflowers all have the potential to be used in cars as well as on salads (see Fig. 1).

The idea of using vegetable oil for fuel is not new. German scientist Rudolf Diesel reportedly used peanut oil to run one of his diesel engines at the Paris Exposition in 1900. In addition, ethyl alcohol from corn has been used widely as a fuel in South America and as a fuel additive in the United States.

Biodiesel, a fuel made by esterifying the fatty acids found in vegetable oil, has some real advantages over regular diesel fuel. Biodiesel produces fewer pollutants such as particulates, carbon monoxide, and complex organic molecules, and since vegetable oils have no sulfur, there is no noxious sulfur dioxide in the exhaust gases. Also, biodiesel can run in existing engines with little modification. In addition, biodiesel is much more biodegradable than petroleum-based fuels, so spills cause less environmental damage.

Biodiesel also has some serious drawbacks, however, one of which is that it currently costs about three times as much as regular diesel fuel. Biodiesel also produces more nitrogen oxides in the exhaust than conventional diesel fuel and is less stable in storage. Biodiesel can leave more gummy deposits in engines and must be "winterized" by removing components that tend to solidify at low temperatures.

The best solution may be to use biodiesel as an additive to regular diesel fuel (similar to adding ethanol made from corn to gasoline). One such fuel is known as B20 because it is 20% biodiesel and 80% conventional diesel fuel. B20 is especially attractive because of the higher lubricating ability of vegetable oils, which reduces diesel engine wear.

Vegetable oils are also being looked at as replacements for motor oils and hydraulic fluids. Tests of a sunflower seed–based engine lubricant have shown satisfactory lubricating ability while lowering particle emissions. In addition, researchers at the University of Northern Iowa have developed BioSOY, a vegetable oil–based hydraulic fluid for use in heavy machinery.

Vegetable oil fuels and lubricants will have a growing market as petroleum supplies wane and as environmental laws become more stringent. In Germany's Black Forest region, for example, environmental protection laws require that farm equipment use only vegetable oil fuels and lubricants. In the near future, there may be vegetable oil in the garage as well as in the kitchen.

Figure 1 Veggie Power! This bus in England is fueled by biodiesel made from vegetable cooking oil.

Chemists have long tried to duplicate the compounds of nature. As the science of chemistry progressed and formulas and basic components became known, chemists were able to synthesize some of these natural compounds by reactions involving the appropriate elements or compounds. During this early trial-and-error period, there were probably as many serendipitous (accidental, but fortunate) as deliberate discoveries.

Attempts to synthesize natural compounds led to the discovery of **synthetics**, materials whose molecules have no duplicates in nature. The first synthetic compound was prepared by Belgian chemist Leo Baekeland in 1907; it was used as a common electrical insulator known commercially as Bakelite.

Baekeland's discovery triggered serious efforts to prepare synthetic materials. Chemists became aware that substituting different atoms or groups in a molecule would change its properties. For example, substituting a chlorine atom for a hydrogen atom in ethane produces chloroethane, which has very different properties from ethane. By knowing the general properties of the substituted groups, a chemist often can tailor a molecule to satisfy a given requirement.

Scientists also determined that molecules with large numbers of atoms having exceedingly high formula masses are often made up of repeating units of smaller molecules that have bonded together to form long, chain-like structures. The fundamental repeating unit is called the **monomer**, and the long chain made up of the repeating units is called the **polymer**.

Because of these discoveries, multitudes of synthetic compounds have been constructed. Probably the best known is the group of synthetic polymers that can be molded and hardened: *plastics*. Plastics have become an integral part of our modern life and are used in clothing, shoes, buildings, autos, sports equipment, art, electrical appliances, toothbrushes, toys, and many other items.

Conceptual Question and Answer

Keep It in Place

Q. Is hair spray a synthetic material?

A. Hair spray is a solution of polymers (long, chain-like molecules) in a very rapidly evaporating solution. Spraying deposits a stiff layer of the polymer on the hair after the solvent evaporates. Some hair sprays use natural polymers and solvents such as vegetable gums dissolved in alcohol.

The two major types of polymers are the addition polymers and the condensation polymers. **Addition polymers** are those formed when molecules of an alkene monomer add to one another. Under proper reaction conditions, often including a catalyst, one bond of the monomer's double bond opens up. This allows the monomer to attach itself by single bonds to two other monomer molecules, then each end attaches to another monomer, and so on. The polymerization of ethylene (ethene) to polyethylene is illustrated below, where the subscripted n means that the unit shown in the brackets is repeated thousands of times, as thousands of the monomer molecules join. (The atoms that eventually terminate the polymer are not usually shown.)

$$\begin{array}{c}
H \\
 \\
H
\end{array}
C=C
\begin{array}{c}
H \\
 \\
H
\end{array}
\xrightarrow{\text{Catalyst}}
\left[
\begin{array}{cc}
H & H \\
| & | \\
C & - C \\
| & | \\
H & H
\end{array}
\right]_n$$

Ethylene (ethene)   Polyethylene

Polyethylene is the simplest synthetic polymer. Because of its chemical inertness, it is used for chemical storage containers and many other packaging applications (● Fig. 14.16). Polyethylene comes in two major varieties. High-density polyethylene (HDPE) is rather rigid (milk jugs, for example). HDPE containers are stamped with these letters and have a 2 inside the triangular recycling symbol. The number 2 indicates that the plastic container can and should be recycled. On the other hand, low-density polyethylene (LDPE) is very flexible (trash bags, for example), bears the number 4, and is rarely recycled. ● Figure 14.17 summarizes the names, formulas, applications, and recycling information about common addition polymers.

Another common polymer that has found many uses, especially in coating cooking utensils, is Teflon, a hard, strong, chemically resistant fluorocarbon resin with a high melting point and low surface friction. Its monomer (tetrafluoroethene) and the polymer structure are

$$\begin{array}{c}
F \\
 \\
F
\end{array}
C=C
\begin{array}{c}
F \\
 \\
F
\end{array}
\xrightarrow{\text{Catalyst}}
\left[
\begin{array}{cc}
F & F \\
| & | \\
C & - C \\
| & | \\
F & F
\end{array}
\right]_n$$

Tetrafluoroethen   Teflon

Figure 14.16 Plastic Containers
Note the recycling symbols, numbers, and letters on these containers. (See text for description.)

Figure 14.17 Some Common Addition Polymers and Recycling Information

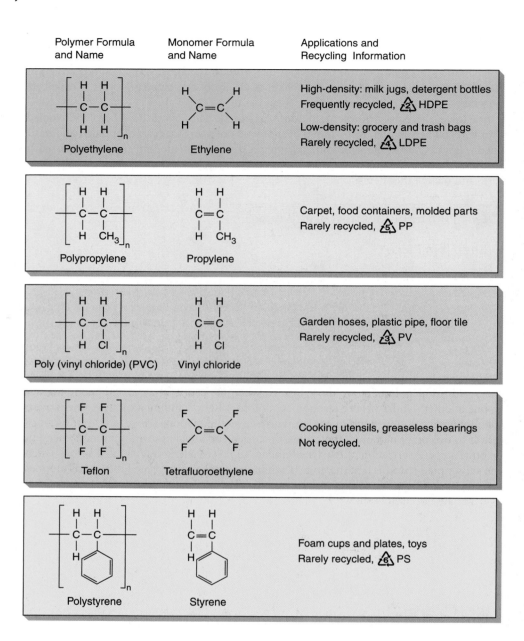

Polymer Formula and Name	Monomer Formula and Name	Applications and Recycling Information
Polyethylene	Ethylene	High-density: milk jugs, detergent bottles. Frequently recycled, 2 HDPE. Low-density: grocery and trash bags. Rarely recycled, 4 LDPE
Polypropylene	Propylene	Carpet, food containers, molded parts. Rarely recycled, 5 PP
Poly (vinyl chloride) (PVC)	Vinyl chloride	Garden hoses, plastic pipe, floor tile. Rarely recycled, 3 PV
Teflon	Tetrafluoroethylene	Cooking utensils, greaseless bearings. Not recycled.
Polystyrene	Styrene	Foam cups and plates, toys. Rarely recycled, 6 PS

EXAMPLE 14.5   Drawing the Structure of an Addition Polymer

An addition polymer can be prepared from vinylidene chloride, $CH_2{=}CCl_2$. Draw the structure of this polymer.

Solution

In forming the polymer, one bond of the double bond opens and adds to another molecule of the monomer on each end. This process repeats as more and more monomers add to each side of the growing chain, so the polymer's structure is shown as

Confidence Exercise 14.5

Draw the structure of the monomer from which this addition polymer was made.

$$\begin{bmatrix} & H & Cl \\ & | & | \\ -C & - & C- \\ & | & | \\ & H & H \end{bmatrix}_n$$

Condensation polymers are constructed from molecules that have two or more reactive groups. Generally, one molecule attaches to another by an ester or amide linkage. Water is the other product; hence the name *condensation polymer*. Of course, if a monoacid reacts with a monoalcohol or monoamine, then the reaction stops with the condensation of the two molecules and there is no chance to form a long-chain polymer.

However, if a diacid reacts with a dialcohol or a diamine, then the reaction can go on and on. An example is the polyester named *polyethylene terephthalate* (PETE), which is formed from thousands of molecules of two monomers. PETE is used to make plastic bottles that are marked with a number 1 and are commonly recycled. This same polymer is called *Dacron* when drawn into fibers and used to make polyester clothing. When Dacron is fashioned into a film rather than fibers, it is called *Mylar*, the polymer used for party balloons and audio and video recording tape. Dacron, a truly versatile polymer, has also been used in synthetic heart valves.

Another condensation polymer is the widely used polyamide *nylon* (● Fig. 14.18; also see the chapter-opening photo), which is formed from a diamine and a diacid. Nylon was an instant hit when it was introduced to the public at the 1939 World's Fair in New York City. In short order, women were buying more than four million pairs of nylon stockings each day, and during World War II when silk was unavailable for stockings, there were reports of "nylon riots" in some department stores.

Velcro is made of two nylon strips, one having thick loops that are slit open to form "hooks" and the other having thin, closed loops that entangle the slit fibers when the sides are pressed together (● Fig. 14.19). The inventor of Velcro got his idea by noticing how cockleburs clung to his clothing when he walked through a field.

Figure 14.18 Nylon, a Polyamide, Was First Synthesized in 1935 at DuPont
Here a strand of nylon is being drawn from the interface (boundary) of the two reactants, where the reaction occurs.

Sean Brady

Did You Learn?

- Polymers are formed from high-mass molecules bonded together in long chains.
- Plastics, such as Teflon nonstick surfaces and polyester clothing, are polymers.

14.6 Biochemistry

Preview Questions

- How are most carbohydrates used?
- What is the composition of soap?

The field of biochemistry emerged in the early 1900s as the study of chemical compounds and reactions in living organisms. Since life as we know it is carbon-based, biochemistry is naturally a part of organic chemistry.

Proteins

Biochemical molecules are natural polymers, unlike the synthetic polymers made in the lab. **Proteins** are biological polymers and are extremely long-chain polyamides formed

Figure 14.19 Velcro
The "hooks" of one nylon surface entangle with the loops of the other in this nylon Velcro fastener.

2003 Paul Silverman, Fundamental Photographs, NYC

Figure 14.20 A Polypeptide Protein
Proteins twist themselves into long curls and then fold onto themselves. This β-polypeptide protein surrounds an iron ion structure called a heme group.

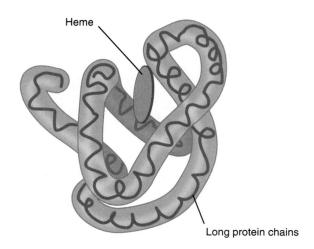

Heme

Long protein chains

by the enzyme-catalyzed condensation of *amino acids* under the direction of nucleic acids in the cell (or the biochemist in the lab). Their formula masses, measured in atomic mass units (u), range from a few thousand (insulin, 6000 u) to millions for the most complex (hemocyamine, 9 million u). Proteins function in living organisms both as structural components, such as muscle fiber, hair, and feathers, and as enzymes (biological catalysts).

Amino acids are organic compounds that contain both an amino group and a carboxyl group. More than 20 natural amino acids exist, 8 of which are essential in the human diet. The simplest amino acids are glycine and alanine. When a molecule of glycine combines with a molecule of alanine, a molecule of water is eliminated and an amide linkage, —CONH—, is formed in the resulting molecule called a *dipeptide*.

$$\underset{\text{Glycine}}{H-\overset{\overset{\displaystyle H}{|}}{N}-\overset{\overset{\displaystyle H}{|}}{\underset{\underset{\displaystyle H}{|}}{C}}-\overset{\overset{\displaystyle O}{||}}{C}-OH} + \underset{\text{Alanine}}{\overset{}{H}-\overset{\overset{\displaystyle H}{|}}{N}-\overset{\overset{\displaystyle H}{|}}{\underset{\underset{\displaystyle CH_3}{|}}{C}}-\overset{\overset{\displaystyle O}{||}}{C}-OH} \longrightarrow$$

$$\underset{\text{Water}}{H_2O} + \underset{\text{A dipeptide}}{H-\overset{\overset{\displaystyle H}{|}}{N}-\overset{\overset{\displaystyle H}{|}}{\underset{\underset{\displaystyle H}{|}}{C}}-\overset{\overset{\displaystyle O}{||}}{C}-\overset{\overset{\displaystyle H}{|}}{N}-\overset{\overset{\displaystyle H}{|}}{\underset{\underset{\displaystyle CH_3}{|}}{C}}-\overset{\overset{\displaystyle O}{||}}{C}-OH}$$

This process can be repeated by linking more amino acid molecules to each end of such a dipeptide, eventually forming a protein. The structures of proteins are complex, resulting in long coils or sheets that fold onto themselves (● Fig. 14.20). This complex folding is fundamental to the unique function of proteins and is necessary for living organisms.

Conceptual Question and Answer

My Twisted Double Helix

Q. Why does the famous DNA molecule have a "twisted" double-helix shape?

A. Hydrogen bonds exist between some of the H atoms and O or N atoms in other parts of the same molecule that have twisted around into close proximity. DNA contains four bases—adenine (A), cytosine (C), guanine (G), and thymine (T)—that attach to each other in pairs (A with T, and C with G). This "base pairing" is due to hydrogen bonding, which allows one DNA strand to link with another and replicate. The famous "unzipping" of DNA molecules is actually a breaking of intramolecular hydrogen bonds.

Carbohydrates

Carbohydrates, an important class of compounds in living matter, contain multiple hydroxyl groups in their molecular structures, and their names end in *-ose*. A primary use of carbohydrates in cells is as an energy source, and the most important ones are sugars, starches, and cellulose. Two important simple sugars are the isomers glucose ($C_6H_{12}O_6$) and fructose ($C_6H_{12}O_6$). When a glucose molecule is bonded to a fructose molecule, a molecule of sucrose (ordinary table sugar) is formed. Conversely, when sucrose is digested, the atoms of a water molecule are added to the structure as it breaks to form a molecule of glucose and a molecule of fructose (● Fig. 14.21).

Figure 14.21 Breaking Up Sucrose
Liquid-center candy uses the enzyme-catalyzed reaction of sucrose with water to form glucose and fructose, sugars that are much more soluble in water than the original sucrose and that dissolve to give the liquid center.

Conceptual Question and Answer

Carbohydrates and You

Q. What does the name carbohydrate imply?

A. Carbohydrates consist only of carbon, hydrogen, and oxygen, and most have a general formula $C_n(H_2O)_n$; thus, they can be viewed as hydrates of carbon. Like water, carbohydrates have a hydrogen-to-oxygen atom ratio of 2 to 1; an example is the sugar isomers of $C_6H_{12}O_6$. The arrangement of atoms in carbohydrates has little to do with water molecules, however. The primary function of carbohydrates is for energy storage (sugars). When we regularly eat too many carbohydrates, some of the excess sugars are converted to fats, giving rise to obesity and health problems.

Fructose (fruit sugar) is the sweetest of all sugars and is present in fruits and honey. If sucrose is given an arbitrary sweetness value of 100, then fructose is rated 173. The artificial sweetener aspartame has a value of 15,000, and saccharin is rated 35,000. No wonder a little artificial sweetener goes a long way.

Glucose, also known as *dextrose*, formed in plants by the action of sunlight and chlorophyll, is found in sweet fruits like grapes and figs and in flowers and honey. Carbohydrates must be digested into glucose for circulation in the blood. Hospitalized patients are sometimes fed intravenously with glucose solutions, because glucose requires no digestion. Glucose is normally present in the blood to the extent of 0.1%, but it occurs in much greater amounts in people with diabetes.

Starch is a natural polymer consisting of long chains of up to 3000 glucose units. It is a noncrystalline substance formed by plants in their seeds, tubers, and fruits. After we eat these plant parts, the digestion process converts the starches back to glucose. Glycogen (animal starch), a smaller and more highly branched polymer of glucose, is stored in the liver and muscles of animals as a reserve food supply that is easily converted to energy.

Cellulose, another polymer of glucose, has the same general formula, $(C_6H_{10}O_5)_n$, as starch, but its structure is slightly different, so it has different properties. Cellulose is the main component of the cell walls of plants; it is the most abundant organic substance found in our environment. Cellulose cannot be digested by humans because our digestive systems do not contain the enzymes (called *cellulases*) that are necessary to break the linkages in the molecular chain. The bacteria in the digestive tracts of termites and herbivores (such as cows and deer) do have the enzymes necessary to obtain nutrition from cellulose. Cellulose is also contained in many commercial products, such as rayon, explosives, and paper.

Fats

Fats and oils are used in the diets of humans and other organisms. In the digestive process, fats are broken down into glycerol and acids, which are absorbed into the bloodstream and oxidized to produce energy that may be used immediately or stored for future use. Fats are also used by the body as insulation to prevent loss of heat and are important components of cell membranes. The metabolism of 1 gram of fat produces 9 kcal of energy, whereas

1 gram of protein or carbohydrate produces only 4 kcal of energy. It is well established that a diet heavy in saturated fats is unhealthy because the fats lead to a buildup of cholesterol, a waxy substance that can clog arteries.

Fats are esters composed of the trialcohol named *glycerol*, $CH_2(OH)CH(OH)CH_2(OH)$, and long-chain carboxylic acids known as *fatty acids*. A typical fatty acid is stearic acid, $C_{17}H_{35}COOH$, a component of beef fat. Stearic acid's structure is that of a long-chain hydrocarbon containing 17 carbon atoms (and their associated hydrogen atoms) attached to a carboxyl group. It is written below in condensed form to save space. When stearic acid is combined with glycerol, the triester *glyceryl tristearate* (a fat) is obtained. The reaction is

$$3\ \underset{\text{Stearic acid}}{C_{17}H_{35}-\overset{\overset{\displaystyle O}{\|}}{C}-OH} + \underset{\text{Glycerol}}{\left(\begin{array}{c} H-O-CH_2 \\ H-O-CH \\ H-O-CH_2 \end{array}\right)} \longrightarrow 3\ H_2O + \underset{\text{Glyceryl tristearate (a fat)}}{\left(\begin{array}{c} C_{17}H_{35}-\overset{\overset{\displaystyle O}{\|}}{C}=O-CH_2 \\ C_{17}H_{35}-\overset{\overset{\displaystyle O}{\|}}{C}=O-CH \\ C_{17}H_{35}-\overset{\overset{\displaystyle O}{\|}}{C}=O-CH_2 \end{array}\right)}$$

Fats that come from mammals and birds are generally solids at room temperature, but those that come from plants and fish are usually liquid (● Fig. 14.22a). These liquid fats are usually referred to as *oils*. Their molecules are composed of hydrocarbon chains with double bonds between some of the carbon atoms; they are unsaturated. These oils can be changed to solid (saturated) fats by a process called *hydrogenation* (Fig. 14.22b). In this process, hydrogen is added to the carbon atoms that have the double bonds (Chapter 14.3), and the hydrocarbon chains become saturated, or nearly so. Thus, liquid fats (oils) are esters of glycerol and unsaturated acids, and solid fats are esters of glycerol and saturated acids. When cottonseed oil (a liquid) is hydrogenated, margarine (a solid) is obtained.

When fats are treated with sodium hydroxide (NaOH), commonly called lye, the ester linkages break to give glycerol and sodium salts of fatty acids. The sodium salts of fatty acids are *soap*. A typical soap is sodium stearate, whose condensed structural formula is

$$\underset{\text{Sodium stearate}}{CH_3(CH_2)_{16}\overset{\overset{\displaystyle O}{\|}}{C}-O^-\ Na^+}$$

Figure 14.22 Fats and Oils
The partial hydrogenation of the double bonds in the molecules of a liquid vegetable oil (a) produces a semisolid substance similar to an animal fat (b).

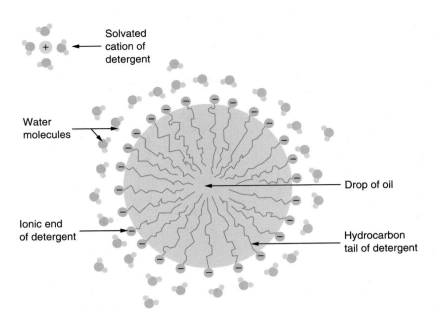

Figure 14.23 Like Dissolves Like
The ionic ends of detergent molecules dissolve in the polar water, and the long nonpolar chains of the detergent molecules dissolve in the grease. The emulsified grease droplets can then be rinsed away

To dissolve stains made by a nonpolar compound such as grease, either a nonpolar solvent or a soap or detergent must be used. One end of the soap or detergent molecule is highly polar and dissolves in the water, whereas the other part of the molecule is a long, nonpolar hydrocarbon chain that dissolves in the grease (● Fig. 14.23). The grease is then emulsified and swept away by rinsing.

Soaps have the disadvantage of forming precipitates when used in acidic solutions or with hard water, which contains ions of calcium, magnesium, or iron. The modern *synthetic detergents*, which are soap substitutes, are effective cleansing agents in hard water. For another interesting use of fats and oils, see the previous **Highlight: Veggie Gasoline**.

Did You Learn?

- Carbohydrates, like sugar, are used in cells as an energy source.
- Soap is a sodium salt of fatty acids.

KEY TERMS

1. organic chemistry
2. hydrocarbons (14.2)
3. aromatic hydrocarbons
4. aliphatic hydrocarbons (14.3)
5. alkanes
6. alkyl group
7. constitutional isomers
8. cycloalkanes
9. alkenes
10. alkynes
11. functional group (14.4)
12. alkyl halide
13. CFCs
14. alcohols
15. amine
16. carboxylic acids
17. ester
18. amides
19. synthetics (14.5)
20. monomer
21. polymer
22. addition polymers
23. condensation polymers
24. proteins (14.6)
25. carbohydrates
26. fats

MATCHING

For each of the following items, fill in the number of the appropriate Key Term from the preceding list. Compare your answers with those at the back of the book.

a. _____ Compounds that contain only the elements carbon and hydrogen

b. _____ A long-chain molecule made up of repeating units called monomers

c. _____ A hydrocarbon derivative that contains a halogen atom

d. _____ Hydrocarbons that contain at least one carbon-to-carbon triple bond

e. _____ Hydrocarbons having no benzene rings

f. _____ Polymers formed from alkene monomers

g. _____ A substituent that contains one less hydrogen atom than the corresponding alkane

h. _____ The branch of chemistry that studies carbon compounds

i. _____ Saturated hydrocarbons that have a ring structure

j. _____ Materials whose molecules have no duplicates in nature

k. _____ Hydrocarbons with at least one carbon-to-carbon double bond

l. _____ A carboxylic acid and an amine react to give this nitrogen-containing compound

m. _____ Generally, the most reactive part of a molecule

n. _____ Ethylene glycol is a common example of this organic compound

o. _____ Important compounds used primarily in cells as an energy source

p. _____ Hydrocarbons that contain only single bonds and no ring structure

q. _____ These compounds come from animals and are generally solid at room temperature

r. _____ Long-chain polyamides formed from amino acids

s. _____ Hydrocarbons that possess one or more benzene rings

t. _____ The fundamental repeating unit of a polymer

u. _____ Nitrogen-containing compounds that usually have strong, foul odors

v. _____ An alkyl group combined with a carboxyl group

w. _____ Polymers formed from molecules having two or more reactive groups

x. _____ Alkyl halides used as refrigerants but harmful to the ozone layer

y. _____ A carboxylic acid and an alcohol react to give this pleasant-smelling compound

z. _____ Compounds having the same molecular formula but different structural formulas

MULTIPLE CHOICE

Compare your answers with those at the back of the book.

1. How many covalent bonds does an oxygen atom form? (14.1)
   (a) 1   (b) 2   (c) 3   (d) 4

2. When a carbon atom bonds to four other atoms, what are the bond angles? (14.1)
   (a) 180°   (b) 120°   (c) 109°   (d) 90°

3. Which of the following is the most common aromatic compound? (14.2)
   (a) benzedrine   (b) ethylene
   (c) butane   (d) benzene

4. How many electrons are shared by all the carbon atoms in a benzene ring? (14.2)
   (a) 2   (b) 4   (c) 6   (d) 8

5. A molecule of propane contains how many carbon atoms? (14.3)
   (a) three   (b) five
   (c) seven   (d) eight

6. Which of the following is an ethyl group? (14.3)
   (a) CH_3- (b) CH_3CH_2-
   (c) CH_2CH_3- (d) CH_2CH_2-

7. Which of the following is a correct formula for an alkyl halide? (14.4)
   (a) CH_3Br (b) C_7H_{16}
   (c) CH_2NH_2 (d) CH_3CH_2OH

8. Alcohols contain which group attached to an alkyl group? (14.4)
   (a) amino   (b) hydroxyl
   (c) halogen   (d) carboxyl

9. Which of these are the best-known synthetic compounds? (14.5)
   (a) plastics
   (b) monomers
   (c) partially hydrogenated vegetable oils
   (d) amines

10. Which of the following is a condensation polymer? (14.5)
    (a) Teflon   (b) PVC
    (c) nylon   (d) polyethylene

11. Which process converts unsaturated fats into saturated fats? (14.6)
    (a) carbonation
    (b) oxidation
    (c) hydrogenation
    (d) combustion

12. The metabolism of which of the following produces the most energy? (14.6)
    (a) carbohydrates
    (b) fats
    (c) proteins
    (d) acids

FILL IN THE BLANK

Compare your answers with those at the back of the book.

1. Organic chemistry studies the compounds of the element named ___. (Intro)

2. In organic compounds, a nitrogen atom forms ___ bonds. (14.1)

3. ___ bonds are the type of bonds in organic compounds. (14.1)

4. Hydrocarbons are first divided into aromatic and ___. (14.2)

5. ___ is the most important aromatic hydrocarbon. (14.2)

6. A compound with the general formula C_nH_{2n} is either an alkene or a(n) ___. (14.3)

7. Hexyne is a member of the ___ class of aliphatic hydrocarbons. (14.3)

8. Organic compounds that contain a hydroxyl group are called ___. (14.4)

9. An amine and a carboxylic acid react to form water and a(n) ___. (14.4)

10. Two classes of synthetic polymers are addition and ___. (14.5)

11. Velcro is made of ___, a widely used type of condensation polymer. (14.5)

12. Sucrose, common table sugar, is part of the ___ class of compounds. (14.6)

SHORT ANSWER

1. Distinguish between organic chemistry and biochemistry.

14.1 Bonding in Organic Compounds

2. Explain the type of bonding used for organic compounds.

3. Tell the number of covalent bonds formed by an atom of each of these common elements in organic compounds: C, H, O, S, N, a halogen.

14.2 Aromatic Hydrocarbons

4. Describe the physical characteristics of benzene. What is its molecular formula?

5. Show the Kekulé representation and the preferred representation of benzene's structural formula.

14.3 Aliphatic Hydrocarbons

6. What structural feature distinguishes an aromatic hydrocarbon from an aliphatic hydrocarbon?

7. Give the general molecular formulas for alkanes, cycloalkanes, alkenes, and alkynes. Write the structural formula for one example of each of these classes.

8. Name the first eight members of the alkane series. What are some useful products and properties of alkanes?

9. Describe the geometry and bond angles when a carbon atom forms four single bonds.

10. Two children build different objects from identical boxes of toy blocks. How is this related to constitutional isomers?

11. What are the basic rules of nomenclature for organic compounds?

12. Use both full and condensed structural formulas to show the difference between methane and a methyl group and between ethane and an ethyl group.

13. What is the structural difference between benzene and cyclohexane?

14. Name the compound represented by a hexagon and give its molecular formula and full structural formula.

15. Both ethene and ethyne are often called by their more common names. Give the common names and draw the structural formula for each compound.

16. Distinguish between saturated and unsaturated hydrocarbons.

17. Describe an addition reaction that alkenes and alkynes undergo. Why can't alkanes undergo addition reactions?

14.4 Derivatives of Hydrocarbons

18. What are the physical characteristics of simple alkyl halides? What are the benefits and detriments to their use?

19. Give the condensed structural formula for methanol.

20. Give the general formula for an alcohol. Name the characteristic group it contains. What suffix is used in the IUPAC names of alcohols?

21. What characteristic group does an amine contain, and what are some applications for these compounds? What property makes the simple amines unpopular?

22. Give the general formula for a carboxylic acid. Name the characteristic group it contains.

23. Why are esters popular? Give examples of where you can find these compounds.

24. What characteristic element is found in amides? What are the uses of amides?

14.5 Synthetic Polymers

25. To form an addition polymer, what structural feature must the monomer possess?

26. To form a condensation polymer, what structural feature is needed in the monomers?

27. Name a well-known synthetic fiber that is a polyester and one that is a polyamide. Name two addition polymers.

14.6 Biochemistry

28. Name the monomers of proteins. Name and write the structural formula for the simplest one of these monomers.

29. How do humans get energy from carbohydrates? Which energy source do plants use to make carbohydrates?

30. What two simpler sugars combine to form sucrose? Which sugar is the monomer of both starch and cellulose?

31. Why can herbivores digest cellulose but humans cannot?

32. What are the physical and chemical differences between fats and oils?

33. Which types of fats can cause health issues if eaten in large quantities?

34. What is the relationship between a fat and a soap?

VISUAL CONNECTION

Visualize the connections and give answers for the blanks. Compare your answers with those at the back of the book.

Classification of Hydrocarbons

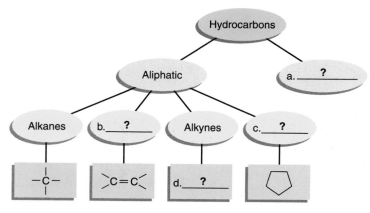

APPLYING YOUR KNOWLEDGE

1. Although life elsewhere in the universe is probably based on carbon, some science fiction writers have speculated that it could be based on another element with similar bonding properties. Which element do you think they choose?

2. You overhear someone comment that a lot of "cat cracking" takes place in oil refineries. Should you call the Society for the Prevention of Cruelty to Animals?

3. Look up the structural formula of aspirin (acetylsalicylic acid) and Tylenol (acetaminophen). What types of organic compounds are they? What alkyl and functional groups do they contain, and how are they similar?

4. As you take out the garbage one morning, you see several plastic items stamped with a 2 inside a triangle. Can these items be recycled?

5. If you get tears in your eyes when you peel onions, then the water-soluble compound thiopropanal S-oxide (CH_3CH_2CHSO) is the culprit. What simple strategy would enable you to peel onions without irritating your eyes?

6. Describe the process in which soap "cleans" a drop of grease off a piece of clothing. Would using soap mixed with alcohol be more effective than using soap and water? Explain.

7. How does hydrogen bonding affect large molecules like proteins? What shapes do proteins take because of this effect?

EXERCISES

14.1 Bonding in Organic Compounds

1. Which of these structural formulas is valid, and which is incorrect?

(a)
```
     H   H
     |   |
 H—C—C—H
     \ /
      N
      |
      H
```

(b)
```
     H       H
     |       |
 H—C—Cl—C—O—H
     |       |
     H       H
```

Answer: (a) is valid. (b) is incorrect because Cl should have one bond, not two.

2. Which of these structural formulas is valid, and which is incorrect?

(a) H—S—C≡C—O—H

(b)
```
     O
     ||
 H—C—Br
```

14.2 Aromatic Hydrocarbons

3. Draw the structural formula for 1-bromo-2-methyl-benzene.

Answer: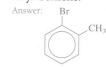

4. Draw the structural formula for 1,4-diethylbenzene.

14.3 Aliphatic Hydrocarbons

5. Classify each of the following hydrocarbon structural formulas as an alkane, cycloalkane, alkene, alkyne, or aromatic hydrocarbon.

(a)
```
 H₃C         CH₃
     \       /
      C = C
     /       \
    H         H
```

(b) cyclopentane with two CH_3 groups

(c) CH₃—C=C—CH₃

(d) CH₃CH₂CH₃

(e) 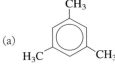CH₂CH₃

6. Classify each of the following hydrocarbon structural formulas as an alkane, cycloalkane, alkene, alkyne, or aromatic hydrocarbon.

(a)
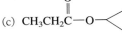

(b) CH₃CH₂CH₂CH₂CH₃

(c) ▽—CH₂CH₃

(d)
$$\underset{H}{\overset{H}{C}}=\underset{H}{\overset{CH_2CH_3}{C}}$$

(e) H—C=C—CH₂CH₃

7. Use each of the following names to classify each hydrocarbon as an alkane, cycloalkane, alkene, alkyne, or aromatic hydrocarbon.

(a) 2-methylbutane

(b) 3-methyl-1-pentyne

(c) 1,1-dimethylcyclobutane

(d) 3-octene

(e) 1,3-dimethylbenzene

8. Use each of the following names to classify each hydrocarbon as an alkane, cycloalkane, alkene, alkyne, or aromatic hydrocarbon.

(a) 2-methyl-2-hexene

(b) methylcyclohexane

(c) 3,3,4-triethylhexane

(d) 2-heptyne

(e) ethylbenzene

9. State whether the structural formulas shown in each case represent the *same compound*, are *constitutional isomers*, or are *neither*.

(a) CH₃CH₂CH₃ and
$$\underset{CH_3}{\overset{CH_2CH_3}{|}}$$

(b) CH₃CH₂OH and CH₃ CH₂CH₂OH

(c) CH₂=CHCH₃ and
$$\underset{CH_2}{\overset{CH_2CH_2}{\diagdown\diagup}}$$

10. State whether the structural formulas shown in each case represent the *same compound*, or are *constitutional isomers*, or are *neither*.

(a) CH₃CH₂NH₂ and CH₃NHCH₂CH₃

(b) HOOCCH₂CH₃ and CH₃CH₂COOH

(c) CH₃CHCH₃ and CH₃CH₂CH₂OH
    |
    OH

11. Given the IUPAC name, draw the structural formula for each alkyl halide.

(a) 1,1-dibromo-2-fluorobutane

(b) 1-bromo-2-methylcyclopentane

12. Given the IUPAC name, draw the structural formula for each alkyl halide.

(a) 1,2,2-trifluorobutane

(b) 1-bromo-2-chlorocyclopropane

13. Two constitutional isomers of continuous-chain butenes exist: 1-butene and 2-butene. How many constitutional isomers of continuous-chain pentenes exist? Name each and draw its structural formula.

14. How many constitutional isomers of continuous-chain heptenes exist? Name each and draw its condensed structural formula.

14.4 Derivatives of Hydrocarbons

15. Identify each structural formula as belonging to an alkyl halide, alcohol, amine, carboxylic acid, ester, or amide.

(a) CH₃CH₂NH₂

(b)
$$\underset{}{\overset{O\ \ H}{CH_3CH_2C-N-CH_3}}$$
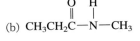

(c)
$$\underset{}{\overset{O}{CH_3CH_2C-O-\triangle}}$$
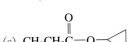

(d)
$$\underset{}{\overset{O}{CH_3CH_2C-O-H}}$$
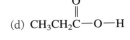

(e) CH₃CHCH₃
        |
        Cl

(f) CH₃CHCH₂CH₃
        |
        OH

16. Identify each structural formula as belonging to an alkyl halide, alcohol, amine, carboxylic acid, ester, or amide.

(a) CH₃CH₂COOH

(b)
$$\underset{}{\overset{O}{CH_3C-O-CH_2CH_3}}$$

(c) CH₃CH₂ CH₂OH

(d) CH₃CH₂NH₂

(e) CF₃CF₃

(f)
$$\underset{}{\overset{O\ \ H}{CH_3C-N-CH_3}}$$

17. Draw the constitutional isomers for (a) C₂H₆O (two) and (b) C₃H₇Cl (two).

18. Draw the constitutional isomers for (a) C₂H₇N (two) and (b) C₃H₈O (three).

14.5 Synthetic Polymers

19. Polystyrene, or Styrofoam, is an addition polymer made from the monomer styrene. Show by means of an equation how styrene polymerizes to polystyrene.

$$\underset{H}{\overset{H}{C}}=\underset{H}{\overset{H}{C}}$$

20. Acrilan is an addition polymer made from the monomer named acrylonitrile (cyanoethene). Show by means of an equation how acrylonitrile polymerizes to Acrilan.

$$\underset{H}{\overset{H}{>}}C=C\underset{CN}{\overset{H}{<}}$$

21. Draw the structure of a portion of the chain of Kevlar, used in bulletproof vests, which is made by the condensation polymerization of the monomers shown.

$$H_2N\!-\!\!\langle\bigcirc\rangle\!\!-\!NH_2 \quad and \quad HO\!-\!\overset{O}{\overset{\|}{C}}\!\!-\!\!\langle\bigcirc\rangle\!\!-\!\overset{O}{\overset{\|}{C}}\!-\!OH$$

Answer:

$$\left[\!-\!HN\!-\!\!\langle\bigcirc\rangle\!\!-\!NH\!-\!\overset{O}{\overset{\|}{C}}\!\!-\!\!\langle\bigcirc\rangle\!\!-\!\overset{O}{\overset{\|}{C}}\!-\!\right]_n$$

22. The polyester formed from lactic acid (shown below) is used for tissue implants and surgical sutures that will dissolve in the body. Draw the structure of a portion of this polymer.

$$HO\!-\!\underset{\underset{CH_3}{|}}{CH}\!-\!\overset{O}{\overset{\|}{C}}\!-\!OH$$

ON THE WEB

1. The Essence of Hydrocarbons

Can you answer the question, "What is an organic compound?" What is the difference between organic and inorganic compounds? What is the basis of the special role that carbon plays in the chemistry of elements? Follow the links on the student website at **www.cengagebrain.com/shop/ISBN/1133104096** to find answers to these questions and then consider how what you've learned about organic compounds is relevant to your own life.

2. Synthesizing What You've Learned about Synthetic Polymers

What is a polymer? Where are polymers seen in nature? What is the effect of polymers on the environment? What effect do they have on your everyday life? Why should you care about polymers, and why learn about them? In what ways are polymers being recycled? What are some major myths about polymers? To further explore synthetic polymers and to answer these and other questions, follow the recommended links on the student website at **www.cengagebrain.com/shop/ISBN/1133104096**.

Place and Time

NASA Earth Observatory

What place have you?
What place have I?
What time have you?
What time have I?
Only here and now
Have you and I.

•

James T. Shipman
(1919–2009)

< *The terminator, shown here run-*
ning through Europe and Africa, is
the line or boundary between illumi-
nated day time and dark night time.
(Notice the lights from the cities on
the dark side.) As the Earth rotates,
the terminator moves westward.

I n physical science, events that take place in our environment are observed and examined. These events occur at different places and at different times. Some occur nearby and are observed immediately, whereas others occur at great distances and are not observed until a later time. For example, a star explodes at some great distance from the Earth, and years later the event is observed as the radiation reaches our planet. On a local level, a distant flash of lightning is seen instantaneously, but the sound of thunder is not heard until later. Thus, the events taking place in our environment are separated in space and time.

In this chapter we will introduce the concept of reference systems. A reference system is particularly important in describing exact locations of places on the Earth's surface. Our senses make it possible for us to know and observe objects

ASTRONOMY FACTS

▶ Common definition of time: Time is the continuous, forward flow of events. It has never been observed to run "backward," as would appear if a movie reel were run backward.

▶ An observer in the Northern Hemisphere can easily

Chapter Outline

determine their latitude on the Earth; it is the measured altitude of the North Star, Polaris.

▶ Marcus Aurelius (121–180 CE), Roman emperor and philosopher, wrote, "Time is sort of a river of passing events, and strong is its current."

▶ If the Earth's axis were not tilted, then there would be 12-hour days everywhere and no seasons. At the poles, the Sun would always be on the horizon.

▶ Assuming no cloudy days, every place on the Earth receives 12 hours of sunlight each day *averaged over a year* and therefore a total of 4380 hours of sunshine each year.

and their places relative to one another. The concept of time, on the other hand, is rather elusive. What is time? The periodic changes of day and night, the phases of the Moon, and the yearly cycle of seasons as a result of the Earth's revolving around the Sun are used to measure *time*. That is, time can be referenced to event changes observed in our environment. This chapter introduces and explains the concept of time that is so important in our daily lives (See also the Conceptual Question and Answer, Time and Time Again in Chapter 1.4).

15.1   Cartesian Coordinates

Preview Questions

● What is a Cartesian coordinate system?

● What do numbers on the *x*- and *y*-axes indicate on a rectangular coordinate system?

Designating the location of an object requires a reference system that has one or more dimensions. A one-dimensional axis system is depicted by the number line shown in ● Fig. 15.1. For a line to represent a coordinate system, an origin needs to be designated, along with a numerical unit scale.

A two-dimensional system is shown in ● Fig. 15.2, in which two number lines are drawn perpendicular to each other and the *origin* is assigned at the point of intersection. Such a two-dimensional system is called a **Cartesian coordinate system**, in honor of its originator, French philosopher and mathematician René Descartes (1596–1650). It is also referred to as a *rectangular coordinate system*.

The horizontal line is normally designated the *x*-axis and the vertical line the *y*-axis, each with a scale unit. In graphing data, it is important to give scale units, such as meters (m) or seconds (s). Every position or point in the *x*–*y* plane has a pair of coordinates (*x*, *y*); an example is the (3 m, 2 m) point shown in Fig. 15.2. Many cities in the United States are laid out in a Cartesian coordinate system. For example, one street may run east and west, corresponding to the *x*-axis, and another street may run north and south, corresponding to the *y*-axis.*

A Cartesian (*x*, *y*) coordinate system is used to determine a position or location in a plane, but determining a location on a spherical curved surface is different. The location of any position on the surface of a sphere can be designated using two

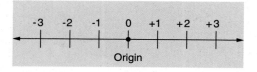

Origin

Figure 15.1 A One-Dimensional Reference System
For a line to represent a coordinate system, an origin needs to be designated along with a numerical unit scale.

Figure 15.2 A Two-Dimensional Reference System
Such a system is known as a Cartesian (*x*, *y*) coordinate system. The intersection of the *x*- and *y*-axes (the origin) is the reference point. The scale unit here is in meters (m) for both axes. The *x*- and *y*-axes may have different units; for example, *y* could be in meters (m) and *x* in seconds (s), which is read as distance versus time (*y* versus *t*).

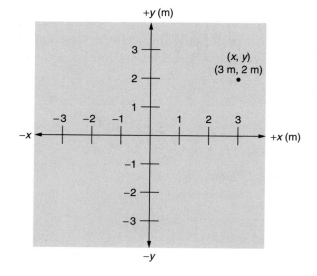

*A Cartesian system may be expanded to three dimensions by use of a *z*-axis perpendicular to the plane of *x* and *y*. For the street analogue, the *z*-axis would correspond to height, or distance above a street.

reference *circles*, analogous to the two Cartesian coordinate axes. The coordinates used to designate a position on the Earth's surface constitute latitude and longitude.

Did You Learn?

- A Cartesian coordinate system is a two-dimensional system in which two numbered lines (with units) are drawn perpendicular to each other.
- The numbers on the *x*- and *y*-axes indicate the scale units of each. Together (*x*, *y*) they locate a point.

15.2 Latitude and Longitude

Preview Questions

- How are locations on the Earth designated?
- How are parallels and meridians related to latitude and longitude?

The location on the surface of the Earth is designated by means of a coordinate system known as *latitude and longitude*. The *geographic poles*, the imaginary points on the surface of the Earth where the axis of rotation projects from the sphere, are used as north–south reference points. East–west coordinates are measured east or west along the equator or along lines parallel to it.

Lines called *parallels* and *meridians* form the latitude–longitude grid. Circles drawn around the Earth parallel to the equator are called **parallels**. Any number of such circles can be drawn, and they become smaller as the distance from the equator becomes greater. When you travel due east or west, you travel along a parallel travel along a parallel (● Fig. 15.3).

Imaginary lines drawn along the surface of the Earth running from the geographic North Pole, perpendicular to the equator, to the geographic South Pole, are known as **meridians**. Meridians are half circles, which are portions of a **great circle**, one whose plane passes through the center of the Earth (Fig. 15.3). The **latitude** of a surface position is defined as the angular measurement in degrees north or south of the equator. The latitude angle is measured from the center of the Earth relative to the equator (see ● Fig. 15.4a for the latitude of Washington, D.C.). Latitude is measured in degrees along a meridian, and parallels are lines of equal latitude. The equator has a latitude of 0°, the North Pole has a latitude of 90° north (90°N), and the South Pole's latitude is 90° south (90°S).

The longitude coordinate runs east and west. **Longitude** is defined as the angular measurement, in degrees, east or west of the reference meridian, known as the **Greenwich (prime) meridian** (Fig. 15.4b). This meridian was chosen as the zero (0°) meridian because England ruled the seas when the coordinate system of latitude and longitude was devised, and a large telescope located at Greenwich, England (a suburb of London), was used to promote celestial navigation.

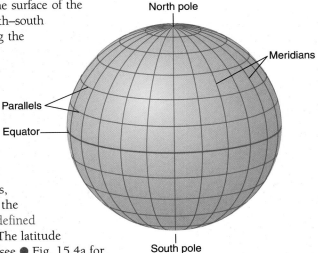

North pole

Meridians

Parallels

Equator

South pole

Figure 15.3 Parallels and Meridians
Parallels are circles drawn around the Earth parallel to the equator. Meridians are lines drawn from the geographic North Pole, perpendicular to the equator, to the geographic South Pole. Meridians are half circles, which are portions of a great circle, one whose plane passes through the center of the Earth. The equator is the only parallel that is a great circle.

Conceptual Question and Answer

Parallels and Perpendiculars

Q. How many parallels are great circles? How many meridians?

A. The center of the Earth is the reference point for latitude and longitude; by definition, the planes of all great circles pass through the Earth's center. Therefore, only one parallel is a great circle: the equator. However, since meridians are perpendicular to the equator, any number of meridians can be drawn, and every meridian is one-half of a great circle.

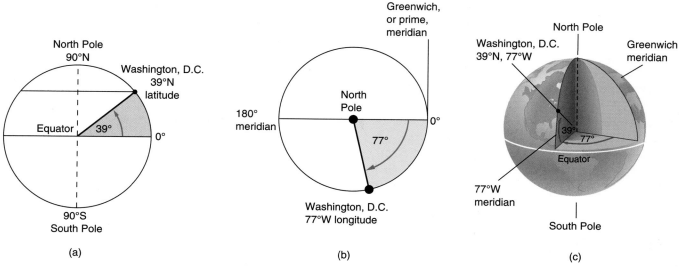

Figure 15.4 Diagrams Showing the Determination of the Latitude and Longitude of Washington, D.C.
(a) Latitude is the angle north or south of the equator. The latitude of Washington, D.C, is 39°N. (b) Longitude is the angle measured east or west from the Greenwich, or prime, meridian (0°). The longitude of Washington, D.C, is 77°W. (c) A cutaway view showing that Washington, D.C, is located at 39°N, 77°W.

Longitude has a minimum value of 0° at the prime meridian and a maximum value of 180° east and west. The longitude of Washington, D.C., is shown in Fig. 15.4b. The combined latitude and longitude of Washington, D.C., are given as 39°N, 77°W, as shown in the cutaway view in Fig. 15.4c.

In an *x–y* Cartesian system, the shortest distance between two points is a straight line. However, the shortest surface distance between any two points on the spherical Earth is the arc length along a great circle. Thus, the distance between two places on the Earth's surface can be determined if the great circle angle between them is known.

One minute of arc of a great circle is equal to one *nautical mile* (n mi), which is longer than the regular (statute) mile; 1 n mi = 1.15 mi. Sixty minutes of arc is equal to one degree, so 60 n mi equals 1°. You probably have heard the term *knot* in the context of a boat's speed, which is given in knots. One knot is is one nautical mile per hour (1 n mi/h).

| EXAMPLE 15.1 | Determining the Distance between Two Locations |

Determine the number of nautical miles between place *A* (10°S, 90°W) and place *B* (60°N, 90°E).

Solution

Step 1

Draw a diagram showing the latitude and longitude of place *A* and place *B*. Because the two places are 180° apart in longitude, they are on the same great circle (running through 90°E, 90°W, and the North Pole). A portion of this great circle is shown here (vertical line), from the perspective of looking down on the North Pole.

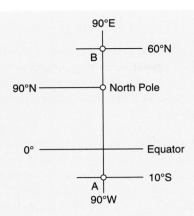

Step 2

Determine the number of degrees between place A and place B. Start at place A and add the latitude angle values from 10°S → North Pole → 60°N:

$$10° + 90° + 30° = 130°$$

Step 3

Calculate the number of nautical miles (n mi) between place A and place B.

$$60 \text{ n mi} = 1°$$

Therefore,

$$130° \times \frac{60 \text{ n mi}}{1°} = 7800 \text{ n mi}$$

Confidence Exercise 15.1

Determine the number of nautical miles between place A (43°N, 84°W) and place B (34°N, 84°W).

Answers to Confidence Exercises may be found at the back of the book.

In 1589, Gerardus Mercator (1512–1594), a Flemish cartographer and geographer, used the word *atlas* to describe a collection of projection maps.* The places on the maps are shown relative to one another with reference lines of latitude and longitude, but the maps are "projected" on flat surfaces, which leads to distortion of the size and shape of areas close to the poles. Mercator projection maps are used chiefly for navigation. Such maps have become an indispensable reference for showing the location of a place on the surface of the Earth.

Most maps are drawn with north at the top, south at the bottom, east on the right, and west on the left. With these references, direction can be determined, and when a scale is provided, the maps can also be used to obtain distance. More will be learned about determining the latitude and longitude of a location in Chapter 15.4. For now, see the **Highlight: Global Positioning System (GPS)**.

Did You Learn?

● Locations on the Earth are designated by longitude and latitude angles (coordinates).

● Lines parallel to the Earth's equator are called parallels, and lines perpendicular to the Earth's equator are called meridians. They form a latitude–longitude grid.

---

*After Atlas, who in Greek mythology is commonly depicted carrying the Earth on his shoulders.

Highlight Global Positioning System (GPS)

The GPS consists of a network of two dozen satellites. These solar-powered satellites circle the Earth at altitudes of about 20,000 km (12,400 mi), making two complete orbits every day. The orbits are arranged so that there are at least four satellites observable at any time, from anywhere on the Earth (Fig. 1).

Originally developed for the U.S. Department of Defense as a military navigation system, the GPS is now available to everyone. All you need is a GPS receiver to find your location anywhere on the Earth, except where the satellite radio signals cannot be received, such as in caves and under water.

GPS receivers are widely used for finding locations in navigation and other applications. They are used by hunters, hikers, and boaters. GPS devices are found in handheld units, cell phones, and automobiles to provide locations for roadside assistance. The accuracy of a receiver location depends on its price. High-end receivers have accuracies down to 1 meter. Really expensive units can come within 1 centimeter!

How does the GPS determine a position (latitude and longitude) on the Earth? The electronics involved are quite complicated, but the general principles of locating a position can be understood. The process involves triangulation. For example, you have probably seen one type of triangulation on TV shows or in movies when police are trying to locate a cell phone signal. One receiver gets a "fix" on the direction of the signal from its position, and a straight line is drawn on a map. Another receiver at another location does the same, and where the two directional lines cross is the location of the signal. Just to make sure, a third receiver is used for a three-line intersection (hence, triangulation).

In the case of the GPS, however, distance rather than direction is plotted. Let's consider a two-dimensional example of finding a location. Suppose you are at a large university and want to find your location on a campus map. You ask a passing student how far it is to the bell tower and learn that this distance is one block. By drawing a circle with a one-block radius with the bell tower at the center, you know that you are somewhere on this circle (Fig. 2a).

That doesn't help much, so you ask another student how far it is to the gym and learn that this second distance is two blocks. By drawing a two-block radius circle with the gym at the center, you know that you are at either point *A* or point *B,* the two points where the circles intersect (Fig. 2b). By doing the same for the campus gate, which you are told is three blocks away, you know that your location is at point *A,* the one point where all three circles intersect (Fig. 2c).

The same idea works in three dimensions on spheres. The satellites send time radio signals to the receiver, and its electronics interpret this information in terms of a satellite's distance. Then, in a manner analogous to our two-dimensional example, the

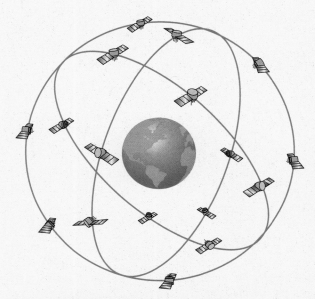

Figure 1 Global Positioning System (GPS)
An artist's conception of the GPS satellites.

15.3 Time

Preview Questions

- Are there a 12 A.M. and a 12 P.M.?
- What is the significance of the International Date Line (IDL)?

The concept of time is an interesting topic. In terms of measurement, the second is the standard unit of time. It was originally defined as a fraction of a solar day, but now is defined in terms of the emitted frequency of the cesium-133 atom. (See Chapter 1.4.)

For everyday purposes, the day is an important unit of time with divisions of hours, minutes, and seconds. The day has been defined in two ways. In one definition, the **solar day** is the elapsed time between two successive crossings of the same meridian by the

Highlight

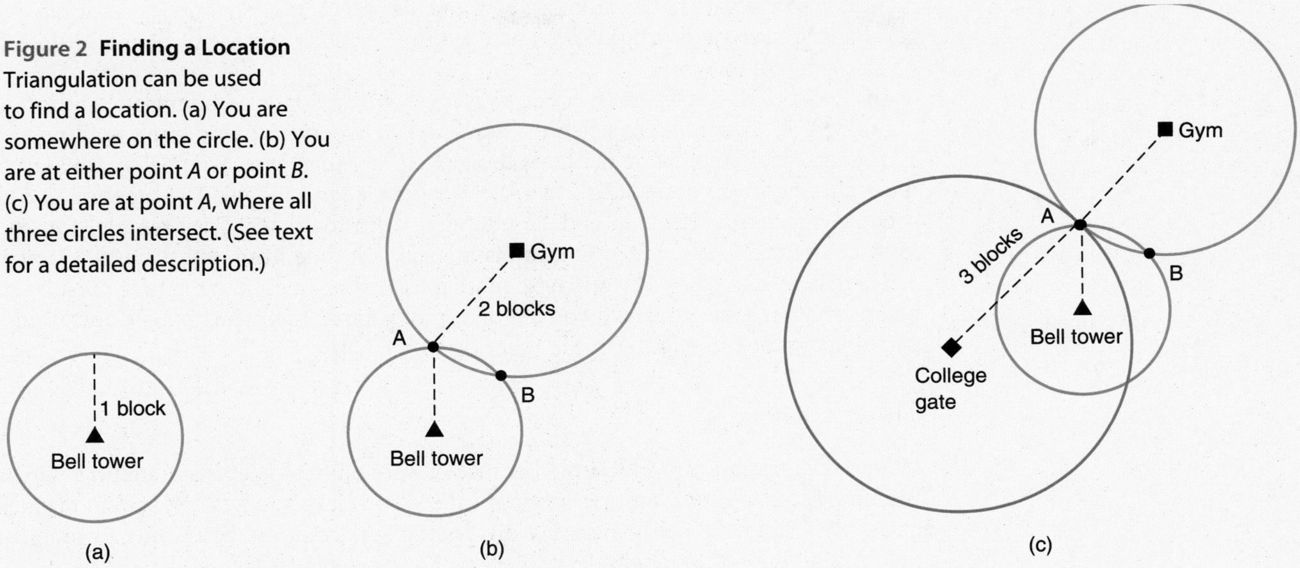

Figure 2 Finding a Location

Triangulation can be used to find a location. (a) You are somewhere on the circle. (b) You are at either point *A* or point *B*. (c) You are at point *A*, where all three circles intersect. (See text for a detailed description.)

(a)

1 block

Bell tower

(b)

Gym

2 blocks

A

B

Bell tower

(c)

Gym

3 blocks

A

B

Bell tower

College gate

positions and distances provide three circles on the globe, and their intersection is the receiver's location (Fig. 3).* Time is the measured quantity for the distance calculation (distance = speed of light × time), so time measurements must be very accurate to give precise locations. Atomic clocks are used (Chapter 1.4).

---

*The receiver's altitude must also be supplied. Adding a fourth satellite's information makes it possible to determine the receiver's latitude, longitude, and altitude.

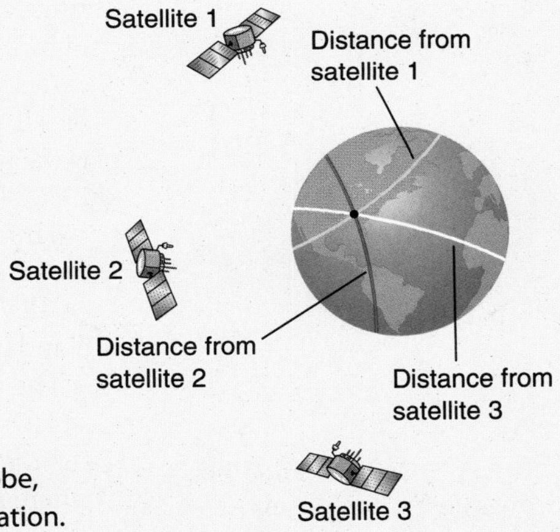

Satellite 1

Distance from satellite 1

Satellite 2

Distance from satellite 2

Distance from satellite 3

Satellite 3

Figure 3 Location on the Earth

Satellite data provide three circles on the globe, the intersection of which is the receiver's location.

Sun. This period is also known as the *apparent solar day*, because that is what appears to happen. Because the Earth travels in an elliptical orbit, its orbital speed is not constant (Chapter 3.7). Therefore, the apparent days are not of the same duration. As a remedy for this discrepancy, an average, or a *mean solar day*, is computed from all the apparent days during one year.

In another definition, a **sidereal day** is the elapsed time between two successive crossings of the same meridian by a star other than the Sun. ● Figure 15.5 shows the difference between a solar day

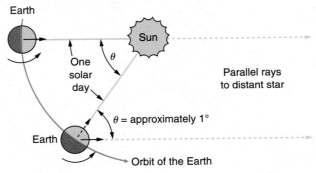

Earth

Sun

One solar day

θ

Parallel rays to distant star

θ = approximately 1°

Earth

Orbit of the Earth

Figure 15.5 The Difference between the Solar Day and the Sidereal Day

One rotation of the Earth on its axis with respect to the Sun is known as one solar day. One rotation of the Earth on its axis with respect to any other star is known as one sidereal day. See the reference arrow and notice that the Earth turns through an angle of 360° for one sidereal day and through an angle of approximately 361° for one solar day. (Drawing not to scale.)

and a sidereal day. One sidereal day is the time interval between the parallel (dashed) arrows pointing toward a distant star (parallel because of the great distance). To complete a solar day, the Earth must rotate a bit farther for the Sun to be overhead on an observer's meridian. The additional rotation is about one degree (1°). Therefore, the solar day is longer than the sidereal day by approximately 4 min, because the Earth rotates 360°/24 h, or 15°/h, or 1°/4 min.

The 24-hour day as we know it begins at midnight and ends 24 hours later at the following midnight. By definition, when the Sun is on an observer's meridian, it is 12 noon *local solar time*. The hours before noon are designated A.M. (*ante meridiem*, before midday), and those after noon are designated P.M. (*post meridiem*, after midday). The times of 12 o'clock should be stated as 12 noon and 12 midnight, with the dates. For example, you should state 12 midnight, December 3–4, to distinguish that time from, say, 12 noon, December 4. There is *no* 12 A.M. or 12 P.M. These times are dividing lines, and the time must be either before or after. Analogously, if you are on top of a fence, then you are on a "dividing line" and aren't on one side (A.M.) or the other (P.M.).

Time Zones

Modern civilization runs efficiently because of our ability to keep accurate time. Since the late nineteenth century, most countries of the world have adopted the system of *standard time zones*. This scheme theoretically divides the surface of the Earth into 24 time zones, each containing 15° of longitude, or 1 h (the Earth rotates 15°/h).

The first zone is referenced to the Greenwich meridian, or prime meridian (0°), which runs through Greenwich, England, and the zone extends 7.5° on each side of the prime meridian. The zones continue east and west from the Greenwich meridian, each zone extending 7.5° on each side of its central meridian. The actual boundaries of the zones vary because of local conditions, but all places within a zone have the same standard time, which is the time of the central meridian of that zone. For example, Washington, D.C., is located at 77°W longitude, which is within 7.5° of the 75°W central meridian (● Fig. 15.6).

Thus, Washington, D.C., is located in the Eastern Standard Time (EST) zone. Alaska, for the most part, is two time zones west (or 2 hours ahead) of the Pacific Standard Time zone. Alaska's Aleutian Islands and Hawaii are three time zones west (3 hours ahead) of Pacific Standard Time. In the conterminous United States, thirteen states have two time zones.*

The time of the first zone, centered on the central prime meridian, or Greenwich meridian (0°), is known as **Greenwich Mean Time (GMT)**. It became an international time standard, and reference was made to it in expressing times, particularly for navigation, where it is referred to as "Zulu." For example, an airline pilot in the United States might say his or her estimated time of arrival (ETA) in Washington, D.C., was 1400 ("fourteen hundred") hours GMT, or 1400 hours Zulu, in 24-h notation. (Fourteen hundred hours corresponds to 2:00 P.M. in 12-h notation.) When a person speaks in terms of GMT, there is only one reference point, and time zone mix-ups cannot occur. Under this time standard, when the Sun is on the prime meridian, it is officially noon, 1200 hours. Midnight is 0000 hours, so there is no confusion between 12 noon and 12 midnight.

However, because of slight variations in the Earth's rotation and revolution, there were variations in time as determined by the noon standard. Therefore, in 1972, the time of a grid of atomic clocks became the official standard of time. These clocks are accurate to within a billionth of a second (Chapter 1.4). Now the international time standard is called **Coordinated Universal Time (UTC)**.† This term is currently used for what was previously referred to as GMT. The terms *GMT* and *Zulu* are still widely used, however.

● Figure 15.7 shows the time and date on the Earth for any Tuesday at 7 A.M. (PST), 8 A.M. (MST), 9 A.M. (CST), and 10 A.M. (EST). As the Earth rotates eastward, the Sun

---

*Conterminous means "contiguous or adjoining," so the conterminous United States is the original 48 states. It is different from the continental United States. Why?

†The UTC abbreviation was a compromise between the English (CUT) and French (TUC) language abbreviations.

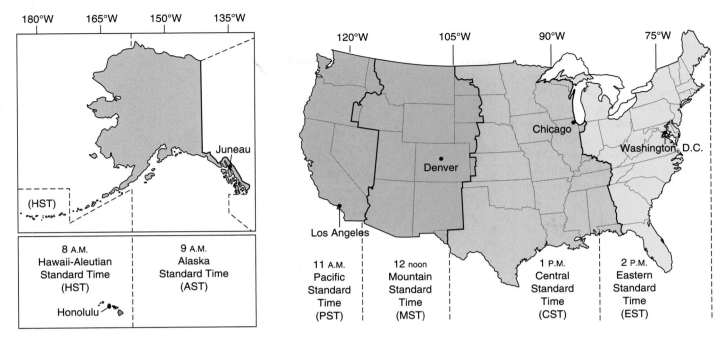

Figure 15.6 Time Zones of the Conterminous United States
The conterminous United States spans four time zones. Thus, there is a 3-hour difference between the East Coast and the West Coast. Alaska and the Aleutian and Hawaiian Islands are in different, earlier time zones.

appears to move westward, taking 12 noon with it. Twelve midnight is 180°, or 12 h from the Sun, and as 12 noon moves westward, 12 midnight follows, bringing the new day.

When traveling west into an adjacent time zone, the time kept on your watch will be 1 hour later than the standard time of the westward zone. Therefore, you would have to set the hour hand of your watch 1 hour earlier, for example, from 6 P.M. to 5 P.M., to have the correct standard time. This process will be necessary again and again as you continue west through additional time zones. A trip all the way around the Earth in a westward direction would mean an apparent loss of 24 h, or one complete day. When you travel east, the

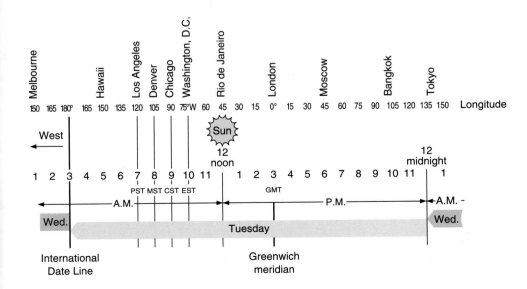

Figure 15.7 Diagram Showing the Times and Dates on the Earth for Any Tuesday at 10 A.M. EST
As time passes, the Sun appears to move westward; thus, 12 noon moves westward, and midnight follows (180° or 12 h) behind, bringing the new day, Wednesday, with it.

Figure 15.8 The International Date Line (IDL)
Theoretically, the 180° meridian is the IDL, but variations are made to accommodate various countries. At the top, the zigzag keeps parts of Siberia from being on different days (dates). It also keeps Alaska and the Aleutian Islands on the same day, with Siberia being one day (date) ahead of Alaska. Other variations occur for the islands in the South Pacific, as shown in the blow-up view.

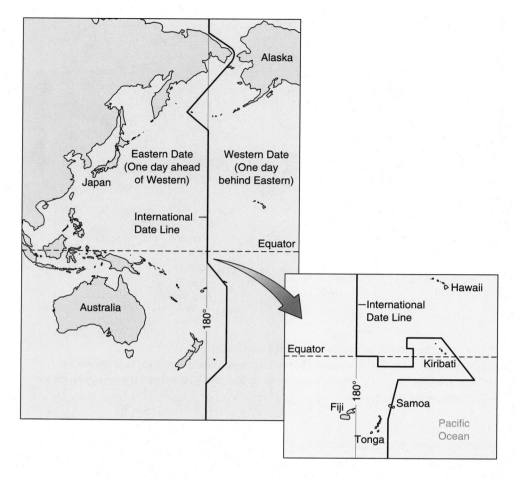

opposite is true; that is, your watch will be 1 h slow for each zone and must repeatedly be set ahead 1 h.*

A better understanding of why a day is apparently lost in traveling around the Earth in a westward direction can be obtained by taking a make-believe trip. Suppose you leave Washington, D.C., by jet plane at exactly 12 noon local solar time on Tuesday and then fly west at a speed equal to the apparent westward speed of the Sun. Because Washington, D.C., is located at 39°N latitude, the plane would have to travel along the 39°N parallel west at about 1300 km/h (800 mi/h).

On take off, you notice that the Sun is out the left window of the plane, or toward the south. One hour after leaving the airport, you can still see the Sun out the left window. Six hours later, with your watch indicating 6 P.M., the Sun still has the same position as observed from the left window of the plane. Because the plane is flying at the same apparent speed as the Sun, the Sun will continue to be observed at the same position out the left window of the plane.

Twenty-four hours later you arrive back in Washington, D.C., with the Sun in the same apparent position. During your 24-h trip, the time remained at 12 noon local solar time (the Sun always on your meridian). The Sun was not seen to rise or set, so it is still 12 noon on Tuesday, but your friends meeting you at the airport will stubbornly insist that it is 12 noon on Wednesday. Something is wrong.

To remedy such situations, the **International Date Line (IDL)** was established on the 180° meridian (● Fig. 15.8). When one crosses the IDL traveling westward, the date is

---

*For historical or political reasons, there are several variations in the 15° time zone designation. Newfoundland, Venezuela, Iran, and parts of India and Australia, for example, are on a half-hour time zone; China, which would span five time zones, has the same time throughout the country, that of Beijing, the capital.

advanced to the next day; when one crosses the IDL traveling eastward, one day is subtracted from the present date.

Note in Fig. 15.8 that the IDL does not follow the 180° meridian line exactly; rather, it zigzags to accommodate the needs of various countries in the Pacific Ocean (much like time zone boundaries). A large deviation is shown in the inset. In 1995, the island nation of Kiribati came into existence and unilaterally relocated the IDL to the east of the country, a change that is now generally accepted. Also note that the IDL separates the islands of Tonga and Samoa. These islands are in the same time zone and so have the same time, but they are one day (date) apart.*

Conceptual Question and Answer

Polar Time

Q. What is the time and date at the North Pole?

A. Since all the time zone boundaries converge at the North and South Poles, many answers would seem possible. To solve this conundrum, it was internationally agreed that Coordinated Universal Time (UTC)—that is, the time and date of Greenwich, England—be used. It is the same time and date at both the top and the bottom of the world.

If the standard time is known in one time zone, then the standard time in another zone can be determined by remembering that there are 15° or 1 hour for each time zone. Should the calculation extend through midnight or the International Date Line be crossed, the date would change.

One problem of practical importance is finding the time and date in a distant city when you know your local standard time and date. This situation may be encountered when you are trying to make a long-distance telephone call and don't want to awaken someone in the middle of his or her night.

EXAMPLE 15.2   Finding the Standard Time and Date at Another Location

It is 6 A.M. on March 21 in Los Angeles (34°N, 118°W). What are the time and date in Perth, Australia (32°S, 115°E)?

Solution

Step 1

Draw a diagram of the Earth's equator (as viewed from above) with the 15° time zones as shown in ● Fig. 15.9. (A recommended procedure follows this example.)

Step 2

You can see that Los Angeles is in the time zone with the 120°W central meridian and Perth is in the time zone with the 120°E central meridian. (A time zone extends 7.5° on each side of a central meridian.) Now simply count around going westward and subtract 1 hour for each time zone. (If it is shorter to count eastward, then add 1 hour for each time zone.)

As can be seen, it is 10 P.M. on March 21 in Perth. Note that it is March 22 in some time zones. As the Sun moves westward, the midnight line does too (on the opposite side of the Earth), bringing with it a new day (date) as it passes into a time zone. (The latitudes of the locations were not considered. Why?)

---

*At the end of 2011, Samoa plans to skip Dec. 31 and jump ahead one day and move to the western side of the IDL, so as to have the same date as major trading partners Australia and New Zealand. American Samoa, just east of these islands, will remain on the eastern side of the IDL.

Figure 15.9 Finding the Time and Date
A diagram for Example 15.2.

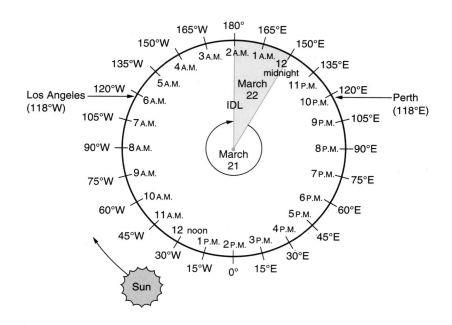

Confidence Exercise 15.2

Determine the standard time and date at 130°W when it is 12 noon on July 4 at 82°W longitude. (The procedure following this question might be helpful.)

Answers to the Confidence Exercises may be found at the back of the book.

A Recommended Procedure for Solving Time and Date Problems

1. Draw a circle and add tick marks to divide it into quarters (for 0°, 90°E, 90°W, and 180°).

2. Divide the four sections into halves with a tick mark, and then add two more dividing marks in each section, giving 24 marks to represent the central meridians of the 24 different 15° time zones. Draw a line from the center of the circle to the 180° mark to indicate the International Date Line (IDL).

3. Label the longitudes of the central meridians around the circle. (*Note:* Analyze the problem first. If the two locations of interest are relatively close together, then you may not need to label the complete circle but only the part with the locations, thus saving a little *time*. Pun intended.)

4. Locate the longitudes of the places of interest and note the nearest time zone central meridians. Using the known time, count around the time zones and find 12 noon. Then draw a midnight line from the center of the circle to the longitude opposite noon to locate where the date changes and which time zones have which date.

   As the Sun (noon) moves westward, the midnight line also moves, bringing the new date to each time zone as it moves into it. The new date appears to "come out" of the 180° line, whereas the old day is "pushed" into it.

5. Then count around to the central meridian of the time zone in which the place of interest is located, and you now have the time and date.

Because most watches and clocks keep standard time, the time shown on these watches and clocks around the world will display the same minute but a different hour. For example, if your watch shows 20 minutes past 10:00 A.M. (10:20 A.M.), then most clocks worldwide will display 20 minutes past some other hour. Many airports and hotels have six or more clocks on a wall, showing the time at certain major cities around the world.

During World Wars I and II, the clocks of many countries were set ahead 1 hour during the summer months to give more daylight hours in the evening, thus conserving fuel used to generate electricity for lighting. This practice has now become standard for all but a few

Highlight   Daylight Saving Time

Every year, most of us go through the tedious process of setting our clocks to Daylight Saving Time (DST; note that the S stands for *Saving*, not *Savings*). We "spring ahead" in the spring, advancing the clocks 1 hour, and then we "fall back" in the fall and set the clocks back 1 hour to return to standard time. (One thing this ritual does is make us realize how many clocks we have and appreciate that atomic clocks, computers, and cell phones set themselves.)

The idea of DST is credited to Ben Franklin, who proposed it in 1784 in an essay entitled "An Economical Project," published in Paris. Franklin called for a tax on every Parisian window shuttered after sunrise to "encourage the economy of using sunlight instead of candles." Not much came of it until many years later. Germany began observing DST during World War I; Great Britain followed shortly thereafter. The United States waited until 1918. There was strong opposition, particularly from farmers who usually worked schedules on sun time and who had cows that didn't like being milked an hour off schedule. After the war, Congress passed a bill to do away with DST.

What, then, are the advantages of Daylight Saving Time? Probably the most important is that it saves energy. (Franklin wanted to save candles.) In going on DST, we move our clocks forward 1 hour. In effect, DST moves 1 hour of daylight from the morning to the evening. Then when coming home from work or school, there is an "extra" hour of daylight to engage in outdoor activities and not turn on inside lights, television sets, computers, and so on.

Another advantage of Daylight Saving Time is that people travel home from work or school in daylight hours, and because visibility is better in daylight, there are fewer traffic accidents. Also, because there are more daylight hours in the evening, DST is also said to reduce crime, which is more common after dark.

Daylight Saving Time was instituted in many countries again during World War II specifically to save energy. In the United States, President Franklin D. Roosevelt instituted year-round "War Time," putting the entire country on DST from 1942 until 1945. In Great Britain, clocks were advanced 2 hours during the summer and 1 hour during the winter.

After World War II, states and localities in the United States had a choice—go on DST or not go on DST—which caused much controversy and confusion. Coordinating railroad, bus, and plane schedules among various cities and states was a problem. A traveler might go in and out of DST zones several times on a trip, which made departure and arrival times confusing.

In 1966, Congress passed the Uniform Time Act. This law established a system of uniform DST throughout the United States and its possessions, exempting only those in which the state governments voted to keep the entire state or possession on standard time. Today, *not* going on DST are Arizona (with the exception of the Navajo Nation, which lies in three states: Arizona, Utah, and New Mexico), Hawaii, Puerto Rico, the U.S. Virgin Islands, Guam, and American Samoa. Arizona gets very hot in the summer, and many people there don't want an extra hour of daylight for the air conditioners to run. For the islands, the reason is different; the length of the day does not vary much over the year. A 1972 revision allowed a state that had two or more time zones to exempt part of the state provided that the other went on DST. There have been exceptions along the way, Indiana being the most notable. Until a few years ago, most of Indiana's counties remained on standard time. The other counties, mostly near large cities in other states, have been using DST since the 1970s. The entire state of Indiana joined DST in 2006, although all counties are not in the same time zone.

The national DST system has also undergone a recent change. No longer does the country spring ahead on the first Sunday in April and fall back on the last Sunday in October, as it did for 40 years. Congress passed an energy bill that included expanding the period for DST by several weeks. Starting in 2007, DST is in effect from the second Sunday in March until the first Sunday in November.

Today, more than one billion people in approximately 70 countries around the world observe DST in some form, although the times for going on and off DST vary greatly.

of the 50 U.S. states. This time during the summer months in the United States is known as **Daylight Saving Time (DST)**. (See the **Highlight: Daylight Saving Time**.)

Did You Learn?

- The times of 12 o'clock are correctly stated as 12 noon and 12 midnight.
- The International Date Line is the meridian (180° longitude) located in the Pacific Ocean where the calendar date changes out of necessity for timekeeping.

15.4   Determining Latitude and Longitude

Preview Questions

- What is meant by the altitude of the Sun?
- How are longitude and latitude determined?

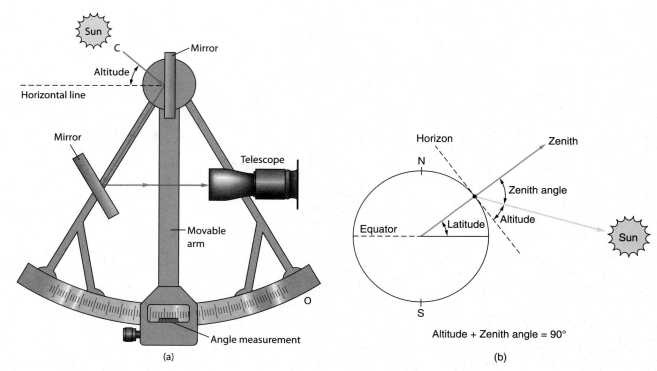

Figure 15.10 Locating the Sun
(a) The sextant used to "shoot the Sun" and determine the altitude angle, the angle of the Sun above the horizon. (b) When one shoots (a line) to the Sun, the zenith is perpendicular to the horizon, so the zenith angle plus the altitude equals 90°.

Imagine being the captain of a three-masted sailing ship in the late 1700s and being somewhere on the ocean with no land in sight. To reach your destination, it is imperative to know where you are, so a reliable position, or "fix," is needed to reasonably approximate your latitude and longitude. How can you determine your position? (There were, of course, no GPS satellites in those days.)

Prior to the 1700s, navigational methods were relatively crude. Compasses were available, and the celestial positions of stars and the Moon were generally known. Without maps, though, early explorers sailed in a particular direction, guessing how far they had gone by "dead reckoning," that is, by estimating their position from the starting point, using the direction and the distance traveled (from estimated speed and time; see Chapter 2.2), with corrections for wind and drift. The fervent hope was that someone in the crow's nest would eventually call out, "Land ho!"

As the world was explored, rather crude maps became available, but a ship's position on a map still needed to be known to navigate efficiently. Fortunately, a couple of major developments enabled navigators to determine the positions of ships at sea.

Latitude

One of these developments was the sextant (● Fig. 15.10a). By "shooting the Sun" with this instrument, it was possible to measure accurately the **altitude** of the Sun (its angle above the horizon, Fig. 15.10b). The **zenith angle** is the complementary angle of the altitude. The zenith is the point directly over an observer's head. The zenith is 90° from the horizon, and the sum of the zenith angle and altitude is 90°.

Knowing the Sun's noon altitude and date, one can determine latitude. The date is important in knowing where (at what latitude) the noonday Sun is overhead at the time. For those who live in northern midlatitudes, you have no doubt noticed that in summer the Sun is high in the sky and in winter it is lower (similar for southern midlatitudes with the reverse of seasons). This difference is a result of the tilt of the Earth's axis, which is $23\frac{1}{2}°$ from the vertical to its orbital plane. As the Earth revolves about the Sun, this tilt causes the noonday Sun to be overhead at different latitudes at different times of the year. (The axis tilt is also responsible for the seasons, which is discussed in the next section.)

● Figure 15.11 shows the noonday overhead latitude positions of the Sun at different times of the year. The farthest north latitude, $23\frac{1}{2}°N$, occurs on June 21 or 22. It is the longest day of the year for the Northern Hemisphere. The farthest overhead south latitude, $23\frac{1}{2}°S$, occurs on December 21 or 22, which is the shortest day of the year for the Northern Hemisphere. Between these dates, the noonday Sun appears to move to intermediate latitudes. Because of the Earth's orbit, the axis tilts toward and away from the Sun, and we see the Sun higher and lower in the sky, respectively.*

Let's illustrate finding latitude graphically. Suppose that the Sun is to the south and its noon altitude is measured to be 50° on June 21. It is known that the noonday Sun is overhead at $23\frac{1}{2}°N$ latitude on this date. These data are represented in the drawing in ● Fig. 15.12. The procedure for making such a drawing is summarized in ● Table 15.1. Because the altitude is 50° and the Sun is to the south, you know that you are in the Northern Hemisphere. (Why?)

The rays of the Sun that reach the Earth are parallel because the Sun is so far away. The zenith angle is defined to be 90° minus the Sun altitude, or 40°. By similar angles relative to the zenith line, the interior angle at the center of the Earth between the Sun and the zenith is also 40°. Adding the two interior angles between the equator and zenith lines gives the latitude of the observer at $63\frac{1}{2}°N$.

By studying Fig. 15.12 and Table 15.1, you can see a relationship.

Latitude = zenith angle + degrees overhead the Sun in your hemisphere

or

Latitude = zenith angle − degrees overhead the Sun in other hemisphere

In simplified form,

$$L = ZA \pm \text{Sun degrees} \qquad 15.1$$

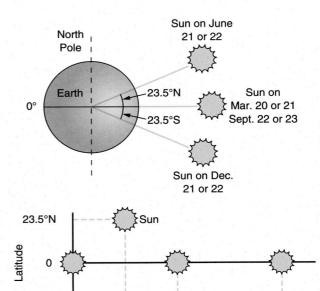

Figure 15.11 Diagrams of the Sun's Position (Degrees Latitude) at Four Different Times of the Year
The upper drawing shows a greatly magnified Earth with respect to the Sun, illustrating the Sun's overhead positions on the dates indicated. The lower graph plots the Sun's position (degrees latitude) versus time (months).

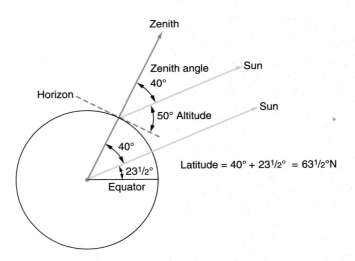

Latitude = 40° + 23¹/2° = 63¹/2°N

Figure 15.12 Determining Latitude
By knowing the date and "shooting the Sun" with a sextant to determine the altitude of the noonday Sun, it is possible to determine the latitude. (See text for description.)

---

*Because of slight variations in the Earth's orbit, the key overhead positions of the Sun can occur on adjacent dates. For our purposes and ease of remembering, the following dates will be used: March 21, June 21, September 22, and December 22. (Two 21s and two 22s.)

| Table 15.1 | Graph Procedure for Determining Latitude |

Example: Determine the latitude when the measured Sun altitude is 60° on a date when the noon Sun is overhead at 15°N.

1. Draw a circle and a horizontal equator line from the center of the Earth.

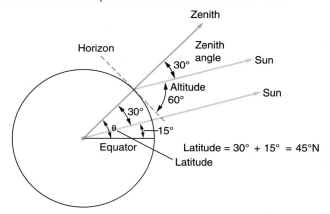

2. Estimate where the observer might be on the circle and draw a line from the center of the circle through this location; this line is the zenith line. From the given data, you should be able to tell which hemisphere the observer is in and to approximate (guess) the latitude.

3. Draw a line perpendicular to the zenith line through the observer's location; this line is the horizon line.

4. Shoot the Sun. (a) Draw a line from the center of the Earth toward the Sun through the latitude at which the Sun is overhead on that date. (b) Draw a line toward the Sun from the observer's position at the measured (given) altitude angle.

5. Use similar angles to determine the observer's latitude, the angle between the equator and the zenith line.

where ZA is the zenith angle. This method can be seen to work in these examples:

Figure 15.12: $L = 40°(N) + 23\frac{1}{2}°(N) = 63\frac{1}{2}°N$

Table 15.1: $L = 30°(N) + 15°(N) = 45°N$

EXAMPLE 15.3   Not in My Hemisphere!

A ship's captain measures the noonday Sun at an altitude of 30° toward the south on December 22. What is the latitude of the ship?

Solution

Let's make a drawing to illustrate this situation (● Fig. 15.13). As can be seen, with an altitude of 30°, the zenith angle is 60°; and on December 22, the noonday Sun is overhead at $23\frac{1}{2}°S$. Noting the two interior similar angles, and because the Sun is overhead in the other, or opposite, hemisphere (negative sign in Eq. 15.1), we have

$$L = ZA - \text{Sun degrees} = 60°(N) - 23\frac{1}{2}°(S) = 36\frac{1}{2}°N$$

which illustrates the use of the minus sign in Eq. 15.1.

Confidence Exercise 15.3

What would be the latitude of the ship in Example 15.3 were the altitude of the Sun $70\frac{1}{2}°$ to the north?

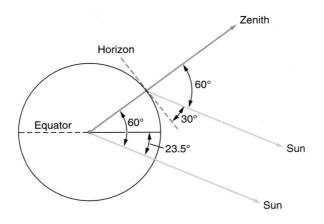

Figure 15.13 Another Case of Computing Latitude
A diagram for Example 15.3.

Latitude = 60° − 23.5° = 36.5°N

Thus, a diagram is not needed if the directional (north or south) altitude of the Sun is measured and the noon overhead position of the Sun is known on that date. How do you get the latter? Easy. We know that the overhead position of the noon Sun "travels" between the equator and $23\frac{1}{2}$°N (and S) between March 21 (September 22) and June 21 (December 22), or about 3 months, or 90 days. So, 23.5°/90 days = 0.26°/day. With this "speed" we can determine the overhead position of the noon Sun north or south of the equator on a particular date, as shown in ● Table 15.2.

For example, 20 days after the Sun is over the equator (April 10 or October 12), the overhead position of the Sun is

$$20 \text{ days } (0.26°/\text{day}) = 5°$$

north or south of the equator, depending on the time of the year.

Longitude

In theory, determining longitude is easy. For example, suppose that a ship leaves England at noon local time (the Sun on the meridian) with the ship's clock set at Greenwich Mean Time (GMT). After the ship sails for a month or so, the noonday Sun is observed when it is 2 P.M. on the clock (still set for GMT). Because the Sun "travels" westward at 15°/h, the ship's longitude position is 15°/h × 2 h = 30°W. [At 1 P.M., the noonday Sun was 15° west of the prime meridian (0°); at 2 P.M., the noonday Sun is at 30° west of 0°, or 30°W.]

This sounds simple, but the problem in the 1700s was that the clocks of the day didn't keep time very well. On long sea voyages, a loss or gain of time would result in faulty longitude determinations that could be quite far off. While the sextant solved the problem of finding accurate latitudes, an accurate clock was needed to provide reliable longitude measurements. In the mid-1760s, a metallic "marine chronometer" (actually a metal pocket watch) was developed. It was tested on a voyage from England to Barbados, and the error was 39 seconds over a voyage of 47 days. With such a watch, longitude could be determined to a good approximation.

EXAMPLE 15.4   Ahoy! Where Are We?

A sailing ship leaves England on the morning tide, bound for a coastal settlement in Virginia. After sailing for a month and a half, on December 22 the captain shoots the noonday Sun with his sextant and measures an altitude of 30° southward. He notes the time on the ship's clock to be 5 P.M. (GMT). What is the ship's position in terms of latitude and longitude?

Table 15.2	Overhead Positions of the Noon Sun at Different Times of the Year	
Times	**Degrees (latitude)**
Mar. 21 Sept. 22	0°
Apr. 10 Oct. 12	5°
Apr. 30 Nov. 1	10°
May 20 Nov. 21	15°
June 7 Dec. 8	20°
June 21 Dec. 22	23.5°

Solution

Wanted: Latitude

We know altitude = 30° and zenith angle = 60°, and on December 22, the noon Sun is overhead at $23\frac{1}{2}°$S. Then, using Eq. 15.1, with the observer and Sun in opposite hemispheres, we get

$$L = ZA - \text{Sun degrees} = 60°(N) - 23\frac{1}{2}°(S) = 36\frac{1}{2}°N$$

Wanted: Longitude

Because it is noon at the ship's position and 5 P.M. GMT, it was noon in England (0° meridian) 5 h earlier. The Sun "travels" westward at 15°/h, so it is now at longitude

$$15°/h \times 5\ h = 75°W$$

Thus, the ship is at (36.5°N, 75°W), which is about 100 mi off the Virginia coast. It may be there in time for Christmas.

Confidence Exercise 15.4

Make drawings illustrating the determination of latitude and longitude for Example 15.4.

Finally, don't forget that it is possible for the noonday Sun to be north of a position in the Northern Hemisphere (or south of a position in the Southern Hemisphere). For example, at latitude 5°N on June 21, the noonday Sun would be to the north. You would know where the Sun is overhead because of the date (Table 15.1), and a large altitude would be measured. Equation 15.1 still applies. One state, Hawaii (general latitude 21°N), has the noonday Sun overhead twice a year and would appear both to the north and to the south. (Why?)

Did You Learn?

- The Sun's altitude is its angle measured above the horizon.
- Latitude and longitude may be determined by "shooting the Sun" with a sextant and keeping track of noon with a preset clock, respectively.

15.5   The Seasons and the Calendar

Preview Questions

- What gives rise to the Earth's seasons?
- Why is our present-day calendar called the Gregorian calendar?

The Seasons

Most of us experience the change of seasons—summer, autumn, winter, and spring—with notable weather changes. The seasons result from the tilt in the Earth's axis in its yearly revolution about the Sun, as illustrated in ● Fig. 15.14. Because the Earth's orbit is slightly elliptical, it is about 5.9 million km (3.7 million mi) farther from the Sun at *aphelion* (the point farthest from the Sun) than at *perihelion* (the point closest to the Sun). To help you keep these definitions straight, note that *ap* goes with "away." This difference in distance is very small relative to the overall Earth–Sun distance and has no effect on the seasons. Furthermore, aphelion and perihelion occur in July and January, respectively, which means that the Earth is closer to the Sun in winter (in the Northern Hemisphere) than in summer. Thus, it is the Earth's tilt that gives rise to the seasons as we know them, not the Earth–Sun distance.

Note in Fig. 15.14 that the Earth's axis is tilted toward the Sun on June 21. As a result, the Northern Hemisphere receives more direct rays of the Sun, and June 21 is taken to be the beginning of summer in the Northern Hemisphere. On June 21, the noon Sun is highest in the sky for observers above $23\frac{1}{2}°$N. The altitude of the noonday Sun on June 21 can be calculated from (90° − observer's latitude) + $23\frac{1}{2}°$. For example, for midnorthern latitudes, about 40°N, the altitude of the Sun on June 21 is 50° + $23\frac{1}{2}°$ = $73\frac{1}{2}°$.*

Thus, on June 21, the Sun has reached its greatest altitude for latitudes above $23\frac{1}{2}°$N; this **summer solstice** (*solstice* meaning "the Sun stands still") marks the beginning of summer. It is the longest day of the year in the Northern Hemisphere. The **winter solstice**, the beginning of winter in the Northern Hemisphere, occurs on December 22. For latitudes above $23\frac{1}{2}°$N, the noonday Sun is at its minimum altitude, and it is the shortest day of the year. In the Southern Hemisphere, however, the noonday Sun has reached its maximum altitude for latitudes greater than $23\frac{1}{2}°$S. The Southern Hemisphere receives the more direct rays of the Sun, and it is "summer" weather there in winter. As pointed out, because the Earth's orbit is slightly elliptical, it is closer to the Sun in the Northern Hemisphere's winter (Fig. 15.14).

Intervening between summer and winter is the **autumnal equinox** (*equinox* meaning "equal nights") on September 22, the beginning of fall. At that time, the noonday Sun is directly overhead at the equator, and there are approximately equal hours of daylight and darkness. Equinox means "equal night." The **vernal equinox**, the beginning of spring, occurs on March 21, when the Sun is again over the equator.

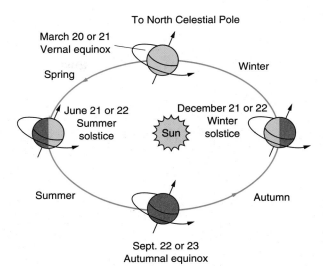

Figure 15.14 The Earth's Positions Relative to the Sun and the Four Seasons
As the Earth revolves around the Sun yearly, its north–south axis remains pointing in the same direction. On March 20 or 21 and September 22 or 23, the Sun is directly above the equator. On June 21 or 22, the noonday Sun is overhead at $23\frac{1}{2}°$N and the Northern Hemisphere has more daylight hours than dark hours. On December 21 or 22, the noonday Sun is overhead at $23\frac{1}{2}°$S and the situation is reversed. The Sun is drawn slightly off center to indicate that the Earth is slightly closer to the Sun during winter in the Northern Hemisphere than during summer.

Conceptual Question and Answer

Equal Days and Nights

Q. Are the days and nights equal on the equinoxes?

A. You often hear that day and night are of equal length on the spring and fall equinoxes, but this is not the case. The Sun is not a small point of light; rather, it is a disk with an angular size of about $\frac{1}{2}°$. Sunrise is defined as the moment when the top edge of the Sun appears on the horizon, and sunset is when the Sun dips below the horizon. In addition, the atmosphere refracts the sunlight (see Chapter 7.2, Fig. 7.10), and when close to the horizon, the rays from the Sun are seen a few minutes ahead or later than the "true" sunrise and sunset, respectively.

The cumulative effects make the day about 14 minutes longer than at night at the equator and still longer toward the North and South Poles. The equality of day and night only happens in places that have seasonal differences in day and night. As winter progresses toward the vernal equinox, the days get longer and the nights shorter (and opposite as summer progresses toward the autumnal equinox). At some point the lengths match, which occurs a few days before March 21 and a few days after September 22.

Because the times of sunrise and sunset vary with an observer's geographic location (latitude and longitude), the equal day–equal night depends on location and does not exist for locations close to the equator. However, an equinox is the precise moment the center of the Sun is over the equator, which is common for all observers on the Earth.

*Because of slight variations in the Earth's orbit, the solstices and equinoxes can occur on different dates; they are the dates of the overhead positions of the Sun considered in the previous section (Chapter 15.4). Similar dates will be taken here: March 21, June 21, September 22, and December 22.

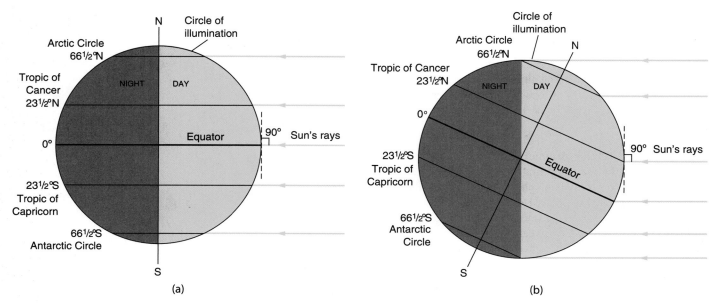

Figure 15.15 Circle of Illumination
(a) At the time of the vernal equinox (March 21), the Sun is directly overhead at the equator and half of the Earth is illuminated from North Pole to South Pole, which is referred to as the circle of illumination. (b) At the time of the summer solstice (June 21), the boundary of the circle of illumination runs from the Arctic Circle to the Antarctic Circle. These two conditions lead to different hours of daylight and darkness at different locations. (See text for description.)

To designate the latitudes of the solstices, the parallel circle of $23\frac{1}{2}°$N is called the *Tropic of Cancer*. *Tropic* means "turning place," and the noonday Sun appears to turn and travel to lower latitudes in the background of the constellation (star pattern) Cancer. Similarly, the parallel $23\frac{1}{2}°$S is called the *Tropic of Capricorn*.*

Other designations occur for the $66\frac{1}{2}°$N and $66\frac{1}{2}°$S parallels. At the time of the summer solstice, all locations above $66\frac{1}{2}°$N experience 24 hours of daylight. This parallel is called the *Arctic Circle*. Similarly, in the Southern Hemisphere at the time of the winter solstice, all locations above $66\frac{1}{2}°$S experience 24 h of daylight, and the $66\frac{1}{2}°$S parallel is called the *Antarctic Circle*.

During the spring and summer months, the North Pole has 24 hours of daylight each day. The Sun never sets. To better understand the lengths of days and nights, consider ● Fig. 15.15a. At the vernal equinox (March 21), the Sun is directly overhead at the equator and illuminates one half of the Earth as usual. The lighted portion is referred to as the *circle of illumination*. (Analogously, the illuminated half of the Moon at full moon appears as an illuminated circle.) With the illuminated and dark portions being equal in each hemisphere, as the Earth makes one rotation there are 12 hours of daylight and 12 hours of darkness for all locations. The situation is the same for the autumnal equinox (September 22).

Now look at Fig. 15.15b, which illustrates the summer solstice (June 21). The Sun is then over the Tropic of Cancer ($23\frac{1}{2}°$N). Notice that the boundary of the circle of illumination between daylight and dark (called the *terminator*; see the chapter-opener photo) runs from the Arctic Circle ($66\frac{1}{2}°$N) to the Antarctic Circle ($66\frac{1}{2}°$S). Then, as the Earth rotates, all locations from $66\frac{1}{2}°$N to 90°N have 24 hours of daylight. The Sun doesn't rise or set, but instead goes up and down in a path around and above the horizon (the Land of the Midnight Sun). Conversely, all locations from $66\frac{1}{2}°$S to 90°S have 24 hours of darkness.

---

*The Tropic of Cancer ($23\frac{1}{2}°$N) runs through Mexico, the Bahamas, Egypt, Saudi Arabia, India, and southern China. The Tropic of Capricorn ($23\frac{1}{2}°$S) runs through countries in southern Africa, Australia, Chile, and southern Brazil (Brazil is the only country through which the equator and a tropic pass).

Comparing Fig. 15.15a and Fig. 15.15b, you can see what happens during spring and summer. After March 21, the top of the terminator of the circle of illumination "rotates" from 90°N to $66\frac{1}{2}$°N, bringing varying degrees of daylight and darkness to locations above the Arctic Circle until it reaches $66\frac{1}{2}$°N on June 21 (Fig. 15.15b). Then there are 24 hours of daylight above the Arctic Circle. After June 21, the terminator "rotates" back, and on September 22 (the autumnal equinox), all locations have 12 hours of daylight and 12 hours of darkness again. Can you tell what happens after that?

Conceptual Question and Answer

Hot and Cold Weather

Q. On June 21, the length of the daytime in Washington, D.C., is about 14 hours, but it is 24 hours near the North Pole. Why isn't the temperature hotter at the North Pole?

A. The angle at which the sunlight hits the ground is the real factor. When the Sun is high in the sky, the Sun's rays are more intense (more energy hits the ground per area per time), thus warming the land and the water. On June 21 at the North Pole, the Sun's altitude is only $23\frac{1}{2}$° and the rays are not very intense. Thus, the land (ice) does not warm up very much, and the temperatures remain cold.

The rotating Earth revolves around the Sun in an orbit that is elliptical, yet nearly circular. When the Earth makes one complete orbit about the Sun, the elapsed time is known as one *year*. We will be concerned with two different definitions of the year. The **tropical year**, or year of seasons, is the time interval from one vernal equinox to the next vernal equinox; in other words, it is the elapsed time between one northward crossing of the Sun above the equator (vernal equinox) and the next northward crossing. With respect to the rotational period of the Earth, the tropical year is 365.2422 mean solar days.*

The **sidereal year** is the time interval for the Earth to make one complete revolution around the Sun with respect to any particular star other than the Sun (sidereal means "pertaining to stars"). The sidereal year is equal to 365.2536 mean solar days and is approximately 20 min longer than the tropical year. The reason for this difference is due to the precession of Earth's axis (Chapter 15.6).

The Calendar

The measurement of time requires the periodic movement of some object as a reference. The first unit for the measurement of time was probably the day, as defined by the periodic apparent motion of the Sun and perhaps measured from sunrise to sunrise. Then there is the periodic change of the Moon's phases, which gives a month as a time unit. On a yearly basis, there is the change in seasons (for some latitudes) and the periodic apparent motion of the constellations of the zodiac, used as an early calendar, the *zodiacal calendar*.

The *zodiac* is a central, circular section of the celestial sphere that is divided into 12 sections (● Fig. 15.16). Each section is identified by a prominent group of stars called a *constellation*. Ancient civilizations named constellations for the figures the stars seemed to form, such as Leo (the Lion) and Taurus (the Bull).

Because of the Earth's revolution, the 12 constellations, or "signs," of the zodiac periodically change in the night sky during the course of a year. Thus, a particular constellation's appearance in the night sky marked a particular time of the year. Perhaps

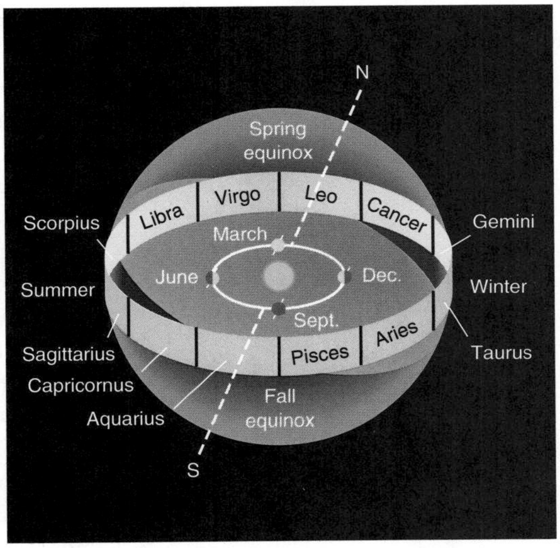

Figure 15.16 Signs of the Zodiac
The drawing illustrates the boundaries of the zodiacal constellations. Each of the 12 sections of the zodiac is 30° wide and 16° high, or 8° above and below the central plane.

---

*During a tropical year, the overhead Sun makes one cycle between the tropics, the Tropic of Cancer ($23\frac{1}{2}$°N) and the Tropic of Capricorn ($23\frac{1}{2}$°S).

Table 15.3	The Days of the Week	
Celestial Object	**English Name**	**Saxon Name**
Sun	Sunday	Sun's day
Moon	Monday	Moon's day
Mars	Tuesday	Tiw's day
Mercury	Wednesday	Woden's day
Jupiter	Thursday	Thor's day
Venus	Friday	Fria's day
Saturn	Saturday	Saturn's day

when Libra appeared, it was time to plant crops, for example. Because of the Earth's precession (slow rotation of the Earth's axis, see Chapter 15.6), however, the 12 constellations change, slowly appearing at different months after many hundreds of years.

The Moon has probably been the greatest contributor to the development of the calendar. The length of our month, which originated from the periodic phases of the Moon, is based on the 29.5 solar days it takes the Moon to orbit the Earth.

The first calendar seems to have originated before 3000 BCE with the Babylonians. This calendar was based on the motion and phases of the Moon, which divided the year into 12 lunar months consisting of 30 days each. Because $30 \times 12 = 360$ days and the year actually has approximately 365.25 days, corrections had to be made to keep the calendar in sync with the seasons. The Babylonians adjusted the length of the months and added an extra month when needed. This Babylonian calendar set the pattern for many of the calendars adopted by ancient civilizations. For more history of the calendar, see the **Highlight: A Brief History of the Calendar**.

On the calendar there are seven days in a week. The origin of the seven-day week is not definitely known, and not all cultures have had it. One possible origin is that it takes approximately seven days for the Moon to go from one phase to the next (for example, from new to first-quarter phase).

Another theory involves the nighttime sky. As the ancients watched the sky night after night, they saw that seven celestial bodies moved relative to the fixed stars. These seven objects visible to the unaided eye are the Sun, the Moon, and five visible planets: Mars, Mercury, Jupiter, Venus, and Saturn.

● Table 15.3 lists the celestial objects and the English and Saxon names for the seven days. Our Tuesday, Wednesday, Thursday, and Friday come from Tiw, Woden, Thor, and Fria, in Nordic mythology. Woden was the principal Nordic god, Tiw and Thor were the gods of law and war, and Fria was the goddess of love. (Apparently, we should be thanking a goddess it's Friday.) Sunday, Monday, and Saturday retain their connection to the Sun, the Moon, and Saturn.

One final note on the calendar: It is common to designate years using BC and AD (for example, 300 BC and AD 450). The BC stands for "before Christ," and AD for *anno Domini* or "in the year of our Lord." The years are designated as being before or after the birth of Jesus Christ. The Gregorian calendar is sometimes also called the Christian calendar.

Now, new designations are used (as we have done in this book), such as 300 BCE and 450 CE. The BCE stands for "before the Common Era" and is identical to BC. The CE stands for "Common Era" and corresponds to AD. The phrase *Common Era* refers to the most commonly used internationally accepted civil calendar system, the Gregorian calendar. Keep in mind that there are other calendars too, such as the Chinese and Muslim calendars. Check them out. It makes for interesting reading.

Did You Learn?

- The Earth's seasons result from the tilted axis of the Earth as it orbits the Sun each year.

- The current international civil calendar is the Gregorian calendar, so named because Pope Gregory XIII instituted a method to keep the calendar more accurate.

Highlight   A Brief History of the Calendar

The ancestor of the calendar we use today originated with the Romans. The early Roman calendar contained only ten months, and the year began with the advent of spring. The months were named March, April, May, June, Quintilis, Sextilis, September, October, November, and December.* The months of January and February did not exist. Around 713 BCE, these months were added to the calendar.

In 46 BCE, Julius Caesar introduced a new solar-based calendar that he had learned about in Egypt. It was a vast improvement over the lunar calendar, which had become very inaccurate over the years. The month of Quintilis, the month of Caesar's birth, was renamed July in his honor.

This so-called Julian calendar had 365 days in a year. In every year evenly divisible by 4 (a leap year), an extra day was added to make up for the approximately 365.25 days the Earth takes to orbit the Sun. For example, 2008 is a leap year and has 366 days with an extra day in February, February 29. The years 2009, 2010, and 2011 are not evenly divisible by 4 and have the usual 365 days. The Julian calendar was fairly accurate and was used for more than 1600 years.

There was an initial discrepancy, however. For several decades, a leap year was observed every 3 years instead of every 4 years. Emperor Augustus, the adopted nephew of Julius Caesar, corrected the problem. He abolished all leap years between 8 BCE and CE 8, which put the calendar back on track. Augustus changed the month Sextilis to August in honor of himself. He also removed one day from February, which he added to August so that his month would have the same number of days as his uncle's month of July. Quite a bit of juggling went on with the number of days in the months, at 29, 30, and 31 days. Augustus settled this issue and established the months the lengths they have today, with a short 28 days for February.

The Julian calendar was still not quite accurate, though, because the tropical year is not 365 days and 6 hours (365.25 days) but rather is approximately 365 days, 5 hours, and 46 seconds (365.2422 days). Therefore, the Julian calendar was 11 minutes and 46 seconds too slow, which added up to a loss of about a full day every 128 years.

After hundreds of years, this discrepancy began showing up and dated events didn't occur at the right time of the year or season. For example, in the sixteenth century, the vernal equinox occurred on March 11 and religious holidays were occurring at the wrong times. By 1582, there was a 10-day discrepancy and adjustments needed to be made. Pope Gregory XIII did just that. The pope decreed that 10 days on the calendar would be skipped to get the calendar back in sync with the seasons; October 4, 1582, would be followed by October 15, 1582.

Pope Gregory XIII also instituted a method to keep the calendar more accurate. To keep it from getting ahead, he decreased the number of leap years. Every 400 years, 3 leap years would be skipped; these years were the century years not evenly divisible by 400. For example, the years 1700, 1800, and 1900 were not leap years, but the year 2000 was. This method took away 3 "leap days" every 400 years, or 1 day per 133.3 years, which is close to the correction of 1 day per 128 years. It also made the calendar accurate to 1 day in 3300 y. Our present-day calendar with these leap-year designations is called the **Gregorian calendar**.

Even though the Gregorian calendar is quite accurate, scientists add a "leap second" to their atomic clocks when needed. As noted in the previous (Chapter 15.2) Highlight: Global Positioning System (GPS), atomic clocks are very important in locating positions on the Earth accurately.

The adoption of the Gregorian calendar was not immediate. At first, only predominantly Catholic countries adopted the new calendar. Great Britain and its colonies made the adoption in 1752, Russia in 1919 (after the revolution), and China in 1949 (after another revolution). In some countries, there was rioting, with people believing that the pope had stolen 10 days of their lives!

There is an interesting sidelight to the calendar change. After the change, dates were written with O.S. (Old Style) or N.S. (New Style) to make it clear whether records and other dated material had a Julian or a Gregorian date. It is sometimes said that George Washington wasn't born on February 22, because prior to the Gregorian calendar, the new year began on March 25, but that was changed to January 1 on the new calendar. George Washington was born on February 11, 1731 (O.S.), but his birthday becomes February 22, 1732 (N.S.), on the Gregorian calendar, with the year of his birthday coming more than a year later.

*After June, the names of the months indicate their numbers. In Latin, they are *quinque*, 5; *sex*, 6; *septum*, 7; *octo*, 8; *novem*, 9; and *decem*, 10, December being the tenth month on the early Roman calendar.

15.6 Precession of the Earth's Axis

Preview Questions

- How long is the Earth's precession period?
- What is the effect of the Earth's precession?

Many of us are acquainted with the action of a toy top that has been set in rapid motion and allowed to spin about its axis. After spinning a few seconds, the top begins to wobble

Figure 15.17 Precession of a Top
The axis of rotation precesses in a circular path.

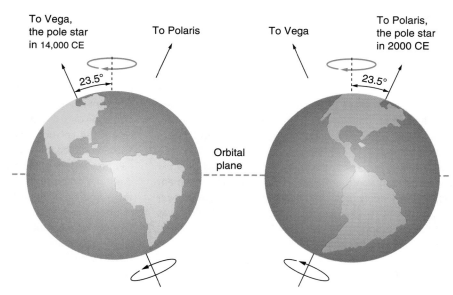

FIGURE 15.18 Precession of the Earth's Axis
Right: The Earth's axis is presently pointing toward the star Polaris, which we call the pole star or North Star. *Left:* In approximately 12,000 years, the axis will again be pointing toward the star Vega in the constellation Lyra.

or do what physicists call *precess* (● Fig. 15.17). The top, a symmetric object, will continue to spin about a vertical axis if the center of gravity remains above the point of support. When the top slows down and the center of gravity is not in a vertical line with the point of support, the axis starts to slowly rotate. This slow rotation of the axis is called **precession**.

Because the Earth is rotating, it bulges at the equator and is not a perfect sphere. The Moon and the Sun apply a gravitational torque on the Earth, and because of this torque, the Earth's axis slowly rotates clockwise as viewed from above (● Fig. 15.18). The period of the precession is 25,800 y; that is, it takes 25,800 y for the axis to precess through 360°.

Currently, the rotational axis (out of the North Pole) points toward Polaris, the pole star or North Star. As the axis precesses, it will point toward different stars, which will then successively become the North Star. The precession is very slow, however, and Polaris will be the North Star for some time to come.

The precession of the Earth's axis has little current effect on the seasons. Today in the Northern Hemisphere, winter occurs when the Earth is closest to the Sun and moving fastest (conservation of angular momentum, Chapter 3.7). As a result, the winter season is several days shorter than the summer. Because of axis precession, over the next 10,000 years the winter in the Northern Hemisphere will gradually lengthen and summer will become shorter. Some scientists believe that this change will create conditions likely to initiate another ice age.

Did You Learn?

● The period of the Earth's precession is 25,800 years.

● Earth's precession gives rise to stars slowly shifting positions in the sky; Polaris will not always be the North Star.

KEY TERMS

1. Cartesian coordinate system (15.1)
2. parallels (15.2)
3. meridians

4. great circle
5. latitude
6. longitude

7. Greenwich (prime) meridian
8. solar day (15.3)
9. sidereal day

10. Greenwich Mean Time (GMT)
11. Coordinated Universal Time (UTC)
12. International Date Line (IDL)
13. Daylight Saving Time (DST)
14. altitude (15.4)

15. zenith angle
16. summer solstice (15.5)
17. winter solstice
18. autumnal equinox
19. vernal equinox

20. tropical year
21. sidereal year
22. Gregorian calendar
23. precession (15.6)

MATCHING

For each of the following items, fill in the number of the appropriate Key Term from the preceding list. Compare your answers with those at the back of the book.

a. _____ Circles parallel to the equator

b. _____ Designates the beginning of winter

c. _____ Time used to save energy

d. _____ The zero meridian

e. _____ Occurs on or near March 21

f. _____ Angular measurement in degrees east or west of the prime meridian

g. _____ 90° − altitude angle

h. _____ The time of one *revolution* of the Earth with respect to a star other than the Sun

i. _____ A rectangular coordinate system

j. _____ The angle of the Sun above the horizon

k. _____ The time of one *rotation* of the Earth with respect to a star other than the Sun

l. _____ The equator, for example

m. _____ Designates the beginning of summer

n. _____ Time referenced to atomic clocks

o. _____ Angular measurement in degrees north and south of the equator

p. _____ Occurs on or near September 22

q. _____ The theoretical 180° meridian

r. _____ The year of the seasons

s. _____ Half circles that are portions of a great circle

t. _____ The time of one rotation of the Earth with respect to the Sun

u. _____ Skips 3 leap years every 400 years

v. _____ The slow coning motion of the Earth's axis

w. _____ Zulu

MULTIPLE CHOICE

Compare your answers with those at the back of the book.

1. What must a coordinate system have? (15.1)
   (a) an indicated origin
   (b) only one dimension
   (c) a scale with units
   (d) both (a) and (c)

2. Which of the following is true? (15.1) A Cartesian coordinate system
   (a) can be a two-dimensional system.
   (b) normally designates the horizontal line known as the *x*-axis.
   (c) normally designates the vertical line known as the *y*-axis.
   (d) All of these are true.

3. Which of the following is true? (15.2) Latitude
   (a) is a linear measurement.
   (b) can have greater numerical values than longitude.
   (c) is measured in a north–south direction.
   (d) can have negative values.

4. Which of the following is true? (15.2) Longitude
   (a) can have a maximum value of 360°.
   (b) is an angular measurement measured east or west.
   (c) is measured in units of kilometers.
   (d) is measured north or south of the equator.

5. Which of the following is true of meridians? (15.2) Meridians
   (a) run east and west.
   (b) are halves of great circles.
   (c) are great circles.
   (d) All of these are true.

6. Which of the following is false? (15.2)
   (a) Parallels become smaller as the distance from the equator becomes greater.
   (b) Parallels are all small circles.
   (c) Parallels run east–west.
   (d) The equator is a parallel.

7. What is a mean solar day? (15.3)
   (a) a day that begins at noon
   (b) a day with a really bad attitude
   (c) a day measured by the positions of distant stars
   (d) the average of all apparent solar days during 1 year

8. When is the Sun overhead on a person's meridian at local time? (15.3)
   (a) 12 midnight
   (b) on June 21
   (c) 12 noon
   (d) all of these

9. How does a solar day compare with a sidereal day? (15.3)
   (a) 4 minutes shorter than the sidereal day
   (b) 2 minutes longer than the sidereal day
   (c) 1 minute shorter than the sidereal day
   (d) 4 minutes longer than the sidereal day

10. During the month of January, how does the number of daylight hours at Washington, D.C., compare with the number of daylight hours at Orlando, Florida? (15.3)
    (a) more than
    (b) less than
    (c) the same as

11. If the altitude of the Sun is measured to be 25°, then what is the zenith angle? (15.4)
    (a) 45°
    (b) 55°
    (c) 65°
    (d) none of these

12. Which of the following is used to measure altitude? (15.4)
    (a) compass
    (b) sextant
    (c) clock
    (d) dead reckoning

13. In early sailing days, what was used to determine longitude? (15.4)
    (a) compass
    (b) sextant
    (c) clock
    (d) GPS

14. Why does the Earth have seasons? (15.5)
    (a) The Earth spins on its axis.
    (b) The Earth orbits the Sun with the rotation axis tilted.
    (c) The Earth is a spherical body.
    (d) The Earth's distance to the Sun changes as it orbits.

15. Which of the following is true for an observer at 20°N? (15.5)
    (a) always looks south to see the Sun
    (b) observes the Sun directly overhead twice a year
    (c) is near the Tropic of Capricorn
    (d) none of these

16. From September 22 to March 21, in what direction must an observer in the Northern Hemisphere look to see sunrise? (15.5)
    (a) east
    (b) northeast
    (c) southeast
    (d) west

17. When does the longest day occur for latitudes greater than $23\frac{1}{2}$°S in the Southern Hemisphere? (15.5)
    (a) on the summer solstice
    (b) on the winter solstice
    (c) on the autumnal equinox
    (d) on the vernal equinox

18. Which of the following is true for the Gregorian calendar? (15.5)
    (a) It is based on lunar months.
    (b) It is accurate to 1 day in 6000 years.
    (c) It has a leap year every century year that is divisible by 4.
    (d) It differs from the Julian calendar only in its name.

19. What can be said about the precession of the Earth's axis? (15.6)
    (a) maintains Polaris as the "North" Star
    (b) changes the angle between the axis and the vertical
    (c) has no important effect on the Earth's seasons
    (d) none of these

20. What causes the precession of the Earth's axis? (15.6)
    (a) the tilt of the axis
    (b) the Earth's equatorial bulge
    (c) the apparent north–south movement of the Sun
    (d) gravitational torque

FILL IN THE BLANK

Compare your answers with those at the back of the book.

1. A Cartesian coordinate system is also called a(n) ___ coordinate system. (15.1)

2. Cartesian coordinates (x, y) are referenced to the ___. (15.1)

3. Circles drawn around the Earth parallel to the equator are called ___. (15.2)

4. A ___ is a circle whose plane passes through the center of the Earth. (15.2)

5. A position north or south of the equator is designated by ___. (15.2)

6. A position east or west of the prime meridian is designated by ___. (15.2)

7. A ___ day is longer than a sidereal day. (15.3)

8. The Earth is divided into ___ time zones. (15.3)

9. The conterminous United States has ___ time zones. (15.3)

10. Crossing the ___ traveling westward, the date is advanced to the next day. (15.3)

11. The angle of the Sun above the horizon is called the Sun's ___. (15.4)

12. Latitude can be determined by knowing the date and measuring the Sun's ___. (15.4)

13. On a sailing ship with its marine chronometer set for GMT, it is noted that the noonday Sun occurs at 10 A.M. The longitude of the ship is then ___. (15.4)

14. The vernal equinox indicates the beginning of ___. (15.5)

15. Most of the United States lies above the Tropic of ___. (15.5)

16. The tropical year is called the year of the ___. (15.5)

17. One of the earliest calendars is based on constellations in the ___. (15.5)

18. For the Gregorian calendar, ___ leap years are skipped every 400 years. (15.5)

19. A year is given as 2010 CE. The CE stands for ___. (15.5)

20. The precession of the Earth's axis is much like that of a toy ___. (15.6)

SHORT ANSWER

15.1 Cartesian Coordinates

1. What common thing is often laid out in a Cartesian coordinate system?
2. What, on a sphere, is analogous to a two-dimensional Cartesian coordinate system?
3. State a reference point and then use a three-dimensional reference system to give the position of your textbook in the room where you are located.

15.2 Latitude and Longitude

4. What are the minimum and maximum values for latitude and longitude?
5. What are the reference points for longitude and latitude?
6. What are the names of lines of equal (a) longitude and (b) latitude?
7. What are the latitude and longitude of the North Pole?
8. At what point on or inside the Earth are longitude/latitude angles measured?
9. Are meridians great circles? Explain.
10. Are any parallels great circles? Explain.

15.3 Time

11. Give the longitude of the central meridian for each of the time zones in the conterminous United States.
12. What do A.M. and P.M. mean? How about GMT and UTC?
13. What is the standard unit of time, and how is the length of a day defined?
14. What are two advantages of Daylight Saving Time? Do all states go on DST? Explain.
15. Distinguish between a solar day and a sidereal day.
16. How does the date change when one crosses the IDL?
17. Explain why the Sun moves 15°/h.

15.4 Determining Latitude and Longitude

18. Describe the apparent north–south motion of the overhead Sun throughout the year. What causes this motion?
19. What is your zenith? Distinguish between altitude and zenith angle.
20. How were the latitude and longitude of a ship determined in early sailing days?
21. At what latitudes can the noonday Sun be seen overhead in (a) the Northern Hemisphere and (b) the Southern Hemisphere?

15.5 The Seasons and the Calendar

22. Distinguish between a tropical year and a sidereal year.
23. Give the times during the year when a place near the North Pole receives (a) the maximum daily amount of solar radiation and (b) the minimum daily amount.
24. Where is the noon Sun directly overhead on (a) the beginning of our summer, (b) the beginning of our winter, (c) the beginning of spring, and (d) the beginning of fall?
25. What is the origin of the month?
26. Distinguish between the Julian and Gregorian calendars.
27. How often is there a leap year in the Gregorian calendar?
28. Explain how the circle of illumination changes over 1 year.
29. Which century year will next be a leap year?

15.6 Precession of the Earth's Axis

30. What effect does the precession of the Earth's axis have with respect to the pole star or North Star?
31. What evidence is there to support precession of the Earth's axis?
32. How long does it take the Earth's axis to precess one cycle?

VISUAL CONNECTION

Visualize the connections and give answers for the blanks. Compare your answers with those at the back of the book.

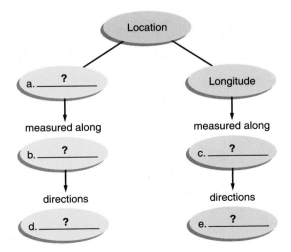

APPLYING YOUR KNOWLEDGE

1. When measuring time using a sundial, are you measuring apparent solar time or mean solar time? Explain.

2. How could an observer in the Northern Hemisphere estimate his or her latitude by observing the sky on a clear night?

3. Is there a place on the Earth's surface where 1° of latitude and 1° of longitude have approximately equal numerical values in nautical miles? Explain.

4. Barrow, Alaska (71°N), gets 24 hours of daylight for several weeks near June 21. Austin, Texas (34°N), only gets 14 hours of sunlight near June 21. Why is it hotter in Texas?

5. October is the tenth month of the year, but *octo* means "eight." Explain.

6. (a) Can you travel continuously eastward and circle the Earth? Why or why not? (b) Can you travel continuously southward and circle the Earth? Why or why not?

7. What is the implication for someone born on February 29?

8. Why does a person's zodiac "sign" change over long periods of time?

IMPORTANT EQUATION

Latitude: L = ZA ± Sun degrees   (15.1)

EXERCISES

15.2 Latitude and Longitude

1. What is the number of latitude degrees between place *A* (50°N, 90°W) and place *B* (60°S, 90°E)?

   Answer: 190°

2. What is the number of latitude degrees between place *A* (20°N, 75°W) and place *B* (30°S, 75°W)?

3. What is the number of nautical miles between place *A* and place *B* in Exercise 1?

   Answer: 11,400 n mi

4. What is the number of nautical miles between place *A* and place *B* in Exercise 2?

5. What are the latitude and longitude of the point on the Earth that is opposite Washington, D.C. (39°N, 77°W)?

   Answer: 39°S, 103°E

6. What are the latitude and longitude of the point on the Earth that is opposite Tokyo (36°N, 140°E)?

15.3 Time

7. What are the standard time and date at (40°N, 118°W) when the standard time at (35°N, 80°W) is 6 P.M. on October 8?

   Answer: 3 P.M., October 8

8. What are the standard time and date at (40°N, 110°W) when the standard time at (30°N, 70°W) is 1 A.M. on October 16?

9. When it is 10 P.M. standard time on November 26 in Moscow (56°N, 38°E), what are the standard time and date in Tokyo (36°N, 140°E)?

   Answer: 4 A.M., November 27

10. When it is 10 A.M. standard time on February 22 in Los Angeles (34°N, 118°W), what are the standard time and date in Hong Kong, China (22°N, 114°E)?

11. When it is 12 noon in London, what is the standard time in Denver, Colorado (40°N, 105°W)?

    Answer: 5 A.M. MST

12. When it is 12 noon in London, what is the standard time in Sydney, Australia (34°S, 151°E)?

13. Some military personnel stationed at a base in Germany (51°N, 10°E) wanted to watch a figure skating event live in Salt Lake City, Utah (41°N, 112°W), that was scheduled for 7 P.M. local time on February 22. At what time and date should they have watched the television to see this event?

    Answer: 3 A.M. on February 23

14. It is 6 A.M. on July 1 in London (51.5°N, 0°). What are the time and date in Cape Town, South Africa (34°S, 18°E)?

15.4 Determining Latitude and Longitude

15. What is the altitude angle of the Sun for someone in Atlanta, Georgia (34°N), on June 21?

    Answer: 79.5°

16. What is the altitude angle of the Sun for someone in Atlanta, Georgia (34°N), on December 22?

17. What is the latitude of someone in the United States who sees the Sun at an altitude of 71.5° on June 21?

    Answer: 42°N

18. What is the latitude of someone in Europe who sees the Sun at an altitude of 65.5° on December 22?

19. Determine the month and day when the Sun is at maximum altitude for an observer in Washington, D.C. (39°N). What is the altitude of the Sun at this time?

    Answer: on or about June 21, 74.5°

20. Determine the month and day when the Sun is at minimum altitude for an observer in Washington, D.C. (39°N). What is the altitude of the Sun at this time?

21. At approximately what latitude is the noonday Sun overhead on April 15?

    Answer: 6.5°N

22. At approximately what latitude is the noonday Sun overhead on October 22?

23. A ship's captain with a sextant measures the altitude of the noon-day Sun to be 40° toward the south on December 21. What is the latitude of the ship? (Solve both graphically and with Eq. 15.1.)

Answer: $26\frac{1}{2}$°N

24. Later in the voyage, on March 21, the captain in Exercise 23 measures the altitude of the noonday Sun to be 60° toward the north. What is the latitude of the ship? (Solve both graphically and using Eq. 15.1.)

25. Shooting the Sun with a sextant, the first mate on a ship measures the altitude of the noonday Sun to be 35° toward the north on May 20. What is the latitude of the ship? (Solve both graphically and using Eq. 15.1.)

Answer: 45°S

26. On December 7, the first mate on a ship measures the altitude of the noonday Sun to be 56° to the south. What is the latitude of the ship? (Solve both graphically and using Eq. 15.1.)

27. The reading taken on a sailing ship shows that on December 21 the noonday Sun has an altitude of $86\frac{1}{2}$° to the south and the marine chronometer reads 7 A.M. What are the latitude and longitude of the ship, and where is it located geographically?

Answer: 20°S, 75°E, Indian Ocean

28. On June 21, the altitude of the noonday Sun is measured to be $66\frac{1}{2}$° to the north and the marine chronometer reads 2 A.M. What are the latitude and longitude of the ship, and where is it located geographically?

ON THE WEB

1. Time and Place

How do we tell time? Would time be different if you lived on another planet? How do you tell someone where you are on the Earth (designate your location)? What system would you use to describe your position on another planet? In the solar system? The Milky Way galaxy? Visit the student website at **www.cengagebrain.com/shop/ISBN/1133104096** to learn more about timekeeping methods and about coordinate systems.

2. How accurate is our current calendar?

In another 10,000 years, on what date will be the first day of spring? What is the date of your birthday using the Old Style (Julian) calendar? Explore answers to these questions through recommended links on the student website at **www.cengagebrain.com/shop/ISBN/1133104096**.

The Solar System

Courtesy of International Astronomical Union. Martin Kornmesser.

> *In my studies of astronomy and philosophy I hold this opinion about the universe, that the Sun remains fixed in the center of a circle of heavenly bodies, without changing its place; and the Earth, turning on itself, moves around the Sun.*
>
> •
> Galileo (1564–1642)

A montage of the eight planets > of the solar system. Notice the relative sizes of the planets.

ASTRONOMY FACTS

▸ All planets and dwarf planets in the solar system, except the Earth, are named after ancient mythological gods or goddesses.

▸ Venus is the hottest planet, even though it is almost twice as far from the Sun as Mercury.

A **stronomy** is the study of the universe, which is the totality of all matter, energy, space, and time. The vastness of the universe staggers the imagination. On a clear night, one might look at the sky and wonder what lies beyond the planets, stars, and galaxies. Less than 200 years ago, it was generally thought that our Milky Way Galaxy constituted the entire universe, but in the last century, our knowledge of the universe has expanded many times over. With modern telescopes at least a billion galaxies can be identified.

The same explosion of knowledge applies to our solar system, which consists of the Sun and everything that orbits it. In a single generation, more has been learned about the solar system than was known for centuries. Humans have walked on the Moon. Space probes visit other planets and the far reaches of the solar system;

Chapter Outline

two of them are on the verge of leaving the solar system. Astronomy is one of the oldest sciences, but it has never been more exciting than it is today.

The study of the universe has advanced at a tremendous rate. Technology has improved to the level where scientists have constructed huge land-based telescopes with sophisticated electronic detectors interfaced with computers that store, analyze, and enhance the images received. Unfortunately, the Earth's atmosphere absorbs much of the incoming radiation. However, telescopes can now be put in satellites that orbit above the atmosphere. One of these, the Hubble Space Telescope (HST) has sent back amazing pictures, as you will see in this chapter.

Other changes have taken place, in particular the redefinition of a planet. This change was made in 2006, and as a result, Pluto has been demoted and is no longer considered a major planet. Rather, Pluto and other objects have been classified as "dwarf planets." In this chapter you will find sections on terrestrial planets (16.4), Jovian planets (16.5), and dwarf planets (16.6).

Chapter 15, Place and Time, was generally limited to the study of phenomena here on the Earth. Chapters 16 through 18 extend this study. First, we'll take a close look at the planets of the solar system (Chapter 16), then our Moon and other satellites (Chapter 17), and finally in Chapter 18 we'll take off into space, exploring stars, galaxies, and cosmology (the structure and evolution of the entire universe).

16.1 The Solar System and Planetary Motion

Preview Questions

● What is the difference between the geocentric model and the heliocentric model?

● What does Kepler's harmonic law describe?

The **solar system** is a complex system of moving masses held together by gravitational forces. At the center of this system is a star called the Sun, which is the dominant mass. Revolving around the Sun are 8 major planets with more than 170 satellites (moons), and currently 4 dwarf planets. In addition there are thousands of asteroids, vast numbers of comets and meteoroids, as well as interplanetary dust particles, gases, and a solar wind composed of charged particles.

The term *planet* originated with ancient observers of stars. These observers of the nighttime sky also viewed, with the unaided eye, star-like objects (Mercury, Venus, Mars, Jupiter, and Saturn) that moved with respect to fixed stars. They called these objects planets, from the Greek word meaning "wanderer."

The rotating and revolving motions of the Earth were concepts not readily accepted at first. In early times, most people were convinced that the Earth was motionless and that the Sun, Moon, planets, and stars revolved around the Earth, which was considered the center of the universe. This concept is called the Earth-centered model, or **geocentric model**, of the universe, and it persisted for some 18 centuries.

Nicolaus Copernicus (1473–1543), a Polish astronomer, developed the Sun-centered model, or **heliocentric model**, of the solar system (● Fig. 16.1). Although Copernicus did not prove that the Earth revolves around the Sun, he did provide elegant mathematical proofs that could be used to predict future positions of the planets.

After the death of Copernicus in 1543, the study of astronomy was continued and improved by several astronomers. Notable among them was Danish astronomer Tycho Brahe ("BRAH-he") (1546–1601), who built an observatory on the island of Hven near Copenhagen and spent most of his life observing and studying the stars and planets (● Fig. 16.2).

Brahe is considered the greatest practical astronomer since the Greeks. His measurements of the planets and stars, all made with the unaided eye (the telescope had not yet been invented), proved to be more accurate than any made previously. Brahe's data, published in 1603, were edited by his colleague Johannes Kepler (1571–1630), a German

Figure 16.1 Nicolaus Copernicus (1473–1543)
Copernicus, a Polish astronomer, developed the heliocentric (Sun-centered) model of the universe.

© Bettmann/CORBIS

Figure 16.2 Tycho Brahe (1546–1601)
The Danish astronomer Tycho Brahe is known for his very accurate observations, made with the unaided eye, of the positions of stars and planets.

mathematician and astronomer who had joined Brahe during the last year of his life (● Fig. 16.3). After Brahe's death, his lifetime's observations were at Kepler's disposal and provided him with the data necessary to formulate the three laws known today as *Kepler's laws of planetary motion.*

Kepler was interested in the irregular motion of the planet Mars. He spent considerable time and energy before coming to the conclusion that the uniform circular orbit proposed by Copernicus was not a true representation of the observed facts. Perhaps because he was a mathematician, Kepler saw that a simple type of geometric figure would fit the observed motions of Mars and the other known planets.

Kepler's first law is known as the **law of elliptical orbits**:

> **All planets move in elliptical orbits around the Sun, with the Sun at one focus of the ellipse.**

An ellipse is a two-dimensional figure that is symmetric about two unequal axes, sort of a flattened circle (● Fig. 16.4). An ellipse can be drawn by using two thumbtacks, a closed loop of string, paper, and pencil. The points where the two tacks are positioned are called the *foci* (singular, focus) of the ellipse. The longer axis, which passes through both foci, is called the *major axis*. Half the major axis is called the *semimajor axis*. The semimajor axis of the Earth's elliptical orbit is taken as the average distance between the Earth and the Sun.

A convenient unit for expressing distances in the solar system is the **astronomical unit** (AU), which is the average distance between the Earth and the Sun. One AU is 1.5×10^8 km (9.3×10^7 mi). Using astronomical units, one can get a relative idea of planet distances from the Sun. For example, if a planet were 3 AU from the Sun, then it would be three times as far from the Sun as is the Earth.

Kepler's first law gives the shape of the orbit but fails to predict when the planet will be at a particular position in the orbit. Aware of this problem, Kepler set out to find a solution from the mountain of data he had at his disposal. After a tremendous amount of work, he discovered what is now known as Kepler's second law, the **law of equal areas**:

> **An imaginary line (radial vector) joining a planet to the Sun sweeps out equal areas in equal periods of time.**

Figure 16.3 Johannes Kepler (1571–1630)
Kepler, a German mathematician and astronomer, formulated the basic quantitative laws that describe planetary motion.

As illustrated in ● Fig. 16.5, the speed of a revolving planet will be greatest when the planet is closest to the Sun because the time periods are equal (conservation of angular momentum; see Chapter 3.7, Fig. 3.20). This closest point in its orbit is called *perihelion* ("per-i-HEE-lee-on") and occurs for the Earth around January 4. The speed of a planet is least when it is farthest from the Sun. This point in the orbit is called *aphelion* ("a-FEE-lee-on") and occurs around July 5 for the Earth. (Remember *ap* with "away.")

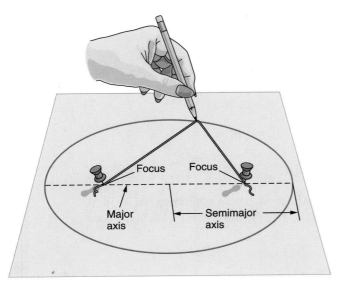

Figure 16.4 Drawing an Ellipse
An ellipse can be drawn using two thumbtacks, a loop of string, a pencil, and a sheet of paper. (See text for description.)

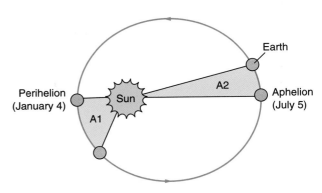

Figure 16.5 Kepler's Law of Equal Areas
An imaginary line joining a planet to the Sun sweeps out equal areas in equal periods of time. Area A1 equals area A2. The Earth has greater orbital speed in January than in July. (Figure exaggerated for illustration. The orbits of all the planets are nearly circular.)

After the publication of his first two laws in 1609, Kepler began a search for a relationship among the motions of the different planets and an explanation to account for these motions. Ten years later he published *De Harmonice Mundi* (*Harmony of the Worlds*), in which he stated his third law of planetary motion, known as the **harmonic law** (or law of harmonic motion):

> **The square of the sidereal period of a planet is proportional to the cube of its semimajor axis (one-half the major axis).**

This law can be written as

$$(\text{period})^2 \propto (\text{semimajor axis})^3$$

or, in equation form,

$$T^2 = kR^3 \qquad\qquad \textbf{16.1}$$

where T is the sidereal period (time of one complete revolution around the Sun), R the length of the semimajor axis, and k a constant of proportionality (the same value for all planets).

If the period of the planet is measured in Earth years and the semimajor axis in astronomical units, then k is conveniently equal to 1 y²/(AU)³.* ● Table 16.1 gives the mean distance of each major planet from the Sun. The mean distance can generally be used as a substitute for the length of the semimajor axis in Eq. 16.1.

EXAMPLE 16.1   Calculating the Period of a Planet

Calculate the period of a planet whose orbit has a semimajor axis of 1.52 AU.

Solution

Step 1

Use Eq. 16.1 and substitute in the values for k and R.

$$T^2 = kR^3 = \left[\frac{1\,y^2}{(AU)^3}\right](1.52\ AU)^3$$

---

*$k = T^2/R^3 = (1\ y)^2/(1\ AU)^3 = 1\ y^2/(AU)^3$ (that is, years squared divided by astronomical units cubed).

Step 2

Cube 1.52 AU and cancel the (AU)^3 units.

$$T^2 = 3.51 \text{ y}^2$$

Step 3

Take the square root of both sides.

$$T = 1.87 \text{ y}$$

Confidence Exercise 16.1

Calculate the period of a planet whose orbit has a semimajor axis of 30 AU.
Answers to Confidence Exercises may be found at the back of the book

---

Table 16.1 The Planets of Our Solar System

	Mean Distance from the Sun (AU)	Orbit Period (y)	Mass (Earth = 1)	Diameter (Earth = 1)
Sun			333,000	
Mercury	0.387	0.24	0.055	0.38
Venus	0.723	0.62	0.81	0.95
Earth	1.000	1.00	1.00	1.00
Mars	1.524	1.88	0.11	0.53
Jupiter	5.203	11.86	318	11.2
Saturn	9.529	29.42	94.3	9.46
Uranus	19.19	83.75	14.54	4.00
Neptune	30.06	163.70	17.1	3.88

Did You Learn?

● The geocentric, or Earth-centered, model considers the Earth at the center of the universe, whereas the heliocentric model is a Sun-centered universe.

● Kepler's third law, or harmonic law, states that the square of the orbital period is proportional to the cube of the semimajor axis. It means that objects that are farther from the Sun have longer orbital periods.

16.2 Major Planet Classifications and Orbits

Preview Questions

● What are the names of the terrestrial planets and the names of the Jovian planets?

● What is the difference between inferior and superior planets?

Galileo Galilei (1564–1642)—an Italian astronomer, mathematician, and physicist who is usually called just Galileo—was one of the greatest scientists of all time. The most important of his many contributions to science were in the field of mechanics. He founded the modern experimental approach to scientific knowledge (Chapter 1.2), and the motion of objects, including the planets, was of prime interest to him. (See the Highlight: Galileo and the Leaning Tower of Pisa, Chapter 2.3.)

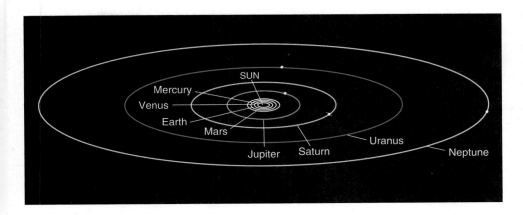

Figure 16.6 The Orbits of the Planets
All the planets revolve counterclockwise around the Sun, as observed from a position in space above the North Pole of the Earth.

In 1609, Galileo became the first person to observe the Moon and planets through a telescope. He also discovered four of Jupiter's moons, thus proving that the Earth is not the center of motion for all objects in the universe, as was thought previously. Equally important was his discovery that the planet Venus went through a change of phase similar to that of our Moon. This observation could be explained by the heliocentric model but ran counter to the geocentric model, which predicted Venus to only have new or crescent phases.

The works of Copernicus, Kepler, and Galileo were integrated by Sir Isaac Newton in 1687 with the publication of the *Principia* (Chapter 3, Introduction). Newton, an English physicist regarded by many as the greatest scientist the world has known, formulated the principles of gravitational attraction between objects (Chapter 3.5).

Newton also established physical laws determining the magnitude and direction of the forces that cause the planets to move in elliptical orbits in accordance with Kepler's laws. He invented calculus and used it to help explain Kepler's first law. Newton also used the law of conservation of angular momentum (Chapter 3.7) to explain Kepler's second law. These explanations of Kepler's laws unified the heliocentric model of the solar system and brought an end to the confusion.

The Sun is the dominant mass of the solar system, possessing 99.87% of the mass of the system. The distribution of the remaining 0.13% of the solar system's mass is shown in Table 16.1. More than half of this 0.13% is the mass of Jupiter.

Mercury, Venus, Earth, and Mars are classified as *inner planets*, or **terrestrial planets**. They are called terrestrial (from the Latin *terra*, meaning "earth") because they have physical and chemical characteristics that resemble the Earth in certain respects.

Jupiter, Saturn, Uranus, and Neptune are classified as *outer planets*, or **Jovian planets**. They are called Jovian after the Roman god Jupiter, who was also called Jove. These planets resemble Jupiter, having gaseous outer layers. Jupiter is by far the largest and most massive planet (Table 16.1).

The relative distances of the planets from the Sun are shown in ● Fig. 16.6. The orbits are all elliptical, but nearly circular, and generally lie in a plane. Note how far Jupiter is from the Sun compared with the distance of Mars. Also note that the distance from Saturn to Neptune is greater than that from the Sun to Saturn (Table 16.1).

When the solar system is viewed from a position in space above the North Pole of the Earth, the planets all *revolve* counterclockwise around the Sun. This revolution is called *direct* or **prograde motion**. The planets also *rotate* with a counterclockwise motion, or prograde motion, when viewed from above the North Pole, with the exception of Venus and Uranus. These two planets have **retrograde rotation**: the rotation is east to west (westward), or clockwise, as viewed from above.

The relative sizes of the planets are shown in ● Fig. 16.7. Notice how huge the Jovian planets are compared with the terrestrial planets and how small the planets are with respect to the Sun. A mnemonic for remembering the order of the planets from the Sun is the following: *My Very Endearing Mother Just Served Us Nachos*, for *Mercury, Venus, Earth, Mars, Jupiter, Saturn, Uranus,* and *Neptune*.

Figure 16.7 The Solar System
The eight major planets and the Sun are shown in size for comparison. The colors are similar to their surface colors.

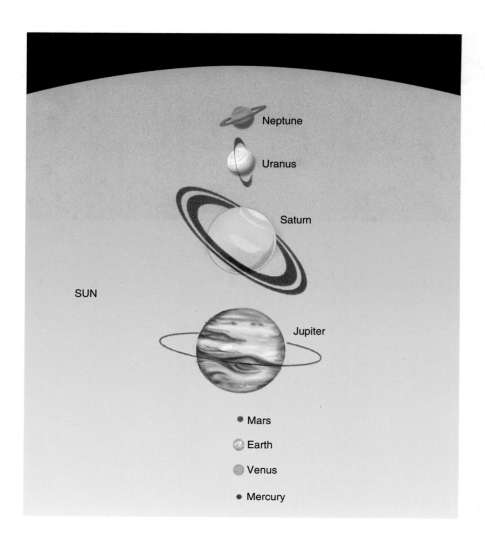

The orbital period is the time it takes a planet to travel one complete orbit around the Sun. This is expressed in terms of its **sidereal period**, the time it takes the planet to make one full orbit around the Sun relative to a fixed star. For example, the sidereal period for the planet Mercury—that is, Mercury's sidereal "year"—is 88 Earth days. The sidereal orbit period for the Earth is 365.25 days (Chapter 15.5).

Planets that have orbits smaller than the Earth's are classified as *inferior* and those with orbits greater than the Earth's as *superior*. A **conjunction** is said to occur when two planets are lined up with respect to the Sun. When the Earth and one of the *inferior planets* (Mercury or Venus) are lined up *on the same side of the Sun*, an *inferior conjunction* is said to occur. When the Earth and one of the inferior planets are lined up *on opposite sides of the Sun*, a *superior conjunction* occurs. These conditions are illustrated in ● Fig. 16.8.

When the Sun and one of the *superior* planets (Mars, Jupiter, etc.) are lined up *on the same side of the Earth*, they are said to be in *conjunction*. When the Sun and one of the superior planets are lined up on *opposite sides of the Earth* (that is, 180° apart), they are said to be in **opposition**. In short, if the Sun and planet as viewed from the Earth are together in the sky, then it is a conjunction. If they are opposite each other in the sky, then it is an opposition.

Did You Learn?

● The terrestrial planets are Mercury, Venus, Earth, and Mars; the Jovian planets are Jupiter, Saturn, Uranus, and Neptune.

● Inferior planets are closer to the Sun than the Earth, whereas superior planets are farther away.

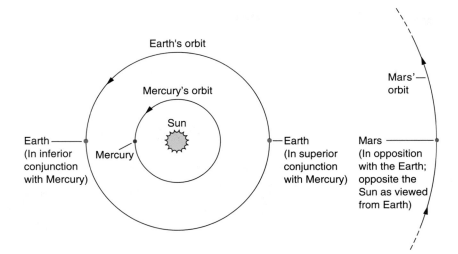

16.3 The Planet Earth

Preview Questions

- Which has the greater albedo, the Earth or the Moon? Why?
- How are the rotation and revolution of the Earth proven?

The planet Earth is a solid, rather spherical, rocky body with oceans and an atmosphere. Among the eight major planets in our solar system, the Earth is unique. It is the only planet with large amounts of surface water, an atmosphere that contains oxygen, a temperate climate, and living organisms (as far as we know).

Because oxygen is a very reactive element, it dominates the chemistry of the planet. In addition to the atmosphere being 21% oxygen (O_2), oxygen is the most abundant element in the Earth's crust. (Nitrogen, N_2, is the most abundant element in the atmosphere, 78%.) Oxygen-containing compounds constitute 90% or more of the Earth's rocks by volume. When oxygen combines with another substance, the process is called *oxidation* (Chapter 13.4). Consequently, we live in an oxidized environment.

The Earth is not a perfect sphere but rather an oblate spheroid, flattened at the poles and bulging at the equator. The difference between the Earth's diameter at the poles and its diameter at the equator (about 43 km, or 27 mi) is very small considering the total average diameter of the Earth, which is about 12,800 km (8000 mi). The ratio of 43 km to 12,800 km is only $\frac{1}{300}$. If the Earth were represented by a basketball with similar proportional dimensions, then the human eye would not be able to detect that it was nonspherical.

The fraction of incident sunlight reflected by a celestial object is called its **albedo** (from the Latin *albus*, meaning "white"). The Earth's albedo is 0.33, and the Moon's albedo is only 0.07. This value indicates that the Moon's surface reflects 7% of the incoming sunlight falling on its surface. The Earth reflects more light (33%) because the clouds and water areas are much better reflecting surfaces than the dull, dark surface of the Moon, which has no atmosphere. The planet Venus, the third brightest object in the sky (only the Sun and Moon are brighter), has an albedo of 0.76 (76%). Although the Moon's albedo is much smaller (0.07), it is much closer to the Earth and so appears brighter than the bright but distant Venus. The full Earth, as viewed from the Moon, appears about 4 times larger in diameter than the full moon viewed from the Earth. Due to the increased area and higher albedo, however, the Earth reflects more than 70 times as much light.

The Earth is undergoing several motions simultaneously. Two that have major influences on our daily lives are explained in this section: (1) the daily rotation of the Earth on its axis and (2) the annual revolution of the Earth around the Sun. A third motion, precession, was discussed in Chapter 15.6.

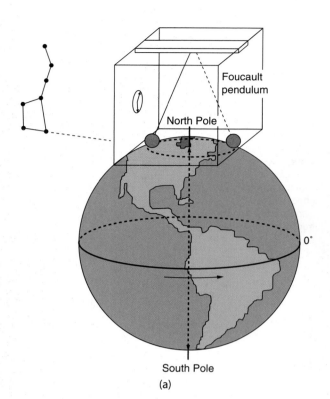

Figure 16.9 Foucault Pendulum
(a) The drawing illustrates a Foucault pendulum positioned in a fictitious room at the North Pole of the Earth. To an observer in the room, the pendulum will appear to change its plane of swing by 360° every 24 h. (See text for description.) (b) The photograph shows a Foucault pendulum at the Science Center in San Diego, CA. As the pendulum swings back and forth, its plane of swing appears to change, as noted by the consecutive knocking over of the markers positioned in a circle.

It is important to know the difference between rotation and revolution. An object is said to be in **rotation** when it rotates on an internal axis. Examples include a spinning toy top and a Ferris wheel at an amusement park. **Revolution** is the movement of one object around another. The Earth revolves around the Sun, and the Moon revolves around the Earth.

The Earth *revolves* eastward around the Sun (counterclockwise as viewed from above), and the revolution of the Earth causes the Sun to appear to move relative to the background stars. The Earth also *rotates* eastward around a central internal axis that is tilted $23\frac{1}{2}°$ from a line perpendicular to its orbital plane. The Earth's rotational period of 24 hours and the axis tilt ($23\frac{1}{2}°$) distribute the solar energy that is radiated onto the planet's surface. Also, as noted in Chapter 15.5, the $23\frac{1}{2}°$ tilt of the axis and the revolution of the Earth around the Sun account for the seasons the Earth experiences annually.

That the Earth rotates on its axis was not generally accepted until the nineteenth century. A few scientists had considered the possibility, but because no definite proof was available to support their beliefs, their ideas were not accepted. In 1851, an experiment demonstrating the rotation of the Earth was performed in Paris by Jean Foucault ("foo-KOH") (1819–1868), a French engineer, using a 61-m (200-ft) pendulum. Today, any pendulum used to demonstrate the rotation of the Earth is called a **Foucault pendulum**. For even more noticeable and better-understood results, imagine that the experiment is performed at the Earth's North Pole.

Imagine a large, one-room building made of glass with a high ceiling located at the North Pole during the winter months for the Northern Hemisphere. The North Pole has 24 hours of darkness during these months, and with clear skies, the stars are always visible (● Fig. 16.9). Fastened to the ceiling precisely above the North Pole is a swivel support having very little friction to which a fine steel wire is attached. Connected to the lower end of the wire is a massive iron ball.

The pendulum is carefully set into motion, initially swinging in a plane that goes through a clock on the wall and in a plane with a fixed star. After a few minutes, the plane of the swinging pendulum *appears* to be rotating clockwise. The swivel support allows the pendulum to change planes—directions—of oscillation (a swivel is found, for example, on the end of many dog leashes). At the end of 1 hour, the plane has rotated 15° clockwise from its original position relative to the clock. When 6 hours have elapsed, the plane of the pendulum appears to have rotated 90° clockwise. The plane of the pendulum continues to rotate, and at the end of 24 hours, it has made an apparent rotation of 360°.*

Did the room rotate, or did the pendulum rotate? A person who believes in a nonrotating Earth might argue that the pendulum *actually* rotated 360°. However, through the transparent walls of the room, it is observed that the pendulum swings in the same plane with the reference star. That is, *the pendulum has not rotated with reference to the fixed star.* No forces have been acting on the pendulum to change its plane of swing. Only the force of gravity acts, and it is vertically downward. Therefore, the pendulum does not really rotate westward (clockwise). Rather, the building and the Earth rotate eastward, or counterclockwise, once during the 24-hour period, as viewed from above. The Foucault pendulum is experimental proof of the Earth's rotation on its axis.

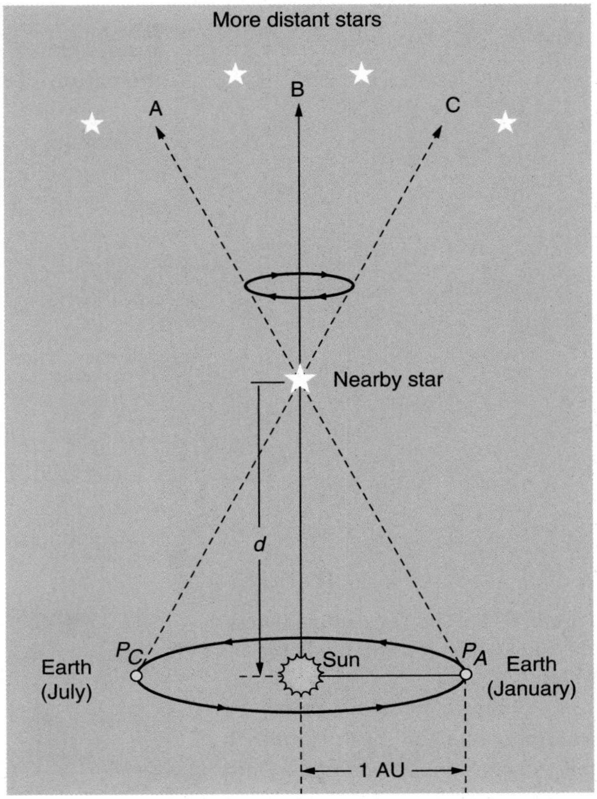

Figure 16.10 Stellar Parallax
The parallax of a star is the apparent displacement of a star that is located fairly close to the Earth with respect to more distant stars. When an observer is at P_A, the star appears in the direction A. As the Earth revolves counterclockwise, the star appears to be displaced and appears in the direction indicated for different positions of the Earth. Positions P_A and P_C are 6 months apart.

Conceptual Question and Answer

Another Foucault Pendulum

Q. Describe the motion of a Foucault pendulum if it is placed at the Earth's equator and swung along an east–west line.

A. At the equator, the Foucault pendulum is swinging in a plane that is perpendicular to the Earth's spin (rotational) axis. As at the North Pole, the pendulum swings in a plane fixed with a distant star. However, the plane of the Earth's equator also remains fixed to the same star. Therefore, the pendulum does not rotate at all; it continues to swing in the same line as it did initially.

Now, what experimental observation would prove that the Earth revolves around the Sun? As the Earth orbits the Sun once a year, the apparent positions of nearby stars change with respect to more distant stars. This effect is called *parallax*. In general, **parallax** is the apparent motion, or shift, that occurs between two fixed objects when the observer changes position. To see parallax for yourself, hold your finger at a fixed position in front of you. Close one eye, move your head from side to side, and note the apparent motion between your finger and some distant object. Note also that the apparent motion becomes less as you move your finger farther away. ● Figure 16.10 illustrates the parallax of a nearby star as measured from the Earth relative to stars that are more distant.

The motion of the Earth as it revolves around the Sun leads to an apparent shift in the positions of the nearby stars with respect to more distant stars. Because the stars are at very great distances from the Earth, the parallax is too small to be seen with the unaided eye. It was first observed with a telescope in 1838 by Friedrich W. Bessel (1784–1846), a German astronomer and mathematician. The observation of parallax was experimental proof that the Earth revolves around the Sun. Today, the measurement of parallax is the best method we have of determining the distances to nearby stars (Chapter 18.1).

---

*In actual practice, energy is supplied to the pendulum by an electric device via the steel wire so that the pendulum will continue to oscillate.

A second proof of the Earth's orbital motion around the Sun is the telescopic observation of a systematic change in the position of all stars annually. The observed effect, called the **aberration of starlight**, is defined as the apparent displacement in the direction of light coming from a star because of the orbital motion of the Earth.

By analogy, you have probably experienced the "aberration of raindrops." Suppose you are sitting in a parked car and it is raining. The rain comes straight down (no wind). As you start driving, it appears as though the raindrops are hitting the windshield at an angle, and the greater your speed, the greater the raindrop angle from the vertical. The rain is coming straight down, but there is an apparent direction change because of the motion of the car. It is analogous to the aberration of starlight because of the motion of the Earth's revolution.

Did You Learn?

- The Earth reflects more sunlight than the Moon. The larger albedo is caused by the more reflective clouds and surface water.

- The Foucault pendulum proves that the Earth rotates; parallax and the aberration of starlight prove that the Earth revolves about the Sun.

Figure 16.11 The Terrestrial Planets, the Jovian Planets, and the Earth's Moon
This montage of photos was taken by NASA spacecraft. At the top (*from left to right*) are Mercury, Venus, the Earth and Moon, and Mars. At the bottom (*from center right to left*) are Jupiter, Saturn, Uranus, and Neptune. The terrestrial planets are roughly to scale with one another, as are the Jovian planets, which are much, much larger than the terrestrial planets (see Fig. 16.7).

NASA/JPL

16.4   The Terrestrial Planets

Preview Questions

● What makes a planet terrestrial, or pertaining to the Earth?

● What is the most abundant molecule in the atmosphere of each terrestrial planet?

As mentioned earlier, Mercury, Venus, Earth, and Mars are called the *terrestrial planets* because in certain respects, the physical and chemical characteristics of these three inner planets resemble those of the Earth. ● Figure 16.11 shows these four planets and the Earth's Moon, along with the Jovian planets. All four terrestrial planets are relatively small in size and mass (for size comparison, see Fig 16.7). They are composed of rocky material and metals (their cores are mostly iron and nickel). All four are relatively dense (their average density is about 5.0 g/cm³, five times that of water) and have solid surfaces.

The orbits of the terrestrial planets, comparatively speaking, are close together and are relatively close to the Sun (all are within about 1.5 AU; Table 16.1). None has a Saturn-like ring system, and only the Earth and Mars have moons. Although the terrestrial planets have some similarities, they are also very different from one another. Only Mercury and the Earth have global magnetic fields, although they are weak (Mars has weak regional fields). The Earth alone has an abundance of surface water and an atmosphere that is 21% oxygen. The other terrestrial planets have no surface water and no free oxygen in their atmospheres.

Let's take a brief look at each terrestrial planet.

Mercury

Mercury is the closest planet to the Sun and has the shortest period of revolution (88 days). The early Greeks named the planet Mercury after the speedy messenger of the gods. It is the fastest-moving of all the planets because of its position closest to the Sun. Mercury can only be seen from the Earth just after sunset or just before sunrise. The surface of Mercury is cratered and crisscrossed by faults that formed as the planet originally cooled and contracted. The *MESSENGER* (*ME*rcury *S*urface *S*pace *EN*vironment *GE*ochemistry and *R*anging) spacecraft revealed 50% of the surface of the planet never seen before (● Fig. 16.12). In 2011, it was the first spacecraft to orbit Mercury, discovering many new surface plains of volcanic origin, long ridges and cracks, and hundreds of craters, some showing fresh, white subsurface material ejected on impact. Mercury has a density almost equal to that of the Earth. This relatively high density indicates that it probably has a core of mostly iron, as does the Earth, giving rise to a magnetic field.

Mercury's rotation period is two-thirds that of its period of revolution. Thus, it rotates three times while circling the Sun twice. As it rotates, the side facing the Sun has temperatures as high as 430°C (806°F), whereas the dark side may be as low as −180°C (−292°F). Because of its high temperature and relatively small mass (that is, small gravitational force), Mercury has virtually no atmosphere; any original gases have escaped.

NASA/Johns Hopkins University Applied Physics Laboratory/Carnegie Institution of Washington

Figure 16.12 Battered Mercury
Mercury is seen up close by the MESSENGER spacecraft on a flyby of altitude 27,600 kilometers (17,100 miles) in 2008. Impressive craters surrounded by white rays and long scars on the surface indicate Mercury's violent history.

Figure 16.13 Crescent Venus
The clouds of Venus are seen in detail in this ultraviolet light image taken by the Hubble Space Telescope.

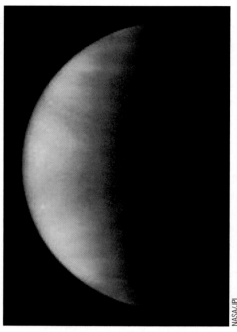

NASA/JPL

Venus

Venus is the Earth's closest planetary neighbor. It is the third-brightest object in the sky, exceeded only by the Sun and our Moon. Venus was named in honor of the Roman goddess of beauty, probably because of its planetary brightness.

Venus and the Earth resemble each other in several ways—they have similar average density, mass, size, and surface gravity—but there the similarities end. Venus is covered with a dense atmosphere that is 96% carbon dioxide, CO_2 (● Fig. 16.13). At the surface of Venus, the atmospheric pressure is a tremendous 90 atm and its high temperature, 467°C (873°F), exceeds that of Mercury. The high temperature is due mainly to the large amount of carbon dioxide in the atmosphere, which produces a large "greenhouse effect" (see Chapter 19.2).

The surface of Venus cannot be seen by an observer on the Earth because of the dense, thick carbon dioxide clouds that cover the planet. However, general surface features have been obtained with *radar* (radio detecting and ranging) imaging that penetrates the clouds. ● Figure 16.14 shows a composite radar image of Venus without clouds from NASA's *Magellan* spacecraft.

The *Magellan* radar images revealed that the surface of Venus is dominated by vast plains and lowlands, with some highlands, and relatively few large craters. These images con-

Figure 16.14 A Radar Image of Venus without Clouds
Radar can penetrate the clouds to give a view of Venus' surface. Vast volcanic plains cover about 80% of the planet's surface. There are also fault lines with high walls and cliffs and mountain chains that extend hundreds of kilometers in length. (Image is shown in false color to highlight different terrains.) Inset: The Surface of Venus taken by the *Venera 13* lander in 1982. The surface soil and rock of Venus are seen in this image. The foot of the spacecraft is also seen along with a color bar.

NASA History Office

NASA/JPL/USGS

firmed the earlier data and photos taken from the Soviet *Venera 13* lander in 1982 (Fig. 16.14 inset). The hot surface is gray-colored and similar in composition to terrestrial basalt.

Other surface features identified by *Magellan* were fractures and fault lines with high walls and cliffs, mountain chains that extend hundreds of kilometers in length, and volcanic plains that cover more than 80% of the planet's surface. No active volcanoes appeared in *Magellan* radar images, but many extinct volcanoes are seen. Most surface rocks appear to be volcanic in origin, indicating that volcanism was the last geologic process to take place on the planet. No surface feature in the images appears older than about 1 billion years; most features appear to be approximately 400 million years old.

Mars

Viewed from the Earth, Mars has a reddish color, so the Romans named it after their bloody god of war. The reddish color is believed to be due to fine-grain iron oxide minerals. The planet rotates once every 24.5 h, very close to an Earth day. However, because Mars is about 1.5 times as far from the Sun as the Earth and because its orbit is larger than that of the Earth, it takes 687 days (about 23 Earth months) for Mars to revolve around the Sun. Thus, a Martian year is almost twice as long as an Earth year.

The axis of Mars is tilted like that of the Earth, and Mars undergoes slow seasonal changes during the course of a Martian year. When Mars is viewed from the Earth, the most obvious feature is its bright polar ice caps (● Fig. 16.15). The caps grow and diminish with the seasons, almost disappearing in the Martian summer. The caps are composed of some water (H_2O) ice, but are mostly frozen carbon dioxide (CO_2, dry ice).

The dark surface features on Mars also change with the seasons. In the nineteenth century, there was fanciful speculation that these changes were due to changes in vegetation and that perhaps Mars was inhabited. To top it off, in 1877, an Italian astronomer, Giovanni Schiaparelli, discovered long, faint lines on Mars, which he called *canali* (Italian for "channels" or "grooves"). In English-speaking countries, *canali* was translated as *canals*. Were these canals dug by Martians for the transportation of water from the polar caps for irrigation? Some scientists of the time thought so, but the answer is no. Pictures taken by probes to Mars in the 1970s showed conclusively that the *canali* are not canals; they are changing surface features due to shifting dust from atmospheric winds.

NASA/ESA/The Hubble Heritage Team (STScI/AURA)

Figure 16.15 Mars at Closest Approach
The full disk of Mars is seen from the Hubble Space Telescope in 2003. The South polar ice cap is clearly visible along with dark surface features that periodically get covered by dust.

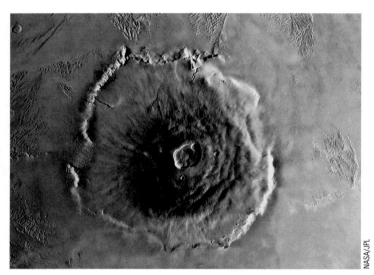

Figure 16.16 Olympus Mons, "Mount Olympus"
This huge Martian volcano (the largest in the solar system) is about 24 km (15 mi) high and 600 km (370 mi) wide at its base. The caldera is 80 km (50 mi) across at the summit.

Mars has 12 or more extinct volcanoes. The largest, Olympus Mons (Mount Olympus), shown in ● Fig. 16.16, is the largest known volcano in the solar system. It rises 24 km (15 mi) above the plain and has a base with a diameter greater than 600 km (370 mi). The volcano is crowned with a crater 80 km (50 mi) wide. The largest volcano on the Earth is Mauna Loa on the island of Hawaii. The base of Mauna Loa rests on the ocean floor about 5 km (3 mi) below the surface of the Pacific Ocean and extends upward another 4.2 km (2.6 mi) above the level of the ocean. Thus, Mauna Loa is about 40% as high as Olympus Mons.

Mars has its own "Grand Canyon" called *Valles Marineris*, or Mariner Valley (● Fig. 16.17). It is about 4000 km (2500 mi) long (roughly the width of the conterminous United States) and 6 km (4 mi) deep (almost four times as deep as our Grand Canyon). This tremendous gash in the surface of Mars is thought to be a gigantic fracture caused by stress within the planet.

Because Mars is the terrestrial planet that most closely resembles the Earth, it has been the focus of space exploration for many decades. The *Viking I* spacecraft landed on the Martian surface in 1976. Since then, a variety of spacecraft have been sent to Mars to explore its characteristics and learn more about the planet. The ultimate goal is someday to send a crewed mission to Mars. But first, uncrewed spacecraft are being sent to Mars to obtain the information needed for a live landing.

The public awareness of Mars was heightened in 1997 when dramatic pictures of the Martian surface were sent back by a mobile, radio-controlled (from the Earth) rover called *Sojourner*. In January 2004, two more rovers, *Spirit* and *Opportunity*, landed on opposite sides of the planet. These rovers have sent pictures of and information about Mars back to the Earth, and much more has been learned about our Martian neighbor. One important confirmation was that Mars once had water on its surface. This finding is based on the rovers' visual, chemical, and mineral analyses, which indicate that water percolated through the rock at some time.

In October 2006, the long-lived *Opportunity* rover reached the edge of Victoria Crater (● Fig. 16.18). The crater is half a mile wide with scalloped edges of jutting ledges and gentler slopes, and it is five times larger than other inspected craters. In 2009 the rover *Spirit* became stuck in soft soil, and after radio contact was lost in 2010, the mission was

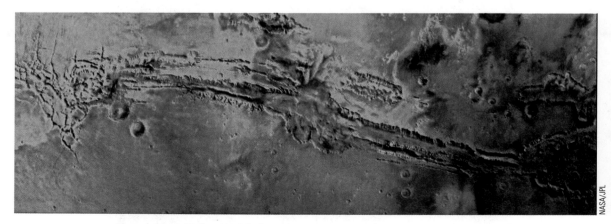

Figure 16.17 Valles Marineris
This enhanced color mosaic shows the great canyon Valles Marineris on Mars. The canyon is 4000 km (2500 mi) in length and 6 km (4 mi) deep. Geologists believe that it is a fracture in the planet's crust caused by internal forces.

Figure 16.18 An Opportunity
The long-lived *Opportunity* reached the edge of the half-mile wide Victoria Crater (the rover *Opportunity* is superimposed at the top center in this image).

declared ended. *Opportunity* marches on, however, and by summer 2011 it had traveled more than 30 km (18 mi) on the Martian surface. Its target is a 22 km (14 mi) wide crater named Endeavour, thought to have old geologic rock for analysis.

The *Mars Reconnaissance Orbiter* (*MRO*), launched in August 2005, had a mission to search for evidence that water persisted on the Martian surface for a long period of time. It found subsurface ice exposed by impacts and calculated the amount of water ice in the polar caps. Although there is evidence that water flowed on the Martian surface, it remains a mystery whether water was around long enough to provide a habitat for life.

Mars is the only terrestrial planet other than the Earth to have a moon. In fact, Mars has two moons, which are discussed in Chapter 17.

Before going on to study the Jovian planets, look at ● Table 16.2, which summarizes information on the terrestrial planets.

Table 16.2 Physical Parameters of the Planets

Terrestrial Planets	Density (kg/m³)	Rotation Period (solar days)	Surface Temperature (°C)	Atmosphere Composition	Magnetic Field (Earth = 1)	Satellites
Mercury	5420	58.65	430 day −180 night	Trace H and He	0.01	0
Venus	5250	243.01 (retrograde)	453	96% CO_2, 3% N_2	<0.001	0
Earth	5520	0.997	20	78% N_2, 21% O_2	1	1
Mars	3940	1.026	27 day −123 night	95% CO_2, 3% N_2	0.001	2
Jovian Planets	**Density (kg/m³)**	**Rotation Period (solar days)**	**Surface Temperature (°C)**	**Atmosphere Composition**	**Magnetic Field (Earth = 1)**	**Satellites**
Jupiter	1314	0.414	−121	90% H_2, 10% He	13.9	63
Saturn	690	0.444	−180	97% H_2, 3% He	0.67	62
Uranus	1220	0.718 (retrograde)	−215	83% H_2, 15% He	0.74	27
Neptune	1640	0.671	−225	74% H_2, 25% He	0.43	13

Did You Learn?

- The terrestrial planets have physical and chemical characteristics that resemble the Earth.

- The most abundant molecule in each terrestrial atmosphere is as follows: Mercury, none; Venus, CO_2; Earth, N_2; and Mars, CO_2 (see Table 16.2).

16.5 The Jovian Planets

Preview Questions

- What are the Jovian planets, and what is their general composition?
- Why do Uranus and Neptune have a blue color?

The four major gaseous planets—Jupiter, Saturn, Uranus, and Neptune—are collectively known as the *Jovian planets* because of their similarity to the planet Jupiter. (In Roman mythology, Jove was another name for the reigning god, Jupiter.) The Jovian planets are large compared with the terrestrial planets (see Table 16.1). They possess strong magnetic fields, have many moons and rings, and are very distant from the Sun, with orbits far apart from one another. Because the Jovian planets are composed mainly of hydrogen and helium gases, they all have relatively low densities. (Table 16.2 gives the average densities for the terrestrial planets and the Jovian planets. Do you notice any distinctions or trends?)

All four planets are believed to have a rocky core with layers of ice above it. Upper layers of molecular and metallic hydrogen* apply high pressure to the ice layers and rock core, producing ice and rock that are much different from those on the Earth. The ices are believed to be methane, ammonia, and water. The Jovian planets are very different from the terrestrial planets.

Around 5 billion years ago when the planets first began to coalesce (Chapter 16.7), the predominant elements were the two least massive ones, hydrogen and helium. The heat from the Sun allowed these two elements to escape from the inner terrestrial planets; that is, the velocities of the atoms of these light elements were sufficient to allow them to escape the planets' gravitational pulls.

The terrestrial planets were left with mostly metal cores and rocky mantles, giving them high densities. The four large Jovian planets were much colder and massive, and they retained their hydrogen and helium, which now surround their ice layers and rocky cores. The Jovian planets consist primarily of hydrogen and helium in various forms, and this composition gives them much lower densities. See Table 16.2 for a summary of physical information about the Jovian planets.

Jupiter

Jupiter, named after the chief Roman god because of its brightness and giant size, is the largest planet of the solar system, both in volume and in mass. Its rotation is faster than that of any other planet; Jupiter takes only about 10 h to make one rotation.

Jupiter's diameter is 11 times as large as the Earth's, and it has 318 times as much mass; its density, however, is much less. Jupiter consists of a rocky core, a layer of ice, a layer of hydrogen in liquid metallic form (because it is at high pressure), and an outer layer of molecular hydrogen. A probe from the *Galileo* spacecraft parachuted into Jupiter in 1995 and confirmed that the outer atmosphere is a thin layer of clouds composed of hydrogen, helium, methane, ammonia, and several other gases. The mean temperature at the top of the clouds is about $-121°C$ ($-186°F$). Interestingly, these temperatures are warmer than

---

*Metallic hydrogen is a phase of atomic hydrogen that occurs at extremely high pressures in which the material transforms into a conducting solid or liquid.

expected; Jupiter is believed to have an internal energy source, probably from slow gravitational contraction of the interior.

Jupiter's clouds have many different patterns: bands, ovals, and light and dark areas in white, yellow, orange, red, and brown, as can be seen in Fig. 16.19. Convection currents are present, and the tops of updrafts (the lighter areas) and downdrafts (the darker areas) are visible.

Conceptual Question and Answer

Gaseous Planet Rotations

Q. How do astronomers measure the rotations of the gaseous planets?

A. All the Jovian planets rotate differentially, that is, different latitude regions rotate at different rates. Cloud features move and change over time, so their average rotation values are not very accurate. Serendipitously, radio emission was discovered coming from Jupiter in 1955. Because radio emission is associated with the magnetic field of the planet and because the magnetic field is generated in the interior of the planet, a rotation period of the interior can be calculated. These radio (internal) rotation values for all the Jovian planets are listed in Table 16.2.

An interesting feature of Jupiter is the Great Red Spot (● Fig. 16.19). The spot exhibits an erratic movement and changes color and shape. The most recent theory of the Great Red Spot postulates that it is a huge counterclockwise storm similar to a hurricane on the Earth but lasting hundreds of years. In 2006, a strange event occurred; a long-observed "white spot" reddened to a color almost identical to that of the Great Red Spot. The new red spot, affectionately dubbed "Red Spot Junior" by some astronomers, is nearly half the size of the Great Red Spot, or on the order of the diameter of the Earth.

Another feature of Jupiter is a faint ring system of particle matter orbiting in the plane of the planet's equator. This ring was discovered by a 1979 *Voyager* mission. The small, dark particles that make up the ring may be collision fragments by asteroid impacts from two small moons in the vicinity of the ring.

Jupiter has many moons (64 at latest count). The four largest moons, discovered by Galileo in 1610, are called the *Galilean moons* of Jupiter. In order of increasing distances from Jupiter, they are Io, Europa, Ganymede (the largest moon in the solar system), and Callisto. These and other moons are discussed in greater depth in Chapter 17.

Saturn

Saturn, named after the Roman god of agriculture, is famous for its stunning array of rings. Although the other Jovian planets have rings, no rings are as spectacular as those of Saturn.

The structure of Saturn itself is similar to that of Jupiter: a small, solid core surrounded by a layer of ice, a layer of metallic hydrogen, and an outer layer of liquid hydrogen and helium. Saturn's density is the lowest of any planet in the solar system, only about 0.70 g/cm^3, so the planet would float in water if there were an ocean big enough to hold it (see Buoyancy in Chapter 3.6). At more than nine times the diameter of the Earth, Saturn is the second-largest planet in the solar system. The temperature at the top of its clouds is approximately −180°C (−292°F); like Jupiter, these temperatures are warmer than expected, partly due to an internal heat source. Because of its distance from the center of the solar system, Saturn takes almost 30 years to orbit the Sun. However, it takes only 10.2 h for the planet to complete one rotation, making for a short Saturn day.

NASA, ESA, I. de Pater and M. Wong (University of California, Berkeley)

Figure 16.19 The Planet Jupiter
Jupiter's clouds, clearly seen in this color-enhanced image, have many different patterns: bands, ovals, and light and dark areas. A main feature is the Great Red Spot (right), which has erratic movement and changes color and shape. Another red spot, dubbed "Red Spot Junior" (left), is nearly half the size of the Great Red Spot, or on the order of the diameter of the Earth.

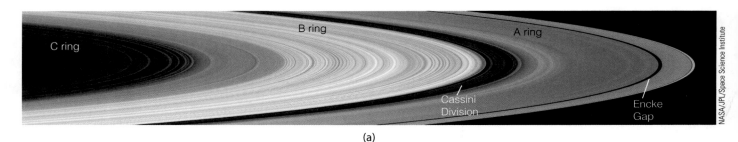

C ring   B ring   A ring

Cassini
Division

Encke
Gap

NASA/JPL/Space Science Institute

(a)

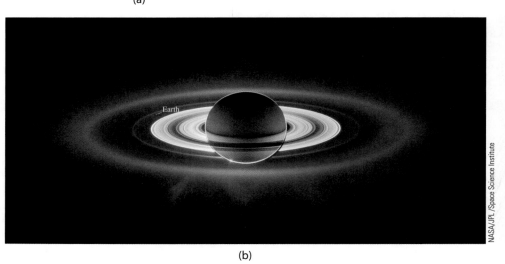

Earth

NASA/JPL /Space Science Institute

(b)

Figure 16.20 Saturn's Rings
(a) This *Cassini* spacecraft image shows the natural color of Saturn's rings in extraordinary detail. (b) The *Cassini* spacecraft flew into Saturn's shadow and captured this magnificent view of the ringed planet eclipsing the Sun. Notice Earth in the background, far away.

The rings of Saturn were first seen by Galileo in 1610 using a telescope. The rings are very wide but relatively flat bands, no more than several hundred meters in thickness. They extend from near the top of Saturn's atmosphere to well beyond the closest of its 62 known moons. The rings are made up mostly of highly reflective water ice particles and of rocks ranging in size from dust grains to huge boulders. These rocks are believed to be pieces of shattered moons, asteroids, or comets (Chapter 17.6) that broke up before reaching the planet (● Fig. 16.20).

The ring system shows structures on many scales, ranging from the divisions of major rings (labeled C, B, and A outward from the planet), down to myriad individual structures. More than 100,000 separate ringlets make up the broad major C, B, and A rings. One notable feature—the apparent separation, or distinct gap, between the B and A rings—is called the Cassini Division, named for Italian–French astronomer Giovanni Cassini, who discovered it in 1675. This division is approximately 4800 km (3000 mi) wide; the division was shown by the *Voyager* probe in 1980 to contain many ringlets within it.*

A wealth of information is being learned about Saturn. The *Cassini-Huygens* spacecraft arrived at the planet in July 2004, navigated through the 22,000-km (14,000-mi) gap between the F and G rings, and went into orbit around the planet. *Cassini* made 76 orbital surveys of the planet system during its 4-year mission, making many discoveries: it updated the radio rotation period, observed an atmospheric storm that lasted a year, detected strong lightning, observed gigantic vortex storms at the poles, discovered eight new moons, and found a new ring. It will have many more opportunities, because the mission has been extended to 2017 to observe an entire seasonal cycle on Saturn.

In addition, the *Cassini* orbiter ejected a probe toward the surface of Saturn's largest moon, Titan. For details about Titan and Saturn's 61 other moons, see Chapter 17.4. To understand how the *Cassini-Huygens* spacecraft made it to Saturn and how spacecraft have reached other planets, see the **Highlight: Solar System Exploration: Gravity Assists**.

---

*Seven rings are distinct enough for labeling. From the planet outward, they are D, C, B, A, F, G, and E. The major ring groups are C, B, and A. Rings B and A and the Cassini Division are easily visible from the Earth with low-power telescopes. The smaller Encke Gap, in honor of German astronomer Johann Encke, requires a high-power telescope.

Highlight   Solar System Exploration: Gravity Assists

After a 7-year, 3.5-billion-km (2.2-billion-mi) journey, the *Cassini-Huygens* spacecraft arrived at Saturn in July 2004, having made two Venus flybys and one Earth flyby (Fig. 1). Why was the spacecraft launched toward Venus, an inner planet, to go to Saturn, an outer planet?

Spacecrafts launched from the Earth have limitations, including a frustrating trade-off between fuel and payload: the more fuel, the smaller the payload. Using rockets alone, spacecraft are limited to visiting planets no farther from the Earth than Mars and Jupiter. More distant planets could not be reached by a spacecraft of reasonable size without taking decades to get there.

How, then, did the *Cassini-Huygens* craft get to Saturn in 7 years after being launched? Through the use of gravity in a clever scheme called *gravity assist*. Using planetary gravity makes missions to all the planets possible. Rocket energy is needed to get a spacecraft to the first planet; after that, the energy for the flight is more or less "free." During a planetary flyby, there is an exchange of energy between the planet and the spacecraft, which enables the spacecraft to increase its speed. (This phenomenon is sometimes called a *slingshot effect*.)

Let's look briefly at this ingenious use of gravity. Imagine *Cassini* making a flyby of Jupiter. When two objects interact, they essentially "collide," even though there may not be direct contact as there is in a billiard ball collision. In such an interaction, there is generally an exchange of momentum and energy. When the spacecraft approaches from "behind" the planet and leaves in "front" (relative to the planet's rotational direction), the gravitational interaction gives rise to a change in momentum (mv). The momentum of the spacecraft has a greater magnitude (speed) afterward and a different direction. The spacecraft leaves with more (kinetic) energy, a greater speed, and a new direction. (If the flyby occurred in the opposite direction, then the spacecraft would be slowed down through a braking effect.)

Recall from the early physics chapters that momentum and energy are conserved. That is, the swing-by planet gets an equal and opposite change in momentum ($\Delta p = m\Delta v$), giving it a retarding effect. Because the planet's mass is much greater than that of the spacecraft, the effect on the planet is negligible.

To help grasp the idea of a gravity assist, consider the analogous roller derby "slingshot maneuver" shown in Fig. 2. Two skaters interact, and energy and momentum are transferred to the "spacecraft" skater when she is being pulled around by the "planet" skater as shown in the figure. As a result, the "spacecraft" skater leaves with a greater speed and different direction.

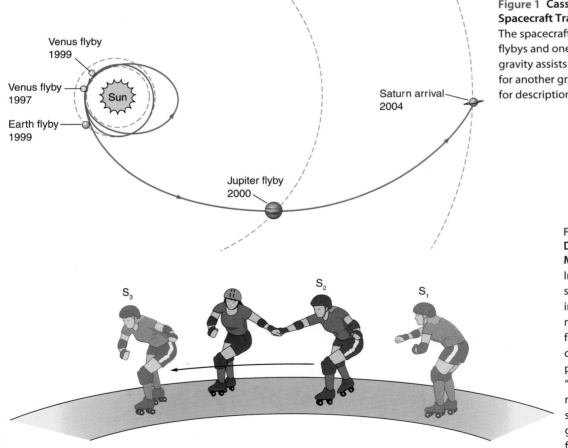

Figure 1 Cassini-Huygens Spacecraft Trajectory
The spacecraft made two Venus flybys and one Earth flyby to get gravity assists on its way to Jupiter for another gravity assist. (See text for description.)

Venus flyby 1999
Venus flyby 1997
Earth flyby 1999
Sun
Saturn arrival 2004
Jupiter flyby 2000

Figure 2 The Roller Derby "Slingshot Maneuver"
In a roller derby slingshot maneuver, skaters interact; energy and momentum are transferred to the "spacecraft" skater (S_2) when pulled around by the "planet" skater. As a result, the "spacecraft" skater leaves with a greater speed and different direction (S_3).

S_3 S_2 S_1

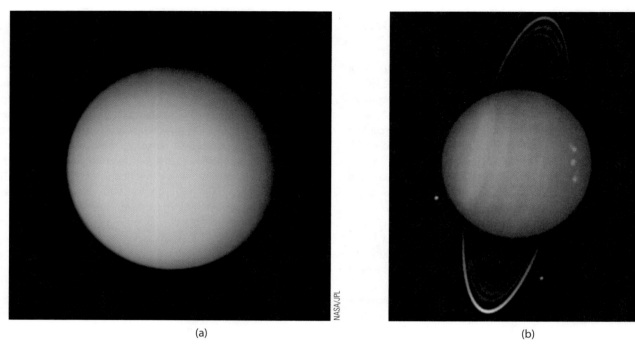

(a) NASA/JPL

(b) NASA and Erich Karkoschka, University of Arizona

Figure 16.21 Uranus in True and False Color
(a) A true-color picture of Uranus. This image shows how the planet would be seen by the human eye from a spacecraft. (b) A false-color image of Uranus reveals the planet's rings and bright cloud patches in red.

Uranus

Uranus ("YUR-en-us") was discovered in 1781 by William Herschel (1738–1822), an English astronomer. The name Uranus was chosen in keeping with the tradition of naming planets for gods of mythology. In Roman mythology, Uranus was the father of the Titans and the grandfather of Jupiter.

● Figure 16.21a shows the true color of Uranus. Its atmosphere is mostly hydrogen and helium, but it appears a blue-green color because it contains methane, a gas that absorbs the red end of the sunlight spectrum. Uranus has a distinct peculiarity. Unlike other planets, where the planet's spin axis is roughly perpendicular to its orbital plane, Uranus' rotation axis lies almost within the plane, tilted 98° from the perpendicular. The planet essentially revolves around the Sun on its side. It is not known why Uranus is tilted this way. Perhaps it occurred as a result of some event, such as a grazing collision between the planet and another planet-sized object. Such a collision event probably disturbed Uranus enough so that it is not producing extra heat from gravitational contraction, resulting in few clouds and bands in the atmosphere. Also, as noted in Chapter 16.2, Uranus has retrograde rotation.

Uranus has a very thin ring system. Figure 16.21b shows Uranus' rings and bright cloud masses. These low-lying clouds move around Uranus in the same direction as the planet's rotation. The rings are composed mainly of dark material, from small grains to particles up to 1 m in diameter. Lacking ices, the rings do not reflect light as well as the rings of Saturn. Along with these 13 rings, some 27 moons orbit the planet.

Neptune

Neptune was first observed in 1846 by Johann G. Galle (1812–1910), a German astronomer. Sharing the credit for the discovery are John Couch Adams and Urbain Le Verrier, English and French mathematicians, respectively. Using Newton's law of gravitation, Adams and Le Verrier made calculations that produced information on where to look for a suspected planet that was disturbing the orbital motion of Uranus.*

In 1989, *Voyager 2* arrived at Neptune and sent back to the Earth a photographic record of Neptune's clouds, storms, large wind systems, 13 known moons, 11 rings, and the Great Dark Spot, similar to Jupiter's Great Red Spot (● Fig. 16.22). Because of internal heating from gravitational contraction, Neptune's atmosphere is actually warmer than that of Uranus, and more clouds and storms are visible. These warmer temperatures also cause less haze in the atmosphere; therefore, Neptune appears bluer than Uranus.

Neptune is sometimes regarded as a twin to Uranus. Not only are the two similar in size and the composition of their atmospheres, but their internal compositions are also thought to be similar. Both planets have a rocky core surrounded by a mantle of water, methane, and ammonia ice. The mantle is surrounded by a layer of gas composed mainly of hydrogen and helium.

Neptune is massive enough to gravitationally affect its surrounding space, especially the small objects farther out from the Sun. These intriguing objects are discussed in the next section.

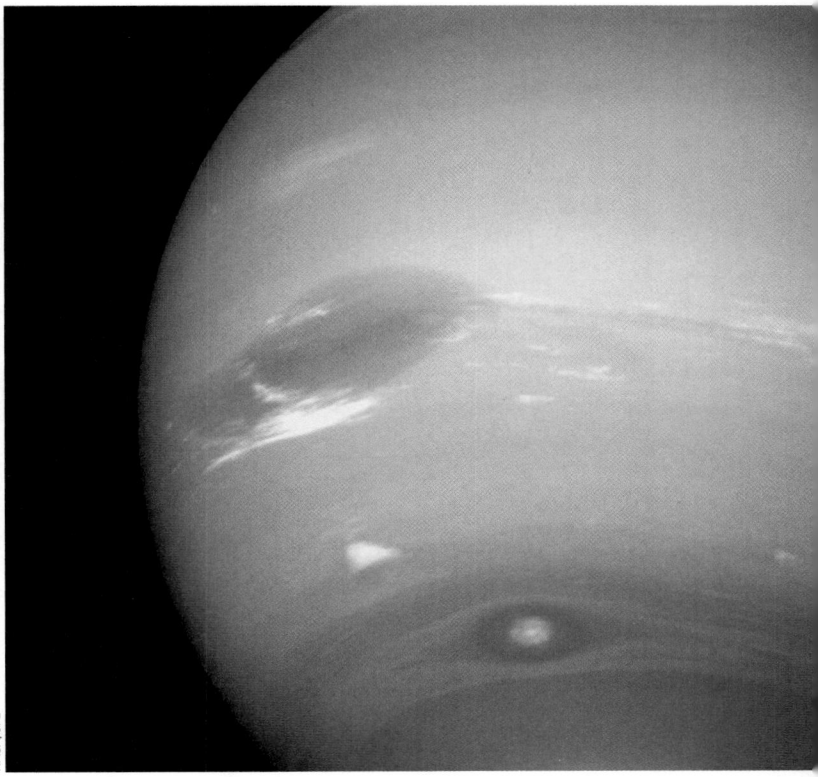

NASA/JPL

Figure 16.22 The Planet Neptune Reconstructed image taken by *Voyager 2* shows the Great Dark Spot (*left*) accompanied by bright, white clouds that undergo rapid changes in appearance. Below the Great Dark Spot is a bright feature that has been nicknamed "Scooter." To the right of this is a feature called "Dark Spot 2," which has a bright core.

Did You Learn?

- Jupiter, Saturn, Uranus, and Neptune are Jovian planets; they are mostly composed of hydrogen gas.

- The bluish color of Uranus and Neptune is caused by a small percent of methane gas in their atmospheres that absorbs the red part of sunlight.

16.6 The Dwarf Planets

Preview Questions

- Which dwarf planet's orbit goes inside that of Neptune's orbit?

- Why is Pluto considered a dwarf planet?

As noted in the chapter introduction, 2006 was a year of change. The definition of a planet was revised by the International Astronomical Union (IAU), an organization of more than 8800 astronomers from 85 countries. The IAU is the internationally recognized authority for assigning designations to celestial bodies and to any surface features of them. In 2006, the IAU adopted the following criteria for a solar system body to be a planet:

1. It must be in orbit about the Sun.

2. It must have sufficient mass for self-gravity to form a nearly round shape.

3. It must be the dominant body within its orbit; that is, it must have cleared the neighborhood around its orbit.

The last criterion disqualifies Pluto from the planet club. The phrase "have cleared the neighborhood around its orbit" means that during formation, the planet became gravitationally dominant, clearing out any debris around its orbit other than its own satellites.

---

*In Roman mythology, Neptune was the god of the sea, the son of Saturn, and the brother of Jupiter and Pluto.

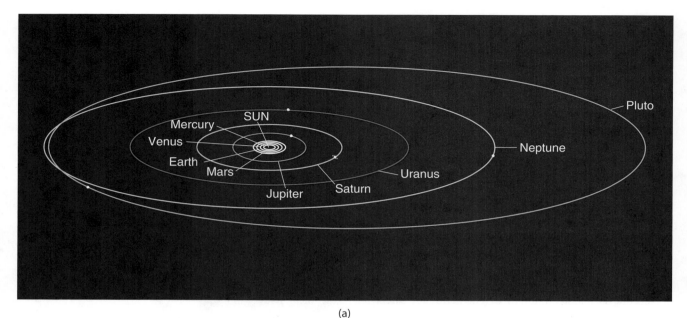

(a)

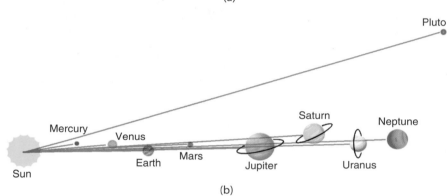

(b)

Figure 16.23 Oddities of Pluto
(a) Pluto comes inside of Neptune's orbit, which eliminates it as a major planet. (b) The orbits of
the major planets lie generally in a plane. Pluto's orbit is greatly inclined to this plane.

Pluto's orbit takes it inside that of Neptune, and so Pluto is not the dominant body of its
orbit (● Fig. 16.23a). Another oddity of Pluto is that its orbit is highly tilted compared
with those of the major planets, which generally lie in a plane (Fig. 16.23b). Because of
its small size and noncircular and tilted orbit, scientists hypothesize that Pluto may be an
escaped moon of one of the Jovian planets.

As more and more objects are discovered beyond Neptune, they are classified and sometimes reclassified. New classes help scientists better study the objects by allowing better
comparisons regarding structure, composition, and origin. In addition to redefining planet
status, IAU established two new categories for objects that orbit the Sun. One was **dwarf
planets**, which is now Pluto's designation, along with *Ceres* (a large asteroid in the asteroid
belt between Mars and Jupiter) and several others beyond Pluto's orbit. The other category
is *small solar-system bodies*, including all objects that are not satellites but that orbit the Sun
(Chapter 17.6). A discussion of the dwarf planets follows.

Ceres

Ceres lies in the asteroid belt between Mars and Jupiter (Chapter 17.6). Discovered in 1801
by Italian astronomer Giuseppe Piazzi, Ceres was named after the Roman goddess of agri-

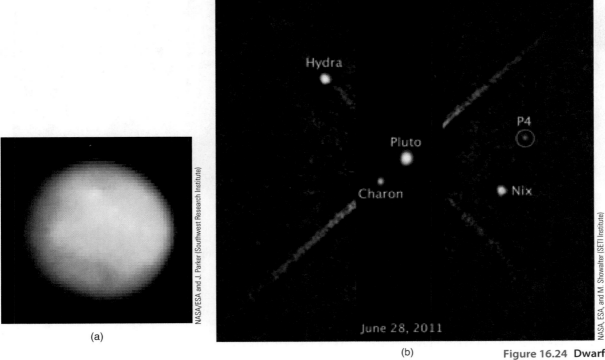

(a)

(b)

June 28, 2011

NASA/ESA and J. Parker (Southwest Research Institute)

NASA, ESA, and M. Showalter (SETI Institute)

Figure 16.24 Dwarf Planets Ceres and Pluto

(a) Ceres, the smallest dwarf planet, is found in the asteroid belt; it orbits the Sun every 4.7 years. (b) This image of Pluto and its four moons was taken by the Hubble Space Telescope in 2011. The moon Charon, discovered in 1978, and Pluto are purposely dimmed in this image to reveal the other moons. Nix and Hydra, discovered in 2005, are actually much fainter than Charon. Even fainter is the moon P4, a discovery in 2011.

culture and fertility. Although Ceres was initially classified as a planet and then for a long time as an asteroid, it is now designated a dwarf planet. Its orbit is not clear of other objects (the asteroid belt), so it does not have planet status.

Ceres is not large. With a diameter of about 940 km (580 mi), it is the smallest dwarf planet. Infrared images suggest that Ceres has a textured surface, and it is theorized still to contain pristine water from when the solar system was formed (● Fig. 16.24a).

Pluto

Small, cold, and distant from the Sun, Pluto was named after the Roman god of the underworld. By the end of the nineteenth century, observations indicated that Neptune's gravitational influence was not sufficient to account for all the irregularities in Uranus' orbital motion. Following the success of the discovery of Neptune, another planet was predicted. Astronomers tried to calculate its location, and chief among them was Percival Lowell (1855–1916), an American astronomer who founded the Lowell Observatory in Arizona. He searched for the anticipated body but was unsuccessful. Not until 14 years later in 1930, after Lowell's death, did C. W. Tombaugh, working at the Lowell Observatory, finally observe Pluto only 6° away from Lowell's predicted position (Fig. 16.24b).

Pluto is so far away that little is known about its physical and chemical makeup. Spectroscopic investigations indicate that the planet is covered with methane and nitrogen ice. Because Pluto is so far from the Sun, its mean surface temperature is about −236°C (−393°F). Recall from Chapter 5.1 that absolute zero is −273°C.

Unlike the major planets, Pluto has not been visited by a space probe. NASA launched *New Horizons* in January 2006, the first space probe mission to a dwarf planet. Using a gravity assist from Jupiter (see the Highlight: Solar System Exploration: Gravity Assists in Chapter 16.5), a flyby of Pluto is expected in July 2015. It should shed some light on the surface features, chemical makeup, and atmosphere of this far-distant dwarf planet.

The *New Horizons* space probe will also inspect the four moons of Pluto: Charon, Hydra, Nix, and the unnamed moon P4. The largest moon Charon ("KERH-on") was discovered in 1978 and named after the mythical boatman who took the dead across the River Styx to

(a)   (b)   (c)

Figure 16.25 Other Dwarf Planets
(a) Eris and its small moon Dysnomia; (b) Haumea and its moon Hi'iaka (above) and Namaka
(below); and (c) Makemake, a small, icy body 46 AU from the Sun.

Hades. Little is known about Charon, which is about half the diameter of Pluto, and less is
known about Hydra and Nix, discovered in 2005, and P4 discovered in 2011.*

Other Dwarf Planets: Eris, Haumea, and Makemake

Previously known as 2003 UB313 (and informally as Xena), this far-away dwarf planet
was officially named Eris, after the Greek goddess of chaos and strife. Eris is slightly larger
than Pluto and about 27% more massive. In its current orbital position, this dwarf planet
is about three times farther away from the Sun than Pluto. Its period of solar revolution is
about 560 Earth years. Eris' orbit takes it through a belt of comets (Chapter 17.5) and its
orbit is not cleared, so it is a dwarf planet.

Eris' orbit is highly elliptical. Currently, it is near aphelion (farthest from the Sun) at a
distance of about 98 AU. In roughly 280 years, it will be near perihelion (closest to the
Sun) at a distance of 38 AU.

The composition of Eris is probably similar to that of Pluto: rocky and methane ice. As
the dwarf planet returns from the outer reaches of its orbit, more studies will be made.
Eris has one satellite, or moon, Dysnomia, named after the daughter of Eris, the spirit of
lawlessness (● Fig. 16.25a).

Discovered in 2004, dwarf planet Haumea ("heu-MAY-a") is named after a Hawaiian
goddess of fertility and childbirth (Fig. 16.25b). This small object has a mass about 30%
that of Pluto, orbits at a distance of 43 AU, and has two small moons, Hi'iaka and Namaka.
Another dwarf planet, named Makemake ("ma-KEE-ma-KEE") after a Rapanui (Polynesian)
god of the creator of humanity, was announced in 2005 (Fig. 16.25c). Its size is about
75% that of Pluto, and it has a semimajor axis of 75 AU from the Sun and a 310-year orbit
period. Not much is known about this dwarf planet; it does not have moons, and it is probably composed of methane and nitrogen ices.

And Beyond?

What lies in the outermost reaches of the solar system? One thing we know about is the
Kuiper ("KY-per") Belt, which extends just beyond the orbit of Neptune to well beyond the
orbit of Pluto and into the space of Eris. This disk-shaped belt consists of millions of comets and of cometary material (Chapter 17.6). The generic name for all small objects orbiting

*Hydra was named after Hydra, the nine-headed serpent who battled Hercules in Greco-Roman mythology,
and Nix was a mythological water sprite.

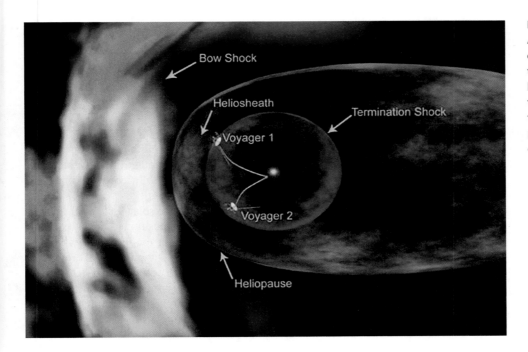

Figure 16.26 The Final Frontier
An artist's conception of the positions of Voyager spacecraft in relation to the structures formed around our Sun by the solar wind. Also illustrated is the *termination shock*, a violent region through which the spacecraft must pass before reaching the outer limits of the solar system.

beyond Neptune, including the Kuiper Belt, is *trans-Neptunian objects* (TNOs). It is hoped that after the spacecraft *New Horizons* rendezvous with Pluto in 2015, it will continue on through the Kuiper Belt, making observations and new discoveries.

The most distant solar system object that has been discovered is Sedna, which was named after the Inuit goddess of the sea and is slightly smaller than Pluto. It was found at a distance 90 times greater than that of the Earth from the Sun. Interestingly, the orbit of Sedna is extremely elliptical, with a perihelion of about 75 AU and an aphelion of about 90 AU. Sedna requires 10,500 years to orbit the Sun, and its orbit takes it well beyond the Kuiper Belt. There are currently more than 200 TNO objects, and with new technology and observations, hundreds more will be discovered and will give clues to the origin and formation of the solar system.

Where is the edge of the solar system? Astronomers have predicted that it is somewhere around 100 AU from the Sun. No one is certain, but there may be more information soon. An important boundary is the *termination shock*. This boundary is where the solar wind (streams of electrically charged particles from the Sun; see Chapter 18.2) collides with the interstellar medium that fills the vast regions between stars. The *Voyager 1* spacecraft first crossed this boundary in 2004, and by 2010, scientists confirmed that *Voyager 1* reached a point of zero solar wind velocity.

Voyager 1 was launched in 1977, and after passing Jupiter and Saturn taking photos, the spacecraft headed toward outer space. *Voyager 1* has traveled about 116 AU (about 17 billion km, or 11 billion mi).* *Voyager 2* is not far behind on a different trajectory, at about 95 AU from the Sun, and is also expected to reach the termination shock boundary. ● Figure 16.26 is an illustration of what the Voyager spacecraft are predicted to encounter at this boundary.

Did You Learn?

- Pluto's orbit carries it inside the orbit of Neptune for about 10 years of its 248-year orbit.

- Pluto is now considered a dwarf planet because it was not gravitationally dominant enough to clear the neighborhood around its orbit.

---

*It is expected that *Voyager 1* will not reach another system for 400,000 years. The spacecraft carries greetings in 55 languages and audiovisual materials depicting life on the Earth.

16.7   The Origin of the Solar System

Preview Questions

- With what did the formation of the solar system begin?
- How old is the solar system?

Any theory that purports to explain the origin and development of the solar system must account for the system as it presently exists. The preceding sections in this chapter have given a general description of the system in its present state, which, according to our best measurements, has lasted for about 4.56 billion years.

Currently, most astronomers believe that the formation of the solar system began with a large, spherical cloud of cold gases and dust, a **solar nebula**, positioned in space among the stars of the Milky Way galaxy. The nebula contracted under the influence of its own gravity, began rotating, and then flattened into a swirling disk of gas and dust. Through the process of condensation, the nebula evolved into the system we observe today. This explanation, which is known as the **condensation theory**, is today supported by such nebulae being observed throughout the universe. What initiated the process of forming the nebula from interstellar matter is not well known but is probably due to a shock wave from other star formations, interstellar cloud collisions, or supernovae explosions (Chapter 18).

Astronomers generally believe that the interstellar dust played a major role in the condensation process by allowing condensation to take place before the gas had a chance to disperse. The collection of particles was slow at first, but it became greater as the central mass increased in size. As the particles moved inward, the rotation of the mass had to increase to conserve angular momentum. Because of the rapid turning, the cloud began to flatten and spread out in the equatorial plane (see ● Fig. 16.27).

Figure 16.27 The Formation of the Solar System
This drawing illustrates the formation of the solar system according to the condensation theory. A cloud of gas and dust (*top*) in gravitational collapse developed into a flattened, rotating disk called the solar nebula. With further contraction, a protosun and protoplanets evolved. (See text for description.)

This motion set up shearing forces, which, coupled with variations in density, enhanced the formation of other masses. These masses moved around the large central portion, the "protosun," sweeping up more material and forming the protoplanets (*proto-* means "earliest form of"). Over a vast expanse of time, the protosun evolved into the current Sun, and the protoplanets evolved into the planets of our present solar system.

Did You Learn?

- The solar system formed out of a large, spherical cloud of gas and dust.
- The solar system is about 4.56 billion years old.

16.8 Other Planetary Systems

Preview Questions

- Why is it difficult to detect other planetary systems?
- What are two methods used to hunt for other planetary systems?

Are there other planetary systems in our galaxy or the universe? For centuries scientists have pondered this age-old question—whether or not planets orbit stars other than our own Sun. Such possible planets are referred to as *extrasolar planets* (or *exoplanets*). With an estimated 100 billion galaxies, the larger of which contain billions of stars, it would seem probable that planetary systems other than our own had formed. Because planets are relatively small, lie close to their parent star, and shine only by reflected light from it, the faint glimmer of a planet is lost in the brightness of its "sun" and cannot be detected. Or can it?

How do we search for other planetary systems? There are some physical methods. Although a planet itself is too faint to be seen, the effects of its gravitational pull on the parent star might be observed. As the planet revolves around the star, its gravitation attracts the parent star, giving it a slight "wobble." The more massive the planet, the more noticeable the effect, and these effects can be measured in several different ways to discover exoplanets. One method uses *astrometry*, the branch of astronomy that deals with the measurement of positions and motions of celestial objects, to measure the wobble. As the star moves through space (galaxies and stars move), the gravitational tugging shows up as tiny deviations from the straight-line path that a star without a planet(s) would follow, and the star is observed to wobble slightly (● Fig. 16.28). Although this method is time consuming (most planets take months or years to orbit a star), it is useful for finding planets at large distances from their parent star.

A second method of detection is the use of the Doppler effect (Chapter 6.5). A massive planet and the star both revolve about their center of mass (a gravitational balance point). The star then alternately moves toward and away from us. As a result, the starlight we observe would be Doppler shifted slightly toward the blue end of the spectrum when approaching and slightly toward the red end when receding. The Doppler shift is extremely small, requiring very sophisticated instruments to measure it. Although this technique has been very productive in discovering exoplanets, it is limited to the most massive planets and to those that orbit closely to their parent star.

A third technique using gravitational wobble is to measure time. Pulsars are ultradense, rapidly spinning stars that emit radio waves at very regular time intervals (Chapter 18.5). By accurately measuring the time of arrival of these pulses, planets can be discovered from variations in these radio signals. The first exoplanet was discovered using this technique in 1992.

One other planet-finding technique is called the transit method. Light from a star is dimmed if a planet passes in front of the star (transit) and causes an eclipse. The amount of the light dimmed and the length of time of the eclipse can give clues to the size, mass, and atmo-

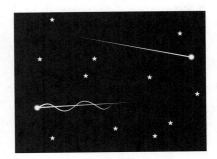

Figure 16.28 A Case of the "Wobbles"
As a star with a planet moves through space, gravitational tugging shows up with tiny deviations (*bottom*) from the straight-line path (*top*) of a star without planets. (The wobbling in this illustration is greatly exaggerated.)

Highlight   The Kepler Mission: Searching for Planets

NASA's Kepler Mission is a space telescope designed to survey our local region of the Milky Way and search for Earth-sized exoplanets, perhaps many orbiting in the habitable zone around its parent star (Fig. 1). Launched in 2009 and named for German astronomer Johannes Kepler, this Earth-orbiting spacecraft is expected to have a 3.5-year lifetime.

During this time, the telescope will observe more than 100,000 stars continuously in the constellations Cygnus and Lyrae and detect any fluctuations in brightness. Many of these fluctuations will be periodic due to an orbiting planet eclipsing the light from the star (Fig. 2). Known as the transit method, when an object crosses in front of a star, the dimmed light can be analyzed for

physical characteristics of the planet: orbit, size, mass, temperature, and atmosphere. One key parameter to know is whether these planets reside in the habitable zone around a star. This zone requires temperature ranges at which water can exist in liquid form, thus giving rise to the possibility of life.

As of 2011, the Kepler Mission had officially discovered 16 new exoplanets and more than 1200 planet candidates, nearly 70 being Earth-sized objects and more than 50 within the habitable zone of the parent star. One of those confirmed detections was a rocky planet 1.4 times the size of the Earth. It is an exciting time for discovery, and more is sure to come.

Figure 1   The Kepler Telescope
An artist's rendition of the Kepler Telescope orbiting above the Earth.

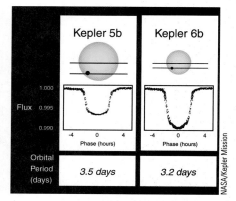

Figure 2   Transit Light Curves
These light curves show the change in flux (light intensity) as a function of time (orbit phase). The light decreases when the star is eclipsed.

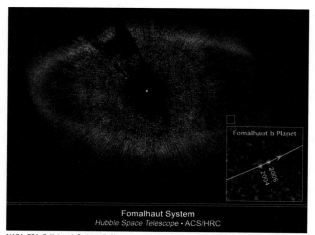

NASA, ESA, P. Kalas, J. Graham, E. Chiang, E. Kite (Univ. California, Berkeley), M. Clampin (NASA/Goddard), M. Fitzgerald (Lawrence Livermore NL), K. Stapelfeldt, J. Krist (NASA/JPL)

Figure 16.29   The First Visible Light Image of an Exoplanet
NASA's Hubble Space Telescope shows a planet at the lower right nearly hidden in the ring of dust and scattered starlight that surrounds the star. The inset image shows a magnified view of the planet in its orbit at two different times. The star Fomalhaut and it's planet are 25 ly from the Earth.

of the light dimmed and the length of time of the eclipse can give clues to the size, mass, and atmosphere of the orbiting planet (see the **Highlight: The Kepler Mission: Searching for Planets**).

Clearly, the search for other planetary systems is difficult, hampered by great distances and faintness of signals. More than 500 exoplanets have been detected by such indirect methods as described here. In 2008, astronomers using the Hubble Space Telescope photographed an exoplanet for the first time in visible light (● Fig. 16.29). This planet, estimated to be about three times the mass of Jupiter, orbits about 100 AU away from the bright star called Fomalhaut in the constellation Piscis Australis.

The first exoplanet orbiting within the habitable zone ("life" zone where liquid water could exist) around a star was discovered in 2010. Other detections have shown planets with atmospheres containing water, oxygen, and organic molecules. These discoveries are significant, but they are just the beginning. As telescopes and technology improve, smaller exoplanets will be discovered and many more details about the characteristics of these planets will become known. Will another Earth-like planet be found? As American astronomer Percival Lowell once said, "When dealing with the far-reaching scientific questions, it can be hard to separate one's science from one's imagination."

Did You Learn?

- Other planetary systems are very difficult to detect because they are very far away and are very faint.

- Planetary systems are best detected by the Doppler effect and the transit (eclipse) method.

KEY TERMS

1. astronomy
2. solar system (16.1)
3. geocentric model
4. heliocentric model
5. law of elliptical orbits
6. astronomical unit
7. law of equal areas
8. harmonic law

9. terrestrial planets (16.2)
10. Jovian planets
11. prograde motion
12. retrograde rotation
13. sidereal period
14. conjunction
15. opposition
16. albedo (16.3)

17. rotation
18. revolution
19. Foucault pendulum
20. parallax
21. aberration of starlight
22. dwarf planets (16.6)
23. solar nebula (16.7)
24. condensation theory

MATCHING

For each of the following items, fill in the number of the appropriate Key Term from the preceding list. Compare your answers with those at the back of the book.

a. _____ Planetary clockwise rotation
b. _____ Considers the Earth to be the center of the universe
c. _____ Earth's motion around the Sun
d. _____ Two planets lined up on the same side of the Sun
e. _____ Planets having Earth-like characteristics
f. _____ A flattened, rotating disk of gas and dust
g. _____ The study of the totality of all matter, energy, space, and time
h. _____ Spinning on an internal axis
i. _____ Apparent shift of positions of two objects when an observer changes position
j. _____ A planet sweeps out equal areas in equal periods of time
k. _____ Proves that the Earth rotates

l. _____ Orbit period with respect to the fixed stars
m. _____ Contains eight planets
n. _____ Counterclockwise revolution about the Sun
o. _____ Fraction of sunlight reflected by a celestial object
p. _____ All planets move in elliptical orbits about the Sun
q. _____ The gaseous planets
r. _____ Apparent change in direction because of orbital motion
s. _____ Considers the Sun to be the center of the solar system
t. _____ Two planets lined up on opposite sides of the Earth
u. _____ Relates the period of a planet to its semimajor axis of orbit
v. _____ Describes evolution of the solar system
w. _____ Average distance between the Earth and the Sun
x. _____ A new class of planets including Pluto

MULTIPLE CHOICE

Compare your answers with those at the back of the book.

1. Astronomy is best defined by which of the following? (Intro)
   (a) The study of the comets of the solar system.
   (b) The attempt to relate solar and planetary positions to human traits.
   (c) A study of Earth-like planets.
   (d) The scientific study of the universe.

2. Which of Kepler's laws gives the most direct indication of the changes in the orbital speed of a planet? (16.1)
   (a) law of elliptical orbits   (b) law of equal areas
   (c) harmonic law

3. Which of Kepler's laws gives an indication of the semimajor axis? (16.1)
   (a) law of elliptical orbits   (b) law of equal areas
   (c) harmonic law

4. One of the greatest scientists of all time, he discovered the true phases of Venus. (16.2)
   (a) Kepler   (b) Galileo   (c) Einstein   (d) Copernicus

5. Which of the following is abundant on the Earth but not on the other seven planets? (16.3)
   (a) oxygen   (b) water   (c) life   (d) all the preceding

6. The Foucault pendulum is an experimental proof of what characteristic of the Earth? (16.3)
   (a) revolution   (b) rotation
   (c) precession   (d) retrograde rotation

7. Terrestrial planets have which of the following physical characteristics? (16.4)
   (a) surfaces of mostly ice
   (b) primarily composed of rocks and metals
   (c) large ring systems
   (d) strong magnetic fields

8. Which of the following statements concerning the terrestrial planets is false? (16.4)
   (a) Mercury and Venus can never be in opposition.
   (b) All have magnetic fields except Venus.
   (c) All rotate clockwise as viewed from above the North Pole.
   (d) They are relatively close to the Sun.

9. Which of the following is *not* a physical characteristic of a terrestrial planet? (16.4)
   (a) small diameter
   (b) solid surface
   (c) relatively low density
   (d) relatively high-temperature environment

10. What are the primary constituents of the Jovian planets? (16.5)
   (a) hydrogen and helium   (b) hydrogen and carbon dioxide
   (c) methane and oxygen   (d) nitrogen and oxygen

11. Which of the following is *not* a physical characteristic of a Jovian planet? (16.5)
   (a) gaseous   (b) relatively high density
   (c) rocky/ice cores   (d) rapid rotation

12. What atmospheric feature was seen on Neptune? (16.5)
   (a) great red spot   (b) great dark spot
   (c) volcanic plumes   (d) large craters

13. Which Jovian planet revolves on its side and has retrograde rotation? (16.5)
   (a) Jupiter   (b) Neptune   (c) Uranus   (d) Saturn

14. Which one of the following criteria disqualifies Pluto as a major planet? (16.6)
   (a) sufficient mass for a round shape
   (b) in orbit about the Sun
   (c) orbit clear of other objects
   (d) none of the preceding

15. Which statement about the dwarf planet Ceres is true? (16.6)
   (a) It is farthest from the Sun.
   (b) It is between Mars and Jupiter.
   (c) It is larger than Pluto.
   (d) It has a thick atmosphere.

16. What is the age of the solar system? (16.7)
   (a) 65 million years   (b) 6000 years
   (c) 4.5 billion years   (d) 13.5 billion years

17. The planets of the solar system evolved most directly from which of the following? (16.7)
   (a) planetary nebula   (b) protosun
   (c) exoplanets   (d) condensation

18. Which of the following is *not* a very useful method for detecting an exoplanet? (16.8)
   (a) the observation of a star's motion
   (b) the observation of Doppler shifts in the spectrum of a star
   (c) the detection of alien electromagnetic signals

FILL IN THE BLANK

Compare your answers with those at the back of the book.

1. ____ is the study of the universe. (Intro)
2. The ____ model considers the Earth to be the center of the universe. (16.1)
3. The long axis of an ellipse is the ____. (16.1)
4. The square of the ____ is related to the cube of its semimajor axis. (16.1)
5. Relative to the Earth, Mars is a(n) ____ planet. (16.2)
6. The counterclockwise revolution of a planet about the Sun (as viewed from above) is called ____ motion. (16.2)
7. When the Earth and Mars are in line on the same side of the Sun, they are said to be in ____. (16.2)
8. The Earth has a(n) ____ of 0.33. (16.3)

9. A(n) ____ is used to prove the Earth rotates on its axis. (16.3)
10. ____ is the smallest terrestrial planet. (16.4)
11. Venus has a high surface temperature due to the ____. (16.4)
12. Mars appears reddish because of ____ minerals. (16.4)
13. The Jovian planet with retrograde rotation is ____. (16.4)
14. ____ is the smallest dwarf planet. (16.6)
15. The dwarf planet most distant from the Sun is ____. (16.6)
16. The explanation that the planets formed from a solar nebula is called the ____ theory. (16.7)
17. Exoplanets revolving around a star might cause tiny ____ from a straight-line path. (16.8)

SHORT ANSWER

16.1 The Solar System and Planetary Motion

1. What objects make up the solar system?
2. What is the main difference between the heliocentric model and the geocentric model?
3. When is the speed of a planet revolving around the Sun the slowest?
4. According to Kepler's harmonic law, how are the period and semimajor axis of a planet related?

16.2 Major Planet Classifications and Orbits

5. Describe the orientation and the shape of the motion of the orbits of the major planets.
6. What's the difference between inferior and superior planets?
7. Can Mars be in inferior conjunction with the Earth? Explain.

16.3 The Planet Earth

8. Is the Earth a perfect sphere? Explain.

9. Approximately how many times greater is the Earth's albedo than that of the Moon?

10. What proofs are there that the Earth is rotating and that it is revolving around the Sun?

16.4 The Terrestrial Planets

11. Describe four characteristics of the terrestrial planets.

12. Why can't we see the surface of Venus?

13. Explain the differences between the Grand Canyon on Earth and *Valles Marineris* on Mars.

16.5 The Jovian Planets

14. What are the major distinctions between the Jovian planets and the terrestrial planets?

15. Explain how a spacecraft can use a planet for acceleration.

16. Which planet's axis of rotation is a peculiarity, and why?

17. What is the mechanism that causes Uranus and Neptune to be blue in color?

16.6 The Dwarf Planets

18. What are dwarf planets?

19. Why is Pluto not considered a major planet, and what other peculiarity distinguishes it from the major planets?

20. What is the Kuiper Belt?

16.7 The Origin of the Solar System

21. What is a solar nebula, and what is its relationship to the condensation theory?

22. What was the major influence in the formation of the solar system?

16.8 Other Planetary Systems

23. What is astrometry?

24. Name two indirect methods by which other planetary systems can be detected.

VISUAL CONNECTION

Visualize the connections and give answers for the blanks.
Compare your answers with those at the back of the book.

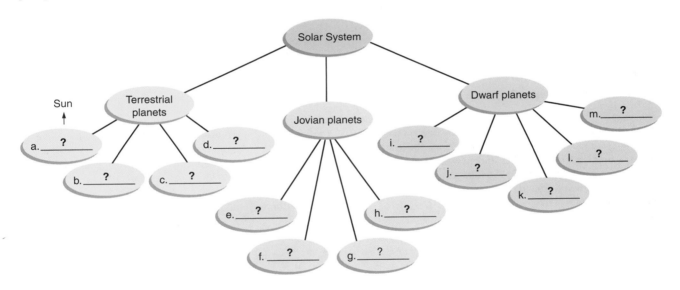

APPLYING YOUR KNOWLEDGE

1. Give some reasons our knowledge of the solar system has increased considerably in the past few years.

2. Why was it thought that there were other planetary systems in the universe before the technology was developed to search for them?

3. A Foucault pendulum suspended from the ceiling of a high tower located at the Earth's equator is put in motion in a north–south plane. Explain what would be observed regarding any apparent deviation, or the lack thereof, of the pendulum from the north–south plane during a 24-hour period.

4. Explain how the scientific method was used to reclassify the planets in the solar system.

5. How does the solar nebula theory explain the orbits of the major planets? Dwarf planets? Does it explain the rotations of the planets? Why or why not?

IMPORTANT EQUATION

Kepler's Third Law: $T^2 = kR^3$, where $k = 1 \ y^2/(AU)^3$ (16.1)

EXERCISES

16.1 The Solar System and Planetary Motion

1. Calculate the period T of a planet whose orbit has a semimajor axis of 5.2 AU.

   Answer: 11.9 y

2. Calculate the period T of a planet whose orbit has a semimajor axis of 19 AU.

3. Calculate the length R of the semimajor axis of a planet whose period is 225 days.

   Answer: 0.724 AU

4. Calculate the length R of the semimajor axis of a planet whose period is 165 years.

5. Determine what the period of revolution of the Earth would be if its distance from the Sun were 2 AU rather than 1 AU. Assume that the mass of the Sun remains the same.

   Answer: 2.8 y

6. Determine what the period of revolution of the Earth would be if its distance from the Sun were 4 AU rather than 1 AU. Assume that the mass of the Sun remains the same.

7. Asteroids are believed to be material that never collected into a planet. The asteroid Ceres has a period of 4.6 years. Assuming Ceres to be at the center of the planet if it had formed, how far would the planet be from the Sun?

   Answer: 2.8 AU

8. Show that the asteroid belt lies between Mars and Jupiter, and indicate to which planet it is closer.

9. Use Kepler's third law to show that the closer a planet is to the Sun, the shorter its period.

   Answer: $\dfrac{T^2}{R^3} = k$ or $T^2 = kR^3$

10. Use Kepler's third law to show that the closer a planet is to the Sun, the greater its speed around the Sun.

11. List the terrestrial planets in order of increasing distance from the Sun.

    Answer: Mercury, Venus, the Earth, Mars

12. List the Jovian planets in order of increasing distance from the Sun.

ON THE WEB

1. Journey through the Universe

How common are planetary systems around other stars (stars besides the Sun), and which might support life? Is there life beyond our solar system? Begin your journey and answer the questions above by following the recommended links on the student website at **www .cengagebrain.com/shop/ISBN/1133104096**. Discover how common planetary systems are around other stars (stars besides the Sun) that might support life.

2. The Search for Terrestrial Planets

Visit the recommended links on the student website at **www .cengagebrain.com/shop/ISBN/1133104096** and join NASA and the Kepler Mission on the search for habitable planets. What are the science objectives of the Kepler Mission? How are planets detected? Why do we care whether life exists elsewhere in the universe? What difference might the existence of such life make to our own lives here on the Earth?

Moons and Small Solar System Bodies

That's one small step for man, one giant leap for mankind.

•

Neil Armstrong, on taking the first step onto the Moon (1969)

< Buzz Aldrin and Neil Armstrong putting the American flag on the Moon.

J ust as planets revolve around the Sun, moons (natural satellites) revolve around most planets. You might be surprised by how many moons there are in the solar system. At present, 176 are known (● Table 17.1). Many moons are small; about 30% of Jupiter's 64 moons are 1 km or less in diameter. Because a moon is defined as a small, natural object in orbit about a planet, these small chunks of rock qualify.

Of course, some moons are large. Jupiter takes the prize for having the largest moon in the solar system, Ganymede. Our Moon is larger than all but four other moons and is even larger than Pluto. This chapter examines some of the larger moons, those that have known features and characteristics. Major attention is given to our own Moon because of our intimate relationship with it. Songs and

ASTRONOMY FACTS

❱ From the Earth, the same side of the Moon is always seen.

❱ The "dark" (unseen) side of the Moon is not dark; the Sun illuminates it too.

❱ Currently, there are 176 known moons in the solar system.

Chapter Outline

Table 17.1 Moons in the Solar System (as of 2011)

	Planet	Number of Known Moons
Terrestrial	Mercury	0
	Venus	0
	Earth	1
	Mars	2
Jovian	Jupiter	64
	Saturn	62
	Uranus	27
	Neptune	13
Dwarf	Ceres	0
	Pluto	4
	Haumea	2
	Makemake	0
	Eris	1
		Total 176

poems are written about the Moon, and much folklore involves the Moon, in particular its changing phases. People living near the ocean or vacationing at the beach observe tides, which are caused principally by the Moon, and an eclipse of the Moon is occasionally observed.

Other orbiting objects in the solar system, classified as *small solar system bodies*, are discussed in this chapter. Most of these objects orbit the Sun beyond Neptune and collectively are called *trans-Neptunian objects* (TNOs); see the Highlight: Trans-Neptunian Objects in Chapter 17.6. There are "minor planets," or *asteroids*, *meteoroids* that become *meteors* in the Earth's atmosphere, and *comets* that can be seen in the night sky when their orbits bring them close to the Sun. Halley's comet is one of the most famous. Finally, we look at *interplanetary dust*.

17.1 Structure, Origin, and Features of the Earth's Moon

Preview Questions

● How is the Moon associated with a month?

● What types of features on the surface of the Moon are the most prominent?

The exact origin of the word *moon* seems to be unknown, but in ancient times the Moon was used to measure time. The length of a month (moonth?) was related to the motion and phases of the Moon. The term *lunar* is used in reference to the Moon, for instance, a lunar month or the lunar surface. The word *lunar* comes from the ancient Roman goddess of the moon, Luna.

Because of its size and relative closeness, the Moon appears to us as the second-brightest object in the sky. (Which is the brightest?) The average distance of the Moon from the Earth is 380,000 km (240,000 mi). Many of the surface features of the Moon can be seen with the unaided eye, and even greater detail can be observed with binoculars or a small telescope. Soviet and U.S. space probes began taking photographs of the Moon in the 1960s, and the first human explorations of the Moon began with the landing of the first crewed U.S. spacecraft on the Moon on July 20, 1969.

The bright full Moon is a wondrous sight as it reflects the Sun's light to the Earth. Our Moon is the fifth-largest moon in the solar system and is nearly spherical, with a diameter of 3476 km (2155 mi), slightly greater than one-fourth the Earth's diameter (● Fig. 17.1). The mass of the Moon is $\frac{1}{81}$ that of the Earth, and its average density is 3.3 g/cm³, which is less than the Earth's (5.5 g/cm³). As a result, the surface gravity of the Moon is one-sixth that of the Earth's. Therefore, your weight on the surface of the Moon would be about one-sixth your weight on the surface of the Earth (but you would still have the same mass, Chapter 3.3).

A model of the Moon's interior is determined from orbiting spacecraft and seismic instruments left on the Moon by the Apollo astronauts. The Moon is thought to be made up of a small, solid, iron-rich inner core surrounded by a soft or liquid outer core. Most of the interior of the Moon is a solid, rocky mantle having a near uniform density. In general, the Moon has fewer heavy elements such as iron and nickel, leading to its lower average density. The mantle is covered by a thick outer crust that varies in thickness across the lunar surface on different sides of the Moon. The top of the crust is covered by a layer of pulverized rock and fine dust, called the regolith, formed because of impacts by meteoroids (see Ch. 17.6) over the Moon's 4.5 billion year history.

Conceptual Question and Answer

No Magnetic Field

Q. Why doesn't the Moon have a magnetic field if it has iron in the interior?

A. A magnetic field (Ch. 8.4) can be generated in the interior of a body if two conditions are met: (1) the body must be rotating, and (2) it must contain a liquid, electrically conducting material in the interior. The Moon may have a liquid iron outer core, which satisfies the second condition, but the Moon is not rotating to any reasonable extent. It rotates once per orbit, every 27.3 days. This rotation is not enough to generate a magnetic field. The planet Venus does not have a magnetic field. Why?

The size, orbit, structure, and composition of the Moon led scientists to the theory that the Moon was formed when a large object impacted the Earth during its formation. This theory is consistent with a model of the early solar system in which many collisions occurred during the formation of the planets. ● Figure 17.2 shows the impact theory in which the early Earth was struck with a glancing blow by a large object. Most of the material fell back to the Earth, but some of it coalesced together to form the Moon. The composition of the Moon's crust and mantle is similar to that of the Earth's mantle, consisting mostly of oxygen, silicon, magnesium, and aluminum.

The Moon's surface features include the prominent *highlands*, *maria*, and *craters*, along with the less prominent *rays*, *mountain ranges*, *faults*, and *rills*. All these features vary in size,

Figure 17.1 The Moon
The full disk of the Moon as seen from the Apollo 11 spacecraft. It is nearly spherical, with a diameter of 3476 km (2155 mi), about one-fourth the Earth's diameter. Note the differences between the dark and light features.

Origin: The Great Impact Theory

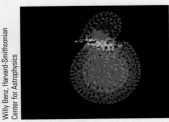
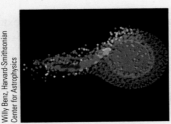
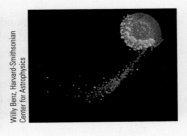
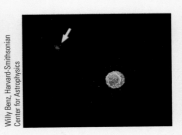

FIGURE 17.2 The Large Impact Theory
According to this theory, a large-sized object struck the Earth with a glancing blow, causing the ejection of matter into orbit to form the Moon.

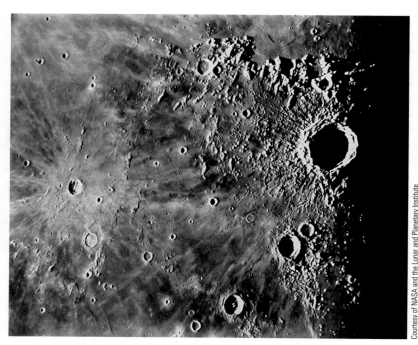

Courtesy of NASA and the Lunar and Planetary Institute

Figure 17.3 Craters on the Moon
Lunar craters are clearly visible, along with maria (plains), mountains, and rays.

shape, and structure. The most outstanding are the craters, which are clearly visible to an Earth observer with binoculars or a low-power telescope.

Highlands

The lunar **highlands** are the light-colored rock surface easily seen on the Moon (see Fig 17.1). Aptly named, the highlands extend to several kilometers above the average elevation on the Moon. On closer inspection the highland regions have more bumps and craters giving direct indication that the highlands are an older lunar crust. Surprisingly, the first photographs of the far side of the Moon taken by spacecraft showed that it was almost entirely covered by highlands. The highlands make up 85% of the lunar surface.

Maria

The lunar surface exhibits large, flat areas called **maria** ("MAH-ree-ah"), which were named by Galileo (● Fig. 17.3). *Maria* is a Latin word meaning "seas". They are believed to be craters formed by the impacts of huge objects from space and later filled with lava. Also called plains, these areas are much lower in elevation than the highlands, and are very dark because they have an albedo similar to black asphalt. The maria can easily be seen with the unaided eye during a full moon.

The Moon's maria and a few of its craters (about 1%) were produced by volcanic eruptions. The maria are composed of black volcanic lava that covered many craters, and most of them are on the near side of the Moon seen from the Earth. That fewer volcanic eruptions occurred on the far side is probably related to the Moon's surface crust being thicker there.

Craters

Craters are the best-known feature of the Moon's surface. The word **crater** comes from the Greek *krater*, meaning "bowl." Approximately 30,000 craters can be seen with an Earth-based telescope. They range in size from very small to hundreds of kilometers in diameter. Craters about 1 km (0.62 mi) in diameter tend to have a smooth, bowl-shaped interior. The floors of larger craters are flattened and form *basins*. Figure 17.3 gives a detailed view of some lunar craters, which are believed to have been formed by the impact of meteoroids hitting the Moon.

Rays, Mountain Ranges, Faults, and Rills

Some craters are surrounded by streaks, or *rays*, that extend outward over the surface. These rays are believed to be pulverized rock that was thrown out when the crater was formed. The pulverized rock reflects light well, and the bright rays are very distinct. The ray system of a crater has an average diameter about 12 times the diameter of the crater. One of the largest ray craters, dubbed Tycho, is shown in ● Fig. 17.4.

The *mountain ranges* on the lunar surface, some with peaks as high as 6 km (4 mi), all seem to be arranged in circular patterns bordering the great maria. This pattern indicates that the mountains were not formed and shaped by the same processes as mountain ranges on the Earth (Chapter 21.6).

A *fault* is a break or fracture in the surface of a planet or moon along which movement has occurred (see Earth faults in Chapter 21.5). The motion of a fault can be vertical,

Courtesy of NASA and the Lunar and Planetary Institute

Figure 17.4 Ray Crater
Tycho is one of the largest ray craters on the Moon. It may be viewed with binoculars or a small telescope during a full moon. You can also see Mare Tranquilitatis (the Sea of Tranquility), the site of the first Apollo moon landing.

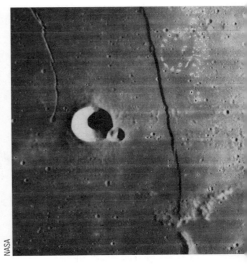

NASA

Figure 17.5 The Straight Wall
This photo shows a unique steep slope on the eastern side of Mare Nubium. The fault wall is about 130 km (80 mi) long and 0.25 km (0.16 mi) high. Rima Birt 1 (*top left*) is an irregular trough or rill. The photo was taken by *Lunar Orbiter V*.

horizontal, or parallel. Several faults are observed on the lunar surface. The very large cliff shown in ● Fig. 17.5 is the result of slippage of the Moon's crust along a fault. This fault, called the Straight Wall, is about 130 km (80 mi) long and 0.25 km (0.16 mi) high.

Another feature on the lunar surface are long, narrow trenches or valleys called *rills*. These trenches vary from a few meters to about 5 km (3 mi) in width and extend for hundreds of kilometers. Rills have steep walls and fairly flat bottoms that are almost 1 km below the lunar surface. Moonquakes are thought to be the cause of rills, which are similar to the separations of the Earth's surface produced by earthquakes.

The first crewed landing on the Moon took place on July 20, 1969, when the landing craft of *Apollo 11* settled on Mare Tranquilitatis (see Fig. 17.4 and chapter-opening photo). Between 1969 and 1972, the United States successfully completed five other lunar landing missions: *Apollo 12, 14, 15, 16,* and *17.** The Apollo astronauts collected and brought back to the Earth 370 kg (814 lb) of lunar material and placed scientific stations on the Moon's surface that collected data for some years.

Lunar rock samples, such as the one shown in ● Fig. 17.6, have enabled scientists to gain a better understanding of the Moon's composition and history. As the Moon formed and cooled, rocks rich in lighter metals such as calcium, silicon, and aluminum rose to the surface and formed the highlands. Samples from the maria, or lowlands, are more iron-rich and more dense so they sank deeper into the mantle. Rocks from the maria have been dated to be considerably younger than those from the highlands. Rocks from the highlands were formed between 3.9 and 4.4 billion years ago, whereas those from the maria have

NASA/Johnson Space Center

Figure 17.6 Scientists Examine a Lunar Sample
The samples were stored in an atmosphere of dry nitrogen, thus isolating them from oxygen and moisture to prevent chemical reactions.

*Apollo 13 was to make a lunar landing, but en route to the Moon an explosion occurred, causing the spacecraft to lose part of its electrical, oxygen, and other systems. The mission was aborted, but the spacecraft could not return directly to the Earth; after the craft had made a single pass around the Moon the *Apollo 13* crew returned safely to the Earth.

ages between 3.1 and 3.9 billion years. No lunar rocks older than 4.4 billion years or younger than 3.1 billion years have been found.

Did You Learn?

● The length of the month is based on the period of the Moon's orbit.

● The highlands, maria, and craters are the most prominent types of features on the Moon.

17.2 Lunar Motion Effects: Phases, Eclipses, and Tides

Preview Questions

● Why is only one side of the Moon seen from the Earth?

● What is the color of the Moon during a total lunar eclipse, and why?

The Moon revolves eastward around the Earth in an elliptical orbit, making one revolution in a little over 29.5 days. This is a **synodic month**, or the time it takes the Moon to go through a complete cycle of phases (a month of phases). A **sidereal month** is a little more than 27.3 days, which is the time of one complete revolution relative to stars. The synodic month is a little more than 2 days longer than the sidereal month for the same reason that a solar day is longer than a sidereal day (Chapter 15.3). Because of the Earth's motion around the Sun, the Moon must complete slightly more than one revolution to return to the same phase in its orbit.

The Moon's orbital plane does not coincide with that of the Earth but rather is tilted at an angle of approximately 5° with respect to the Earth's orbital plane (● Fig. 17.7). The 5° tilt allows the Moon to be overhead at any latitude between $28\frac{1}{2}°$ N and $28\frac{1}{2}°$ S. The 5° tilt accounts for the Moon's being higher and lower in the sky at different times of the year. The Moon also rotates as it revolves, making one rotation on its axis during one revolution around the Earth. This movement is described as a *synchronous* rotation. Because the Moon's periods of rotation and revolution are the same, on the Earth we see only one side of the Moon (see the **Highlight: Seeing Only One Side of the Moon**).

To an observer on the Earth, the Moon rises in the east and sets in the west each day. This apparent motion of the Moon is due to the Earth completing completing one daily rotation.

Figure 17.7 The Relative Motions of the Moon and the Earth
The top diagram is an angled view from above the Earth's orbital plane; the lower diagram is a view from within the Earth's orbital plane. Notice the Moon's tilted orbit; this causes the Moon to appear higher and lower in the sky at different times as seen from the Earth.

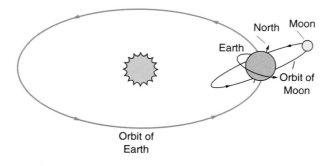

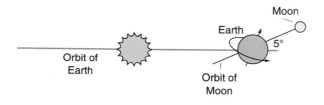

Highlight   Seeing Only One Side of the Moon

From the Earth, we see one side of the Moon and only that side. The scientific explanation is that the Moon's period of rotation is the same as its period of revolution. To better understand this explanation, consider the following analogy.

Visualize yourself on a circle facing a lamp in the center (Fig. 1). Then imagine moving sideways (sidestepping) around the circle, still always facing the lamp. When you return to the starting point, you will have made one *revolution* around the lamp.

Then consider the directions you faced while making the revolution. At the starting point (Fig. 1), you were facing east. After sidestepping to the right a quarter of a revolution, you were facing north; after half a revolution, west; after three-quarters of a revolution, south; and after one complete revolution, east again. Hence, you made one complete *rotation* around the axis of your body.

Without making the revolution around the lamp, you could have made these same changes in direction simply by standing in place and making one rotation (turning in place). Thus, always facing the lamp while you revolve requires a simultaneous rotation, with the same period as that of the revolution. Such is the case with the Moon; consequently, the same side of the Moon is always toward the Earth.

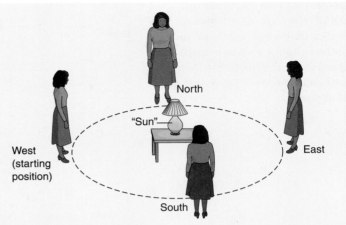

Figure 1 Synchronous Rotation
(See text for description.)

Actually, a little more than the same side of the Moon is seen because of variations in the Moon's orbit. Observations of the far, or "dark," side of the Moon were not available until picture-taking satellites orbited the Moon.

The lunar clock was important in early civilizations, but it receives little attention in our modern society. Ask a few people what time the Moon rises and compare the answers. You will probably get replies such as "about 7 P.M." or "near 9 P.M.," because people consider the Moon to be a nighttime object because of the bright nocturnal full moon. Actually, the Moon rises at various times during the day.

Because the Moon revolves around the Earth (360°) in 27.3 days, it must travel 360°/27.3 days = 13.2°/day. So, the Earth has to turn an extra 13.2° for the Moon to be in the same position in the sky as the previous night. For example, if last night the Moon was on the eastern horizon (rising) at a certain time, then tonight at the same time it will be 13.2° below the horizon.

How long does it take the Earth to turn 13.2°? In 24 hours the Earth turns 360°, so (13.2°/360°) × 24 h = 0.88 h, or 52.8 min. Hence, on the average, the Moon rises about 50 minutes later each day and is seen in the daytime as well as at night.*

Phases of the Moon

Probably the most familiar feature of the Moon to an Earth observer is the periodic variation in how much of the Moon is illuminated, known as *phases* of the Moon. One-half of the Moon is always reflecting light from the Sun, but only once during the lunar month does an Earth observer see all the illuminated half (a full moon). The starting point for the Moon's month of phases is arbitrarily taken to be its new phase. The new phase of the Moon occurs when the Moon is positioned between the Earth and the Sun (● Fig. 17.8). As can be seen in the figure, the side of the Moon toward the Earth is not illuminated, and the Moon cannot be seen. Because the Sun is on an observer's meridian with the Moon, the new moon occurs at 12 noon local solar time. During the rest of the month, the Moon exhibits different phases (illuminated portions) as illustrated in ● Fig. 17.9.

---

*Because of the Moon's orbit, the delay time varies, particularly at the equinoxes, as much as 10 min at the autumnal equinox and 1.5 h at the vernal equinox (see Chapter 15.5).

Figure 17.8 The Moon Relative to the Earth and the Sun
Diagram showing the position of the Moon relative to the Earth and the Sun during one lunar month as observed from a position in space above the Earth's North Pole. During the month, the Moon passes through all its phases.

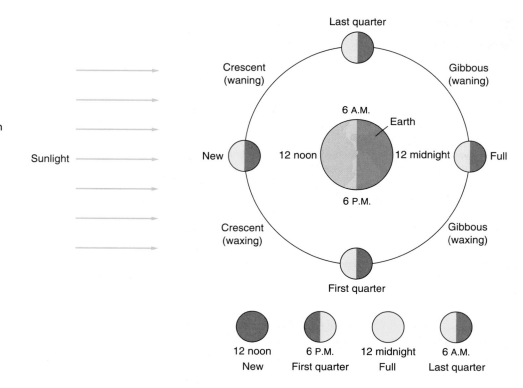

The **new moon** actually occurs for only an instant—the instant it is on the same meridian as the Sun. The Moon revolves eastward from the new-phase position, and for 7.38 solar days (one-fourth of 29.5 days) it is seen as a *waxing crescent moon*. The term *waxing phase* means that the illuminated portion of the Moon is getting larger each day. The term *waning phase* means that the illuminated portion of the Moon is getting smaller each day. A **crescent moon** has less than one-half of its observed surface illuminated. A **gibbous moon** occurs when more than one-half of the Moon's observed surface appears illuminated. (The Latin root of the word *gibbous* means "hump" and thus reflects the shape of the illuminated portion of the Moon in its gibbous phases.)

From new moon to first-quarter phase, the Moon is in *waxing crescent phase*, and the illuminated portion of the Moon waxes (or grows larger). The Moon is in **first-quarter phase** when it is 90° east of the Sun and appears as a quarter moon on an observer's meridian at

Figure 17.9 The Sequence of Lunar Phases
Diagram illustrating the phases of the Moon, as observed from any latitude north of $28\frac{1}{2}$°N. The observer is looking south, so east is on the left. The Sun's position can be determined by noting the local solar time at which the Moon is on the overhead meridian. The time period represented in the drawing is 29.5 days. Compare this drawing with Figure 17.10.

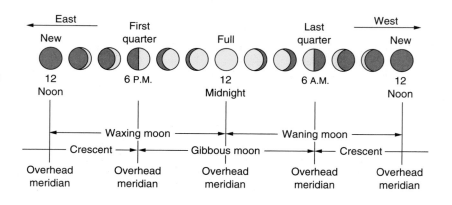

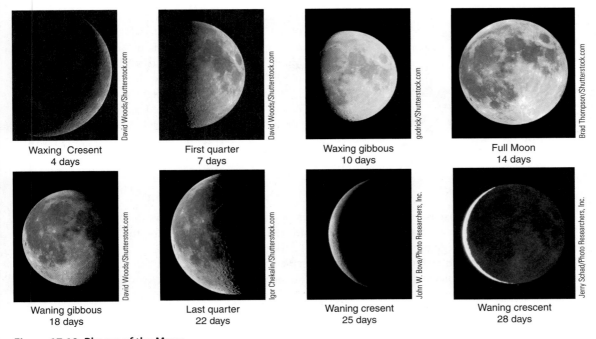

Waxing Cresent	First quarter	Waxing gibbous	Full Moon
4 days	7 days	10 days	14 days
Waning gibbous	Last quarter	Waning cresent	Waning crescent
18 days	22 days	25 days	28 days

Figure 17.10 Phases of the Moon
These eight photographs show the Moon at different times during the lunar month. (Why isn't a picture of a new moon shown?) They are arranged so that the phases are the way they appear in sequence. Compare these photos with Figure 17.9. Notice the "Earthshine" on the Waning crescent Moon caused by reflected sunlight from the Earth (see text for description).

6 P.M. (see Fig. 17.8 and 17.9). Like the new moon, the first-quarter moon occurs only for an instant because the Moon is 90° east of the Sun for only an instant.*

From the first-quarter phase, the Moon enters the *waxing gibbous phase* for 7.38 days. During this phase, the illuminated portion is larger than a quarter moon but less than a full moon. When the Moon is 180° east of the Sun, it will be in full phase and will appear as a **full moon**. The full moon appears on an observer's meridian at 12 midnight local solar time.

And so the sequence continues. The Moon enters the *waning gibbous phase* for 7.38 solar days, and the illuminated portion gets smaller. When the Moon is 270° east of the Sun, it is in **last-quarter phase** (sometimes called the *third-quarter phase*) and is on an observer's meridian at 6 A.M. local solar time. From the last-quarter moon, the Moon enters the *waning crescent phase* and the illuminated portion decreases. Then the moon is back to new moon, and the month of phases begins again. ● Figure 17.10 shows photographs of the phase sequence. ● Table 17.2 summarizes the times for the various phases of the Moon to rise, to be overhead (on meridian), and to set.

Table 17.2 Times for the Various Phases of the Moon to Rise, to Be Overhead, and to Set

Phase	Rising Time	Time Overhead	Setting Time
New moon	6 A.M.	Noon	6 P.M.
First-quarter moon	Noon	6 P.M.	Midnight
Full moon	6 P.M.	Midnight	6 A.M.
Last-quarter moon	Midnight	6 A.M.	Noon

*The quarter-moon phase does not refer to a fraction of illumination but to the period in the phase cycle, 7.38 days/29.3 days = 1/4. The Moon has moved one quarter of its orbit. Similarly, for the third-quarter (or last-quarter) phase, the Moon has 1/4 cycle to go to complete a sequence of phases.

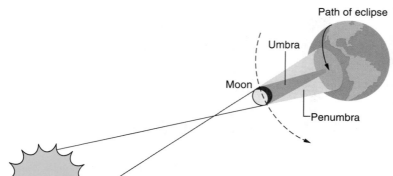

Figure 17.11 A Total Solar Eclipse
This diagram shows the positions of the Sun, the Moon, and the Earth during a total solar eclipse. The umbra and penumbra are, respectively, the dark and semidark shadows cast by the Moon on the surface of the Earth. Observers in the umbra see a total eclipse, and those in the penumbra see a partial eclipse.

At times during crescent moons, the dark portion of the Moon is seen to be faintly illuminated. During a crescent moon, the Sun is on the other side of the Moon from the Earth, and sunlight reflected from the Earth gives the dark portion of the Moon a faint glow. This faint, reflected light is called "Earthshine." Leonardo da Vinci explained this phenomenon some 500 years ago. The sight is sometimes referred to as "the old moon in the new moon's arms" or as the Moon's "ashen glow."

Eclipses

An **eclipse** is the blocking of the light of one celestial body by another. Consider the Moon to be between the Earth and the Sun (● Fig. 17.11). Examining the shadow cast by the Moon, regions of different degrees of darkness are observed. The darker and smaller region is known as the **umbra**. For an observer located in the umbra, the Sun's surface is completely blocked and the observer experiences a **total solar eclipse** (● Fig. 17.12). The other, semidark region of the Moon's shadow is called the **penumbra**, and an observer in this region during an eclipse sees only a portion of the Sun darkened in a *partial solar eclipse*.

Conceptual Question and Answer

A Phase for Every Eclipse

Q. Which phase must the Moon be in for a solar eclipse? A lunar eclipse?

A. A solar eclipse occurs when the Moon is at or near its new-phase position (Fig. 17.11). That is the only time all three bodies line up such that the Moon's shadow can be on the Earth. Similarly, for a lunar eclipse, the Moon must be at or near full phase (Fig. 17.14). Otherwise the Earth's shadow could not cover the Moon. Can you see why the lunar phases could *never* be caused by the shadow of the Earth? For example, at first-quarter phase, the Sun, the Earth, and the Moon are at right angles to one another. Can any shadow make a right-angle turn?

The length of the Moon's shadow varies as the Moon's distance from the Earth varies. The average length of the Moon's umbra is 373,000 km (231,000 mi), which is slightly less than the mean distance between the Earth and the Moon. Because the length of the umbra

Figure 17.12 The Solar Corona and Umbra of a Total Solar Eclipse
(a) During a total solar eclipse, the solar corona can be photographed. The corona is composed of hot gases that extend millions of miles into space. This photo shows only the brightest inner part of the solar corona. (b) The umbra during a total solar eclipse as seen from space. Within this shadow, which measures about 160 km (100 mi) across, it is dark for a few minutes while the Sun is totally blocked by the Moon.

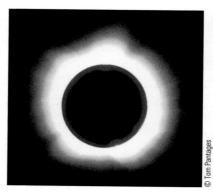

(a)

(b)

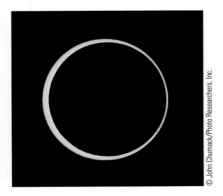

Figure 17.13 Annular Eclipse of the Sun
When the umbra of the Moon's shadow does not reach all the way to the surface of the Earth, an annular eclipse of the Sun is observed.

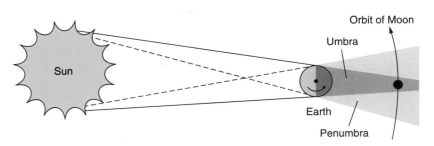

Figure 17.14 A Lunar Eclipse
This diagram shows the positions of the Sun, the Earth, and the Moon during an eclipse of the Moon.

is shorter, an eclipse of the Sun can occur in which the umbra fails to reach the Earth. An observer then sees the Moon's disk projected against the Sun but not covering its total surface, and a bright ring, or *annulus*, appears outside the dark Moon. This condition is called an *annular eclipse* (● Fig. 17.13).

The motions of the Moon and the Earth are such that the shadow of the Moon moves generally eastward during the time of the eclipse with a speed of about 1600 km/h (990 mi/h). Only those observers over whom the umbra shadow passes see a total eclipse. A total eclipse does not remain very long. The greatest possible length of time is 7.5 min, and the average time is about 3 or 4 min.

A *lunar eclipse* occurs when the Sun, the Earth, and the Moon are positioned in nearly a straight line, with the Earth between the Sun and the Moon (● Fig. 17.14). The Earth's shadow is long enough to reach the Moon, and occasionally the entire face of the Moon is obscured in a **total lunar eclipse**. Usually, however, the alignment is imperfect, so the shadow never completely covers the Moon and a *partial lunar eclipse* occurs. A total lunar eclipse may last for 1.5 hours; partial eclipses can last as long as 3 h 40 min. During a lunar eclipse, the Moon may appear to have a darkened reddish or copper color (● Fig. 17.15).

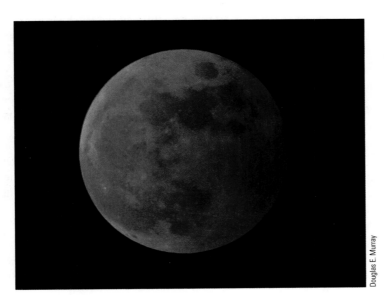

Figure 17.15 Lunar Eclipse
During an eclipse, the Moon takes on a darkened reddish or copper color because the sunlight that reaches the Moon is refracted (bent) and scattered by the Earth's atmosphere.

Conceptual Question and Answer

Copper Moon

Q. If the Moon is in a total lunar eclipse, then how can we see the Moon at all, and why is it reddish or copper in color?

A. During a total lunar eclipse, the Sun's rays that skim the edge of the Earth are refracted (bent) by the Earth's atmosphere and reach the Moon during the eclipse so it can be seen. However, in going through the atmosphere, blue light is scattered more (Rayleigh scattering) than red light, some of which reaches the Moon (see the Chapter 19.2 Highlight: Blue Skies and Red Sunsets). On reflection, the Moon appears reddish or copper-colored to an Earth observer.

Two to four solar eclipses may occur per year. But the land area on the Earth covered by the Moon's umbra during a solar eclipse is only on the order of 160 km (100 mi) wide, which makes solar eclipses quite infrequent for any particular location. Lunar eclipses are actually less numerous than solar eclipses, but everyone on the night side of the planet generally has the opportunity to see a total or partial lunar eclipse.

The frequency with which eclipses occur depends on the positions of the Earth, the Sun, and the Moon; the three bodies must be aligned during new or full moons. Also, because the orbital plane of the Moon is tilted approximately 5° with respect to the Earth's orbital plane, eclipses can occur only at a *node*, a point at which the Moon is crossing the Earth's orbital plane (see Fig. 17.7). As a result, two lunar eclipses occur in most calendar years. There can be up to three lunar eclipses in a given year, or there can be none at all. Solar eclipses occur two to five times a year, five being exceptional. In any *one* location on the Earth, a total solar eclipse is observed only once in three to four centuries.

The last total solar eclipse that could be seen from the conterminous United States occurred on February 26, 1979, and was visible only in the northwestern part of the country. The next total solar eclipse will be on August 21, 2017. The path of totality will cross the country from northwest to southeast. The last total solar eclipse for Alaska was on July 10, 1972. Alaskans have a long wait for the next one, which is due to occur in 2033. The last total solar eclipse for Hawaii was on July 16, 1991. Hawaiians will have a really long wait for the next one, which is scheduled for 2106.

Ocean Tides

At an ocean, it is observed that the water is sometimes far up on the beach, leaving only a small band of sand. At other times, the water is much farther down the beach, exposing a wide swath of sand. These changes are due to the alternating rise and fall of the ocean's water and are known as **tides**.

Ocean tides are due to the gravitational attraction that the Sun and the Moon exert on the Earth, chiefly that of the Moon. Specifically, the tides are due to the differences in the gravitational attraction at different places on the Earth. Because the Moon is much closer to the Earth than is the Sun, its differential gravitational attraction is much larger and has a greater influence on the tides.*

The Moon's gravitational attraction to the water on the side of the Earth closer to it causes a tidal "bulge," or high tide, on that side of the Earth. However, another bulge—or high tide—is formed at the same time on the opposite side of the Earth (● Fig. 17.16). The reason for this opposite tidal bulge is that the Earth itself is attracted toward the

---

*The tidal force is a *differential force* because it measures the difference in the gravitational attraction from two points on a body. The tidal force is an inverse-cube relationship with distance; therefore, the Moon has a greater effect on the Earth's tides than the Sun.

Moon but with more force than the water on its opposite side. The Moon's gravitational force therefore pulls the Earth away from the water, resulting in a second and opposite tidal bulge. As the Earth rotates, these two bulges, or *high tides*, "travel" around the Earth daily. The intervening water depressions form two *low tides*.

Thus, at any one location on the Earth, two high tides and two low tides generally occur daily because of the Moon's gravitational attraction to the Earth, the rotation of the Earth, and the revolution of the Moon around the Earth. Although the period of the Earth's rotation with respect to the Sun is 24 h, its period of rotation with respect to the Moon is 24 h 50 min. The two daily high tides are therefore about 12 h 25 min apart.

When the Sun, the Earth, and the Moon are positioned in nearly a straight line, the gravitational forces of the Moon and the Sun combine to produce higher high tides and lower low tides than usual. The variations between high and low tides are greatest at this time. These tides of greatest variations are called *spring tides*, and they occur at the new and full phases of the Moon (● Fig. 17.17). When the Moon is at the first-quarter or last-quarter phase, the Sun and the Moon are at angles of 90° with respect to the Earth. At these times, the gravitational forces of the Moon and the Sun tend to cancel each other, and there is a minimum difference in the height of the tides. Such tides are known as *neap tides*. Because of the tilted orbit of the Moon, the height of the tide also varies with latitude.

The time of high tide does not correspond to the time of the meridian crossing of the Moon. Because of the Earth's rotation, the tidal bulge is always a little ahead (eastward) of the Moon. The Earth rotates faster than the Moon revolves, and the Earth carries the tidal bulge forward in the direction it is rotating.

The action of tides also produces a retarding effect on the Earth's rotation, slowing it and lengthening the solar day about 0.002 second per century. Because the conservation of

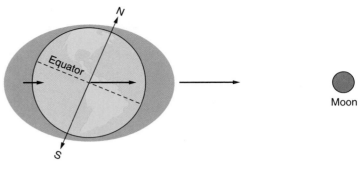

Figure 17.16 Tides
There are two tidal bulges, or two high tides, caused by the differences in the gravitational attractions of the Moon at different distances (indicated by the lengths of the arrows in the figure). The rotation of the Earth and revolution of the Moon generally give rise to two high tides per day (about 12 hours and 25 minutes apart).

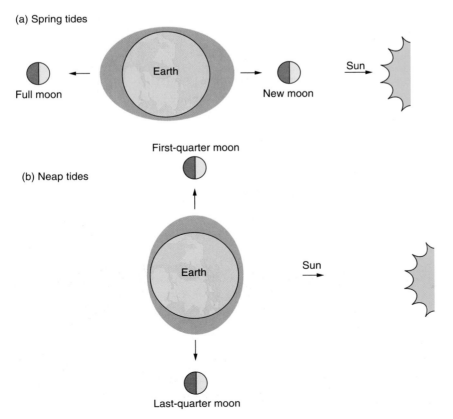

Figure 17.17 Spring and Neap Tides
This diagram shows the relative positions of the Earth, the Moon, and the Sun at the times of (a) spring and (b) neap tides. The tidal bulges are highly exaggerated.

angular momentum applies (Chapter 3.7), the decrease in the Earth's angular momentum must appear as an increase in the Moon's angular momentum. Thus, with constant rotational speed of the Moon, the distance between the Earth and the Moon must increase, and the Moon is gradually receding from the Earth. Measurements of the Moon's orbit show that its semimajor axis is increasing by about 3.8 cm/y.

Did You Learn?

● The same side of the Moon faces the Earth because the periods of its rotation and revolution are the same.

● The light that reaches the Moon during a lunar eclipse is reddish or copper in color due to the refraction and scattering of the Sun's light rays through the Earth's atmosphere.

17.3 Moons of the Terrestrial Planets

Preview Questions

● Which terrestrial planets have moons?

● What is thought to be the origin of the Martian moons?

Unlike Mercury and Venus, which have no moons, Mars has two small moons. Named Phobos ("Fear") and Deimos ("Panic") after the horses that drew the chariot of the Roman war god Mars, on a cosmic scale they are little more than large rocks (● Fig. 17.18). Both moons are irregularly shaped and heavily cratered. They have quite dark surfaces, with albedos no more than 0.06 (6%), which makes them difficult to observe from the Earth with the unaided eye. They can, however, be seen with a small telescope.

The larger of the two moons, Phobos, is about 28 km (17 mi) long and 20 km (12 mi) wide. The Martian moons orbit relatively close to the planet, only 9400 km (5800 mi) and 23,000 km (14,300 mi) for Phobos and Deimos, respectively, compared with 384,000 km (238,000 mi) for the distance of the Moon's orbit from the Earth. Phobos and Deimos orbit Mars in circular, equatorial orbits. And they are synchronous, always keeping the same side toward Mars, just as the Moon does toward the Earth.

The composition and density of Mars's two moons are quite unlike those of the planet. Astronomers believe that Phobos and Deimos did not form along with Mars but instead are asteroids that were slowed down and captured by Martian gravity early in the planet's history. (Asteroids are rocky objects that usually orbit the Sun between Mars and Jupiter; see Chapter 17.6.)

Did You Learn?

● Only the terrestrial planets Mars and the Earth have moons.

● The Martian moons, Phobos and Deimos, are believed to be captured bodies from the asteroid belt.

Figure 17.18 The Moons of Mars
On a cosmic scale, Phobos ("Fear") and Deimos ("Panic") are little more than large rocks.

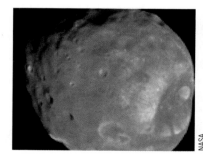 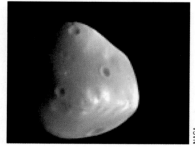

17.4 Moons of the Jovian Planets

Preview Questions

- Which moons of the Jovian planets are volcanically active?
- Which planet(s) do Titan and Triton orbit?

Unlike the terrestrial planets, the Jovian planets have numerous moons. As noted earlier, by definition a moon is any natural object in orbit about a planet, and some of the moons of the Jovian planets are quite small. For example, a couple of Jupiter's moons are only 1 km (0.62 mi) wide. Only the largest of the Jovian moons will be discussed here.

Jupiter's Galilean Moons: Io, Europa, Ganymede, and Callisto

The planet Jupiter has 64 known moons. Its four largest moons were discovered by Galileo in 1610 and so are referred to as the *Galilean moons*. In order of increasing distance from the planet, they are Io, Europa, Ganymede, and Callisto (● Fig. 17.19). Some physical data for the Galilean moons are given in ● Table 17.3.

Io and Europa are the smallest of the Galilean moons and are relatively close to the planet (● Fig. 17.20). Io is unique in composition, containing mostly sulfur and rock, whereas Europa, Ganymede, and Callisto have little rocky material and are made up of lighter substances, such as water ice. The moon densities are given in Table 17.3.

NASA/JPL

Figure 17.19 The Galilean Moons of Jupiter
A montage of the Galilean moons photographed by *Voyager 1* from a distance of about 1 million km (0.62 million mi). Clockwise from upper left are Io, Europa, Callisto, and Ganymede.

Table 17.3 The Large Moons of the Outer Planets

Moon Name	Mean Distance from Planet	Orbit Period (days)	Diameter	Mass (Earth Moon masses)	Density (kg/m³)
Io (Jupiter)	421,800 km (261,500 mi)	1.77	3643 km (2259 mi)	1.22	3500
Europa (Jupiter)	671,100 km (416,100 mi)	3.55	3122 km (1940 mi)	0.65	3000
Ganymede (Jupiter)	1,070,400 km (663,600 mi)	7.15	5262 km (3262 mi)	2.02	1900
Callisto (Jupiter)	1,882,700 km (1,167,300 mi)	16.7	4800 km (3000 mi)	1.46	1830
Titan (Saturn)	1,221,900 km (757,600 mi)	15.95	5150 km (3190 mi)	1.82	1880
Miranda (Uranus)	129,900 km (80,538 mi)	1.41	471 km (292 mi)	0.00090	1100
Triton (Neptune)	354,800 km (220,000 mi)	5.88 retrograde	2700 km (1680 mi)	0.29	2060
Comparison to Earth Moon Data					
Moon	384,400 km from the Earth (239,000 mi from the Earth)	29.5	3476 km (2155 mi)	1.0	3340

Io is the densest of the Galilean moons consisting mostly of silicate rock. In size, mass, and density, it is similar to our own Moon, but there are major dissimilarities. Io's surface is a colorful collage of yellows, reds, and dark browns due to the presence of sulfur and sulfur compounds (Fig. 17.19). Perhaps its most striking difference from our Moon and the other Galilean moons is that Io is volcanically active, with more than 400 active volcanoes. Also in contrast to our Moon and the other Galilean moons, Io's surface is not overly cratered. Its surface is relatively smooth, apparently from molten volcanic matter filling in the craters.

But what is the energy source for Io's volcanic activity? The moon is too small to have a radioactive internal source as the Earth does (Chapter 24.4). Like our Moon, Io should be long dead. Scientists believe that the source of Io's energy is Jupiter's gravity. Jupiter's enormous gravitational force on Io causes stresses that continually flex the moon's interior. Just as bending (flexing) a metal wire back and forth heats the wire through friction, the constant flexing of Io's interior produces heat that is released in volcanic eruptions.

Europa is quite different from Io. It has craters but only a few, which suggests that Europa's crust is relatively young. Also, lines crisscross bright, clear fields of water ice. Close-up spacecraft images show flat chunks of ice that have broken apart, like Europan "icebergs." Scientists speculate that underneath Europa's ice surface is an ocean of liquid water. This surface ice may be several kilometers thick, and the ocean is believed to be

Figure 17.20 Jupiter and Closely Orbiting Moons
Io (dot on the left) and Europa (dot on the right) orbit relatively close to the planet. Note that Io is orbiting over the Great Red Spot.

100 km (62 mi) deep below it. The liquid interior is also believed to result from Jupiter's tidal forces, and scientists are intrigued about the possibilities of life in this Europan ocean.

Ganymede and Callisto, the two outermost Galilean moons, are similar in size, mass, and density (Table 17.3). The densities suggest that the moons are composed of about a 50–50 mixture of ices and rock.

Ganymede, the largest moon in the solar system, is 35% larger than the Earth's Moon and is also larger than the planet Mercury. Its interior is believed to have a core, a mantle, and crust structure and, like Europa, is thought to have a subsurface liquid water ocean. Its surface has many impact craters, fault lines, and flat areas, or "maria." Unlike our Moon's maria, which originated from huge craters being filled in with lava, Ganymede's craters were filled in by water that solidified into ice. Ganymede's surface also has fault lines of lateral (sideways) displacement, similar to the San Andreas Fault in California (Chapter 21.5). Ganymede is the only place besides the Earth where such faults have been found. In 1996, Ganymede was found to have its own magnetic field, further making it a very unique moon.

In several ways, *Callisto* is similar to Ganymede. For example, Callisto is covered by rock and ice and is heavily cratered. However, it has many more craters indicating that its surface is the oldest of the Galilean moons. The craters are no wider than 50 km (31 mi), which suggests that Callisto's surface is not very firm.

A distinctive surface feature of Callisto is a huge bull's-eye formation of concentric rings or ridges that surrounds a large basin some 3000 km (1860 mi) across (● Fig. 17.21). These rings probably resulted from a huge impact, the ridges being analogous to the water ripples made when a stone is thrown into a pond. Perhaps the impact caused the icy surface to melt and move outward, resolidifying before the ridge ripples had time to dissipate.

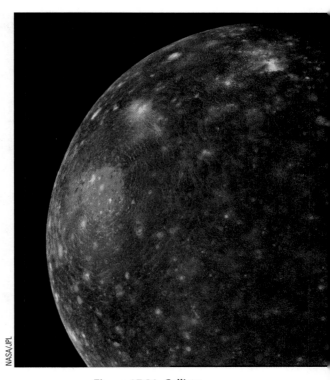

NASA/JPL

Figure 17.21 Callisto
Callisto, the outermost Galilean moon of Jupiter, is similar to Ganymede in composition but is more heavily cratered. The large series of concentric ridges visible on the left of the image is known as Valhalla. Extending nearly 1500 km (930 mi) from the basin center, these ridges probably formed when "ripples" from a large meteoritic impact froze before they could disperse completely. The image was taken by *Voyager 2*.

Saturn's Moon Titan

Saturn has 62 known moons. Six are of medium size, with diameters between 400 km (250 mi) and 1600 km (990 mi), but they are dwarfed by Saturn's largest moon, Titan. *Titan* is the second-largest moon in the solar system (after Jupiter's Ganymede). It is the only moon in the solar system to have a dense atmosphere, denser than the Earth's atmosphere. Also like the Earth, Titan's atmosphere is rich in nitrogen. Spacecraft data have shown a nitrogen content of about 90%, along with argon and methane. These elements and compounds provide a rich laboratory for chemical reactions, with much of the energy provided by sunlight. The presence of methane, however, has presented somewhat of a quandary. On the geologic time scale (Chapter 24.5), methane has a relatively short lifetime. Before being broken up by sunlight, a molecule of methane (CH_4) may last only 10 million years or so. With Titan being billions of years old, there was the question of how the gas was renewed.

After flybys of Venus, Earth, and Jupiter, the *Cassini-Huygens* spacecraft arrived at Saturn in July 2004 (Chapter 16.5), and the surface detail of Titan is slowly being revealed. Infrared images penetrating through the clouds reveal dark and light surface features (● Fig. 17.22). These features, including ridges, cracks, and large sand-like dunes, suggest geologic activity on the surface, and other surface features are confirmed lakes of methane. In January 2005, the *Cassini-Huygens* probe parachuted through Titan's thick atmosphere, gathering data on temperature, pressure, and composition. The spacecraft's camera revealed a rugged landscape (Fig. 17.22b). On descent, images showed an almost Earth-like "shoreline" along a lake (most likely methane, not water). This finding would answer the question on how methane in the atmosphere is renewed. Several dark, meandering channels were seen that appear to drain into the lake.

Titan's composition of rock and ice is like that of Jupiter's Ganymede and Callisto because of similar physical properties (compare in Table 17.3). The internal structure is probably a large, rocky core surrounded by a thick mantle of water ice with the possibility of a subsurface ocean.

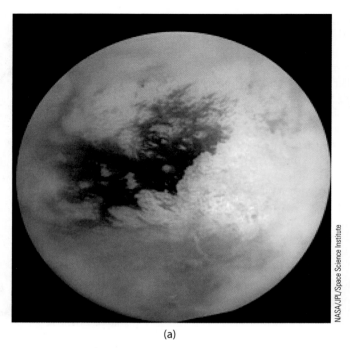

(a)   (b)

Figure 17.22 Titan
(a) A 16-image mosaic of Titan taken by the Cassini spacecraft. Dark and light surface features as well as some of the thick clouds in the atmosphere are seen. (b) A color-enhanced image taken by the *Huygens* probe camera after successfully landing on Titan. Most of the rocks in the image are roughly the size of pebbles, the larger ones with a size of about 15 cm (6 in).

Another Saturnian moon, called Enceladus, has recently been shown to have not only a thin atmosphere, but also volcanic geysers. The material coming from them is believed to be water vapor and may be a source of material that populates the E ring of Saturn. Enceladus is a small, icy moon about 500 km (300 mi) in size.

Uranus' Moon Miranda

Unlike Jupiter and Saturn, Uranus has no large moons. Of the 27 known moons of Uranus, the five major moons are relatively medium-sized. They are, in order of increasing distance from the planet, Miranda, Ariel, Umbriel, Titania, and Oberon.* Most of these moons have dark surfaces and are heavily cratered, with little evidence of geologic activity. One exception is Ariel, which shows some surface cracks, probably due to gravitational stresses. The other exception is Miranda; its physical data are shown in Table 17.3.

Miranda, the smallest and innermost of the five major moons, is a strange one, displaying a variety of surface features (● Fig. 17.23). There are large, curved regions of grooves and ridges. The regions that appear chevron-shaped have large faults and other rather unusual geologic features. Scientists hypothesize that this moon was broken into two large pieces and re-formed later under the influence of gravity.

Neptune's Moon Triton

Like Saturn's moon Titan, there is one major moon among Neptune's 13 known moons. Its name is *Triton* after the son of Poseidon, the Greek god of the sea. Triton is the only large moon to have a retrograde orbit; it orbits in the opposite direction from most planets and moons (Chapter 16.1). Compare its physical data in Table 17.3.

Figure 17.23 Uranus' Moon Miranda
Uranus' innermost large moon, Miranda, is roughly 480 km (300 mi) in diameter and exhibits a variety of geologic forms, some of the most bizarre in the solar system. Chevron-shaped regions and folded ridges in circular racetrack patterns are visible on the satellite's surface. There are large scarps, or cliffs, ranging up to 5 km (3 mi) in height; they are clearly visible in the lower right part of the photo. Next to them is a deep canyon approximately 50 km (30 mi) wide.

*Miranda and Ariel were characters in Shakespeare's play *The Tempest*. Umbriel comes from Alexander Pope's poem *The Rape of the Lock*, and Titania and Oberon were characters in Shakespeare's play *A Midsummer Night's Dream*.

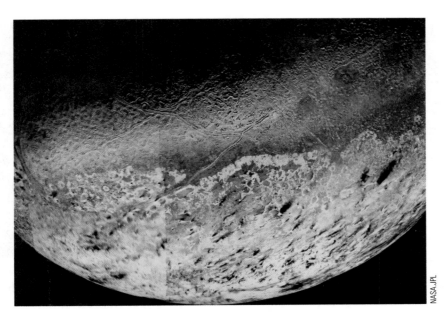

NASA.JPL

Figure 17.24 Triton's South Polar Cap
Neptune's largest satellite is a primarily white object with a pinkish cast in some areas. The pinkish color is probably due to frozen nitrogen. The land areas are strange and complex, and a scarcity of craters indicates that the surface may have been melted or flooded by icy "slush." A number of high-resolution photographs were combined to produce this image of Triton's south polar region.

Because of Neptune's thick cloud cover, the surface features of the planet itself are not revealed in photographs. However, excellent photographs of Triton have been obtained; some of its surface details are shown in ● Fig. 17.24. There are complex landforms and a polar ice cap of frozen nitrogen. These high-resolution photographs also show that Triton is volcanically active, with small geysers of nitrogen spewing from its icy surface. Because of Triton's retrograde orbit, astronomers speculate that Triton did not form with the planet but rather was captured by Neptune's gravitational field in the not-too-distant past.

Did You Learn?

- Currently known volcanically active moons are Jupiter's moon Io, Saturn's moon Enceladus, and Neptune's moon Triton.

- The large moon Titan orbits Saturn; Triton orbits Neptune.

17.5 Moons of the Dwarf Planets

Preview Questions

- What is unique about the orbit of Pluto's largest moon, Charon?
- What is the composition of the moons of the dwarf planets?

Pluto's Moon Charon

In 1978, Pluto was discovered to have a moon. It was named Charon ("KEHR-on"), after the mythical boatman who ferried the dead across the River Styx to the underworld, which was Pluto's domain. ● Figure 17.25 was taken by the Hubble Space Telescope and shows Charon as separate from Pluto.

Charon's diameter is about half that of Pluto, which makes it the largest satellite in relation to its parent planet. Charon's mass is about one-sixth that of Pluto, and its surface is thought to be mostly water ice. When the *New Horizon* spacecraft reaches Pluto in 2015 (Chapter 16.6), more will be learned about Charon. One odd thing is known, however: Pluto and Charon are in synchronous orbit. Charon has a rotation period of 6.39 days and an orbital period of 6.39 days. They are the same as Pluto's rotation period, which means

Dr. R. Albrecht, ESA/ESO Space Telescope European Coordinating Facility; NASA

Figure 17.25 Pluto and Its Moon, Charon
This image of Pluto and its moon was taken by the Hubble Space Telescope when the planet was 4.4 billion km (2.6 billion mi) from the Earth, or nearly 30 times the separation between the Earth and the Sun. The different colors suggest that the bodies have different surface composition and structure. The bright highlight of Pluto suggests that it has a relatively smooth reflecting surface layer.

that one side of Charon is always toward Pluto and one side of Pluto is always toward Charon. The two bodies are like a spinning dumbbell rotating in space.

In 2005, two small moons of Pluto were discovered beyond Charon and were christened Nix and Hydra. These names are associated with Charon and Pluto in Greek mythology. Nix was the goddess of the night and mother of Charon. Hydra was the nine-headed monster that guarded an entrance to Pluto's underworld (see Fig. 16.24b).

Another small moon was discovered in 2011 and is tentatively named P4 (Fig 16.24b). Little is known about these distant moons. Nix is the closest of them, with a diameter about 80 km (50 mi) and an orbital period of 25 days. P4 is about 25 km (15 mi) in size with a period of 31 days, and Hydra is about 100 km (62 mi) across with a period of 38 days.

Eris' Moon Dysnomia and Haumea's Moons Hi'iaka and Namaka

In 2005, Eris was discovered to have a small moon (see Fig. 16.25a). It was given the name Dysnomia ("diss-NOH-mee-uh") after the daughter of Eris, the spirit of lawlessness.

Little is known about Dysnomia. Its diameter is estimated to be about eight times smaller than that of Eris, but this moon is probably too small to have formed into a spherical shape. Its orbital is circular with a period of about 16 days. Astronomers still do not know its composition or mass. (For more information, see the Highlight: Trans-Neptunian Objects).

The other dwarf planet Haumea, the Hawaiian goddess of childbirth, has two small moons, Hi'iaka ("hee-ee-AH-kah"), and Namaka ("nah-MA-kah"), appropriately named after the two daughters of Haumea (see Fig. 16.25b). Both discovered in 2005, Hi'iaka, the larger of the two moons, is about 300 km in diameter and orbits Haumea every 49 days. Namaka, about one-tenth the size of Hi'iaka, orbits eccentrically with a period of about 18 days. Because the moons show signatures of water ice, they are believed to be fragments of Haumea from an ancient collision.

Did You Learn?

- Pluto and Charon are synchronous; that is, they always have the same face toward each other.

- The moons of the dwarf planets are believed to consist mostly of water ice.

17.6 Small Solar System Bodies: Asteroids, Meteoroids, Comets, and Interplanetary Dust

Preview Questions

- What is the difference between a meteor and a meteorite?
- Where are comets believed to come from?

Thus far, this chapter has dealt with the moons of the planets. Still more objects inhabit the solar system.

Asteroids

There are more than 20,000 named and numbered objects that orbit the Sun in a belt about midway between the orbits of Mars and Jupiter (● Fig. 17.26). These objects are called **asteroids**, and scientists sometimes refer to them as *minor planets*.* The asteroid belt contains millions of asteroids, and an estimated 100,000 are bright enough to be photographed by Earth-based telescopes.

The first of these solar-orbiting bodies between Mars and Jupiter was discovered in 1801 by Giuseppi Piazzi, an Italian astronomer. This small body, 940 km (580 mi) in diameter, was named Ceres after the Roman goddess of agriculture. For more than 150 years, Ceres was classified as an asteroid, but now it is considered to be a dwarf planet (Chapter 16.6).

The origin of the asteroid belt is sometimes attributed incorrectly to the breakup of a former planet. To prove the point, estimates of the total mass of the asteroids are much less than the mass of an ordinary planet. A more plausible explanation is that the debris never formed into a large object in the first place; asteroids are simply leftover debris from the formation of the solar system. As discussed in Chapter 16.7, it is believed that the planets gravitationally condensed from swirling debris. Calculations show that once Jupiter had formed, its immense gravitational force would have stirred up debris near its orbit, giving the debris particles speeds too great to allow them to condense or stick together.

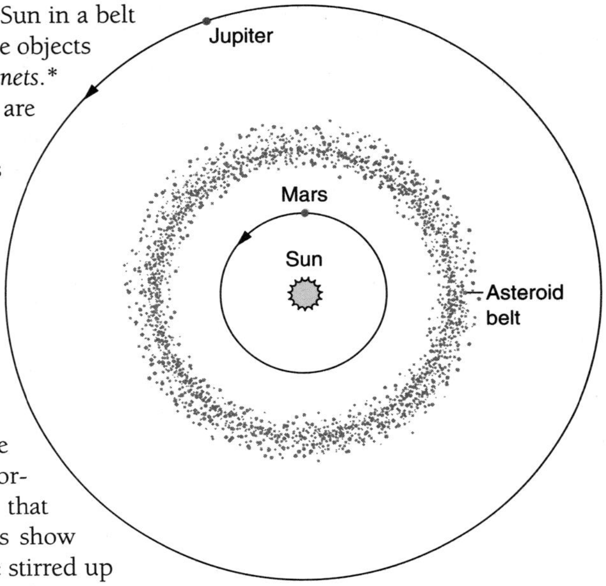

With Ceres now being a dwarf planet, the two largest asteroids are now Pallas (diameter 580 km, or 360 mi) and Vesta (540 km, or 340 mi). Pallas has a very low albedo, and only Vesta, with a relatively high albedo, can be seen with the unaided eye.

The first close-up image of an asteroid was taken in 1991 by the Jupiter-bound *Galileo* spacecraft. ● Figure 17.27 shows the irregularly shaped asteroid named Gaspra, which is about 11 km wide and 19 km long. Gaspra is classified as a stony asteroid.

The diameters of the known asteroids range from that of Pallas (580 km, or 360 mi) down to only a few kilometers. Most asteroids are probably less than a few kilometers in diameter. There are perhaps many thousands the size of boulders, marbles, and grains of sand. Only the largest asteroids are roughly spherical; the others are irregular in shape. The asteroid Eros, which approaches the Earth to within about 1 AU, is about the size and shape of the island of Manhattan. In 2001, a spacecraft called NEAR actually orbited Eros, sending back pictures of its rocky, pitted barren surface and close-up data.[†]

Figure 17.26 Asteroid Belt
The asteroid belt lies between the orbits of Mars and Jupiter and contains millions of asteroids.

---

*Because of their star-like appearance, British astronomer William Herschel in 1802 began calling these objects asteroids, after *aster*, the Greek word for "stars."

[†]For those interested in name origins, Pallas (or Pallas Athena) was the daughter of Zeus in Greek mythology and the goddess of many areas, including wisdom and war; Vesta was a household deity, the Roman goddess of the hearth; Gaspra was named by its discoverer, Grigoriy Neujmin, a Ukrainian astronomer, for a resort on the Crimean peninsula; and Eros is the Greek god of love, the counterpart of Cupid in Roman mythology.

Figure 17.27 Asteroid Gaspra
An image of the stony asteroid Gaspra taken by the Jupiter-bound *Galileo* spacecraft in October 1991, from a distance of about 53,000 km (33,000 mi). The Sun is shining from the right. Note the surface craters.

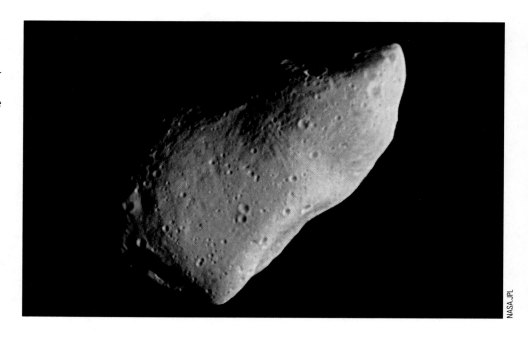

NASA/JPL

Thus, asteroids are believed to be early solar system material that never collected into a single planet because of Jupiter's gravitation. One piece of evidence supporting this view is that there seem to be several different kinds of asteroids. Those at the inner edge of the asteroid belt appear to be stony, whereas those farther out are darker, indicating more carbon content. A third group may be composed mostly of iron and nickel.

Like the planets, asteroids revolve counterclockwise around the Sun. Although most asteroids move in an orbit between Mars and Jupiter, some have orbits that range beyond Saturn and inside the orbit of Mercury.

Meteoroids, Meteors, and Meteorites

Meteoroids are small interplanetary metallic and stony objects that range in size from a fraction of a millimeter to about a hundred meters. They are probably the remains of comets and fragments of shattered asteroids. Meteoroids circle the Sun in elliptical orbits and sometimes strike the Earth from all directions at very high speeds. Their high speed, which is increased by the Earth's gravitational force, produces great frictional heating when the meteoroids enter the Earth's atmosphere.

A meteoroid is called a **meteor**, or "shooting star," when it enters the Earth's atmosphere and becomes luminous because of the tremendous heat generated by friction with the air. Spectacular displays of meteors, or *meteor showers*, occur when the Earth passes through a debris zone and hits a dust trail of a previous comet, thereby sparking a yearly meteor shower. Historically, the *Perseid* (mid-August) and *Leonid* (mid-November) *meteor showers* give the best numbers of meteors per hour. In some years, as many as 1000 "shooting stars" may be observed per hour. The number observed can vary a great amount depending on Earth's path through the comet trail, as well as on such conditions as the phase of the Moon and local clouds.

Most meteors are vaporized in the atmosphere, but some larger ones survive the flight through the atmosphere and strike the Earth's surface. They then become known as **meteorites**. When a large meteorite strikes the Earth's surface, a large crater is formed. ● Figure 17.28 is a photograph of a sizable meteorite crater near Winslow, Arizona, which scientists estimate to be about 50,000 years old.

The largest known meteorite, with a mass of more than 55,000 kg (121,000 lb), fell in southwest Africa. The largest meteorite known to have struck North America, with a mass of about 36,000 kg (79,200 lb), was found near Cape York, Greenland, in 1895. It is on display at the Hayden Planetarium in New York City (● Fig. 17.29a). A "ring" meteorite on display at the Smithsonian Institution in Washington, D.C., is shown in Fig. 17.29b.

Meteor Crater Enterprises, Inc.

Figure 17.28 The Barringer Meteorite Crater near Winslow, Arizona
The crater is 1300 m (4300 ft) across and 180 m (590 ft) deep, and its rim is 45 m above the surrounding land.

American Museum of Natural History

© Chip Clark

(a)

(b)

Figure 17.29 Large Meteorites
(a) The largest known meteorite to strike North America had a mass of 36,000 kg (79,200 lb) and was discovered in Greenland in 1895. (b) The Tucson ring meteorite. This 623-kg (1370-lb) meteorite was found by the first Spanish explorers near Tucson, Arizona, in 1851. Native Americans had known of it 300 years prior to that time. The Smithsonian acquired the meteorite early in the twentieth century.

Comets

The term *comet* derives from the Greek *aster kometes*, meaning "long-haired star." Comets are the solar system members that may periodically appear in our sky for a few weeks or months and then disappear. A **comet** is a relatively small object that is composed of dust and ice and that revolves about the Sun in a highly elliptical orbit. As a comet comes near the Sun, some of the surface vaporizes to form a gaseous head and a long tail.

A comet consists of four parts, as shown in ● Fig. 17.30: (1) the *nucleus*, typically a few kilometers in diameter and composed of rocky or metallic material, as well as solid ices of water, ammonia, methane, and carbon dioxide; (2) the head, or *coma*, which surrounds the nucleus, can be as much as several hundred kilometers in diameter, and is formed from the nucleus as it approaches within about 5 astronomical units (AU) of the Sun; (3) the long, voluminous, and magnificent *tail*, which is composed of ionized molecules, dust, or a combination of both and can be millions of kilometers in length; and (4) a spherical

Figure 17.30 The Principal Parts of a Comet
(See text for description.)

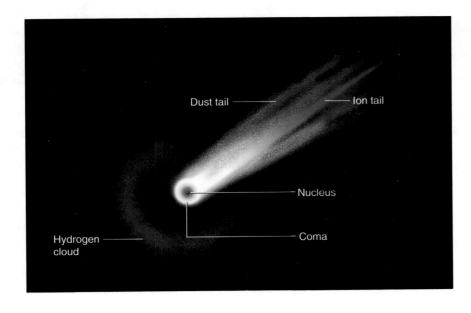

Dust tail —————— ———— Ion tail

Nucleus

Coma

Hydrogen cloud

Highlight   Trans-Neptunian Objects (TNOs)

The redefinition of a planet in the solar system in 2006 has caused some controversy both inside and outside the astronomical community. Although not a perfect definition, scientists classify objects by similar properties so that they can better study their physical characteristics, origins, and evolution. As more and more objects are discovered beyond the orbit of Neptune, astronomers have collectively called these objects trans-Neptunian objects, or TNOs.

Some of the largest TNOs are given the distinction of being dwarf planets (Pluto, Eris, Haumea, and Makemake). More than 1000 other TNOs have been discovered, most notably Varuna in 2000, Quaoar ("KWAH-whar") in 2002, Sedna in 2003, and Orcus in 2004. Their physical shapes and sizes are given in Fig. 1 and are easily compared with the size of Pluto and the Earth. Some of these large TNOs might later be classified as dwarf planets.

Scientists also classify the TNOs by their orbits and other physical characteristics, such as color and composition. Many of these objects have a blue-gray-white color indicative of a surface of ices, probably some combination of water, methane, and nitrogen ices. A large number of TNOs are reddish, however. It is unknown what causes this coloration, but scientists believe that the two groups of different colors might have different origins.

Orbital characteristics can lead to further understanding of the objects. The orbits of several TNOs are shown in Fig. 2, along with the relative locations of the Kuiper belt and the Oort cloud (see Chapters 16.6 and 17.6). TNOs like Pluto and others that are perturbed gravitationally by Neptune are called resonant objects. Objects like Quaoar and Makemake are unperturbed by Neptune, have nearly circular orbits, and lie in the plane of the solar system. They are classic Kuiper belt objects that have distances of 30 to 55 AU from the Sun and orbit periods ranging from 160 to 400 years.

Objects that have irregular orbits—more elliptical and more tilted orbit planes—are the scattered disk part of the Kuiper belt. These objects are farther away from the Sun on average; Eris is an example having this type of orbit.

Finally, there are estimated to be 1 trillion to 2 trillion small, icy objects that are much farther away, and collectively they are found in a region called the Oort cloud. The Oort cloud lies approximately 5000 to 100,000 AU from the Sun (Fig. 2). Astronomers are learning more about this cloud since the discovery of Sedna in 2003. Sedna has an orbit period of about 12,000 years, and its elongated orbit takes it out to about 1,000 AU, near the inner part of the Oort cloud. The outer edge of the Oort cloud is considered to be the limit to the gravitational influence of the Sun.

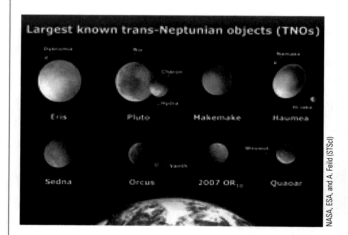

Figure 1 The Largest Known TNOs
An artist's drawing of the eight largest trans-Neptunian objects (TNOs). Known moons are labeled and their sizes are drawn to scale with the Earth as a reference shown at the bottom.

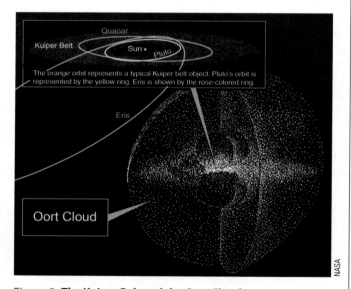

Figure 2 The Kuiper Belt and the Oort Cloud
An artist's drawing of the Kuiper belt and the Oort cloud. The flattened Kuiper belt is where the small solar system bodies are being discovered. The spherical Oort cloud is believed to be the source of long-period comets. The orbits of several TNOs are shown in the inset.

hydrogen cloud surrounding the coma, believed to be formed from the dissociation of water molecules in the nucleus. The sphere of hydrogen in some comets may have a diameter exceeding that of the Sun.

Halley's comet, named after the British astronomer Edmond Halley (1656–1742), is one of the brightest and best-known comets (● Fig. 17.31). The first proven observation of the comet was made by the Chinese in 240 BCE. It was described as a "broom" star that "appeared in the east and was seen in the north." Halley was the first to suggest and predict the periodic appearance of the comet (he did not discover it). Halley observed the comet that bears his name in 1682 and, using Newton's laws of motion and gravitation, correctly predicted its return in 76 years (in 1758). Halley did not live to see the return of his comet. It has continued to reappear every 76 years, including 1910 and 1986. If the comet remains on its regular schedule, then it will be back in the year 2062.*

Comets are visible by reflected sunlight and the fluorescence of some of the molecules making up the comet. As a comet approaches the Sun and moves around it, the amount of material in the coma and tail gets larger and longer and then diminishes as the comet recedes (● Fig. 17.32). This increase in size is evidently caused by the Sun's heating a thin outer shell of the comet's nucleus.

Scientists believe that comets originate and evolve from dirty, icy objects that were part of the primordial debris thrown outward into interstellar space when the solar system was formed. These objects are believed to be more dirt than ice and are sometimes described as "frozen mud balls" or "dirty snowballs."

The majority of comets take hundreds of thousands of years to complete their highly elliptical orbits around the Sun. Only a small number of these *long-period* comets have orbits that lie within the inner solar system. We may see such a comet once and then never see it again. Astronomers believe that there must be a huge "cloud" far beyond the orbit of Pluto and completely surrounding the Sun. This region is called the **Oort cloud**, after Dutch astronomer Jan Oort (1900–1992), who first suggested the possibility of a vast reservoir of frozen comets. The Oort cloud is thought to contain billions of comets in far-distant orbits around the Sun. (For a diagram of the Oort cloud, see the **Highlight: Trans-Neptunian Objects (TNOs)**.)

Figure 17.31 Halley's Comet
This famous comet is named after British astronomer Edmond Halley (name rhymes with *valley*), who was a contemporary of Sir Isaac Newton. Here the comet is shown on its last swing around the Earth in 1986. Halley predicted the periodic appearance of the comet using Newton's laws of motion and gravitation, but he did not discover the comet, which is mentioned in historical documents spanning a period of 22 centuries.

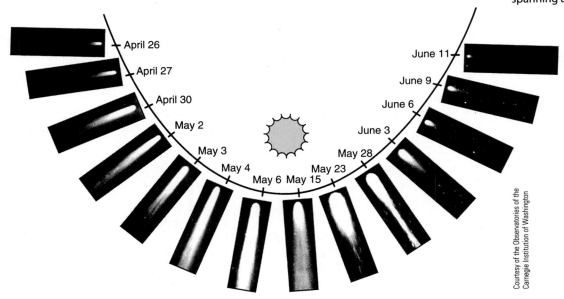

Figure 17.32 The Tail of a Comet
These 14 views of Halley's comet were taken between April 26 and June 11, 1910. Note the change in size of the coma and the tail as well as the tail's direction.

---

*Some people have a unique relationship with Halley's comet. Mark Twain was one of them. He told his biographer, "I came in with Halley's Comet in 1835. It is coming again next year, and I expect to go out with it. The Almighty has said, no doubt: 'Now here are these two unaccountable freaks; they came in together, they must go out together.'" Mark Twain died on April 21, 1910, just as the comet began its pass within sight of the Earth.

However, there are some *short-period* comets that have orbital periods of less than 200 years. These short-term comets—for example, Halley's comet—return to perihelion in relatively short times. Astronomers believe that the source region for short-term comets is the **Kuiper Belt** ("KY-per"), named after Dutch astronomer Gerard Kuiper (1905–1992), who predicted its existence. The Kuiper Belt lies beyond the orbit of Neptune and extends to well beyond the orbit of Eris (Chapter 16.6). It contains thousands of comets moving in orbits that never take them inside the Jovian planets. However, occasionally a comet leaves the belt, probably because of the gravitational influence of an outer planet, and goes into an elliptical orbit that passes through the inner solar system.

Interplanetary Dust

In addition to the planets and other large bodies discussed thus far, a tremendous volume of the solar system's space is occupied by very small solid particles known as *micrometeoroids*, or **interplanetary dust**. Two celestial phenomena, which can be observed with the unaided eye and photographed, show that the dust particles do exist.

On a very clear, dark night, just after sunset in the western sky, you may have observed the first of these phenomena, *zodiacal light*, a faint band of light along the zodiac (Chapter 15.5). The band of light can also be seen just before sunrise. The faint glow is due to sunlight reflected from the dust particles.

The other phenomenon is called the *Gegenshein* (German for "counterglow") and is also due to sunlight reflected from dust particles. This faint glow is observed exactly opposite the Sun. Appearing as a diffuse, oval spot, it is more difficult to observe than zodiacal light.

Did You Learn?

- A meteor burns up in the Earth's atmosphere, producing a streak of light (shooting star); a meteorite is a space-related object that reaches the Earth.

- Comets come from either the Kuiper Belt beyond the orbit of Neptune or from the Oort cloud far beyond the orbit of Pluto.

KEY TERMS

1. highlands (17.1)
2. maria
3. crater
4. synodic month (17.2)
5. sidereal month
6. new moon
7. crescent moon
8. gibbous moon
9. first-quarter phase
10. full moon
11. last-quarter phase
12. eclipse
13. umbra
14. total solar eclipse
15. penumbra
16. total lunar eclipse
17. tides
18. asteroids (17.5)
19. meteoroids
20. meteor
21. meteorites
22. comet
23. Oort cloud
24. Kuiper Belt
25. interplanetary dust

MATCHING

For each of the following items, fill in the number of the appropriate Key Term from the preceding list. Compare your answers with those at the back of the book.

a. _____ Bowl-shaped feature on the Moon

b. _____ Phase of the Moon occurring at 6 A.M.

c. _____ Not burned up in the Earth's atmosphere

d. _____ Phase of the Moon occurring at 6 P.M.

e. _____ Region from which partial solar eclipses are seen

f. _____ Entire face of the Moon obscured

g. _____ Less than half of the Moon's observed surface illuminated

h. _____ Incorrectly called "shooting stars"

i. _____ Month of phases

j. _____ Light-colored and heavily cratered lunar crust

k. _____ Darkest, smallest region of the Moon's shadow

l. _____ Predominantly found between Mars and Jupiter

m. _____ Sun's surface is completely obscured

n. _____ Occurs at 12 noon local solar time

o. _____ More than half of the Moon's observed surface illuminated

p. _____ Also known as micrometeoroids

q. _____ Phase of the Moon occurring at 12 midnight

r. _____ Stellar month

s. _____ Rise and fall of the ocean's surface

t. _____ Characterized by "tails"

u. _____ Source region of short-term comets

v. _____ A vast reservoir of long-term comets

w. _____ Interplanetary metallic and stony objects

x. _____ Lunar lowland regions

y. _____ Blocking of light of one celestial body by another

MULTIPLE CHOICE

Compare your answers with those at the back of the book.

1. Which of the following is *not* a general physical feature of the Moon's surface? (17.1)
   - (a) craters
   - (b) volcanoes
   - (c) plains
   - (d) rays

2. Which one of the following statements is true? (17.1)
   - (a) The Moon's surface gravity is $\frac{1}{81}$ that of the Earth.
   - (b) The Moon has an appreciable magnetic field.
   - (c) The Moon is the second-brightest object in the sky.
   - (d) The Moon revolves around the Earth in 31 days.

3. Most craters on the surface of the Moon are believed to be caused by which of the following? (17.1)
   - (a) faults
   - (b) meteoroids
   - (c) volcanoes
   - (d) asteroids

4. What is the oldest type of surface on the Moon? (17.1)
   - (a) astronaut footprints
   - (b) highlands
   - (c) rays
   - (d) maria

5. Which of the following statements is false? (17.2)
   - (a) The Moon rotates and revolves westward.
   - (b) The orbital plane of the Moon is tilted about 5° to the Earth's orbital plane.
   - (c) The difference between the sidereal and synodic months is about 2 days.
   - (d) The Moon revolves in an elliptical orbit.

6. The rising of the Moon in the east and its setting in the west are due to which of the following? (17.2)
   - (a) the orbital motion of the Moon
   - (b) the rotational motion of the Moon
   - (c) the Earth's rotation
   - (d) none of the preceding

7. During 1 month, the Moon passes through how many different phases? (17.2)
   - (a) four
   - (b) six
   - (c) eight
   - (d) none of the preceding

8. What is the approximate time between new and full moons? (17.2)
   - (a) 1 week
   - (b) 2 weeks
   - (c) 1 sidereal month
   - (d) 1 synodic month

9. The first-quarter moon is overhead at what time? (17.2)
   - (a) 6 A.M.
   - (b) noon
   - (c) 6 P.M.
   - (d) midnight

10. A lunar eclipse occurs during what phase of the Moon? (17.2)
    - (a) new
    - (b) first-quarter
    - (c) last-quarter
    - (d) full

11. Which of the following is *not* a contributing factor in causing eclipses? (17.2)
    - (a) the rotation of the Earth about its axis
    - (b) the inclination of the Moon's orbit
    - (c) the varying distance between the Earth and the Moon
    - (d) the varying distance between the Earth and the Sun

12. During a total eclipse of the Sun, the center of totality goes across the surface of the Earth in which direction? (17.2)
    - (a) east to west
    - (b) north to south
    - (c) west to east
    - (d) south to north

13. The two daily high tides are due mainly to which of the following? (17.2)
    - (a) the Moon's gravitational force lifting the ocean water away from the solid Earth
    - (b) the differential gravitational attraction of the Moon because of the inverse-square relationship
    - (c) gravitational forces between the Sun and the Earth
    - (d) none of the preceding

14. When do spring tides on the Earth take place? (17.2)
    - (a) only during the spring season
    - (b) only during the times of new moons
    - (c) near the times of full and new moons
    - (d) near the times of first-quarter and last-quarter moons

15. Which of the following planets has no moon? (17.3)
    - (a) Venus
    - (b) the Earth
    - (c) Mars
    - (d) None of the preceding; they all have moons.

16. Which of the following planets has the most moons? (17.3)
    - (a) Mercury
    - (b) Venus
    - (c) the Earth
    - (d) Mars

17. How many Galilean moons are there? (17.4)
    - (a) three
    - (b) four
    - (c) five
    - (d) six

18. Which of the following is the largest moon in the solar system? (17.4)
    - (a) Callisto
    - (b) Titan
    - (c) Europa
    - (d) Ganymede

19. Which of the following is the second-largest moon in the solar system? (17.4)
    - (a) the Earth's Moon
    - (b) Callisto
    - (c) Ganymede
    - (d) Titan

20. Which are the four Galilean moons of Jupiter? (17.4)
    (a) Europa, Titan, Ganymede, and Callisto
    (b) Io, Ganymede, Callisto, and Titan
    (c) Io, Europa, Ganymede, and Callisto
    (d) Europa, Ganymede, Io, and Triton

21. The *Cassini-Huygens* probe successfully landed on which of the following? (17.4)
    (a) Mars   (b) Saturn
    (c) Titan   (d) Triton

22. Which dwarf planet has no moons? (17.5)
    (a) Ceres   (b) Pluto
    (c) Charon   (d) Eris

23. Which dwarf planet moon has a rotation period and a revolution period that are the same as the rotation period of its parent planet? (17.5)
    (a) Charon
    (b) Nix
    (c) Hydra
    (d) none of the preceding

24. Which of the following is *not* true of asteroids? (17.6)
    (a) They are believed to be initial solar system material that never collected into a single planet.
    (b) They are located mainly in orbits around the Sun between the Earth and Mars.
    (c) They range in size from hundreds of kilometers down to the size of sand grains.
    (d) They are generally irregular in shape.

25. Which of the following is *not* true of comets? (17.6)
    (a) They are composed of dust and ice.
    (b) They revolve around the Sun in highly elliptical orbits.
    (c) They can be observed on entering the solar system.
    (d) They usually have a long tail when they are close to the Sun.

26. Which of the following is (are) true of meteoroids? (17.6)
    (a) They are small, solid, interplanetary metallic and stony objects.
    (b) They are usually smaller than a kilometer.
    (c) They are known as meteors when they enter the Earth's atmosphere.
    (d) They are known as meteorites when they strike the Earth's surface.
    (e) All the preceding are true.

FILL IN THE BLANK

Compare your answers with those at the back of the book.

1. Smooth, dark regions called ___ cover about 15% of the lunar crust. (17.1)

2. The crater Tycho is surrounded by streaks of powder called ___. (17.1)

3. The Moon rises approximately 50 minutes ___ each day. (17.2)

4. The last-quarter moon is overhead at ___. (17.2)

5. The ___ Moon rises at 6 p.m. (17.2)

6. The phase of the Moon preceding the last-quarter moon is a(n) ___ gibbous phase. (17.2)

7. The Earth is between the Sun and the Moon during a(n) ___ eclipse. (17.2)

8. To see a total solar eclipse, an observer must be in the ___ of the Moon's shadow. (17.2)

9. The greatest tidal variation occurs during ___ tides. (17.2)

10. The two daily high tides occur approximately ___ hours apart. (17.2)

11. Venus has ___ moons. (17.3)

12. Mars has ___ moons. (17.3)

13. The largest moon of Jupiter is ___. (17.4)

14. The Galilean moon with volcanic activity is ___. (17.4)

15. The Saturnian moon that has an atmosphere is ___. (17.4)

16. Pluto has three named moons: ___, ___, and ___. (17.5)

17. ___ is the primary composition of the moons of the dwarf planets. (17.5)

18. A belt of ___ is found between Mars and Jupiter. (17.6)

19. A meteoroid that strikes the Earth is called a(n) ___. (17.6)

20. The head of a comet surrounding the nucleus is called the ___. (17.6)

SHORT ANSWER

17.1 Structure, Origin, and Features of the Earth's Moon

1. Name and define three major surface features of the Moon.

2. What are lunar rays, and how are they formed?

3. In what year was the first crewed landing on the Moon, and how many crewed landings were there after that?

4. How old are the rocks brought back from the Moon?

17.2 Lunar Motion Effects: Phases, Eclipses, and Tides

5. What is the Moon's orbital period? What is its rotation period?

6. Why does the Moon rise 50 minutes later each day?

7. What is the difference between a waxing phase and a waning phase of the Moon?

8. What is the difference between a gibbous moon and a crescent moon?

9. What is a synchronous orbit?

10. What is the difference between an umbra and a penumbra?

11. Relative to the Earth, what are the positions of the Moon and the Sun at the first-quarter and last-quarter phases of the Moon?

12. If an observer sees a partial eclipse of the Sun, then what can be said about the observer's position?

13. If an observer sees a partial eclipse of the Moon, then what can be said about the observer's position?

14. What are the relative positions of the Earth, the Sun, and the Moon during a lunar eclipse? During a solar eclipse?

15. The Moon attracts the ocean on the side nearest it, creating a bulge. What causes the bulge on the opposite side of the Earth that results in two daily high tides?

16. What is the difference between a spring tide and a neap tide?

17.3 Moons of the Terrestrial Planets

17. What is the total number of moons of the terrestrial planets?

18. How many moons does Mars have, and what are their names?

17.4 Moons of the Jovian Planets

19. How many Galilean moons are there, and why are they called Galilean?

20. Which Galilean moon has volcanic activity, and what is believed to be the cause?

21. Which moon is the largest in the solar system, and what is its parent planet?

22. Which moon is the second largest in the solar system, and what is its parent planet?

23. Which planet has more moons, Uranus or Neptune?

24. The moons Titan, Titania, and Triton are often confused. Tell which planet each orbits.

17.5 The Moons of the Dwarf Planets

25. What does it mean to say that Pluto and Charon have synchronous orbits?

26. Which of Pluto's moons is farthest from the planet?

27. How big is Dysnomia compared with its parent planet?

28. Which dwarf planet has two moons?

17.6 Small Solar System Bodies: Asteroids, Meteoroids, Comets, and Interplanetary Dust

29. Where can one find the most asteroids?

30. When is a meteoroid termed a meteor? When is it termed a meteorite?

31. What is the composition of comets?

32. Where do short-period comets come from? What about long-period comets?

33. How often does Halley's comet reappear?

34. What is zodiacal light? Is it related to the *Gegenshein*?

VISUAL CONNECTION

Visualize the connections and give answers for the blanks. Compare your answers with those at the back of the book.

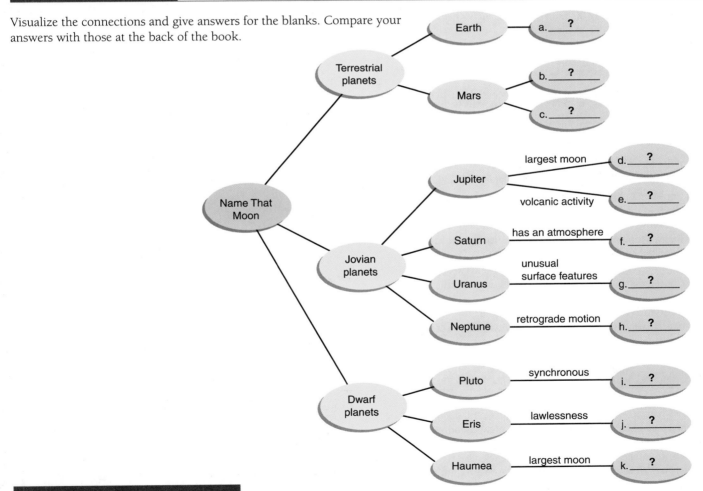

APPLYING YOUR KNOWLEDGE

1. Suppose you are on the Moon facing the Earth. Will you observe phases of the Earth? If so, then describe them.

2. The Moon rises and sets. For example, a new moon that is overhead at 12 noon rises at 6 A.M. and sets at 6 P.M. Suppose an

astronaut on the Moon saw the Earth overhead at 12 noon. At what time would the Earth set?

3. ● Figure 17.33 shows a lunar situation. What is wrong with the picture? (The figure is based on a stamp issued in 1988 by the East African country Tanzania.)

4. The apparent diameter of the full moon is observed to be about the same as the apparent diameter of the Sun, but the actual diameter of the Sun is about 400 times that of the Moon. Why are the apparent diameters about the same?

5. What kind of phases would the satellite Charon have if you lived on the dwarf planet Pluto?

Figure 17.33
See Applying Your Knowledge Question 3.

EXERCISES

17.1 Structure, Origin, and Features of the Earth's Moon

1. If a person weighs 800 N on the Earth, then what is the person's weight (in newtons) on the Moon?

Answer: 133 N

2. If a person weighs 160 lb on the Earth, then what is the person's weight (in pounds) on the Moon?

17.2 Lunar Motion Effects: Phases, Eclipses, and Tides

3. How many days are there in 12 lunar months (synodic months)?

Answer: Approximately 354

4. How many days are there in 12 lunar months (sidereal months)?

5. How many days is the Moon in (a) the waxing phase and (b) the waning phase?

Answer: 14.75 days for both

6. If the Moon rises at 6 P.M. on a particular day, then approximately what time will it rise 30 days later?

7. Consider a person in the United States who sees the first-quarter phase of the Moon.
   (a) Which side of the Moon is illuminated, east or west?
   (b) What phase does an observer in Australia see at the same time, and which side is bright?

Answer: (a) west (b) first-quarter phase; west (left) side

8. Consider a person in the United States who sees the last-quarter phase of the Moon.
   (a) Which side of the Moon is illuminated?
   (b) What phase does an observer in Australia see at the same time, and which side is bright?

9. The Moon is halfway through the waxing crescent phase. In approximately how many days will it be in the last-quarter phase?

Answer: 18 days

10. The Moon is just entering the waning gibbous phase. In approximately how many days will it enter the waning crescent phase?

11. Draw a diagram illustrating a total solar eclipse. Include the orbital paths of the Earth and the Moon and indicate the approximate time of day at which the eclipse is taking place.

Answer:

Sun   Moon   M   noon — Earth

12. Draw a diagram illustrating a total lunar eclipse. Include the orbital paths of the Earth and the Moon and indicate the approximate time of day at which the eclipse is taking place.

13. A high tide is occurring at Charleston, South Carolina (33°N, 84°W).
    (a) What other longitude is also experiencing a high tide?
    (b) What two longitudes are experiencing low tides?

Answer: (a) 96°E (b) 6°E, 174°W

14. A low tide is occurring at Galveston, Texas (29°N, 95°W).
    (a) What other longitude is also experiencing a low tide?
    (b) What two longitudes are experiencing high tides?

ON THE WEB

1. "Skywatcher's Guide to the Moon"

What are the theories about the formation of the Moon? What evidence is there for these theories? What problems did Alan Shepard and his crew encounter on *Apollo 14*? Why do some scientists think the Moon is a planet? What effect does the Moon have on the Earth's oceans and our tides? Finally, is the Moon moving closer to or farther from the Earth, or is it remaining at a constant distance? How do we know? Visit the student website at **www.cengagebrain.com/shop/ISBN/1133104096** to discover more about the Moon and to answer these questions.

2. The Great Moon Hoax: Can You Really Believe We've Been There?

Have you ever wondered whether U.S. astronauts really did go to the Moon, walk on its surface, and bring back specimens of rock and other artifacts? Apparently, some *have* wondered. Follow the recommended links on the student website at **www.cengagebrain.com/shop/ISBN/1133104096** to learn about the "great Moon hoax." Summarize the "evidence" presented by both sides to suggest either a "conspiracy" or a true event. What fallacies might you point out in either argument?

The Universe

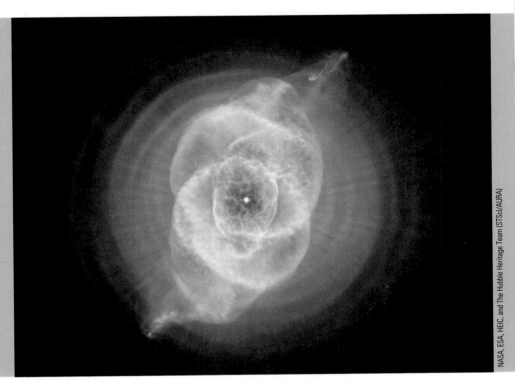

NASA, ESA, HEIC, and The Hubble Heritage Team (STScI/AURA)

*We shall not cease
from exploration
And the end of all our
exploring
Will be to arrive where
we started
And know the place
for the first time.*

•

T. S. Eliot (1888–1965)

< The beautiful Cat's Eye Nebula (NGC 6543), a planetary nebula, which is a dying star's gas and dust ejected into space.

The **universe** is the totality of all matter, energy, and space. In this chapter, the *celestial sphere* is defined and used to help describe the locations of *stars* (the Sun in particular) and *galaxies* (giant assemblages of stars and other matter). Techniques used to measure distances to astronomical objects are discussed in the Highlight: Determining Astronomical Distances. After a discussion of stellar life cycles and properties of galaxies, the chapter concludes with a section on *cosmology*, the branch of astronomy that studies the structure and evolution of the universe. The Highlight: The Age of the Universe discusses the recent determination of the age and composition of the universe.

Stars and galaxies emit all types of electromagnetic radiation, including radio waves, microwaves, infrared rays, visible light, ultraviolet rays, X-rays, and gamma

ASTRONOMY FACTS

▶ Recent evidence suggests that the number of stars in the universe is three times higher than previously thought. The latest data estimate about 500 billion stars per galaxy on average and about 500 billion galaxies in the universe, giving a staggering total of about 250 sextillion (2.5×10^{23}) stars in the universe.

▶ The Hubble Space Telescope (HST) has had more effect on astronomy than any other object, person, or event in history, even more than Galileo, who used the newly invented telescope to study the skies in 1610. The HST can point to celestial objects with amazing precision, comparable to pointing a laser pointer on the tip of a pencil that is 1.6 km (1 mi) away!

Chapter Outline

rays (Chapter 6.3). Until 1931, astronomers had to rely on information gathered from only one type of electromagnetic radiation, visible light. The advent of radio telescopes led to the discovery of celestial objects such as quasars and pulsars, as well as to clues about the size and structure of our galaxy, the Milky Way.

Most of the radiation of the electromagnetic spectrum, from infrared to gamma rays, is largely absorbed by our atmosphere. Only when balloons—and then satellites and other spacecraft—were able to rise above our atmosphere did these regions of the spectrum reveal abundant new information about the universe. The result has been that a multitude of discoveries and advances are presently being made in astronomy unequalled in most other sciences.

Some people confuse the science of astronomy with the pseudoscience called *astrology*. Astrologers contend that the positions of the planets and the Sun in the sky at the time of a person's birth affect the individual's personality or future. There are no known forces that could cause such effects. Scientific studies have thoroughly debunked astrology, yet some people think astrology is valid.

18.1 The Celestial Sphere

Preview Questions

● How is the position of a star designated in the sky?

● What is the simplest way to measure the distance to a nearby star?

A view of the stars on a clear, dark night makes a deep impression on an observer. Stars appear as bright points of light on a huge overhead dome. As the night passes, the dome of stars appears to move westward as part of a great rotating sphere (● Fig. 18.1). The unaided eye is not able to detect any relative motion of the stars on the apparent sphere or to perceive their relative distances from the Earth. The stars all appear to be "mounted" on a very large sphere with the Earth at its center. Astronomers call this huge, *imaginary* sphere the **celestial sphere** (● Fig.18.2). Actually, the daily *apparent* east-to-west motion of the dome of stars is just a reflection of the Earth's *real* daily west-to-east rotation (Chapter 16.3).

In the Northern Hemisphere, the stars seem to rotate around a point on the celestial sphere called the *North celestial pole* (NCP). The NCP is the point on the celestial sphere that is the extension of the northern end of the Earth's axis of rotation (Fig. 18.2). At present, a fairly bright star named Polaris lies close to the NCP. Therefore, Polaris is known as the North Star or pole star. Because of the Earth's 26,000-year cycle of precession of its axis (Chapter 15.6), the pole star slowly becomes a different star, and often there is no pole star. The *South celestial pole* (SCP) is the point on the celestial sphere that is the extension

of the southern end of the Earth's axis of rotation (Fig.18.2). At present there is no "south pole star."

Figure 18.2 shows the *celestial equator*, which is simply the extension of the Earth's equator onto the celestial sphere. Also shown in Fig. 18.2 is the **ecliptic**, which is the apparent path the Sun traces annually along the celestial sphere. Of course, the Sun does not really move around the Earth, but vice versa, so the ecliptic is really a projection of Earth's orbital plane onto the celestial sphere.

Note that the plane of the ecliptic and the plane of the celestial equator intersect each other at an angle (● Fig. 18.3). The angle is 23.5°, which is the tilt of the Earth's axis of rotation with respect to its orbital plane. The ecliptic and the celestial equator intersect on the celestial sphere at two points

Figure 18.1 Star Trails
This time-exposure photograph shows the circular pattern that stars seem to make around the South celestial pole. Actually, the pattern is due to the Earth's rotation during the time of exposure.

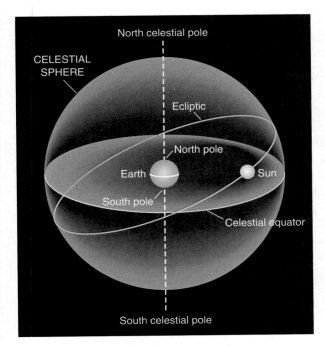

Figure 18.2 The Celestial Sphere
Key locations on the celestial sphere are the North and South celestial poles, the celestial equator, and the ecliptic (the apparent annual path of the Sun on the celestial sphere).

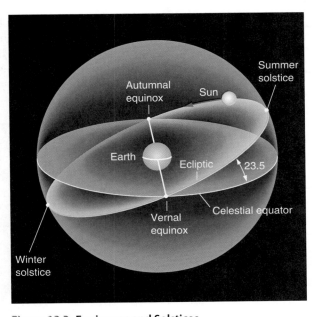

Figure 18.3 Equinoxes and Solstices
Because of the tilt of the Earth's axis of rotation, the ecliptic is inclined 23.5° to the celestial equator. The vernal equinox is the time and position when and where the Sun reaches the intersection of the ecliptic and the celestial equator as it moves northward. The summer solstice is the northern-most overhead position that the Sun reaches each year.

called *equinoxes*. When the Sun reaches either point, it is directly over the Earth's equator and there are approximately 12 hours of daytime and 12 hours of nighttime at all locations on the Earth (see Conceptual Question and Answer: Equal Days and Nights in Chapter 15.5).

As the Sun is observed traveling on the ecliptic, it intersects the celestial equator each year approximately on September 22, moving southward (Fig. 18.3). This point of intersection is called the *autumnal equinox*. Six months later, approximately on March 21, the Sun again crosses the celestial equator, but this time moving northward. This point of intersection is termed the *vernal equinox*. Two other significant locations lie along the ecliptic between the equinoxes. The point on the ecliptic where the Sun is farthest north (approximately on June 21) is termed the *summer solstice*, and the point where the Sun is farthest south (approximately on December 22) is called the *winter solstice* (Fig. 18.3).

The half-circle that passes through the vernal equinox, the NCP, and the SCP defines the *celestial prime meridian* (● Fig. 18.4). It is analogous to the Greenwich prime meridian defining the reference for designating longitude on the surface of the Earth (Chapter 15.2). The celestial prime meridian currently passes through the constellation Pisces in the sky, but its position is changing very slowly due to precession (see Chapter 15.6).

The *celestial longitude* of a star or galaxy is the angle measured eastward along the celestial equator from the celestial prime meridian. It is called the **right ascension** (RA) of the star or galaxy, and because time is more conveniently measured than angle, RA is not expressed in degrees but rather in units of hours and minutes and seconds.

What about a star's latitude? *Celestial latitude* is known as **declination** (DEC) and is simply the angular measure in degrees, minutes ('), and

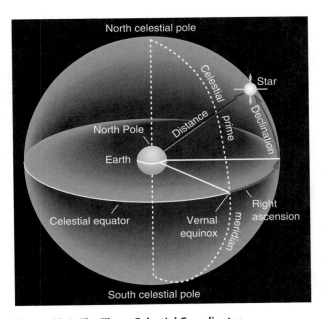

Figure 18.4 The Three Celestial Coordinates
A star's position is described using three celestial coordinates. Right ascension is measured eastward along the celestial equator from the celestial prime meridian. Declination is measured northward or southward from the celestial equator. Distance is measured in a straight line from the Earth to the star.

seconds ('') north or south of the celestial equator (Fig. 18.4). The celestial equator can be found in the sky by locating the plane that is perpendicular to the celestial poles. This plane currently passes through several constellations, including Orion and Virgo.

For example, the brightest star in the night sky is Sirius, which has celestial coordinates RA = 6 h 45 min and DEC = −16°43'. Thus, to locate Sirius in the night sky, you would use its RA and DEC to specify its location on the celestial sphere.

Conceptual Question and Answer

Celestial Coordinates

Q. What are the minimum and maximum values for RA and DEC?

A. Right ascension (RA) has a value of beginning at 0 h at the celestial prime meridian and continues in an eastward circle to a maximum of 24 h (Fig. 18.4). Declination (DEC) has a minimum value of zero at the celestial equator, and its value increases to a maximum of +90° at the NCP and −90° at the SCP, similar to latitude measurement on the Earth.

But what about distance? The distance coordinate is usually measured in astronomical units, light-years, or parsecs. The *astronomical unit* (AU) was defined in Chapter 16.1 as the mean distance of the Earth from the Sun (1.5×10^8 km, or 93 million mi). A *light-year* (ly) is the distance traveled by light in 1 year (9.5×10^{12} km, or 6 trillion mi). It is calculated by multiplying the speed of light (3.00×10^5 km/s) by the number of seconds in a year (3.16×10^7 s/y). One **parsec** (pc) is defined as the distance to a star when the star exhibits a parallax of 1 second of arc, where 1 second of arc is defined to be $\frac{1}{3,600}$ of 1° (● Fig. 18.5). A parsec is related to a light-year by the following:

$$1 \text{ pc} = 3.26 \text{ ly}$$

The nearby star in Fig. 18.5, observed from two positions, appears to move back and forth against the background of more distant stars. This apparent motion is called parallax

Figure 18.5 Annual Parallax of a Nearby Star
Over the course of a year, a nearby star seems to move back and forth against a background of more distant stars. This parallax effect is due to the viewer's position changing as the Earth moves around the Sun. The shaded angle *p*, measured in arc seconds, represents the annual parallax of the nearby star.

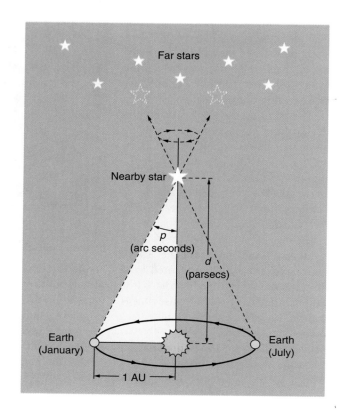

(see Chapter 16.3). The angle *p* measures the parallax in seconds of arc (arc seconds, or arcsec). Defining a parsec this way provides an easy method for determining the distance to a celestial object because taking the reciprocal of the angle *p*, measured in arc seconds, gives the distance in parsecs.

$$d = \frac{1}{p}$$

18.1

where *d* is the distance in parsecs and *p* is the parallax angle in arc seconds.

EXAMPLE 18.1   **Calculating the Distance to a Star**

Calculate the distance (in parsecs and light-years) to Proxima Centauri, the nearest star to the Earth (other than the Sun). The annual parallax of the star is 0.762 arcsec.

Solution

$$d = \frac{1}{p} = \frac{1}{0.762} = 1.31 \text{ pc} \ (= 4.26 \text{ ly})$$

Confidence Exercise 18.1

Calculate the distance (in parsecs and light-years) to Sirius, the brightest star in the night sky. Its annual parallax is 0.376 arcsec.

Answers to Confidence Exercises may be found at the back of the book

Because Aristotle and others of his time observed no stellar parallax, they rejected the idea that the Earth moved around the Sun. It was not realized that even the closest stars were much too far away for their parallaxes to be detected with the unaided eye.

It was not until 1838 that Friedrich Bessel, a German astronomer, was able to observe stellar parallax using a telescope. It is easy to appreciate why parallax is so difficult to observe. The parallax angle of even the closest star is equivalent to the diameter of a dime seen at a distance of 6 km (3.7 mi). Parallax measurements can determine the distances only to stars that are within about 300 ly of the Earth.

Did You Learn?

- Star positions are given by two angles: right ascension (RA) and declination (DEC).
- The simplest way to measure a nearby star's distance is by parallax.

18.2   The Sun: Our Closest Star

Preview Questions

- What is the composition of the Sun?
- Why does the Sun "shine"?

The Sun is a star, a self-luminous sphere of hot gas held together by gravity and energized by nuclear reactions in its core. Our closest star is quite large, with a diameter about 100 times the size of the Earth. General information about several of the Sun's characteristics is given in ● Table 18.1.

A cross-sectional view of the Sun is shown in ● Fig. 18.6. Let's begin our discussion with the visible surface of the Sun, move out to the Sun's atmosphere, and then delve into the Sun's interior. (*Note:* Never look directly at the Sun. Doing so may damage your eyesight.)

The bright, visible "surface" of the Sun is called the **photosphere**. The temperature of the photosphere is about 6000 K. Analysis of the Sun's spectrum shows that its photosphere is

Table 18.1 Characteristics of the Sun	
Diameter	1.39×10^6 km (8.63×10^5 mi)
Mass	2.0×10^{30} kg
Density (mean)	1.4 g/cm^3
Period of rotation at equator	25 Earth days (longer at higher latitudes)
Distance (mean) from the Earth	1.5×10^8 km (9.3×10^7 mi)

composed, by mass, of about 75% hydrogen, 25% helium, and less than 1% heavier elements, the most abundant being oxygen, carbon, neon, and iron.

A distinct feature of the Sun's surface is the periodic occurrence of sunspots. *Sunspots* are huge regions of cooler material on the surface of the Sun where the magnetic field is very intense. A close-up view of a sunspot shows that the photosphere, viewed through a telescope with appropriate filters, has a granular appearance (● Fig. 18.7). The granules are hot spots (some 100 K hotter than the surrounding surface) that are about the width of the state of Texas (1000 km or 620 mi) and last only a few minutes. These granules are actually convection cells that bring thermal energy from the interior to the surface. ● Figure 18.8 is an image of the whole solar disk; it shows large groups of sunspots. A given sunspot generally lasts for several weeks before disappearing.

The number of sunspots appearing on the Sun varies over an 11-year *sunspot cycle* (● Fig. 18.9). A cycle begins with the appearance of a few sunspots near 30° latitude in both hemispheres of the Sun. The number slowly increases, and a maximum (generally between 100 and 200) occurs in the middle of the cycle, at around 15° latitude. Then, the number of spots slowly decreases until, at the end of the 11 years, only a few are observed near 8° latitude.

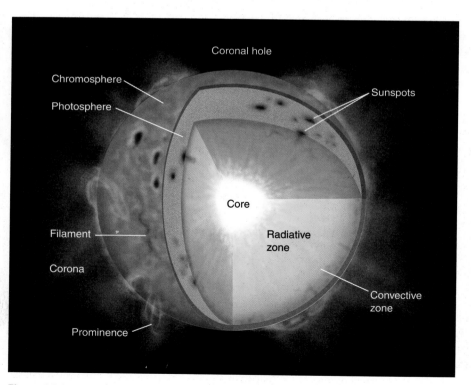

Figure 18.6 A Cutaway View of the Interior and Atmosphere of the Sun
The various layers of the Sun are shown. The interior, the photosphere, and the corona are all gaseous regions with different densities. The boundary between two layers is not sharply defined.

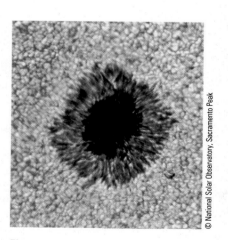

© National Solar Observatory, Sacramento Peak

Figure 18.7 A Sunspot
This dark region on the Sun is a typical sunspot. The granular nature of the photosphere can be seen all around the sunspot.

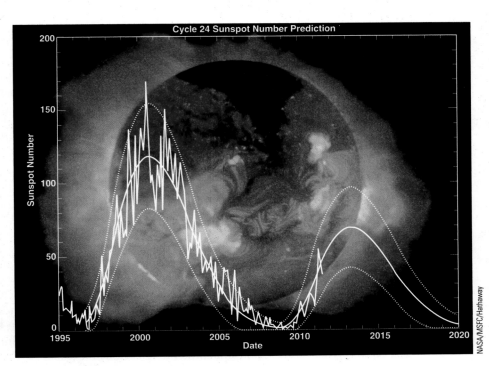

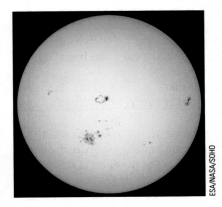

Figure 18.8 Sunspots
An incredible number of large sunspot groups from October 2003.

Figure 18.9 The Most Recent Solar Cycle
Sunspot number is graphed versus time, showing the average 11-year period. The sunspot minimum occurred in 2009.

The cycle appears to repeat itself, but there is a noticeable difference. Sunspots have magnetic polarity. If a sunspot has a north magnetic pole during the initial increase and decrease, then the next 11-year cycle will show a south magnetic pole associated with the sunspot. Thus, the 11-year sunspot cycle appears to be an aspect of a more fundamental 22-year *magnetic cycle.*

Other distinct features of the Sun's surface are solar flares and prominences. *Flares* are sudden, bright, explosive events originating on the Sun's surface. They are releases of energy associated with the Sun's magnetic field. *Prominences* are enormous filaments of excited gas arching over the surface and usually extending hundreds of thousands of kilometers outward. They are very evident to astronomers during solar eclipses, at which time they appear as great eruptions at the edge of the obscured solar disk (see Chapter 17.2). Prominences have an associated magnetic field and may appear as streamers, loops, twisted columns, or fountains. An extraordinarily large prominence and a number of flares are shown in ● Fig. 18.10.

Above the photosphere is a layer of hotter gas (mainly hydrogen) known as the *chromosphere* ("color" + "sphere"). No distinct boundary exists between the photosphere and the chromosphere. The chromosphere (temperature range 6000 K to 25,000 K) can be studied during a total solar eclipse, when it can be seen as a thin, red crescent for only the few seconds during which the photosphere is concealed from view.

At the time of a total solar eclipse, the photosphere is hidden by the Moon, and the Sun's outer solar atmosphere, the *corona* (Latin for "crown"), can be seen as a beautiful, tenuous, white halo extending far beyond the solar disk (see Fig. 17.12a). The temperature of the corona is about 1 million K. Astronomers think that the high temperatures of both the chromosphere and the corona are due to the transport and release of energy from intense magnetic fields that permeate the gases, coupled with the magnetic fields under the photosphere.

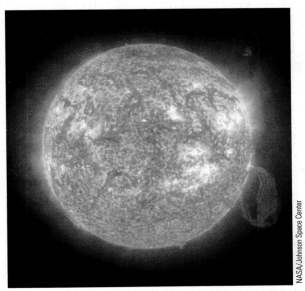

Figure 18.10 The Sun during a Major Solar Eruption
This ultraviolet photograph, showing several flares (the bright yellow areas) and a large prominence (*right*), was taken from the Solar and Heliospheric Observatory (SOHO) spacecraft in 2001.

The extreme temperature of the corona is sufficient to give protons, electrons, and ions enough energy to escape the Sun's atmosphere. The charged particles are projected into space, giving rise to the *solar wind*. The particles move quickly away from the Sun, reaching the Earth in 3 to 5 days. The solar wind is controlled by the Sun's magnetic field. Measurements made by the *Voyager 1* spacecraft have confirmed that the solar wind and the accompanying magnetic field extend outward to at least 100 AU from the Sun's surface. This volume of space over which the solar wind extends is called the *heliosphere*.

What about the Sun's interior? The interior of the Sun is so hot that neutral atoms do not exist because high-speed collisions continually knock the electrons loose from the atomic nuclei. The interior is composed of nuclei and electrons rapidly moving about more or less independently, similar to a gas. The charged nuclei (ions) and electrons exist as a "fourth phase of matter" called *plasma* (see Chapter 5.5). The plasma has an average density of 1.4 g/cm³, whereas at the very core of the Sun this density is 150 g/cm³, which is 150 times greater than the density of water (1.00 g/cm³).

For us, one of the most important features of the Sun is that it radiates energy. However, not until 1938 did scientists understand that the Sun radiates energy because of nuclear fusion of hydrogen into helium in its core (Chapter 10.6).

The Sun is made up mostly of hydrogen nuclei, or protons or, in nuclear notation, $^1_1$H. At the 15 million-K temperature of the core, these protons are moving at very high speeds, and they occasionally fuse together, as shown in ● Fig. 18.11a. The products of this nuclear fusion reaction are a deuteron (a proton and a neutron together, or $^2_1$H), a positron $(^0_{+1}e)$, and a neutrino (designated by the symbol ν, the Greek letter nu). A *neutrino* is an elementary particle that has no electric charge, travels at nearly the speed of light, and hardly ever interacts with other particles such as electrons and protons. Only recently was it determined that neutrinos have mass.

As Fig. 18.11b shows, once the deuteron is formed, it quickly reacts with a proton to form a helium-3 nucleus $(^3_2$He$)$, with the emission of a neutrino (ν) and a gamma ray (γ). Finally, two helium-3 nuclei fuse to form the more common helium-4 nucleus $(^4_2$He$)$ and two protons (Fig. 18.11c).

In each of these three fusion reactions, energy is liberated by the conversion of mass. These three reactions are called the *proton–proton chain* and can be written as shown in Fig. 18.11. If the first two reactions are multiplied by 2 (that is, let each happen *twice* to produce the two $^3_2$He nuclei needed for the third reaction to happen *once*) and added together, then the equation for all three reactions in a net reaction is

$$4\,^1_1\text{H} \rightarrow\, ^4_2\text{He} + 2(^0_{+1}e + \nu + \gamma) + \text{energy}$$

In the net reaction, four protons form a helium nucleus, two positrons, two neutrinos, and two high-energy gamma rays. The "energy" factor on the right-hand side of the equa-

Figure 18.11 The Proton–Proton Chain for Nuclear Fusion

In the Sun and similar stars, hydrogen is fused into helium by a series of three nuclear reactions shown here in order as (a), (b), and (c).

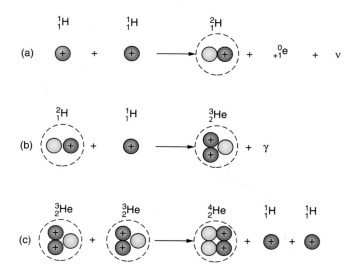

tion simply means that the particles on the right possess more kinetic and radiant energy than the particles on the left. In the reaction, mass has been converted into energy in conformity with Einstein's equation, $E = mc^2$ (Chapter 10.6).

The Sun contains about 10^{30} kg of hydrogen and has been fusing it into helium for about 5 billion years. Every second in the Sun's interior, about 6.0×10^{11} kg of hydrogen is converted into 5.96×10^{11} kg of helium and 4.0×10^{26} J of energy. Even at this rate, the Sun is expected to radiate energy from hydrogen fusion for another 5 billion years.

The amount of energy that reaches the Earth from the Sun is about 1400 joules per second per square meter ($J/s \cdot m^2$). This amount of energy is called the *solar constant*. Nearly all life on planet Earth is dependent on energy from the Sun because this energy controls our total environment: atmosphere, food supplies, water sources, and so on. A variation in the solar constant of as little as 0.5% would have catastrophic effects on our ability to stay alive.

Did You Learn?

- The Sun is composed of hydrogen and helium and traces of many other elements.
- The Sun shines because of energy released from nuclear fusion in its core.

18.3 Classifying Stars

Preview Questions

- Are all stars the same size?
- Why are stars different colors?

Like the Sun, most stars are huge balls of plasma, composed largely of hydrogen along with some helium and a very small percentage of other elements. Nuclear fusion of the hydrogen into helium produces the enormous energy that stars emit. Their surface temperatures range from about 1500 K to 50,000 K. Stars vary in mass from about 0.08 to 100 solar masses, but most have a mass between 0.1 and 5 solar masses. (One *solar mass* is the mass of our Sun.)

A star's size (volume) represents a balance between the inward force of gravity and the outward force of thermal and radiation pressure. This condition is called *hydrostatic equilibrium*. If the two forces did not balance, then the star would either shrink (if inward gravity dominated) or expand (if outward pressure dominated). The sizes of stars vary greatly, and a given star can have different sizes (and masses) during its lifetime, as we will discuss later.

Unlike the Sun, most stars are not single stars but rather are part of multiple-star systems. A *binary star* consists of two stars orbiting a common center of mass. Systems of three or more stars also exist, but they are not nearly as abundant as binary stars.

Prominent groups of stars in the night sky appear to an Earth observer as distinct patterns. Many of these groups, or *constellations*, have names that can be traced back to early Babylonian and Greek civilizations, but nearly all cultures had different names for the constellations. Although the constellations have no physical significance, modern astronomers find them useful in referring to certain areas of the sky. In 1927, astronomers set specified boundaries for the 88 constellations so as to encompass the complete celestial sphere. Appendix IX shows the positions of some stars and constellations in the night sky around March and September.

The Sun has an apparent daily westward motion across the sky, and when the Moon is visible, it also has an apparent westward motion. When one observes the stars for an hour or two, the constellations also appear to move westward across the sky. These apparent daily motions are due to the eastward rotation of the Earth (see Chapter 16.3).

In addition to their daily motion, the constellations have an annual motion that results from the Earth's revolution about the Sun. For example, currently from the Northern Hemisphere, the constellations Pisces (the Fish) and Capricornus (the Goat) are observed in the autumn night sky. In winter, Orion (the Hunter) and Taurus (the Bull) are seen.

Figure 18.12 Constellations Ursa Major and Ursa Minor
The asterisms called the Big Dipper and the Little Dipper are parts of the constellations Ursa Major and Ursa Minor, respectively. Polaris, the North Star, is at the tail of the Little Dipper. Can you find the constellation Leo (the Lion) at the bottom of the figure?

Leo (the Lion) is prominent in spring, and Cygnus (the Swan) is a summer constellation.* Constellations with high values of DEC, such as Ursa Major (the Big Bear) with declination +50°, can be seen throughout the year from the United States (● Fig. 18.12).

Constellations often have a story, or myth, associated with them. For example, the "bears" in Ursa Major and Ursa Minor, shown in Fig. 18.12, were said to have such long tails because Zeus (king of the gods) in a fit of rage slung them around his head several times by their tails before launching them into the sky.

Some familiar groups of stars, called *asterisms*, are part of a constellation or parts of different constellations. The Big Dipper, which is part of Ursa Major, is an asterism (find the Big Dipper in Fig. 18.12). Another example is the Summer Triangle, which is formed by the bright stars Altair, Deneb, and Vega. These stars are in three different constellations: Altair is in Aquila (the Eagle), Deneb is in Cygnus (the Swan), and Vega is in Lyra (the Lyre). (See Appendix IX, September.)

Celestial Magnitude

Hipparchus of Nicaea, a Greek astronomer and mathematician, was antiquity's greatest observer of the stars. He measured the celestial locations of more than 800 stars and compiled the first star catalog, which was completed in 129 BCE. Hipparchus assigned the stars to six groups of brightness called magnitudes. The apparent brightness, or **apparent magnitude**, of a star (or other celestial object) is its brightness as observed from the Earth. The brightest ones were listed as stars of the first magnitude, those not quite as bright were stars of the second magnitude, and so on, down to the sixth magnitude, which includes stars barely visible to the unaided eye.

An extended and modified version of Hipparchus' scale is used today. Although estimates of stellar brightness can be made with the unaided eye, stellar brightness is now measured more accurately by using instruments. When a comparison was made between Hipparchus' first-magnitude stars and his sixth-magnitude stars, each first-magnitude star was measured to be about 100 times brighter than each sixth-magnitude star. When this scale is used for stellar magnitude, the greater the magnitude number, the dimmer the star, whereas the smaller the magnitude number, the brighter the star.

On the magnitude scale, the brightest star, Sirius, has an apparent magnitude of about −1. (Magnitudes are actually measured to several decimal places, but for simplicity rounded values are given here.) Sirius, which is 8.7 ly distant, is the closest star to the Earth (except for the Sun) that can be seen by any observer in the United States.

The three stars of the Summer Triangle asterism have the following apparent magnitudes: Vega (0), Altair (1), and Deneb (1). Six of the seven stars of the Big Dipper have apparent magnitudes of about 2 (as does Polaris, our North Star), and the seventh (the one at the junction of the "handle" and the "bowl") has a magnitude of about 3 (Fig. 18.12). From an urban area with light pollution, stars of magnitude 3 to 6 are not easily seen; to see these with the unaided eye, a good, dark site is needed.

Alpha Centauri A (magnitude 0) and its close companion Alpha Centauri B (magnitude 1) revolve around each other. To the unaided eye, they appear as a single bright star and, having a DEC of −61°, can only be seen from the Southern Hemisphere. Close to Alpha Centauri A and Alpha Centauri B is a faint, red star called Proxima Centauri. At a distance of 4.3 ly, these three stars are the closest ones to the Earth, Proxima Centauri being slightly closer to the Earth than the other two.

Thus far, we have been discussing apparent magnitude, or how bright the star looks from the Earth. The distances of various stars from us vary over a wide range, however, and their distances away have a tremendous effect on how bright they appear to us. The **absolute magnitude** of a star is defined as the brightness it would have if it were placed 10 pc (32.6 ly) from the Earth. For example, the *absolute* magnitude of the Sun is about

---

*As pointed out in Chapter 15.5, the motion of the constellations of the zodiac were used as an early calendar. However, because of the Earth's precession (Chapter 15.6), the seasonal visibility of the constellations has changed slowly over many thousands of years.

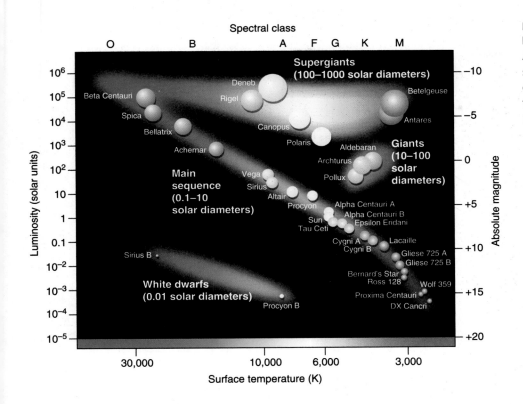

Figure 18.13 Hertzsprung-Russell Diagram
An H-R diagram shows the luminosity, spectral class, temperature, and color of the various classes of stars. Note that the temperature increases from right to left along the horizontal axis.

+5, whereas, because it is so close to us, its *apparent* magnitude is about −27. If the annual parallax (from which distance is calculated) and the apparent magnitude of a star can be measured, then the absolute magnitude can be calculated.

The Hertzsprung-Russell Diagram

When the absolute magnitudes of stars are plotted versus the temperature of their photospheres, the **H-R diagram** (Hertzsprung-Russell diagram) is obtained. This tool is named after Ejnar Hertzsprung, a Danish astronomer, and Henry Norris Russell, an American astronomer, who first created diagrams in 1911 and 1913, respectively. An H-R diagram for stars is shown in ● Fig. 18.13. Note that the temperature axis is reversed; the temperature increases to the left instead of to the right.

Most stars on an H-R diagram follow a trend in which a brighter star has a higher temperature. These stars form the **main sequence**, a narrow band going from the lower right to the upper left. As Fig. 18.13 shows, the color of a star has a direct correlation with the temperature of its photosphere. The hottest stars on the main sequence are blue, whereas the coolest stars are red. Between these extremes are stars that are white, yellow, and orange. Stars above and to the right of the main sequence are cool, yet very bright. Because they must be unusually large to be so bright, these stars are called **red giants**, and the very brightest are known as *red supergiants*. Stars below the main sequence that are hot yet very dim must be small, and they are called **white dwarfs**.

The elements that make up a star can be identified from a study of the star's spectrum (see Chapters 7.2 and 9.3). Even though the compositions of most stars are about the same, the spectra of stars vary considerably, a variation due almost entirely to a star's photospheric temperature. Therefore, the pattern of the absorption lines in a star's spectrum can be used to determine its temperature, as well as its composition. Analyses of stellar spectra show that even the most distant stars contain the same elements found in our solar system.

Stars are placed in seven different spectral classes. From highest temperature to lowest temperature, the order of the classes is O, B, A, F, G, K, and M. (This order may be recalled by using the mnemonic *Oh, Be A Fine Girl (Guy), Kiss Me*; or *Oh Boy, A Failing Grade Kills Me*.)

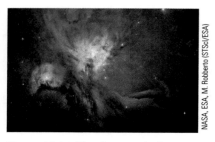

Figure 18.14 The Great Nebula in Orion
The Orion nebula is a bright emission nebula, some 25 ly in diameter and about 1300 ly distant. It can be seen with the unaided eye in the constellation Orion. Many stars are forming in this nebula.

NASA, ESA, M. Robberto (STScI/ESA)

As Fig. 18.13 shows, the horizontal axis in an H-R diagram can be labeled using spectral class instead of temperature. The Sun is a class G star, and Sirius is a class A star. The majority of stars are small, cool, red, class M stars called *red dwarfs*. For example, Proxima Centauri is a red dwarf.

Did You Learn?

- Stars come in a variety of sizes, ten times smaller to thousands of times larger than the Sun.
- Stars have different colors because they have different temperatures.

18.4   The Life Cycle of Low-Mass Stars

Preview Questions

- Why do stars burn out?
- Where do stars originate, and how are they formed?

Astronomers know that stars are born, radiate energy, expand, possibly explode, and then die. In general, that is a star's life cycle, but the exact details depend on a star's mass and, in a minor way, on its composition. The greater the mass of a star, the faster it moves through its life cycle.

Gas and dust distributed among the stars are known as the *interstellar medium*. The gas consists of about 75% hydrogen by mass and about 24% helium, along with a trace of heavier elements. About 1% of the interstellar medium is dust particles about the size of those in smoke. The dust consists primarily of compounds of carbon, iron, oxygen, and silicon. The gas and dust do not seem to be distributed uniformly, but instead form cool, dense clouds called **nebulae** (Latin for "clouds").

There are two major types of nebulae, *bright nebulae* and *dark nebulae*. Bright nebulae can be *emission nebulae*, in which the emission comes from hydrogen gas ionized by the energy from nearby stars. An example of an emission nebula is the Great Nebula in Orion, one of the brightest in the night sky and the closest such nebula to us (● Fig. 18.14). On the other hand, some bright nebulae are *reflection nebulae*, in which the energy of the nearby stars is not sufficient to ionize the hydrogen, and the dust just reflects and scatters the starlight. Reflection nebulae have a characteristic blue color for the same reason that the Earth's atmo-

Figure 18.15 The Horsehead Nebula
The Horsehead Nebula, a dark nebula, projects against a bright emission nebula behind it. The Horsehead Nebula cannot be seen with the unaided eye.

© Anglo-Australian Observatory/Royal Observatory, Edinburgh

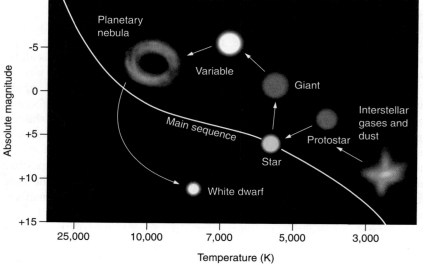

Figure 18.16 Stellar Evolution on the H-R Diagram
The stages of evolution of a star like the Sun are shown. (See text for description.)

sphere makes the sky blue (see the Highlight: Blue Skies and Red Sunsets in Chapter 19.2).

Dark nebulae are produced when starlight is obscured by a relatively dense cloud of interstellar dust. By far the most famous of the dark nebulae is the Horsehead Nebula, also located in the constellation Orion but not visible to the unaided eye (● Fig. 18.15). The gigantic dark cloud is framed against the light-emitting region behind it. Its name comes from its shape, which resembles that of the neck and head of a horse.

The general evolution of a low-mass star like our Sun, and of stars with even lower mass, is plotted on the H-R diagram shown in ● Fig. 18.16. The birth of a star begins with the accretion (gathering) of interstellar material (mostly hydrogen) in large pockets of cold gas and dust within enormous, interstellar, molecular clouds. The accretion is due to gravitational attraction between the interstellar material, radiation pressure from nearby stars, and shock waves from exploding stars. The Great Nebula in Orion is a classic example of a place where stars are being born.

The size of the star formed in a nebula depends on the total mass available, which also determines the rate of contraction. As the interstellar mass condenses and loses gravitational potential energy, the temperature rises and the material gains thermal energy. It becomes what is known as a *protostar*.

As the protostar continues to decrease in size, the temperature continues to increase and a thermonuclear (fusion) reaction begins in which hydrogen is converted into helium, as discussed in Chapter 18.1. At this time, a protostar becomes a star as it moves onto the main sequence at a position determined by its temperature and brightness, both of which depend ultimately on its mass.

The stars spend 90% of their life on the main sequence of the H-R diagram, continuing to fuse hydrogen into helium. The lifetime on the main sequence for a star such as the Sun is about 10 billion years, and the Sun has been there for about 5 billion years so far. For a very low-mass star, the lifetime may be up to a few trillion years. A high-mass star fuses its nuclear fuel at such a great rate that its time on the main sequence may be only a few million years.

As hydrogen in a star's core is converted into helium, the core begins to contract and heat up. This process heats the surrounding shell of hydrogen and causes the fusion of the hydrogen in the shell to proceed more rapidly. The rapid release of energy upsets the pressure–gravity force balance (hydrostatic equilibrium) and causes the star to expand, cool down, and enter the red-giant phase of its evolution. It now moves off the main sequence, as shown in Fig. 18.16. Because the Sun has the fuel supply of a one-solar-mass star, it is expected to become a red giant in about 5 billion years. The Sun will become so large that it will engulf the orbits of Mercury, Venus, and probably the Earth. (Betelgeuse, a well-known red supergiant in the constellation Orion, is so large that were it placed where the Sun is located, it would extend out nearly to the orbit of Jupiter!)

Eventually, the core of a red giant gets so hot that helium can fuse into carbon, and if the star's mass is great enough, then other nuclear reactions soon occur in which heavier nuclei may be created. The creation of the nuclei of elements inside stars is known as *nucleosynthesis*. If the star is a high-mass red supergiant, then nucleosynthesis can continue all the way up to the element iron, which is the most stable nucleus. During the red-giant phase, a star varies in temperature and brightness.* It becomes a *variable star* for a relatively short time, and its position on the H-R diagram moves to the left (Fig. 18.16). The star becomes very unstable and the outer layers are blown off, forming a beautiful **planetary nebula**, such as the Eskimo Nebula (● Fig. 18.17). The chapter opening photo is also a planetary nebula appropriately nicknamed the Cat's Eye Nebula. It is 3000 ly from the Earth and 0.5 ly in diameter. Despite its name, a planetary nebula has absolutely nothing to do with planets. The name came about because the first fuzzy photographic images of planetary nebulae looked to astronomers something like faint, distant planets.

Figure 18.16 shows the position of a star expelling a planetary nebula on the H-R diagram. The shell of expelled matter diffuses into space over a period of about 50,000 years

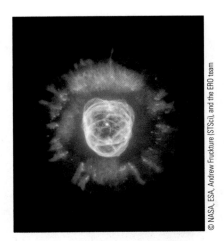

Figure 18.17 A Planetary Nebula
This photo taken by the Hubble Space Telescope in 1999 shows the Eskimo Nebula, a planetary nebula about 5000 ly distant. Much of the star's matter was ejected, leaving behind the white dwarf that is visible in the center of the nebula.

© NASA, ESA, Andrew Fruchture (STScI), and the ERO team

---

*Actually, a star with a mass similar to that of our Sun undergoes two red-giant stages. The first stage begins after all the hydrogen in the core is fused into helium. The second red-giant stage begins after all the helium in the core has been fused into carbon and oxygen.

and provides material from which new stars may eventually form. Our Sun is thought to be a second- or third-generation star, being composed partially of elements ejected into space by stars that lived and died before the birth of the Sun.

The remaining core of the planetary nebula is now a *white dwarf*, in which fusion no longer occurs, and the star slowly radiates its residual energy into space as it cools. When a star becomes a white dwarf, it gets very small. It has gravitationally collapsed to the smallest possible size at which the force of gravity is offset by the repulsive force of the electrons in each atom. The star is about the size of the Earth and is so dense that a single teaspoon of matter may weigh 5 tons. Because it is not very bright, it is low on the H-R diagram (see Fig. 18.13). Many white dwarfs in our galaxy have been identified, and one of the best known is related to the bright star Sirius. Actually, Sirius is a binary star system composed of a class A main-sequence star (Sirius A) and a companion white dwarf, Sirius B (● Fig. 18.18). Sirius B is only 85% of the size of the Earth but has the mass of the Sun. Its photosphere temperature is about 25,000 K (compared with 6000 K for the Sun).

What if a protostar does not have enough mass for sustained fusion to occur? Astronomers theorized for years that many "failed stars," named *brown dwarfs*, must exist. They realized that brown dwarfs would be very dim and hard to find, probably showing up only by the infrared radiation they emit. The brown dwarf Gleise 229B was the first to be found in 1995. Since then, more than 600 brown dwarfs have been identified, and astronomers think that galaxies might be teeming with more.

Could Jupiter have become a star, forming a binary star system with the Sun? No; Jupiter's mass is only 1/1000 that of the Sun, and the gas giant would have needed at least 80 times more mass before it could have become a star.

Such is the life cycle of low-mass stars: protostar, main-sequence star, red giant, planetary nebula, and (finally) white dwarf (Fig. 18.16). With the universe thought to be only about 14 billion years old, not many of the lowest-mass stars have had time to evolve into white dwarfs. High-mass stars have a different life cycle, as we will see in Section 18.5.

Figure 18.18 Sirius A and Sirius B
Sirius is a binary star. It consists of Sirius A, which is a main-sequence star, and its small, dim companion, Sirius B (the white dot at 5 o'clock), the first white dwarf to be discovered (1862).

Did You Learn?

● Stars burn out because the nuclear fuel in the core runs out.

● Stars originate in interstellar gas and dust clouds; they form when conditions allow gravity to pull material together.

18.5   The Life Cycle of High-Mass Stars

Preview Questions

● Why do some stars explode?

● How does a black hole form?

Many stars appear dim and insignificant but suddenly, in a matter of hours, increase in brightness by a factor of 100 to millions. A star undergoing such a drastic increase in brightness is called a **nova**, or "new" star. A nova is the result of a nuclear explosion on the surface of a white dwarf: an explosion caused by relatively small amounts of matter falling onto its surface from the atmosphere of a larger binary companion. A nova is not a new star but rather a faint white dwarf that temporarily increases in brightness.

Sometimes, stars explode catastrophically and throw off large amounts of material and radiation. Such a gigantic explosion is known as a **supernova**. A *Type I supernova* results from the destruction of a white dwarf with a carbon–oxygen core. It is the outcome of the accretion (capture) of enough material from a companion star to put the white dwarf's mass over the limit of stability for a white dwarf (1.4 times the Sun's mass).

A *Type II supernova* results from the collapse of the iron core of a massive red supergiant. This type of supernova leaves behind either a neutron star or a black hole, depending

Figure 18.19 The Crab Nebula
The Crab Nebula is the remnant of the supernova explosion recorded in 1054 CE. The nebula is about 2000 pc (6500 ly) away. At its center is a pulsar (neutron star) that spins 30 times per second.

on the remaining mass of the star. Both types of supernovae scatter much of their original matter into space and can be distinguished by their spectra. A Type II supernova shows hydrogen spectral lines from the red supergiant, and a Type I supernova shows carbon spectral lines from the white dwarf.

Fewer than ten supernovae have been recorded in the Milky Way galaxy (more about galaxies in Chapter 18.6). The best known is the Crab Nebula in the constellation Taurus (● Fig. 18.19). By knowing the Crab Nebula's average angular radius and expansion rate, the original time of the explosion can be calculated. The result agrees closely with Chinese and Japanese records that report the appearance of a bright new star in the constellation Taurus in 1054 CE. Reportedly, the supernova was so bright that for 2 weeks it was possible to read at night by its light.

In 1987, a supernova was observed in the Large Magellanic Cloud (LMC). It is the first supernova to have been observed in the LMC, which, at 180,000 ly away, is the third-closest galaxy to our Milky Way. The supernova was designated 1987A. The discovery of supernova 1987A was important to astronomers because they could at last observe a stellar explosion from the beginning; they can now observe the remnant star and track what happens to the debris coming from the explosion. Since 1987, valuable information has been obtained from supernova 1987A, especially by the Hubble Space Telescope.

Neutron Stars

As you have learned, Type II supernovae result when high-mass stars develop into supergiants. When a supergiant's nuclear fuel is depleted enough, the interior can undergo a sudden, catastrophic implosion to form a small-diameter **neutron star**. During the implosion, the outer layers of the star bounce off the rigid inner core and explode into space, giving rise to the supernova and leaving behind the small core. During the explosion of a supernova, the energy and neutrons emitted cause the nucleosynthesis of elements heavier than iron. Hundreds of neutron stars have been identified since the first one was discovered in 1967 by Jocelyn Bell, a graduate student at Cambridge University in England.

Except for the primordial ("present from the start") hydrogen and helium atoms (and some lithium atoms), all the elements up to and including iron are thought to be made during the normal fusion processes in stars of various masses. All elements past iron are thought to have originated in supernovae explosions. Thus, the elements scattered into space during explosions of supernovae, novae, and planetary nebulae provide the material for future generations of stars, planets, moons, and so forth. Carl Sagan (1934–1996), a famous American astronomer, was fond of saying that we are made of "star stuff," and scientists think he was right.

The core of a Type II supernova may collapse to approximately 10 km in diameter, the size of a small city. The electrons and protons in this superdense star combine to form neutrons—hence the name *neutron star*—that have enormous density (equal to that of a nucleus of an atom). If you think a white dwarf is dense at 5 tons a teaspoon, then consider that a teaspoon of neutron star might weigh about 1 billion tons!

Because the angular momentum of the original star must be conserved (Chapter 3.7), the small size of a neutron star dictates that it must be spinning rapidly. Rapidly rotating neutron stars seem to emit radio wave pulses. These stars are called *pulsars* when the pulses are detected and measured on the Earth by large radio telescopes, such as the Very Large Array (VLA) (● Fig. 18.20).

Figure 18.20 The Very Large Array (VLA) in New Mexico
Nine movable radio telescopes 25 m (82 ft) in diameter are mounted on each of the three 21-km (13-mi) railroad tracks of this Y-shaped array. This arrangement allows the system to operate as one huge radio telescope with an equivalent diameter of up to 32 km (20 mi). The VLA produces very sharp radio wave images of distant celestial objects.

© Photo Researchers, Inc.

*A pulsar's period very slowly but steadily decreases as the pulsar ages.

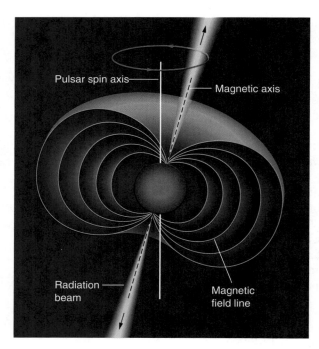

Figure 18.21 A Model of a Rotating Neutron Star

A pulsar is a rapidly rotating neutron star, the magnetic axis of which is tilted relative to its axis of rotation. Under these conditions, two oppositely directed beams of radiation sweep across the sky. If the Earth is in the path of one of these beams, then a pulsar is detected.

The waves from a specific pulsar have a near "constant" period that may be between 0.002 s and 4 s.* One of the most rapidly spinning pulsars, or neutron stars, is located at the middle of the Crab Nebula, the remnant of the supernova of 1054 CE. Its period is 0.03 s, and it emits pulses not only of radio waves but also of visible light (Fig. 18.19). Photographs show that it seems to blink on and off about 30 times a second. The apparent pulsation is observed because the magnetic axis of the spinning star is not the same as its rotational axis. This apparent pulsation can periodically send a beam of radiation along the magnetic axis toward the Earth, somewhat similar to a lighthouse beacon sweeping across a nearby ship's deck (● Fig. 18.21). Thus, the periodic flashing of pulsars is actually due to rotation, not pulsation.

Black Holes

As dense as neutron stars are, they do not represent the ultimate in compression. If the core remaining after a supernova explosion is greater than about three times the mass of the Sun, then theory predicts that it will end up as a gravitationally collapsed object even smaller and denser than a neutron star. An object called a **black hole** is so dense that even light cannot escape from its surface because of the object's intense gravitational field. By Einstein's theory of general relativity, gravity affects the propagation of light (see the Highlight: Albert Einstein in Chapter 9.2). A black hole represents gravity's final victory over all other forces.

In a black hole, the star's matter continues to contract until the original volume becomes a fantastically dense point called a *singularity*. The singularity is surrounded (at a distance at which the escape velocity equals the speed of light) by an invisible spherical boundary known as the *event horizon*. Any matter or radiation within the event horizon cannot escape the influence of the singularity's gravitational pull. The event horizon defines the size of the black hole and is a one-way boundary; matter and radiation can enter, but they cannot leave.*

● Figure 18.22 is a three-dimensional sketch of the configuration of a nonrotating black hole. The value R, called the *Schwarzschild radius* (named after German astronomer Karl Schwarzschild, who first calculated it), is the radial distance the event horizon is located from the singularity. Calculation of the event horizon for a black hole with a mass equal to that of the Sun yields a Schwarzschild radius of only about 2.9 km (1.8 mi). Because most stars rotate, most black holes probably rotate, which would lead to them being more elliptical than spherical.

One method of identifying a black hole is the detection of X-rays coming from its vicinity. Captured hot gases form an *accretion disk* just outside the event horizon and emit X-rays as they orbit and spiral into the black hole.

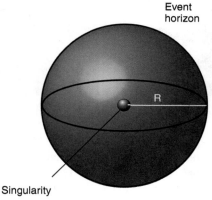

Figure 18.22 Configuration of a Nonrotating Black Hole

The black dot represents the singularity at the center of a black hole. The Schwarzschild radius R is the distance from the singularity to the event horizon.

Conceptual Question and Answer

Black Hole Sun

Q. Will the Sun turn into a black hole?

A. No; the Sun does not have enough mass (gravity) to ever become a black hole. Only stars with masses larger than about 8 solar masses have enough gravity to compress the core into a black hole. Our Sun has a less violent fate: it will shed most of its exterior layers as a planetary nebula, and the core will become a white dwarf.

---

*A quantum effect known as virtual-pair production indicates that black holes can ever-so-slowly evaporate as one particle of the pair is formed outside of the event horizon. In 2004, British astrophysicist Stephen Hawking, who pioneered the understanding of black holes, explained that black holes do not destroy everything they consume but instead emit radiation and information "in a mangled form" known as Hawking radiation.

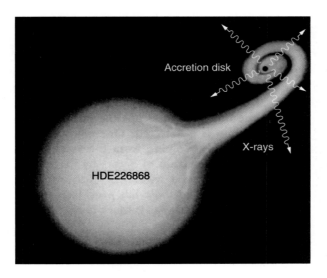

Figure 18.23 Cygnus X-1, a Black Hole
Cygnus X-1, a black hole, captures material from its giant companion star. The captured material forms an accretion disk around the black hole's event horizon and emits X-rays. Cygnus X-1's mass is about ten times the mass of the Sun, which is much too large for it to be a white dwarf or a neutron star.

For example, a class O supergiant star that is located some 8000 ly from the Earth has a companion called Cygnus X-1 that cannot be seen directly. However, X-rays have been detected coming from the unseen companion, indicating that it may be a black hole. The X-rays are apparently generated in the accretion disk by gases captured from the supergiant by Cygnus X-1 (● Fig. 18.23). Calculations from experimental data indicate that Cygnus X-1 is smaller than the Earth and probably more than ten times as massive as the Sun. This object is too massive to be a white dwarf or a neutron star, so it is quite likely that Cygnus X-1 is a black hole.

Satellites such as the Chandra X-Ray Observatory launched in 1999 have been placed in orbit around the Earth for the specific purpose of detecting celestial objects that emit X-rays. Chandra has already revealed hundreds of new black holes and made estimates that 30 million black holes existed in the early universe.

High-mass stars are formed in the same manner as low-mass stars, but they are hotter and brighter and so move on to the main sequence at higher points on the H-R diagram. High-mass stars fuse their nuclear fuel more rapidly than do low-mass stars and hence do not stay on the main sequence as long.

When high-mass stars leave the main sequence, they become supergiants and eventually explode as Type II supernovae, scattering much of their material into space and leaving behind a neutron star or, in the case of the most massive stars, a black hole. ● Figure 18.24 summarizes the life cycle of high-mass stars.

As has been noted, the initial mass that a celestial object manages to gain during its formation determines its fate. However, the final mass after ejection of any stellar material is the conclusive determinant. ● Table 18.2 summarizes this relationship.

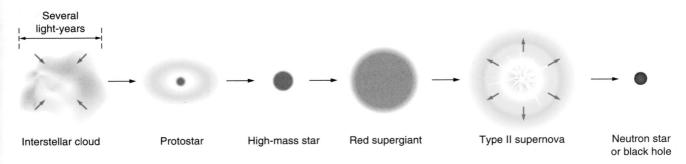

Several light-years					
Interstellar cloud	Protostar	High-mass star	Red supergiant	Type II supernova	Neutron star or black hole

Figure 18.24 Evolution of a High-Mass Star
High-mass stars form in the same way as all stars, but they become red supergiants after they move off the main sequence, and they eventually explode as a Type II supernova, leaving behind either a neutron star or a black hole.

Table 18.2 The Fate of Celestial Objects	
Final Mass of Object	**Fate of Object**
Less than 0.01 solar masses	Planet or moon
0.01 to 0.07 solar masses	Brown dwarf
0.08 to 1.4 solar masses	White dwarf
1.5 to 3.0 solar masses	Neutron star
Greater than 3.0 solar masses	Black hole

Did You Learn?

- High-mass stars can explode when nuclear reactions cease and the star violently collapses on itself.

- A black hole forms when a massive star explodes, causing the star core to gravitationally collapse.

18.6 Galaxies

Preview Questions

- How big is the Milky Way galaxy?
- What is dark matter?

A **galaxy** is an extremely large collection of stars bound together by mutual gravitational attraction. Galaxies are the fundamental components for the structure of the universe. Our solar system occupies a relatively small volume of space in a very large system of about 100 billion to 200 billion stars known as the Milky Way galaxy.

The Milky Way Galaxy

The Milky Way galaxy is so named because ancient peoples thought that the broad band of stars across the dark night sky resembled a trail of spilled milk. This broad band is observed when the viewer is looking into the plane of the galactic disk and thus sees an abundance of stars too small to be seen individually.

The Milky Way contains an estimated 100 billion to 200 billion stars and has three basic parts: (1) the crowded center called the nuclear bulge, (2) the thin plane containing the spiral arms called the disk, and (3) the spherical distribution of star clusters called the halo. The nuclear bulge of the Milky Way cannot be seen with the unaided eye because the interstellar dust between the nuclear bulge and the observer blocks most of the visible light. In 1989, the galactic disk and nuclear bulge were imaged for the first time, in infrared light by a satellite named COBE (*Cosmic Background Explorer*) in orbit above the Earth's atmosphere (● Fig. 18.25). The nuclear bulge and the spiral arms have also been imaged via radio telescopes.

In the halo around the Milky Way lie about 200 **globular clusters**, each globular cluster being a spherical collection of hundreds of thousands of stars (● Fig. 18.26). In 1917, American astronomer Harlow Shapley mapped the locations of the 93 then-known globular clusters. He found that they were centered not about our solar system, but about a point some 28,000 ly away, toward the constellation Sagittarius. Shapley correctly inferred that the globular clusters were orbiting the nuclear bulge and that our solar system was located in the "suburbs" in a spiral arm. For the first time, the true size and extent of the Milky Way—and our position in it—were established.

We now know that the Milky Way is about 100,000 ly in diameter, the disk is about 2000 ly thick, and the nuclear bulge is about 13,000 ly thick (● Fig. 18.27). Our solar system is located about 28,000 ly from the center of the Milky Way, in a spiral arm known as the Orion arm, which takes about 230 million years to make one orbit.

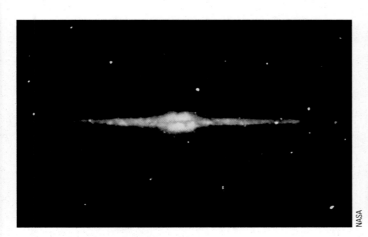

Figure 18.25 The Milky Way Galaxy
Although not easily seen in visible light, the nuclear bulge and part of the disk can be imaged in infrared light. This image of the Milky Way was obtained by the Cosmic Background Explorer (COBE) satellite.

Figure 18.26 The Hercules Globular Cluster
About 200 globular clusters form a halo around the Milky Way. Each globular cluster contains hundreds of thousands of stars. The Hercules globular cluster is one of three that can be seen with the unaided eye.

New evidence strongly suggests that a supermassive black hole, containing perhaps 2.6 million solar masses, exists in the core of the Milky Way's nuclear bulge. Such a hypothesis explains the high velocity of stars orbiting close to the core and the tremendous amount of energy that the core is generating. In fact, observations indicate that supermassive black holes inhabit the cores of nearly all galaxies.

Classification of Galaxies

American astronomer Edwin P. Hubble (1889–1953), for whom the Hubble Space Telescope is named, established a system for classifying galaxies that is based on how they look in photographs. Hubble's system classifies galaxies into three types: elliptical, normal or barred spiral, and irregular.

In an *elliptical galaxy*, the stars are bunched in a spherical or elliptical shape. Such a galaxy has no curved "arms." Examples include M84 (NGC 4374)* and M86 (NGC 4406), in the Virgo Cluster (● Fig. 18.28).

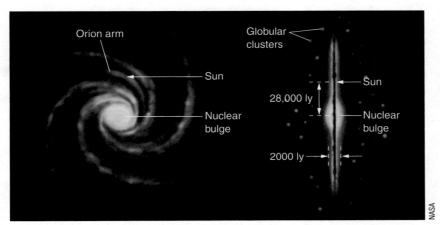

Figure 18.27 The Milky Way Galaxy
Schematic face-on and edge-on views of the Milky Way are shown. The Sun is located in the Orion arm, about 28,000 ly from the center of our galaxy.

*M84 is a catalogue number, named after the famous French astronomer Charles Messier (1730–1817). NGC refers to the New General Catalogue of astronomical objects first compiled in the 1880s.

Figure 18.28 Elliptical Galaxies in the Virgo Cluster
The center of the Virgo Cluster is dominated by two giant elliptical galaxies, M84 (*right*) and M86 (*center*). The Virgo Cluster is about 48 million ly away and contains more than 2000 galaxies.

In a *normal spiral galaxy*, many stars are gathered into a *nuclear bulge*, but many other stars are located in a disk consisting of arms that curve ("spiral") outward from the nuclear bulge. Examples of normal spiral galaxies include the Andromeda galaxy, the Whirlpool galaxy, and the Sombrero galaxy (● Fig. 18.29). Also note in Fig. 18.29a that the Andromeda galaxy has two small elliptical-galaxy companions.

In a *barred spiral galaxy*, a broad bar extends outward from opposite sides of the nuclear bulge before arms start to curve from the outer ends of the bars. An example is NGC 1300 (● Fig. 18.30).

(a)

(c)

(b)

Figure 18.29 Normal Spiral Galaxies
(a) The Andromeda galaxy (M31) is a normal spiral galaxy about 2.5 million ly away. It has two small, elliptical-galaxy companions, NGC 205 (*lower right*) and M32 (*upper left*). (b) The Whirlpool galaxy (M51) is a normal spiral galaxy seen face-on to us. It is interacting with the smaller galaxy (NGC 5195) shown below it. (c) The Sombrero galaxy (M104) is a normal spiral galaxy seen edge-on to us. The dark band across the galaxy's center is composed of gas and dust.

An *irregular galaxy*, as the name implies, has no regular geometric shape. Examples include the Large Magellanic Cloud (LMC) and the Small Magellanic Cloud (SMC) (● Fig. 18.31). These two galaxies were named in honor of explorer Ferdinand Magellan.*

Several hundred thousand galaxies have been identified. Using the latest ultradeep-field images from the Hubble Space Telescope, astronomers estimate that the universe may contain upwards of 500 billion galaxies (which might give you some idea of the vastness of the universe). About half of galaxies are the elliptical type and half are the spiral type, and only about 3% of galaxies are irregulars. Spiral galaxies are generally brighter than the others, accounting for about 75% of the brightest galaxies observed from the Earth.

As mentioned, the LMC and SMC can be seen with the unaided eye from the Southern Hemisphere. At only 180,000 ly away, the LMC is the third-closest galaxy to us. The SMC, about 210,000 ly distant, is the fourth closest. From the Northern Hemisphere, the Andromeda galaxy can be seen with the unaided eye, at a distance of about 2.2 million ly. Except for the LMC, the SMC, and the Andromeda galaxy, everything seen with the unaided eye is part of our own Milky Way galaxy. (A galaxy's distance is determined by several methods; see the **Highlight: Determining Astronomical Distances**.)

What is the closest galaxy to the Milky Way? That honor belongs to the Canis Major Dwarf, which is about 42,000 ly away and was discovered in 2003. The second-closest galaxy is the Sagittarius Dwarf, discovered in 1994 and about 50,000 ly distant. The gravitational forces exerted by the more massive Milky Way are tearing apart both of these dwarf galaxies, and their stars will ultimately be captured by the Milky Way.

Figure 18.30 A Barred Spiral Galaxy (NGC 1300)
The spiral arms of the barred spiral galaxy NGC 1300 curve from the ends of a bar extending from each side of the nuclear bulge.

Quasars

In 1963, astronomers using new radio telescopes with high *resolution* (the ability to distinguish the separation of two points) began detecting extremely strong radio waves from

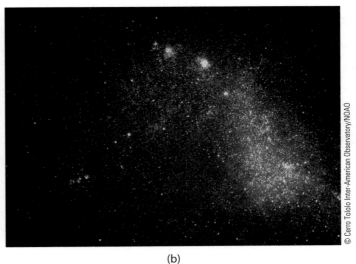

(a)   (b)

Figure 18.31 The Large and Small Magellanic Clouds
(a) The Large Magellanic Cloud (LMC) is an irregular galaxy about 180,000 ly away. The huge bright region at the left is the Tarantula Nebula. (b) The Small Magellanic Cloud (SMC), also an irregular galaxy, is about 210,000 ly away. Both galaxies are easily seen from the Southern Hemisphere.

*In 1519, Magellan left Spain with five ships and 240 men. He and his men were the first Europeans to view these galaxies because they can be seen only from the Southern Hemisphere. Three years later, one of the ships and 17 men (not including Magellan, who was killed by natives in the Philippines) returned to Spain after sailing completely around the Earth and proving, once and for all, that it is round.

Highlight   Determining Astronomical Distances

There is no tape measure long enough to measure even the nearest objects to the Earth. How do astronomers make accurate determinations of distances? They have developed various techniques to measure distances.

To measure distances to the nearest solar system objects, scientists have developed powerful radar to bounce signals off Venus, Mars, Mercury, and even the Sun. For more distant objects, astronomers use an age-old technique called geometric *parallax* that was first devised by the Greeks in 300 BCE. Figure 18.5 shows a diagram of the technique. If any two angles and a baseline are known, then the other sides of the triangle can be computed. Because stellar distances are so great, this parallax technique is accurate only to distances of about 325 ly.

The distances to stars can also be determined by knowing their absolute and apparent magnitudes (Chapter 18.3). This technique is not easy because the distance to the star must be known to determine the absolute magnitude of the star. Fortunately, most stars can be classified by their spectrum and placed on the H-R diagram, which can then be used to determine absolute magnitudes. This technique is useful for measuring stellar distances out to 35,000 ly. Careful application of the scientific method, checking and rechecking distances to the same star with different methods, significantly reduces errors.

Before 1915, the techniques of accurately measuring distances were limited to objects within the Milky Way galaxy. A breakthrough came when American astronomer Henrietta Leavitt discovered that certain unstable red giant stars, called *Cepheid variable stars*, could be used to accurately measure distances.

Leavitt found that the periods of variability of the unstable stars were directly related to their brightness; therefore, the distances to the stars could be determined. Because the Cepheid variable stars are very bright, they are seen to great distances, and greater distances could therefore be measured. With this important discovery, the distances to the nearest galaxies could be accurately determined. The use of Cepheid variable stars (Fig. 1) allows accurate measurements to 50 million light-years.

How are the distances to more distant galaxies found? Astronomers in the 1970s discovered that brighter spiral galaxies (those having more stars) rotate faster. Measuring rotation via the Doppler shift of light allows one to determine the absolute brightness of the galaxy. Distances out to 600 million light-years can be determined by comparing the absolute and apparent brightness of these galaxies

Supernovae (see Chapter 18.5) are stars that explode and become incredibly bright objects, sometimes even brighter than an entire galaxy of stars, although only for a short time. Specifically, certain types (Type I) of supernovae can be used to measure distances.

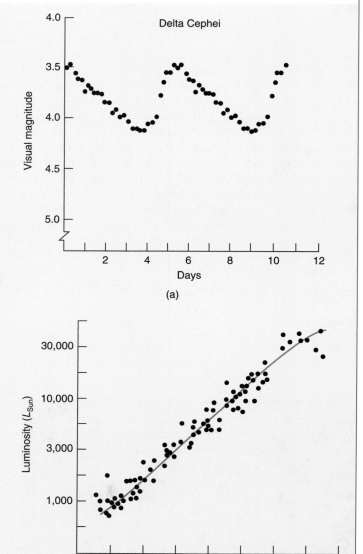

Figure 1 Cepheid Variable Stars
(a) The brightness of the variable star is plotted as a function of time. Because the variability of the star is very periodic, the period of the "pulsations" of the star is easily determined. (b) The period of variation is directly related to the brightness of the star, and the distance can be determined.

Highlight

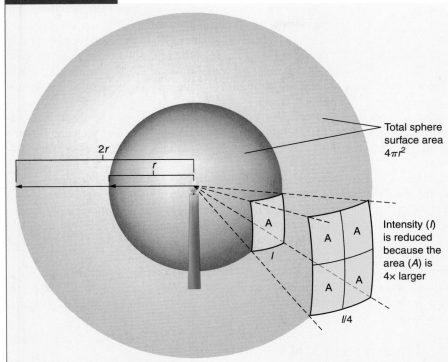

2r
r
A
A A
A
A A
I
I/4

Total sphere
surface area
$4\pi r^2$

Intensity (I)
is reduced
because the
area (A) is
4× larger

Figure 2 The Standard Candle Technique
As the candle is moved farther away, the brightness decreases. If the distance to the candle doubles, then the brightness of the candle decreases by a factor of 4.

The idea is simple: All type I supernovae are caused by similar explosion events, so their absolute magnitudes will be the same (that is, they are a benchmark). Astronomers can measure their apparent magnitudes, which differ for each supernova because they are all at different distances. Using the absolute and apparent brightness of supernovae, accurate distances are determined and are reliable out to about 650 million light-years.

Many of these techniques use a general principle of physics, the inverse-square law of intensity ($I \propto 1/r^2$, intensity proportional to $1/r^2$; see Fig. 6.11). When astronomers use this principle to measure distances, they refer to the method as standard candles. Because the brightness of the candle decreases as the square of the distance increases, the distance to the candle is determined. Figure 2 shows the simple idea for this technique.

Finally, Hubble's law (Chapter 18.7) can be used to determine distances to the farthest known objects in the universe. Figure 18.35 shows that there is a linear relationship between the velocity of a galaxy and its distance. The velocities of many distant

galaxies are measured with large telescopes by observing their spectra (Chapter 9.3). The absorption lines in the spectra are redshifted in direct relation to the overall velocity of the galaxy (the well-known Doppler effect; see Chapter 6.5). Hubble's law can be written in equation form as

$$V_r = H \times d$$

where V_r is the recessional velocity, H is the Hubble's constant (currently estimated to be 73 km/s/Mpc), and d is the distance to the galaxy.

Thus, if a galaxy's velocity is measured to be 10,000 km/s, then

$$d = \frac{V_r}{H} = \frac{10{,}000 \text{ km/s}}{73 \text{ km/s/Mpc}} = 137 \text{ Mpc} \approx 446 \text{ Mly}$$

The astute reader will note that to determine distances to the most distant objects in the universe, a very good determination of the Hubble constant is needed. The Hubble Space Telescope does that very task.

Figure 18.32 Quasar 3C 273
This Hubble Space Telescope photo shows one of the first quasars to be found, 3C 273, which was detected in the Virgo constellation in 1963. The redshift of its spectral lines indicates a distance of about 2 billion ly.

sources with small angular dimensions. These sources were named *quasars*, a shortened form for *quasi-stellar radio sources* (● Fig. 18.32).

Tens of thousands of quasars have now been detected, and they have two important characteristics: They are generally the most distant objects observed in the universe, and they emit tremendous amounts of electromagnetic radiation (energy) at all wavelengths. In some cases, they emit 1000 times the energy of the entire Milky Way. The closest known quasar is 800 million ly away in the radio galaxy Cygnus A. The record for the farthest known galaxy keeps being broken, but as of 2011, it is held by a compact blue galaxy that is thought to be about 13.2 billion ly away.

For a time, quasars were mysterious objects. Astronomers are now convinced that quasars are the cores of galaxies that were forming when the universe was young and that a quasar's energy comes from material as it orbits, and falls into, a supermassive black hole at its center.

Photographs of distant stars and galaxies show the way the objects appeared at the moment that the light radiated from them. Objects that are seen in the night sky, or photographed with telescope cameras, enable us to look back in time. If a photograph is taken of a quasar that is 13 billion light-years away, then that quasar is seen as it was 13 billion years ago. Because astronomers think that the universe is 13.7 billion years old, the quasar is seen as it looked when the universe was very young.

In other words, the picture is of the quasar's past, not its present. Even when the Sun is photographed, the surface is seen as it existed about 8 minutes ago, the time it takes sunlight to reach the Earth. Be aware that the farther one looks into space, the farther one is looking back in time.

The Local Group and Local Supercluster

The Milky Way and the Andromeda galaxies are the two largest members of a group of almost 40 galaxies that are gravitationally bound in what is called the *Local Group* (● Fig. 18.33). Most of the galaxies in the Local Group are within 3 million ly of the Earth, and most are small, dim ellipticals and irregulars.

The galaxies that astronomers observe throughout the vast volume of the universe are lumped together in *clusters* that vary in size and number. The clusters range from 3 million

Figure 18.33 The Local Group of Galaxies
The Local Group of about 40 galaxies is dominated by three large spiral galaxies: the Andromeda galaxy, the Triangulum galaxy (M33), and the Milky Way. Most of the other Local Group galaxies are small ellipticals and irregulars.

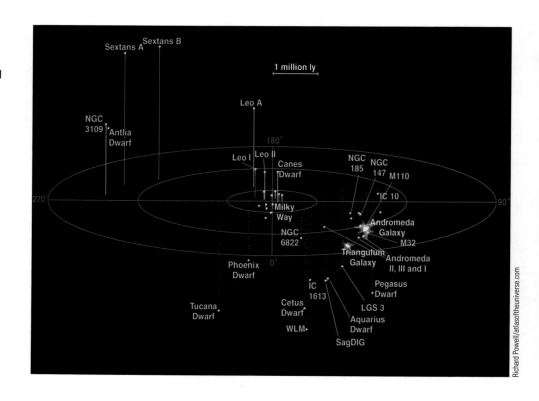

to 15 million ly in diameter. Some, such as the Local Group, contain a few galaxies. Others, such as the Virgo Cluster, consist of thousands of galaxies (see Fig. 18.28).

The clusters of galaxies are themselves bunched in what are called *superclusters*, or clusters of clusters. These superclusters have diameters as large as 300 million ly and masses equal to or greater than 10^{15} solar masses. The Local Group and adjoining clusters such as the Virgo Cluster form what is known as the *Local Supercluster*. The Virgo Cluster seems to be close to the center of the Local Supercluster, whereas the Local Group lies near its outer boundary.

Dark Matter

Analysis of the orbital speed of stars in the Milky Way indicates that it contains much more mass than is accounted for by the stars and gas in it. This problem recurs throughout the universe. Almost everywhere astronomers look—whether in the Milky Way, in other galaxies, or in clusters of galaxies—most of the matter needed to cause the galaxy's gravitational behavior seems to be unseen by light, radio waves, X-rays, or any other electromagnetic radiation. Oddly enough, it appears that only about 5% of the universe is accounted for as normal matter. The unseen matter was originally referred to as the *missing mass*, but a better term, **dark matter**, is now used. Apparently, the disk of the Milky Way is embedded in a "halo" of dark matter that extends perhaps 150,000 ly from the Milky Way's center.

Dark matter is thought to exist in two basic forms: ordinary matter and exotic matter. The category of ordinary matter is given the name MACHOs (*massive compact halo objects*) and contains brown dwarfs, dim stars such as white dwarfs and neutron stars, black holes, and even dark galaxies. Many of these MACHOs have been detected by gravitational interactions with visible matter, but so far their numbers are too few to explain all the dark matter. A discovery of many faint, low-mass stars in 2011 tripled the number of stars in the universe (see Astronomy Facts), but that still cannot explain the discrepancy. The exotic matter is made up of subatomic particles such as neutrinos or other "theoretical" particles called WIMPs (*weakly interacting massive particles*). Neutrinos have been detected but in too few numbers. The WIMP particles are difficult to detect because they interact weakly with other matter, but scientists are optimistic that they are the answer and are still searching. See the **Highlight: The Age of the Universe**.

Did You Learn?

- The Milky Way is about 100,000 light-years across.
- Dark matter is unknown material only detected by its gravitational presence.

18.7 Cosmology

Preview Questions

- Is the universe expanding?
- What does the universe look like on a large scale?

The Structure of the Universe

Cosmology is the branch of astronomy that is the study of the structure and evolution of the universe. Astronomers determine the structure of the universe by detecting and analyzing electromagnetic waves that come from the galaxies, the "building blocks" of the universe. From the data thus collected, they determine the way in which these galaxies are distributed throughout the vast volume of the universe. Astronomers estimate that tens of billions of galaxies are within range of optical telescopes.

Figure 18.34 The Distribution of Galaxies
More than 200,000 galaxies are represented in this map. The galaxies extend to 2 billion ly away. The Milky Way is located at the center of the map on the left. The expanded view on the right shows the two-dimensional view of the deep sky.

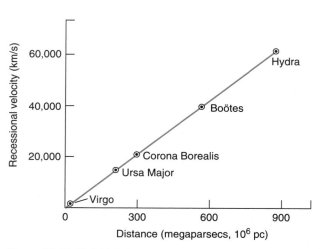

Figure 18.35 Hubble's Law
This graph shows the relationship between how fast a cluster of galaxies is moving away (its recessional velocity) and its distance from the Milky Way.

Even if they were inclined to do so, astronomers would not have the time to observe the tremendous volume of the universe to obtain information about all the galaxies. Instead, the astronomers' model of the structure of the universe is based on the sampling of different regions.

In 2003, a team of astronomers calculated the distances to about 200,000 galaxies, entered the coordinates (DECs, RAs, and distances) into a computer, and obtained an illustration showing the location and relative positions of these galaxies (● Fig. 18.34). Nearly every galaxy in the sample belongs to either a two-dimensional "sheet" or a one-dimensional "thread" that is millions of light-years in length.

The data also revealed large voids, millions of light-years across, that contained very few galaxies. In other words, the two-dimensional sheets were separated by vast volumes of nearly empty space. Such a structure is reminiscent of the structure of a bunch of soap bubbles attached to one another. Think of the galaxies as being located on the surface and intersections of the soap bubbles, with very few galaxies being inside a given bubble. Like the galaxies, dark matter does not seem to be present to any significant extent in the voids. It seems that galaxies are not evenly distributed in space.

Hubble's Law

When Edwin Hubble began looking at spectrum shifts of galaxies, he found nothing surprising at first. In the Local Group, the shifts were small. Some shifts were blue and some were red, a blue one indicating that the galaxy is moving toward us and a red one meaning the galaxy is moving away. However, as Hubble looked at galaxies farther and farther away, he found only redshifts. In fact, the farther away the galaxy is, the larger the redshift. Hubble concluded that the universe is expanding.

In addition to these redshift measurements, the distances to remote galaxies can also be determined independently. Hubble converted the observed redshift of a galaxy to radial velocity and then plotted the velocity versus distance (● Fig. 18.35). Hubble's discovery is now known as **Hubble's law**:

The greater the recessional velocity of a galaxy, the farther away the galaxy.

Consequences of Hubble's Law

If galaxies (actually galactic clusters) are receding from us in all directions, then does that mean we are at the center of the universe? The answer is *no*. Any observer anywhere in an

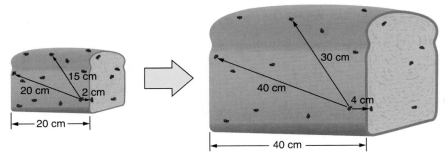

Figure 18.36 Analogy of the Expanding Universe
The expanding universe is somewhat analogous to the expanding dough in a loaf of raisin bread.
The raisins (galaxies) are carried along as the dough (space) expands. The numbers show that
the farther an "observed" raisin is from the "observing" raisin, the faster the former is moving. For
example, the most distant raisin has moved 20 cm in the same time that the closest raisin has
moved only 2 cm. The view from any raisin would show the other raisins receding from it.

expanding universe would make the same observations and come to the same conclusion.
This concept can be illustrated with an analogy.

Consider baking a loaf of raisin bread (● Fig. 18.36). Imagine that the raisins represent
the galaxies and the dough represents space. As the loaf bakes and expands, each raisin
remains the same size but moves away from every other raisin. No matter which raisin
an observer might be "riding," the other raisins would move away. The greater the initial
distance a specific raisin was from the observer's raisin, the faster and farther the observed
raisin would recede (study Fig. 18.36). Note that the raisins (galaxies) stay the same size,
but they are carried along by the expansion of the dough (space). Essentially, the raisins
behave in accordance with Hubble's law.

Conceptual Question and Answer

The Expanding Universe

Q. Into what is the universe expanding?

A. This question is common, and it is a misconception to think the universe is
expanding into something. The answer is that the universe is not expanding into
anything. Galaxies are not flying apart from one another from some explosion;
rather, galaxies are part of space that expands (like expanding dough in baking
bread) and carries the galaxies with it. If the universe is infinite (many scientists
think it is), then all space is expanding and there is no "outside" of the universe
in which to expand.

Astronomers can determine the age of the universe using Hubble's law (see Fig. 18.35).
The slope of the graph represents the rate at which the universe is expanding; therefore,
astronomers can calculate how long the universe has been expanding, and the result of that
calculation is the age of the universe. If the slope of the line is H (the Hubble constant),
then the age of the universe is

$$\text{age (in years)} = \frac{1 \times 10^{12}}{H} \qquad \text{18.2}$$

where H is the Hubble constant (currently estimated to be 73 km/s/Mpc).

The number 1×10^{12} is a conversion factor for the time and distance units used in the
problem. The best estimate of the constant H is about 73 km/s/Mpc (1 Mpc = 1 *mega*-parsec,
1 million parsecs). This calculation gives about 13.7 billion years for the age of the universe.

EXAMPLE 18.2   Calculating the Age of the Universe

Calculate the age of the universe if the constant rate of expansion is 73 km/s/Mpc.

Solution

$$\text{age (in years)} = \frac{1 \times 10^{12}}{H} = \frac{1 \times 10^{12}}{73} = 1.37 \times 10^{10} \text{ years}$$

$$= 13.7 \text{ billion years}$$

Confidence Exercise 18.2

Calculate the age of the universe if the constant rate of expansion is 50 km/s/Mpc.
Answers to Confidence Questions may be found at the back of the book.

Astronomers agree that the universe is expanding. Will the universe expand forever? Recent evidence strongly indicates that we live in what cosmologists call a *flat universe* that will continue to expand forever. Contrary to earlier ideas that gravity would slow down the expansion, recent discoveries have confirmed that the expansion of the universe is actually accelerating. The acceleration is thought to be due to the repulsive effect of a mysterious **dark energy** that seems to make up 72% of the mass–energy of the universe ($E = mc^2$). The remaining mass–energy appears to be 5% ordinary matter and 23% dark matter (see the Highlight: The Age of the Universe). One of the current mysteries in astronomy is whether dark energy really exists, and, if so, then what is it?

The Big Bang

If galaxies are moving away from one another at the present time, then the galaxies must have been closer to one another in the past. That is, the universe must have been more compressed. Carrying this idea to its logical conclusion, most astronomers think that the universe began in a small, hot, dense state, the rapid expansion of which is called the **Big Bang**.

The hypothesis called the Big Bang was first proposed in 1927 by Georges Lemaitre, a Belgian Catholic priest and cosmologist.* Since then, scientists have focused great efforts on investigating the Big Bang model. Our knowledge of the structure of the universe and of subatomic particles is able to take us back to within 10^{-43} second of the Big Bang. Before that time, the universe was so compressed and opaque that our present understanding of the laws of relativity and quantum mechanics is inadequate.

Scientists think that by 10^{-4} second after the beginning, the universe was filled with photons having a high temperature and density. Within the first 4 seconds, some of the photons formed protons, neutrons, and electrons. The universe continued to expand and cool. By the time the universe was 3 minutes old, some of the protons had formed deuterium and helium nuclei.

When the universe was 30 minutes old, it had cooled so much that nuclear reactions ceased; the matter in the universe consisted, by mass, of about 25% helium nuclei and 75% hydrogen nuclei. By the time the universe was about 500,000 years old, it was cool enough for the helium and hydrogen nuclei to capture electrons and become neutral

---

*The geometry and mathematics of an expanding universe were first formulated in 1922 by Alexander Friedmann, a Russian mathematician. Five years later, Lemaitre developed a new cosmology theory of an expanding universe and postulated an explosive beginning from a "Primeval Atom." Having heard Hubble lecture at Harvard University, Lemaitre knew of Hubble's evidence concerning the cosmological redshift of galaxies. The term *big bang* was first applied disparagingly by Fred Hoyle, a British astrophysicist and cosmologist, who in the 1950s helped develop a rival theory called the steady state model, which no longer receives serious consideration.

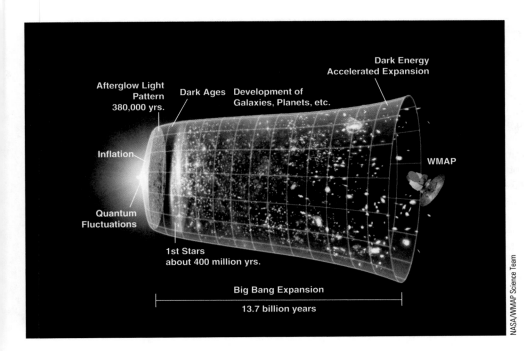

Figure 18.37 A History of the Universe
This timeline shows the history of the universe according to the inflationary model of the Big Bang. The vertical axis shows the size of the universe, and the horizontal axis shows the time. The period of inflation occurs between 10^{-35} second and 10^{-30} second after the Big Bang, followed by a slow, steady expansion.

atoms. Photons were now able to move freely throughout space, and matter was able to be influenced by gravity and begin forming galaxies and stars.

The standard Big Bang model has received broad acceptance because experimental evidence supports it in three major areas:

1. Astronomers observe galaxies that show a shift in their spectrum lines toward the low-frequency (red) end of the electromagnetic spectrum. This redshift is known as the *cosmological redshift*. It is not really a Doppler shift (Chapter 6.5), but rather a lengthening in the wavelength as a consequence of the expansion of space. The galaxies aren't moving *through* space. Instead, space itself is expanding and carrying the galaxies along.

2. Astronomers detect a cosmic background radiation in all directions coming from space. The presence of this **cosmic microwave background** was predicted by the Big Bang theory (see Fig. 1 of the Highlight: The Age of the Universe). In 1965, the background radiation was detected by two scientists at Bell Telephone Laboratories, and it had exactly the properties predicted by the Big Bang model. This residual microwave radiation is the greatly redshifted radiation of the extremely hot universe that existed about 400,000 years after the Big Bang.

3. Astronomers observe a mass ratio of hydrogen to helium of 3 to 1 in stars and interstellar matter, a ratio predicted by the Big Bang model.

However, what is called the *standard model* of the Big Bang was unable to answer certain questions. This led to a modification known as the *inflationary model* of the Big Bang, advanced in 1980 by Alan Guth, an astrophysicist at the Massachusetts Institute of Technology (M.I.T.). The inflationary model fits a flat universe (indeed, recent experiments indicate that our universe is exactly flat).* The model proposes that at 10^{-35} second after the Big Bang, the universe had cooled enough that some of the four fundamental forces (gravitational, electromagnetic, strong nuclear, and weak nuclear) separated from a single, unified force and, in doing so, released tremendous energy.

This energy release caused an outward pressure that inflated the universe to at least 10^{50} times its original tiny size by the time the universe was 10^{-30} second old (● Fig. 18.37).

---

*An *open universe* expands forever. A *closed universe* stops expanding and then collapses. *A flat universe* is so finely tuned that the expansion speed becomes zero when the universe has reached infinite size.

Highlight   The Age of the Universe

A recent top science story was the determination of the age of the universe, which turned out to be 13.7 billion years old (plus or minus 1%). Prior to the first detailed all-sky map of the early universe, astronomers could give only approximate answers to some fundamental questions: How old is the universe? Of what is it composed? What is its shape?

The map (Fig. 1) was compiled from data collected by the Wilkinson Microwave Anisotropy Probe (WMAP), an orbiting space probe that was launched by NASA in 2001. WMAP, named after American cosmologist David Wilkinson, has given us an unprecedented overview of the universe as it was 380,000 years after the Big Bang. It was a time immediately after an opaque "soup" of atom fragments combined for the first time to form actual atoms. This formation of neutral atoms allowed the first glow of radiation, known as the cosmic microwave background, to fill the entire universe.

From the data, the WMAP research team calculated that the universe is 13.7 billion years old (plus or minus 1%), an astounding degree of accuracy given that most prior estimates could give only a range of 12 to 20 billion years. The WMAP team was also able to determine that the first stars appeared 400 million years after the Big Bang, which is earlier than most previous estimates.

The data also showed the exact proportions of the universe: 5% normal matter, 23% dark matter, and 72% dark energy (Fig. 2). This proportion fits only for a universe that is flat (rather than spherical, or hyperbolic like the surface of a saddle), and it predicts that the universe will continue to expand forever rather than someday contracting in a Big Crunch.

WMAP's lead scientist, astrophysicist Charles Bennett of Goddard Space Flight Center, states, "The WMAP results are a turning point. Now we need to ask a whole new set of questions, like what happened in the first moments of inflation, and what is dark matter?"

Source: From Kathy Svitil, "Probe Reveals Age, Composition, and Shape of the Cosmos," *Discover Magazine,* January 2004, p. 37.

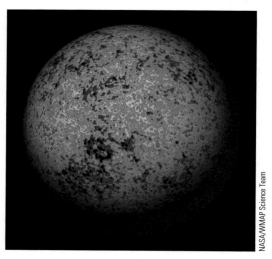

NASA/WMAP Science Team

Figure 1   WMAP's All-Sky Map
This map shows temperature fluctuations from relatively warm (red) to cool (blue) in the microwave light that bathed the universe some 13 billion years ago.

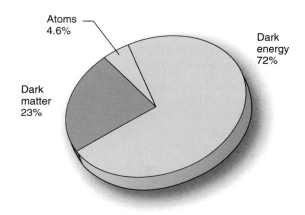

Figure 2   The Composition of the Universe as Determined by WMAP
Data from WMAP indicate that the universe is composed of about 5% normal "ordinary" matter, 23% dark matter, and 72% dark energy.

The model indicates that the universe grew from about the size of an atom to the size of a grape during that brief period of inflation. After the first 10^{-30} second, no differences exist between the inflationary model and the standard model. The timeline in Fig. 18.37 shows the cosmic background radiation followed by a period during which stars had yet to form (dark ages). The expansion rate of the universe was nearly constant for 13 billion years, and now we are detecting an increase in the expansion due to dark energy.

This model of the universe means that the universe is infinite in extent. There are no centers or "edges" in the universe; the Big Bang and inflation occurred everywhere in the universe. Because the universe is not infinitely old, however, only part of it is able to be seen. What is seen is called the observable universe; beyond the observable limit is impossible to see at the present time because light has not had enough time to travel to our telescopes.

The inflationary model of the Big Bang is astronomers' best theory of the evolution of our universe. Of course, our current state of knowledge is incomplete and scientists do not know what initially caused the Big Bang. It is expected that in the future, there will be improvements in the theory of how the universe formed, evolved, and will end.

In this chapter, our exposition of the Big Bang theory has been severely limited. Lack of space (no pun intended) prohibits a more thorough discussion of this fascinating, complex topic. For readers who wish to explore further, almost any astronomy textbook has a full chapter on cosmology.

Rest assured that there will be new astronomical discoveries throughout your lifetime. New, giant telescopes of all kinds have recently been built, are being built, or are in the planning stage, including both ground-based and space-based instruments. For the latest in astronomical discoveries and for a discussion of astronomy topics on a level appropriate to educated nonscientists, the reader is referred to *Astronomy* magazine and to its excellent website, www.astronomy.com.

Did You Learn?

- Observational evidence shows that the universe is expanding according to Hubble's law.
- On a large scale, giant clusters of galaxies are distributed in clumps and sheets around great voids in space.

KEY TERMS

1. universe
2. celestial sphere (18.1)
3. ecliptic
4. right ascension
5. declination
6. parsec
7. photosphere (18.2)
8. apparent magnitude (18.3)
9. absolute magnitude
10. H-R diagram
11. main sequence
12. red giants
13. white dwarfs
14. nebulae (18.4)
15. planetary nebula
16. nova (18.5)
17. supernova
18. neutron star
19. black hole
20. galaxy (18.6)
21. globular clusters
22. dark matter
23. cosmology (18.7)
24. Hubble's law (18.7)
25. dark energy
26. Big Bang
27. cosmic microwave background

MATCHING

For each of the following items, fill in the number of the appropriate Key Term from the preceding list. Compare your answers with those at the back of the book.

a. _____ The brightness a star would have at a distance of 10 pc

b. _____ An object so dense that it has a singularity at the center

c. _____ Our best theory to explain the temperature, composition, and expansion of the universe

d. _____ The graph used to plot the temperature and absolute magnitude of stars

e. _____ Emission, reflection, or dark clouds of gas and dust in space

f. _____ The relationship between the distance of a galaxy and its recessional velocity

g. _____ The visible layer of the Sun in which sunspots are found

h. _____ The huge, imaginary dome on which the stars appear to be mounted

i. _____ The branch of astronomy that studies the structure and evolution of the universe

j. _____ The event when a star completely blows itself up

k. _____ Large, cool stars to the upper right on the H-R diagram

l. _____ The latitude coordinate on the celestial sphere

m. _____ The brightness of a celestial object as observed from the Earth

n. _____ The mysterious, unseen matter that accounts for about 23% of the mass of the universe

o. _____ Primordial radiation that resulted from the Big Bang and fills the entire universe

p. _____ The longitude coordinate on the celestial sphere

q. _____ A sudden brightening of a white dwarf star due to the explosion of accreting material

r. _____ An extremely large collection of stars held together by gravity

s. _____ The narrow band of stars going from upper left to lower right on an H-R diagram

t. _____ A shell of matter ejected from a variable star late in its life

u. _____ Astronomers use this distance unit when measuring the parallax of a star

v. _____ A mysterious energy that seems to cause the expansion of the universe to accelerate

w. _____ A rapidly rotating, superdense core left over at the end of the life of a high-mass star

x. _____ These small, hot, dense stars are the fate of low-mass stars

y. _____ The name given to all matter, energy, and space

z. _____ Large groups of stars found in a spherical halo around the Milky Way

aa. _____ The apparent path that the Sun traces annually on the celestial sphere

MULTIPLE CHOICE

Compare your answers with those at the back of the book.

1. What is the point on the celestial sphere representing the extension of the axis of Earth's North Pole? (18.1)
   (a) North celestial pole   (b) core
   (c) vernal equinox   (d) celestial equator

2. What is the name of the path on the sky that the Sun appears to follow as it moves among the stars? (18.1)
   (a) the ecliptic   (b) the meridian
   (c) the equator   (d) the equinox

3. The angular measure in degrees north or south of the celestial equator is called which of the following? (18.1)
   (a) latitude   (b) longitude
   (c) right ascension   (d) declination

4. Which of the following is part of the Sun's structure? (18.2)
   (a) an accretion disk   (b) the corona
   (c) a globular cluster   (d) a singularity

5. Why are sunspots darker than the regions of the Sun around them? (18.2)
   (a) They consist of different elements than the rest of the Sun.
   (b) They are located far out into the Sun's atmosphere (corona).
   (c) They are hotter than the material around them.
   (d) They are cooler than the surrounding material.

6. Where in the Sun does fusion of hydrogen occur? (18.2)
   (a) only in the core
   (b) only near the photosphere (near the surface layer)
   (c) pretty much anywhere throughout the Sun
   (d) in the sunspot regions

7. The part of the Sun's structure that we see in visible light is which of the following? (18.2)
   (a) the photosphere   (b) the chromosphere
   (c) the core   (d) the corona

8. What force keeps the all stars from flying apart? (18.3)
   (a) nuclear force   (b) gravitational force
   (c) radiation pressure   (d) electrical force

9. Which property classifies the star based on its temperature? (18.3)
   (a) its apparent magnitude   (b) its mass
   (c) its radius   (d) its spectral type

10. A relatively cool star is which of the following? (18.3)
    (a) red   (b) yellow
    (c) white   (d) blue

11. At the end of the Sun's life, what kind of object will remain? (18.4)
    (a) blue giant star   (b) white dwarf star
    (c) red giant star   (d) red dwarf star

12. The famous Horsehead Nebula is which of the following? (18.4)
    (a) an emission nebula
    (b) a reflection nebula
    (c) a planetary nebula
    (d) a dark nebula

13. A white dwarf has about the same diameter as which of the following? (18.4)
    (a) Texas   (b) the Earth
    (c) Jupiter   (d) the Sun

14. What is the fundamental quantity of a star that indicates its ultimate fate? (18.5)
    (a) color   (b) mass
    (c) size   (d) temperature

15. What event is believed to create elements heavier than iron? (18.5)
    (a) a supernova explosion
    (b) a nova outburst
    (c) the formation of a neutron star
    (d) the accretion of matter into a black hole

16. Black holes appear black for what basic reason? (18.5)
    (a) They are invisible and can never be detected by astronomers.
    (b) They appear in starless, dark areas of the Milky Way.
    (c) They "shine" by blackbody radiation and are easily detected.
    (d) Light cannot escape them.

17. The Milky Way is classified as what kind of galaxy? (18.6)
    (a) elliptical   (b) spiral
    (c) irregular   (d) collisional

18. How much of the universe appears to be normal ordinary matter? (18.6)
    (a) 5%   (b) 20%
    (c) 50%   (d) 95%

19. What key observation led Hubble to conclude that the universe is expanding? (18.7)
    (a) the distribution of the galaxies in the universe
    (b) the classification of different types of galaxies
    (c) the redshift of light from distant galaxies
    (d) the ratio of hydrogen to helium in a galaxy

20. What is the cosmic background radiation? (18.7)
    (a) the light-energy remnants of the explosion in which the universe was born
    (b) the radio noise from hot gas in rich galactic clusters
    (c) the faint glow of light across the night sky
    (d) the light emitted by supernova explosions

FILL IN THE BLANK

1. The apparent change of the position of a star due to the Earth's orbiting the Sun is called ___. (18.1)

2. The extension of the Earth's equator onto the celestial sphere is called the ___. (18.1)

3. The longitude of a place on the Earth's surface is comparable to the ___ of a star on the celestial sphere. (18.1)

4. The angle between the Earth's orbit plane and the plane of the equator projected on the celestial sphere is ___. (18.1)

5. The ___ cycle is 11 years long. (18.2)

6. The Sun's energy comes from the fusion of hydrogen into ____. (18.2)

7. ___ are sudden, bright, explosive events originating from the Sun's surface. (18.2)

8. The celestial sphere is divided into 88 ___. (18.3)

9. Type M stars are ___ in color. (18.3)

10. Analysis of the ___ of a star reveals the composition of the star. (18.3)

11. The magnitude scale is really a scale of ___. (18.3)

12. The diagonal group of stars on an H-R diagram is called the ___. (18.3)

13. The ____ phase is the *next* phase of the Sun's life. (18.4)

14. When a star runs out of nuclear fuel, the core begins to ___. (18.4)

15. Stars are born in giant clouds of gas and dust, such as the Great ___ in Orion. (18.4)

16. A "new" star is called a(n) ___. (18.5)

17. A neutron star that emits radio waves in a periodic fashion is a(n) ___. (18.5)

18. The most catastrophic of stellar explosions is called a(n) ___. (18.5)

19. Astronomers think that a(n) ___ black hole lies at the center of the Milky Way. (18.6)

20. Astronomers think that the universe is ___ billion years old. (18.7)

SHORT ANSWER

1. Define the term *universe*.

18.1 The Celestial Sphere

2. Name the three celestial coordinates.

3. At what angle is the ecliptic tilted with respect to the celestial equator? What is the physical reason for this angle?

4. What is the vernal equinox, and what does it have to do with the celestial prime meridian?

5. Which is the larger unit of distance, the light-year or the parsec? How much larger is one than the other?

6. Why is the measurement of distance limited by using the parallax technique? What is this distance limit for stars?

18.2 The Sun: Our Closest Star

7. What is the approximate temperature of the Sun at its surface?

8. What happens to the latitude of the sunspots over the course of one solar cycle?

9. When is the Sun's corona visible?

10. Approximately how many times larger than the Earth is the Sun?

11. Describe the various layers of the Sun beginning from the center.

12. Give the net reaction of the proton–proton chain.

18.3 Classifying Stars

13. What two forces are balanced inside a star to keep it stable?

14. What is the difference between the apparent magnitude and the absolute magnitude of a star?

15. Describe the motions of the Moon, stars, and constellations in the sky over one day.

16. How far away is the star closest to the Sun? What is its name?

17. Give seven spectral classes of stars, in order from hottest to coolest photospheres.

18. Sketch an H-R diagram and show the position of each of the following: (a) the Sun, (b) the main sequence, (c) red giants, (d) white dwarfs, and (e) red supergiants.

18.4 The Life Cycle of Low-Mass Stars

19. What is the difference between an emission nebula and a reflection nebula?

20. When the Sun first moves off the main sequence, what type of star will it become?

21. Define the term *nucleosynthesis*.

22. Describe how the mass of a star plays a role in the lifetime of a star.

23. What is a brown dwarf? Have any brown dwarfs been found?

24. State the order in which the following possible stages of a star occur: main-sequence star, planetary nebula, white dwarf, protostar, red giant.

18.5 The Life Cycle of High-Mass Stars

25. What is the difference between a nova and a supernova?

26. What is the difference between Type I and Type II supernovae?

27. Compare a pulsar with a lighthouse.

28. Sketch a nonrotating black hole, showing the singularity, the Schwarzschild radius, and the event horizon.

29. How do astronomers "detect" a black hole?

18.6 Galaxies

30. Name the three major classifications of galaxies.

31. What is a quasar?

32. What are two types of spiral galaxy, and how do they differ?
33. What is the significance of (a) dark matter and (b) dark energy?
34. State the structure and dimensions of the Milky Way galaxy.
35. What are globular clusters, where are they found, and what conclusion was reached by examining their distribution?
36. What two basic groups of objects are thought to make up dark matter?
37. Name the two galaxies that can be seen with the unaided eye in the Southern Hemisphere.

18.7 Cosmology

38. What is Hubble's law, and what conclusion about the universe do astronomers draw from it?
39. Briefly, how do galaxies seem to be distributed in the universe?
40. State three experimental findings that support the Big Bang model of the universe.
41. Name the most recent modification of the Big Bang model and briefly explain how it differs from the standard model.
42. About how many years after the Big Bang did gravity start to shape matter into stars and galaxies?

VISUAL CONNECTION

Visualize the connections and give answers for the blanks. Compare your answers with those at the back of the book.

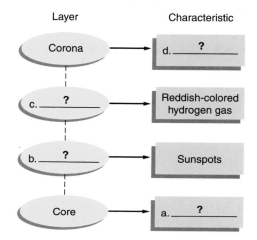

APPLYING YOUR KNOWLEDGE

1. Why do you think it took so long for astronomers to measure the distances to stars accurately?
2. What might happen if the bright star Sirius exploded in a supernova? How long would it be before we found out that it had happened?
3. Consider three main sequence stars: Star A is 12,000 K, Star B is 6000 K, and Star C is 3000 K. What are the colors of each of these stars? If these stars are the same distance from the Earth, then which star is the brightest? (See Chapter 9.2 for help.)
4. If you went outside on a clear night to locate Polaris, then how would you find it? (*Hint*: See Appendix IX.)
5. How are the light from emission nebulae and the light from reflection nebulae analogous to the light from the Sun and the Moon, respectively?
6. Do you need to worry that the Sun will swell to a red giant and destroy the Earth any time soon?
7. Everyone realizes how dependent we Earthlings are on the Sun, but how are we also indebted to massive, long-dead stars?

IMPORTANT EQUATIONS

Stellar distance (in parsecs): $d = \dfrac{1}{p}$ (18.1)

Age of the Universe: age (in years) $= \dfrac{1 \times 10^{12}}{H}$ (18.2)

EXERCISES

18.1 The Celestial Sphere

1. Find the distance in parsecs to the star Altair, which has an annual parallax of 0.20 arcsec.

   Answer: 5.0 pc

2. The bright star Sirius has a parallax angle of 0.38 arcsec. Find the distance in parsecs and in light-years.

3. Calculate the number of seconds in a year (365 days). Express your answer in standard exponential notation, with three significant figures.

   Answer: 3.15×10^7 s/y

4. Calculate the number of miles in a light-year, using 1.86×10^5 mi/s as the speed of light. (*Hint:* The answer to Exercise 3 will be useful.)

5. The red supergiant star in Orion is called Betelgeuse, and its distance is 130 pc. What is the parallax of this star?

   Answer: 0.008 arcsec

6. The famous Pleiades star cluster is about 130 parsecs distant. How many light-years is that?

18.3 Classifying Stars

7. How many times brighter is a Cepheid variable of absolute magnitude −3 than a white dwarf of absolute magnitude +7?

   Answer: 10,000 times

8. How many times dimmer is a star with magnitude +1 compared with the planet Venus with a magnitude of −4?

9. A galaxy cluster in Ursa Major has a recessional velocity of 15,000 km/s. Using the best estimate for Hubble's constant, find the distance to the galaxy cluster (see the Highlight: Determining Astronomical Distances).

   Answer: $d = V_r/H = 205$ Mpc

10. The Hydra supercluster has a distance of about 800 Mpc. Using Hubble's law, determine the recessional velocity to this cluster.

18.7 Cosmology

11. Suppose that the Hubble constant had a value of $H = 100$ km/s/Mpc. What would be the age of the universe? What if H had a value of 50 km/s/Mpc?

    Answer: Age = $(1 \times 10^{12})/H = (1 \times 10^{12})/100 = 10$ billion years; age = $(1 \times 10^{12})/H = (1 \times 10^{12})/50 = 20$ billion years

12. If Hubble's constant had a value of 75 km/s/Mpc, then what would be the age of the universe?

ON THE WEB

1. Exploring the Universe

How did the universe begin, and how did it come to be in its present state? What is the "Big Bang"? How do the different observations explain the origins of the universe? How does the universe continue to expand today? To find answers to these questions, follow the recommended links on the student website at **www.cengagebrain.com/shop/ISBN/1133104096** and take a ride back in time, forward into the future, and around the cosmos to explore the past, present, and ultimate fate of our universe.

2. Black Holes Aren't Just a Matter for Science Fiction

Just what are black holes? Why are they called "black" holes? Why study them? How do they form? What are they made of? Where do they begin and end? Visit the recommended links on the student website at **www.cengagebrain.com/shop/ISBN/1133104096** to explore answers to these questions.

3. The Wonders of the Hubble Space Telescope

Where do I find all those great images from space? How does the Hubble Space Telescope work? Are the images in true color? Visit **www.hubblesite.org** to find all the HST images, explanations of Hubble's workings, and educational materials.

*. . . this most excellent
canopy, the air.*

•

William Shakespeare
(1564–1616)

The setting Moon over the >
Earth's horizon and troposphere.
(Can you identify the phase of
the Moon?)

NASA

ATMOSPHERIC FACTS

▶ If the Earth were the size of
a basketball, then the atmosphere would be about 1 mm
thick.

▶ Baseball records may be set in
Denver, Colorado, known as
the "Mile-High City." The air
there is about 15% less dense
than at sea level, resulting in

Earth science is a collective term that involves all aspects of our planet:
land, sea, air, and even its history. This chapter and the next will be
concerned with one aspect of Earth science: the composition and phenomena of the atmosphere.

Our atmosphere (from the Greek *atmos*, "vapor," and *sphaira*, "sphere") is the
gaseous shell or envelope of air that surrounds the Earth. Just as certain sea creatures live at the bottom of the ocean, we humans live at the bottom of this vast
atmospheric sea of gases.

In recent years, the study of the atmosphere has expanded because of advances
in technology. Every aspect of the atmosphere is now investigated in what is called
atmospheric science. An older term, **meteorology** (from the Greek *meteora*, "the

Chapter Outline

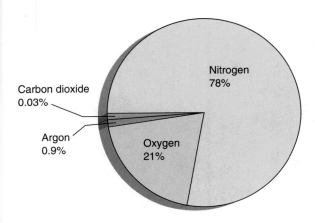

Figure 19.1 Composition of Dry Air
A graphical representation of the volume composition of the major constituents of air. (The CO_2 and Ar percentages are shown larger than scale for clarity.)

air"), is now more commonly applied to the study of the lower atmosphere. The conditions of the lower atmosphere are what we call *weather*. Lower atmospheric conditions are monitored daily, and changing patterns are studied to help predict future conditions. The resulting conditions and weather forecasts are broadcast on TV, printed in newspapers, and posted to the Internet.

With increasing awareness of environmental issues, interest in the study of the atmosphere has grown. Because the air we breathe is such an integral part of our environment, knowledge of the atmosphere's constituents, properties, and workings is needed to understand, appreciate, and solve environmental problems. In this chapter the normal atmospheric properties and conditions are discussed. In Chapter 20 atmospheric effects and some environmental problems are presented.

19.1 Composition and Structure

Preview Questions

- What are the four major gaseous constituents of the atmosphere and their percentages?
- What are the divisions of the atmosphere based on temperature?

The air of the atmosphere is a mixture of many gases. In addition, the air holds many suspended liquid droplets and particulate matter. However, only two gases make up about 99% of the volume of air near the Earth. ● Figure 19.1 and ● Table 19.1 show that air is composed primarily of nitrogen (78%) and oxygen (21%), with nitrogen being almost four

Table 19.1 Composition of Air

Nitrogen	N_2	78% (by volume)
Oxygen	O_2	21%
Argon	Ar	0.9%
Carbon dioxide	CO_2	0.03%
Others (traces)		*Others (variable)*
Neon	Ne	Water vapor (H_2O) 0 to 4%
Helium	He	Carbon monoxide (CO)
Methane	CH_4	Ammonia (NH_3)
Nitrous oxide	N_2O	Solid particles—dust, pollen, etc.
Hydrogen	H_2	

Figure 19.2 Photosynthesis
Energy from the Sun is necessary to the photosynthesis process whereby plants produce carbohydrates and oxygen from water and carbon dioxide. The "shafts" of light are called *crepuscular rays*, which are caused by the scattering of sunlight by particles in the air.

Gregory G. Dimijian, M.D./Photo Researchers, Inc.

times as abundant as oxygen. Note that atmospheric nitrogen and oxygen are diatomic (two-atom) molecules, N_2 and O_2. The other main constituents of air are argon (Ar, 0.9%) and carbon dioxide (CO_2, 0.03%).

Minute quantities of other gases are found in the atmosphere. Some of these gases, especially water vapor and carbon monoxide, vary in concentration depending on conditions and locality. The amount of water vapor in the air depends to a large extent on temperature, as will be discussed in Chapter 19.3. Carbon monoxide (CO) is a product of incomplete combustion (Chapters 13.2 and 20.4). A sample of air taken near a busy freeway would contain a concentration of CO considerably higher than that normally found in the atmosphere.

In general, the relative amounts of the major constituents of the atmosphere remain fairly constant. Nitrogen, oxygen, and carbon dioxide are involved in the life processes of plants and animals. These gases are continuously taken from the air and replenished as by-products of various processes. For example, nitrogen is taken in by some plants and released during organic decay. Animals inhale oxygen and exhale carbon dioxide, whereas plants convert carbon dioxide to oxygen.

Plants produce oxygen by **photosynthesis**, the process by which CO_2 and H_2O are converted into carbohydrates (needed for plant life) and O_2, using energy from the Sun (● Fig. 19.2). The key to photosynthesis is the ability of *chlorophyll*, the green pigment in plants, to convert sunlight into chemical energy. Billions of tons of CO_2 are removed from the atmosphere annually and replaced with billions of tons of oxygen. More than half of all photosynthesis takes place in the oceans, which contain many forms of green plants.

Note the "shafts" of light in Fig. 19.2. These *crepuscular rays* result from the scattering of sunlight by particles in the air. The effect is sometimes said to be "the Sun drawing water," but of course, that is not the case. Crepuscular rays are also seen emanating from clouds. (See Fig. 1b in the Chapter 19.2 Highlight: Blue Skies and Red Sunsets.)

The Earth's atmosphere evolved into its present condition over billions of years. The atmospheres of other planets evolved in different fashions and contain different constituents. However, the planet Mercury and our Moon have virtually no atmosphere. Gravitational attraction was not strong enough to hold the energetic gas molecules of their early atmospheres, which escaped into space.

The gravitational attraction between the Earth and the atmosphere is greatest near the planet's surface. (Why?) As a result, the density of air is greatest near the Earth's surface and decreases with increasing altitude. Because of gravitational attraction, more than half the mass of the atmosphere lies below an altitude of 11 km (7 mi) and almost 99% lies below an altitude of 30 km (19 mi). At an altitude of 320 km (200 mi), the density of the atmosphere is such that a gas molecule may travel a distance of 1.6 km (1 mi) before encountering another gas molecule. There is no clearly defined upper limit of the Earth's atmosphere. It simply becomes more and more tenuous, merging into the interplanetary gases, which may be thought of as part of the extensive "atmosphere" of the Sun.

To distinguish different regions of the atmosphere, we look for physical properties that vary with altitude. Such changes in physical properties can be used to define vertical divisions. Atmospheric density decreases continuously with altitude, so this property provides no distinction. Two atmospheric properties that *do* show vertical variations are (1) temperature and (2) ozone and ion concentrations.

Temperature

Plotting the temperature of the atmosphere versus altitude reveals distinctions, and these distinctions lead to major divisions of the atmosphere based on temperature variations. Near the Earth's surface, the temperature of the atmosphere decreases with increasing altitude at an average rate of about $6\frac{1}{2}$°C/km (or $3\frac{1}{2}$°F/1000 ft) up to about 16 km (10 mi). This region is called the **troposphere** (from the Greek *tropo*, meaning "change"; ● Fig. 19.3).

The troposphere contains about 80% of the atmospheric mass and virtually all the clouds and water vapor. There is continuous mixing and a great deal of change in this region. The atmospheric conditions of the lower troposphere are referred to collectively as **weather**. Generally, jet aircraft fly in the upper troposphere to avoid turbulent weather. Changes in the weather reflect the local variations of the atmosphere near the Earth's surface. The lower troposphere was the only region of the atmosphere investigated (via hot-air balloons) before the twentieth century. At the top of the troposphere, the temperature falls to −45° to −50°C (about −50° to −60°F).

Above the troposphere, the temperature of the atmosphere increases nonuniformly up to an altitude of about 50 km (30 mi; see Fig. 19.3). This region of the atmosphere, from approximately 16 to 50 km (10 to 30 mi) in altitude, is called the **stratosphere** (from the Greek *stratum*, "covering layer"). Together, the troposphere and stratosphere account for about 99.9% of the atmospheric mass.

Above the stratosphere the temperature decreases rather uniformly with altitude to a value of about −95°C (−140°F) at an altitude of 80 km (50 mi). This region, between about 50 and 80 km (30 and 50 mi) in altitude, is called the **mesosphere** (from the Greek *meso*, "middle"). It is the coldest region of the atmosphere.

Above the mesosphere, the thin atmosphere is heated intensely by the Sun's rays and the temperature climbs to more than 1000°C (about 1800°F). This region, which contains less than 0.01% of all the air of the atmosphere and extends to its outer reaches, is called the **thermosphere** (from the Greek *therme*, "heat"). The temperature of the thermosphere varies considerably with solar activity. Most satellites orbit the Earth in this region. The International Space Station is in orbit in the middle of the thermosphere, at an altitude of about 350 km (220 mi).

Another "sphere" is the *exosphere*, which starts at the top of the thermosphere and continues until it merges with interplanetary gases or outer space. Hydrogen and helium are the major components of the exosphere and are present in extremely low densities. Atoms and molecules may exit the exosphere into space.

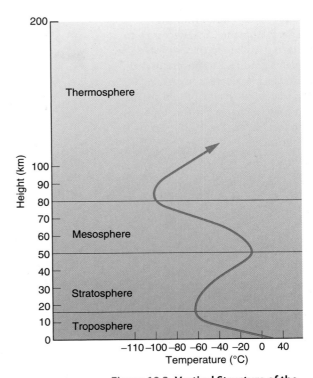

Figure 19.3 Vertical Structure of the Atmosphere
Divisions of the atmosphere are based on variations in physical properties such as temperature, as shown here.

Ozone and Ion Concentrations

The atmosphere also may be divided into two parts based on regions of concentration of ozone and ions. The ozone region lies below the ion region. **Ozone** (O_3) is formed by the dissociation of molecular oxygen (O_2) and the combining of atomic oxygen (O) with molecular oxygen:

$$O_2 + energy \rightarrow O + O$$
$$O + O_2 \rightarrow O_3 \ (ozone)$$

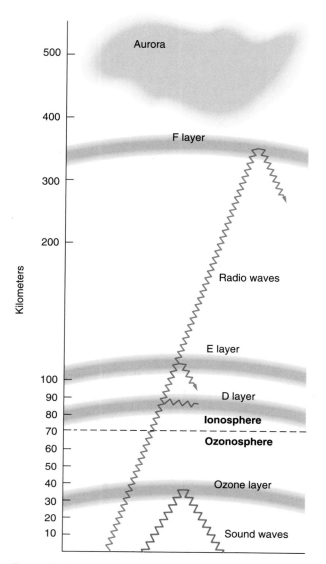

Figure 19.4 Ozonosphere and Ionosphere
These atmospheric regions are based on ozone and ion concentrations. A warm-air layer reflects sound waves and was first investigated by using this property. The upper ion layers reflect radio waves.

At high altitudes, energetic ultraviolet (UV) radiation from the Sun provides the energy necessary to dissociate the molecular oxygen. Oxygen is less abundant at higher altitudes, so the production and concentration of ozone depend on the appropriate balance of UV radiation and oxygen molecules. The optimum conditions occur at an altitude of about 30 km (20 mi), and in this region the central concentration of the ozone layer is found, as illustrated in ● Fig. 19.4. The ozone layer is a warm, broad band of gas that extends through nearly all the stratosphere.

The ozone concentration becomes less with increasing altitude up to an altitude of about 70 km (45 mi). The region of the atmosphere below this altitude is referred to as the **ozonosphere**. Ozone is unstable in the presence of sunlight, and it dissociates into atomic and molecular oxygen. When an oxygen atom meets an ozone molecule, they combine to form two ordinary oxygen molecules $(O + O_3 \rightarrow O_2 + O_2)$ and thus destroy the ozone. This process and the formation of ozone go on simultaneously, producing a balance in the concentration of the ozone layer.

Relatively little ozone is present naturally near the Earth's surface, but you may have experienced ozone when it is formed by electrical sparking discharges. The gas is easily detected by the distinct, pungent smell from which it derives its name (from the Greek *ozein*, "to smell"). In some areas—for example, Los Angeles—ozone is classified as a pollutant. It is found in relatively high concentrations resulting from photochemical reactions of air pollutants. Such reactions give rise to photochemical smog, which is discussed in Chapter 20.4.

The ozone layer in the stratosphere acts as an umbrella that shields life from harmful ultraviolet radiation from the Sun by absorbing most of the short wavelengths of this radiation. The portion of the UV radiation that gets through the ozone layer burns and tans our skin in the summer (and may cause skin cancer). Were it not for the ozone, we would be badly burned and find the sunlight intolerable.

Because the ozone layer absorbs energetic ultraviolet radiation, one can expect an increase in temperature in the ozonosphere. A comparison of Figs. 19.3 and 19.4 shows that the ozone layer lies in the stratosphere. Hence, the ozone absorption of UV radiation explains the temperature increase in the stratosphere, which contrasts with the continuously decreasing temperature that occurs with increasing altitude in the neighboring troposphere and mesosphere.

In the upper atmosphere above the ozonosphere, energetic particles from the Sun cause the ionization of gas molecules. For example, for a nitrogen molecule,

$$N_2 + energy \rightarrow N_2^+ + e^-$$

The electrically charged ions and electrons are trapped in the Earth's magnetic field and form ion layers in the upper region of the atmosphere, which is called the **ionosphere**.

Variations in the ion density with altitude give rise to the labeling of three regions or layers, D, E, and F (Fig. 19.4). The D layer strongly absorbs radio waves below a certain frequency. Radio waves with frequencies above this value pass through the D layer but are reflected by the E and F layers, up to a limiting frequency. Thus, the ionosphere provides global radio communications via the reflection of waves from ion layers, as illustrated in ● Fig. 19.5. Solar disturbances, which produce a shower of incoming energetic particles, may disrupt the uniformity of the layers, causing communication problems. Today satellites are used to relay communications.

Solar disturbances are also associated with the beautiful displays of light in the upper atmosphere, predominantly in the northern regions of Canada and Alaska. In the Northern Hemisphere, they are called *northern lights*, or *aurora borealis* (from the Latin *aurora*,

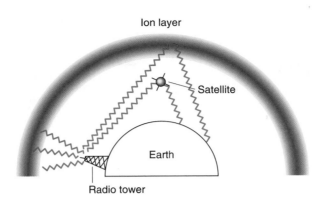

Ion layer

Satellite

Earth

Radio tower

Figure 19.5 **Global Radio Transmission**
Radio waves travel in straight lines and are reflected around the curvature of the Earth by ion layers. Ionic disturbances from solar activity may affect ion density and disrupt communications. Transmission via low-Earth orbiting satellites is not as severely affected by ion disturbances.

"dawn," and *boreas*, "northern wind"; ● Fig. 19.6). The Southern Hemisphere tends to be forgotten by people living north of the equator. However, light displays of equal beauty occur in the southern polar atmosphere ("southern lights") and are called *aurora australis* (from the Latin *auster*, "southern wind").

In general, the ions and electrons trapped in the Earth's magnetic field are deflected toward the Earth's magnetic poles (polar regions), over which the majority of the auroras occur. However, auroras are sometimes observed at lower latitudes. The emission of light is associated with the recombination of ions and electrons; energy is needed to ionize, and on recombining, energy is emitted in the form of visible light, or radiation.

Because of the movement of the magnetic north pole, the aurora seen over Canada and Alaska may be lost some day. (See the Highlight: Magnetic North Pole in Chapter 8.4).

Did You Learn?

● The four major gases of the atmosphere (with percentages by volume) are nitrogen, N_2 (78%); oxygen, O_2 (21%); argon, Ar (0.9%); and carbon dioxide, CO_2 (0.03%). Note that nitrogen and oxygen make up 99% of the atmospheric gases.

● With increasing altitude, the divisions of the atmosphere based on temperature are the troposphere, stratosphere, mesosphere, and thermosphere.

George Lepp/Getty Images

Figure 19.6 Aurora
The aurora borealis, or northern lights, as seen over Denali National Park, Alaska.

Figure 19.7 Insolation Distribution
An illustration of how the incoming solar radiation is distributed. The percentages vary somewhat, depending on atmospheric conditions.

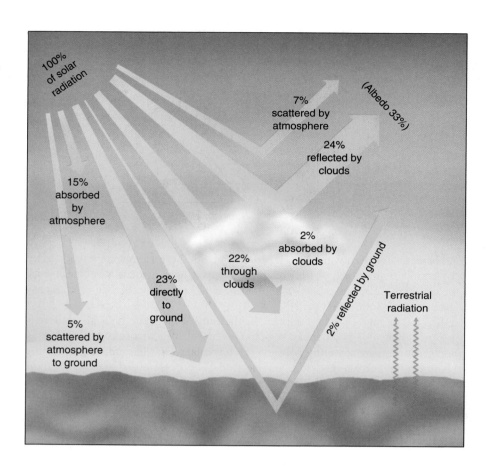

19.2 Atmospheric Energy Content

Preview Questions

- What is insolation, and how much reaches the Earth's surface?
- What gives rise to the greenhouse effect?

The Sun is by far the most important source of energy for the Earth and its atmosphere. At an average distance of 93 million miles from the Sun, the Earth intercepts only a small portion of the vast amount of solar energy emitted. This energy traverses space in the form of radiation, and the portion incident on the Earth's atmosphere is called **insolation** (which stands for *incoming solar radiation*). Because the Earth's axis is tilted $23\frac{1}{2}°$ with respect to a line normal (perpendicular) to the plane of its orbit, the insolation is not evenly distributed over the Earth's surface. This tilt, coupled with the Earth's revolution around the Sun, gives rise to the seasons (Chapter 15.5).

The solar radiation (energy) output fluctuates, but the Earth receives a relatively constant average intensity at the top of the atmosphere. However, depending on atmospheric conditions, only about 50% or less of the insolation reaches the Earth's surface. In considering the energy content of the atmosphere, one might think that it comes directly from insolation. Surprisingly enough, most of the direct heating of the atmosphere comes not from the Sun but from the Earth. To understand why, we need to examine the distribution and disposal of the insolation (● Fig. 19.7).

About 33% of the insolation received is returned to space as a result of reflection by clouds, scattering by particles in the atmosphere, and reflection from terrestrial surfaces such as water, ice, and ground. The reflectivity, or the amount of light a body reflects, is known as its *albedo* (from the Latin *albus*, "white") and is expressed as a fraction or percentage of solar radiation received.

Figure 19.8 Planet Earth
As seen from space, the brightness of the Earth depends on its size and on the amount of sunlight reflected. Clouds and water are also important factors. The Earth has an albedo of 0.33 or 33%. The distant Moon (upper left), has an albedo of of 7% and appears dimmer because of its dark surface and size.

NASA Earth Observatory

For example, the Earth has an albedo of 0.33, or 33%. It reflects about one-third of the incident sunlight. The brightness of the Earth as viewed from space depends on its size and on the amount of sunlight it reflects; clouds and water play important roles (● Fig. 19.8). In comparison, the Moon, with a dark surface and no atmosphere, has an albedo of only 0.07; that is, it reflects only 7% of the insolation. With its larger size and larger albedo, the Earth is a much more impressive sight than the Moon when viewed from space. Some typical albedos of Earth surfaces are given in ● Table 19.2.

Conceptual Question and Answer

Hot Time

Q. The hottest part of a summer day is around 2:00 or 3:00 P.M. Is that when the maximum insolation is received?

A. No. The maximum insolation is received when the Sun is highest in the sky, or around 12 noon. The warming continues after this time. By analogy, think of slowly turning a gas flame up and down under a pan of water. The maximum heating occurs when the flame is highest, but as it is lowered, the water continues to heat and temperature increases.

In the atmosphere, scattering of insolation occurs from gas molecules of the air, dust particles, water droplets, and so on. (*Scattering* is the absorption of incident light and its reradiation in all directions.) As shown in Fig. 19.7, some of the scattered radiation is dispersed back into space and some is scattered toward the Earth's surface.

An important type of scattering is **Rayleigh scattering**, named after Lord Rayleigh (1842–1919), a British physicist who developed the theory. Lord Rayleigh showed that the amount of scattering by particles of a given molecular size was proportional to $1/\lambda^4$, where λ is the wavelength of the incident light (Chapter 6.2). That is, the longer the wavelength, the less the scattering. It is this scattering that gives rise to the blue color of the sky. (See the **Highlight: Blue Skies and Red Sunsets**.)

About 15% of the insolation is absorbed directly by the atmosphere. Most of this absorption is accomplished by ozone, which removes the ultraviolet radiation, and by water vapor, which absorbs strongly in the infrared region of the spectrum. A major portion of the solar spectrum lies in the narrow visible region (Chapter 6.3). The atmosphere is practically transparent to (absorbs little) visible radiation, so most of the visible light reaches the Earth's surface.

Table 19.2	Typical Albedos of Earth Surfaces
Surface Type	**Albedo (%)**
Water*	3–24
Polar ice	30–40
Fresh snow	75–95
Dry sand	20–30
Soil	5–15
Thin cloud	35–50
Thick cloud	70–90

*Depends on solar elevation. When the Sun is high in the sky, less radiation is reflected, whereas when the Sun is low in the sky, a water surface acts rather like a mirror.

Highlight   Blue Skies and Red Sunsets

If we lived on the Moon, which has no atmosphere, then the sky would appear black except in the vicinity of the Sun. However, our sky is blue as a result of the scattering of sunlight in the atmosphere. The gas molecules of the air account for most of the scattering in the visible region of the spectrum. In this region the wavelengths increase from violet to red, and scattering is greater for shorter wavelengths. The blue end is therefore scattered more than the red end. (The colors of the visible spectrum and the rainbow may be remembered with the help of the acronym name ROY G. BIV for red, orange, yellow, green, blue, indigo, and violet.)

As sunlight passes through the atmosphere, the blue end of the spectrum is preferentially scattered. Some of this scattered light reaches the Earth, where we see it as blue sky light (Fig. 1a). Keep in mind that all colors are present in sky light, but the dominant wavelength or color lies in the blue. You may have noticed that the sky is more blue directly overhead or high in the sky and less blue toward the horizon, becoming almost white just above the horizon. You see these effects because there are fewer scatterers along a path through the atmosphere overhead than toward the horizon. Extensive scattering along the horizon path mixes the colors, giving the white appearance.

The scattering of sunlight by the atmospheric gases *and* small particles gives rise to red sunsets. One might think that because sunlight travels a greater distance through the atmosphere to an observer at sunset, most of the shorter wavelengths would be scattered from the sunlight and that only light in the red end of the spectrum would reach the observer. However, the dominant color of this light, were it due solely to molecular scattering, would be orange. Hence, additional scattering by small particles in the atmosphere shifts the light from the setting (or rising) Sun toward the red.

Foreign particles (natural or pollutants) in the atmosphere are not necessary to give a blue sky and may even detract from it, yet they are necessary for deep red sunsets and sunrises. The beauty of red sunrises and sunsets is often made more spectacular by layers of pink-colored clouds. The cloud color is due to the reflection of red light (Fig. 1b).

A popular saying is associated with red sunsets: "Red sky at night, sailors delight. Red sky in the morning, sailors take warning." It is reasonably accurate for the westerlies wind zone (Chapter 19.4). If the sky (or Sun) is red in the evening, then much scattering of sunlight has occurred and there is likely to be a high-pressure area to the west (denser air, more scattering). High-pressure areas are associated with good weather, so as this good weather moves west, sailors in its path are delighted.

Conversely, a red sky in the morning indicates that the high is now to the east and will be followed by a low with bad weather. (Interestingly, there is a biblical reference to this phenomenon in Matthew 16:1–4.)

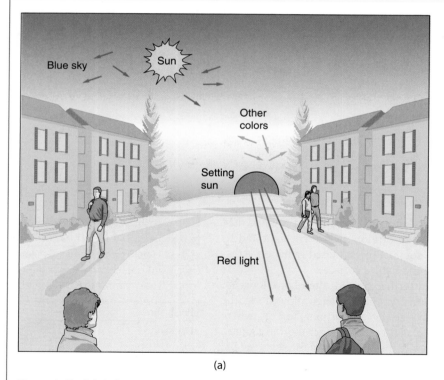

(a)

(b)

Figure 1 Rayleigh Scattering
(a) The preferential scattering by air molecules causes the sky to look blue, which in turn, together with the scattering caused by small particles, produces red sunsets. (See text for description.) (b) Scattering in action. Blue sky and red sunset. Note the fan-like crepuscular rays. (See Fig. 19.2.)

After reflection, scattering, and direct absorption, about 50% of the total incoming solar radiation reaches the Earth's surface. This radiation goes into terrestrial surface heating, primarily through the absorption of visible radiation. As noted previously, the atmosphere—in particular, the troposphere—derives most of its energy content directly from the Earth. This absorption of energy is accomplished in three main ways, listed here in order of decreasing contribution:

1. Absorption of terrestrial radiation
2. Latent heat of condensation
3. Conduction from the Earth's surface

Absorption of Terrestrial Radiation

The Earth, like any warm body, radiates energy that may be subsequently absorbed by the atmosphere. The wavelength of the radiation emitted by the Earth depends on its temperature. From the wavelength relationship of the energy of a photon and the Kelvin temperature (Chapter 5.1) of the emitting source, it can be shown that

$$\lambda \propto \frac{1}{T}$$

The wavelength of radiation emitted by a source is inversely proportional to its temperature.

The Earth's temperature is such that it radiates energy primarily in the long-wavelength infrared region. Water vapor and carbon dioxide (CO_2) are the primary absorbers of infrared radiation in the atmosphere, water vapor being the more important. These gases are referred to as *selective absorbers* because they absorb certain wavelengths and transmit others. This selective absorption gives rise to the **greenhouse effect**, about which the public now hears a great deal in terms of the greenhouse gases. (See the **Highlight: The Greenhouse Effect**.)

Because the wavelength of the radiation is dependent on the temperature of the source, infrared images taken from satellites are used to study temperature variations in the Earth's crust arising from such things as underground rivers and volcanic or geothermal activity (● Fig. 19.9).

Latent Heat of Condensation

Approximately 70% of the Earth's surface is covered by water. Consequently, a great deal of evaporation occurs because of the insolation reaching the Earth's surface. Recall from

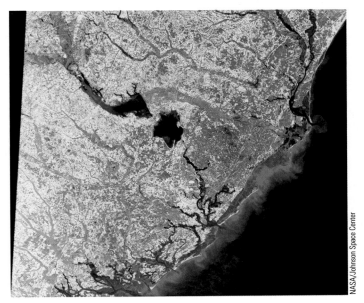

Figure 19.9 Infrared Analysis
An infrared photo taken from a satellite shows the variations in temperature of the Earth's surface. Shown here is most of the coast of South Carolina, from Myrtle Beach (near upper right corner) south almost to Parris Island. Inland lakes and rivers are clearly visible.

NASA/Johnson Space Center

Highlight   The Greenhouse Effect

The absorption of terrestrial radiation—primarily by water vapor (H_2O), carbon dioxide (CO_2), and methane (CH_4)—adds to the energy content of the atmosphere. The heat-retaining process of such gases is referred to as the *greenhouse effect* because a similar effect occurs in greenhouses. The absorption and transmission properties of regular glass are similar to those of the atmospheric "greenhouse gases"; in general, visible radiation is transmitted and infrared radiation is absorbed (Fig. 1).

We have all observed the warming effect of sunlight passing through glass, such as in a closed car on a sunny but cold day. In a greenhouse the objects inside become warm and reradiate long-wavelength infrared radiation, which is absorbed and reradiated by the glass. Thus, the air inside a greenhouse heats up and is quite warm on a sunny day, even in the winter.

Actually, in this case the maintained warmth is primarily due to the glass enclosure, which prevents the escape of warm air. The temperature of the greenhouse in the summer is controlled by painting the glass panels white to reflect sunlight and by opening windows to allow the hot air to escape.

The greenhouse effect in the atmosphere is quite noticeable at night, particularly on cloudy nights. With a cloud and water vapor cover to absorb the terrestrial radiation, the night air is relatively warm. Without this insulating effect, the night is usually "cold and clear" because the energy from the daytime insolation is quickly lost.

Despite the daily and seasonal gain and loss of heat, the *average* temperature of the Earth has remained fairly constant. The Earth must lose, or reradiate, as much energy as it receives. If it did not, then the continual gain of energy would cause the Earth's average temperature to rise. The selective absorption of atmospheric gases provides a thermostatic, or heat-regulating, process for the planet.*

To illustrate, suppose the Earth's temperature were such that it emitted radiation with wavelengths that were absorbed by the

*There is now concern that greenhouse gases being released into the atmosphere from human activities could cause a rise in the Earth's temperature and "global warming." See Chapter 20.5.

Figure 1 The Greenhouse Effect
The gases of the lower atmosphere transmit most of the visible portion of the sunlight, as does the glass of a greenhouse. The warmed Earth emits infrared radiation, which is selectively absorbed by atmospheric gases, the absorption spectrum of which is similar to that of glass. The absorbed energy heats the atmosphere and helps maintain the Earth's average temperature.

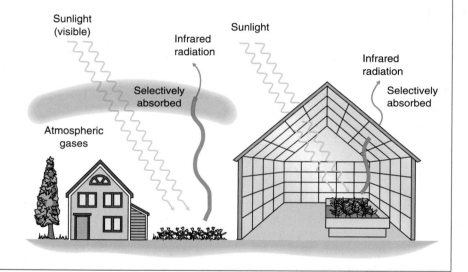

Chapter 5.3 that the latent heat of vaporization for water is 540 kcal/kg; 540 kcal of heat energy is required to change 1 kg of water into the gaseous state. This is at the boiling temperature, but it is also a reasonable approximation for evaporation.

Thus, with condensation, a large amount of energy is transferred to the atmosphere in the form of latent heat. This energy is released during the formation of clouds, fog, rain, dew, and so on.

Conduction from the Earth's Surface

A comparatively smaller but significant amount of heat is transferred to the atmosphere by conduction from the Earth's surface. Because the air is a relatively poor conductor of heat, this process is restricted to the layer of air in direct contact with the Earth's surface. The heated air is then transferred aloft by convection. As a result, the temperature of the air tends to be greater near the surface of the Earth and decreases gradually with altitude.

Highlight

atmospheric gases. As a result of this absorption, the lower atmosphere would become warmer and would effectively hold in the heat, thus insulating the Earth. With additional insolation, the Earth would become warmer and its average temperature would rise. However, according to the previously mentioned relationship between temperature and wavelength ($\lambda \propto 1/T$), the greater the temperature, the shorter the wavelength of the emitted radiation. Thus, the wavelength of the higher-temperature terrestrial radiation is shifted to a shorter wavelength.

The wavelength would eventually be shifted to a "window" in the absorption spectrum where little or no absorption takes place.

The terrestrial radiation then passes through the atmosphere into space. The Earth loses energy, and its temperature decreases. But with a temperature decrease, the terrestrial radiation returns to a longer wavelength, which would then be absorbed by the atmosphere. In effect, there is a turning on and off of absorption, similar to the action of a thermostat (Fig. 2).

Averaged over that total spectrum, the selective absorption of atmospheric gases plays an important role in maintaining the Earth's average temperature. Recall from Chapter 16.3 that a relatively large amount of atmospheric CO_2 and the resulting greenhouse effect keep the surface of Venus very hot (453°C).

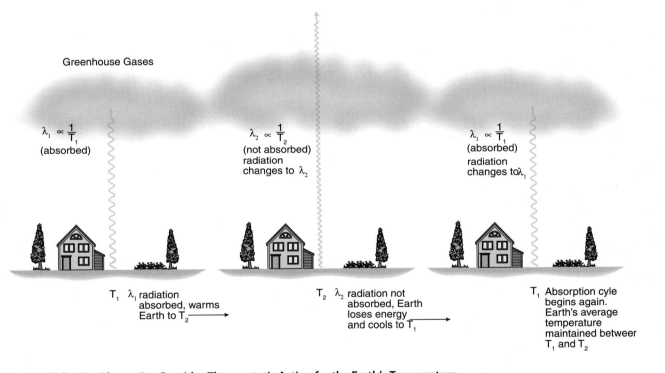

Figure 2 Selective Absorption Provides Thermostatic Action for the Earth's Temperature

Conceptual Question and Answer

Violet Sky

Q. If Rayleigh scattering of visible light by atmospheric gases is inversely proportional to the wavelength λ, then why isn't the sky violet since it has the shortest wavelength?

A. The wavelengths of visible light in ROY G. BIV decrease from left to right, and violet light does have a shorter wavelength than blue light and is scattered more. The sky appears blue for a couple of reasons. First, the eye is more sensitive to blue light than violet light. Second, sunlight contains more blue light than violet light. (The greatest color component of sunlight is yellow-green.)

19.3   Atmospheric Measurements and Observations

Preview Questions

● What is one standard atmosphere of pressure?

● What is meant by a relative humidity of 50%, and what is the dew point?

Measurements of the atmosphere's properties and characteristics are important in the study and analysis of the atmosphere. These properties are measured daily, and records have been compiled over many years. Meteorologists study such records with the hope of observing cycles and trends in atmospheric behavior so as to understand and predict its changes better.

Daily atmospheric readings are made to obtain a qualitative picture of the conditions for that day in the region and around the country. Fundamental atmospheric measurements include (1) temperature, (2) pressure, (3) humidity, (4) wind speed and direction, and (5) precipitation.

Temperature

Having discussed temperature measurements in Chapter 5.1, we will not dwell on this property. Keep in mind that it is the *air temperature* that is being measured. Heat transfer to a thermometer by radiation (sunlight) may result in a higher temperature. A truer air temperature in the summer is often expressed as being so many degrees "in the shade," This implies that when making a temperature measurement, the thermometer should not be exposed to the direct rays of the Sun.

Pressure

Pressure is defined as the force per unit area ($p = F/A$, Chapter 5.6). Living at the bottom of the atmosphere, we experience the weight of the gases above us. Because this weight is experienced before and after birth as a part of our natural environment, little thought is given to the fact that every square inch of our bodies sustains an average weight of 14.7 lb at sea level, that is, a pressure of 14.7 lb/in². This pressure of 14.7 lb/in² is referred as *one standard atmosphere of pressure.*

One of the first investigations of atmospheric pressure was initiated by Galileo. In attempting to pipe water to elevated heights by evacuating air from a tube, he found that it was impossible to sustain a column of water taller than about 10 m (33 ft). Evangelista Torricelli (1608–1647), who was Galileo's successor as professor of mathematics in Florence, pointed out the difficulty through the invention of a device that showed the height of a liquid column in a tube to be dependent on the atmospheric pressure.

A glass tube filled with mercury (Hg) was inverted into a pool of mercury. Although some mercury ran out of the tube, a column of mercury 76 cm (30 in.) high was left in the tube, as illustrated in ● Fig. 19.10. Such a device, called a **barometer** (from the Greek *baros,* "weight"), is still used to measure atmospheric pressure.

A mercury barometer is shown in ● Fig. 19.11. Because the column of mercury has weight, a force must hold up the column. The only available force is that of atmospheric

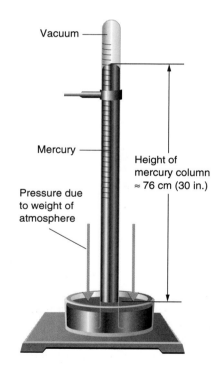

Vacuum

Mercury

Pressure due to weight of atmosphere

Height of mercury column ≈ 76 cm (30 in.)

Figure 19.10 Principle of the Mercury Barometer
The external air pressure, which is due to the weight of the atmosphere on the surface of a pool of mercury, supports the mercury column in the inverted tube. The height of the column depends on the atmospheric pressure and provides a means of measuring it. Normally, the column would be about 76 cm (30 in.) high.

pressure on the surface of the mercury pool. Thus, the greater the height (h), the greater the pressure (p), or $p \propto h$. The equation relationship is $p = \rho g h$, where ρ (the Greek letter rho) is the density of the liquid and g is the acceleration due to gravity. Therefore, the less dense the liquid, the greater the height of the column for a given pressure.

Conceptual Question and Answer

Not Dense Enough

Q. Why could Galileo not get a sustained column of water taller than about 10 m?

A. Water has a density of 1.00 g/cm³, whereas mercury has a density of 13.6 g/cm³. So, a water column would need to be 13.6 times taller than a mercury column for a given pressure. As was noted, one atmosphere supports a mercury column 76 cm tall, so a water barometer would have a column of 76 cm × 13.6 = 1034 cm, which is a little more than 10 m. Hence, the atmosphere generally does not have sufficient pressure to support a column of water much over 10 m, as Galileo found.

The standard units of pressure ($p = F/A$) are N/m² and lb/in² in the SI and the British system, respectively. However, these units are not used for the barometric readings given on radio and TV weather reports. Instead, these readings are expressed in length units, or so many "inches" (of mercury as related to the barometer). For weather phenomena, we are primarily interested in changes in pressure, and because $p \propto h$, the variation in column height gives this directly.

Because of Torricelli's barometric work, a pressure unit has been named after him. A height of one millimeter of mercury is called a *torr* (1 mm Hg = 1 torr), so one atmosphere of pressure is 76 cm = 760 mm = 760 torr. Summarizing the atmospheric pressure units,

$$1 \text{ atmosphere (atm)} = 76 \text{ cm Hg} = 760 \text{ mm Hg}$$
$$= 760 \text{ torr}$$
$$= 30 \text{ in. Hg}$$
$$= 14.7 \text{ lb/in}^2$$

In the metric system, one atmosphere (atm) has a pressure of 1.013×10^5 N/m², or approximately 10^5 N/m². Meteorologists use yet another unit called the *millibar* (mb). A *bar* is defined to be 10^5 N/m², and because 1 bar = 1000 mb,

$$1 \text{ atm} \approx 1 \text{ bar} = 1000 \text{ mb}$$

With 1000 units, finer changes in atmospheric pressure can be measured.

Because mercury vapor is toxic and a column of mercury 30 inches tall is awkward to handle, another type of barometer, called the *aneroid* ("without fluid") *barometer*, is commonly used. This is a mechanical device having a metal diaphragm that is sensitive to pressure, much like a drumhead. A pointer on the dial face of the barometer is used to indicate the pressure changes on the diaphragm. Aneroid barometers with dial faces are found in many homes and are usually inlaid in wood, along with dial thermometers, and used as decorative wall displays (● Fig. 19.12).

Atmospheric pressure quickly becomes evident when sudden pressure changes occur. A relatively small change in altitude will cause our ears to "pop" because the pressure in the inner ear does not equalize quickly, which puts pressure on the eardrum. When the pressure equalizes (swallowing helps), the ears pop.

Airplanes are equipped with pressurized cabins that maintain normal atmospheric pressure on passengers' bodies. The internal pressure of the human body is accustomed to an external pressure of 14.7 lb/in². Should this pressure be reduced, the excess internal pressure may be evidenced in the form of a nosebleed. Also, altitude and pressure have an effect on the boiling points of liquids, as was seen in Chapter 5.3.

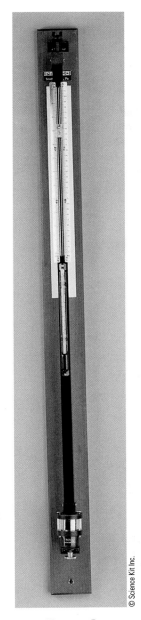

© Science Kit Inc.

Figure 19.11 Mercury Barometer
A typical mercury barometer as used in the laboratory for measuring atmospheric pressure. The mercury pool is seen at the bottom, and the height of the column is measured using the scale on the mounting board.

Figure 19.12 Aneroid Barometer
Changes in the atmospheric pressure on a sensitive metal diaphragm are reflected on the barometer's dial face. For reasons discussed in Chapter 20.2, fair weather is generally associated with high barometric pressure and rainy weather is associated with low barometric pressure. (What is the pressure reading of the barometer in inches and millibars?)

Conceptual Question and Answer

Slurp It Up

Q. What is the principle of drinking through a straw?

A. It is commonly thought that sucking on a straw draws the liquid up through it. Actually, the sucking action reduces the pressure in the straw, and the atmospheric pressure on the liquid's surface then "pushes" the liquid up the straw. (Put a small hole near the top of the straw and see what happens.)

Humidity

Humidity is a measure of the moisture, or water vapor, in the air. It affects our comfort and indirectly our "energy" and state of mind. In summer, many homes use dehumidifiers to remove moisture from the air. In winter, humidifiers or strategically placed exposed pans of water allow water to evaporate into the air.

Humidity can be expressed in several ways. *Absolute humidity* is simply the amount of water vapor in a given volume of air. In the United States, humidity is commonly measured in the nonstandard unit of grains per cubic foot. The grain (gr) is a small weight unit, with 1 lb equal to 7000 gr. An average value of humidity is approximately 4.5 gr/ft³.

The most common method of expressing the water vapor content of the air is in terms of relative humidity. **Relative humidity** is the ratio of the actual moisture content to the maximum moisture capacity of a volume of air at a given temperature. This ratio is commonly expressed as a percentage:

$$(\%)\, RH = \frac{AC}{MC} (\times 100\%) \qquad\qquad 19.1$$

where *RH* is the relative humidity, *AC* the actual moisture content, and *MC* the maximum moisture capacity.

The actual moisture content (*AC*) is just the absolute humidity, or the actual amount of water vapor in a given volume of air. The maximum moisture capacity (*MC*) is the maximum amount of water vapor that the volume of air can hold *at a given temperature*. For normal household temperatures, a relative humidity of about 45% to 50% is comfortable.

In summer when it is "hot and humid," the high humidity interferes with the evaporation of perspiration and the cooling of our bodies (Chapter 5.3). As a result, we feel that

National Weather Service
Heat Index

Temperature (°F)

		80	82	84	86	88	90	92	94	96	98	100	102	104	106	108	110
	40	80	81	83	85	88	91	94	97	101	105	109	114	119	124	130	136
	45	80	82	84	87	89	93	96	100	104	109	114	119	124	130	137	
	50	81	83	85	88	91	95	99	103	108	113	118	124	131	137		
	55	81	84	86	89	93	97	101	106	112	117	124	130	137			
	60	82	84	88	91	95	100	105	110	116	123	129	137				
	65	82	85	89	93	98	103	108	114	121	126	130					
Relative Humidity (%)	70	83	86	90	95	100	105	112	119	126	134						
	75	84	88	92	97	103	109	116	124	132							
	80	84	89	94	100	106	113	121	129								
	85	85	90	96	102	110	117	126	135								
	90	86	91	98	105	113	122	131									
	95	86	93	100	108	117	127										
	100	87	95	103	112	121	132										

Likelihood of Heat Disorders with Prolonged Exposure or Strenuous Activity

■ Caution   ■ Extreme Caution   ■ Danger   ■ Extreme Danger

Figure 19.13 Heat Index
High humidity interferes with the evaporation of perspiration, which is a cooling mechanism of our bodies. Taking temperature and humidity into account, the temperature we "feel" is given by a *heat index table*. Find the relative humidity on the left vertical row and the temperature on the top horizontal row; where they intersect gives the "real feel." For example, for 88°F and 70% relative humidity, the "real feel" is 100°F. The heat and humidity conditions that generate a high heat index are associated with the likelihood of heat disorders.

the temperature is greater than the measured air temperature. This feeling may be not only uncomfortable but also dangerous. A *heat index*, which takes temperature *and* humidity into account, has been devised (● Fig. 19.13). This index is often given in weather reports as the "real feel" for the high temperature of the day.

Relative humidity is essentially a measure of how "full" of moisture a volume of air is at a given temperature. For example, if the relative humidity is 0.50, or 50%, then a volume of air is "half full" or contains half as much water as it is capable of holding *at that temperature*.

To understand better how the water vapor content of air varies with temperature, consider an analogy with a saltwater solution. Just as a given amount of water at a certain temperature can dissolve only so much salt, a volume of air at a given temperature can hold only so much water vapor. When the maximum amount of salt is dissolved in solution, the solution is *saturated* (Chapter 11.1). This condition is analogous to a volume of air being at its maximum moisture capacity.

The addition of more salt to a saturated solution results in salt on the bottom of the container. However, more salt may be dissolved if the water is heated and the temperature raised. Similarly, when air is heated, it can hold more water vapor; that is, warm air has a greater capacity for water vapor than does cold air.

Conversely, when the temperature of a nearly saturated salt solution is lowered, the solution will become saturated at a lower temperature. Any additional lowering of temperature will cause salt to crystallize and come out of solution because otherwise the solution would be supersaturated. Analogously, when the temperature of a sample of air is lowered, it will become saturated at a certain temperature.

The temperature to which a sample of air must be cooled to become saturated is called the **dew point** (temperature). Hence, at the dew point, the relative humidity is 100%. (Why?) Cooling below this point causes supersaturation and generally results in condensation and loss of moisture in the form of precipitation.

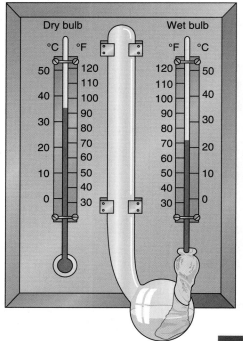

Figure 19.14 Psychrometer and Relative Humidity
The dry bulb of a psychrometer records the air temperature, which is greater than that of the wet bulb (evaporation cools the wet bulb). The lower the humidity, the greater the evaporation and the greater the difference between the two temperature readings. Thus, the psychrometer provides a means for measuring relative humidity.

Humidity may be measured by several means. One of the most common methods is the use of a **psychrometer** (*psychro-* is Greek for "cold"). This instrument consists of two thermometers, one of which measures the air temperature, while the other has its bulb surrounded by a cloth wick that keeps it wet. The thermometers are referred to as the *dry bulb* and the *wet bulb*, respectively. They may be simply mounted, as shown in ● Fig. 19.14.

The dry bulb measures the air temperature, and the wet bulb has a lower reading that is a function of the amount of moisture in the air. This reading occurs because of the evaporation of water from the wick around the wet bulb. The evaporation removes latent heat from the wet thermometer bulb. If the humidity is high and the air contains a lot of water vapor, then little water is evaporated and the wet bulb is only slightly cooled. Consequently, the temperature of the wet bulb is only slightly lower than the temperature of the dry bulb; that is, the wet-bulb reading is slightly "depressed."

However, if the humidity is low, a great deal of evaporation will occur, accompanied by cooling, and the wet-bulb reading will be considerably depressed. Hence, the temperature difference between the thermometers—that is, the depression of the wet-bulb reading—is a measure of relative humidity.

Using the air (dry-bulb) temperature and the wet-bulb depression, one can read the relative humidity, the maximum moisture capacity, and the dew point directly from the tables in Appendix VIII. An example of how this is done follows.

EXAMPLE 19.1   **Using Psychrometric Tables**

The dry bulb and wet bulb of a psychrometer have readings of 80°F and 73°F, respectively. Find the following for the air: (a) relative humidity, (b) maximum moisture capacity, (c) actual moisture content, and (d) dew point.

Solution

Given: 80°F (dry-bulb reading)
        73°F (wet-bulb reading)
Wanted:
(a) *RH* (relative humidity)
(b) *MC* (maximum capacity)
(c) *AC* (actual content)
(d) dew point (temperature)

First, we find the wet-bulb depression—in other words, the difference between the dry-bulb and wet-bulb readings:

$$\Delta T = 80°F - 73°F = 7°F$$

(a) Then, using Table 1 in Appendix VIII to determine the relative humidity, we find the dry-bulb temperature in the first column and then locate the wet-bulb depression in the top row of the table. Move down the column under the wet-bulb depression to the row that corresponds to the dry-bulb temperature reading. The intersection of this row and column gives the value of the relative humidity. (It may be helpful to move one finger down and another one across to find the intersection number.) The relative humidity is 72% for this depression.

(b) The maximum moisture content (*MC*) is read directly from the table. Find the dry-bulb temperature in the first column; the *MC* for that temperature is given in the adjacent column. In this case, $MC = 10.9 \text{ gr/ft}^3$.

(c) Knowing *RH* and *MC*, the actual moisture content can be found from Eq. 19.1, $RH = AC/MC$. Rearranging yields

$$AC = RH \times MC = 0.72 \times 10.9 \text{ gr/ft}^3$$
$$= 7.8 \text{ gr/ft}^3$$

Note that the relative humidity is used in decimal (not percentage) form in calculations.

(d) The dew point is found by using Table 2 in Appendix VIII in the same manner as the relative humidity in part (a). The intersection value of the appropriate row and column is 70°F.

Hence, if the air is cooled to 70°F, then it will be saturated and the relative humidity will be 100%, with $AC = MC$. Note from Table 1 in Appendix VIII that MC for 70°F is 7.8 gr/ft³, which is the value found in part (c). The value of the actual content at 80°F should correspond to the maximum capacity at the dew point (70°F).

Confidence Exercise 19.1

On a particular day, the dry-bulb and wet-bulb readings of a psychrometer are 70°F and 66°F, respectively, (a) What is the relative humidity? (b) How many degrees would the air temperature have to be lowered for precipitation to be likely?

Answers to Confidence Exercises may be found at the back of the book.

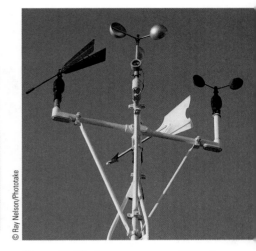

Figure 19.15 Anemometer and Wind Vanes
An array of two anemometers and two wind vanes is shown. The anemometer cups catch the wind to measure its speed. The shapes of the wind vanes cause them to point in the direction from which the wind is coming. That is, pointing north indicates a north wind.

© Ray Nelson/Phototake

It could be raining, but psychrometer readings give a relative humidity of less than 100%. Why? The cloud in which the precipitation formed was certainly saturated, but rain falls to the ground, where the humidity can be well below 100%.

Wind Speed and Direction

Wind speed is measured with an **anemometer**. This instrument consists of three or four cups attached to a rod that is free to rotate, much like a pinwheel. The cups catch the wind, and the greater the wind speed, the faster the anemometer rotates (● Fig. 19.15).

A **wind vane** (often called a weather vane) indicates the direction from which the wind is blowing. This instrument is simply a free-rotating indicator that, because of its shape, lines up with the wind (Fig. 19.15). Wind direction is reported as the direction *from which the wind is coming*. For example, a north wind comes from the north, blowing from north to south.

Precipitation

The major forms of precipitation are rain and snow. Rainfall is measured by a **rain gauge**. This device in simple form may be an open container with a vertical inch or centimeter scale on the side that is placed outside in an open area. After a rainfall, the rain gauge is read and the amount of precipitation is reported as so many inches.* The assumption is that this much rainfall is distributed relatively evenly over the surrounding area.

If precipitation is in the form of snow, then the depth of the snow (where it is not drifted) is reported in inches. The actual amount of water received depends on the density of the snow. Typically, 10 inches of snow yield about 1 inch of water. To obtain this reading, a rain gauge is sprayed with a chemical that melts the snow and the actual amount of water is recorded. More elaborate rain-measuring instruments automatically measure and record rainfall and snowfall.

Weather Observations

Technology has extended our means of making weather observations well beyond our direct visual capabilities. Two such means will be considered here: radar and satellites.

Radar (*ra*dio *d*etecting *a*nd *r*anging) can be used to detect and monitor precipitation, especially that of severe storms. Radar operates on the principle of reflected electromagnetic waves. Radar installations were initially located mainly in the tornado belt of the midwestern states and along the Atlantic and Gulf coasts where hurricanes are most probable. Additional radar information is obtained from air traffic control systems at many airports.

---

*In the United States, meteorologists commonly report precipitation in inches. Nearly all other countries report precipitation in centimeters.

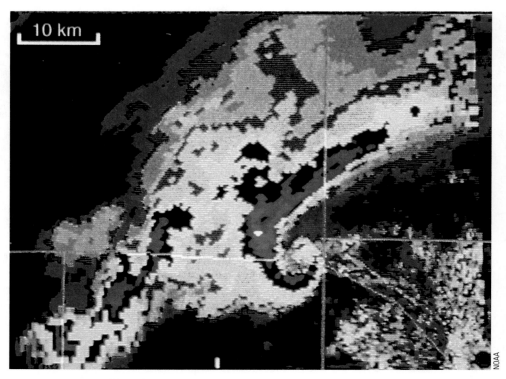

Figure 19.16 Doppler Radar
A relative-velocity image. The colors indicate wind directions. Note the characteristic hooked signature of a tornado. This image was taken in May 1999 during a tornado outbreak in Oklahoma and Kansas, which involved 70 to 80 tornadoes. Doppler radar has greatly increased the tornado warning lead time.

A more advanced radar system, called **Doppler radar**, is now used, and a network for this purpose is in place across the United States. Like conventional radar, Doppler radar measures the distribution and intensity of precipitation over a broad area. However, Doppler radar also has the ability to measure wind speeds. It is based on the Doppler effect (Chapter 6.5), the same principle used in police radar to measure the speeds of automobiles. Doppler radar scans are commonly seen on TV weather reports.

Radar waves are reflected from raindrops in storms. The direction of a storm's wind-driven rain, and hence a wind "field" of the storm region, can be mapped. This map provides strong clues, or signatures, of developing tornadoes (● Fig. 19.16).

Conventional radar can detect the hooked signature of a tornado only after the storm is well developed. With a wind-field map, a developing tornado signature can be detected much earlier. Using Doppler radar, forecasters are able to predict tornadoes as much as 20 min before they touch down, compared with just over 2 min for conventional radar. Doppler radar saves many lives with this increased advance warning time. Currently, there are Doppler radar sites in every state, along with Guam and Puerto Rico, for a total of more than 160 sites.

Probably the greatest progress in general weather observation came with the advent of the weather satellite. Before satellites, weather observations were unavailable for more than 80% of the globe. The first weather picture was sent back from space in 1961. The first fully operational weather satellite system was in place by 1966. These early pole-orbiting satellites, traveling from pole to pole at altitudes of several hundred miles, monitored only limited areas below their orbital paths. It took almost three orbits to photograph the entire conterminous United States.

Today, a fleet of GOES (*Geostationary Orbiting Environmental Satellites*), which orbit at fixed points, provide an almost continuous picture of weather patterns all over the globe.

At an altitude of about 36,800 km (23,000 mi), GOES orbiters have the same orbital period as the Earth's rotation and hence are "stationary" over a particular location.

At this altitude, GOES can send back pictures of large portions of the Earth's surface. Geographic boundaries and grids are prepared by computer and electronically combined with the picture signal so that the areas of particular weather disturbances can be easily identified (● Fig. 19.17).

With satellite photographs, meteorologists have a panoramic view of weather conditions. The dominant feature of such photographs is cloud coverage. However, with the aid of radar, which uses wavelengths that pick up only precipitation, the storm areas are easily differentiated from regular cloud coverage.

Figure 19.17 Satellite Image
In addition to regular cloud coverage photos, GOES can take infrared (IR) images, which indicate the temperature of the clouds. The orange and yellow regions are cold, high clouds moving across the southeastern United States.

Did You Learn?

- The pressure of one atmosphere (in various units) is 14.7 lb/in^2 = 76 cm Hg = 760 torr = 30 in. Hg.

- A relative humidity of 50% means that a volume of air is "half full" or contains 50% of the water vapor it is capable of holding at a given temperature. The dew point is the temperature to which a sample of air must be cooled to become saturated. At the dew point, the relative humidity is 100%.

19.4 Air Motion

Preview Questions

- What causes wind and air currents?
- How do winds rotate around high- and low-pressure regions (in the Northern Hemisphere as viewed from above)?

If the air in the troposphere were static, then there would be little change in the local atmospheric conditions that constitute weather. Air motion is important in many processes, even biological ones, such as carrying scents and distributing pollen. As air moves into a region, it brings with it the temperature and humidity that are mementos of its travels.

Wind is the horizontal movement of air or air motion along the Earth's surface. Vertical air motions are referred to as *updrafts* and *downdrafts* or collectively as **air currents**. For winds and air currents to exist, the air must be in motion. What causes the air to move? As in all dynamic situations, forces are necessary to produce motion and changes in motion. The gases of the atmosphere are subject to two primary forces: (1) gravity and (2) pressure differences due to temperature variations.

The force of gravity is vertically downward and acts on each gas molecule of the air. Although this force is often overwhelmed by forces in other directions, the downward gravity component is ever present and accounts for the greater density of air near the Earth's surface.

Because the air is a mixture of gases, its behavior is governed by the gas laws discussed in Chapter 5.6 and by other physical principles. The pressure of a gas is directly proportional to its Kelvin temperature ($p \propto T$), so if there is temperature variation, then there will be a pressure difference ($\Delta p \propto \Delta T$). Pressure is the force per unit area, so a pressure difference corresponds to an unbalanced force. When there is a pressure difference, air moves from a high-pressure to a low-pressure region.

The pressures of a region may be mapped by taking barometric readings at different locations. A line drawn through the locations (points) of equal pressure is called an **isobar.** Because all points on an isobar are of equal pressure, there will be no air movement along

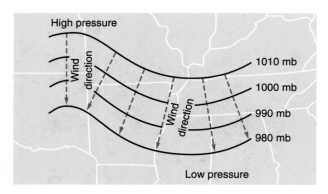

Figure 19.18 Isobars
Isobars are lines drawn through locations having equal atmospheric pressure. In the absence of other forces, the air motion or wind direction is perpendicular to the isobars from a region of greater pressure (greater mb values) to a region of lower pressure (lower mb values).

an isobar. The wind direction will be at right angles to the isobar in the direction of the low-pressure region, as illustrated in ● Fig. 19.18. (This is an idealized situation. Other forces, which are discussed shortly, may cause deflections.)

The pressure *and* volume of a gas are directly proportional to its Kelvin temperature, $pV \propto T$ (Chapter 5.6). Thus, a change in temperature causes a change in the pressure and/or volume of a gas. With a change in volume, there is also a change in density ($\rho = m/V$). For example, if the air is heated and expands, then the air density decreases. As a result of this relationship, localized heating sets up air motion in a **convection cycle**, which gives rise to *thermal circulation*.

Thermal circulations due to geologic features give rise to local winds. Land areas heat up more quickly during the day than do water areas, and the warm, buoyant (less dense) air over the land rises. As air flows horizontally into this region, the rising air cools and falls, and a convection cycle is set up. As a result, when the land is warmer than the water during the day, a *sea breeze* is experienced, as shown in ● Fig. 19.19a. You may have noticed daytime sea breezes at an ocean beach.

At night, the land loses its heat more quickly than the water does, so the air over the water is warmer than the air over the land. The convection cycle is then reversed, and at night a *land breeze* blows (Fig. 19.19b). Sea and land breezes are sometimes referred to as *onshore* and *offshore winds*, respectively.

In southern Asia, a giant seasonal convection cycle brings heavy, summer monsoon rains in from the ocean. Most of the land mass is in the Northern Hemisphere, whereas most of the ocean is in the Southern Hemisphere. In summer, the Northern Hemisphere receives the more direct rays of the Sun because of the Earth's tilt (Chapter 15.5). The heat is absorbed by the land mass, warming the air above it. The warmed air rises, and cooler ocean air from the Southern Hemisphere moves in to replace it. The ocean air carries moisture, which is released overland as the rainy *summer monsoon.**

Six months later, the Southern Hemisphere receives the more direct solar rays. During this season, the northern continental land mass is cooler than the water. The cycle reverses circulation, with warmer air rising over the ocean and cooler land air replacing it. The cooler air warms and picks up moisture as it travels over tropical waters and releases it over Indonesia and northern Australia as a *winter monsoon*.

The monsoons are analogous to the day–night thermal circulations at an ocean beach.

*Monsoon translates from other languages as "season."

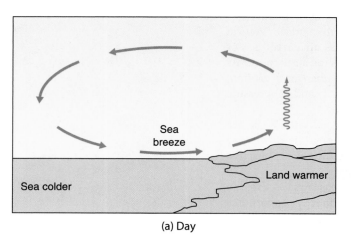

(a) Day

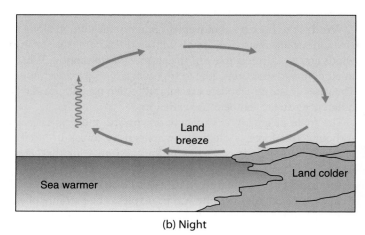

(b) Night

Figure 19.19 Daily Convection Cycles over Land and Water
(a) During the day, the land surface heats up more quickly than a large body of water, setting up a convection cycle in which the surface winds are from the water (a sea breeze). (b) At night, the land cools more quickly than water. The convection cycle is then reversed, with surface winds coming from the land (a land breeze).

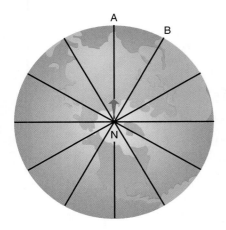

| Projectile fired south from | Earth rotates beneath projectile, | Earthbound observer sees projectile |
| North Pole along meridian A | which lands to the west at B | deflected to right toward B |

Figure 19.20 The Coriolis Force
(*Left*) Imagine someone firing a projectile from the North Pole toward position A. (*Center*) The Earth turns while the projectile is in flight, and the projectile lands to the right of A at location B. That is the situation you would see if you were viewing the Earth from space over the North Pole. (*Right*) For an observer on the Earth, who does not perceive the Earth's rotation, if the projectile lands at B, then it seems that it must have been deflected by some force as required by Newton's laws. The "Coriolis force" was invented to account for this apparent deflection.

Once the air has been set into motion, velocity-dependent forces act. These secondary forces are (1) the Coriolis force and (2) friction.

The **Coriolis force**, named after the French engineer who first described it, results because an observer on the Earth is in a rotating frame of reference. This force is sometimes referred to as a *pseudoforce*, or false force, because it is introduced to account for the effect of the Earth's rotation.

We humans tend to consider ourselves to be motionless, even though we are on a rotating Earth that has a surface speed of about 1600 km/h (1000 mi/h) near the equator. Rotating with the Earth, we are in an accelerating reference frame. (How is it accelerating?) Newton's laws of motion apply to nonaccelerating reference frames and may be used for ordinary motions on the Earth without correction. However, for high speeds or huge masses such as the atmospheric gases, the correction for the Earth's rotation becomes important.

To help understand this effect, imagine a high-speed projectile being fired from the North Pole southward along a meridian (● Fig. 19.20). While the projectile travels southward, the Earth rotates beneath it and lands to the west of the original meridian. To an observer at the North Pole looking southward along the meridian, it appears that the projectile is deflected to the right. By Newton's laws, this deflection requires a force, even though no such "force" really exists. Hence, the Coriolis force is a pseudoforce that is invented so that the effect will be consistent with the laws of motion.

In general, projectiles or particles moving in the Northern Hemisphere are apparently deflected to the right. By similar reasoning, objects in the Southern Hemisphere appear to be deflected to the left. We say that because of the Coriolis force, moving objects are deflected to the right in the Northern Hemisphere and to the left in the Southern Hemisphere, as observed in the direction of motion. The Coriolis force is at a right angle to the direction of motion of an object, and its magnitude varies with latitude; it is zero at the equator and increases toward the poles.

Consider this effect on wind motion. Initially, air moves toward a low-pressure region (a "low") and away from a high-pressure region (a "high"). Winds are deflected because of the Coriolis force, and in the Northern Hemisphere, winds tend to rotate counterclockwise around a low and clockwise around a high, as viewed from above (● Fig. 19.21). These

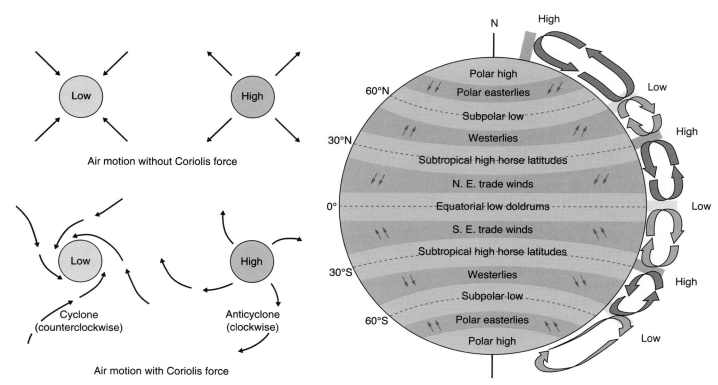

Figure 19.21 Effects of the Coriolis Force on Air Motion
In the Northern Hemisphere, the Coriolis deflection to the right produces counterclockwise air motion around a low and clockwise rotation around a high (as viewed from above).

Figure 19.22 The Earth's General Circulation Structure
For rather complicated reasons, the Earth's general circulation pattern has six large convection cycles, or convection cells. Most of the conterminous United States lies in the westerlies zone. (The small blue arrows show the general directions of the prevailing winds.)

disturbances are referred to as *cyclones* and *anticyclones*, respectively. The rotational directions are reversed in the Southern Hemisphere. The opposite rotations in the different hemispheres are evident from satellite photos. Water motion and currents in the oceans are also affected by the Coriolis force.

Friction, or *drag*, can also cause the retardation or deflection of air movements. Moving air molecules experience frictional interactions (collisions) among themselves and with terrestrial surfaces. The opposing frictional force along a surface is in the opposite direction of the air motion. Thus, winds moving into a cyclonic disturbance may be deflected differently because of the sum of the forces.

Air motion changes locally with altitude, geographic features, and the seasons. However, the air near the Earth's surface does exhibit a general circulation pattern. Because of the Coriolis force, land and sea variations, and other complicated factors, the hemispheric circulation is broken into six general convection cycles, or pressure cells. The Earth's general circulation structure is shown in ● Fig. 19.22.

Many local variations occur within the cells, which shift seasonally in latitude because of variations in insolation. The prevailing winds of this semipermanent circulation structure are important in influencing general weather movement around the world.

The conterminous United States lies generally between the latitudes of 30°N and 50°N.* This is predominately in the westerlies wind zone. As a result, our weather conditions generally move from west to east across the country.

You commonly hear about other high-altitude winds on TV weather reports, particularly in winter. In the upper troposphere are fast-moving "rivers" of air called **jet streams**. They

*Conterminous means "contiguous or adjoining," so the conterminous United States is the original 48 states. This is different from the continental United States. (Why?)

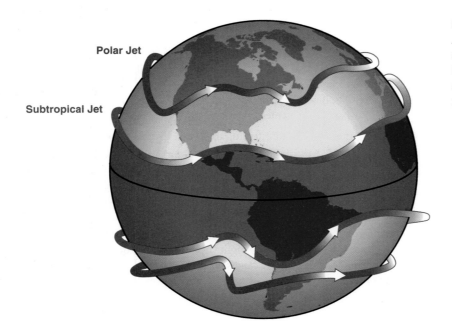

Figure 19.23 Jet Streams
Jet streams meander like rivers of air around each hemisphere. A *polar jet* is located in the 50° to 60° latitudes, and a *subtropical jet* is located in the 30° latitudes. (See text for description.)

were first noted in the 1930s but did not receive much attention until World War II, when high-flying aircraft encountered them.

Jet streams meander like rivers around each hemisphere. They occur generally where the circulation cells meet, namely latitudes 30° N (and S) and 50° to 60° N (and S). The *polar jet* is located in the 50° to 60° latitudes and the *subtropical jet* in the 30° latitudes (● Fig. 19.23). These relatively narrow bands of strong wind, flowing west to east with speeds on the order of 170 km/h (275 mi/h), are usually found somewhere between 10 km and 15 km (6 mi and 9 mi) above the Earth's surface. They meander around the globe, dipping and rising in altitude and latitude, and even disappearing altogether to appear somewhere else. The locations of the jet streams shift throughout the year, and they are said to "follow the Sun" since they generally move north with warm weather and south during cold weather.

Jet streams form along the upper boundaries of warm and cold air masses. Because these warm and cold boundaries are more pronounced in winter, jet streams are strongest for both the Northern and Southern Hemispheres during the winter season. The polar jet stream is most prominent for the conterminous United States. Normally at 50° to 60° N latitude, it can shift southward in the winter, indicating an invasion of a cold polar air mass.

Did You Learn?

- The primary forces producing air motion are (1) gravity and (2) pressure differences due to temperature variations.

- In the Northern Hemisphere, winds rotate clockwise around a high-pressure region and counterclockwise around a low-pressure region (as viewed from above).

19.5 Clouds

Preview Questions

- What are the four cloud families?
- How are clouds generally formed?

Classification of Clouds

Clouds are both a common sight and an important atmospheric consideration. **Clouds** are buoyant masses of visible water droplets or ice crystals. The size, shape, and behavior of clouds are useful keys to the weather.

Clouds are classified according to their *shape*, *appearance* (light or dark), and *altitude*. There are four basic root names:

Cirrus, meaning "curl" and referring to wispy, fibrous forms

Cumulus, meaning "heap" and referring to billowy, rounded forms

Stratus, meaning "layer" and referring to stratified, or layered, forms

Nimbus, meaning "rain cloud" and referring to a cloud from which precipitation is occurring or threatens to occur

These root forms are then combined to describe the types of clouds and precipitation potential.

When classified according to height, clouds are separated into four families: (1) *high clouds*, (2) *middle clouds*, (3) *low clouds*, and (4) *clouds of vertical development*. See the pictorial **Highlight: Cloud Families and Types**, which gives the approximate heights and cloud types that belong to each family. Brief descriptions or common names are also given, along with a collection of illustrative photographs. Study these so that you will be able to identify the clouds when you see them next time.

Cloud Formation

To be visible as droplets, the water vapor in the air must condense. Condensation requires a certain temperature, the dew point temperature. Hence, if moist air is cooled to the dew point, then the water vapor contained therein will generally condense into fine droplets and form a cloud.

The air is continuously in motion, and when an air mass moves into a cooler region, clouds may form. Because the temperature of the troposphere decreases with height, cloud formation is associated with the vertical movement of air. In general, clouds are formed in vertical air motion (currents) and are shaped and moved about by horizontal air motion (winds).

Air may rise as a result of heating or wind motion along an elevating surface, such as up the side of a mountain or the boundary of a front (Chapter 20.2). Here we consider cloud formation resulting from the rising of warm, buoyant air (clouds of vertical development). As the warm air ascends, it becomes cooler because it expands and because its internal energy is used to do the work of expansion against the surrounding stationary air. With less energy, it becomes cooler.

The temperature of the air in the troposphere decreases with altitude, and the rate at which this temperature decreases with height is called the **lapse rate**. The normal lapse rate in stationary air in the troposphere is about $6\frac{1}{2}$°C /km (or $3\frac{1}{2}$°F/1000 ft).

Because energy is used in the expansion of a warm air mass, the rising air cools more quickly and has a greater lapse rate. When the rising air mass cools to the same temperature as the surrounding stationary air, the densities of the two air masses become equal. The rising air mass then loses its buoyancy and is said to be in a *stable condition*. A heated air mass rises until stability is reached, and this portion of the atmosphere is referred to as a *stable layer*, a layer of air of uniform temperature and density.

Clouds are formed when water vapor in the rising air condenses into droplets and can be seen. If the rising air reaches its dew point before becoming stable, then condensation occurs and the rising air carries the condensed droplets upward, forming a cloud.

Highlight Cloud Families and Types

High Clouds

[above 6 km (4 mi)]. All are composed of ice crystals.

Cirrus: Wispy and curling. Known as "artist's brushes" or "mares' tails" (a).

Cirrocumulus: Layered patches. Known as "mackerel scales" (a).

Cirrostratus: Thin veil of ice crystals. Scattering from crystals gives rise to solar and lunar halos (b).

Middle Clouds

[1.8 to 6 km (1.1 to 4 mi)]. All names have the prefix *alto*.

Altostratus: Layered forms of varying thickness. May hide the Sun or the Moon and cast a shadow (c).

Altocumulus: Woolly patches or rolled, flattened layers (d).

Geraldine Buckley/Alamy

(a) Cirrus and cirrocumulus clouds. Artist's-brush cirrus are to the top, and mackerel-scale cirrocumulus are at the bottom.

Trevor McDonald/Photo Researchers, Inc.

(c) Altostratus clouds. Thick, gray altostratus clouds.

© Gianni Muratore/Alamy

(b) Cirrostratus clouds. The clouds cover the sky and are evidenced by a solar halo.

© Wayne Decker/Fundamental Photographs, NYC

(d) Altocumulus clouds. These clouds are often rolled and arranged in flattened layers by moving air.

Highlight Cloud Families and Types (continued)

Low Clouds

[ground level to 1.8 km (1.1 mi)].

Stratus: Thin layers of water droplets. May appear dark; common in winter. Fog may be thought of as low-lying stratus clouds (e).

Stratocumulus: Long layers of cotton-like masses, sometimes with a wavy appearance (f).

Nimbostratus: Dark, low clouds given to precipitation (g).

Advection fog: Forms when moist air moving over a colder surface is cooled below dew point. Advection fogs "roll in"(h).

Radiation fog: Condensation in stationary air overlying a surface that cools. Typically occurs in valleys and is called a "valley fog" (i).

Clouds with Vertical Development

[5 to 18 km (3 to 11 mi)]. Formed by updrafts.

Cumulus: Commonly seen on a clear day (j).

Cumulonimbus: Darkened cumulus cloud, referred to as a "thunderhead"(k).

(e) Stratus clouds. Low-lying stratus clouds are sometimes called high fogs.

(g) Nimbostratus clouds. Dark nimbostratus clouds are given to precipitation.

(f) Stratocumulus clouds. Appears as long layers of cotton-like masses.

(h) Advection fog. Formed in moist air moving over a cool surface, an advection fog rolls in around San Francisco Bay's Golden Gate Bridge.

Highlight

(i) Radiation fog. Commonly formed overnight in valleys when radiational heat loss cools the ground. The nearby air is then cooled, leading to condensation and fog.

(j) Cumulus clouds. These billowy, white clouds are commonly seen on a clear day.

(k) Cumulonimbus clouds. This huge cloud of vertical development has a dark nimbus lower portion, from which precipitation is occurring or threatening to occur. Such dark clouds are sometimes called thunderheads.

Family	Types	Illustration
High clouds [above 6 km (4 mi)]	Cirrus (Ci)	(a)
	Cirrocumulus (Cc)	(a)
	Cirrostratus (Cs)	(b)
Middle clouds [1.8 to 6 km (1.1 to 4 mi)]	Altostratus (As)	(c)
	Altocumulus (Ac)	(d)
Low clouds [ground level to 1.8 km (1.1 mi)]	Stratus (St)	(e)
	Stratocumulus (Sc)	(f)
	Nimbostratus (Ns)	(g)
	Advection fog	(h)
	Radiation fog	(i)
Clouds with vertical development [5 to 18 km (3 to 11 mi); see Chapter 19.5]	Cumulus (Cu)	(j)
	Cumulonimbus (Cb)	(k)

DHuss/iStockphoto.com

Figure 19.24 Vertical Cloud Development
The cloud begins to form at the elevation at which the rising air reaches its dew point and condensation occurs. The vertical development continues until the rising air stabilizes. Horizontal air motion in the upper regions gives this cloud an anvil shape, hence the name anvil cumulus.

The height at which condensation occurs is the height of the base of the cloud. The vertical distance between the level where condensation begins and the level where stability is reached is the height (vertical thickness) of the cloud. For purposes of this discussion, we assume that no precipitation occurs. Once the clouds are formed, the wind shapes them. They may break up into smaller clouds or assume massive forms, as shown for an anvil cumulus in ● Fig. 19.24.

Did You Learn?

● The cloud families are (1) high clouds, (2) middle clouds, (3) low clouds, and (4) clouds of vertical development.

● Clouds are formed when water vapor in rising air condenses into droplets that are visible.

KEY TERMS

1. atmospheric science
2. meteorology
3. photosynthesis (19.1)
4. troposphere
5. weather
6. stratosphere
7. mesosphere
8. thermosphere
9. ozone (O_3)
10. ozonosphere

11. ionosphere
12. insolation (19.2)
13. Rayleigh scattering
14. greenhouse effect
15. barometer (19.3)
16. relative humidity
17. dew point
18. psychrometer
19. anemometer
20. wind vane

21. rain gauge
22. Doppler radar
23. wind (19.4)
24. air currents
25. isobar
26. convection cycle
27. Coriolis force
28. jet streams
29. clouds (19.5)
30. lapse rate

MATCHING

For each of the following items, fill in the number of the appropriate Key Term from the preceding list. Compare your answers with those at the back of the book.

a. _____ Atmospheric conditions of the lower atmosphere
b. _____ Radiation that comes from the Sun
c. _____ Thermal circulation
d. _____ Region of atmosphere with highest temperature
e. _____ Used to measure wind speed
f. _____ Horizontal movement of air
g. _____ Lowest region of atmosphere
h. _____ A pseudoforce
i. _____ Region of atmosphere characterized by O_3 content
j. _____ "Rivers" of air
k. _____ Investigates every aspect of the atmosphere
l. _____ Rate of temperature decrease with height
m. _____ Used to measure air pressure
n. _____ Together with troposphere, accounts for 99.9% of atmospheric mass

o. _____ A line of equal pressure
p. _____ A measure of the moisture content of a volume of air at a given temperature
q. _____ Molecule with three atoms of oxygen
r. _____ Temperature at which air becomes saturated
s. _____ Used to measure precipitation
t. _____ Study of the lower atmosphere
u. _____ Indicates wind direction
v. _____ Region where layers reflect radio waves
w. _____ Vertical movements of air
x. _____ Causes the sky to be blue
y. _____ Buoyant masses of visible water droplets or ice crystals
z. _____ Process that replenishes atmospheric oxygen
aa. _____ Gives early warnings of tornadoes

bb. _____ Used to measure relative humidity

cc. _____ Provides the Earth with a thermostatic effect

dd. _____ Upper region of the atmosphere characterized by decreasing temperature with height

MULTIPLE CHOICE

Compare your answers with those at the back of the book.

1. Which is the second-most-abundant gas in the atmosphere? (19.1)
   (a) oxygen
   (b) carbon dioxide
   (c) nitrogen
   (d) argon

2. In what region does the ozone layer lie? (19.1)
   (a) thermosphere
   (b) troposphere
   (c) stratosphere
   (d) mesosphere

3. Photosynthesis is responsible for the atmospheric production of which of the following? (19.1)
   (a) carbon dioxide
   (b) oxygen
   (c) nitrogen
   (d) carbon monoxide

4. What regulates the Earth's average temperature? (19.2)
   (a) Rayleigh scattering
   (b) the greenhouse effect
   (c) atmospheric pressure
   (d) photosynthesis

5. Approximately what percentage of insolation reaches the Earth's surface? (19.2)
   (a) 33%
   (b) 40%
   (c) 50%
   (d) 75%

6. What instrument is used to measure relative humidity? (19.3)
   (a) an anemometer
   (b) a barometer
   (c) a wind vane
   (d) a psychrometer

7. With what instrument is atmospheric pressure measured? (19.3)
   (a) an anemometer
   (b) a barometer
   (c) a wind vane
   (d) a psychrometer

8. Near a large body of water, which wind is the predominant wind during the day? (19.4)
   (a) a sea breeze
   (b) a land breeze
   (c) an updraft
   (d) a jet stream

9. What is the direction of rotation around a cyclone in the Northern Hemisphere as viewed from above? (19.4)
   (a) clockwise
   (b) counterclockwise
   (c) sometimes clockwise, sometimes counterclockwise

10. Convection cycles give rise to which of the following? (19.4)
    (a) land breeze
    (b) sea breeze
    (c) air currents
    (d) all the preceding

11. What is the cloud root name that means "heap"? (19.5)
    (a) stratus
    (b) cirrus
    (c) nimbus
    (d) cumulus

12. The altostratus cloud is a member of which family? (19.5)
    (a) high clouds
    (b) middle clouds
    (c) low clouds
    (d) vertical development

FILL IN THE BLANK

Compare your answers with those at the back of the book.

1. ___ is the study of the lower atmosphere. (Intro)

2. The ozone layer causes a temperature increase in the ___. (19.1)

3. The International Space Station is in orbit in the ___. (19.1)

4. The primary absorbers of infrared radiation in the atmosphere are water vapor and ___. (19.2)

5. One standard atmosphere of pressure is ___ lb². (19.3)

6. One atmosphere is equal to ___ cm or ___ in. of mercury. (19.3)

7. Relative humidity is the ratio of the actual moisture content to the ___ moisture content (19.3)

8. The arrow of a wind vane (or weather vane) points in the ___ direction from which the wind is blowing. (19.3)

9. A line drawn through points of equal pressure is called a(n) ___. (19.4)

10. A(n) ___ breeze blows at the ocean coast during the evening. (19.4)

11. In the Southern Hemisphere, the wind rotates ___ around a high, as viewed from above. (19.5)

12. Nimbostratus clouds belong to the ___ cloud family. (19.5)

SHORT ANSWER

19.1 Composition and Structure

1. Other than moisture, what are the three major components of the air you breathe?

2. Humans and other animals inhale oxygen and exhale carbon dioxide. With our large population, wouldn't this reduce the atmospheric oxygen level over a period of time? Explain.

3. What key compound makes photosynthesis in green plants possible?

4. Describe how the temperature of the atmosphere varies in each of the following regions:
   (a) the mesosphere
   (b) the stratosphere
   (c) the troposphere
   (d) the thermosphere

5. Of what importance is the atmospheric ozone layer?

6. What causes the displays of lights called auroras?

19.2 Atmospheric Energy Content

7. What does it mean to say that the Earth has an albedo of 33%? How does it compare with the Moon's albedo?

8. From what source does the atmosphere receive most of its *direct* heating, and how is the overall heating accomplished?

9. Why is the sky blue overhead and more white toward the horizon?

10. In terms of Rayleigh scattering, why is it advantageous to have amber fog lights and red taillights on cars?

11. Is the atmospheric "greenhouse effect" the same as that in an actual greenhouse? Explain.

12. How does the selective absorption of atmospheric gases provide a thermostatic effect for the Earth?

19.3 Atmospheric Measurements and Observations

13. What are the four fundamental atmospheric measurements discussed in this chapter, and with what instrument is each measured?

14. What is the principle of the liquid barometer? What is the height of a mercury barometer column for one atmosphere of pressure?

15. Why does water condense on the outside of a glass containing an iced drink?

16. When is the relative humidity 100%? It may be raining, but a sheltered psychrometer doesn't read 100%. Why?

17. Which way, relative to the wind direction, does a wind vane (or weather vane) point and why?

18. At small airports, a wind sock (a tapered bag that pivots on a pole) acts as a wind vane and gives some indication of the wind speed. Explain its operation.

19. What information does Doppler radar give that conventional radar cannot?

19.4 Air Motion

20. What is a convection cycle, and what are the related effects near a large body of water such as the ocean?

21. Indicate any differences between cyclones and anticyclones in the Northern and Southern Hemispheres and explain.

22. Why does weather generally move from west to east in the conterminous United States?

23. Generally speaking, on which side of town would it be best to build a house in the United States so as to avoid smoke and other air pollutants generated in the town?

24. Should the prevailing wind direction be considered when designing the heating plan and insulation of a house?

19.5 Clouds

25. Name the cloud family for each of the following:
    (a) nimbostratus   (d) stratus
    (b) cirrostratus   (e) cumulonimbus
    (c) altostratus

26. Name the cloud type associated with each of the following:
    (a) mackerel sky   (c) thunderhead
    (b) solar or lunar halo

27. What is "lapse rate," and how is it involved in cloud formation?

28. What happens if the dew point of a rising air mass is not reached before stability?

VISUAL CONNECTION

Visualize the connections and give answers for the blanks. Compare your answers with those at the back of the book.

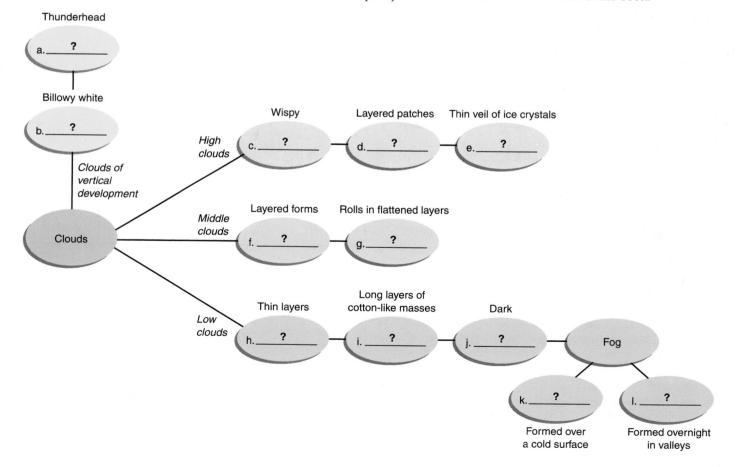

APPLYING YOUR KNOWLEDGE

1. (a) Why does the land lose heat more quickly at night than a body of water? (b) Deserts are very hot during the day and cold at night. Why is there such a large nocturnal temperature drop?

2. How do the summer and winter monsoons compare to the daily thermal circulations experienced at an ocean beach?

3. In summer, sudden rain showers can occur over a limited area. Why?

4. Describe what would happen if a projectile were fired directly north from the equator. (*Hint*: See Fig. 19.20.)

5. Water leaks from a cup with a hole, as shown in ● Fig. 19.25. What would happen if you placed the palm of your hand tightly over the mouth of the cup? Why? (*Hint*: What happens to the pressure of the air above the water with respect to the pressure outside the cup?)

Figure 19.25 Stop That Leak
See Applying Your Knowledge
Question 5.

IMPORTANT EQUATION

Relative Humidity: $(\%) \, RH = \dfrac{AC}{MC} \, (\times \, 100\%)$ (19.1)

EXERCISES

19.1 Composition and Structure

1. Express the approximate thicknesses of the (a) stratosphere, (b) mesosphere, and (c) thermosphere in terms of the thickness of the troposphere (at the equator).
   Answer: (a) 2.1 (b) 1.9 (c) 7.5

2. On a vertical scale of altitude in kilometers and miles above sea level, compare the heights of the following (the heights not listed can be found in the chapter).
   (a) the top of Pikes Peak, 14,000 ft
   (b) the top of Mt. Everest, 29,000 ft
   (c) commercial airline flight, 35,000 ft
   (d) supersonic transport flight (SST), 65,000 ft
   (e) communications satellite, 400 mi
   (f) the E and F ion layers
   (g) aurora displays
   (h) syncom satellite, 23,000 mi (satellite with period synchronous to the Earth's rotation, so it stays over one location)

   (*Note*: Conversion factors are found on the inside back cover.)

3. If the air temperature is 70°F at sea level, then what is the temperature at the top of Pikes Peak (elevation 14,000 ft)? (*Hint*: The temperature decreases rather uniformly in the troposphere.)
   Answer: 21°F

4. If the air temperature is 20°C at sea level, then what is the temperature outside a jet aircraft flying at an altitude of 10,000 m?

19.3 Atmospheric Measurements and Observations

5. On a day when the air temperature is 85°F, the wet-bulb reading of a psychrometer is 70°F. Find each of the following: (a) relative humidity, (b) dew point, (c) maximum moisture capacity of the air, and (d) actual moisture content of the air.
   Answer: (a) 46% (b) 62°F (c) 12.7 gr/ft³ (d) 5.8 gr/ft³

6. A psychrometer has a dry-bulb reading of 90°F and a wet-bulb reading of 80°F. Find each of the quantities asked for in Exercise 5.

7. On a very hot day with an air temperature of 105°F, the wet-bulb thermometer of a psychrometer records 100°F. (a) What is the actual moisture content of the air? (b) How many degrees would the air temperature have to be lowered for the relative humidity to be 100%?
   Answer: (a) 21 gr/ft³ (b) 4°F

8. On a winter day, a psychrometer has a dry-bulb reading of 35°F and a wet-bulb reading of 29°F. (a) What is the actual moisture content of the air? (b) Would the water in the wick of the wet bulb freeze? Explain.

9. The dry-bulb and wet-bulb thermometers of a psychrometer read 75°F and 65°F, respectively. What are (a) the actual moisture content of the air and (b) the temperature at which the relative humidity would be 100%?
   Answer: (a) 5.5 gr/ft³ (b) 59°F

10. On a day when the air temperature is 80°F, the relative humidity is measured to be 79%. How many degrees would the air temperature have to be lowered for the relative humidity to be 100%?

11. On a day when the air temperature is 70°F, a fellow student reads the psychrometer in the physical science lab and tells you that with a wet-bulb depression of 3°F, the relative humidity is 83%. Would you agree? Explain.
    Answer: No

12. On another day with the same air temperature (70°F), you read the psychrometer and determine the relative humidity to be 68%. A sudden weather change (that does not affect the actual moisture content of the air) reduces the relative humidity to 63%. What was the temperature change?

ON THE WEB

1. A Trip to the NASA Lab

Do people have an effect on clouds and radiation, or are these phenomena beyond our control? Take a trip to the NASA Earth Observatory and find out. Follow the links on the student website at **www.cengagebrain.com/shop/ISBN/1133104096** to study clouds and investigate cloud forcing, high clouds, and low clouds.

2. Recreating the Greenhouse Effect

Why has the "greenhouse effect" become a major political issue around the globe? Visit the student website at **www.cengagebrain.com/shop/ISBN/1133104096** to try an experiment. What could happen if this greenhouse effect changed the Earth's climate?

Atmospheric Effects

Joel Santore/Getty Images

And pleas'd the Almighty's orders to perform Rides in the whirlwind and directs the storm

•

Joseph Addison, English poet (1672–1719)

< Lightning: a most spectacular atmospheric effect, but it can be dangerous.

R eadily available weather reports and forecasts help us decide such things as how we should dress for the day, when to take an umbrella along, or whether a weekend picnic should be canceled. The weather changes frequently because the lower atmosphere is a very dynamic place.

The air you now breathe may have been far out over the Pacific Ocean a week ago. As air moves into a region, it brings with it the temperature and humidity of previous locations. Cold, dry arctic air may cause a sudden drop in the temperature of the regions in its path. Warm, moist air from the Gulf of Mexico may bring heat and humidity, creating uncomfortable summer conditions.

Moving air transports the physical characteristics that influence the weather and cause changes in it. A mass can influence a region's weather for a considerable

ATMOSPHERIC FACTS

▶ The community of Tamarack, California (on the west slope of the Sierra Nevada), holds the national snowfall record for 1 month. In January 1911, Tamarack received 390 in. (32.5 ft) of snow.

▶ A tornado in March 1925 was believed to have stayed

Chapter Outline

period of time or have only a brief effect. The movement of air masses depends largely on the Earth's air circulation structure and seasonal variations.

When air masses meet, variations of their properties may trigger storms along their common boundary. The types of storms depend on the properties of the air masses involved. Also, variations within a single air mass can give rise to storms locally. The violence and destruction of some storms demonstrate the vast amount of energy contained in the atmosphere. As you will learn in this chapter, the variations of our weather are closely associated with air masses and their movements and interactions.

An unfortunate issue that arises in atmospheric science is pollution. Various pollutants are being released into the atmosphere, affecting health, living conditions, and the environment. Climate also may be affected by pollution. For example, we hear about the ozone "hole" over the South Pole and are warned of climate change brought about by emissions of greenhouse gases, giving rise to global warming. These and other topics are considered in this chapter.

20.1   Condensation and Precipitation

Preview Questions

● What are the three essentials of the Bergeron process?

● Is frost frozen dew?

How are visible droplets of water formed? You might think that the collision and coalescing of water molecules would form a droplet, but this event would require the collision of millions of molecules. Moreover, only after a small droplet has reached a critical size does it have sufficient binding force to retain additional molecules. The probability of a droplet forming by molecular collisions is quite remote.

In Chapter 19.5 on cloud formation, it was noted that condensation occurs in an air mass when the dew point is reached. However, it is quite possible for an air mass containing water vapor to be cooled below the dew point without condensation occurring. In this state the air is said to be *supersaturated*, or *supercooled*.

Water droplets form from supercooled vapor condensing on microscopic foreign particles, called *hygroscopic nuclei*, which are already present in the air. These particles are in the form of dust, combustion residue (smoke and soot), salt from seawater evaporation, and so on. Because foreign particles initiate the formation of droplets that eventually fall as precipitation, condensation provides a mechanism for cleansing the atmosphere.

Also, liquid water may be cooled below the freezing point (supercooled) without the formation of ice if it does not contain the proper type of foreign particles to act as ice nuclei. For many years, scientists believed that ice nuclei could be just about anything, such as dust. However, research has shown that "clean" dust—that is, dust without biological materials from plants or bacteria—will not act as ice nuclei. This discovery is important because precipitation sometimes involves ice crystals, as will be discussed.

Because cooling and condensation occur in updrafts, the formed droplets are readily suspended in the air as a cloud. For precipitation, larger droplets or drops must form. This condition may be brought about by two processes: (1) coalescence and/or (2) the Bergeron process.

Coalescence

Coalescence is the formation of drops by the collision of droplets, the result being that larger droplets grow at the expense of smaller ones. The efficiency of this process depends on the variation in the size of the droplets.

Raindrops vary in size, reaching a maximum diameter of approximately 7 mm. A drop 1 mm in diameter would require the coalescing of a million droplets 10 μm (micrometers) in diameter, but only 1000 droplets 100 μm in diameter. Larger droplets are necessary for the coalescence process.

Bergeron Process

Named after Swedish meteorologist Tor Bergeron (1891–1977), the **Bergeron process** is the more important process for the initiation of precipitation. This process involves clouds that contain ice crystals in their upper portions and have become supercooled in their lower portions (● Fig. 20.1).

Mixing or agitation within such a cloud allows the ice crystals to come into contact with the supercooled vapor. Acting as nuclei, the ice crystals grow larger as the vapor condenses on them. The ice crystals melt into large droplets in the warmer, lower portion of the cloud and coalesce to fall as precipitation. Air currents are the normal mixing agents. Note that there are three essentials in the Bergeron process: (1) ice crystals, (2) supercooled vapor, and (3) mixing.

Rainmaking is based on the essentials of the Bergeron process. The early rainmakers were mostly charlatans. With much ceremony, they would beat on drums or fire cannons and rockets into the air. Explosives may sometimes have supplied the agitation or mixing for rainmaking, assuming that the other two essentials of the Bergeron process were present. However, modern rainmakers use a different approach. There are usually adequate air currents present for mixing, but the ice crystal nuclei may be lacking. To correct this deficiency, clouds are "seeded" with silver iodide (AgI) crystals or dry-ice pellets (solid CO_2).*

Silver iodide crystals have a structure similar to that of ice and act as a substitute for ice crystals. Silver iodide crystals are produced by a burning process. The burning may be done on the ground, with the iodide crystals being carried aloft by the rising warm air, or the burner may be attached to an airplane and seeded in a cloud.

Dry-ice pellets are also seeded into a cloud from an airplane. The pellets do not act as nuclei but rather serve another purpose. The temperature of solid dry ice is $-79°C$ ($-110°F$), and it quickly sublimes, or goes directly from the solid to the gaseous phase. Rapid cooling associated with the sublimation triggers the conversion of supercooled cloud droplets into ice crystals. Precipitation may then occur if this part of the Bergeron process has been absent. Also, the latent heat released by formation of the ice crystals is available to set up convection cycles for mixing.

China has had an active program for rainmaking. It has been reported that cloud seeding with silver iodide produced ample rainfall in various dry areas of China. Seeding is conducted by aircraft and with outdated anti-aircraft guns. Farmers shoot shells containing silver iodide at passing clouds. Another rainmaking project in China was to seed any clouds that were approaching Beijing during the 2008 Olympic games so that it wouldn't rain during the games.

More recently, silver iodide cloud seeding has been done in Russia near Moscow. Here it involves "snowmaking." The idea is to have the clouds release their precipitation before reaching the capital, because the cost of seeding is less than the usual street snow removal.

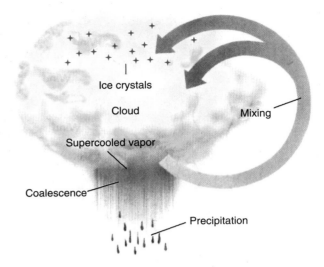

Figure 20.1 The Bergeron Process The essence of the Bergeron process is the mixing of ice crystals and supercooled vapor, which produces water droplets and initiates precipitation.

Types of Precipitation

Precipitation can occur in the form of rain, snow, sleet, hail, dew, or fog, depending on atmospheric conditions. *Rain* is the most common form of precipitation in the lower and middle latitudes. You learned about coalescence, or the formation of large water drops that fall as rain, in the previous section.

If the dew point is below 0°C, then the water vapor freezes upon condensing and the resulting ice crystals fall as *snow*. In cold regions these ice crystals may fall individually, but

---

*An interesting side note: In the early 1890s, Congress appropriated about $20,000 (a sizable sum in those days) to finance the rainmaking efforts of a General Dryenforth. It had been observed that rain frequently occurred during a battle with cannon fire. The general carried out experiments near San Antonio, Texas, using cannon and explosives carried aloft by balloons. After many firings and explosions, a little rain did occur (which may have been natural), but subsequent attempts failed. To the dismay of the local residents, the general called in more cannon, without success.

Figure 20.2 Hailstones
The successive vertical ascents of ice pellets into super-cooled air and regions of condensation produce large, layered "stones" of ice. The layered structure can be seen in these cross sections.

Figure 20.3 Frost
Frost forms when water vapor in the air is cooled and changes directly into ice on objects, as shown here on a flower and buds.

in warmer regions the ice crystals stick together, forming a snowflake that may be 1 or 2 centimeters across. Because ice crystallizes in a hexagonal (six-sided) pattern, snowflakes are hexagonal (see Fig. 2 in the Highlight: Freezing from the Top Down in Chapter 5.2). On average, 10 in. of snow would produce 1 in. of water when melted.

Sleet, which is frozen rain or pellets of ice, occurs when rain falls through a cold surface layer of air and freezes or, more often, when the ice pellets fall directly from the cloud without melting before striking the ground. Large pellets of ice, or *hail*, result from successive vertical descents and ascents in vigorous convection cycles associated with thunderstorms. Additional condensation on successive cycles into supercooled regions that are below freezing may produce layered-structure hailstones the size of golf balls and baseballs. When hailstones are cut in two, the layers of ice can be observed, much like the rings in a section of tree (● Fig. 20.2).

Dew is formed when atmospheric water vapor condenses on various surfaces. The land cools quickly at night, particularly with no cloud cover, and the temperature may fall below the dew point. Water vapor then condenses on available surfaces, such as blades of grass, giving rise to the "early morning dew."

If the dew point is below freezing, then the water vapor condenses in the form of ice crystals as *frost* (● Fig. 20.3). Frost is *not* frozen dew but rather results from the direct change of water vapor into ice (the reverse of sublimation, called *deposition*). Ideal conditions for frost formation are a night with clear skies, light or no wind, and a temperature near or below freezing. Much of the water vapor that goes into frost and dew actually comes from the soil and plants. The air temperature does not necessarily have to be below freezing for frost to form. Many surfaces, such as grass and plants, can cool several degrees below the air temperature.

Frost undeservingly gets a bad name in terms of "frostbite." This condition really has nothing to do with frost but can occur with exposure to extremely cold weather. The body's survival mechanisms are activated to protect the inner vital organs by cutting back on blood circulation to the extremities: the feet, hands, and nose. Deprived of a warming blood flow, these parts freeze when exposed to extreme cold.

Did You Learn?

● The three essentials of the Bergeron process are (1) ice crystals, (2) supercooled vapor, and (3) mixing.

● Frost is not frozen dew but rather results from the direct change of water vapor into ice (deposition).

20.2 Air Masses

Preview Questions

- What are the main distinguishing characteristics of air masses?
- What is the distinction between a warm front and a cold front?

The weather continually changes with time, but we often have several days of relatively uniform weather conditions. Our general weather conditions depend in large part on vast air masses that move across the conterminous United States. When a large body of air takes on physical characteristics that distinguish it from the surrounding air, it is referred to as an **air mass**. The main distinguishing characteristics are *temperature* and *moisture content*.

A mass of air remaining for some time over a particular region, such as a large body of land or water, takes on the physical characteristics of the surface of the region. The region from which an air mass derives its characteristics is called its **source region**.

An air mass eventually moves from its source region, bringing that region's characteristics to regions in its path and thus changing the weather. As an air mass travels, its properties may become modified because of local variations. For example, if Canadian polar air masses did not become warmer as they traveled southward, then Florida would experience some extremely cold temperatures.

Whether an air mass is termed *cold* or *warm* depends on whether it is colder or warmer than the surface over which it moves. Quite logically, if an air mass is warmer than the land surface, then it is referred to as a *warm air mass*. If the air is colder than the surface, it is a *cold air mass*. Remember, though, that these terms are relative. *Warm* and *cold* do not always imply warm and cold weather. A "warm" air mass in winter may not raise the temperature above freezing.

Air masses are classified according to the surface and general latitude of their source regions:

Surface	*Latitude*	
Maritime (m)	Arctic (A)	Tropical (T)
Continental (c)	Polar (P)	Equatorial (E)

The surface of the source region, abbreviated by a lowercase letter, gives an indication of the moisture content of an air mass. An air mass forming over a body of water (maritime, m) would naturally be expected to have greater moisture content than one forming over land (continental, c).

The general latitude of a source region, abbreviated by an uppercase letter, gives an indication of the temperature of an air mass. For example, "mT" designates a maritime (m) tropical (T) air mass, which would be expected to be a warm, moist one. The air masses that affect the weather in the conterminous United States are listed in ● Table 20.1 along with their source regions and are illustrated in ● Fig. 20.4.

The movement of air masses is influenced to a great extent by the Earth's general circulation patterns (Chapter 19.4). Because the conterminous United States lies predominantly

Table 20.1 Air Masses That Affect the Weather of the United States

Classification	Symbol	Source Region
Maritime arctic	mA	Arctic regions
Continental arctic	cA	Greenland
Maritime polar	mP	Northern Atlantic and Pacific Oceans
Continental polar	cP	Alaska and Canada
Maritime tropical	mT	Caribbean Sea, Gulf of Mexico, and Pacific Ocean
Continental tropical	cT	Northern Mexico, southwestern United States

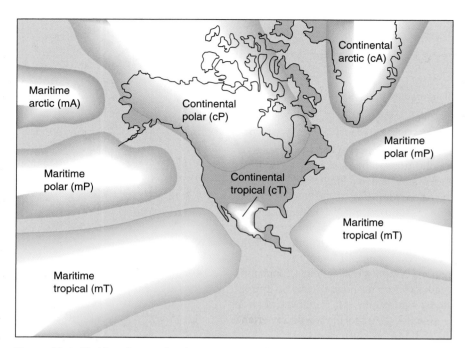

Figure 20.4 Air-Mass Source Regions
The map shows the source regions for the air masses of North America.

in the westerlies wind zone, the general movement of air masses—and hence the weather—is from west to east across the country. Global circulation zones vary to some extent in latitude with the seasons, and the polar easterlies may also move air masses into the eastern United States during the winter.

Let's not forget Alaska (The Land of the Midnight Sun) and Hawaii (The Aloha State). As a result of arctic air masses that form over the Arctic Ocean and northern Canada and Siberia, Alaska experiences cold weather and storms. Annual variations depend in large part on shifts in the jet streams. In general, there are cold winters and relatively warm summers (excess hours of daylight and solar radiation – the land of the midnight sun).

In Hawaii, northeast trade winds blow rather consistently on the islands. In the warm "summer" season (May through September), the Sun is more directly overhead. (At a latitude of about 21°N, Hawaii lies below the Tropic of Cancer; see Chapter 15.5.) In the "winter" or cooler season (October through April), the Sun is lower and the winds variable, with more rainfall.

Fronts

The boundary between two air masses is called a **front**. A *warm front* is the boundary of a warm air mass advancing over a colder surface, and a *cold front* is the boundary of a cold air mass moving over a warmer surface. These boundaries, called *frontal zones*, may vary in width from a few miles to more than 160 km (100 mi). Along fronts, which divide air masses of different physical characteristics, drastic changes in weather may occur. Turbulent weather and storms usually characterize a front.

The degree and rate of weather change depend on the difference in temperature of the air masses and on the degree of vertical slope of a front. A cold front moving into a warmer region causes the lighter (less dense), warm air to be displaced upward over the front. The lighter air of an advancing warm front cannot displace the heavier, colder air as readily, and generally it moves slowly up and over the colder air giving rise to different cloud formations (● Fig. 20.5).

In winter, this process sets the stage for a snowstorm. Moist air moving up over the cold front may form ice clouds that give rise to precipitation in the form of snow. Falling through the cold air with freezing temperatures at ground level, a snowstorm accumulation may occur, and with wind, conditions could worsen to a blizzard (Chapter 20.3).

Heavier, colder air is associated with high pressure, and downward divergent air flow in a high-pressure region generally gives a cold front greater speed than a warm front. A cold front may have an average speed of 30 to 40 km/h (20 to 25 mi/h), whereas a warm front averages about 15 to 25 km/h (10 to 15 mi/h).

Cold fronts have sharper vertical boundaries than warm fronts, and warm air is displaced upward more rapidly by an advancing cold front. As a result, cold fronts are accompanied by more violent or sudden changes in weather. The sudden decrease in temperature is often described as a "cold snap." Dark altocumulus clouds often mark a cold front's approach. The sudden cooling and the rising warm air may set off rainstorm or snowstorm activity along the front.

A warm front may be characterized by precipitation and storms. The more gradual slope of a warm front is usually heralded by a period of lowering clouds. Cirrus and mackerel

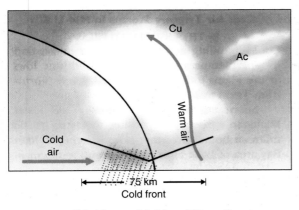 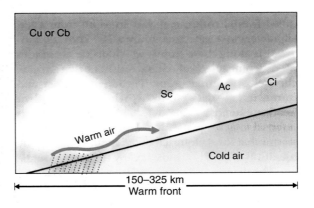

Figure 20.5 Side Views of Cold and Warm Fronts
Note in the left diagram the sharp, steep boundary that is characteristic of cold fronts. The boundary of a warm front, as shown in the right diagram, is less steep. As a result, different cloud types are associated with the approach of the two types of fronts. (See the Highlight: Cloud Families and Types in Chapter 19.5.)

scale (cirrocumulus) clouds drift ahead of the front, followed by alto clouds, and as the front approaches, cumulus or cumulonimbus clouds resulting from the rising air produce precipitation and storms. Most precipitation occurs before the front passes.

The graphical symbols for fronts are given in (● Fig. 20.6a). A cold front has sharp triangles and a warm front rounded ones. The side of the line with the symbol indicates the direction of advance.

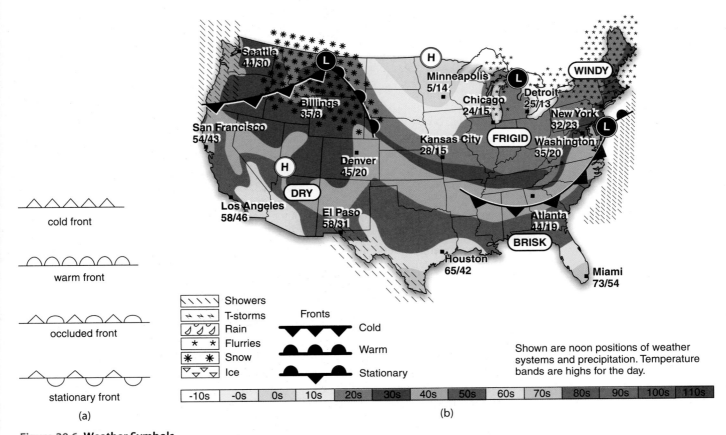

Figure 20.6 Weather Symbols
(a) Graphical symbols for fronts. The side of the line with the symbol indicates the direction of advance. (See text for description.) (b) Typical weather map for the conterminous United States showing fronts and other weather conditions.

As a faster-moving cold front advances, it may overtake a warm air mass and push it upward. The boundary between these two air masses is called an *occluded front* and is indicated by sharp and rounded symbols on the same side of the front line. The cold front occludes, or cuts off, the warm air from the ground along the occluded front. When a cold front advances under a warm front, a *cold front occlusion* results. When a warm front advances up and over a cold front, the air ahead is colder than the advancing air, and the occluded front is referred to as a *warm front occlusion*.

Sometimes fronts traveling in opposite directions meet. The opposing fronts may balance each other so that no movement occurs. This case is referred to as a *stationary front* and is indicated by sharp and rounded symbols on opposite sides of the front line. These symbols indicate fronts on a surface weather map (Fig. 20.6).

Air masses and fronts move across the country, bringing changes in weather. Dynamic situations give rise to cyclonic disturbances around low-pressure regions ("lows") and high-pressure regions ("highs"). As learned in Chapter 19.4, these disturbances are called *cyclones* and *anticyclones*, respectively (see Fig. 19.21).

As a "low" (cyclone) moves, it carries with it rising air currents, clouds, possibly precipitation, and generally bad weather. Hence, lows, or cyclones, are usually associated with poor weather, whereas highs, or anticyclones, are usually associated with good weather. The lack of rising air and cloud formation in highs gives clear skies and fair weather. The movements of highs and lows are closely monitored because of their influence on the weather.

Winds and ocean currents can also have major effects on regional weather. See the **Highlight: El Niño (the Little Boy) and La Niña (the Little Girl)**.

Did You Learn?

● Temperature and moisture content are the main distinguishing characteristics of air masses.

● A warm front is the boundary of a warm air mass advancing over a cold surface, and a cold front is the boundary of a cold air mass advancing over a warm surface.

20.3 Storms

Preview Questions

● What different types of lightning discharges are there, and which is the most dangerous?

● What is the most violent of storms? Is it the most energetic?

Storms are atmospheric disturbances that may develop locally within a single air mass or may be due to frontal activity along the boundary of air masses. Several types of storms, distinguished in terms of their intensity and violence, are discussed in the following sections. They are divided generally into local storms and tropical storms.

Local Storms

There are several types of local storms. A heavy downpour is commonly referred to as a *rainstorm*. Storms with rainfalls of 1 to 3 in./h are common.

A *thunderstorm* is a rainstorm distinguished by thunder and lightning, and sometimes hail. **Lightning** associated with a thunderstorm is a huge discharge of electrical energy. In the turmoil of a thundercloud, or "thunderhead," there is a separation of charge associated with the breaking up and movement of water droplets, which gives rise to an electric potential (Chapter 8.2). When that potential is of sufficient magnitude, a lightning discharge occurs.

Highlight El Niño (the Little Boy) and La Niña (the Little Girl)

A source region is usually thought of as being relatively hot or cold in terms of difference in latitude, but significant variations can occur within particular latitudes, giving rise to abnormal weather conditions. A good example is *El Niño*, an occasional disruption of the ocean–atmosphere system in the tropical Pacific Ocean, which has important weather consequences for many parts of the world. Originally recognized by fishermen off the Pacific coast of South America, the appearance at irregular intervals of unusually warm water near the beginning of the year was named El Niño, which means "the little boy" in Spanish or "the Christ child," reflecting the tendency of El Niño to arrive at Christmastime.

In normal conditions, the trade winds generally blow from east to west across the tropical Pacific (see Fig. 19.22). These winds pile up warm surface water in the western Pacific (Fig. 1a). The sea surface temperature is about 8°C higher in the west, with cool temperatures off South America because of an upwelling of cold water from deeper levels. The cold water is nutrient-rich and supports an abundance of fish and the fishing industry. The warm water in the western Pacific gives rise to convection cycles that produce wet weather for the countries in this region, whereas the Pacific coast of South America is relatively dry.

At irregular intervals, typically every 3 to 5 years, the normal trade winds relax and the warm pool of water in the western Pacific is free to move back eastward toward the South American continent (Fig. 1b). The shift of convection and precipitation to the central and eastern Pacific usually results in heavier-than-normal rainfall for the west coast of South America, particularly Ecuador, Peru, and Chile. This shift means that the mechanism for precipitation for Indonesia and Australia is no longer available over those regions, and the area often experiences drought conditions during an El Niño.

More than 20 El Niños have occurred since 1900. The cause of this relatively frequent event, which lasts about 18 months, is not known. The 1982–1983 El Niño was one of the worst, accounting for the loss of 2000 lives and the displacement of thousands of people from their homes. Indonesia and Australia suffered droughts and bush fires. Peru was hit with heavy rainfall, 11 ft in some areas where 6 in. is normal.

El Niño has a twin sister, La Niña ("the little girl"). The effects of La Niña tend to be nearly opposite of El Niño. For instance, the eastern Pacific surface waters are colder than normal and extend farther westward than usual. Strong La Niñas cause unusual weather conditions around the United States. For example, winter temperatures are usually warmer than normal in the Southeast and cooler than normal in the Northwest, and precipitation is below normal in California and the Southeast. Also, La Niña conditions are thought to have contributed to the huge 2011 tornado outbreak in the United States (Ch. 20.3). The cycle from El Niño conditions to La Niña takes about 4 years.

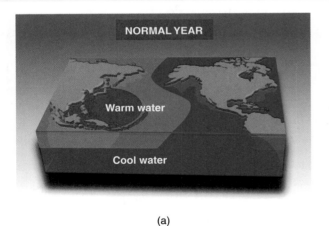

(a)

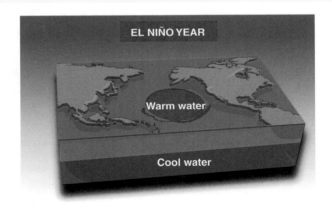

(b)

Figure 1 El Niño
(a) Normally, east-to-west winds pile up warm water in the western Pacific, while cold water from deep in the ocean rises to the surface along the South American coast. (b) Every few years the trade winds change, allowing the pool of warm water to move to the east where it blocks the rising cold water. The changes help trigger global weather patterns associated with El Niño.

Lightning can occur entirely within a cloud (intracloud or cloud discharges), between two clouds (cloud-to-cloud discharges), between a cloud and the Earth (cloud-to-ground or ground discharges), or between a cloud and the surrounding air (air discharges). See ● Fig. 20.7. People are killed every year by lightning strikes. Fortunately, about 80% of the discharges occur within the atmosphere and never strike the Earth. The 20% cloud-to-ground discharges are the most hazardous.

Jhaz Photography/Shutterstock.com

Figure 20.7 Lightning
Lightning discharges can occur between a cloud and the Earth, between clouds, and within a cloud.

Lightning can be seen in clear skies, giving rise to the expression "a bolt from the blue." A bolt can on occasion jump 10 or more miles out from the parent storm cloud and strike in a region with blue skies overhead. When lightning occurs below the horizon or behind clouds, it often illuminates the clouds with flickering flashes. Such lightning commonly occurs on a still summer night and is known as *heat lightning*.

Although the most frequently occurring form of lightning is the intracloud discharge, of greatest concern is lightning between a cloud and the Earth. The shorter the distance from a cloud to the ground, the more easily the electric discharge takes place. For this reason, lightning often strikes trees and tall buildings. It is very inadvisable to take shelter in a thunderstorm under a tree. If lightning strikes and heats the sap, then the tree could explode.

A person in the vicinity of a lightning strike may experience an electric shock that causes breathing to fail. In such a case, breathing resuscitation should be given immediately, and the person should be kept warm as a treatment for shock. (See the Lightning Safety box.)

Lightning Safety

If you are outside during a thunderstorm and feel an electric charge, as evidenced by your hair standing on end or skin tingling, then what should you do? *Fall to the ground fast!* Lightning may be about to strike.

Statistics show that lightning kills, on average, 100 people a year in the United States and injures another 300. Most deaths and injuries occur at home. Indoor casualties occur most frequently when people are talking on the telephone, working on computers, working in the kitchen, doing laundry, or watching TV. During severe lightning activity, the following safety rules are recommended:

Stay indoors away from open windows, fireplaces, and electrical conductors such as sinks and stoves.

Avoid using the telephone. Lightning may strike the telephone lines outside.

Do not use electrical plug-in equipment such as computers, radios, TVs, and lamps. (It is good to have a voltage surge protector on electronic equipment for protection from damage that may result from a lightning voltage surge coming in on a line.)

Should you be caught outside, seek shelter in a building. If no buildings are available, then seek protection in a ditch or ravine. Getting wet is a lot better than being struck by lightning.

A lightning stroke's sudden release of energy explosively heats the air to about 30,000°C, producing sound compressions that are heard as **thunder**. When heard at a distance of about 100 m (330 ft) or less from the discharge channel, thunder consists of one loud bang, or "clap." When heard at a distance of 1 km (0.62 mi) from the discharge channel, thunder generally consists of a rumbling sound punctuated by several large claps. In general, thunder cannot be heard at distances of more than 25 km (16 mi) from the discharge channel.

Conceptual Question and Answer

What a Thundersnow!

Q. Is it possible to have lightning and thunder in a snowstorm?

A. Yes, a *thunder snowstorm*, or *thundersnow*, is a rare kind of thunderstorm with snow falling as precipitation instead of rain or, with extreme cold, ice pellets.

The ingredients for thundersnow are a mass of cold air on top of warm and moist air close to the ground. Although rare, thundersnows are most common with lake-effect snow in the Great Lakes region, the midwestern United States, and the Great Salt Lake in Utah. In the United States, only about 6 thundersnows are reported per year on average, with March being the dominant month.

An interesting aspect of the lightning-induced thunder of a thundersnow is that the snow fall acts as an acoustic suppressor of the thunder. Although thunder from a typical thunderstorm may be heard many miles away, the thunder from a thundersnow can be heard only within 2 to 3 miles, so the effects are somewhat localized.

Because lightning strokes generally occur near the storm center, the resulting thunder provides a method of approximating the distance to the storm. Light travels at about 300,000 km/s (186,000 mi/s), so the lightning flash is seen instantaneously. Sound, however, travels at approximately $\frac{1}{3}$ km/s ($\frac{1}{5}$ mi/s), so a time lapse occurs between an observer seeing the lightning flash and hearing the thunder. By counting the seconds between seeing the lightning and hearing the thunder (by saying, "one-thousand-one, one-thousand-two," and so on), you can estimate your distance from the lightning stroke or the storm.

EXAMPLE 20.1   Estimating the Distance of a Thunderstorm

Suppose some campers notice an approaching thunderstorm in the distance. Lightning is seen, and the thunder is heard 5.0 s later. Approximately how far away is the storm center in (a) kilometers and (b) miles?

Solution

The approximate or average speed of sound is $\bar{v} = \frac{1}{3}$ km/s $= \frac{1}{5}$ mi/s. Then the distance (d) that the sound travels in a time (t) is given by Eq. 2.1 ($d = \bar{v}t$).

(a) Using the metric speed,
$$d = \bar{v}t = \left(\tfrac{1}{3} \text{ km/s}\right)(5.0 \text{ s}) = 1.6 \text{ km}$$

(b) You could convert the distance in (a) to miles (and, in fact, you may recognize the conversion right away), but let's compute it.
$$d = \bar{v}t = \left(\tfrac{1}{5} \text{ mi/s}\right)(5.0 \text{ s}) = 1.0 \text{ mi}$$

(Recall that 1 mi = 1.6 km.)

Confidence Exercise 20.1

If thunder is heard 3.0 s after a flash of lightning is seen, then approximately how far away, in kilometers and miles, is the lightning?

The answers to Confidence Exercises may be found at the back of the book.

If the temperature of the Earth's surface is below 0°C (32°F) and raindrops do not freeze before striking the ground, then the rain will freeze when it strikes cold surface objects. Such an **ice storm** builds up a layer of ice on objects exposed to the freezing rain. The ice layer may build up to more than half an inch in thickness, depending on the magnitude of the rainfall. Viewed in sunlight, this glaze produces beautiful winter scenes as the ice-coated landscape glistens in the Sun. However, damage to trees and power lines often detracts from the beauty.

Snow is made of ice crystals that fall from ice clouds. A **snowstorm** is an appreciable accumulation of snow. What may be considered a severe snowstorm in some regions may be thought of as a light snowfall in areas where snow is more prevalent. When a snowstorm is accompanied by high winds and low temperatures, the storm is referred to as a *blizzard*. The

Highlight Wind Chill Temperature Index

In 2001, the National Weather Service implemented an updated Wind Chill Temperature (WCT) Index. The update improved on the WCT Index developed in 1945.

The WCT, or, more commonly, the "wind chill," is how cold people *feel* when they are outside. Wind chill is based on the rate of heat loss from exposed skin caused by wind and cold (Fig. 1).

As the wind increases, more heat is drawn from the body, decreasing the skin temperature and eventually the internal body temperature. Therefore, wind can make it *feel* much colder than it actually is. If the temperature is 0°F and the wind is blowing 15 mi/h, then the wind chill is −19°F (see Table 1). At this wind chill temperature, exposed skin can freeze (frostbite occurs) in 30 minutes.

During severe cold spells in the winter, the wind chill can be as low as −40° (take your choice of Fahrenheit or Celsius; at that particular temperature, the Fahrenheit and Celsius scale readings are the same as discussed in Chapter 5.1).

Figure 1 Wind Chill

The wind chill temperature reflects how cold people *feel* when they are outside under specific conditions of temperature and wind speed. It is based on the rate at which wind and cold remove heat from exposed skin.

Julien/Shutterstock.com

Table 1 Wind Chill Temperature Index

Find the temperature in the top row and the wind speed in the left column. Their intersection shows the wind chill. For example, when the temperature is 0°F and the wind speed is 15 mi/h, the wind chill is −19°F.

Wind Chill Chart

Temperature (°F)

Wind (mph)	40	35	30	25	20	15	10	5	0	−5	−10	−15	−20	−25	−30	−35	−40	−45
5	36	31	25	19	13	7	1	−5	−11	−16	−22	−28	−34	−40	−46	−52	−57	−63
10	34	27	21	15	9	3	−4	−10	−16	−22	−28	−35	−41	−47	−53	−59	−66	−72
15	32	25	19	13	6	0	−7	−13	−19	−26	−32	−39	−45	−51	−58	−64	−71	−77
20	30	24	17	11	4	−2	−9	−15	−22	−29	−35	−42	−48	−55	−61	−68	−74	−81
25	29	23	16	9	3	−4	−11	−17	−24	−31	−37	−44	−51	−58	−64	−71	−78	−84
30	28	22	15	8	1	−5	−12	−19	−26	−33	−39	−46	−53	−60	−67	−73	−80	−87
35	28	21	14	7	0	−7	−14	−21	−27	−34	−41	−48	−55	−62	−69	−76	−82	−89
40	27	20	13	6	−1	−8	−15	−22	−29	−36	−43	−50	−57	−64	−71	−78	−84	−91
45	26	19	12	5	−2	−9	−16	−23	−30	−37	−44	−51	−58	−65	−72	−79	−86	−93
50	26	19	12	4	−3	−10	−17	−24	−31	−38	−45	−52	−60	−67	−74	−81	−88	−95
55	25	18	11	4	−3	−11	−18	−25	−32	−39	−46	−54	−61	−68	−75	−82	−89	−97
60	25	17	10	3	−4	−11	−19	−26	−33	−40	−48	−55	−62	−69	−76	−84	−91	−98

Frostbite times: ☐ 30 minutes ☐ 10 minutes ☐ 5 minutes

winds whip the fallen snow into blinding swirls. Visibility may be reduced to a few inches. For this reason, a blizzard is often called a *blinding* snowstorm, and with violent gusts of wind, it is called a snow *squall*. (Rain squalls also occur, particularly on large bodies of water.)

Swirling snow may cause disorientation, and people have gotten lost only a few feet from their homes. The wind may blow the snow across level terrain, forming huge drifts against some obstructing object. Drifting is common on the flat prairies of the western United States.

Even without snow, wind affects how cold you feel. The combined effect of wind and cold is described by wind chill temperatures. (See the **Highlight: Wind Chill Temperature Index**.)

Conceptual Question and Answer

Snowy Cold

Q. Does it ever get too cold to snow?

A. You might have heard someone say, "It is too cold to snow," but that statement is somewhat inaccurate. Cold clouds contain tiny ice or "snow" crystals. A snowflake is an aggregate of snow crystals. Snowflakes near freezing will be relatively large, and accumulations can occur. Below 20°F (-7°C), the formation of snowflake aggregation decreases with temperature and the snowflakes become smaller. Below around -20°F (-29°C), snowflakes do not form, but snow may be in the form of ice or snow crystals, which are sometimes called "diamond dust" from sparkling in the sunlight. So, with moisture present, it isn't too cold to snow but too cold for snowflakes to form.

Severe rainstorms and snowstorms can result from atmospheric rivers. An *atmospheric river* is a stream of concentrated moisture in the atmosphere. These rivers are typically several thousand miles long and a few hundred miles wide. One river can carry more water than the Mississippi River. Over 90% of the global north–south water transport depends on atmospheric rivers. Their existence has only been known for about a decade.

When strong atmospheric rivers strike land, they produce flooding rains that create mud slides and cause catastrophic damage, sometimes with the loss of life. When these rivers move up mountain slopes, heavy snows can result.

A well-known example of a strong atmospheric river is the "Pineapple Express," which every few years brings moisture from the tropics near Hawaii to the West Coast of the United States. It is believed to be driven by a southern branch of the polar jet stream. In December 2010, a Pineapple Express system ravaged southern California with as much as 2 ft of rain in some locations and more than 5 ft of snow in the Sierra Nevada.

The **tornado** is the most violent of storms. Although a tornado may have less *total* energy than some other storms (for example, hurricanes), the concentration of its energy in a relatively small region gives the tornado great destructive potential. Characterized by a whirling, funnel-shaped cloud that hangs from a dark cloud mass, the tornado is commonly referred to as a *twister* (● Fig. 20.8).

(a)

(b)

Figure 20.8 Tornado Destruction
(a) A tornado funnel touching down with a town in its path. (b) The high-speed winds of a tornado can push in the windward wall of a house, lift off the roof, push the other walls outward, and mangle structures, as shown here. Imagine all the debris flying around during a tornado.

Alan R. Moller/Getty Images

AP Photo/Dave Martin

Table 20.2 Enhanced Fujita Tornado Scale (EF-Scale)		
EF-Scale Number	Wind Speed	Effects
EF0	65–85 mi/h (105–137 km/h)	Some damage to roofs and chimneys, branches broken off trees, shallow-rooted trees pushed over, sign boards damaged. Moderate damage.
EF1	86–110 mi/h (138–178 km/h)	Roofs severely stripped, mobile homes over-turned or badly damaged, windows and other glass broken. Considerable damage.
EF2	111–135 mi/h (179–218 km/h)	Roofs torn off frame houses, mobile homes demolished, large trees snapped or uprooted, light-object missiles generated, cars lifted off ground. Severe damage.
EF3	136–165 mi/h (219–266 km/h)	Entire stories of well-constructed houses destroyed, trains overturned, trees debarked, heavy cars lifted off the ground and thrown. Devastating damage.
EF4	166–200 mi/h (267–322 km/h)	Well-constructed houses and whole frame houses completely leveled, cars thrown and small missiles generated. Total destruction.
EF5	> 200 mi/h (> 322 km/h)	Strong frame houses lifted off foundations and swept away, automobile-sized missiles fly through the air in excess of 300 ft (100 m), steel-reinforced concrete structures badly damaged. Incredible phenomena will occur.

As shown in Fig. 20.8b, tornadoes can be very damaging. Tornadoes are classified using the Enhanced Fujita Tornado Scale, or EF-Scale.* The EF-Scale uses a set of wind estimates (not measurements) on the amount of damage caused and classifies twisters into six categories of estimated wind speed. (See ● Table 20.2.)

Tornadoes occur around the world, with the United States having the highest average annual number. Outside the United States, Australia ranks second in tornado frequency. In an average year, approximately 1000 tornadoes are reported across the United States. Most are small, but others are devastating, resulting annually in an average of 180 deaths, more than 1500 injuries, and millions of dollars of damage. Damage paths can be in excess of 1 mile wide and 50 miles long. The average tornado path is about 5 miles. Tornadoes have been spotted in every state of the conterminous United States, but most tornadoes occur in the Midwest, in the South, and in the broad, relatively flat basin between the Rockies and the Appalachians, known as "Tornado Alley" (● Fig. 20.9). The tornado season is generally considered to be March through August, and the months of peak tornado activity are April, May, and June. The southern states are usually hit hardest in the spring, the northern states in the summer. A typical time of occurrence is between 3:00 and 7:00 P.M. on an unseasonably warm, sultry spring afternoon, but tornadoes have occurred in every month and at all times of day and night.

Most tornadoes travel from southwest to northeast, but the direction of travel can be erratic and may change suddenly. Tornadoes usually travel at an average speed of 48 km/h (30 mi/h). The wind speed of a major tornado may vary from 160 to 480 km/h (from

*The original Fujita Tornado Scale, or F-Scale, was introduced in 1971 by Professor T. Theodore Fujita of the University of Chicago. Researchers have since developed an Enhanced Fujita Scale, or EF-Scale, that better reflects a tornado's strength. The EF-Scale came into use in 2007.

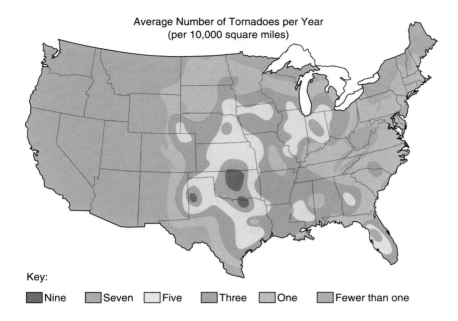

Average Number of Tornadoes per Year
(per 10,000 square miles)

Key:

■ Nine ■ Seven □ Five ■ Three □ One ■ Fewer than one

Figure 20.9 Average Number of Tornadoes per Year
Note that the most tornadoes occur in the flat, Midwest basin between the Rockies and the Appalachians (known as Tornado Alley) and in the southern states.

100 to 300 mi/h). The wind speed of a devastating 1999 Oklahoma tornado was measured by Doppler radar to be 502 km/h (312 mi/h), the highest ever recorded.

Because of many variables, the complete mechanism of tornado formation is not known. One essential component, however, is rising air, which occurs in thunderstorm formation and in the collision of cold and warm air masses. As the ascending air cools, clouds are formed that are swept to the outer portions of the cyclonic motion and outline its funnel form. Because clouds form at certain heights (Chapter 19.5), the outlined funnel may appear well above the ground. Under the right conditions, a full-fledged tornado develops. Winds increase, and the air pressure near the center of the vortex is reduced as the air swirls upward. When the funnel is well developed, it may "touch down," or be seen extending up from the ground, as a result of dust and debris picked up by the swirling winds.

At times, there is a rush or "outbreak" of tornadoes. The spring outbreak of 2011 was one of the largest. During May and June, more than 300 tornadoes were reported in the southern and midwestern United States, with particularly devastating ones in Alabama, Mississippi, and Missouri. Many were EF4 tornadoes, and some were EF5, on the Fujita Scale (Table 20.2). There were thousands of injuries, and hundreds lost their lives. Damages were in the billions of dollars.

The system for a tornado alert has two phases. A **tornado watch** is issued when atmospheric conditions indicate that tornadoes may form. A **tornado warning** is issued when a tornado has actually been sighted or indicated on radar. The similarity between the terms *watch* and *warning* can be confusing. Remember that you should *watch* for a tornado when the conditions are right, but when you are given a *warning*, the situation is dangerous and critical; no more watching.

Care should be taken after a tornado has passed. There may be downed power lines, escaping natural gas, and dangerous debris. (See the Tornado Safety box.)

Tornado Safety

Knowing what to do in the event of a tornado is critically important. If a tornado is sighted, if the ominous roar of one is heard at night, or if a tornado warning is issued for your particular locality, then *seek shelter fast*!

The basement of a home or building is one of the safest places to seek shelter.

Avoid chimneys and windows, because there is great danger from debris and flying glass.

Get under a sturdy piece of furniture, such as an overturned couch, or into a stairwell or closet, and *cover your head*.

In a house or building without a basement, seek the lowest level in the central portion of the structure and the shelter of a closet or hallway.

If you live in a mobile home, then evacuate it. Seek shelter elsewhere.

Tropical Storms

The term *tropical storm* (or *tropical cyclone*) refers to the massive disturbances that form over tropical oceanic regions. A tropical storm becomes a **hurricane** when its wind speed reaches 119 km/h (74 mi/h or 64 knots).

The hurricane is known by different names in different parts of the world. For example, in Southeast Asia it is called a *typhoon*; in the Indian Ocean, a *cyclone*; and in Australia, a *willy-willy* (● Fig. 20.10a). The only places on the Earth not affected by hurricanes are Antarctica and latitudes near the equator. As Fig. 20.10b shows, the three most active hurricane months are August, September, and October.

Figure 20.10 Tropical Storm Regions of the World and U.S. Hurricanes
(a) Tropical storms are known by different names in different parts of the world. The arrows show the general paths. (b) The average number of hurricanes by month and the number making landfall in the United States for the period 1851–2004. August, September, and October are the most active months.

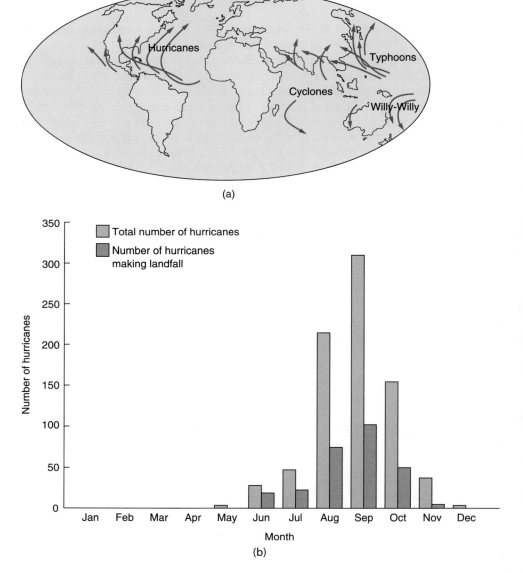

(a)

(b)

Regardless of the name, this type of storm is characterized by high-speed rotating winds, with energy spread over a large area. A hurricane may be 480 to 960 km (300 to 600 mi) in diameter and may have wind speeds of 119 to 320 km/h (74 to 200 mi/h).

Hurricanes form over tropical oceanic regions where the Sun heats huge masses of moist air, and an ascending spiral of rising air results. When the moisture of the rising air condenses, the latent heat provides additional energy and more air rises up the column. This latent heat is the chief source of a hurricane's energy and is readily available from the condensation of the evaporated moisture from its source region.

Unlike a tornado, a hurricane gains energy from its source region. As more and more air rises, the hurricane grows, with clouds and increasing winds that blow in a large spiral around a relatively calm low-pressure center, called the *eye* of the hurricane (● Fig. 20.11). The eye may be 32 to 48 km (20 to 30 mi) wide, and ships sailing into this area have found that it is usually calm and clear, with no indication of the surrounding storm. The air pressure is generally reduced 6% to 8% near the eye. Hurricanes move rather slowly, at only a few kilometers or miles per hour.

(a)

Figure 20.11 The Eye of a Hurricane (a) Hurricane Katrina bears down on the coasts of Louisiana and Mississippi in 2005. The eye is clearly visible. (b) A radar profile of a hurricane. Note the generally clear, low-pressure eye and the vertical buildup of clouds near the sides.

NASA/Goddard Space Flight Center Scientific Visualization Studio

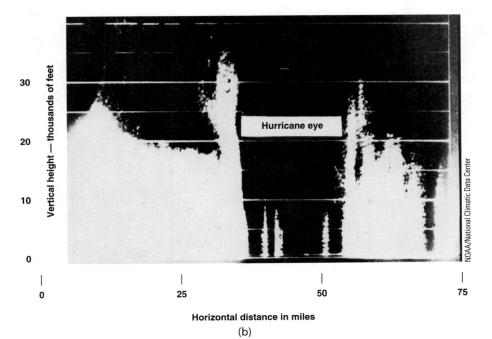

(b)

NOAA/National Climatic Data Center

Although the tornado is labeled the most violent of storms, the hurricane is the most energetic. One method for estimating the energy production of a hurricane uses the amount of rain produced within an average hurricane per day and the latent heat of condensation released (Chapter 5.3). This method gives an energy production of 6.0×10^{11} kWh per day, which is equivalent to about 200 times the worldwide electrical generating capacity. That's an incredible amount of energy!

With such tremendous amounts of energy, hurricanes can be particularly destructive. Hurricane winds do much damage, but drowning is the greatest cause of hurricane deaths. As the eye of a hurricane comes ashore, or *makes landfall*, a great dome of water called a **storm surge**, often more than 80 km (50 mi) wide, comes sweeping across the coastline. It brings huge waves that may reach more than 5 m (17 ft) above normal (● Fig. 20.12). Should this surge occur at high tide, it can be even higher. The storm surge comes suddenly, often flooding coastal lowlands. Nine out of ten casualties are caused by the storm surge. The torrential rains that accompany the hurricane commonly produce flooding as the storm moves inland. As its winds diminish, floods constitute a hurricane's greatest threat.

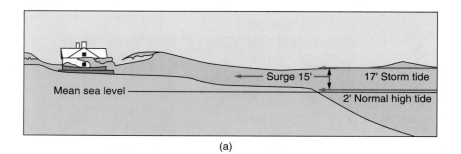

(a)

Figure 20.12 Hurricane Storm Surge and Damage
(a) The drawing illustrates a storm surge coming ashore at high tide.
(b) A close-up of hurricane damage.

Table 20.3   Saffir-Simpson Hurricane Scale

Saffir-Simpson Category	Maximum Sustained Wind Speed	Minimum Surface Pressure	Storm Surge (Height)	
	mi/h	mb	ft	m
1	74–95	Greater than 980	3–5	1.0–1.7
2	96–110	979–965	6–8	1.8–2.6
3	111–130	964–945	9–12	2.7–3.8
4	131–155	944–920	13–18	3.9–5.6
5	156+	Less than 920	19+	5.7+

Category	Winds	Effects
1	74–95 mi/h	No real damage to building structures. Damage primarily to unanchored mobile homes, shrubbery, and trees. Some coastal road flooding and minor pier damage.
2	96–110 mi/h	Some roofing material, door, and window damage to buildings. Considerable damage to vegetation, mobile homes, and piers. Coastal and low-lying escape routes flood 2–4 hours before arrival of center. Small craft in unprotected anchorages break moorings.
3	111–130 mi/h	Some structural damage to small residences and utility buildings, with a minor amount of curtainwall (an exterior wall having no structural function) failures. Mobile homes are destroyed. Flooding near the coast destroys smaller structures, and larger structures are damaged by floating debris. Terrain continuously lower than 5 ft above sea level (ASL) may be flooded inland 8 mi or more.
4	131–155 mi/h	More extensive curtainwall failures, with some complete roof structure failure on small residences. Major erosion of beach. Major damage to lower floors of structures near the shore. Terrain continuously lower than 10 ft ASL may be flooded, requiring massive evacuation of residential areas inland as far as 6 mi.
5	greater than 155 mi/h	Complete roof failure on many residences and industrial buildings. Some complete building failures, with small utility buildings blown over or away. Major damage to lower floors of all structures located less than 15 ft ASL and within 500 yd of the shoreline. Massive evacuation of residential areas on low ground within 5–10 mi of the shoreline may be required.

Hurricanes are classified into categories, such as a Category 2 hurricane or a Category 4 hurricane. These categories come from the Saffir-Simpson Hurricane Scale (● Table 20.3).*
Note the maximum speed, minimum pressure, and height of storm surge for each category.

Once cut off from the warm ocean, the storm dies, starved for moisture and heat energy and dragged apart by friction as it moves over the land. Even though a hurricane weakens rapidly as it moves inland, the remnants of the storm can bring 6 to 12 in. or more of rain for hundreds of miles.

The breeding grounds of the hurricanes that generally affect the United States are in the warm waters off the west coast of Africa. As hurricanes form, they move westward with the

---

*The Saffir-Simpson Hurricane Scale was developed in 1969 by Herbert Saffir, a consulting engineer, and Robert Simpson of the National Hurricane Center.

Highlight   Naming Hurricanes

For several hundred years, many hurricanes in the West Indies were named after the saint's day on which they occurred. In 1953, the National Weather Service began to use women's names for tropical storms and hurricanes. This practice had begun in World War II when military personnel named typhoons in the western Pacific after their wives and girlfriends, using alphabetical order. The first hurricane of the season received a name beginning with A, the second one was given a name beginning with B, and so on.

Table 1 The 6-Year List of Names for Atlantic, Gulf of Mexico, and Caribbean Sea Storms*

2010	2011	2012	2013	2014	2015
Alex	Arlene	Alberto	Andrea	Arthur	Ana
Bonnie	Bret	Beryl	Barry	Bertha	Bill
Colin	Cindy	Chris	Chantal	Cristobal	Claudette
Danielle	Don	Debby	Dorian	Dolly	Danny
Earl	Emily	Ernesto	Erin	Edouardo	Erika
Fiona	Franklin	Florence	Fernand	Fay	Fred
Gaston	Gert	Gordon	Gabrielle	Gonzalo	Grace
Hermine	Harvey	Helena	Humberto	Hanna	Henri
Igor	Irene	Isaac	Ingrid	Isaias	Ida
Julia	Jose	Joyce	Jerry	Josephine	Joaquin
Karl	Katia	Kirk	Karen	Kyle	Kate
Lisa	Lee	Leslie	Lorenzo	Laura	Larry
Matthew	Maria	Michael	Melissa	Marco	Mindy
Nicole	Nate	Nadine	Nestor	Nana	Nicholas
Otto	Ophelia	Oscar	Olga	Omar	Odette
Paula	Philippe	Patty	Pablo	Paulette	Peter
Richard	Rina	Rafael	Rebekah	Rene	Rose
Shary	Sean	Sandy	Sebastien	Sally	Sam
Tomas	Tammy	Tony	Tanya	Teddy	Teresa
Virginie	Vince	Valerie	Van	Vicky	Victor
Walter	Whitney	William	Wendy	Wilfred	Wanda

*The 2016 names will be the same as the list for 2010 and so on.

trade winds (Chapter 19.4), usually making landfall in the United States along the Gulf Coast and the Atlantic Coast of the southern United States. During the hurricane season, the area of their formation is constantly monitored by satellite. When a tropical storm is detected and becomes a hurricane, radar-equipped airplanes, or "hurricane hunters," track the storm and make local measurements to help predict its path. The hurricane season in the Atlantic Ocean is officially from June 1 to November 30. During these 6 months, 97% of the hurricanes usually occur.

As for tornadoes, the hurricane alerting system has two phases. A **hurricane watch** is issued for coastal areas when there is a threat of hurricane conditions within 24 to 36 hours. A **hurricane warning** indicates that hurricane conditions are expected within 24 hours (winds of 74 mi/h or greater, or dangerously high water and rough seas). A memory device to differentiate between a hurricane watch and a hurricane warning is similar to that for tornadoes. A *watch* is an alert to watch out for a possible coming hurricane; a *warning* is

Highlight

The practice of naming hurricanes solely after women was changed in 1979 when men's names were first included in the lists. A 6-year list of names for Atlantic storms is given in Table 1. A similar list is available for Pacific storms. Names beginning with the letters Q, U, X, Y, and Z are excluded because of their scarcity. The lists are recycled every 6 years. For example, the 2012 list is used again in 2018.

A record-setting year for hurricanes was 2005. In that year, there were 28 named tropical storms, 15 of which became hurricanes; of those, four were Category 5 hurricanes, and four were intense hurricanes making landfall in the United States.

Hurricane Katrina in 2005 was one of the most deadly and costly hurricanes in the United States. Its storm surge breached the levees that protect New Orleans and flooded the city (Fig. 1). Some 1300 lives were lost. The damage caused by Katrina is esti-mated to have been in the tens of billions of dollars. In addition to the personal and public destruction, harm done to offshore oil rigs and coastal oil refineries damaged the U.S. economy. Hurricane damage has increased over the years due to more industry developing and more people living along the coastline.

When a hurricane is particularly destructive, its name is retired from the list so as not to cause confusion with another storm of the same name in the next 6-year cycle. For example, the name Andrew, which had been applied to a destructive 1992 storm that hit Florida, was replaced with Alex. Alex has behaved quite nicely, surviving two 6-year cycles, and is listed again for 2010 (Table 1). Since 1954, approximately 70 hurricane names have been retired. For 2005, a record five names were retired: Dennis, Katrina, Rita, Wilma, and Stan. These names were all replaced in the 2011 list.

(a)

(b)

Figure 1   Hurricane Katrina Damage
(a) In New Orleans, 53 levees of the flood protection system were breached and 80% of the city was flooded. Almost half the city lies below sea level. (b) Displaced people sought shelter in the Superdome. The waterproof membrane of its roof was essentially peeled off by the storm.

notice that a hurricane is imminent and that you should batten down the hatches or evacuate (head for the hills or higher ground; no more watching).

As you are probably aware, tropical storms and hurricanes are given names. How that is done is explained and an account of a very active hurricane year is given in the **Highlight: Naming Hurricanes**.

Did You Learn?

- Lightning can occur within a cloud (intracloud discharge), between two clouds (cloud-to-cloud discharge), between a cloud and the ground (cloud-to-ground discharge), or between a cloud and the surrounding air (air discharge). A cloud-to-ground discharge is the most dangerous for people and property.

- The tornado is the most violent of storms because of its high energy density (more concentrated energy), whereas the hurricane is the most energetic.

Conceptual Question and Answer

There She Blows

Q. A tropical storm becomes a hurricane when its wind speed reaches 74 mi/h (119 km/h, or 64 knots). Why 74 mi/h?

A. In the early 1800s, Commander Francis Beaufort of the British Royal Navy devised a descriptive wind scale based on the state and behavior of a "well-conditioned man-of-war [sailing ship]." This was a numbered "force" scale with no mention of wind speed. There were 12 forces ranging from 0 to 12—Force 0 being calm up to Force 12, a hurricane.

With the development of accurate anemometers, which measure wind speeds (see Chapter 19.3), wind speed ranges were assigned to the force numbers. For example, Force 4 is 13–19 mi/h (moderate breeze); Force 10 is 56–64 mi/h (storm); Force 11 is 65–73 mi/h (violent storm); and Force 12 is 74+ mi/h (hurricane). Thus, this somewhat arbitrary wind speed of 74 mi/h was taken to be the initial wind speed of a hurricane.

20.4   Atmospheric Pollution

Preview Questions

- What is a temperature inversion, and what problem can it cause?
- What is pollution, and what is the major source of atmospheric pollution?

At the beginning of Chapter 19, the quotation by William Shakespeare describes the atmosphere as "this most excellent canopy, the air." This quotation was taken out of context; in full, it reads: "this most excellent canopy, the air, look you, this brave o'erhanging firmament, this majestical roof fretted with golden fire, why, it appears no other thing to me but a foul and pestilent congregation of vapours" (*Hamlet*, act 2, scene 2).

Even Shakespeare in the early seventeenth century made reference to atmospheric pollution, an unfortunate topic of earth science. **Pollution** is any atypical contribution to the environment resulting from the activities of humans. Of course, gases and particulate matter are spewed into the air from volcanic eruptions and lightning-initiated forest fires, but those are natural phenomena over which there is little or no control.

Air pollution results primarily from the products of combustion and industrial processes that are released into the atmosphere. It has been a common practice to vent these wastes, and the resulting problems are not new, particularly in areas of high human population.

Smoke and soot from the burning of coal plagued England more than 700 years ago. London recorded air pollution problems in the late 1200s, and the use of particularly smoky types of coal was taxed and even banned, but the problem was not alleviated. In the middle 1600s, King Charles II commissioned one of the outstanding scholars of the day, Sir John Evelyn, to make a study of the situation. The degree of London's air pollution at that time is described in the following passage from his report, *Fumifugium* (translated from the Latin as *On Dispelling of Smoke*):

> . . . the inhabitants breathe nothing but impure thick mist, accompanied with a fuliginous and filthy vapor, corrupting the lungs. Coughs and consumption rage more in this one city (London) than in the whole world. When in all other places the [air] is most serene and pure, it is here eclipsed with such a cloud . . . as the sun itself is hardly able to penetrate. The traveler, at miles distance, sooner smells than sees the City.

However, the Industrial Revolution was about to begin, and Sir John's report was ignored and left to gather dust (and soot).

As a result of such air pollution, London has experienced several disasters involving the loss of life. Thick fogs are quite common in this island nation, and the combination of smoke and fog forms a particularly noxious mixture known as **smog**, the term itself a contraction of *smoke* and *fog*.

The presence of fog indicates that the temperature of the air near the ground is at the dew point, and with the release of latent heat, there is the possibility of a *temperature inversion*. As learned in Chapter 19.1, the atmospheric temperature decreases with increasing altitude in the troposphere (with a lapse rate of about $6\frac{1}{2}°C/km$). Hot combustive gases generally rise. However, under certain conditions, such as rapid radiational cooling near the ground surface, the temperature may locally *increase* with increasing altitude. The lapse rate is then said to be *inverted*, giving rise to a **temperature inversion** (● Fig. 20.13).

Most common are radiation and subsidence inversions. A *radiation temperature inversion*, which is associated with the Earth's radiational heat loss, may occur daily. The ground is heated by insolation during the day, and at night it cools by radiating heat back into the atmosphere (Chapter 19.2). If it is a clear night, then the land surface and the nearby air cool quickly. The air some distance above the surface, however, remains relatively warm, thus giving rise to a temperature inversion. *Radiation fogs* provide common evidence of this cooling effect in valleys. [See Fig (i) in Chapter 19 Highlight: Cloud Families and Types.]

A *subsidence temperature inversion* occurs when a high-pressure air mass moves over a region and becomes stationary. As the dense air settles, it becomes compressed and heated. If the temperature of the descending air exceeds that of the air below it, then the lapse rate is inverted, similar to that shown in Fig. 20.13.

With a temperature inversion, emitted gases and smoke cannot rise and are held near the ground. Continued combustion causes the air to become polluted, creating particularly hazardous conditions for people with heart and lung ailments. Smog episodes in various parts of the world have contributed to numerous deaths (● Fig. 20.14).

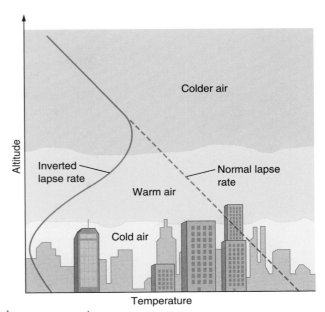

Figure 20.13 Temperature Inversion Normally, the lapse rate near the Earth's surface decreases uniformly with increasing altitude. However, radiational cooling of the ground can cause the lapse rate to become inverted, and the temperature increases with increasing altitude (usually below 1 mi). A similar condition may come about as a result of the subsidence of a high-pressure air mass.

Standard Oil Co./Carnegie Library, Pittsburgh

Figure 20.14 Smog Episode Waiting to Happen Shown here is a picture of a 1940s steel mill on the Monongahela River near Pittsburgh, Pennsylvania. A similar scene in Donora, 20 miles down the river from Pittsburgh, gave rise to a 5-day smog episode in which hundreds of people became ill and at least 20 died.

The major source of air pollution is the combustion of *fossil fuels*: coal, gas, and oil (petroleum). Technically, combustion (burning) is the chemical combination of certain substances with oxygen (Chapter 13.2). Fossil fuels are the remains of plant and animal life and are composed chiefly of hydrocarbons (compounds of hydrogen and carbon).

When a fuel is pure and combustion is complete, the products released are CO_2 and H_2O. For example, when carbon, C (as in coal), ormethane, CH_4 (as in natural gas), is burned completely, the reactions are

$$C + O_2 \rightarrow CO_2$$

$$CH_4 + 2O_2 \rightarrow CO_2 + 2H_2O$$

The vast amounts of carbon dioxide (CO_2) being vented to the atmosphere can contribute to greenhouse warming (Chapter 19.2).

However, if fuel combustion is incomplete, the products released may also include carbon (soot), various hydrocarbons, and carbon monoxide (CO). *Carbon monoxide* results from the incomplete combustion (oxidation) of carbon:

$$2C + O_2 \rightarrow 2CO$$

Carbon monoxide is a noxious gas and can cause problems in closed areas. (Automobile exhausts contain carbon monoxide). Fuel impurities can add a variety of pollutants.

Increased concentrations of CO_2 can affect our environment. Carbon dioxide combines with water in the atmosphere to form carbonic acid, a mild acid that many of us drink in the form of carbonated beverages (carbonated water):

$$CO_2 + H_2O \rightarrow H_2CO_3$$

Carbonic acid is a natural agent of chemical weathering in geologic processes (Chapter 23.1), but as a product of air pollution, increased concentrations may also exacerbate the corrosion of metals and react with certain materials, causing decomposition (● Fig. 20.15).

Oddly enough, some by-products of complete combustion contribute to air pollution. For example, **nitrogen oxides** (NO_x) are formed when combustion temperatures are high enough to cause a reaction of the nitrogen and oxygen in the air. This reaction typically occurs when combustion is nearly complete, a condition that produces high temperatures, or when combustion takes place at high pressure, such as in the cylinders of automobile engines.

These oxides, normally NO (nitric oxide) and NO_2 (nitrogen dioxide), can combine with water vapor in the air to form nitric acid (HNO_3), which is very corrosive. Also, this acid contributes to acid rain, which is discussed shortly. Nitrogen dioxide (NO_2, "laughing gas") is a colorless gas with a slightly sweet odor. In polluted aerosol form it has a yellow-brown color. During peak rush-hour traffic in some large cities, it is evident as a whiskey-brown haze.

Conceptual Question and Answer

A Laughing Matter

Q. Why is NO_2 called "laughing gas"?

A. Nitrous oxide (NO_2) is used as an anesthetic, primarily in dentistry. Inhalation of NO_2 leads to euphoria, numbness, and eventually loss of consciousness. Breathing air after inhaling the gas sometimes leads to hysterical laughter, hence the name "laughing gas," which was the first artificial anesthetic.

Nitrogen oxides also can cause lung irritation and are key substances in the chemical reactions that produce what is known as "Los Angeles smog," where it was first identified. This smog is not the classic London smoke–fog variety but rather a **photochemical smog** that results from the chemical reactions of hydrocarbons with oxygen in the air and other pollutants in the presence of sunlight. The sunlight supplies the energy for chemical reactions that take place in the air.

(a)

(b)

Figure 20.15 Acid Weathering
The damage inflicted on this statue by pollution and acid rain is evident. The photos are of a decorative statute on the Field Museum in Chicago; the first was taken about 1920 and the second in 1990.

(a)   (b)

Figure 20.16 The Effect of Smog on Visibility
(a) A clear day in Los Angeles. (b) A day of heavy smog in Los Angeles.

More than 18 million people live in the Los Angeles area, which is in the form of a basin, with the Pacific Ocean to the west and mountains to the east. This topography makes air pollution and temperature inversions a particularly hazardous combination. A temperature inversion essentially puts a "lid" on the city, which then becomes engulfed in its own fumes and exhaust wastes (● Fig. 20.16).

Los Angeles has more than its share of temperature inversions, which may occur as frequently as 320 days per year. These inversions, a generous amount of air pollution, and an abundance of sunshine set the stage for the production of photochemical smog. In comparison with the smoke–fogs of London, photochemical smog contains many more dangerous contaminants. They include organic compounds, some of which may be *carcinogens* (substances that cause cancer).

One of the best indicators of photochemical reactions, and a pollutant itself, is **ozone** (O_3), which is found in relatively large quantities in photochemically polluted air.

Fuel Impurities

Fuel impurities occur in a variety of forms. Probably the most common impurity in fossil fuels, and the most critical to air pollution, is *sulfur*. Sulfur is present in various fossil fuels in different concentrations. A low-sulfur fuel has less than 1% sulfur content, and a high-sulfur fuel has a sulfur content greater than 2%. When fuels containing sulfur are burned, the sulfur combines with oxygen to form sulfur oxides (SO_x), the most common of which is **sulfur dioxide** (SO_2):

$$S + O_2 \rightarrow SO_2$$

A majority of SO_2 emissions comes from the burning of coal, and an appreciable amount comes from the burning of fuel oils. Coal and oil are the major fuels used in generating electricity. Almost half the SO_2 pollution in the United States occurs in seven northeastern industrial states.

Sulfur dioxide in the presence of oxygen and water can react chemically to produce sulfurous and sulfuric acids. Sulfurous acid (H_2SO_3) is mildly corrosive and is used as an industrial bleaching agent. Sulfuric acid (H_2SO_4), a very corrosive acid, is a widely used industrial chemical. In the atmosphere, these sulfur compounds can cause a great deal of damage to practically all forms of life and property. Anyone familiar with sulfuric acid, the electrolyte used in car batteries, can appreciate its undesirability as an air pollutant.

The sulfur pollution problem has received considerable attention because of the occurrence of **acid rain**. Rain is normally slightly acidic as a result of carbon dioxide combining with water vapor to form carbonic acid (H_2CO_3). However, sulfur oxide and nitrogen oxide pollutants cause precipitation from contaminated clouds to be even more acidic, giving rise to acid rain (and also to acid snow, sleet, fog, and hail; ● Fig. 20.17).

Figure 20.17 Formation and Effect of Acid Rain

(a) Sulfur dioxide and nitrogen oxide emissions react with water vapor in the atmosphere to form acid compounds. The acids are deposited in rain or snow and also may join dry airborne particles and fall to the Earth as dry depositions. (b) Forest damaged by acid rain in the Great Smoky Mountains between Tennessee and North Carolina. Fortunately, government regulations are reducing sulfur emissions and acid rain.

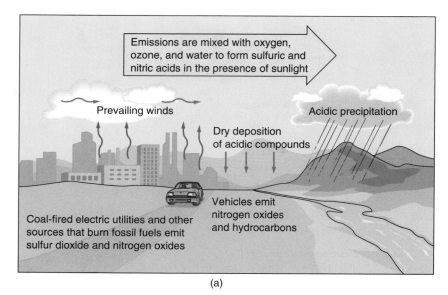

Emissions are mixed with oxygen, ozone, and water to form sulfuric and nitric acids in the presence of sunlight

Prevailing winds

Acidic precipitation

Dry deposition of acidic compounds

Vehicles emit nitrogen oxides and hydrocarbons

Coal-fired electric utilities and other sources that burn fossil fuels emit sulfur dioxide and nitrogen oxides

(a)

(b)

The U.S. government has imposed limits on the levels of sulfur emissions, and acid rain has been greatly reduced. However, before the regulations were in place, rainfall with a pH of 1.4 had been recorded in the northeastern United States. This value surpasses the pH of lemon juice (pH 2.2). Canada had monthly rainfalls with an average pH of 3.5, which is as acidic as tomato juice. The yearly average pH of the rain in the affected regions was about 4.2 to 4.4. Recall that a neutral solution has a pH of 7.0. (See Chapter 13.3 for a discussion of pH.)

In addition to acid rain, there are acid snows. Over the course of a winter, acid precipitations build up in snowpacks. During the spring thaw and resulting runoff, the sudden release of these acids gives streams and lakes an "acid shock."

Air pollution can be quite insidious and can consist of a great deal more than the common particulate matter (smoke, soot, and fly ash) that blackens the outside of buildings. Pollutants may be in the form of mists and aerosols. Some pollutants are metals, such as lead and arsenic. Approximately 100 atmospheric pollutants have been identified, 20 of which are metals that come primarily from industrial processes.

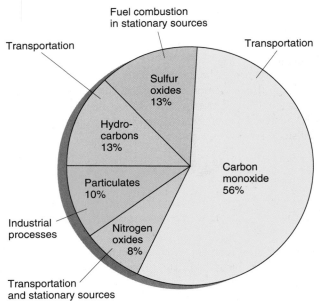

Fuel combustion
in stationary sources

Transportation

Transportation

Sulfur
oxides
13%

Hydro-
carbons
13%

Carbon
monoxide
56%

Particulates
10%

Industrial
processes

Nitrogen
oxides
8%

Transportation
and stationary sources

Figure 20.18 Air Pollutants and Their Major Sources

Figure 20.19 Transportation Air Pollution
Vast amounts of air pollution are generated by cars, trucks, and aircraft.

● Figure 20.18 shows the sources and relative magnitudes of the various atmospheric pollutants. Transportation is clearly the major source of total air pollution. The United States is a mobile society with more than 250 million registered vehicles powered by internal combustion engines (● Fig. 20.19).

The other major sources of air pollution are stationary sources and industrial processes. The term *stationary source* refers mainly to plants that generate electricity. These sources account for the majority of the sulfur oxide (SO_x) pollution. It results primarily from the burning of coal, which always has some sulfur content.

Awareness of the problem of air pollution is readily evident today in the air quality reports in the media. How good the air is to breathe (or how polluted) is rated in newspapers, on the radio and TV, and on the Internet. The Environmental Protection Agency (EPA) now reports how good or bad the air is by rating it on a color-coded Air Quality Index (AQI), as shown in ● Fig. 20.20. Notice that in the reports shown in the figure, ozone is the main pollutant, which is particularly true in summer, when photochemical processes give rise to an "ozone season." Many people check the AQI to determine whether

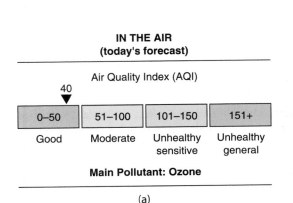

IN THE AIR
(today's forecast)

Air Quality Index (AQI)

40

0–50	51–100	101–150	151+
Good	Moderate	Unhealthy sensitive	Unhealthy general

Main Pollutant: Ozone

(a)

(b)

Figure 20.20 Pollution Reports
(a) A typical color-coded Air Quality Index (AQI) as found in newspapers. This day is a good one, with an index of 40. Note that the main pollutant is ozone. (b) A mobile board showing the AQI color report. What is the category of the AQI on this day?

they should jog or plan other outdoor activities. Elevated concentrations of ozone can reduce lung capacity, causing a shortness of breath; they can also cause headaches, nausea, and eye irritation, particularly for the elderly and for people with asthma and other lung ailments.

Yes, air pollution exists, but there are ways to make the air cleaner. As you might guess, cleaner air can be accomplished through the regulation of emissions from transportation, industry, and other pollution sources, and regulation works. During the 1990s, the air quality improved in more than 200 metropolitan areas. Los Angeles, for example, reports that smog has been greatly reduced since 1985.

Did You Learn?

- A temperature inversion occurs with an inverted lapse rate, and temperature increases with increasing altitude. A temperature inversion can keep pollutants from rising and traps them near the Earth.

- Pollution is any atypical contribution to the environment resulting from human activity. The major atmospheric source is the incomplete combustion of fossil fuels: coal, gas, and oil.

20.5 Climate and Pollution

Preview Questions

- What is climate?
- What is the major concern about carbon dioxide and methane pollution?

It is generally believed that changes in the global climate are brought about by atmospheric pollution. **Climate** is the long-term average weather conditions of a region. Some regions are identified by their climates. For example, when someone mentions Florida or California, one usually thinks of a warm climate; Arizona is known for its dryness and low humidity. Because of such favorable conditions, the climate of a region often attracts people to live there, and the distribution of population (and pollution) is affected.

Dramatic changes in climate have occurred throughout the Earth's history. Probably the most familiar of these changes are the *ice ages*, periods when glacial ice sheets advanced southward over the world's northern continents. The most recent ice age ended some 10,000 years ago, after glaciers came as far south as the midwestern conterminous United States.

Climatic fluctuations are continually occurring on a smaller scale. For example, in the past several decades there has been a noticeable southern shift of warm climate, which has produced drought conditions in some areas. The question being asked today is whether air pollution may be responsible for some of the observed climate variations.

There have also been shifts in the frost and ice boundaries, a weakening of zonal wind circulation, and marked variations in the world's rainfall pattern. However, this pattern appears to be changing, possibly as a result of the depletion of the ozone layer (discussed shortly).

Global climate is sensitive to atmospheric contributions that affect the radiation balance of the atmosphere. These contributions include the concentration of CO_2 and other "greenhouse" gases (Chapter 19.2), the particulate concentration, and the extent of cloud cover, all of which affect the Earth's albedo (the fraction of insolation reflected back into space). See ● Table 20.4 for data on CO_2 emissions and notice the two major industrialized countries that contribute the most.

Air pollution and other human activities do contribute to changes in climatic conditions. Scientists are now trying to understand climate changes by using various models. These models of the workings of the Earth's atmosphere and oceans enable scientists to compare theories, using historical data on climate changes. However, specific data are scant.

Table 20.4 Carbon Dioxide Emission

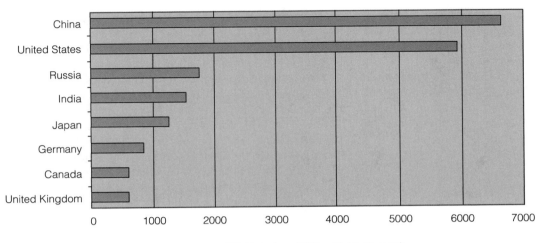

Carbon Dioxide Emissions in 2008

CO$_2$ Emissions (Million Metric Tonnes)

Particulate pollution could contribute to changes in the Earth's thermal balance by decreasing the transparency of the atmosphere to insolation. We know that this effect occurs from observing the results of fires and volcanic eruptions. The particulate matter that is emitted could cause changes in the Earth's albedo.

Recent widespread fires have occurred in the western United States. Large amounts of particulate matter go into the atmosphere (● Fig. 20.21), along with significant amounts of carbon dioxide (Chapter 22.4).

In 1991, Mount Pinatubo in the Philippines erupted. Debris was sent more than 15 mi (24 km) into the atmosphere, and over a foot of volcanic ash piled up in surrounding regions. Measurements indicate that Pinatubo was probably the largest volcanic eruption of the twentieth century, belching out tons of debris and sulfur dioxide (SO$_2$). The particulate

Figure 20.21 A Result of Wildfire
A halo around the Sun and reddish sky caused by heavy particulate smoke over San Diego, California, in 2003.

UPI/Earl S. Cryer/Landov

Highlight   The Ozone Hole and Global Warming

In 1974, scientists in California warned that chlorofluorocarbon gases, or CFCs, might seriously damage the ozone layer through depletion. Observations generally supported this prediction, and the United States banned the use of these gases as propellants in aerosol spray cans and as refrigerants for cooling.

When released, these gases slowly rise into the stratosphere, a process that takes 20 to 30 years. In the stratosphere, the CFC molecules are broken apart by ultraviolet radiation, with the release of reactive chlorine atoms. These atoms in turn react with and destroy ozone molecules in the repeating cycle

$$CFC \rightarrow Cl$$

$$Cl + O_3 \rightarrow ClO + O_2$$
$$\text{(ozone)}$$

$$ClO + O \rightarrow Cl + O_2$$

Note that the chlorine atom is again available for reaction after the process. These atoms may remain in the atmosphere for a year or two. During this time, a single Cl atom may destroy as many as 100,000 ozone molecules.

In 1985, scientists announced the discovery of an ozone "hole" over Antarctica (Fig. 1). Investigations have shown that this polar hole in the ozone layer opens up annually during the southern springtime months of September and October. It is thought that the seasonal hole began forming in the late 1970s because of increasing concentrations of ozone-destroying chlorine pollutants in the stratosphere.

The worldwide depletion of the ultraviolet-absorbing ozone layer will have some undesirable effects. Experts estimate that the number of cases of skin cancer will increase by 60% and that there will be many additional cases of cataracts. In 1994, the National Weather Service began issuing a "UV index" forecast for many large cities. On a scale of 0 to 15, the index gives a relative indication of the amount of UV light that will be received at the Earth's surface at noontime the next day; the higher the number, the greater the amount of UV. The scale is based on upper-atmosphere ozone levels and clouds.

International concern over ozone depletion prompted industrialized nations to agree to reduce the production of CFCs and eventually phase them out, replacing them with newly developed, environmentally friendly compounds. As a result, the use of CFCs has been greatly reduced. Even so, the ozone hole over Antarctica is now even larger, with an average area of 17 million km² (10.6 million mi²).

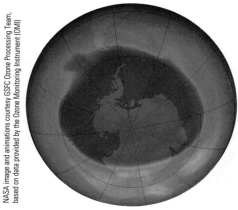

NASA image and animations courtesy GSFC Ozone Processing Team, based on data provided by the Ozone Monitoring Instrument (OMI)

Figure 1   The Ozone Hole
A computer map of the South Pole region showing the total stratospheric ozone concentrations. The ozone hole, or region of minimal concentration, is shown in blue and purple. Note the outline of Antarctica.

matter caused beautiful sunrises and sunsets around the world during the following year. The sulfur dioxide gas reacted with oxygen and water to form tiny droplets, or aerosols, of sulfuric acid, which can stay aloft for several years before falling back to the ground (See Fig. 22.18.)

The classic example is the eruption of Mount Tambora on Sumbawa, an Indonesian island just east of Java. Mount Tambora erupted for several days in April 1815, and an estimated 145 km³ of solid debris was thrown into the air. The volcanic dust darkened the sky and circulated around the globe. The Sun's rays were partially blocked, and 1816 was unseasonably cold. It was known as "the year without summer."

Another environmental concern is the thinning of the Earth's protective ozone layer, thus allowing more ultraviolet radiation to reach the surface and increase the temperature. More insight to these major problems is given in the **Highlight: The Ozone Hole and Global Warming**.

Highlight

CFCs are also greenhouse gases and can contribute to an increase in the Earth's temperature, or global warming. (See the Highlight: The Greenhouse Effect in Chapter 19.2.) However, the main concern arises from other greenhouse gases: carbon dioxide, methane, and nitrogen oxides. CO_2 emissions arise chiefly from the combustion of fossil fuels, which is essential to industry and our society. With a world population of about 7 billion, CO_2 emissions are substantial. The United States, with a population of more than 300 million, contributes significantly. According to *Consumer Reports* (January 2007), "A car that travels 15,000 miles a year and averages 25 miles per gallon exhales about 5.5 metric tons of CO_2."

Many scientists agree that these tons of gaseous emissions will cause the Earth's average temperature to rise, producing *global warming*. In the extreme, this temperature increase is expected to lead to the melting of polar ice caps, rising sea levels, and more disastrous weather events, such as storms and floods. Increased warming of the ocean could fuel more hurricanes.

There is already evidence of global warming. Satellite photos show the melting of Greenland's ice sheet and other seasonal differences are noted (Fig. 2). Continued melting could raise the ocean levels with an inundation of coastal areas. Calls are made to limit and reduce the emission of greenhouse gases. However, some scientists cast doubt on global warming extreme changes and see only moderate climate changes, which are perhaps natural.

How will this debate play out? Time will tell. There are many unanswered questions about the effects of pollution—which has occurred over a relatively short time—on the natural interacting cycles of the atmosphere and biosphere, cycles that have taken millions of years to become established.*

*A great deal of information on global warming can be found at www.epa.gov and search "climate change."

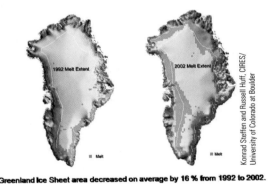

Greenland Ice Sheet area decreased on average by 16 % from 1992 to 2002.

Konrad Steffen and Russell Huff, CIRES/ University of Colorado at Boulder

Figure 2 Global Warming
Global warming is evident from satellite photos showing the increased melting of Greenland's ice sheet.

Aircraft have a cause for concern because of particulate and gaseous emissions. In April 2010, an ash plume from a volcano erupting in Iceland curtailed jet aircraft flights and closed airports in Europe. In the troposphere, precipitation processes "wash" out particulate matter and gaseous pollutants, but there is no snow or rain washout mechanism in the stratosphere. Furthermore, the stratosphere is a region of high chemical activity, and chemical pollutants (such as NO_x and hydrocarbons) might possibly give rise to climate-changing reactions.

Temperature increases may also result from pollution. Vast amounts of carbon dioxide (CO_2) and methane (CH_4) are expelled into the atmosphere. As discussed in Chapter 19.2, CO_2, methane, and water vapor play important roles in the Earth's energy balance in the greenhouse effect. An increase in the atmospheric greenhouse gases could alter the amount of radiation absorbed from the planet's surface and thus prompt an increase in the Earth's average temperature and give rise to *global warming*.

Conceptual Question and Answer

Ruminating Up Some CH₄

Q. Methane (CH_4) is a greenhouse gas. How does it get into the atmosphere?

A. Some methane emissions are associated with natural gas and coal mining industries. However, the largest emissions in the United States come from livestock digestive processes (enteric fermentation) and landfills. Some domesticated animals, such as cattle and sheep, produce methane as part of their normal digestive fermentation processes. (These animals have two stomachs and ruminate, or chew cud.) The methane produced is belched or exhaled by the animals. Considering the number of cattle and sheep in the United States, it adds up to a lot of CH_4.

Landfills are the second largest source of methane. The gas is generated as waste decomposes in anaerobic (without oxygen) conditions.

Did You Learn?

- Climate is the long-term average weather conditions of a region.

- An increase in atmospheric carbon dioxide (CO_2) and methane (CH_4) greenhouse gases could cause an increase in the Earth's average temperature and global warming.

KEY TERMS

1. coalescence (20.1)
2. Bergeron process
3. air mass (20.2)
4. source region
5. front
6. lightning (20.3)
7. thunder
8. ice storm
9. snowstorm
10. tornado
11. tornado watch
12. tornado warning
13. hurricane
14. storm surge
15. hurricane watch
16. hurricane warning
17. pollution (20.4)
18. smog
19. temperature inversion
20. nitrogen oxides (NO_x)
21. photochemical smog
22. ozone (O_3)
23. sulfur dioxide (SO_2)
24. acid rain
25. climate (20.5)

MATCHING

For each of the following items, fill in the number of the appropriate Key Term from the preceding list. Compare your answers with those at the back of the book.

a. _____ Boundary between two air masses

b. _____ Combination of smoke and fog

c. _____ Associated with freezing rain

d. _____ Threat of hurricane conditions within 24 to 36 hours

e. _____ Describes essentials for precipitation

f. _____ Has wind speed exceeding 74 mi/h

g. _____ Results primarily from nitrogen oxides

h. _____ Distinguished from surrounding air by physical characteristics

i. _____ Hurricane conditions expected within 24 hours

j. _____ Associated with lightning

k. _____ Long-term, average weather conditions

l. _____ Indicates that a tornado may form

m. _____ Increasing temperature with height in the troposphere

n. _____ Formation of drops by collision of droplets

o. _____ Causes most hurricane causalities

p. _____ Good indicator of photochemical pollution

q. _____ An atmospheric electrical discharge

r. _____ Result of SO_2 pollution

s. _____ Issued when a tornado has been sighted or indicated on radar

t. _____ Pollutant formed with high combustion temperatures

u. _____ Where an air mass derives its characteristics

v. _____ Atypical contributions to the environment by humans

w. _____ Most violent of storms

x. _____ Results from impurity in fossil fuels

y. _____ An appreciable accumulation of snow

MULTIPLE CHOICE

Compare your answers with those at the back of the book.

1. When the temperature of the air is below the dew point without precipitation, the air is said to be what? (20.1)
   (a) stable (b) supercooled
   (c) sublimed (d) coalesced

2. Which of the following is *not* essential to the Bergeron process? (20.1)
   (a) silver iodide (b) mixing
   (c) supercooled vapor (d) ice crystals

3. Which of the following is the result of deposition? (20.1)
   (a) snow (b) rain
   (c) dew (d) frost

4. Which type of air mass is mainly responsible for mostly warm weather in the conterminous United States? (20.2)
   (a) mP (b) cA
   (c) mT (d) cT

5. Which of the following air masses would be expected to be cold and dry for the conterminous United States? (20.2)
   (a) cP (b) mA
   (c) cT (d) mE

6. What is a cold front advancing under a warm front called? (20.2)
   (a) a warm front
   (b) a stationary front
   (c) a cold front occlusion
   (d) a warm front occlusion

7. What is the critical alert for a tornado? (20.3)
   (a) tornado alert
   (b) tornado warning
   (c) tornado watch
   (d) tornado prediction

8. The greatest number of hurricane casualties is caused by which of the following? (20.3)
   (a) high winds
   (b) low pressure
   (c) flying debris
   (d) storm surge

9. A subsidence temperature inversion is caused by which of the following? (20.4)
   (a) a high-pressure air mass
   (b) acid rain
   (c) radiational cooling
   (d) subcritical air pressure

10. Which is a major source of air pollution? (20.4)
    (a) nuclear electrical generation
    (b) incomplete combustion
    (c) temperature inversion
    (d) acid rain

11. A change in the Earth's albedo could result from which of the following? (20.5)
    (a) nitrogen oxides
    (b) acid rain
    (c) photochemical smog
    (d) particulate matter

12. Major concern about global warming arises from increased concentrations of which of the following? (20.5)
    (a) sulfur oxides
    (b) nitrogen oxides
    (c) greenhouse gases
    (d) photochemical smog

FILL IN THE BLANK

Compare your answers with those at the back of the book.

1. The two processes of raindrop formation are the Bergeron process and ___. (20.1)
2. Sleet is frozen _____. (20.1)
3. The main distinguishing characteristics of air masses are ___ and moisture content. (20.2)
4. The type of air mass with a source region in Canada is a(n) ___. (20.2)
5. A(n) ___ front is the boundary of an advancing warm air mass over a colder surface. (20.2)
6. When a cold front advances under a warm front, a cold front ___ occurs. (20.2)
7. Lightning that occurs below the horizon or behind clouds is called ___ lightning. (20.3)
8. When a tornado has been sighted or indicated by radar, a tornado ___ is issued. (20.3)
9. When a hurricane warning is announced, hurricane conditions are expected within _____ hours. (20.3)
10. When a high-pressure air mass moves over a region and becomes stationary, a(n) ___ temperature inversion could result. (20.4)
11. "Los Angeles smog" is ___ smog. (20.4)
12. The chief cause of the "ozone hole" is believed to be ___. (20.5)

SHORT ANSWER

20.1 Condensation and Precipitation

1. Describe the three essentials of the Bergeron process and explain how they are related to methods of modern rainmaking.

2. (a) Is frost frozen dew? Explain. (b) How are large hailstones formed?

20.2 Air Masses

3. How are air masses classified? Explain the relationship between the characteristics of air masses and source regions.

4. Give the source region(s) of the air mass(es) that affect the weather in your area.

5. What is a front? List the meteorological symbols for four types of fronts.

6. Describe the weather associated with warm fronts and with cold fronts. What is the significance of the sharpness of these fronts' vertical boundaries?

20.3 Storms

7. (a) Where can lightning take place? (Where can it begin and end?) (b) Describe what is meant by heat lightning and "a bolt from the blue."

8. What type of first aid should be given to someone suffering from the shock of a lightning stroke?

9. An ice storm is likely to result along what type of front? Explain.

10. What is the most violent of storms and why?

11. A tornado is sighted nearby. What should you do?

12. What is the major source of energy for a tropical storm? When does a tropical storm become a hurricane?

13. Distinguish between a hurricane watch and a hurricane warning.

14. What three months are the peak season in the United States (a) for hurricanes? (b) For tornadoes?

20.4 Atmospheric Pollution

15. Define the term *air pollution*.

16. Is air pollution a relatively new problem? Explain.

17. What are the two types of temperature inversions, and how does a temperature inversion affect atmospheric pollution in an area?

18. What are the products of complete combustion of hydrocarbons? What are the products of incomplete combustion?

19. How are nitrogen oxides (NO_x) formed, and what role do they play in air pollution?

20. Distinguish between classical smog and photochemical smog. What is the prime indicator of the latter?

21. You have a big bonfire going. The smoke rises to a point and then spreads out in a flat layer. What does this indicate?

22. What is the major fossil-fuel impurity?

23. What are the causes and effects of acid rain? In which areas is acid rain a major problem, and why?

24. Name the major sources of the following pollutants:
    (a) carbon monoxide
    (b) sulfur dioxide
    (c) particulate matter
    (d) nitrogen oxides
    (e) ozone

20.5 Climate and Pollution

25. Define *climate*.

26. What effect might each of the following have on the Earth's climate?
    (a) CO_2
    (b) particulate pollution
    (c) aircraft in the stratosphere

27. What is a direct effect on humans that increases with atmospheric ozone depletion?

28. What effects could CFCs have on the Earth's climate?

29. What concern has been raised about air pollution in the stratosphere?

30. What are the concerns about global warming?

VISUAL CONNECTION

Visualize the connections and give answers for the blanks. Compare your answers with those at the back of the book.

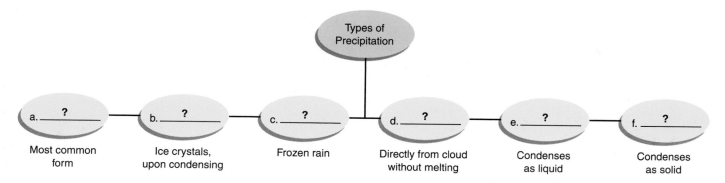

Types of Precipitation

a. ? — Most common form

b. ? — Ice crystals, upon condensing

c. ? — Frozen rain

d. ? — Directly from cloud without melting

e. ? — Condenses as liquid

f. ? — Condenses as solid

APPLYING YOUR KNOWLEDGE

1. Why do household barometers often have descriptive adjectives such as *rain* and *fair* on their faces, along with the direct pressure readings? (See Fig. 19.12.)

2. How could CO_2 pollution be decreased while our energy needs are still being met?

3. Assuming that the name has not been retired, what will be the name of the third Atlantic tropical storm in 2018?

4. Which is hotter, the air in the vicinity of a lightning stroke or the surface (protosphere, Chapter 18.2) of the Sun? How many times hotter?

EXERCISES

20.2 Air Masses

1. Locate the source regions for the following air masses that affect the conterminous United States:
   (a) cA   (d) mT
   (b) mP   (e) cP
   (c) cT

   Answer: (a) Greenland (b) northern Atlantic Ocean and Pacific Ocean (c) Mexico (d) middle Atlantic Ocean and Pacific Ocean (e) Canada

2. What would be the classifications of the air masses that form over the following source regions?
   (a) Sahara Desert   (d) mid-Pacific Ocean
   (b) Antarctic Ocean   (e) Siberia
   (c) Greenland

3. On average, how far do (a) a cold front and (b) a warm front travel in 24 h?

   Answer: (a) 840 km (521 mi) (b) 480 km (298 mi)

4. How long does it take (a) a cold front and (b) a warm front to travel from west to east across your home state?

5. If thunder is heard 10 s after a lightning flash is observed, then approximately how far away is the storm? Give your answer in kilometers and in miles.

   Answers: 3.3 km (2.0 mi)

6. While picnicking on a summer day, you hear thunder 6 s after seeing a lightning flash from an approaching storm. Approximately how far away, in miles, is the storm?

ON THE WEB

1. Stormy Weather

Go to the student website at **www.cengagebrain.com/shop/ ISBN/1133104096** to learn more about El Niño and to answer the following questions. What role does El Niño play in our weather patterns? What other theories have been advanced to explain the causes of El Niño, and what is their common denominator? What is the Global Weather Machine? How accurate are weather forecasters?

2. Ozone: A Real Health Concern

What exactly is the ozone hole? How did scientists first become aware of it? More fundamentally, what is ozone and how is it formed? How is it destroyed? Why is it important? What is being done? What are TOMS satellite measurements? What steps might you take to reduce the risks posed by depletion of the ozone layer? Explore answers to these questions by following the recommended links on the student website at **www.cengagebrain.com/shop/ISBN/1133104096**.

CHAPTER

21 Structural Geology and Plate Tectonics

*The face of places and
their forms decay;
And that is solid earth
that once was sea;
Seas, in their turn,
retreating from
the shore,
Make solid land what
ocean was before.*

•

Ovid (43 BCE–18 CE)

A man walks in a snowstorm >
over land that once was a flat
plateau; this significant land
movement was caused by an
earthquake that struck Afghani-
stan in 1998. The earthquake
destroyed five villages and killed
about 5000 people.

AP Photo/Zaheeruddin Abdullah

GEOLOGY FACTS

▶ The deepest hole drilled into
the Earth's crust had a depth of
12.3 km (7.63 mi). It was drilled
in Russia from 1970 to 1994.
This depth is only 0.2% of the
radius of the Earth.

▶ As a consequence of plate
tectonic motion, the Americas
are moving away from Europe

Geology is the study of planet Earth: its composition, structure, processes, and history. Geology also plays a role in the study of the Moon and other solar system objects, but the study of these objects is primarily the province of astronomy (Chapter 17). This chapter and the following three introduce the geologic concepts necessary for understanding the physical nature of our planet.

In this chapter you will learn about the basic structure of the Earth's interior and why plate tectonics is the primary mover of the Earth's outer shell, the lithosphere. After a look at Earth's interior, plate tectonics will be discussed, along with the theories of continental drift and seafloor spreading. Finally, three manifestations of plate movement are examined: earthquakes, crustal deformation, and mountain building (including volcanoes).

Chapter Outline

21.1 The Earth's Interior Structure

Preview Questions

● What are the four regions of the Earth's interior?

● What type of seismic waves travel through the Earth?

The Earth's Interior

None of the remarkable scientific advances of the twentieth century revealed with certainty the composition and structure of the Earth's interior. Our ideas about it rest on indirect evidence provided by (1) earthquake body waves, their speed and direction identifying the types of materials through which they move; (2) meteorites, whose composition is believed to be similar to the Earth's; (3) spacecraft measurements of gravity and magnetic variations; and (4) laboratory experiments performed on rocks under very high temperature and pressure.

Earthquakes, despite their potential for destruction, can be an aid to science. By monitoring earthquake waves at different locations and applying knowledge of wave properties, such as speed and refraction (bending) in various types of materials, scientists obtain information about the Earth's interior structure.

Two general types of **seismic waves** result from the vibrations produced by earthquakes: *surface waves,* which travel along the outer layer of the Earth, and *body waves,* which travel through the Earth. Surface waves cause most earthquake damage.

There are two types of body waves: P (for primary) waves, or compressional waves, and S (for secondary) waves, or shear waves. *P waves* are longitudinal compressional waves that are propagated by particles moving longitudinally back and forth in the same direction the wave is traveling (Chapter 6.2). In *S waves,* the particles move at right angles to the direction of wave travel; hence, they are transverse waves.

The P and S waves have two other important differences: the material in which they propagate and their speed. S (transverse) waves can only travel through solids. Liquids and gases cannot support a shear (right angle) stress; therefore, their particles will not oscillate in the direction of a shearing force. For example, little or no resistance is felt when one shears (slices) a knife through a liquid or gas. The compressional P waves, on the other hand, can travel through any kind of material, be it solid, liquid, or gas.

The other important difference is in the speed of the waves. P waves are called *primary* because they travel faster in any particular solid material and arrive earlier at a seismic station than the *secondary* S waves. Both the type and speed of the waves allow seismologists to determine the location of an earthquake and to learn about the Earth's internal structure.

The speed of body waves depends on the density of the material in which they propagate, and these densities generally increase with depth. As a result, the waves are curved, or refracted. Also, the waves are refracted when they cross a boundary, or discontinuity, between different media. (This refraction is analogous to the refraction of light waves; see Chapter 7.2.) The refraction of seismic waves, and the fact that S waves cannot travel through liquid media, provide our present view of the Earth's structure, as shown in ● Fig. 21.1.

From these indirect observations, scientists believe that the Earth is made up of four regions: (1) the inner core, (2) the outer core, (3) the mantle, and (4) the crust. These regions are characterized by the different compositions and physical properties shown in ● Fig. 21.2.

The density of the inner and outer cores suggests a metallic composition, which is believed to be chiefly iron (85%) and nickel, similar to the composition of metallic meteorites (Chapter 17.6). The **outer core,** some 2240 km (1390 mi) thick, is believed to be viscous liquid, whereas the **inner core** is thought to be a solid ball with a radius of approximately 1230 km (763 mi). Evidence for a liquid outer core and a solid inner core is found in the behavior of P and S waves. P waves slow down on entering the outer core and speed

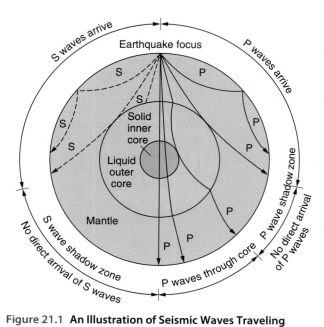

Figure 21.1 An Illustration of Seismic Waves Traveling through the Earth's Interior
Because of the S and P wave shadow zones (where no waves arrive), the Earth is believed to have a liquid outer core. For clarity, the S wave behavior is shown only on the left and the P wave behavior only on the right.

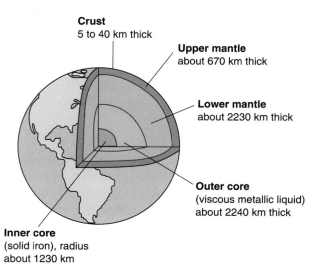

Figure 21.2 Interior Structure of the Earth
The radius of the Earth is about 6400 km (4000 mi). Together, the two cores have a radius of about 3500 km (2200 mi), and the upper and lower mantles together are about 2900 km (1800 mi) thick. Comparatively the crust is very thin, less than the width of the black lines in the figure.

up again on entering the inner core. S waves stop altogether at the edge of the outer core. As shown in Fig. 21.1, this behavior of P and S waves creates a *shadow zone* on each side of the core.

Conceptual Question and Answer

The Earth's Interior Boundaries

Q. If earthquakes generated only S waves, then what would we know about the Earth's interior?

A. Because S waves can only travel through solids, the size of the Earth's core could still be determined by shadow zones, but seismologists would not know that Earth has a solid inner core and a liquid outer core. We need the P waves to show this boundary.

Around the outer core is the **mantle**, which averages about 2900 km (1800 mi) thick. The mantle is generally divided into two parts, the upper mantle and the lower mantle. In the upper mantle, the temperature of the rock is near the melting point. The molten rock extruded by some volcanoes originates in this region. The lower mantle contains rocky materials at high temperatures and pressures. The composition of the rocky mantle differs sharply from that of the metallic core, and their boundary is distinct.

The mantle is surrounded by a thin, rocky, outer layer on which we live, called the **crust**. It ranges in thickness from about 5 to 11 km (3 to 7 mi) beneath the ocean basins to about 19 to 40 km (12 to 25 mi) under the continents. The overall radius of the Earth is 6400 km (4000 mi). Thus, the crust represents less than 1% of the Earth's radius, the mantle about 45%, and the core about 55%.

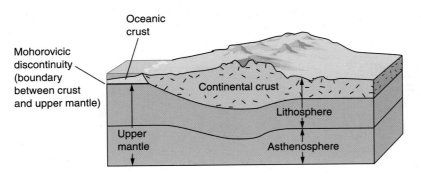

Figure 21.3 Lithosphere and Asthenosphere
The crust ranges from 5 to 40 km (3 to 25 mi) in depth, and the lithosphere, which includes the crust, to 70 km (43 mi). The asthenosphere extends to a depth of about 700 km (434 mi). The boundary of the lithosphere and asthenosphere is in the upper mantle.

An abrupt change in the behavior of body waves as they travel toward the Earth's interior reveals the existence of the *Mohorovicic* ("MOE-HOE-ROE-ve-chich") *discontinuity*, a sharply defined boundary that separates the crust from the upper mantle (● Fig. 21.3).* Both scientists and students prefer the shortened term, *Moho*, for this important boundary.

If the crust and upper mantle are viewed in terms of behavior rather than composition, then they can be divided somewhat differently into what might be called zones. The outer zone, called the **lithosphere**, extends to a depth of approximately 70 km (43 mi) and includes all the crust and a thin part of the upper mantle (Fig. 21.3). The lithosphere is rigid, brittle, and relatively resistant to deformation. Faults and earthquakes are largely restricted to this layer.

The **asthenosphere**, the part of the mantle that lies beneath the lithosphere, is essentially solid rock, but it is so close to its melting temperature that it contains pockets of thick, molten rock and is relatively plastic. Therefore, it is more easily deformed than the lithosphere. It extends to a depth of roughly 700 km (434 mi) below the Earth's surface, and the interface between the lithosphere and the asthenosphere is significant in terms of internal geologic processes. For this reason, the asthenosphere plays an essential role in tectonic plate movement, as described in the next two sections.

Did You Learn?

● The Earth has four interior regions: inner core, outer core, mantle, and crust.

● Seismic P waves (compressional waves travel completely through the Earth's interior, whereas S (shear) waves do not.

21.2 Continental Drift and Seafloor Spreading

Preview Questions

● What is Pangaea?

● Why are the continents believed to be in motion?

When one looks at a map of the continents of the world, it is tempting to speculate that the Atlantic coasts of Africa and those of North America and South America could fit nicely together, as though they were pieces of a jigsaw puzzle. This observation has led scientists at various times to suggest that these continents, and perhaps the other continents, were once a single, giant supercontinent that broke into large fragments, called *plates*, and then drifted apart. However, scientists had no evidence to support the idea of such an event other than the shapes of the continents.

---

*The Earth's mantle-crust boundary is named in honor of Andrija Mohorovičić (1857–1936), a Croatian geophysicist and meteorologist who discovered it in 1909.

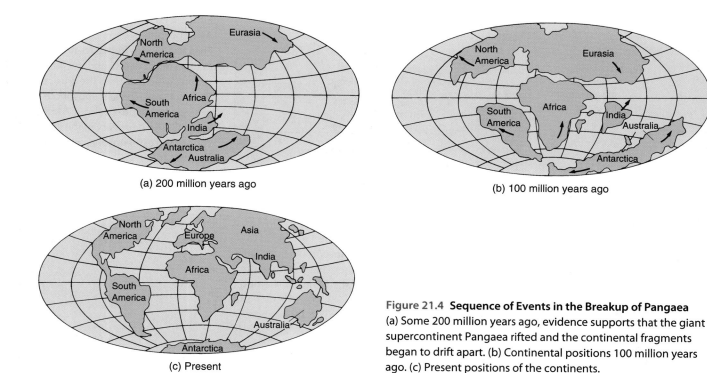

(a) 200 million years ago

(b) 100 million years ago

(c) Present

Figure 21.4 Sequence of Events in the Breakup of Pangaea
(a) Some 200 million years ago, evidence supports that the giant supercontinent Pangaea rifted and the continental fragments began to drift apart. (b) Continental positions 100 million years ago. (c) Present positions of the continents.

Continental Drift

In the early 1900s, Alfred Wegener (1880–1930), a German meteorologist and geophysicist, revived the idea of plates and their movement and assembled various types of geologic evidence supporting the model known as **continental drift**. Wegener's hypothesis gave rise to considerable controversy, and only in the last few decades has conclusive evidence been found that supports some aspects of it. Today continental drift is a well-known and well-tested theory.

Wegener's assumption was that the continents were once part of a single giant continent, which he called **Pangaea** (from the Greek, meaning "all lands," and pronounced "pan-JEE-ah"). He proposed that this hypothetical supercontinent rifted (broke apart) about 200 million years ago and that its sections somehow drifted to their present positions and became today's continents (● Fig. 21.4).

The scientific evidence supporting Wegener's hypothesis that the continents once formed a single land mass is of several different types. Let's briefly consider the three most prominent.

1. *Biological evidence.* Similarities in biological species and fossils found on distant continents that are today separated by oceans strongly suggest that these land masses were at one time together as Pangaea. For example, a certain variety of garden snail is found only in the western part of Europe and the eastern part of North America. It would not be expected that this species could traverse the present-day oceans. Similarly, fossils of identical reptiles have been found in South America and Africa, and identical plant fossils have been found in South America, Africa, India, Australia, and Antarctica.

2. *Continuity of geologic features.* The roughly interlocking shapes of the coastlines of the African and American continents inspired the theory of continental drift. Imagine the continental shapes to be interlocking jigsaw puzzle pieces. When the pieces are put back together, the fitted pieces display a continuity of geologic features, as shown in ● Fig. 21.5.

   If indeed the continents rifted and drifted apart, then it might be expected that there would be evidence of geologic features in common on different continents.

For example, if the continents were put back together, then the Cape Mountain Range in southern Africa would line up with the Sierra Range near Buenos Aires. These mountains are strikingly similar in geologic structure and rock composition (Fig. 21.5). In the Northern Hemisphere, the Hebrides Mountains in northern Scotland match up with similar formations in Labrador, and the Caledonian Mountains in Norway and Sweden have a logical extension in the Canadian Appalachians.

3. *Glacial evidence.* Solid geologic evidence confirms that a glacial ice sheet covered the southern parts of South America, Africa, India, and Australia about 300 million years ago, an ice sheet similar to the one that covers Antarctica today. A reasonable conclusion is that the southern portions of these continents were under the influence of a polar climate at that time.

Wegener's hypothesis suggests an answer and derives support from these observations. The direction of the glacier flow is easily determined by marks of erosion on rock floors. If the continents were once grouped together, then the glaciations area was common to the various continents, as illustrated in ● Fig. 21.6. Although this evidence supported Wegener's theory, it was not generally accepted, primarily because he could not explain how continental crust could move through much denser oceanic crust.

Seafloor Spreading

The mechanism behind continental drift was suggested in 1960 by Harry H. Hess, an American geologist. Geologists knew at the time that a *mid-ocean ridge* system stretches through the major oceans of the world (● Fig. 21.7). In particular, the *Mid-Atlantic Ridge* runs along the center of the Atlantic Ocean between the continents. The East Pacific Rise runs along the Pacific Ocean floor just west of the North and South American continents. Geologists also knew about the deep, narrow depressions that ring the Indian and Pacific oceans. Called *deep-sea trenches,* they are the lowest places on the Earth's surface.

Hess suggested a theory of **seafloor spreading**, where the seafloor slowly spreads and moves sideways away from the mid-ocean ridges. Molten rock then wells up into the gap and cools to form new seafloor rock. Thus, the seafloor moves away from the ridges, conveyor-belt style, cooling and contracting as it moves. When it reaches the trenches, it descends back into the mantle.

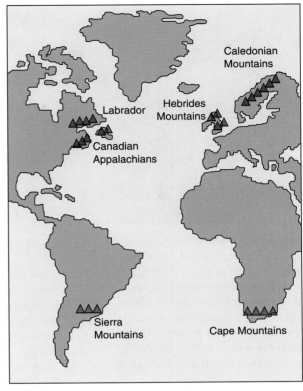

Figure 21.5 Continuity of Geologic Features Supporting the Theory of Continental Drift
If the continents were fitted together, then various mountain ranges of similar structure and rock composition on the different continents would line up. (See text for description.)

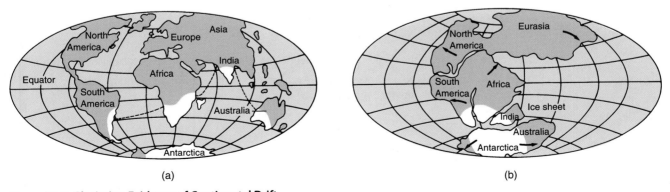

(a)   (b)

Figure 21.6 Glaciation Evidence of Continental Drift
(a) Geologic evidence shows that a glacial ice sheet covered parts of South America, Africa, India, and Australia some 300 million years ago, but there is no evidence of this glaciation in Europe and North America. (b) This glaciation is readily explained if a single ice sheet covered the southern polar region of Pangaea and the continents subsequently rifted and drifted apart.

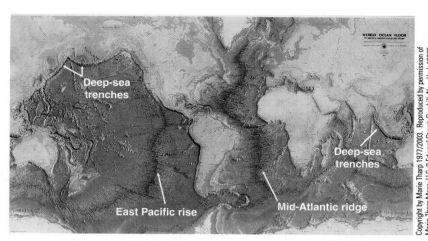

Figure 21.7 The Seafloor Surfaces
The seafloor surfaces are as irregular as the surfaces of the continents. A system of mid-ocean ridges extends throughout the ocean basins. Notice also the deep-sea trenches rimming the Pacific Ocean. Title (World Ocean Floor Panorama), Authors (Bruce C. Heezen and Marie Tharp), Date (1977)

Support for this theory has come from studies of remanent magnetism and from determination of the ages of the rocks on each side of the mid-ocean ridges. **Remanent magnetism** is the magnetization of rocks that form in the presence of an external magnetic field. When molten material containing the mineral magnetite (Fe_3O_4) is extruded upward from the mantle—for example, in a volcanic eruption—and solidifies in the Earth's magnetic field (Chapter 8.4), it becomes permanently magnetized. The direction of the magnetization indicates the direction of the Earth's magnetic field at the time. Thus, the newly formed seafloor rock preserves the magnetic record at that time.

Another example of remanent magnetism occurs when solidified rock that contains magnetite is worn down by erosion. The fragments are carried away by water and eventually settle in bodies of water, where they become layers in future sedimentary rock (Chapter 23.2). In the settling process, the magnetized particles, which are in fact small magnets, become generally aligned with the Earth's magnetic field. These magnetite-rich sediments, similar to the new seafloor rock described above, create a long-term record of the past magnetism of the Earth.

Measurements of the remanent magnetism of rock on the ocean floor revealed long, narrow, symmetric bands of *magnetic anomalies* on both sides of the Mid-Atlantic Ridge. That is, the direction of the magnetization was reversed in adjacent parallel regions (● Fig. 21.8). Along with data on land rock, the magnetic anomalies indicate that the Earth's magnetic field has reversed polarity fairly frequently and regularly throughout recent geologic time. The most recent reversal was about 700,000 years ago. Why this and other reversals should occur is not well understood, but recent data show that the Earth's inner and outer cores rotate at slightly different rates. This effect may cause the Earth's magnetic field to change slowly over a few thousand years and then reverse polarity. The symmetry of the magnetic anomaly bands on either side of the ridge indicates movement away from the ridge at the rate of several centimeters per year and provides evidence for seafloor spreading.

These observations support the idea of seafloor spreading as a mechanism for continental drift, an idea that has culminated in the modern theory of plate tectonics, discussed in the following section. Instead of continental crust plowing through oceanic crust, it is now known that both are carried across the upper mantle as part of the *lithosphere*.

Did You Learn?

- Pangaea is the single, giant continent believed to have broken apart 200 million years ago.

- Continents are in motion because molten rock wells up because of energy released by the Earth's interior causing seafloor spreading.

21.3 Plate Tectonics

Preview Questions

- How are the plates of the Earth able to move?

- What geologic features are created at plate boundaries?

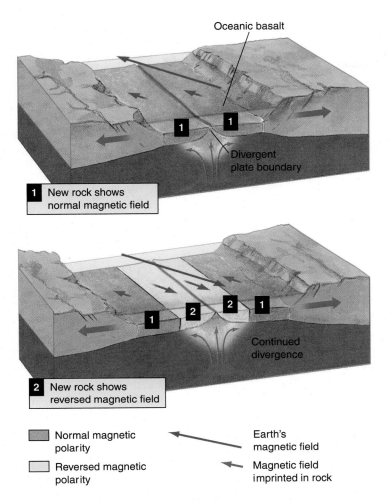

Oceanic basalt

1 New rock shows normal magnetic field

Divergent plate boundary

2 New rock shows reversed magnetic field

Continued divergence

■ Normal magnetic polarity

□ Reversed magnetic polarity

← Earth's magnetic field

← Magnetic field imprinted in rock

Figure 21.8 Marine Magnetic Anomalies Showing Reversals in the Earth's Magnetic Field
As molten rock cools and solidifies at divergent boundaries along mid-ocean ridges, the magnetized crystals within the rock become aligned with the prevailing direction of the Earth's magnetic field. Each resulting strip of rock has either normal magnetism (magnetism like today's magnetic field) or reversed magnetism (magnetism opposite to today's field). Refer to the Earth's magnetic field in Chapter 8.4.

The view of ocean basins in a process of continual self-renewal has led to acceptance of the theory of **plate tectonics**.* The lithosphere is now viewed not as one solid rock but as a series of solid sections or segments called *plates* that are constantly interacting with one another in very slow motion. The surface of the globe is segmented into about 20 plates. Some are very large, and some are small. The major plates and their rates of drift are illustrated in ● Fig. 21.9. For example, Hawaii (on the Pacific plate) is moving toward Japan (on the Eurasian plate) at a rate of about 7 cm per year. North America and Europe are moving apart at the rate of about 3 cm per year. The oceanic plates generally move the fastest (up to 10 cm per year), whereas the continental plates move more slowly (about 2 cm per year).

The most active, restless parts of the Earth's crust are located at the plate boundaries. A **divergent boundary** along the mid-ocean ridges is where one plate is moving away from another and new oceanic rock is formed. An example of a divergent boundary is at the Red Sea, where the African and Arabian plates are separating. Where plates are driven together is a **convergent boundary**, and rock is consumed. An example is the Himalayan Mountains, where the Indian plate is colliding with the Eurasian plate, forming a mountain range. In still other parts of the Earth's surface, the abutting edges of neighboring plates slide horizontally past one another in a **transform boundary**, and rock is neither produced nor destroyed. The San Andreas Fault in California is an example. A summary of these boundary types is illustrated in ● Figure 21.10.

---

*Tectonics (from the Greek *tekto*, for "builder") is the study of the Earth's general structural features and their changes.

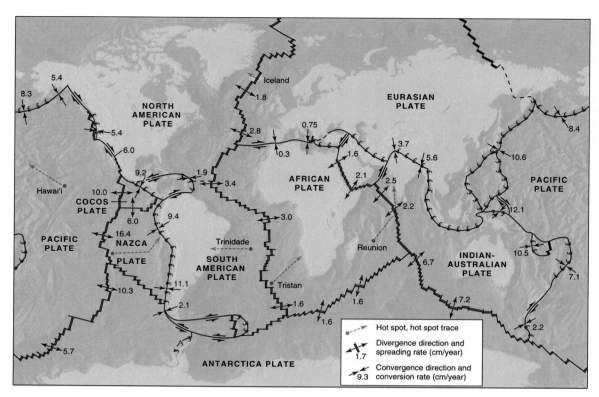

Figure 21.9 Plate Movement
The directions and rates at which the Earth's plates move, calculated from marine magnetic anomalies, offset rocks along transform faults, and island distances relative to hot spots. The Hawaiian Islands are examples of the work of a volcanic hot spot (see Chapter 22.4).

The movement of plastic rock during structural adjustments takes place essentially within the asthenosphere. The lithosphere essentially "floats" on the asthenosphere. The continents float higher because of their lower density. This state of buoyancy is called **isostasy** (● Fig. 21.11).

Isostatic balance can be maintained only because rock is plastic deep below the surface within the asthenosphere, where pressures and temperatures are very large. Hence, rock, although still solid, has the capacity to flow and therefore to adjust itself to the distribution of mass above it.

The difference in density in the crustal rock accounts for the differences in elevation between ocean basins and continental masses. There are large differences in height within a continent; high mountain chains and low river valleys are both in isostatic balance. How is that possible? (See the following Conceptual Question and Answer.)

Conceptual Question and Answer

Continents in Balance

Q. How do towering mountain chains and low elevation plains both stand in isostatic balance if they are basically the same type of rock?

A. This arrangement is possible because the continental masses vary markedly in thickness. Just as the top of a massive iceberg floats higher above water than an ice cube, so does a thick mass of crustal rock stand high as a mountain, whereas a thin mass of crustal rock is only slightly above sea level (see Fig. 21.11).

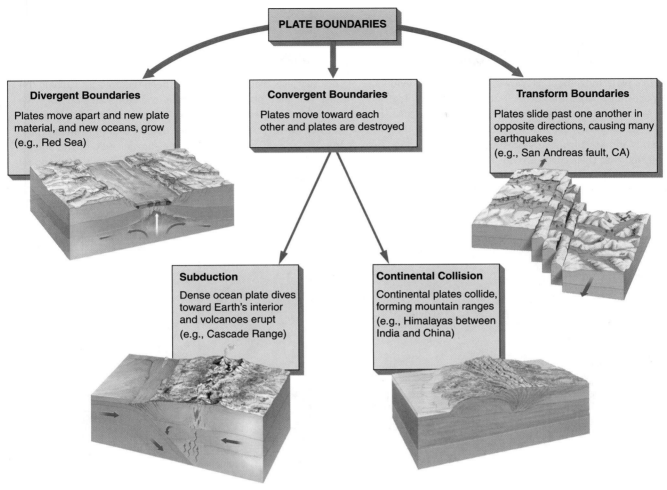

Figure 21.10 **A Summary of Plate Boundaries**
(See text for description.)

PLATE BOUNDARIES

Divergent Boundaries
Plates move apart and new plate material, and new oceans, grow (e.g., Red Sea)

Convergent Boundaries
Plates move toward each other and plates are destroyed

Transform Boundaries
Plates slide past one another in opposite directions, causing many earthquakes (e.g., San Andreas fault, CA)

Subduction
Dense ocean plate dives toward Earth's interior and volcanoes erupt (e.g., Cascade Range)

Continental Collision
Continental plates collide, forming mountain ranges (e.g., Himalayas between India and China)

Figure 21.11 **Isostasy**
The principle of isostasy states that the depth to which a floating object sinks into underlying material depends on the object's density and thickness. (a) All floating blocks of ice have the same density; thus, they sink in water so that the same proportion of their volume, about 90% (actually 89%), becomes submerged. The thicker the block of ice, the greater depth it reaches below the surface. (b) Continental lithosphere, because it is of lower density than oceanic lithosphere, is more buoyant and has a larger proportion of its volume above the asthenosphere. Because it is so thick compared with oceanic lithosphere, however, continental lithosphere extends farther into the asthenosphere (that is, it has a deep "root").

90% of larger iceberg submerged

90% of smaller iceberg submerged

Continental crust

Mountains

Oceanic crust

Lithosphere

Deep "root"

Asthenosphere

(a)

(b)

Highlight   Crustal Motion on Other Planets and Moons

Do other planets, their moons, and the Earth's Moon show plate tectonic motion? The simple answer at this time is no, not like the Earth's plate movement. However, there is ample evidence from spacecraft images that other worlds are also geologically active. Mercury is not believed to be geologically active, but crustal deformations show that it probably shrunk in the past. Both Venus and Mars have very obvious ancient volcanoes, but it is believed that their crusts are not currently able to support tectonic motion (Fig. 1). The crust of Venus might be too thick to support plate motions, or too hot and soft, or both. Mars' crust

was also believed to be too thick, but recent analysis of crustal features and the discovery of crust with magnetically aligned rock suggests that Mars may have been tectonically active in the past. The giant planets are gaseous; therefore, they do not have a solid surface (see Chapter 16.5). Their moons do show signs of activity, however. Figure 2 shows Jupiter's moons Io and Europa; Io is volcanically active, and Europa shows ice floes. The energy needed to form these features comes from Jupiter's gravitational tidal forces that create internal heat within the moons. The source for the energy may be different than that in the Earth, but the

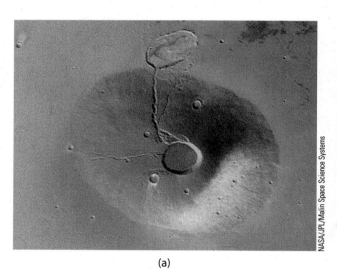

(a)

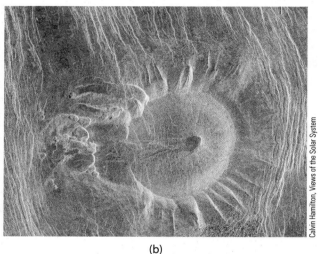

(b)

Figure 1 Mars and Venus
(a) The dormant volcano Ceraunius Tholus on Mars. (b) The "tick" volcano on Venus.

The forces that move the plates—one against another, one away from another, or one past another—are also found within the asthenosphere. The plates, which are segments of the lithosphere, are actually passive bodies driven into motion by the drag of the more active asthenosphere beneath them.

Geologists, although they differ in the explanations of details, view this motion in terms of convection cells that form in all or part of the Earth's mantle, perhaps in the *asthenosphere* (● Fig. 21.12). As an analogy, consider a large pot of thick tar sitting over a fire. The tar at the bottom of the pot absorbs heat, expands, becomes less dense, and rises to the surface through the colder tar. At the surface, the tar spreads, loses heat to the air, becomes denser, and eventually sinks back to the bottom of the pot, where reheating occurs. This type of heat movement is called convection (Chapter 5.4), and the cyclic motion of the material constitutes a *convection cell*.

The heat that causes the convection cells to form is thought to come principally from radioactive decay within the mantle. The lower mantle's contact with the hot, outer core provides additional heat.

Is there plate tectonic activity on other planets? With the prevalence of more sophisticated technology and spacecraft, scientists are finding out more about our planetary neighbors. There is evidence of bulk geologic activity on other planets and moons in the solar system that is similar to activity on the Earth. For more details, see the **Highlight: Crustal Motion on Other Planets and Moons.**

Highlight

manifestations (volcanoes, ice floes) are similar. Saturn's moon Enceladus and Neptune's moon Triton are also known to be geologically "active" (see Chapter 17.4).

What about our Moon? Its crust is too thick and its mantle is too solid and cold to generate enough energy to cause plate motion.

However, have you ever heard of moonquakes? Several different types of moonquakes do exist, but they are extremely small in scale. They are probably caused by gravitational forces from the Earth or extreme changes in temperature.

(a)

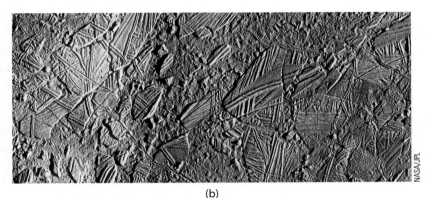

(b)

Figure 2 Moons of Jupiter
(a) A volcanic plume on Io, seen over the limb of the moon in June 1997. (b) The icy surface of Europa, showing evidence of ice floes similar to polar ice floes on Earth.

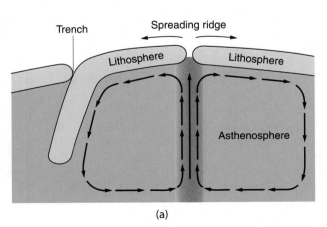

(a)

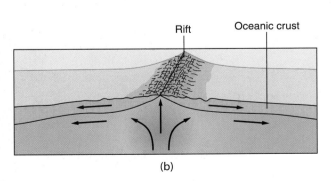

(b)

Figure 21.12 Convection Cells
(a) Unequal distribution of temperature within the Earth causes hot, less dense material to rise and cooler, denser material to sink, generating convection currents in the asthenosphere. The lithospheric plates, resting on the asthenosphere, are put in motion by the driving force of the convection cells. (b) As two plates move apart at divergent boundaries, molten rock rises into the rift, cools, and forms new oceanic crust.

Plate Motion at Divergent Boundaries

Let's examine the role of convection cells in plate tectonics by focusing attention first on divergent boundaries, where the plates are being driven away from one another (Fig. 21.12). Beneath these spreading mid-ocean ridges, convection currents lift material from the hot asthenosphere up into a region of lower pressure, where it begins to melt. The material melts only partially, however, forming plastic rock that can flow.

As more mantle material wells up from below, the upper-mantle material is forced to both sides and moves slowly in a horizontal direction beneath the lithospheric plates. It is the drag of the mobile asthenosphere against the bottom of the lithospheric plates that keeps the plates in motion.

A lithospheric plate cools as it moves slowly away from the hot, spreading ridge. Therefore, the plate loses volume. As a result, the top of the plate gradually subsides and causes the oceans to grow progressively deeper away from the spreading ridges. The slope of denser, cooler rock away from the mid-ocean ridges also contributes to the motion of a plate at convergent boundaries.

Plate Motion at Convergent Boundaries

When two plates collide, what happens next depends on whether the colliding margins are oceanic or continental crust. Three combinations are possible: (1) oceanic–oceanic, (2) oceanic–continental, and (3) continental–continental.

1. *Oceanic–oceanic convergence.* When two oceanic plates collide, they buckle, and one of them, ever so slowly, slides beneath the other. The place, or zone, where the plate descends into the asthenosphere is called a *subduction zone*, and the process is called **subduction**. The descending oceanic plate, now in contact with the asthenosphere, begins to melt. The molten material begins to rise, and a series of volcanoes develops in an arc shape on the overriding plate (● Fig. 21.13). Deep-sea trenches lie in front of the arc system, marking the places where the plates are being subducted. The deepest trench known is the Marianas Trench in the western Pacific Ocean, which is 11 km (7 mi) below sea level.

2. *Oceanic–continental convergence.* Whenever oceanic crust collides with lower-density continental crust, it is always subducted beneath the continental crust (● Fig. 21.14).

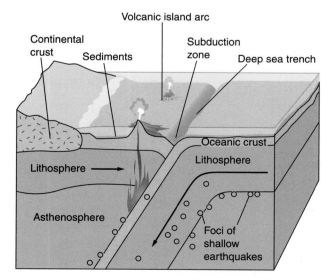

Figure 21.13 The Convergence of Two Oceanic Plates
The subduction zone is the region in which one lithospheric plate plunges beneath another. Destructive earthquakes occur most often in the subduction zones.

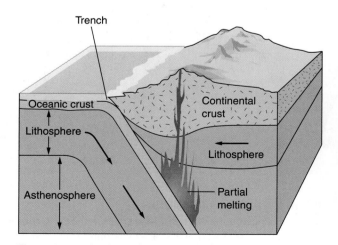

Figure 21.14 An Oceanic–Continental Convergence
Examples are the Andes Mountains of South America and the Cascade Mountains of North America.

A trench will develop at the point where the oceanic plate is being subducted, but it is never as deep as the trench formed in the oceanic–oceanic convergence. The oceanic plate begins to melt as it descends into the asthenosphere. Molten rock then moves up into the overriding plate, causing large igneous intrusions and often volcanic mountains at the surface. Examples include the Andes Mountains of South America and the Cascade Mountains of Washington and Oregon in North America.

3. *Continental–continental convergence.* The continental plates have about the same density, so when they collide with one another, the boundary of rock between the continental edges is pushed and crumpled intensely to form fold-mountain belt systems (● Fig. 21.15). In this manner, continents grow in size by "suturing" themselves together along fold-mountain belt systems. Examples include the Alps, the Appalachians, and the Himalayas. As will be seen later in this chapter, geologists believe that the Himalayas formed in this manner when the Indian plate collided with the Eurasian plate.

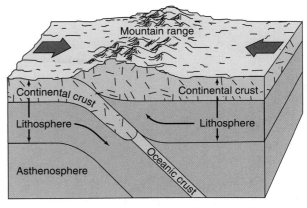

Figure 21.15 A Continental–Continental Convergence The boundary of rock between the continental edges is pushed and crumpled to form fold-mountain systems. The Appalachian Mountain Range in the eastern United States is an example.

Plate Motion at Transform Boundaries

A zone of shear faults, or transform faults, is a boundary at which adjacent plates slide past each other with no gain or loss in surface area. This zone occurs along faults that mark the plate boundaries. Movements and the resulting release of energy along these boundaries give rise to earthquakes. Examples of such fault zones are the San Andreas Fault in California and the Anatolian Fault in Turkey.

Did You Learn?

- The Earth's plates, part of the lithosphere, actually "float" on the asthenosphere, allowing them to move.
- Common features at plate boundaries are volcanoes, mountains, fault lines, and seas.

21.4 Plate Motion and Volcanoes

Preview Questions

- What is the Ring of Fire?
- Why do volcanoes occur near the boundaries of ocean and continental plates?

A **volcano** can mean either a vent from which hot molten rock (lava), ash, and gases escape from deep below the Earth's surface, or it can refer to the mountain or elevation created by solidified lava and volcanic debris that accumulates near the vent. (The word *volcano* comes from Vulcan, the Roman god of fire.) It is common to visualize a volcano as an erupting mountain, but that is not always the case. (See Chapter 22.4 for more information about volcanoes.) The occurrence of volcanic activity is for the most part unpredictable. New volcanoes may form unexpectedly, and existing volcanoes lying dormant may suddenly erupt with practically no warning. However, the locations of eruptions and potential eruptions are known.

The map in ● Fig. 21.16 shows where most of the active volcanoes of the world are located. Volcanoes are so numerous along the margins of the Pacific Ocean that the region has been dubbed the *Ring of Fire*. The theory of *plate tectonics* explains not only the existence of the Ring of Fire but also the locations of volcanoes the world over. It is mainly along convergent plate boundaries that energy is released in the form of volcanoes and earthquakes. As Fig. 21.16 shows, a large majority of the world's volcanoes occur where

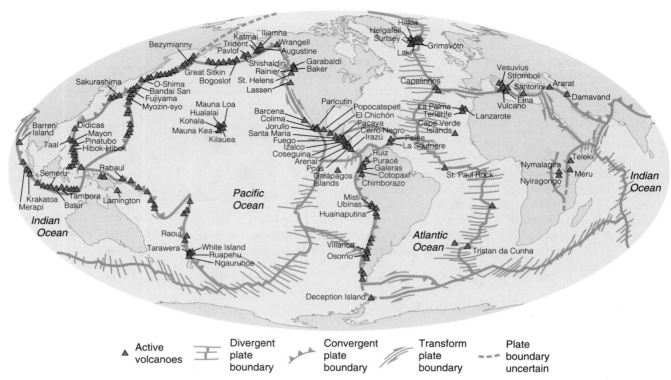

| Active volcanoes | Divergent plate boundary | Convergent plate boundary | Transform plate boundary | Plate boundary uncertain |

Figure 21.16 Active Volcanoes of the World
This map shows the three types of plate boundaries: divergent, convergent, and transform.
Notice that the great majority of active volcanoes lie along these boundaries.

plates collide with one another. Note there are very few volcanoes in Africa; the African continent is surrounded mostly by divergent plate boundaries.

As pointed out previously, when two plates collide and both consist of oceanic crust, subduction occurs. During this process, rock just above the subducting plate margin melts and the molten rock rises to the surface to form volcanic islands (● Fig. 21.17). Various segments of the Pacific's Ring of Fire, such as Indonesia, the Philippines, Japan, and the Aleutians, are volcanic island arcs.

Figure 21.17 Subduction of Oceanic Lithosphere Forms a Volcanic Island Arc
This model shows the tectonic activity in the region of the islands of Japan. For the big picture, look for this region on the western edge of the Pacific Ocean in the world map shown in Fig. 21.16.

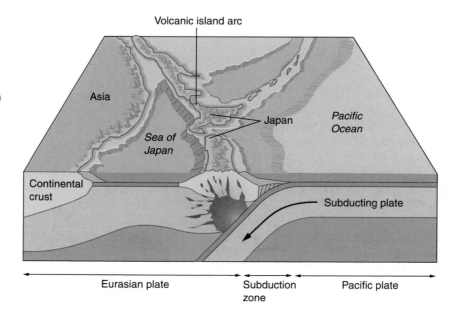

Where the oceanic crust of one plate collides with a continent on a neighboring plate, the oceanic crust subducts because it is denser and the rising molten rock forms a volcanic mountain chain along the margin of the continent. The Andes Mountains of western South America and the Cascade Mountains of western North America are volcanic mountain chains (Fig. 21.14).

Did You Learn?

- The Ring of Fire is the geologically active region around the rim of the Pacific Ocean.
- Volcanoes occur where one plate subducts under another, which allows hot magma to rise to the surface.

21.5 Earthquakes

Preview Questions

- What causes an earthquake?
- On what crustal plate is Los Angeles located and in which direction is the plate moving?

The study of earthquakes is called *seismology*. An **earthquake** involves a tremendous release of energy accompanying the rupture or repositioning of underground rock and is manifested by the vibrating and sometimes violent movement of the Earth's surface. Earthquakes rattle our globe perhaps a million times each year, but the vast majority are so mild that they can only be detected with sensitive instruments. A very powerful quake, however, can lay waste to a large area. The chapter-opening photo shows the power that earthquakes have to alter the surface of the Earth.

The Causes of Earthquakes

Earthquakes can be caused by explosive volcanic eruptions or even explosions detonated by humans, but the great majority are associated with the movement of lithospheric plates. These movements form large crustal features called **faults**. In a fault, the rock on one side of the fracture has moved relative to the rock on the other side of the fracture (● Fig. 21.18). The optimum place for movement is at the plate boundaries; thus, the major earthquake belts of the world are observed in these regions (● Fig. 21.19). Note how the Pacific Ocean earthquake region is similar to that of the volcanic Ring of Fire (see Fig. 21.16).

Movement of neighboring plates exerts stress on the rock formations along the plate margins. Because rocks possess elastic properties, energy is stored until the stresses acting on the rocks are great enough to overcome the force of friction. Then the fault walls move suddenly, and the energy stored in the stressed rocks is released, causing an earthquake. After a major earthquake, the rocks may continue to adjust to their new positions, causing additional vibrations called *aftershocks*. In some cases, the aftershocks are quite large. After the tremendous 9.0 magnitude earthquake off the northeast coast of Japan in 2011, hundreds of aftershocks were recorded, some as large as magnitude 7.4 (earthquake magnitudes are discussed below).

Many large horizontal faults associated with plate boundaries are transform faults. An example of a transform fault is the famous San Andreas Fault in California. It is the master fault of an intricate network of faults that runs along the coastal regions of California (● Fig. 21.20). This huge fracture in the Earth's crust is more than 960 km (595 mi) long and at least 32 km (20 mi) deep. Over much of its length, a linear trough of narrow ridges reveals the fault's presence.

As Fig. 21.20 shows, the San Andreas Fault system lies on the boundary of the Pacific plate and the North American plate. Movement along the transform fault arises from the relative motion between these plates.

University of Washington Libraries, Special Collections, John Shelton Collection, KGR318

Figure 21.18 Faults
Faulted rocks are those that have moved relative to one another. The fault in this photo is evident in the displacement of the various colored rock layers.

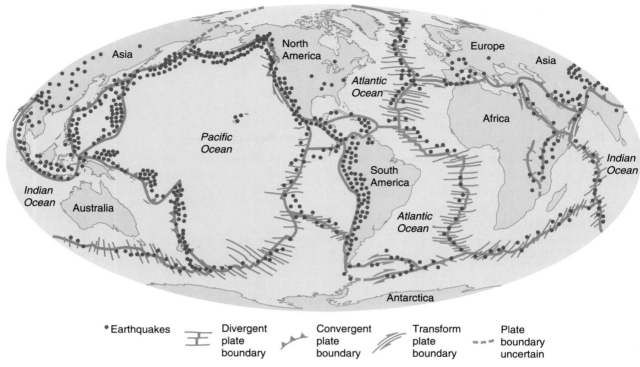

Figure 21.19 World Map of Recorded Earthquakes
The great majority of earthquakes occur at plate boundaries.

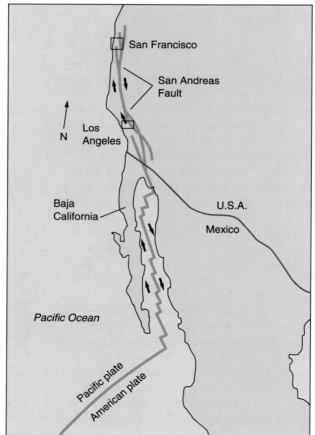

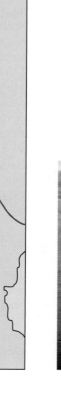

Figure 21.20 The San Andreas Fault
Left: A map showing the boundary, marked by the fault, between the Pacific and North American plates, which are moving in opposite directions. This fault is a classic example of a transform fault. *Right*: An aerial view of the fault on the Carrizo Plain, California.

Conceptual Question and Answer

Los Angeles Meets San Francisco

Q. Due to plate tectonics, will Los Angeles split off of the West Coast and move out into the ocean?

A. The San Andreas fault is a *transform boundary*; therefore, the two plates are moving along the boundary, not separating. The Pacific plate is moving northward relative to the North American plate at a rate of several centimeters per year. At this rate and direction, in about 10 million years Los Angeles will have moved northward to the same latitude as San Francisco.

The horizontal movement along the San Andreas Fault is cause for considerable concern in the densely populated San Francisco Bay area through which it runs. Many earthquakes, both mild and strong, have occurred along the San Andreas. The famous San Francisco earthquake of April 18, 1906, resulted in the destruction of most of the city. In 1989, another major earthquake shook the San Francisco Bay area, damaging bridges, buildings, and highways. Little can be done about the San Andreas Fault except to study it and live with it. Building codes in the area are strict. However, increased population has pushed housing construction right onto some fault lines. People live in these houses with the knowledge of the fracture in the crust below. (See the **Highlight: Earthquake Risk in North America** for an indication of the earthquake risk in your area.)

When an earthquake does occur, the point of the initial energy release or slippage is called its **focus**. A focus generally lies at considerable depth, from a few miles to several hundred miles. Consequently, geologists designate the location on the Earth's surface directly above the focus as the **epicenter** of the quake (● Fig. 21.21). The epicenter is the point on the surface that receives the greatest impact from the quake. You can remember the distinction between the focus and the epicenter if you recall that skin, the outer layer of the body, is called the epidermis.

The energy released at the focus of an earthquake propagates outward as *seismic waves* (S and P waves; Chapter 21.1). The seismic waves of an earthquake are monitored by an instrument called a *seismograph*, the principle of which is illustrated in ● Fig. 21.22. The greater the energy of the earthquake, the greater the amplitude of the traces on the recorded *seismogram*.

The severity of earthquakes is represented on different scales. The most common is the *Richter scale*. A reading on this scale is an absolute measure of the energy released; it is found by calculating the energy of seismic waves at a standard distance from the epicenter.

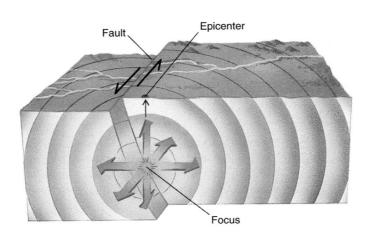

Figure 21.21 Focus and Epicenter
The focus of an earthquake is the point of initial rupture, where seismic waves originate. The epicenter is the surface projection of the focus.

Highlight   Earthquake Risk in North America

Many dangerous geologic effects occur during an earthquake. They include large-scale shifting of the landscape, landslides and mudflows, liquefaction of soil into a kind of quicksand, and tsunamis. For example, in 1692 liquefaction and landsliding sent the city of Port Royal, Jamaica, sliding into the ocean.

In the United States, we are all aware that California is earthquake-prone. Other areas of the country have also experienced violent quakes, and not all quakes occur at present plate boundaries. In late 1811 and early 1812, the area around New Madrid, Missouri, was hit by a series of earthquakes. It was reported that the last major shock broke windows as far away as Washington, D.C.,

and even caused church bells to toll in Boston, Massachusetts. At times the Mississippi River flowed backwards, and Reelfoot Lake in northern Tennessee was formed when a vast area of land subsided during the quakes. Fortunately, the area was sparsely populated in 1811–1812, but that is not the case today. Major cities such as Memphis and St. Louis still lie under the influence of the New Madrid Fault.

Do earthquakes pose a significant risk to you in your present locality? Should you purchase earthquake insurance? Examining Fig. 1 may help you answer such questions.

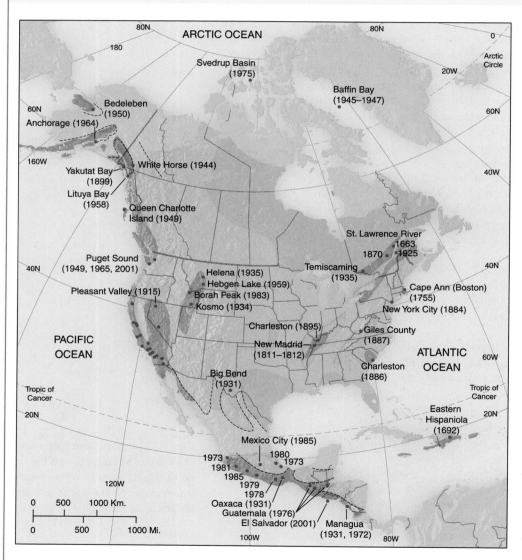

	Major damage
	Moderate damage
	Minor damage
	No damage

• Magnitude > 6 earthquake epicenter

------ Seismic-risk zone boundary poorly defined

Figure 1 A Seismic-Risk Map of North America
Significant earthquake activity does not occur just at plate boundaries, as this seismic-risk map of North America shows. Also marked on the map are the locations and dates of some of the continent's major historical earthquakes.

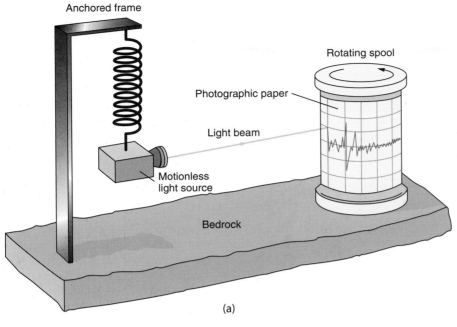

Anchored frame

Rotating spool

Photographic paper

Light beam

Motionless
light source

Bedrock

(a)

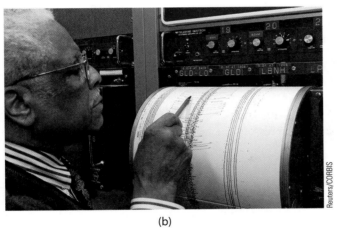

Reuters/CORBIS

(b)

Figure 21.22 An Illustration of the Principle of the Seismograph
(a) The rotating spool (anchored in bedrock) vibrates during the quake. A light beam from the relatively motionless source on a spring traces out a record, or seismogram, of the earthquake's energy on light-sensitive photographic paper. (b) A geologist studies a modern seismograph record.

The Richter scale, developed in 1935 by Charles Richter of the California Institute of Technology, has magnitudes expressed in whole numbers and decimals, usually between 3 and 9. It is a logarithmic scale, however; that is, each whole-number step represents about 31 times more energy than the preceding whole-number step. For example, when an earthquake registers a magnitude of 5.5 on the Richter scale, the measuring seismograph has received about 31 times more energy as it would from a magnitude 4.5 earthquake. An earthquake with a magnitude of 2 is the smallest tremor felt by humans. The largest recorded earthquakes range from 8.7 to 9.5 in magnitude.

The Richter scale gives no indication of the damage caused by an earthquake, just its potential for damage (● Table 21.1). Damage depends not only on the magnitude of the quake but also on the location of its focus and epicenter and on the environment of that region, specifically the local geologic conditions, the density of population, and the construction designs of buildings.*

*Another earthquake scale, the *Mercalli scale*, originally developed by Italian volcanologist Giuseppi Mercalli in 1902, describes the severity of an earthquake in terms of its observed damage effects. A useful scale for geologists is the moment magnitude scale, which gives a measure of the total amount of energy released by an earthquake. This moment magnitude scale is set by the length, depth, and the amount of slippage of the rupture.

Table 21.1 Earthquake Severity

Richter Magnitude	Effects
Less than 3.5	Generally not felt, but recorded.
3.5–5.4	Often felt, but rarely causes damage.
Under 6.0	At most, slight damage to well-designed buildings. Can cause major damage to poorly constructed buildings over small regions.
6.1–6.9	Can be destructive in areas up to about 100 km from the epicenter.
7.0–7.9	Major earthquake. Can cause serious damage over larger areas.
8.0 or greater	Great earthquake. Can cause serious damage in areas several hundred kilometers across.

An earthquake can be low on the Richter scale and still cause great damage, whereas another can be high on the Richter scale and cause relatively little damage. For example, an earthquake of magnitude 6.6 that struck southeastern Iran in 2003 killed more than 26,000 people, but in 1992 a California earthquake of magnitude 7.3—more than ten times stronger than the Iranian quake—killed only one person, in the town of Landers. The reason is clear. Landers is a small town in a sparsely populated region of the Mojave Desert. There are only a few buildings in the town, and most have been constructed according to strict state building codes. The Iranian quake was centered 10 km (6 mi) outside of Bam, Iran, an ancient city with a population of 78,000. Seventy percent of the city's mud-brick buildings collapsed on their inhabitants almost as soon as the shaking began.

Conceptual Question and Answer

The 2010 Big Shake in Haiti

Q. What are two reasons the magnitude 7.0 earthquake in Haiti in 2010 was so deadly?

A. The Haiti earthquake was devastating. More than 300,000 people were killed and another 1 million were left homeless. Not only was it a strong earthquake causing incredible damage, but Haiti is a poor country with few resources for earthquake-proof buildings. The combination of a strong earthquake and poor construction combined to make this earthquake an exceptional tragedy.

Earthquake damage may result directly from the vibrational tremors or indirectly from landslides and subsidence, as in the magnitude 8.4 earthquake that struck Anchorage, Alaska, in 1964 (● Fig. 21.23). In populated areas, a great deal of the property damage is caused by fires because of a lack of ability to fight them as a result of disrupted water mains and roads.

Earthquakes that occur at subduction zones can be particularly devastating because they pack a one-two punch: movement of land *and* water. Subduction zones occur near the edges of continents, so not only can the earthquake cause land damage, but the energy release from the ocean floor sometimes pushes an incredible amount of water upwards and causes huge waves called **tsunamis**. These devastating waves sometimes cause more damage than the earthquake, such as with the 2011 earthquake near Japan and the 2004 earthquake off the coast of Indonesia. (See the **Highlight: Deadly Tsunamis**.)

U.S. Geological Survey

Figure 21.23 Subsidence
Subsidence damage from the 1964 Alaska earthquake was extensive. In downtown Anchorage, seen here, the subsidence was more than 3 m (10 ft).

Did You Learn?

● Most earthquakes result from the movement of lithospheric plates.

● Los Angeles is located on the Pacific plate, and the plate is moving northward.

21.6   Crustal Deformation and Mountain Building

Preview Questions

- What is the primary cause of mountain building?
- How were the Himalaya Mountains formed?

The forces that build up in the vicinity of plate boundaries can buckle and fracture rocks or break them and shift their positions. Plate edges can be ruptured into huge displaced blocks or squeezed together and uplifted into great folds. In this section we will look briefly at the ways in which plate movements cause crustal deformation and create mountain ranges.

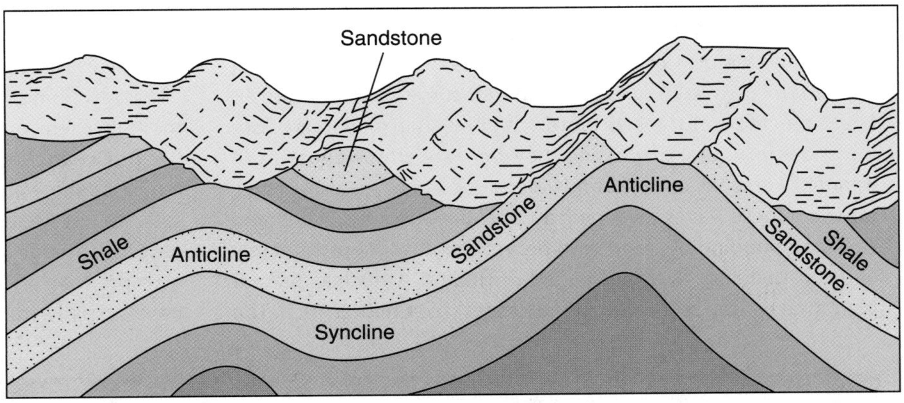

Crustal Deformation

Two major types of slow structural deformations are folding and faulting. *Folding* of the Earth's crust takes place when extreme pressure is exerted horizontally or vertically. Rock can be compressed only a limited amount before it begins to buckle and fold. The folded rock layers, or **fold**, can form an arch (*anticline*) or a trough (*syncline*). A good way to remember the distinction is that both *arch* and *anticline* start with the letter *a*. ● Figure 21.24 illustrates the principal types of folded rocks. The folding occurs during the early stages of mountain formation.

Faulting begins with fracturing. Fractures are breaks in rock caused by stress; faults are fractures that display evidence that one side of the fracture has moved relative to the other. The stresses may be vertical and produce uplifts, they may be horizontal and produce compressions, or they may cause tension and bring about a lengthening of the crust. ● Figure 21.25 illustrates a few essential terms needed to describe fault geometry. The *fault plane* is

Figure 21.24 Types of Folds
Here are some of the more common relationships between anticlines and synclines and the landforms above them.

Figure 21.25 Some Common Types of Faults and Their Associated Terminology
Relative motion of the opposing blocks on either side of the fault plane determines whether a fault is classified as normal, reverse, or transform (a transform fault is also called a strike-slip fault). A thrust fault is a low-angle reverse fault. (See text for description.)

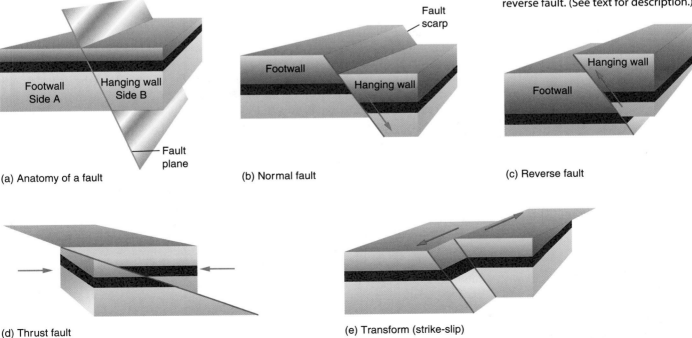

(a) Anatomy of a fault

(b) Normal fault

(c) Reverse fault

(d) Thrust fault

(e) Transform (strike-slip)

Highlight Deadly Tsunamis

On March 11, 2011, at 2:46 P.M. local time, a massive submarine earthquake of magnitude 9.0 struck 130 km (80 mi) off the northeast coast of Japan. The event occurred at a subduction zone at a depth of 32 km (20 mi) where the Pacific plate slipped under Japan and the Eurasian plate. This quake, the fifth largest quake in modern history, moved the seafloor upwards approximately 30–40 m (100–130 ft) causing tremendous tsunami waves.

With only minutes of warning, the tsunami struck the coast of Japan with a wave 10 m (33 ft) high that surged inland up to 10 km (6 mi). Most of the coastal towns in northern Japan are fishing and agricultural towns, and they were literally swept away in minutes. Personal accounts and videos document the destructive power of these waves, sweeping away entire buildings, trains, and cars, and flooding the land with debris and mud. Approximately 100,000 buildings were damaged, including a nuclear reactor plant that caused some radiation releases (see Chapter 10.5). The

human cost was also high: more than 15,000 deaths and another 17,000 people missing. The devastation is seen in the before-and-after photos of Japan shown in Fig. 1.

These recent events are all too familiar to people in Indonesia and surrounding nations who suffered incredible tsunamis after an earthquake in 2004. In fourteen countries, the total death toll was more than 230,000, making that tsunami one of the worst natural disasters in recorded history.

Most tsunamis are generated by submarine earthquakes, and most of them originate in the subduction zones that rim the Pacific Ocean. The vibrations set off on the ocean floor are transferred to the water and radiate outward across the ocean in a series of low waves. On the open sea, a tsunami wave is difficult to detect. Although the wave is many kilometers in length, it may be no more than a 50-cm-high bulge on the surface of the water. It travels rapidly, though—at speeds of up to 960 km/h (595 mi/h)—

AFP PHOTO/HO /GEODAC, National Cheng-Kung University - National Space Organization, Taiwan

Figure 1 Before-and-After Photos of the 2011 Japanese Tsunami
These images show the same part of the Japanese coastline before March 11, 2011 (left), and again after the tsunami (right). The force of the waves destroyed a vast number of buildings in the photograph. Much of the coastline was flooded, as well as some agricultural lands up to 10 km (6 mi) inland.

the actual surface itself, the *hanging wall* is the rock on the upper side of the inclined fault plane, and the *footwall* is the rock on the underside (Fig. 21.25a). The upthrown side has moved up relative to the downthrown side, as indicated by the arrows in Fig 21.25b–d.

The three types of faults are *normal*, *reverse*, and *transform*. Faults may occur in any type of rock, and they may be vertical, horizontal, or inclined at an angle. Most faults occur at an angle.

A *normal fault* occurs as the result of expansive forces that cause its overlying side to move downward relative to the side beneath it (Fig. 21.25b). In this case, the stress forces are in opposite directions and the faulting tends to pull the crust apart. A *reverse fault* occurs as the result of compressional stress forces that cause the overlying side of the fault to move upward relative to the side beneath it (Fig. 21.25c). A special case of reverse faulting, called *thrust faulting*, describes the faulting when the fault plane is at less than a 45° angle to the horizontal (Fig. 21.25d). As might be expected, this type of faulting occurs in subduction zones, where one plate slides over a descending plate, that is, along convergent plate boundaries.

Finally, *transform*, or *strike-slip, faulting* occurs when the stresses are parallel to the fault boundary such that the fault slip is horizontal (Fig. 21.25e). This type of faulting takes

Highlight

and can travel for great distances, even across the Pacific Ocean, about 14,500 km (9000 mi). When it reaches the shallow coast, friction slows the wave down at the same time as it causes it to roll up into a 5- to 30-m-high wall of water that crashes down on the shoreline with immense force (Fig. 2). About 10 to 11 hours after the Japanese quake in 2011, a 2-m (7-ft) tsunami wave struck the West Coast of Oregon and California, causing serious damage to some harbors, docks, and boats.

A tsunami coming ashore is commonly referred to as a *tidal wave*, but these deadly waves are not related to tides. Scientists have adopted the more appropriate Japanese name *tsunami*, which means "harbor wave." The effects of a tsunami are intensified in the confined spaces of bays, where most harbors are located. As a tsunami approaches a shallower harbor and rolls up to its full height, the water that was in the harbor is drawn out and into the oncoming wave. Coastal residents of Japan know that when a harbor is suddenly emptied, it is the signal to drop everything

and run to high ground, because a tsunami is only seconds away.

The coastlines of the Pacific Rim (Japan, Hawaii, Asia, Indonesia, and the Americas) are especially vulnerable. Japanese history records at least 15 major tsunami disasters, some of which killed hundreds of thousands of people.

Fortunately for countries on the Pacific Rim, a tsunami warning system is now in place. This warning system consists of technologically advanced hardware, including seismometers and undersea and floating ocean monitoring stations as well as Earth-orbiting relay satellites. Furthermore, a coordinated effort of scientists and civil authorities from many countries is needed to alert populations in threatened areas. These efforts and warnings probably saved thousands of lives in Japan, a country that routinely practices tsunami evacuation drills. Unfortunately, that was not the case for the 2004 Indonesian tsunami, which caused much more loss of human life. As a result of that devastation, an Indian Ocean tsunami warning system was put in place in 2008.

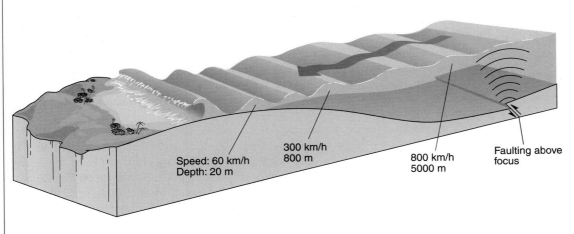

Speed: 60 km/h
Depth: 20 m

300 km/h
800 m

800 km/h
5000 m

Faulting above focus

Figure 2 Behavior of a Tsunami
As a tsunami approaches shore, the wave is shortened in length while its height is increased. Notice that the harbor (*left*) empties out just before the wave strikes.

place along the transform boundary of two plates that *strike* and *slip* by each other without an appreciable gain or loss of surface area.

Mountain Building

Mountain building occurs primarily because of converging plate boundaries. Mountains are generally classified on the basis of characteristic features into three principal kinds: (1) volcanic, (2) fault-block, and (3) fold.

Volcanic mountains, as the name implies, have been built by material ejected during volcanic eruptions. As discussed in Chapter 21.4, most volcanoes, and hence volcanic mountains, are located above the subduction zones of plate boundaries. If the colliding plates are both oceanic, then volcanic mountain chains are formed on the ocean floor of the overlying plate. Chains of oceanic volcanic mountains are manifested above sea level by island arcs; Japan and the islands of the West Indies are good examples.

Continental volcanic mountain chains occur when the overlying plate of the subduction zone is continental crust. The Andes Mountains along the western coast of South

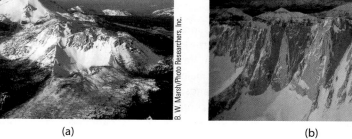

(a)   (b)   (c)

Figure 21.26 Three Types of Mountain Building
(a) Volcanic mountains. The Cascades in Washington and Oregon are an example of a continental volcanic mountain chain. (b) Fault-block mountains. Mount Whitney, in California, is the highest peak in the Sierra Nevada Range. (c) Fold mountains in Alberta, Canada. Notice the arch (anticline) in the upper right.

America are an example of such a continental mountain range. The Cascade Mountains in Washington and Oregon are the only active volcanic mountains in the conterminous United States (● Fig. 21.26a).

Fault-block mountains are believed to have been built by normal faulting in which giant pieces of the Earth's crust were faulted and uplifted at the same time (Fig. 21.26b). These mountains give evidence of the great stresses within the Earth's lithospheric plates. Fault-block mountains rise sharply above the surrounding plains and are usually steep on the faulted side but gently sloping on the opposite side. The Sierra Nevada Range in California, the Grand Teton Mountains in Wyoming, and the Wasatch Range in Utah are examples of fault-block mountains in the United States.

A **fold mountain**, as the name implies, is characterized by folded rock strata. Examples of fold mountains include the Alps, the Himalayas, and the Appalachian Mountains (Fig. 21.26c).

Although folding is their main feature, these complex structures also contain external evidence of faulting and internal evidence of high temperature and pressure changes. Fold mountains are also characterized by exceptionally thick sedimentary strata, which indicate that the material of these mountains was once at the bottom of an ocean basin. Indeed, marine fossils have been found at high elevations in the Himalayas and other fold-mountain systems.

Within the framework of plate tectonics, consider the formation of the grandest fold mountains of all, the Himalayas. These mountains are believed to have formed after the supercontinent Pangaea broke up some 200 million years ago. In short, India broke away from Africa and ran into Asia, and the collision resulted in the Himalayas (● Fig. 21.27).

In more detail, the northward movement of the Indian plate after its separation from Africa resulted in the loss of oceanic lithosphere formerly separating India and Asia, with the eventual collision of these two continental masses. The Indian plate descended under the Eurasian plate, and the depositional basins along the Eurasian plate were folded into mountains, as illustrated in ● Fig. 21.28. When the two continental plates met, the edge of the Eurasian plate was lifted, the spectacular result being one of the highest mountain ranges on the Earth. Present-day analyses of plate movements indicate that the Indian plate is still moving northward relative to the Eurasian plate, causing the slow, continuing uplift of the Himalayas.

Did You Learn?

● Mountain building occurs primarily from converging plate boundaries.

● The Indian plate collided with the Eurasian plate, forming the Himalayas.

KEY TERMS

1. geology
2. seismic waves (21.1)
3. outer core
4. inner core
5. mantle

6. crust
7. lithosphere
8. asthenosphere
9. continental drift (21.2)
10. Pangaea

11. seafloor spreading
12. remanent magnetism
13. plate tectonics (21.3)
14. divergent boundary
15. convergent boundary

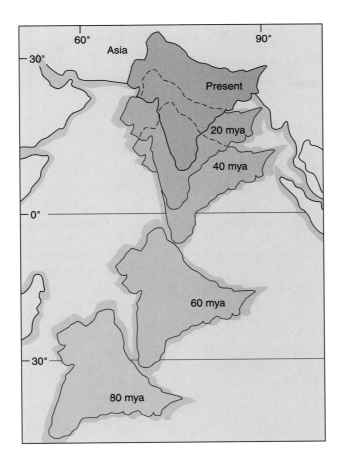

Figure 21.27 Collision of India with Asia
Following the breakup of Pangaea, the plate carrying the Indian continent moved northward until it collided with the Eurasian plate. (mya = million years ago.)

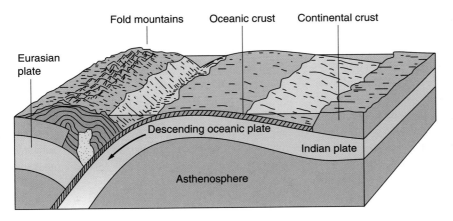

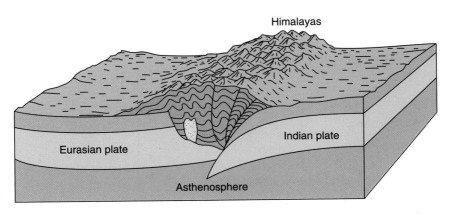

Figure 21.28 An Illustration of the Formation of the Himalayas
(a) As the northward-moving Indian plate descended under the Eurasian plate, the depositional basins along the Eurasian plate were folded into mountains. (b) When the two continental crusts met, the edge of the Eurasian plate was lifted, giving rise to the lofty Himalayas, including Mount Everest.

16. transform boundary
17. isostasy
18. subduction
19. volcano (21.4)
20. earthquake (21.5)

21. faults
22. focus
23. epicenter
24. tsunamis

25. fold (21.6)
26. volcanic mountain
27. fault-block mountain
28. fold mountain

MATCHING

For each of the following items, fill in the number of the appropriate Key Term from the preceding list. Compare your answers with those at the back of the book.

a. _____ A tremendous release of energy generally caused by the movements of lithospheric plates

b. _____ The solid, innermost region of the Earth; composed primarily of iron

c. _____ Large crustal fractures where rock on one side has moved relative to rock on the other

d. _____ The point on the Earth's surface that lies directly above the focus of an earthquake

e. _____ Sometimes-devastating sea waves that travel thousands of miles and are caused by earthquakes under the seafloor

f. _____ The largest interior layer of the Earth, about 2500 km (1550 mi) thick

g. _____ A region along the mid-ocean ridges where one plate is pulling away from another

h. _____ A mountain built by normal faulting, in which giant pieces of the Earth's crust are uplifted

i. _____ The giant supercontinent that is thought to have existed about 200 million years ago

j. _____ A mountain characterized by exceptionally thick sedimentary strata sometimes containing fossilized marine life

k. _____ The rigid, rocky outer layer of the Earth that includes the crust and part of the upper mantle

l. _____ The well-tested theory that Earth's surface is made of plates that move

m. _____ The layer of rock in the mantle that is close to its melting point and therefore is plastic and movable

n. _____ A region where two lithospheric plates are driven together and rock is consumed

o. _____ A process in which two plates collide and the denser plate is deflected downward beneath the other plate

p. _____ A mountain that has been built from material ejected during volcanic eruptions

q. _____ The point of the initial energy release or slippage during an earthquake

r. _____ The theory that the Earth's lithosphere is made up of giant, rigid plates that move relative to one another

s. _____ The state of buoyancy between the lithosphere and the asthenosphere

t. _____ Vibrations produced in and on the Earth during an earthquake

u. _____ An internal layer of the Earth that is liquid metal iron and nickel

v. _____ The outermost layer of the Earth

w. _____ Two lithospheric plates sliding past one another with no production or destruction of rock

x. _____ The magnetism retained in iron-containing rocks after they solidify in the Earth's magnetic field

y. _____ The process that produces new seafloor between two diverging plates

z. _____ A rock layer buckled into arches and troughs as a result of extreme horizontal pressure

aa. _____ A mountain or elevation formed by molten rock and debris ejected through a vent in the Earth's surface

bb. _____ The study of the planet Earth: its composition, structure, processes, and history

MULTIPLE CHOICE

Compare your answers with those at the back of the book.

1. About how many plates exist on the Earth? (21.1)
   (a) 3   (b) 20
   (c) 250   (d) 1000

2. Which of the following correctly describes the lithosphere? (21.1)
   (a) It is the same as the Earth's crust.
   (b) It is the same as the Earth's upper mantle.
   (c) It is the part of the mantle below the asthenosphere.
   (d) It consists of the crust and part of the upper mantle.

3. What layer is close to the melting point of rock and is relatively plastic? (21.1)
   (a) mantle
   (b) lithosphere
   (c) asthenosphere
   (d) Moho

4. Which of the following geologic evidence does *not* support continental drift? (21.2)
   (a) similarities in biological species and fossils found on distant continents

(b) continuity of geologic structures such as mountain ranges

(c) ancient glaciation in the Southern Hemisphere

(d) similar types of rivers on every continent

5. Wegener proposed what name for the ancient, giant supercontinent he envisioned? (21.2)

(a) Pangaea

(b) Rhodinia

(c) Moho

(d) Alfredia

6. Which of the following is a primary cause of volcanoes, earthquakes, and mountain building? (21.3)

(a) solar radiation

(b) remanent magnetism

(c) hot spots

(d) plate tectonics

7. Which of the following is *not* a type of plate boundary? (21.3)

(a) divergent

(b) convergent

(c) reverse

(d) transform

8. The principle of isostasy states that the depth to which a floating object sinks into underlying material depends on which two properties? (21.3)

(a) density and composition

(b) temperature and the state of matter

(c) density and thickness

(d) mass and temperature

9. What process during plate collisions leads to the formation of volcanic islands arcs? (21.4)

(a) remanent magnetism

(b) seismic waves

(c) subduction

(d) isostasy

10. On the Richter scale, a magnitude 7.0 earthquake is how many times more powerful than a magnitude 6.0 earthquake? (21.5)

(a) 1   (b) 25

(c) 31   (d) 100

11. By what name is an upward pointing arch of a fold known? (21.6)

(a) anticline

(b) syncline

(c) hot spot

(d) hanging wall

12. Which of the following is *not* a mountain-building mechanism? (21.6)

(a) volcanic

(b) reverse-slip

(c) fault-block

(d) fold

FILL IN THE BLANK

Compare your answers with those at the back of the book.

1. ___ is the study of the Earth's general structural features and their changes. (21.1)

2. The ___ travel along the outer layer of the Earth. (21.1)

3. Directly below the lithosphere lies the ___. (21.1)

4. The ___ is the part of the Earth's interior that is believed to be made up largely of molten iron. (21.1)

5. The ___ are the lowest places on the Earth's surface. (21.2)

6. A(n) ___ ridge system stretches through the major oceans of the world. (21.2)

7. The lithosphere essentially "floats" on the ___. (21.3)

8. Lithospheric plate boundaries can be divergent, convergent, or ___. (21.3)

9. Dubbed the ___, it is the geologically active region along the margins of the Pacific Ocean. (21.4)

10. The majority of ___ are caused by the movement of lithospheric plates. (21.5)

11. The three principal types of mountains are fold, fault-block, and ___. (21.6)

12. The three major types of faults are normal, ___, and transform. (21.6)

SHORT ANSWER

21.1 The Earth's Interior Structure

1. What is the radius of the Earth, in kilometers and in miles?

2. Name the four major regions of the Earth, from "outermost" to "innermost."

3. About how thick, in kilometers, is the Earth's crust under the continents? Under the ocean basins?

4. Why is the asthenosphere so unique and important?

21.2 Continental Drift and Seafloor Spreading

5. What is meant by the term *Pangaea*?

6. What is the Mid-Atlantic Ridge?

7. Briefly state three types of evidence that support the concept of continental drift.

8. What are magnetic anomalies on the ocean floor?

21.3 Plate Tectonics

9. What, in the context of plate tectonics, is a plate?

10. What and where are the lithosphere and the asthenosphere?

11. Explain how continental crust and oceanic crust are like icebergs in water.

12. Give an example of a feature at each of the three general types of plate boundaries.

13. What is a subduction zone?

14. At what type of plate boundary is rock consumed? Produced? Neither consumed nor produced?

15. When an oceanic plate collides with a continental plate at a convergent boundary, which one is subducted? Why?

21.4 Plate Motion and Volcanoes

16. Briefly, how are volcanoes explained in the theory of plate tectonics?

17. Where on the Earth are volcanoes not found?

18. Give three examples of mountains formed by volcanic activity.

21.5 Earthquakes

19. Briefly, how are earthquakes explained in the theory of plate tectonics?

20. Name the two plates that border the San Andreas Fault. How are the plates moving relative to each other?

21. Distinguish between the focus and the epicenter of an earthquake.

22. Describe the deadly events that can occur because of undersea earthquakes.

23. What is measured by the Richter scale? Describe the effects of each whole-number magnitude.

24. On what three factors does the damage caused by an earthquake depend?

21.6 Crustal Deformation and Mountain Building

25. Make a simple sketch of a fold and show the difference between a syncline and an anticline.

26. Distinguish among a normal fault, a reverse fault, and a transform fault.

27. Describe the process by which a volcanic mountain is built.

28. To which of the three principal types of mountains does each of the following belong: (a) the Grand Tetons, (b) the Appalachians, (c) the Cascades, and (d) the Himalayas?

VISUAL CONNECTION

Visualize the connection and give answers for the blanks. Compare your answers with those at the back of the book.

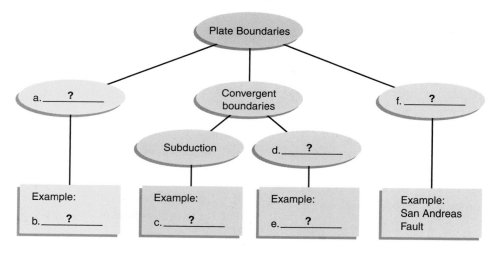

APPLYING YOUR KNOWLEDGE

1. State two similarities between the methods used by geologists to solve geologic problems and those used by police detectives to solve crimes.

2. How do we know that the Earth's interior is made up of heavier elements such as metals? Calculate the average density of the Earth (in g/cm³) and compare this answer with the density of water, which is equal to 1 g/cm³.

3. Which slow-moving tectonic plate are *you* riding on?

4. Oil and gas deposits have been found off the coast of Ghana in central Africa, sparking exploration for oil and gas off the eastern coast of South America. Understanding the concept of plate tectonics, what is the reason for exploring this area?

5. The Highlight: Earthquake Risk in North America maps earthquake zones according to the damage expected from earthquake

activity. What is the zone number in which you are located? What area(s) of North America has the highest seismic risk?

6. Refer to Figures 21.16 and 21.19. In terms of volcanoes and earthquakes, which two continents on the Earth are the safest? On which would you choose to live? Why?

7. You and two friends go out to sea on your boat and are motoring back into the harbor when you hear a radio warning that a tsunami will reach the area in 15 minutes. The first friend says, "Let's jump overboard and swim for shore." The other friend argues, "No, let's stay with the boat and head for the dock as quickly as possible." Being the captain, you decide that the best course of action is to turn around and head back out to sea. Who picked the correct course of action? Why?

ON THE WEB

1. Hang on to Your Hat, the Earth's on the Move

What does the term *plate tectonics* mean? Why is the Earth so restless, and what causes earthquakes, volcanoes, and the rise of mountains? How does the Earth's internal structure influence plate tectonics? How did the theories about continental drift and plate tectonics develop? What are the four major scientific developments that have spurred the formulation of the theory of plate tectonics? How does the ocean floor get mapped? Visit the student website at **www.cengagebrain.com/ shop/ISBN/1133104096** to discover more about plate tectonics and the recycling of the Earth's crust.

2. Shake, Rattle, and Roll

You've just experienced an earthquake. It may be jarring, it may be rolling; we hope you were not at the epicenter. What caused this event? Why is the seismic moment important? What are some practical ways to estimate the magnitude of an earthquake? What sources of information do seismologists use to make intensity ratings? Why do they make these ratings? To learn more about earthquakes and to answer these questions, visit the student website at **www .cengagebrain.com/shop/ISBN/1133104096**.

Minerals, Rocks, and Volcanoes

Touch the earth, love the earth ,honour the earth, her plains, her valleys, her hills and her seas; rest your spirit in her solitary places.

•

Henry Beston,
American author
(1888–1968)

Beautiful colors and crystal > patterns of several minerals. Purple amethyst (a form of quartz) is surrounded by brown selenite gypsum, blue azurite, green tourmaline, metallic stibnite, and iron pyrite (fool's gold).

Courtesy Chuck Higgins

GEOLOGY FACTS

▶ Igneous rock makes up to 80% of the Earth's crust, while the most common minerals in the crust are feldspars.

▶ About 30 different minerals are required to make a computer; a telephone needs 40 different minerals, and an automobile about 15.

▶ There are on average 15 to 20 volcanoes currently erupting on the Earth. However, by scientific estimates there are about 1500 active volcanoes around the world, active being defined as erupting in the last 10,000 years.

▶ The tallest volcano in the world is Mauna Kea, Hawaii, which stands more than 10 km (6.3 mi, or 33,465 ft) tall if measured from the ocean floor to its summit.

In Chapter 21 the Earth's structure and its tectonic processes, such as continental drift and earthquakes, were discussed. In this chapter rocks and minerals are considered and analyzed. Briefly, minerals are naturally occurring inorganic solids that formed within the Earth; rocks are naturally formed aggregates of one or more minerals. The processes that formed these different types of rocks and minerals are not small in scale. They are directly related to the bulk forces that changed the Earth on large scales, and they include the volcanic processes that were introduced in Chapter 21. First, minerals are defined and their formation processes studied, and the methods of identification are discussed. Second, the three basic types of rocks are introduced, their properties

Chapter Outline

are compared, and then each type of rock is presented in detail including formation and changes over time. An in-depth study of igneous activity and volcanoes is included.

22.1 Minerals

Preview Questions

- What are the two major chemical elements that make up minerals?
- Without trying it, how would you know that a diamond will scratch glass?

The study of minerals is called *mineralogy*. A **mineral** is a naturally occurring, crystalline, inorganic element or compound that possesses a fairly definite chemical composition and a distinctive set of physical properties.

Everywhere around us there are minerals. Some have high monetary value, whereas others are worthless. For example, precious stones, such as diamonds, emeralds, and rubies, are valuable minerals. Also, a nation's mineral wealth is defined by its natural raw materials, such as ores containing iron, gold, silver, and copper. However, the minerals of common rocks, such as sandstone, have little value.

The term *mineral* also has taken on popular meanings. For example, foods are said to contain vitamins and "minerals." In this case, *mineral* refers to compounds in food that contain elements required in small quantities by the human body, such as iron, iodine, and potassium. The names of minerals, like those of chemical elements, have historical connotations and many reflect the names of localities where they were formed. For example, the mineral galaxite was named after Galax, Virginia, where it is found.

For the most part, minerals are composed of eight elements. These elements, along with their relative abundance in the Earth's crust, are shown in Fig. 11.9. Seventy-five percent of the Earth's crust is made up of two of these eight elements—oxygen and silicon. More than 2000 minerals have been found in the Earth's crust. Approximately 20 of them are common, and fewer than 10 account for more than 90% of the crust by mass.

The Silicates

Solid, cohesive, natural aggregates of minerals are called *rocks* (Chapter 22.2). Most rock-forming minerals are composed mainly of oxygen and silicon. The fundamental silicon–oxygen compound is silicon dioxide, or *silica*, which has the chemical formula SiO_2. Quartz, a hard and brittle solid, is an example (see chapter-opening photo and ● Fig. 22.1). Because carbon and silicon are in the same chemical group, one might think

(a)   (b)

Figure 22.1 Two Mineral Samples of Quartz
(a) Citrine and (b) amethyst. Both are quartz composed of silicon dioxide (SiO_2), but their colors differ because they contain minute traces of different impurities.

Figure 22.2 The Silicon–Oxygen (Covalently Bonded) Tetrahedron and the Structure of Quartz (SiO_2)
(a) The silicon–oxygen tetrahedron is the fundamental building block for all silicate minerals. (b) The structure of quartz is based on interlocking SiO_4 tetrahedra, in which each oxygen atom (blue) is shared by two silicon atoms (orange).

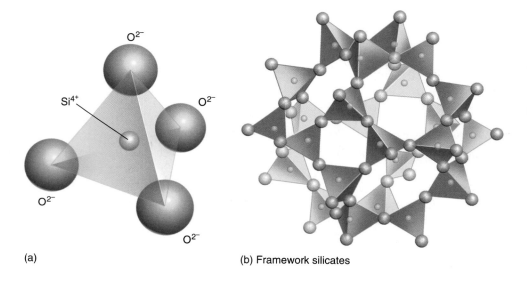

(a)

(b) Framework silicates

that SiO_2 would be a molecular gas similar to carbon dioxide, CO_2, but the bonding of the two compounds is very different. (See Chapter 12.4 and 12.5 for a review of chemical bonds.) Rather than being composed of SiO_2 molecules, the silicon–oxygen structure of quartz is based on a network of SiO_4 tetrahedra with shared oxygen atoms (● Fig. 22.2).

In silica, the oxygen-to-silicon ratio is 2 to 1. However, the oxygen-to-silicon ratios in the group of **silicates**, the most common types of minerals, are greater than 2 to 1 and can vary significantly. The variation occurs because the silicon–oxygen tetrahedra may exist as separate independent units or may share oxygen atoms at corners, edges, or faces in many different ways. Thus, the structures of the silicate minerals are determined by the way the SiO_4 tetrahedra are arranged.

In addition to oxygen and silicon, most rocks contain aluminum and at least one other common element, typically calcium, sodium, or potassium. *Feldspars* are the most abundant minerals in the Earth's crust. There are two main types of feldspar:

1. Plagioclase feldspar contains ions of oxygen, silicon, aluminum, and calcium or sodium.

2. Potassium (orthoclase) feldspar is composed of ions of oxygen, silicon, aluminum, and potassium.

Some photographs of typical minerals along with their silicate structures are shown in ● Fig. 22.3. Olivine is a mineral with independent SiO_4^{4-} tetrahedra that are bonded with two Mg^{2+} and/or Fe^{2+} ions to remain electrically neutral. These metal ions can substitute for each other, depending on the conditions of mineral formation.

The silicon–oxygen tetrahedra also can form single- and double-chain structures, as shown in Fig. 22.3, that share the oxygen ions to remain neutral. Examples of minerals with these structures are *pyroxene* and *hornblende*. In the pyroxene chain, each tetrahedron shares two oxygens; in the double hornblende chain, half the tetrahedra share two oxygens and the other half share three oxygens. These structures have various metallic-ion components. Mica has a two-dimensional continuous sheet structure of tetrahedra, and quartz shows a three-dimensional framework structure (Fig. 22.3).

Nonsilicate Minerals

Nonsilicates constitute less than 10% of the mass of the Earth's crust. They include valuable "pure" elements, such as gold and silver, and gemstones, such as diamonds and sapphires. Their ores are also sources of useful metals, such as iron, copper, nickel, and tin. The most common nonsilicate groups are the carbonates, the oxides, and the sulfides.

Carbonate minerals are formed when the carbonate ion (CO_3^{2-}) bonds with other ions. For example, carbonate ions commonly bond with calcium ions (Ca^{2+}) to form the min-

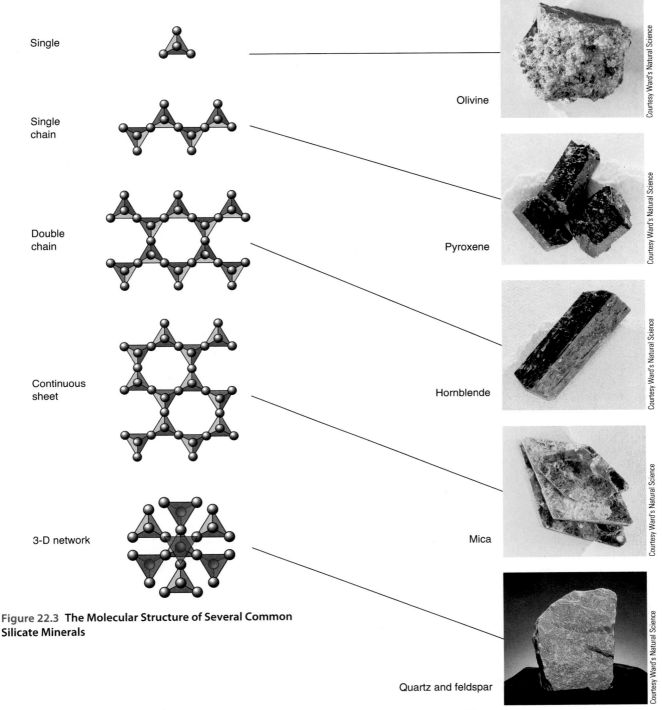

Single

Single chain

Double chain

Continuous sheet

3-D network

Olivine

Pyroxene

Hornblende

Mica

Quartz and feldspar

Courtesy Ward's Natural Science

Figure 22.3 The Molecular Structure of Several Common Silicate Minerals

eral calcite ($CaCO_3$). Calcite is a soft mineral, and like all carbonates, it dissolves readily in acidic water (see Chapter 13.3). This property of calcite, a key component of limestone, contributes to the formation of limestone caves discussed in Chapter 23.1.

Oxide minerals are produced when oxygen ions bond with metallic ions. Oxide ores—such as the iron ore hematite (Fe_2O_3), the tin ore cassiterite (SnO_2), and the uranium ore uraninite (UO_2)—are sources of valuable metals.

Ions of sulfur bond with various positive ions to produce the *sulfide minerals*. Sulfides such as chalcopyrite ($CuFeS_2$), which contains copper and iron, and galena (PbS), which contains lead, are valuable metal ores.

Identifying Minerals

A classification of minerals based on the physical and chemical properties of substances is advantageous because it distinguishes between different forms of minerals composed of the same element or compound. For example, graphite—a soft, black, slippery substance commonly used as a lubricant—and diamond are both composed of carbon (Chapter 11.3). However, because of different crystalline structures, their properties are quite different. (Graphite mixed with other substances, such as clay, to obtain various degrees of hardness is the "lead" in lead pencils.)

Minerals can be identified by chemical analysis, but most of these methods are detailed and costly and are not available to the average person. More commonly, distinctive physical properties are used as the key to mineral identification. These properties are well known to all serious rock and mineral collectors. Some of the properties used to identify minerals are described in the following paragraphs.

Crystal form is the size and shape assumed by the crystal faces when a crystal has time and space to grow. It represents the results of the interaction of the atomic structure with the environment in which it grows. Many minerals have such characteristic crystal forms that they often can be identified by this property alone (see the chapter-opening photo and Fig. 22.1).

All crystalline substances crystallize in one of seven major geometric patterns. When a mineral grows in unrestricted space, it develops the external shape of its crystal form. However, during the growth of most crystals, the space is restricted, resulting in an intergrown mass that does not exhibit its crystal form. The crystalline forms are studied in detail by means of X-ray analysis (Chapter 9.6).

Hardness is a comparative property that reflects the ability of a mineral to resist scratching. The degrees of hardness are represented on the **Mohs scale**, which runs from 1 to 10, soft to hard.* This arbitrary scale is expressed by the ten minerals listed in ● Table 22.1. Talc is the softest, and diamond is the hardest. A particular mineral on the scale is harder

Table 22.1 The Mohs Hardness Scale

Mineral	Hardness	Common Examples
Diamond	10	
Corundum	9	Ruby, Sapphire (9)
Topaz	8	Emerald (8)
		Zircon (7.5)
Quartz	7	Ceramic Tile (7)
		Steel File (6.5)
Orthoclase (potassium feldspar)	6	
		Glass (5-6)
Apatite	5	Tooth Enamel (5)
		Platinum (4.3)
Fluorite	4	Iron, Nickel (4)
		Copper Penny (3.5)
Calcite	3	Gold (2.8)
		Human Fingernail (2.5)
Gypsum	2	Lead (1.5)
Talc	1	Sodium (0.5)
	0	

*Named after Friedrich Mohs, a German mineralogist, who devised this scale in 1812.

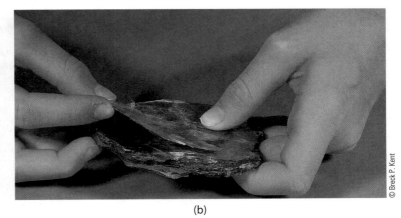

(a)

(b)

Figure 22.4 Calcite and Mica Have Distinctive Cleavages
(a) Calcite's three cleavage planes are not mutually perpendicular. Calcite cleavage produces a geometric shape called a rhombohedron. (b) Mica has one perfect cleavage plane, forming sheets.

than (can scratch) all those with lower numbers. Using these minerals as standards, one finds the following on the hardness scale: fingernail, 2.5; penny, 3.5; window glass or knife blade, 5–6; steel file, 6.5.

Cleavage is the tendency of some minerals to break along definite smooth planes. The mineral may exhibit distinct cleavage along one or more planes, or it may exhibit indistinct cleavage or no cleavage. The degree of cleavage is a clue to the identity of the mineral (● Fig. 22.4).

Fracture refers to the way in which a mineral breaks. It may break into splinters; into rough, irregularly surfaced pieces; or into shell-shaped forms known as conchoidal fractures (● Fig. 22.5).

Figure 22.5 A Conchoidal Fracture on Quartz
Quartz, with equally strong covalent bonds in all directions, has no planes of weakness. It therefore fractures irregularly instead of cleaving.

Conceptual Question and Answer

Identifying Diamonds

Q. A street vendor wants to sell you a diamond ring to match the one you are wearing. It looks like a diamond, but you are not sure; it could be a zircon that resembles a diamond. The vendor assures you that it is a diamond and scratches a glass plate to prove it. Still, how could you make sure that it is a diamond?

A. Diamond is the hardest on the Mohs hardness scale (number 10). To see if the other ring is a diamond, try to scratch the stone with your genuine diamond. If you get a scratch, then it is an imitation. (Zircon hardness, 7.5.)

Figure 22.6 Azurite Crystals
Azurite, a copper mineral, is one of the few minerals that come in only one distinctive color. Quartz, on the other hand, comes in a variety of colors (two are seen in Fig. 22.1).

Color is the property of reflecting light of one or more wavelengths. Although the color of a mineral may be impressive, it is not a reliable property for identifying the mineral because the presence of small amounts of impurities may cause drastic changes in the color. Some minerals, such as quartz, occur in a variety of colors (Fig. 22.1). Blue azurite is one of the few minerals that are found in only one color (● Fig. 22.6).

Streak is the color of the powder of a mineral. A mineral may exhibit an appearance of several colors, but it will always show the same color streak. A mineral rubbed (streaked) across the surface of an unglazed porcelain tile will be powdered, showing its true color (● Fig. 22.7).

Luster is the appearance of the mineral's surface in reflected light. Mineral surfaces have either a metallic or a nonmetallic luster. A *metallic luster* has the appearance of polished metal, such as pyrite

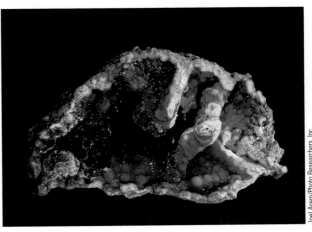

Figure 22.7 Testing Hematite's Streak
Although many samples of hematite (Fe_2O_3) are steel gray, the streaks of all samples are reddish brown.

(a)

(b)

Figure 22.8 Metallic and Nonmetallic Lusters
The pyrite sample (*left*) displays metallic luster. The limonite sample (*right*) displays an earthy or dull luster. Both are iron minerals.

(● Fig. 22.8). A nonmetallic appearance may be of various lusters, such as glassy-looking or *vitreous* (topaz), *adamantine* (diamond), *pearly* (opal), *greasy* (talc), or *earthy/dull* (clay).

Specific gravity is the ratio of a mineral sample's mass to the mass of an equal volume of water (Chapter 1.6). Each mineral has a characteristic specific gravity. For example, fluorite (CaF_2) has a specific gravity of 3.2, whereas galena (PbS) has a specific gravity of 7.6.

A summary of properties to identify minerals is shown in ● Table 22.2. Gems incorporate many of these physical properties. A *gem* is any mineral or other precious or semiprecious stone valued for its beauty and monetary worth. Most gems have a crystalline structure, and when they are shaped and polished, their appearance is enhanced. Their beauty depends on the special characteristics of brilliancy, color, prismatic fire, luster, optical effects, and durability.

Did You Learn?

● Minerals are mostly composed of the elements oxygen and silicon.

● A diamond can scratch glass because diamond has a greater degree of hardness than glass (Mohs scale; see Table 22.1).

Table 22.2 Properties to Identify Minerals

Identifying Characteristic	Examples
Crystal form	Size and shape of crystal faces: tetrahedron, rhombohedron, etc.
Hardness	Ability to resist scratches: scale from 1 to 10
Cleavage	Tendency to break along definite, smooth planes
Fracture	The way the mineral breaks: splinters, irregular pieces, or shell-shaped forms
Color	Reflected wavelength(s) of light
Streak	Color of the powder of the mineral
Luster	Surface appearance in reflected light: metallic or nonmetallic
Specific gravity	Ratio of sample mass to the mass of an equal volume of water

22.2 Rocks

Preview Questions

- How many classes of rocks are there, and what are they?
- By what processes are rocks formed?

As mentioned previously, **rock** is a solid, cohesive, natural aggregate of one or more minerals. Rock is a natural and substantial part of the Earth's crust. When one looks at a mountain cliff, one sees rock rather than individual minerals. The Colorado River has carved the Grand Canyon through layers of rock, and the continents and ocean basins are composed of rock. Rocks are classified into three major categories on the basis of the way they originated.

Igneous rocks are formed by the solidification of magma. *Magma* is molten rock material that originates far beneath the Earth's surface, where the temperatures are thousands of degrees. Magma that reaches the Earth's surface (by way of a volcanic eruption, for example) is called *lava*. Thus, igneous rocks may be formed deep inside the crust or at the Earth's surface.

Sedimentary rocks are formed at the Earth's surface by compaction and cementation of layers of sediment. They are aggregates of the following:

1. Rock fragments derived from the wearing away of older rocks
2. Minerals precipitated from a solution
3. Altered remains of plants or animals

Metamorphic rocks are formed by the alteration of preexisting rock in response to the effects of pressure, temperature, or the gain or loss of chemical components. *Metamorphism* occurs well below the surface but above the depths at which rock is molten.

The Rock Cycle

Most eighteenth-century scientists believed that the Earth's structure was caused by catastrophic events and that all rocks on its surface had been deposited by a great flood that took place in the relatively recent past. This doctrine was known as *catastrophism*.

Geology became a recognized science early in the nineteenth century when scientists first came to understand that the geologic processes that today form and change rocks on the Earth's surface and within its interior are the same processes that have been at work throughout the very long history of the Earth. Thus, ancient rocks were formed in the same way as modern rocks and can be interpreted in the same manner. This concept, known as **uniformitarianism**, is the very foundation on which the science of geology rests; that is,

Figure 22.9 Portrait of James Hutton (1726–1797)
Hutton developed the concept of uniformitarianism and is known as "The Father of Geology."

the present is the key to the past. James Hutton (1726–1797), a Scottish physician, gentleman farmer, and part-time geologist, is credited with developing the concept (● Fig. 22.9). The founding of modern geology is known as the Huttonian revolution, and Hutton is called "The Father of Geology."

Hutton recognized that rocks are continuously being formed, broken down, and re-formed as a result of igneous, sedimentary, and metamorphic processes. His model, called the *rock cycle*, reflects the manner in which internal heat, solar energy, water, and gravity act on and transform the materials of the Earth's crust (● Fig. 22.10). Specifically, the **rock cycle** is a series of events through which a rock changes over time between igneous, sedimentary, and metamorphic forms. Over long periods of time, each of the three types of rock may be transformed into another type of rock or a different form of the same type. Rocks eroded at the Earth's surface are then deposited as new sediment, and these new sediments may become sedimentary rock. Sedimentary rocks may also become so deeply buried in the Earth's hot interior that they are changed into metamorphic rock or melted to become igneous rock. Under heat and pressure, igneous rocks can also become metamorphic rocks. There is no prescribed sequence to the rock cycle. A given rock's evolution may be altered at any time by a change in the geologic conditions around it.

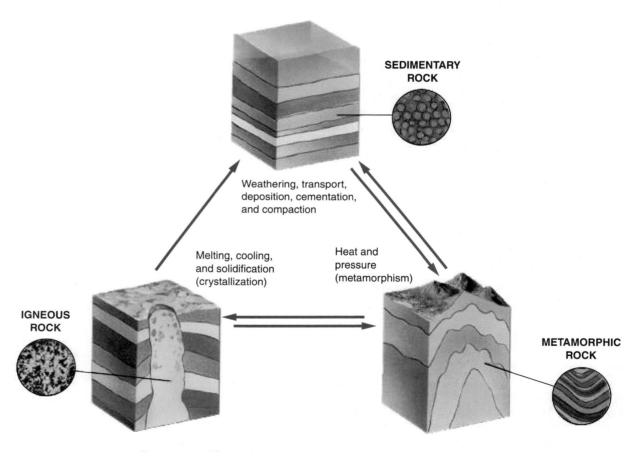

Figure 22.10 The Rock Cycle
This simplified scheme illustrates the variety of ways in which the Earth's rocks may evolve into other types of rocks. For example, an igneous rock may weather away and its particles eventually consolidate to become a sedimentary rock. The same igneous rock may remain buried deep beneath the Earth's surface, where heat and pressure might convert it into a metamorphic rock. The same igneous rock, if it is buried even deeper, may melt to become magma—which may eventually cool again and solidify to form a new igneous rock.

Conceptual Question and Answer

Energy for the Rock Cycle

Q. What is one global Earth process we have studied that drives the rock cycle?

A. The rock cycle is driven by plate tectonics, which creates pressure and heat at plate boundaries and volcanic regions (Chapter 21.4). This cycle makes igneous and metamorphic rock change over time, creating and consuming rock in the process. Another process, to be studied in Chapter 23, is the Earth's hydrologic cycle, which causes weathering, transport, and deposition to form sedimentary rocks.

Hutton realized that the processes of the rock cycle require a great deal of time. He also considered that much of the rock material seen at the Earth's surface today has been through that cycle many times over. Thus, he hypothesized that the Earth must be very old, far older than his fellow eighteenth-century scientists had ever imagined and far, far older than the age calculated by biblical scholars. Along with Hutton's principle of uniformitarianism, the concept of the rock cycle and the recognition of the Earth's great age remain central themes of modern geology.

Did You Learn?

● Rocks occur in three classes: igneous, sedimentary, and metamorphic.

● Rocks are formed by compaction of sediment, heat and pressure changes, and solidification of magma.

22.3 Igneous Rocks

Preview Questions

● How are igneous rocks formed?

● Where on the Earth is most of the igneous rock located?

Igneous rock forms when molten material from far beneath the Earth's surface cools and solidifies. Accumulations of solid particles thrown from a volcano during an explosive eruption are also considered igneous rocks. **Magma** is the molten material beneath the Earth's surface, but is called **lava** if it flows on the surface. The term *lava* may be a bit confusing, because it can be used to mean either the hot molten material or the resulting solidified igneous rock.

Assuming that the Earth was originally molten, the very first rocks of the continents and ocean basins must have been igneous. Geologic processes, however, long ago obscured any recognizable remnant of these most ancient materials. Nevertheless, igneous activity has continued throughout the history of the Earth, and igneous rocks are by far the most abundant type. They are estimated to constitute as much as 80% of the Earth's crust.

The eruption of a volcano is a spectacular geologic phenomenon. Any igneous rock that cools from extruded molten lava is described as *extrusive rock*. The most common occurrence of this type of rock is the formation of new oceanic crust. As tectonic plates continue to move apart, magma continues to well up, cool, and move aside (Chapter 21.2). The entire oceanic crust has been formed in this manner and consists mostly of the extrusive igneous rock called *basalt*. Few people realize that the vast majority of magma (molten rock) never finds its way to the surface but rather cools to solid rock somewhere within the Earth's interior as *intrusive rock*. We observe intrusive rock only where erosion has stripped away the overlying material or where movement of the Earth's crust has brought it to the

(a)

Figure 22.11 Volcanic Rocks
(a) Volcanic rocks, such as this basalt, typically have small-grain textures, because (b) they solidify rapidly above ground.

(b)

(a)

(b)

Igneous Rock Texture and Composition

surface. The prime example of this process occurs at the subduction zone regions (Chapter 21.3). Much of the magma generated at subduction zones does not rise to the surface but cools inside the crust to form intrusive igneous rock. Therefore, because most of the igneous activity occurs at plate boundaries, most of the igneous rock is produced near these regions (mid-ocean ridges, mountain building zones, subduction zones, etc.).

Intrusive and extrusive igneous rocks are classified in terms of their *texture*, or physical appearance, and their mineral composition. The texture of an igneous rock is dictated primarily by the size of its mineral grains. The mineral crystals of the basalt sample shown in ● Fig. 22.11 are tiny, resulting in a fine-grained texture, whereas the mineral grains of the granite sample are large enough to be visible to the unaided eye. Thus, granite is a coarse-grained igneous rock (● Fig. 22.12).

Grain size is determined primarily by the rate at which molten rock cools. An igneous body must cool slowly if its mineral grains are to grow large. Rapid cooling invariably yields small grains. The basalt sample in Fig. 22.11, for example, was probably brought to the surface as molten lava. Lava exposed to the cool atmosphere loses heat quickly and therefore develops only small grains. Globs of lava shot into the air during an explosive eruption cool so quickly that the lava solidifies without crystallizing. The results are rocks with a glassy texture, such as obsidian, or pumice, which has many air-filled cavities (● Fig. 22.13).

Magma deep within the Earth loses heat very slowly because the cover of rock that overlies it is a very poor conductor of heat. Igneous rocks formed at these great depths are almost invariably coarse grained. ● Table 22.3 shows how the location of cooling affects the rate of cooling, which in turn affects the texture of an igneous rock.

Igneous rocks can be roughly divided according to their composition into those rich in silica (SiO_2) and those relatively low in silica. Granite rocks are rich in silica and contain minerals with abundant silicon, sodium, and potassium. These

Figure 22.12 Granite Rocks
(a) Rocks that solidify slowly underground, as did this granite, have coarse-grained textures. (b) This sample is typical of the granite found in Yosemite National Park in California.

Table 22.3 Effects of Cooling on the Textures of Igneous Rocks				
Type of Igneous Rock	Location of Cooling	Rate of Cooling	Texture	Example
Intrusive	At depth	Slow	Coarse	Granite
Extrusive	At surface	Rapid	Fine	Basalt
Extrusive	In the air or water	Very rapid	Glassy	Obsidian

(a)

© Doug Sokell/Visuals Unlimited

minerals are mostly light in color. Basalt rocks are low in silica and rich in iron, magnesium, and calcium. These minerals make low-silica rocks both darker and denser than silica-rich rocks. This density difference explains why the high-silica granitic continents stand higher than the low-silica basaltic ocean basins (see Chapter 21.3).

Most volcanic islands and many continental mountain chains are built of andesite, an extrusive igneous rock of medium-silica content named after the Andes Mountains of South America, where it was first discovered. See ● Table 22.4 for the extrusive and intrusive forms of igneous rocks high, medium, and low in silica content.

(b)

Eric Schrempp/Photo Researchers, Inc.

Figure 22.13 Glassy Volcanic Rocks
(a) Obsidian and (b) pumice contain no crystals because they solidify instantaneously. Pumice, which forms from lava foam, commonly has so many tiny air-filled cavities that it can float in water.

Did You Learn?

● Igneous rocks are formed by the solidification of magma.

● Most of the igneous rock is located at plate boundaries.

22.4 Igneous Activity and Volcanoes

Preview Questions

● What erupts from a volcano?

● What are hot spots?

Plutons

Intrusive igneous rocks, formed below the surface of the Earth by solidification of magma, are known as *plutons*. **Plutons** are classified according to the size and shape of the intrusive bodies and according to their relationship to the surrounding rock they penetrate. A pluton is *discordant* if it cuts across the grain of the surrounding rock and is *concordant* if it is parallel to the grain. The most important discordant igneous rock body is a *batholith* because of its enormous size (● Fig 22.14). A batholith, by definition, must have a surface area (when exposed) of at least 100 km² (40 mi²), but many batholiths are vastly larger than that. For example, the Coast Range Batholith in western Canada is more than 1600 km (990 mi) long and in places more than 160 km (99 mi) wide.

All surface exposures indicate that batholiths grow larger with depth, but the nature of their bottoms remains uncertain because no canyons or mines have penetrated that deep. A batholith forms when viscous magma crystallizes deep within the Earth's crust; its

Table 22.4 Common Igneous Rocks Organized by Composition				
Silica Content	Mineral Colors	Extrusive Rocks (small grains, fine texture)	Intrusive Rocks (large grains, coarse texture)	Primarily Found
High (> 70%)	Mostly light	Rhyolite	Granite	Continental crust
Medium (55–65%)	Mostly dark	Andesite	Diorite	Volcanic island arcs, some continental mountain chains
Low (40–50%)	Dark	Basalt	Gabbro	Oceanic crust

Figure 22.14 An Illustration of Plutonic Bodies
Magma solidifying within the Earth forms intrusive igneous bodies. The batholith is the largest intrusive body. Sills and laccoliths are concordant bodies that lie parallel to existing rock formations. Dikes, which are discordant bodies, cut across existing rock formations.

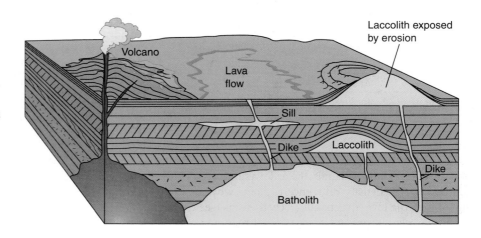

formation is an intimate part of the complex process of mountain building. Batholiths are exposed at the Earth's surface by uplift or where mountain chains have been deeply scarred by erosion. Figure 22.14 illustrates their shape and their relationship to the intruded rock.

Dikes are discordant plutons formed from magma that has filled fractures that are vertical or nearly so; therefore, their shape is tabular—thin in one dimension and extensive in the other two—as illustrated in Fig. 22.14. The sizes of dikes are just as variable as the sizes of the fractures they fill. Dikes are quite common and have been recognized in many different kinds of geologic environments.

A *sill* has the same shape as a dike but is concordant rather than discordant (Fig. 22.14). A *laccolith*, also shown in Fig. 22.14, is a concordant pluton formed from a blister-like intrusion that has pushed up the overlying rock layers.

Products of Volcanic Eruptions

As defined in Chapter 21.4, a volcano is a mountain or elevation formed from lava and debris ejected through a vent in the Earth's surface. What erupts from a volcano? People who know that volcanoes produce lava may not be aware that lava is but one of three products of volcanic eruptions.

1. *Gas.* The expulsion of gas is the most widespread general characteristic of all volcanoes. A volcano may expel gases in its earliest infancy, at the height of its activity, and during its final dying moments when all other signs of activity are gone. Steam (H_2O) may account for as much as 90% of these gases, but carbon dioxide (CO_2) and hydrogen sulfide (H_2S) are also present. Volcanoes can also bring valuable materials to the surface (see the **Highlight: Making Gold in an Active Volcano**).

2. *Lava.* Volcanoes may eject lava in variable quantities under various conditions. Some volcanoes produce vast amounts of very fluid lava, which flows easily and quietly with little explosive violence. If much gas escapes at the same time, then minor explosions may hurl incandescent lava into the air (● Fig. 22.15). These lava fountains, impressive enough by day, create magnificent fireworks at night. Clots of lava sometimes harden in midair and strike the ground as spindle-shaped volcanic projectiles of various sizes. Other volcanoes eject relatively small volumes of lava so stiff and viscous that it barely flows at all.

3. *Solids.* Some volcanoes spew enormous volumes of solids that can range in size from fine dust to huge boulders. Such particles are known collectively as **pyroclastics** or *tephra* and include not only fragments of rock but also gas-laden material that the volcano ejects in molten form but that hits the ground as a solid. Such debris is blasted into the air by violently explosive volcanoes. When the amount of pyroclastic material expelled by a volcano is so great that gravity almost immediately pulls it down onto the volcano slope, this material rushes downslope as *a pyroclastic flow* (● Fig. 22.16).

D. Griggs/USGS

Figure 22.15 A Volcanic Eruption
The force of gas escaping from the underground reservoir of magma throws molten lava high into the air over the crater of Hawaii's Kilauea volcano.

Highlight Making Gold in an Active Volcano

At the foot of volcano Galeras lies the peaceful city of Pasto, Columbia, with a population of nearly 400,000 residents. One of the most active volcanoes in the world (another is Nevada del Ruiz; see Fig 22.22), it has erupted many dozens of times since 1988 after a period of dormancy (Fig. 1). Tragically, Galeras erupted unexpectedly in 1993 killing six scientists and three tourists while they were at the summit. Geologists regularly probe the volcano's crater, collecting gases, water from hot springs, and samples of lava from past eruptions for study and to help predict future activity.

In 1994, a team from New Mexico's Los Alamos National Laboratory visited Galeras to sample its volcanic rock and escaping gas and came away with a valuable discovery: The rocks of the Galeras volcano are accumulating gold. Laboratory tests of rock samples collected at several sites at the base of the volcano showed some tiny flakes of gold in the rock. Geologists think that these rocks might be part of veins of old volcanic deposits containing as much as 7 ounces of gold per ton of rock. Compared with many active gold mines in the world, which produce less than 1 ounce per ton of rock, it is an impressive concentration.

The volcano's gold, like the gold in Alaska, the Colorado Rockies, and California's Sierra Nevada Range, was probably formed from heated water and gases that dissolve metals from the surrounding rock. When the heated waters rise from the volcano's subterranean magma chamber and seep into cooler rocks at shallow depths, the metallic minerals precipitate out to form veins of mineral-rich ores.

Compared with other volcanoes around the world, Galeras appears to be one of the richest for gold content. Geologists estimate that the Galeras volcano could make up to 45 pounds of gold each year in the rocks that line its crater. Another rich gold deposit is found within the extinct Luise volcano on Lihir Island, a tiny section of Papua New Guinea (next to Indonesia) in the southwestern Pacific Ocean.

Already making plans for your next gold mining trip? Don't. Mining companies have already laid their claims to the extinct volcanoes, and active volcanoes are far too dangerous.

Figure 1 Colombia's Galeras Volcano
Galeras, one of the world's most active volcanoes, is seen pouring hot lava and venting gases in 2010. Surprisingly, it is a veritable factory of gold production. Any enthusiasm for searching for and mining this treasure, however, is tempered by Galeras' recent history of activity.

Eruptive Style

From the preceding description of the products of volcanoes, it should be obvious that there are two basic styles of volcanic eruptions: *peaceful* and *explosive*. The manner in which a volcano erupts—peacefully or explosively—depends by and large on the **viscosity** of its magma. Low viscosity allows magma to flow fluidly and therefore peacefully. Highly viscous magma is thick and stiff and moves only when subjected to great force. Magma viscosity in turn depends on the temperature and silica content of the magma. High temperature and low silica content result in low viscosity; low temperature and high silica content result in high viscosity (● Table 22.5).

In peaceful eruptions, lava flows fluidly out of the volcanic vent and is accompanied by the occasional fireworks caused by the expulsion of gases. Many visitors to Hawaii's Kilauea volcano experience the thrill of observing such eruptions from the safety of observation points a few hundred meters from spewing vents. Peaceful volcanic eruptions

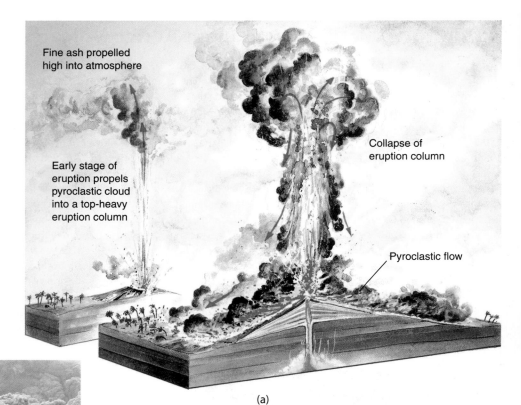

Fine ash propelled
high into atmosphere

Collapse of
eruption column

Early stage of
eruption propels
pyroclastic cloud
into a top-heavy
eruption column

Pyroclastic flow

(a)

(b)

Alberto Garcia/Corbis

Figure 22.16 Pyroclastic Flows
(a) Pyroclastic flows are produced when a massive amount of airborne pyroclastic material is pulled to the Earth by gravity and rushes downslope. (b) A pyroclastic flow from the May 1991 eruption of Mount Pinatubo in the Philippines. (Better step on the gas!)

involve basaltic magma. Because it originates deep beneath the crust, basaltic magma is hot. It flows fluidly because of its high temperature and low silica content.

Basaltic eruptions occur in two settings: along divergent plate boundaries, where magma wells up from beneath the lithosphere to fill the rift between spreading plates, and in plate interiors, where the plate is riding over a hot spot. A *hot spot* is the surface expression of magma that originates in the asthenosphere and deeper (Chapter 21.1). Hot-spot magma rises as a plume and pierces the overlying lithospheric plate, forming volcanic holes as the plate rides over it. For the past 70 million years, the Emperor Seamount chain and all the volcanic islands of Hawaii formed as the Pacific plate rode over one such hot spot (● Fig. 22.17). At present, a new seamount called

Table 22.5	Magma Viscosity and Eruptive Style			
Magma	Temperature	Silica Content	Style of Eruption	Tectonic Setting
Rhyolitic	Low	High (> 70%)	Violent	Oceanic/continental subduction zones
Andesitic	Low to moderate	Medium (55–65%)	Moderately violent to violent	Oceanic/continental subduction zones, volcanic island arcs
Basaltic	High	Low (40–50%)	Peaceful	Divergent boundaries, hot spots

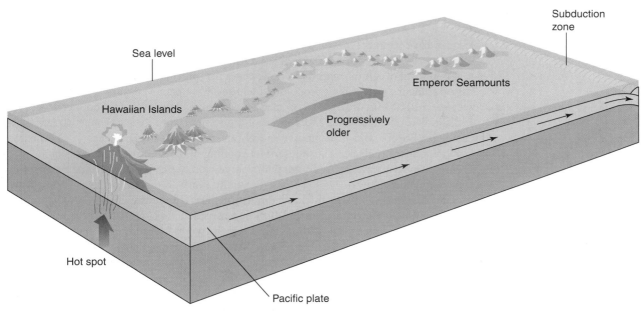

Figure 22.17 Formation of the Hawaiian Islands and Emperor Seamount Chain
Magma rising from a stationary hot spot deep in the Earth's interior "punches out" volcanic holes in the Pacific plate as it moves in a north–northwesterly direction over the hot spot. For the bigger picture, locate the Hawaiian Islands on the world map in Fig. 21.16.

Loihi is forming over the hot spot. Someday it may become another island in the Hawaiian chain.

Volcanoes that erupt explosively are another matter altogether. Unpredictable in both timing and strength, explosive eruptions can cause sudden, massive devastation. For example, early in 1980, Mount Saint Helens, a moderate-size dormant volcano located in southwestern Washington about 60 km (37 mi) from Portland, Oregon, showed signs that it was going to erupt. The immediate area was evacuated, and geologists set up stations around the volcano to monitor its activity. For months the scientists watched and waited. Then, on May 18, 1980, Mount Saint Helens exploded (● Fig. 22.18). Triggered by a magnitude 5.1 earthquake, most of the peak of the mountain slid away in a gigantic avalanche that raced down the mountain. With the pressure released within the mountain, a major eruption of the volcano followed. The eruption devastated an area of more than 400 km² (150 mi²) and left more than 60 people dead or missing. A column of ash rose to an altitude of more than 20 km (12 mi). Ash blanketed nearby cities, and a light dusting fell as far away as 1450 km (900 mi) to the east. Huge mudflows from the ash caused flooding and silting in the rivers near the volcano.

Explosive eruptions occur in subduction zones (for example, along the Pacific's Ring of Fire; Chapter 21.4). Because subduction-zone magmas do not originate as deep inside the Earth as basaltic magmas, they are cooler. They are also higher in silica content. For these reasons, they are highly viscous and, instead of flowing out, tend to form a resistant plug in the volcanic vent. Gases that accompany fluid magmas are able to escape the vent gradually, but the gases that accompany viscous magmas are kept bottled up under high pressure beneath the plugged vent. When the plug rises and the weight of overlying material is lowered, the gases explode suddenly and violently, blowing the plug. In the case of Mount Saint Helens, much of the mountain itself was blown to pieces, propelling the debris great distances.

USGS/Cascades Volcano Observatory

Figure 22.18 The Mount Saint Helens Eruption in 1980
On May 18, 1980, Mount Saint Helens in the state of Washington exploded with a blast that removed most of the peak of the mountain. Volcanic ash was deposited up to 800 km (500 mi) away and caused near blackout conditions in many cities.

Volcanic Structures

Volcanoes have various types of structures depending on the geologic environment. Some lava reaches the surface through long fractures in the surface rocks. More commonly, however, it pours out of central vents or volcanic cones that previous eruptions have constructed.

Fissure eruptions, which issue from long fractures, have been very rare in human history. However, in 2010 a fissure opened up in the volcano Eyjafjallajökull (pronounced "AYE-ya fyah-dla jow-kudl") in Iceland and sent ash 9 km (5.6 mi) into the air. The jet stream blew the ash over Europe and disrupted air travel for several weeks. The volcano also melted part of the mountain glacier, causing serious flooding in the southern part of Iceland.

Fissure eruptions have poured remarkably large volumes of basalt onto the Earth's surface at various times in the geologic past. Virtually every continent has its own extensive area of *flood basalts*, as these thick accumulations are commonly called. An example is the great Columbia Plateau in North America, a conspicuous landform over several of the states in the northwestern U.S. Although it is not the world's largest, the Columbia Plateau basalt covers an area of 576,000 km^2 (220,000 mi^2) to an average depth of 150 m (490 ft). The ocean basins are floored with even greater floods of basalt, estimated to be as much as 5 km (3 mi) thick.

The basaltic lavas that came from fissure eruptions were so extremely fluid that they flowed many miles over the Earth's surface without constructing volcanoes. Individual lava flows were not thick, however, because they extended for many miles over the surface. As one lava flow followed another over millions of years, the entire landscape was eventually drowned in a sea of solid basalt, with perhaps a high peak forming an island in the black and desolate ocean of rock. The Blue Mountains of Oregon are examples of such islands within the Columbia Plateau.

Not quite as hot and fluid as the ancient basalts that built the Columbia Plateau, the frequently repeated flows of modern-day basalts form gently sloping, low-profile *shield volcanoes*. The classic example of a shield volcano is Mauna Loa in the Hawaiian chain. In fact, Mauna Loa of Hawaii is a huge volcanic mountain, the largest single mountain on the Earth in sheer bulk. Although not as tall as Mount Everest, Mauna Loa rises 4.7 km (2.9 mi) from the ocean floor to sea level and protrudes an additional 4.2 km (2.6 mi) above sea level for a total height of about 8.9 km (5.5 mi). This partially submerged mountain has a base almost 160 km (100 mi) in diameter.

Volcanic eruptions of both lava and tephra form a more steeply sloping, layered composite cone that is called a *stratovolcano* or *composite volcano*. The lava of stratovolcanoes has a relatively high viscosity, and eruptions are more violent and generally less frequent than those of shield volcanoes. Many stratovolcanoes have an accumulation of material up to 1.8 to 2.4 km (1.1 to 1.5 mi) above their base and exhibit a characteristic symmetric profile. Mount Saint Helens is a stratovolcano. Dormant stratovolcanoes include Mount Fuji in Japan (● Fig. 22.19) and Mount Shasta, Mount Hood, and Mount Rainier in the Cascade Mountains.

A volcanic eruption may also consist primarily of tephra. In this case, *cinder cones* are formed with steep slopes, which rarely exceed 300 m (980 ft) in height. A cinder cone rising above a previous lava flow is shown in ● Fig. 22.20.

Activity is usually not confined to the region of the central vent of a volcano. Fractures may split the cone, with volcanic material emitted along the flanks of the cone. Also, material and gases may emerge from small auxiliary vents, forming small cones on the slope of the main central vent. Near the summit of most volcanoes is a funnel-shaped depression called a *volcanic crater* from which material and gases are ejected.

Many volcanoes are marked by a much larger depression called a **caldera**. These roughly circular, steep-walled depressions may be up to several miles in diameter. Calderas result primarily from the collapse of the chamber at the volcano's summit from which lava and ash were emitted. The weight of the ejected material on the partially empty chamber causes its roof to collapse, much like the collapse of the snow-laden roof of a building. Crater Lake on top of Mount Mazama in Oregon occupies the caldera formed by the collapse of the volcanic chamber of this once-active stratovolcano (● Fig. 22.21).

Figure 22.19 Mount Fuji in Japan
A dormant stratovolcano.

Figure 22.20 Volcano Izalco in El Salvador
A cinder cone volcano.

Figure 22.21 Crater Lake
Crater Lake on top of Mount Mazama in Oregon occupies a caldera about 10 km (6.2 mi) in diameter.

Table 22.6 Historic Volcanic Eruptions

Name	Date of Eruption	Location	Summary
Tambora	1815	Indonesia	Largest in recorded history; 145 km^3 (35 mi^3) of solids were ejected in a gigantic blast that was heard thousands of miles away. The ash caused several days of darkness and an uncommonly cold year in 1816. An estimated 71,000 people were killed, mostly from starvation.
Krakatoa	1883	Indonesia	Blast was four times the most powerful atomic bomb ever detonated. Hot ash rained down on the nearby islands, and the pyroclastic flows triggered tsunamis, a combined effect that killed approximately 36,000 people.
Pelée	1902	Martinique, Caribbean Sea	Incandescent cloud-like mixtures of superheated gas and tephra were released. The pyroclastics swept through the nearby city of St. Pierre, destroying nearly all life and property; nearly 30,000 people perished. One of only two survivors was a convicted murderer awaiting execution in a dungeon, where he was shielded from the blast.
Nevado del Ruiz	1985	Colombia	This volcano released sulfur-rich gases and ejected tephra 30 km (19 mi) into the air. Much of the ice-covered crater melted and sent rivers of mud swiftly down the mountain, killing about 23,000 people.
Vesuvius	79 CE	Italy	Buried 2000 people in the city of Pompeii under a thick, burning blanket of ash, with a total of 16,000 people killed in the region. Firsthand accounts recorded the effects of the devastating pyroclastic flows, choking gases, and raining hot pumice rocks.

Historic Eruptions

The largest known volcanic eruption occurred about 28 million years ago in what is now southern Colorado. It created a crater about 30 km (20 mi) wide and 80 km (50 mi) long. The blast was an estimated 1000 times as powerful as that of Mount Pinatubo and 10,000 times as powerful as that of Mount Saint Helens. In terms of the human cost, a list of the worst volcanic eruptions in recorded history is given in ● Table 22.6. One of these, the Nevado del Ruiz volcano, is shown in ● Figure 22.22. The satellite image clearly shows the mud and lava flows that caused most of the devastation.

ISS Crew Earth Observations experiment and Image Science & Analysis Laboratory, Johnson Space Center/NASA

Figure 22.22 Nevado del Ruiz Volcano in Colombia
This NASA satellite image shows the volcano in 2010. The 1-km (0.6-mi) caldera is seen in the upper right of the glacier-covered volcano. The long mudflow channels are seen in almost every direction, and the lava flow is seen in the lower right of the image. The image width is 23 km (14 mi).

Leonard Von Matt/Photo Researchers, Inc.

Figure 22.23 Pompeii Victims
When Mount Vesuvius erupted in 79 CE, the people in nearby Pompeii were trapped and suffocated beneath a layer of volcanic ash as much as 6 m (20 ft) thick. Archeologists who excavated Pompeii found cavities lined with imprints of the decomposed bodies. By pouring plaster into the cavities, they were able to make casts that displayed the victims' agonized facial expressions and in some cases even the folds in their clothing.

Throughout human history, volcanoes have been viewed with awe and fear, yet in many areas of the world, large populations continue to live in the danger zones around volcanoes. Take southern Italy, for example. The Mount Vesuvius eruption buried 2000 residents of the city of Pompeii (● Fig. 22.23). Between then and 1944, Vesuvius erupted some 83 times, yet the population of the region continued to grow, and the city of Naples, located only a few kilometers from the volcano, became southern Italy's largest metropolis. An estimated 3 million people live close to Mount Vesuvius.

The reason is based on a simple economic fact: Volcanic ash forms rich, fertile soil. The region around Naples is one of the world's most agriculturally productive. However, as learned in Chapter 21, eruptions are not the only volcanic hazards. The soil formed from volcanic ash is rich, but it is also fine and loose, forming masses of mud when wet. In early 1998, record rainfall in southern Italy resulted in the burial alive of 200 people in the region around Naples, this time under rivers of mud 4 m (13 ft) thick that flowed downslope more rapidly than people could get out of the way.

Did You Learn?

● Volcanoes can eject gas, lava, and tephra (pyroclastics).

● A hot spot is the surface expression of magma that originates below the crust.

22.5   Sedimentary Rocks

Preview Questions

● How are sedimentary rocks classified?

● What features do sedimentary rocks have in common?

As discussed earlier, James Hutton's model of the rock cycle was derived from his observations that rocks are continuously being formed, broken down, and re-formed as a result of igneous, sedimentary, and metamorphic processes. His conceptualization began with an understanding of sedimentary processes.

In his field trips around Scotland, Hutton observed the everyday effects of rain and wind on rocks and soil. He saw mud, sand and rock fragments, large and small, carried by streams from the Scottish highlands to the sea. These particles were deposited as **sediment** along rivers, in lakes, and on the seafloor.

Hutton also observed rocks in the highlands that were made of cemented sand and rock fragments and concluded that they were *sedimentary rocks*, rocks whose components had been derived from the wearing away of older rocks. But how did sediments deposited on the seafloor end up as highland rocks?

Hutton knew that sediments are deposited on the seafloor in horizontal layers, or *strata*, and that each successive layer is deposited on top of the previous, and older, layer. This simple principle, called *superposition* (Chapter 24.2), was established long before Hutton. He realized that as layer upon layer was piled on top, the older layers were compacted by the weight above them and converted into rock. Subsequently, he concluded, the rocks were uplifted by powerful forces from within the Earth to form a mountain range. But even as the new mountains were rising, rain and wind were working to wear them down and streams were transporting their sediments to the sea.

Sediments and sedimentary rocks make up only about 5% of the Earth's crust, but their importance entirely outweighs their relatively limited abundance. One reason is that they

(a)   (b)   (c)

Figure 22.24 Detrital Sedimentary Rocks
Note the differences in grain size. (a) Shale is very fine grained, (b) sandstone is coarse, and (c) conglomerate is very coarse with pebble-size grains.

cover about 75% of the surfaces of continents and even more of the ocean basins. In a sense, they are a bit like clothing; they cover only the surface and are conspicuous. The North American continental interior is composed of a foundation of igneous and metamorphic rocks with a sedimentary veneer only a few kilometers thick. Residents of many of the interior states of the United States may never have seen any rocks other than sedimentary ones. Our modern way of life owes much to sedimentary rocks, because they contain abundant petroleum, coal, metal deposits, and many of the materials essential to modern industry.

Sedimentary rocks, with their many varieties, commonly form fabulous landscapes. For example, the beauty and variety of the Colorado River's Grand Canyon, with its staircase-like, brightly colored slopes and cliffs of sedimentary rock, are famous around the world (see Chapter 23 opening photo).

The Origins of Sedimentary Rocks

The transformation of sediment into a sedimentary rock is a process called **lithification**. During this process, the loose, solid particles (sediment) are compacted by the weight of overlying material and are eventually cemented together. Common cementing agents are silica (SiO_2), calcium carbonate ($CaCO_3$), and iron oxides, which are dissolved in groundwater that permeates the sediment. An example is the formation of shale. When fine-grained mud is subjected to pressure from overlying rock material, water is forced out and the clay minerals begin to compact (consolidate). As groundwater moves through the compacted sediment, materials dissolved in the water precipitate around the individual small mud particles and cementation of the particles occurs. A sedimentary rock is the result.

Sediments are classified according to the source of their constituents into two main groups: *detrital* and *chemical*. *Detrital sediments* are composed of solid fragments, or *detritus*, derived from preexisting rock (● Fig. 22.24). Detrital rocks are classified according to the size of their components (● Table 22.7). *Shale* is composed of the very fine particles that make up mud. *Sandstone* consists of sand-size grains in a silica or calcium carbonate cement. *Conglomerate* is made from rounded pebbles of varying sizes embedded in silica,

Table 22.7	Classification of Detrital Sedimentary Rocks	
Sediment	Grain Size (mm)	Rock Name
Gravel	> 2	Conglomerate (rounded pebbles) Breccia (angular pebbles)
Sand	$\frac{1}{16}$ to 2	Sandstone
Mud	$< \frac{1}{16}$	Shale

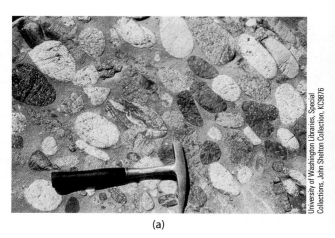

(a)

University of Washington Libraries, Special Collections, John Shelton Collection, KC9876

© Breck P. Kent

(b)

Figure 22.25 Conglomerates and Breccias
The grains in these coarse sedimentary rocks reveal much about their history. (a) The roundness of the grains in conglomerates suggests long-distance transport by vigorously moving water. (b) The angularity of the grains in breccias suggests short-distance transport.

calcium carbonate, or iron oxide cement (● Fig. 22.25a). Rock similar to conglomerate but with angular, not rounded, pebbles is called *breccia* (Fig. 22.25b).

Chemical sediments are composed of minerals that were transported to the sea in solution. ● Table 22.8 shows that there are two types of chemical sedimentary rocks: *organic* and *inorganic*. Organic chemical sedimentary rocks are composed of minerals that were transported in solution but were subsequently acted on by marine organisms. *Organic limestone*, for example, consists of calcite ($CaCO_3$). The calcite dissolved in seawater is extracted by microscopic marine organisms and converted into calcium carbonate skeletal and shell matter. When these organisms die, they fall to the seafloor, where their hard parts collect and eventually lithify into *chalk*. England's famous White Cliffs of Dover is an ancient chalk deposit composed of the carbonate shells of microorganisms (● Fig. 22.26).

Coal is classified as an organic chemical sedimentary rock. Although it does not meet the requirement of having been derived from minerals transported in solution, it is included under this heading because it is derived from the lithified remains of plant matter.

Inorganic chemical rocks (evaporites) are formed when the evaporation of water containing dissolved materials leaves behind residues of chemical sediment such as sodium

Table 22.8 Chemical Sedimentary Rocks

	Rock Name	Typical Composition
Inorganic	Inorganic limestone	Calcite ($CaCO_3$)
	Evaporites	Halite (NaCl), gypsum ($CaSO_4 \cdot H_2O$)
	Dolostone	Dolomite [$CaMg(CO_3)_2$]
	Inorganic chert	Chemically precipitated silica (SiO_2)
Organic	Biogenic limestone	Calcium carbonate remains of marine organisms (e.g., algae, foraminifera)
	Biogenic chert	Silica-based remains of marine organisms (e.g., radiolaria, diatoms, sponges)
	Coal	Compressed remains of terrestrial plants

Source: Chernicoff, Stanley, and Haydn A. Fox, *Essentials of Geology*, Second Edition. Copyright © 2000 by Houghton Mifflin Company. Used with permission.

© Cheryl Hogue/Visuals Unlimited

Figure 22.26 The Formation of Chalk
Chalk is a shallow-marine limestone formed from the calcium carbonate remains of microscopic organisms. Accumulation of chalk can eventually produce deposits of impressive size, such as the famed White Cliffs of Dover in England. The cliffs are composed mainly of the skeletons of microscopic marine plants and animals that accumulated about 100 million years ago, when the global sea level was apparently higher and coastal England was under water.

chloride (halite, or or rock salt; see ● Fig. 22.27) and calcium sulfate (gypsum). Another example of inorganic chemical sedimentary rock is cave dripstone, which is formed primarily by calcium carbonate precipitated from dripping water. Dripstone takes a variety of forms, but most common are the icicle-shaped *stalactites* and cone-shaped *stalagmites* (● Fig. 22.28). Stalactites extend down from the ceiling (*c* for ceiling), whereas stalagmites protrude up from the ground (*g* for ground).

Pgiam/istockphoto.com

Figure 22.27 Evaporite Deposits at the Bonneville Salt Flats
The modern-day Great Salt Lake is a small remnant of Lake Bonneville, a vast lake that existed in Utah about 15,000 years ago, when the local climate was cooler, cloudier, and more humid than it is today. The salt deposits formed when most of Lake Bonneville evaporated under the modern climate, which is warm, clear, and dry.

Jupiter Images/Agence Images/Alamy

Figure 22.28 Dripstone
Stalactites (*c* for ceiling) and stalagmites (*g* for ground) in a limestone cavern.

Figure 22.29 Bedding (Stratification) in Limestone

Sedimentary Characteristics and Structures

Geologists gain much information from sedimentary rocks because their characteristics reveal so much about their origin and history. Color, rounding, sorting, bedding, fossil content, ripple marks, mud cracks, footprints, and even raindrop prints are common characteristics of sedimentary rocks that indicate under what conditions the rocks formed.

The *color* of a sedimentary rock is especially conspicuous in drier parts of the world where little soil or vegetation covers the surface. The bright, variegated colors of the Painted Desert in Arizona draw crowds of tourists, but even rocks with dull colors are conspicuous in dry landscapes. Gray, the most common sedimentary rock color, reflects the rock's origin in shallow, well-aerated marine water. Those few sedimentary rocks that are dark gray to black contain carbon from the organic matter that accumulated in the stagnant water where the sediment was deposited. Rocks deposited above sea level in the presence of abundant oxygen are usually colored by iron oxides and are red-brown or yellow-brown, colors that are so striking in many of the western states.

The shape, or *rounding*, of the sediment gives clues to its place of origin. The larger detrital sediments tend to be worn as the currents and waves that move them grind the particles against the stream bottom and against one another. Angular fragments within the rocks inform geologists that these sediments were not carried far and that they were dropped quickly. The rounder grains of quartz sandstone tell of their long journey downstream and of the many hours they were shifted and rolled by the waves and currents of the sea.

A current or wave that can keep small grains in motion may be unable to move larger grains. Because wind and moving water transport some particles more easily than others, the particles tend to become *sorted* or separated according to size. The greater the distance of transportation, the more effectively the grains are sorted. In a lake or ocean, the largest particles come to rest near shore, but the finer grains are transported into the quieter water farther from shore, where they eventually settle to the bottom.

Bedding, or *stratification*, is the layering that develops at the time the sediment is deposited. The bedding may be vague, or it may be conspicuous as shown in ● Fig. 22.29. It may be in either thick or very thin layers. Because most sediment comes to rest on a level surface, most bedding is horizontal. In many environments, however, sediments accumulate in tilted layers known as *cross-bedding*. Much research and many pages have been devoted to the geometry and origin of the many kinds of cross-bedding. In every case, the surface is sloping where the sediments come to rest.

Cross-bedding is especially common where a river empties into a lake, where sediments fill in depressions scoured by floods along river channels and where wind drapes sand down the flanks of a dune (● Fig. 22.30). By recognizing the type of cross-bedding, a geologist can make an observation that helps unravel the origin and history of the rock.

The most distinctive and interesting characteristic of sedimentary rock is the fossils it sometimes contains (● Fig. 22.31). Although a mystery to those geologists who lived several centuries ago, a *fossil* is now known to be the remains or traces of a prehistoric organism. Geologists have collected so many fossils and have refined the techniques of examining them so effectively that they can determine the relative age of a sedimentary rock from its fossil content. The role of fossils in determinations of geologic age will be discussed in Chapter 24.

Figure 22.30 Cross-bedding in Sandstone

Did You Learn?

● SSedimentary rocks are classified by the source of the constituents that make up their sediment: either detrital (solid fragments) or chemical (minerals in solution).

● Sedimentary rocks have common features: color, shape, bedding, and fossil content, among others.

Lee Boltin/Bridgeman Art Library

Figure 22.31 Fossils Are Often Found in Sedimentary Rocks Most fossils, such as this fish cast, consist only of the impression left in the rock by the organism after the remains have decomposed or dissolved.

22.6 Metamorphic Rocks

Preview Questions

- How are metamorphic rocks formed?
- What percentage of the Earth's crust is metamorphic rock?

As discussed previously, magmas that cool to form igneous rocks are created by high temperatures deep inside the Earth's interior, and sedimentary rocks are formed when sediments are buried and lithified just below the Earth's surface. Metamorphic rocks are created by the conditions that exist in a zone in the Earth's interior between 10 and 30 km in depth, below the region where rocks are lithified and above the region where they are melted.

About 15% of the Earth's crust is metamorphic rock. **Metamorphism** is the process by which the structure, mineral content, or both of a rock is changed while the rock remains a solid. All igneous and sedimentary rocks can be metamorphosed, and any metamorphic rock can be subjected to further metamorphism (see Fig. 22.10). The agents of metamorphism are heat, pressure, and chemically reactive hot-water solutions. The resulting metamorphic rock depends on these influences as well as on the composition of the *parent rock* being metamorphosed (● Table 22.9).

Temperature increases with crustal depth at the rate of about 30°C/km, so the temperature at 15 km (9 mi) is about 450°C. Pressure also increases with depth, and the pressure at 15 km (9 mi) is about 4000 times as great as the pressure at the Earth's surface. Pressure can be either *confining pressure* (equal in all directions) or *directed pressure* (strongest in a certain direction). The texture, the mineral composition, or both can change as a rock is metamorphosed. For example, in the metamorphism of organic limestone to *marble*, the texture of the rock is changed, but not its mineral content (● Fig. 22.32). Organic limestone is made of fossils and a cement matrix, both of which are composed of small calcite crystals. When the limestone undergoes temperature and pressure changes, the calcite crystals grow large. All traces of fossiliferous (formed of fossils) limestone texture are obliterated in the marble, but the calcite content remains the same.

In general, if the parent rock contains only one mineral, then the metamorphic rock will be composed of that mineral alone (for example, calcite, in the case of limestone);

Table 22.9 Classification of Some Common Metamorphic Rocks

Foliated Rocks			
Parent Rock	Metamorphic Rock	Key Minerals	Characteristics
Shale	Slate	Clay, quartz, mica, chlorite	Fine grains, slaty cleavage
Shale, basalt, slate	Schist	Chlorite, plagioclase, mica, garnet	Coarse grains, well-foliated
Shale, granite, slate, schist	Gneiss	Plagioclase, garnet, kyanite, sillimanite	Coarse grains, light- and dark-colored bands
Nonfoliated Rocks			
Parent Rock	Metamorphic Rock	Key Minerals	Characteristics
Limestone	Marble	Calcite	Coarse, interlocking calcite grains
Sandstone	Quartzite	Quartz	Fine to coarse interlocking quartz grains
Shale, basalt, or any fine-grained rock	Hornfels	Mica, quartz	Fine grains, variable composition

however, if the parent rock contains several minerals, then metamorphism will create new and different minerals. For example, shale is commonly composed of clay quartz, mica, and chlorite. ● Figure 22.33 shows that as shale undergoes increasing pressure and temperature changes, it metamorphoses first to slate, then to schist, and finally to gneiss (pronounced "nice"). As it progresses though each stage, both its texture and its composition change. If the temperature and pressure are high enough, then the gneiss will melt and become magma.

Although it is difficult to make clear-cut separations, we recognize several kinds of metamorphism, including contact, shear, regional, and hydrothermal metamorphism. **Contact metamorphism** is change brought about primarily by heat, with very little pressure

Figure 22.32 Fossiliferous Limestone (*left*) and Marble (*right*)
When fossiliferous limestone is metamorphosed into marble, the texture of the rock is radically changed but the calcite mineral content of the rock remains the same. The sculpture is a replica of the statue of David by Michelangelo. Like the original, it was created from a large block of white marble.

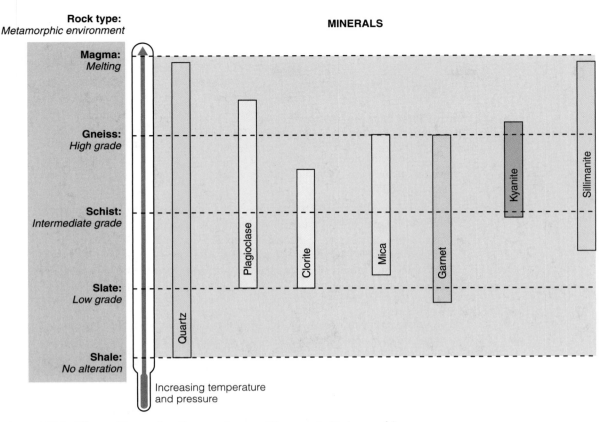

Figure 22.33 Effects of Increasing Temperature and Pressure in Metamorphism
As sedimentary shale is subjected to increasingly intense temperatures and pressures, it becomes metamorphic slate, then schist, and then gneiss. Finally, the rock melts and becomes magma. Also shown are some of the minerals found in each type of rock.

involved. Such a change commonly occurs in shallow bedrock when it is subjected to the heat of a molten body of magma moving up from greater depths. Contact metamorphism is most obvious at such a shallow depth because bedrock near the Earth's surface is normally cool, and the effects of great temperature changes are therefore quite pronounced. The rock immediately next to the molten magma experiences intense metamorphism and may be coarse grained, but it grades out into finer-grained rock that has been less severely transformed by the heat. This dark, fine-grained rock, containing recrystallized minerals with random crystal orientation, is known as *hornfels*.

Rocks changed more by pressure than by temperature are said to have undergone the effects of **shear (cataclastic) metamorphism**, which is most common in active fault zones where one rock unit slides past another. Mechanical deformation shatters the grains or changes their shapes plastically. Recrystallization accompanies the more intense forms of shear metamorphism, but the significant physical effects are more obvious.

Most metamorphic rocks have been affected by *both* high temperature and high pressure and have therefore experienced both mechanical deformation and chemical recrystallization. **Regional metamorphism**, as this type of change is known, gets its name from the extremely vast area it generally affects. The widely exposed metamorphic rocks found in central Canada are of this sort. Although much remains to be learned about regional metamorphism, it appears to affect rocks undergoing intense deformation during mountain building.

In rocks subjected to regional metamorphism, the mineral grains (most often mica) are flattened, elongated, and aligned perpendicular to the direction of the directed pressure.

The flattened mineral grains become aligned parallel to one another, and this arrangement of mineral grains results in a pronounced layering of the rock called **foliation**. Rocks that are metamorphosed solely by the heat of contact metamorphism do not develop foliation.

Conceptual Question and Answer

Name That Metamorphic Process

Q. If one rock type forms deep underneath a volcano, a second type forms by magma moving near the surface, and a third type forms in a subduction zone where one tectonic plate slides under another, what type of metamorphism is experienced in each?

A. Rock deep under a volcano exists in a high-temperature and high-pressure environment; therefore, it is altered by regional metamorphism. High-temperature magma cooling near the surface is altered in a low-pressure environment defined by contact metamorphism. Near a subduction zone, the pressures are extremely high but the temperatures are not, and the rock is therefore altered by shear metamorphism.

The progressive metamorphism of shale is a good illustration of changes that occur as a sedimentary rock is subjected to more and more intense regional metamorphism. Shale, after relatively mild metamorphism, is transformed into *slate*, a fine-grained metamorphic rock similar in many respects to shale but different fundamentally in its excellent slaty cleavage (● Fig. 22.34). A rock exhibiting *slaty cleavage* breaks apart very easily along the planes of its thin, smooth layers. If the shale is subjected to more intense heat and pressure, then it will change into *schist*, a foliated metamorphic rock whose grains are visible to the unaided eye (● Fig. 22.35). Very intense regional metamorphism produces *gneiss*, an even coarser-grained rock with rough foliation characterized by distinct banding (● Fig. 22.36). Higher grades of metamorphism therefore produce larger grains but rougher foliation.

Metamorphic rocks may be either *foliated* or *nonfoliated*. Metamorphic rocks that lack foliation include such well-known examples as marble, which is formed from limestone, and *quartzite*, which is metamorphosed sandstone.

Hydrothermal metamorphism is the chemical alteration of preexisting rocks by chemically reactive, hot-water solutions, which dissolve some ions from the original minerals and replace them with other ions, thus changing the mineral composition of the rock. Most hydrothermal metamorphism takes place at divergent plate boundaries on the ocean floor, where both heat and water are abundant. *Verde* (pronounced "verd") *antique* is a hydrothermally metamorphosed rock composed of serpentinite, a hydrous magnesium silicate.

Figure 22.34 Slate Bed
Slates are fine-grained, metamorphic rocks that exhibit a type of foliation known as slaty cleavage.

Figure 22.35 Schist
This mica schist is a product of intermediate-grade metamorphism.

Figure 22.36 Gneiss
Feldspar and quartz are the chief minerals of gneiss, a metamorphic rock contorted under high temperature and pressure.

Courtesy of Marble Modes, Inc.

Figure 22.37 Verde Antique
Verde antique is a highly prized decorative stone formed by hydrothermal metamorphism.

Verde antique is a decorative stone that is highly prized by architects because of its dark green color, with white and greenish-white accents (● Fig. 22.37). Mined in Vermont, its beauty and durability make it suitable for flooring, countertops, monuments, and building exteriors. This rock does not stain and keeps its shine when exposed to the weather.

Did You Learn?

- Metamorphic rocks are formed when rock undergoes heat, pressure, or chemical changes.

- About 15% of the Earth's crust is metamorphic rock.

KEY TERMS

1. mineral (22.1)
2. silicates
3. Mohs scale
4. cleavage
5. rock (22.2)
6. igneous rocks
7. sedimentary rocks
8. metamorphic rocks
9. uniformitarianism
10. rock cycle
11. magma (22.3)
12. lava
13. plutons (22.4)
14. pyroclastics (tephra)
15. viscosity
16. caldera
17. sediment (22.5)
18. lithification
19. bedding
20. metamorphism (22.6)
21. contact metamorphism
22. shear metamorphism
23. regional metamorphism
24. foliation
25. hydrothermal metamorphism

MATCHING

For each of the following items, fill in the number of the appropriate Key Term from the preceding list. Compare your answers with those at the back of the book.

a. _____ Chemical alteration of preexisting rocks by hot solutions

b. _____ The breaking of a mineral crystal along distinctive planes

c. _____ General term for large bodies of intrusive igneous rock

d. _____ Hot, molten material below the Earth's surface

e. _____ Rocks formed by alteration of preexisting rocks by pressure, temperature, or chemistry

f. _____ Used to describe mineral hardness

g. _____ Metamorphism over a vast area; brought about by both heat and pressure

h. _____ A solid, cohesive, aggregate of one or more minerals

i. _____ Solids, liquids, and gases that are ejected into the air during a volcanic eruption

j. _____ Rocks formed from the solidification of magma or lava

k. _____ The transformation of materials of the Earth's crust over long periods of time

l. _____ Hot, molten material flowing on the Earth's surface, or the rock this material forms

m. _____ Rocks formed by compaction and cementation of sediment

n. _____ The process of changing the structure or mineral content of a rock while the rock remains solid

o. _____ Metamorphism brought about mostly by pressure

p. _____ A naturally occurring crystalline, inorganic element or compound

q. _____ The layering that develops at the time the sediment is deposited

r. _____ The process by which sediment is compacted and cemented together

s. _____ Minerals that contain silicon, oxygen, aluminum, and usually one other element

t. _____ The basic understanding that ancient rocks were formed in the same way as modern rocks

u. _____ Sand, mud, precipitated material, or rock fragments that have been transported by water or air

v. _____ Metamorphism brought about mostly by heat

w. _____ A large depression in the top of a volcano; usually formed by the collapse of the chamber

x. _____ The internal property of a substance that offers resistance to flow

y. _____ The parallel arrangement of mineral grains that can occur in a metamorphic rock as a consequence of pressure

MULTIPLE CHOICE

Compare your answers with those at the back of the book.

1. Which of the following is *not* a characteristic property of a mineral? (22.1)
   (a) naturally occurring
   (b) crystalline
   (c) definite chemical composition
   (d) low boiling point

2. Density is related most closely with which property of identifying minerals? (22.1)
   (a) hardness   (b) specific gravity
   (c) color   (d) fracture

3. The simple, common method of classifying minerals depends on which general property? (22.1)
   (a) physical   (b) chemical
   (c) nuclear   (d) optical

4. What does the Mohs scale measure? (22.1)
   (a) streak   (b) luster
   (c) hardness   (d) cleavage

5. Which of the following scientists is generally designated "The Father of Geology"? (22.2)
   (a) Friedrich Mohs   (b) James Hutton
   (c) Alfred Wegener   (d) Edwin Hubble

6. Which of the following is *not* one of the three basic classes of rocks? (22.2)
   (a) igneous   (b) pyroclastic
   (c) metamorphic   (d) sedimentary

7. Coal is the compacted remains of plant matter. What basic type of rock is coal? (22.2)
   (a) igneous   (b) sedimentary
   (c) metamorphic   (d) magma

8. Which of the following does the rock cycle describe? (22.2)
   (a) the interrelationships between rock-producing processes
   (b) the chemical composition of basaltic rocks
   (c) subduction of lithospheric plates
   (d) volcanic eruptions and magma viscosity

9. Igneous rocks are classified according to their combination of chemical composition and what else? (22.3)
   (a) age   (b) texture
   (c) permeability   (d) density

10. What does the grain size of an igneous rock tells a scientist about that rock? (22.3)
    (a) the color of the rock
    (b) the rate at which the molten rock cooled
    (c) whether the rock will float in water
    (d) the overall size of the rock

11. Which of the following is *not* an igneous rock? (22.3)
    (a) basalt   (b) granite
    (c) sandstone   (d) obsidian

12. What are plutons? (22.4)
    (a) bodies of intrusive igneous rocks
    (b) explosive volcanoes
    (c) undersea metamorphic rocks
    (d) pyroclastic flows

13. Which of the following statements is true? (22.4)
    (a) Dikes are concordant rock bodies.
    (b) Dikes are formed from lava.
    (c) Dikes are spherical in shape.
    (d) Dikes are intrusive igneous rock.

14. What type of volcano forms a layered, composite cone? (22.4)
    (a) fumarole   (b) cinder cone
    (c) stratovolcano   (d) shield volcano

15. How does sedimentary rock from the ocean floor sometimes end up in highland and mountainous regions on the Earth? (22.5)
    (a) Sedimentary rock is commonly made inside volcanoes.
    (b) Wind and water carry the sediment to the mountaintops.
    (c) It is believed that asteroid impacts probably caused the sedimentary rock to move great distances.
    (d) The sedimentary rock was uplifted by powerful forces to form mountain chains.

16. Which of these is a common color of sedimentary rock? (22.5)
    (a) yellow-green   (b) blue-green
    (c) purple-gray   (d) red-brown
17. What is the process of transforming sediment into sedimentary rock called? (22.5)
    (a) lithification   (b) striation
    (c) metamorphism   (d) hardening
18. To which rock does the metamorphosis of limestone lead? (22.6)
    (a) marble   (b) slate
    (c) schist   (d) gneiss

19. Metamorphism occurs for which class of rocks? (22.6)
    (a) metamorphic
    (b) sedimentary
    (c) igneous
    (d) all the preceding
20. Which of these is *not* a common metamorphic rock? (22.6)
    (a) gneiss   (b) granite
    (c) slate   (d) schist

FILL IN THE BLANK

Compare your answers with those at the back of the book.

1. The two most abundant elements in the Earth's crust are oxygen and ___. (22.1)
2. The most abundant minerals in the Earth's crust are ___. (22.1)
3. The metallic appearance of some minerals is called ___. (22.1)
4. A(n) ___ is any mineral or other precious or semiprecious stone valued for its beauty. (22.1)
5. Rocks solidified from molten material, either below or above ground, are classified as ___ rocks. (22.2)
6. The concept that ancient rocks were formed in the same way as modern rocks is called ___. (22.2)
7. Weathering and compaction of igneous or metamorphic rock form ___ rock. (22.3)
8. The common igneous rock that has a fine-grained texture and low silica content (dark color) is called ___. (22.3)
9. Dripstone formations that extend downward from a cavern ceiling are called ___. (22.3)
10. The manner in which a volcano erupts depends primarily on the ___ of its magma. (22.4)
11. A(n) ___ is a steep-sloped formation from a volcano that ejected tephra. (22.4)

12. Intrusive igneous rock formations that lie more or less parallel to older formations are said to be ___. (22.4)
13. The Hawaiian Islands and Emperor Seamount chain were created by a(n) ___ in the Earth's crust. (22.4)
14. The general name for an intrusive igneous rock formation is ___. (22.4)
15. Plutons are classified as either concordant or ___. (22.4)
16. A common energy source, ___ is classified as an organic chemical sedimentary rock. (22.5)
17. Stratification, or ___, is the layering that develops at the time of sediment deposition. (22.5)
18. The sedimentary rock called shale metamorphoses into the rock called ___. (22.6)
19. ___ is a pronounced layering of rock caused by regional metamorphism. (22.6)
20. Metamorphism that results from the circulation of hot solutions is called ___ metamorphism. (22.6)

SHORT ANSWER

22.1 Minerals

1. What is a mineral, and what is the study of minerals called?
2. Sketch the structure of the silicon–oxygen tetrahedron.
3. How does the oxygen-to-silicon ratio differ between silica and other silicates?
4. Give four examples of nonsilicate materials.
5. What are the limits of the Mohs scale? Give an example mineral at each limit.
6. Name six physical characteristics that are used to identify minerals.
7. Define the mineralogical terms *luster* and *streak*.
8. What is the characteristic cleavage of mica?

22.2 Rocks

9. Define the term *rock*. What is the study of rocks called?

10. Name the three basic types of rocks and briefly state the process by which each type is formed.
11. What is the basis of uniformitarianism? How does it relate to modern geology?
12. Briefly describe the rock cycle.

22.3 Igneous Rocks

13. Distinguish between magma and lava.
14. How are intrusive and extrusive igneous rocks classified?
15. Name a common igneous rock that is an example of each texture.
16. Use Table 22.4 and compare gabbro and rhyolite with regard to grain size (large or small), color (light or dark), and silica content (high or low).

22.4 Igneous Activity and Volcanoes

17. Distinguish between discordant and concordant igneous rock bodies.

18. What are three products of a volcano?

19. What two factors determine the viscosity of magma and lava?

20. How did the Hawaiian Islands form?

21. Describe how flood basalts have changed the landscape of many continents.

22. Distinguish among (a) shield volcanoes, (b) stratovolcanoes, and (c) cinder cones.

23. What are calderas, and how are they formed? Name a famous one in the United States.

22.5 Sedimentary Rocks

24. Why are sedimentary rocks the most important rocks for humans? Distinguish between detrital and chemical sedimentary rocks.

25. Distinguish between a stalactite and a stalagmite. (How can one remember?)

26. List several physical characteristics of sedimentary rock.

27. What is often the most interesting characteristic of sedimentary rocks?

22.6 Metamorphic Rocks

28. How do temperature and pressure change with depth in the Earth? What kind of rock is produced at great depths?

29. Distinguish among contact, shear, and hydrothermal metamorphism.

30. Define the term *foliation*.

31. Describe how shale is progressively metamorphosed into three types of metamorphic rock.

VISUAL CONNECTION

Visualize the connection and give answers for the blanks. Compare your answers with those at the back of the book.

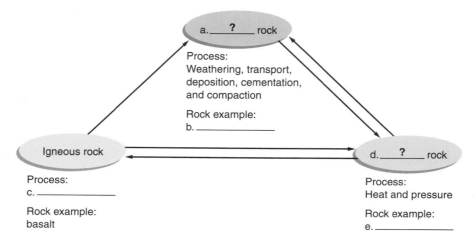

APPLYING YOUR KNOWLEDGE

1. While on a geologic expedition, you unearth a large, dark stone with an excellent print of a fossil embedded within the rock. Because of the dark color, your friend automatically assumes that it is an igneous rock. Explain why it is uncommon to find fossils in igneous rock. In what kind of rock are most fossils found?

2. While in Hawaii, you want to visit Mauna Loa. Your friend, however, is afraid to go anywhere near the volcano because it might

   explode like Mount Saint Helens. Explain to your friend why Mauna Loa, which erupts regularly, can be expected to erupt peacefully. Then explain how geologists knew in advance that if and when Mount Saint Helens blew, it would be a dangerously explosive eruption.

3. You are given five transparent objects: a calcite crystal, a diamond, a piece of window glass, a sample of quartz, and a piece

of zircon. How would you go about identifying each sample? How would they rank on the Mohs hardness scale?

4. In lab, your instructor hands you a steel-gray mineral and asks you to identify it. You find that the mineral gives a red-brown streak. What is the mineral's probable identity?

5. The lab instructor hands you an igneous rock to identify. You notice that the rock has large grain size and contains only dark-colored minerals. What is the rock's probable identity?

6. As a student in geology, you have three favorite mountains: Mount Fuji, Mount Kilimanjaro, and Mount Vesuvius. Where is each of these mountains located, and what kind of mountain is each? Be specific in your classification.

7. You are leading a scout troop on a field trip when you are asked to identify the igneous rock formation shown in ● Fig. 22.38. Give your troop the correct answer.

Figure 22.38 An Igneous Feature
See Applying Your Knowledge Question 7.

Fred Hirschmann

ON THE WEB

1. A Gem of an Idea

How common are gems? Where are they found? What are the two major characteristics of gemstones? How are they categorized? Which would be your favorite and why? If you want to go out looking for your own gemstones, where can you find them? To learn more about gems and to answer the above questions, visit the links on the student website at **www.cengagebrain.com/shop/ISBN/1133104096** and follow the recommended links.

2. Journey to the Center of the Earth: Looking Inside Volcanoes

Explore the benefits and hazards of volcanoes as well as their types and natures by following the recommended links on the student website at **www.cengagebrain.com/shop/ISBN/1133104096**. Imagine yourself a volcanologist exploring a volcano in Hawaii, Greece, or some other part of the world. What do you need to know about planning and logistics, operations, and the equipment you'll be using? What safety precautions might be important?

Surface Processes

Running water labors continually to reduce the whole of the land to the level of the sea.

•

Charles Lyell,
British geologist
(1797–1875)

Arizona's Grand Canyon is a > vivid example of erosion.

David Muench/Getty Images

GEOLOGY FACTS

▶ The largest known cave in the world was discovered in Vietnam in 2009. The Son Doong cave is 80 m by 80 m (260 ft by 260 ft) in most places and is at least 4.5 km (2.8 mi) in length. In some places the cave is 240 m (800 ft) tall, the size of a 40-story building.

▶ The driest desert in the world is the Atacama in Chile and Peru. In some places it rains only once in 20 years, because the tall mountains on either side shield the desert from potential rain or snow. Fog is the main source of water. Some of the riverbeds have been dry for 120,000 years.

▶ The wettest place on the Earth is Cherrapunji, India, in the northeast part of the country, with an average rainfall amount of nearly 13 m (43 ft).

I n the context of geologic time, nothing is permanent on the Earth's surface. Since early times, people have erected edifices and monuments as memorials to persons and peoples and as symbols of various cultures. The most durable rock materials were used for this purpose; however, even recent structures show the inevitable signs of deterioration. Buildings, statues, and tombstones are eventually worn away when exposed to the elements, and on a larger scale, nature's mountains are continually being leveled.

An integral part of changing the Earth's crustal rocks is decomposition by weathering. As soon as rock is exposed at the Earth's surface, its destruction begins. Many of the resulting particles of rock are washed away by surface water and transported

Chapter Outline

as sediment toward the oceans in streams and rivers. Over geologic time, this process results in a leveling of the Earth by wearing away high places and transporting sediment to lower elevations, a process known as *gradation*. (See the chapter-opening quotation.)

Important geologic processes take place on the ocean floor, which accounts for about 70% of the Earth's surface. Therefore, waves and ocean currents along with shoreline and seafloor topography are also considered in this chapter.

23.1 Weathering

Preview Questions

- What are the two basic types of weathering processes?
- How is chemistry associated with weathering?

Weathering is the process of breaking down rock on or near the Earth's surface. The rate of weathering depends on a number of factors, such as the type of rock, moisture, temperature, and overall climate.

Mechanical weathering involves the physical disintegration or fracture of rock, primarily as a result of pressure. For example, a common type of mechanical weathering in some regions is *frost wedging*. When rock solidifies or is altered by deposits, heat, or pressure, internal stresses are produced in it. One common result of these stresses is cracks or crevices in the rock, called *joints*. Joints provide an access route for water to penetrate rock. If the water freezes, then the expanding ice exerts strong pressure on the surrounding rock. Just as freezing water in a glass container may crack the glass, the rock may break apart, as illustrated in ● Fig. 23.1. Frost wedging is most effective in cold regions where freezing and thawing occur daily.

In cold upper latitudes, the subsurface soil may remain frozen permanently, creating a **permafrost** layer. During a few weeks in summer, the topsoil may thaw to a depth of a few inches to a few feet. The frozen subsurface prevents the melted water from draining. As a result, the ground surface becomes wet and spongy (● Fig. 23.2a). The permafrost in Alaska caused many problems during construction of the Alaskan oil pipeline. Because the oil is heated in the pipeline to make it flow more easily, the pipeline is elevated above the permafrost. In some areas, special cooling fins are used to keep the ground frozen (Fig. 23.2b).

Plants and animals play a relatively small role in mechanical weathering. Most notable is the fracture of rock by plant root systems when they invade and grow in rock crevices.

Figure 23.1 Frost Wedging
The expansive forces of freezing water in rock joints cause the rock to fracture.

P. Carrara/USGS

Figure 23.2 Permafrost in Alaska
(a) This "roller-coaster" railroad near Strelna, Alaska, was caused by differential subsidence resulting from the thawing of topsoil over subsurface permafrost. (b) The Alaskan oil pipeline shown is elevated to circumvent the permafrost. The cooling fins are visible at the top of the posts. (See text for description.)

(a)

(b)

Burrowing animals—earthworms in particular—loosen and bring soil to the surface. This action promotes aeration and access to moisture. The activities of humans also give rise to weathering and erosion that are often unwanted, as will be discussed later in the chapter.

In all cases of mechanical weathering, the disintegrated rock still has the same composition. *Chemical weathering*, however, involves a chemical change in the rock's composition. Because heat and moisture are two important factors in chemical reactions, this type of weathering is most prevalent in hot, moist climates. One of the most common types of chemical weathering involves limestone, which is made up of the mineral calcite ($CaCO_3$). Rain can absorb and combine with carbon dioxide (CO_2) in the atmosphere to form a weak solution of carbonic acid:

$$H_2O + CO_2 \rightarrow H_2CO_3$$
Carbonic acid

Also, as water moves downward through soil, it can take up even more carbon dioxide that is released by soil bacteria involved in plant decay. Recall that carbonic acid (carbonated water) is the weak acid in carbonated drinks (Chapter 13.3).

Figure 23.3 A Sinkhole in Winter Park, Florida
A home, a car dealership, part of a swimming pool, and part of the road were destroyed when the ceiling of a cavern collapsed, creating a sinkhole.

When carbonic acid comes in contact with limestone, it reacts with the limestone to produce calcium hydrogen carbonate, or calcium bicarbonate:

$$H_2CO_3 + CaCO_3 \rightarrow Ca(HCO_3)_2$$

Carbonic acid   Limestone   Calcium bicarbonate

Calcium bicarbonate dissolves readily in water and is carried away in solution. Because limestone is generally impermeable to water (and dilute carbonic acid), this type of chemical weathering acts primarily on the surfaces of limestone rock over which water flows.

Water flowing through underground limestone formations can carve out large caverns over millions of years. The cavern ceilings may collapse, causing depressions called **sinkholes** to appear on the land surface (● Fig. 23.3).

The rate of chemical weathering depends primarily on climate and on the mineral content of the rock; humidity and temperature are the chief climate controls. Chemical weathering is more rapid in hot, humid climates than in cold or dry regions. Egyptian pyramids and statues have stood for millennia with relatively little chemical weathering.

Some rock minerals are more susceptible to chemical weathering than others, as is apparent in the decomposition of gravestones (● Fig. 23.4). Marble gravestones, which consist of relatively soluble calcite, may show a great deal of chemical weathering. Granite, because of its hardness, firmness, and durability, is the rock most suitable for gravestones.

Figure 23.4 Differential Weathering
Marble and granite gravestones from the same cemetery in Williamstown, Massachusetts, show differential weathering of rock types. The marble gravestone that reads 1870 (*left*) has been exposed to the same climate as the granite gravestone (*right*). The marble stone was erected 50 years later, but has suffered more weathering damage. Marble, composed predominantly of chemically reactive calcite, is much more susceptible to chemical weathering than is granite, a rock composed of very stable minerals.

Marble

Granite

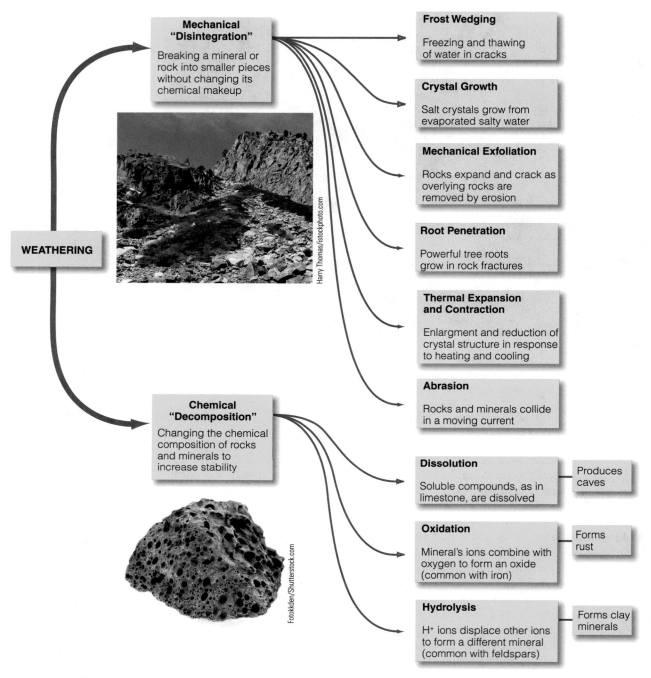

Mechanical "Disintegration"

Breaking a mineral or rock into smaller pieces without changing its chemical makeup

Frost Wedging

Freezing and thawing of water in cracks

Crystal Growth

Salt crystals grow from evaporated salty water

Mechanical Exfoliation

Rocks expand and crack as overlying rocks are removed by erosion

Root Penetration

Powerful tree roots grow in rock fractures

Thermal Expansion and Contraction

Enlargment and reduction of crystal structure in response to heating and cooling

Abrasion

Rocks and minerals collide in a moving current

WEATHERING

Harry Thomas/istockphoto.com

Chemical "Decomposition"

Changing the chemical composition of rocks and minerals to increase stability

Fotokkden/Shutterstock.com

Dissolution

Soluble compounds, as in limestone, are dissolved

Produces caves

Oxidation

Mineral's ions combine with oxygen to form an oxide (common with iron)

Forms rust

Hydrolysis

H^+ ions displace other ions to form a different mineral (common with feldspars)

Forms clay minerals

Figure 23.5 A Summary of Weathering
The six types of mechanical and the three types of chemical weathering are summarized here. (See text for description.) Top: mechanical weathering caused visible piles of rock at the base of this mountain; Bottom: chemical weathering caused holes to form in this rock.

There are six general forms of mechanical weathering, and three forms of chemical weathering. These processes are listed and summarized in ● Figure 23.5.

Conceptual Question and Answer

Moon Weathering

Q. Is there weathering on the Moon?

A. Because of the lack of atmosphere and surface water, the Moon has no weathering as we know it. However, there is "space weathering" on the surface due to the impact of meteoroids (sizes range from dust particles to large boulders) and high-energy particles from the Sun over billions of years. As a result, the lunar surface is covered with a layer of dust called the regolith. You may have seen the images of the footprints of the Apollo astronauts in the powdery lunar soil (see Chapter 17 opening photo). An example of big lunar space weathering is the formation of rays, which are radial streaks thrown outward during the formation of an impact crater (see Fig. 17.4).

Did You Learn?

- Weathering occurs in two basic forms: mechanical and chemical.
- Chemical weathering changes the chemical composition of the rock.

23.2 Erosion

Preview Questions

- How do glaciers erode the land?
- What are the three main agents of erosion?

Erosion is the downslope movement of soil and rock fragments under the influence of gravity (mass wasting) or by agents such as streams, glaciers, wind, and waves.

Streams

The term *running water* refers primarily to the waters of streams and rivers that erode the land surface and then transport and deposit eroded materials. Rainfall that is not returned to the atmosphere by evaporation either sinks into the soil as groundwater or flows over the surface as runoff. *Runoff* occurs whenever rainfall exceeds the amount of water that can be immediately absorbed into the ground. It can move surprisingly large amounts of sediment, particularly from slopes without vegetation and from cultivated fields (● Fig. 23.6). Runoff usually occurs only for short distances before the water ends up in a stream.

Geologists define a **stream** as any flow of water occurring between well-defined banks. The term is applied to channeled flows of any size, from meter-wide mountain brooks to kilometer-wide rivers. The material transported by a stream is referred to as *stream load*. The load of a stream varies from dissolved minerals and fine particles to large rocks. The transportation process, as well as the degree to which the stream can erode its channel, depends on the volume and swiftness (discharge) of the stream's current. A stream's load is divided into three components: dissolved load, suspended load, and bed load.

The *dissolved load* consists of water-soluble minerals that are carried along by a stream in solution. As much as 20% of the material reaching the oceans is transported in solution. Fine particles not heavy enough to sink to the bottom are carried along in suspension. This *suspended load* is quite evident when the stream appears muddy after a heavy rain. Coarse particles and rocks along or close to the bed of the stream constitute its *bed load*. These particles and rocks are rolled and bounced along by the current.

The action of the flowing water in a river causes its bed to erode. (See the chapter-opening photo.) The principal landform resulting from a stream's

Figure 23.6 Runoff
This runoff is beginning to carve out a channel. If its supply of water continues, it may become a stream.

Stacey Lynn Payne/Shutterstock.com

Glenn Oliver/Visuals Unlimited

Figure 23.7 A Meandering Stream in Northern Alaska
This meandering stream has oxbow lakes created by erosion. Note the sandy sediment deposited along the riverbanks. Can you find at least three oxbow lakes?

erosive power is a V-shaped valley. The depth of an eroded valley is limited by the level of the body of water into which the river flows. The limiting level below which a stream cannot erode the land is called its *base level*. In general, the ultimate base level for rivers is sea level.

Once a stream has eroded its channel close to base level, downward erosion slows and the stream begins to expend its energy by eroding its bed from side to side. The valley floor is widened as one bank is eroded and then the other. Periodic flooding is caused by heavy rains or spring thaw in the mountains upstream. During flooding, the water overflows its banks and deposits sediment across the low adjacent land, or **floodplain**. As a consequence of this natural process, the floodplains of large rivers offer some of the most fertile soil on the Earth.

That rich soil is important for agriculture and is the reason floodplains become heavily populated. However, the people who come to live on a floodplain want the soil, but not the floods that go with it. Therefore, they often build artificial levees to hold the river in its channel.

As a stream flows overland, it is influenced by gravity and the rotating Earth (see Chapter 19.4). Therefore, it naturally forms twists and turns, following the path of least resistance. A loop-like bend in a stream channel is called a **meander** (● Fig. 23.7). Meanders shift or migrate because of greater erosion on the outside of the curved loops. The speed of the streamflow is greater in this region than near the inside bank of the meander, where sediment is deposited. When the erosion along two sharp meanders causes the stream to meet itself, it may abandon the water-filled meander, which is then known as an *oxbow lake* (Fig. 23.7).

The amount of material eroded and transported by streams is enormous. The oceans receive billions of tons of sediment each year as a result of the action of running water. The Mississippi River alone discharges approximately 500 million tons of sediment yearly into the Gulf of Mexico. When a stream enters a lake or ocean, its suspended load and bed load may accumulate at its mouth and form a **delta**, such as the Mississippi River Delta (● Fig. 23.8). The rich sediment makes delta regions important for agriculture.

However, some sediment deposits are not always beneficial. Many harbors must be dredged to remove sediment deposits to keep the harbors navigable, and the buildup behind dams may pose operational problems. Human activities that remove erosion-preventing vegetation from the land may give rise to sediment pollution in rivers and streams and thus cause environmental problems.

Glaciers

Parts of the Earth are covered with large, thick masses of "permanent" ice; these large masses are called **glaciers**. Large areas of Greenland and Antarctica are presently covered with glacial ice sheets, or *continental glaciers*, similar in size to those that covered Europe and North America during the last ice age more than 10,000 years ago. Due to natural and human causes, these glaciers are changing due to increased air and ocean temperatures (see the Chapter 20 Highlight: The Ozone Hole and Global Warming).

Glaciers are formed when, over a number of years, more snow falls than melts. As the snow accumulates and becomes deeper, it is compacted into ice by its own weight. When enough ice accumulates, the glacier "flows" downhill under the influence of gravity or, if it rests on a flat region, flows out from its center. The icebergs commonly found in the North Atlantic and Antarctic oceans are huge chunks of ice that have broken off from the edges of the glacial ice sheets of Greenland and Antarctica, respectively.

Are there any glaciers in the United States today? The answer is yes. In fact, you may be surprised to learn that there are more than 1100 glaciers in the western part of the conterminous United States and that 3% of the land area of Alaska is covered by glaciers.

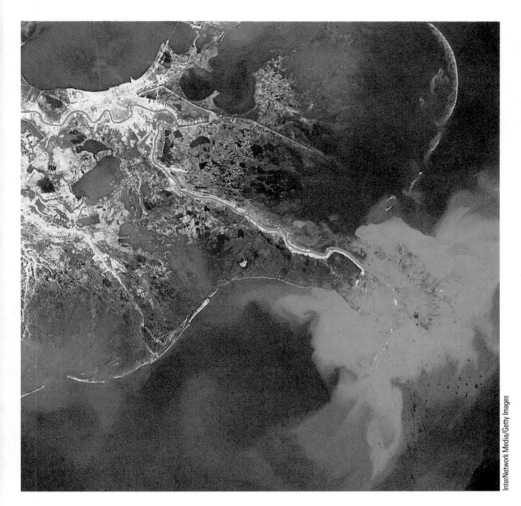

Figure 23.8 Delta of the Mississippi River
Deltas form when streams deposit sediment upon entering a large body of water. The rich sediment makes delta regions important for agriculture.

InterNetwork Media/Getty Images

Small glaciers, called *cirque* (pronounced "sirk") *glaciers*, form along mountains in hollow depressions that are protected from the Sun. The majority of the glaciers in the western United States are of this variety. The ice movement further erodes the land and forms an amphitheater-like depression called a *cirque*. When the ice melts, these glacier-eroded cirques often become lakes.

When snow accumulates in a valley, the valley floor may be covered with compressed glacial ice, and a *valley glacier* is formed that flows down the valley (● Fig. 23.9). At lower and warmer elevations, the ice melts, and where the melting rate equals the glacier's flow rate, the glacier becomes stationary. The flow rate may be anywhere from a few inches to more than 30 m (100 ft) per day, depending on the glacier's size and on other conditions.

The erosive action of a valley glacier is similar to that of a stream. As the ice flows, it loosens and carries away materials, or bed load, that will be ground fine by abrasion. Although it moves much more slowly than a stream, a glacier can pick up huge boulders and gouge deep holes in the valley. The paths of vigorous mountain glaciers are well marked by the deep U-shaped valleys they leave.

Glaciers, like streams, deposit the material they carry. Rock material that is transported and deposited by ice is called **glacial drift**. As shown in Fig. 23.9, the glacial drift may form large ridges known as **moraines** along the sides of a glacier (a *lateral moraine*), in the middle of the glacier (a *medial moraine*), or at its end (a *terminal moraine*). The terminal moraine marks the farthest advance of the glacier. Terminal moraines give us an indication of the extent and advance of the glacial ice sheet in North America, which retreated about 10,000 years ago. These moraines lie as far south as Indiana, Ohio, and Long Island, New York. When these large ice sheets retreated and melted, they carved out and filled the Great Lakes region of the United States and Canada.

Photo

Figure 23.9 A Valley Glacier Showing Moraines
(Sketch) Medial moraines form when adjacent glaciers converge, causing the lateral moraines at their edges to run together. End moraines form as the glacier recedes. (Photo) The Kennicott Glacier in southeastern Alaska shows medial and lateral moraines.

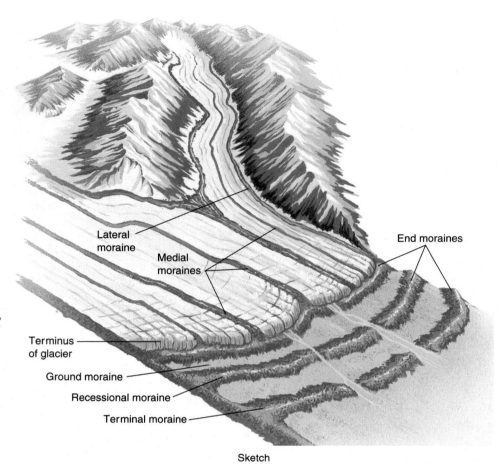

Sketch

Wind

Erosion by wind is a slow process, but wind contributes significantly to the leveling of the land surface, especially in deserts. Most people think of a desert as a hot, dry, sandy place, but to a geologist, the term **desert** is defined by a severe lack of precipitation, not by temperature. Occurring in cold as well as torrid regions, deserts account for one-fifth of the Earth's land surface. Because vegetation is sparse in deserts, the occasional rainfall wreaks erosive havoc on the land surface. However, from day to day, wind is the prime mover of the land, transporting it grain by grain from high places to low.

A region of arid and semiarid climate in western North America extends from northern Mexico to eastern Washington State and includes such well-known deserts as the Sonoran Desert on the U.S.–Mexican border, the Mojave Desert in California, and the Great Basin centered in Nevada.

Dust particles that are small enough are transported great distances by the wind, and larger particles are moved short distances by rolling or bouncing along the surface. Wind action can lead to remarkable surface features in areas with large quantities of loose, weathered debris (● Fig. 23.10). Dust storms may darken the sky and be of such intensity that visibility is reduced almost to zero. During the 1930s in the Midwest, drought conditions created areas known as *dust bowls* in which layers of fertile topsoil were blown away by the wind.

Waves

As wind blows across the ocean, friction on the water surface creates waves. The wind-driven water waves erode the land along the shorelines. This erosion is most evident where the ocean surf pounds the shore. Some coasts are rocky and jagged, which show that only

Figure 23.10 Wind Erosion in Utah
In desert areas, rocks may perch precariously on narrow pedestals that have been eroded by wind-blown sand.

Figure 23.11 A Massive Rockslide
This rockslide occurred in 1997 along Highway 140 close to Yosemite National Park in California. Note the relative sizes of the people and the largest rock.

the hardest materials can withstand the unrelenting wave action over long periods of time. Along other coastlines, cliffs are formed, terraced by the eroding action of waves. Waves are discussed further in Chapter 23.4.

Mass Wasting

Wherever the ground slopes, debris consisting of soil and rock fragments that have been loosened by weathering—or shaken loose by an earthquake or volcanic eruption—is pulled downslope by gravity. **Mass wasting** is the general geologic term for the downslope movement of soil and rock under the influence of gravity. To start the debris moving, the force of gravity must overcome friction, the force that tends to keep the material stationary (Chapter 4.2).

Both the forces of gravity and friction are influenced by the presence of water. Water adds mass to the debris, thus increasing its weight, but it also acts as a lubricant, thereby lessening the effects of friction. For this reason, mass-wasting disasters occur most frequently during or after a heavy snow or rainfall.

Mass movement may be fast or slow, depending on the steepness of the slope. Furthermore, the amount of material involved in a single mass-wasting episode may be from a few pebbles to an entire mountainside. Thus, the effect of an episode on the local environment and on any people living there can range from unnoticeable to devastating.

Two important types of *fast mass wasting* are landslides and mudflows. *Landslides* involve the downslope movement of large blocks of weathered materials. Spectacular landslides occur in mountainous areas when large quantities of rock break off and move rapidly down the steep slopes. This type of landslide is termed a *rockslide* (● Fig. 23.11).

A comparatively slower form of landslide is a *slump*, the downslope movement of an unbroken block of rock or soil, which leaves a curved depression on the slope. Slumps are

Gary Phelps/Ventura County Star

Figure 23.12 January 2005 Mudflow Disaster in Southern California
A mudslide in La Conchita crashed down on the town, killing 10 people and destroying 18 homes.

commonly accompanied by debris flows that consist of a mixture of rock fragments, mud, and water flowing downslope as a viscous liquid. Small slumps are commonly observed on the bare slopes of new road construction.

Mudflows are the movements of large masses of soil that have accumulated on steep slopes and become unstable through the absorption of large quantities of water from melting snows or heavy rains. Vegetation hinders mass movement. Consequently, mudflows are common in hilly and mountainous regions where land is cleared for development without regard to soil conservation.

In January 2005, the seaside community of La Conchita, California, felt the wrath of nature in the form of a mudflow. Nearly 60 cm (24 in.) of rain fell over five days. Combined with mountain snowmelt, the rain caused the entire mountainside to race down into the town (● Fig. 23.12). Nearly 9 m (30 ft) of mud covered part of the town, killing 10 people and destroying 18 homes.

Fast mass wasting is quite dramatic, but *slow mass wasting* is a more effective geologic transport process. In contrast to the rates of fast mass wasting, the rates of slow mass wasting are generally imperceptible. The slowest type, called **creep**, is the slow movement of weathered debris down a slope, taking place year after year. It cannot be seen happening, but the manifestations of creep are evident (● Fig. 23.13). Although spectacular landslides and slumps involve large quantities of mass movement, they represent only a fraction of the cumulative total of mass movement by creep over a period of time.

Did You Learn?

● Glaciers transport rock and sediment as they advance and retreat.

● The three main agents of erosion are water, glaciers (ice), and wind.

23.3 Groundwater

Preview Questions

- What is the hydrologic cycle?
- How much of the Earth's water is fresh water, and where is the majority located?

Water is often referred to as the basis of life. The human body is composed of 55–60% water by mass, and water is necessary to maintain our body functions. This common chemical compound is an essential part of our physical environment, not only in life processes but also in agriculture, industry, sanitation, firefighting, and even religious ceremonies. Early civilizations developed in valleys where water was abundant, and even today the distribution of water is a critical issue.

Our water supply is a reusable resource that is constantly being redistributed over the Earth. Many factors enter into this redistribution, but in general it is a movement of moisture from large reservoirs of water, such as oceans and seas, to the atmosphere, to the land, and back to the oceans. This perpetual cyclic movement of the Earth's water supply is known as the **hydrologic cycle** (● Fig. 23.14).

Moisture evaporated from the oceans moves over the continents through atmospheric processes and falls as precipitation. Some of this water evaporates and returns to the

Dr. John D. Cunningham/Corbis

Figure 23.13 Creep
The downhill bends in these trees are evidence of creep in the surface beneath them.

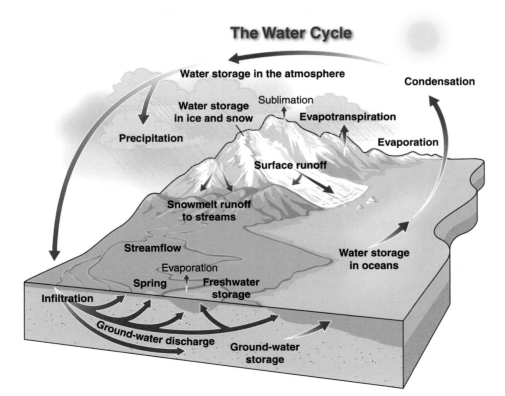

The Water Cycle

Figure 23.14 An Illustration of the Hydrologic Cycle
Evaporated moisture from oceans, rivers, lakes, and soil is distributed by atmospheric processes. It eventually falls as precipitation and returns to the soil and bodies of water.

Highlight   The Earth's Largest Crystals

In 2000, two miners who were surveying for a tunnel in the Naica Mountains, a region southwest of the city of Chihuahua, Mexico, stumbled across a cave containing the largest crystal structures ever seen. Silver in the Naica Mountains was first discovered in 1794, but not until about 1900 did the mining increase because of the discovery of the important elements lead and zinc.

The newly discovered cave is called Cueva de los Cristales (Cave of Crystals) because of the spectacular crystals located there. It is about 300 m (1000 ft) below the surface and is the size of a small basketball court, about 300 m² (3230 ft²). Huge beams of crystals have grown from floor to ceiling and out of the walls of

the cave (Fig. 1). The largest crystals are on the order of 11 m (36 ft) in length and 2 m (6.6 ft) wide. The cave is horseshoe-shaped, and crystal blocks cover the entire floor (Fig. 2). The antechamber was initially discovered by the miners, who called it the Eye of the Queen because of its small entrance in the shape of an eye (Fig. 3).

According to geologists, volcanic activity created the Naica Mountains about 26 million years ago and filled the cave with the mineral anhydrite. As the surrounding rock cooled, the minerals began to dissolve and thus enriched the subterranean waters. The crystals are made of calcium sulfate in the form of selenite

Figure 1 A geologist is seen standing on the world's largest known crystals. They are made of gypsum and are up to 11 m (36 ft) in length.

Figure 2 Another scientist photographed in the "cave of ice." Near-perfect conditions of hot water and minerals of calcium sulfate allowed the crystals to grow to their enormous size.

atmosphere and some of it becomes runoff, but a large part of it soaks into the soil and down into the subsurface, where it collects as **groundwater**. For an example of how groundwater created some amazing geologic forms, see the **Highlight: The Earth's Largest Crystals**.

Conceptual Question and Answer

Powering the Hydrologic Cycle

Q. What powers the hydrologic cycle?

A. The hydrologic cycle is powered primarily by the Sun and gravity. Solar insolation (Ch. 19.2) warms the atmosphere, which causes evaporation, and the moisture is carried upwards to form clouds. After the Sun has done its job, gravity causes the moisture to fall back to the Earth as precipitation. Gravity also causes water and ice to move downhill via stream, glacier, and groundwater flow. These flows fill our lakes and oceans only to be evaporated (in part) again, and the cycle repeats.

Highlight

gypsum. Gypsum is the same material used to make plaster of Paris. It was calculated that the crystals grew in near perfect conditions of 100% humidity at 58°C (136°F) for hundreds of thousands of years. The temperature was kept at such a narrow range that the crystals were able to grow to their enormous sizes.

The cave was originally flooded with mineral water kept hot from volcanism, but it is currently being pumped out for surveying, mining, and research. There are 55,000 liters (14,500 gallons) of water being pumped out of the mine every minute. Because the crystals are as delicate as glass, the cave has been sealed with a special door to keep it separate from the industrial mines.

Because of the high humidity and temperature, initial photographs of the crystals were difficult. In fact, the extreme environment will only allow a human to survive in the cave for several minutes. Someone did break into the mine and try to steal some crystals, but he was unprepared for the heat and suffocated in a very short time. Now, however, the mine and cave are now ventilated and the Cave of Crystals is open on a limited basis. Currently, only scientists, geologists, mineralogists, and crystal admirers are allowed to see this magnificent structure in person.

Javier Trueba/MSF/Photo Researchers, Inc.

Figure 3 "The Eye of the Queen" is the name given to the small entrance to the first cave discovered in this complex. This cave is smaller than the main cave and is only 10 m (33 ft) by 20 m (66 ft).

Groundwater Mechanics

The Earth's water supply, some 1.25×10^{18} m³ (4.4×10^{19} ft³), may seem inexhaustible because it is one of our most abundant natural resources. Approximately 70% of the Earth's surface is covered with water. However, about 97% of that water is salt water, and only about 3% is fresh water. Most of the fresh water is frozen in the glacial ice sheets of Greenland and Antarctica. A mere 0.6% of fresh water is groundwater, yet it is the major source of our drinking water and the water used in agriculture and industry (● Fig. 23.15).

The rate of human use of groundwater increases every year, and the rate of yearly rainfall is not enough to replenish the groundwater reserves that took thousands of years to accumulate. In addition, groundwater supplies in many places have been contaminated by human, agricultural, and industrial wastes. The use and abuse of groundwater have become key environmental issues facing all nations of the world, including the United States, where bottled water is now a standard commodity sold in supermarkets.

The movement of groundwater is controlled by the physical properties of soil and rocks. *Porosity* is the percentage volume of unoccupied space in the total volume of a substance. The porosity of rocks and soil near the Earth's surface determines the ground's capacity to store water. The rock that most commonly serves to store water is sandstone.

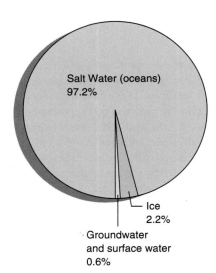

Figure 23.15 Distribution of the Earth's Water
Only about 3% of the Earth's water is fresh water (ice, groundwater, and fresh surface water), and most of it is in the form of ice.

The factor that determines the availability of groundwater is the permeability of the subsurface soil and rock. **Permeability** is a material's capacity to transmit fluids, which is a function of the porosity of the material. Of course, loosely packed soil components, such as sand and gravel, permit greater movement of water. Clay, on the other hand, has fine openings and relatively low permeability.

Under the influence of gravity, water percolates downward through the soil until at some level the ground becomes saturated. The upper boundary of this *zone of saturation* is called the **water table**. The unsaturated zone above the water table is called the *zone of aeration* (● Fig. 23.16).

In the zone of aeration, the pores of the soil and rocks are partially or completely filled with air. In the zone of saturation, all the voids are saturated with groundwater, forming a reservoir from which we obtain part of our water supply by drilling wells to depths below the surface of the water table (see Fig. 23.16). Lakes, rivers, and springs occur where the water table intersects the surface. The level of the water table shows seasonal variations, and shallow wells may go dry in late summer.

A body of permeable rock that both stores and transports groundwater is called an **aquifer** (Latin for "water carrier"). Sand, gravel, and loose sedimentary rock are good aquifer materials. Aquifers are found under more than half the area of the conterminous United States.

Generally, water must be pumped out of *water-table wells*, but a special geometry of impermeable rock layers makes possible what are called *artesian wells*, wells in which water under pressure rises to the surface without the aid of pumping. As illustrated in ● Fig. 23.17, this geometry can occur when an aquifer is sandwiched between sloping, impermeable rock strata (*confined aquifers*) and the higher end is exposed to the surface so that it can receive water to replenish the aquifer. Gravity can cause water to spurt or bubble onto the surface above the aquifer. The name *artesian* comes from the French province of Artois, where such wells and springs are common.

Groundwater Depletion and Contamination

Groundwater reserves are not inexhaustible. When the rate of extraction is greater than the rate of *recharge*—that is, the rate at which the water is replaced—an aquifer can be seriously depleted. The Ogallala aquifer, which extends from South Dakota to the Texas Panhandle, provides an example. Most of the water in the aquifer was added during the wetter climate of the last ice age. Today, the aquifer provides water for agricultural irriga-

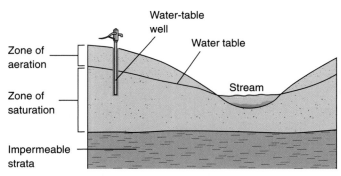

Figure 23.16 Zone of Aeration and Zone of Saturation
The boundary between the zone of aeration and the zone of saturation is called the water table. Wells are drilled below the water table into the zone of saturation, from which water may be pumped.

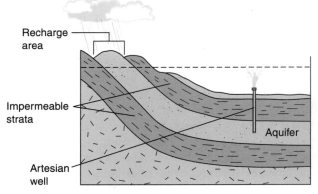

Figure 23.17 Confined Aquifer, Recharge Area, and Artesian Well
As long as the top of the well is below the height of the recharge area (dashed line), water will rise out of the well without the aid of pumping.

tion in a semiarid region covering 65,000 km² (25,000 mi²). More than 150,000 wells are currently tapping the Ogallala at ten times the rate of recharge, and the zone of saturation is shrinking drastically. Eventually, this essentially nonrenewable aquifer will be depleted. With no other source of water available to replace it, the agricultural economy of the region will surely fail.

Besides depletion, excessive groundwater extraction can lead to a variety of problems, one of which is land *subsidence*. Because the water in an aquifer helps support the weight of the overlying strata, excessive extraction of water may cause the land surface to sink, or subside. ● Figure 23.18 shows the drastic 9-m (30-ft) subsidence of the land in California's San Joaquin Valley between 1925 and 1977. Land subsidence is accelerated if the surface strata are supporting additional weight, such as buildings, roads, and other structures. Subsidence may cause water and drainage pipes to rupture, building foundations to shift, cracks to develop suddenly in roads, and roadbeds to sag.

Another problem of excessive groundwater extraction that occurs in coastal regions is saltwater contamination of wells. Because salt water is denser than fresh water, fresh groundwater in the zone of saturation floats on salt water that has seeped in from the ocean and penetrated the strata. As pumping lifts fresh water up to the surface through the well, the salt water directly beneath the well rises to fill the gap (● Fig. 23.19). Thus, excessive pumping eventually brings undesirable salt water up into the well.

Other forms of contamination also threaten the supply of usable groundwater. The quality of water is generally expressed in terms of the amount of dissolved chemical substances present as a concentration in parts per thousand (ppt), parts per million (ppm), or parts per billion (ppb). For example, when a water sample contains 1 ppm salt, it contains 1 gram of salt per 1 million grams of water.

Also important with respect to quality are the types of chemicals that water contains. Chemicals range from nontoxic to extremely toxic; concentrations range from very high to very minute values. For example, water quality can be very good with a fairly high concentration of calcium carbonate, but it would be very toxic with minute amounts of some other substances, such as methylmercury. Specifically for drinking water, common levels of calcium carbonate are 100–300 ppm. However, the safety standard for methylmercury in drinking water is only 0.002 ppm, which is about 100,000 times less.*

Our best natural source of fresh water is rainwater, but even that contains a variety of dissolved chemicals due to air pollution. Acid rain, which is caused mainly by the burning

---

*Methylmercury is a poisonous compound. Mercury is released into the environment through the burning of coal, from volcanoes, and from volcanoes and forest fires. The mercury is transformed to methylmercury by bacteria that live in lakes, rivers, and soil.

Figure 23.18 Land Subsidence Due to Groundwater Depletion
The signs on the utility pole show that between 1925 and 1977, this part of the San Joaquin Valley subsided almost 9 m (30 ft) because of the withdrawal of groundwater and the resulting compaction of sediments.

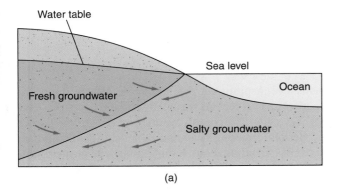

(a)

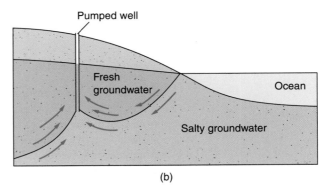

(b)

Figure 23.19 Saltwater Contamination of a Coastal Water-Table Well
(a) Salt water seeps into the ground beneath the aquifer, but because it is denser, it stays beneath the fresh water. (b) Overdrawing the fresh water causes the salt water to rise up into the well.

of sulfur-containing fossil fuels, has been a problem (see Chapter 20.4). Rain falling on the Earth's surface enters the ground, where it takes on more chemicals when it reacts with soil, rock, and organic material. Water quality is reduced further, sometimes drastically, when the groundwater is contaminated by phosphates (found in laundry detergents and fertilizers), sewage and other waste materials, pesticides (organic compounds used to kill insects), herbicides (organic compounds used to kill unwanted plant life), and industrial wastes.

Municipal water supplies are treated by chlorination to eliminate bacteria and viruses and thus increase the quality of our drinking water. However, this process does not remove organic compounds such as acetone, benzene, carbon tetrachloride, or chloroform that may enter the water supply by way of industrial wastes. Some of these compounds can be removed by using charcoal filters.

In some cases, water containing bicarbonates (hydrogen carbonates) and chlorides is sold commercially as "mineral water." However, dissolved minerals in "hard" water have undesirable effects. *Hard water* is so called because of its high content of dissolved calcium, iron, and magnesium salts (bicarbonates, chlorides, and sulfates).

Such dissolved minerals not only affect the taste of the water but also cause usage problems. For example, the minerals can form a scale that clogs pipes. Also, the salts combine with the organic acids in soaps to form insoluble compounds, thereby reducing the lathering and cleansing qualities of the soap (see Chapter 14.6). Because of these insoluble compounds, white clothing becomes tarnished during washing, and rings form around the bathtub. Water softening, which is the removal of these insoluble compounds, is an active business in many parts of the United States.

Because water makes up 55–60% of the human body mass, water of good quality is extremely important for good health. Therefore, it is vital for all individuals, wherever they live, to protect their freshwater supply.

Did You Learn?

- The hydrologic cycle is the perpetual cyclic movement of the Earth's water supply.

- About 3% of the Earth's water is fresh water, and most of that is in the form of ice in Antarctica and Greenland. (Only about 0.6% of the fresh water is groundwater.).

23.4   Shoreline and Seafloor Topography

Preview Questions

- How do waves affect shorelines?
- Where are underwater volcanoes located?

Waves, Currents, and Tides

The vastness of the restless oceans is awe-inspiring to most observers. Where the ocean meets the land, we observe the beauty of coastal regions shaped by surface waves that vary from broad, low beaches to steep, rocky cliffs. The oceans cover about 70% of the Earth's surface. The five major oceans in order of decreasing size are the Pacific, Atlantic, Indian, Antarctic, and Arctic oceans. Their average depth is about 4 km (2.5 mi); the greatest measured depth is about 11 km (7 mi) in the Marianas Trench in the western Pacific.

Ocean water is in constant motion. Three types of seawater movement are waves, currents, and tides. Ocean waves continually lap the shore (● Fig. 23.20). Although the form of a wave moves directly toward the shore, the water "particles" move in more or less cir-

Figure 23.20 Ocean Waves Erode the Land
The surf crashes against the rocky beach along the coast of central Oregon. For scale, note the people at the right of the photo.

Gary Braasch/Corbis

cular paths, as illustrated in ● Fig. 23.21. This circular pattern is the reason debris bobs up and down as the waveforms pass underneath.

As a wave approaches shallower water near the shore, the water particles are forced into more elliptical paths. The surface wave then grows higher and steeper. Finally, when the depth becomes too shallow, the water particles can no longer move through the bottom part of their paths and the wave breaks, with the crest of the wave falling forward to form *surf*.

At the beach, you may have noticed that unless a piece of bobbing debris comes close enough to be caught in the surf and thrown up on shore, it appears to move steadily in a direction parallel to the shoreline. This parallel movement is an indication of a **long-shore current** flowing along the shore. This current arises from incoming ocean waves that break at an angle to the shore. The component of water motion along the shore causes a current in that direction. Waves and their resulting long-shore currents are important agents of erosion along coastlines.

Tides, the periodic rise and fall of the water level along the shores of large bodies of water, are also quite evident at the beach. Tides result from the two tidal bulges that "move" around the Earth daily as a result of the gravitational attractions of the Moon and the Sun and the rotation of the Earth (see Chapter 17.2). Because of the geography at some

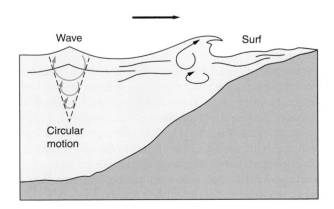

Figure 23.21 An Illustration of a Surface Wave Approaching the Shore
(See text for description.)

Highlight   The Highest Tides in the World

At certain places on the Earth, the tidal changes are enormous. The place with the largest variation in tides is the Bay of Fundy, located between the Canadian provinces of New Brunswick and Nova Scotia (Fig. 1). This bay was created millions of years ago when continental plates drifted apart and formed a rift valley.

Because of the ocean currents and the unique shape and location of the bay, there is an effect called resonance, which causes the tidal changes to be large (for more on resonance, see Chapter 6.6). The resonance occurs because the time it takes water to move in and out of the bay is almost the same as the time between high and low tides, which is about 6 hours (Chapter 17.2).

The difference between the high and low tides at certain places within the Bay of Fundy is generally about 12 to 15 m (40 to 50 ft) but can reach as high as 17 m (56 ft). Figure 2 shows a fishing dock where the water depth routinely changes 9 m (30 ft) because of the tides. Compare that change with the average change in ocean heights of about 1 m (3 ft) due to tidal forces. Those who pilot fishing boats in the Bay of Fundy have to be aware of the changing water depth due to these tides or their boats may get stuck out in the bay for many hours.

There are some important consequences of these large tides. The first is that the rivers that flow into the bay sometimes reverse

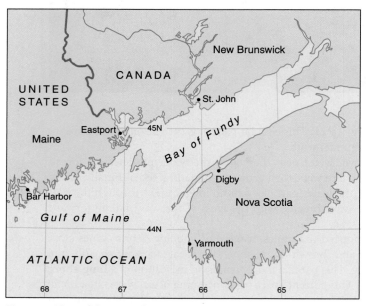

Figure 1 Map of the Bay of Fundy

locations, such as the Bay of Fundy in New Brunswick, Canada, the water level may rise as much as 17 m (56 ft) at high tide (see the **Highlight: The Highest Tides in the World**).

Shoreline Topography

When incoming waves attack an irregular stretch of coastline, the parts of the shoreline that jut out are subjected to the most erosion. If the land along the coast is elevated, then wave action cuts into the base of the slopes below the high-tide mark. As shown in ● Fig. 23.22, when the overlying rock loses support, it collapses, leaving behind a nearly vertical *wave-cut cliff*. Hollowed-out portions of the wave-cut cliff form *sea caves*. Isolated remnants of resistant rock remain as *sea stacks* and *sea arches* (● Fig. 23.23).

Along with the sediments transported to the sea by rivers, the fragments eroded from elevated stretches of coastline are transported by long-shore currents to bays and to calmer stretches of the coast where the land is at a lower elevation. Here beaches and other features of a depositional coastline are seen.

Highlight

direction. The inflowing tidal water creates a wave, called a tidal bore, that travels up the river. As the river narrows, the wave can reach up to 1 m (3 ft) in height and travels at speeds of up to 15 km/h (10 mi/h). Witnesses say that the in-rushing wave sounds like an approaching freight train. People in boats, canoes, and rafts can sometimes ride this wave up the river.

Second, because the cold water is in nearly constant motion, it stirs up nutrients on the bottom of the bay, creating a vast feeding ground for many whales and porpoises within the bay. Sometimes, however, whales get stuck in the bay during times of low tide and must wait 6 hours until the water rises again.

Third, the tides also create impressive wave-cut cliffs and sea stacks (see Figure 23.22). The most famous are the Hopewell

Rocks, which look like giant flower pots with trees on top (Fig. 3). Underneath one of these rocks is a sea arch (called Lover's Arch) where many weddings have taken place. These rocks and sea stacks have been created by the water pounding the shoreline every 6 hours.

Finally, engineers are looking at ways to harness the energy created by the tidal waters; perhaps a hydroelectric plant may be built in the future (similar to the Rance River tidal power station in France; see Chapter 4.6). The amount of water that sloshes in and out of the bay every tidal period is about 100 km³ (24 mi³), which is equal to the volume that all the freshwater rivers in the world discharge into the oceans. The bay is potentially a nearly limitless source of energy without pollution.

Figure 2 The Bay of Fundy at High Tide and Low Tide

Figure 3 Hopewell Rocks at the Bay of Fundy
Note the sea stacks, wave-cut cliffs, and sea arch in this photo.

As shown in ● Fig. 23.24, some of the more common depositional features are the following:

1. *Pocket beaches* form in the low-energy wave environment between headlands.

2. *Barrier islands* extend more or less parallel to the mainland. New York's Fire Island is an example. Between a barrier island and the mainland is a protected body of water called a *lagoon*.

3. *Spits* are narrow, curved projections of beach that extend into the sea, elongating the shoreline. Cape Cod, Massachusetts, is a spit.

Seafloor Topography

Scientists once thought that the surface features, or topography, of the ocean basins consisted of an occasional volcanic island arc on a relatively smooth, sediment-covered floor. This incorrect view resulted from a lack of direct observation. The surface of the oceanic

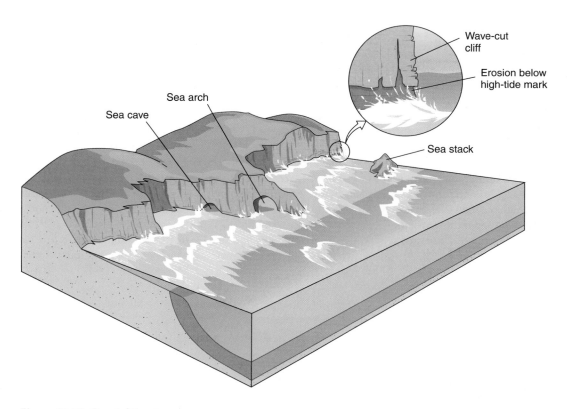

Figure 23.22 Coastal Erosion
The action of waves on the coastline create features of erosion shown here. (See text for description.)

Figure 23.23 A Sea Stack and a Sea Arch
A sea stack (*left*) and a sea arch (*right*) are seen at the tip of Mexico's Baja Peninsula.

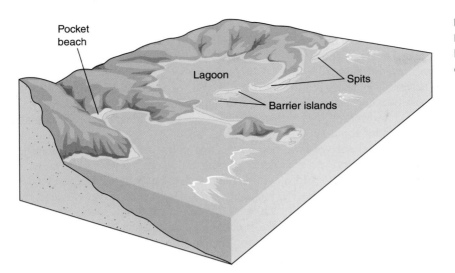

Figure 23.24 Coastal Deposition
Features include pocket beaches, barrier islands, and spits. (See text for description.)

crust was not explored in great detail until after World War II. With the advent of modern technology, sounding and drilling operations revealed that the ocean floor is about as irregular as the surfaces of the continents, if not more so (see Fig. 21.7).

We now know that the seafloor has a system of mid-ocean ridges: rocky, submarine mountain chains that mark divergent plate boundaries. Volcanic island arcs rim the Pacific and Indian oceans. Large volcanic mountains also rise from the ocean floor away from plate boundaries, marking the place where the plate has ridden over a mantle hot spot (Chapter 22.4).

Many isolated, submarine, volcanic mountains have also been discovered. Known as **seamounts**, these individual mountains may extend to heights of more than 1.6 km (1 mi) above the seafloor. Some seamounts have flat tops and are given the special name of *guyots* ("GEE-ohs").* Their shapes suggest that the tops were once islands that were eroded away by wave action. However, many of the guyot tops are several thousand feet below sea level. The eroded seamounts subsided and sank below sea level as the oceanic crust moved away from a spreading ridge.

Other notable features of seafloor topography are *trenches*, which mark the locations of the deep-sea subduction zones. These trenches are as much as 240 km (150 mi) in width, 24,000 km (15,000 mi) or more in length, and 11 km (7 mi) in depth.

The huge volumes of sediment flowing into the oceans from continental regions do have an effect on seafloor topography. Distributed by ocean currents, sediment accumulates in some regions such that a layer covers and masks the irregular features of the rocky ocean floor. The resulting large, flat areas on the deep-ocean floor are called *abyssal plains*. Abyssal plains are most common near the continents, which supply the sediment that forms them.

Continental shelves are gently sloping, relatively shallowly submerged areas that border continental land masses (● Fig. 23.25). Even though these shelves are submerged, they

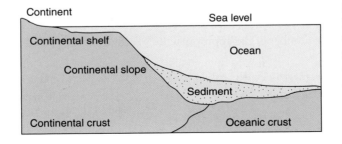

Figure 23.25 A Cross-Sectional Illustration of a Continental Shelf
Continental shelves are relatively shallow submerged areas that border continental land masses. (See text for description.)

---

*Named in honor of Arnold Guyot, the first geologist at Princeton University, by Professor Harry Hess, a geologist at Princeton who discovered the first flat-topped seamounts in the 1950s.

are not part of the oceanic crust, but rather of the continental crust. The continental crust makes up 35% of the Earth's surface area, which is 30% land mass and 5% submerged continental shelves. Thus, the oceanic crust basins account for only about 65% of the surface area, even though 70% of the Earth's surface is covered with water. The widths of these shelves vary greatly but average on the order of 64 to 80 km (40 to 50 mi). The Pacific coast of South America has almost no continental shelf, only a relatively sharp, abrupt slope. However, off the north coast of Siberia, the continental shelf extends outward into the ocean for about 1280 km (800 mi).

Continental shelves have become a focal point of international interest and dispute, because the majority of commercial fishing is done in the waters above the shelves. Also, the shelves are the locations of oil deposits that are now being tapped by offshore drilling. As a result, many countries, including the United States, have extended their territorial claims to an offshore limit of 320 km (200 mi).

Beyond a continental shelf, the surface of the continental land mass slopes steeply downward to the floor of the ocean basin. The **continental slopes** define the true edges of the continental land masses. Erosion along these slopes gives rise to deep, submarine canyons that extend downward toward the ocean basins. Near the edges of the ocean basins, the sediment collects in depositional basins.

Undersea exploration is a relatively new phase of scientific investigation. Indeed, a great deal remains to be learned about this vast region. As advances in technology provide more data on these previously inaccessible depths, it is expected that our knowledge of the Earth and its geologic processes will grow.

Did You Learn?

- Shorelines are eroded by waves causing movement of sand and formations such as sea stacks and sea arches.

- Underwater volcanoes are located near plate boundaries and over hot spots.

KEY TERMS

1. weathering (23.1)
2. permafrost
3. sinkholes
4. erosion (23.2)
5. stream
6. floodplain
7. meander
8. delta
9. glaciers
10. glacial drift
11. moraines
12. desert
13. mass wasting
14. creep
15. hydrologic cycle (23.3)
16. groundwater
17. permeability
18. water table
19. aquifer
20. long-shore current (23.4)
21. tides
22. seamounts
23. continental shelves
24. continental slopes

MATCHING

For each of the following items, fill in the number of the appropriate Key Term from the preceding list. Compare your answers with those at the back of the book.

a. _____ Any flow of water occurring between well-defined banks

b. _____ Dry regions of the Earth, accounting for one-fifth of the Earth's land

c. _____ Isolated, submarine, volcanic mountains

d. _____ Permanently frozen, subsurface soil

e. _____ A body of permeable rock that both stores and supports water

f. _____ The constant, cyclical redistribution of the water supply over the Earth

g. _____ A loop-like bend in a stream channel

h. _____ The downslope movement of soil and rock fragments under the direct influence of gravity alone

i. _____ Large ridges of glacial drift at the ends, sides, or middle of a glacier

j. _____ Circular depressions that appear on the land's surface as a consequence of the collapse of cavern ceilings

k. _____ The periodic rise and fall of the water level along the shores of large bodies of water

l. _____ Downslope movement of rock and soil under the influence of gravity or of water, ice, or wind

m. _____ The steep, submerged surface of the continental land mass beyond the continental shelves

n. _____ A large accumulation of sediment where a stream enters a lake or ocean

o. _____ The breaking down of rock at or near the Earth's surface

p. _____ Rock material that is transported and deposited by ice

q. _____ Water that soaks into the soil and down into the subsurface

r. _____ Gently sloping, shallowly submerged areas that border the continental land masses

s. _____ Large, thick masses of ice that remain year-round

t. _____ The upper boundary of the zone of water saturation

u. _____ Occurs when water overflows its banks and deposits sediments across the low, adjacent land

v. _____ An ocean current that flows almost parallel to the shoreline

w. _____ Slow, continual, particle-by-particle downslope movement of weathered debris

x. _____ A material's capacity to transmit a fluid

MULTIPLE CHOICE

Compare your answers with those at the back of the book.

1. What is the general wearing away of high places and transporting of material to lower places called? (Intro)
   - (a) gradation
   - (b) weathering
   - (c) depositing
   - (d) glaciation

2. Which type of mechanical disintegration is due to rocks and minerals colliding in a moving current? (23.1)
   - (a) frost wedging
   - (b) root penetration
   - (c) abrasion
   - (d) crystal growth

3. Which two primary factors does the rate of weathering of rock depend on? (23.1)
   - (a) temperature and pressure
   - (b) humidity and mineral content
   - (c) temperature and humidity
   - (d) wind and geographic location

4. Chemical weathering can be determined by analyzing what property of a rock? (23.1)
   - (a) temperature
   - (b) density
   - (c) composition
   - (d) mass

5. Which of the following are agents that cause erosion? (23.2)
   - (a) streams
   - (b) glaciers
   - (c) waves
   - (d) all the preceding

6. In a desert, what is the prime mover of the land material? (23.2)
   - (a) rain
   - (b) wind
   - (c) gravity
   - (d) creep

7. What term refers to the downslope movement of soil and rock fragments that is caused solely by gravity? (23.2)
   - (a) sheet erosion
   - (b) mass wasting
   - (c) meandering
   - (d) mechanical weathering

8. Which of the following best describes the total amount of water on the Earth? (23.3)
   - (a) increasing
   - (b) decreasing
   - (c) remaining constant

9. About what percent of Earth's surface is fresh water? (23.3)
   - (a) 50%
   - (b) 32%
   - (c) 5%
   - (d) 3%

10. What energy source powers the Earth's hydrologic cycle? (23.3)
    - (a) ocean currents
    - (b) solar insolation
    - (c) Earth's rotational energy
    - (d) geothermal power

11. At a depth of 11 km (7 mi), what are the deepest regions below the ocean called? (23.4)
    - (a) seamounts
    - (b) trenches
    - (c) continental shelves
    - (d) abyssal plains

12. Which one of the following is a feature of coastal deposition? (23.4)
    - (a) spit
    - (b) sea stack
    - (c) wave-cut cliff
    - (d) sea arch

FILL IN THE BLANK

Compare your answers with those at the back of the book.

1. Frost wedging is a common type of ___ weathering. (23.1)
2. Permanently frozen subsurface soil is called ___. (23.1)
3. Heat and ___ are important factors in chemical weathering. (23.1)
4. Any flow of water between well-defined banks is called a(n) ___. (23.2)
5. A stream's ___ load makes up the largest-sized particles. (23.2)
6. Greenland and Antarctica are covered with ice sheets called ___. (23.2)
7. Two important types of fast mass wasting are landslides and ___. (23.2)
8. Large ridges of glacial drift are called ___. (23.2)
9. ___ occurs when land sinks as a consequence of the excessive extraction of water. (23.3)
10. A(n) ___ is the body of permeable rock that stores and transports groundwater. (23.3)
11. Three types of seawater movement are waves, currents, and ___. (23.4)
12. The true edges of the continental land masses are the ___. (23.4)

SHORT ANSWER

23.1 Weathering

1. Give three examples each of mechanical weathering and chemical weathering.
2. What is frost wedging?
3. How do plants and animals contribute to weathering?
4. What is permafrost?
5. What common type of rock is most susceptible to chemical weathering?
6. What causes sinkholes?

23.2 Erosion

7. What are the natural influences and agents of erosion?
8. Name and describe the three components of a stream's load.
9. What causes a river to meander? What kind of lakes can form from this?
10. What are the pros and cons of living on a floodplain?
11. Distinguish among continental glaciers, cirque glaciers, and valley glaciers.
12. What is glacial drift?
13. Distinguish among terminal moraines, lateral moraines, and medial moraines.
14. Describe each of the following and state whether it is a fast or a slow type of mass wasting: (a) rockslide, (b) creep, and (c) slump.

23.3 Groundwater

15. What is the term for the process that returns moisture to the air from the Earth's surface? What is the term for the process that returns water to the surface from the air?
16. How does *porosity* differ from *permeability*?
17. How does a *zone of saturation* differ from a *zone of aeration*?
18. What is an aquifer? How common are aquifers in the United States?
19. What are problems associated with groundwater extraction?
20. Describe three ways in which groundwater may become contaminated.

23.4 Shoreline and Seafloor Topography

21. Draw a simple sketch that shows the differences among a *pocket beach,* a *barrier island,* and a *spit.*
22. Define the term *seamount.* What are seamounts with flat tops called?
23. What are tides, and which solar system body has the most influence on the Earth's tides?
24. How do continental shelves differ from continental slopes? Which are the true edges of continents?

VISUAL CONNECTION

Visualize the connection and give answers for the blanks. Compare your answers with those at the back of the book.

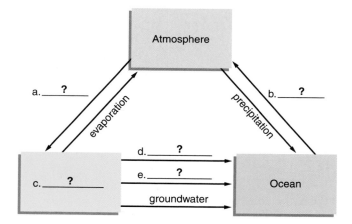

APPLYING YOUR KNOWLEDGE

1. The Moon has neither an atmosphere nor surface water. Can mass wasting occur on the Moon? Explain.
2. State two procedures that can reduce the hazards of landslides.
3. ● Figure 23.26a is a photograph of Cleopatra's Needle as it appeared for 3500 years in Egypt. Figure 23.26b is the same structure after spending 100 years in New York City. Describe the various weathering processes that caused the disintegration.

4. How does the Earth's water supply change due to melting of the glaciers?

5. Describe several ways in which you can protect your drinking water.

6. Suppose that you collected a bucket of water from a stream. What part of the stream's load would settle to the bottom of the bucket? What part would not settle? What part would probably not even be in the bucket?

Figure 23.26 Cleopatra's Needle
These photographs show Cleopatra's Needle (left) before and (right) after standing in New York City since 1879. See Question 3.

ON THE WEB

1. Erosion and Soil Degradation

During the 1930s, the United States and Canada experienced extremely harsh economic times, not just because of the Great Depression but also because of drought and soil erosion. We are clearly dependent on weather conditions to maintain the cycles we need to produce the food we eat, but is it merely something left to nature, or do we have a role to play? Can you answer the following questions? What effect has increased human population had on the world's soil systems? How does water erosion become a problem? How is this problem reduced? What suggestions can be made to stop soil erosion or even enhance the soil? Why do many farmers choose not to use these techniques? You can explore the role that you can play in affecting the mechanisms of soil degradation and erosion by following the recommended links on the student website at **www.cengagebrain.com/shop/ISBN/1133104096**.

2. Demonstration Erosion

Here's a chance to consider yourself an instructor as you try a classroom demonstration. Go to the student website at **www.cengagebrain.com/shop/ISBN/1133104096**, gather the equipment you'll need, and then follow the instructions (or improvise as you see fit). What results did you get from each of these demonstrations? Were your results the same as those predicted? For additional information on how to use techniques that will reduce erosion or enhance soil quality, download the Nebraska Conservation Planning Sheets at the end of the website.

Geologic Time

*Lives of great men all remind us
We can make our lives sublime,
And, departing, leave behind us
Footprints on the sands of time.*

•

Henry Wadsworth Longfellow (1807–1882)

A well-preserved fossil > discovered in 1990, this *Tyrannosaurus rex* (Latin for "tyrant lizard king"), called "Sue," walked the Earth in the Cretaceous Period about 70 million years ago.

GEOLOGY FACTS

▶ Sponge-like fossils dating back 650 million years may be the oldest animal fossils yet discovered. The fossils were discovered in Australia in 2010 and are thought to be part of an ancient ocean reef.

▶ The warmest time of the Earth's history was the Cretaceous period, a time so temperate that forests thrived in Antarctica. *Tyrannosaurus rex* and other well-known dinosaurs lived during the Cretaceous period.

▶ The Texas Pterosaur is one of the most famous finds in the history of paleontology. The bones of the largest creature ever to fly were found in 1971 at Big Bend National Park in West Texas. Its wingspan of 12 m (40 ft) surpasses that

Geology is indeed a broad topic, encompassing the composition, structure, and processes of both the surface and the interior of the Earth. Geologists not only study the Earth as it exists today but also look back at the Earth's long history, a span called **geologic time**.

Perhaps geology's major contribution to human knowledge is the finding that the Earth is very old, about 4.56 billion years; and humans (the genus *Homo*) have been around for only a minute part, about 2 million years of that immense span of time. This recognition has revolutionized our view of reality, comparable to the finding in astronomy that the Earth is like a mere speck of dust, far from the center of the Milky Way galaxy, one of hundreds of billions of galaxies in the universe.

Chapter Outline

In science, evidence is what matters; what people wish to be true must give way to scientific observation. This perspective on our place in time and space should heighten both our sense of responsibility as the present-day custodians of the Earth and our appreciation of the magnitude and complexity of nature and what the human mind can discover and comprehend.

This final chapter starts with a discussion of what *fossils* are, how they were formed, and what part they play in our understanding of geologic time. Second, *relative geologic time* is discussed and determined by placing rocks and associated geologic events in chronologic order. Third, *radiometric dating* techniques are described and then shown how they are used to determine *absolute geologic time*, the actual ages of rocks and geologic events, including the age of the Earth. Finally, the absolute *geologic time scale* is constructed using all this information.

24.1 Fossils

Preview Questions

- Are dinosaur footprints fossils?
- How do fossils help oil companies find oil?

The fossil record is vital to the understanding of geologic time. A **fossil** is any remnant or indication of prehistoric life preserved in rock. The study of fossils is called **paleontology**, an area of interest to both biologists and geologists. Evidence of ancient plants and animals can be preserved in several ways.

1. *Original remains.* Ancient insects have been preserved by the sticky tree resin in which they were trapped. The hardened resin, called **amber** and often used for jewelry, is found globally but most is extracted in eastern Europe and the Dominican Republic (● Fig. 24.1a).

   The entire bodies of woolly mammoths have been found frozen in the permafrost of Alaska and Siberia. More often, only the hardest parts of organisms are preserved, such as bones. Shark teeth and the shells of shallow-water marine organisms that were buried in ancient sediment are common types of original remains.

2. *Replaced remains.* The hard parts (bone, shell, etc.) of a buried organism can be slowly replaced by minerals such as silica (SiO_2), calcite ($CaCO_3$), and pyrite (FeS_2) in circulating groundwater. The result is a copy of the original plant or animal material. Petrified wood, such as the beautiful samples from Arizona's Petrified National Forest, is a common type of **replacement fossil** (Fig. 24.1b).

   The replacement process of *carbonization* occurs when plant remains are decomposed by bacteria under anaerobic (no oxygen) conditions. The hydrogen, nitrogen, and oxygen are driven off, leaving a carbon residue that may retain many of the features of the original plant (Fig. 24.1c). That is how coal was formed.

3. *Molds and casts of remains.* When an embedded shell or bone is dissolved completely out of a rock, it leaves a hollow depression called a **mold**. When new mineral material fills the mold and hardens, it forms a **cast** of the original shell or bone. Molds and casts can show only the original shape of the remains (Fig. 24.1d). For more about this process, see the **Highlight: How Fossils Were Formed.**

4. *Trace fossils.* A fossil imprint made by the movement of an animal is called a **trace fossil**. Examples include tracks (including those of dinosaurs), borings, and burrows (Fig. 24.1e).

The earliest evidence of ancient life is a fossil of blue-green algae, or cyanobacteria, which are single-celled organisms. The oldest algal fossils are found in Australia and date to about 3.5 billion years ago (● Fig. 24.2). The fossil record shows that as time passed, larger and more complex life forms developed.

Figure 24.1 Some Modes of Fossil Formation
(a) *Original remains*. About 40 million years ago, this insect was trapped in tree resin that later hardened to amber. (b) *Replaced remains*. About 200 million years ago, the organic matter in this log from Arizona was replaced by silica, bit by bit, to form petrified wood. (c) *Replaced remains*. About 350 million years ago, this seed fern was changed to carbon, but its form was preserved. (d) *Molds and casts*. This near-perfect cast of *Archaeopteryx*, one of the earliest birds (note the feather impressions), was found in Germany and dates from 145 million years ago. It had teeth, claws, and a bony tail. (e) *Trace fossils*. About 200 million years ago in Arizona's Painted Desert, a dinosaur left these tracks in what was then a mudflat.

Highlight How Fossils Were Formed

Fossils typically formed when all or part of an organism was buried in layers of sediment. As the organism decomposed the surrounding rock preserved their shapes. Geologists have discovered large numbers of fossils with hard shells, skeletons, or bones, such as clam shells, fish skeletons, and dinosaur bones. However, they have found far fewer remains of worms, slugs, and jellyfish because the soft organisms (or parts of organisms) decompose quickly, before they are buried and preserved. The more resilient shells, skeletons, and bones can also decompose after they are buried if they come in contact with underground water afterward. For this reason, most fossils are typically composed of some type of replacement material rather than an organism's original biological substance.

As mentioned in this section, if an embedded shell or bone is dissolved completely out of a rock, then it leaves a hollow depression in the sediment called a mold. At a later time a new mineral material such as ion-rich groundwater may fill the mold and harden. This process creates a solid cast of the mold, which is what is typically found as a fossil (Fig. 1). The solid cast shows the organism's size and shape, and in some cases the external part of the cast may reveal remarkable detail about the texture of the organism (Fig. 2a and Fig. 24.1d).

If the deceased organism decomposes very slowly, it may undergo a cell-by-cell replacement of its original material by secondary substances such as calcium carbonate or other substances dissolved in groundwater. During this slow process the internal structure of the organism may be retained and would therefore give fantastic clues about the organism and perhaps its environment. These clues are very exciting to paleontologists.

For the original parts of the organism to be preserved, the environmental conditions must be extraordinarily unusual. An example is the infamous La Brea tar pits beneath downtown Los Angeles, where thousands of mammoths, mastodons, saber-toothed cats, and other late Ice Age mammals were preserved in the viscous tar. These and other animals and birds became trapped in the tar about 44,000 years ago.

Another example of an unusual environment for preserving animals is the arctic permafrost. Intact woolly mammoths have been dug up from the tundras of both Alaska and Siberia. Also mentioned previously, original bodies of entire insects have been recovered from hardened tree sap, or amber, in which they became stuck millions of years ago (see Fig. 24.1a).

Recently, it was discovered that fine volcanic ash preserved the oldest "undeniably male" fossil, a 425-million year old sea creature called an *ostracode* (Fig. 2b). The volcanic ash allowed for the preservation of all its soft external parts. In each of these cases, the tar, the amber, the permafrost, or the ash did not permit the circulation of water or air, which prevents the organism's original material from decomposing.

As a final example of fossil formation, in 2005 paleontologists discovered *Tyrannosaurus rex* bones in Montana that still contained *T. rex* bone marrow and other soft tissues. Scientists now have the opportunity to study the actual DNA of this long extinct dinosaur. This amazing process of fossilization allows us to learn in part about the Earth's incredible history.

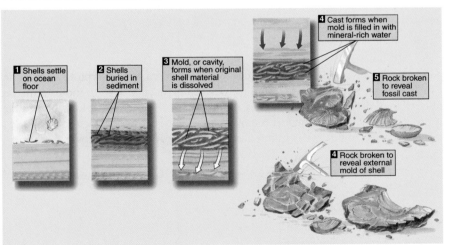

Figure 1 The Process of Fossilization
An organism that has been buried in sediment may be preserved in the form of a mold or a cast in the resulting rock

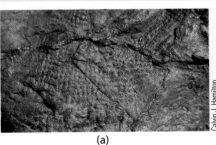

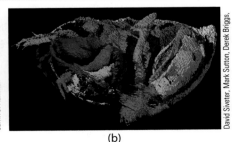

(a)   (b)

Figure 2 Various Fossils
(a) The skin texture of a dinosaur.
(b) A 425-million-year-old male ostracode with soft tissues shown in different colors.

Calvin J. Hamilton

David Siveter, Mark Sutton, Derek Briggs, Derek Siveter

(a)

(b)

Figure 24.2 Fossil and Modern Algae
(a) The fossils in this rock are *stromatolites*, structures created by an early form of algae about 3.5 billion years ago. (b) Modern stromatolites in Shark Bay, Western Australia.

Fossils not only help geologists determine the relative ages of rocks, but they also reveal something about past climatic conditions. For example, the finding of a coral reef in a farmer's cornfield indicates that the area was once a warm, shallow sea.

Conceptual Question and Answer

Fossilized Jellyfish

Q. Why don't scientists commonly find fossils of ancient jellyfish?

A. Jellyfish are made of soft tissue; therefore, the organic material decomposed too quickly before it was preserved in rock. It is much more common to find fossils of animals that contain hard shells, skeletons, or bones because they took much longer to decay.

Certain microfossils, such as some species of foraminifera, which typically have linear, spiral, or concentric shells and are found in rock layers, have proven to indicate the presence of nearby oil deposits (● Fig. 24.3). Companies involved in oil exploration drill deep underground and examine the cores removed in search of these characteristic fossils. As our discussion of geologic time continues, other fossil creatures will become evident.

Did You Learn?

● Dinosaur footprints are one kind of fossil called a trace fossil.

● Certain fossil types found in cores drilled deep underground indicate oil deposits.

24.2 Relative Geologic Time

Preview Questions

● How are the relative ages of rocks and geologic events determined?

● What is the current geologic eon, era, and period?

Relative geologic time is obtained when rocks and the geologic events that they record are placed in chronologic order without regard to actual dates. As an analogy, you might find out that a friend of yours graduated from college, then served in the military, and then got married. You know the *relative time*, or the *order* in which the events occurred, but what if you knew the actual date of each of these events in your friend's life? Then you would know what geologists call *absolute time*, or *numerical time* (Chapter 24.3).

The three principles used to determine the relative ages of rocks in one locality are listed and discussed here:

▶ The **principle of original horizontality** states that sediments and lava flows are deposited as horizontal layers (● Fig. 24.4). Rock layers that are found in nonhorizontal positions were originally horizontal and then altered later by other forces.

▶ The **principle of superposition** states that in a sequence of undisturbed sedimentary rocks, lavas, or ash, each layer is younger than the layer beneath it and older than the layer above it (● Fig. 24.5). Sometimes layers are disturbed by folding or faulting. A geologist must look for any evidence of such events and, if found, take this disturbance into account.

▶ The **principle of cross-cutting relationships** states that an igneous rock is younger than the rock layers it has intruded (cut into or across). The same principle applies to faults; a fault must be younger than any of the rocks it has affected (● Fig. 24.6).

In a given locality, sediments are not deposited continuously over time. Breaks called **unconformities** occur in the rock record. The missing layers of rocks may never have been deposited, or they may have been deposited and then eroded away after the rock surface was uplifted by a geologic event. If the area was then resubmerged, then new layers may have been deposited. Unconformities represent gaps in the geologic record, and usually it is impossible to determine how much time is represented by an unconformity.

Figure 24.3 Foraminifera from About 350 Million Years Ago
The needle's eye shows how tiny were the aquatic creatures named *foraminifera* but usually referred to as *forams*. These microfossils indicate the presence of nearby oil deposits.

Figure 24.4 Principle of Original Horizontality
Most layers of sediment are deposited in a nearly horizontal position. Thus, when rock layers are seen that are folded or tilted, it can be assumed that they must have been moved into that position by crustal disturbances *after* their deposition. These folded layers are exposed in the Namib Desert of southwestern Africa.

Figure 24.5 The Principle of Superposition
The principle of superposition tells us that the limestone layers at the bottom are older than the thick layers of brown sandstone and thin layers of gray shale above it. Older layers are beneath younger ones.

Figure 24.6 The Principle of Cross-Cutting Relationships
In this diagram, the dike of igneous rock has cut across the sedimentary strata and so is younger than the strata (shown for clarity as layers of sandstone and brick-like rock). The fault has displaced the dike (as well as the strata) and so is younger than the dike.

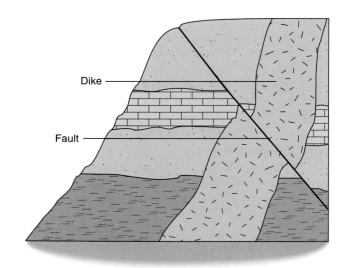

EXAMPLE 24.1   Applying Principles of Relative Dating

Using the principles of relative dating, analyze ● Fig. 24.7 and put the rocks marked 1 through 5 in order from youngest to oldest.

Solution

The principle of superposition indicates that the topmost layer, rock 5, is the youngest, followed by rock 4, rock 3, and rock 1. The principle of cross-cutting relationships shows that rock 2 is younger than rock 1 but older than rock 3. Thus, the correct order from youngest to oldest is 5, 4, 3, 2, 1.

Confidence Exercise 24.1

Figure 24.7 shows an unconformity. Where is it, and when (relatively) must it have been formed?

   Answers to Confidence Exercises may be found at the back of the book.

Figure 24.7 Applying Principles of Relative Dating
See Example 24.1.

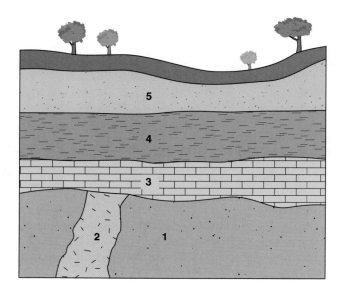

Figure 24.8 Trilobites
The trilobites *Modicia* (large) and *Ptychagnostus* (small) are
index fossils of the Cambrian period. Trilobites were early
marine arthropods with hard exoskeletons, whose name
refers to the "three lobes" that run lengthwise down the
body. The first organisms with eyes, they ruled the early
Paleozoic seas but were wiped out by the extinction event
that ended the Paleozoic era (about 250 million years ago).

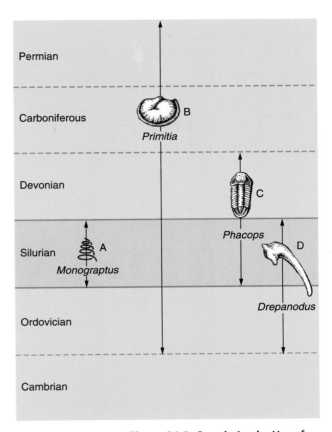

**Figure 24.9 Correlation by Use of
Index Fossils**
See Example 24.2.

Geologists use the principles just discussed to determine the relative ages of rocks in a
specific locality. **Correlation** is the process of matching rock layers in different localities
by the use of fossils or other means. If the age of rock in locality A is known and rock in
locality B is *correlated* with A, then the rock in locality B is thereby determined to be the
same age as that in locality A.

Certain fossils, called *index fossils*, are a major aid in correlation. **Index fossils** are those
that are widespread, numerous, easily identified, and typical of a particular limited time
segment of the Earth's history. For example, many species of trilobites are important index
fossils (● Fig. 24.8). After an index fossil has been thoroughly established, geologists know
that any newly investigated rock layer in which it is found is the same age as previously
known layers in which that index fossil was found. Let's look at an example.

EXAMPLE 24.2   **Using the Process of Correlation**

The Cambrian is the oldest (earliest) and the Permian the youngest (latest) of the six
geologic periods named in ● Fig. 24.9. Four fossils, labeled A through D, are shown,
along with their time ranges. (a) Which fossil would be the most useful as an index fossil? (b) If a rock layer from a certain locality contains both fossil C and fossil D, what can
be said about the period of the rock?

Solution

(a) Fossil A would be the best index fossil, because of the narrow range of time in
which it lived.
(b) Rock that contains both fossil C and fossil D would have to be from the Silurian
period, because only during this period did *both* of these fossils live.

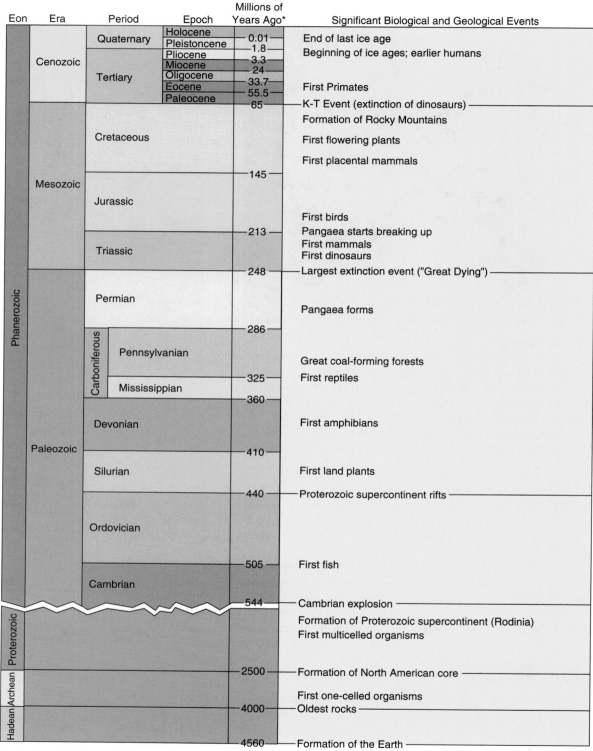

Figure 24.10 The Geologic Time Scale
The time units are arranged with the oldest at the bottom and successively younger ones above
it. The absolute geologic time scale is also listed and includes significant geologic events. Dates
from U.S. Geological Survey.

By correlating rocks over large areas, geologists have been able to determine the relative ages of most of the rocks on the surface of the Earth and establish a relative time scale for the Earth's history. Refer to ● Fig. 24.10 during this discussion of some features of the relative geologic time scale.

The largest units of geologic time are the **eons**. We live in the *Phanerozoic* ("evident life") *eon*. The time before that is often collectively called *Precambrian time*, because it immediately precedes the Cambrian period. Eons are divided into **eras**, and the oldest era in our eon is the Paleozoic era (the "age of ancient life"). Then come the Mesozoic era (the "age of reptiles") and the Cenozoic era (the "age of mammals").

In turn, eras are divided into smaller time units called **periods**. The Paleozoic era is divided into seven periods. (In Europe, the Mississippian and Pennsylvanian periods are combined and called the *Carboniferous period*.) Three periods make up the Mesozoic era (you have probably heard of the middle one). The Cenozoic has only two periods, the Tertiary and Quaternary ("qua-TUR-nay-ry") periods. We owe a debt to the plants that lived during the Pennsylvanian period, because much of our energy comes from the coal that was eventually formed from them (● Fig. 24.11). Whereas beds of coal come from plant life in ancient swamps, the Earth's deposits of petroleum and natural gas result from ancient marine life, such as plankton and algae.

In Chapter 24.5 and referring to Figure 24.10, some absolute ages will be assigned to these and other geologic time divisions. The major events that caused these divisions will be discussed briefly.

Figure 24.11 A Pennsylvanian-Period Swamp
This reconstruction of a swamp of 300 million years ago shows conifers and ferns that were buried and slowly changed to coal in the oxygen-depleted layers.

Did You Learn?

- Relative ages of rocks are determined by three major principles: original horizontality, superposition, and cross-cutting relationships.

- The current geologic time is the Phanerozoic eon, the Cenozoic era, and the Quaternary period (in which we live).

24.3 Radiometric Dating

Preview Questions

- How does radioactivity help us date rocks?
- How is carbon-14 dating used in geology?

The establishment of the relative geologic time scale was a significant scientific achievement, but one that left geologists with a sense of frustration. Not only did they need to know the *order* in which geologic events occurred, but also *how long ago* these events occurred. That is, they needed absolute ages as well as relative ages. The need to measure **absolute (numerical) geologic time**—the actual age of geologic events—became apparent in 1785 when James Hutton's concept of uniformitarianism indicated that the Earth is very ancient (Chapter 22.2). It also became crucial in 1859 when Charles Darwin's theory of organic evolution redoubled interest in the early history of our planet.

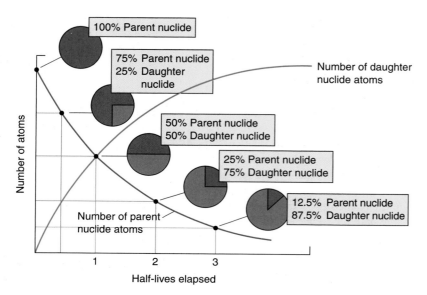

Figure 24.12 Half-Life and Radiometric Dating

As a parent radioactive isotope decays to its daughter isotope, the proportion of the parent decreases (purple line), while the proportion of the daughter increases (blue-green line). By measuring the proportion of parent to daughter in a sample, it is possible to obtain the number of half-lives since "time zero" (100% parent). Multiplying the number of half-lives by the half-life of the isotope gives the age of the rock.

Radiometric dating, the determination of age by using radioactivity, has become geology's best tool for establishing absolute geologic time. Recall from Chapter 10.3 that atomic nuclei that decay of their own accord are said to be *radioactive*. The decay product—or *daughter nucleus*, as it is commonly called—may be stable. If so, then the transformation reaches completion in a single step. However, the daughter nuclei of many naturally occurring isotopes are themselves radioactive and hence undergo further decay. The radioactive series can be long and complex (see Fig. 10.6 for the decay of uranium-238 to stable lead-206).

Scientists express the rate of decay of a radioactive isotope in terms of *half-life*, the span of time required for half of the parent nuclei in a sample to decay (Chapter 10.3). Because the rate of decay for a given isotope remains constant—unaffected by temperature, pressure, or chemical environment—radioactivity can be used as a "clock" to measure the march of geologic time. The older the rock, the smaller the ratio of the amount of radioactive parent to the stable daughter (● Fig. 24.12). For example, uranium-238 has a half-life of 4.46 billion years for its decay to lead-206. For example, for every 1000 atoms of uranium-238 present in a rock 4.46 billion years ago, only 500 atoms of uranium-238 are still there today (1 half-life has elapsed), and 500 atoms of lead-206 have been formed (ratio: 50%/50% = 1). Looking ahead, 4.46 billion years in the future, only 250 atoms of uranium-238 will have survived, with 750 atoms of lead-206 having been formed (ratio: 25%/75% = $\frac{3}{4}$, getting smaller).

Therefore, the rate of decay of an isotope in a rock can make it possible to assign a date to the rock. Ideally, an isotope within a rock can be used to tell the age of the rock under the following conditions:

1. No addition or subtraction of the parent or daughter has occurred over the lifetime of the rock, other than that caused by radioactive decay.

2. The age of the rock does not differ too much from the half-life of the parent isotope.

3. None of the daughter element was present in the rock when it formed, or if present, it is known.

Satisfying condition 1 is usually not difficult. However, in certain cases the decay of potassium-40 to argon-40 cannot be used because the daughter product occurs in the form of a gas that sometimes leaks out of the rock being dated.

Condition 2 must be satisfied because if the rock's age is too much greater than the half-life of the isotope, then too many half-lives go by and so little of the parent isotope remains that its amount cannot be measured accurately. On the other hand, if the rock has an age very much less than the half-life, then so little of the daughter nuclei may have formed that it may be hard to measure accurately. Because most rocks that geologists are interested in dating have ages of hundreds of millions, or even billions, of years, only parent isotopes of similarly long half-lives can be used. Of course, many rocks do not contain appropriate isotopes and thus cannot be dated by radiometric methods. Radioactive isotopes that are commonly used to date rocks are listed in ● Table 24.1.

Satisfying condition 3 is sometimes a problem, as can be illustrated by the *uranium-lead dating* of rocks. (Uranium-lead dating means that uranium is the parent isotope and lead is the final daughter product.) Lead that comes from radioactive decay is called *radiogenic lead*, whereas lead that does *not* come from radioactive decay is termed *primordial lead*.

All primordial elements on the Earth occur in a fixed ratio of isotopes. Therefore, primordial lead always has the same proportion of isotopes lead-204, -206, -207, and -208. If any primordial lead is in the rock that is being dated, then not only will the other lead isotopes be present, but also lead-204, *which never comes from radioactive decay*. For every

Table 24.1	The Major Radioactive Isotopes Used for Radiometric Dating					
Method	Parent Isotope	Daughter Isotope	Half-Life of Parent (years)	Effective Dating Range (years)	Materials Commonly Dated	Comments
Rubidium–strontium	Rb-87	Sr-87	47 billion	10 million–4.6 billion	Potassium-rich minerals; volcanic and metamorphic rocks (whole-rock analysis)	Useful for dating the Earth's oldest metamorphic and plutonic rocks
Uranium–lead	U-238	Pb-206	4.5 billion	10 million–4.6 billion	Zircons, uraninite, and uranium ore; igneous and metamorphic rock (whole-rock analysis)	Uranium isotopes usually coexist in minerals such as zircon. Multiple dating schemes enable geologists to cross-check dating results.
Uranium–lead	U-235	Pb-207	704 million	10 million–4.6 billion		
Thorium–lead	Th-232	Pb-208	14.1 billion	10 million–4.6 billion	Zircons, uraninite	Thorium coexists with uranium isotopes in minerals such as zircon.
Potassium–argon	K-40	Ar-40	1.3 billion	100,000–4.6 billion	Potassium-rich minerals; volcanic rocks (whole-rock analysis)	High-grade metamorphic and plutonic igneous rocks may have been heated sufficiently to allow Ar-40 gas to escape.
Carbon-14	C-14	N-14	5730	100–70,000	Any carbon-bearing material, such as bones, wood, shells, charcoal, cloth, paper, animal droppings; also water, ice, cave deposits	Commonly used to date archeological sites, recent glacial events, evidence of recent climate change, environmental effects of human activity

Source: Chernicoff, Stanley, and Haydn A. Fox, *Essentials of Geology*, Third Edition. Copyright © 2003 by Houghton Mifflin Company. Used with permission.

gram of lead-204 that is present in a rock, the geologist knows that from the fixed ratio there is 17.2 g of lead-206, 15.8 g of lead-207, and 37.4 g of lead-208. Thus, the geologist can subtract these amounts from the total of each lead isotope to determine the amount of radiogenic lead.

For example, if a rock has 1.0 g of lead-204 and a total amount of lead-206 equal to 46.0 g, then the amount of radiogenic lead-206 is 46.0 g minus 17.2 g, or 28.8 g. The value 28.8 g would be the amount of lead-206 used to help date the rock (● Fig. 24.13).

As stated previously, uranium-238, the most abundant isotope of uranium, has a half-life of 4.46 billion years and decays to lead-206. Uranium-235 and thorium-237 decay into different isotopes of lead. Hence, it may be possible to date the rock by three independent methods. If all three

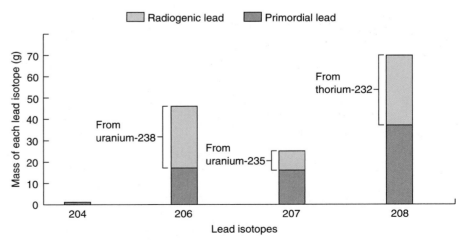

Figure 24.13 Primordial and Radiogenic Lead
Because lead-204 is *never* radiogenic, if any lead-204 is found in a rock, then it is known that primordial 206, 207, and 208 isotopes are there as well. By knowing the constant proportion of the primordial lead isotopes (blue bars), geologists can tell how much of the total amount of each isotope is actually radiogenic (green bars).

give basically the same age, then geologists can be confident that the rock has been dated correctly.

Uranium-238, uranium-235, and thorium-232 are clocks that have timed the birth of many rocks on the Earth. Most of the early measurements used uranium-bearing minerals such as uraninite, but this mineral is rare. Fortunately, advanced procedures permit the measurement of small traces of uranium and lead in zircon ($ZrSiO_4$), a much more abundant and representative mineral.

Potassium, one of the most abundant and widely occurring elements, contains a very small percentage (0.012%) of radioactive potassium-40. This isotope is found in many rocks that do not contain measurable quantities of uranium, so it has become a useful tool in determining where a rock falls on the calendar of geologic time.

Potassium-40 (half-life 1.25 billion years) undergoes beta decay to argon-40, so this type of radiometric dating is referred to as *potassium–argon dating*. A potassium–argon date may merely reveal the last time the rock was heated rather than its true age. Therefore, geologists tend to regard potassium–argon dates as the *minimum* ages of the rocks and to require that dates derived this way be consistent with other geologic evidence before being fully accepted as the correct ages.

Rubidium-87 (half-life of 49 billion years) is more abundant than potassium-40 and commonly occurs in the same minerals that contain potassium. It has the further advantage of decaying to daughter nuclei that is a solid (not of a gas), but a disadvantage is that a lot of the daughter product is primordial, and so a correction must be made. Geologists often use *rubidium–strontium dating* to compare with potassium–argon determinations from the same rock.

EXAMPLE 24.3 Using Radiometric Dating

Analysis of samples of a certain igneous rock layer shows that the ratio of uranium-235 to its daughter, radiogenic lead-207, is 1.0 to 3.0. That is, only 25% of its original uranium-235 (half-life 704×10^6 years) remains. How old is the rock?

Solution

To decay from 100% to 25% would take two half-lives, as shown in Fig. 24.12 or found by

$$100\% \rightarrow 50\% \rightarrow 25\%$$

Now, to find the time in years, multiply the 2.0 half-lives by the half-life of the isotope:

$$(2.0 \text{ half-lives}) (704 \times 10^6 \text{ years/half-life}) = 1.41 \times 10^9 \text{ years, or } 1.41 \text{ billion years}$$

Confidence Exercise 24.3

Analysis of samples of another igneous rock layer shows that the ratio of uranium-235 to its daughter, radiogenic lead-207, is 1.0 to 7.0. That is, only 12.5% of its original uranium-235 (half-life 704×10^6 years) remains. How old is the rock?

Carbon-14 Dating

A dating technique developed in 1950 by Willard Libby, an American chemist, is the radiometric method that is used to find the age of ancient, once-living remains such as charcoal, parchment, or bones. Carbon-14 (^{14}C) is a radioactive isotope whose relatively short half-life of 5730 years puts a "second hand" on the radioactive time clock. **Carbon-14 dating** dates organic remains by measuring the amount of ^{14}C in an ancient sample and comparing it with the amount in present-day organic matter. (See also carbon-14 dating in Chapter 10.3.)

Throughout history, ^{14}C has been produced in the atmosphere by the action of neutrons on atmospheric nitrogen (● Fig. 24.14). The newly formed ^{14}C reacts with oxygen in the

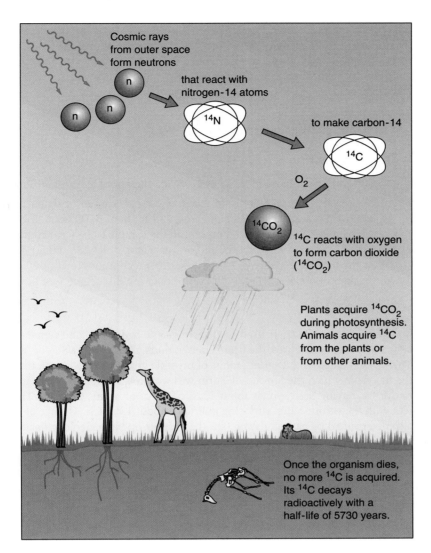

Figure 24.14 Carbon Dating
This illustration shows how carbon-14 forms in the atmosphere and enters the biosphere. (See text for further description.)

air to form radioactive carbon dioxide, $^{14}CO_2$, which, along with ordinary $^{12}CO_2$, is used by plants in photosynthesis. About one out of every one trillion (10^{12}) carbon atoms in plants is ^{14}C. Animals that eat the plants incorporate the radioactive ^{14}C into their cells, as do animals that eat the animals that ate the plants.

Thus, all living matter has about the same level of radioactivity due to ^{14}C, an activity of about 16 counts per minute per gram of total carbon. Once an organism dies, it ceases to take in ^{14}C, but the ^{14}C in its remains continues to undergo radioactive decay. Therefore, the longer the organism has been dead, the lower the radioactivity of each gram of carbon in its remains.

The newest method of carbon dating relies on a method that separates and counts both the ^{14}C atoms and the ^{12}C atoms in a sample. By comparing the ratio of the isotopes in the specimen to the ratio in living matter, it is possible to calculate the time since the specimen's death. This method uses tiny samples and can date specimens as old as 75,000 years. Beyond that age, only about 0.02% of the original ^{14}C is still undecayed, which is too little to measure accurately.

Carbon dating assumes that the amount of ^{14}C in the atmosphere (and hence in the biosphere) has been the same throughout the past 75,000 years. However, because of changes in solar activity and in the Earth's magnetic field, that amount apparently has varied by as much as 5% above or below normal. California's bristlecone pines, which live as long as

5000 years, allowed geologists to correct for the slight changes in the abundance of ^{14}C. By studying the ^{14}C activity of samples taken from the annual growth rings in both dead and living trees, geologists have developed a calibration curve for ^{14}C dates as far back as about 5000 BCE. Therefore, carbon dating is most reliable for specimens no more than 7000 years old.

Conceptual Question and Answer

Dinosaur Dating

Q. Can carbon-14 be used to accurately date the age of a dinosaur bone?

A. No. Material that is older than about 7000 years cannot be dated using carbon-14. Because the half-life of carbon-14 is about 5700 years, too little of the parent isotope would be present in the sample to be measured accurately. Dinosaurs became extinct 65 million years ago, much too old for carbon-14 dating to work reliably. Scientists use other techniques to accurately determine the ages of rocks where the dinosaur bones are discovered (see how the ages are "bracketed" in Example 24.4).

Carbon dating has been used to determine the age of organic remains such as bones, charcoal from ancient fires, the beams in pyramids, the Dead Sea Scrolls, and the Shroud of Turin. Such dates have a certain range of possible error associated with them. For example, the flax from which the Shroud of Turin was woven was dated by three independent laboratories as having been grown about 1325 CE, plus or minus 65 years. This date is consistent with the shroud's first surfacing in France about 1357 CE. The age of "the Iceman" was also determined by carbon dating (see the chapter-opening Geology Facts).

Did You Learn?

- Radioactivity helps us date rocks because the ratio of parent-daughter isotopes can be measured.

- Carbon-14 dating is useful for dating items to about 7000 years old.

24.4   The Age of the Earth

Preview Questions

- How old is the Earth?

- What is the age of other solar system objects such as the Moon?

In the middle and late 1800s, the distinguished physicist Lord Kelvin (associated with the absolute temperature scale; Chapter 5.1) attempted to determine the Earth's absolute age from the rate of heat loss from its interior. Kelvin assumed that the Earth began as a hot, molten body that became solid as it cooled and continued to lose its residual heat from its still-hot interior. From a measurement of the present rate of heat loss, he calculated that the Earth became solid between 20 and 40 million years ago. This estimate, although based on actual measurements and supported by Kelvin's considerable prestige, was at odds with estimates of both geologists and biologists, who thought the present slow pace of geologic processes and organic evolution hinted at a much greater age.

After the discovery of radioactivity in 1896 (Chapter 10.2), it became evident that Kelvin's calculation was badly in error because of the incorrectness of one of his basic assumptions: that all the heat in the Earth's interior was residual. It is now known that most of this heat

Figure 24.15 An Outcrop of Rock in Western Australia's Jack Hills
Grains of the mineral zircon that have been extracted from this rock have been dated at 4.4 billion years, making them some of the few surviving remnants of the Hadean eon, the earliest of the eons (see Figure 24.10).

E.B. Watson, Rensselaer Polytechnic Institute

is actually from radioactive decay, which is a continuous process. Therefore, much more heat is available to flow out over a much longer time than Kelvin reckoned. Ironically, the phenomenon of radioactivity, which torpedoed his calculation, became geology's best tool for establishing absolute geologic time.

Kelvin did the best he could with the knowledge at that time, but this story brings home three important aspects about the scientific method (Chapter 1.2): (a) the necessity of examining basic assumptions, (b) the realization that scientific results are always subject to reinterpretation as new evidence accumulates, and (c) the wisdom of not getting overly concerned when scientists disagree. As research continues on a scientific problem, usually the correct answer finally emerges.

Fortunately, geologists are confident that they now have a reasonably accurate value for the Earth's age: about 4.56 billion years. Three major pieces of evidence support this date.

1. *The age of Earth rocks.* Using radiometric dating, it is possible to put absolute dates on many igneous and metamorphic rocks. The longevity record is held by zircon crystals discovered in Western Australia in the 1980s, which were dated as 4.4 billion years old (● Fig 24.15). Other ancient deposits include 4.0-billion-year-old metamorphic and igneous rocks in Canada, 3.8-billion-year-old metamorphic rocks in Minnesota, 3.8-billion-year-old granites in southwestern Greenland, and 3.4-billion-year-old granites in South Africa.

2. *The age of meteorites.* Meteorites from the asteroid belt (Chapter 17.6), which presumably formed about the same time as the Earth, date at about 4.56 billion years. A few special meteorites are probably rock from Mars, and they also show an age of 4.5 billion years. This age is given by both uranium–lead and rubidium–strontium methods.

3. *The age of Moon rocks.* The rocks from the lunar highlands are the oldest materials brought back from the Moon, and they yield dates of 4.45 billion years.

Thus, compelling evidence exists that the planets, moons, and asteroids of the solar system all formed about 4.56 billion years ago. It is unlikely that rocks formed on the Earth will be found that are quite 4.56 billion years old because the Earth's surface was probably molten for several hundred million years. Weathering and subduction due to plate tectonics undoubtedly have destroyed many ancient rocks, so we are fortunate to find as many old rock formations as we do.

One final point about the age of the Earth needs to be addressed, and a statement by American geologist Anatole Dolgoff is well-crafted:

> Some people argue on religious grounds that the Earth is only 5000 to 10,000 years old, and geologists are often drawn into public debate as advocates of their scientific estimates. Sometimes, the position of geologists in this debate is misunderstood—for in the final analysis, geologists have no stake in how old the Earth is. They simply want to know how old it is! If the Earth is only 5000 years old, as some who take a literal interpretation of the Scriptures claim, so be it. However, the evidence points overwhelmingly to the contrary.*

Did You Learn?

- The age of the Earth is about 4.56 billion years.
- Other solar system bodies, including the Moon, have an age of about 4.56 billion years.

24.5   The Geologic Time Scale

Preview Questions

- What was the Cambrian explosion?
- When was the first indication of humans (*Homo sapiens*)?

Relative geologic time and absolute geologic time are combined to give the **geologic time scale** shown in Figure 24.10. Time in millions of years, as determined by radiometric dating methods, is shown on the right side of the relative time scale, along with a list of some major geologic and biological events that took place at given times.

Although radiometric dating has provided geologists with numerous essential dates for the ages of rocks, many of the dates on the geologic time scale are estimated values and are subject to minor changes as new evidence is found.

Geologists have principally used sedimentary rocks to establish the relative time scale, whereas most of their radiometric determinations have been made on igneous rocks. Even though sedimentary rocks often contain radioactive isotopes, the sediments that form the rocks have been weathered from rocks of different, and older, ages. Thus, the age of the sedimentary rock is not the same as the ages of its constituents.

Reasonable values for the absolute dates of sedimentary rock layers are frequently determined by relating them to igneous rocks, as shown in Example 24.4.

EXAMPLE 24.4   Using Igneous Rocks to Date Sedimentary Rocks

● Figure 24.16 shows the intrusion of two igneous dikes X and Y, which were radiometrically dated at 400 my (my means "megayears," or million years) and 350 my, respectively. The dikes cross strata whose relative ages were determined by the fossils they contained. What can be said about the age of the Devonian stratum labeled B?

Solution

Igneous dike X intruded the Silurian strata, and then its top and part of the Silurian strata were eroded. Stratum B was then deposited at the unconformity, so stratum B must be younger than 400 my (the date of X). Because of the intrusion of dike Y through stratum B, stratum B must be older than 350 my (the date of Y). Therefore, radiometric dating tells us that the Devonian stratum is between 350 my and 400 my old. Geologists say that the age of the stratum has been *bracketed*.

Confidence Exercise 24.4

Refer to Fig. 24.16. What can be said about the absolute age of the sedimentary rock layer A from the Mississippian period?

*Anatole Dolgoff, *Essentials of Physical Geology* (Boston: Houghton Mifflin, 1998).

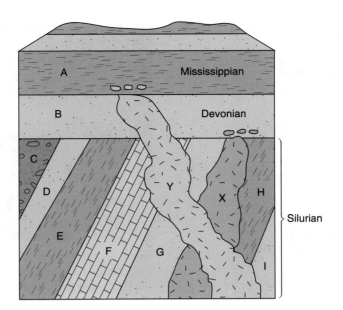

Figure 24.16 Using Igneous Rock to Date Sedimentary Rock
See Example 24.4.

Space does not permit a discussion of all the information in the geologic time scale shown in Fig. 24.10, so please examine the feature closely. The Hadean eon (at the bottom of the scale) ends and the Archean eon begins about 4000 mya (mega- or million years ago), the date of the earliest known Earth rocks. The Proterozoic eon begins 2500 mya, when the core rocks of what is now North America came together. ● Figure 24.17 shows a reconstruction of the huge supercontinent, called *Rodinia*, that formed during the Proterozoic and broke up during the early Paleozoic.

The Phanerozoic eon (our present eon) and the Paleozoic era began about 544 mya when the hard-shelled marine invertebrate fossils first became abundant. At this time, an extinction event occurred, followed by a great proliferation of life forms that is sometimes referred to as the **Cambrian explosion**. Rather suddenly, oceans that had previously held nothing more complicated than burrowing worms teemed with complex animal life.

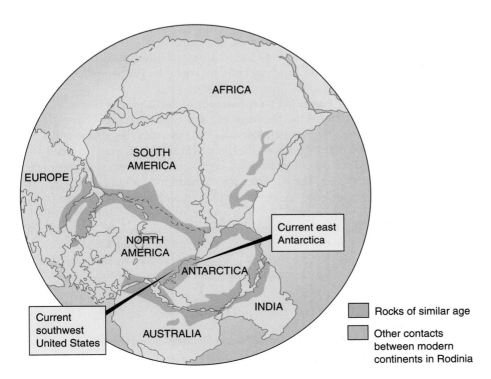

Figure 24.17 Approximate Configuration of the Proterozoic Supercontinent
A supercontinent called Rodinia formed in the late Proterozoic eon. It began to rift (break apart) in the early Paleozoic era. However, in the late Paleozoic era, the pieces reassembled into the plate tectonics supercontinent called *Pangaea*, which itself began to rift in the middle of the Mesozoic era (see Ch 21.2).

Jim Jurica/istockphoto.com

Figure 24.18 Crinoid Fossils of the Mississippian Period, Found in Indiana
Crinoids are marine invertebrate animals that look like plants. They were abundant (about 5000 species) during most of the Paleozoic era, and a few species (commonly called "sea lilies") exist today. Flat, circular, or star-shaped segments of crinoid columns are common fossils, sometimes called (inaccurately) "Indian money."

Walter Hodge/Peter Arnold, Inc./PhotoLibrary

Figure 24.19 Brachiopods from the Paleozoic Era
Brachiopods are hard-shelled marine invertebrate animals that were abundant during the Paleozoic era. Casts are shown on the left, molds on the right. Unlike mussels, whose two shell valves are similar but opposite, brachiopod valves are unlike and unequal. A few species of brachiopods (called "lamp shells") exist today.

The Paleozoic era ended and the Mesozoic era began about 248 mya when the most devastating extinction known to geologists—the Permian event sometimes called the **Great Dying**—occurred. An estimated 90% of ocean species, including the trilobites and most of the crinoids and brachiopods, and 70% of land species were wiped out (● Fig. 24.18 and ● Fig. 24.19). What caused the Great Dying, or Permian event? Hypotheses range from the explosion of a nearby star to greenhouse warming to ice age cooling. In 2001, in locations such as Hungary, China, and Japan, scientists found a layer of 244-million-year-old rock that contains buckyballs (Chapter 11.3) that have argon gas trapped within their cage-like structures. The ratio of argon isotopes was characteristic not of terrestrial argon but of asteroids. Perhaps a large asteroid strike triggered dramatic climatic shifts that led to the "mother of all extinctions." Recent evidence indicates that it did.

In the Mesozoic era, Pangaea broke into the familiar continents of today (Chapter 21.2). The Mesozoic climate was mild. Coral grew in what is now Europe, and the poles were free of glacial ice. Dinosaurs lived on all the continents, but the numerous fossils found in the western United States and Canada indicate that these localities provided a particularly good environment for dinosaurs.

The **Highlight: The K-T Event: The Disappearance of the Dinosaurs** discusses the extinction event that 65 mya ended the Mesozoic era and began the Cenozoic era, sometimes called the *age of mammals*. Our present period, the Quaternary, began about 1.8 mya with the appearance of the oldest fossils of the genus *Homo*, which was preceded for about another 2 million years by other hominids (all great apes), who were members of the genus *Australopithecus*. Our species, *Homo sapiens*, has been around for only about 100,000 years.

The periods of the Cenozoic era are subdivided into **epochs**, the names of which end in *-cene*, a suffix meaning "recent." The Pleistocene epoch is also known as the *ice ages*. Our present epoch, the Holocene, begins with the last retreat of the glaciers from North America and Europe about 10,000 years ago. ● Figure 24.20 is a calendar that helps us realize how long geologic time really is and how short is the part we call "recorded history." Each day on the calendar represents about 12.5 million years of history.

Did You Learn?

● The Cambrian explosion is a period of a great proliferation of life forms.

● The first evidence of humans (species *Homo sapiens*) dates about 100,000 years ago.

Highlight   The K-T Event: The Disappearance of the Dinosaurs

What killed about 70% of all the world's plant and animal species, including all the dinosaurs, 65 million years ago? The answer began to emerge in the late 1970s, when a team of scientists from the University of California at Berkeley began investigating a thin layer of clay that marks the boundary between the Mesozoic and Cenozoic eras. The Cretaceous (symbolized K) is the latest period of the Mesozoic era, whereas the earliest period of the Cenozoic era is the Tertiary (symbolized T). Where the two periods meet is called the *K-T boundary*, and the event that marks the division is often referred to as the **K-T event.**

In Cretaceous rocks are found fossils of thousands of marine invertebrate species, dinosaurs, and other organisms, none of which survived to leave fossils in Tertiary rock strata. No doubt exists that a gigantic mass extinction occurred. In the Tertiary strata are found, for the first time, fossils of modern types of plants and fossils of mammals larger than rodents.

The team, led by American geologist Walter Alvarez and his father, Nobel laureate physicist Luis Alvarez, was investigating the clay layer in the cliffs near Gubbio in northern Italy. First, they found that the clay layer contained a concentration of the rare element *iridium* (Ir) hundreds of times greater than that found in normal clay or in the Earth's crust as a whole. However, the concentration closely matched the iridium content of some meteorites. Second, the clay layer contained distinctive glassy beads called *spherules* that were once molten droplets like those associated with meteorite impact craters on the Moon and the Earth. Third, the clay contained grains of *shocked quartz*, a type of quartz formed from meteorite impacts. Fourth, particles of soot were found, apparently from extensive fires.

The Alvarez team hypothesized that a massive meteorite struck the Earth 65 million years ago, causing widespread fires and driving dust, ash, and other debris (about 100 trillion tons!) high into the stratosphere. The cloud girdled the globe, blocking out sunlight for perhaps a decade, causing acid rain, and catastrophically disrupting the food chain on land and in the oceans.

Subsequently, the same materials (iridium, spherules, shocked quartz, and soot) have been identified at all the 95 thin K-T boundary layer locations scattered throughout the world. Apparently, when the distinctive debris from the impact slowly settled back to Earth, it formed the thin layer of sediment that is now found intact at only certain localities because plate tectonic movement and erosion have wiped it out in most places.

A search was launched for an impact crater that was of the appropriate date (65 million years old) and size (many miles wide) to do such damage, and it was found. Satellite images detected such a filled-in crater about 200 km (125 mi) in diameter in the Caribbean on the Yucatan peninsula of Mexico, a site called Chicxulub ("CHEEK-shoe-lube") (Figs. 1a and b). The rock strata at the location contain a lot of sulfate rock, which would have formed a deadly amount of acid rain in the fallout.

Cores drilled in 1996 and 1997 in the Atlantic Ocean provide evidence of a fossil-rich layer topped with an iron-rich band and glass spherules. The clincher seems to have come in 1998, when a 65-million-year-old core layer of Pacific Ocean seafloor yielded specks of a carbon–iron meteorite that may be part of the killer asteroid itself (Fig. 1c). As research continues, the evidence grows that the dinosaurs and many other species met their demise at the hands of a visitor from space that was 10 km (6 mi) wide.

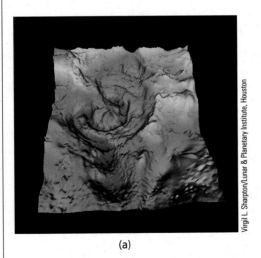
(a)

Virgil L. Sharpton/Lunar & Planetary Institute, Houston

(b)

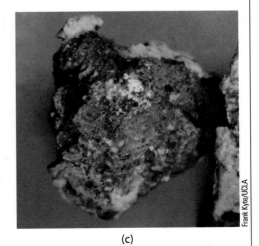
(c)

Frank Kyte/UCLA

Figure 1 The Chicxulub Crater and Meteorite Fragment
(a) This radar image shows the circular outline of the enormous Chicxulub crater, 200 km (125 mi) in diameter. (b) Chicxulub crater is located on the Yucatan peninsula of Mexico. (c) A 2.5-mm (0.1-in.) fragment of a 65-million-year-old meteorite, retrieved from the Pacific Ocean floor.

January	February	March
1st Earth forms (4.56 billion years ago) 26th Oceans form	17th Oldest rock on Earth (3.96 billion years ago)	4th First single-celled organisms (3.77 billion years ago) 29th Oldest stromatolites
April	**May**	**June**
Earth's high surface temperatures gradually fall Early life continues to develop	Early life continues to develop	2nd First prokaryotes and eukaryotes 14th First convincing microfossils (2.5 billion years ago) 26th First snowball Earth
July	**August**	**September**
25th Widespread microbial life in the oceans (2 billion years ago)	Atmospheric oxygen levels continue to increase	21st First multicellular fossils (1.2 billion years ago)
October	**November**	**December**
8th Supercontinent Rodinia exists (1.1 billion years ago)	19th Cambrian Explosion (525 mya) 21st First fish (505 mya) 26th First land plants 27th First terrestrial animals	7th First reptiles (310 mya) 9th Great Dying extinction event 12th First dinosaurs (230 mya) 26th Extinction of dinosaurs (K-T event, 65 mya) 31st 10am Earliest human relatives 31st 7pm Earliest stone tools 31st 11:59pm First civilizations (8,000 years ago) 31st 11:59:46 (1 AD) 31st 11:59:59 Declaration of Independence signed (1776 AD)

Figure 24.20 Geologic Time in Perspective
A calendar helps to put geologic time in perspective. Each day corresponds to about 12.5 million years of Earth history (each hour approximately equals 500,000 years of time). Only for about 6000 years have we had recorded, or written, history.

KEY TERMS

1. geologic time
2. fossil (24.1)
3. paleontology
4. amber
5. replacement fossil
6. mold
7. cast
8. trace fossil
9. relative geologic time (24.2)

10. principle of original horizontality
11. principle of superposition
12. principle of cross-cutting relationships
13. unconformities
14. correlation
15. index fossils
16. eons
17. eras
18. periods

19. absolute (numerical) geologic time (24.3)
20. radiometric dating
21. carbon-14 dating
22. geologic time scale (24.5)
23. Cambrian explosion
24. Great Dying
25. epochs
26. K-T event

MATCHING

For each of the following items, fill in the number of the appropriate Key Term from the preceding list. Compare your answers with those at the back of the book.

a. _____ A fault must be younger than any of the rocks it has affected.

b. _____ Recent periods of time, names of which end in -cene

c. _____ The process of matching rock layers in different localities

d. _____ In a sequence of undisturbed sedimentary rocks, each layer is younger than the layer beneath it.

e. _____ The science that studies fossils

f. _____ The time spans into which eras are divided

g. _____ The general term for the determination of age by use of radioactivity

h. _____ The total span of the Earth's history

i. _____ The event between the Cretaceous and Tertiary periods

j. _____ The actual age of geologic events

k. _____ A hollow depression formed when a bone or shell is dissolved out of rock

l. _____ The great proliferation of life forms that occurred near the start of the Paleozoic era

m. _____ Fossil formed when new material fills a mold and hardens

n. _____ Fossil formed by slow replacement of the hard parts of a buried organism

o. _____ A hardened tree resin in which ancient insects have sometimes been preserved

p. _____ Sediments and lava flows are laid down flat on the surface on which they are deposited.

q. _____ Breaks in the rock record due to nondeposition or erosion

r. _____ The largest units of geologic time

s. _____ The time spans into which eons are divided

t. _____ Any remnant or indication of prehistoric life preserved in a rock

u. _____ Time obtained when geologic events are arranged in chronologic order without absolute dates

v. _____ A fossil formed by an animal track

w. _____ The dating of organic remains by measuring their carbon-14 content

x. _____ The combination of relative and absolute geologic time

y. _____ The most devastating extinction that has ever occurred

z. _____ Widespread, easily identified fossils that are typical of a particular time segment of the Earth's history

MULTIPLE CHOICE

Compare your answers with those at the back of the book.

1. Which statement is *not* true? (24.1)
   (a) Fossils are indications of prehistoric life.
   (b) Fossils help determine the relative ages of rocks.
   (c) Fossils can be tracks imprinted in rocks.
   (d) Fossils are irrelevant to our understanding of geologic time.

2. What is the name for the branch of science that specifically studies fossils? (24.1)
   (a) petrology   (b) mineralogy
   (c) paleontology   (d) fossiligraphy

3. What type of remain is a shark tooth? (24.1)
   (a) replacement   (b) index
   (c) original   (d) trace

4. What is the name for the type of fossil formed when mineral material fills a rock's hollow depression that once contained an embedded bone? (24.1)
   (a) mold   (b) cast
   (c) trace fossil   (d) nodule

5. What geologic principle tells us that each sediment layer is younger than the layer beneath it? (24.2)
   (a) superposition   (b) original horizontality
   (c) cross-cutting relationships (d) faunal succession

6. What is obtained when rocks and geologic events are put into chronologic order without regard to the actual dates? (24.2)
   (a) absolute (numerical) geologic time
   (b) relative geologic time
   (c) a geologic formation
   (d) a correlation period

7. Into what time spans are eras next divided? (24.2)
   (a) eons   (b) ages
   (c) epochs   (d) periods

8. What is the name of the time period when plants lived that eventually became coal? (24.2)
   (a) Devonian
   (b) Pennsylvanian
   (c) Cambrian
   (d) Triassic

9. What kind of fossil is widespread and easily identifiable? (24.2)
   (a) index   (b) trace
   (c) replacement   (d) amber

10. What radioactive element gradually decays to lead and can be used to date ancient rocks? (24.3)
    (a) rubidium   (b) potassium
    (c) carbon   (d) uranium

11. If the half-life of a radioactive specimen is 100 years, then how long will it take for the specimen to reach 25% of its original amount? (24.3)
    (a) 25 years   (b) 50 years
    (c) 100 years   (d) 200 years

12. Radiometric dating is used to determine which of the following? (24.3)
    (a) absolute (numerical) geologic time
    (b) relative geologic time
    (c) a geologic formation
    (d) a correlation period

13. Which radiometric dating method is most useful for dating bones and wood? (24.3)
    (a) uranium–lead   (b) potassium–argon
    (c) carbon-14   (d) rubidium–strontium

14. What is the age of the Earth as determined by radiometric dating? (24.4)
    (a) 150 million years   (b) 1200.2 years
    (c) 4.56 billion years   (d) 320 million years

15. Which of these rocks are useful in helping us determine the age of the Earth? (24.4)
    (a) Earth rocks   (b) Moon rocks
    (c) meteorite rocks   (d) all the preceding

16. Which became the dominant life form in the Ordovician period? (24.5)
    (a) dinosaurs   (b) plants
    (c) mammals   (d) fish

17. Fossils from Precambrian time could include which of the following? (24.5)
    (a) human skulls
    (b) dinosaur bones
    (c) algae, bacteria, and sea worms
    (d) leaf impressions and shark teeth

18. In what era were dinosaurs common? (24.5)
    (a) Mesozoic   (b) Paleozoic
    (c) Cambrian   (d) Cenozoic

19. In what era were trilobites common? (24.5)
    (a) Mesozoic   (b) Paleozoic
    (c) Cambrian   (d) Cenozoic

20. When did the K-T extinction event take place? (24.5)
    (a) at the end of the Paleozoic era
    (b) at the start of the Paleozoic era
    (c) at the end of the Mesozoic era
    (d) at the start of the Mesozoic era

FILL IN THE BLANK

Compare your answers with those at the back of the book.

1. The study of fossils is called ___. (24.1)

2. Petrified wood is a common type of ___. (24.1)

3. The earliest evidence of ancient life is fossil blue-green ___.(24.1)

4. We live in the ___ era. (24.2)

5. A break or gap in the rock record is called a(n) ___. (24.2)

6. A dike that intrudes a sedimentary rock bed is ___ than the sedimentary rock. (24.2)

7. The order in which events occur is called ___. (24.2)

8. ___ is the process of matching rock layers in different localities. (24.2)

9. The longest spans of geologic time are the ___. (24.2)

10. An example of a very important index fossil is a ___. (24.2)

11. To determine the relative age of a layer of sedimentary rock and a layer of ash deposited on it, a geologist would use the principle of ___. (24.2)

12. Carbon dating uses the isotope of carbon known as ___. (24.3)

13. The younger the rock, the ___ the ratio of the radioactive parent to the stable daughter. (24.3)

14. Lead-204 is found only in ___ lead. (24.3)

15. Radiometric dating indicates that the age of the Earth is about ___ billion years. (24.4)

16. Earth rock is a little ___ than Moon rock and meteorite rock, because Earth rock is subject to weathering and tectonic activity. (24.4)

17. The huge supercontinent called ___ broke up in the Paleozoic era. (24.5)

18. The extinction that occurred at the end of the Mesozoic era is called the ___. (24.5)

19. The great proliferation of life forms that occurred at the beginning of the Paleozoic era is called the Cambrian ___. (24.5)

20. The suffix of the name indicates that the Oligocene is a(n) ___. (24.5)

SHORT ANSWER

1. What is meant by the phrase *geologic time*?

24.1 Fossils

2. Why is amber important for paleontology?

3. Why are the most abundant fossils those of marine creatures?

4. Give some examples of replaced remains.

5. Arizona is famous for what variety of fossil?

6. How is coal formed?

7. Explain the difference between a *mold* and a *cast*.

8. What type of organisms are the oldest for which geologists can find evidence? About how long ago did the earliest of these apparently live?

9. How do fossils aid in oil exploration?

10. Give three examples of trace fossils.

24.2 Relative Geologic Time

11. How might the principle of superposition be violated?

12. State the principle of original horizontality.

13. Give two examples that could demonstrate the principle of cross-cutting relationships.

14. State how much time is represented by an unconformity.

15. What is meant by correlation?

16. How do index fossils aid correlation?

17. What four features characterize the best index fossils?

18. Use events in your daily life as an analogy to relative geologic time.

19. Name the three eras of the Phanerozoic eon in order of oldest to youngest.

20. Into how many periods is the Paleozoic era divided? Name the latest (youngest) of these periods.

21. In which period of the Mesozoic era did dinosaurs thrive?

22. Name the two periods of the Cenozoic era, with the older first.

23. The Mississippian and Pennsylvanian periods in North America go by what name in Europe?

24. What is meant by Precambrian time?

24.3 Radiometric Dating

25. What is the daughter product of potassium-40?

26. What type of rock is commonly dated with rubidium–strontium dating?

27. Carbon-14 dating measures the ratio of carbon-14 *not* to its daughter, nitrogen-14, but instead to what isotope?

28. What are the three conditions for using an isotope in a rock to date that rock?

29. Distinguish between primordial lead and radiogenic lead.

30. Give examples of objects dated with carbon-14.

24.4 The Age of the Earth

31. Give evidence that the Earth is about 4.56 billion years old. Include in your answer the age of the Sun (see Chapter 18.2).

32. About how old are the oldest rocks found on Earth? On the Moon?

24.5 The Geologic Time Scale (and Highlight: The K-T Event)

33. Briefly, how can the absolute age of a sedimentary rock be determined?

34. On the basis of what geologic or biological events is Precambrian time split into three eons? Name the three eons.

35. Which era on the geologic time scale encompasses the shortest time span?

36. In what era do you live? What period? What epoch?

37. What are some of the explanations for the Great Dying?

38. What geologic event separates the Pleistocene epoch from the Holocene epoch?

39. What biologic event is used by geologists to separate the Tertiary period from the Quaternary period?

40. What is the biological event that started the Paleozoic era called?

41. What probably caused the demise of the dinosaurs and other species?

42. To what was the extinction at the end of the Mesozoic probably due? State some evidence for that explanation.

43. What evidence points to the K-T event being astronomical in origin?

44. What are the starting and ending dates of the Paleozoic era?

45. Briefly, what effect would a supercontinent, such as Pangaea, have on the distribution of land plants and animals of the time?

VISUAL CONNECTION

Visualize the connections and give answers for the blanks. Compare your answers with those given at the back of the book

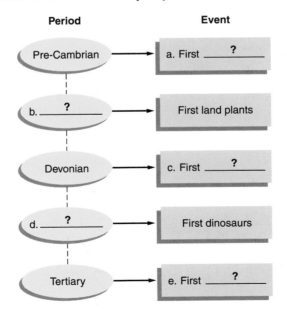

APPLYING YOUR KNOWLEDGE

1. While on a nature hike, you come upon a rock quarry whose shale rock contains fossils of crinoids, brachiopods, and trilobites. Your companion asks you what era the rocks are from and how old the fossils are. In light of this assemblage of fossils, what are your answers?

2. Your friend shows you a cow bone dug up from a field on his farm and declares the bone to be a fossil. Do you agree? Why or why not?

3. You see a comic strip in which cavemen and dinosaurs are shown coexisting. Why do you realize that artistic license is being taken?

4. At a jewelry show, you overhear a person looking at an amber brooch say that she wishes it contained an insect so that she would have a fossil. What error is she making?

5. While standing in line at the grocery store, you see a magazine headline stating that carbon dating has been used to find the age of a dinosaur bone. Why are you skeptical of that claim?

EXERCISES

24.2 Relative Geologic Time

1. ● Table 24.2 shows, in color, the range in the rock record of six different fossils of the Paleozoic era. Along the top of the chart is a letter for each period of the era (C for Cambrian, P for Pennsylvanian, PR for Permian. etc.).
   (a) What is the range of geologic periods for the crinoid *Platycrinites*?
   (b) To what period does rock belong that contains the brachiopod *Zygospira* and the trilobite *Phacops*?
   (c) List the fossils shown that might be found in rock of the Pennsylvanian period.
   (d) Considering only the time range, indicate which of the fossils would be the best index fossil.

   Answer: (a) Mississippian, Pennsylvanian, Permian (b) Silurian
   (c) *Platycrinites* and *Lingula* (d) *Elrathia*

2. Refer to Table 24.2.
   (a) What is the range of geologic periods for the brachiopod *Zygospira*?
   (b) To what period does rock belong that contains the crinoid *Taxocrinus* and the trilobite *Phacops*?
   (c) List the fossils shown that might be found in rock of the Silurian period.
   (d) Why could neither of the two trilobites listed be used to identify rock of the Ordovician period?
   (e) Considering only the time range, indicate which fossil would be the *worst* index fossil.

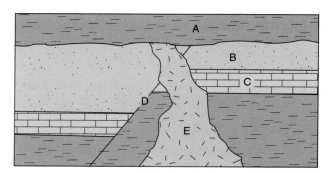

Figure 24.21 Relative Dating
See Exercises 3 and 4.

3. Refer to ● Fig. 24.21.
   (a) Which rock stratum is younger, A or C? What geologic principle did you use?
   (b) Which is younger, the rock stratum marked C or the igneous intrusion marked E? What geologic principle did you use?

   Answer: (a) A is the younger; superposition
   (b) E is the younger; cross-cutting relationships

4. Refer to Fig. 24.21.
   (a) Which is younger, the rock stratum C or the fault marked D? What geologic principle did you use?
   (b) Which is younger, the fault marked D or the igneous intrusion marked E? What geologic principle did you use?

24.3 Radiometric Dating

5. Charcoal from an ancient campfire has a ratio of ^{14}C to ^{12}C that is one-fourth that of new wood. About how old is the charcoal? The half-life of ^{14}C is 5730 years.

   Answer: About 11,460 y

6. Carbon obtained from a sample of frozen skin in a glacier is found to have one-eighth the ^{14}C-to-^{12}C ratio of present-day carbon. About how old is the skin? The half-life of ^{14}C is 5730 years.

Table 24.2 Some Fossils and Their Periods

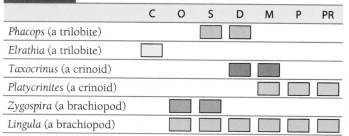

	C	O	S	D	M	P	PR
Phacops (a trilobite)			■	■			
Elrathia (a trilobite)	■						
Taxocrinus (a crinoid)			■	■			
Platycrinites (a crinoid)					■	■	■
Zygospira (a brachiopod)		■	■				
Lingula (a brachiopod)	■	■	■	■	■	■	■

24.5 The Geologic Time Scale

7. Refer to ● Fig. 24.22, which shows a lava flow radiometrically dated at 245 my and an igneous dike dated at 210 my. What can be said about the absolute age of the sandstone layer?

   Answer: Older than 210 my but younger than 245 my

8. Suppose that one species of index fossil lived between 410 and 380 mya and another lived between 440 and 350 mya. What can be said about the age of a rock that contains fossils *of both* species?

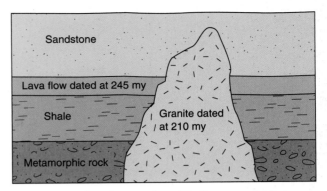

Figure 24.22 Absolute Dating of Sedimentary Rocks
See Exercise 7.

ON THE WEB

1. Foraging for Fossils

What are the many reasons we study our Earth? How does recorded history differ from Earth history? What is the geologic time scale? How are events put in order? How and where was the geologic time scale developed? What are fossils, and how do they tell us the age of rocks? What are some of the challenges of studying fossils? What do you believe to be the most important reason for studying our fossil records? Explore answers to these questions by following the recommended links on the student website at **www.cengagebrain.com/shop/ISBN/1133104096**.

2. Techniques of Science

What are the scientific principles used to determine the ages of dinosaurs? Were humans and dinosaurs ever living at the same time? How are the ages of dinosaurs "bracketed"? What are the errors associated with dating the ages of dinosaurs? How is the scientific method used to estimate the ages? What are the strengths and weaknesses of the radiometric dating method? Explore answers to these questions by following the recommended links on the student website at **www.cengagebrain.com/shop/ISBN/1133104096**.

Appendixes

Appendix I The Seven Base Units of the International System of Units (SI)

1. **meter, m (length):** The meter is defined in reference to the standard unit of time. One meter is the length of the path traveled by light in a vacuum during a time interval of 1/299,792,458 of a second. That is, the speed of light is a universal constant of nature whose value is defined to be 299,792,458 meters per second.
2. **kilogram, kg (mass):** The kilogram is a cylinder of platinum-iridium alloy kept by the International Bureau of Weights and Measures in Paris. A duplicate in the custody of the National Institute of Standards and Technology serves as the mass standard for the United States. This is the only base unit still defined by an artifact.
3. **second, s (time):** The second is defined as the duration of 9,192,631,770 cycles of the radiation associated with a specified transition of the cesium-133 atom.
4. **ampere, A (electric current):** The ampere is defined as that current that, if maintained in each of two long parallel wires separated by one meter in free space, would produce a force between the two wires (due to their magnetic fields) of 2×10^{-7} newtons for each meter of length.
5. **kelvin, K (temperature):** The kelvin is defined as the fraction 1/273.16 of the thermodynamic temperature of the triple point of water. The temperature 0 K is called *absolute zero*.

Table I.1 Prefixes Representing Powers of 10*

Multiple	Prefix	Abbreviation
10^{18}	exa-	E
10^{15}	peta-	P
10^{12}	tera-	T
10^{9}	giga-	G
10^{6}	mega-	M
10^{3}	kilo-	k
10^{2}	hecto-	h
10	deka-	da
10^{-1}	deci-	d
10^{-2}	centi-	c
10^{-3}	milli-	m
10^{-6}	micro-	μ**
10^{-9}	nano-	n
10^{-12}	pico-	p
10^{-15}	femto-	f
10^{-18}	atto-	a

*The most commonly used prefixes are highlighted in color.

**This is the Greek letter mu.

6. **mole, mol (amount of substance):** The mole is the amount of substance of a system that contains as many elementary entities as there are atoms in 0.012 kilogram of carbon-12.
7. **candela, cd (luminous intensity):** The candela is defined as the luminous intensity of 1/600,000 of a square meter of a black body at the temperature of freezing platinum (2045 K).

Appendix II   Solving Mathematical Problems in Science

When taking a science course, some students worry unduly about their ability to handle exercises involving mathematics. Any such apprehension should be replaced with a sense of confidence that they can handle *any* exercise presented in this textbook if they will just learn, *early on,* the few fundamental skills presented in Chapter 1 and these appendixes.

   Mathematics is fundamental in the physical sciences, and it is difficult to understand or appreciate these sciences without certain basic mathematical abilities. Dealing with the quantitative side of science gives the student a chance to review and gain confidence in the use of basic mathematical skills, to learn to analyze problems and reason them through, and to see the importance and power of a systematic approach to problems. We recommend the following approach to solving mathematical problems in science.

1. Read the exercise and list what you are given and what is wanted (unknown). Include all units—not just the numbers. Sometimes, making a rough sketch of the situation is helpful.
2. From that information, decide the type of problem with which you are dealing, and select the appropriate equation. (A brief list of **Important Equations** is provided near the end of any chapter that contains mathematical exercises.) *Occasionally,* you may need to use conversion factors (Chapter 1.6) to change, say, 66 g to kg. In general, all quantities should be in the same system of units (generally, the mks system).
3. If necessary, rearrange the equation for the unknown. (Appendix III discusses equation rearrangement.)
4. Substitute the known numbers *and their units* into the rearranged equation.
5. See whether the units combine to give you the appropriate unit for the unknown. (Appendix IV discusses the analysis of units.)
6. Once you have determined that the units are correct, do the math, being sure to express the answer to the proper number of significant figures and to include the unit. (Appendixes V, VI, and VII describe how to use positive and negative numbers, powers-of-10 notation, and significant figures.)
7. Evaluate the answer for reasonableness. (For example, it would not be *reasonable* for a car to be traveling 800 mi/h.)

Appendix III   Equation Rearrangement

An important skill for solving mathematical problems in science is the ability to rearrange an equation for the unknown quantity. Basically, we want to get the unknown

1. into the numerator
2. positive in sign
3. to the first power (not squared, cubed, etc.)
4. alone on one side of the equals sign (=)

Addition or Subtraction to Both Sides of an Equation

The equation $X - 4 = 12$ states that the numbers $X - 4$ and 12 are equal. To solve for X, use this rule: Whatever is added to (or subtracted from) one side can be added to (or subtracted from) the other side, and equality will be maintained. We want to get X, the unknown in this case, alone on one side of the equation, so we add 4 to both sides (so that $-4 + 4 = 0$ on the left side). Example III.1 illustrates this procedure.

EXAMPLE III.1

$$X - 4 = 12$$
$$X - 4 + 4 = 12 + 4 \quad \text{(4 added to both sides)}$$
$$X = 16$$

EXAMPLE III.2

Suppose in the scientific equation $T_K = T_C + 273$ we wish to solve for T_C. All we need to do is to get T_C alone on one side, so we subtract 273 from each side. The procedure is

$$T_K = T_C + 273$$
$$T_K - 273 = T_C + 273 - 273 \quad \text{(273 subtracted from both sides)}$$
$$T_K - 273 = T_C \text{ (or } T_C = T_K - 273\text{)}$$

Shortcut After you understand the principle of adding or subtracting the same number or symbol from each side, you may wish to use a shortcut for this type of problem: To move a number or symbol added to (or subtracted from) the unknown, take it to the other side but change its sign. See how Examples III.1 and III.2 are solved using the shortcut.

Change sign

$$X - 4 = 12 \quad \text{gives } X = 12 + 4$$

Change sign

$$T_K = T_C + 273 \quad \text{gives } T_K - 273 = T_C$$

PRACTICE PROBLEMS

Solve each equation for the unknown. As you proceed, check your answers against those given at the end of this appendix.

(a) $X + 9 = 11; X = ?$

(b) $T_f - T_i = \Delta T; T_f = ?$

(c) $\lambda f = E_i - E_f; E_i = ?$

(d) $H = \Delta E_i + W; W = ?$

(e) $\dfrac{\Delta E_p}{mg} = h_2 - h_1; h_2 = ?$

(f) $H = E_p - \dfrac{mv^2}{2}; E_p = ?$

Multiplication or Division to Both Sides of an Equation

For equations such as $\frac{X}{4} = 6$, the rule is: Multiply (or divide) both sides by the number or symbol that will leave the unknown alone on one side of the equation. Thus, in the equation $\frac{X}{4} = 6$, we multiply both sides by 4, as shown in Example III.3.

EXAMPLE III.3

$$\frac{X}{4} = 6$$

$$\frac{4X}{4} = 6 \times 4 \quad \text{(both sides multiplied by 4)}$$

$$X = 24$$

EXAMPLE III.4

In the scientific equation $F = ma$, solve for a. The unknown is already in the numerator, so all we do is move the m by dividing both sides by m.

$$F = ma$$

$$\frac{F}{m} = \frac{ma}{m} \quad \text{(both sides divided by m)}$$

$$\frac{F}{m} = a \left(\text{or } a = \frac{F}{m} \right)$$

EXAMPLE III.5

Suppose the unknown is in the denominator to start with; for example, solving for t in $P = \frac{W}{t}$. Multiply both sides by t to get t in the numerator, and then divide both sides by P to move P to the other side. The procedure is

$$P = \frac{W}{t}$$

$$tP = \frac{tW}{t} \quad \text{(both sides multiplied by } t)$$

$$tP = W$$

$$\frac{tP}{P} = \frac{W}{P} \quad \text{(both sides divided by } P)$$

$$t = \frac{W}{P}$$

Shortcut After you understand the principle of multiplying or dividing both sides of the equation by the same number or symbol, you may wish to use a shortcut to move the number or symbol: Whatever is multiplying (or dividing) the whole of one side winds up dividing (or multiplying) the whole other side. See how Examples III.3, III.4, and III.5 are solved using the shortcut.

$$\frac{X}{4} = 6 \quad \text{gives } X = 4 \times 6 \text{ (or 24)}$$

$$F = ma \quad \text{gives } \frac{F}{m} = a$$

$$P = \frac{W}{t} \quad \text{gives } t = \frac{W}{P}$$

(The shortcut saves many steps when an unknown is in the denominator.)

PRACTICE PROBLEMS

Solve for the unknown. The answers are given at the end of this appendix.

(g) $7X = 21$; $X = ?$

(l) $\lambda = \dfrac{h}{mf}$; $m = ?$

(h) $\dfrac{X}{3} = 2$; $X = ?$

(m) $E_k = \dfrac{mv^2}{2}$; $m = ?$

(i) $w = mg$; $m = ?$

(n) $v^2 = \dfrac{3kT}{m}$; $T = ?$

(j) $\lambda = \dfrac{c}{f}$; $c = ?$

(o) $pV = nRT$; $R = ?$

(k) $\lambda = \dfrac{c}{f}$; $f = ?$

Multiple Operations to Both Sides of an Equation

To solve some problems requires the use of both the addition/subtraction and the multiplication/division rules. For example, solving $\dfrac{X - 4}{3} = 8$ and $\dfrac{X}{3} - 4 = 8$ requires both rules. These two equations are similar but not identical. We will follow a different order of application of the rules as we solve each.

A good way to proceed is to follow the principle: If the *whole side* on which the unknown is found is multiplied and/or divided by a number or symbol, get that number or symbol to the other side first. *Then* move any numbers or symbols that are added to or subtracted from the unknown. (This is the case in Example III.6.) On the other hand, if only the unknown (and not the whole side) is multiplied or divided by a number or symbol, first move any number or symbol added or subtracted. (This is the case in Example III.7.) The successive steps for each example are shown.

EXAMPLE III.6

$$\frac{(X - 4)}{3} = 8$$

$$\frac{3 \times (X - 4)}{3} = 3 \times 8 \qquad \text{(both sides multiplied by 3)}$$

$$X - 4 = 24$$

$$X - 4 + 4 = 24 + 4 \qquad \text{(4 added to both sides)}$$

$$X = 28$$

EXAMPLE III.7

$$\frac{X}{3} - 4 = 8$$

$$\frac{X}{3} - 4 + 4 = 8 + 4 \qquad \text{(4 added to both sides)}$$

$$\frac{X}{3} = 12$$

$$\frac{3X}{3} = 12 \times 3 \qquad \text{(both sides multiplied by 3)}$$

$$X = 36$$

Note that in Example III.6 both sides were first multiplied by 3, and then 4 was added to both sides. But in Example III.7, 4 was added to both sides first, and then both sides were divided by 3. We used different strategies because in Example III.6 the 3 was dividing the whole side that X was on, whereas in Example III.7 it wasn't. Other strategies may be applied to problems like these, but things generally work out better when the recommended principle is used.

The rules for rearranging scientific equations are exactly the same. Before substituting numbers and units for the various letters in a scientific equation, most experts advise that the equation first be rearranged so that the unknown is positive, to the first power, and alone in the numerator on one side. It is easier and more accurate to move several letters than to move a bunch of numbers and units, as you would have to do if you substituted first.

EXAMPLE III.8

How would you proceed in solving the equation $a = \dfrac{v_f - v_i}{t}$ for v_f? The v_i and the t need to be moved, and t should be moved first because it is dividing the whole side that the unknown, v_f, is on. The successive steps are

$$a = \frac{v_f - v_i}{t}$$

$$at = \frac{t(v_f - v_i)}{t} \qquad \text{(both sides multiplied by } t)$$

$$at = v_f - v_i$$

$$at + v_i = v_f - v_i + v_i \qquad (v_i \text{ added to both sides})$$

$$at + v_i = v_f \,(\text{or } v_f = at + v_i)$$

Shortcut Of course, the same shortcuts can be used in these more complicated equations. Just be sure you use them in the correct order. For Examples III.6, III.7, and III.8, the circled numbers show the correct order for the shortcuts:

$$\frac{(X - 4)}{3} = 8 \qquad \text{gives } X = (3 \times 8) + 4 = 28$$

$$\frac{X}{3} - 4 = 8 \qquad \text{gives } X = (8 + 4) \times 3 = 36$$

$$a = \frac{v_f - v_i}{t} \qquad \text{gives } at + v_i = v_f$$

PRACTICE PROBLEMS

Solve for the unknown. The answers are given at the end of this appendix.

(p) $3X + 2 = -;\ 7X = ?$

(q) $\dfrac{X + 2}{4} = 6;\ X = ?$

(r) $a = \dfrac{F_2 - F_1}{m};\ F_2 = ?$

(s) $H = \dfrac{B}{U} - M;\ B = ?$

(t) $T_F = 1.8T_C + 32;\ T_C = ?$

(u) $H = mc_w(T_f - T_i);\ T_f = ?$

(v) $F = \dfrac{m(v - v_o)}{t};\ v = ?$

Squared or Negative Unknowns

You will encounter some equations in which the unknown is squared (raised to the second power). For example, solve for v in $a = \dfrac{v^2}{r}$. In such cases, solve for the squared form first; then, either before or after substituting the numbers and units, take the square root of both sides.

EXAMPLE III.9

$$a = \frac{v^2}{r}$$

$$ar = \frac{rv^2}{r} \qquad \text{(both sides multiplied by } r\text{)}$$

$$ar = v^2$$

$$\sqrt{ar} = \sqrt{v^2} \qquad \text{(the square root of both sides taken)}$$

$$\sqrt{ar} = v \ (\text{or } v = \sqrt{ar})$$

Rarely, you may need to solve for an unknown that is negative at the start. For example, find T_i in $\Delta T = T_f - T_i$. In such cases, multiply both numerators by -1, and then proceed as usual.

EXAMPLE III.10

$$\Delta T = T_f - T_i$$

$$-\Delta T = -T_f + T_i \qquad \text{(both sides multiplied by } -1\text{)}$$

$$T_f - \Delta T = -T_f + T_i + T_f \qquad (T_f \text{ added to both sides})$$

$$T_f - \Delta T = T_i$$

PRACTICE PROBLEMS

Solve for the unknown. The answers are given at the end of this appendix.

(w) $E_k = \dfrac{mv^2}{2}$; $v = ?$ (x) $a = \dfrac{v_f - v_i}{t}$; $v_i = ?$ (y) $F_G = \dfrac{Gm_1m_2}{r^2}$; $r = ?$

ANSWERS

(a) $X = 2$

(b) $T_f = \Delta T + T_i$

(c) $E_i = \lambda f + E_f$

(d) $W = H - \Delta E_i$

(e) $h_2 = \dfrac{\Delta E_p}{mg} + h_i$

(f) $E_p = H + \dfrac{mv^2}{2}$

(g) $X = 3$

(h) $X = 6$

(i) $m = \dfrac{w}{g}$

(j) $c = \lambda f$

(k) $f = \dfrac{c}{\lambda}$

(l) $m = \dfrac{h}{\lambda f}$

(m) $m = \dfrac{2E_k}{v^2}$

(n) $T = \dfrac{v^2 m}{3k}$

(o) $R = \dfrac{pV}{nT}$

(p) $X = -3$

(q) $X = 22$

(r) $F_2 = ma + F_1$

(s) $B = U(H + M)$

(t) $T_C = \dfrac{T_F - 32}{1.8}$

(u) $T_f = \dfrac{H}{mc_w} + T_i$

(v) $v = \dfrac{Ft}{m} + v_o$

(w) $v = \sqrt{\dfrac{2E_k}{m}}$

(x) $v_i = v_f - at$

(y) $r = \sqrt{\dfrac{Gm_1m_2}{F_G}}$

Appendix IV Analysis of Units

Measured quantities always have dimensions, or *units*—for example, 17 *grams*, 1.4 *meters*, 25 *seconds*, and so forth. Analyzing the units is important when dealing with scientific equations because the units can often show you whether you have rearranged the equation and put in your data correctly. For example, if the unknown for which you are solving is a distance, yet the units show you coming out with m/s^2 (a combination of units characteristic of an acceleration), you have done something wrong (probably you rearranged the equation incorrectly). Units follow the same rules as numbers. Only a few basic situations are encountered when analyzing units, and most are illustrated in the problems given in this appendix.

You have often heard that you cannot add apples to oranges; neither can you add, say, grams to meters. When adding and subtracting, the units must be the same.

EXAMPLE IV.1 **(Two Different Cases)**

$8 \text{ mL} + 2 \text{ mL} = 10 \text{ mL}$ (that is, mL + mL gives mL)
$15 \text{ g} - 7 \text{ g} = 8 \text{ g}$ (that is, g − g = g)

When multiplying and dividing, a unit in the numerator will cancel its counterpart in the denominator. A unit multipled by the same unit gives the unit squared.

EXAMPLE IV.2 **(Three Different Cases)**

$$\frac{8 \text{ m}}{2 \text{ m}} = 4 \qquad 5.0 \frac{\text{m}}{\text{s}^2} \times 3.0 \text{ s} = 15 \frac{\text{m}}{\text{s}}$$
$$8 \text{ m} \times 2 \text{ m} = 16 \text{ m}^2$$

What about a problem where $\frac{\text{m}}{\text{s}}$ is divided by $\frac{\text{m}}{\text{s}^2}$? Recall how fractions are divided. To divide $\frac{3}{4}$ by $\frac{1}{4}$ you *invert* (turn over) the fraction in the denominator and then multiply by the inverted fraction. Units are handled the same way. Be sure to practice this trickiest part of analyzing units. Do not omit the intermediate step; this is no place to take shortcuts.

EXAMPLE IV.3 **(Two Different Cases)**

$$\frac{\frac{3}{4}}{\frac{1}{4}} = \frac{3}{4} \times \frac{4}{1} = 3 \qquad \frac{\frac{\text{m}}{\text{s}}}{\frac{\text{m}}{\text{s}^2}} = \frac{\text{m}}{\text{s}} \times \frac{\text{s}^2}{\text{m}} = \text{s}$$

PRACTICE PROBLEMS

Analyze these units. As you proceed, check your answers against those given at the end of this appendix.

(a) $\dfrac{\text{g}}{\text{mol}} \times \text{mol}$

(b) $\dfrac{\frac{\text{g}}{\text{mol}}}{\text{mol}}$

(c) $\dfrac{\text{cm}}{\text{s}} + \dfrac{\text{cm}}{\text{s}}$

(d) $\dfrac{\text{g}}{\frac{\text{g}}{\text{mol}}}$

(e) $\dfrac{\frac{\text{g}}{\text{mol}}}{\text{g}}$

(f) $\dfrac{\text{m}}{\text{s}} \times \text{m}$

(g) $\dfrac{\frac{\text{m}}{\text{s}}}{\text{m}}$

(h) $\dfrac{\text{J}}{\text{cal}} \times \text{J}$

(i) $\dfrac{\text{J}}{\text{cal}} \times \text{cal}$

(j) $\text{kg} - \text{kg}$

(k) $\dfrac{\frac{\text{g}}{\text{mL}}}{\text{mL}}$

(l) $\text{kg} \times \dfrac{\text{m}}{\text{s}^2}$

Chapter 1 discusses the use of conversion factors, a procedure in which the proper analysis of units is crucial. Using conversion factors, try the following practice problems. In parts (o) through (t), put in the units of the conversion factor first to make sure they will work out correctly; then insert the proper numbers. Check your answers against those given at the end of this appendix.

PRACTICE PROBLEMS

(m) If 1 L = 1.06 qt, then how many $\dfrac{qt}{L}$ are there? How many $\dfrac{L}{qt}$?

(n) If there are 1.61 $\dfrac{km}{mi}$, then how many $\dfrac{mi}{km}$ are there?

(o) 15.0 m = ? yd, if 1 m = 1.09 yd

(p) 52.0 L = ? qt, if 1 L = 1.06 qt

(q) 46.0 km = ? mi, if 1 mi = 1.61 km

(r) 93.0 lb = ? kg, if 1 kg = 2.20 lb (at the Earth's surface)

(s) 87 kg = ? g, if 1 kg = 1000 g

(t) 49 cm = ? m, if 1 m = 100 cm

ANSWERS

(a) g

(b) $\dfrac{g}{mol^2}$

(c) $\dfrac{cm}{s}$

(d) mol

(e) $\dfrac{1}{mol}$

(f) $\dfrac{m^2}{s}$

(g) $\dfrac{m^2}{s}$

(h) $\dfrac{J^2}{cal}$

(i) J

(j) kg

(k) $\dfrac{g}{mL^2}$

(l) $\dfrac{kg \cdot m}{s^2}$

(m) $1.06 \dfrac{qt}{L}; \dfrac{1}{1.06} \dfrac{L}{qt}$

(n) $\dfrac{1}{1.61} \dfrac{mi}{km}$

(o) $15.0 \text{ m } (1.09) \dfrac{yd}{m} = 16.4 \text{ yd}$

(p) $52.0 \text{ L } (1.06) \dfrac{qt}{L} = 55.1 \text{ qt}$

(q) $46.0 \text{ km} \left(\dfrac{1}{1.61}\right) \dfrac{mi}{km} = 28.6 \text{ mi}$

(r) $93.0 \text{ lb} \left(\dfrac{1}{2.20}\right) \dfrac{kg}{lb} = 42.3 \text{ kg}$

(s) $87 \text{ kg } (1000) \dfrac{g}{kg} = 87 \times 10^3 \text{ g}$

(t) $49 \text{ cm} \left(\dfrac{1}{100}\right) \dfrac{m}{cm} = 0.49 \text{ m}$

Appendix V Positive and Negative Numbers

Multiplying and Dividing

To multiply and divide positive and negative numbers, follow these simple rules: *If both* numbers are positive or *both* numbers are negative, the result is positive. If one number is positive and the other is negative, the result is negative.

EXAMPLE V.1 (Three Different Cases)

$$3 \times 4 = 12$$
$$-20 \div -5 = 4$$
$$(-3) \times 4 = -12$$

PRACTICE PROBLEMS

Perform the designated operations. The answers are given at the end of this appendix.

(a) $14 \div 2$
(b) $-30 \div 6$
(c) 5×8
(d) $-7 \times (-6)$
(e) $8 \times (-2)$
(f) $-40 \div -5$

Algebraic Addition

Algebraic addition of positive and negative numbers is illustrated in Example V.2. The numbers may be grouped and added or subtracted in any sequence without affecting the result.

EXAMPLE V.2 (Six Different Cases)

$$4 + 5 = 9$$
$$4 + (-5) = 4 - 5 = -1$$
$$4 - 5 = -1$$
$$4 - (-5) = 4 + 5 = 9$$
$$-4 - 5 = -9$$
$$-5 + 4 - 6 + 8 = -11 + 12 = 1$$

PRACTICE PROBLEMS

Perform the designated operations. Check your answers.

(g) $7 + 6$
(h) $-7 + 3 - 6$
(i) $-8 - 7$
(j) $14 - 5 + 8$
(k) $18 - 10$
(l) $8 - (-6)$

ANSWERS

(a) 7
(b) -5
(c) 40
(d) 42
(e) -16
(f) 8
(g) 13
(h) -10
(i) -15
(j) 17
(k) 8
(l) 14

Appendix VI Powers-of-10 Notation

Changing Between Decimal Form and Powers-of-10 Form

In physical science, many numbers are very big or very small. To express them, we frequently use **powers-of-10 (scientific) notation.** When the number 10 is squared or cubed, we get

$$10^2 = 10 \times 10 = 100$$
$$10^3 = 10 \times 10 \times 10 = 1000$$

You can see that the number of zeros in the answers above is just equal to the exponent, or power of 10. As an example, 10^{23} is a 1 followed by 23 zeros.

Negative powers of 10 also can be used. For example,

$$10^{-2} = \frac{1}{10^2} = \frac{1}{100} = 0.01$$

We see that if a number has a negative exponent, we shift the decimal place to the left once for each power of 10. For example, 1 centimeter (cm) is 1/100 m or 10^{-2} m, which is 0.01 m.

We also can multiply numbers by powers of 10. Table VI.1 shows examples of various large and small numbers expressed in powers-of-10 notation.

We can represent a number in powers-of-10 notation in many different ways—all correct. For example, the distance from the Earth to the Sun is 93 million miles. This value can be represented as 93,000,000 miles, or 93×10^6 miles, or 9.3×10^7 miles, or 0.93×10^8 miles. Any of the given representations of 93 million miles is correct, although 9.3×10^7 is preferred. (In expressing powers-of-10 notation, it is customary to have one digit to the left of the decimal point. This is called *conventional* or *standard form*.)

TABLE VI.1 Numbers Expressed in Powers-of-10 Notation

Number	Powers-of-10 Notation*
247	2.47×10^2
186,000	1.86×10^5
4,705,000	4.705×10^6
0.025	2.5×10^{-2}
0.0000408	4.08×10^{-5}
0.00000010	1.0×10^{-7}

*Note: The exponent (power of 10) is increased by 1 for every place the decimal point is shifted to the left and is decreased by 1 for every place the decimal point is shifted to the right.

Thus it can be seen that the exponent, or power of 10, changes when the decimal point of the prefix number is shifted. General rules for this are as follows:

1. The exponent, or power of 10, is *increased* by 1 for every place the decimal point is shifted to the *left*.
2. The exponent, or power of 10, is *decreased* by 1 for every place the decimal point is shifted to the *right*.

This is simply a way of saying that if the coefficient (prefix number) gets smaller, the exponent gets correspondingly larger, and vice versa. Overall, the number is the same.

Try the following practice problems.

PRACTICE PROBLEMS

Put parts (a) through (d) in standard powers-of-10 form, and parts (e) through (h) in decimal form. The answers are given at the end of this appendix.

(a) 2500
(b) 870,000
(c) 0.0000008
(d) 0.0357

(e) 6×10^4
(f) 5.6×10^3
(g) 5.6×10^{-6}
(h) 7.9×10^{-2}

Changing Between Powers-of-10 Forms

When changing from one powers-of-10 form to another, you must ensure that the final number is equal to the number with which you started. Thus, if the exponential is made larger, the decimal in the prefix becomes correspondingly smaller, and vice versa.

EXAMPLE VI.1

Change 83×10^5 to the 10^6 form.
Going from 10^5 to 10^6 is an *increase* by a factor of 10. Therefore, the prefix part must *decrease* by a factor of 10. Because 83 divided by 10 is 8.3, the answer is 8.3×10^6.

Change 4.5×10^{-9} to the 10^{-10} form.
Going from 10^{-9} to 10^{-10} is a decrease by a factor of 10. Therefore, the decimal prefix must *increase* by a factor of 10. Because 4.5 multipled by 10 is 45, the answer is 45×10^{-10}.

PRACTICE PROBLEMS

Determine the value required in place of the question mark for the equation to be true. Check your answers.

(i) $3.02 \times 10^7 = ? \times 10^6$
(j) $126 \times 10^{-3} = ? \times 10^{-2}$
(k) $896 \times 10^4 = ? \times 10^6$
(l) $32.7 \times 10^5 = 3.27 \times 10^?$

Addition and Subtraction of Powers of 10

In addition or subtraction, the exponents of 10 must be the same value.

EXAMPLE VI.2

$$\begin{array}{r} 4.6 \times 10^{-8} \\ + 1.2 \times 10^{-8} \\ \hline 5.8 \times 10^{-8} \end{array} \quad \text{and} \quad \begin{array}{r} 4.8 \times 10^7 \\ -2.5 \times 10^7 \\ \hline 2.3 \times 10^7 \end{array}$$

PRACTICE PROBLEMS

Perform the designated arithmetical operations. Check your answers.

(m) $\begin{array}{r} 4.5 \times 10^5 \\ + 3.2 \times 10^5 \\ \hline ? \end{array}$ (n) $\begin{array}{r} 5.66 \times 10^{-3} \\ -3.24 \times 10^{-3} \\ \hline ? \end{array}$

Multiplication of Powers of 10

In multiplication, the exponents are added.

EXAMPLE VI.3

$(2 \times 10^4)(4 \times 10^3) = 8 \times 10^7$ and $(1.2 \times 10^{-2})(3 \times 10^6) = 3.6 \times 10^4$

PRACTICE PROBLEMS

Perform the designated arithmetical operations.

(o) $(7 \times 10^5)(3 \times 10^4) = ?$
(p) $(2 \times 10^{-3})(4 \times 10^6) = ?$

Division of Powers of 10

In division, the exponents are subtracted.

EXAMPLE VI.4

$$\frac{4.8 \times 10^8}{2.4 \times 10^2} = 2.0 \times 10^6 \qquad \text{and} \qquad \frac{3.4 \times 10^{-8}}{1.7 \times 10^{-2}} = 2.0 \times 10^{-6}$$

An alternative method for division is to transfer all powers of 10 from the denominator to the numerator by changing the sign of the exponent. Then, the exponents of the powers of 10 may be added, because they are now multiplying. The decimal parts are not transferred; they are divided in the usual manner. This method requires an additional step, but many students find it leads to the correct answer more consistently. Thus,

$$\frac{4.8 \times 10^8}{2.4 \times 10^2} = \frac{4.8 \times 10^8 \times 10^{-2}}{2.4} = 2.0 \times 10^6$$

PRACTICE PROBLEMS

Perform the designated arithmetical operations.

(q) $\dfrac{18 \times 10^7}{3 \times 10^4}$

(r) $\dfrac{(3 \times 10^{17})(4 \times 10^{-8})}{6 \times 10^{-11}} = \,?$

Squaring Powers of 10

When squaring exponential numbers, multiply the exponent by 2. The decimal part is multiplied by itself.

EXAMPLE VI.5

$$(3 \times 10^4)^2 = 9 \times 10^8$$
$$(4 \times 10^{-7})^2 = 16 \times 10^{-14}$$

PRACTICE PROBLEMS

Perform the designated algebraic operations. Check your answers.

(s) $(8 \times 10^{-5})^2$

(t) $(4 \times 10^3)^2$

(u) $(3 \times 10^{-8})^2$

Finding the Square Root of Powers of 10

To find the square root of an exponential number, follow the rule $\sqrt{10^a} = 10^{\left(\frac{a}{2}\right)}$. Note that the exponent must be an even number. If it is not, change to a power-of-10 form that gives an even exponent. Find the square root of the decimal part by determining what number multiplied by itself gives that number.

EXAMPLE VI.6   (Two Examples)

$$\sqrt{9 \times 10^8} = 3 \times 10^4$$
$$\sqrt{2.5 \times 10^{-17}} = \sqrt{25 \times 10^{-18}} = 5 \times 10^{-9}$$

PRACTICE PROBLEMS

Perform the designated algebraic operations. Check your answers.

(v) $\sqrt{4 \times 10^8}$

(w) $\sqrt{16 \times 10^{-10}}$

(x) $\sqrt{78 \times 10^{-11}}$

ANSWERS

(a) 2.5×10^3	(m) 7.7×10^5
(b) 8.7×10^5	(n) 2.42×10^{-3}
(c) 8×10^{-7}	(o) 21×10^9
(d) 3.57×10^{-2}	(p) 8×10^3
(e) 60,000	(q) 6×10^3
(f) 5600	(r) 2×10^{20}
(g) 0.0000056	(s) 64×10^{-10}
(h) 0.079	(t) 16×10^6
(i) 30.2×10^6	(u) 9×10^{-16}
(j) 12.6×10^{-2}	(v) 2×10^4
(k) 8.96×10^6	(w) 4×10^{-5}
(l) 3.27×10^6	(x) 2.8×10^{-5}

Appendix VII   Significant Figures

Chapter 1 introduced the concept of significant figures (Chapter 1.7). Let's expand on what you learned there and get some practice.

In scientific work, most numbers are *measured* quantities and thus are not exact. All measured quantities are limited in significant figures (abbreviated s.f.) by the precision of the instrument used to make the measurement. The measurement must be recorded in such a way as to show the degree of precision to which it was made—no more, no less. Furthermore, calculations based on the measured quantities can have no more (or no less) precision than the measurements themselves. Thus the answers to the calculations must be recorded to the proper number of significant figures. To do otherwise is misleading and improper.

Counting Significant Figures

Measured Quantities

Rule 1: **Nonzero integers** are always significant (for example, both 23.4 g and 234 g have 3 s.f.).

Rule 2: **Captive zeros,** those bounded on both sides by nonzero integers, are always significant (e.g., 20.05 g has 4 s.f.; 407 g has 3 s.f.).

Rule 3: **Leading zeros,** those *not bounded* on the *left* by nonzero integers, are never significant (e.g., 0.04 g has 1 s.f.; 0.00035 has 2 s.f.). Such zeros just set the decimal point; they always disappear if the number is converted to powers-of-10 notation.

Rule 4: **Trailing zeros,** those bounded *only on the left* by nonzero integers, are probably not significant *unless a decimal point is shown,* in which case they are always significant. For example, 45.0 L has 3 s.f. but 450 L probably has only 2 s.f.; 21.00 kg has 4 s.f. but 2100 kg probably has only 2 s.f.; 55.20 mm has 4 s.f.; 151.10 cal has 5 s.f.; 3.0×10^4 J has 2 s.f. If you wish to show *for sure* that, say, 150 m is to be interpreted as having 3 s.f., change to powers-of-10 notation and show it as 1.50×10^2 m.

Exact Numbers

Rule 5: Exact numbers are those obtained not by measurement but by definition or by counting small numbers of objects. They are assumed to have an unlimited number of significant figures. For example, in the equation $c = 2\pi r$, the "2" is a defined quantity, not a measured one, so it has no effect on the number of significant figures to which the answer can be reported. In counting, say, 15 pennies, you can see that the number is exact because you cannot have 14.9 pennies or 15.13 pennies. The $\frac{9}{5}$ and 1.8 found in temperature conversion equations are exact numbers based on definitions.

PRACTICE PROBLEMS

Using Rules 1–5, determine the number of significant figures in the following measurements. As you proceed, check your answers against those given at the end of the appendix.

(a) 4853 g
(b) 36.200 km
(c) 0.088 s
(d) 30.003 J
(e) 6 dogs
(f) 74.0 m
(g) 340 cm

(h) 40 mi
(i) 8.9 L
(j) 1.30×10^2 cal
(k) 0.002710 ft
(l) 4000 mi
(m) 0.0507 mL
(n) the 2 in $E_k = \dfrac{mv^2}{2}$

Multiplication and Division Involving Significant Figures

Rule 6: In calculations involving only multiplication and/or division of measured quantities, the answer shall have the same number of significant figures as the *fewest* possessed by any measured quantity in the calculation.

EXAMPLE VII.1

A calculator gives 4572.768 cm³ when 130.8 cm is multiplied by 15.2 cm and then by 2.3 cm. However, this answer would be rounded and reported as 4.6×10^3 cm³, because 2.3 cm has the fewest significant figures (two). The reasoning behind this is that the measured 2.3 cm could easily be wrong by 0.1 cm.

Suppose it were really 2.4 cm. What difference would this make in the answer on the calculator? You would get 4771.584 cm³! Comparing this to what the calculator originally gave, you see that the uncertainty in the answer is in the hundreds place, so the answer is properly reported only to the hundreds place—that is, to 2 s.f.

The measured quantity with the fewest significant figures will have the greatest effect on the answer because of percentage effects (a miss of 1 out of 23 is more damaging than a miss of 1 out of 1308, for example).

PRACTICE PROBLEMS

Rewrite the following calculator-given answers so that the proper number of significant figures is shown in each case. When necessary, use exponential notation to avoid ambiguity in the answer. Refer to the rounding rules given in Section 1.7 of the textbook.

(o) $7.7 \dfrac{m}{s^2} \times 3.222 \text{ s} \times 2.4423 \text{ s} = 60.59199762 \text{ m}$

(p) $93.0067 \text{ g} \div 35 \text{ mL} = 2.65733428571 \dfrac{g}{mL}$

(q) $7.43 \dfrac{kg}{L} \times 15 \text{ L} = 111.45 \text{ kg}$

(r) $5766 \dfrac{m}{s} \times 322 \text{ s} = 1{,}856{,}652 \text{ m}$

Addition and Subtraction Involving Significant Figures

Rule 7:   When adding or subtracting measured quantities, report the same number of *decimal places* as there are in the quantity with the fewest decimal places.

That is, in calculations where measured quantities are added or subtracted, the final answer can have only one "uncertain" figure, so it stops at the place on the right where any of the data first stops.

Carry the calculation one place further, and then round up the answer if justified. (If the last figure is less than 5, drop it; if it is 5 or greater, round the preceding figure up one.)

EXAMPLE VII.2

In each of these examples, the vertical dashed line shows how far over to the right the answer can go. Note that each answer has been calculated to one place further than will be reported. This is so we can see whether rounding up is necessary. The final answers are shown below the double underline.

$$
\begin{array}{ll}
46.6 \ \text{m} & \\
+ \ 5.72 \ \text{m} & \\
\hline
52.32 \ \text{m} & \leftarrow \text{initial answer} \rightarrow \\
52.3 \ \text{m} & \leftarrow \text{final answer} \rightarrow
\end{array}
\qquad
\begin{array}{ll}
38 \quad \ \text{cm} \\
- \ 7.44 \ \text{cm} \\
\hline
30.6 \ \text{cm} \\
31 \quad \ \text{cm}
\end{array}
$$

PRACTICE PROBLEMS

Perform the designated arithmetic operations, being careful to retain the proper number of significant figures.

(s)   0.0012 m
    + 1.334 m
    ——————
       ?

(t)   879 g
    − 79.9 g
    ——————
       ?

(u)   6.788 cm
    + 5.6 cm
    ——————
       ?

(v)   67.4 kg
    − 0.06 kg
    ——————
       ?

(w) 54.09×10^4 g
    + 3 $\times 10^4$ g
    ——————
       ?

ANSWERS

(a) 4 (Rule 1)
(b) 5 (Rule 4)
(c) 2 (Rule 3)
(d) 5 (Rule 2)
(e) Unlimited (Rule 5)
(f) 3 (Rule 4)
(g) 2 or 3 (Rule 4)
(h) 1 or 2 (Rule 4)
(i) 2 (Rule 1)
(j) 3 (Rule 4)
(k) 4 (Rules 3, 4)

(m) 3 (Rules 2, 3)
(n) Unlimited (Rule 5)
(o) 61 m
(p) 2.7 $\frac{\text{g}}{\text{mL}}$
(q) 1.1×10^2 kg
(r) 1.86×10^6 m
(s) 1.335 m
(t) 799 g
(u) 12.4 cm
(v) 67.3 kg
(w) 57×10^4 g

Appendix VIII Psychrometric Tables (pressure: 30 in. of Hg)

Table VIII.1 Relative Humidity (%) and Maximum Moisture Capacity

Air Temp. (°F) (Dry Bulb)	Max. Moisture Capacity (gr/ft³)	Degrees Depression of Wet-Bulb Thermometer (°F)													
		1	2	3	4	5	6	7	8	9	10	15	20	25	30
25	1.6	87	74	62	49	37	25	13	1						
30	1.9	89	78	67	56	46	36	26	16	6					
35	2.4	91	81	72	63	54	45	36	27	19	10				
40	2.8	92	83	75	68	60	52	45	37	29	22				
45	3.4	93	86	78	71	64	57	51	44	38	31				
50	4.1	93	87	80	74	67	61	55	49	43	38	10			
55	4.8	94	88	82	76	70	65	59	54	49	43	19			
60	5.7	94	89	83	78	73	68	63	58	53	48	26	5		
65	6.8	95	90	85	80	75	70	66	61	56	52	31	12		
70	7.8	95	90	86	81	77	72	68	64	59	55	36	19	3	
75	9.4	96	91	86	82	78	74	70	66	62	58	40	24	9	
80	10.9	96	91	87	83	79	75	72	68	64	61	44	29	15	3
85	12.7	96	92	88	84	80	76	73	69	66	62	46	32	20	8
90	14.8	96	92	89	85	81	78	74	71	68	65	49	36	24	13
95	17.1	96	93	89	85	82	79	75	72	69	66	51	38	27	17
100	19.8	96	93	89	86	83	80	77	73	70	68	54	41	30	21
105	23.4	97	93	90	87	83	80	77	74	71	69	55	43	33	23
110	26.0	97	93	90	87	84	81	78	75	73	70	57	46	36	26

Note: To use the table, determine the air temperature with a dry-bulb thermometer and degrees depressed on the wet-bulb thermometer. Read the relative humidity (in percent) opposite and below these values. Read the maximum capacity directly.

Table VIII.2 Dew Point (°F)

Air Temp. (°F) (Dry Bulb)	Degrees Depression of Wet-Bulb Thermometer (°F)													
	1	2	3	4	5	6	7	8	9	10	15	20	25	30
25	22	19	15	10	5	−3	−15	−51						
30	27	25	21	18	14	8	2	−7	−25					
35	33	30	28	25	21	17	13	7	0	0				
40	38	35	33	30	28	25	21	18	13	15				
45	43	41	38	36	34	31	28	25	22	25				
50	48	46	44	42	40	37	34	32	29	34	0			
55	53	51	50	48	45	43	41	38	36	42	15			
60	58	57	55	53	51	49	47	45	43	49	25	−8		
65	63	62	60	59	57	55	53	51	49	56	34	14		
70	69	67	65	64	62	61	59	57	55	62	42	26	−11	
75	74	72	71	69	68	66	64	63	61	69	49	36	15	
80	79	77	76	74	73	72	70	68	67	74	56	44	28	−7
85	84	82	81	80	78	77	75	74	72	80	62	52	39	19
90	89	87	86	85	83	82	81	79	78	86	69	59	48	32
95	94	93	91	90	89	87	86	85	83	91	74	66	56	43
100	99	98	96	95	94	93	91	90	89	80	80	72	63	52
105	104	103	101	100	99	98	96	95	94	86	86	78	70	61
110	109	108	106	105	104	103	102	100	99	91	91	84	77	68

Note: To use the table, determine the air temperature with a dry-bulb thermometer and degrees depressed on the wet-bulb thermometer. Find the dew point opposite and below these values.

Appendix IX   Seasonal Star Charts

Chart time (local standard time):	
10 P.M.	First of month
9 P.M.	Middle of month
8 P.M.	Last of month

Latitude of chart is 40°N, but it is practical throughout the conterminous United States and other places in northern mid-latitudes. The relative brightness of the stars is indicated by the size of the white dot on the chart. Non-stellar objects like star clusters and nebulae are indicated by the orange * symbol and galaxies are indicated by the orange ovals. The Messier catalog numbers are also given (for example, M42 in Orion).

To use: Hold chart vertically and turn it so the direction you are facing shows at the bottom.

The Evening Sky in March (Adapted from telescope.com)

Latitude of chart is 40°N, but it is practical throughout the conterminous United States and other places in northern mid-latitudes. The relative brightness of the stars is indicated by the size of the white dot on the chart. Non-stellar objects like star clusters and nebulae are indicated by the orange * symbol and galaxies are indicated by the orange ovals. The Messier catalog numbers are also given (for example, M42 in Orion).

To use: Hold chart vertically and turn it so the direction you are facing shows at the bottom.

Chart time (local standard time):	
10 P.M.	First of month
9 P.M.	Middle of month
8 P.M.	Last of month

The Evening Sky in September (Adapted from telescope.com)

Answers to Confidence Exercises

CHAPTER 1 ANSWERS TO CONFIDENCE EXERCISES

1.1 $\rho = m/V$ or $V = m/\rho = 452$ g$/(19.3$ g/cm$^3) = 23.4$ cm^3 (gold), compared to 20.0 cm^3 (osmium).

1.2 10 yd (0.914 m/yd) = 9.1 m ("First and 9.1")

1.3 1 day$\left(\dfrac{24 \text{ h}}{\text{day}}\right)\left(\dfrac{60 \text{ min}}{\text{h}}\right)\left(\dfrac{60 \text{ s}}{\text{min}}\right) = 86{,}400$ s

1.4 8.01 (to three significant figures)

CHAPTER 2 ANSWERS TO CONFIDENCE EXERCISES

2.1 $d = vt = (20$ m/s$)(10$ s$) = 2.0 \times 10^2$ m

2.2 $t = \dfrac{d}{v} = \dfrac{3.56 \times 10^7 m}{3.00 \times 10^8 \, m/s} = 0.119$ s

(Now you know why people on TV seem to pause when they are asked a question on satellite interviews.)

2.3 $\bar{v} = \dfrac{d}{t} = \dfrac{2\pi R_E}{t} = \dfrac{2(3.14)(4.00 \times 10^3 \, m)}{24.0 \, h}$

$= 1.05 \times 10^3 \, mi/h$

Pretty fast—over 1000 mi/h

2.4 $v_f = v_o + at = 0 + (3.6$ m/s$^2)(10$ s$) = 36$ m/s

2.5 Using Eq. 2.3a, with $a = g$, $v = gt = (9.80$ m/s$^2)(1.50$ s$) = 14.7$ m/s

2.6 $a_c = \dfrac{v^2}{r} = \dfrac{(3.0 \times 10^4 \text{ m/s})^2}{1.5 \times 10^{11} \text{ m}}$

$= 6.0 \times 10^{-3}$ m/s^2

CHAPTER 3 ANSWERS TO CONFIDENCE EXERCISES

3.1 $a = \dfrac{F_{net}}{m_1 + m_2} = \dfrac{6.0 \text{ N} - 9.0 \text{ N}}{2.0 \text{ kg}} = -1.5$ m/s^2

(in the direction opposite to that in Example 3.1)

3.2 $w_M = mg_M = m(0.39)g_E = (1.0$ kg$)(0.39)(9.8$ m/s$^2) = 3.8$ N

3.3 $\dfrac{F_2}{F_1} = \dfrac{r_1^2}{r_2^2}$ or $\dfrac{F_2}{F_1} = \left(\dfrac{r_1}{r_2}\right)^2$, and with $r_2 = 3r_1$ or $\dfrac{r_1}{r_2} = \dfrac{1}{3}$, then

$F_2 = \left(\dfrac{1}{3}\right)^2 F_1 = \dfrac{F_1}{9}$ $\left(\text{reduced to } \dfrac{1}{9} \text{ of original value}\right)$

3.4 $\dfrac{m_1}{m_2} = \dfrac{m}{3m} = \dfrac{1}{3}$ and $v_2 = -\left(\dfrac{m_2}{m_1}\right)v_1 = -\left(\dfrac{1}{3}\right)(1.8$ m/s$) = -0.60$ m/s

3.5 Greater speed at closer distance.

$mv_1r_1 = mv_2r_2$, and with r_1 the closer distance,

$v_1 = \left(\dfrac{r_2}{r_1}\right)v_2 = \left(\dfrac{1.52 \times 10^8 \text{ km}}{1.47 \times 10^8 \text{ km}}\right)v_2 = (1.03)v_2$, or 3% greater

CHAPTER 4 ANSWERS TO CONFIDENCE EXERCISES

4.1 No. $\Delta v = v_2 - v_1 = 8.0$ m/s $- 4.0$ m/s $= 4.0$ m/s, and $\frac{1}{2} m(\Delta v)^2 = \frac{1}{2}(1.0$ kg$)(4.0$ m/s$)^2 = 8.0$ J. Wrong! $(v_2 - v_1)^2 \neq (v^2_2 - v^2_1)$. E_k must be computed for each speed.

4.2 Because the stone has fallen halfway, half of the original E_p, or 4.9 J, is converted to E_k. (4.9 J).

4.3 (a) $W = Pt = (7.5$ W$)(1.0$ s$) = 7.5$ J
(b) $F = W/d = 7.5$ J$/0.50$ m $= 15$ N

4.4 30 days = 720 h

$E = Pt = (2.00$ kW$)(720$ h$) = 1440$ kWh

1440 kWh ($0.08/kWh) = $115.20

CHAPTER 5 ANSWERS TO CONFIDENCE EXERCISES

5.1 $T_F = \frac{9}{5} T_C + 32 = \frac{9}{5}(-40) + 32 = -40°$F. The two scales have the same value at $-40°$.

5.2 $H = mc\Delta T = (1.0$ kg$)(1.0$ kcal/kg-°C$)(5°$C $- 20°$C$) = -15$ kcal

5.3 $H_1 = mc\Delta T = (0.20$ kg$)(1.0$ kcal/kg-°C$)(0°$C $- 10°$C$) = -2.0$ kcal
$H_2 = mL_f = (0.20$ kg$)(-80$ kcal/kg$) = -16$ kcal (minus because removed)
$H_T = H_1 + H_2 = (-2.0$ kcal$) + (-16$ kcal$) = -18$ kcal

5.4 $p_2 = 2 p_1$ and $T_2 = (p_2/p_1)T_1 = (2)(293$ K$) = 586$ K

CHAPTER 6 ANSWERS TO CONFIDENCE EXERCISES

6.1 $f = \dfrac{v}{\lambda} = \dfrac{344 \text{ m/s}}{0.500 \text{ m}} = 688$ Hz

6.2 $\lambda = \dfrac{v}{f} = \dfrac{3.00 \times 10^8 \text{ m/s}}{9.00 \times 10^7 \text{ Hz}} = 0.333 \times 10^1$ m $= 3.33$ m

6.3 $\lambda = \dfrac{v}{f} = \dfrac{3.44 \text{ m/s}}{10.0 \text{ Hz}} = 34.4$ m (infrasound wavelength is much longer)

CHAPTER 7 ANSWERS TO CONFIDENCE EXERCISES

7.1 (a) $c_m = \dfrac{c}{n} = \dfrac{c}{1.000} = c$ (or 3.00×10^8 m/s)

(b) $c_m = \dfrac{c}{n} = \dfrac{3.00 \times 10^8 \text{ m/s}}{1.00029} = 2.999 \times 10^8$ m/s

(significant figures ignored)
Essentially the same for normal calculations.

7.2 Draw a ray diagram. Image distance ≈ 10 cm (behind mirror), and the image is virtual, upright, and magnified.

7.3 Draw a ray diagram. Image distance ≈ 24 cm, and the image is virtual, upright, and reduced.

CHAPTER 8 ANSWERS TO CONFIDENCE EXERCISES

8.1 $P = IV = (10$ A$)(120$ V$) = 1200$ W or 1200 J/s

8.2 Greater resistance, less current

$I = \dfrac{V}{R_s} = \dfrac{12 \text{ V}}{6.0 \text{ }\Omega + 6.0 \text{ }\Omega + 3.0 \text{ }\Omega} = 0.80$ A

8.3 $V_2 = \left(\dfrac{N_2}{N_1}\right)V_1 = \left(\dfrac{500}{25}\right)100$ V $= 2000$ V

CHAPTER 9 ANSWERS TO CONFIDENCE EXERCISES

9.1 $E = hf = (6.63 \times 10^{-34}$ J·s$)(7.50 \times 10^{14}$ 1/s$) = 49.7 \times 10^{-20}$ J $= 4.97 \times 10^{-19}$ J

9.2 $r_n = 0.053 \, n^2$ nm, or $r_2 = 0.053 \, (2)^2$ nm $= 0.212$ nm. Similarly for $n = 3$, $r_3 = 4.77$ nm.

9.3 $E_2 = \dfrac{-13.60}{2^2}$ eV $= \dfrac{-13.60}{4}$ eV $= -3.40$ eV Similarly for $n = 3$, $E_3 = -1.51$ eV.

9.4 Table 9.1 shows values of -13.60 eV for the $n = 1$ level and -1.51 eV for $n = 3$. Thus $E_{photon} = E_1 - E_3 = -13.60$ eV $- (-1.51$ eV$) = -12.09$ eV (the negative value indicating that the photon is absorbed).

9.5 $\lambda = \dfrac{h}{mv} = \dfrac{6.63 \times 10^{-34} \text{ J} \cdot \text{s}}{(10^3 \text{ kg})(25 \text{ m/s})} = 2.7 \times 10^{-38}$ m

CHAPTER 10 ANSWERS TO CONFIDENCE EXERCISES

10.1 53 protons, 53 electrons, 78 neutrons (131 − 53 = 78)

10.2 $^{226}_{88}Ra \rightarrow ^{222}_{86}Rn + ^{4}_{2}He$

10.3 $^{232}_{90}Th$ (above $Z = 83$) and $^{40}_{19}K$ (odd-odd nuclide)

10.4 $\frac{58\ y}{29\ y/\text{half-life}}$ 2 half-lives, $N_o \rightarrow \frac{N_o}{2} \rightarrow \frac{N_o}{4}$, o, or one-fourth will remain

10.5 First, see how many half-lives are needed to get to one-sixteenth of the original activity.

$$N_o \rightarrow \frac{N_o}{2} \rightarrow \frac{N_o}{4} \rightarrow \frac{N_o}{8} \rightarrow \frac{N_o}{16}$$

Four half-lives (count the arrows) are needed, so (4 half-lives) (6.0 h/half-life) = 24 h

10.6 Number of half-lives from 16 cpm to 1 cpm is 4, and $4 \times 5730\ y \approx 23{,}000\ y$

10.7 $^{2}_{1}H + ^{27}_{13}Al \rightarrow ^{25}_{12}Mg + ^{4}_{2}He$

10.8 $^{236}_{92}U \rightarrow ^{90}_{38}Sr + ^{144}_{54}Xe + 2^{1}_{0}n$

10.9 (2.0140 u + 3.0161 u) − (4.0026 u + 1.0087 u) = 5.0301 u − 5.0113 u = 0.0188 u of mass defect (0.0188 u) (931 MeV/u) = 17.5 MeV of energy released

CHAPTER 11 ANSWERS TO CONFIDENCE EXERCISES

11.1 (a) $Z = 15$; atomic mass is 31.0 u
(b) representative; nonmetal
(c) Period 3; Group 5A
(d) 15 electrons; 15 protons
(e) valence electrons (same as group number)
(f) shells (same as period number)
(g) 2, 8, 5

11.2 (a) AsF_5 is a binary compound of two nonmetals, so the *-ide* suffix will be used for F, and because fluorine has more than one atom showing in the formula, Greek prefixes are needed for it. The answer is *arsenic pentafluoride.* (b) $CaCl_2$ is a binary compound of a metal and a nonmetal. Name the metal, and then use the *-ide* ending for the nonmetal. The answer is *calcium chloride.*

11.3 Potassium, sodium, and lithium are all in Group 1A. Because sodium chloride is NaCl and potassium chloride is KCl, lithium chloride's formula is LiCl. Hydroxide has a chemical formula OH^-. Since potassium is Group 1A then the formula for potassium hydroxide is KOH.

CHAPTER 12 ANSWERS TO CONFIDENCE EXERCISES

12.1 The answer is 71.0 g of chlorine. Because the compound is composed of only calcium and chlorine, the difference in mass between 111.1 g of compound and 40.1 g of calcium must be the mass of the chlorine.

12.2 The answer is 34.1 u. FM = (2 × 1.0 u for H) + 32.1 u for S = 34.1 u.

12.3 The answer is 52.9% Al and 47.1% O. Al_2 is 54 u, and O_3 is 48 u.

$$\%Al = \frac{54\ u}{102.0\ u} \times 100\% = 52.9\%$$

$$\%O = 100.0\% - 52.9\% = 47.1\%$$

12.4

	X^-	$X2^-$	$X3^-$
M^+	MX	M_2X	M_3X
M^{2+}	MX_2	MX	M_3X_2
M^{3+}	MX_3	M_2X_3	MX

12.5 The Stock system names are copper(I) fluoride and copper(II) fluoride, respectively. Cu is the symbol for the metal copper, and F is the symbol for the nonmetal fluorine. The name of the nonmetal changes to its *-ide* form, so both compounds are copper fluorides. Because the ionic charge of F is always 1−, copper's ionic charge must be 1+ in the first compound and 2+ in the second.

12.6 O (Group 6A) forms two bonds, and H forms one bond. Only the O atoms can connect to two atoms. Thus the structure of H_2O_2 must be

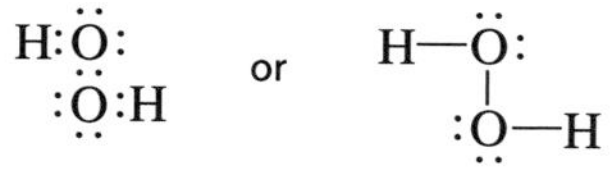

12.7 PCl_3 is *covalent* (two nonmetals). MgF_2 is *ionic* (Group 2A metal and a nonmetal).

12.8 Nitrogen (EN = 3.0) is more electronegative than hydrogen (EN = 2.1), so the arrows denoting the polarity of the bonds would point to the nitrogen atom as shown here.

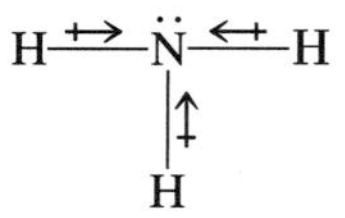

12.9 In order for the three bond dipoles to cancel exactly, the BF_3 molecule must be flat, with bond angles at 120°. Only in this way can the center of positive charge and the center of negative charge be at the same place. (See the SO_3 case in Table 12.6.)

CHAPTER 13 ANSWERS TO CONFIDENCE EXERCISES

13.1 $2\ H_2O \rightarrow 2\ H_2 + O_2$; a decomposition reaction

13.2 $C_3H_8 + 5\ O_2 \rightarrow 3\ CO_2 + 4\ H_2O$

13.3 $3\ HCl + Al(OH)_3 \rightarrow 3\ H_2O + AlCl_3$

13.4 $2\ HNO_3 + Na_2CO_3 \rightarrow CO_2 + H_2O + 2\ NaNO_3$

13.5 $Na_2SO_4(aq) + BaCl_2(aq) \rightarrow 2\ NaCl(aq) + BaSO_4(s)$

13.6 Examination of the activity series (Table 13.5) shows that Al is higher than Cu, so a reaction takes place only in the first beaker. It may be written

$$2\ Al + 3\ CuSO_4(aq) \rightarrow 3\ Cu + Al_2(SO_4)_3(aq)$$

13.7 1 K (39.10 g) + 1 Mn (54.93 g) + 4 O (16.0 g) = 158.03 g/mol. Then, $1.70\ \text{mol KMnO}_4 \times \frac{158.03\ g}{\text{mol}} = 268.65\ g$; $25.0\ g\ NaCl \times \frac{1\ \text{mol}}{58.44\ g} = 0.43$ moles

13.8 $M = \frac{1.60\ \text{mol}}{2.00\ L} = 0.800\ M$

CHAPTER 14 ANSWERS TO CONFIDENCE EXERCISES

14.1 Model (a) is correct. Model (b) is not correct. Model (b) is not correct because fluorine shows 2 bonds (it should only have 1).

14.2 Attach a chlorine atom to the benzene ring. That carbon automatically becomes the 1 position. The carbon next to it must now be the 2 position, so attach a fluorine atom there. The final structure is

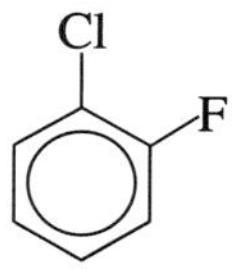

14.3 The compound 2,2,4-trimethylpentane has five carbons in a chain connected by single bonds. Two methyl groups are attached to carbon 2, and one methyl group is on carbon 4. Hydrogen atoms are at the end of all remaining bonds necessary to give each carbon atom four bonds. The final structure is

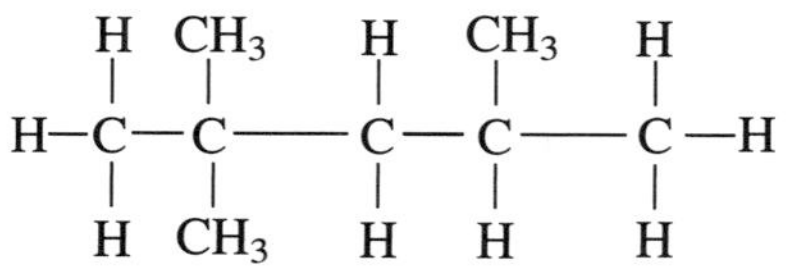

14.4 When the three carbon atoms are connected by single bonds, enough bonds remain to connect eight singly bonded atoms. Seven hydrogen atoms and one fluorine atom fill that requirement, so the only question is how the fluorine atom can be attached to give the two isomers. The result is 1-fluoropropane and 2-fluoropropane, as shown. (Putting the F on one end of the chain rather than on the other would make no difference; that is, "3-fluoropropane" is really 1-fluoropropane.)

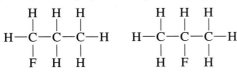

1-Fluoropropane   2-Fluoropropane

14.5 The monomer must be the alkene that corresponds to the basic unit shown in the polymer. That is, the monomer must have the structure

$$\underset{H}{\overset{H}{\diagdown}}C=C\underset{H}{\overset{Cl}{\diagup}}$$

CHAPTER 15 ANSWERS TO CONFIDENCE EXERCISES

15.1 $43°N - 34°N = 9°$, and $\dfrac{9° \times 60 \text{ n mi}}{1°} = 540$ n mi

15.2 8 A.M. on July 4.

15.3 43°S

15.4 Drawing

CHAPTER 16 ANSWER TO CONFIDENCE EXERCISE

16.1 $T_2 = \left(\dfrac{1y^2}{AU^3}\right)(30 \text{ AU})^3$

$T^2 = 27{,}000 \ y^2$

$T = 164 \ y$

CHAPTER 18 ANSWERS TO CONFIDENCE EXERCISES

18.1 Calculate the distance (d) using Eq. 18.1.

$d = \dfrac{1}{p} = \dfrac{1}{0.376} = 2.66$ pc

18.2 If Hubble's constant H is a smaller value than the currently accepted value, then the age of the Universe will be larger. Calculate the age using Eq. 18.2.

Age (in billions of years) $= \dfrac{1 \times 10^{12}}{H} = \dfrac{1 \times 10^{12}}{50} = 2.0 \times 10^{10}$ years

$= 20$ billion years

CHAPTER 19 ANSWER TO CONFIDENCE EXERCISE

19.1 (a) 81%

(b) 10°F

CHAPTER 20 ANSWER TO CONFIDENCE EXERCISE

20.1 $d = \bar{v}t = \frac{1}{3}$ km/s $\times$ 3 s $= 1$ km

$= \frac{1}{5}$ mi/s $\times$ 3 s $= 0.6$ mi

CHAPTER 24 ANSWERS TO CONFIDENCE EXERCISES

24.1 The unconformity is represented by the wavy border between rock 1 and rock 3, where rock 2 has been truncated by erosion. The unconformity must be younger than rock 2 and older than rock 3.

24.2 (a) Fossil B has such a wide time range that it would be of limited use as an index fossil.

(b) Rock that contains fossil C must be from *either* the Silurian or the Devonian, because these are the only two periods in which this fossil has ever been found. Additional evidence would be needed to say exactly to which of the two periods the rock belongs.

(c) *Phacops* is a trilobite. (See Fig. 24.8.)

24.3 To decay from 100% to 12.5% would take three half-lives, as shown in Fig. 24.12 or found by

$$100\% \rightarrow 50.0\% \rightarrow 25.0\% \rightarrow 12.5\%$$

Find the time in years by multiplying the 3.00 half-lives by the half-life of the radioisotope. Thus (3.00 half-lives) (704 $\times$ 10^6 years/half-life) = 2.11 $\times$ 10^9 years, or 2.11 billion years.

24.4 The principle of cross-cutting relationships tells us that A is younger than dike Y, because dike Y was eroded before A was deposited. Hence A must be younger than 350 my.

Answers to Selected Questions

CHAPTER 1 ANSWERS

Answers to Matching Questions

a. 5, b. 17, c. 23, d. 7, e. 22, f. 16, g. 10, h. 2, i. 12, j. 1, k. 9, l. 4, m. 18, n. 6, o. 11, p. 3, q. 20, r. 8, s. 15, t. 14, u. 22, v. 13. w. 19

Answers to Multiple-Choice Questions

1. c, 2. b, 3. c, 4. b, 5. b, 6. b, 7. d, 8. c, 9. d, 10. c, 11. d, 12. d. 13. a

Answers to Fill-in-the-Blank Questions

1. biological 2. experiment 3. scientific method 4. sight 5. limitations 6. greater 7. shorter 8. fundamental 9. time or second 10. 10^6 or 1,000,000 (million) 11. liter 12. mass

Answers to Visual Connection

a. kilogram (kg) b. meter (m) c. second (s) d. pound (lb) e. foot (ft) f. second (s)

CHAPTER 2 ANSWERS

Answers to Matching Questions

a. 9, b. 4, c. 11, d. 17, e. 7, f. 10, g. 3, h. 12, i. 5, j. 15, k. 18, l. 8, m. 2, n. 14, o. 6, p. 1, q. 13, r. 16

Answers to Multiple-Choice Questions

1. d, 2. c, 3. d, 4. c, 5. d, 6. b, 7. c, 8. d, 9. d, 10. d, 11. c, 12. c

Answers to Fill-in-the-Blank Questions

1. position 2. scalar 3. vector 4. distance 5. speed 6. constant or uniform 7. time t^2 8. free-fall 9. m/s^2 10. speed 11. 4 12. acceleration (due to gravity)

Answers to Visual Connection

a. acceleration b. distance c. displacement d. velocity e. m/s f. m/s g. m/s^2

CHAPTER 3 ANSWERS

Answers to Matching Questions

a. 3, b. 18, c. 8, d. 13, e. 17, f. 4, g. 15, h. 6, i. 19, j. 10, k. 1, l. 12, m. 7, n. 9, o. 2, p. 16, q. 11, r. 5, s. 14

Answers to Multiple-Choice Questions

1. d, 2. d, 3. c, 4. b, 5. a, 6. d, 7. d, 8. b, 9. c, 10. a, 11. c, 12. a, 13. b, 14. d, 15. c

Answers to Fill-in-the-Blank Questions

1. capable 2. vector 3. could 4. net or unbalanced 5. mass 6. inversely 7. $kg \cdot m/s^2$ 8. static, kinetic or sliding 9. different 10. everywhere 11. greater 12. more 13. net or unbalanced 14. torque

Answers to Visual Connection

a. action b. inertia c. acceleration

CHAPTER 4 ANSWERS

Answers to Matching Questions

a. 4, b. 10, c. 2, d. 12, e. 7, f. 6, g. 1, h. 3, i. 15, j. 11, k. 9, l. 5, m. 13, n. 8, o. 14

Answers to Multiple-Choice Questions

1. a, 2. d, 3. d, 4. c, 5. b, 6. d, 7. c, 8. a, 9. b, 10. a, 11. a, 12. c, 13. b, 14. c

Answers to Fill-in-the-Blank Questions

1. parallel 2. scalar 3. joule 4. work 5. kinetic 6. square 7. transferring 8. isolated 9. work 10. 0.75 11. energy 12. coal 13. exhausted 14. ethanol (alcohol)

Answers to Visual Connection

a. power b. kinetic energy c. potential energy d. conservation of mechanical energy

CHAPTER 5 ANSWERS

Answers to Matching Questions

a. 10, b. 8, c. 14, d. 23, e. 12, f. 1, g. 15, h. 25, i. 4, j. 11, k. 2, l. 16, m. 7, n. 5, o. 13, p. 20, q. 18, r. 3, s. 24, t. 9, u. 17, v. 22, w. 6, x. 19, y. 21

Answers to Multiple-Choice Questions

1. b, 2. a, 3. a, 4. c, 5. b, 6. c, 7. a, 8. b, 9. d, 10. c, 11. c, 12. b

Answers to Fill-in-the-Blank Questions

1. smaller 2. temperature 3. 1000 4. $J/kg \cdot °C$ 5. seven 6. pressure 7. conduction 8. gas 9. area 10. inversely 11. direction 12. pump

Answers to Visual Connection

a. melting b. freezing c. vaporization d. deposition e. condensation

CHAPTER 6 ANSWERS

Answers to Matching Questions

a. 2, b. 17, c. 10, d. 5, e. 19, f. 8, g. 20, h. 12, i. 1, j. 21, k. 14, l. 6, m. 16, n. 11, o. 3, p. 9, q. 13, r. 15, s. 18, t. 7, u. 4

Answers to Multiple-Choice Questions

1. b, 2. a, 3. a, 4. a, 5. d, 6. d, 7. c, 8. a, 9. d, 10. a, 11. b, 12. a, 13. d

Answers to Fill-in-the-Blank Questions

1. energy 2. perpendicular 3. wavelength 4. frequency 5. light or 3.00×10^8 m/s 6. electromagnetic 7. longitudinal 8. 20 9. intensity 10. 3 11. higher 12. approaching 13. natural or characteristic

Answers to Visual Connection

a. transverse b. longitudinal c. light or electromagnetic d. sound

CHAPTER 7 ANSWERS

Answers to Matching Questions

a. 4, b. 14, c. 12, d. 21, e. 13, f. 6, g. 20, h. 1, i. 11, j. 22, k. 5, l. 16, m. 10, n. 8, o. 17, p. 2, q. 19, r. 7, s. 9, t. 18, u. 3, v. 15

Answers to Multiple-Choice Questions

1. d, 2. d, 3. b, 4. a, 5. b, 6. c, 7. b, 8. c, 9. c, 10. a, 11. a, 12. a

Answers to Fill-in-the-Blank Questions

1. geometrical or ray 2. diffuse 3. vacuum 4. toward 5. total internal 6. converging 7. cannot 8. thinner 9. concave or diverging 10. transverse 11. greater 12. principle of superposition

Answers to Visual Connection

a. reflection b. total internal reflection c. dispersion

CHAPTER 8 ANSWERS

Answers to Matching Questions

a. 3, b. 21, c. 12, d. 4, e. 22, f. 1, g. 25, h. 11, i. 23, j. 2, k. 18, l. 5, m. 19, n. 14, o. 7, p. 17, q. 9, r. 15, s. 10, t. 20, u. 13, v. 24, w. 8, x. 16, y. 6

Answers to Multiple-Choice Questions

1. d, 2. c, 3. a, 4. b, 5. d, 6. a, 7. a, 8. b, 9. d, 10. d, 11. c, 12. b, 13. c, 14. b

Answers to Fill-in-the-Blank Questions

1. positively 2. ampere (amp) 3. Semiconductors 4. charge 5. open 6. ohm 7. I^2R 8. direct or dc 9. smallest 10. Curie 11. geographic 12. secondary

Answers to Visual Connection

a. electrons b. voltage (potential difference) c. resistance d. current e. I^2R or joule heat

CHAPTER 9 ANSWERS

Answers to Matching Questions

a. 10, b. 7, c. 3, d. 14, e. 8, f. 18, g. 1, h. 16, i. 4, j. 2, k. 13, l. 6, m. 9, n. 17, o. 5, p. 15, q. 11

Answers to Multiple-Choice Questions

1. c, 2. a, 3. a, 4. d, 5. c, 6. c, 7. b, 8. d, 9. d, 10. a, 11. b, 12. b, 13. b, 14. d

Answers to Fill-in-the-Blank Questions

1. electron 2. Rutherford 3. Planck's constant 4. photon 5. increases 6. continuous 7. higher 8. water 9. unknown 10. uncertainty 11. wave 12. probability

Answers to Visual Connection

a. Dalton b. Plum pudding c. Rutherford d. Planetary e. Schrödinger

CHAPTER 10 ANSWERS

Answers to Matching Questions

a. 12, b. 23, c. 4, d. 14, e. 1, f. 16, g. 13, h. 9, i. 2, j. 20, k. 19, l. 3, m. 24, n. 6, o. 22, p. 8, q. 11, r. 15, s. 7, t. 18, u. 21, v. 5, w. 10, x. 25, y. 17, z. 26

Answers to Multiple-Choice Questions

1. b, 2. d, 3. c, 4. d, 5. c, 6. d, 7. b, 8. c, 9. a, 10. c, 11. c, 12. a, 13. c, 14. b, 15. d

Answers to Fill-in-the-Blank Questions

1. potassium 2. ^{12}C 3. nucleons 4. isotopes 5. 83 6. beta particle (electron) 7. three 8. one more 9. mass number 10. critical 11. deuteron (or deuterium)12. 82%, 13. quarks

Answers to Visual Connection

a. electrons b. neutrons c. + 1 d. 0 (zero)

CHAPTER 11 ANSWERS

Answers to Matching Questions

a. 12, b. 14 c. 13, d. 21, e. 18, f. 1, g. 7, h. 10, i. 2, j. 3, k. 23, l. 25, m. 24, n. 22, o. 4, p. 8, q. 16, r. 15, s. 6, t. 5, u. 19, v. 11, w. 9, x. 17, y. 20

Answers to Multiple-Choice Questions

1. b, 2. c, 3. a, 4. b, 5. d, 6. c, 7. c, 8. c, 9. b, 10. d, 11. a, 12. c

Answers to Fill-in-the-Blank Questions

1. Organic chemistry 2. solvent 3. more 4. experimentation 5. molecules 6. aluminum 7. O_3 8. Mendeleev 9. four 10. increases 11. nitrate 12. sodium hydroxide

Answers to Visual Connection

a. Li b. K c. Br d. I e. Zn f. V g. Mg h. Ca i. Ne j. Ar

CHAPTER 12 ANSWERS

Answers to Matching Questions

a. 10, b. 18, c. 6, d. 14, e. 4, f. 17, g. 15, h. 8, i. 11, j. 19, k. 1, l. 5, m. 12 n. 3, o. 13, p. 9, q. 2, r. 16, s. 7

Answers to Multiple-Choice Questions

1. b, 2. a, 3. d, 4. c, 5. c, 6. d 7. c, 8. c, 9. d, 10. a, 11. c, 12. b 13. a 14. b

Answers to Fill-in-the-Blank Questions

1. mass 2. CO_2 3. definite proportions 4. chemistry 5. CuO 6. anion 7. two 8. M_2X 9. ionic 10. double 11. decreases 12. nitrogen

Answers to Visual Connection

a. electron sharing b. ionic bonding c. unequal sharing d. nonpolar covalent bonding

CHAPTER 13 ANSWERS

Answers to Matching Questions

a. 13, b. 9, c. 1, d. 14, e. 22, f. 19, g. 4, h. 20, i. 6, j. 23, k. 3, l. 21, m. 8 n. 16, o. 10, p. 25, q. 17, r. 11, s. 5, t. 2, u. 7, v. 12, w. 15, x. 18, y. 24, z. 26

Answers to Multiple-Choice Questions

1. d, 2. c, 3. b, 4. b, 5. d, 6. a, 7. c, 8. a, 9. c, 10. a, 11. a, 12. d

Answers to Fill-in-the-Blank Questions

1. physical, 2. the same, 3. *AB*, 4. enzymes, 5. lower, 6. catalyst, 7. OH^-, 8. neutral, 9. acid–carbonate, 10. redox, 11. will not, 12. two

Answers to Visual Connection

a. combination, b. $AB \rightarrow A + B$, c. single replacement, d. $AB + CD \rightarrow AD + CB$, e. hydrocarbon combustion

CHAPTER 14 ANSWERS

Answers to Matching Questions

a. 2, b. 21, c. 12, d. 9, e. 4, f. 22, g. 6, h. 1, i. 8, j. 19, k. 9, l. 18, m. 11, n. 14, o. 25, p. 5, q. 26, r. 24, s. 3, t. 20, u. 15, v. 16, w. 23, x. 13, y. 17, z. 7

Answers to Multiple-Choice Questions

1. b, 2. c, 3. d, 4. c, 5. b, 6. b, 7. a, 8. b, 9. a, 10. c, 11. c, 12. b

Answers to Fill-in-the-Blank Questions

1. carbon	7. alkyne
2. three	8. alcohols
3. Covalent	9. amide
4. aliphatic	10. condensation
5. Benzene	11. nylon
6. cycloalkane	12. carbohydrate

Answers to Visual Connection

a. aromatic

b. alkenes

c. cycloalkanes

d. $-C \equiv C-$

CHAPTER 15 ANSWERS

Answers to Matching Questions

a. 2, b. 17, c. 13, d. 7, e. 19, f. 6, g. 15, h. 21, i. 1, j. 14, k. 9, l. 4, m. 16, n. 11, o. 5, p. 18, q. 12, r. 20, s. 3, t. 8, u. 22, v. 23, w. 10

Answers to Multiple-Choice Questions

1. d, 2. d, 3. c, 4. b, 5. b, 6. b, 7. d, 8. c, 9. d, 10. b, 11. c, 12. b 13. c, 14. b, 15. b, 16. c, 17. b, 18. c, 19. c, 20. d

Answers to Fill-in-the-Blank Questions

1. rectangular 2. origin 3. parallels 4. great circle 5. latitude 6. longitude 7. solar 8. twenty-four 9. four 10. International Date Line 11. altitude 12. date 13. 30°E 14. spring 15. Cancer 16. seasons 17. zodiac 18. three 19. Common Era 20. top

Answers to Visual Connection

a. latitude b. meridians c. parallels d. north–south e. east–west

CHAPTER 16 ANSWERS

Answers to Matching Questions

a. 12, b. 3, c. 18, d. 14, e. 9, f. 23, g. 1, h. 17, i. 20, j. 7, k. 19, l. 13, m. 2, n. 11, o. 16, p. 5, q. 10, r. 21, s. 4, t. 15, u. 8, v. 24, w. 6, x. 22

Answers to Multiple-Choice Questions

1. d, 2. b, 3. c, 4. b, 5. d, 6. b, 7. b, 8. c, 9. c, 10. a, 11. b, 12. b, 13. c, 14. c, 15. b, 16. c, 17. d, 18. c

Answers to Fill-in-the-Blank Questions

1. Astronomy	10. Mercury
2. geocentric	11. greenhouse effect
3. major axis	12. iron oxide
4. period	13. Uranus
5. superior	14. Ceres
6. prograde	15. Eris
7. opposition	16. condensation
8. albedo	17. deviations (wobbles)
9. Foucault pendulum	

Answers to Visual Connection
a. Mercury   h. Neptune
b. Venus   i. Ceres
c. Earth   j. Pluto
d. Mars   k. Eris
e. Jupiter   l. Haumea
f. Saturn   m. Makemake
g. Uranus

CHAPTER 17 ANSWERS

Answers to Matching Questions
a. 3, b. 11, c. 21, d. 9, e. 15, f. 14, g. 7, h. 20, i. 4, j. 1, k. 13, l. 18, m. 14, n. 6, o. 8, p. 25, q. 10, r. 5, s. 17, t. 22, u. 24, v. 23, w. 19, x. 2, y. 12

Answers to Multiple-Choice Questions
1. b, 2. c, 3. b, 4. b, 5. a, 6. c, 7. a, 8. b, 9. c, 10. d, 11. a, 12. c, 13. b, 14. c, 15. a, 16. d, 17. b, 18. d, 19. d, 20. c, 21. c, 22. a, 23. a, 24. b, 25. c, 26. e

Answers to Fill-in-the-Blank Questions
1. maria, 2. rays, 3. later, 4. 6 A.M., 5. full, 6. waning, 7. lunar, 8. umbra, 9. spring, 10. twelve, 11. zero, 12. two, 13. Ganymede, 14. Io, 15. Titan, 16. Charon, Nix, and Hydra, 17. Water, 18. asteroids, 19. meteorite, 20. coma

Answers to Visual Connection
a. Moon, b. Phobos (Deimos), c. Deimos (Phobos), d. Ganymede, e. Io, f. Titan, g. Miranda, h. Triton, i. Charon, j. Dysnomia, k. Hi'iaka, j. Namaka

CHAPTER 18 ANSWERS

Answers to Matching Questions
a. 9, b. 19, c. 26, d. 10, e. 14, f. 24, g. 7, h. 2, i. 23, j. 17, k. 12, l. 5, m. 8, n. 22, o. 27, p. 4, q. 16, r. 20, s. 11, t. 15, u. 6, v. 25, w. 18, x. 13, y. 1, z. 21, aa. 3

Answers to Multiple-Choice Questions
1. a, 2. a, 3. d, 4. b, 5. d, 6. a, 7. a, 8. b, 9. d, 10. a, 11. b, 12. d, 13. b, 14. b, 15. a, 16. d, 17. b, 18. a, 19. c, 20. a

Answers to Fill-in-the-Blank Questions
1. parallax 2. celestial equator 3. right ascension 4. 23.5° 5. sunspot 6. helium 7. Flares 8. constellations 9. red 10. spectrum 11. brightness 12. main sequence 13. red giant 14. contract 15. Nebula 16. nova 17. pulsar 18. supernova 19. supermassive 20. 13.7

Answers to Visual Connection
a. Nuclear reactions b. Photosphere c. Chromosphere d. Solar wind

CHAPTER 19 ANSWERS

Answers to Matching Questions
a. 5, b. 12, c. 26, d. 8, e. 19, f. 23, g. 4, h. 27, i. 10, j. 28, k. 1, l. 30, m. 15, n. 6, o. 25, p. 16, q. 9, r. 17, s. 21, t. 2, u. 20, v. 11, w. 24, x. 13, y. 29, z. 3, aa. 22, bb. 18, cc. 14, dd. 7

Answers to Multiple-Choice Questions
1. a, 2. c, 3. a, 4. b, 5. c, 6. d, 7. b, 8. a, 9. b, 10. d, 11. d, 12. b

Answers to Fill-in-the-Blank Questions
1. Meteorology 2. stratosphere 3. stratosphere 4. carbon dioxide, CO_2 5. 14.7 6. 76 cm, 30 in. 7. maximum 8. opposite 9. isobar 10. land 11. counterclockwise 12. low

Answers to Visual Connection
a. cumulonimbus b. cumulus c. cirrus d. cirrocumulus e. cirrostratus f. altostratus g. altocumulus h. stratus i. stratocumulus j. nimbostratus k. advection l. radiation

CHAPTER 20 ANSWERS

Answers to Matching Questions
a. 5, b. 18, c. 8, d. 15, e. 2, f. 13, g. 21, h. 3, i. 16, j. 7, k. 25, l. 11, m.19, n. 1, o. 14, p. 22, q. 6, r. 24, s. 12, t. 20, u. 4, v. 17, w. 10, x. 23, y. 9

Answers to Multiple-Choice Questions
1. b, 2. a, 3. d, 4. d, 5. a, 6. c, 7. b, 8. d, 9. a, 10. b, 11. d, 12. c

Answers to Fill-in-the-Blank Questions
1. coalescence 2. rain 3. temperature 4. cP 5. warm 6. occlusion 7. heat 8. warning 9. 24 10. subsidence 11. photochemical 12. CFCs

Answers to Visual Connection
a. rain b. snow c. sleet d. hail e. dew f. frost

CHAPTER 21 ANSWERS

Answers to Matching Questions
a. 20, b. 4, c. 21, d. 23, e. 24, f. 5, g. 14, h. 27, i. 10, j. 28, k. 7, l. 9, m. 8, n. 15, o. 18, p. 26, q. 22, r. 13, s. 17, t. 2, u. 3, v. 6, w. 16, x. 12, y. 11, z. 25, aa. 19, bb. 1

Answers to Multiple-Choice Questions
1. b, 2. d, 3. c, 4. d, 5. a, 6. d, 7. c, 8. c, 9. c, 10. c, 11. a, 12. b

Answers to Fill-in-the-Blank Questions
1. Geology 2. surface waves 3. asthenosphere 4. outer core 5. deep sea trenches 6. mid-ocean 7. asthenosphere 8. transform 9. Ring of Fire 10. earthquakes 11. volcanic 12. reverse

Answers to Visual Connection
a. Divergent Boundaries b. Red Sea c. Cascade Range d. Continental Collision e. Himalayas f. Transform Boundaries

CHAPTER 22 ANSWERS

Answers to Matching Questions
a. 25, b. 4, c. 13, d. 11, e. 8, f. 3, g. 23, h. 5, i. 14, j. 6, k. 10, l. 12, m. 7, n. 20, o. 22, p. 1, q. 19, r. 18, s. 2, t. 9, u. 17, v. 21, w. 16, x. 15, y. 24

Answers to Multiple-Choice Questions
1. d, 2. b, 3. a, 4. c, 5. b, 6. b, 7. b, 8. a, 9. b, 10. b 11. c, 12. a, 13. d, 14. c, 15. d, 16. d, 17. a, 18. a, 19. d, 20. b

Answers to Fill-in-the-Blank Questions
1. silicon 2. feldspars 3. luster 4. gem 5. igneous 6. uniformitarianism 7. sedimentary 8. basalt 9. stalactites 10. viscosity 11. cinder cone 12. concordant 13. hot spot 14. pluton 15. discordant 16. coal 17. bedding 18. slate 19. Foliation 20. hydrothermal

Answers to Visual Connection
a. Sedimentary b. sandstone, limestone, etc. c. Melting, cooling, solidification d. Metamorphic e. marble, gneiss, etc.

CHAPTER 23 ANSWERS

Answers to Matching Questions
a. 5, b. 12, c. 22, d. 2, e. 19, f. 15, g. 7, h. 13, i. 11, j. 3, k. 21, l. 4, m. 24, n. 8, o. 1, p. 10, q. 16, r. 23, s. 9, t. 18, u. 6, v. 20, w. 14, x. 17

Answers to Multiple-Choice Questions
1. a, 2. c, 3. c, 4. c, 5. d, 6. b, 7. b, 8. c, 9. d, 10. b, 11. b, 12. a

Answers to Fill-in-the-Blank Questions
1. mechanical 2. permafrost 3. moisture 4. stream 5. bed 6. continental glaciers 7. mudflows 8. moraines 9. Subsidence 10. aquifer 11. tides 12. continental slopes

Answers to Visual Connection
a. precipitation b. evaporation c. Land d. streams e. glacier flow

CHAPTER 24 ANSWERS

Answers to Matching Questions
a. 12, b. 25, c. 14, d. 11, e. 3, f. 18, g. 20, h. 1, i. 26, j. 19, k. 6, l. 23, m. 7, n. 5, o. 4, p. 10, q. 13, r. 16, s. 17, t. 2, u. 9, v. 8, w. 21, x. 22, y. 24, z.15

Answers to Multiple-Choice Questions
1. d, 2. c, 3. c, 4. b, 5. a, 6. b, 7. d, 8. b, 9. a, 10. d, 11. d, 12. a 13. c, 14. c, 15. d, 16. d, 17. c, 18. a, 19. b, 20. c

Answers to Fill-in-the-Blank Questions
1. paleontology 2. replacement fossil 3. algae 4. Cenozoic 5. unconformity 6. younger 7. relative time 8. Correlation 9. eons 10. trilobite 11. superposition 12. carbon-14 13. larger 14. primordial 15. 4.56 16. younger 17. Rhodinia 18. K-T event 19. explosion 20. epoch

Answers to Visual Connection
a. one-celled (or multicelled) organisms b. Silurian c. amphibians d. Triassic e. primates

Glossary

The number in parentheses following each definition refers to the section in which the term is discussed.

aberration of starlight The apparent displacement in the direction of light coming from a star because of the orbital motion of the Earth. (16.3)

absolute (numerical) geologic time The actual age of geologic events, established on the basis of the radioactive decay of certain atomic nuclei. (24.3)

absolute magnitude The brightness a star would have if it were placed 10 pc (32.6 ly) from the Earth. (18.3)

acceleration The time rate of change of velocity; $a = \Delta v/\Delta t$ (2.3)

acceleration due to gravity Usually given as the symbol g; equal to 9.80 m/s^2, or 32 ft/s^2. (2.3, 3.5)

acid A substance that gives hydrogen ions, H$^+$ (or hydronium ions, H$_3$O$^+$) in water (Arrhenius definition). (13.3)

acid-base reaction The H$^+$ of an acid unites with the OH$^-$ of a base to form water, while the cation of the base combines with the anion of the acid to form a salt. (13.3)

acid-carbonate reaction An acid and a carbonate (or hydrogen carbonate) react to give carbon dioxide, water, and a salt. (13.3)

acid rain Rain that has a relatively low pH (i.e., relatively high acidity) because of air pollution. (20.4)

activation energy The energy necessary to start a chemical reaction; a measure of the minimum kinetic energy that colliding molecules must possess in order to react chemically. (13.2)

activity series A list of elements in order of relative tendency to lose electrons to ions of another metal or to hydrogen ions.. (13.4)

addition polymers Polymers formed when molecules of an alkene monomer add to one another. (14.5)

air current Vertical air motions. (19.4)

air mass A large body of air with physical characteristics that distinguish it from the surrounding air. (20.2)

albedo The fraction of incident sunlight reflected by a celestial object. (16.3)

alcohols Organic compounds containing a hydroxyl group, —OH, attached to an alkyl group. The general formula for an alcohol is R—OH, or just ROH, and the IUPAC names of alcohols end in -ol. (14.4)

aliphatic hydrocarbon Hydrocarbons having no benzene rings. (14.3)

alkali metals The elements in Group 1A of the periodic table, except for hydrogen (Li, Na, K, Rb, Cs, Fr). (11.6)

alkaline earth metals The elements in Group 2A of the periodic table (Be, Mg, Ca, Sr, Ba, Ra). (11.6)

alkanes Hydrocarbons that contain only single bonds; general formula, C_nH_{2n+2}. (14.3)

alkenes Hydrocarbons that have a double bond between two carbon atoms; general formula, C_nH_{2n}. (14.3)

alkyl group A substituent that contains one less hydrogen atom than the corresponding alkane; given the general symbol R. (14.3)

alkyl halide An alkane derivative in which one or more of the hydrogen atoms have been replaced by halogen atoms; general formula, RX, where X is a halogen atom and R is an alkyl group. (14.4)

alkynes Hydrocarbons that have a triple bond between two carbon atoms; general formula, C_nH_{2n-2}. (14.3)

allotropes Two or more forms of the same element that have different bonding structures in the same physical phase. (11.3)

alpha decay The disintegration of a nucleus into a nucleus of another element with the emission of an *alpha particle*, which is a helium nucleus. (10.3)

alternating current (ac) Electric current produced by a constantly changing (alternating) voltage from positive (+) to negative (−) to positive (+), and so on. (8.2)

alternative energy sources Energy sources that are not based on the burning of fossil fuels and nuclear processes. (4.6)

altitude The angle measured from the horizon to a celestial object. (15.4)

amber Fossilized tree resin. (24.1)

amides Nitrogen-containing organic compounds that have the general formula RCONHR′. (14.4)

amine Organic compound that contains nitrogen and is basic (alkaline); general formula, R—NH$_2$, or just RNH$_2$. (14.4)

amino acids Organic compounds that contain both an amino group and a carboxyl group. (14.6)

ampere (A) The unit of electric current. (8.2)

amplitude The maximum displacement of any part of the wave (or wave particle) from its equilibrium position. (6.2)

anemometer An instrument used to measure wind speed. (19.3)

angular momentum mvr for a mass m going at a speed v in a circle of radius r. (3.7)

anions Negative ions; so called because they move toward the anode (the positive electrode) of an electrochemical cell. (12.4)

annular eclipse A solar eclipse in which the Moon blocks out all of the Sun except for a ring around the Sun's outer edge. (17.2)

ante meridiem (A.M.) Pertaining to time from 12 midnight to 12 noon. (15.3)

aphelion The point when Earth (or another orbiting object) is farthest from the Sun. (15.5)

apparent magnitude The brightness of a star (or other celestial object) as observed from the Earth. (18.3)

aquifer A body of permeable rock that both stores and transports groundwater. (23.3)

Archimedes' principle An object immersed wholly or partially in a fluid experiences a buoyant force equal in magnitude to the weight of the *volume of fluid* that is displaced. (3.6)

aromatic hydrocarbon A hydrocarbon that possesses one or more benzene rings. (14.2)

asteroids Large and small chunks of matter that orbit the Sun (usually between Mars and Jupiter), sometimes called minor planets. (17.6)

asthenosphere The part of the mantle that lies beneath the lithosphere and is essentially solid rock, but is so close to its melting temperature that it contains pockets of thick, molten rock and is relatively plastic. (21.1)

astronomical unit (AU) The average distance between the Earth and the Sun, which is 1.5×10^8 km (93 million miles). (16.1)

astronomy The scientific study of the universe, which is the totality of all matter, energy, space, and time. (16.Intro)

atmospheric science The investigation of every aspect of the atmosphere. (19.Intro)

atom The smallest particle of an element that can enter into a chemical combination. (9.1)

atomic mass The weighed average mass of an atom of the element in naturally occurring samples; given under its symbol in the period table (in *atomic mass units*, symbolized u). (10.2)

atomic number Symbolized by the letter Z, it is equal to the number of protons in the nucleus of each atom of that element. (10.2)

autumnal equinox The point where the Sun crosses the celestial equator from north to south, around September 22. The beginning of fall. (15.5)

average acceleration The change in velocity divided by the time for the change to occur. (2.3)

average speed The total distance traveled divided by the time spent in traveling the total distance; $\bar{v} = \Delta d/\Delta t$. (2.2)

average velocity The displacement divided by the total travel time. (2.2)

Avogadro's number 6.02×10^{23}, symbolized N_A; the number of entities in a mole. (13.5)

barometer A device used to measure atmospheric pressure. (19.3)

base A substance that produces hydroxide ions, OH^-, in water (Arrhenius definition). (13.3)

bedding The layering that develops at the time sediment is deposited; stratification of sedimentary rock formations. (22.5)

Bergeron process The process by which precipitation is formed in clouds. (20.1)

beta decay The disintegration of a nucleus into a nucleus of another element with the emission of a *beta particle*, which is an electron. (10.3)

Big Bang Theory of the beginning of the universe that states that the known universe was smaller, hotter, and denser in the past, and that it began rapidly expanding 13.7 billion years ago. (18.7)

black hole An object so dense that even light cannot escape from its surface because of the object's intense gravitational field. (18.5)

British system The system of units still often employed in the United States, wherein the foot, pound, and second are the standards of length, weight, and time, respectively. The system is sometimes referred to as the *fps* (foot-pound-second) *system*. (1.4)

brown dwarfs Low-mass objects that are larger than a typical planet but do not have enough mass to begin fusion in their cores. Also called "failed stars." (18.4)

Btu (British thermal unit) The amount of heat necessary to raise one *pound* of water one Fahrenheit degree at normal atmospheric pressure. (5.2)

buoyant force The upward force resulting from an object being wholly or partially immersed in a fluid. (3.6)

caldera A roughly circular, steep-walled depression formed primarily from the collapse of the chamber at a volcano's summit. (22.4)

calorie (cal) The amount of heat necessary to raise one *gram* of pure water one Celsius degree at normal atmospheric pressure. (5.2)

Cambrian explosion The great proliferation of life forms that followed the extinction event at the beginning of the Paleozoic era. (24.5)

carbohydrates An important class of compounds that contain multiple hydroxyl groups in their molecular structures. (14.6)

carbon-14 dating A procedure used to establish the age of ancient organic remains by measuring the amount of ^{14}C in an ancient sample. (10.3, 24.3)

carboxylic acids A class of organic compounds that contain a *carboxyl group* and have thegeneral formula, RCOOH. (14.4)

carcinogen A cancer-causing agent. (14.2)

Cartesian coordinate system A two-dimensional coordinate system in which two number lines (x, y) are drawn perpendicular to each other and the *origin* is assigned at the point of intersection. A third dimension may be taken in the z direction. (15.1)

cast Fossil formed when new mineral material fills a mold and hardens. (24.1)

catalyst A substance that increases the rate of reaction but is not itself consumed in the reaction. (13.2)

cations Positive ions; so called because they move toward the cathode (the negative electrode) of an electrochemical cell. (12.4)

celestial prime meridian An imaginary half-circle running from the north celestial pole to the south celestial pole and crossing perpendicular to the celestial equator at the point of the vernal equinox. (18.1)

celestial sphere The apparent sphere of the sky on which all the stars seem to appear. (18.1)

Celsius scale A temperature scale based on an ice point of 0° and a steam point of 100° with 100 equal units or divisions between these points. (5.1)

centi- The metric prefix meaning 1/100, or 0.01. (1.5)

centripetal acceleration The "center-seeking" acceleration necessary for circular motion; $a = v^2/r$. (2.4)

centripetal force The "center-seeking" force that causes an object to travel in a circle. (3.3)

CFCs Chlorofluorocarbons, such as dichlorodifluoromethane (Freon-12), which have been used commonly in air conditioners, refrigerators, and heat pumps, and which helped deplete the ozone layer. (14.4, 20.5)

chain reaction Occurs when each fission event causes at least one or more fission events. (10.5)

charges, law of Like charges repel; unlike charges attract. (8.1)

chemical properties Characteristics that describe the chemical reactivity of a substance—that is, its ability to transform into another substance. (13.1)

chemical reaction A change that alters the chemical composition of a substance and hence forms one or more new substances. (13.1)

chemistry The study of the composition and structure of matter (anything that has mass) and the chemical reactions by which substances are changed into other substances. (11.Intro)

cleavage The tendency of some minerals to break along definite smooth planes. (22.1)

climate The long-term average weather conditions of a region. (20.5)

cloud A buoyant mass of visible droplets of water and ice crystals. (19.5)

coalescence The formation of drops by the collision of droplets, the result being that larger droplets grow at the expense of smaller ones. (20.1)

combination reaction A reaction in which at least two reactants combine to form just one product: $A + B = AB$. (13.1)

combustion reaction A reaction in which a substance reacts with oxygen to burst into flame and form an oxide. (13.2)

comet A relatively small object that is composed of dust and ice and that revolves about the Sun in a highly elliptical orbit. (17.6)

compound A pure substance composed of two or more elements chemically bonded in a definite, fixed ratio by mass. (11.1)

concave mirror A mirror shaped like the inside (concave side) of a small section of a sphere. (7.3)

condensation polymers Large molecules constructed from smaller molecules that have two or more reactive groups. Generally, one molecule attaches to another by an ester or amide linkage, and water is the other product. (14.5)

condensation theory A process of solar system formation in which interstellar dust grains act as condensation nuclei. (16.7)

conduction (thermal) The transfer of heat by molecular collisions. (5.4)

conjunction When two planets are lined up with respect to the Sun. (16.2)

conservation of angular momentum, law of The angular momentum of an object remains constant if there is no external, unbalanced torque acting on it. (3.7)

conservation of linear momentum, law of The total linear momentum of an isolated system remains the same if there is no external, unbalanced force acting on the system. (3.7)

conservation of mass, law of No detectable change in the total mass occurs during a chemical reaction. (12.1)

conservation of mechanical energy, law of In an ideal system, the sum of the kinetic and potential energies is constant: $E_k + E_p = E$ (a constant). (4.3)

conservation of total energy, law of The total energy of an isolated system remains constant. (4.3)

constellation Prominent groups of stars appearing as patterns in each section of the night sky. (15.5, 18.3)

constitutional (structural) isomers Compounds that have the same *molecular* formula but different *structural* formulas. (14.3)

constructive interference A superposition of waves for which the combined waveform has a greater amplitude. (7.6)

contact metamorphism A change in rock brought about primarily by heat, with very little than pressure involved. (22.6)

continental drift The theory that continents move, drifting apart or together. (21.2)

continental shelf A gently sloping, relatively shallow submerged area that borders a continental landmass. (23.4)

continental slope The seaward slope beyond the continental shelf. It extends downward to the ocean basin. (23.4)

convection The transfer of heat by the movement of a substance, or mass, from one place to another. (5.4)

convection cycle The cyclic movement of matter (such as air) as a result of localized heating and convectional heat transfer. (19.4)

convergent boundary A region where moving plates of the lithosphere are driven together, causing one of the plates to be consumed into the mantle as it descends beneath an overriding plate. (21.3)

converging lens A lens that is thicker at the center than at the edges. (7.4)

conversion factor An equivalence statement expressed as a ratio. (1.6)

convex mirror A mirror shaped like the outside (convex side) of a spherical section. (7.3)

Coordinated Universal Time (UTC) The international time standard based on time kept by atomic clocks. (15.3)

Coriolis force A pseudoforce that results because an observer on the Earth is in a rotating frame of reference. (19.4)

correlation The process of matching rock layers in different localities by the use of fossils or other means. (24.2)

cosmic microwave background The microwave radiation that fills all space and is believed to be the redshifted glow from the Big Bang. (18.7)

cosmological redshift The shift toward longer wavelengths caused by the expansion of the universe. (18.7)

cosmology The branch of astronomy that is the study of the structure and evolution of the universe. (18.7)

coulomb (C) The unit of electric charge, equal to one ampere-second (A·s). (8.1)

Coulomb's law The force of attraction or repulsion between two charged bodies is directly proportional to the product of the two charges and inversely proportional to the square of the distance between them. (8.1)

covalent bond The force of attraction caused by a pair of electrons being shared by two atoms. (12.5)

covalent compounds Those in which the atoms share pairs of electrons to form molecules. (12.5)

crater (lunar) A circular depression on the surface of the Moon caused by the impact of a meteoroid. (17.1)

creep A type of slow mass wasting that involves the particle-by-particle movement of weathered debris down a slope. (23.2)

crescent moon The Moon viewed when less than one-half of its observed surface is illuminated. (17.2)

critical mass The minimum amount of fissionable material necessary to sustain a chain reaction. (10.5)

cross-cutting relationships, principle of An igneous rock or fault is younger than the rock layers it has intruded (cut into or across). (24.2)

crust The thin, rocky, outer layer of the Earth. (21.1)

Curie temperature The temperature above which a material ceases to be ferromagnetic. (8.4)

current (electrical) The time rate of flow of electric charge; $I = q/t$. (8.2)

cycloalkanes Members of a series of saturated hydrocarbons that have the general molecular formula C_nH_{2n} and possess rings of carbon atoms, each carbon atom bonded to a total of four carbon or hydrogen atoms. (14.3)

dark energy A mysterious energy that seems to be causing the expansion of the universe to accelerate. (18.7)

dark matter The as-yet-unidentified nonluminous matter in the universe. (18.6)

Daylight Saving Time (DST) Time advanced one hour from standard time, adopted during the spring and summer months to take advantage of longer evening daylight hours and save electricity. (15.3)

decibel (dB) A unit of sound intensity level; one-tenth of a bel (B). (6.4)

declination The angular measure in degrees, minutes, and seconds north or south of the celestial equator. (18.1)

decomposition reaction One in which only one reactant is present and decomposes into two (or more) products: $AB = A + B$. (13.1)

definite proportions, law of Different samples of a pure compound always contain the same elements in the same proportion by mass. (12.2)

delta The accumulation of sediment formed where running water enters a large body of water such as a lake or ocean. (23.2)

density A measure of the compactness of the matter or mass of a substance using a ratio of mass to volume; $\rho = m/V$. (1.6)

derived units Multiples or combinations of units. (1.6)

desert An area on the Earth's surface that has a severe lack of precipitation. (23.2)

destructive interference A superposition of waves for which the combined waveform has a smaller amplitude. (7.6)

dew point The temperature to which a sample of air must be cooled to become saturated—that is, has a relative humidity of 100%. (19.3)

diffraction The bending of waves as they go through relatively small slits or pass by the corners of objects. (7.6)

direct current (dc) Electric current in which the electrons flow directionally from the negative ($-$) terminal toward the positive ($+$) terminal. (8.2)

dispersion Different frequencies of light refracted at slightly different angles, giving rise to a spectrum. (7.2)

displacement The straight-line distance between the initial and final positions, with direction toward the final position, and is a vector quantity. (2.2)

distance The actual length of the path that is traveled. (2.2)

divergent boundary A region where one plate of the lithosphere is moving away from one another and new oceanic rock is formed. (21.3)

diverging lens A lens that is thinner at the center than at the edges. (7.4)

Doppler effect The apparent change in frequency of a moving sound source. (6.5)

Doppler radar Radar that uses the Doppler effect on water droplets in clouds to measure the wind speed and direction. (19.3)

double-replacement reactions Reactions in which the positive and negative components of the two compounds "change partners." The general format is $AB + CD \rightarrow AD + CB$. (13.3)

dual nature of light Light must be described sometimes as a wave and sometimes as a particle. (9.2)

dwarf planet A new class of planets including Pluto, Ceres, Eris Haumea, and Makemake. (16.6)

earthquake The tremendous release of energy accompanying the rupture or repositioning of underground rock and is manifested by the vibrating and sometimes violent movement of the Earth's surface. (21.5)

eclipse The blocking of the light of one celestial body by another. (17.2)

ecliptic The apparent path the Sun traces annually along the celestial sphere. (18.1)

electric charge A fundamental property of matter that can be either positive or negative and gives rise to electric forces. (8.1)

electric field A force field of imaginary lines surrounding a charge representing the electrical effect a positive unit charge would experience. (8.1)

electric potential energy The potential energy that results from work done in separating electric charges. (8.2)

electric power The expenditure of electrical work divided by time; $P = W/t = IV$. (8.2)

electromagnetic wave A transverse wave consisting of oscillating electric and magnetic fields. (8.5)

electromagnetism The interaction of electrical and magnetic effects. (8.5)

electron configuration The order of electrons in the energy levels of their atoms. (11.4)

electronegativity A measure of the ability of an atom in a molecule to draw bonding electrons to itself. (12.5)

electrons Negatively charged subatomic particles. (8.1, 9.1, 10.2)

element A substance in which all the atoms have the same number of protons (the same atomic number, Z). (10.2, 11.1)

elliptical orbits, law of (Kepler's first law) All planets move in elliptical orbits around the Sun, with the Sun at one focus of the ellipse. (16.1)

endothermic reaction A reaction that causes a net absorption of energy from the surroundings to occur. (13.2)

energy The ability to do work. (4.2)

entropy A mathematical quantity; thermodynamically speaking, its change tells whether or not a process can take place naturally. (5.7)

eon The largest unit of geologic time. Eons are divided into eras. (24.2)

epicenter The point on the surface of the Earth directly above the focus of an earthquake. (21.5)

epoch An interval of geologic time that is a subdivision of a period. (24.5)

equal areas, law of (Kepler's second law) An imaginary line (radial vector) joining a planet to the Sun sweeps out equal areas in equal periods of time. (16.1)

equilibrium In chemistry, a dynamic process in which the reactants are combining to form the products at the same rate at which the products are combining to form the reactants. (13.3)

equinox The two points where the ecliptic and the celestial equator intersect (18.1)

era An interval of geologic time that is a subdivision of an eon and is made up of periods and epochs. (24.2)

erosion The downslope movement of soil and rock fragments under the influence of gravity (mass wasting) or by agents such as streams, glaciers, wind, and waves. (23.2)

ester An organic compound that has the general formula RCOOR′ where R and R′ are any alkyl groups. (14.4)

excess reactant A starting material that is only partially used up in a chemical reaction. (12.2)

excited states The energy levels above the ground state in an atom. *See ground state.* (9.3)

exoplanets (extrasolar planets) Planets orbiting stars other than our own Sun. (16.8)

exothermic reaction A reaction that has a net release of energy to the surroundings. (13.2)

experiment The testing of a hypothesis under controlled conditions to see whether the test results confirm the hypothetical assumptions, can be duplicated, and are consistent. (1.2)

Fahrenheit scale A temperature scale with and ice point of 32° and a steam point of 212° with 180 equal units or divisions between these points. (5.1)

fats Esters composed of the trialcohol named glycerol, $C_3H_5(OH)_3$, and long-chain carboxylic acids known as fatty acids. (14.4)

fault A break or fracture in the surface of a planet or moon along which movement has occurred. (17.1, 21.5)

    normal The result of expansive forces that cause the fault's overlying side to move downward relative to the side beneath. (21.6)

    reverse The result of compressional stress forces that cause the overlying side of the fault to move upward relative to the side beneath. (21.6)

    transform (or strike-slip) The result of stresses that are parallel to the fault boundary such that the fault slip is horizontal. (21.6)

fault-block mountains Mountains that are built by normal faulting in which giant pieces of the Earth's crust were faulted and uplifted at the same time. (21.6)

ferromagnetic Characteristic of substances such as iron, nickel, and cobalt that exhibit the ability to acquire high magnetization. (8.4)

first law of thermodynamics The heat added to a system must go into increasing the internal energy of the system, or any work done by the system, or both. The law, which is based on the conservation of energy, also states that heat energy removed from a system must produce a decrease in the internal energy of the system, or any work done on the system, or both. (5.7)

first-quarter phase The Moon when it is 90° east of the Sun and appears as a quarter moon on an observer's meridian at 6 P.M. (17.2)

fission The process in which a large nucleus "splits" (fissions) into two intermediate-size nuclei, with the emission of neutrons and the conversion of mass into energy. (10.5)

floodplain The low adjacent land to a river or stream that can become inundated when the river or stream overflows. (23.2)

focal length The distance from the vertex of a mirror or lens to the focal point. (7.3)

focus (earthquake) The point within the Earth at which the initial energy release or slippage of an earthquake occurs. (21.5)

fold A folded rock layer that can form an arch (anticline) or a trough (syncline) as a result of compressional forces. (21.6)

fold mountains Mountains characterized by folded rock strata, with external evidence of faulting and internal evidence of high temperature and pressure changes. Fold mountains are believed to be formed at convergent plate boundaries. (21.6)

foliation The parallel alignment of minerals characteristic of some metamorphic rocks that results from directional pressures during transformation. (22.6)

foot-pound (ft·lb) The unit of work (and energy) in the British system. (4.1)

force A vector quantity capable of producing motion or a change in motion, that is, a change in velocity or an acceleration. (3.1)

formula mass The sum of the atomic masses given in the formula of a substance. (12.2)

formula unit The smallest combination of ions that gives the formula of the compound. (12.4)

fossil Any remnant or indication of prehistoric life preserved in rock. (24.1)

Foucault pendulum A pendulum that is used to demonstrate the rotation of the Earth. (16.3)

free fall A state of motion solely under the influence of gravity. (2.3)

frequency The number of oscillations or cycles of a wave that occur during a given period of time, usually one second. (6.2)

friction The ever-present resistance to relative motion that occurs whenever two materials are in contact with each other, whether they are solids, liquids, or gases. (3.3)

front The boundary between two air masses. (20.2)

full moon The phase of the Moon that occurs when the Moon is 180° east of the Sun and appears on the observer's meridian at 12 midnight local solar time. (17.2)

functional group Any atom, group of atoms, or organization of bonds that determines specific properties of a molecule. (14.4)

fusion The process in which smaller nuclei combine to form larger ones with the release of energy. (10.6)

G The universal gravitational constant; $G = 6.67 \times 10^{-11}$ N·m²/kg². (3.5)

galaxy An extremely large collection of stars bound together by mutual gravitational attraction. Galaxies have a spiral, elliptical, or irregular structure. (18.6)

gamma decay An event in which a nucleus emits a *gamma ray* and becomes a less energetic form of the same nucleus. (10.3)

gas Matter that is made up of rapidly moving molecules and assumes the shape and size of its container; has no definite volume or shape. (5.5)

generator A device that converts mechanical work or energy into electrical energy. (8.5)

geocentric model The old false theory of the solar system, which placed the Earth at its center. (16.1)

geologic time The time span that covers the long history of the Earth. (24.Intro)

geologic time scale A relative time scale based on the fossil contents of rock strata and the principles of superposition and cross-cutting relationships. (24.5)

geology The study of the planet Earth: its composition, structure, and history. Also, the study of the chemical and physical properties of other solar system bodies. (21.Intro)

gibbous moon The Moon viewed when more than one-half of its illuminated surface is observed from the Earth. (17.2)

glacial drift General term for rock material that is transported and deposited by ice. (23.2)

glacier A large, thick mass of "permanent" ice that consists of recrystallized snow and that flows on a land surface under the influence of gravity. (23.2)

globular cluster A spherical collection of hundreds of thousands of gravitationally bound stars, usually found in the outlying regions of a galaxy. (18.6)

gravitational potential energy The potential energy resulting from an object's position in a gravitational field—in other words, the stored energy that comes from doing work against gravity. (4.2)

great circle Any circle on the surface of a sphere whose center is at the center of the sphere. It applies especially to imaginary circles on the Earth's surface that pass through both the North Pole and the South Pole. (15.2)

Great Dying The most devastating extinction known to geologists; it marked the end of the Paleozoic era and the beginning of the Mesozoic era. (24.5)

greenhouse effect The heat-retaining process of atmospheric gases, such as water vapor (H_2O) and carbon dioxide (CO_2), that results from the selective absorption of terrestrial radiation. (19.2)

Greenwich Mean Time (GMT) The time at the central prime meridian, or Greenwich meridian (0° longitude). (15.3)

Greenwich (prime) meridian The reference meridian of longitude, which passes through the old Royal Greenwich Observatory near London. (15.2)

Gregorian calendar The reformed Julian calendar—our present-day calendar. (15.5)

ground state The lowest energy level of an atom. (9.3)

groundwater Water that soaks into the soil down into the subsurface. (23.3)

groups The vertical columns in the periodic table. (11.4)

half-life The time it takes for half the nuclei in a given radioactive sample to decay. (10.3)

halogens The elements in Group 7A of the periodic table (F_2, Cl_2, Br_2, I_2). (11.6)

harmonic law (Kepler's third law) The square of the sidereal period of a planet is proportional to the cube of its semimajor axis (one-half the major axis). (16.1)

heat The net energy transferred from one object to another because of a temperature difference; energy in transit because of a temperature difference. (5.2)

heat engine A device that uses converts heat into work. (5.7)

heat pump A device that uses work input to transfer heat from a low-temperature reservoir to a high-temperature reservoir. (5.7)

Heisenberg's uncertainty principle It is impossible to know a particle's exact position and velocity simultaneously. (9.5)

heliocentric model The model of the solar system that places the Sun at its center. (16.1)

hertz (Hz) One cycle per second or 1/s. The SI unit of frequency. (6.2)

horsepower (hp) A unit of power equal to 550 f·lb/s. (746 W); commonly used to rate the power of motors and engines. (4.4)

H-R diagram A plot of the absolute magnitude of stars versus the temperature of their photospheres. (18.3)

Hubble's law The greater the recessional velocity of a galaxy, the farther away the galaxy, $V_r = H \times d$. (18.7)

Hubble's constant The proportionality constant for Hubble's law, currently believed to be 73 km/s/Mpc; also the average rate of expansion of the universe. (18.7)

humidity A measure of the moisture, or water vapor, in the air. (19.3)

hurricane A tropical storm with winds of 119km/h (74 mi/h or 64 knots) or greater. (20.3)

hurricane warning An alert that hurricane conditions are expected within 24 hours. (20.3)

hurricane watch An advisory alert that hurricane conditions are a threat within 24 to 36 hours. (20.3)

hydrocarbons The simplest organic compounds which contain only carbon and hydrogen. (14.2)

hydrogen bond A special kind of dipole–dipole interaction that can occur whenever a compound contains hydrogen atoms covalently bonded to small, highly electronegative atoms (only O, F, and N meet these criteria). (12.6)

hydrothermal metamorphism The chemical alteration of preexisting rocks by chemically reactive, hot-water solutions, which dissolve some ions from the original minerals and replace them with other ions, thus changing the mineral composition of the rock. (22.6)

hydrologic cycle The cyclic movement of the Earth's water supply from the oceans to the mountains and back again to the oceans. (23.3)

hypothesis A possible explanation for observations; tentative answer or an educated guess. (1.2)

ice storm A storm with accumulations of ice as a result of the surface temperature being below the freezing point. (20.3)

ideal gas law Relates the pressure, volume, and absolute temperature of a gas; $p_1V_1/T_1 = p_2V_2/T_2$. (5.6)

igneous rock Rock formed by the solidification of magma. (22.2)

index fossil A fossil that is widespread, numerous, easily identified, and typical of a particular limited time segment of the Earth's history. (24.2)

index of refraction The ratio of the speed of light in a vacuum to the speed of light in a medium. (7.2)

inertia The natural tendency of an object to remain in a state of rest or in uniform motion in a straight line. (3.2)

inner core The innermost region of the Earth, which is solid and probably composed of about 85% iron and 15% nickel. (21.1)

inner transition elements The *lanthanides* and *actinides,* the two rows at the bottom of the periodic table, make up the inner transition elements. (11.4)

insolation The solar radiation received by the Earth and its atmosphere; incoming solar radiation. (19.2)

instantaneous speed An object's speed at a particular instant of time. (2.2)

instantaneous velocity The velocity at a particular instant of time. (2.2)

intensity (of sound wave) The rate of sound energy transfer through a given area, with units of watts per square meter (W/m^2). (6.4)

interference, constructive A superposition of waves for which the combined waveform has a greater amplitude. (7.6)

interference, destructive A superposition of waves for which the combined waveform has a smaller amplitude. (7.6)

International Date Line (IDL) The meridian that is 180° E or W of the prime meridian. (15.3)

interplanetary dust Very small solid particles known as *micrometeoroids* that exist in the space between the planets. (17.6)

ion An atom, or chemical combination of atoms, that has a net electric charge because of a gain or loss of electrons. (11.5)

ionic bonds Electrical forces that hold the ions together in the crystal lattice of an ionic compound. (12.4)

ionic compounds Compounds formed by an electron transfer process in which one or more atoms lose their valence electrons, and other atoms gain these same electrons to achieve noble gas configurations. (12.4)

ionization energy The amount of energy it takes to remove an electron from an atom. (11.5)

ionosphere The region of the atmosphere between about 70 km (43 mi) and several hundred kilometers in altitude. It is characterized by a high concentration of ions. (19.1)

isobar A line on a weather map drawn through the locations (points) of equal pressure. (19.4)

isostasy The depth to which a floating object sinks into underlying material depends on the objects density and thickness; the state of buoyancy between the Earth's lithosphere and asthenosphere. (21.3)

isotopes Forms of nuclei of an element that have the same numbers of protons but differ in their numbers of neutrons. (10.2)

jet streams Rapidly moving "rivers" of air in the upper troposphere. (19.4)

joule (J) A unit of energy equivalent to 1 N·m or 1 kg·m²/s². (4.1)

Jovian planets The four outer planets—Saturn, Jupiter, Uranus, and Neptune. All have characteristics resembling those of Jupiter, having gaseous outer layers. (16.2)

kelvin (K) The unit of temperature on the Kelvin (absolute) temperature scale. A kelvin is equal in magnitude to a degree Celsius. (5.1)

Kelvin scale The "absolute" temperature scale that takes absolute zero as 0 K. (5.1)

kilo- Metric prefix that means 10^3, or one thousand. (1.5)

kilocalorie (kcal) The amount of heat necessary to raise the temperature of one *kilogram* of water one Celsius degree. (5.2)

kilogram (kg) The standard metric unit of mass; 1 kilogram has an equivalent weight of 2.2 pounds. (1.4)

kilowatt-hour (kWh) A unit of energy (power × time); $P = E/t$, and $E = Pt$. (4.4)

kinetic energy Energy of an object's motion, equal to $\frac{1}{2}mv^2$. (4.2)

kinetic theory A gas consists of molecules moving independently in all directions at high speeds (the higher the temperature, the higher the average speed), colliding with one another and the walls of the container, and having a distance between molecules that is large, on average, compared with the size of the molecules themselves. (5.6)

K-T Event The extinction episode that marks the transition from the Cretaceous period (K) to the Tertiary period (T). (24.5)

Kuiper belt A doughnut-shaped ring of space around the Sun beyond the orbit of Neptune and that extends to well beyond the orbit of Pluto and Eris, containing many short-period comets. (17.5)

land breeze A local wind from land to sea resulting from a convection cycle. (19.4)

lapse rate The rate at which the temperature of the air in the troposphere decreases with altitude. The normal lapse rate is −6.5 C°/km, or −3.5 F°/1000 ft. (19.5)

laser An acronym for light amplification by stimulated emission of radiation; it produces coherent, monochromatic light. (9.4)

last-quarter (third quarter) phase The phase that occurs when the Moon is 270° east of the Sun and appears on the observer's meridian at 6 A.M. local solar time. (17.2)

latent heat The heat associated with a phase change. (5.3)
  of fusion The amount of heat necessary to change one kilogram of a solid into a liquid. (5.3)
  of vaporization The amount of heat required to change one kilogram of a liquid to a gas . (5.3)

latitude The angular measurement in degrees north or south of the equator for a point on the surface of the Earth. (15.2)

lava Magma that reaches the Earth's surface through a volcanic vent. (22.3)

law A concise statement in words or a mathematical equation that describes a fundamental relationship of nature. (1.2)

law of:
  charges Like charges repel; unlike charges attract. (8.1)
  elliptical orbits All planets move in elliptical orbits around the Sun, with the Sun at one focus of the ellipse. (16.1)
  equal areas An imaginary line (radial vector) joining a planet to the Sun sweeps out equal areas in equal periods of time. (16.1)
  poles (magnetic) Like poles repel; unlike poles attract. (8.4)
  reflection The angle of incidence is equal to the angle of reflection, $\theta_i = \theta_r$, as measured relative to the normal, a line perpendicular to the reflecting surface. (7.1)

length The measurement of space in any direction. (1.4)

Lewis structures "Electron dot" symbols used to show valence electrons in molecules and ions of compounds. (12.4)

Lewis symbol The element's symbol represents the nucleus and inner electrons of an atom, and the valence electrons are shown as dots arranged around the symbol. (12.4)

lightning A huge discharge of electrical energy in the atmosphere. (20.3)

light-year The distance traveled by light in one year (9.5×10^{12} km or 6 trillion mi). (18.1)

limiting reactant A starting material that is used up completely in a chemical reaction. (12.2)

line absorption spectrum A set of dark spectral lines of certain frequencies or wavelengths, formed by dispersion of light that has come from an incandescent source and has then passed through a sample of cool gas. (9.3)

line emission spectrum A set of bright spectral lines of certain frequencies or wavelengths formed by dispersion of light from a gas discharge tube. Each element gives a different set of lines. (9.3)

linear momentum The product of an object's mass and its velocity. (3.7)

linearly polarized light The condition of transverse light waves that vibrate in only one plane. (7.5)

liquid An arrangement of molecules that may move and assume the shape of the container; has a definite volume but no definite shape. (5.5)

liter (L) A metric unit of volume or capacity; 1 L = 1000 cm^3. (1.5)

lithification The process of transforming a sediment into a sedimentary rock; also called consolidation. (22.5)

lithosphere The outermost solid portion of the Earth, which includes the crust and part of the upper mantle. (21.1)

Local Group The cluster of galaxies that includes our own Milky Way. (18.6)

longitude The angular measurement in degrees east or west of the reference meridian, known as the Greenwich (prime) meridian for a point on the surface of the Earth. (15.2)

longitudinal wave A wave in which the particle motion and the wave velocity are parallel to each other. (6.2)

long-shore current A current along a shore that results from waves that break at an angle to the shoreline. (23.4)

lunar eclipse An eclipse of the Moon caused by the Earth's blocking the Sun's rays to the Moon. (17.4)

   partial lunar eclipse The Earth's umbral (dark) shadow does not completely cover the Moon. (17.4)

   total lunar eclipse The Earth's umbral (dark) shadow completely covers the Moon. (17.4)

luster The appearance of a mineral's surface in reflected light. (22.1)

magma The molten material beneath the Earth's surface. (22.3)

magnetic declination The angle between geographic (true) north and magnetic north. (8.4)

magnetic domains Local regions of alignment of the magnetic fields of numerous atoms in ferromagnetic materials. (8.4)

magnetic field A magnetic force field represented by a set of imaginary lines that indicate the direction in which a small compass needle would point if it were placed near a magnet. (8.4)

magnitude (absolute) The brightness that a star would have if it were placed 10 parsecs from the Earth. (18.3)

magnitude (apparent) The brightness of a star as observed from the Earth. (18.3)

main sequence The narrow band going from the lower right to the upper left on the H-R diagram; most stars fall into this category. (18.3)

mantle The interior region of the Earth between the core and the crust. (21.1)

maria Large, dark, flat areas on the Moon believed to be craters formed by large impacts from space that then filled with volcanic lava. (17.1)

mass A quantity of matter and a measure of the amount of inertia that an object possesses. (1.4, 3.2)

mass defect The decrease in mass in a nuclear reaction. (10.6)

mass number The number of protons plus neutrons in a nucleus; the total number of nucleons. (10.2)

mass wasting The general geologic term for the downslope movement of soil and rock under the influence of gravity. (23.2)

matter (de Broglie) waves The waves associated with moving particles. (9.6)

meander A loop-like bend in a stream channel influenced by gravity and the rotating Earth. (23.2)

measurement A quantitative observation, one involving numbers. (1.2)

mega- Prefix that means 10^6, or one million. (1.5)

meridians Imaginary lines drawn along the surface of the Earth running from the geographic North Pole to the geographic South Pole perpendicular to the equator. (15.2)

mesosphere The region of the Earth's atmosphere that lies between approximately 50 and 80 km (30 and 50 mi) in altitude. (19.1)

metal An element whose atoms tend to lose valence electrons during chemical reactions. (11.4)

metamorphic rock Rock that is formed by the alteration of preexisting rock in response to the effects of pressure, temperature, or the gain or loss of chemical components. (22.2)

metamorphism The process by which the structure, mineral content, or both of a rock is changed while the rock remains a solid. (22.6)

meteor A metallic or stony object that burns up as it passes through the Earth's atmosphere and appears to be a "shooting star." (17.6)

meteorite A metallic or stony object from the solar system that strikes the Earth's surface. (17.6)

meteoroids Small, interplanetary metallic and stony objects in space before they encounter the Earth. (17.6)

meteorology The study of the lower atmosphere. (19.Intro)

meter (m) The standard unit of length in the metric system. It is equal to 39.37 inches, or 3.28 feet. (1.4)

metric system The decimal (base-10) system of units employed predominantly throughout the world. (1.4)

mid-ocean ridge A series of mountain ranges on the ocean floor, more than 84,000 km (52,000 mi) in length, extending through the North and South Atlantic, the Indian Ocean, and the South Pacific. (21.2)

milli- The metric prefix that means 10^{-3}, or 1 one-thousandth. (1.5)

mineral A naturally occurring, crystalline inorganic element or compound that possesses a fairly definite chemical composition and a distinctive set of physical properties. (22.1)

mixture A type of matter composed of varying proportions of two or more substances that are just physically mixed, *not* chemically bonded. (11.1)

mks system The metric system that has the *meter*, *kilogram*, *second*, and *coulomb* as the standard units of length, mass, time, and electric charge, respectively. (1.4)

Moho (Mohorovicic discontinuity) The sharply defined boundary that separates the Earth's crust from the upper mantle. (21.1)

Mohs scale A list of 10 minerals used to measure the hardness of other minerals. (22.1)

molarity (M) A measure of solution concentration in terms of moles of solute per liter of solution. (13.5)

mold A hollow depression formed when an embedded shell or bone is dissolved out of a rock. (24.1)

mole (mol) The quantity of a substance that contains as many elementary units as there are atoms in exactly 12 g of carbon-12; 6.02×10^{23} formula units. (13.5)

molecule An electrically neutral particle composed of two or more atoms chemically bonded. (11.3)

monomer A fundamental repeating unit of a polymer (14.5)

moraine A ridge of glacial drift. (23.2)

motion The undergoing of a continuous change in position. (2.1)

motor A device that converts electrical energy into mechanical energy. (8.5)

mountain range A series of mountains. (21.6)

neap tides Moderate tides with the least variation between high and low. They occur at the first- and last-quarter phases of the Moon. (17.2)

nebulae Cool, dense clouds of interstellar gas and dust. (18.4)

neutron number N, the number of neutrons in the nucleus of an atom. (10.2)

neutrons Neutral particles found in the nuclei of atoms. (10.2)

neutron star An extremely high-density star composed almost entirely of neutrons. (18.5)

new moon The phase of the Moon that occurs when the Moon is on the same meridian as the Sun at 12 noon local solar time. (17.2)

newton (N) The unit of force in the metric system; 1 kg·m/s^2. (3.3)

Newton's first law of motion An object will remain at rest or in uniform motion in a straight line unless acted on by an external, unbalanced force. (3.2)

Newton's law of universal gravitation Every particle in the universe attracts every other particle with a force that is directly proportional to the product of their masses and inversely proportional to the square of the distance between them; $F = Gm_1m_2/r^2$. (3.5)

Newton's second law of motion The acceleration of an object is equal to the net force on the object divided by the mass of the object; $a = F/m$. (3.3)

Newton's third law of motion For every action there is an equal and opposite reaction; for every force there is an equal and opposite force, acting on different bodies. (3.4)

nitrogen oxides (NO$_x$) Chemical combinations of nitrogen and oxygen, such as NO and NO_2. (20.4)

noble gases The elements of Group 8A of the periodic table (He, Ne, Ar, Kr, Xe, Rn). (11.6)

nonmetal An element whose atoms tend to gain (or share) electrons during chemical reactions. (11.4)

nova A white dwarf star that suddenly increases dramatically in brightness for a brief period of time. (18.5)

nucleons A collective term for neutrons and protons (particles in the nucleus). (10.2)

nucleosynthesis The creation of the nuclei of elements inside stars. (18.4)

nucleus The central core of an atom; composed of protons and neutrons. (10.2)

nuclide A particular species or isotope of any element, characterized by a definite atomic number and mass number. (10.3)

octet rule In forming compounds, atoms tend to gain, lose, or share valence electrons to achieve electron configurations of the noble gases; that is, they tend to get eight electrons (an octet) in the outer shell. Hydrogen is the main exception; it tends to get two electrons in the outer shell, like the configuration of the noble gas helium. (12.4)

ohm (Ω) The unit of resistance; equal to one volt per ampere. (8.2)

Ohm's law The voltage across two points is equal to the current flowing between the points times the resistance between the points; $V = IR$. (8.2)

Oort cloud The cloud of cometary objects believed to be orbiting the Sun far beyond the orbit of Neptune at 50,000 astronomical units and from which the majority of comets originate. (17.6)

opposition The time at which one of the superior planets is on the opposite side of the Earth from the Sun. (16.2)

organic chemistry The study of carbon compounds. (14.Intro)

original horizontality, principle of The principle that sediments and lava flows are deposited as horizontal layers. (24.2)

outer core Part of the innermost region of the Earth, which is composed of iron and nickel in two parts: a solid inner core and a molten, highly viscous outer core. (21.1)

oxidation Occurs when oxygen combines with a substance or when an atom or ion loses electrons. (13.4)

ozone O_3, a form of oxygen found naturally in the atmosphere in the ozonosphere. It is also a constituent of photochemical smog. (19.1, 20.4)

ozonosphere A region of the atmosphere, below 70 km (45 mi) in altitude, characterized by ozone concentration. (19.1)

paleontology The systematic study of fossils and prehistoric life forms. (24.1)

Pangaea The giant supercontinent that is believed to have existed over 200 million years ago. (21.2)

parallax The apparent motion, or shift, that occurs between two fixed objects when the observer changes position. (16.3, 18.1)

parallel circuit A circuit in which the voltage across each resistance is the same, but the current through each resistance may vary (different resistances, different currents). (8.3)

parallels Imaginary lines encircling the Earth parallel to the plane of the equator. (15.2)

parsec (pc) The distance to a star when the star exhibits a parallax of one second of arc, where 1 second of arc is defined to be 1/3,600 of 1°. This distance is equal to 3.26 light-years or 206,265 astronomical units. (18.1)

partial lunar eclipse The Earth's umbral (dark) shadow does not completely cover the Moon. (17.2)

partial solar eclipse Partial blocking of the Sun, seen by an observer in the penumbra. (17.2)

penumbra A semidark region of the Moon's shadow. During an eclipse, an observer in the penumbra sees only a partial eclipse. (17.2)

perihelion The point when Earth (or another orbiting object) is closest to the Sun. (15.5)

period In physics, the time for a complete cycle of motion. In chemistry, one of the seven horizontal rows of the periodic table. In geology, an interval of geologic time that is a subdivision of an era and is made up of epochs. (6.2, 11.4, 24.2)

periodic law The properties of elements are periodic functions of their atomic numbers. (11.4)

periodic table Organization of the elements on the basis of atomic numbers into seven rows called periods. (11.4)

permafrost Ground that is permanently frozen. (23.1)

permeability A material's capacity to transmit fluids. (23.3)

pH A measure (on a logarithmic scale) of the hydrogen ion (or hydronium ion) concentration in a solution. (13.3)

phases of matter The physical forms of matter—most commonly, solid, liquid, and gas. (5.5)

photochemical smog Air pollution resulting from the chemical reactions of hydrocarbons with oxygen in the air and other pollutants in the presence of sunlight. (20.4)

photoelectric effect The emission of electrons that occurs when certain metals are exposed to light. (9.2)

photon A "particle" of electromagnetic energy. (9.2)

photosphere The bright, visible "surface" of the Sun. (18.2)

photosynthesis The process by which plants convert CO_2 and H_2O into carbohydrates (needed for plant life) and O_2, using energy from the Sun. (19.1)

physics The most fundamental physical science; concerned with the basic principles and concepts that describe the workings of the universe. It deals with matter, motion, force, and energy. (2.Intro)

planetary nebula A luminous shell of gas ejected from an old, low-mass star. (18.4)

plasma A high-temperature gas of free electrons and positively charged ions. (10.6)

plate tectonics The theory that the Earth's lithosphere is made up of a series of solid sections or segments called *plates* that are constantly interating with one another in very slow motion. (21.3)

plutons Intrusive igneous rocks, formed below the surface of the Earth by solidification of magma. (22.4)

polar covalent bond One in which the pair of bonding electrons is unequally shared, leading to the bond's having a slightly positive end and a slightly negative end. (12.5)

polar molecule A molecule that has a positive end and a negative end—that is, one that has a dipole. (12.5)

polarization The preferential orientation of the electric vector of a light wave. (7.5)

poles, law of (magnetic) Like poles repel; unlike poles attract. (8.4)

pollution Any atypical contribution to the environment resulting from the activities of humans. (20.4)

polymer A compound of very high molecular mass whose chain-like molecules are made up of repeating units called monomers. (14.5)

position The location of an object with respect to a reference point. (2.1)

post meridiem (P.M.) Pertaining to time from 12 noon to 12 midnight. (15.3)

potential energy The energy an object has because of its position or location; the energy of position. (4.2)

power The time rate of doing work. (4.4)

powers-of-10 notation Notation in which numbers are expressed by a coefficient and a power of 10; for example, $2500 = 2.5 \times 10^3$. Also called *scientific notation*. (1.7)

precession The slow rotation of the axis of spin of the Earth around an axis perpendicular to the ecliptic plane. The rotation is clockwise as observed from the north celestial pole. (15.6)

precipitate An insoluble solid that appears when two liquids (usually aqueous solutions) are mixed. (13.3)

pressure The force per unit area; $p = F/A$. (5.6)

principal quantum number The numbers $n = 1, 2, 3, \ldots$ used to designate the various principal energy levels that an electron may occupy in a hydrogen atom. (9.3)

principle of

cross-cutting relationships An igneous rock is younger than the rock layers it has intruded (cut into or across). (24.2)

original horizontality Sediments and lava flows are deposited as horizontal layers. (24.2)

superposition (geology) The principle that in a sequence of undisturbed sedimentary rocks, lavas, or ash, each layer is younger than the layer beneath it and older than the layer above it. (24.2)

superposition (wave) At any time, the combined waveform of two or more interfering waves is given by the sum of the displacements of the individual waves at each point in the medium. (7.6)

products The substances formed during a chemical reaction. (13.1)

prograde motion Orbital or rotational motion in the forward direction. In the solar system, this is west-to-east, or counterclockwise, as viewed from above the Earth's North Pole. (16.2)

projectile motion The motion of a projected or thrown object under the influence of gravity. (2.5)

proteins Biological polymers and extremely long-chain polyamides formed by the enzyme-catalyzed condensation of *amino acids* under the direction of nucleic acids in the cell (or the biochemist in the lab). (14.6)

proton-proton chain A series of stellar nuclear reactions in which four hydrogen nuclei (protons) combine to form one helium nucleus and release energy. (18.2)

protons Positively charged particles in the nuclei of atoms. (8.1, 10.2)

psychrometer An instrument used to measure relative humidity. (19.3)

pyroclastics (tephra) Solid material emitted by volcanoes; range in size from fine dust to large boulders. (22.4)

quantum A discrete amount of energy. (9.2)

quantum mechanics The branch of physics that replaced the classical-mechanical view (that everything moved according to exact laws of nature) with the concept of probability. Schrödinger's equation forms the basis of quantum wave mechanics. (9.7)

quasar Shortened from quasi-stellar radio sources, these are extremely distant objects emitting a tremendous amount of energy. (18.6)

radar An instrument that sends out electromagnetic (radio) waves, monitors the returning waves that are reflected by some object, and thereby locates the object. Radar stands for **ra**dio **d**etecting **a**nd **r**anging. Radar is used to detect and monitor precipitation and severe storms. (19.3)

radiation A method of heat transfer by means of electromagnetic waves. (5.4)

radioactive isotope A nuclide whose nucleotide undergoes spontaneous decay (disintegration). (10.3)

radioactivity The spontaneous process of nuclei undergoing a change by the emitting particles or rays. (10.3)

radiometric dating The determination of age by using radioactivity; geology's best tool for establishing absolute geologic time. (24.3)

radionuclides Types of nuclei that undergo radioactive decay. (10.3)

rain gauge An open, calibrated container used to measure amounts of precipitation. (19.3)

ray A straight line that represents the path of light with a directional arrowhead. (7.1)

Rayleigh scattering The preferential scattering of light by air molecules and particles that accounts for the blueness of the sky. The scattering is proportional to $1/\lambda^4$. (19.2)

rays (lunar) Streaks of light-colored material extending outward from craters on the Moon. (17.1)

reactants The original substances in a chemical reaction. (13.1)

real image An image from a mirror or lens for which the light rays converge so that an image can be formed on a screen. (7.3)

red giant A relatively cool, very bright star that has a diameter much larger than average. (18.3)

redshift A Doppler effect caused when a light source, such as a galaxy, moves away from the observer and shifts the light frequencies lower, or toward the red end of the electromagnetic spectrum. (6.5)

reduction Occurs when oxygen is removed from a compound or when an atom or ion gains electrons. (13.4)

reflection The change in the direction of a wave when it strikes and rebounds from a surface or the boundary between two media. (7.1)

reflection

diffuse Reflection from a rough surface in which reflected rays are not parallel but scattered. (7.1)

law of The angle of reflection equals the angle of incidence, $\theta_i = \theta_r$ as measured relative to the normal, a line perpendicular to the reflecting surface. (7.1)

specular Reflection from very smooth (mirror) surfaces in which the reflected rays are parallel. (7.1)

total internal A phenomenon in which light is totally reflected in a medium because of refraction. (7.2)

refraction The deviation of light from its original path caused by a change in speed in the second medium. (7.2)

regional metamorphism A change in rock over a large area, brought about by both heat and pressure. (22.6)

relative geologic time A time scale obtained when rocks and the geologic events they record are placed in chronologic order without regard to actual dates. (24.2)

relative humidity The ratio of the actual moisture content to the maximum moisture capacity of a volume of air at a given temperature. (19.3)

remanent magnetism The magnetization of rocks that form in the presence of an external magnetic field. (21.2)

renewable energy sources Energy sources from natural processes that are constantly replenished, such as wind and hydro power (4.6)

replacement fossil Fossil formed when a mineral slowly replaces parts of a buried organism. (24.1)

representative elements Those in Groups 1A through 8A in the periodic table. (11.4)

resistance (electrical) The opposition to the flow of electric charge. (8.2)

resonance A wave effect that occurs when an object has a natural frequency that corresponds to an external frequency. (6.6)

retrograde rotation Orbital or rotational motion in the backward direction. In the solar system, this is east to west, or clockwise, as viewed from above the Earth's North Pole. (16.2)

revolution The movement of one object around another. (16.3)

Richter scale The severity of an earthquake as an absolute measure of the energy released. (21.5)

right ascension A coordinate for measuring the east-west positions of celestial objects. The angle is measured eastward from the vernal equinox in hours, minutes, and seconds. (18.1)

rill A narrow trench or valley on the Moon. (17.1)

Ring of Fire The plate boundaries of the Pacific Ocean where volcanoes and earthquakes are common. (21.4)

rock A solid, cohesive natural aggregate of one or more minerals. (22.2)

rock cycle A series of events through which a rock changes over time between igneous, sedimentary, and metamorphic forms. (22.2)

rotation The turning (spinning) of an object about an internal axis. (16.3)

salt An ionic compound composed of any cation except H^+ and any anion except OH^-. (13.3)

saturated solution A solution that has the maximum amount of solute dissolved in the solvent at a given temperature. (11.1)

scalar A quantity that has a magnitude but has no direction associated with it. (2.2)

science An organized body of knowledge about the natural universe and the processes by which that knowledge is acquired and tested. (1.1)

scientific method An investigative process that holds that no concept or model of nature is valid unless the predictions it generates agree with experimental results. That is, all hypotheses should be based on as much relevant data as possible and then should be tested and verified. (1.2)

sea breeze A local wind blowing from the sea to land as a result of a convection cycle. (19.4)

seafloor spreading The theory that the seafloor slowly spreads and moves sideways away from mid-ocean ridges. The spreading is believed to be due to convection cycles of subterranean molten material that cause the formation of the ridges and a surface motion in a lateral direction from the ridges. (21.2)

seamount An isolated submarine volcanic mountain that may extend to heights of more than 1.6 km (1 mi) above the seafloor. (23.4)

second The standard unit of time. It is now defined in terms of the frequency of a certain transition in the cesium atom. (1.4)

second law of thermodynamics It is impossible for heat to flow spontaneously from a colder body to a hotter body. (5.7)

sedimentary rock Rock formed at the Earth's surface by compaction of layers of sediment. (22.2)

sediment Sand, mud, precipitated material, or rock fragments that have been transported or deposited by water, air, or ice. (22.5)

seismic waves The waves generated by the energy release of an earthquake. (21.1)

series circuit A circuit in which an entering current flows individually through all the circuit elements. (8.3)

shear metamorphism A change in rock brought about primarily by pressure rather than heat. (22.6)

SI (International System of Units) A modernized version of the metric system that contains seven base units. (1.4)

sidereal day The elapsed time between two successive crossings of the same meridian by a star other than the Sun. One sidereal day is 23 h, 56 min, 4.091 s. (15.3)

sidereal month The time it takes for the Moon to make one complete cycle relative to the stars, 27.3 days. (17.2)

sidereal period The time it takes a planet (or other object) to make one full orbit around the Sun relative to a fixed star. (16.2)

sidereal year The time interval for the Earth to make one complete revolution around the Sun with respect to any particular star other than the Sun. (15.5)

significant figures A method of expressing measured numbers properly; involves the accuracy of measurement and mathematical operations. (1.7)

silicate Any one of numerous minerals that have the oxygen and silicon tetrahedron as their basic structure. (22.1)

single-replacement reactions Reactions in which one element replaces another that is in a compound: $A + BC \rightarrow B + AC$. (13.4)

sinkhole A depression on the land surface where soluble rock (limestone) has been removed by groundwater. (23.1)

snowstorm An appreciable accumulation of snow. When accompanied by high winds and low temperatures, it is referred to as a *blizzard*. (20.3)

smog A contraction of **smoke-fog**, used to describe the combination of these conditions. (20.4)

solar day The elapsed time between two successive crossings of the same meridian by the Sun. (15.3)

solar eclipse An eclipse of the Sun caused by the Moon blocking the Sun's rays to an observer on the Earth. (17.2)

    partial Partial blocking of the Sun, seen by an observer in the penumbra. (17.2)

    total Complete blocking of the Sun, seen by an observer in the umbra. (17.2)

solar nebula A large, spherical cloud of cold gases and dust that contracted under the influence of its own gravity, began rotating, and then flattened into a swirling disk of gas and dust. (16.7)

solar system A complex system of moving masses held together by gravitational forces, consisting of the Sun, eight major planets and their satellites, dwarf planets, the asteroids, comets, meteoroids, and interplanetary dust. (16.1)

solid Matter that has relatively fixed molecules and a definite shape and volume. (5.5)

solstice The point where the Sun is farthest north (or south) from the equator, approximately on June 21 (or December 22). (18.1)

solubility The amount of solute that will dissolve in a specified volume or mass of solvent (at a given temperature) to produce a saturated solution. (11.1)

solute The substance(s) present in a smaller amount in a solution. (11.1)

solution A mixture that is uniform throughout, also called a homogeneous mixture. (11.1)

solvent The substance present in the larger amount in a solution. (11.1)

sound The propagation of *longitudinal* waves through matter. (6.4)

sound spectrum An ordered arrangement of various frequencies or wavelengths of sound. The three main regions of the sound spectrum are the infrasonic, the audible, and the ultrasonic. (6.4)

source region The region from which an air mass derives its physical characteristics. (20.2)

specific gravity The ratio of a sample?s mass to the mass of an equal volume of water. (1.6, 22.1)

specific heat The amount of heat energy in kilocalories necessary to raise the temperature of one kilogram of the substance one Celsius degree. (5.3)

speed, average The total distance traveled divided by the time spent in traveling the total distance. (2.2)

speed, instantaneous An object's speed at a particular instant of time (Δt becoming extremely small). (2.2)

speed of light (c) How fast light travels. In air or a vacuum, $c = 3.00 \times 10^8$ m/s, or 186,000 mi/s. (6.3)

speed of sound How fast sound travels in a medium; for example, $v_s = 344$ m/s (770 mi/h) in air at room temperature. (6.4)

spring tides The tides of greatest variation between high and low. They occur at the new and full phases of the Moon. (17.2)

standard time zones The division of the surface of the Earth into 24 time zones, each containing about 15° of longitude. (15.3)

standard unit A fixed and reproducible value for the purpose of taking accurate measurements. (1.4)

standing wave A "stationary" waveform arising from the interference of waves traveling in opposite directions. (6.6)

steam point The temperature at which water boils at normal atmospheric pressure, 100 °C or 212 °F. (5.1)

stimulated emission Process in which an excited atom is caused to emit a photon. (9.4)

Stock system A system of nomenclature for compounds of metals that form more than one ion. A Roman numeral placed in parentheses directly after the name of the metal denotes its ionic charge in the compound being named. (12.4)

storm surge The great dome of water associated with a hurricane when it makes landfall. (20.3)

stratosphere The region of the Earth's atmosphere from approximately 16 to 50 km (10 to 30 mi) in altitude. (19.1)

streak The color of the powder of a mineral on a streak plate (unglazed porcelain). (22.1)

stream Any flow of water occurring between well-defined banks. (23.2)

strong nuclear force The short-range force of attraction that acts between two nucleons and holds the nucleus together. (10.2)

structural formula A graphical representation of the way the atoms are connected to one another in a molecule. (14.3)

subduction The process in which one plate is deflected downward beneath another plate into the asthenosphere. (21.3, 22.3)

subsidence The sinking of the land surface due to the excessive extraction of groundwater. (23.3)

sulfur dioxide (SO_2) An atmospheric pollutant formed by the oxidation of sulfur; it contributes to acid rain. (20.4)

summer solstice The farthest point of the Sun's latitude north of the equator (for the Northern Hemisphere), around June 21. The beginning of summer. (15.5)

sunspot Huge regions of cooler gas on the surface of the Sun where the magnetic field is very strong. (18.2)

sunspot cycle The 11-year variation of the number and location of the sunspots on the Sun. (18.2)

supernova An exploding star. (18.5)

superposition, principle of (geology) In a sequence of undisturbed sedimentary rocks, lavas, or ash, each layer is younger than the layer beneath it and older than the layer above it. (24.2)

superposition, principle of (wave) At any time, the combined waveform of two or more interfering waves is given by the sum of the displacements of the individual waves at each point in the medium. (7.6)

supersaturated solution A solution that contains more than the normal maximum amount of dissolved solute at a given temperature and hence is unstable. (11.1)

synodic month The time it takes the Moon to go through one complete cycle of phases (a month of phases), 29.5 days. (17.2)

synthetics Materials whose molecules have no duplicates in nature. (14.5)

system of units A group of standard units and their combinations. The two major systems of units in use today are the metric system and the British system. (1.4)

temperature A measure of the average kinetic energy of the molecules of a substance. (5.1)

temperature inversion A condition characterized by an inverted lapse rate. (20.4)

terminal velocity The maximum velocity reached by a falling object because of air resistance. (2.3)

terminator The boundary of the circle of illumination between daylight and dark. (15.5)

terrestrial planets The four inner planets—Mercury, Venus, Earth, and Mars. All are similar to the Earth in general chemical and physical properties. (16.2)

theory A well-tested explanation of observed natural phenomena. (1.2)

thermodynamics The science dealing with the production of heat, the flow of heat, and the conversion of heat to work. (5.7)

thermodynamics, first law of The heat energy added to a system must go into increasing the internal energy of the system, or any work done by the system, or both. The law, which is based on the conservation of energy, also states that heat energy removed from a system must produce a decrease in the internal energy of the system, or any work done on the system, or both. (5.7)

thermodynamics, second law of It is impossible for heat to flow spontaneously from a colder body to a hotter body. (5.7)

thermodynamics, third law of It is impossible to attain a temperature of absolute zero. (5.7)

thermosphere The region of the Earth's atmosphere extending from about 80 km (50 mi) in altitude to the outer reaches of the atmosphere. (19.1)

thunder The sound associated with lightning; it arises from the explosive release of electrical energy. (20.3)

tides The periodic rise and fall of the water level along the shores of large bodies of water. (17.2, 23.4)

time The continuous forward flow of events. (1.4, 15.3)

tornado The most violent of storms, characterized by a whirling, funnel-shaped cloud and high winds. (20.3)

tornado warning The alert issued when a tornado has actually been sighted or is indicated on radar. (20.3)

tornado watch The alert issued when atmospheric conditions indicate that tornadoes may form. (20.3)

torque A twisting action that produces rotational motion or a change in rotational motion. (3.7)

total internal reflection A phenomenon in which light is totally reflected in a medium because of refraction. (7.2)

total lunar eclipse The Earth's umbral (dark) shadow completely covers the Moon. (17.2)

total solar eclipse Complete blocking of the Sun, seen by an observer in the umbra. (17.2)

trace fossil A fossil imprint made by the movement of an animal. (24.1)

transform boundary A region of the lithosphere where a moving plate slides along one side of another without creating or destroying rock. (21.3)

transformer A device based on electromagnetic induction that increases or decreases the voltage or alternating current. (8.5)

transition elements The B group of elements in the periodic table. (11.4)

trans-Neptunian object (TNO) Large collection of comet-like objects orbiting the Sun beyond the orbit of Neptune. (17.6)

transverse wave A wave in which the particle motion is perpendicular to the direction of the wave velocity. (6.2)

tropical year The time interval from one vernal equinox to the next; the elapsed time between two successive northward crossings of the Sun above the equator (vernal equinox). (15.5)

troposphere The region of the Earth's atmosphere from the ground up to about 16 km (10 mi). (19.1)

tsunami A Japanese word for a "harbor" wave—an unusually large sea wave produced by a seaquake or undersea volcanic eruption. (21.5)

ultrasound Sound with frequency greater than 20,000 Hz or 20 kHz. (6.4)

umbra A region of total darkness in a shadow. During an eclipse, an observer in the umbra sees a total eclipse. (17.2)

unbalanced (net) force The sum of vector forces with a nonzero result. A force capable of producing motion. (3.1)

unconformity A break in the geologic rock record. (24.2)

uniformitarianism The principle that the same processes operate on and within the Earth today as in the past. Hence the present is considered the key to the past. (22.2)

universe The totality of all matter, energy, and space. (18.Intro)

unsaturated solution A solution in which more solute can be dissolved at the same temperature. (11.1)

valence electrons The electrons that are involved in bond formation, usually those in an atom's outer shell. (11.4)

valence shell An atom's outer shell, which contains the valence electrons. (11.4)

vector A quantity that has both magnitude and direction. (2.1)

velocity, average The displacement divided by the total travel time; $\bar{v} = \Delta d/\Delta t$. (2.2)

velocity, instantaneous The velocity at any instant of time. (2.2)

velocity, terminal The maximum velocity reached by a falling object because of air resistance. (2.3)

vernal equinox The point where the Sun crosses the celestial equator from south to north, around March 21. The beginning of spring. (15.5)

virtual image An image from a lens or mirror for which the light rays diverge and cannot be formed on a screen. (7.3)

viscosity The internal property of a substance that offers resistance to flow. (22.4)

volcanic mountains Mountains that have been built by material ejected during volcanic eruptions. (21.6)

volcano A vent from which hot molten rock (lava), ash, and gases escape from deep below the Earth's surface, or the mountain or elevation created by solidified lava and volcanic debris that accumulates near the vent. (21.4, 22.4)

volt (V) The unit of voltage equal to one joule per coulomb. (8.2)

voltage The amount of work it would take to move a charge between two points divided by the value of the charge—that is, work per unit charge $V = W/q$ or the electric potential energy per unit charge. (8.2)

waning phase The illuminated portion of the Moon is getting smaller each day as observed from the Earth. (17.2)

water table The boundary between the zone of aeration and the zone of saturation. (23.3)

watt (W) A unit of power equivalent to 1 kg·m^2/s^3, or 1 J/s. (4.4)

wave The propagation of energy in media from a disturbance. (6.1)

wavelength The distance from any point on a wave to an adjacent point with similar oscillation: the distance of one complete "wave," or where it starts to repeat itself. (6.2)

wave speed The distance a wave travels divided by the time of travel. (6.2)

waxing phase The illuminated portion of the Moon is getting larger each day as observed from the Earth. (17.2)

weather The conditions of the lower atmosphere. (19.1)

weathering The process of breaking down rock on or near the Earth's surface. (23.1)

weight A measure of the force due to gravitational attraction ($w = mg$, on the Earth's surface). (3.3)

white dwarf A hot white star that has a much smaller diameter and much higher density than average. It is believed to be the final stage of a low-mass star. (18.3)

wind The horizontal movement of air or air motion along the Earth's surface. (19.4)

wind vane A freely-rotating device that, because of its shape, lines up with the wind and indicates the direction from which the wind is blowing. (19.3)

winter solstice The farthest point of the Sun's latitude south of the equator (for the Northern Hemisphere), around December 22. The beginning of winter. (15.5)

work The product of the magnitude of a force (or a parallel component of a force) and the parallel distance through which the object moves while the force is applied. (4.1)

X-rays High-frequency, high-energy electromagnetic radiation formed when high-speed electrons strike a metallic target. (9.4)

zenith The position directly overhead for an observer on the Earth. (15.4)

zenith angle The complementary angle of the altitude; it is the angle between the zenith and an object in the sky (the Sun for example). (15.4)

Index

Note: Information in figures and table is indicated by *f* and *t*.